VIRTUAL PRINCIPLES IN AIRCRAFT STRUCTURES
Volume 2: Design, Plates, Finite Elements

MECHANICS OF STRUCTURAL SYSTEMS
Editors: J.S. Przemieniecki and G.Æ. Oravas

L. Fryba, Vibration of solids and structures under moving loads. 1973
ISBN 90-01-32420-7

K. Marguerre and H. Wölfel, Mechanics of vibrations. 1979
ISBN 90-286-0086-6

E.B. Magrab, Vibrations of elastic structural members. 1979
ISBN 90-286-0207-0

R.T. Haftka and M.P. Kamat, Elements of structural optimization. 1985
ISBN 90-247-2950-5(hardbound) ISBN 90-247-3062-7(paperback)

J.R. Vinson and R.L. Sierakowski, The behavior of structures composed of composite materials. 1986
ISBN 90-247-3125-9(hardbound) ISBN 90-247-3578-5(paperback)

B.E. Gatewood, Virtual Principles in Aircraft Structures Volume 1. 1989.
ISBN 90-247-3754-0

B.E. Gatewood, Virtual Principles in Aircraft Structures Volume 2. 1989.
ISBN 90-247-3755-9
ISBN 90-247-3753-2 (set).

Volume 7

B.E. GATEWOOD

*Professor Emeritus, Dept. of Aeronautical and Astronautical Engineering,
The Ohio State University, Columbus, Ohio, U.S.A.*

Virtual Principles in Aircraft Structures

Volume 2: Design, Plates, Finite Elements

Springer-Science+Business Media, B.V.

Library of Congress Cataloging-in-Publication Data

Gatewood, Burford Echols, 1913-
  Virtual principles of aircraft structures.

  (Mechanics of structural systems)
  Includes bibliographies and indexes.
  Contents: v. 1. Analysis -- v. 2. Design, plates,
finite elements.
  1. Airframes. 2. Structures, Theory of.
3. Strength of materials. I. Title. II. Series.
TL671.6.G37 1989   629.134'1   88-13303

ISBN 978-94-010-7018-8   ISBN 978-94-009-1165-9 (eBook)
DOI 10.1007/978-94-009-1165-9

printed on acid free paper

All Rights Reserved
©**1989 by** Springer Science+Business Media Dordrecht
Originally published by Kluwer Academic Publishers in 1989
Softcover reprint of the hardcover 1st edition 1989

No part of the material protected by this copyright notice may be reproduced or
utilized in any form or by any means, electronic or mechanical,
including photocopying, recording or by any information storage and
retrieval system, without written permission from the copyright owner

Contents
Volume 2: Design, Plates, Finite Elements

Preface	xiii
List of selected equations in Volume 1 referred to in Volume 2	xv
Chapter 1 / Allowable stresses of flight vehicle materials	1
1.1 Introduction	1
1.2 Tension, shear, and bearing allowable stresses	2
1.3 Temperature effects on allowable stresses	6
1.4 Allowable compression stresses	11
1.5 Allowable combined stresses	22
1.6 Creep effects on allowable stresses	31
1.7 Room temperature fatigue effects upon allowable stresses	35
1.8 Temperature effects upon allowable fatigue stresses	38
1.9 Crack effects upon allowable fatigue stresses	40
1.10 Problems	41
References	42
Chapter 2 / Analysis and design of joints and splices	44
2.1 Introduction	44
2.2 Analysis of plate splices with axial tension forces	44
2.3 Multi-row tension splices	49
2.4 Joints with eccentric loading	54
2.5 Minimum weight design of splice for beam with rectangular cross section	57
2.6 Design of splices for I-beams and thin shear webs	65
2.7 Deflection effects on load distribution in splices	69
2.8 Temperature and inelastic effects on load distribution in splices	74
2.9 Welded joints	77
2.10 Bonded joints	80
2.11 Problems	86
References	87

Contents of Volume 2: Design, Plates, Finite Elements

Chapter 3 / Structural design of aircraft components   89

3.1	Introduction	89
3.2	Design of minimum weight columns without local buckling	91
3.3	Design of minimum weight sections with local buckling and crippling	96
3.4	Design of minimum weight columns with local buckling	100
3.5	Minimum weight design for stiffened panels in compression	105
3.6	Effective areas for stiffened panels	112
3.7	Effect of load intensity on wing design	114
3.8	Design of box beam cross sections with four spar caps	117
3.9	Analysis of diagonal tension beams	121
3.10	Problems	123
	References	125

Chapter 4 / Analysis and design of pressurized structures   126

4.1	Introduction	126
4.2	Membrane stresses in thin shells	126
4.3	Cut-outs in thin shells with membrane stresses	129
4.4	Bending in circular cylindrical shells with axially symmetric loading	132
4.5	Bending in pressurized aircraft fuselages from stringers and frames	139
4.6	Bending of non-circular cross sections with internal pressure	142
4.7	Bending of non-circular fuselage rings with internal pressure	143
4.8	Bending of non-circular fuselage rings with point loads	146
4.9	Effect of internal pressure on buckling of cylindrical shells	150
4.10	Pressure stabilized structures	152
4.11	Problems	155
	References	156

Chapter 5 / Approximate solutions using the virtual principles   157

5.1	Introduction	157
5.2	Approximate solutions for beams using the principle of virtual displacements	160
5.3	Approximate solutions for columns	166
5.4	The tapered cantilever beam with numerical integration	169
5.5	Tapered beam finite element matrices for columns	173
5.6	The unit load theorem and numerical integration	175
5.7	Approximate solutions for beams using the mixed virtual principle	178
5.8	Problems	184
	References	186

Chapter 6 / Dynamics of simple beams   188

6.1	Introduction	188
6.2	Bending vibrations of simple beams	188

Contents of Volume 2: Design, Plates, Finite Elements

6.3	Forced motion of uniform beam	192
6.4	Approximate solutions for frequencies and mode shapes	194
6.5	Torsional vibrations of simple beams	197
6.6	Finite element matrices for beam frequencies	200
6.7	Flutter of wing segment with one degree of freedom	201
6.8	Flutter of wing segment with two degrees of freedom	204
6.9	Dynamic loads on beams	207
6.10	Problems	209
	References	209

Chapter 7 / The plate equations — 211

7.1	Introduction	211
7.2	The plate inplane case using the principle of virtual displacements	213
7.3	The plate inplane case using the principle of virtual forces	218
7.4	The plate inplane case using the mixed virtual principle	225
7.5	The plate bending case using the principle of virtual displacements	229
7.6	The plate bending case using the principle of virtual forces	233
7.7	The plate bending case using the mixed virtual principle	236
7.8	Combined inplane and lateral forces	239
7.9	Combined forces with large bending deflections	240
7.10	Buckling of plates	243
7.11	Plate vibrations	245
7.12	Problems	250
	References	250

Chapter 8 / Approximate matrix equations for plate finite elements — 252

8.1	Introduction	252
8.2	The point unknowns for the matrices	253
8.3	The methods to obtain the matrix equations	257
8.4	Inplane plate element matrices from the principle of virtual displacements	261
8.5	Inplane plate element matrices from the principle of virtual forces	266
8.6	Inplane plate element matrices from the mixed virtual principle	275
8.7	Bending plate element matrices from the principle of virtual displacements	280
8.8	Bending plate element matrices from the principle of virtual forces	285
8.9	Bending plate element matrices from the mixed virtual principle	294
8.10	Matrices for constant stress triangular elements	299
8.11	Problems	300

Chapter 9 / Matrix structural analysis using finite elements — 303

9.1	Introduction	303
9.2	General beam elements in local coordinates	305

9.3	General beam elements in datum coordinates	311
9.4	Triangular plate elements with inplane forces	315
9.5	Assembly of finite elements by the virtual principles	322
	References	327

Chapter 10 / Composite Materials   329

10.1	Introduction	329
10.2	Stress-strain equations for nonisotropic materials	331
10.3	Stress-strain equations for plane stress in an orthotropic material	332
10.4	Forces and moments in laminated plates	335
10.5	Stresses in laminated plates	340
10.6	Allowable stresses for laminated plates	343
10.7	Interlamina stresses	348
10.8	Joints in laminated plates	352
10.9	Bending deflections of laminated plates	353
10.10	Buckling loads for laminated plates	354
10.11	Vibrations of laminated plates	355
10.12	Problems	356
	References	357

Index   358

Contents
Volume 1: Analysis

Preface

Chapter 1 / The basic three, two, and one dimensional equations in structural analysis

1.1 Introduction
1.2 Three dimensional equations
1.3 The displacement method of solution
1.4 The stress method of solution
1.5 The combined method of solution
1.6 Two dimensional equations
1.7 Saint Venant's principle
1.8 One dimensional beam equations
1.9 No shear stresses in the beam
1.10 Beam cross section of a thin plate with one shear stress
1.11 Thin web beams with large flange areas and one shear stress
1.12 Torsion of circular cross section and thin wall closed box
1.13 Thin web box beam with general loading
1.14 Inelastic effects in beams with temperature
1.15 Example of inelastic axial stresses and strains with temperature
1.16 Sequence loading and thermal cycling in beams
1.17 Load-strain design curves for beams
1.18 Problems
     References

Chapter 2 / Virtual displacement and virtual force methods in structural analysis

2.1 Introduction
2.2 The principle of virtual displacements
2.3 The unit displacement theorem
2.4 The principle of virtual forces

2.5 The unit load theorem
2.6 The principle of mixed virtual stresses and virtual displacements
2.7 The mixed unit displacement and unit load theorem
2.8 Two dimensional form of the virtual principles
2.9 One dimensional forms of the virtual principles
2.10 The one dimensional virtual principles with temperature, inelastic and large displacement effects
2.11 Matrix forms of the virtual principles
2.12 Problems
References

Chapter 3 / The virtual principles for pin-jointed trusses

3.1 Introduction
3.2 The unit displacement theorem for trusses
3.3 The unit load theorem for trusses
3.4 Inelastic effects with temperature changes in trusses
3.5 Matrix equations for trusses from the unit displacement theorem
3.6 Matrix equations for trusses from the unit load theorem
3.7 Matrix equations for trusses from the mixed unit displacement and unit load theorem
3.8 Problems
References

Chapter 4 / The virtual principles for simple beams

4.1 Introduction
4.2 Principle of virtual displacements for beams
4.3 Point values for beam elements by the principle of virtual displacements
4.4 Principle of virtual forces for beams
4.5 Point values for beam elements by the unit load theorem
4.6 Principle of mixed virtual stresses and virtual displacements for beams
4.7 Inelastic and temperature effects in simple beams
4.8 Matrix equations for beams from the unit displacement theorem
4.9 Matrix equations for beams from the unit load theorem
4.10 Matrix equations for beams from the mixed unit displacement and unit load theorem
4.11 The beam column equations
4.12 Problems
References

Chapter 5 / Box beam shear stresses and deflections

5.1 Introduction
5.2 Shear stresses in beams
5.3 Torsional shear stresses in beams

- 5.4 Shear flows in open box beams
- 5.5 Shear flows in single cell box beams
- 5.6 Shear flows in multi-cell box beams
- 5.7 Shear center for closed box beams
- 5.8 Shear flows in tapered box beams
- 5.9 Inelastic and buckling shear stresses in beams
- 5.10 Axial and bending deflections of box beams with inelastic and temperature effects
- 5.11 Shear deflections of beams
- 5.12 Torsional rotation of beams
- 5.13 Rotation of swept wings
- 5.14 Spanwise airload distribution and static wing divergence under rotation
- 5.15 Static aileron effectiveness and reversal speed under wing rotation
- 5.16 Problems
  References

Chapter 6 / Shear lag in thin web structures

- 6.1 Introduction
  Part 1. Solutions for determinate cases
- 6.2 Shear flows due to concentrated loads into thin webs
- 6.3 Shear flows around cut-outs in thin web beams
- 6.4 Cut-outs in box beams
- 6.5 Shear flows in ribs and bulkheads
- 6.6 Forces on ribs due to airloads and taper effects
  Part 2. Solutions for redundant cases
- 6.7 Restraint effects in thin web structures
- 6.8 Shear flows in redundant beams in one plane
- 6.9 Deflections of thin web structures
- 6.10 Flexibility matrices for shear web elements and stiffener elements
- 6.11 Matrix solutions for thin web beams in one plane
- 6.12 Matrix solutions for box beams
- 6.13 Load redistribution in swept back wings
- 6.14 Problems
  References

Appendix A / Notes on matrix algebra

- A.1 Definition of matrices
- A.2 Addition, subtraction, multiplication of matrices
- A.3 Determinants
- A.4 Matrix inversion
- A.5 Solution of systems of simultaneous equations by matrices
- A.6 Solution of systems of simultaneous equations by tri-diagonal matrices

A.7 Solution of systems of equations by Jordan successive transformations
A.8 Matrix representations
A.9 Orthogonal matrices
A.10 Eigenvalues and eigenvectors of matrices
A.11 Note on matrix notation
References

Appendix B / External forces on flight vehicles

B.1 Introduction
B.2 Inertial forces for rigid body translation and rotation in a vertical plane
B.3 Air forces on airplane wing
B.4 Airplane equilibrium equations in flight. Load factors
B.5 Velocity-load factor $(V\text{-}n)$ diagram for design
B.6 Wing spanwise lift coefficient distribution
B.7 Spanwise lift coefficient distribution on twisted wings
B.8 Spanwise airload, shear, and moment distributions on wings
B.9 Distribution of inertia forces on wing and fuselage
B.10 Forces and moments on landing gear structures
B.11 Thermal forces
B.12 Miscellaneous forces
B.13 Deflection effects on the external forces
B.14 Criteria for the structure to support the external forces
B.15 Problems
References

Appendix C / Derivation of the strain energy theorems from the virtual principles

C.1 Work and strain energy
C.2 Maximum and minimum strain energy and total potential energy
C.3 Theorem of minimum total potential energy
C.4 Theorem of minimum strain energy
C.5 Castigliano's theorem (Part I)
C.6 Hamilton's principle
C.7 Theorem of minimum total complementary potential theory
C.8 Theorem of minimum complementary strain energy
C.9 Castigliano's theorem (Part II)
C.10 Reissner's variational principle
C.11 Comparison of the virtual principles and the strain energy theorems
References

Index

Preface

The basic partial differential equations for the stresses and displacements in classical three dimensional elasticity theory can be set up in three ways: (1) to solve for the displacements first and then the stresses; (2) to solve for the stresses first and then the displacements; and (3) to solve for both stresses and displacements simultaneously. These three methods are identified in the literature as (1) the displacement method, (2) the stress or force method, and (3) the combined or mixed method. Closed form solutions of the partial differential equations with their complicated boundary conditions for any of these three methods have been obtained only in special cases.

In order to obtain solutions, various special methods have been developed to determine the stresses and displacements in structures. The equations have been reduced to two and one dimensional forms for plates, beams, and trusses. By neglecting the local effects at the edges and ends, satisfactory solutions can be obtained for many cases. The procedures for reducing the three dimensional equations to two and one dimensional equations are described in Chapter 1, Volume 1, where the various approximations are pointed out.

Integral transform methods, energy methods, Rayleigh-Ritz and Galerkin approximation methods, virtual principles, and finite element methods have been developed to aid in solving more complicated structural problems. In recent years (see Chapter 2, Volume 1, References 1 and 2 and Introduction) it has been shown that three virtual principles, which correspond to the three methods of solution described above, give a rational basis for the energy methods, Rayleigh-Ritz methods, and finite element methods.

By using integral transform methods, it is possible to convert the three sets of three dimensional elasticity equations described above to integral forms involving stresses, strains, and/or displacements directly. These integral forms can be identified as (1) the principle of virtual displacements, (2) the principle of virtual forces, and (3) the principle of mixed virtual stresses and virtual displacements. The principles can be used directly to obtain exact solutions for simple structures, or they can be put into matrix forms and used directly for finite elements. Assumed functions can be put into the virtual principles to obtain a system of simultaneous equations, which correspond exactly to the Rayleigh-Ritz system of simultaneous equations.

The three virtual principles are derived in Chapter 2, Volume 1, and used throughout the text. Temperature and inelastic effects are included in the one dimensional forms. Comparisons of the three methods are made to show that for a given structure one method may be much simpler to use than the others.

The virtual principles are used to obtain differential equations and boundary conditions for beams and beam columns in Chapter 4, Volume 1, for vibration of beams in Chapter 6, Volume 2, and for plates in Chapter 7, Volume 2. They are used to obtain stresses and displacements in determinate and redundant trusses and beams in Chapter 3 and 4, Volume 1, respectively; to get deflections of box beam aircraft type structures in Chapter 5, Volume 1; to solve redundant shear lag problems in Chapter 6, Volume 1, and Chapter 2, Volume 2; to obtain approximate solutions using assumed functions in Chapters 5, 6, 7, and 8, Volume 2; to solve complex structures with finite elements and matrix equations in Chapters 3, 4, and 6, in Volume 1, and Chapters 5, 6, 8, and 9 in Volume 2; to assemble finite element matrices in Chapters 3, 4, and 6 in Volume 1 and Chapter 9 in Volume 2.

All the energy theorems are derived from the virtual principles in Appendix C, Volume 1.

Although the external forces on airplanes are described in Appendix B, Volume 1, and aircraft type structures are emphasized in the text, the virtual principles and the procedures given apply to all types of structures.

There are 230 solved examples in the two volumes so that they can be used not only as textbooks but also as reference books by practicing engineers in structural analysis. A pocket calculator was used to solve the examples and only a pocket calculator is needed to solve the 540 problems in the two volumes. Of course, computers can be used to solve the matrix equations for more complex structures with assumed functions or finite elements.

The material in Volume 1 together with Chapters 1 to 6 in Volume 2 originated from lecture notes used by the author in aircraft structures courses at the Ohio State University. The more advanced Chapters 7-10 (Volume 2) on plates, finite elements, and composite materials have been added. It is assumed that students have a knowledge of differential equations and mechanics of materials. Although not necessary, some knowledge of integral transforms (such as Laplace Transforms or Fourier Transforms) would help in understanding the derivation of the virtual principles in Chapter 2, Volume 1. The necessary matrix algebra is given in the Appendix A in the same volume.

As may be gathered from the above, the book is published in two volumes. Volume 1 contains subject matter usually covered in the two undergraduate courses in analysis of aircraft structures. Volume 2 contains the subject matter for the usual undergraduate design course as well as chapters on plates, dynamics, finite elements, and composite materials for a more advanced course.

B.E. GATEWOOD

List of Selected Equations in Volume 1 Referred to in Volume 2

In Volume 2, references are made to chapters, sections, figures, and equations in Volume 1. For the convenience of those using Volume 2, a number of the referenced equations least affected by being out of context are listed here.

$$\begin{aligned}\sigma_{xx,x} + \sigma_{xy,y} + \sigma_{xz,z} + X_x = 0, \\ \sigma_{xy,x} + \sigma_{yy,y} + \sigma_{yz,z} + X_y = 0, \\ \sigma_{xz,x} + \sigma_{yz,y} + \sigma_{zz,z} + X_z = 0,\end{aligned} \quad (1.4)$$

$$\sigma_{xx,x} = \frac{\partial \sigma_{xx}}{\partial x}, \quad \text{etc. for notation,}$$

$$\begin{aligned}\sigma_{xx}l + \sigma_{xy}m + \sigma_{xz}n = S_x, \\ \sigma_{xy}l + \sigma_{yy}m + \sigma_{yz}n = S_y, \\ \sigma_{xz}l + \sigma_{yz}m + \sigma_{zz}n = S_z,\end{aligned} \quad (1.5)$$

$$\begin{aligned}e_{xx} = u_{x,x}, \quad e_{yy} = u_{y,y}, \quad e_{zz} = u_{z,z}, \\ e_{xy} = u_{y,x} + u_{x,y}, \quad e_{xz} = u_{z,x} + u_{x,z}, \\ e_{yz} = u_{z,y} + u_{y,z}.\end{aligned} \quad (1.7)$$

$$\begin{aligned}u_x = u_{x0} - y\theta_z + z\theta_y, \quad u_y = u_{y0} - z\theta_x + x\theta_z, \\ u_z = u_{z0} - x\theta_y + y\theta_x.\end{aligned} \quad (1.10)$$

$$\begin{aligned}e_{xx} = (1/E)[\sigma_{xx} - \nu(\sigma_{yy} + \sigma_{zz})] + \alpha T, \\ e_{yy} = (1/E)[\sigma_{yy} - \nu(\sigma_{xx} + \sigma_{zz})] + \alpha T, \\ e_{zz} = (1/E)[\sigma_{zz} - \nu(\sigma_{xx} + \sigma_{yy})] + \alpha T, \\ e_{xy} = \sigma_{xy}/G, \quad e_{xz} = \sigma_{xz}/G, \quad e_{yz} = \sigma_{yz}/G.\end{aligned} \quad (1.11)$$

$$G = \frac{E}{2(1+\nu)}. \quad (1.12)$$

$$\sigma_{xx,x} + \sigma_{xz,z} + X_x = 0, \quad \sigma_{xz,x} + \sigma_{zz,z} + X_z = 0, \quad (1.21)$$

$$\sigma_{xx}l + \sigma_{xz}n = S_x, \quad \sigma_{xz}l + \sigma_{zz}n = S_z. \quad (1.22)$$

$$\sigma_{xx} = V + \phi_{,zz}, \quad \sigma_{zz} = V + \phi_{,xx}, \quad \sigma_{xz} = -\phi_{,xz}. \quad (1.24)$$

$$\sigma_{yz} = \frac{P_z}{2I_{zz}}(h^2 - z^2), \quad I_{zz} = t(2h)^3/12,$$

$$\sigma_{yy} = C_0 + C_3 z + (P_z/I_{zz})yz - E\alpha T(z),$$
(1.44)

$$\sigma_{yy} = (E/E_R)[-E_R e_E + (P_y + P_{yE})/A_E - (zI_{zzE} - xI_{zzE}) \times \\ \times (M_z + M_{zE})/H + (xI_{zzE} - zI_{zzE})(M_z + M_{zE})/H]. \quad (1.90)$$

$$\sigma_{yy} = (E/E_R)[-E_R e_E + (P_y + P_{yE})/A_E - (M_z + M_{zE})(z/I_{zzE}) + \\ + (M_z + M_{zE})(x/I_{zzE})]. \quad (1.91)$$

$$\sigma_{yy} = (E/E_R)[-E_R e_E + (P_y + P_{yE})/A_E - (M_z + M_{zE})(z/I_{zzE})]. \quad (1.92)$$

$$\sigma_{yy} = -M_z(z/I_{zz}). \quad (1.93)$$

$$E_{yy} e^*_{yy}/F_y = (\sigma_{yy}/F_y) + (3/7)(\sigma_{yy}/F_y)^n \\ = (\sigma_{yy}/F_y) + (E_{yy} e_p/F_y), \\ e_p = (3F_y/7E_{yy})(\sigma_{yy}/F_y)^n, \quad (1.94)$$

$$\int_V (R_{,x} S_{,x} + R_{,y} S_{,y} + R_{,z} S_{,z}) \, dV = \int_S R(lS_{,x} + mS_{,y} + nS_{,z}) \, dS - \\ - \int_V R(S_{,xx} + S_{,yy} + S_{,zz}) \, dV. \quad (2.2)$$

$$u_x = u_{xa} + z\theta_y, \quad u_y = u_{ya} - z\theta_x + x\theta_z, \quad u_z = u_{za} - x\theta_y, \quad (2.30)$$

$$\theta_z = \frac{du_{za}}{dy} = -\frac{du_y}{dz}, \quad \theta_z = -\frac{du_{za}}{dy} = \frac{du_y}{dx}. \quad (2.31)$$

$$(e_{xy})_g = u_{x,y}, \quad (e_{yx})_g = u_{x,y}. \quad (2.53a)$$

$$e_E = e_I + e_T + e_p \mp e_d, \quad (2.57)$$

$$\{e\} = [J] + \{e_E\}, \quad [J] = \frac{1}{E}\begin{bmatrix} 1 & -\nu & -\nu & 0 & 0 & 0 \\ -\nu & 1 & -\nu & 0 & 0 & 0 \\ -\nu & -\nu & 1 & 0 & 0 & 0 \\ 0 & 0 & 0 & \frac{E}{G} & 0 & 0 \\ 0 & 0 & 0 & 0 & \frac{E}{G} & 0 \\ 0 & 0 & 0 & 0 & 0 & \frac{E}{G} \end{bmatrix}. \quad (2.68)$$

$$\{P\} = [k]\{u\} - \{P_E\}, \quad (2.69)$$

$$[k] = \int_V [e^1][\sigma_1] \, dV, \quad \{P_E\} = \int_V [e^1][J]^{-1}\{e_E\} \, dV. \quad (2.70)$$

$$[k] = \int_V [\sigma_1]^T [J][\sigma_1] \, dV. \quad (2.71)$$

List of Selected Equations

$$\{u\} = [C]\{P\} + \{u_E\}, \tag{2.76}$$

$$[C] = \int_V [\sigma^1][e_1]\,\mathrm{d}V, \quad \{u_E\} = \int_V [\sigma^1]\{e_E\}\,\mathrm{d}V. \tag{2.77}$$

$$[M]\begin{Bmatrix} S \\ u \end{Bmatrix} = \begin{bmatrix} -[C_F] & [s] \\ [s]^T & [0] \end{bmatrix}\begin{Bmatrix} S \\ u \end{Bmatrix} = \begin{Bmatrix} v_E \\ P \end{Bmatrix}. \tag{2.86}$$

$$\{P\} = [s]^T\{S\}. \tag{3.38b}$$

$$[M]^{-1} = \begin{bmatrix} -k_S + k_S s_P k^{-1} s_P^T k_S & k_S s_P k^{-1} \\ k^{-1} s_P^T k_S & k^{-1} \end{bmatrix}, \tag{3.108}$$

$$[k] = [s]_P^T [k_S][s]_P, \quad [k_S] = [C_F]^{-1},$$

$$e_{yy}^V = -z\frac{\mathrm{d}^2 u_{zb}^V}{\mathrm{d}y^2}, \quad \theta_x^V = \frac{\mathrm{d}u_{zb}^V}{\mathrm{d}y}, \quad e_{yz}^V = \frac{\mathrm{d}u_{zs}^V}{\mathrm{d}y},$$
$$u_{zb}^V = u_{zm}^1 u_{zm}^V, \quad u_{zs}^V = u_{zsm}^1 u_{zm}^V, \tag{4.3}$$

$$\sigma_{yy}^V = -\frac{M_x^V z}{I_{xx}}, \quad \sigma_{yz}^V = \frac{V_z^V}{A_s}, \quad M_x^V = m_{xm}^1 P_{zm}^V, \quad V_z^V = p_{zm}^1 P_{zm}^V. \tag{4.4}$$

$$\int_0^L \int_A (\sigma_{yy} e_{yy}^V + \sigma_{yz} e_{yz}^V)\,\mathrm{d}A\,\mathrm{d}y - \int_0^L p_z u_z^V\,\mathrm{d}y - P_{zL} u_{zL}^V -$$
$$- M_{xL} \theta_{xL}^V - P_{z0} u_{z0}^V - M_{x0}\theta_{x0}^V - \sum_m P_{zm} u_{zm}^V = 0. \tag{4.5}$$

$$\frac{\mathrm{d}^2}{\mathrm{d}y^2}\left(EI_{xx}\frac{\mathrm{d}^2 u_{zb}}{\mathrm{d}y^2}\right) - p_z = 0, \tag{4.6}$$

$$\left(EI\frac{\mathrm{d}^2 u_{zb}}{\mathrm{d}y^2} - M_{xL}\right)_L = 0, \quad \left(EI\frac{\mathrm{d}^2 u_{zb}}{\mathrm{d}y^2} + M_{x0}\right)_0 = 0,$$
$$\left[\frac{\mathrm{d}}{\mathrm{d}y}\left(EI\frac{\mathrm{d}^2 u_{zb}}{\mathrm{d}y^2}\right) + P_{zL}\right]_L = 0, \quad \left[\frac{\mathrm{d}}{\mathrm{d}y}\left(EI\frac{\mathrm{d}^2 u_{zb}}{\mathrm{d}y^2}\right) - P_{z0}\right]_0 = 0, \tag{4.7}$$

$$u_z = (3u_{zL} - L\theta_{xL})(\tfrac{y}{L})^2 + (-2u_{zL} + L\theta_{xL})(\tfrac{y}{L})^3 +$$
$$+ (3u_{z0} + L\theta_{x0})(1 - \tfrac{y}{L})^2 + (-2u_{z0} - L\theta_{x0})(1 - \tfrac{y}{L})^3. \tag{4.15}$$

$$\int_0^L p_z u_z^V\,\mathrm{d}y + \sum_m P_{zm} u_{zm}^V + \sum_n M_{xn}\theta_{xn}^V$$
$$= P_{zLd} u_{zL}^V + M_{xLd}\theta_{xL}^V + P_{z0d} u_{z0}^V + M_{x0d}\theta_{x0}^V, \tag{4.18}$$

$$u_{zm} = \int_0^L \left(\frac{M_x m_{xm}^1}{EI_{xx}} + \frac{V_z p_{zm}^1}{GA_s}\right)\mathrm{d}y,$$
$$\theta_{xn} = \int_0^L \left(\frac{M_x m_{xn}^1}{EI_{xx}} + \frac{V_z p_{zn}^1}{GA_s}\right)\mathrm{d}y. \tag{4.39}$$

$$e_{total} = (\sigma_{yy}/E_{yy}) + e_E = -z\frac{d^2 u_{zb}}{dy^2},$$

$$\sigma_{yy} = -E_{yy}\left(z\frac{d^2 u_{zb}}{dy^2} + e_E\right) = \frac{E_{yy}}{E_R}[-E_R e_E - (M_x + M_{xE})\frac{z}{I_{xx}}]. \quad (4.62b)$$

$$\{P\} = [k]\{u\} - \{P_E\}, \quad (4.90)$$
$$[k] = [s]^T [k_S][s], \quad \{P_E\} = [s]^T \{S_E\},$$

$$\begin{Bmatrix} P_1 + P_{1d} + P_{1E} \\ P_2 + P_{2d} + P_{2E} \\ P_3 + P_{3d} + P_{3E} \\ P_4 + P_{4d} + P_{4E} \end{Bmatrix} = \left(\frac{EI}{L^3}\right)_1 \begin{bmatrix} 12 & -6L & -12 & -6L \\ -6L & 4L^2 & 6L & 2L^2 \\ -12 & 6L & 12 & 6L \\ -6L & 2L^2 & 6L & 4L^2 \end{bmatrix}_1 \begin{Bmatrix} u_1 \\ u_2 \\ u_3 \\ u_4 \end{Bmatrix}, \quad (4.91)$$

$$\{P\}_1 = [k]_1 \{u\}_1.$$

$$M_{xc} = -P_c u_{zb}, \quad \text{columns.} \quad (4.136)$$

$$P_{c,cr} = \pi^2 EI_{xx}/L^2 = \pi^2 EA/(L/\rho)^2, \quad \text{simple support.} \quad (4.141)$$

$$P_{c,cr} = \pi^2 EI_{xx}/4L^2, \quad \text{cantilever beam,}$$
$$P_{c,cr} = 4\pi^2 EI_{xx}/L^2, \quad \text{ends clamped.} \quad (4.143)$$

$$\int_{A_s} \sigma_{yzc}(e^V_{yzc})_g dA_s = -\int_{A_s} (P_c/A_s) u_{zb,y} u^V_{zb,y} dA_s$$
$$= -P_c u_{zb,y} u^V_{zb,y} \quad (4.147)$$

$$\left(EI u_{zb,yy}\right)_{,yy} + P_c u_{zb,yy} - p_z = 0. \quad (4.150)$$

$$\sigma_{zas,y} = V_z/GA_w. \quad (5.83)$$

$$\frac{d\theta_y}{dy} = M_y/GJ, \quad 1/GJ = (1/4A_c^2)\sum_{k=1}^{N}(s/Gt)_k. \quad (5.106)$$

The two basic criterions for the material in aircraft design are:

(A) The stress $k_L \sigma_L$, where σ_L is produced by the largest external forces expected during the life of the vehicle, shall not exceed the yield stress F_y of the material. k_L is a nondimensional factor to be specified for the design of a particular vehicle ($k_L = 1$ in many cases). This condition is expressed as a margin of safety (M.S.),

$$\text{M.S.} = (F_y/k_L \sigma_L) - 1 \geq 0.00, \quad (A). \quad (B.74)$$

The condition (A) is a condition to prevent permanent set of the structure during its life, where the stress σ_L is denoted as a limit stress and may be produced by the worst points on the V-n diagram.

List of Selected Equations

(B) The stress $k_u \sigma_L$ shall not exceed the ultimate failing stress F_u of the material during the life of the vehicle, where k_u is a specified factor for the design. The stress σ_L is the same as in condition (A) and $k_u = 1.50$ for many airplanes, but k_u may vary from 1.20 for some guided missiles to $k_u = 2.00$ for some critical types of external forces. The condition is expressed as

$$\text{M.S.} = (F_u/k_u \sigma_L) - 1 \geq 0.00, \quad \text{(B)}. \tag{B.75}$$

The condition (B) is a condition to prevent failure during the life of the airplane.

Both conditions (A) and (B) must be satisfied at every point in the structure of the vehicle. For any particular material with F_y and F_u known and k_L and k_u specified, it is evident that one of the conditions will be more critical than the other, giving a smaller M.S., or

$$\begin{aligned}(F_u/F_y) < (k_u/k_L), \quad \text{use (B)},\\ (F_u/F_y) > (k_u/k_L), \quad \text{use (A)}.\end{aligned} \tag{B.76}$$

1

Allowable stresses of flight vehicle materials

1.1. Introduction

In Volume 1, the primary concern has been the determination of the applied stresses in a given structure with given applied external forces. To complete the analysis it is necessary to compare the applied stresses to the stresses that the materials can take at yield and at failure. This chapter is concerned with the allowable stresses for typical metallic isotropic materials used in flight vehicle structures. Some discussion of allowable stresses for composite materials is given in Chapter 10. The primary source for many of the allowable stresses in flight vehicle metallic materials is Reference 1, Mil-HDBK-5c, "Metallic Materials and Elements for Aerospace Vehicle Structures". This standardization handbook is maintained as a joint effort of the Department of Defence and the Federal Aviation Agency. Selected tables of the allowable stresses from the 1962 edition of Reference 1 are given in Reference 2.

As given in Volume 1, the primary stresses are tension, compression and shear. Also, a bearing stress can be defined for use in joint analysis and design (see the next chapter). For each type of allowable stress for each material alloy, the tables in References 1 and 2 give F_{tu} (ultimate allowable tension stress), F_{ty} (yield allowable tension stress), F_{cy} (yield allowable compression stress), F_{su} (ultimate allowable shear stress), F_{bru} (ultimate allowable bearing stress), F_{bry} (yield allowable bearing stress). The ultimate allowable stress is the stress at failure of the material. The yield allowable stress is taken from the stress-strain curve as the interesection point of the 0.002 strain off-set line of slope E (tension) or E_c (compression) with the curve. The tables also give the moduli of elasticity E, E_c, G (shear), as well as the per cent elongation at failure and the weight density of the alloy material. Some of these tables from Reference 1 are given in Section 1.2 below for the aluminum alloys 2024 and 7075. Since the allowable stresses of the materials vary with temperature, various curves of allowable stresses plotted against temperature are given in References 1 and 2. Some of these curves for the aluminum alloys 2024 and 7075 are given in Section 1.3.

There are four allowable compression stresses, F_{cy}, F_{cr}, F_{cc}, and F_c, three of which depend upon the geometry and restraints of the compression structural member. Only the compression yield stress F_{cy} is given in the tables. The allowable compression local buckling stress F_{cr}, the allowable compression local crippling stress F_{cc}, and the allowable compressive column stress F_c must be calculated from the cross section dimensions, edge and end restraints, and the length of the structural member in compression. Formulas and curves for making these calculations of the various compression allowable stresses are given in Section 1.4.

Since in many cases several different types of stresses may be acting upon a structure at the same location and at the same time, it is necessary to calculate the allowable stresses or determine design curves for combined stresses. This combined case is considered in Section 1.5.

The allowable stresses are affected by various factors during the life of the structure. Besides the direct temperature effects mentioned above, a steady temperature can cause a strain increase by creep and failure at lower stress levels. Applied load cycles can cause fatigue of the material and reduced allowable stresses. Temperature also affects the fatigue stresses and may produce a combined creep and fatigue problem. Cracks in the material have large effects upon the allowable fatigue stresses. These effects are discussed in Sections 1.6 through 1.9.

1.2. Tension, shear, and bearing allowable stresses

The allowable stresses F_{tu}, F_{ty}, F_{cy}, F_{su}, F_{bru}, and F_{bry} can be obtained directly from test data. However, there are small variations in these stresses with the thickness, cross section area, width, and direction of loading. Also, the allowable stresses for any material alloy depend upon the processing and form of the material. All these items are included in the tables for allowable stresses in References 1 and 2. In the tables, L indicates the length direction of extrusions or direction of rolling of plates, LT indicates the long transverse direction, ST is the short transverse direction (thickness direction in plates). Values are also shown for the design basis, where basis S is a minimum specified value, basis A is the value above which at least 99% of test values should fall with a confidence of 95%, and basis B is the value above which at least 90% of the test values should fall with a confidence of 95%. The B basis values may be used if there are several load paths in the structure or several elements of the same material are used together, such as a group of stringers on the wing or fuselage.

It should be noted that the tables give no yield stress for shear. This is due to the fact that no satisfactory yield stress can be obtained in shear tests. Also, the yield stress F_{cy} in compression is used only as a reference stress in the calculation of the allowable compressive stresses to be given in Section 1.4.

Since the aluminum alloys 2024 and 7075 are extensively used in flight vehicle structures, a few tables for these alloys have been taken from the 1966 edition of Reference 1, and are given below. Data from these Tables 1.1 through 1.6 will be used in examples and problems in this and later chapters on design. Of course, on the job values from the latest edition should be used. Here, any values can be used to demonstrate the analysis and design procedures. Also, the values in the Tables are in ksi = 1000 psi = 1000 lb/in² and since these values are based upon tests of plates with standard thicknesses in inches, it is necessary to use lb and in. units in Chapter 1.

Table 1.1. Aluminum alloy 2024 plate.

Condition	-T4	-T4	-T42	-T42	-T42	-T851	
Thickness (in)	0.010-0.020	0.021-0.249	0.010-0.449	0.500-1.000	1.001-2.000	0.250-0.499	
Basis	S	A	B	S	S	S	A	B
F_{tu},ksi,L	62	62	66	62	61	60	67	68
F_{tu}, LT	62	62	66	62	61	60	67	68
F_{ty},ksi,L	40	40	41	38	38	38	59	61
F_{ty}, LT	40	40	41	38	38	38	58	60
F_{cy},ksi	40	40	41	38	38	38	59	61
F_{su},ksi	37	37	40	37	37	36	38	39
F_{bru},ksi								
$e/D = 1.5$	93	93	99	93	92	90	102	104
$e/D = 2.0$	118	118	126	118	116	114	131	132
F_{bry},ksi								
$e/D = 1.5$	56	56	57	53	53	53	87	90
$e/D = 2.0$	64	64	66	61	61	61	102	105
e,percent	12	15	-	-	8	-	5	-
E_t,psi		10,500,000						
E_c,psi		10,700,000						
G,psi		4,000,000						
ν		0.33						
Density, lb/in³		0.100						

Table 1.2. Aluminum alloy clad 2024 plate.

Condition	-T3	-T3		-T3		-T36		-T36		-T42	
Thickness (in)	0.008-0.009	0.010-0.062		0.063-0.249		0.020-0.062		0.063-0.499		0.010-0.062	
Basis	S	A	B	A	B	A	B	A	B	A	B
F_{tu},ksi,L	59	60	62	63	65	63	66	67	69	57	59
F_{tu}, LT	58	59	61	62	64	62	65	66	68	57	59
F_{ty},ksi,L	45	45	47	46	48	55	58	58	60	34	35
F_{ty}, LT	39	39	41	40	42	48	50	50	52	34	35
F_{cy},ksi,L	37	37	39	38	40	46	48	48	49	34	35
F_{cy}, LT	42	42	44	43	45	51	54	54	56	34	35
F_{su},ksi	37	38	39	40	41	39	41	41	42	34	35
F_{bru},ksi											
$e/D = 1.5$	89	90	93	95	98	95	99	101	104	86	89
$e/D = 2.0$	112	114	118	120	124	122	125	127	131	108	112
F_{bry},ksi											
$e/D = 1.5$	64	64	67	64	69	77	81	81	84	48	49
$e/D = 2.0$	73	73	76	74	78	88	93	93	96	54	56
e,percent	10	-	-	15	-	8	-	9	-	-	-
E_t,psi, primary		10,500,000									
E_t,psi, secondary		9,500,000									
E_c,psi, primary		10,700,000									
E_c,psi, secondary		9,700,000									
G, ν, density (see table 1.1)											

Table 1.3. Aluminum alloy 2024 extrusions.

Condition	-T6		-T6		-T6		-T6		-T6		-T42
Thickness (in)	0.050–0.249		0.250–0.499		0.500–0.749		0.750–1.499		1.500–2.999		all
Basis	A	B	A	B	A	B	A	B	A	B	S
F_{tu},ksi,L	57	61	60	62	60	62	65	70	70	74	57
F_{tu},ksi,T	57	61	60	62	62	62	58	61	54	57	50
F_{ty},ksi,L	42	47	44	47	44	47	46	54	52	54	38
F_{ty},ksi,T	42	46	43	46	42	45	41	44	38	41	36
F_{cy},ksi,L	38	41	39	42	39	42	44	52	50	52	38
F_{cy},ksi,T	38	41	39	42	39	42	42	48	42	44	38
F_{su},ksi	30	32	32	33	32	33	34	38	38	40	30
F_{bru},ksi											
$e/D = 1.5$	85	91	85	91	85	91	85	91	85	91	85
$e/D = 2.0$	108	114	108	114	108	114	108	114	108	114	108
F_{bry},ksi											
$e/D = 1.5$	59	66	60	66	60	66	61	66	62	66	53
$e/D = 2.0$	67	75	69	75	69	75	71	75	73	75	61
e,percent,											
L	12	-	12	-	12	-	10	-	10	-	-
T	-	-	6	-	6	-	5	-	2	-	-
E_t, psi	10,500,000										
E_c, psi	10,700,000										
G, psi	4,000,000										
ν	0.33										
Density, lb/in^3	0.100										

Table 1.4. Aluminum alloy 7075 plate.

Condition	-T6		-T6		-T651		-T651		-T651	
Thickness (in)	0.015–0.039		0.040–0.249		0.250–0.499		0.500–1.000		1.001–2.000	
Basis	A	B	A	B	A	B	A	B	A	B
F_{tu},ksi,L	76	78	77	79	76	78	76	78	76	78
F_{tu}, LT	76	78	77	79	77	79	77	79	77	79
F_{ty},ksi,L	66	69	67	70	68	70	68	70	68	70
F_{ty}, LT	65	68	66	69	66	68	66	68	66	68
F_{cy},ksi,L	67	70	68	71	67	69	66	68	65	67
F_{cy}, LT	70	73	71	74	70	72	70	72	70	72
F_{su},ksi	46	47	46	47	43	44	44	45	45	46
F_{bru},ksi										
$e/D=1.5$	114	117	116	119	117	120	117	120	117	120
$e/D=2.0$	144	148	146	150	144	148	144	148	144	148
F_{bry},ksi										
$e/D=1.5$	92	97	94	98	97	100	98	101	100	103
$e/D=2.0$	106	110	107	112	114	117	115	118	117	121
e,percent										
L	7	-	8	-	8	-	6	-	5	-
LT	7	-	8	-	8	-	6	-	4	-
E_t,psi	10,300,000									
E_c,psi	10,500,000									
G, psi	3,900,000									
ν	0.33									
Density,lb/in^3	0.101									

Table 1.5. Aluminum Alloy clad 7075 plate.

Condition	-T6		-T6		-T6		-T6		-T651		-T651	
Thickness (in)	0.015-0.039		0.040-0.062		0.063-0.187		0.188-0.249		0.250-0.499		0.500-1.000	
Basis	A	B	A	B	A	B	A	B	A	B	A	B
F_{tu},ksi,L	70	73	72	74	73	75	75	77	74	76	74	76
F_{tu}, LT	70	73	72	74	75	75	75	77	75	77	75	77
F_{ty},ksi,L	61	64	63	65	64	66	65	67	66	68	66	68
F_{ty}, LT	60	63	62	64	63	65	64	66	64	66	64	66
F_{cy},ksi,L	62	65	64	66	65	67	66	68	65	67	64	66
F_{cy}, LT	64	67	66	68	67	69	68	70	68	70	68	70
F_{su},ksi	42	44	43	44	44	45	45	46	42	43	43	44
F_{bru},ksi												
$e/D=1.5$	105	110	108	111	110	112	112	116	114	117	114	117
$e/D=2.0$	133	139	137	141	139	142	142	146	140	144	140	144
F_{bry},ksi												
$e/D=1.5$	85	90	88	91	90	92	91	94	94	97	95	98
$e/D=2.0$	98	102	101	104	102	106	104	107	110	114	111	115
e,percent												
L	7	-	8	-	8	-	8	-	6	-	5	-
LT	7	-	8	-	8	-	8	-	6	-	4	-
E_t,psi,primary					10.300.000							
E_t,psi,secondary					9,700,000							
E_c,psi,primary					10,500,000							
E_c,psi,secondary					9,900,000							
G, ν, density (see Table 1.4)												

Table 1.6. Aluminum alloy 7075 extrusions.

Condition	-T6		-T6		-T6		-T6		-T6	
Thickness (in)	up to 0.249		0.250-0.499		0.500-0.749		0.750-1.499		1.500-2.999	
Basis	A	B	A	B	A	B	A	B	A	B
F_{tu},ksi,L	78	82	81	85	81	85	81	85	81	85
F_{tu}, LT	76	78	77	79	73	75	72	74	66	68
F_{ty},ksi,L	70	74	73	77	72	76	72	76	72	76
F_{ty}, LT	64	67	66	68	63	66	62	64	57	59
F_{cy},ksi,L	71	75	74	78	73	77	72	76	72	77
F_{cy}, LT	71	75	74	78	73	77	72	75	69	72
F_{su},ksi	43	45	45	47	45	47	45	47	45	47
F_{bru},ksi										
$e/D=1.5$	101	107	97	102	97	102	97	102	97	102
$e/D=2.0$	125	131	130	136	130	136	130	136	130	136
F_{bry},ksi										
$e/D=1.5$	91	96	80	85	79	84	79	84	79	84
$e/D=2.0$	98	104	102	108	101	106	101	106	101	106
e,percent										
L	7	8	7	8	7	8	7	8	7	8
LT	5	-	5	-	4	-	3	-	1	-
E_t,psi			10,300,000							
E_c,psi			10,500,000							
G, psi			3,900,000							
Density,(lb/in^3)			0.101							

1.3. Temperature effects on allowable stresses

The allowable stresses for metallic materials decrease with increase in temperature. Since large increases in temperature occur in high speed flight in the atmosphere and in the engines which propel the vehicle, it is necessary to test the materials at various temperatures to obtain the allowable stresses. Curves of the allowable stresses as a per cent of the room temperature allowable stresses with temperature as the abscissa are given in References 1 and 2 for various alloys. The allowable stresses also vary with the time for which the material is held at a given temperature, or the exposure time. This exposure time is shown as a parameter on the curves. Since the material recovers part of its strength at room temperature after being exposed for a certain time at elevated temperature, curves are also given for these recovery allowable stresses. Selected curves for some of the allowable stresses for the aluminum alloys 2024 and 7075 have been taken from the 1966 edition of Reference 1, and are given in Figures 1.1 through 1.13. Data from these curves will be used in examples and problems.

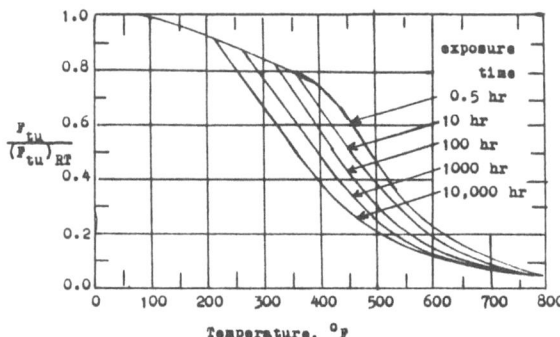

Fig. 1.1. Elevated temperature F_{tu} (ultimate) for 2024-T3 and 2024-T4 aluminum alloy.

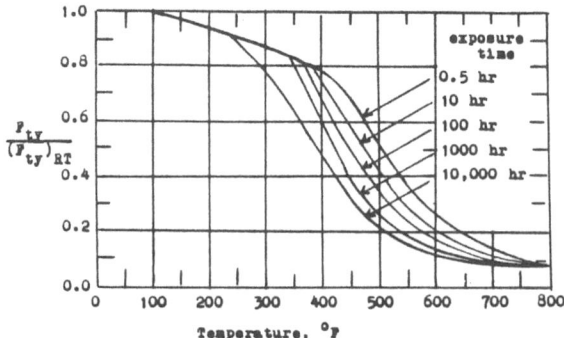

Fig. 1.2. Elevated temperature F_{ty} (yield) for 2024-T3 and 2024-T4 aluminum alloy.

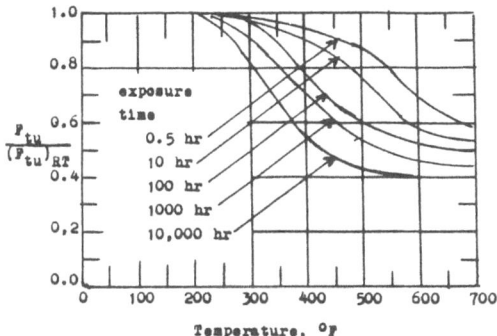

Fig. 1.3. Room temperature F_{tu} after temperature exposure for 2024-T3 and 2024-T4 aluminum alloy.

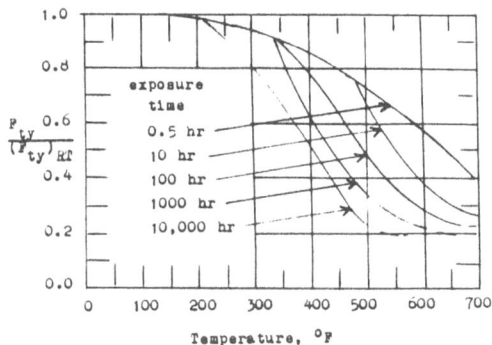

Fig. 1.4. Room temperature F_{ty} after temperature exposure for 2024-T3 and 2024-T4 aluminum alloy.

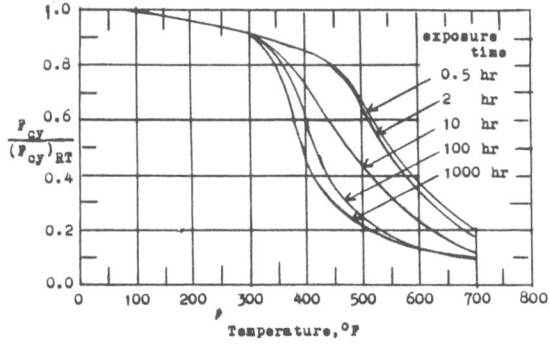

Fig. 1.5. Elevated temperature F_{cy} (yield) for clad 2024-T3 and clad 2024-T4 sheet.

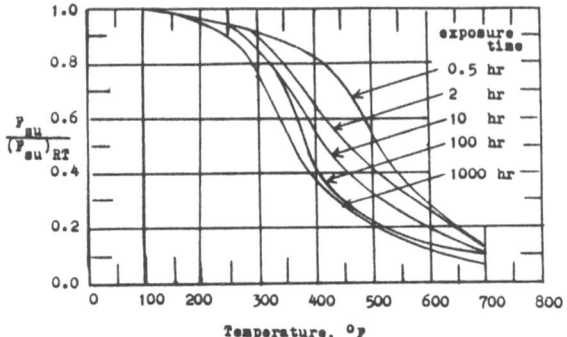

Fig. 1.6. Elevated temperature F_{su} (ultimate) for clad 2024-T3 and clad 2024-T4 sheet.

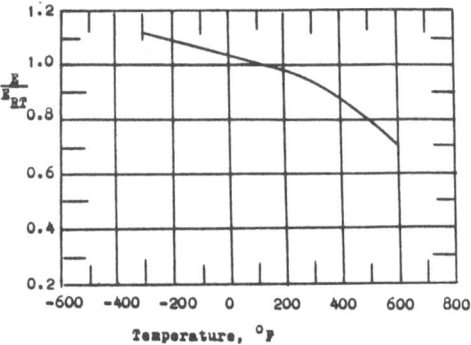

Fig. 1.7. Temperature effects on E_t and E_c for 2024 aluminum alloy.

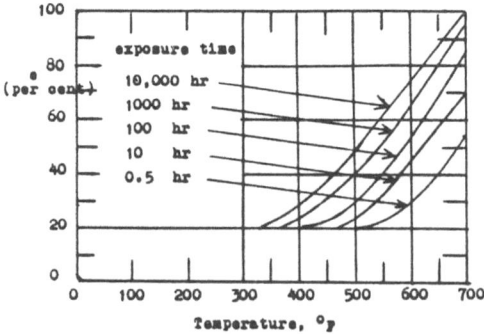

Fig. 1.8. Elevated temperature elongation e for 2024-T3 and 2024-T4 aluminum alloy, except extrusions.

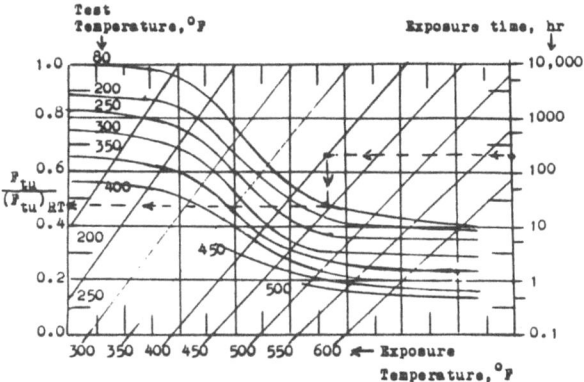

Fig. 1.9. Temperature effects on F_{tu} (ultimate) for 7075-T6 aluminum alloy.

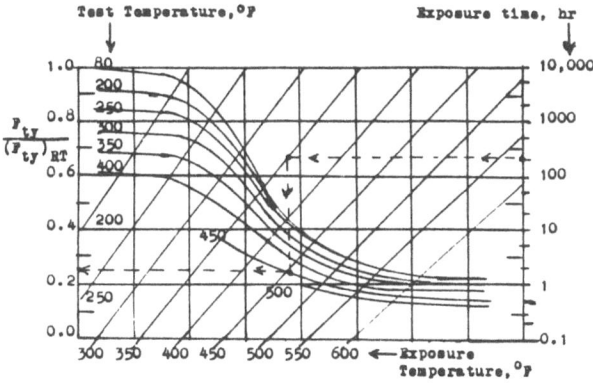

Fig.1.10. Temperature effects on F_{ty} (yield) for 7075-T6 aluminum alloy.

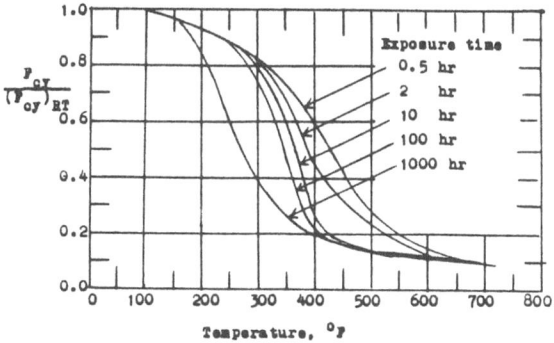

Fig. 1.11. Elevated temperature F_{cy} (yield) for 7075-T6 aluminum alloy.

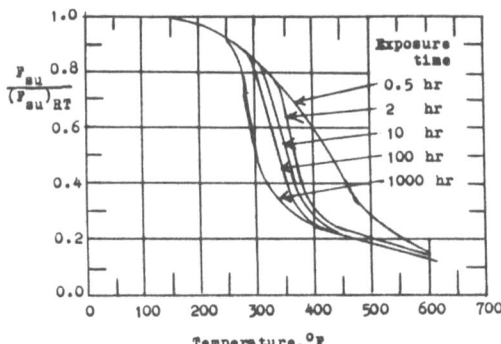

Fig. 1.12. Elevated temperature F_{su} (ultimate) for 7075-T6 aluminum alloy.

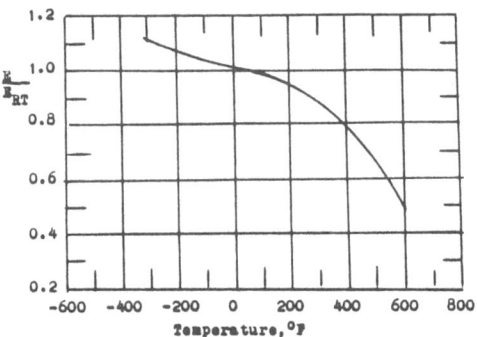

Fig. 1.13. Temperature effects on E and E_c for 7075-T6 aluminum alloy.

Example 1.1. Find the allowable stresses F_{tu}, F_{ty}, F_{su}, and F_{cy} for a clad 2024-T3 aluminum alloy plate of thickness $t = 0.150$ in. at $T = 300°F$. Use B basis and L direction with 10 hour exposure time at 300°F.

Solution. From Table 1.2 at room temperature the allowable stresses are $(F_{tu})_{R.T.} = 65,000$ psi, $(F_{ty})_{R.T.} = 48,000$ psi, $(F_{su})_{R.T.} = 41,000$ psi, $(F_{cy})_{R.T.} = 40,000$ psi. From Figure 1.1 at 300°F and 10 hour exposure time, $F_{tu}/(F_{tu})_{R.T.} = 0.84$ or $F_{tu} = 0.84(65,000) = 54,600$ psi. From Figure 1.2 at 300°F and 10 hour exposure, $F_{ty}/(F_{ty})_{R.T.} = 0.87$ or $F_{ty} = 0.87(48,000) = 41,800$ psi. From Figure 1.6 at 300°F and 10 hour exposure, $F_{su}/(F_{su})_{R.T.} = 0.86$ or $F_{su} = 0.86(41,000) = 35,300$ psi. From Figure 1.5 at 300°F and 10 hour exposure, $F_{cy}/(F_{cy})_{R.T.} = 0.92$ or $F_{cy} = 0.92(40,000) = 36,800$ psi.

Example 1.2. Repeat Example 1.1 for a clad 7075-T6 sheet or plate under the same conditions.

Solution. From Table 1.5 the room temperature values are $(F_{tu})_{R.T.} = 75,000$ psi, $(F_{ty})_{R.T.} = 66,000$ psi, $(F_{su})_{R.T.} = 45,000$ psi, $(F_{cy})_{R.T.} = 67,000$ psi, which are much larger than the corresponding clad 2024-T3 values. From Figure 1.9 at 300°F and 10 hour exposure, $F_{tu}/(F_{tu})_{R.T.} = 0.70$ (go horizontal on 10 hour line to 300°F exposure line and up to 300°F test curve), or $F_{tu} = 0.70(75,000) = 52,500$ psi, which is less than the 2024-T3 value. From Figure 1.10 at 300°F and 10 hour exposure, $F_{ty}/(F_{ty})_{R.T.} = 0.76$, or $F_{ty} = 0.76(66,000) = 50,200$ psi. From Figure 1.12 at 300°F and 10 hour exposure, $F_{su}/(F_{su})_{R.T.} = 0.85$, or $F_{su} = 0.85(45,000) = 38,200$ psi. From Figure 1.11 at 300°F and 10 hour exposure, $F_{cy}/(F_{cy})_{R.T.} = 0.79$, or $F_{cy} = 0.79(67,000) = 52,900$ psi. Note that the F_{cy} compressive yield stress is as large as the ultimate tension stress for this case.

1.4. Allowable compression stresses

The allowable compressive stresses F_{cr} (local buckling stress), F_{cc} (local crippling stress), and F_c (column stress) depend upon a large displacement stability analysis involving the interaction between the applied compression force and the lateral deflection of the structural member in compression. The derivation and solution of the differential equations for the allowable elastic local buckling stress F_{cr} and the elastic column stress F_c are considered later in Chapter 7. These elastic formulas from Chapter 7 will be modified to account for the inelastic effects by using theory and test data. The allowable crippling stress F_{cc} will be based on an empirical formula derived from test data.

To simplify the calculations and to reduce the number of graphs needed to determine the allowable compression stresses, it is necessary to use a non-dimensional form for the stresses. The yield stress F_{cy}, as given in Tables 1.1 through 1.6, can be used as the reference stress so that

$$\begin{aligned} S_{cr} &= F_{cr}/F_{cy}, \quad S_{cc} = F_{cc}/F_{cy}, \quad S_c = F_c/F_{cy}, \\ S_{crE} &= F_{crE}/F_{cy}, \quad S_{cE} = F_{cE}/F_{cy}, \end{aligned} \tag{1.1}$$

will be used for the non-dimensional allowable compression stresses. Here, F_{crE} and F_{cE} are the elastic formula values for F_{cr} and F_c, and all the non-dimensional stresses S_{cr}, S_{cc}, S_c will be related to S_{crE} and S_{cE} by graphs and equations.

(a) *Allowable local buckling stresses* F_{cr} *or* S_{cr}

If the cross section is a thin flat plate or has thin plate elements such as angle or channel sections, then under a compressive force the plate elements tend to buckle in waves, giving a local allowable buckling stress F_{cr} practically independent of the length L. From Equation (7.167), Section 7.10, Examples 7.8 and 7.9, the formula for calculating the S_{crE} and F_{crE} values is

$$S_{crE} = F_{crE}/F_{cy} = (KE_c/F_{cy})(t/b)^2, \tag{1.2}$$

where t is the thickness and b the width of a selected flat element on the cross section and K depends on the shape of the cross section, the length in some cases, and the selected t and b values. E_c and F_{cy} are given in tables 1.1 through 1.6.

Procedures for the calculation of K for various types of cross sections are given in References 3, 4, and 5. Figures 1.14 through 1.18 show curves for K values for plates, angle, channels, and stiffened panels. Other cases, as well as the ones shown here, can be found in various publications such as References 2, 7-9, and 11. Also, further references are given in References 9 and 11. Thus, for a given material and a given cross section for which K can be obtained, the elastic buckling stress S_{crE} can be calculated directly from Equation (1.2).

To obtain S_{cr} in both the elastic and inelastic regions of the material, take

$$S_{cr} = R_{cr}S_{crE}, \tag{1.3}$$

where R_{cr} is 1.00 in the elastic range and varies in the inelastic range.

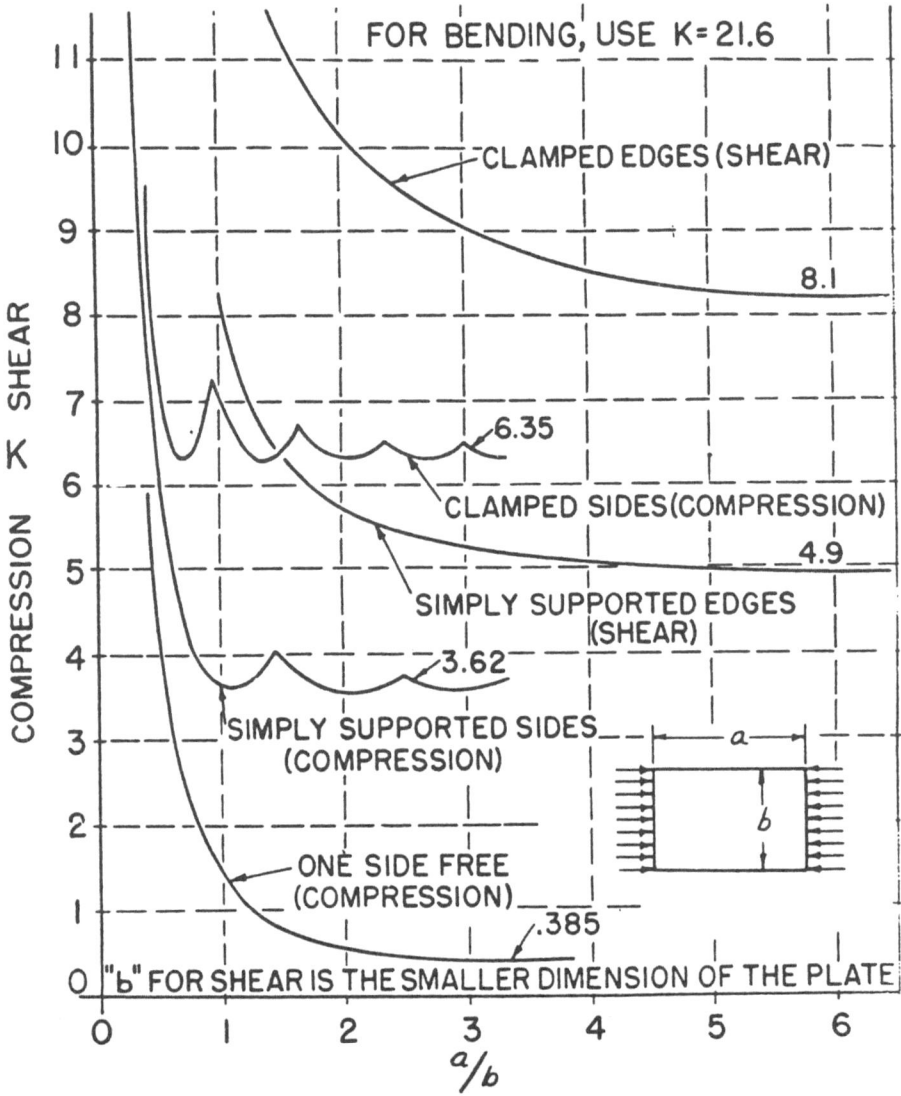

Fig. 1.14. K for compression, shear, and bending of flat plates.

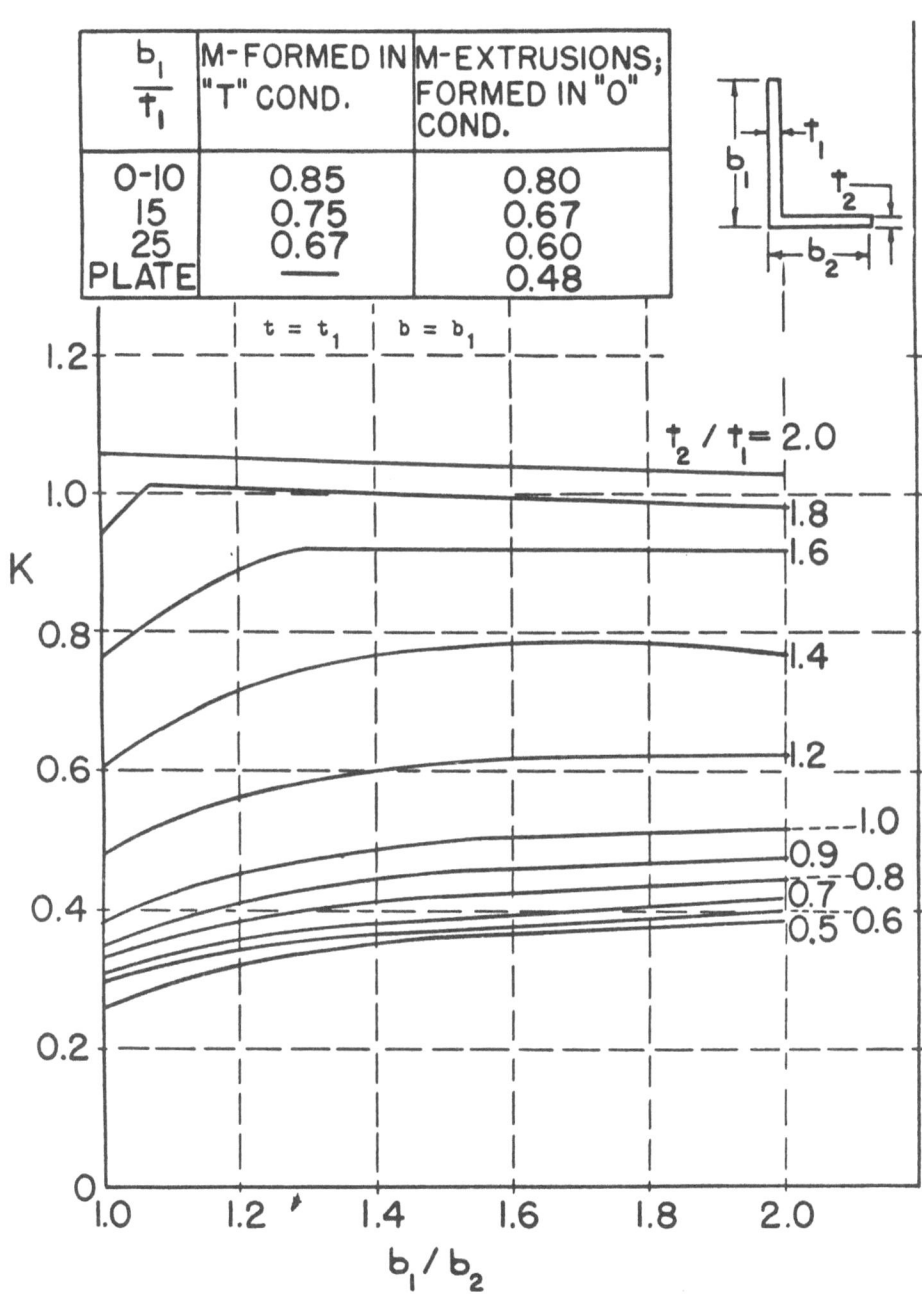

Fig. 1.15. *K* and *M* values for angles in compression.

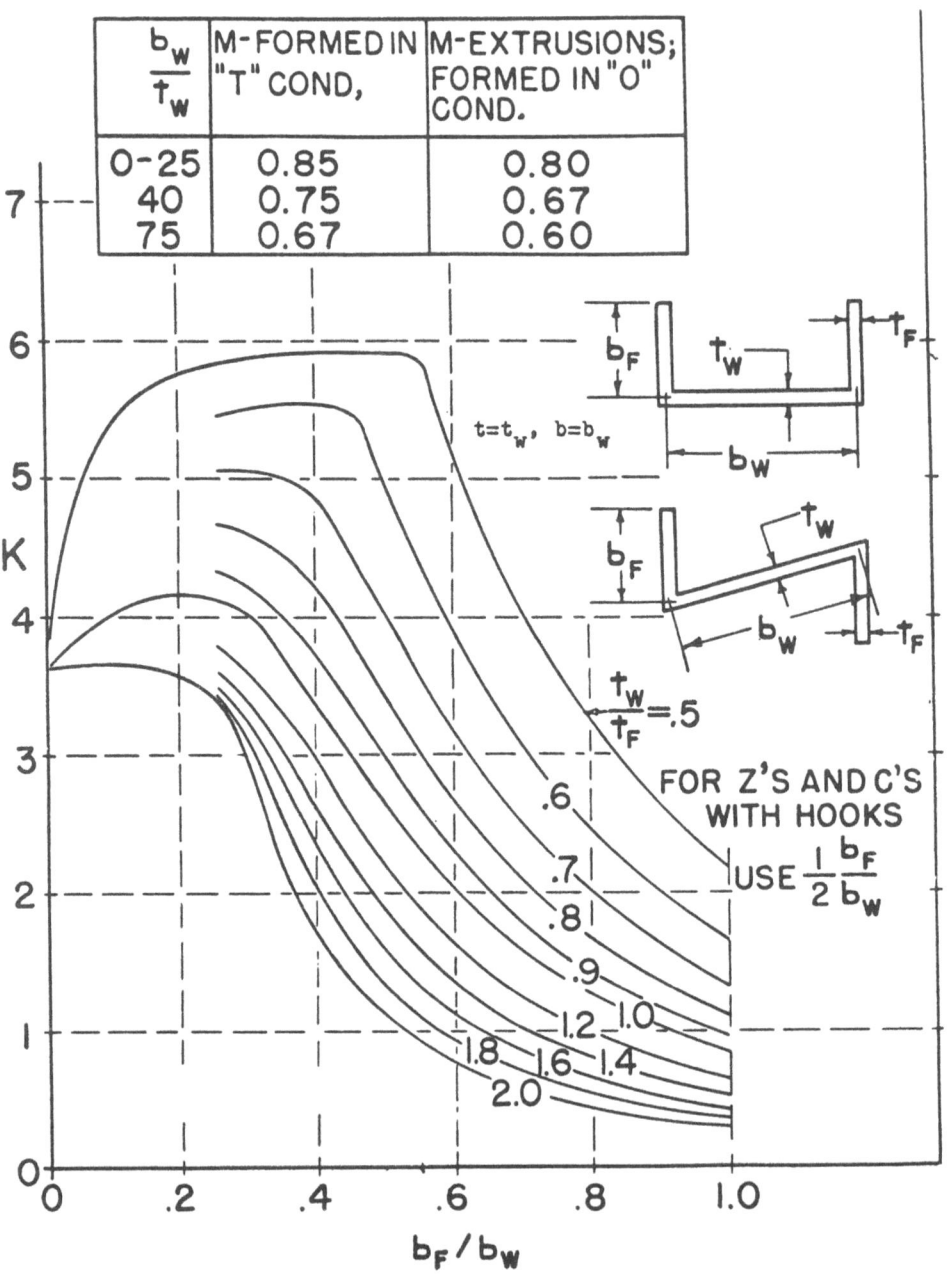

Fig. 1.16. *K* and *M* values for channels and zees in compression.

Allowable stresses of flight vehicle materials

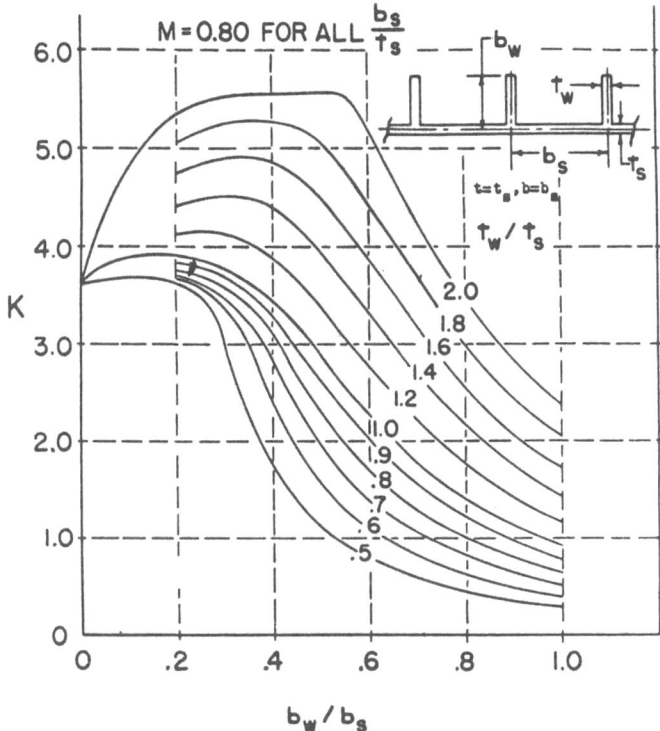

Fig. 1.17. K and M values for angle-stiffened panels in compression.

Various expressions for R_{cr} are given in References 6, 10, and 11 for different types of cross sections. Only one of these expressions, which checks the test data for most practical cross sections, will be used here, or

$$R_{cr} = \frac{E_{cs}}{E_c}\left[0.50 + 0.25\left(1 + 3\frac{E_{ct}}{E_{cs}}\right)^{1/2}\right]. \quad (1.4)$$

E_{ct} is the tangent modulus of elasticity or the slope of the stress-strain curve at any point, and E_{cs} is the secant modulus of elasticity or the slope of the line from the origin to any point on the stress-strain curve. In the elastic range up to the proportional limit, $E_{cs} = E_{ct} = E_c$.

Typical stress-strain and E_{ct} curves for some of the alloys are given in Reference 1 so that E_{cs}, E_{ct}, and R_{cr} can be calculated for a given S_{cr}. However, in a given case, S_{crE} is known and S_{cr} is unknown so that it is necessary to assume various S_{cr}, calculate R_{cr} and S_{crE}, and draw a S_{cr} against S_{crE} curve from which S_{cr} can be read for the given S_{crE}. Rather than to construct curves for each alloy at each temperature, it is simpler to use the Ramberg-Osgood non-dimensional stress-strain curves and construct the S_{cr} against S_{crE} curves for a few cases which can be used for all the alloys and temperature cases.

From Equation (1.94, Vol. 1), the Ramberg-Osgood stress-strain equation can

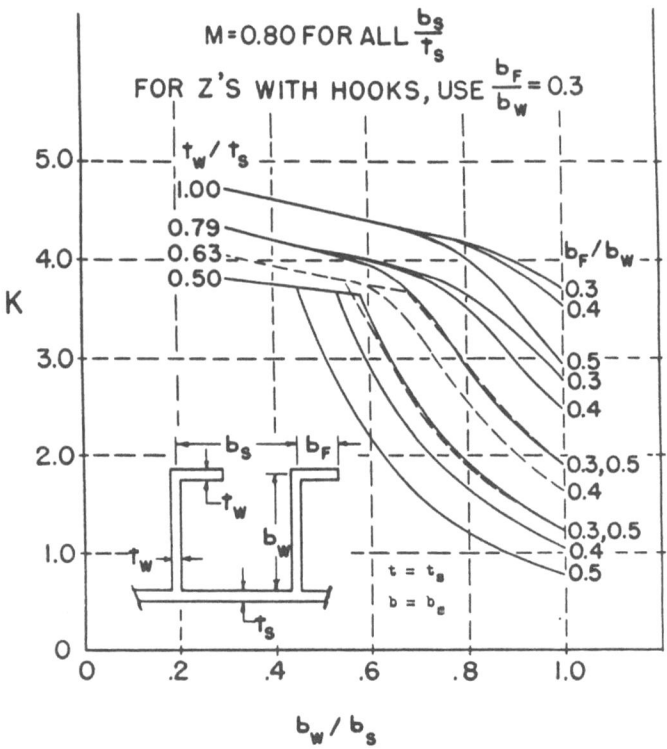

Fig. 1.18. K and M values for Z-stiffened panels in compression.

be written as

$$e_c E_c / F_{cy} = S_{cr} + (3/7)(S_{cr})^n, \tag{1.5}$$

whence

$$\frac{E_{cs}}{E_c} = \frac{F_{cr}}{e_c E_c} = \frac{F_{cy} S_{cr}}{e_c E_c} = \frac{1}{1 + (3/7)(S_{cr})^{n-1}}, \tag{1.6}$$

$$\frac{E_{ct}}{E_c} = \frac{1}{E_c} \frac{\mathrm{d} F_{cr}}{\mathrm{d} e_c} = \frac{1}{E_c} \left(\frac{\mathrm{d} e_c}{\mathrm{d} F_{cr}} \right)^{-1} = \frac{1}{1 + (3/7)n(S_{cr})^{n-1}}. \tag{1.7}$$

For a given n, a range of S_{cr} values can be assumed and E_{cs}, E_{ct}, R_{cr}, and S_{crE} calculated directly from Equations (1.6), (1.7), (1.4), and (1.3). Since it is necessary to read S_{cr} from the curves in the inelastic region for a calculated S_{crE}, it is better to graph S_{cr} against a parameter B_{cr} rather than directly against S_{crE} in Equation (1.3). Rewrite Equation (1.3) as

$$S_{cr} = R_{cr}/B_{cr}^2, \qquad B_{cr} = (S_{crE})^{-1/2}. \tag{1.8}$$

Figure 1.19 shows curves of S_{cr} against B_{cr} for a range of n values that covers most alloys at room and elevated temperatures. Interpolation can be made for n values between the ones shown. Typical values of n for the alloys in Tables 1.1 through 1.6 are given in Equation (1.9).

Allowable stresses of flight vehicle materials

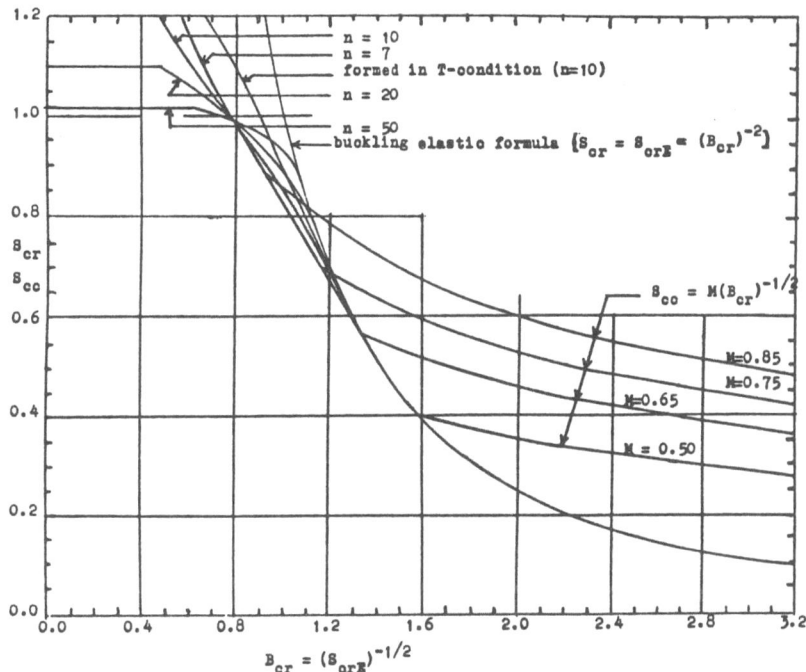

Fig. 1.19. Allowable buckling and crippling stresses.

It should be noted that the yield stress F_{cy} used in the above discussion is assumed to be constant on the cross section. This is true for extrusions and for plates formed in the annealed, or 0-condition, and then heat treated. However, for plates formed in the heat treated, or T-condition, F_{cy} increases in the corners so that the average F_{cy} is larger than the Table values. Test data on 2024 alloys indicates higher buckling stresses in the inelastic range for sections formed in the T-condition. Rather than try to evaluate the average F_{cy}, it is simpler to use a special curve on Figure 1.19 for this case. See Figure 1.19 for the curve marked "formed in T-condition $(n = 10)$".

Test data shows that the local buckling stress F_{cr} reaches a maximum at or slightly above the yield stress, depending upon the alloy and the n value for the stress-strain curve. The cut-offs are indicated on Figure 1.19 for $n = 10$, 20, and 50. The values of n for the alloys given in Tables 1.1 through 1.6 are approximately

$$n = 10 \text{ in Tables } 1.1, 1.2,$$
$$n = 15 \text{ in Tables } 1.3, 1.4, 1.5, \tag{1.9}$$
$$n = 20 \text{ in Table } 1.6.$$

It should be noted that the curves in Figure 1.19 show that the buckling stresses S_{cr} change slowly with the parameter n so that only approximate values for n are needed. The approximate n for each alloy at each temperature can be determined from the stress-strain curves in Reference 1 by plotting the non-dimensional Ramberg-Osgood terms on a set of non-dimensional stress-strain curves with various n values.

Example 1.3. Find the buckling stress F_{cr} for a clad 2024-T3 channel section in Figure 1.16 formed in the T-condition with L-direction, basis A, $t_w = t_F = 0.080$ in., $b_w = 2.00$ in., $b_F = 1.00$ in.

Solution. From Table 1.2, $F_{cy} = 38,000$ psi and $E_c = 9.7(10)^6$ psi (use the secondary modulus for all cases). On Figure 1.16, $b_F/b_w = 0.50$, $t_w/t_F = 1.00$, whence $K = 2.7$. Thus, Equations (1.2) and (1.8) give

$$S_{crE} = \frac{2.7(9.7)(10^6)}{38,000}\left(\frac{0.080}{2.00}\right)^2 = 1.12,$$

$$B_{cr} = (S_{crE})^{-1/2} = 0.94.$$

From Equation (1.9), take $n = 10$ and read S_{cr} on Figure 1.19 as

$$S_{cr} = 0.89, \quad F_{cr} = (0.89)(38,000) = 33,800 \text{ psi}.$$

(b) *Allowable local crippling stresses* F_{cc} *or* S_{cc}

After a plate element buckles, the edges (restrained to remain straight) and the corners of angle sections may take additional load until they reach the yield stress or beyond. As the load is increased the stress distribution may have a variable shape as shown in Figure 1.20. When the edges or corners fail, the cross section fails and the average stress at failure is the local F_{cc} crippling stress.

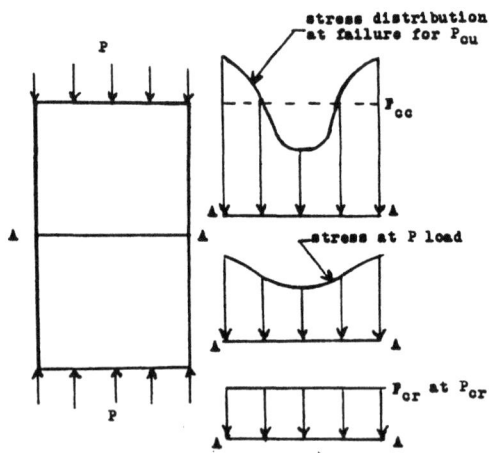

Fig. 1.20. Stress distributions beyond buckling.

Considerable work, References 15 and 16, has been done to calculate F_{cc} from theory but only very simple cases have been solved. To cover various types of cross sections various empirical formulas have been developed for F_{cc} or S_{cc}.

Allowable stresses of flight vehicle materials

From References 7-9 and 13, an approximate formula based on nearly 2000 tests on various cross sections and various materials is

$$S_{cc} = F_{cc}/F_{cy} = M(S_{crE})^{1/4} = M(B_{cr})^{-1/2}, \quad S_{cc} \geq S_{cr},$$
$$S_{cc} = S_{cr}, \quad \text{otherwise,} \tag{1.10}$$

where M depends on the geometry of the cross section. M is shown on Figures 1.15 through 1.18 for the particular cross sections on those figures. The S_{cc} curves from Equation (1.10) are shown on Figure 1.19 for several values of M covering the M range. When these S_{cc} curves intersect the S_{cr} curves on Figure 1.19, then $S_{cc} = S_{cr}$ for smaller B_{cr} or larger S_{crE}. Although the $S_{cc} > S_{cr}$ case can be read from the curves, it is simpler to calculate S_{cc} from Equation (1.10) for this case.

For stiffened panels, such as in Figures 1.17 and 1.18, it should be noted that the rivet size and spacing in the attachment of the stiffeners to the plate can affect the F_{cc} crippling stress. See References 8 and 14 for correction factors to use in Equation (1.10). For Equation (1.10) to apply directly the rivets should be large and closely spaced.

Example 1.4. Find the buckling F_{cr} and crippling F_{cc} stresses for a 7075-T6 angle extrusion with L-direction, B basis, $b_1 = 1.50$ in., $b_2 = 1.00$ in., $t_1 = 0.060$ in., $t_2 = 0.040$ in.

Solution. From Table 1.6, $F_{cy} = 75,000$ psi, $E_c = 10.5(10^6)$ psi. On Figure 1.15, $b_1/b_2 = 1.50$, $t_2/t_1 = 0.67$, whence $K = 0.38$. Thus Equation (1.3) gives

$$S_{crE} = \frac{0.38(10.5)(10^6)}{75,000}\left(\frac{0.060}{1.50}\right)^2 = 0.085.$$

From Figure 1.19, $S_{cr} = S_{crE} = 0.085$ independent of n, whence

$$F_{cr} = 0.085(75,000) = 6400 \text{ psi.}$$

On Figure 1.15, $b_1/t_1 = 1.50/0.060 = 25$, whence $M = 0.60$ for extrusions. Thus Equation (1.10) gives

$$S_{cc} = 0.60(0.085)^{1/4} = 0.32, \quad F_{cc} = 0.32(75,000) = 24,300 \text{ psi.}$$

Here F_{cc} is nearly four times F_{cr}.

(c) Allowable column stresses F_c or S_c

The Euler's column formula for the elastic allowable column stress F_{cE} is given in textbooks on "Strength of Materials", books on "Elasticity Theory", and in Equations (4.141, Vol. 1) and (4.143, Volume 1), or

$$S_{cE} = \frac{F_{cE}}{F_{cy}} = \frac{c\pi^2 E_c I}{AL^2 F_{cy}} = \frac{c\pi^2 E_c}{F_{cy}}\left(\frac{\rho}{L}\right)^2 = \frac{1}{B_c^2}, \quad B_c = (S_{cE})^{-1/2}, \tag{1.11}$$

where c is the end fixity of the column ($c = 1$ for pinned ends, $c = 4$ for fixed ends, etc.), L is the length of the column, A is the cross section area, I is the area moment of inertia about the bending neutral axis, and $\rho^2 = I/A$ is the radius of gyration. ρ is the minimum value for the cross section if the column is not constrained to bend about a given axis. E_c and F_{cy} are given in Tables 1.1 through 1.6.

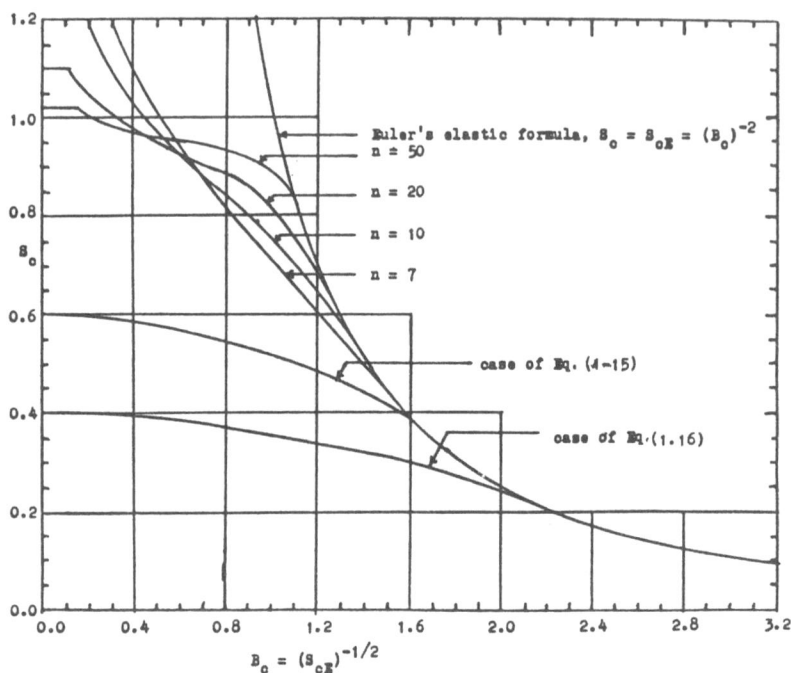

Fig. 1.21. Allowable column stresses.

Theory and test data indicate that S_c in the inelastic range can be taken as

$$S_c = F_c/F_{cy} = (E_{ct}/E_c)S_{cE} = (E_{ct}/E_c B_c^2), \tag{1.12}$$

where E_{ct} is the tangent modulus of elasticity. For the Ramberg-Osgood nondimensional stress-strain curves, E_{ct}/E_c is given by Equation (1.7). Figure 1.21 shows graphs of Equation (1.12) for the same n values in the Ramberg-Osgood equation as used in Figure 1.19 for S_{cr} local buckling.

Test data shows that S_c, as given by Equation (1.12) for columns with no local buckling (such as solid circular and thick rectangular sections with $b/t < 10$), has cut-offs slightly above the yield stress F_{cy}. These approximate cut-offs are shown on Figure 1.21 for the indicated n values, and are the same as those on Figure 1.19 for the buckling and crippling cut-offs. From Reference 1, an empirical formula for $(S_c)_{max}$ for aluminum alloys is

$$(S_c)_{max} = 1 + g(F_{cy})^{1/2}, \tag{1.13}$$

where g for the alloys in Tables 1.1 through 1.6 has the approximate values

$$\begin{aligned} g &= 0.001 \text{ in Tables 1.1, 1.2, with } n = 10, \\ g &= 0.00075 \text{ in Tables 1.3, 1.4, 1.5, with } n = 15, \\ g &= 0.0005 \text{ in Table 1.6, with } n = 20. \end{aligned} \tag{1.14}$$

For columns with local buckling, the cut-off stress is F_{cc} or S_{cc}, which is the

Allowable stresses of flight vehicle materials

local failing stress. For cases with $S_{cc} = S_{cr}$ in Figure 1.19, the curves in Figure 1.21 can be used up to the cut-off at $S_c = S_{cc} = S_{cr}$. For cases with $S_{cr} < S_{cc}$, a transition region occurs between the S_{cr} point on the column curves and the S_{cc} cut-off line. Test data indicates that this transition can be represented by a parabola in the B_c parameter. Since very little inelastic effect will occur before the S_c stress reaches $(1/2)S_{cc}$ crippling stress, the transition can be represented in two parts, or

$$S_c = S_{cc} - (S_{cc} - S_{cr})S_{cr}B_c^2, \quad S_{cr} \leq S_{cc} \leq 2S_{cr}, \tag{1.15}$$
$$S_c = S_{cc} - (1/4)S_{cc}^2 B_c^2, \quad S_{cc} \geq 2S_{cr}. \tag{1.16}$$

Note in Figure 1.21 that these curves are applicable only to the left of the column curves. When $S_{cc} > 2S_{cr}$, follow Euler's curve from S_{cr} up to $S_{cc}/2$ and then follow the parabola as B_c decreases. Note also in Equation (1.15) that $S_{cc} = S_{cr}$ gives $S_c = S_{cc}$ on the cut-off as indicated above. Since only particular cases of Equations (1.15) and (1.16) are shown on Figure 1.21, it is necessary to calculate S_c for the cases for which Equations (1.15) and (1.16) are applicable.

For stiffened panels, the rivet attachments between the stiffeners and the plate as well as the location of the neutral axis may affect the column stresses in Equations (1.15) and (1.16). See References 8 and 14 for correction factors to use. For equations (1.15) and (1.16) to apply directly, the rivets should be large with round heads and closely spaced while the neutral axis should be close to the plate-stiffener joint.

Example 1.5. Find the allowable compressive stresses F_{cr}, F_{cc}, F_c for a 7075-T6 extrusion channel section (*L*-direction, *B* basis) in Figure 1.16 with $t_w = 0.080$ in., $b_F = 1.50$ in., $b_w = 3.00$ in. Take $L = 80$ in., $c = 2$, and bending about an axis parallel to the flanges for the column action.
Solution. From Table 1.6, $F_{cy} = 75,000$ psi and $E_c = 10.5(10^6)$ psi. On Figure 1.16, $b_F/b_w = 0.50$, $t_w/t_F = 0.80$, whence $K = 3.30$. Thus Equations (1.2) and (1.8) give

$$S_{crE} = \frac{3.30(10.5)(10^6)}{75,000}\left(\frac{0.080}{3.00}\right)^2 = 0.329,$$
$$B_{cr} = (0.329)^{-1/2} = 1.745.$$

Since B_{cr} shows that S_{cr} on Figure 1.19 is in the elastic range,

$$S_{cr} = S_{crE} = 0.329, \quad F_{cr} = 0.329(75,000) = 24,700 \text{ psi.}$$

On Figure 1.19 the above B_{cr} shows that $S_{cc} > S_{cr}$, whence in Figure 1.16, $b_w/t_w = 37.5$ and $M = 0.69$. Thus, Equation (1.10) gives

$$S_{cc} = 0.69(0.329)^{1/4} = 0.523,$$
$$F_{cc} = 0.523(75,000) = 39,200 \text{ psi.}$$

The radius of gyration ρ for a symmetrical channel, about a centroid axis parallel to the flanges, can be approximated as

$$\rho = b_w\left[\left(1 + 6\frac{b_F t_F}{b_w t_w}\right) \Big/ 12\left(1 + 2\frac{b_F t_F}{b_w t_w}\right)\right]^{1/2}, \tag{1.17}$$

$$\rho = 3.00\left[\frac{1 + 6(0.50)(1.25)}{12[1 + 2(0.50)(1.25)]}\right]^{1/2} = 1.256 \text{ in.}$$

From Equation (1.11),

$$S_{cE} = \frac{2\pi^2(10.5)(10^6)}{75,000}\left(\frac{1.256}{80}\right)^2 = 0.684,$$

$$B_c = (0.684)^{-1/2} = 1.209.$$

From the above S_{cr} and S_{cc} values, Equation (1.15) applies so that

$$S_c = 0.523 - (0.523 - 0.329)(0.329)(1.209)^2 = 0.430,$$

$$F_c = 0.430(75,000) = 32,200 \text{ psi}.$$

1.5. Allowable combined stresses

In the previous sections of this Chapter 1 the allowable stresses given in the tables or calculated in Section 1.4 apply for the particular stress acting alone. In the thin web beam and the box beam analysis given in Volume 1, the axial stresses in the stringers and stiffeners act alone while the shear stresses in the webs are assumed to act alone. Thus, in truss and thin web beam analysis and design, the applied stresses can be compared directly to the allowable stresses given in Sections 1.2, 1.3 and 1.4.

However, in two and three dimensional structures the applied stresses, Equations (1.21, Vol.1) and (1.4, Vol.1), interact with each other. This interaction affects the allowable stresses for the materials in the structure so that combinations of the allowable stresses must be used to define yield and failure for the structure. Since the allowable stresses must be compared in some manner to the maximum applied stresses, it is necessary to determine the applied stresses in any direction in the material. That is, the maximum applied axial stresses may not be along the x, y, z axes for σ_{xx}, σ_{yy}, σ_{zz}, but may be along some other axes.

Rotation of axes

If the applied stresses σ_{xx}, σ_{yy}, σ_{zz}, σ_{xy}, σ_{xz}, σ_{yz} are assumed to be known in the three or two dimensional cases, then it is possible to calculate the applied stresses for any other directions in the body by rotating the x, y, z axes to the x_1, y_1, z_1 axes. This rotation of the axes for the case of orthogonal rectangular coordinate axes can be done by the procedure discussed in Section A.9 (Vol.1). In order to use this procedure, assemble the three equilibrium equations in Equation (1.5, Vol.1) into a matrix form by multiplying by the area element ΔA and using Equation (A.66, Vol.1), or

$$\begin{Bmatrix} S_x \Delta A \\ S_y \Delta A \\ S_z \Delta A \end{Bmatrix} = \begin{bmatrix} \sigma_{xx} & \sigma_{xy} & \sigma_{xz} \\ \sigma_{xy} & \sigma_{yy} & \sigma_{yz} \\ \sigma_{xz} & \sigma_{yz} & \sigma_{zz} \end{bmatrix} \begin{Bmatrix} l\Delta A \\ m\Delta A \\ n\Delta A \end{Bmatrix}. \tag{1.18}$$

This matrix equation can be expressed in the form of Equation (A.95, Vol.1), or

$$Q_c = AP_c, \tag{1.19}$$

where the stress matrix A transforms the components of the area vector P_c into the components of the applied force vector Q_c.

If the x, y, z coordinate system is rotated to the x_1, y_1, z_1 system, Figure A.1, (Vol.1), by the rotation matrix R in Equation (A.101, Vol.1), then the components of both the area vector and the force vector in Equation (1.19) will be changed in the x_1, y_1, z_1 system to P_{c1} and Q_{c1}. The stress matrix A_1 in the new system is given by Equation (A.104, Vol.1). For rotation through the angle θ about the $z = z_1$ axis, Figure A.2 (Vol.1), the stresses in A_1 are given by Equation (A.107, Vol.1) for three dimensions and by Equation (A.110, Vol.1) for two dimensions. For the two dimensional case,

$$\begin{aligned} 2\sigma_{xx1} &= (\sigma_{xx} + \sigma_{yy}) + (\sigma_{xx} - \sigma_{yy})\cos 2\theta + 2\sigma_{xy}\sin 2\theta, \\ 2\sigma_{yy1} &= (\sigma_{xx} + \sigma_{yy}) - (\sigma_{xx} - \sigma_{yy})\cos 2\theta - 2\sigma_{xy}\sin 2\theta, \\ 2\sigma_{xy1} &= -(\sigma_{xx} - \sigma_{yy})\sin 2\theta + 2\sigma_{xy}\cos 2\theta. \end{aligned} \qquad (1.20)$$

Mohr's circle

As shown in Equation (A.111) and Figure A.3, (Vol.1), elimination of θ from Equation (1.20) gives a circle relation among the stresses, with σ_{xx1}, σ_{yy1} on the abscissa and σ_{xy1} on the ordinate. This is the *Mohr's circle* for the stresses in a plane. The maximum and minimum values of σ_{xx1}, σ_{yy1} on their axis are the *principle stresses*. The angle θ_d from the original axes to determine these values is given by Equation (A.112, Vol.1) and is shown in Figure A.3, (Vol.1) or

$$\tan 2\theta_d = \frac{2\sigma_{xy}}{\sigma_{xx} - \sigma_{yy}}. \qquad (1.21)$$

Note that the radius a of the circle, which also equals the maximum shear stress, is

$$a^2 = \sigma_{xy,\max}^2 = \tfrac{1}{4}(\sigma_{xx} - \sigma_{yy})^2 + \sigma_{xy}^2, \qquad (1.22)$$

whence the principle stresses are

$$\sigma_p = \tfrac{1}{2}(\sigma_{xx} + \sigma_{yy}) \pm a. \qquad (1.23)$$

For the case of no applied shear stress σ_{xy}, Equation (1.21) shows $\theta_d = 0$ so that the applied σ_{xx} and σ_{yy} stresses are the principle stresses and the maximum shear stress is $(\sigma_{xx} - \sigma_{yy})/2$.

The relationships among the stresses in the plane in Equation (1.20) or on Mohr's circle hold for the *applied stresses* produced in the plate by the applied axial forces and by the applied shear forces. Note that one axial force alone produces both axial and shear stresses in the plate in other directions. Also, a shear force alone produces both axial and shear stresses for other directions.

Since the three dimensional applied stresses can be obtained only in special cases and since there is little test data on the allowable stresses in three dimensions, the following discussion of the allowable stresses will be limited to the two dimensional case.

Allowable biaxial tension stresses

In Reference 1 test data for a few material alloys are given for biaxial loading of plates in which σ_{xx} and σ_{yy} tension stresses are applied to the plate for a specified ratio between the two stresses, or

$$B_a = \sigma_{yy}/\sigma_{xx}, \quad B_s = \sigma_{xy}/\sigma_{xx} = 0. \tag{1.24}$$

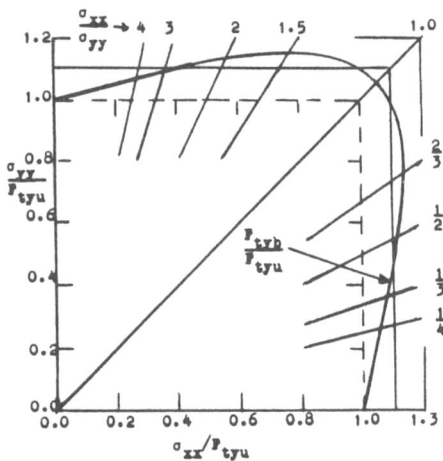

Fig. 1.22. Biaxial non-dimensional yield stress envelope for AISI alloy steel (cylindrical specimens).

Biaxial stress-strain curves can be drawn for each value of B_a by plotting the maximum principle stress against the strain in the same direction. These curves may be different for each B_a ratio so that the biaxial modulus of elasticity, the biaxial yield stress, and the biaxial ultimate stress may vary with B_a. A non-dimensional cross plot of the data from these stress-strain curves can be made with B_a as a parameter. Figure 1.22 shows such a graph of biaxial non-dimensional yield stress for AISI 4340 alloy steel with uniaxial reference yield stress F_{ty} taken in the σ_{xx} hoop direction (Reference 1), on the thin wall cylinder.

In Figure 1.22 the allowable biaxial yield stress F_{tyb} is greater than or equal to the uniaxial yield stress F_{tyu} for all values of the B_a ratios, having a maximum value of $1.14 F_{tyu}$ at $B_a = 2/3$ and 1.5 approximately. This suggests that when biaxial data for a material is not available, then it may be conservative to use the biaxial yield stress the same as the uniaxial yield stress in the direction of the maximum principle stress. Curves for the biaxial ultimate stress are similar to Figure 1.22, and design can be made by using the uniaxial ultimate stress equal to the biaxial value.

Allowable stresses of flight vehicle materials

Yield and ultimate theories for allowable combined stresses

The above discussion applied to the case of both σ_{xx} and σ_{yy} as tension stresses and $\sigma_{xy} = 0$. To allow for compression and shear stresses, consider cases of Mohr's circle in Figure A.3 (Vol.1) and add the allowable axial and shear uniaxial stresses, as shown in Figure 1.23. Assume yielding or failure (ultimate) to occur when the σ_{xx}, σ_{yy}, σ_{xy} stresses are such that Mohr's circle touches the rectangle boundary defined by the allowable uniaxial stresses and assume

$$F_{\text{all},uy} = F_{\text{all},ux}, \quad F_{\text{all},us} = k_s F_{\text{all},ux}. \tag{1.25}$$

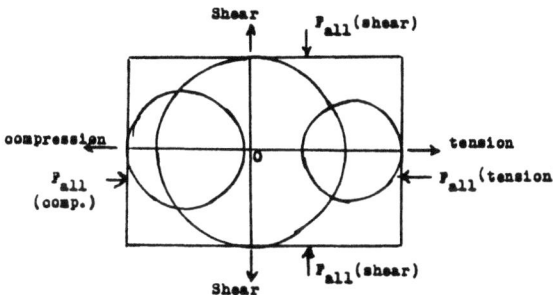

Fig. 1.23. Allowable biaxial maximum principle stresses and maximum shear stresses.

From Equations (1.22)-(1.25),

$$2F_{\text{all},ux}/F_{\text{all},bx} = (1 + B_a) + \left[(1 - B_a)^2 + 4B_s^2\right]^{1/2} \tag{1.26}$$

for the case of *maximum principle stress theory* with yielding or failure at the right or left boundary, and

$$2k_s F_{\text{all},ux}/F_{\text{all},bx} = \left[(1 - B_a)^2 + 4B_s^2\right]^{1/2} \tag{1.27}$$

for the case of *maximum shear stress theory* at the top or bottom boundary. Here $F_{\text{all},bx}$ is the allowable biaxial stress in the x-direction with the corresponding stresses given by Equation (1.25).

Various yield and failure theories for plates are discussed in Chapter 3 of Reference 17, including the maximum principle stress theory and the maximum shear stress theory described by Equations (1.26) and (1.27). The *distortion energy theory* or the *octohedral shear stress theory*, which has the same form, in which

$$F_{\text{all},ux}/F_{\text{all},bx} = \left[1 - B_a + B_a^2 + 3B_s^2\right]^{1/2}, \tag{1.28}$$

checks test data quite well and is recommended for both yield and failure biaxial allowable stresses for ductile materials, without buckling. The theories in Equations (1.26) and (1.27) are recommended for brittle materials. Figure 1.24

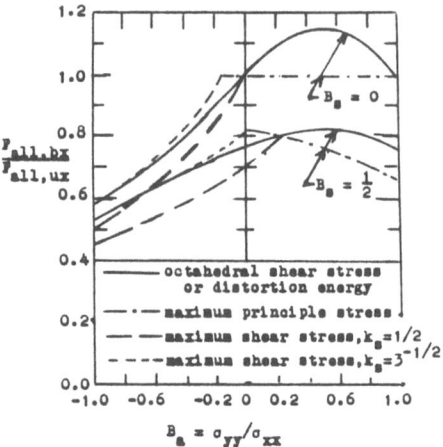

Fig. 1.24. Biaxial stress theories.

shows curves for all three theories for the design stress ratio $F_{all,bx}/F_{all,ux}$ plotted against the ratio B_a as variable and B_s as a parameter. From Reference 17, $k_s = 1/2$ in Equations (1.25) and (1.27) and $k_s = 3^{-1/2}$ in the distortion energy theory. However, from Tables 1.1 through 1.6, the ratio of F_{su}/F_{tu} is approximately $3^{-1/2}$ so that the case of $k_s = 3^{-1/2}$ in Equation (1.27) is also shown on Figure 1.24. Note that only the smaller value of Equations (1.26) and (1.27) is plotted in Figure 1.24.

From Figure 1.24 the maximum principle stress and maximum shear stress with $k_s = 1/2$ give lower allowable stresses for all B_a and B_s stress ratios than the distortion energy theory. With $k_s = 3^{-1/2}$, the maximum shear stress curve is much closer to the distortion energy curve. For B_a positive with σ_{xx} and σ_{yy} tension, $B_s = 0$, the distortion energy curve in Figure 1.24 is very similar to that in Figure 1.22 with approximately the same maximum value.

Example 1.6. Find the allowable biaxial ultimate stress $F_{all,bx}$ for the 7075-T6 sheet in Figure 1.25 by (a) the maximum principle stress theory or maximum shear stress theory, and (b) the distortion energy theory. (c) Repeat for the case of $\sigma_{yy} = -\sigma_{xx}$.

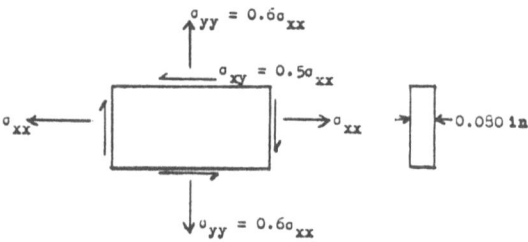

Fig. 1.25. Biaxially loaded plate with shear stress.

Solution. (a) From Table 1.4, the allowable uniaxial ultimate stress is $F_{all,uz} = F_{tu} = 76,000$ psi in the L or z direction for basis (A). From Figure 1.25, $B_a = 0.6$, $B_s = 0.5$, so that from Equations (1.26) and (1.28) or Figure 1.24,

(a) $F_{all,bz} = 0.74 F_{all,uz} = 56,000$ psi,

(b) $F_{all,bz} = 0.81 F_{all,uz} = 62,000$ psi.

(c) For $B_a = -1$, Equations (1.27) and (1.28) or Figure 1.24 give

(a) $F_{all,bz} = 0.45 F_{all,uz} = 34,000$ psi for $k_s = 1/2$,

(a) $F_{all,bz} = 0.52 F_{all,uz} = 40,000$ psi for $k_s = 0.58$,

(b) $F_{all,bz} = 0.52 F_{all,uz} = 40,000$ psi for distortion energy.

From Equation (1.22) the maximum applied shear stress for $B_a = -1$ and $B_s = 1/2$ is 45,000 psi while the allowable maximum shear stress is $0.58(76,000) = 44,000$ psi for $k_s = 0.58$, and $F_{su} = 46,000$ psi from Table 1.4. Thus, the distortion energy theory and the $k_s = 0.58$ case for the maximum shear stress theory check the actual ultimate material properties nearly exactly in this case.

If the B_a and B_s stress ratios are assumed to stay constant for changes in the applied loads and stresses, then the margin of safety for the biaxial stress can be calculated from Equation (B.75, Vol.1) as

$$\text{M.S.} = \frac{F_{all,bz}}{k_u \sigma_{zzL}} - 1, \qquad (1.29)$$

where σ_{zzL} is the applied stress for limit load.

Yield and ultimate combined allowable stresses from stress ratios

In the above yield and ultimate stress theories for the plate in Equations (1.25)-(1.29), it is assumed that the allowable stresses for the axial and shear stresses are related to each other so that one of the stresses can be used for the biaxial design. In many materials and combined loading cases, particularly in the inelastic range, the allowable stresses vary so that no simple formulas relating them can be used. In such cases the ratio of the applied stress to the allowable stress for each type of stress, or

$$R = \frac{f}{F} = \frac{\text{applied}}{\text{allowable}} = \frac{P_{app}}{P_{all}} = \frac{M_{app}}{M_{all}}, \qquad (1.30)$$

can be used, and combinations of the ratios made to define yield and failure, and hence margin of safety. For one stress alone

$$\text{M.S.} = \frac{1}{R} - 1. \qquad (1.31)$$

If two stresses are in the same direction, such as axial stress and bending stress on a beam, but in the inelastic region the allowable stresses are different, then

$$R_a + R_b = 1, \quad \text{M.S.} = \frac{1}{R_a + R_b} - 1. \qquad (1.32)$$

This implies that each stress affects the deformation of the beam in such a way that failure occurs when the sum of the ratios is one. This may happen in a brittle material, but in a ductile material in the inelastic range, strain analysis shows that the sum is not necessarily one. Also, test data shows the sum to be greater than one. See Reference 26 for the load-strain design curves for this case.

When the stresses are at right angles to each other such as σ_{xx}, σ_{yy}, and σ_{xy} in the plate, then Equations (1.26)-(1.28) suggest a sum of the squares of the ratios, or

$$R_{xx}^2 + R_{yy}^2 + R_{xy}^2 = 1, \quad \text{M.S.} = \left(R_{xx}^2 + R_{yy}^2 + R_{xy}^2\right)^{-1/2} - 1. \tag{1.33}$$

It should be pointed out that the distortion energy theory in Equation (1.28) can be modified to give

$$R_{xx}^2 + R_{yy}^2 + R_{xy}^2 \mp R_{xx}R_{yy} = 1,$$
$$\text{M.S.} = \left(R_{xx}^2 + R_{yy}^2 + R_{xy}^2 \mp R_{xx}R_{yy}\right)^{-1/2} - 1, \tag{1.34}$$

where the minus sign applies for σ_{xx} and σ_{yy} with the same sign, and the plus sign applies for σ_{xx} and σ_{yy} with opposite signs. For no buckling of a material meeting the conditions of the distortion energy theory, it is evident that Equation (1.33) is conservative for σ_{xx} and σ_{yy} having the same sign, and unconservative for σ_{xx} and σ_{yy} having different signs. Thus, for materials such as many of the aluminum aloys in which $F_{\text{all},y}$ is approximately $F_{\text{all},x}$ and $F_{\text{all},s}$ is approximately $0.58 F_{\text{all},x}$, it appears that Equation (1.34) will give better results than Equation (1.33) for plates without buckling. See Example 1.6 above.

Allowable combined buckling stresses

When both axial stresses σ_{xx} and σ_{yy} are compression with $\sigma_{xy} = 0$ and buckling occurs, then from Reference 18 with some rearrangement,

$$\frac{4m^2 p^2}{(m^2 p^2 + 1)^2} R_{xx} + \frac{(p^2 + 1)^2}{(m^2 p^2 + 1)^2} R_{yy} = 1, \quad p = b/a,$$
$$a \geq b, \quad \text{Figure 1.26}, \quad R_{xx} = \sigma_{xx}/F_{crxx}, \quad R_{yy} = \sigma_{yy}/F_{cryy}, \tag{1.35}$$
$$m = \text{number of half wave lengths in } x\text{-direction.}$$

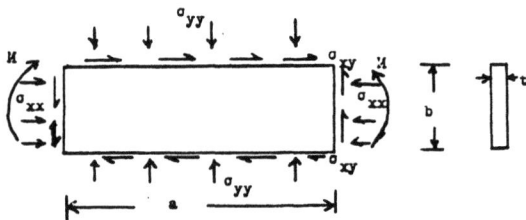

Fig. 1.26. Biaxial buckling of rectangular plate, $a \geq b$.

From Equation (1.2) and Reference 18,

$$F_{crxx} = KE_c(t/b)^2, \quad F_{cryy} = \tfrac{1}{4}(1+p^2)^2 F_{crxx}. \tag{1.36}$$

where K is given by Figure 1.14.

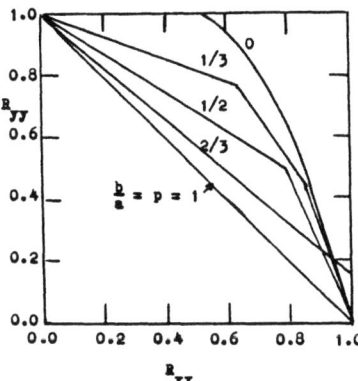

Fig. 1.27. Biaxial compression buckling interaction curves.

In Equation (1.35), for each value of p the values of m must be determined to give the smallest values of R_{xx} and R_{yy}. For $p = 1$, the proper m is $m = 1$ for the *square plate*, or

$$R_{xx} + R_{yy} = 1, \quad \text{M.S.} = (R_{xx} + R_{yy})^{-1} - 1. \tag{1.37}$$

For $p < 1$, it is necessary to consider several values of $m = 1, 2, \cdots$ up to mp of order one. Then the envelope of the straight lines in Equation (1.35) for the different m and specified p gives the interaction curve for any p. Figure 1.27 shows the interaction envelope on the lines for several values of $p = b/a$.

The crippling stresses and the inelastic buckling stresses may not follow Equation (1.35) or Figure 1.27. Beyond buckling, the long edges of the plate at the supports carry the additional x-axial load so that σ_{xx} can increase while σ_{yy} remains fixed. Both σ_{xx} and σ_{yy} can increase for the square plate, probably following the maximum principle stress theory.

The margin of safety on Figure 1.27 can be calculated from the equation for the given p and the proper value of m as

$$\text{M.S.} = \frac{(m^2 p^2 + 1)^2}{m^2 p^2 R_{xx} + (p^2 + 1)^2 R_{yy}} - 1. \tag{1.38}$$

From references 19 and 23, the interaction between σ_{xx} and σ_{xy} with $\sigma_{yy} = 0$ in Figure 1.26 is

$$R_{xy}^2 + R_{xx} = 1, \quad \text{M.S.} = \frac{2}{R_{xx} + (R_{xx}^2 + 4R_{xy}^2)^{1/2}} - 1, \tag{1.39}$$

for the elastic case. For the inelastic case, Reference 23,

$$R_{xy}^2 + R_{xx}^2 = 1, \quad \text{M.S.} = (R_{xy}^2 + R_{xx}^2)^{-1/2} - 1. \tag{1.40}$$

From curves given in Reference 19 for the interaction between σ_{yy} and σ_{xy} with $\sigma_{xx} = 0$ in Figure 1.26, approximate elastic equations are

$$R_{xy}^{(3p+1)/2p} + R_{yy} = 1, \quad 0.20 \le p \le 1,$$
$$R_{xy} + 0.25 R_{yy} = 1, \quad 0 \le R_{yy} \le 1, \quad p = 0. \tag{1.41}$$

The margin of safety for this case can be obtained by assuming $R_{xy} = r_1 R_{yy}$ with r_1 specified, whence $R_{yy,\text{all}}$ for given p can be determined by trial or graphing from

$$\left(r_1 R_{yy,\text{all}}\right)^{(3p+1)/2p} + R_{yy,\text{all}} = 1. \tag{1.42}$$

Thus, it follows that

$$\text{M.S.} = \left(R_{yy,\text{all}}/R_{yy}\right) - 1. \tag{1.43}$$

From References 10 and 23, the interaction between bending σ_{bx} and shear σ_{xy} in Figure 1.26 is

$$R_{bx}^2 + R_{xy}^2 = 1, \quad \text{M.S.} = \left(R_{bx}^2 + R_{xy}^2\right)^{-1/2} - 1. \tag{1.44}$$

From curves in Reference 23 the interaction between bending σ_{bx} and compression σ_{xx} can be approximated by

$$R_{bx}^2 + R_{xx} = 1, \tag{1.45}$$

with the M.S. as in Equation (1.39). The approximation for the interaction between σ_{bx} and σ_{yy} is

$$R_{bx}^{2/p} + R_{yy} = 1, \quad 0.50 \le p \le 1,$$
$$R_{bx} + 0.15 R_{yy} = 1, \quad 0 \le R_{yy} \le 1, \quad p = 1/3,$$
$$R_{bx} + 0.10 R_{yy} = 1, \quad 0 \le R_{yy} \le 1, \quad p = 0. \tag{1.46}$$

For the M.S. use the same procedure as in Equations (1.42) and (1.43).

The interactions among three stresses are more complicated and usually require families of curves for each selected value of p. In Reference 23 curves are given for R_{bx} against R_{yy} with R_{xx} as parameter for several values of p. Also, in Reference 23 curves for $p = 0$ are given for interactions among R_{bx}, R_{xy}, and R_{yy}. From Equations (1.37), (1.39), and (1.41), it appears that the interaction

$$R_{xy}^2 + R_{xx} + R_{yy} = 1,$$
$$\text{M.S.} = \frac{2}{R_{xx} + R_{yy} + \left[(R_{xx} + R_{yy})^2 + 4R_{xy}^2\right]^{1/2}} - 1, \tag{1.47}$$

can be used for the square plate with $p = 1$.

The combined crippling stresses and inelastic buckling stresses are quite complicated in this case of shear and axial loads. After buckling the additional shear load tends to be taken by tension at an angle of approximately 45° (see Section 3.9), while the additional compressive axial load is taken on the edges. The diagonal tension load also produces compression on the edges of the plate. Probably, the maximum principle stress theory will give the best results.

From Reference 1, for tubes in bending, torsion, and compression without local buckling, the compression load tends to increase both the bending moment and the torsional moment so that

$$\left(R_b^2 + R_s^2\right)^{1/2} + R_c = 1, \quad \text{M.S.} = \frac{1}{R_c + \left(R_b^2 + R_s^2\right)^{1/2}} - 1. \tag{1.48}$$

The allowable stress for the column is F_c in Equation (1.12), or $R_c = S_c$ in Equation (1.12) and Figure 1.21. Equation (1.48) probably applies for both the elastic and inelastic cases.

Example 1.7. In Figure 1.26 take $a = b = 3.00$ in., $t = 0.063$ in., $\sigma_{xx} = 6000$ psi, $\sigma_{yy} = 7200$ psi, $\sigma_{xy} = 6000$ psi for the simply supported plate. If the plate is 2024-T3 clad aluminum alloy in Table 1.2, find the margin of safety and the applied stresses to produced buckling.

Solution. From Equation (1.2), Figure 1.14, and Table 1.2 the uniaxial elastic buckling stresses are

$$F_{crxxu} = F_{cryyu} = 3.62(10.2)(10^6)(0.063/3)^2 = 16,300 \text{ psi},$$
$$F_{crxyu} = 8.1(10.2)(10^6)(0.063/3)^2 = 36,500 \text{ psi},$$

whence $R_{xx} = 6000/16,300 = 0.368$, $R_{yy} = 0.442$, $R_{xy} = 0.164$. From Equation (1.47)

$$\text{M.S.} = \frac{2}{0.810 + (0.656 + 0.108)^{1/2}} - 1 = 0.19.$$

The applied stresses to produce buckling, which must maintain the same ratio to each other as the given stresses, are

$$\sigma_{xxb} = 1.19(6000) = 7100 \text{ psi},$$
$$\sigma_{yyb} = 1.19(7200) = 8600 \text{ psi},$$
$$\sigma_{xyb} = 1.19(6000) = 7100 \text{ psi}.$$

1.6. Creep effects on allowable stresses

The materials used in flight structures tend to creep at elevated temperatures so that the structure may have large deformations if the load is carried for a long time. Thus, it is necessary to use an allowable stress that permits a specified amount of deformation in a specified time. To obtain these allowable stresses, tests of the material at a given temperature and a given load or stress can be made by measuring strains at various time intervals as the material creeps over the time of the test. Figure 1.28 shows the type of curves obtained in creep tests for various stress levels at a constant temperature. See References 1, 9 and 17. Cross plots of the data in Figure 1.28 can be made to give a graph of stress against time for a fixed strain or fixed total deformation. Also, a cross plot of the stress against time at failure can be made. Such data of allowable stress against time are given for some materials in Reference 1 for a few selected temperatures.

The curves in Reference 1 for 7075-T6 aluminum alloy in tension at $T = 350°F$ can be represented by simple equations for the range of the test data as

$$R_u = F_{tc}/F_{tuRT} = 0.465 - 0.123 \log_{10} t, \text{ rupture},$$
$$R_u = 0.458 - 0.123 \log_{10} t, \text{ 5\% total deformation},$$

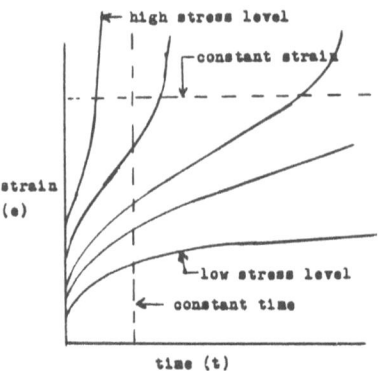

Fig. 1.28. Creep at constant elevated temperature.

$$R_u = 0.420 - 0.110 \log_{10} t, \; 1\% \text{ total deformation},\tag{1.49}$$
$$R_u = 0.340 - 0.083 \log_{10} t, \; 0.5\% \text{ total deformation},$$
$$R_u = 0.150 - 0.015 \log_{10} t, \; 0.2\% \text{ total deformation},$$

where t is in hours, F_{tuRT} is the ultimate tension stress at room temperature in Tables 1.4, 1.5, and 1.6, F_{tc} is the allowable tension creep stress at 375°F for the specified strain conditions. The range of the test data for Equation (1.49) is approximately 0.5 hour to 140 hours.

A cross plot of stress against strain at a constant time can be made from Figure 1.28 to give an apparent stress-strain curve or an isochronous stress-strain curve. A family of these iso-stress-strain curves can be constructed at a given temperature with time as a parameter. In many cases, these iso-stress-strain curves can be used as if they are real stress-strain curves to obtain a yield stress and an ultimate stress. In Section 6.6 of Reference 9 it is shown that the allowable crippling and column stresses calculated by the procedures of Section 1.4 above by using the F_{cy} and E_c from the iso-stress-strain curves check the column test data for the creep lifetime of 2024-T3 aluminum alloy skin-stringer panels at elevated temperatures.

Because of the large number of tests required to obtain the stress and strain data for each material at various temperatures and creep times, efforts have been made to combine the temperature and time variables into one parameter. From References 20 and 9, the Larson-Miller parameter

$$LM = 10^{-3}(T + 460)(C + \log_{10} t)\tag{1.50}$$

checks test data over a wide range of times and temperatures for many steels and aluminum alloys. Although $C = 20$ checks all the data reasonably well, a better check can be obtained by using a different C for each alloy. In Equation (1.50), T is the temperature in °F and t is the time in hours.

Since any iso-stress-strain curve described above corresponds to a given temperature and a given time, it corresponds to a given value of the Larson-Miller

Allowable stresses of flight vehicle materials 33

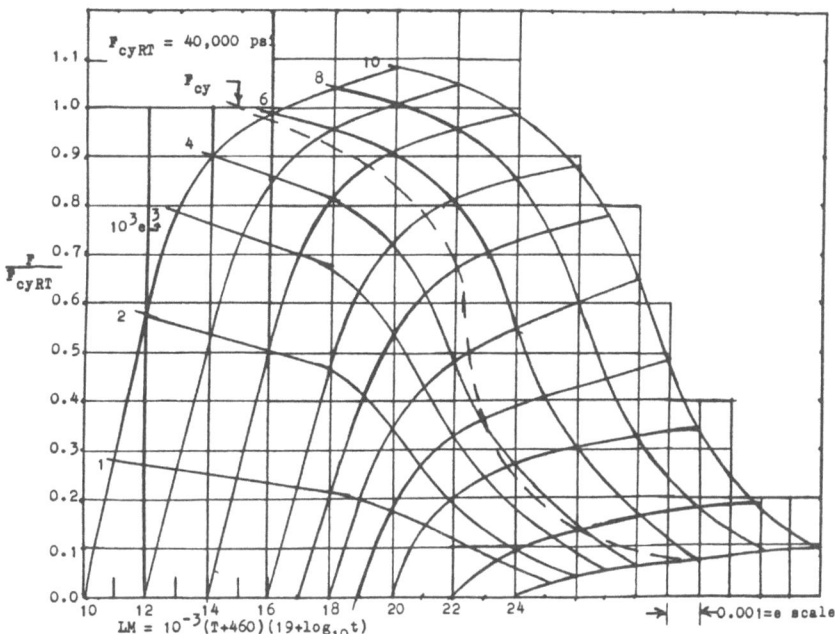

Fig. 1.29. Clad aluminum alloy sheet, 2024-T3, compression creep curves.

parameter LM by Equation (1.50) so that a family of iso-stress-strain curves for a given material alloy can be drawn with LM as the abscissa and stress as the ordinate. Also, the constant strain lines can be drawn on the family to show the creep curves. Figure 1.29 shows such a family of compression master creep curves for 2024-T3 clad aluminum alloy sheet. The stress-strain curves are based upon short-time elevated temperature tests after 1/2 hour soaking time with $t = 0.1$ in the LM parameter.

Example 1.8. Find the allowable tension stress F_{tc} for a 7075-T6 extruded spar cap at 375°F with an allowable total deformation of 1% after 50 hours of constant load.

Solution. From Table 1.6, the ultimate tension stress at room temperature for a spar cap in the 0.500 - 0.749 in. thickness range for basis B is $F_{tuRT} = 85,000$ psi. From Equation (1.49) at 1% total creep deformation

$$F_{tc} = 85,000(0.420 - 0.110\log_{10} 50) = 19,800 \text{ psi.}$$

Note that from Equation (1.49) the allowable rupture stress is 21,800 psi, which compares to $F_{tu} = 26,000$ psi on Figure 1.9 for 50 hours soaking time at 375°F and test at 375°F. Thus, most of the effect on F_{tu} is due to the temperature, while the creep effect reduced F_{tu} from 26,000 psi to 21,800 psi.

Example 1.9. Find the allowable compressive stresses F_{cr}, F_{cc}, F_c for a clad 2024-T3 formed channel section (formed in the T-condition) in Figure 1.16 with $t_w = 0.080$ in., $t_F = 0.100$ in., $b_F = 1.50$ in., $b_w = 3.00$ in., $L = 80$ in., $c = 1$, and bending about an axis parallel to the flanges for the column action. The temperature is 400°F and the creep life is 100 hours.

Solution. The Larson-Miller parameter is

$$LM = 10^{-3}(400 + 460)(19 + \log_{10} 100) = 18.06,$$

whence Figure 1.29 gives $F_{cy} = 0.48F_{cyRT} = 19,200$ psi. From Table 1.2 and Figure 1.7, $E_c = 10.2(10^6)(0.89) = 9.08(10^6)$ psi. On Figure 1.16, $b_F/b_w = 0.50$, $t_w/t_F = 0.80$, whence $K = 3.30$. Thus, Equations (1.2) and (1.8) give

$$S_{crE} = \frac{3.30(9.08)(10^6)}{19,200}\left(\frac{0.080}{3.00}\right)^2 = 1.11,$$

$$B_{cr} = (1.11)^{-1/2} = 0.95.$$

Assume $n = 10$ in Figure 1.19 to get $S_{cr} = 0.95$, $F_{cr} = 18,200$ psi. In this case, $S_{cc} = S_{cr}$ so that $F_{cc} = 18,200$ psi.

From Equation (1.12), $\rho = 1.26$, and From Equation (1.11)

$$S_{cE} = \frac{\pi^2(9.08)(10^6)}{19,200}\left(\frac{1.26}{80}\right)^2 = 1.16, \qquad B_c = (1.16)^{-1/2} = 0.93.$$

Assume $n = 10$ in Figure 1.21 to get $S_c = 0.79$, $F_c = 14,400$ psi.

In the above discussion of designing for a temperature and creep time, it was assumed that only one constant stress level was acting. In flight vehicle structures operating under high temperature conditions, various stress levels acting for various time intervals may occur as the vehicle flys various missions. Among the several cumulative-creep laws described in Reference 21, the *life fraction law* appears to give the best results. This law states that creep failure or creep to a specified total strain is determined by

$$\sum_{i=1}^{N}(t_i/t_{Ri}) = 1, \qquad (1.51)$$

where t_i is the time at a given temperature T_i and stress σ_i and t_{Ri} is the calculated reference lifetime at the same temperature T_i and stress σ_i. Failure occurs when the sum of these lifetimes t_i/t_{Ri} is one for N different temperature and stress conditions.

Let each time t_i be expressed as $t_i = p_i t_b$, where t_b is the unknown creep lifetime and p_i is the average fraction of time for condition t_i, T_i, σ_i in one hour of operation. The p_i values can be estimated from a basic block of missions for the vehicle with the sum of all p_i being one. From Equation (1.51) the life is

$$t_b = 1 \bigg/ \sum_{i=1}^{N}(p_i/t_{Ri}). \qquad (1.52)$$

For the case of *tension creep*, the time t_{Ri} can be obtained from tension creep curves similar to Figure 1.29, or from stress time creep curves in Reference 1. Use the σ_i stress and read the $(LM)_i$ value for the specified creep strain or rupture requirement. Use the specified T_i value and the $(LM)_i$ in Equation (1.50) to get

$$\log_{10} t_{Ri} = \frac{10^3 (LM)_i}{T_i + 460} - 19. \qquad (1.53)$$

Use the t_{Ri} values in Equation (1.52) to get the life t_b. If t_b is too small for the required design value, the stress levels must be reduced and the calculations repeated.

For the case of *compression creep* without local buckling or column action, the same procedure as for tension creep above can be used. However, the crippling

Allowable stresses of flight vehicle materials

and column creep cases are more involved. One procedure is to construct a crippling or column curve against time for each temperature T_i. For each given T_i, use the procedure of Example 1.8 for various selected times so that a curve of allowable crippling or column stress against time can be drawn for the particular structural element. With such a curve available for each temperature T_i, the σ_i stresses can be used to read the t_{Ri} times directly. Equation (1.52) then gives the life of the component.

Example 1.10. A 2024-T3 clad aluminum alloy structure without local buckling and column effects is subjected to (1) $T_1 = 300°F$, $\sigma_1 = 30{,}000$ psi, $p_1 = 0.40$; (2) $T_2 = 400°F$, $\sigma_2 = 20{,}000$ psi, $p_2 = 0.50$; (3) $T_3 = 500°F$, $\sigma_3 = 10{,}000$ psi, $p_3 = 0.10$. Use Figure 1.29 and find the life t_b for $e_{\max} = 0.010$.

Solution. Use the given stresses on Figure 1.29 and read $(LM)_1 = 17.3$, $(LM)_2 = 18.9$, $(LM)_3 = 21.0$. From Equations (1.53) and (1.52), $t_{R1} = 5800$ hours, $t_{R2} = 950$ hours, $t_{R3} = 750$ hours,

$$\frac{1}{t_b} = \frac{0.40}{5800} + \frac{0.50}{950} + \frac{0.10}{750}, \quad t_b = 1372 \text{ hours.}$$

Note that 72% of the creep effect comes from the case (2) conditions. In fact, if σ_2 is reduced to 18,000 psi, then the life will be doubled.

1.7. Room temperature fatigue effects on allowable stresses

In its lifetime the flight vehicle is subjected to varying stresses, repeated stresses, and cyclic stresses due to gusts and vibrations of the structure. These repeated stress cycles tend to *fatigue* the material and to reduce the allowable stresses. The fatigue data for each material alloy must be obtained from tests which are conducted by cycling a given load on the material specimen until failure occurs.

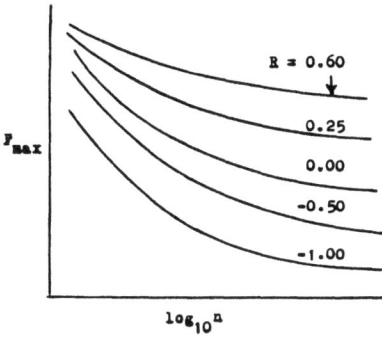

Fig. 1.30. S-N fatigue curves for a material at various stress ratios (R).

The maximum test stress $F_{\max}$ is plotted against the number of cycles n (usually $\log_{10} n$) with one of the expressions

$$R = F_{\min}/F_{\max}, \quad \text{or} \quad A = F_a/F_m, \quad R = \frac{1-A}{1+A}, \quad A = \frac{1-R}{1+R}, \quad (1.54)$$

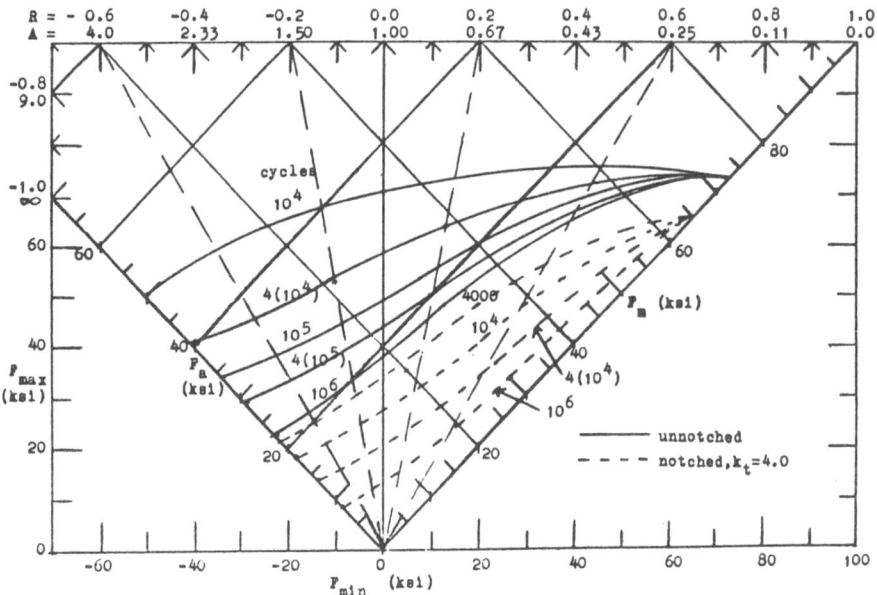

Fig. 1.31. Constant-life fatigue diagram for 2024-T3 aluminum alloy.

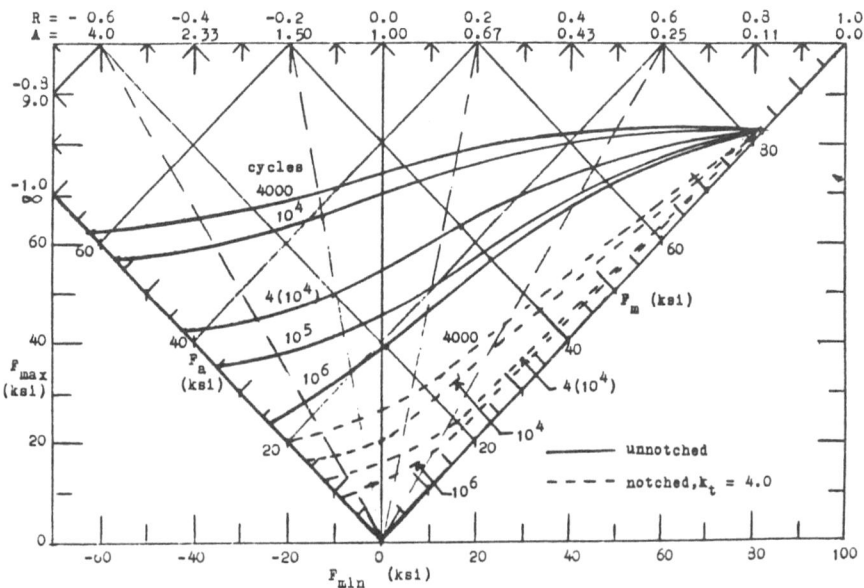

Fig. 1.32. Constant-life fatigue diagram for 7075-T6 aluminum alloy.

used as a parameter for each curve. Here $F_{\max}$ is the largest stress in the stress cycle (algebraically), $F_{\min}$ is the smallest algebraic value in the stress cycle, $F_a = (F_{\max} - F_{\min})/2$ is the alternating stress amplitude of the cycle, and the mean stress $F_m = (F_{\max} + F_{\min})/2$ is the steady stress in the cycle. Figure 1.30 shows the form of the $S - N$ or stress-cycle fatigue curves from test data.

Usually the fatigue data is cross plotted from Figure 1.30 for constant number of cycles to give constant-life diagrams. Figures 1.31 and 1.32 are from Reference 1 and show the typical constant-life diagrams given in Reference 1 for many of the alloys. Note the large effect of the notch upon the fatigue life of the material in Figures 1.31 and 1.32, where the reduction in F_a for given F_m is approximately F_a/k_t for the same life. The four stresses and two ratios R and A shown in Equation (1.54) have scales on the diagrams so that any combination of values can be used.

The diagrams in Figures 1.31 and 1.32 give the cycle life to failure at one fixed combination of stresses. The flight vehicle is subjected to cycling at various stresses and various number of cycles during its life. Among the various cumulative-fatigue laws described in the literature, some of which are considered in Reference 22, Miner's law or life fraction rule, which is similar to Equation (1.51) for creep, states that failure is determined by

$$\sum_{i=1}^{M}(n_i/N_{Ri}) = 1. \qquad (1.55)$$

Here n_i is the number of cycles at a given alternating stress F_{ai} and mean stress F_{mi} and N_{Ri} is the number of cycles from the constant-life diagram that causes failure at the same F_{ai} and F_{mi}. For simplicity, only Miner's law is considered here.

Let $n_i = q_i t_f$, where t_f is the unknown fatigue lifetime and q_i is the average number of cycles at conditions F_{ai} and F_{mi} in one hour of operation. From Equation (1.55),

$$1/t_f = \sum_{i=1}^{M}(q_i/N_{Ri}). \qquad (1.56)$$

Example 1.11. A critical point on a structure at room temperature with $k_t = 4.0$ is subjected to (1) $F_{m1} = 15$ ksi, $F_{a1} = 5$ ksi, $q_1 = 20$ cycles per hour; (2) $F_{m2} = 15$ ksi, $F_{a2} = 10$ ksi, $q_2 = 4$ cycles per hour; (3) $F_{m3} = 15$ ksi, $F_{a3} = 15$ ksi, $q_3 = 0.8$ cycles per hour; (4) $F_{m4} = 15$ ksi, $F_{a4} = 20$ ksi, $q_4 = 0.1$ cycles per hour. Use Equation (1.56) and Figures 1.31 and 1.32 to find the life t_f for (a) 2024-T3 aluminum alloy material, and (b) 7075-T6 aluminum alloy material. (c) compare results.

Solution. (a) From Figure 1.31 for the given stresses, $N_{R1} = 5(10^5)$ cycles, $N_{R2} = 3(10^4)$, $N_{R3} = 6000$, $N_{R4} = 2000$, and

$$t_f = \left(\frac{20}{500,000} + \frac{4}{30,000} + \frac{0.8}{6000} + \frac{0.1}{2000}\right)^{-1} = 2800 \text{ hours}.$$

(b) From Figure 1.32 for 7075-T6, $N_{R1} = 40,000$ cycles, $N_{R2} = 10,000$, $N_{R3} = 3800$, $N_{R4} = 1000$, whence

$$t_f = \left(\frac{20}{40,000} + \frac{4}{10,000} + \frac{0.8}{3800} + \frac{0.1}{1000}\right)^{-1} = 826 \text{ hours}.$$

(c) Although the static strength of 7075-T6 is much larger than 2024-T3, its fatigue strength with $k_t = 4.0$ is smaller for the same cyclic conditions.

1.8. Temperature effects upon allowable fatigue stresses

The temperature not only affects the alternating stress F_a directly but also affects the mean stress F_m by causing creep. If the mean stress is large, the structure may fail from creep with little effect from the cycling stresses. Thus, both creep and fatigue effects are present for any given elevated temperature. If a relationship between cycles and time is specified (q_i in Equation 1.56), then a constant life diagram similar to that for room temperature fatigue in Figures 1.31 and 1.32 can be constructed from cycling tests with creep measurements. The diagrams can be drawn for a specified creep strain or for rupture. The F_m mean stress values for $F_a = 0$ are taken from creep tests while the F_a alternating stress values for $F_m = 0$ are taken from alternating tests at the same

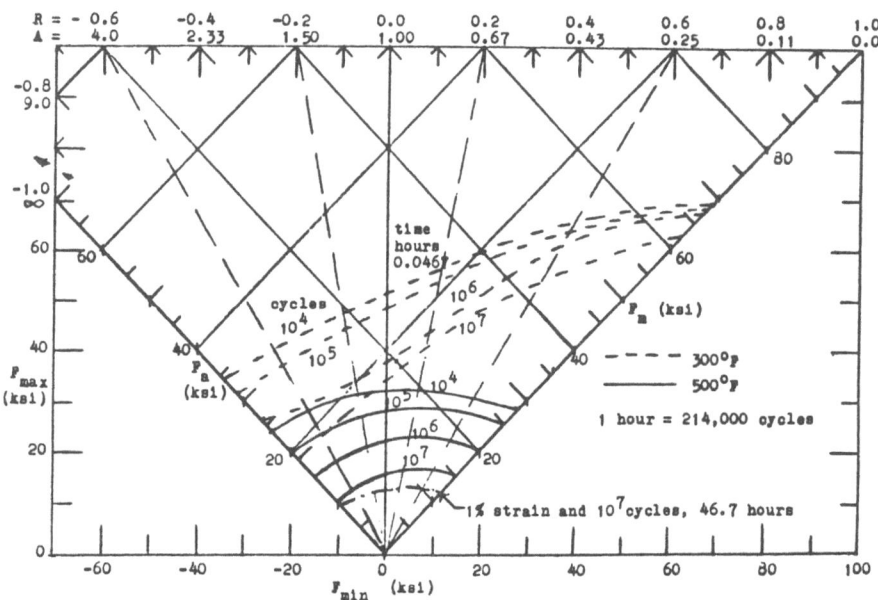

Fig. 1.33. Constant-life fatigue diagram for 2024-T4 rolled aluminum alloy at 300°F and 500°F.

temperature. Figure 1.33 shows constant life rupture curves and one 1% creep strain curve for 2024-T4 aluminum alloy at 300°F and 500°F, Reference 24. The time for these curves is short with $N_i = qt_i$ and $q = 214,000$ cycles per hour.

If the time is longer or q is smaller, the constant life curves will be different due to the creep effect being larger. With q a variable as well as temperature and time, it is evident that the number of tests to get the fatigue creep data for

Allowable stresses of flight vehicle materials

various materials is prohibitive. In order to reduce the number of required tests, interaction curves between R_a and R_m, where

$$R_a = F_a/F_{aR}, \quad R_m = F_m/F_{mR},$$
$$F_{aR} = \text{cyclic stress at } F_m = 0 \text{ and given } T, \quad (1.57)$$
$$F_{mR} = \text{creep stress at } F_a = 0 \text{ and given } T,$$

were used in Reference 24. If the interaction equations are known, then only the fatigue cyclic tests and the creep tests are needed. Also, the Larson-Miller parameter can be used for the creep data, Figure 1.29.

The form of the interaction equation is

$$R_a^g + R_m^h = 1, \quad (1.58)$$

where g and h may depend upon temperature, frequency, time and material. The 500°F curves in Figure 1.33 can be represented reasonably well by $g = h = 2$, or a circle in the non-dimensional form of Equations (1.57) and (1.58). Some of the curves on Figures 1.31 and 1.32 can have $g = 2$, $h = 1$ and others can be approximated by $g = h = 1$, such as the 300°F curves on Figure 1.33. The life of the structure under several different conditions can be obtained by using Equation (1.56), as the following example demonstrates.

Example 1.12. A 2024-T3 clad aluminum alloy compression structure without local buckling and column effects is subjected to (1) $T_1 = 300°F$, $F_{a1} = 15{,}000$ psi, $F_{m1} = 15{,}000$ psi, $p_1 = 0.40$, (2) $T_2 = 500°F$, $F_{a2} = 8000$ psi, $F_{m2} = 6000$ psi, $p_2 = 0.40$, (3) $T_3 = 500°F$, $F_{a3} = 15{,}000$ psi, $F_{m3} = 4000$ psi, $p_3 = 0.20$. Find the life of the structure for a maximum strain of $e_{max} = 0.010$. Assume the frequency to be 1000 cycles per hour and take F_{aR} from the ordinate of Figure 1.33 at $F_m = 0$.

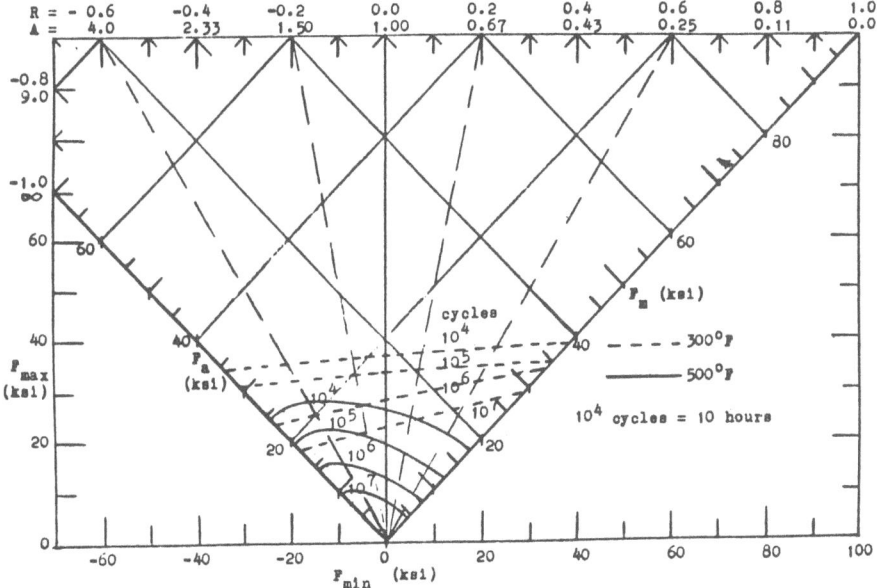

Fig. 1.34. Constant life fatigue diagram for 2024-T3 aluminum alloy for Example 1.12.

Solution. Assume 10^4 cycles = 10 hours and read $F_{aR1} = 33,500$ psi, $F_{aR2} = F_{aR3} = 24,000$ psi from Figure 1.33. Calculate $(LM)_1 = 760(19+1)/1000 = 15.2$, $(LM)_2 = (LM)_3 = 19.2$, and read $F_{m1} = 37,000$ psi, $F_{m2} = F_{m3} = 18,000$ psi from Figure 1.29 on the 0.010 strain curve. Assume the interaction curves to be $R_a + R_m = 1$ at 300°F and $R_a^2 + R_m^2 = 1$ at 500°F, and draw the 10 hour = 10^4 cycle interaction curves on Figure 1.34 using the above reference values. Repeat the procedure for 10^5 cycles = 100 hours, 10^6 cycles = 1000 hours, 10^7 cycles = 10,000 hours, and draw the curves on Figure 1.34. This Figure 1.34 represents the design curves for selected stress combinations within the conditions and range of Figure 1.34. For the given stress conditions interpolate logarithmically on Figure 1.34 to get the reference time t_{Ri}, or $t_{R1} = 300$ hours, $t_{R2} = 1000$ hours, and $t_{R3} = 200$ hours. From Equation (1.52) or (1.56),

$$t_b = t_f = \left(\frac{0.40}{300} + \frac{0.40}{1000} + \frac{0.20}{200}\right)^{-1} = 366 \text{ hours}.$$

Up to this point in this Section 1.8 it has been assumed that the temperature in the unrestrained structure is uniform or changes slowly to a different temperature so that no thermal stresses are present. If the structure is restrained or the temperature is nonuniform steady state or nonuniform cyclic, then thermal stresses are produced which will change the creep and fatigue effects. Although there may be no applied mean stress and no applied alternating stress, the cycling thermal stresses produced by the cycling temperature may cause thermal fatigue and creep of elements in the structure. See References 23 and 24 for further discussion of these thermal stress problems. Also, if a mean stress is present, then a nonuniform cyclic temperature may produce strain growth in the structure which may cause rupture in less than 100 cycles. See Section 1.16 (Vol.1).

1.9. Crack effects upon allowable fatigue stresses

Figures 1.31 and 1.32 show the large effect of notches upon the fatigue life of the structure. Cracks in the structure have similar effects on the fatigue life of the structure, and cracks may be present from flaws in the material manufacture, from the fabrication processes, and from scratches in assembly procedures and in service. The fracture strength of a structural component containing a crack depends upon crack length $2a$, the component geometry, and the material property "fracture toughness" K_{Scr}. The K_{Scr} property is the critical value of the stress intensity factor K_S, where

$$K_S = \sigma(aH)^{1/2}, \text{ (ksi-in}^{1/2}\text{)}, \tag{1.59}$$

with σ the applied stress in ksi, a the half crack length in inches, and H a nondimensional factor involving geometry and a. In flat plates, $H = \pi$ for uniform tension and for uniform shear stresses.

Since the smallest values of K_{Scr} for a given material occur for plate type components and since the H factor in Equation (1.59) can be evaluated for various types of cracks and loadings in plates (Reference 25), the values of the fracture toughness K_{Scr} can be obtained from tests using the stress F_{Scr} and the crack length $2a_{cr}$ at failure in Equation (1.59). These values are given for some of the material alloys in Reference 1. They vary with plate thickness and crack direction, where L-T means load in the L-direction of the material and T

is the transverse crack direction. Some values from Reference 1 for 7075-T6 clad aluminum alloy plates for the *L-T* case are

$$K_{Scr} = 54.7 \text{ ksi-(in)}^{1/2}, \quad t = 0.063",$$
$$K_{Scr} = 60.6 \text{ ksi-(in)}^{1/2}, \quad t = 0.080", \quad (1.60)$$
$$K_{Scr} = 61.4 \text{ ksi-(in)}^{1/2}, \quad t = 0.090".$$

The special clad aluminum alloy 7475-T61 plates have K_{Scr} approximately 80 ksi-(in)$^{1/2}$. Thus, if K_{Scr} and a are known for a plate, then Equation (1.59) gives the allowable fracture stress as

$$F_{Scr} = K_{Scr}(Ha)^{-1/2}. \quad (1.61)$$

Tests shows that under cyclic or fatigue loading the crack grows in a stable manner to some value of a at which it becomes unstable and causes fracture of the material. The rate of growth of the crack depends upon the stress intensity factor K_s, whence

$$da/dN = C(\Delta K_s)^n, \quad (1.62)$$

where C and n depend upon test data, da/dN is the growth of a per cycle, and

$$\Delta K_S = 2F_a(aH)^{1/2}. \quad (1.63)$$

For uniform constant amplitude cycling without mean stress, a_{cr} is given by Equation (1.61) with $F_{Scr} = F_{max}$, and Equation (1.62) can be integrated to give the number of cycles to fracture, or

$$N_{cr} = \int_{a_i}^{a_{cr}} da/(\Delta K_S)^n C, \quad (1.64)$$

where a_i is half the initial crack length. When mean stress is present and the cycles vary in magnitude, the determination of N_{cr} becomes more involved. Also, tests show that yielding around the crack tip may retard the crack growth so that the life may be longer in ductile materials. Various retardation or relaxation models have been developed to account for these effects in the life calculations.

Example 1.13. A 7075-T6 clad aluminum alloy 0.080 in. thick plate is subjected to a constant amplitude uniform cycling with $F_a = 15$ ksi, $F_m = 0$. If $C = 2(10^{-9})$ and $n = 3.43$ for 7075-T6 alloys, use Equation (1.64) to find the fatigue life N_{cr}. Use $a_i = 0.1$ in. and compare the result to the $k_t = 4$ notched case in Figure 1.32.

Solution. From Equations (1.63), (1.61), and (1.60) with $H = \pi$, $\Delta K_S = 2(15)(\pi a)^{1/2} = 53.17 a^{1/2}$, $a_{cr} = K_{Scr}^2/HF_{max}^2 = \frac{1}{\pi}\left(\frac{60.6}{15}\right)^2 = 5.2$ in., whence Equation (1.64) becomes

$$N_{cr} = 0.5(10^9)(53.17)^{-3.43} \int_{0.1}^{5.2} da/a^{1.715} = 4100 \text{ cycles.}$$

This compares to about 12,000 cycles on Figure 1.32 for $k_t = 4.0$.

1.10. Problems

1.1. Solve Example 1.3 for the case of $T = 400°F$ with 10 hour exposure at $T = 400°F$.

1.2. Find the allowable stresses F_{tu} and F_{ty} for a clad 7075-T6 plate or sheet 0.200 in. thick (L-direction, A basis) at $T = 300°F$ after 100 hour exposure at $T = 500°F$.

1.3. In Example 1.4 take $t_1 = 0.100$ in. and determine F_{cr} and F_{cc} at $T = 400°F$ after 10 hour exposure at $T = 400°F$.

1.4. Solve Example 1.5 for the cases of $b_w = 1.00$ in., 1.50 in., 2.00, 3.00, 4.00, and 5.00 in. Graph the results as stress against b_w on the abscissa and comment on the results.

1.5. Repeat Problem 1.4 for $T = 300°F$ after 1000 hour exposure at 300°F.

1.6. In Figure 1.18 take $b_F = 1.00$, $b_w = 2.00$, $b_s = 6.00$ in., $t_w = 0.100$, $t_s = 0.080$ in., clad 2024-T3 material (L-direction, B basis), and find the allowable compressive stresses F_{cr}, F_{cc}, and F_c. For the column use $L = 60$ in., $c = 1.5$, and assume symmetrical channels are properly riveted to the plate so that the axis of bending is parallel to the plate. Use 6.00 in. of skin or plate with one stiffener to calculate the radius of gyration ρ.

1.7. Draw Mohr's circle to scale for the case of $\sigma_{xx} = 40$ ksi, $\sigma_{yy} = 15$ ksi, $\sigma_{xy} = 15$ ksi. (a) Read the principle stresses and the maximum shear stress from the circle, and check by Equations (1.22) and (1.23). (b) Read the $2\theta_d$ angle from the circle and check by Equation (1.21).

1.8. Repeat Problem 1.7 for $\sigma_{xx} = 40$ ksi, $\sigma_{yy} = -20$ ksi, $\sigma_{xy} = 15$ ksi.

1.9. On the Mohr's circle in Problem 1.7 measure the angle $2\theta = 90°$ to the x_1, y_1 axes and read the stresses $\sigma_{xx1}, \sigma_{yy1}, \sigma_{xy1}$. Check by Equation (1.20).

1.10. Construct curves on Figure 1.24 for $B_s = 1/4$, $B_s = 3/4$, and $B_s = 1$.

1.11. Repeat Example 1.6 for $B_a = 0.6$, $B_s = 0$.

1.12. Repeat Example 1.6 for $B_a = -0.6$, $B_s = 0.5$.

1.13. Repeat Example 1.6 for 2024-T3 sheet with L as the x-direction and basis (A).

1.14. Derive the curve for $p = 1/2$ on Figure 1.27.

1.15. Repeat Example 1.7 for $a = 4.00$ in., $b = 2.00$ in.

1.16. Repeat Example 1.7 for $\sigma_{xx} = 10,000$ psi, $\sigma_{yy} = 8000$ psi, $\sigma_{xy} = 0$.

1.17. Solve Example 1.8 for an allowable total deformation of 0.5% after 200 hours of constant load.

1.18. Solve Example 1.9 for a creep life of 500 hours.

1.19. Solve Example 1.9 for a channel section with $t_w = t_f = 0.080$ in., $b_F = 1.00$ in., $b_w = 2.20$ in., $L = 100$ in., $c = 1$.

1.20. Solve Example 1.10 with the σ_2 stress changed to $\sigma_2 = 15,000$ psi.

1.21. Solve Example 1.10 if the maximum allowable strain is $e_{max} = 0.005$.

1.22. Solve Example 1.11 for all the F_{mi} stresses zero.

1.23. Solve Example 1.11 for all the F_{mi} stresses = 25 ksi.

1.24. Solve Example 1.12 for a frequency of 200 cycles per hour.

1.25. Solve Example 1.12 for a frequency of 5000 cycles per hour.

1.26. Solve Example 1.13 for $F_a = 25$ ksi, $F_m = 0$.

1.27. Solve Example 1.13 for $F_a = 10$ ksi, $F_m = 0$.

References

Chapter 1

1. Mil-HDBK-5c, *Metallic Materials and Elements of Aerospace Vehicle Structures*, 2 vols., U.S. Government Printing Office, Washington, D.C. 20013. P.O. Box 1533, Sept. (1976).
2. E.F. Bruhn: *Analysis and Design of Flight Vehicle Structures*, Tri-State Offset Co., Cincinnati, Ohio, 45202.
3. E.E. Lundquist, E.Z. Stowell and E.H. Schuette: *Principles of Moment Distribution Applied to Stability of Structures Composed of Bars and Plates*, NACA TR 809 (1945).
4. E.H. Schuette and J.C. McCulloch: *Charts for Minimum Weight Design of Multiweb Wings in Bending*, NACA TN 1323 (1947).
5. W.D. Kroll: *Tables of Stiffness and Carry-over Factors for Flat Rectangular Plates under Compression*, NACA WRL-398 (ARR 3K27) (1943).
6. E.Z. Stowell: *A Unified Theory of Plastic Buckling of Columns and Plates*, NACA TR 898 (TN 1556) (1948).
7. B.E. Gatewood and E.L. Williams: Allowable compressive stresses in aircraft structures, *J. Aeronautical Sciences* 18, No. 10, pp. 657-664 (1951).
8. B.E. Gatewood and D.W. Breuer: Allowable stresses for channels and zees in bending, *J.*

Aeronautical Sciences **21**, No. 5, p.349 (1954).
9. B.E. Gatewood: *Thermal Stresses*, McGraw-Hill Book Co., New York (1957).
10. G. Gerard and H. Becker: *Handbook of Structural Stability, Part I, Buckling of Flat Plates*, NACA TN 3781 (1957).
11. H. Becker: *Handbook of Structural Stability, Part II, Buckling of Composite Elements*, NACA TN 3782 (1957).
12. G. Gerard and H. Becker: *Handbook of Structural Stability, Part III, Buckling of Curved Plates and Shells*, NACA TN 3783(1957).
13. G. Gerard: *Handbook of Structural Stability, Part IV, Failure of Plates and Composite Elements*, NACA TN 3784 (1957).
14. G. Gerard: *Handbook of Structural Stability, Part V, Compressive Strengths of Flat Stiffened Panels*, NACA TN 3785 (1957).
15. E.Z. Stowell: Compressive Strength of Flanges, NACA TN 2020, 1950.
16. J. Mayers and B. Budiansky: *Analysis of Behavior of Simply Supported Flat Plates Compressed Beyond the Buckling Load into the Plastic Range*, NACA TN 3368 (1955).
17. J. Marin: *Mechanical Behavior of Engineering Materials*, Prentice-Hall, New York (1962).
18. S. Timoshenko: *Theory of Elastic Stability*, McGraw-Hill Book Co., New York (1936).
19. S.B. Batdorf and M. Stein: *Critical Combinations of Shear and Direct Stress for Simply Supported Rectangular Flat Plates*, NACA TN 1223 (1947).
20. F.R. Larson and J.M. Miller: A time-temperature relationship for rupture and creep stresses, *Trans. ASME* **74**, pp. 765-775 (1952).
21. S.S. Manson: *Thermal Stress and Low-Cycle Fatigue*, McGraw-Hill Book Co., New York, 1966.
22. B.E. Gatewood and J.P. Honaker: On S-N curves for fatigue analysis, *Jour. of Aero. Sciences*, **23** (1956).
23. G.E. Maddux, L.A. Vorst, F.J. Giessler and T. Moritz: *Stress Analysis Manual*, Technical Report AFFDL-TR-69-42, Wright-Patterson AFB, Ohio (1970).
24. J. Padlog and A. Schnitt: *A Study of Creep, Creep-fatigue, and Thermal-Stress-Fatigue in Airframes Subject to Aerodynamic Heating*, Wright Air Development Center TR-58-294, Wright-Patterson AFB, Ohio (1958).
25. D.P. Wilhelm: *Fracture Mechanics Guidelines for Aircraft Structural Applications*, AFFDL-TR-69-111, Wright-Patterson AFB, Ohio (1970).
26. B.E. Gatewood and R.W. Gehring: Allowable axial loads and bending moments for inelastic structures under nonuniform temperature distribution, *J. of Aerospace Sciences* **29**, No. 5, May (1962).

References for additional reading

27. W.N. Findley, J.S. Lai and K. Onaran: *Creep and Relaxation of Nonlinear Viscoelastic Materials*, North Holland, Amsterdam (1976).
28. G.P. Cherepanov: *Mechanics of Brittle Fracture*, McGraw-Hill Book Co., New York (1979).
29. R.P. Skelton (Editor): *High Temperature Fatigue*, Elsevier Applied Science Publishers, New York (1987).

2

Analysis and design of joints and splices

2.1. Introduction

In all assembled structures there exists the problem of joining the component parts. The joints, or splices, or attachments, must carry the same loads as the components they join together, and they must be efficient in that they add very little weight to the structure. Ideally, a joint should be as strong as the structure it joins together. Actually, in most cases it is very difficult and expensive to make a joint as strong as the structure. This means that ordinarily the joint is designed on the basis of the loads that the structure carries and not on the basis of the total strength of the structure. Because of strain concentrations and strain redistributions required in a joint, the joint may cause large reductions in the allowable loads in the materials being joined, particularly in materials with low ductility. In all too many cases, the joint becomes the weak link in the structure, with the majority of structural failures occurring at joints and splices. Thus, in the design of flight vehicles, much effort and time are spent on the analysis and design of joints.

Since joints involve rivets and bolts with standard diameters, holes with standard diameters, plates with standard thicknesses, and correction factors based on tests using these standard values, it is necessary to use in this chapter the units for which standard value data is available. Until such time as standards are set up and test data is obtained in SI units, it will be necessary to use the available standard data in the usual units of pounds for force and inches for length in joint analysis and design. The correction factors k_s in Table 2.1 must be obtained from tests using standard plates and rivets.

The primary emphasis in this chapter is on riveted and bolted joints. In Sections 2.2 through 2.6 the joints are regarded as determinate with the materials being sufficiently ductile to adjust to the assumed load distribution before failure. Deflection, temperature, and inelastic effects on the joints are considered in Sections 2.7 and 2.8. Some discussion of bonded and welded attachments is given in Sections 2.9 and 2.10.

2.2. Analysis of plate splices with axial tension forces

Consider the simple lap joint shown in Figure 2.1, where the plate has an axial

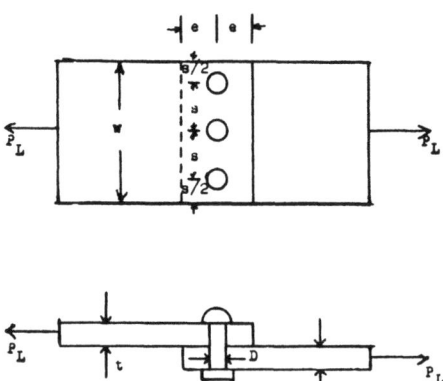

Fig. 2.1. Simple lap joint with tension force.

limit applied force P_L. Assume that the force acts through the centroid of the rivet pattern, that the rivets all have the same diameter D and the same material, and that each rivet takes the same load ($P_L/3$ in Figure 2.1.). To analyze the joint for the yield criterion (A), Equation (B.74, Vol.1),

Margin of Safety = M.S. = $(F_y/k_L\sigma_L) - 1 \geq 0.00$, (A), \hfill (B.74)

and for the failure criterion (B), Equation (B.75, Vol.1),

M.S. = $(F_u/k_u\sigma_L) - 1 \geq 0.00$, (B), \hfill (B.75)

it is necessary to identify the possible modes of failure of the joint.

There are four possible ways a joint can fail:

(1) Rivet shear failure,
(2) Plate tension failure,
(3) Plate bearing failure,
(4) Plate tear out (shear) failure.

It is necessary to calculate the allowable load for the rivet or plate in each of these cases for either the yield criterion (A) or the failure criterion (B), whichever is critical, and compare it to the applied load $k_L P_L$ for (A) or $k_u P_L$ for (B). For given materials and given factors k_L and k_u, Equation (B.76, Vol.1),

$F_u/F_y < k_u/k_L$, use (B); $\quad F_u/F_y > k_u/k_L$, use (A), \hfill (B.76)

identifies whether criterion (A) or (B) is critical (gives the smaller margin of safety). Since each rivet has the same load, these calculations can be made on the basis of one rivet as unit values or on the basis of the total load (for all three rivets in Figure 2.1).

(1) *Rivet Shear.* Since no yield stress is given in the Tables of Reference 1 and in Tables 1.1 through 1.6 for shear, rivet shear is based on criterion (B). The theoretical allowable ultimate or failue load for a rivet is

$$P_{su} = (\pi D^2/4) F_{su}, \hfill (2.1)$$

where D is the rivet diameter and F_{su} is the ultimate shear stress for the rivet material, Table 2.1. However, two corrections to Equation (2.1) must be made

Table 2.1. Data for round head rivets.

Shear strength of solid rivets, (lb), $(P_{su})_{Ref}$

D (in.)		1/16	3/32	1/8	5/32	3/16	1/4	5/16	3/8
F_{su},(ksi)									
5056,	28	99	203	363	556	802	1450	2290	3280
2117-T3,	30	106	217	388	596	862	1550	2460	3510
2017-T31,	34	120	247	442	675	977	1760	2790	3970
2017-T3,	38	135	275	494	755	1090	1970	3110	4450
2024-T31,	41	145	296	531	815	1180	2120	3360	4800
A286,	90	317	651	1170	1790	2580	4670	7370	10500
Monel	49	173	355	635	973	1400	2540	4020	5730

Single-shear rivet strength factors, k_s

t(in.)	1/16	3/32	1/8	5/32	3/16	1/4	5/16	3/8
0.016	0.96	-	-	-	-	-	-	-
0.018	0.98	0.91	-	-	-	-	-	-
0.020	0.99	0.93	-	-	-	-	-	-
0.025	1.00	0.97	0.92	-	-	-	-	-
0.032	-	1.00	0.96	0.92	-	-	-	-
0.040	-	-	0.99	0.96	0.93	-	-	-
0.050	-	-	-	0.99	0.97	0.92	-	-
0.063	-	-	-	1.00	1.00	0.96	0.92	-
0.071	-	-	-	-	-	0.98	0.94	0.91
0.080	-	-	-	-	-	0.99	0.96	0.93
0.090	-	-	-	-	-	1.00	0.98	0.95
0.100	-	-	-	-	-	-	0.99	0.97
0.125	-	-	-	-	-	-	1.00	1.00

Double-shear rivet strength factors, k_s

t (in.)	1/16	3/32	1/8	5/32	3/16	1/4	5/16	3/8
0.016	0.69	-	-	-	-	-	-	-
0.018	0.74	0.52	-	-	-	-	-	-
0.020	0.79	0.58	-	-	-	-	-	-
0.025	0.87	0.71	0.54	-	-	-	-	-
0.032	0.94	0.81	0.69	0.56	-	-	-	-
0.040	0.99	0.89	0.79	0.69	0.58	-	-	-
0.050	1.00	0.95	0.87	0.79	0.69	0.54	-	-
0.063	-	1.00	0.94	0.87	0.81	0.68	0.55	-
0.071	-	-	0.97	0.91	0.85	0.74	0.62	0.51
0.080	-	-	0.99	0.94	0.89	0.79	0.69	0.58
0.090	-	-	1.00	0.97	0.92	0.83	0.74	0.65
0.100	-	-	-	0.99	0.95	0.87	0.79	0.70
0.125	-	-	-	1.00	1.00	0.94	0.87	0.80
0.160	-	-	-	-	-	0.99	0.94	0.89
0.190	-	-	-	-	-	1.00	0.98	0.94
0.250	-	-	-	-	-	-	1.00	1.00

to obtain the true allowable load for the rivet. (1) Since the rivet fills the hole if properly driven, D in Equation (2.1) should be the hole diameter rather than the rivet diameter. Table 2.1 or Reference 1 gives the $(P_{su})_{Ref}$ values as calculated from the standard hole diameters and the specified material F_{su}. (2) Since a thin plate acts more as a knife to cut the rivet rather than to shear it, Table 2.1 shows correction factors k_s, as given in References 1 and 2, for the various plate standard thicknesses and the standard rivet sizes. Thus, with these two corrections, the allowable load for rivet shear can be expressed as

$$P_{su1} = k_s(P_{su})_{\text{Ref}}, \quad \text{single shear},$$
$$P_{su1} = 2k_s(P_{su})_{\text{Ref}}, \quad \text{double shear},$$
(2.2)

as based on Table 2.1. The margin of safety for rivet shear is

$$\text{M.S.} = \frac{P_{su1}}{k_u(P_L/N)} - 1, \quad (B),$$
(2.3)

for N equally loaded rivets.

(2) *Plate Tension.* Since the tension load in the plate must cross the net area region where the rivet holes are located before it can bear on the rivet, the allowable total load P_{tu} or P_{ty} in the plate is

$$P_{ty} = t(w - ND)F_{ty}, \quad P_{tu} = t(w - ND)F_{tu},$$
(2.4)

for N rivets in one row ($N = 3$ in Figure 2.1). If the edge distance on the side is $s/2$ with s the equal rivet spacing in a row, then the allowable unit loading per rivet is

$$P_{ty1} = t(s - D)F_{ty}, \quad P_{tu1} = t(s - D)F_{tu}, \quad \text{per rivet}.$$
(2.5)

The margin of safety for plate tension is

$$\text{M.S.} = \frac{P_{ty}}{k_L P_L} - 1, \quad \text{or} \quad \text{M.S.} = \frac{P_{ty1}}{k_L(P_L/N)} - 1, \quad (A),$$
$$\text{M.S.} = \frac{P_{tu}}{k_u P_L} - 1, \quad \text{or} \quad \text{M.S.} = \frac{P_{tu1}}{k_u(P_L/N)} - 1, \quad (B),$$
(2.6)

where (A) or (B) is used according to Equation (B.76). The F_{ty} yield values and the F_{tu} ultimate values are given in Tables in Section 1.2, or in References 1 and 2. Note that for a compression load, the load bears directly on the rivets and the total area of the plate can be used so that the joint has little effect on the plate compression design.

(3) *Plate Bearing.* When the rivet bears on the plate it tends to elongate the rivet hole and possibly loosen the rivet under repeated loading. The F_{bry} and F_{bru} values in the material property tables in Section 1.2 and in References 1 and 2 are based on the test loads at specified per cent elongation of the hole (2% for F_{bry} and actual failure for F_{bru}). Since this bearing elongation is more critical near the edge, the tables show values for $e/D = 2.0$ and $e/D = 1.5$. For $e/D \geq 2.0$, use the $e/D = 2.0$ values; for $1.5 \leq e/D \leq 2.0$, use straight line interpolation between the given values. Tests are required for any $e/D < 1.5$. The allowable loads for bearing per rivet are

$$P_{bry1} = tDF_{bry}, \quad P_{bru1} = tDF_{bru},$$
(2.7)

where t is the plate thickness and D is the rivet diameter. The margin of safety for plate bearing is

$$\text{M.S.} = \frac{P_{bry1}}{k_L(P_L/N)} - 1, \quad (A),$$
$$\text{M.S.} = \frac{P_{bru1}}{k_u(P_L/N)} - 1, \quad (B),$$
(2.8)

where (A) or (B) is used according to Equation (B.76) with the bearing allowables.

(4) *Plate Tear-out.* If the edge distance e in Figure 2.1 is too small the rivet can shear the plate out through the edge for tension load. See Figure 2.2. The shear out occurs on two faces forcing a plug out. The net area is taken as $2t(e - \frac{D}{2})$ in Figure 2.2, whence the allowable tear-out load per rivet is

$$P_{tou1} = 2t(e - \frac{D}{2})F_{su}, \quad (B),$$
(2.9)

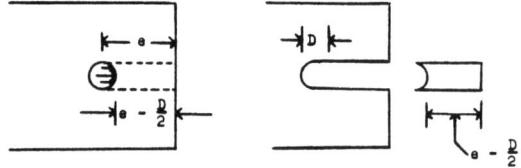

Fig. 2.2. Plate tear-out by shearing plug out of plate.

where only the (B) criterion applies. See Tables in Section 1.2 for F_{su}. If $e \geq 2D$, tear-out is not critical and no checks are made. If $1.5D \leq e \leq 2D$, then the check must be made. If $e \leq 1.5D$, the design is unsatisfactory. The margin of safety for plate tear-out per rivet is

$$\text{M.S.} = \frac{P_{tou1}}{k_u(P_L/N)} - 1, \quad (B).$$
(2.10)

Example 2.1. In Figure 2.1, take $P_L = 800$ lb, $s = 0.50$ in., $e = 0.25$ in., $D = \frac{5}{32}$ in., 2117-T3 solid rivet, $t = 0.040$ in. clad 2024-T3 aluminum alloy plate (L-direction, A basis), $k_L = 1.00$, $k_u = 1.50$. (a) Find the M.S. for each of the four modes of failure. (b) What maximum load P_L will the splice take?

Solution. (a) From Table 1.2, the plate material properties are $F_{tu} = 60,000$ psi, $F_{ty} = 45,000$ psi, $F_{su} = 38,000$ psi, $F_{bru}(e/D = 1.5) = 90,000$ psi, $F_{bru}(e/D = 2.0) = 114,000$ psi, $F_{bry}(e/D = 1.5) = 64,000$ psi, $F_{bry}(e/D = 2.0) = 73,000$ psi. From Table 2.1 the rivet material values are $(P_{su})_{\text{Ref}} = 596$ lb, $k_s = 0.964$ for the single shear joint. From Equations (2.2) and (2.3) for rivet shear

$$\text{M.S.} = \frac{3(0.964)(596)}{(1.50)(800)} - 1 = 0.44 \quad (B), \quad \text{rivet shear.}$$

From Equation (B.76) for plate tension

$$\frac{60,000}{45,000} < \frac{1.50}{1.00}, \quad \text{use (B) criterion,}$$

whence Equations (2.5) and (2.6) give

$$P_{tu1} = 0.040(0.50 - 0.156)(60,000) = 825 \text{ lb},$$
$$\text{M.S.} = \frac{3(825)}{1.50(800)} - 1 = 1.06 \quad (B), \quad \text{plate tension}.$$

For plate bearing, $e/D = 0.25/(5/32) = 1.60$, whence

$$F_{bru} = 90,000 + \frac{1.60 - 1.50}{2.00 - 1.50}(114,000 - 90,000) = 94,800 \text{ psi},$$
$$F_{bry} = 64,000 + (0.10/0.50)(73,000 - 64,000) = 65,800 \text{ psi},$$
$$(94,800/65,800) < (1.50/1.00), \quad \text{use (B) criterion}.$$

From Equations (2.7) and (2.8),

$$P_{bru1} = 0.040(5/32)(94,800) = 592 \text{ lb},$$
$$\text{M.S.} = \frac{3(592)}{1.50(800)} - 1 = 0.48 \quad (B), \quad \text{bearing}.$$

For plate tear-out, Equations (2.9) and (2.10) give

$$P_{tou1} = 2(0.040)(0.25 - 0.078)(38,000) = 523 \text{ lb}$$
$$\text{M.S.} = \frac{3(523)}{1.50(800)} - 1 = 0.31 \quad (B), \quad \text{plate tear-out}.$$

(b) Since the smallest M.S. is 0.31, the joint can take more load, or for M.S. = 0.00 in the critical case of plate tear-out,

$$(P_L)_{\max} = (1.31)(800) = 1048 \text{ lb}.$$

2.3. Multi-row tension splices

Assume the rivets are the same size and material and that each rivet takes the same load (see Section 2.8 for discussion of deflection effects on the loads in the rivets in each row). The analysis of rivet shear and plate bearing is the same as for the single row case in Section 2.2. Since plate tear-out cannot occur directly in a multi-row splice, it is necessary to specify that $e \geq 2D$ in the splice and make no check of tear-out. Thus, the analysis for plate tension is the only mode of failure that is different from the single row case. From Figures 2.3 and 2.4, it is necessary to check plate tension on the net area at the rivet row where the plate load is largest. That is, load diagrams must be drawn to give the information for the analysis of the plates and any splice plate, Figures 2.3 and 2.4.

For the plate tension case use the net cross section area at the point of the highest load (on occasion, the net area may be smaller at some other point so that the load and net area at that point should also be checked), whence

$$P_{ty} = t(w - qD)F_{ty}, \qquad P_{tu} = t(w - qD)F_{tu}, \tag{2.11}$$

where q is the number of rivets in the row at the point being checked. In this case the margin of safety is based on the highest load at the cross section being checked, or

$$\begin{aligned}\text{M.S.} &= (P_{ty}/k_L P_L)_{\text{s.c.}} - 1, \quad (A), \\ \text{M.S.} &= (P_{tu}/k_u P_L)_{\text{s.c.}} - 1, \quad (B),\end{aligned} \tag{2.12}$$

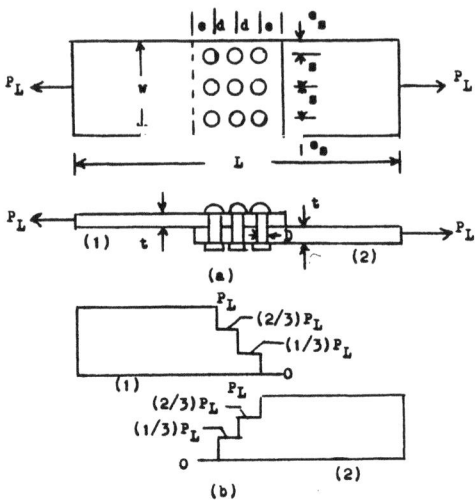

Fig. 2.3. (a). Multi-row lap tension splice. (b). Load diagrams for plates.

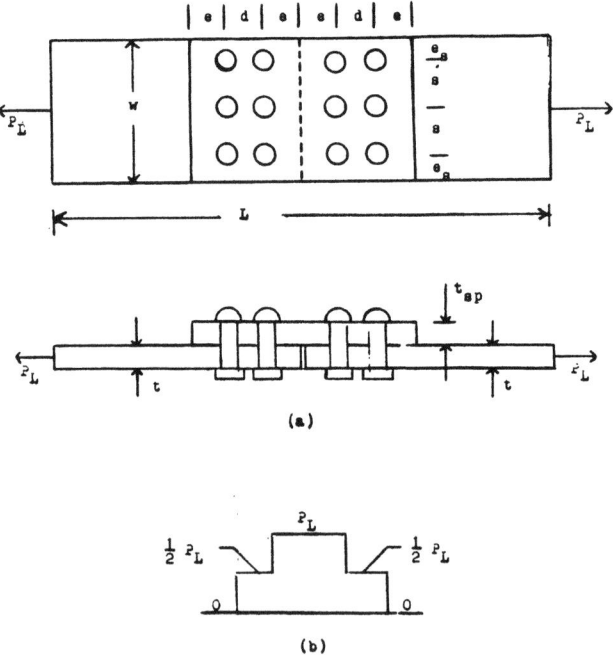

Fig. 2.4. (a). Multi-row butt tension splice. (b). Load diagram for splice plate.

at the section checked (s.c.).

In butt splices with the width w restricted, the number of rows, the rivet diameter, the splice plate width and thickness are variables that give a minimum weight design problem for the splice plates. However, the variables involved are not continuous so that the determination of the minimum weight is not a simple calculus problem. The conditions, factors, and practical limitations involved in a simple butt joint as in Figure 2.4 are (1) $d \geq 3D$, (2) $s \geq 3D$, (3) $e \geq 2D$, (4) $e_s \geq 1.5D$, (5) D has a few standard values, Table 2.1, (6) t has a few standard values, Table 2.1, (7) N = integer for number of rivets, (8) r = integer for number of rows of rivets, (9) q = integer for number of rivets in a row, (10) $N = rq$ = integer, (11) D, N must make M.S. ≥ 0 in Equation (2.3), (12) D, t in Equation (2.7) must make M.S. ≥ 0 in Equation (2.8), (13) w, t, q, D in Equation (2.11) must make M.S. ≥ 0 in Equation (2.12), (14) w is usually specified or restricted to a small range.

The weight of the splice plate in Figure 2.4 can be expressed as

$$W_{s.p.} = \rho_w L t w = \rho_w t w [4e + 2(r-1)d]$$
$$= \rho_w t w (2D + 6rD) = 2D\rho_w t w (1 + 3r), \qquad (2.13)$$

where $e = 2D$ and $d = 3D$ have been used, and ρ_w = weight density of the material. The variables D, t, w, r are involved directly in the weight in Equation (2.13) with N and q involved indirectly.

To get an idea of what is involved in obtaining a minimum weight for the splice plate, assume in Figure 2.4 that w and t are the same for both the splice plates and the original plates and that M.S. = 0.00 in Equation (2.12) for $s = 3D$, $e_s = 1.5D$, which makes $q \leq w/3D$. When $q < w/3D$, the M.S. for tension will be positive. Consider the weight expression from Equation (2.13) in the form

$$Q_w = W_{s.p.}/2\rho_w t w = D(1 + 3r), \qquad (2.14)$$

and assume $k_u P_L$ = 6000 lb and 12,000 lb for two cases, w = 2.00 in., 2024-T31 rivets with D = (1/8)in., (5/32) in., (3/16)in., (1/4)in., and (5/16) in. If k_s = 1.00 in Equation (2.3), then from Equation (2.3)

$$N \geq k_u P_L / (P_{su})_{\text{Ref}} = 6000/(P_{su})_{\text{Ref}} \text{ and } 12,000/(P_{su})_{\text{Ref}}.$$

From Equations (2.7) and (2.8) the bearing area is NtD. Table 2.2 shows the calculations for Q_w and for the bearing area for the two cases of $k_u P_L$ = 6000 lb and 12,000 lb. From Equation (2.14) it is obvious that the smallest r should be selected within the limitations on N and q. This has been done in Table 2.2.

From Table 2.2 the weight expression Q_w for the splice plate has an erratic variation, being a minimum for the 5/32 rivet for the 6000 lb load, but a minimum for the 1/8 rivet for the 12,000 lb load. Also, the weight expression Q_w for all the rivets increases when the load is doubled so that the weight $W_{s.p.}$ in Equation (2.14) more than doubles due to the fact that the thickness t must approximately double to maintain M.S. = 0.00 in tension (w = 2.00 in. is fixed). The bearing area is larger for the smaller rivets so that both the minimum weight and the bearing area indicate that the smaller rivets may be best in a splice similar to Figure 2.4.

Table 2.2. Variation of splice weight Q_w with D.

D (in.)	$\dfrac{6000}{(P_{su})_{Ref}}$	$\dfrac{2.00}{3D}$	N	q	r	Q_w	NtD
1/8	11.3	5.3	12	4	3	1.250	1.500t
5/32	7.4	4.3	8	4	2	1.094	1.250t
3/16	5.1	3.6	6	3	2	1.312	1.125t
1/4	2.8	2.7	4	2	2	1.750	1.000t
5/16	1.8	2.1	2	2	1	1.250	0.625t
	$\dfrac{12.000}{(P_{su})_{Ref}}$						
1/8	22.6	5.3	25	5	5	2.000	3.125t
5/32	14.7	4.3	16	4	4	2.031	2.500t
3/16	10.2	3.6	12	3	4	2.438	2.250t
1/4	5.7	2.7	6	2	3	2.500	1.500t
5/16	3.6	2.1	4	2	2	2.188	1.250t

Example 2.2. The comparison for Q_w and Ntd in Table 2.2 hold for a constant t in each load case. Calculate the actual t required for M.S. = 0.00 in both bearing and tension for the cases in Table 2.2 using the same plate material as in Example 2.1. Also, give the standard t_{st} for each case.

Solution. For the clad 2024-T3 material in Example 2.1, criterion (B) applies for tension and criterion (A) applies for bearing at $e/D = 2.0$. From Equations (2.7) and (2.8) with M.S. = 0.00, the required t_{br} for bearing is

$$t_{br} = \frac{k_L P_L}{NDF_{bry}} = \frac{4000}{73,000ND} \quad \text{and} \quad \frac{8000}{73,000ND}.$$

From Equations (2.11) and (2.12) with M.S. = 0.00, the required t_t for tension is

$$t_t = \frac{k_u P_L}{(w-qD)F_{tu}} = \frac{6000}{60,000(2.00-qD)} \quad \text{and} \quad \frac{12,000}{60,000(2.00-qD)}.$$

Table 2.3 shows the values for t_{br} and t_t as obtained from the D, N, and q values in Table 2.2. The t_{st} values are obtained from Table 2.1. Although the t_{br} and t_t values vary, the t_{st} values are essentially constant for each load case. The actual minimum weight $W_{s.p.}$ based on t_{st} still occurs for the same rivet sizes as for the Q_w weight expression.

Table 2.3. Results for Example 2.2.

	6000 lb load case			12,000 lb load case		
D (in.)	t_{br} (in.)	t_t (in.)	t_{st} (in.)	t_{br} (in.)	t_t (in.)	t_{st} (in.)
1/8	0.0365	0.0667	0.071	0.0351	0.1455	0.160
5/32	0.0438	0.0727	0.080	0.0438	0.1455	0.160
3/16	0.0487	0.0696	0.071	0.0487	0.1391	0.160
1/4	0.0548	0.0667	0.071	0.0731	0.1333	0.160
5/16	0.0878	0.0593	0.090	0.0877	0.1455	0.160

A different situation from that in Tables 2.2 and 2.3 arises in the splice of the T-section shown in Figure 2.5, where the width w and the thickness t will vary depending upon how much the splice plates can extend over the edges (limited by centering the rivets in the plates beneath the flange and on the web). Because of limitations on the rivets in a cross row, the larger rivets usually give the shortest length and hence least weight splice.

When it is necessary to put a concentrated force into a shear web, bearing plates can be used to transfer the load into the stiffener, which then transfers it to the shear web. To save weight, the bearing plates should be as short as

Analysis and design of joints and splices

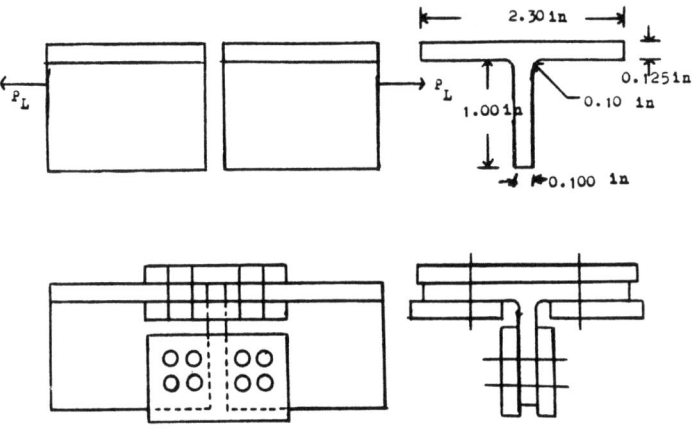

Fig. 2.5. Tension splice in T-section.

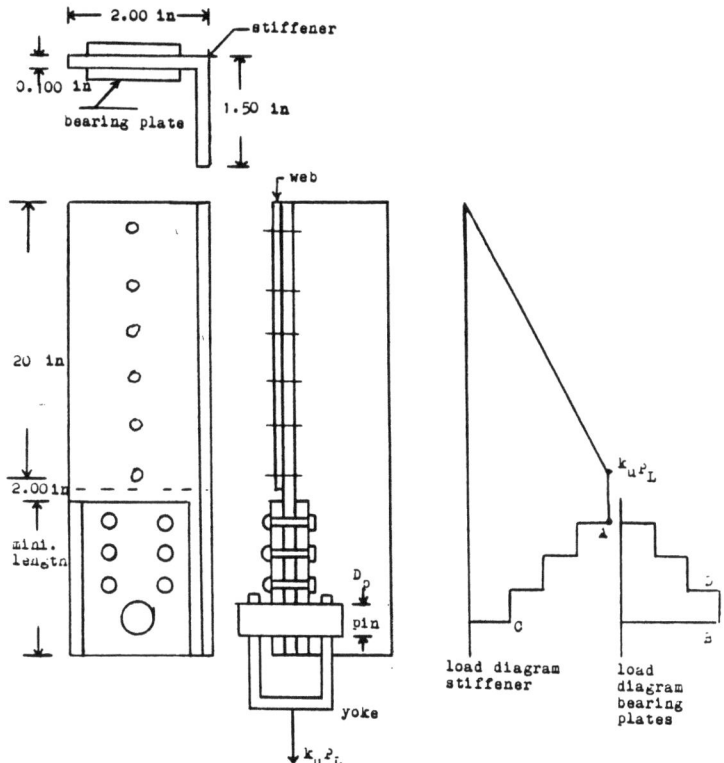

Fig. 2.6. Bearing plates for concentrated load into web.

possible with multi-rows of rivets to put load into the stiffener, Figure 2.6. The pin bringing the load in from a yoke bears on the stiffener and bearing plates so that

$$k_u P_l = D_p[2(tF_{bru})_{bp} + (tF_{bru})_{st}]. \tag{2.15}$$

The t_{st} is determined by tension in the stiffener at point (B) on the stiffener load diagram, the diameter D_p of the steel pin or bolt is determined from the double shear load from the yoke, whence t_{bp} for the bearing plates is determined from Equation (2.15). Equation (2.15) also gives the load in the bearing plates and in the stiffener at the bolt so that point (B) on the bearing plates load diagram is known, as well as point (C) on the stiffener load diagram. The next step is to design the rivet pattern for the load in the bearing plates to transfer this load into the stiffener. The size, number, and arrangement of the rivets can be made to give the shortest length of the bearing plates, where the width of the bearing plates depends upon the tension load at (B) and the net area at (B) or (D).

2.4. Joints with eccentric loading

In Sections 2.2 and 2.3, it was assumed that the applied load acted through the centroid of the rivet pattern and produced equal loads on equal area rivets. In many joints the load is eccentric to the centroid so that it produces a moment on the rivet pattern giving unequal loads on the rivets. Consider Figure 2.7, where the centroid of the rivet pattern as located from the reference axes (x_R, z_R) is

$$\bar{x} = \sum_{i=1}^{N} A_i x_{Ri} \Big/ \sum_{i=1}^{N} A_i, \quad \bar{z} = \sum_{i=1}^{N} A_i z_{Ri} \Big/ \sum_{i=1}^{N} A_i, \tag{2.16}$$

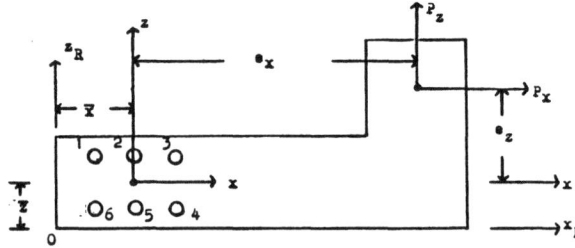

Fig. 2.7. Eccentric loads on rivet pattern.

with A_i the cross section area of rivet i. Usually the rivets are the same size so that $A_i = A$ and

$$\bar{x} = (1/N) \sum_{i=1}^{N} x_{Ri}, \quad \bar{z} = (1/N) \sum_{i=1}^{N} z_{Ri}. \tag{2.17}$$

The eccentric load components P_x and P_z in Figure 2.7 produce the same

Fig. 2.8. Direction of shear stresses on joint with eccentric loads.

component loads at the centroid plus a moment M_y about the centroid, or

$$M_y = e_z P_x - e_x P_z, \qquad (2.18)$$

where the applied moment is plus clockwise and acts as a torsional moment in producing shear stresses on the rivets proportional to the distance of the rivet from the centroid and perpendicular to the line from the centroid to the rivet. Figure 2.8 shows the directions of the applied shear stresses produced on each rivet by P_x, P_z, and M_y.

The shear stresses in rivet i are

$$\sigma_{siox} = P_x \Big/ \sum_{i=1}^{N} A_i, \quad \sigma_{sioz} = P_z \Big/ \sum_{i=1}^{N} A_i,$$
$$\sigma_{sie} = F r_i = F(x_i^2 + z_i^2)^{1/2}, \qquad (2.19)$$

whence the shear forces on rivet i are

$$P_{i0x} = A_i P_x \Big/ \sum_{i=1}^{N} A_i, \quad P_{i0z} = A_i P_z \Big/ \sum_{i=1}^{N} A_i, \quad P_{ie} = F A_i r_i. \qquad (2.20)$$

For equal areas, $A_i = A$,

$$P_{i0x} = P_x/N, \quad P_{i0z} = P_z/N, \quad P_{ie} = F A r_i. \qquad (2.21)$$

To determine the constant F, take equilibrium of moments about the centroid, or

$$M_y = e_z P_x - e_x P_z = \sum_{i=1}^{N} r_i P_{ie} = F \sum_{i=1}^{N} A_i r_i^2,$$

$$F = M_y \Big/ \sum_{i=1}^{N} A_i r_i^2 = M_y \Big/ \sum_{i=1}^{N} A_i (x_i^2 + z_i^2). \qquad (2.22)$$

Thus,

$$P_{ie} = M_y A_i r_i \Big/ \sum_{i=1}^{N} A_i r_i^2, \qquad (2.23)$$

or in component form

$$P_{iex} = FA_i z_i = M_y A_i z_i \bigg/ \sum_{i=1}^{N} A_i(x_i^2 + z_i^2),$$

$$P_{iez} = -FA_i x_i = -M_y A_i x_i \bigg/ \sum_{i=1}^{N} A_i(x_i^2 + z_i^2). \qquad (2.24)$$

The final component combined values on each rivet are

$$P_{ix} = P_{i0x} + P_{iex} = \frac{A_i P_x}{\sum_{i=1}^{N} A_i} + \frac{M_y A_i z_i}{\sum_{i=1}^{N} A_i(x_i^2 + z_i^2)},$$

$$P_{iz} = P_{i0z} + P_{iez} = \frac{A_i P_z}{\sum_{i=1}^{N} A_i} - \frac{M_y A_i x_i}{\sum_{i=1}^{N} A_i(x_i^2 + z_i^2)}, \qquad (2.25)$$

or for equal areas

$$P_{ix} = (P_x/N) + \left[M z_i \bigg/ \sum_{i=1}^{N}(x_i^2 + z_i^2)\right],$$

$$P_{iz} = (P_z/N) - \left[M x_i \bigg/ \sum_{i=1}^{N}(x_i^2 + z_i^2)\right]. \qquad (2.26)$$

The final resultant applied load on each rivet and its direction are given by

$$P_i = \left(P_{ix}^2 + P_{iz}^2\right)^{1/2}, \quad \tan\theta_i = P_{iz}/P_{ix}. \qquad (2.27)$$

This load P_i must be used for the design of the rivet in shear, or

$$\text{M.S.} = \frac{k_s (P_{su})_{\text{ref}}}{k_u P_i} - 1, \quad \text{rivet } i, \qquad (2.28)$$

for the rivet with the largest load $(P_i)_{\max}$. This largest $(P_i)_{\max}$ must also be used to check bearing on the plate and splice plates. However, the plate and splice plates are taking axial load, shear, and moments so that they must be checked for the tension, compression, and shear stresses at the proper locations in the splice. The analysis of these stresses is given in the next section 2.5.

Example 2.3. In Figure 2.7, let $P_z = -900$ lb, $P_x = 1000$ lb, $e_x = 8.00$ in., $e_z = 1.50$ in., $(x_i, z_i) = (-2, 1.5), (0, 1.5), (2, 1.5), (2, -1.5), (0, -1.5), (-2, -1.5)$ for equal area rivets. (a) Find the load and direction on each rivet. (b) If $k_s = 1.00$, $k_u = 1.50$, what size 2024-T31 rivets are needed?

Solution. (a) From Equations (2.18) and (2.26),

$$M_y = (1.50)(-900) - (8.00)(1000) = -9350 \text{ in-lb},$$

$$\sum_{i=1}^{6}(x_i^2 + z_i^2) = 4(2^2 + 1.5^2) + 2(1.5)^2 = 29.50 \text{ in}^2,$$

$$P_{ix} = -\tfrac{900}{6} + \tfrac{(-9350)z_i}{29.50} = -150 - 316.95 z_i,$$

$$P_{iz} = \tfrac{1000}{6} - \tfrac{(-9350)x_i}{29.50} = 167 + 316.95 x_i,$$

whence in sequence

$$(P_{ix}, P_{iz}) = (-625, -467), (-625, 167), (-625, 801),$$
$$(325, 801), (325, 167), (325, -467).$$

Analysis and design of joints and splices

With θ_i measured counterclockwise from the positive x-axis, in sequence Equation (2.27) gives

P_i = 781 lb, 647 lb, 1015 lb, 863 lb, 365 lb, 570 lb,

θ_i = 217°, 165°, 128°, 68°, 27°, 305°.

(b) Rivet (3) has the largest load with $k_u P_3 = 1523$ lb, so that from Table 2.1 the smallest permissible 2024-T31 rivet is $D = (1/4)$ in., or

M.S. = $\frac{2120}{1523} - 1 = 0.39$, rivet(3).

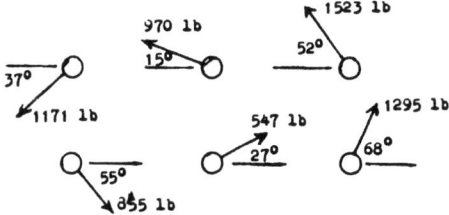

Fig. 2.9. Resultant ultimate rivet loads and directions for Example 2.3.

All the other rivets have larger M.S. Figure 2.9 shows the resultant applied ultimate loads $k_u P_i$ and directions.

2.5. Minimum weight design of splice for beam with rectangular cross section

The splice of any beam must transfer both the beam shear force and the beam moment at the splice. Thus, the analysis of joints with eccentric loads in Section 2.4 can be used for beam splices.

Consider the simple rectangular cantilever beam in Figure 2.10, which is to be spliced at the cut line. Use two splice plates as shown in Figure 2.11 and assume that the rivet arrangements are known. Usually, the same rivet arrangements are used on both sides of the cut, and the analysis is made for the side with the largest moment. The component loads on the rivets in the joint are calculated by Equation (2.25).

However, in design the size, number, and arrangement of the rivets is not known. Thus, to obtain a minimum weight design of the splice plates, it is necessary to select and arrange the rivets to give the shortest possible length of the splice plates. To do this, assume h in Figure 2.10 and 2.11 is given, assume the rivet diameter D, and arrange the maximum number of rivets in the h-direction consistent with the restrictions in Figure 2.11. The determination of the number and spacing of the rivets lengthwise can be made by a trial and error procedure based on the highest loaded rivet. However, with the rivets located in the h-direction, it is possible to construct a graph of the allowable moment against the location of the rivets lengthwise.

As a demonstration of the procedure to get the shortest length of the splice plates, consider the simple case of Figure 2.12, where $b \geq 3D/2$ is fixed and d

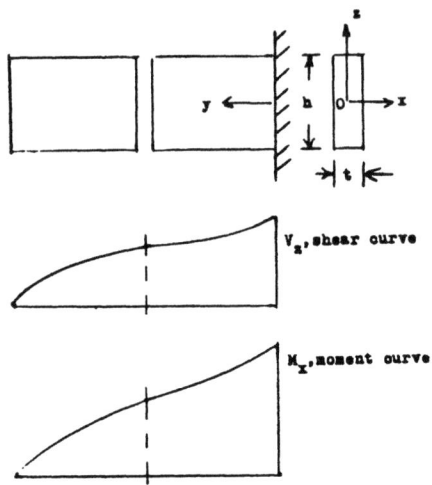

Fig. 2.10. Rectangular beam to be spliced.

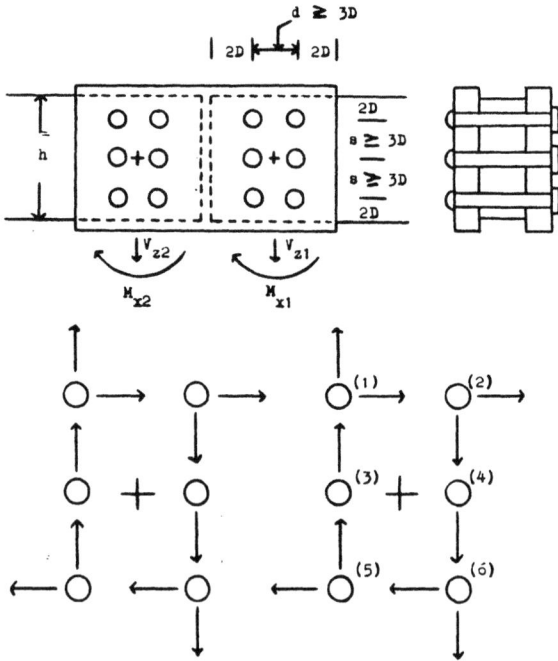

Fig. 2.11. Splice plates and rivet component loads for beam splice.

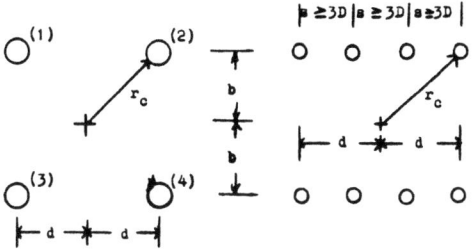

Fig. 2.12. Cases of two rivets in each vertical row.

is variable. Assume no shear force and consider the load on the highest loaded corner rivets due to the moment $M_u = k_u M_L$ about the centroid of the rivet pattern. From Equation (2.23) with M.S. = 0.00 for the highest loaded rivet, take

$$P_{ceu} = M_u r_c \Big/ \sum_{i=1}^{N} r_i^2 = k_s(P_{su})_{\text{Ref}} = P_{su}, \tag{2.29}$$

$$P_{su} = 2k_s(P_{su})_{\text{Ref}} \quad \text{for double shear,}$$

where P_{su} is known for a selected rivet size and material in Table 2.1, and r_c is the distance from the centroid to the highest loaded rivet. Thus

$$M_{\text{all}} = M_u = k_u M_L = (P_{su}/r_c) \sum_{i=1}^{N} r_i^2. \tag{2.30}$$

For the one row case, $d = 0$, $r_1 = r_2 = b = r_c$, $\sum r_i^2 = 2b^2$,

$$M_{\text{all}}/bP_{su} = 2, \quad 0 \le d/b \le 3D/2b. \tag{2.31}$$

For two rows, $r_i = (b^2 + d^2)^{1/2} = r_c$, $\sum r_i^2 = 4(b^2 + d^2)$,

$$M_{\text{all}}/bP_{su} = 4\left[1 + (d/b)^2\right]^{1/2}, \quad \frac{3D}{2b} \le \frac{d}{b} \le \frac{3D}{b}. \tag{2.32}$$

For three rows, $r_c = (b^2 + d^2)^{1/2}$, $\sum r_i^2 = 2b^2 + 4(b^2 + d^2)$,

$$M_{\text{all}}/bP_{su} = \frac{6 + 4(d/b)^2}{\left[1 + (d/b)^2\right]^{1/2}}, \quad \frac{3D}{b} \le \frac{d}{b} \le \frac{9D}{2b}. \tag{2.33}$$

For four rows, $r_c = (b^2 + d^2)^{1/2}$, $\sum r_i^2 = 8b^2 + (40/9)(d/b)^2$,

$$M_{\text{all}}/bP_{su} = \frac{8 + (40/9)(d/b)^2}{\left[1 + (d/b)^2\right]^{1/2}}, \quad \frac{9D}{2b} \le \frac{d}{b} \le \frac{6D}{b}. \tag{2.34}$$

Note that the increase in the number of required rows occurs at $3D/2b$ intervals on the d/b variable.

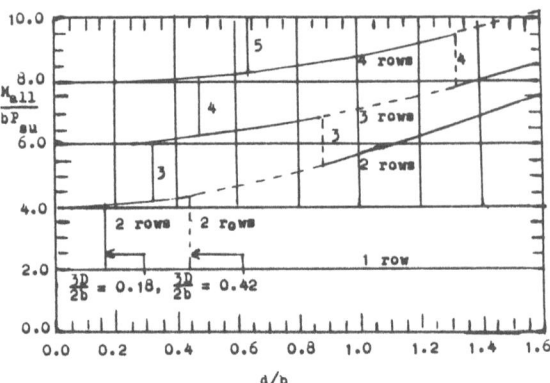

Fig. 2.13. Design curves for beam splice with two rivets in each row, Fig. 2.12.

Figure 2.13 shows graphs of the Equations (2.31)-(2.34) together with typical design curves for $3D/2b = 0.18$ and 0.42. The jumps between the curves can be marked for any given value of $3D/2b$ to obtain a design curve for the particular rivet arrangements used in Figures 2.12 and 2.13. For any given moment M_u, any given b, and a given rivet size D for a specified rivet material, the number of rows and spacing can be obtained for the least length L of the splice plates. The equation for the best L for the full length of the splice plates is

$$L = 4b\left(\frac{d}{b} + \frac{2D}{b}\right). \tag{2.35}$$

Since D^2 occurs in P_{su}, M_{all}/bP_{su} is smaller for the larger rivets so that on Figure 2.13 the least L is usually given by the largest permissible rivets. However, there may be exceptions when the larger rivet is slightly above the one row case. Whenever the design point occurs on a jump, the M.S. will be positive since the maximum allowable value is at the top of the jump.

Example 2.4. In Figures 2.12 and 2.13, let $b = 0.50$ in., $M_u = 1.50$, $M_L = 4500$ in-lb $= M_{\text{all}}$. Find the number of rows, minimum length L for the splice plates, and M.S. for rivet shear for 1/8, 5/32, 3/16, and 1/4 in. diameter 2024-T31 rivets. Use $k_s = 1.00$ with the rivets in double shear.

Solution. From Table 2.1 and Equation (2.29), with the rivets in the order 1/8, 5/32, 3/16, 1/4, it follows that

$$P_{su} = 1062 \text{ lb}, \quad 1630 \text{ lb}, \quad 2360 \text{ lb}, \quad 4240 \text{ lb},$$
$$\frac{M_{\text{all}}}{bP_{su}} = 8.47, \quad 5.52, \quad 3.81, \quad 2.12,$$
$$\frac{3D}{2b} = 0.375, \quad 0.469, \quad 0.563, \quad 0.750.$$

From Figure 2.13 with these $3D/2b$ intervals and the M_{all}/bP_{su} values, in sequence,

$$\text{rows} = 4, \quad 3, \quad 2, \quad 2,$$
$$d/b = 1.125, \quad 0.938, \quad 0.563, \quad 0.750,$$
$$L = 3.25 \text{ in.}, \quad 3.13 \text{ in.}, \quad 2.63 \text{ in.}, \quad 3.50 \text{ in.},$$
$$\text{M.S.} = 0.07, \quad 0.26, \quad 0.20, \quad 1.36,$$

and the (3/16) rivet gives the least length. Note that the large (1/4) rivet is slightly above the one rowe case (it has $L = 2.00$ in. and M.S.$= -0.06$ for 1 row) and thus gives the largest L in this case.

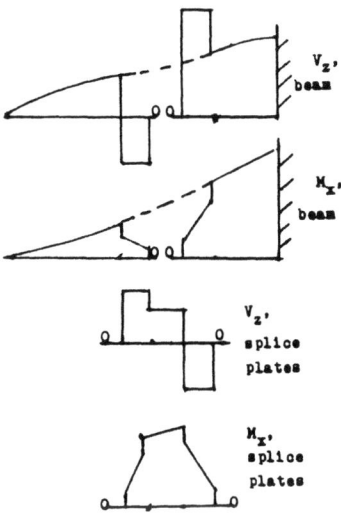

Fig. 2.14. Shear and moment curves in beam splice.

Once the best rivet arrangement has been obtained for the splice, it is necessary to calculate the applied component loads on each rivet so that the shear and moment curves for the beam and splice plates can be drawn. At each row of rivets in Figure 2.11, the vertical components add to produce a point shear force on the beam and splice plates, while the horizontal components produce a point moment couple about the centroid at the cross section of the row. Figure 2.14 shows a sketch of the shear and moment curves for the beam and splice plates for a case similar to Figures 2.10 and 2.11. It is evident that the beam and splice plates must be analyzed for the shear, tension, and compression stresses produced by the maximum values of these shear forces and moments.

For the rectangular beam under consideration, Figures 2.10 and 2.11, the applied stresses are given by Equations (1.93, Vol.1) and (1.44, Vol.1), or

$$\sigma_{yy} = -\frac{M_{xL} z}{I_{xx}}, \quad \sigma_{yz} = \frac{V_{zL}}{2I_{xx}}\left(\frac{h^2}{4} - z^2\right), \tag{2.36}$$

whence the maximum values are

$$(\sigma_{yy})_{\max} = \pm\frac{(M_{xL})_{\max} h}{2I_{xx}}, \quad (\sigma_{yz})_{\max} = \frac{(V_{zL})_{\max} h^2}{8I_{xx}}. \tag{2.37}$$

To allow for the rivet holes in the cross section, the simplest procedure is to change I_{xx} to

$$I_{xx} = (th^3/12) - 2tD\sum z_i^2, \tag{2.38}$$

where z_i is the distance from the centroid to rivet i on one side for the symmetrical case. These formulas can be used to find the M.S. for the beam in shear (B criterion), tension (A or B criteria), compression (B criterion) from bending, which may be taken as F_{tu} for small h/t ratio. Also, the bearing of the highest loaded rivet on the beam must be checked.

The next step is to design the splice plates for minimum cross section area wt, where w is the width of the plates (approximately equal to h of the beam) and t is a standard thickness (Table 2.1). For two plates, each of standard t, Equations (2.38) and (2.36) become

$$I_{sp} = (2tw^3/12) - 4tD\sum z_i^2,$$
$$\sigma_{tu} = w(M_u)_{sp}/2I_{sp}, \quad (B); \quad \sigma_{ty} = w(M_L)_{sp}/2I_{sp}, \quad (A). \tag{2.39}$$

For M.S. = 0.00 in tension, Equation (2.39) gives

$$t = \frac{w(M_u)_{sp,\max}}{4F_{tu}((w^3/12) - 2D\sum z_i^2)}, \quad (B), \tag{2.40}$$

or $(M_L)_{sp,\max}$ and F_{ty} for the (A) criterion. Assume w and get t; then, vary w within its limitations around h to get a standard value of t. With w and t selected for least weight on the basis of tension, check the splice plates for shear and bearing M.S. If these are positive, the plate design is satisfactory. If either shear or bearing has a negative M.S., increase t and reduce w as much as possible to get all M.S. zero or positive.

Example 2.5. Design a minimum weight splice for the 7075-T6 (*L*-direction, *B* basis) extruded rectangular beam shown in Figure 2.15 with M_L = constant = 32,000 in-lb, $k_L = 1.00$, $k_u = 1.50$, $k_s = 1.00$. Use (1/4) 2024-T31 rivets and two standard thickness clad 2024-T42 (*L*-direction, *B*-basis) splice plates with $e = 2D$, $s \geq 3D$, $d \geq 3D$ between rows, and equal number of rivets in each row.

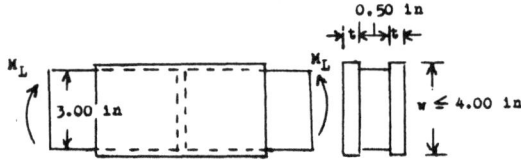

Fig. 2.15. Beam for Example 2.5.

Solution. Since the beam has $h = 3.00$ in., only three (1/4) in. rivets can be put in one row, whence with $e = 2D = 0.50$ in., $b = 1.00$ in. puts the rivets as far from the center as possible. Thus, with $(P_{su})_{\text{Ref}} = 2120$ lb from Table 2.1,

$$\frac{M_u}{bP_{su}} = \frac{(1.50)(32,000)}{(1.00)(2)(2120)} = 11.32.$$

Although Figure 2.13 applies for only two rivets in a row, it suggests that four rows might be needed for this example. Thus, for four rows, with three rivets in a row, Equation (2.30) becomes

$$\frac{M_{\text{all}}}{bP_{su}} = \frac{8b^2 + (20/3)d^2}{b(b^2+d^2)^{1/2}} = \frac{8+(20/3)(d/b)^2}{(1+(d/b)^2)^{1/2}},$$

or with $b = 1.00$, $M_{\text{all}}/bP_s u = 11.32$ for M.S. = 0.00, it follows that $d = 1.20$ in., or the row spacing is 0.80 in. which is greater than $3D = 0.75$. Thus, this point is on the curve between

Analysis and design of joints and splices

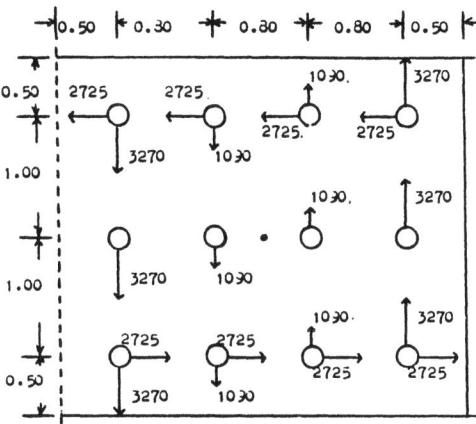

Fig. 2.16. Component loads on one side of beam splice for Example 2.5, (in. and lb).

the jump from 3 to 4 rows at $d = 1.125$ and from 4 to 5 rows at $d = 1.50$. With these values, Equation (2.35) gives the minimum length of the splice plates as

$$L = 4(1.20 + \tfrac{2}{4}) = 6.80 \text{ in.}$$

The next step is to calculate the component loads on the rivets by using Equation (2.26). The results for the ultimate component loads are shown in Figure 2.16 for one side of the splice. These values can be used to draw the shear and moment curves for the beam and splice plates. For this constant moment case, the numerical maximum values are the same in the beam and in the splice plates so that only the shear and moment curves for the splice plates are shown in Figure 2.17.

To check the margins of safety for the beam, get F_{tu}, F_{ty}, F_{su}, F_{bru}, and F_{bry} from Table 1.6, or $F_{tu} = 85{,}000$ psi, $F_{ty} = 76{,}000$ psi, whence criterion (B) is applicable, $F_{su} = 47{,}000$ psi, $F_{bru} = 136{,}000$ psi, $F_{bry} = 106{,}000$ psi for $e/D = 2.00$, whence criterion (B) is applicable. Since the largest shear and largest moment occur at a cross section with the three rivet holes,

$$I = t(3.00)^3/12 - 2(0.25t)(1)^2 = 1.75t = 0.875 \text{ in}^4.$$

Thus, for tension with Equation (2.37),

$$\text{M.S.} = \frac{85{,}000}{(48{,}000)(1.50)/(0.875)} - 1 = 0.03 \quad (B),$$

and for shear,

$$\text{M.S.} = \frac{47{,}000}{(13{,}080)(3.00)^2/(8)(0.875)} - 1 = 1.80 \quad (B).$$

Since the largest rivet load is $[(3270)^2 + (2725)^2]^{1/2} = 4250$ lb, the margin of safety for bearing on the beam is

$$\text{M.S.} = \frac{0.500(136{,}000)(0.25)}{4250} - 1 = 3.03 \quad (B).$$

To find the t and w for the splice plates, get the material properties from Table 1.2 by assuming a t in the range 0.063 - 0.249, whence $F_{tu} = 62{,}000$ psi, $F_{ty} = 38{,}000$ psi so that criterion (A) applies for tension, $F_{su} = 37{,}000$ psi, $F_{bru} = 118{,}000$ psi, $F_{bry} = 61{,}000$ psi so that criterion (A) applies for bearing. Assume $w = 4.00$ in. so that

$$I_{sp} = \frac{2t(4.00)^3}{12} - 4(0.25t)(1.00)^2 = 9.66t,$$

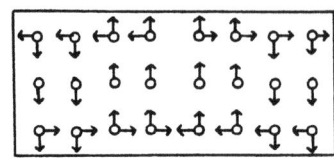

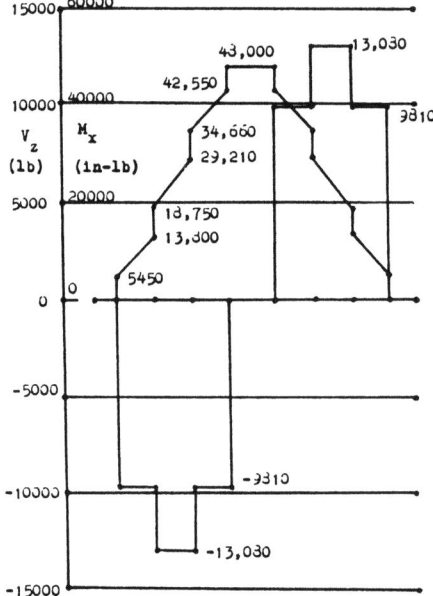

Fig. 2.17. Shear and moment curves for splice plates in Example 2.5.

$\sigma_L = M_L w/2I = (32,000)(4.00)/(2)(9.66t) = 6625/t$,

M.S. $= 0.00 = \dfrac{38,000}{6625/t} - 1$, for tension (A),

$t = 0.174$ in.

From Table 2.1, use standard thickness t as $t = 0.190$ in. and reduce w to make M.S. $= 0.00$. Thus,

$\sigma_L = 38,000 = \dfrac{(32,000)(w/2)}{0.190[(w^3/6) - 1]} = \dfrac{84,200w}{(w^3/6) - 1}$,

for which $w = 3.85$ by trial. With $t = 0.190$ in. and $w = 3.85$ in.

$I_{sp} = [2(0.190)(3.85)^3/12] - 0.190 = 1.617$ in^4,

whence for shear stress,

M.S. $= [(37,000)(8)(1.617)/(13,080)(3.85)^2] - 1 = 1.47$ (B).

For bearing of the splice plates,

M.S. $= \dfrac{2(0.190)(0.25)(61,000)}{4250} - 1 = 1.04$ (A).

Analysis and design of joints and splices

Of course, the w and t were selected to give the tension

M.S. = 0.00, (A).

The weight of the two splice plates is

weight = $2(0.100)(0.190)(3.85)(6.80) = 0.997$ lb.

2.6. Design of splices for I-beams and thin shear webs

Several types of joint or attachment problems arise in I-beams, depending on whether the I-beam is rolled or extruded as a unit, or whether it is assembled by

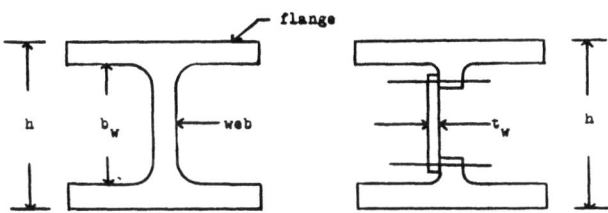

Fig. 2.18. I-beam cross sections.

riveting or bolting the web to the flanges. See Figure 2.18. Also, the attachments depend upon whether the web is thick with no buckling or a thin sheet that buckles in bending and shear. If the web is thin and buckles, then it is assumed to take shear only and to take no moment. In such a case the shear flow in the web is constant and the flanges take the moment as a couple

$$P = \pm M/h. \tag{2.41}$$

However, if the web is thick and takes moment, then the moment must be divided between the flanges and the web.

In the *thick web case* assume that the moment M is divided as

$$M = M_F + M_w. \tag{2.42}$$

Since the bending stress is $\sigma_b = Mz/I_b$, it follows that for the web

$$M_w = 2\int_0^{b_w/2} \sigma_b z\, dA = (2Mt_w/I_b) \int_0^{b_w/2} z^2\, dz$$
$$= Mt_w b_w^3/12I_b = M(I_w/I_b), \quad M_F = M - M_w. \tag{2.43}$$

That is, the moment divides according to the moments of inertia. Thus, the web can be spliced for M_w by using the procedures of Section 2.5, while the flanges can be spliced for M_F using Equation (2.41) and the procedures of Section 2.3 (see Figure 2.5).

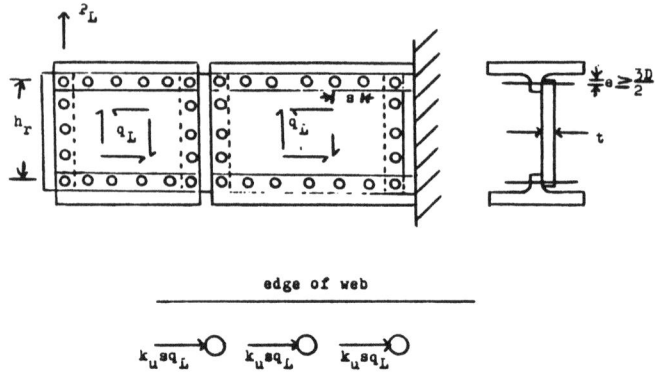

Fig. 2.19. Shear web attachments.

In the *thin web case*, the flanges can be spliced for the total moment M using Equation (2.41) and the procedures of Section 2.3. Since the shear flow is constant in the web, the attachments of the web to the flanges, to the end stiffeners, and for any web splices parallel to the sides of the rectangular web will be the same. The following discussion of shear web joints with constant shear flow neglects any buckling effects.

Since the shear flow is parallel to the edges of the rectangular shear web in Figure 2.19, there is no tear out problem so that the edge distance e for shear and the rivet spacing s may be taken as

$$e \geq 3D/2, \quad s \geq 3D \tag{2.44}$$

to meet restrictions on installing the rivets. As the attachments for the shear web are the same on all sides and at splices, it is simpler to use unit values and take the ultimate applied load per rivet as

$$(P_{su})_{\text{applied}} = k_u s q_L, \quad \text{per rivet}, \tag{2.45}$$

which acts parallel to the rivet line, Figure 2.19.

If buckling of the shear web is neglected there are three modes of failure for the shear attachments,
    (1). Rivet shear,
    (2). Web shear,
    (3). Web bearing.
For rivet shear, Equations (2.2), (2.3), and (2.45) give

$$\text{M.S.} = \frac{k_s (P_{su})_{\text{Ref}}}{k_u s q_L} - 1, \quad \text{rivet shear}, \tag{2.46}$$

where Table 2.1 gives k_s and $(P_{su})_{\text{Ref}}$. For web shear on the net area of $t(s - D)$

Analysis and design of joints and splices

per rivet (Figure 2.19),

$$\text{M.S.} = \frac{t(s-D)F_{su}}{k_u s q_L} - 1, \quad \text{web shear,} \tag{2.47}$$

where Tables 1.1 through 1.6 give the F_{su} allowable shear stress. For web bearing per rivet,

$$\text{M.S.} = \frac{tDF_{bry}}{k_L s q_L} - 1, \quad \text{web bearing (A),}$$
$$\text{M.S.} = \frac{tDF_{bru}}{k_u s q_L} - 1, \quad \text{web bearing (B),} \tag{2.48}$$

where F_{bry} and F_{bru} are given in Tables 1.1 through 1.6 and Equation (B.76, Vol.1) can be used to determine which case, (A) or (B), applies for design.

The Equations (2.46)-(2.48) can be used to analyze a given shear web joint. However, the design of the joint requires the determination of the number of rows in the joint (one row is desired if possible), the diameter D of the rivets (the smallest rivets possible, second only to the one row condition), and the spacing s of the rivets. If the margin of safety is set equal to zero in Equations (2.46)-(2.48) and Equation (2.44) is used, then four conditions on the term s/D are obtained for one row of rivets, or

(1) $s/D \geq 3$,

(2) $s/D \geq 1/(1-R), \quad R = k_u q_L/tF_{su}$,

(3) $s/D \leq \frac{1}{R}\left(\frac{F_{bru}}{F_{su}}\right)$, (B); $\quad s/D \leq \frac{1}{R}\left(\frac{k_u F_{bry}}{k_L F_{su}}\right)$, (A), $\tag{2.49}$

(4) $s/D \leq \frac{1}{R}\left[\frac{k_s(P_{su})_{\text{Ref}}}{tDF_{su}}\right]$,

where the parameter R represents the non-dimensional ratio of the applied ultimate load $k_u q_L$ per inch to the allowable ultimate load tF_{su} per inch (without the rivet hole). Thus, $R < 1$ for a possible design. Also, the values of s/D in conditions (3) and (4) must be greater than the values in conditions (1) and (2) for a possible design. If this condition cannot be met for the largest permissible rivet size, then it is necessary to try two rows for which the values in conditions (3) and (4) are doubled.

In order to construct design curves for the shear web joint to meet the conditions in Equation (2.49), take cases (3) and (4) in the form

$$\frac{s}{D} \leq \frac{H}{R}, \tag{2.50}$$

$$H = N_r\left(\frac{F_{bru}}{F_{su}}\right) \quad \text{for (3), (B),}$$
$$H = N_r\left(\frac{k_u F_{bry}}{k_L F_{su}}\right) \quad \text{for (3), (A),} \tag{2.51}$$
$$H = N_r\left[\frac{k_s(P_{su})_{\text{Ref}}}{tDF_{su}}\right] \quad \text{for (4),}$$

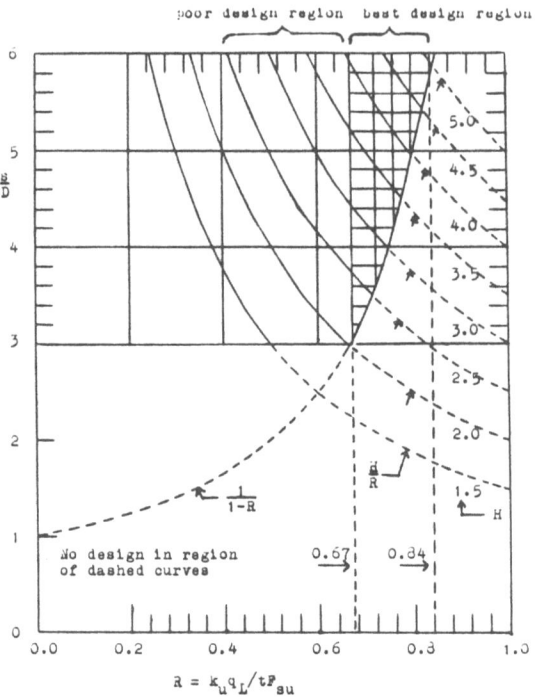

Fig. 2.20. Design regions for shear web attachments.

where N_r is the number of rows of rivets in the joint. Figure 2.20 shows the graphs of Equation (2.49), parts (1) and (2), and Equation (2.50) with R as the abscissa. Since R represents the efficiency of the web in shear, it should be as large as possible for a standard value of t. In practical design, the range of R is usually

$$0.67 \leq R \leq 0.84, \tag{2.52}$$

with R approximately 0.75 for many designs. Once the t is selected for the web for a given material, which gives R, Equation (2.51) and Figure 2.20 can be used to select the standard rivet diameter D and the number of rows N_r. Note in Figure 2.20 that a small change in R in the region of Equation (2.52) corresponds to a large change in H and gives a small best design region in Figure 2.20.

Example 2.6.. In Figure 2.19 take the web as $t = 0.071$ in. clad 2024-T3 aluminum alloy (B-basis) and the rivets as 2117-T3 with $D = (1/8), (5/32), (3/16),$ or $(1/4)$ as needed for best design. Use $k_L = 1.00$, $k_u = 1.50$, and $q_L = 1500$ lb/in. (a) Find the N_r, D, and maximum s for the best design. (b) Give the M.S. for rivet shear, web shear, and web bearing.

Solution. (a) From Tables 1.2 and 2.1, the material values for clad 2024-T3 are $F_{su} = 41,000$ psi, $F_{bru} = 124,000$ psi, $F_{bry} = 78,000$ psi, whence the (A) criterion is proper for bearing, and for the 2117-T3 rivets $(P_{su})_{\text{Ref}} = 388$ lb, 596 lb, 862 lb, and 1550 lb in order with $k_s = 1.00$,

1.00, 1.00, and 0.979 in order. With these values Equation (2.49) gives

$$R = \frac{1.50(1500)}{(0.071)(41,000)} = 0.773,$$

$$\frac{s}{D} \geq \frac{1}{1 - 0.773} = 4.41 \quad \text{for web shear,}$$

whence Figure 2.20 and Equation (2.51) give

$H = 3.45$, required for web shear,

$$H = N_r \frac{(1.50)(78,000)}{(1.00)(41,000)} = 2.85 N_r \quad \text{for bearing,}$$

$N_r = 2$, required for web bearing,

$$H = \frac{2(1.00)(596)}{(0.071)(41,000)(5/32)} = 2.62 \quad \text{for 5/32 rivets,}$$

$$H = \frac{2(1.00)(862)}{(0.071)(41,000)(3/16)} = 3.16 \quad \text{for 3/16 rivets,}$$

$$H = \frac{2(0.979)(1550)}{(0.071)(41,000)(1/4)} = 4.17 \quad \text{for 1/4 rivets.}$$

Thus, two rows of (1/4) rivets are needed to obtain an H greater than the required 3.45, whence Equation (2.50) gives

$$\frac{s}{D} \leq \frac{4.17}{0.773} = 5.39, \quad s_{\max} = (5.39)(1/4) = 1.35 \text{ in.}$$

(b) From $k_u s q_L = (1.50)(1.35)(1500) = 3038$ lb, Equations (2.46) through (2.48) give

$$\text{M.S.} = \frac{(0.979)(1550)(2)}{3038} - 1 = 0.00 \quad \text{(B)}, \quad \text{rivet shear,}$$

$$\text{M.S.} = \frac{(0.071)(1.35 - 0.25)(41,000)}{3038} - 1 = 0.05 \quad \text{(B)}, \quad \text{web shear,}$$

$$\text{M.S.} = \frac{(2)(0.071)(1/4)(78,000)}{2025} - 1 = 0.37 \quad \text{(A)}, \quad \text{web bearing.}$$

Note that a small change in R in Figure 2.20 can give a large change in the required values for H. For example, if q_L is 1400 lb/in. instead of 1500 lb/in. in this problem, then $R = 0.721$, $s/D \geq 3.59$ for shear, $H = 2.6$, and two rows of (5/32) in. rivets would be satisfactory (one row is satisfactory for bearing, but one row of 1/4 rivets gives $H = 2.08$, which is unsatisfactory).

2.7. Deflection effects on load distribution in splices

In the previous sections of this chapter, it has been assumed that the joints were determinate with the materials being sufficiently ductile to adjust to the assumed load distribution before failure. In particular, it was assumed that if all the rivets in a joint without bending were the same size, then each rivet took the same shear load. This neglects any temperature effects upon the rivet loads and assumes that the rivets or bolts are very flexible in bending or bearing.

When the rivet or bolt is loaded in a joint, Figure 2.21, it tends to deflect in bending, shear, and bearing. Also, if the bolt does not have a tight fit, then relative deflections can occur between bolts. This deflection d_{Ri} of the rivet i in the joint can be represented as

$$d_{Ri} = S_i/K, \tag{2.53}$$

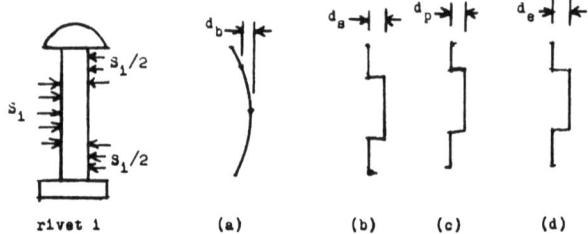

Fig. 2.21. Joint deflections: (a) rivet bending, (b) rivet shear and bearing, (c) plate bearing, (d) loose fitting bolt.

where S_i is the rivet load and K measures the stiffness of the joint, which must be obtained from tests of typical joints. It is the slope of the elastic portion of the load-deformation curve of the test joint, and may be regarded as a material property. Some tests of single lap joints with 0.19 in. 7075-T6 aluminum alloy plates and $\frac{3}{8}$ in. hi-shear bolts show that $K = 500,000$ lb/in. approximately, Reference 3.

Consider the three rivet joint in Figure 2.22, (which may be three rows of rivets), where S_i is the double shear load in rivet i. Use Equation (2.53) and the load diagrams in Figure 2.22 to write the one equilibrium equation and the two deflection equations between rivet rows (1) and (2), (2) and (3), as

$$S_1 + S_2 + S_3 = P,$$
$$s\alpha_1 T_1 + s\left(\frac{P - S_1}{E_1 A_1}\right) + \frac{S_2 - S_1}{K} = s\alpha_2 T_2 + \frac{sS_1}{E_2 A_2}, \quad (2.54)$$
$$s\alpha_1 T_1 + s\left(\frac{S_3}{E_1 A_1}\right) + \frac{S_3 - S_2}{K} = s\alpha_2 T_2 + s\left(\frac{P - S_3}{E_2 A_2}\right).$$

The solution for the rivet shear loads S_1, S_2, S_3 is

$$S_1 = (1 + RH)^{-1}\left[RP_T + \left(R + \frac{1}{3 + RH}\right)P\right], \quad (2.55a)$$
$$S_2 = P/(3 + RH), \quad (2.55b)$$
$$S_3 = (1 + RH)^{-1}\left[-RP_T + (H - 1)RP + \frac{P}{3 + RH}\right], \quad (2.55c)$$
$$R = \frac{sK}{E_1 A_1}, \quad H = 1 + \frac{E_1 A_1}{E_2 A_2}, \quad P_T = E_1 A_1(\alpha_1 T_1 - \alpha_2 T_2). \quad (2.55d)$$

The factor R is a comparison of the joint stiffness sK to the plate (1) stiffness $E_1 A_1$. The limits for R very large for the very stiff joint, and for R very small for the very flexible joint, give

$$R \to \infty, \quad HS_1 = P + P_T, \quad S_2 = 0, \quad HS_3 = -P_T + (H - 1)P, \quad (2.56)$$
$$R \to 0, \quad S_1 = S_2 = S_3 = P/3. \quad (2.57)$$

Thus, Equation (2.57) corresponds to the assumptions in the previous sections of this chapter.

Analysis and design of joints and splices

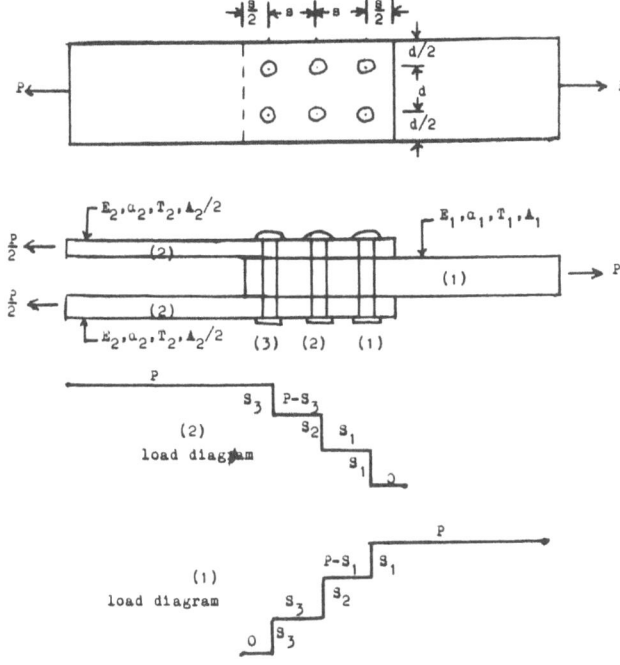

Fig. 2.22. Three row symmetrical joint with load diagrams.

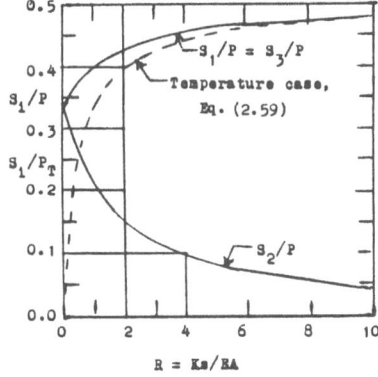

Fig. 2.23. Rivet loads in joint with three rows of rivets, Fig. 2.22.

If the temperature is omitted and $E_1 A_1 = E_2 A_2$, then $H = 2$ and Equation (2.55) reduces to

$$S_1/P = S_3/P = (1+R)/(3+2R), \qquad S_2/P = 1/(3+2R). \qquad (2.58)$$

Figure 2.23 shows the graph of these non-dimensional rivet loads against the stiffness parameter R. From Figure 2.23, it is evident that R must be small and less than about 0.2 in order for the loads to be approximately equal in this elastic case.

If the temperature is considered alone in the elastic case with $H = 2$, then Equation (2.55) gives

$$S_1/P_T = -S_3/P_T = R/(1+2R), \quad S_2 = 0, \qquad (2.59)$$

which is shown in Figure 2.23. The thermal load in rivets (1) and (3) can be large unless R is very small.

It should be emphasized that the trend of the curves in Figure 2.23 holds for any number of rows of rivets. For N rows instead of 3 rows in Figure 2.22, the N equations have the form

$$\sum_{i=1}^{N} S_i = P, \qquad (2.60a)$$

$$R(P + P_T) - RHS_1 + S_2 - S_1 = 0, \qquad (2.60b)$$

$$R(P + P_T) - RH \sum_{i=1}^{n} S_i + S_{n+1} - S_n = 0, \quad n = 2, 3, \cdots, N-2, \qquad (2.60c)$$

$$R(P + P_T) - RHP + RHS_N + S_N - S_{N-1} = 0. \qquad (2.60d)$$

The limits for R small and R large are

$$R \to 0, \quad S_1 = S_2 = \cdots = S_N = P/N, \qquad (2.61)$$

$$R \to \infty, \quad HS_1 = P + P_T, \quad HS_N = -(P + P_T) + HP,$$

$$S_2 = \cdots = S_{N-1} = 0. \qquad (2.62)$$

Thus, in a stiff joint with R large, the end rivets take both the applied and thermal loads, while all rivets in between take practically no load. This means that in joint design, $R = sK/EA$ must be kept as small as possible, and that the non-linear and inelastic form of the joint load deformation curve should be used. Of course the plate and rivet materials must be ductile in order to reduce R. The inelastic effects of both the plates and the joints are considered in the next Section 2.8.

In the above discussion only the effect of the temperature in the load direction was considered. If there is more than one rivet in the rows perpendicular to the load direction, as in a tension skin splice, then the temperature will produce shear loads in the rivets in the row direction. If the rivet spacing in the rows is d, Figure 2.22, then the R_d factor for rivet stiffness in the row direction can be approximated as

$$R_d = Kd/E_1 t_1 Ns. \qquad (2.63)$$

If there are M rivets in each row, then Equations (2.60)-(2.62) apply in the row direction with $P = 0$, R_d for R, M for N, and H_d for H, where

$$H_d = 1 + (E_1 t_1/E_2 t_2). \qquad (2.64)$$

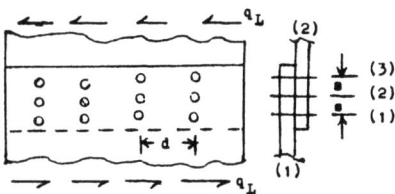

Fig. 2.24. Shear splice.

If R_d is large, the rivets at the end of the rows take the thermal load, while all rivets between take little load. The maximum rivet loads will be a vector combination of the component loads. See Example 2.8 below.

Consider the shear splice in Figure 2.24. In this case the stiffness factors are

$$R = \frac{Ks}{G_1 t_1 d}, \quad R_d = \frac{Kd}{G_1 t_1 s}, \quad H = H_d = 1 + \frac{G_1 t_1}{G_2 t_2}, \tag{2.65}$$

and all the above equations with temperature omitted apply with $k_u q_L d$ for P, $G_1 t_1 d$ for $E_1 A_1$. Since the shear loads on the rivets are in the long direction or q_L direction of the joint, but the component thermal loads on the rivets are the same as for the tension load case, the vector combination of the applied shear load and the thermal load components will be interchanged from that for the applied tension load case.

Example 2.7. In Figure 2.22 take $P = 10,000$ lb, $A_1 = 0.20$ in^2, $E_1 = 9(10^6)$ psi, $T_1 = 300°F$ change in temperature, $\alpha_1 = 13(10^{-6})/°F$, $A_2 = 0.10$ in^2, $E_2 = 3(10^7)$ psi, $T_2 = 0$, $\alpha_2 = 7(10^{-6})/°F$, $K = 800,000$ lb/in. Find the shear loads on the rivets for (a) the P load alone, (b) the thermal load alone, (c) both P and T_1 acting. (d) Comment on the results.

Solution. (a) From Equation (2.55),

$$R = \frac{(1)(800,000)}{0.20(9)(10^6)} = 0.44, \quad H = 1 + \frac{0.20(9)(10^6)}{0.10(3)(10^7)} = 1.60,$$

$$S_1 = \left(0.44 + \frac{1}{3.70}\right)\left(\frac{10,000}{1.70}\right) = 4200 \text{ lb},$$

$$S_2 = \frac{10,000}{3.70} = 2700 \text{ lb}, \quad S_3 = 10,000 - 4200 - 2700 = 3100 \text{ lb}.$$

(b) From Equation (2.55), $P_T = 9(10^6)(0.20)[13(10^{-6})(300) - 0] = 7020$ lb, whence $S_1 = (0.44)(7020)/1.70 = 1820$ lb $= -S_3$, $S_2 = 0$.

(c) In this elastic case add the results or parts (a) and (b) to get $S_1 = 6000$ lb, $S_2 = 2700$ lb, $S_3 = 1300$ lb.

(d) The equal load case gives 3333 lb on each rivet so that the critical rivet (1) takes 1.26 times this load in the applied P load case alone, and 1.80 times this load in the combined load and temperature case. Thus, if the rivets are designed for the 3333 lb shear load, then rivet (1) will fail unless there are sufficient inelastic effects to reduce its load to 3333 lb. See the next Section 2.8 for discussion of the inelastic effects.

Example 2.8. Suppose in Example 2.7 that there are three rivets in each row with $d = 0.75$ in. (a) Find the thermal loads on the rivets in the row direction. (b) Combine these loads with the loads in Example 2.7 to get the maximum load on the highest loaded rivet.

Solution. (a) From Equation (2.63) the assumed effective area is $t_1 Ns = 3t_1 = 3A_1/3d = 0.20/0.75 = 0.27$ in^2. Thus,

$$R_d = \frac{(0.75)(800,000)}{0.27(9)(10,000,000)} = 0.25, \quad H_d = 1.60, \quad R_d H_d = 0.40,$$

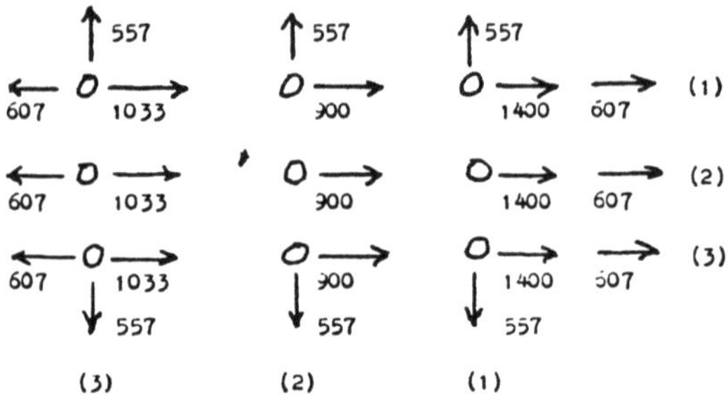

```
          ↑ 557          ↑ 557         ↑ 557
       ←─ O ─→        O ─→          O ─→      ─→    (1)
       607   1033         900            1400   607

       ←─ O ─→        O ─→          O ─→      ─→    (2)
       607   1033         900            1400   607

       ←─ O ─→        O ─→          O ─→      ─→    (3)
       607 │ 1033         900            1400   607
           ↓ 557        ↓ 557         ↓ 557

         (3)             (2)            (1)
```

Fig. 2.25. Component loads (lb) for Example 2.8.

$P_T = (9)(10^6)(0.27)(13)(10^{-6})(300) = 9360 \text{ lb},$
$S_1 = (0.25)(9360)/1.40 = 1670 \text{ lb} = -S_3, \quad S_2 = 0.$

(b) The component loads on the rivets are shown in Figure 2.25, from which the load on the worst rivets is

$$S_{11} = S_{13} = \left[(2007)^2 + (557)^2\right]^{1/2} = 2083 \text{ lb}.$$

The highest applied load alone is 1400 lb, and the highest thermal load alone is 824 lb. Besides the effective area approximation, there are two other effects being approximated or neglected in the above thermal load calculations. Since the thermal loads in the plate are two-dimensional, there is a Poisson's ratio effect, that tends to increase the thermal stresses. On the other hand, the thermal strains are restrained by the concentrated rivet loads so that the thermal stresses vary in the plate with the average being smaller than that used in the above calculations. The two effects tend to cancel each other so that the results given above are probably within the tolerance of the test results to get the stiffnesses K and K_d of the joint.

2.8. Temperature and inelastic effects on load distribution in splices

Consider the joint in Figure 2.26 with N rows of rivets and variable temperature. Use T_{1n} and T_{2n} as the average temperature changes in plates (1) and (2)

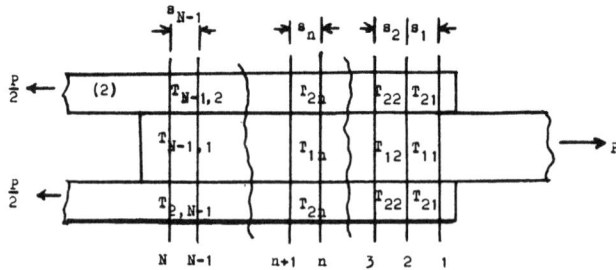

Fig. 2.26. N-row joint with variable temperature, rivet spacing, and material properties.

between rivet rows n and $n+1$. Take the strains e_{E1n} and e_{E2n} between rows n and $n+1$ to represent the thermal strains, the inelastic strains in the plates, and the non-linear and inelastic strains in each rivet row, or

$$e_{E1n} = (\alpha T)_{1n} + e_{p1n} + e_{f,n+1} - e_{fn},$$
$$e_{E2n} = (\alpha T)_{2n} + e_{p2n}. \tag{2.66}$$

From Equations (2.54), (2.60), (2.66), and Figure 2.26, the N equations for the N rivet shear loads S_i are

$$\sum_{i=1}^{N} S_i = P, \tag{2.67a}$$

$$e_{E1n} + (EA)_{1n}^{-1}\left(P - \sum_{i=1}^{n} S_i\right) + \frac{S_{n+1}}{s_n K_{n+1}} - \frac{S_n}{s_n K_n}$$
$$= e_{E2n} + (EA)_{2n}^{-1} \sum_{i=1}^{n} S_i, \quad n = 1, 2, \cdots, N-2, \tag{2.67b}$$

$$e_{E1,N-1} + (EA)_{1,N-1}^{-1} S_N + \frac{S_N}{s_{N-1} K_N} - \frac{S_{N-1}}{s_{N-1} K_{N-1}}$$
$$= e_{E2,N-1} + (EA)_{2,N-1}^{-1}(P - S_N). \tag{2.67c}$$

Use the definitions

$$P_{En} = (EA)_{1n}(e_{E1n} - e_{E2n}), \quad H_n = 1 + \frac{(EA)_{1n}}{(EA)_{2n}},$$
$$R_n = s_n K_n/(EA)_{1n}, \quad Q_n = s_{n-1} K_n/(EA)_{1,n-1}, \tag{2.68}$$

and rewrite Equations (2.67b) and (2.67c) in the form

$$\frac{P_{En} + P}{H_n} - \sum_{i=1}^{n} S_i + \frac{S_{n+1}}{H_n Q_{n+1}} - \frac{S_n}{H_n R_n} = 0, \quad n = 1, 2, \cdots, N-2, \tag{2.69}$$

$$S_{N-1} = -R_{N-1}(H_{N-1} - 1)P + R_{N-1} P_{E,N-1} +$$
$$+ \frac{R_{N-1}}{Q_N}(1 + Q_N H_{N-1}) S_N. \tag{2.70}$$

Subtract the number n equation from the $n+1$ equation in Equation (2.69) to get the recursion formula

$$S_n = \left(H_n R_n + \frac{R_n}{Q_{n+1}} + \frac{H_n R_n}{H_{n+1} R_{n+1}}\right) S_{n+1} - \frac{H_n R_n}{H_{n+1} Q_{n+2}} S_{n+2} +$$
$$+ H_n R_n \left(\frac{P_{En} + P}{H_n} - \frac{P_{E,n+1} + P}{H_{n+1}}\right), \quad n = N-2, N-3, \cdots, 1. \tag{2.71}$$

Since Equation (2.70) gives S_{N-1} in terms of S_N, then Equation (2.71) gives S_{N-2} in terms S_N, etc. for all the S_n by backward recursion in Equation (2.71). The results can be put into Equation (2.67a) to get S_N, whence all the S_n can be obtained. The inelastic strains e_{p1n}, e_{p2n}, e_{fn}, $e_{f,n+1}$ in Equation (2.66) can be obtained from the Ramberg-Osgood equation, or from Equation (1.94, Vol.1),

$$e_{pn} = \tfrac{3}{7}\left(\tfrac{F_{yn}}{E_{yyn}}\right)\left(\tfrac{P_n}{AF_{yn}}\right)^m, \qquad e_{fn} = \tfrac{3}{7}\left(\tfrac{P_{yn}}{sK_n}\right)\left(\tfrac{S_n}{P_{yn}}\right)^m, \tag{2.72}$$

where m may be of the order of 10 for the e_{pn} but may be as low as 2 or 3 for the e_{fn}. Here, P_{yn} is defined for the joint at the intersection of the $0.7K_n$ slope line with the load-deformation test curve. Also, P_n is the load in the plate between rivet n and rivet $n + 1$. Since these inelastic values are not known until the loads are known, it is necessary to iterate for the values of e_{pn} and e_{fn} starting from zero, or to use an incremental loading with straight line segments on the stress-strain curves. If only one rivet row or only one plate element is inelastic then it is possible to use Equation (2.72) and solve directly for the load.

Example 2.9. Solve Example 2.7 for the inelastic case with rivet (1) having $P_{y1} = 2500$ lb and $m = 3$ in Equation (2.72).

Solution. In this case $H_n = H$, $R_n = Q_n = R$, and $N = 3$ so that Equations (2.67a), (2.70), and (2.71) can be solved to give

$$S_1 = \frac{1}{1+RH}\left[R(P+P_{E1}) + \frac{P+R(P_{E2}-P_{E1})}{3+RH}\right], \tag{2.73a}$$

$$S_2 = \frac{P+R(P_{E2}-P_{E1})}{3+RH}, \tag{2.73b}$$

$$S_3 = \frac{1}{1+RH}\left[(H-1)RP - RP_{E2} + \frac{P+R(P_{E2}-P_{E1})}{3+RH}\right], \tag{2.73c}$$

which checks Equation (2.55) when $P_{E1} = P_{E2} = P_T$. However, with rivet (1) inelastic, Equations (2.66), (2.68), and (2.72) give

$$P_{E1} = (EA)_1\big[(\alpha T)_1 - (\alpha T)_2 - e_{f1}\big] = P_T - (EA)_1\left(\tfrac{3}{7}\right)\left(\tfrac{P_{y1}}{sK}\right)\left(\tfrac{S_1}{P_{y1}}\right)^3$$

$$= P_T - \tfrac{3}{7}(P_{y1}/R)(S_1/P_{y1})^3, \tag{2.74a}$$

$$P_{E2} = P_T. \tag{2.74b}$$

Use the data in Example 2.7 to get $P_{E2} = 7020$ lb, $P_{E1} = 7020 - 2435(S_1/2500)^3$,

$$S_1 = \binom{4200}{6000} - 460(S_1/2500)^3, \qquad S_2 = 2700 + 290(S_1/2500)^3,$$

$$S_3 = \binom{3100}{1300} + 170(S_1/2500)^3,$$

where parts (a) and (c) of Example 2.7 are included. The solutions are

no temperature, $\quad S_1 = 3200, \quad S_2 = 3300, \quad S_3 = 3500$ lb,
with temperature, $\quad S_1 = 4050, \quad S_2 = 3930, \quad S_3 = 2020$ lb.

In this case, the non-linear effect in rivet (1) makes the loads approximately equal for the applied load case alone, and makes a large reduction in the rivet (1) load for the combined applied load and temperature case. The ±1820 lb thermal load on rivets (1) and (3) in the elastic case in Example 2.7 is reduced and changed to -1480 lb on rivet (3), 630 lb on rivet (2), and 850 lb on rivet (1).

Example 2.10. Solve Examples 2.7 and 2.9 for $s = 50$ in. and discuss the results.

Solution. From the data in Examples 2.7 and 2.9, $R = 22$, $H = 1.60$, $RH = 35$, $P_{E2} = P_T = 7020$ lb, and

$$P_{E1} = 7020 - 49(S_1/2500)^3.$$

From Equation (2.55) for the elastic case with the $P = 10,000$ lb applied load alone, $S_1 = 6120$ lb, $S_2 = 260$ lb and $S_3 = 3620$ lb. From Equation (2.55) for the elastic case with temperature alone, $P_T = 7029$, $S_1 = 4290$, $S_2 = 0$, $S_3 = -4290$ lb. Thus, the combined elastic solution is $S_1 = 10,410$ lb, $S_2 = 260$ lb, and $S_3 = -670$ lb. For both the applied load case alone and the combined with rivet (1) inelastic, Equation (2.73) gives

$$S_1 = \binom{6120}{10,410} - 30(S_1/2500)^3, \quad S_2 = 260 + 29(S_1/2500)^3,$$

$$S_3 = \binom{3620}{-670} + 1.0(S_1/2500)^3.$$

Since rivet (1) will usually fail if $S_1/2500$ exceeds 2.0, it is evident that the inelastic effect in rivet (1) can do little to reduce the S_1 load before failure. This effect is due to the long joint in which the inelastic deflections are small compared to the total deflection in the joint. Although the joint is short for the applied tension load splice, the joints are long in the spanwise and chordwise skin splices on the wing and in shear web attachments. The thermal loads on the rivets can produce failure in these long joints. The end rivets fail first with the rivet failures proceeding toward the middle of the joint until the length is short enough to allow the inelastic deflections to reduce the loads below the failing values. This length is of the order of 10 inches in ductile materials such as the aluminum, steel, and titanium alloys. See Reference 3 for further discussion of this length effect upon the thermal loads in joints.

In the above analysis, the rivet loads were obtained by solving the deflection equations by recursion. The loads can also be obtained by using the principle of virtual forces. Since all the rivet loads except one are redundant, cut all the rivets except the last one and apply unknown redundant loads. Use unit values and put the equations into the matrix form of Equation (3.70) in Section 3.6 in Volume 1. Follow the analysis in Section 3.6 to get the solution for the rivet loads $S_i = Q_i$ in Equation (3.72). This procedure is used in Reference 4 to solve a complicated airplane joint problem with temperature, mixed materials, and inelastic effects.

2.9. Welded joints

In welding, two or more pieces of metal are joined by applying heat or pressure or both, with or without a filler material, to produce a localized union through fusion or recrystallization in the joint interface. Some welding processes used in flight vehicle structures are *fusion* welding with electric arc-welding using a tungsten or other electrode in inert gas, *resistance* welding with an intense electric current at spots or along a seam, and electric *flash* welding under pressure in direct joining of steel tubes. Since the heat produced in the welding process tends to destroy the heat treatment of the alloy locally, the allowable stresses must be reduced unless it is possible to heat treat the structural part after welding. Allowable stresses for both of these cases and for all three welding processes are given in Reference 1 for some material alloys. In particular, values are given for F_{su} and F_{tu} for fusion welded joints of various steel alloys. Values for the ultimate tension stress and bending modulus of rupture in flash welding of steel tubes are given in terms of the F_{tu} of the orginal material. Tables and curves show the shear strength of spot welds and the efficiency of the plates being spot welded for both steel and aluminum alloys.

Table 2.4. Spot weld allowable ultimate shear loads P_{sus} (lb) for bare and clad aluminum alloys.

F_{tu}, ksi,	>56	35-56	19.5-35	<19.5
t, in., thin plate				
	P_{sus}	P_{sus}	P_{sus}	P_{sus}
0.010	48	40		
0.016	88	80	56	40
0.020	112	108	80	64
0.025	148	140	116	88
0.032	208	188	168	132
0.040	276	248	240	180
0.050	372	344	320	236
0.063	536	488	456	316
0.071	660	576	516	360
0.080	820	684	612	420
0.090	1004	800	696	476
0.100	1192	936	752	540
0.125	1696	1300	840	628
0.160	2496	1952		
0.190	3228	2592		
0.250	5880	5120		

Just as for plate riveted joints in tension and shear, the strength of the spot welded plate must be reduced to allow for the decreased strength at the spot. In the riveted joints, the diameter of the rivet hole was used to give an effective net area of the plate, but in spot welded joints the reduced strength of the plate must be based upon test data for each plate thickness. Also, the shear strength of the spot welds must be based upon test data for each plate thickness. The data is based upon the thinnest plate in the joint with the ratio of the thickest plate to the thinnest plate ≤ 4. Table 2.4 from Reference 1 shows the shear strength of the spot welds in aluminum alloys for various plate thicknesses and various ultimate tension stress ranges for the alloys. At elevated temperatures and under creep conditions, the shear strength of the spot weld is based upon the reduced F_{tu} of the alloy material.

Comparisons of Tables 2.1 and 2.4 show that in most cases the shear strength of the spot weld in a given plate thickness is much smaller than the shear strength of the proper rivet for the plate. For example, in an 0.063 in. 2024-T3 plate at room temperature, the spot weld takes 536 lb while a 1/4 in. 2024-T31 rivet takes $(0.935)(2120) = 1982$ lb.

It is possible to space the spot welds closer than the rivets, but this weakens the plate and reduces its allowable tension or shear load. In the design of the tension or shear splice for the plate, the best design occurs when the spot weld spacing is such that the allowable shear load of the spot welds equals the allowable plate load, or

$$sR_t t F_{tu} = NP_{sus} = sk_u p_L, \tag{2.75}$$

where s is the spacing (in.), N the number of rows, p_L the applied load (lb/in.), and R_t the efficiency of the plate, $(P_{tu})_{\text{all}}/tF_{tu}$. When Equation (2.75) is sat-

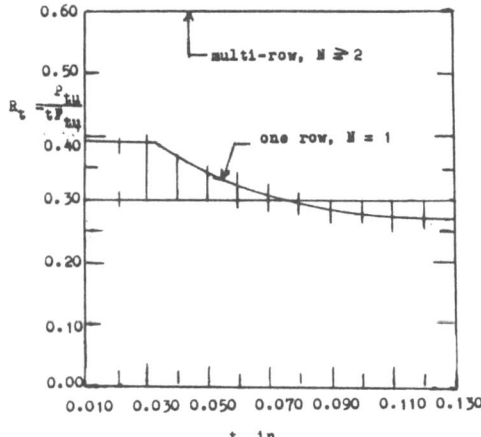

Fig. 2.27. Efficiency of spot welded aluminum alloy plates in tension at $sP_{tu} = NP_{sus}$.

isfied, the ultimate margin of safety M.S. is zero for both spot weld shear and plate tension. Based upon test data in Reference 1, the value of R_t is shown in Figure 2.27 for the case in which Equation (2.75) holds.

It is evident from Figure 2.27 that optimum designed spot welded plates will be less efficient and heavier than optimum designed riveted or bolted joints. With the minimum $s = 3D$ spacing for rivets, the plate efficiency is $R_t = 0.667$. It is 0.75 efficient for $s = 4D$. In Figure 2.20, the best design region for shear web splices with rivets is $0.67 < R_t < 0.84$. In some of the examples in the above sections on rivets, the plate efficiency was about $R_t = 0.80$. The maximum $R_t = 0.60$ in Figure 2.27 for spot welded joints means a larger t to take the applied load.

The values of R_t in Figure 2.27 also apply for seam welds as well as for

$$s \leq \frac{NP_{sus}}{R_t t F_{tu}} \tag{2.76}$$

from Equation (2.75). From Reference 1 the minimum edge distance varies approximately as a straight line from 3/16 in. for $t = 0.016$ in. to 5/8 in. for $t = 0.160$ in. The minimum distance between rows is about $8t$ for spot welds.

Example 2.11. Design the splice used in Example 2.2 and Tables 2.2 and 2.3 for spot welds and compare the plate thicknesses and splice plate weights.
Solution. From Equation (2.75)

$$t_t = k_u p_L / R_t F_{tu} = 3000/(0.60)(60,000) = 0.083 \text{ in}$$

for the 6000 lb load case and $t_t = 0.167$ in. for the 12,000 lb load case. Use the standard thicknesses as $t_{st} = 0.090$ in. and $t_{st} = 0.190$ in. With these thicknesses Table 2.4 and Equation (2.76) give for two rows of spot welds

$s \leq 2(1004)/3000 \leq 0.67$ in. for 6000 lb load case,

$s \leq 2(3228)/6000 \leq 1.08$ in. for 12,000 lb load case.

Three spots can be used in each row in the 6000 lb load case for the 2.00 in. wide plate, and two spots in each row in the 12,000 lb load case. The lengths of the splice plates are $L = 4e + 16t = 4(0.41) + 16(0.090) = 3.08$ in. for the 6000 lb case and $L = 4(0.71) + 16(0.190) = 5.88$ in. for the 12,000 lb case. The weights of the splice plates are $(3.08)(2.00)(0.090)(0.100) = 0.055$ lb for the 6000 lb case and 0.223 lb for the 12,000 lb case.

The comparisons of the best riveted splices to the spot welded splices show 0.071 in. to 0.090 in. and 0.040 lb to 0.055 lb for the 6000 lb case, and 0.160 in. to 0.190 in. and 0.128 lb to 0.223 lb for the 12,000 lb case. Thus, the riveted joint is the better design.

When spot welds are used to attach the skin to the ribs or fuselage rings and no splice is present, then the efficiency R_t of the skin in tension perpendicular

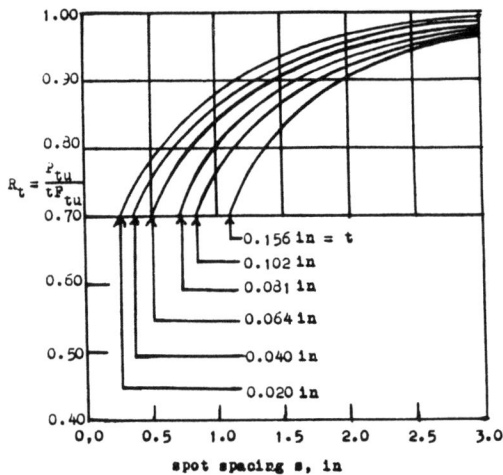

Fig. 2.28. Plate tension efficiency for spot welded aluminum alloys.

to the row of spot welds depends upon the spacing of the spots. Figure 2.28, from Reference 1, shows the values of R_t for various plate thicknesses as plotted against the spacing s.

The above tables and curves apply to tension splices and to tension loads in the plates. For shear loads in the plates, replace F_{tu} by F_{su} in Equations (2.75) and (2.76) and p_L by the shear flow q_L in Equation (2.75). The efficiency values R_t in Figures 2.27 and 2.28 can be used for plate shear.

Fatigue S-N test curves for spot welded joints in aluminum alloys are given in Reference 1. The scatter band in the tests is very wide, varying from $0.07F_{tu}$ to $0.19F_{tu}$ at 10^5 cycles for failure in the plate. The multi-row tests are not much better than the single row tests.

2.10. Bonded joints

In bonded joints the plates are joined by using an organic adhesive or glue as shown in Figure 2.29, where t_b is the thickness of the adhesive and $2s$ the length

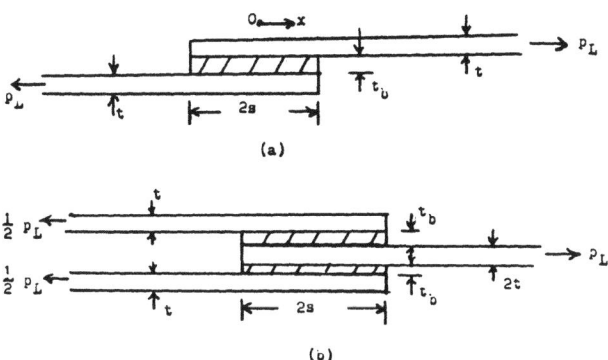

Fig. 2.29. Bonded joints, (a) single lap joint, (b) double lap joint.

of the joint. The applied limit load p_L is in (lb/in.) and a unit width of the joint is used in the calculations. Theoretical analysis of the stress distributions in bonded joints has been made by many investigators, some of which are given in References 5-10. Tests of bonded joints have been made to determine adhesive strengths and modes of failure. The primary problem in bonded joints is the wide variation in the adhesive material properties. Not only do the adhesive allowable tension and shear stresses and the tension and shear moduli of elasticity vary at room temperature, but also the properties vary with temperature, time, and environmental service conditions. Some adhesives creep at room temperature. Thus, many tests must be made on each adhesive in order to establish lower bound trends for the design material properties. Although these tests are made, there is no way to determine whether the life of the bonded joint even approaches the ten to twenty year expected life of many airplane structures.

Bonded joints can be made in both metal and composite structures. Since additional modes of failure can occur in the composite at the joint, the discussion of composite bonded joints will be omitted. See Chapter 10. Bonded joints for metal plates will be considered here, where the modes of failure may be tension in the plates, shear in the adhesive, or tension in the adhesive. The tension stress in the adhesive is due to the bending in the joint and is called the *peel* stress. The theoretical distributions of both the adhesive shear stress and the adhesive peel stress are described in References 5-10. The particular expressions given here are based upon References 5 and 8 for the single lap case in Figure 2.29. The average ultimate shear stress σ_{xya} in the adhesive and the average ultimate tension stress $\sigma_{xx\,max}$ in the plates, or

$$\sigma_{xya} = k_u p_L/2s, \quad \sigma_{xx\,max} = k_u p_L/t, \quad \text{single lap,} \tag{2.77}$$

will be used as the reference stresses in the non-dimensional equations.

In Reference 5 for the single lap joint, the elastic shear stress σ_{xy} is given in the non-dimensional form

$$\sigma_{xy}/\sigma_{xya} = C_1 \cosh R_b R_1(x/s) + C_2,$$
$$C_1 = 2R_b R_2/(R_1 \sinh R_b R_1), \quad C_2 = 1 - (2R_2/R_1^2),$$
$$R_1^2 = 2 + 6(1+R_3)^2, \quad R_3 = t_b/t, \tag{2.78}$$
$$R_b^2 = G_b s^2 (1-\nu^2)/(Ett_b),$$
$$R_2 = 1 + 3(1+R_3)^2 R_4, \quad R_4 = 2M_0/[k_u p_L t(1+R_3)],$$

where M_0 is the induced plate moment at the ends of the joint. From Reference 8, the equation for R_4 is

$$1/R_4 = 1 + s(k_u p_L/D)^{1/2} + (s^2/6)(k_u p_L/D), \quad D = \frac{Et^3}{12(1-\nu^2)}. \tag{2.79}$$

In Equation (2.78) the maximum elastic shear stress occurs at $x = s$, the end of the joint, and is

$$\frac{\sigma_{xy\,\max}}{\sigma_{xya}} = \frac{2R_b R_2}{R_1} \coth R_b R_1 + 1 - \frac{2R_2}{R_1^2}. \tag{2.80}$$

The most important parameter in this Equation (2.80) is R_b, which corresponds to the R stiffness ratio in Figure 2.23 for the riveted joints. From Reference 5, select the typical values $R_3 = 0.10$, $R_4 = 0.43$, whence $R_1 = 3.04$, $R_2 = 2.56$,

$$\sigma_{xy\,\max}/\sigma_{xya} = 1.68 R_b \coth 3.04 R_b + 0.55. \tag{2.81}$$

For $R_b = 0.2$, $\sigma_{xy\,\max}/\sigma_{xya} = 1.17$, but for $R_b = 2.0$, the stress ratio is 3.91, and the maximum elastic stress is very large for large R_b. If the adhesive is ductile and yields locally at the ends, then $\sigma_{xy\,\max}$ is reduced and the maximum shear strain must be used, Reference 8. Elastic test data and finite element solutions given in Reference 6 are in the low range of 0.1 to 0.4 for R_b, and the results check Equation (2.80).

The tension stress or peel stress σ_p acts on the adhesive due to the effect of the bending moment at the ends of the joint. Different formulas for this stress as a function of x are given in References 5 and 8, with the maximum stress $\sigma_{p\,\max}$ occurring at the ends of the joint. In Reference 8, the parameters affecting $\sigma_{p\,\max}$ are R_4, R_3, and

$$R_5^2 = \frac{3E_b t(1-\nu^2)}{2Et_b}. \tag{2.82}$$

In Reference 5, R_b also affects the stress in a complicated manner. From Reference 8, a simple formula for $\sigma_{p\,\max}/(k_u p_L/t)$ is

$$\sigma_{p\,\max}/(k_u p_L/t) = (1+R_3) R_4 R_5, \tag{2.83}$$

which will be used below.

The plate can fail from the combination of axial load and bending at the edge of the joint. From Reference 8, the expression for the maximum elastic stress $\sigma_{xx\,\max}$ in the plate is

$$\sigma_{xx\,\max}/\sigma_{xxa} = 1 + 3(1+R_3)R_4, \tag{2.84}$$

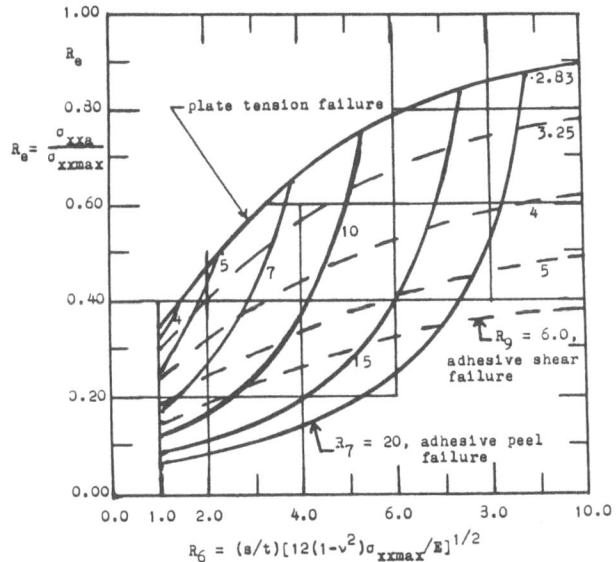

Fig. 2.30. Design curves for single lap bonded joints for metal plates.

where R_4 is given in Equation (2.79). Since R_4 involves the average applied plate stress $\sigma_{xxa} = k_u p_L/t$, it is necessary to solve Equation (2.84) for the efficiency ratio $\sigma_{xxa}/\sigma_{xx\,max}$, where $\sigma_{xx\,max} = F_{ty}$ for a zero margin of safety in the elastic case. Since this leads to a cubic equation, it is simpler to define

$$R_6^2 = \frac{12\sigma_{xx\,max}(1-\nu^2)}{E}\left(\frac{s}{t}\right)^2, \quad R_e = \frac{\sigma_{xxa}}{\sigma_{xx\,max}}, \tag{2.85}$$

and solve Equation (2.84) for R_6 in terms of the unknown efficiency ratio R_e. The result is

$$R_6 = (R_e)^{-1/2}\left[-3 + \left\{3 + \frac{18(1+R_3)R_e}{1-R_e}\right\}^{1/2}\right] \tag{2.86}$$

for plate tension. Assume values of R_e and calculate R_6 to draw a graph of R_e against R_6 in Figure 2.30, with $R_3 = 0$. The practical values of $R_3 < 0.1$ have little effect on the curve in Figure 2.30. It is evident from Figure 2.30 that the efficiency R_e of the joint in tension is very small unless s/t in R_6 is large, which usually means a long joint of length $2s$.

Since the term R_4 in Equation (2.79) is also in Equation (2.83) for the peel stress, the same definitions in Equation (2.85) can be used and Equation (2.83) can be solved for R_6, or

$$R_6 = R_e^{-1/2}\left[-3 + \left(3 + 7.35 R_7 R_e\right)^{1/2}\right], \tag{2.87a}$$

$$R_7 = \left(1 + \frac{t_b}{t}\right)\left(\frac{t}{t_b}\right)^{1/2}\left(\frac{\sigma_{xx\,max}}{\sigma_{p\,max}}\right)\left[(1-\nu^2)\frac{E_b}{E}\right]^{1/2}, \tag{2.87b}$$

for the adhesive peel stress. This Equation (2.87) is graphed on Figure 2.30 with R_7 as a parameter. In this case for the peel stress R_7 should be small and R_6 large to get a reasonably efficient joint. Note for given plate and adhesive properties, R_7 increases with t/t_b.

Since the R_2 term in the maximum shear stress in Equation (2.80) involves the same R_4 term in Equation (2.79), the solution for the R_e efficiency for the adhesive shear stress can be made by the same procedure as for the plate tension and adhesive peel stresses. Rewrite Equation (2.80) in the form

$$\frac{2s}{t}\frac{1}{R_e}\frac{\sigma_{xy\,\max}}{\sigma_{xx\,\max}} - 1 = \frac{2}{R_1}[1 + 3(1+R_3)^2 R_4]\left(R_b \coth R_b R_1 - \frac{1}{R_1}\right). \quad (2.88)$$

Put R_6 and R_e from Equation (2.85) into R_4, whence Equation (2.88) can be solved for R_6 as

$$R_6 = R_e^{-1/2}[-3 + (3+R_8)^{1/2}], \quad (2.89a)$$
$$R_8 = 18(1+R_3)^2/(R_1 B - 1), \quad (2.89b)$$
$$B = \left[\frac{s}{t}\frac{1}{R_e}\frac{\sigma_{xy\,\max}}{\sigma_{xx\,\max}} - 0.50\right]\Big/\left[R_b \coth R_1 R_b - \frac{1}{R_1}\right]. \quad (2.89c)$$

Assume $R_3 = 0$ so that $R_1 = 8^{1/2} = 2.83$ and use R_b in the form

$$R_b = \frac{s}{t}\left(\frac{t}{t_b}\right)^{1/2}[(1-\nu^2)G_b/E]^{1/2}. \quad (2.90)$$

For $R_6 > 1$, s/t will usually be sufficiently large to make $\coth 2.83 R_b$ close to one, whence B in Equation (2.89c) can be approximated as

$$B = 1/R_9 R_e, \quad R_9 = \frac{\sigma_{xx\,\max}}{\sigma_{xy\,\max}}\left(\frac{t}{t_b}\right)^{1/2}[(1-\nu^2)G_b/E]^{1/2}. \quad (2.91)$$

Thus, Equation (2.89a) becomes

$$R_6 = R_e^{-1/2}\left[-3 + \left(3 + \frac{18 R_9 R_e}{2.83 - R_9 R_e}\right)^{1/2}\right], \quad (2.92)$$

which is similar to Equation (2.86) for the plate tension case. In fact, for $R_9 = 2.83$, Equation (2.92) is the same as Equation (2.86), whence on Figure 2.30 values of $R_9 < 2.83$ give curves above the plate tension curve, while values of $R_9 > 2.83$ give elastic adhesive shear curves below the plate tension curve.

All the curves in Figure 2.30 for the three modes of failure in the single lap bonded joint are based upon elastic material properties of the plates (E, $\sigma_{xx\,\max} = F_{ty}$) and the adhesive ($E_b$, $\sigma_{p\,\max} = F_{tub}$, G_b, $\sigma_{xy\,\max} = F_{sub}$), where the adhesive stress-strain curves are approximated as horizontal in the inelastic range with $F_{tub} = F_{tyb}$, $F_{sub} = F_{syb}$. In order to obtain a reasonably efficient bonded joint on Figure 2.30, it is necessary to select an adhesive with properties that make the R_7 and R_9 parameters small. That is, $\sigma_{p\,\max}$ and $\sigma_{xy\,\max}$ should be large but E_b and G_b should be small. Also, a large adhesive thickness t_b and a small plate thickness t make both R_7 and R_9 smaller. If the adhesive is ductile in both shear and tension and fails at large strains, then the inelastic effect can be approximated by E_{beff} and G_{beff} effective values. In some

Analysis and design of joints and splices

adhesives these values may be as low as $0.05G_b$ and $0.15E_b$. Thus, all the shear curves on Figure 2.30 could shift above the plate tension curve so that shear failure would not occur. The $R_7 = 20$ peel stress curve could move to the left to about the $R_7 = 7$ curve. It is evident that the adhesive inelastic effects and the adhesive thickness t_b are important in the design of the bonded joint.

Since the peel stress curves in Figure 2.30 are very steep near the plate tension curve and since there is a large scatter in the adhesive properties, the design of the joint should be made for plate tension as far to the right of the peel stress R_7 curve as possible. If a design is made at the intersection of the peel stress and plate tension curves, then a few per cent increase in R_7 for any reason could produce a peel stress failure as low as one-half the design load.

If the plate becomes inelastic in the region of the ends of the joint, then an effective E_{eff} increases R_6 on Figure 2.30 so that both t and s could be reduced for the more efficient joint. However, the E_{eff} also increases both R_7 and R_9 so that the adhesive margins of safety for the shear and peel stresses may be reduced. The inelastic plate also increases the shear strain in the adhesive so that a shear failure in the adhesive could be produced by the inelastic plate.

It is evident that there are many problems in trying to design an efficient reliable single lap bonded joint. One way to improve the efficiency is to support the joint against bending. This support can be provided by a rib cap or stringer member at the splice. Also, if possible a double lap joint can be used, which is more efficient than a single lap joint. See Reference 7 for analysis of the double lap joint. A scarf joint is more efficient than a single lap joint. See Reference 9 for the analysis of a scarf joint. Also, see Reference 10 for special joints in flight vehicle structures.

Example 2.12. Assume the material properties for a single lap joint bonded with a ductile adhesive to be $\sigma_{xy\,\text{max}} = 5400$ psi, $\sigma_{p\,\text{max}} = 5200$ psi, $G_b = 84,000$ psi, $G_{b\text{eff}} = 5000$ psi, $E_b = 300,000$ psi, $E_{b\text{eff}} = 50,000$ psi, $\sigma_{zz\,\text{max}} = 70,000 = F_{ty}$ for 7075-T6 plates, $E = 10^7$ psi, $\nu = 0.3$. (a) Express the parameters R_6, R_7 and R_9 on Figure 2.30 in terms of s, t, and t_b. (b) Take the applied load as $k_u p_L = 3000$ lb/in. and design the bonded joint for $t_b = 0.010$ in. and $R_e = 0.60$ efficiency. (c) Repeat part (b) with $t_b = 0.015$ in. (d) Repeat part (b) with an efficiency of $R_e = 0.70$.

Solution. (a) Put the given data into Equations (2.85), (2.87b), and (2.91) to get $R_6 = 0.276(s/t)$,

$$R_7 = \begin{Bmatrix} 2.22 \\ 0.91 \end{Bmatrix} \left(1 + \frac{t_b}{t}\right)\left(\frac{t}{t_b}\right)^{1/2}, \quad \text{elastic and inelastic cases,}$$

$$R_9 = \begin{Bmatrix} 1.13 \\ 0.28 \end{Bmatrix} (t/t_b)^{1/2}, \quad \text{elastic and inelastic cases.}$$

(b) For the given load and efficiency, $t = 3000/(0.60)(70,000) = 0.071$ in. With this $t = 0.071$ in. and $t_b = 0.010$ in., $R_7 = 6.75$ and 2.77, $R_9 = 3.01$ and 0.75 for the elastic and inelastic cases. On Figure 2.30, $R_6 = 3.20$ from the plate tension curve for $R_e = 0.60$. The adhesive shear and peel stress curves for the elastic case are to the right of this plate tension point, whence elastic design would require R_6 to be larger or at least $R_6 = 3.80$. The adhesive shear and peel stress curves for the inelastic case are far to the left of the plate tension point so that this point can be used for design. Thus the splice length is $2s = 2R_6(0.071)/0.276 = 1.96$ in. for the elastic case in adhesive shear, and $2s = 1.65$ in. for the inelastic case with plate tension.

(c) With $t = 0.071$ in. and $t_b = 0.015$ in., $R_7 = 5.85$ and 2.40, $R_9 = 2.46$ and 0.61. The adhesive shear and peel stress curves are to the left of the plate tension point for both the elastic and inelastic cases so that the $2s = 1.65$ in. in part (b) applies.

(d) In this case, $t = 3000/(0.70)(70,000) = 0.061$ in. and $R_6 = 4.43$ on the plate tension curve. With $t = 0.061$ in. and $t_b = 0.010$ in., it follows that $R_7 = 6.28$ and 2.56, $R_9 = 2.80$

and 0.69. On Figure 2.30, the adhesive shear and peel stress curves are to the left of this plate tension point for both the elastic and inelastic cases. Thus, $2s = 2(4.43)(0.061)/0.276 = 1.96$ in. Not only does this more efficient joint require thinner plates but also the weight of the overlap is practically the same as in part (b).

2.11. Problems

2.1. Repeat Example 2.1 using $D = \frac{1}{8}$ in. 2117-T3 rivets.

2.2. In Figure 2.1 assume five rivets in the one row with $s = 0.70$ in., $e = 0.25$ in., $D = 1/8$ in. 2017-T3 rivets, $t = 0.032$ in. clad 2024-T3 (L-direction, B basis), $k_L = 1.00$, $k_u = 1.50$, and $P_L = 1300$ lb. (a) Find the M.S. for each of the four modes of failure. (b) What maximum load P_L will the splice take?

2.3. In Figure 2.3 take $s = d = 3D$, $e = e_s = 2D$, $D = (5/32)$ in. 2017-T3 rivets, $t = 0.045$ in. clad 2024-T42 (L-direction, B basis), $k_L = 1.00$, $k_u = 1.50$. How much load P_L will the plate take in (a) rivet shear, (b) plate tension, and (c) plate bearing?

2.4. Recalculate Tables 2.2 and 2.3 for $w = 3.00$ in. and compare results to those in Tables 2.2 and 2.3.

2.5. In Figure 2.3 let $e = 2D$, $d = 3D$, and assume r rows of rivets. Show that the weight of the assembly W_A can be expressed by

$$Q = \frac{W_A F_{bru}}{\rho_w k_u P_L} = \frac{F_{bru}}{k_u P_L}(wDt_{st})\left(1 + 3r + \frac{L}{D}\right), \qquad (2.93)$$

where t_{st} is the standard thickness of the plate.

2.6. In Figure 2.3 let $e = 2D$, $d = 3D$, assume q rivets in each of r rows of rivets, and assume L and P_L are given. With w, t, r, and $N = qr$ regarded as variables, use M.S. $= 0.00$ for rivet shear, plate bearing, and plate tension to eliminate the N, t, and w variables so that an approximate $(W_A)_{\text{app}}$ can be derived for t and r as continuous variables. Thus, show that

$$Q_{\text{app}} = \frac{(W_A)_{\text{app}} F_{bru}}{\rho_w k_u P_L} = k_4 D\left(k_5 + \frac{1}{r}\right)(1 + 3r + \frac{L}{D}), \qquad (2.94)$$

where k_4 and k_5 depend upon the material and the (A) and (B) criterions. Their values are $k_4 = 1$, $k_5 = F_{bru}/F_{tu}$ for tension (B) and bearing (B); $k_4 = 1$, $k_5 = k_L F_{bru}/k_u F_{ty}$ for tension (A) and bearing (B); $k_4 = k_L F_{bru}/k_u F_{bry}$, $k_5 = k_u F_{bry}/k_L F_{tu}$ for tension (B) and bearing (A); $k_4 = k_L F_{bru}/k_u F_{bry}$, $k_5 = F_{bry}/F_{ty}$ for tension (A) and bearing (A).

2.7. In Equation (2.94) in Problem 2.6 above graph Q_{app} against r ($1 \leq r \leq 6$) for $L = 10$ in., $k_L = 1.00$, $k_u = 1.50$, clad 2024-T3 (L-direction, B basis), $0.010 \leq t \leq 0.062$, and for $D = (1/8)$, $(5/32)$, $(3/16)$, and $(1/4)$ in. rivets. Comment on the results, reading r_{opt} and $(Q_{\text{app}})_{\text{min}}$ values from the curves.

2.8. Verify the r_{opt} and $(Q_{\text{app}})_{\text{min}}$ in Problem 2.7 by differentiation of Equation (2.94) to get

$$r_{\text{opt}} = \left[\frac{1 + (L/D)}{3k_5}\right]^{1/2}, \quad (Q_{\text{app}})_{\text{min}} = 3k_4 D(1 + k_5 r_{\text{opt}})^2. \qquad (2.95)$$

2.9. Suppose the T-section in Figure 2.5 is 7075-T6 extrusion (L-direction, B basis) with $P_L = 16000$ lb, $k_L = 1.00$, $k_u = 1.50$. (a) Design a minimum weight splice using the three flange plates only with $(1/4)$ in. 2024-T31 rivets and clad 2024-T42 (L-direction, B basis), standard t_{st}, splice plates. Neglect bending due to transferring the web load through the flange splice plates. (b) Give all M.S. for the extrusion and the splice plates, and the weight of the splice plates.

2.10. Repeat Problem 2.9 using five splice plates.

2.11. Make the splice in Problem 2.9 by using two formed angles, one on either side of the web and attached to both the web and the flange.

2.12. In Figure 2.6 assume $P_L = 12{,}000$ lb, $k_L = 1.00$, $k_u = 1.50$, $D_p = (1/2)$ in. stiffener and bearing plates as clad 2024-T3 (L-direction, A basis), rivets as 2017-T3. Find the thickness of the bearing plates and the maximum load in the bearing plates. Select rivets to make the bearing plates as short as possible within a permissible width, and draw the load diagrams for the stiffener and the bearing plates.

Analysis and design of joints and splices

2.13. Repeat Example 2.3 for the case in which $P_z = 800$ lb and $P_x = 1200$ lb.

2.14. In Example 2.3 let the location of the rivets be $(x_i, z_i) = (-a, 1.50)$, $(0,1.50)$, $(a, 1.50)$, $(a, -1.50 \text{ in.})$, $(0,-1.50)$, $(-a, -1.50)$. Use other data in Example 2.3, except $e_z = 6.00 + a$, and graph the ultimate applied load P_{3u} on rivet (3) against a as abscissa $(0.50 \le a \le 6.00)$. What are the theoretical limit values of P_{3u} at $a = 0$ and $a = \infty$?

2.15. Solve Example 2.4 for the cases of (a) $M_u = 1.50 M_L = 5000$ in-lb (b) $M_u = 4000$ in-lb.

2.16. Solve Example 2.5 for the cases of (a) $M_L = 34{,}000$ in-lb, (b) $M_L = 24{,}000$ in-lb.

2.17. Solve Example 2.5 for the case of $M_L = 24{,}000$ in-lb and $D = 3/16$ in. rivets. Compare results to part (b) in Problem 2.16.

2.18. Solve Example 2.6 for the cases of (a) $t = 0.080$ in., (b) $t = 0.090$ in.

2.19. In Example 2.6, assume $q_L = 800$ lb/in. What values of t_{st} would give a design in the best region of Figure 2.20? Make the design for the rivets, etc., as in Example 2.6 for the best t_{st}.

2.20. (a) For the case of 4 rivets in Figure 2.22 and $H = 2$ in Equation (2.55d) show that

$$\begin{Bmatrix} S_1 \\ S_4 \end{Bmatrix} = \frac{1+2R}{4(1+R)}P \pm \frac{R(3+2R)P_T}{2(1+4R+2R^2)},$$
$$\begin{Bmatrix} S_2 \\ S_3 \end{Bmatrix} = \frac{P}{4(1+R)} \pm \frac{RP_T}{2(1+4R+2R^2)}. \tag{2.96}$$

(b) Graph the non-dimensional form of these rivet loads similar to Figure 2.23.

2.21. Solve Examples 2.7 and 2.8 for the case in which $T_1 = T_2 = 400°F$ change in temperature.

2.22. Solve Example 2.7 for $s = 2.5$ in.

2.23. Solve Example 2.9 for $P_{y1} = 3000$ lb.

2.24. In Equations (2.73) and (2.74) take $R = 2$, $H = 2$, $P = 10{,}000$ lb, $P_T = 7000$ lb, $P_{y1} = 2500$ lb, and solve for the rivet loads S_1, S_2, S_3 with and without the inelastic effect.

2.25. Repeat Problem 2.24 for $R = 0.50$.

2.26. Solve Example 2.10 for $s = 6.0$ in. and compare results to those in Example 2.10.

2.27. Use Equation (2.54) to determine the elastic deflections u_{12} and u_{23} between rivets (1) and (2) and rivets (2) and (3) in Examples 2.7 and 2.10.

2.28. In Figure 2.3 take $w = 3.00$ in., $t = 0.050$ in. clad 2024-T42 aluminum alloy (L-direction, B basis), $k_L = 1.00$, $k_u = 1.50$, and find the load P_L for a spot welded joint.

2.29. Repeat Example 2.11 for $k_u p_L = 4000$ lb/in.

2.30. From the $t = 0.064$ in. curve on Figure 2.28 determine the closest rivet size that takes out an area equivalent to the efficiency of the spot welded tension joint.

2.31. Repeat Example 2.12 for 2024-T6 plate with $\sigma_{zz\,max} = 40{,}000$ psi $= F_{ty}$, $E = 10^7$ psi, $\nu = 0.3$.

2.32. Repeat part (b) of Example 2.12 for $t_b = 0.005$ in.

2.33. Solve Example 2.12 for a brittle adhesive with $\sigma_{xy\,max} = 8500$ psi, $\sigma_{p\,max} = 16{,}000$ psi, $G_b = 220{,}000$ psi, $G_{b\,eff} = 100{,}000$ psi, $E_b = 1{,}500{,}000$ psi, $E_{b\,eff} = 800{,}000$ psi.

2.34. Repeat parts (b) and (c) of Example 2.12 for $k_u p_L = 5000$ lb/in.

2.35. Repeat part (b) of Example 2.12 for an efficiency of $R_e = 0.80$.

References

Chapter 2

1. Mil-HDBK-5c, *Metallic Materials and Elements of Aerospace Vehicle Structures*, Sept. 1976, 2 Vols., U.S. Government Printing Office, Washington, D.C., 20402.
2. E.F. Bruhn: *Analysis and Design of Flight Vehicle Structures*, Tri-State Offset Co., Cincinnati, Ohio 45202.
3. B.E. Gatewood: *Thermal Stresses*, McGraw-Hill Book Co., New York (1957).
4. B.E. Gatewood and R.W. Gehring: Inelastic mechanical joint analysis method with temperature and mixed materials, *AMMRC MS 74-8*, Sept. 1974, pp. 193-210, Army Materials and Mechanics Research Center, Watertown, Massachusetts 02172.
5. I.U. Ojalvo and H.L. Eidinoff: Bond thickness effects upon stresses in single-lap adhesive joints, *AIAA Journal* 16, pp. 204-211, March (1978).

6. W.N. Sharpe, Jr. and T.J. Muha, Jr.: Comparison of theoretical and experimental shear stress in the adhesive layer of a lap joint model, *AMMRC MS 74-8*, pp. 23-44, Sept. 1974.
7. L.J. Hart-Smith: *Adhesive-Bonded Double-Lap Joints*, NASA CR 112235, Langley Research Center, Hampton, Va. 23366 (1973).
8. L.J. Hart-Smith: *Adhesive-Bonded Single-Lap Joints*, NASA CR 112236, Langley Research Center, Hampton, Va. 23366 (1973).
9. L.J. Hart-Smith: *Adhesive-Bonded Scarf and Stepped-Lap Joints*, NASA CR 112237, Langley Research Center, Hampton, Va. 23366 (1973).
10. L.J. Hart-Smith: *Non-Classical Adhesive-Bonded Joints in Practical Aerospace Construction*, NASA CR 112238, Langley Research Center, Hampton, Va. 23366 (1973).

References for additional reading

11. D.J. Peery and J.J. Azar: *Aircraft Structures*, 2nd Ed., McGraw-Hill Book Co., New York, Chapter 12 (1982).
12. J.G. Hicks: *Welded Joint Design*, Halstead Press, N.J. (1979).

3

Structural design of flight vehicle components

3.1. Introduction

Since the flight vehicle structure must be lifted from the earth's surface and transported on every flight during the life of the vehicle, the primary concern in the design of the structure is weight. Every component of the structure must be designed to have a minimum practical weight and still be strong enough to take all the applied forces and temperatures expected during the vehicle life. As discussed in Section B.14,(Vol.1), it is necessary to use the highest strength low density materials available and to use optimum design procedures for materials and structure types to obtain a minimum weight structure.

Since the most efficient structure to take a given load is the one with the least weight, the parameter for comparing the efficiencies of materials is $F/\rho_w =$ allowable stress/weight density. The least weight structure has the largest value of F/ρ_w. However, as discussed in Chapter 1, there are many allowable stresses: tension, ultimate F_{tu} and yield F_{ty}; shear ultimate F_{su}; compression, yield F_{cy}, local buckling F_{cr}, local crippling F_{cc}, and column F_c; bearing, ultimate F_{bru} and yield F_{bry}. The best material for one type of stress at a given temperature may not be the best for a different type of stress at a different temperature. The data given in Reference 1 can be used to draw curves of F/ρ_w against temperature for some of the allowable stresses, F_{tu}, F_{ty}, F_{cy}, and F_{su} so that the various materials can be compared directly at any temperature. Figures 3.1 and 3.2 show such a comparison for F_{tu}/ρ_w and F_{cy}/ρ_w for two aluminum alloys clad 2024-T3 and 7075-T6 extrusions, one titanium alloy Ti-4Al-3Mo-1V, and one nickel-steel alloy inconel X-750. The values used are for 1/2 hour exposure time. Figure 3.3 shows the E_c modulus of elasticity comparison for the same four materials. From the curves in Figures 3.1 and 3.2, it is evident that for these allowable stresses the aluminum alloys can be used up to about 400°F, the titanium alloys up to about 1000°F, and the nickel-steel alloys up to about 1500°F.

As shown in Figures 1.19 and 1.21, the allowable compression stresses F_{cr}, F_{cc}, and F_c depend upon the shape of the cross section, the length of the member, and the compressive yield stress F_{cy}. Also, as will be seen below, the optimum values of F_{cr}, F_{cc}, and F_c at M.S. = 0.00 depend upon the applied loads. Thus,

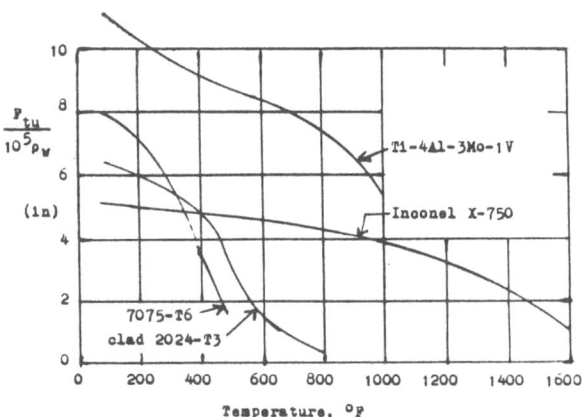

Fig. 3.1. Temperature effects on ultimate tension stress-density ratio.

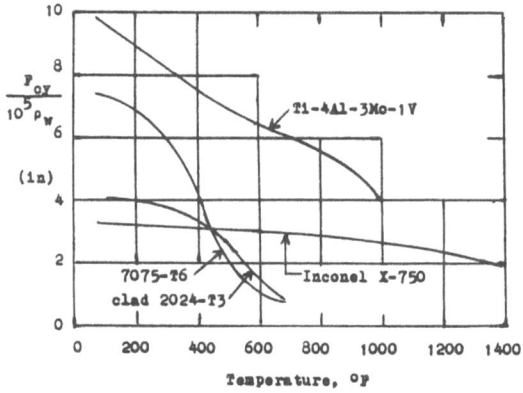

Fig. 3.2. Temperature effects on yield compressive stress-density ratio.

the comparison of the materials for these compressive stresses is complicated and can be made only for similar cross sections using the best design for each material. Also, insofar as possible, the types of cross sections used should give large allowable stresses (small B_{cr} and B_c on Figures 1.19 and 1.21, respectively) in order to have the least weight.

Since in most cases for tension members, it is a simple matter to select the cross section area to make the applied stress equal to the allowable tension stress (within the joint limitations and standard thickness limitations discussed in Chapter 2), most of the effort in this chapter will be devoted to the optimum design of compression members for minimum weight. The cases of minimum weight design for columns without buckling, for sections with local buckling

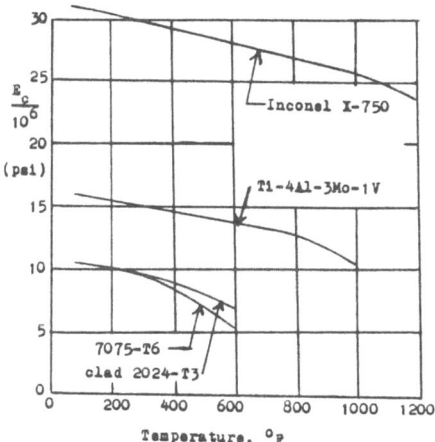

Fig. 3.3. Temperature effects on compressive modulus.

and local crippling, for columns with local buckling, and for stiffened panels in compression are considered in Sections 3.2 through 3.5. The procedure of using effective areas for stiffened panels is described in Section 3.6. The effect of load intensity on wing design and box beam design are considered in Sections 3.7 and 3.8, respectively. Diagonal tension beams are described in Section 3.9 and references made to other books for the analysis and design details.

3.2. Design of minimum weight columns without local buckling

The allowable non-dimensional column stress $S_c = F_c/F_{cy}$ is given by Equations (1.12) and (1.13) and in Figure 1.21 in terms of the parameter B_c, where

$$B_c^2 = (L/\rho)^2 (F_{cy}/c\pi^2 E_c), \quad I = A\rho^2. \tag{3.1}$$

This parameter B_c involves the material properties F_{cy} and E_c and the geometric properties L, ρ, and c. The non-dimensional applied compression stress can be expressed as

$$s_c = \sigma_c/F_{cy} = k_u P_L/AF_{cy}, \tag{3.2}$$

which also involves the material and geometry.

The optimum design of the structure for minimum weight occurs for

$$\text{M.S.} = (S_c/s_c) - 1 = 0.00, \quad s_c = S_c. \tag{3.3}$$

Thus, for any point on the column curves in Figure 1.21, $s_c = S_c$ gives a value for B_c in Equation (3.1) and a value for $s_c = S_c$ in Equation (3.2), or

$$\rho^2 = (L^2 F_{cy})/(c\pi^2 E_c B_c^2), \quad k_u P_L = AS_c F_{cy}. \tag{3.4}$$

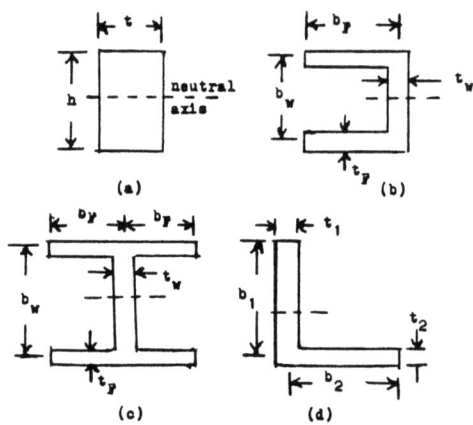

Fig. 3.4. (a). Rectangular Section. (b). Channel or Zee Section. (c). I-Section. (d). Angle Section. See Table 3.1.

Table 3.1. Expressions for Figure 3.4 and Equation (3.6).

Section (Figure 3.4)	h	t	k_0	k_c
(a)	h	t	1	1
(b)	b_w	t_w	$1 + 2\left(\frac{t_F b_F}{t_w b_w}\right)$	$(3k_0 - 2)/k_0$
(c)	b_w	t_w	$1 + 4\left(\frac{t_F b_F}{t_w b_w}\right)$	$(3k_0 - 2)/k_0$
(d)	b_1	t_1	$1 + \left(\frac{b_2 t_2}{b_1 t_1}\right)$	$(4k_0 - 3)/k_0^2$

If the cross section of the column has one variable dimension h so that $\rho = f(h)$ and $A = g(h)$, then Equation (3.4) gives a value for h and for the applied load $k_u P_L$ at any point on the column curves. This gives a procedure to construct a minimum weight design curve for a given material by graphing

$$F_c/\rho_w = S_c F_{cy}/\rho_w \tag{3.5}$$

against a load parameter for the column. It also allows materials to be compared directly for any value of the temperature.

Various type of cross sections, Figure 3.4, can be represented in terms of a variable h in the form

$$A = k_0 t h, \quad \rho^2 = k_c h^2/12, \tag{3.6}$$

where k_0 and k_c are non-dimensional parameters involving the geometry, h is the variable dimension and t is to be selected to prevent local buckling. Table 3.1 and Figure 3.4 show expressions and definitions of k_0, k_c, t, and h for several cross sections.

Structural design of flight vehicle components 93

From Equations (3.4) and (3.6),

$$h^2 = 12\rho^2/k_c = (12L^2 F_{cy})/(ck_c\pi^2 E_c B_c^2), \tag{3.7}$$

$$k_u P_L = AS_c F_{cy} = k_0 th S_c F_{cy} = k_0 th F_c,$$

$$F_c = k_u P_L/k_0 th = (\text{L.P.})_c \left(\pi^2 E_c B_c^2/12 F_{cy}\right)^{1/2},$$

$$(\text{L.P.})_c = \left(k_u P_L/k_0 t\right)\left(ck_c/L^2\right)^{1/2}, \tag{3.8a}$$

where $(\text{L.P.})_c$ is defined as the *load parameter* for columns without local buckling. For any given material with F_{cy}, E_c, n, and ρ_w known, a design column curve of F_c/ρ_w against $(\text{L.P.})_c$ can be constructed. To obtain the curve from Figure 1.21, rewrite Equation (3.8a) in the form

$$(\text{L.P.})_c = \left(k_0 th S_c F_{cy}/k_0 t\right)\left(ck_c/L^2\right)^{1/2}$$

$$= (S_c/B_c)\left(12 F_{cy}^3/\pi^2 E_c\right)^{1/2}. \tag{3.8b}$$

To construct the curves, select B_c values on Figure 1.21 and read S_c on the proper curve. Put these values in Equation (3.8b) to get $(\text{L.P.})_c$ and use S_c to get F_c/ρ_w. Various materials can be compared directly for the same type cross section with the same load parameter.

In constructing the design curves, the points on Euler's curve in Figure 1.21 have $S_c = 1/B_c^2$ so that in this region the equation for the design curve can be written from Equation (3.8b) as

$$F_c/\rho_w = S_c F_{cy}/\rho_w = H_c(\text{L.P.})_c^{2/3}, \text{ region of Euler's curve,}$$

$$H_c = (1/\rho_w)\left(\pi^2 E_c/12\right)^{1/3}, \ (\text{L.P.})_c \text{ from Equation (3.8a)}, \tag{3.9}$$

which simplifies the calculations for the Euler's region.

The following example demonstrates the procedure for constructing the design curves using Figure 1.21 and Equation (3.8b). Note that to get the optimum value of h for a given case, calculate the $(\text{L.P.})_c$ and read F_c/ρ_w from the design curve, which gives $F_{c,\max}$. Since $A = k_u P_L/F_c$, Equation (3.6) gives

$$h_{\text{opt}} = A_{\min}/k_0 t = k_u P_L/(k_0 t F_{c,\max}). \tag{3.10}$$

Example 3.1. Construct room temperature column design curves for cross sections in Equation (3.6) and Figure 3.4 for the four materials in Figures 3.2 and 3.3. (1) For the clad 2024-T3 aluminum alloy use $F_{cy} = 40,000$ psi, $E_c = 10^7$ psi, $\rho_w = 0.100$ lb/in^3, and $n = 10$. (2) For the 7075-T6 aluminum alloy extrusions use $F_{cy} = 75,000$ psi, $E_c = 10.5(10^6)$ psi, $\rho_w = 0.101$ lb/in^3, and $n = 20$. (3) For the Ti-4Al-3Mo-1V titanium alloy use $F_{cy} = 161,000$ psi, $E_c = 16(10^6)$ psi, $\rho_w = 0.163$ lb/in^3, and $n = 20$. (4) For the inconel X-750 nickel-steel alloy use $F_{cy} = 100,000$ psi, $E_c = 31(10^6)$ psi, $\rho_w = 0.300$ lb/in^3, and $n = 10$.

Solution. For the four materials in the sequence (1), (2), (3), and (4), the Equations (3.5), (3.8), and (3.9) give

$$F_c/10^5 \rho_w = 4.00 S_c, \ 7.43 S_c, \ 9.88 S_c, \text{ and } 3.33 S_c,$$

$$\left(12 F_{cy}^3/\pi^2 E_c\right)^{1/2} = 2790, \ 6989, \ 17,808, \text{ and } 6263,$$

$$H_c = 2018, \ 2033, \ 1449, \text{ and } 980.$$

On Figure 1.21, B_c values are selected to cover the S_c range from 0.1 to cut-off for the materials, with the S_c being calculated or read in each case, and the results together with the above values used in Equations (3.8b) and (3.9). Figure 3.5 shows the design curves for the four materials, where a logarithmic scale is used for the load parameter in order to cover the load range up to stress cut-off on the materials.

For the room temperature case on Figure 3.5 it is evident that the aluminum alloys are best for low loads and that the titanium alloy is best for medium and high loads. The curves show that the minimum weight design column stresses increase with the load parameter $(k_u P_L/k_0 t)(ck_c/L^2)^{1/2}$. For a given applied load $k_u P_L$, the parameter is larger for larger values of end fixity c and radius of gyration factor k_c, and for smaller values of the area factor k_0, the thickness t, and the length L. In design, supports can be provided to reduce L and channel and I sections can be used to increase k_c. Also, t can be reduced as much as possible to avoid local buckling. The best adjustment of all these factors will give the largest possible load parameter for a specified load and hence the least possible weight.

Design curves similar to Figure 3.5 can be constructed at any temperature by using the data in Figures 3.2 and 3.3 and in Reference 1. However, for any known load parameter (L.P.), the (F_c/ρ_w) at elevated temperature can be obtained directly from the room temperature values on Figure 3.5 and the E_c and F_{cy} values at the elevated temperature, provided the n value for the material does not change appreciably with temperature. The procedure is to calculate an equivalent room temperature load parameter $(\text{L.P.})_{RT}$ from Equation (3.8) and use it on Figure 3.5 to get an equivalent $(F_c/\rho_w)_{RT}$ to use in Equation (3.5). In summary,

$$\text{Calculate } (\text{L.P.})_{RT} = \left(\frac{F_{cy,RT}}{F_{cy,ET}}\right)^{3/2} \left(\frac{E_{c,ET}}{E_{c,RT}}\right)^{1/2} (\text{L.P.})_c,$$

$$\text{Read } \left(F_c/10^5 \rho_w\right)_{RT} \text{ on Figure 3.5 for } (\text{L.P.})_{RT}, \qquad (3.11)$$

$$\text{Calculate } \left(F_c/10^5 \rho_w\right)_{ET} = \frac{F_{cy,ET}}{F_{cy,RT}} \left(F_c/10^5 \rho_w\right)_{RT}.$$

In the Euler range on Figure 3.5 for the $(F_c/10^5 \rho_w)_{RT}$, the calculations can be further simplified by using Equation (3.9), or

$$\left(F_c/\rho_w\right)_{ET} = \left(\frac{E_{c,ET}}{E_{c,RT}}\right)^{1/3} H_c (\text{L.P.})_c^{2/3}. \qquad (3.12)$$

Example 3.2. Suppose a rectangular column 2.00 in. by h in. with $c = 1$ and $L = 10$ in. has a compression load of $k_u P_L = 40{,}000$ lb. Find the optimum value of h for least weight and find the column stress F_c for the four materials in Example 3.1 and on Figure 3.5.

Solution. From Equation (3.8) and Figure 3.4, $k_0 = 1$, $k_c = 1$, and $t = 2.00$ in. so that the load parameter is

$$(\text{L.P.})_c = (1)(40{,}000)/(1)(2.00)(10) = 2000 \text{ psi},$$

whence from Figure 3.5 and Example 3.1 data

$$\begin{aligned} F_c/10^5 \rho_w &= 3.00, \qquad\quad 3.227, \qquad\quad 2.300, \qquad\quad 1.556, \\ F_c &= 30{,}000 \text{ psi}, \quad 32{,}600 \text{ psi}, \quad 37{,}500 \text{ psi}, \quad 46{,}700 \text{ psi}, \end{aligned}$$

in sequence for clad 2024-T3, 7075-T6 extrusion, Ti-4Al-3Mo-1V titanium, and inconel X-750. Thus, from Equation (3.10), h_{opt} and the minimum weight are, in sequence,

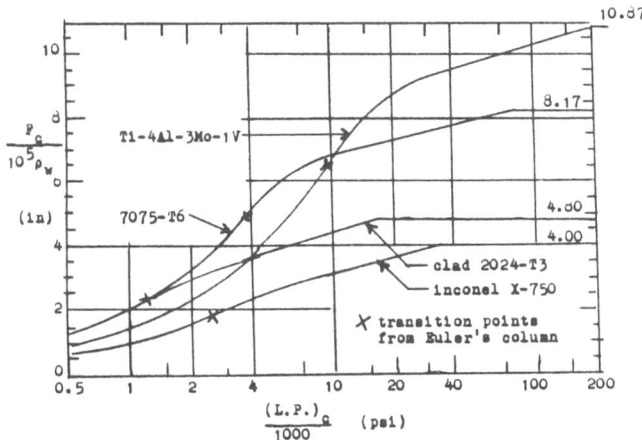

Fig. 3.5. Column design curves without local buckling.

$$h_{opt} = k_u P_L/(2.00 F_c) = \quad 0.67, \quad 0.61, \quad 0.53, \quad 0.43 \text{ in.},$$
$$\text{weight} = 2 h_{opt} \rho_w L \quad = \quad 1.33, \quad 1.23, \quad 1.73, \quad 2.57 \text{ lb},$$

and the 7075-T6 extrusion is the best material in this case. Note that a better design would be obtained by reducing the $t = 2.00$ in. dimension, so long as local buckling does not occur. See Sections 3.3 and 3.4.

Example 3.3. (a) Repeat Example 3.2 for a channel section with $t_w = 0.200$ in., $t_F/t_w = 1.00$, $b_F/b_w = 0.50$, $c = 1$, $L = 10$ in., $k_u P_L = 40,000$ lb. (b) Compare the results to Example 3.2.
Solution. (a) From Equation (3.8) and Figure 3.4 the load parameter is, with $k_0 = 2.00$ and $k_c = 2.00$,

$$(\text{L.P.})_c = (2.00)^{1/2}(40,000)/(2.00)(0.200)(10) = 14,140,$$

whence from Figure 3.5 and Example 3.1 data

$$F_c/(10^5 \rho_w) = \quad 4.65, \quad 7.10, \quad 7.90, \quad 3.35,$$
$$F_c = \quad \quad 46{,}500 \text{ psi}, \quad 71{,}700 \text{ psi}, \quad 128{,}800 \text{ psi}, \quad 101{,}000 \text{ psi},$$

in sequence for clad 2024-T3, 7075-T6 extrusions, Ti-4Al-3Mo-1V, and inconel X-750. Thus, h_{opt} and the minimum weight are

$$h_{opt} = k_u P_L/(0.40 F_c) = \quad 10^5/F_c = \quad 2.15, \quad 1.39, \quad 0.78, \quad 0.99 \text{ in.},$$
$$\text{weight} = 0.40 h_{opt} \rho_w L = \quad \quad \quad \quad 0.86, \quad 0.56, \quad 0.51, \quad 1.19 \text{ lb},$$

in sequence, and the Ti-4Al-3Mo-1V is the least weight material in this case.

(b) As might be expected, all the materials have less weight for the channel section as compared to a rectangular section for this case of $L = 10$ in., $k_u P_L = 40,000$ lb. For the rectangle, the 7075-T6 extrusion weighs much less than the titanium alloy, but for the channel section the titanium weighs slightly less than the 7075-T6. Thus, the comparison of column stress for materials not only depends upon the applied load and temperature, but also depends upon the type of cross section. The $h_{opt} = 0.78$ in. for the titanium points up another factor in the comparisons. With $t_w = 0.20$ in., it is impractical to form the channel from a flat plate. Also, $h_{opt} = 0.78$ in. is probably too small for the section to be extruded. Thus, to be more realistic, the $t_w = 0.20$ in. should be reduced, which increases the $(\text{L.P.})_c$ and from Equation (3.5) gives a further advantage to the titanium. However, if t_w is too small so that $h_{opt}/t_w > 12$, then local buckling can occur and the column equations may change. See Sections 3.3 and 3.4 for the buckling cases.

Example 3.4. Repeat Example 3.2 at $T = 400°F$ for the four materials.

Solution. From Figures 3.2 and 3.3 or Tables in Reference 1, $F_{cy,ET} = 34,300$ psi, 44,000 psi, 122,000 psi, 90,000 psi, in sequence for clad 2024-T3, 7075-T6 extrusion, Ti-4Al-3Mo-1V, and inconel X-750, and $E_{c,ET} = 8.98(10^6)$ psi, $8.40(10^6)$ psi, $14.51(10^6)$ psi, and $29.17(10^6)$ psi. Use these values and the room temperature values in Example 3.1 in Equation (3.11) with $(L.P.)_c = 2000$ psi to get

$(L.P.)_{RT} =$	2420 psi,	3980 psi,	2890 psi,	2270 psi,
$(F_c/10^5 \rho_w)_{RT} =$	3.20,	5.00,	2.94,	1.70,
$(F_c/10^5 \rho_w)_{ET} =$	2.72,	2.93,	2.23,	1.53,
$F_c =$	27,200 psi,	29,600 psi,	36,300 psi,	45,900 psi,
$h_{opt} = 20,000/F_c =$	0.74,	0.68 in.	0.55 in.	0.44 in.
weight$= 20\rho_w h_{opt} =$	1.48,	1.36 lb	1.79 lb	2.61 lb

whence the 7075-T6 extrusion with 1/2 hour exposure time still gives the least weight at 400°F for the 40,000 lb load with $L = 10$ in. At larger loads or higher temperatures, the titanium and inconel will show better results.

3.3. Design of minium weight sections with local buckling and crippling

The allowable non-dimensional local buckling stress $S_{cr} = F_{cr}/F_{cy}$ and the allowable local crippling stress $S_{cc} = F_{cc}/F_{cy}$ are given in Equations (1.8) and (1.10) and on Figure 1.19 in terms of the parameter B_{cr}, where Equations (1.2) and (1.3) give

$$B_{cr}^2 = (b/t)^2 (F_{cy}/KE_c). \tag{3.13}$$

The non-dimensional applied compressive stresses can be expressed as

$$s_{cr} = \sigma_{cr}/F_{cy} = k_b P_L/AF_{cy}, \qquad s_{cc} = \sigma_{cc}/F_{cy} = k_u P_L/AF_{cy}, \tag{3.14}$$

where k_b represents any specified buckling criterion such as $k_b = k_L$ for the outside skin on some airplanes to prevent control problems up to limit load. In most cases there is no criterion for buckling so that only the k_u or (B) failure criterion applies. Of course, in the region of Figure 1.19 where $S_{cr} = S_{cc}$, it follows that $k_b = k_u$.

The optimum design of the structure for minimum weight occurs for

$$\begin{aligned} \text{M.S.} &= (S_{cr}/s_{cr}) - 1 = 0.00, \quad \text{or} \quad s_{cr} = S_{cr}, \\ \text{M.S.} &= (S_{cc}/s_{cc}) - 1 = 0.00, \quad \text{or} \quad s_{cc} = S_{cc}. \end{aligned} \tag{3.15}$$

Thus, for any point on either the buckling or crippling curves on Figure 1.19, $s_{cr} = S_{cr}$ or $s_{cc} = S_{cc}$ gives a value for B_{cr} in Equation (3.13) and a value for $s_{cr} = S_{cr}$ or $s_{cc} = S_{cc}$ in Equation (3.14), or

$$K(t/b)^2 = F_{cy}/(E_c B_{cr}^2), \qquad k_b P_L = AS_{cr}F_{cy}, \qquad k_u P_L = AS_{cc}F_{cy}. \tag{3.16}$$

If the cross section has one variable dimension t so that $A = g(t)$, then Equation (3.16) gives a value for t and for the load $k_b P_L$ or $k_u P_L$ at any point on the curves on Figure 1.19. This gives a procedure to construct minimum weight design curves for a given material by graphing

$$(F_{cr}/\rho_w) = S_{cr}F_{cy}/\rho_w, \quad \text{or} \quad (F_{cc}/\rho_w) = S_{cc}F_{cy}/\rho_w, \tag{3.17}$$

against a load parameter (L.P.) for the cross section.

Consider the types of cross sections shown in Figure 3.4 and used in Section 3.2 for the column, and take t as the variable in

$$A = k_0 t h, \tag{3.18}$$

where $h = b$ is a parameter to be selected to give practical cross sections and k_0, t, and h are defined in Figure 3.4 for each type of cross section. From Equation (3.16) and Equation (3.8)

$$t^2 = (F_{cy}h^2)/(KE_c B_{cr}^2), \tag{3.19}$$

$$(\text{L.P.})_{cr} = (K^{1/2}/k_0)(k_b P_L/h^2) = (S_{cr}/B_{cr})(F_{cy}^3/E_c)^{1/2}, \tag{3.20}$$

$$(\text{L.P.})_{cc} = (K^{1/2}/k_0)(k_u P_L/h^2) = (S_{cc}/B_{cr})(F_{cy}^3/E_c)^{1/2}, \tag{3.21}$$

where L.P. is the load parameter. Note that the constant K is given in Figures 1.14 through 1.16 in terms of the same constant ratios t_w/t_F and b_F/b_w as in Figure 3.4.

Just as for the column case in Section 3.2, design buckling and crippling curves can be constructed for a given material at a given temperature by graphing F_{cr}/ρ_w and F_{cc}/ρ_w against the load parameter L.P. Various materials can be compared directly for the same type cross section for the same load parameter. For a selected B_{cr}, the values S_{cr} and S_{cc} can be read from the proper curve on Figure 1.19 and F_{cr}/ρ_w, F_{cc}/ρ_w, $(\text{L.P.})_{cr}$, and $(\text{L.P.})_{cc}$ calculated from Equations (3.17), (3.20), and (3.21). The calculations can be simplified for cases where $S_{cr} = 1/B_{cr}^2$ for buckling and $S_{cc} = M/(B_{cr})^{1/2}$ for crippling on Figure 1.19. In these cases

$$F_{cr}/\rho_w = H_{cr}(\text{L.P.})_{cr}^{2/3}, \qquad H_{cr} = E_c^{1/3}/\rho_w, \tag{3.22}$$

where $(\text{L.P.})_{cr}$ is defined in Equation (3.20), and

$$F_{cc}/\rho_w = H_{cc}(\text{L.P.})_{cc}^{1/3}, \qquad H_{cc} = (M^4 F_{cy}^3 E_c)^{1/6}/\rho_w, \tag{3.23}$$

where $(\text{L.P.})_{cc}$ is defined in Equation (3.21).

Example 3.5. Construct room temperature local buckling and crippling design curves for cross sections in Equation (3.18) and Figure 3.4 for the four materials in Figures 3.2 and 3.3 and in Example 3.1, which gives the room temperature material property data.

Solution. For the four materials (1) clad 2024-T3 aluminum alloy, (2) 7075-T6 extruded aluminum alloy, (3) Ti-4Al-3Mo-1V titanium alloy, (4) inconel X-750 alloy in sequence, Equations (3.17), (3.20)-(3.23) give

$$\frac{F_{cr}}{10^5 \rho_w} = 4.00 S_{cr}, \quad 7.43 S_{cr}, \quad 9.88 S_{cr}, \quad 3.33 S_{cr},$$

$$\frac{F_{cc}}{10^5 \rho_w} = 4.00 S_{cc}, \quad 7.43 S_{cc}, \quad 9.88 S_{cc}, \quad 3.33 S_{cc},$$

$$\frac{K^{1/2} k_b P_L}{k_0 h^2} = 2530 \frac{S_{cr}}{B_{cr}}, \quad 6340 \frac{S_{cr}}{B_{cr}}, \quad 16,150 \frac{S_{cr}}{B_{cr}}, \quad 5680 \frac{S_{cr}}{B_{cr}},$$

$$\frac{K^{1/2} k_u P_L}{k_0 h^2} = 2530 \frac{S_{cc}}{B_{cr}}, \quad 6340 \frac{S_{cc}}{B_{cr}}, \quad 16,150 \frac{S_{cc}}{B_{cr}}, \quad 5680 \frac{S_{cc}}{B_{cr}},$$

$$H_{cr} = 2154, \quad 2168, \quad 1546, \quad 1047,$$

$$H_{cc} = 29,360 M^{2/3}, \quad 40,170 M^{2/3}, \quad 39,080 M^{2/3}, \quad 18,680 M^{2/3}.$$

On Figure 1.19, select B_{cr} values from 3.2 to 0.5 to cover the range from small stresses to stress cut-off. The corresponding S_{cr} or S_{cc} values can be read from Figure 1.19 for each curve on Figure 1.19. Also, Equations (3.22) and (3.23) can be used to calculate many of the points on the design curves. Figure 3.6 shows the design curves for the four materials, where a logarithmic scale is used for the load parameter in the same manner as for the column curves in Figure 3.5. Only the $M = 0.85$ curves for crippling are shown in order to improve the clarity of Figure 3.6. Any particular value of M can be used in Equation (3.23) to obtain values directly for any case. Note that in Equation (3.23) and on Figure 3.6 the 7075-T6 extrusions and the Ti-4Al-3Mo-1V materials for crippling are practically the same, with the 7075-T6 being about three per cent larger.

For the room temperature cases in Example 3.5 and Figure 3.6, it is evident that for the buckling stress F_{cr} the aluminum alloys are best for low loads and that the titanium alloy is best for high loads. However, for the crippling stress F_{cc}, the 7075-T6 and titanium alloys are practically the same for low loads (at the same M value), the 7075-T6 alloy is better in the range 5000 to 10,000 for the load parameter, and the titanium alloy is better for higher loads. In actual design, the M values are not necessarily the same at the same load for the two materials so that the comparison may be different. See examples below. In order to increase the load parameter for a given load in order to get a higher allowable stress, the proportions on the cross section should be adjusted to give the constant K as large as practical, the constant k_0 as small as possible consistent with K, and the parameter h as small as practical.

Just as for the column case in Equations (3.11) and (3.12), $(F_{cr}/10^5\rho_w)_{ET}$ and $(F_{cc}/10^5\rho_w)_{ET}$ can be calculated for a given (L.P.) by using Equations (3.20) and (3.21) and Figure 3.6, or Equations (3.22) and (3.23). The procedure is

$$\text{Calculate (L.P.)}_{RT} = \left(\frac{F_{cy,RT}}{F_{cy,ET}}\right)^{3/2}\left(\frac{E_{c,ET}}{E_{c,RT}}\right)^{1/2}(\text{L.P.}),$$

$$\text{Read }\left(\frac{F_{cr}}{10^5\rho_w}\right)_{RT}\text{ or }\left(\frac{F_{cc}}{10^5\rho_w}\right)_{RT}\text{ on Figure 3.6 at (L.P.)}_{RT},$$

$$\text{Calculate }\left(\frac{F_{cr}}{10^5\rho_w}\right)_{ET} = \frac{F_{cy,ET}}{F_{cy,RT}}\left(\frac{F_{cr}}{10^5\rho_w}\right)_{RT},$$

$$\text{Calculate }\left(\frac{F_{cc}}{10^5\rho_w}\right)_{ET} = \frac{F_{cy,ET}}{F_{cy,RT}}\left(\frac{F_{cc}}{10^5\rho_w}\right)_{RT},$$

(3.24)

where it is assumed that the n values for the materials do not change with temperature. For the cases in Equations (3.22) and (3.23) with (L.P.)$_{RT}$ in Equation (3.24) in the Euler range,

$$\text{Calculate }\left(\frac{F_{cr}}{10^5\rho_w}\right)_{RT}, \left(\frac{F_{cc}}{10^5\rho_w}\right)_{RT}\text{ from Equations (3.22), (3.23)},$$

$$\text{Calculate }\left(\frac{F_{cr}}{10^5\rho_w}\right)_{ET} = \left(\frac{E_{c,ET}}{E_{c,RT}}\right)^{1/3}\left(\frac{F_{cr}}{10^5\rho_w}\right)_{RT},$$

(3.25)

$$\text{Calculate }\left(\frac{F_{cc}}{10^5\rho_w}\right)_{ET} = \left(\frac{F_{cy,ET}}{F_{cy,RT}}\right)^{1/2}\left(\frac{E_{c,ET}}{E_{c,RT}}\right)^{1/6}\left(\frac{F_{cc}}{10^5\rho_w}\right)_{RT}.$$

It should be noted that the constant M in H_{cc} in Equation (3.23) is not known until $t_w = t$ has been obtained (see Figures 1.15 and 1.16). This requires M to be assumed and one or two iterations made to get the proper M.

Example 3.6. Suppose a channel section with $h = b_w = 2.00$ in., $b_F/b_w = 0.3$, $t_w/t_F = 1.00$, has a compression load of $k_u P_L = 8000$ lb. Find the optimum value of $t = t_w$ for least weight per unit length of the cross section for the crippling stress F_{cc} for the four materials in Figure

Structural design of flight vehicle components

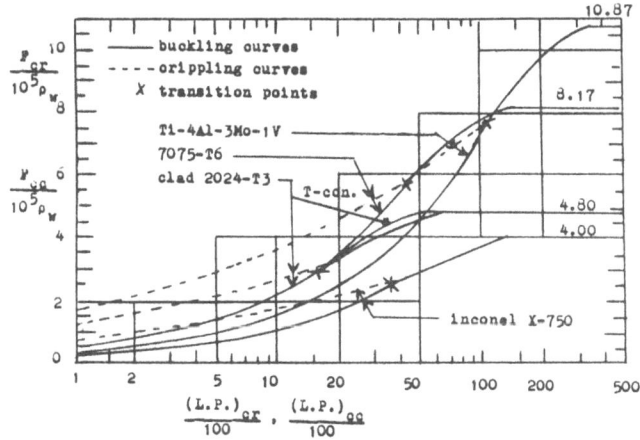

Fig. 3.6. Local buckling and crippling design curves.

3.6. Give F_{cc}, F_{cr}, t_w, $A\rho_w$, $(t_w)_{stand}$, and $A\rho_w$ with standard thickness t_w, and comment on the results.

Solution. From Figure 3.4 for the channel section, $k_0 = 1 + 2(1.00)(0.30) = 1.60$. From Figure 1.16, $K = 4.00$ so that the load parameter is

$$\text{L.P.} = \frac{(4.00)^{1/2}(8000)}{(1.60)(2.00)^2} = 2500 \text{ psi.}$$

On Figure 3.6 for this L.P. it can be seen that for the clad 2024-T3 alloy $F_{cr}/10^5\rho_w = F_{cc}/10^5\rho_w = 3.50$ for the case of formed in the 0-condition and heat treated, and $= 3.80$ for the case of formed in the T-condition, while Equation (3.23) may apply for the other three alloys, depending upon the value of M. From Figure 1.16 assume $M = 0.80$ and use values in Example 3.5 for H_{cc} to calculate F_{cc}/ρ_w from Equation (3.23), or $F_{cc}/10^5\rho_w = 4.70, 4.57$, and 2.19 for 7075-T6, Ti-4Al-3Mo-1V, and inconel X-750 in sequence. This gives $F_{cc} = 47,500$ psi, 74,500 psi, and 65,700 psi, whence

$$b_w/t_w = (2.00)(3.2/A) = 6.4F_{cc}/k_u P_L = 0.00080F_{cc}$$
$$= 38, \quad 60, \quad 53, \quad \text{in sequence.}$$

From Figure 1.16, use the extrusion and formed in the 0-condition for the three materials and get $M = 0.68, 0.63$, and 0.65 from the b_w/t_w values, whence Equation (3.23) now gives

$$F_{cc}/10^5\rho_w = 4.22, \quad 3.90, \quad 1.91,$$
$$F_{cc} = 42,600 \text{ psi}, \quad 63,600 \text{ psi}, \quad 57,300 \text{ psi},$$
$$b_w/t_w = 34, \quad 51, \quad 46.$$

New M values can be determined, but the change is small and will not be done here. Use Figure 3.6 to get F_{cr}, whence the results for the four materials are as follows, in sequence for clad 2024-T3 formed in 0-cond., clad 2024-T3 formed in T-cond., 7075-T6 extrusion, Ti-4Al-3Mo-1V formed in 0-cond., and inconel X-750 formed in 0-cond.,

F_{cc} (psi)	=	35,000	38,000	42,600	63,600	57,300
F_{cr} (psi)	=	35,000	38,000	40,300	46,400	57,300
t_w (in)	=	0.071	0.066	0.059	0.039	0.044
$A\rho_w$ (lb/in)	=	0.023	0.021	0.019	0.021	0.042
$t_{w,st}$ (in)	=	0.071	0.071	0.063	0.040	0.045
$(A\rho_w)_{st}$ (lb/in)	=	0.023	0.023	0.020	0.021	0.043

In this case, the 7075-T6 extrusion is the best material, being slightly better than the titanium alloy. Note that the standard thickness requirement may change the comparison in some cases, depending upon the actual required values. However, it may be possible to change $h = b_w$ slightly to make M.S. = 0.00 for the standard t_w. A large decrease in h increases the load parameter and may favor the titanium alloy. On the other hand h has a practical lower limit and also h enters into the column problem when column action is present. This combination of local buckling and crippling combined with the column is considered in Section 3.4.

Example 3.7. Assume $K = 4.0$ for the stiffener in Example 3.6 with $h = b_w = 2.00$ in., $b_F/b_w = 0.3$, and $t_w/t_F = 1.00$. What are the required loads and required thicknesses t_w at the cut-off points for the four materials on Figure 3.6?

Solution. From the load parameter and area expressions

$$k_u P_L = (1.6)(2.00)^2 (\text{L.P.})_{c.o.}/(4.0)^{1/2} = 3.2(\text{L.P.})_{c.o.},$$
$$t_w = A/(1.60)(2.00) = k_u P_L/3.2 F_{cr} = (\text{L.P.})_{c.o.}/F_{cr},$$

or in sequence for the four materials 2024-T3, 7075-T6, Ti-4Al-3Mo-1V, and inconel X-750 on Figure 3.6

$k_u P_L =$ 19,500,   45,000,   115,000,   and 43,000 lb,
$t_w\ \ =$ 0.127,   0.170,   0.203,   and 0.113 in.

Thus, for sufficiently large loads, it is feasible to design to the cut-offs for buckling and crippling stresses.

3.4. Design of minimum weight columns with local buckling

As discussed in Section 1.4, part (c), the allowable crippling stress F_{cc} for local failure of the cross section is a cut-off stress for the allowable column stress F_c. Since the column allowable stress F_c is unaffected up to the buckling stress F_{cr}, it follows that for cases in which $S_{cc} = S_{cr}$ in Figure 1.19, the design condition for minimum weight is

$$\begin{aligned} S_{cc} &= S_{cr} = S_c = s_c, \quad \text{or} \\ F_{cc} &= F_{cr} = F_c = k_u P_L/A = k_u P_L/k_0 t h, \end{aligned} \quad (3.26)$$

for cross sections of the type in Figure 3.4. Since two conditions are available, both variables t and h can be determined and design curves constructed using a load parameter without t and h.

The procedure to construct the design curves for the region of Equation (3.26) is to select a material on Figures 3.5 and 3.6 and assume various values of $F_{cr}/10^5 \rho_w = F_c/10^5 \rho_w$ and read the corresponding load parameters

$$(\text{L.P.})_{cr} = \frac{K^{1/2} k_u P_L}{k_0 h^2}, \quad (\text{L.P.})_c = \frac{(ck_c)^{1/2} k_u P_L}{k_0 t L}. \quad (3.27)$$

This gives

$$h^2 = \frac{K^{1/2} k_u P_L}{k_0 (\text{L.P.})_{cr}}, \quad t = \frac{(ck_c)^{1/2} k_u P_L}{k_0 L (\text{L.P.})_c}, \quad (3.28)$$

whence from Equation (3.26)

$$F_c = \frac{k_u P_L}{k_0 t h} = \left[\frac{k_0 L^2}{c k_c K^{1/2} k_u P_L} \right]^{1/2} (\text{L.P.})_c (\text{L.P.})_{cr}^{1/2}. \quad (3.29)$$

Thus, define a load parameter $(L.P.)_{cb}$ and solve for it from Equation (3.29), or

$$(L.P.)_{cb} = \frac{ck_c K^{1/2} k_u P_L}{k_0 L^2} = \frac{(L.P.)_c^2 (L.P.)_{cr}}{(10^5 \rho_w)^2 (F_c/10^5 \rho_w)^2}. \tag{3.30}$$

The design curve is the graph of $F_c/10^5 \rho_w$ against $(L.P.)_{cb}$ for any given material.

In the region of Figures 3.6 and 1.19 where $S_{cr} < S_{cc}$, Equations (1.15) and (1.16) show transition equations for S_c, where the column allowable stress S_c is between S_{cr} and S_{cc}. See Figure 1.21. To determine the minimum weight design points for this region, assume t is fixed and express h in terms of the parameters B_c and B_{cr} on Figures 1.21 and 1.19, or

$$B_c^2 = \left(\frac{L}{h}\right)^2 \frac{12 F_{cy}}{\pi^2 c k_c E_c}, \quad B_{cr}^2 = \left(\frac{h}{t}\right)^2 \frac{F_{cy}}{K E_c}, \tag{3.31}$$

$$h = \frac{L}{B_c} \left(\frac{12}{\pi^2} \frac{F_{cy}}{c k_c E_c}\right)^{1/2}, \quad h = t B_{cr} \left(\frac{K E_c}{F_{cy}}\right)^{1/2}. \tag{3.32}$$

This gives the applied stress s_c as

$$s_c = \frac{k_u P_L}{k_0 h t F_{cy}} = \frac{k_u P_L}{tL} \left(\frac{\pi^2 c k_c E_c}{12 k_0^2 F_{cy}^3}\right)^{1/2} B_c, \tag{3.33}$$

$$s_c = \frac{k_u P_L}{t^2} \left(k_0^2 K E_c F_{cy}\right)^{-1/2} \left(\frac{1}{B_{cr}}\right). \tag{3.34}$$

The applied stress s_c in Equation (3.33) graphs as a straight line on Figure 1.21 with positive slope. Thus, the design point for maximum stress occurs at $s_c = S_c$, when S_c is on the regular column curves and not on any of the transition curves between S_{cr} and S_{cc}, Equations (1.15) and (1.16). This point corresponds to a $S_{cr} = S_c$ value from which a transition curve starts. Also, if s_c in Equation (3.34) is graphed on Figure (1.19), it crosses the buckling curves for S_{cr} at larger values than on the S_{cc} crossing points. It follows that the maximum stress design points, and hence minimum weight points, are given by

$$S_c = S_{cr} = s_c \tag{3.35}$$

for the region where $S_{cr} < S_{cc}$. Of course, if the design is made at this point for a specified L, and later L is reduced then S_c will be on the transition curve starting from the design point.

Since Equation (3.35) is the same as Equation (3.26) with S_{cc} omitted, the design curves in the region $S_{cr} < S_{cc}$ could be constructed as described above in Equations (3.26)-(3.30). However, since the equations for S_c and S_{cr} are known in this region of the Euler curve and the elastic buckling curve, formulas can be obtained for $F_c/10^5 \rho_w$ in terms of the load parameter $(L.P.)_{cb}$ defined in Equation (3.30). Equation (3.35) becomes

$$1/B_c^2 = 1/B_{cr}^2 = k_u P_L/(k_0 h t F_{cy}) \tag{3.36}$$

with B_c and B_{cr} in Equation (3.31), whence

$$h = \left(\frac{k_u P_L}{k_0 E_c}\right)^{1/5} \left(\frac{12 L^2 K^{1/3}}{\pi^2 c k_c}\right)^{3/10}, \tag{3.37}$$

$$A = k_0 t h = \left(\frac{k_u P_L}{E_c}\right)^{3/5} \left(\frac{12 k_0 L^2}{\pi^2 c k_c K^{1/2}}\right)^{2/5}, \tag{3.38}$$

$$\frac{F_c}{10^5 \rho_w} = \frac{k_u P_L}{10^5 A \rho_w} = H_{cb}(\text{L.P.})_{cb}^{2/5}, \tag{3.39}$$

$$H_{cb} = \left(\frac{\pi^2}{12}\right)^{2/5} \frac{E^{3/5}}{10^5 \rho_w}, \qquad (\text{L.P.})_{cb} = \frac{c k_c K^{1/2} k_u P_L}{k_0 L^2}. \tag{3.40}$$

Example 3.8. Construct room temperature design curves for columns with local buckling for cross sections in Figure 3.4 and for the same four materials in Figures 3.2, 3.3, 3.5, and 3.6. Use the basic material property data from Example 3.1.

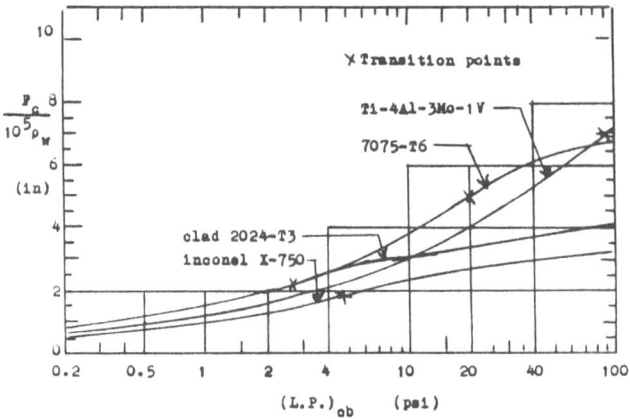

Fig. 3.7. Column design curves with local buckling (0.2-100 range).

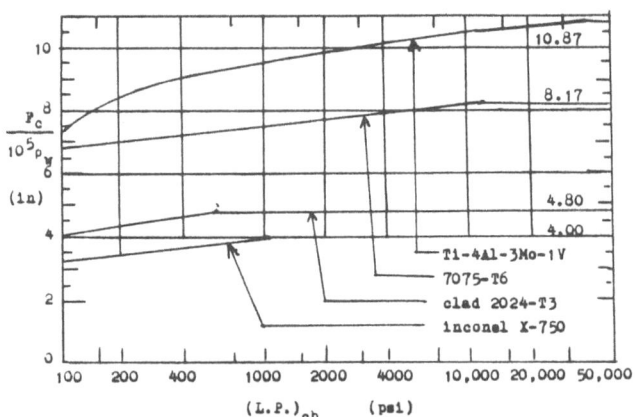

Fig. 3.7. (continued) (100-50,000 range).

Solution. For the four materials (1) clad 2024-T3 aluminum alloy, (2) 7075-T6 extruded aluminum alloy, (3) Ti-4Al-3Mo-1V titanium alloy, and (4) inconel X-750 nickel-steel alloy, Equation (3.39) can be used from $F_c/10^5 \rho_w = 0$ and $(\text{L.P.})_{cb} = 0$ up to the limits, in sequence, or

$F_c/10^5 \rho_w =$ 2.20, 5.00, 6.60, 1.80,
$H_{cb} =$ 1.469, 1.494, 1.192, 0.963,
$(\text{L.P.})_{cb} =$ 2.74, 20.48, 72.14, 4.77,

where Figures 3.5 and 3.6 and Equation (3.40) have been used. From these limit values of $F_c/10^5\rho_w$ up to their cut-off values on Figures 3.5 and 3.6, select various values and read the corresponding load parameters to use in Equation (3.30). Figure 3.7 shows the design curves for the four materials, where a logarithmic scale has been used for the load parameter in order to cover a wide load range up to the cut-off stresses. As in Figure 3.5 for the column without buckling, the aluminum alloys give the least weight for low loads while the titanium alloy is best for high loads.

The effects of temperature can be included by putting Equations (3.11) and (3.24) into Equation (3.30) and using Figure 3.7. Thus, from Equation (3.30) with $S_c = F_c/F_{cy}$ fixed, the procedure is

$$\text{Calculate } (\text{L.P.})_{cb,RT} = \left[\frac{(\text{L.P.})_{RT}}{(\text{L.P.})_{ET}}\right]_c^2 \left[\frac{(\text{L.P.})_{RT}}{(\text{L.P.})_{ET}}\right]_{cr} \left[\frac{F_{cy,ET}}{F_{cy,RT}}\right]^2 (\text{L.P.})_{cb}$$

$$= \left[\frac{F_{cy,RT}}{F_{cy,ET}}\right]^{5/2} \left[\frac{E_{c,ET}}{E_{c,RT}}\right]^{3/2} (\text{L.P.})_{cb},$$

Read $(F_c/10^5\rho_w)_{RT}$ on Figure 3.7 from $(\text{L.P.})_{cb,RT}$, (3.41)

$$\text{Calculate } \left(\frac{F_c}{10^5\rho_w}\right)_{ET} = \left[\frac{F_{cy,ET}}{F_{cy,RT}}\right] \left(\frac{F_c}{10^5\rho_w}\right)_{RT}.$$

If the $(\text{L.P.})_{cb,RT}$ in Equation (3.41) is in the Euler range on Figure 3.7, then the calculations can be simplified by using Equation (3.39),

Calculate $(F_c/10^5\rho_w)_{RT}$ in Equation (3.39),

$$\text{Calculate } \left(\frac{F_c}{10^5\rho_w}\right)_{ET} = \left[\frac{E_{c,ET}}{E_{c,RT}}\right]^{3/5} \left(\frac{F_c}{10^5\rho_w}\right)_{RT}. \tag{3.42}$$

Example 3.9. Redesign the channel section in Example 1.5 to take the same load with minimum weight. Use the same $b_F/b_w = 0.50$, $t_w/t_F = 0.80$ ratios.
  Solution. The area of the channel cross section in Example 1.5 is $(3.00)(0.080+0.100)=0.54$ in² so that the load supported is

$$k_u P_L = 0.54 F_c = 0.54(32,200) = 17,390 \text{ lb}.$$

Also, from Figure 3.4, $k_0 = 1 + 2(0.50)(1.25) = 2.25$, $k_c = [3(2.25) - 1]/2.25 = 2.11$. From Figure 1.16, $K = 3.30$, whence Equation (3.40) gives

$$(\text{L.P.})_{cb} = \frac{2(2.11)(3.30)^{1/2}(17,390)}{2.25(80)^2} = 9.26.$$

From Example 3.8 and Figure 3.7 this $(\text{L.P.})_{cb}$ is in the Euler range for the 7075-T6 extrusion so that Equation (3.39) applies and

$$F_c = (10^5)(0.101)(1.494)(9.26)^{2/5} = 36,800 \text{ psi},$$

which is larger than the result in Example 1.5. From Equations (3.37) and (3.38)

$$h = b_w = \left[\frac{17,390}{2.25(10.5)(10^6)}\right]^{0.2} \left[\frac{12(80)^2(3.30)^{1/3}}{\pi^2(2)(2.11)}\right]^{0.3} = 2.54 \text{ in.},$$

$A = 17,390/36,800 = 0.47 = 2.25(2.54)t_w,$
$t_w = 0.083 \text{ in.}, \quad t_F = 1.25(0.083) = 0.103 \text{ in.}$

Thus, the minimum design weight is $100(0.54 - 0.47)/0.54 = 13\%$ less than the weight in Example 1.5. Use the optimum $b_w = 2.54$ in. and $t_w = 0.083$ in. to calculate

$$B_{cr} = \left(\frac{2.54}{0.083}\right)\left[\frac{75,000}{3.30(10.5)(10^6)}\right]^{0.5} = 1.42,$$

$$F_{cr} = 75,000/(1.42)^2 = 37,000 \text{ psi},$$

$$F_{cc} = (0.76)(75,000)/(1.42)^{1/2} = 47,800 \text{ psi}.$$

Note that F_{cr} is the same as F_c, as should be expected, and that both F_{cr} and F_{cc} are much larger than in Example 1.5.

Example 3.10. Repeat Example 3.9 at a temperature of 400°F.
Solution. From Figures 3.2 and 3.3 for 7075-T6 extrusions $F_{cy,RT}/F_{cy,ET} = 7.42/4.42 = 1.679$, $E_{c,ET}/E_{c,RT} = 8.4/10.5 = 0.800$, $F_{cy,ET} = 75,000/1.679 = 44,700$ psi, whence Equation (3.41) gives

$$(\text{L.P.})_{cb,RT} = (1.679)^{5/2}(0.800)^{3/2}(9.26) = 24.2,$$

which is outside the Euler range for the 7075-T6 extrusion. From Figure 3.7 for this load parameter $(F_c/10^5\rho_w)_{RT} = 5.30$, whence $(F_c/10^5\rho_w)_{ET} = 5.30/1.679 = 3.16$, $F_c = 31,900$ psi, $A = 17,390/31,900 = 0.55$ in$^2 = 2.25 b_w t_w$, and $t_w = 0.244/b_w$. From Equation (3.37) and b_w in Example 3.9,

$$b_w = 2.54(1/0.80)^{0.2} = 2.66 \text{ in.},$$

$$t_w = 0.244/2.66 = 0.092 \text{ in.}, \quad t_F = 0.115 \text{ in.}$$

To determine F_{cr} and F_{cc}, calculate

$$B_{cr} = \left(\frac{2.66}{0.092}\right)\left[\frac{44,700}{3.30(8.4)(10^6)}\right]^{1/2} = 1.16,$$

whence from Figure 1.19, $S_{cr} = 0.73$, $F_{cr} = 0.73(44.700) = 32,600$ psi, which is close to F_c. With $M = 0.77$, $F_{cc} = F_{cr}$. Thus, the temperature has the largest effect on the crippling stress in this case, reducing it from 47,800 psi to 32,000 psi, while reducing the column and buckling stresses from 37,000 psi to 32,000 psi.

It should be pointed out that, although Example 3.7 shows the feasibility of designing to the cut-off stresses for buckling and crippling, the situation is quite different in the column case. For the particular channel section geometry considered for the column in Example 3.9, the required loads at cut-off on Figure 3.7 for the four materials clad, 2024-T3, 7075-T6, Ti-4Al-3Mo-1V, and inconel X-750 in order are

$$k_u P_L = 210L^2, \quad 3500L^2, \quad 11,200L^2, \quad 350L^2. \tag{3.43}$$

Thus, either L must be very small or $k_u P_L$ very large in order to reach the cut-off stress in design. To reach the required buckling loads at cut-off in Example 3.7, L must be, in sequence for this special case,

$$L = 9.6, \quad 3.6, \quad 3.2, \quad 11.1 \text{ in.} \tag{3.44}$$

Since practical values of L on airplanes may be 20-40 in., unless the applied load is very large, it is not feasible to design to the column cut-off stress. However, due to the small slopes on Figure 3.7 above $(\text{L.P.})_{cb} = 100$, the design can usually be made at stresses within 10% to 15% of the cut-off values.

Structural design of flight vehicle components

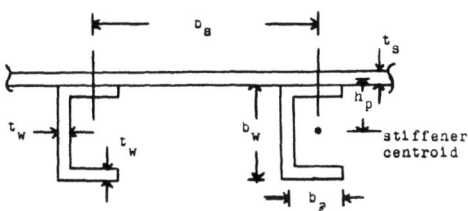

Fig. 3.8. Section of flat plate stiffened with equal spaced channel sections.

3.5. Minimum weight design for stiffened panels in compression

Consider a flat plate stiffened with equally spaced stiffeners of cross sections similar to those in Figure 3.4. Figure 3.8 shows a section of the panel with channel stiffeners while Figures 1.17 and 1.18 show idealized panels for determination of the K values for buckling of the panel. The panel will be considered to be sufficiently wide so that the unit case of one stiffener with area $A_{st} = (k_0 h t)_{st}$ together with its associated area $b_s t_s$ of the plate can be used to represent the panel. Let $k_u p_L = k_u P_L / w =$ the applied ultimate load per unit width of the panel, where w is the panel width. Then the average area per unit width is

$$A = (A_{st}/b_s) + t_s, \text{ per unit width,} \tag{3.45}$$

whence the applied non-dimensional stresses are

$$s_{cr} = \frac{b_s k_b p_L}{(A_{st} + b_s t_s) F_{cy}} \text{ for buckling,}$$

$$s_{cc} = \frac{b_s k_u p_L}{(A_{st} + b_s t_s) F_{cy}} \text{ for failure.} \tag{3.46}$$

The three cases of minimum weight design for local buckling, local crippling, and column with local buckling will be considered.

For the cases of *local buckling and local crippling* on Figure 1.19, it appears that any point on the curves up to the cut-off can be represented by

$$S_{cr} = G_{cr}/B_{cr}^q, \quad S_{cc} = G_{cc}/B_{cr}^r, \quad B_{cr} = \frac{b_s}{t_s}\left(\frac{F_{cy}}{KE_c}\right)^{1/2}, \tag{3.47}$$

where B_{cr} in Equation (3.31) has this form for panels in Figures 1.17 and 1.18. Now, in Equation (3.18) and on Figure 3.6, decreasing values of the parameter h increases s_{cr}, s_{cc} and S_{cr}, S_{cc}, but here a decreasing $h = b_s$ decreases s_{cr} and s_{cc} but increases S_{cr} and S_{cc}. Thus, there is an optimum value of b_s so that both b_s and t_s must be regarded as variables to be determined from

$$s_{cr} = S_{cr}, \quad s_{cc} = S_{cc},$$
$$\frac{ds_{cr}}{db_s} = \frac{dS_{cr}}{db_s} = 0, \quad \frac{ds_{cc}}{db_s} = \frac{dS_{cc}}{db_s} = 0. \tag{3.48}$$

From both $ds_{cr}/db_s = 0$ and $ds_{cc}/db_s = 0$ in Equation (3.46), with A_{st} assumed constant, it follows that

$$\frac{dt_s}{db_s} = \frac{A_{st}}{b_s^2}. \tag{3.49}$$

Also, from both $dS_{cr}/db_s = 0$ and $dS_{cc}/db_s = 0$ in Equation (3.47), with G_{cr}, G_{cc}, and K as constants and for any given material, it follows that

$$\frac{dt_s}{db_s} = \frac{t_s}{b_s}. \tag{3.50}$$

Since at $s_{cr} = S_{cr}$ and $s_{cc} = S_{cc}$ the derivatives dt_s/db_s must be the same, there results

$$A_{st}/b_s^2 = t_s/b_s, \quad \text{or} \quad b_s t_s = A_{st}. \tag{3.51}$$

Thus, at the maximum stress or minimum area the area of the skin $b_s t_s$ associated with the stiffener should equal the area of the stiffener. In other words, on a stiffened panel the optimum design requires one half the area in the stiffeners and one half the area in the skin.

The relation between b_s and t_s in Equation (3.51) can be used in $S_{cr} = s_{cr}$ and $S_{cc} = s_{cc}$ to get the optimum values of $b_{s,\text{opt}}$ and hence the maximum stresses in terms of a load parameter. From Equation (3.46)

$$\begin{aligned} b_{s,cr} &= (2A_{st}S_{cr,\max}F_{cy})/k_b p_L, \\ b_{s,cc} &= (2A_{st}S_{cc,\max}F_{cy})/k_u p_L, \end{aligned} \tag{3.52}$$

which can be put into B_{cr} in Equation (3.47) to get

$$\begin{aligned} B_{cr} &= \frac{4A_{st}^2 S_{cr}^2 F_{cy}^2}{A_{st}(k_b p_L)^2} \left(\frac{F_{cy}}{KE_c}\right)^{1/2} \quad \text{for buckling,} \\ B_{cr} &= \frac{4(A_{st}S_{cc}F_{cy})^2}{A_{st}(k_u p_L)^2} \left(\frac{F_{cy}}{KE_c}\right)^{1/2} \quad \text{for crippling,} \end{aligned} \tag{3.53}$$

where $t_s = A_{st}/b_s$ in Equation (3.51) has been used. This Equation (3.53) can be expressed in terms of load parameters

$$\begin{aligned} (\text{L.P.})_{cr} &= \frac{K^{1/4} k_b p_L}{2A_{st}^{1/2}} = \frac{S_{cr}}{(B_{cr})^{1/2}} \left(\frac{F_{cy}^5}{E_c}\right)^{1/4}, \\ (\text{L.P.})_{cc} &= \frac{K^{1/4} k_u p_L}{2A_{st}^{1/2}} = \frac{S_{cc}}{(B_{cr})^{1/2}} \left(\frac{F_{cy}^5}{E_c}\right)^{1/4}, \end{aligned} \tag{3.54}$$

from which minimum weight design curves can be constructed in the same manner as in Sections 3.2, 3.3, and 3.4.

The calculations can be simplified for the cases where $S_{cr} = B_{cr}^{-2}$, and $S_{cc} = MB_{cr}^{-1/2}$. In these cases

$$F_{cr}/\rho_w = H_{cr}(\text{L.P.})_{cr}^{4/5}, \quad H_{cr} = E_c^{1/5}/\rho_w, \tag{3.55}$$

$$F_{cc}/\rho_w = H_{cc}(\text{L.P.})_{cc}^{1/2}, \quad H_{cc} = \left(M^4 F_{cy}^3 E_c\right)^{1/8}/\rho_w, \tag{3.56}$$

where $(\text{L.P.})_{cr}$ and $(\text{L.P.})_{cc}$ are defined in Equation (3.54).

Example 3.11. Construct room temperature minimum weight design curves for stiffened panels with constant area stiffeners for which the parameter K can be obtained for the four materials used in the previous sections 3.2, 3.3, 3.4.

Solution. For the four materials in the order (1) clad 2024-T3, (2) 7075-T6 extrusion, (3) Ti-4Al-3Mo-1V, and (4) inconel X-750, the multiplying factors in Equations (3.54)-(3.56) above are

$$
\begin{aligned}
F_{cy}^{5/4}/E_c^{1/4} &= 10{,}060, & 21{,}800, & & 51{,}000, & & 23{,}800, \\
H_{cr} &= 251, & 251, & & 169, & & 105, \\
H_{cc} &= 3567, & 4498, & & 3912, & & 1931,
\end{aligned}
$$

where $M = 0.80$ for stiffened panels in Figures 1.17 and 1.18. Proceed as in Example 3.5 using Figure 1.19 and Equations (3.54)-(3.56) to obtain the points to draw the design curves shown on Figure 3.9.

The elevated temperature cases can be solved in the same manner as in Section 3.3. From Equation (3.54) and Figure 3.9, the procedure is

$$
\begin{aligned}
&\text{Calculate } (\text{L.P.})_{RT} = \left[\frac{F_{cy,RT}}{F_{cy,ET}}\right]^{5/4} \left[\frac{E_{c,ET}}{E_{c,RT}}\right]^{1/4} (\text{L.P.}), \\
&\text{Read } \left(\frac{F_{cr}}{10^5 \rho_w}\right)_{RT} \text{ or } \left(\frac{F_{cc}}{10^5 \rho_w}\right)_{RT} \text{ on Figure 3.9 at } (\text{L.P.})_{RT}, \\
&\text{Calculate } \left(\frac{F_{cr}}{10^5 \rho_w}\right)_{ET} = \frac{F_{cy,ET}}{F_{cy,RT}} \left(\frac{F_{cr}}{10^5 \rho_w}\right)_{RT}, \\
&\text{Calculate } \left(\frac{F_{cc}}{10^5 \rho_w}\right)_{ET} = \frac{F_{cy,ET}}{F_{cy,RT}} \left(\frac{F_{cc}}{10^5 \rho_w}\right)_{RT}.
\end{aligned}
\tag{3.57}
$$

For the cases in Equations (3.55) and (3.56),

$$
\begin{aligned}
&\text{Calculate } \left(\frac{F_{cr}}{10^5 \rho_w}\right)_{RT}, \quad \left(\frac{F_{cc}}{10^5 \rho_w}\right)_{RT} \text{ from Equations (3.55), (3.56),} \\
&\text{Calculate } \left(\frac{F_{cr}}{10^5 \rho_w}\right)_{ET} = \left[\frac{E_{c,ET}}{E_{c,RT}}\right]^{1/5} \left(\frac{F_{cr}}{10^5 \rho_w}\right)_{RT}, \\
&\text{Calculate } \left(\frac{F_{cc}}{10^5 \rho_w}\right)_{ET} = \left[\frac{F_{cy,ET}}{F_{cy,RT}}\right]^{3/8} \left[\frac{E_{c,ET}}{E_{c,RT}}\right]^{1/8} \left(\frac{F_{cc}}{10^5 \rho_w}\right)_{RT}.
\end{aligned}
\tag{3.58}
$$

It should be noted that in the above analysis the constant K in the load parameter is not known until b_s and t_s are known (see Figures 1.17 and 1.18). This requires K to be assumed and one or two iterations made to get the proper K. Since $K^{1/4}$ is involved in the (L.P.), and hence $K^{1/5}$ in F_{cr} and $K^{1/8}$ in F_{cc}, small variations in K have little effect on the design stresses.

Note also that for the real panel of finite width w with $m = w/b_s$ equal spaces, the optimum b_s must be varied to make m and integer. However, this variation has only a small effect. If b_s/t_s is assumed constant so that F_{cr} and F_{cc} remain constant, then Equations (3.43), (3.49)-(3.51) can be used to get

$$
\frac{\Delta A}{A_{\text{opt}}} = (1/2)\left(\frac{\Delta b_s}{b_{s,\text{opt}}}\right)^2.
\tag{3.59}
$$

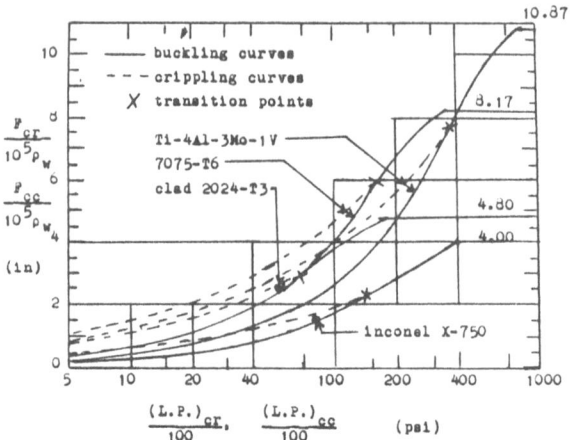

Fig. 3.9. Local buckling and crippling design curves for stiffened panels.

Thus, by varying t_s and b_s together, the stiffener spacing on the panel can be changed up to 20% with only about 2% increase in weight. Of course, the K variation must also be considered in the changes.

Example 3.12. Suppose a 7075-T6 extruded panel shown in Figure 1.17 has $b_w = 2.50$ in., $t_w = 0.160$ in. and $k_u p_L = 7000$ lb/in. compression. (a) Find the room temperature maximum crippling stress $F_{cc} = \sigma_{cc}$ and the optimum dimensions $b_{s,opt}$ and $t_{s,opt}$. (b) Find the corresponding buckling stress F_{cr} for the panel. (c) Design the panel for buckling at $k_b p_L = 7000/1.5 = 4667$ lb/in. (d) Increase $b_{s,opt}$ and $t_{s,opt}$ in part (a) by 20% and find the new F_{cc} and σ_{cc} stresses and the M.S.

Solution. (a) The stiffener area is $A_{st} = 2.50(0.160) = 0.40$ in^2. Assume $K = 4.0$, whence the load parameter and H_{cc} from Equation (3.56) are

$$(\text{L.P.})_{cc} = \frac{(4.0)^{1/4}(7000)}{2(0.40)^{1/2}} = 7826,$$

$$H_{cc} = \frac{(.80)^{1/2}(75,000)^{3/8}(10,500,000)^{1/8}}{0.101} = 4498.$$

On Figure 3.9, this $(\text{L.P.})_{cc}$ for the 7075-T6 curve is in the range of Equation (3.56) so that

$$F_{cc} = 0.101(4498)(7826)^{1/2} = 40,200 \text{ psi}.$$

From Equations (3.52) and (3.51), $b_{s,opt} = 2(0.40)(40,200)/7000 = 4.59$ in., $t_{s,opt} = 0.40/4.59 = 0.087$ in. To check K, calculate $b_w/b_s = 2.50/4.59 = 0.54$, $t_w/t_s = 0.160/0.087 = 1.84$, whence Figure 1.17 gives $K = 5.0$. With this K the new $(\text{L.P.})_{cc} = 8275$, and

$$F_{cc} = 41,300 \text{ psi}, \quad b_{s,opt} = 4.72 \text{ in.}, \quad t_{s,opt} = 0.085 \text{ in.}$$

The new K is about 5.2 so that no further iterations will be made.

(b) To obtain the buckling stress F_{cr} for this panel with $b_s = 4.72$ in. and $t_s = 0.085$ in., use Figure 1.19 with

$$B_{cr} = \frac{4.72}{0.085}\left[\frac{75,000}{5.00(10,500,000)}\right]^{1/2} = 2.10.$$

This is in the elastic range so that

$$F_{cr} = F_{cy}/B_{cr}^2 = 75,000/(2.10)^2 = 17,000 \text{ psi}.$$

(c) If the design requires no buckling up to limit load with $k_b = k_L = 1.00$, $k_u = 1.50$, then the design for buckling from Equation (3.55) gives ($K = 4.0$)

$$F_{cr} = (10,500,000)^{1/5}(7826/1.50)^{4/5} = 23,900 \text{ psi},$$

whence $b_{s,\text{opt}} = 4.10$ in. and $t_{s,\text{opt}} = 0.098$ in. To check K, $b_w/b_s = 0.61$, $t_w/t_s = 1.63$, whence $K = 3.9$, which is close enough to the assumed value. To obtain the crippling stress for the panel

$$B_{cr} = \left(\frac{4.10}{0.098}\right)\left[\frac{75,000}{4.0(10,500,000)}\right]^{1/2} = 1.77,$$

$$F_{cc} = (0.80)(75,000)/(1.77)^{1/2} = 45,100 \text{ psi},$$

which gives the M.S. for failure as

$$\text{M.S.} = \frac{45,100}{1.5(23,900)} - 1 = 0.26.$$

(d) From Equation (3.45) $A = (0.40/4.72) + 0.085 = 0.170 \text{ in}^2/\text{in}$. With a 20% increase in both b_s and t_s, $A = (0.085)/(1.20) + (1.20)(0.085) = 0.173 \text{ in}^2/\text{in}$, which decreases the applied stress by 2%, as Equation (3.59) predicts. Thus, $\sigma_{cc} = 7000/0.173 = 40,500$ psi, and the K for $b_s = (1.20)(4.72) = 5.30$ in. and $t_s = (1.20)(0.085) = 0.102$ becomes $K = 4.3$. This gives $B_{cr} = 2.26$, $F_{cc} = 0.80(70,000)/(2.26)^{1/2} = 39,900$ psi. The margin of safety is

$$M.S = \frac{39,900}{40,500} - 1 = -0.01,$$

which is within the usual tolerance of -0.03. Note that this possible change in b_s and t_s with little change in weight allows t_s to be changed to a standard thickness for the panel.

It should be pointed out that in order to design to the cut-off stresses on Figure 3.9 the required unit loading $k_u p_L$ must be large or the stiffener area A_{st} must be small. Since a small A_{st} means closer stiffener spacing, there must be a compromise in the design. If $k_u p_L = 30,000$ lb/in. in the above Example 3.12, then the cut-off is reached ($K = 4.0$) and $F_{cc} = F_{cr} = 82,500$ psi with $b_{s,\text{opt}} = 2.20$ in. and $t_{s,\text{opt}} = 0.182$ in. But, in Figure 1.17, $b_w/b_s = 2.50/2.20 = 1.14$ and K is less than one so that $(\text{L.P.})_{cc}$ is below the cut-off. Thus, a larger stiffener area is needed and hence a larger $k_u p_L$ in order to reach the cut-off.

For the *column case* it is necessary to have the actual dimensions of the stiffener on the panel in order to get the radius of gyration of the panel. Use stiffeners of the types in Figure 3.4 with

$$A_{st} = (k_0 th)_{st} \tag{3.60}$$

and use the relation in Equation (3.51) as

$$b_s t_s = k_{0,st} t_{st} h_{st} \tag{3.61}$$

in order to have the maximum buckling stress of the panel. Use Equation (3.35) for the minimum weight design of the column with buckling, or

$$F_c = F_{cr} = b_s k_u p_L / 2(k_0 th)_{st}. \tag{3.62}$$

If t_{st}/t_s is regarded as a parameter, then it is possible to solve Equations (3.61) and (3.62) for the optimum values of the variables b_s, h_{st}, and t_{st}.

From Equation (3.61),

$$\frac{b_s}{t_s} = k_{0,st}\left(\frac{t_{st}}{t_s}\right)^2 \left(\frac{h}{t}\right)_{st}, \tag{3.63}$$

which gives B_{cr} in Equation (3.31) for the panel as

$$B_{cr}^2 = k_{0,st}^2 \left(\frac{t_{st}}{t_s}\right)^4 \left(\frac{h}{t}\right)_{st}^2 \left(\frac{F_{cy}}{KE_c}\right). \tag{3.64}$$

If $h_p = k_p h_{st}$ is the distance from the center of the skin to the centroid of the stiffener, Figure 3.8, then for the types of sections in Figure 3.4 the radius of gyration ρ_p for the panel is

$$\rho_p^2 = k_{cp} h_{st}^2 / 12, \qquad k_{cp} = (k_{c,st}/2) + 3k_p^2. \tag{3.65}$$

Thus B_c in Equation (3.31) for the panel is

$$B_c^2 = \left(\frac{L}{h_{st}}\right)^2 \left[\frac{12 F_{cy}}{\pi^2 c k_{cp} E_c}\right]. \tag{3.66}$$

Since B_{cr} and B_c in Equations (3.64) and (3.66) are expressed in terms of the stiffener values alone, the procedure of Section 3.4 for columns with local buckling can be used to construct minimum weight design curves for the stiffened panel as a column. Since the load on the stiffener is $b_s k_u p_L / 2$ with $b_s = k_{0,st}(t_w/t_s)h_{st}$, the load parameters in Equation (3.27) for the panel become

$$(\text{L.P.})_{cr} = \frac{K^{1/2} k_u p_L}{2 k_{0,st}(t_{st}/t_s) h_{st}}, \qquad (\text{L.P.})_c = \frac{(c k_{cp})^{1/2}(t_{st}/t_s) h_{st}(k_u p_L)}{2 t_{st} L} \tag{3.67}$$

This gives

$$h_{st} = \frac{K^{1/2} k_u p_L}{2 k_{0,st}(t_{st}/t_s)(\text{L.P.})_{cr}}, \qquad t_{st} = \frac{(K c k_{cp})^{1/2}(k_u p_L)^2}{4 k_{0,st} L (\text{L.P.})_{cr} (\text{L.P.})_c}, \tag{3.68}$$

whence Equation (3.62) becomes

$$\frac{F_c}{10^5 \rho_w} = \left(\frac{1}{10^5 \rho_w}\right) \left[\frac{2 k_{0,st}(t_{st}/t_s) L}{(K c k_{cp})^{1/2} k_u p_L}\right] (\text{L.P.})_{cr}(\text{L.P.})_c. \tag{3.69}$$

The load parameter for the panel can be defined as

$$(\text{L.P.})_{cp} = \frac{(K c k_{cp})^{1/2} k_u p_L}{2 k_{0,st}(t_{st}/t_s) L} = \frac{(\text{L.P.})_{cr}(\text{L.P.})_c}{(10^5 \rho_w)(F_c/10^5 \rho_w)}. \tag{3.70}$$

In the region of the Euler column curve and the elastic buckling curve, Equation (3.36) can be used to obtain for the panel

$$h_{st} = \left[\frac{k_u p_L}{E_c}\right]^{1/4} \left[\frac{12 L^2 K^{1/3}}{\pi^2 c k_{cp}(2 k_{0,st} t_{st}/t_s)^{2/3}}\right]^{3/8}, \tag{3.71}$$

$$F_c/10^5 \rho_w = H_{cp}(\text{L.P.})_{cp}^{1/2}, \qquad H_{cp} = \left(\frac{\pi^2}{12}\right)^{1/4} \left(\frac{E_c^{1/2}}{10^5 \rho_w}\right), \tag{3.72}$$

where $(\text{L.P.})_{cp}$ is defined in Equation (3.70).

Figure 3.10 shows the column design curves for the same four materials as in Figure 3.7, and as obtained from Equations (3.72) and (3.70).

Example 3.13. (a) Find the maximum column stress and dimensions of the channel stiffened panel in Figure 3.8 for $k_u p_L = 12{,}000$ lb/in., $c = 2$, $L = 40$ in., $t_w/t_s = 1.00$, $b_F/b_w = 0.40$.

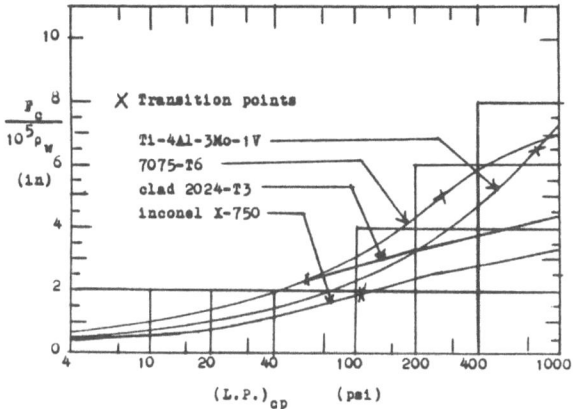

Fig. 3.10. Column design curves for stiffened panels (4-1000 range).

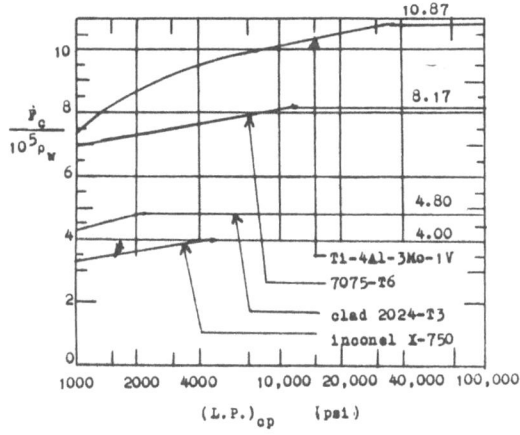

Fig. 3.10. (continued) (1000-100,000 range).

Use clad 2024-T3 material at room temperature. (b) Comment on the effects of the parameter (t_w/t_s).

Solution. (a) From Figure 3.4 for the channel stiffener $h_{st} = b_w$, $t_{st} = t_w$, $k_{0,st} = 1.00 + 2(1.00)(0.40) = 1.80$, $k_{c,st} = [3(1.8) - 2]/1.8 = 1.89$. In Figure 3.8 take $h_p = b_w/2$ so that in Equation (3.65) $k_p = 0.50$, $k_{cp} = (1.89/2) + 3(0.50)^2 = 1.70$. From Equation (3.59), $(h_{st}/b_s) = (t_s/t_{st})(1/k_{0,st}) = 0.56$, whence on Figure 1.18, $K = 4.45$. Equation (3.70) now gives

$$(\text{L.P.})_{cp} = \frac{[2(1.70)(4.45)]^{1/2}(12,000)}{2(1.80)(1.00)(40)} = 324,$$

whence Figure 3.10 gives

$$F_c/10^5 \rho_w = 3.65, \quad F_c = 36,500 \text{ psi}.$$

With this $F_c/10^5\rho_w$, read $(\text{L.P.})_{cr} = 2800$ from Figure 3.6 so that Equation (3.68) gives

$$h_{st} = \frac{(4.45)^{1/2}(12,000)}{2(1.80)(1)(2800)} = 2.51 \text{ in}.$$

The value of t_s is $t_s = (k_u p_L)/(2F_c) = 0.164 = t_{st}$. Also, $b_s = (1.80)(1)(2.51) = 4.52$. If t_s and t_{st} are decreased to the standard thickness of 0.160, the area decreases while the allowable column stress does not change so that the margin of safety is M.S. $= (0.160/0.163) - 1 = -0.02$.

(b) The parameter $(t_w/t_s) = (t_{st}/t_s)$ not only affects the load parameter in Equation (3.68) directly, but it also affects K in Figure 1.18. For the four t_w/t_s values on Figure 1.18 with $b_w/b_s = 1/(1.80 t_w/t_s)$,

$$
\begin{aligned}
t_w/t_s &= 1.00,\ 0.79,\ 0.63,\ 0.50, \\
b_w/b_s &= 0.56,\ 0.70,\ 0.88,\ 1.11, \\
K &= 4.45,\ 3.80,\ 2.00,\ 0.90, \\
K^{1/2}/(t_w/t_s) &= 2.11,\ 2.47,\ 2.24,\ 1.90.
\end{aligned}
$$

Thus the load parameter and maximum stress reach a maximum at $t_w/t_s = 0.79$. However, for this case the b_w/b_s makes b_s smaller than for $t_w/t_s = 1.00$ so that the stiffeners will have less area and will be closer together.

3.6. Effective areas for stiffened panels

The minimum weight design of stiffened panels discussed in the previous Section 3.5 applies to flat panels with uniform compressive unit loading, and is most efficient for large values of the load parameter. Although on airplane wings the stiffened skin panels are curved and may have non-uniform unit compression loads, the above procedures can be adapted to the actual wing panels. For small and medium values of the load parameter, it may be more practical and more efficient to use an effective skin area with the stiffener and then use the combined area as a point area in the analysis of the wing or fuselage cross section. After the skin buckles, the stress distribution in the skin becomes non-uniform as shown in Figures 1.20 and 3.11, and in Reference 2. This effective skin area can be determined from the fact that at the stiffener, the strain in the skin and stiffener must be the same.

From Equation (1.2) and Figure 1.18 the buckling strain of the skin is

$$e_{cr} = F_{cr}/E_c = K(t_s/b_s)^2. \tag{3.73}$$

Assume the skin to be simply supported at the stiffener so that $K = 3.62$ from Figure 1.14. Since the skin is attached to the stiffener, the strain e_s in the skin at the stiffener must equal the strain e_{st} in the stiffener, or

$$e_{st} = e_s = 3.62(t_s/w_c)^2, \tag{3.74}$$

where w_c is the effective width of the skin to produce a buckling strain equal to the stiffener strain. Solve Equation (3.74) for w_c to get

$$w_c = 1.90 t_s \left(e_{st}\right)^{-1/2}, \tag{3.75}$$

$$w_c = 1.90 t_s \left(E_c/F_c\right)_{st}^{1/2}, \quad \text{elastic case.} \tag{3.76}$$

In actual structures K is probably larger and e_{st} is larger for inelastic effects so that more appropriate simple values based on test data are

$$w_c = 1.70 t_s \left(E_c/F_c\right)_{st}^{1/2}, \quad w_{ce} = 0.62 t_s \left(E_c/F_c\right)_{st}^{1/2}, \tag{3.77}$$

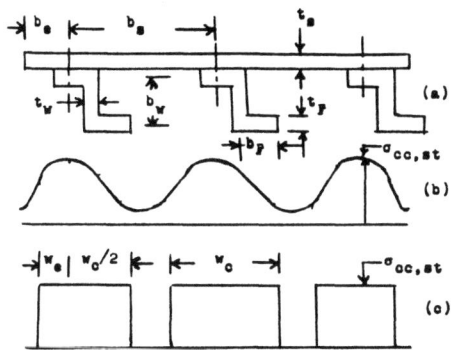

Fig. 3.11. (a). Stiffened panel. (b). Skin stress distribution after buckling. (c). Equivalent skin stress distribution for same load as in (b), and effective skin widths at stiffener strain level.

which are used in the aircraft industry. Note that F_{cc} replaces F_c for no column effects. The effective skin area associated with each stiffener is

$$A_{s,\text{eff}} = w_c t_s, \quad w_c \leq b_s, \quad \text{interior stiffener,}$$
$$A_{s,\text{eff}} = \left(w_e + \frac{w_c}{2}\right)t_s, \quad w_c \leq b_s, \quad w_e \leq b_e, \quad \text{edge.} \quad (3.78)$$

If the stiffener and skin are the same material, then

$$A_{st,\text{eff}} = A_{st} + A_{s,\text{eff}}, \quad \text{same material.} \quad (3.79)$$

If the skin and stiffener are different materials, then

$$A_{st,\text{eff}} = A_{st} + (E_{c,s}/E_{c,st})A_{s,\text{eff}}, \quad \text{elastic case,}$$
$$A_{st,\text{eff}} = A_{st} + (F_{cy,s}/F_{cy,st})A_{s,\text{eff}}, \quad \text{inelastic case.} \quad (3.80)$$

When the panel is in tension, effective areas must also be used to allow for rivet holes of spacing s to attach a rib to the skin and to allow for mixed materials. Thus

$$A_{s,\text{eff}} = b_s\left(\frac{s-D}{s}\right)t_s,$$
$$A_{st,\text{eff}} = A_{st} + (E_s/E_{st})A_{s,\text{eff}}, \quad \text{elastic case,} \quad (3.81)$$
$$A_{st,\text{eff}} = A_{st} + (F_{tu,s}/F_{tu,st})A_{s,\text{eff}}, \quad \text{inelastic case.}$$

Example 3.14. In Figure 3.11 assume $b_w = 1.80$ in., $b_F = 1.00$ in., $t_w = t_F = 0.100$ in., $b_s = 6.00$ in., $t_s = 0.063$ in. If the stiffener material has $F_{cy} = 80,000$ psi, $E_c = 10^7$ psi, and the skin has $F_{cy} = 40,000$ psi, $E_c = 10^7$ psi, find the F_{cc} crippling stress, the effective area for one middle stiffener, and the crippling load for one stiffener.

Solution. To find F_{cc} for the stiffener, calculate K and M from Figure 1.16 as $K = 2.30$, $M = 0.80$ for the extrusion, whence

$$B_{cr} = \left(\frac{1.80}{0.100}\right)\left[\frac{80,000}{23,000,000}\right]^{1/2} = 1.06,$$
$$S_{cc} = 0.84 \text{ on Figure 1.19 for } n = 20, \quad F_{cc} = 67,000 \text{ psi.}$$

From Equation (3.77), $w_c = 1.70(0.063)(10^7/67,000)^{1/2} = 1.31$ in., which is less than b_s. From Equations (3.78) and (3.80),

$$A_{s,\text{eff}} = (1.31)(0.063) = 0.083 \text{ in}^2,$$

$$A_{st,\text{eff}} = (3.80)(0.100) + \frac{40,000}{80,000}(0.083) = 0.42 \text{ in}^2,$$

and the effective skin area is about 10% of the stiffener area. The crippling load is $P_{cc} = (0.42)(67,000) = 28,100$ lb.

3.7. Effect of load intensity on wing design

In the discussion of the design compressive stresses for stiffened panels in Section 3.5, the applied load is represented by the load intensity or unit loading $k_u p_L$ (lb/in.) on the panel. This load intensity is the load term in the load parameter in Figures 3.9 and 3.10 for minimum weight design of the panel. An approximate value of the load intensity $k_u p_L$ can be calculated in the top and bottom skin panels at any wing cross section from the primary wing bending moment M_{xL} or

$$k_u p_L = \pm(k_u M_{xL}/c_s h_s), \tag{3.82}$$

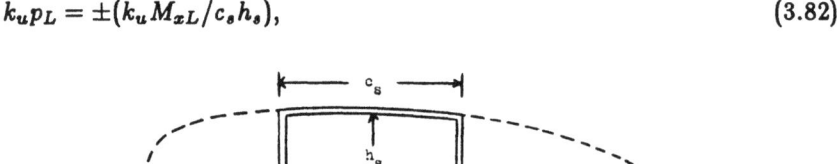

Fig. 3.12. Wing structural box cross section.

where c_s is the width of the panel and h_s is an average height of the wing structural cross section, Figure 3.12. In general, M_{xL}, c_s, and h_s vary along the wing so that $k_u p_L$ is small toward the wing tip and large at the wing root.

To get an idea of the variation of $k_u p_L$ along the wing, assume the section lift coefficient and the distributed wing weight coefficient to be constant, whence the $k_u p_L$ can be derived from the spanwise lift distribution

$$p_{zL} = nc(W/S) = nc_R(W/S)[1 - (1-\lambda)(2y/b_s)], \tag{3.83}$$

where n is the effective wing load factor for a given flight condition corrected for tail load and wing weight, W/S is the wing loading (lb/ft^2), c is the wing chord (ft) with c_R the root chord, $\lambda = c_T/c_R$ with c_T the tip chord for a trapezoidal wing, b_s is the wing structural span (ft), and y is the distance along the span from the fuselage center line. The shear force V_{zL} and bending moment M_{xL} can be obtained by integration of p_{zL}, or with $s = 2y/b_s$,

$$V_{zL} = (b_s/2)\int_s^1 p_{zL}\,ds = \frac{nb_s c_R}{2}\frac{W}{S}\int_s^1 [1-(1-\lambda)s]ds,$$

Structural design of flight vehicle components

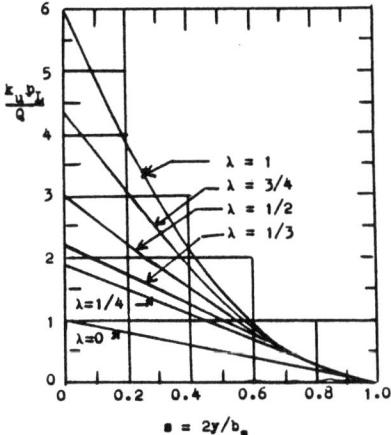

Fig. 3.13. Load intensity variation on wings.

$$M_{xL} = (b_s/2) \int_s^1 V_{zL}\, ds = \frac{nb_s^2 c_R}{24}\frac{W}{S}(1-s)^2[1+2\lambda-(1-\lambda)s]. \qquad (3.84)$$

From Equation (3.82) $k_u p_L$ can now be expressed in (lb/in.) as

$$k_u p_L = \frac{k_u n b_s (W/S)(\text{A.R.})_s(1+\lambda)(1-s)^2[1+2\lambda-(1-\lambda)s]}{576(c_s/c)(h_s/c)[1-(1-\lambda)s]^2}, \qquad (3.85)$$

where $(\text{A.R.})_s$ = wing structural aspect ratio = b_s^2/S, c_s/c = effective structural chord ratio, h_s/c = effective structural thickness ratio, the relation $S/bc_R = (1+\lambda)/2$ has been used, and $s = 2y/b_s$ is the non-dimensional variable along the span from the fuselage center line.

In order to graph Equation (3.85), rewrite it in a non-dimensional form with the taper ratio λ as a parameter, or

$$\frac{k_u p_L}{Q} = \frac{(1+\lambda)(1-s)^2[1+2\lambda-(1-\lambda)s]}{[1-(1-\lambda)s]^2}, \qquad (3.86)$$

$$Q = \frac{k_u n b_s (W/S)(\text{A.R.})_s}{576(c_s/c)(h_s/c)}. \qquad (3.87)$$

Figure 3.13 shows the graph of $k_u p_L$ against $s = 2y/b_s$ for several taper ratios λ. The large variations in $k_u p_L$ along the wing are quite evident and $k_u p_L$ may be very large near the wing root for large values of load factor, wing loading, and aspect ratio in Equation (3.87). However, a small taper ratio λ reduces $k_u p_L$ and gives a smaller variation along the span. If Q in Equation (3.87) is calculated for a particular airplane, then Figure 3.13 can be used to get an approximate value of $k_u p_L$ at any cross section of the wing.

The above discussion of the load intensity included only the case of bending about the x-axis from the M_{xL} moment. The M_{zL} moment, although much smaller than M_{xL}, produces a variable load intensity on the top and bottom surfaces of the wing. With M_{xL} and M_{zL} known at a cross section, the approximate combined load intensity is (+ for tension and minus for compression)

$$k_u p_L = -\frac{k_u M_{xL}}{c_s h_s} + \frac{12(k_u M_{zL})x}{c_s^3}, \quad \text{top surface,}$$
$$k_u p_L = +\frac{k_u M_{xL}}{c_s h_s} - \frac{12(k_u M_{zL})x}{c_s^3}, \quad \text{bottom surface,} \qquad (3.88)$$

where x is from the centroid of the cross section, $+M_{xL}$ produces compression in the top surface, and $+M_{zL}$ produces tension for $+x$.

Since the $k_u p_L$ load intensity for the design flight conditions produces the largest compression loads on the top surface of the wing and the largest tension loads on the bottom surface, the primary design problem is the type of structure to use on the top surface to get the largest possible allowable compression stresses. If $k_u p_L$ is large, it is possible to approach the cut-off stresses in Figures 3.9 and 3.10 and obtain a minimum weight design. If $k_u p_L$ is small, it may be possible to get a better design by using effective areas with the stringers and use large spar caps to take most of the compression load. Since in Figure 3.13, $k_u p_L$ varies rapidly along the wing, it is necessary to vary the area taking the load. The area also must be varied on the tension side in order to keep the design stresses close to F_{tu} ultimate.

Figure 3.14 shows several designs for the wing compression surface. Each of these designs fit a certain range of load intensity and can vary the area along the span. In particular, design (a) is used for large $k_u p_L$ with machine taper to

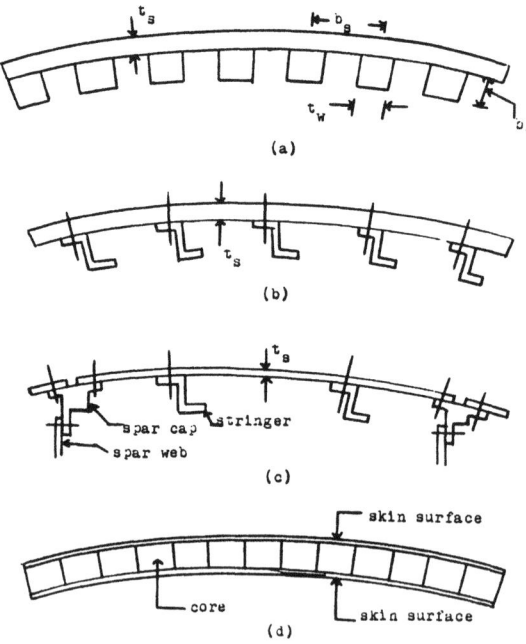

Fig. 3.14. Various designs for wing compression surface.

vary t_s, t_w, b_s, and b_w so that it is very efficient except near the wing tip where $k_u p_L$ becomes small. It can also be used on the tension side, where the skin and integral stiffeners can be made thicker at splices so that the splice efficiency is 100%. However, it is difficult to attach the ribs to the stringers in this design (a).

Design (b) is also used for large $k_u p_L$ on the compression and tension sides with the skin t_s varied by machining. The stringers may vary by machining or they may be constant in steps with the size, area, and number being changed at splices. The skin tension splices can be made 100% efficient. Usually, the lower surface has fewer stringers than the upper surface, since the compression loads on the lower surface are normally less than one-half those on the upper surface. The Z-stringers give good column action and can be attached to the rib without difficulty. The skin thickness can be varied chordwise to allow for the M_{zL} moment in Equation (3.88).

Design (c) is used on many airplanes with medium and small load intensity. The skin t_s is a constant standard thickness, which may change at splices, and the t_s is usually determined from the shear flows in the skin. The stringers are constant with more on top than on the bottom, and are used to support the skin between the ribs as well as to take some of the tension and compression loads with effective skin areas. The spar caps take most of the loads from the bending moments and vary in area along the wing. The spar caps have large areas so that for compression they can be designed to the cut-off stresses. This offsets the ineffective skin in compression and gives a satisfactory average stress for the cross section. See Section 3.8 following for the design procedure for this case (c).

Design (d) is a sandwich panel with thin skins and a low density core, such as honeycomb. It is efficient in compression for small $k_u p_L$. It may be possible to vary the thickness of the skins, or else to use steps in the skin by bonding additional thin skins to the orginal ones.

The choice of which of the four designs to use on a given airplane not only depends upon the $k_u p_L$ load intensity and taper ratio but also upon the cost of building the structure. Design (c) in Figure 3.14 is the simplest to build, since only the spar caps have to be machine tapered. However, it is less efficient than design (b) for large $k_u p_L$ in which the skin is tapered and sculptured by machining and chemical milling. Either design (b) or (c) or variations of them will be found on many airplanes. If Q in Equation (3.87) is greater than 10,000 lb/in., then design (b) would probably be favored. Note that Example 3.13 shows an efficient and reasonable design of the type (b) for $k_u p_L = 12,000$ lb/in. On the 747 commercial jet transport, which has a type (b) design, approximately $Q = 25,000$ lb/in. and $\lambda = 0.36$, whence Figure 3.13 shows that about 75% of the wing span has $k_u p_L$ greater than 10,000 lb/in. The procedure for making the type (c) design is discussed in the following section 3.8.

3.8. Design of box beam cross sections with four spar caps

Consider the type (c) design in Figure 3.14, where the bottom surface of the wing box is similar to the top surface. The analysis of the box beam cross section under the plane strain assumption has been given in Chapter 1, Vol.1. For specified element area, materials, and temperature, the general elastic stresses are given in Equation (1.90, Vol.1). The simple case of tension or compression on beam elements for the inelastic case in Section 1.15, and on Figure 1.15 (Vol.1) shows that in ductile materials inelastic failure does not occur until all elements in tension or in compression fail. Although some elements may be inelastic, they continue to increase in strain at nearly constant stress while the elastic elements take the additional applied loads.

The stringers in Figure 3.14, part (c), are further from the neutral axis than the spar caps so that they must have a larger strain than the spar caps. Also, the compression stringers will buckle or cripple or reach their allowable column stress at a lower stress level than the large spar caps, which may reach the cut-off allowable stress. Once the compression stringers reach their allowable stress or load, they tend to hold the load at approximately the same value as the spar caps take the additional load until they reach their failing load. Of course, the stringers must be ductile on both the tension and compression surfaces in order to allow the continued increase in strain as the spar caps strains increase.

Since the design of the cross section must be made for the ultimate moments $k_u M_{xL}$, $k_u M_{zL}$, and ultimate load $k_u P_{yL}$ at the cross section, the above discussion suggests a procedure to determine the spar cap areas so that M.S. = 0.00 for all four spar caps for a given flight condition. The design procedure is as follows for a given material, given web thicknesses for the shear flows, and a given stringer spacing necessary to support the skin:

(1) Design optimum stringers close to the cut-off (see Examples 3.7 and 3.9 and Equation (3.43)) for compression. Use similar stringers at wider spacing on the tension side.

(2) Calculate effective skin areas for the stringers and spar caps (use cut-off stresses for the spar caps).

(3) Put the applied $k_u P_{yL}$ load at the origin of the reference axes (x_R, z_R) and write the three equilibrium equations in terms of the four unknown spar cap areas, or

$$\sum_{i=1}^{N}(F_i)_{\text{all}}(A_i)_{\text{eff}} = k_u P_{yL},$$

$$\sum_{i=1}^{N}(F_i)_{\text{all}}(A_i)_{\text{eff}} x_{Ri} = k_u M_{zL}, \qquad (3.89)$$

$$\sum_{i=1}^{N}(F_i)_{\text{all}}(A_i)_{\text{eff}} z_{Ri} = -k_u M_{xL}.$$

(4) Assume $\bar{x}$ is specified (midpoint between the spars or closer to the spar with the larger height), or in terms of the four unknown areas

$$\sum_{i=1}^{N}(E_i/E_R)(A_i)_{\text{eff}} x_{Ri} = \bar{x}\sum_{i=1}^{N}(E_i/E_R)(A_i)_{\text{eff}}. \qquad (3.90)$$

Structural design of flight vehicle components   119

(5) Solve Equations (3.89) and (3.90) for the four cap areas.
(6) Repeat the solution for the areas for other flight conditions (particularly with a large change in M_{zL}).
(7) Take the final spar cap areas as the largest area obtained for each cap.

Example 3.15. Design the box beam cross section shown in Figure 3.15 for $k_u M_{zL} = 4(10^6)$ in-lb, $k_u M_{xL} = 800{,}000$ in-lb, $k_u P_{yL} = 0$ lb. Use $t_s = 0.080$ in. clad 2024-T3 (L-direction, B basis) for all the skins and webs. Use 7075-T6 extrusions (L-direction, B-basis) for elements (1)-(4) and 2024-T4 extrusions (L-direction, B-basis) for elements (5)-(8). Take $s = 4D$ rivet spacing for the rib attachments. Take the location of the elements in sequence from the reference axes as (x_{Ri}, z_{Ri}) (in.) = (0,10), (12,12), (24,12), (36,9), (36,0), (24,-1), (12,-1), (0,0). Use rib spacing of $L = 25$ with $c = 3.50$ end fixity for stringers (2) and (3). Take $b_F/b_w = 0.50$, $t_w/t_F = 0.80$ for stringers (2), (3), (6), and (7).

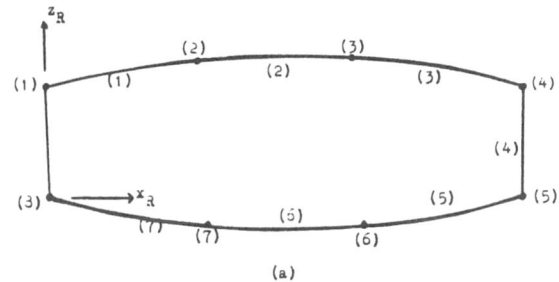

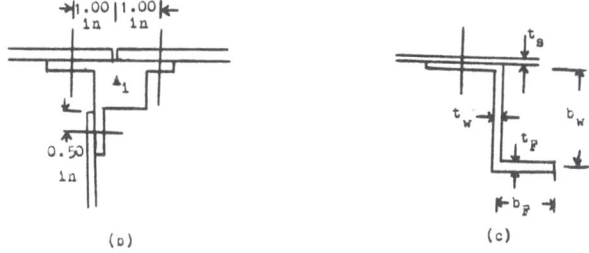

Fig. 3.15. (a). Simple box beam cross section with point areas. (b). Typical spar cap cross section. (c) Typical stringer cross section.

Solution. Follow the steps listed above. (1) For the compression stringers (2) and (3) assume each stringer takes 25,000 lb. From Figure 1.16, $K = 3.30$, and from Figure 3.4, $k_0 = 1.00 + 2(0.50)(1.25) = 2.25$ and $k_c = 2.11$. The load parameter in Equation (3.30) is

$$(\text{L.P.})_{cb} = \frac{(3.50)(2.11)(3.30)^{1/2}(25{,}000)}{(2.25)(25)^2} = 238,$$

whence from Figure 3.7 for 7075-T6 extrusion, $F_c/10^5 \rho_w = 7.07$ and $F_c = 71{,}400$ psi. From Figure 3.6 for this $F_{cr} = F_c$, $(\text{L.P.})_{cr} = 6600$, whence Equation (3.28) gives

$$h = b_w = \left[\frac{(3.30)^{1/2}(25{,}000)}{(2.25)(6600)}\right]^{1/2} = 1.75 \text{ in.}$$

Thus the area and dimensions of the stiffeners are $A_{st} = 25{,}000/71{,}400 = 0.35$ in^2, $t_w = 0.35/(1.75)(2.25) = 0.089$, $t_F = (1.25)(0.089) = 0.111$, $b_w = 1.75$, $b_F = (0.50)(1.75) = 0.88$. Hence, the assumption of 25,000 lb compression load for the stringer gives a stringer with practical dimensions and a column stress close to the yield stress. The $b_F = 0.88$ gives sufficient space for a row of rivets. The same extrusion of 2024-T4 material with area 0.35 in² will be used for stringers (6) and (7) on the tension side.

(2) To calculate the effective skin areas from Equations (3.77)-(3.81), various material property values are needed from Tables 1.2, 1.3, and 1.6. For the compression members (1)-(4) in sequence,

$$10^{-6}E_{c,st} = 10.5, \quad 10.5, \quad 10.5, \quad 10.5,$$
$$10^{-3}F_{cy,st} = 75.0, \quad 75.0, \quad 75.0, \quad 75.0,$$
$$10^{-3}F_{cy,s} = 40.0, \quad 40.0, \quad 40.0, \quad 40.0,$$
$$10^{-3}F_{c,st} = 82.5, \quad 71.4, \quad 71.4, \quad 82.5.$$

For the tension members (5)-(8), in sequence,

$$10^{-3}F_{tu,st} = 70.0, \quad 61.0, \quad 61.0, \quad 70.0,$$
$$10^{-3}F_{tu,s} = 65.0, \quad 65.0, \quad 65.0, \quad 65.0.$$

From Equation (3.77) and Figure 3.15,

$$w_c = 1.70(0.080)(10,500/71.4)^{1/2} = 1.65 \text{ for } (2), (3),$$
$$w_c = 1.70(0.080)(10,500/82.5)^{1/2} = 1.53 \text{ for } (1), (4),$$
$$w_{ce} = (0.62/1.70)(1.53) = 0.56 \text{ for edge for } (1), (4),$$
$$(w_c)_{\text{total}} = 2(1.53) + 0.5(1.53) + 0.50 = 4.33 \text{ for } (1), (4).$$

From Equation (3.80),

$$A_{st,\text{eff}} = 0.35 + (40/75)(1.65)(0.080) = 0.42 \text{ in}^2 \text{ for } (2), (3),$$
$$A_{st,\text{eff}} = A_1 + (40/75)(4.33)(0.080) = A_1 + 0.18 \text{ for } (1),$$
$$A_{st,\text{eff}} = A_4 + 0.18 \text{ for } (4).$$

From Equation (3.81),

$$A_{s,\text{eff}} = 12(3/4)(0.080) = 0.72 \text{ in}^2 \text{ for } (6), (7),$$
$$A_{s,\text{eff}} = (3/4)(6 + \tfrac{10}{4})(0.080) = 0.51 \text{ in}^2 \text{ for } (8),$$
$$A_{s,\text{eff}} = (3/4)(6 + \tfrac{9}{4})(0.080) = 0.50 \text{ in}^2 \text{ for } (5),$$

where 1/4 of the vertical web is used due to the stress variation from tension to compression. For the stiffeners

$$A_{st,\text{eff}} = 0.35 + (65/61)(0.72) = 1.12 \text{ in}^2 \text{ for } (6), (7),$$
$$A_{st,\text{eff}} = A_5 + (65/70)(0.50) = A_5 + 0.46 \text{ for } (5),$$
$$A_{st,\text{eff}} = A_8 + (65/70)(0.51) = A_8 + 0.47 \text{ for } (8).$$

(3) and (4). Assume $\bar{z} = 18$ so that Equations (3.89) and (3.90) become

$$(A_1 + 0.18)(-82.5) + 2(0.42)(-71.4) + (A_4 + 0.18)(-82.5)+$$
$$+(A_5 + 0.46)(70.0) + 2(1.12)(61.0) + (A_8 + 0.47)(70.0) = 10^{-3}k_u P_{yL} = H_y,$$
$$(0.42)(-71.4)(12 + 24) + (A_4 + 0.18)(-82.5)(36) + (A_5 + 0.46)(70.0)(36)+$$
$$+(1.12)(61.0)(12 + 24) = 10^{-3}k_u M_{xL} = H_x,$$
$$(A_1 + 0.18)(-82.5)(10) + 2(0.42)(-71.4)(12) + (A_4 + 0.18)(-82.5)(9)+$$
$$+2(1.12)(61.0)(-1) = -10^{-3}k_u M_{zL} = -H_z,$$
$$(0.42)(12 + 24) + (A_4 + 0.18)(36) + (A_5 + 0.46)(36) + (1.12)(12 + 24)$$
$$= 18(A_1 + A_4 + A_5 + A_8 + 0.36 + 0.84 + 0.93 + 2.24),$$

whence

$$-82.5(A_1 + A_4) + 70(A_5 + A_8) = H_y - 112.1,$$
$$-2970 A_4 + 2520 A_5 = H_x - 2005,$$
$$-825 A_1 - 742.5 A_4 = -H_z + 1138,$$
$$-A_1 + A_4 + A_5 - A_8 = 0.$$

(5) With A_4 as the reference area, the general solution for the four areas is

$$A_4 = -0.73 + \frac{k_u P_{yL}}{290,000} - \frac{k_u M_{xL}}{5,217,000} + \frac{k_u M_{zL}}{1,568,000},$$
$$A_1 = -0.90 A_4 - 1.38 + (k_u M_{zL}/825,000),$$
$$A_5 = 1.18 A_4 - 0.80 + (k_u M_{zL}/2,520,000),$$
$$A_8 = 3.08 A_4 + 0.58 + (k_u M_{zL}/2,520,000) - (k_u M_{zL}/825,000).$$
(3.91)

For the given values $k_u P_{yL} = 0$, $k_u M_{xL} = 800,000$, $k_u M_{zL} = 4,000,000$, the areas are

$$A_4 = 1.67 \text{ in}^2, \quad A_1 = 1.97 \text{ in}^2, \quad A_5 = 1.49 \text{ in}^2, \quad A_8 = 1.19 \text{ in}^2.$$

(6) From Equation (3.91), it is a simple procedure to calculate the areas for any flight condition for the given cross section. Since $k_u M_{xL}$ is nearly constant along the upper line of the V-n diagram from HAA flight condition to LAA flight condition, the calculations can be made for M_{zL} at HAA (the above case) and for LAA, where M_{zL} may be nagative. Note that the areas have a linear variation between HAA and LAA. Since the effective areas will be different for compression in the bottom surface and tension in the top surface, Equation (3.91) will be different for the - HAA and - LAA flight conditions.

3.9. Analysis of diagonal tension beams

As described above, the failure or crippling stress F_{cc} of various cross sections in compression may be much larger than the buckling stress F_{cr} (see Figure 1.19). In a similar manner, a thin web beam may fail at a much larger shear stress than that at which the web buckles in shear. If there are stiffeners on the web at spacing d, then when the web buckles at the stress

$$F_{scr} = q_{cr}/t = K_s E_c (t/d)^2,$$
(3.92)

where K_s is given on Figure 1.14, it forms folds and acts as a set of tension diagonals, while the stiffeners act as compression posts. In other words, the web and stiffeners behave like a truss after buckling, and continue to take additional load until the web fails in tension, or the stiffeners fail in compression, or the rivets fail.

Consider the cantilever beam in Figure 3.16 and the Mohr's circle for an element of the thin web in Figure 3.17. With the web in shear only, the tension and compression stresses at $\pm 45°$ for no buckling are $q_t = q_c = q_L$. If the web buckles for $F_{scr} = q_{cr}/t = 0$, approximately, then $q_c = 0$ and $q_t = 2q_L$ and the web acts in tension provided the stiffeners and flanges support it on the edges. The actual situation can be represented by the diagonal tension factor k,

$$kq_L = \text{shear flow supported in tension},$$
$$(1-k)q_L = \text{shear flow supported in shear}.$$
(3.93)

In Figure 3.18 it is evident that the component forces of the diagonal tension produce compression loads in the stiffeners and flanges, and bending moments in the flanges and the end stiffeners. For the case of pure diagonal tension with

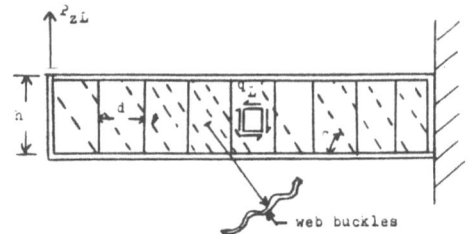

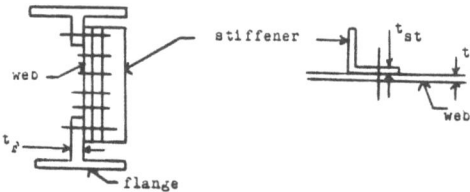

Fig. 3.16. Diagonal tension beam and components.

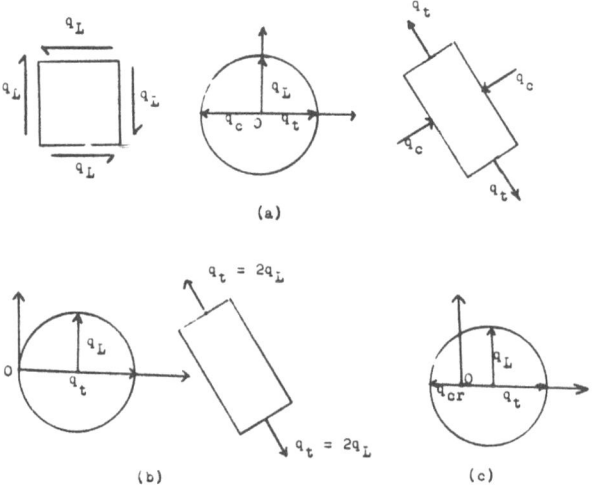

Fig. 3.17. (a). Mohr's circle, no buckling. (b). Mohr's circle, $q_c = 0$. (c). Mohr's circle, $q_c = q_{cr}$ = buckling shear flow.

$k = 1$, the loads and moments on a panel are

$$\begin{aligned}
P_{st} &= q_L d \tan\alpha = \text{ stiffener load,} \\
P_{Fd} &= q_L(h/2)\cot\alpha = \text{ additional flange load,} \\
M_{F,\max} &= q_L(d^2/12)\tan\alpha = \text{ max. moment on flange,} \\
M_{st,\max} &= q_L(h^2/8)\cot\alpha = \text{ max. moment on stiffener.}
\end{aligned} \qquad (3.94)$$

Structural design of flight vehicle components 123

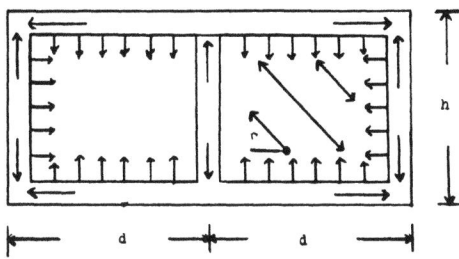

Fig. 3.18. Component forces from diagonal tension.

Because of the interactions among the web, stiffeners, and flanges, the evaluation of the diagonal tension factor k in Equation (3.93) and the determination of the allowable and applied stresses in the web, stiffeners, and flanges must be based upon experimental data and the empirical formulas derived from the test data on various beams with thin webs. This experimental work was done by NACA (now NASA) and various aircraft companies, and the resulting analysis methods are described in References 2-7. Chapter 3 of Reference 3 gives the analysis and design procedures for both plane webs and curved webs, and Chapter 12 of Reference 3 gives the experimental data and modes of failure for diagonal tension beams. Section 1.3.1 of Reference 6 gives analysis and design curves for diagonal tension beams. A discussion of the analysis procedure for plane web diagonal tension beams is given on pages 373-387 of Reference 7.

3.10. Problems

Chapter 3

3.1. Use the procedures of Example 3.1 and construct a curve for 17-7 PH Stainless Steel on Figure 3.5 with $F_{cy} = 212,000$ psi, $E_c = 30,000,000$ psi, $\rho_w = 0.276$ lb/in^3, and $n = 20$.

3.2. Rewrite Equations (3.6)-(3.9) for the solid circular cross section with the diameter as the variable and construct design curves similar to Figure 3.5 for this case, for which the load parameter is $c^{1/2} k_u P_L / L^2$.

3.3. For the thick wall circular tube with D_o and D_i the outer and inner diameters, take D_o/D_i as a parameter and D_o as the variable, and thus modify the load parameter in Problem 3.2 to include this case.

3.4. For the rectangular cross section $k_0 h$ by h, take k_0 as a specifed value and h as the variable, and thus modify the load parameter in Problem 3.2 to include this case.

3.5. Repeat Example 3.2 for the case of $t = 0.40$ in. and compare the results to Example 3.2.

3.6. In Example 3.3, reduce the t_w to $t_w = 0.100$ in. and rework the titanium and inconel cases.

3.7. Repeat Example 3.2 at $T = 600°$F for the four materials.

3.8. Repeat Example 3.2 at $T = 1000°$F for the titanium and inconel alloys.

3.9. Use the procedure of Example 3.5 to construct curves on Figure 3.6 for the 17-7 PH Stainless Steel alloy. Use the properties given in Problem 3.1 above.

3.10. Repeat Example 3.6 at $T = 400°$F for the four materials.

3.11. Suppose an angle section in Figure 3.4 with $h = b_1 = 1.70$ in., $b_1/b_2 = 2.00$, $t_2/t_1 = 2.00$, has a compression load of $k_u P_L = 6000$ lb. Find the optimum value of $t = t_1$ for least weight per unit length of the cross section for the crippling stress F_{cc} for the four materials in Figure 3.6.

3.12. For the angle section in Problem 3.11, what are the required loads and required thicknesses t_1 at the cut-off points for the four materials in Figure 3.6?

3.13. Repeat Example 3.7 at $T = 500°F$ for the four materials.

3.14. Use the procedure of Example 3.8 to construct a curve on Figure 3.7 for the 17-7 PH stainless steel alloy. Use the properties given in Problem 3.1 above.

3.15. Repeat the design in Example 3.9 for $L = 30$ in.

3.16. Suppose an extruded 7075-T6 angle section with $b_1/b_2 = 2.00$, $t_2/t_1 = 2.00$, $L = 30$ in., has a compression load of $k_u P_L = 6000$ lb. Find the optimum values of b_1 and t_1 for least weight for (a) room temperature and (b) $T = 400°F$.

3.17. In Problem 3.16, what are the required loads and required dimensions at the cut-off points?

3.18. Use the procedure of Example 3.11 to construct curves on Figure 3.9 for the 17-7 PH stainless steel alloy. Use the properties given in Problem 3.1 above.

3.19. Repeat Example 3.12 for $k_u p_L = 15,000$ lb/in.

3.20. Repeat Example 3.12 at $T = 400°F$.

3.21. Suppose the stiffened panel in Figure 3.8 has $b_w = 2.00$ in., $b_F = 0.80$, $t_w = 0.125$, 7075-T6 extruded material, and a compression load of $k_u p_L = 10,000$ lb/in. Use Figure 1.18 for K and find the maximum crippling stress $F_{cc} = \sigma_{cc}$ and the optimum dimensions $b_{s,opt}$ and $t_{s,opt}$ at (a) room temperature and (b) $T = 400°F$.

3.22. In Equation (3.52), the b_s can be regarded as given and the A_{st} as the variable to be eliminated. Show that for this case the curves on Figure 3.6 apply with the load parameters

$$(\text{L.P.})_{cr} = \frac{K^{1/2} k_b p_L}{2 b_s}, \quad (\text{L.P.})_{cc} = \frac{K^{1/2} k_u p_L}{2 b_s}.$$

For the crippling curves $M = 0.80$ and Equation (3.23) can be used with this $(\text{L.P.})_{cc}$.

3.23. Check the formulas derived in Problem 3.22 by using the b_s answers in Example 3.12 and determining the stresses and A_{st}.

3.24. If $b_s = 6.00$ in. in Example 3.12, find the stress and required A_{st} in part (a) by the procedure in Problem 3.22.

3.25. Solve Example 3.13 for the case of 7075-T6 material.

3.26. Show that the effects of temperature on the stiffened panel column case can be obtained from Equations (3.70) and (3.72) by the procedure,

Calculate $(\text{L.P.})_{cp,RT} = \left[\dfrac{F_{cy,RT}}{F_{cy,ET}}\right]^2 \left[\dfrac{E_{c,ET}}{E_{c,RT}}\right] (\text{L.P.})_{cp}$,

Read $(F_c/10^5 \rho_w)_{RT}$ on Figure 3.10 from $(\text{L.P.})_{cp,RT}$,

Calculate $(F_c)_{ET} = (F_{cy,ET}/F_{cy,RT})(F_c)_{RT}$,

and for the Euler case,

Calculate $(F_c/10^5 \rho_w)_{RT}$ in Equation (3.72),

Calculate $(F_c)_{ET} = (E_{c,ET}/E_{c,RT})^{1/2} (F_c)_{RT}$.

3.27. Use the procedure of Problem 3.26 and solve Example 3.13 at $T = 400°F$.

3.28. Solve part (a) of Example 3.13 with $L = 20$ in. and $t_w/t_s = 0.79$.

3.29. Repeat Example 3.14 for $b_w = 2.20$ in.

3.30. Calculate the F_c column stress of the channel stiffener that was obtained in Example 3.13, using $L = 40$ in. With this F_c calculate the effective area of the 0.164 in. skin in Example 3.13, and compare the total load using the effective area to that of $4.52(12,000) = 54,000$ lb supported per stiffener in Example 3.13.

3.31. Verify the $\lambda = 1/2$ curve on Figure 3.13.

3.32. To show the effect of the size of the structural wing box upon the load intensity $k_u p_L$, graph the equation

$$Q_s = \frac{576 Q}{k_u n b_s (W/S)(\text{A.R.})_s} = \frac{1}{(c_s/c)(h_s/c)},$$

with Q_s on ordinate against $(c_s/c)(h_s/c)$ for the range $0.025 \leq (c_s/c)(h_s/c) \leq 0.150$.

3.33. An airplane has the aerodynamic values of $b = 200$ feet and A.R. $= 7.00$. What is the percent increase in Q in Equation (3.87) for sweepback angles of (a) $15°$, (b) $30°$, (c) $45°$?

3.34. Calculate the spar cap areas in Equation (3.91) with $k_u M_{zL} = 6,000,000$ in-lb, $k_u P_{yL} = 80,000$ lb, and for the cases of (a) $k_u M_{zL} = 1,000,000$ in-lb, (b) $k_u M_{zL} = 0$ in-lb, (c) $k_u M_{zL} = -500,000$ in-lb.

3.35. In Example 3.15, with the material unchanged, repeat the calculations for compression in the bottom surface and tension in the top surface and thus obtain a new Equation (3.91) for this case. Use your results to get the areas for the case of $k_u P_{yL} = 0$, $k_u M_{zL} = 400,000$ in-lb, $k_u M_{zL} = -2,000,000$ in-lb.

3.36. Repeat Example 3.15 for all elements at $T = 400°F$.

3.37. Repeat Example 3.15 for the four spar caps at room temperature but the four stringers and the skin near the stringers at $400°F$.

References

Chapter 3

1. Mil-HDBK-5c, *Metallic Materials and Elements of Aerospace Vehicle Structures*, Sept. 1976, 2 vols., U.S. Government Printing Office, Washington, D.C., 20013, P.O. Box 1533.
2. E.F. Bruhn: *Analysis and Design of Flight Vehicle Structures*, Tri-State Offset Co., Cincinnati, Ohio, 45202, 1965.
3. P. Kuhn: Stresses in Aircraft and Shell Structures, McGraw-Hill Book Co., New York (1956).
4. P. Kuhn, J.P. Peterson and L.R. Levin: *A Summary of Diagonal Tension, Part I, Methods of Analysis*, NACA TN 2661, May (1952).
5. P. Kuhn, J.P. Peterson and L.R. Levin: *A Summary of Diagonal Tension, Part II, Experimental Evidence*, NACA TN 2662, May (1952).

References for Additional Reading

6. G.E. Maddux, L.A. Vorst, F.J. Giessler and T. Moritz: *Stress Analysis Manual*, AFFDL - TR-69-42, Wright-Patterson AFB, Ohio, 45433, 1970.
7. D.J. Peery and J.J. Azar: *Aircraft Structures*, McGraw-Hill Book Co., New York, 2nd Ed. (1982).

4

Analysis and design of pressurized structures

4.1. Introduction

In airplanes, missiles, and space vehicles, much of the structure must be designed to take internal pressure loads. The cockpits and cabins of all airplanes that fly at high altitudes must be pressurized. Man carrying space vehicles must be pressurized. Fuel tanks, oxidizer tanks, combustion chambers, hydraulic systems, many boxes for electronic components, etc., are pressurized.

One possible way to reduce compression stresses that produce buckling in thin wall structures is to use internal pressure to produce tension stresses which cancel or reduce the compression stresses from external loads. Such pressure stabilized structures are balloons, airships, blimps, air mattresses, inflated tires, and various inflated fabric structures. By using drop threads to form the fabric structure into a beam shape, it is possible to use internal pressure to support moments, torque, shear, and axial loads just as any beam can, even though the allowable stress in compression is zero.

The basic problems of membrane stresses in thin shells are considered in Sections 4.2 and 4.3. The bending in circular cylindrical shells with axially symmetric loading is examined in Section 4.4. The bending in pressurized aircraft fuselages from stringers and frames is discussed in Section 4.5. Section 4.6 considers the problem of non-circular cross sections with internal pressure. The bending of non-circular fuselage rings with internal pressure and point loads is analyzed in Sections 4.7 and 4.8, respectively. The effect of internal pressure upon buckling of cylindrical shells is considered in Section 4.9. The case of pressure stabilized structures is examined in Section 4.10.

4.2. Membrane stresses in thin shells

The main characteristics of pressure vessels are thin walls and shell of revolution design that supports the internal pressure by tension membrane stresses. In Figure 4.1 revolve the curve $y = f(x)$ about the x-axis to generate the pressure vessel. Thus, cross section perpendicular to the x-axis is a circle of radius $r = y$. In Figure 4.1, r_1 and r_2 are the principal radii of curvature, $N\theta = t\sigma_\theta$ = internal

Analysis and design of pressurized structures

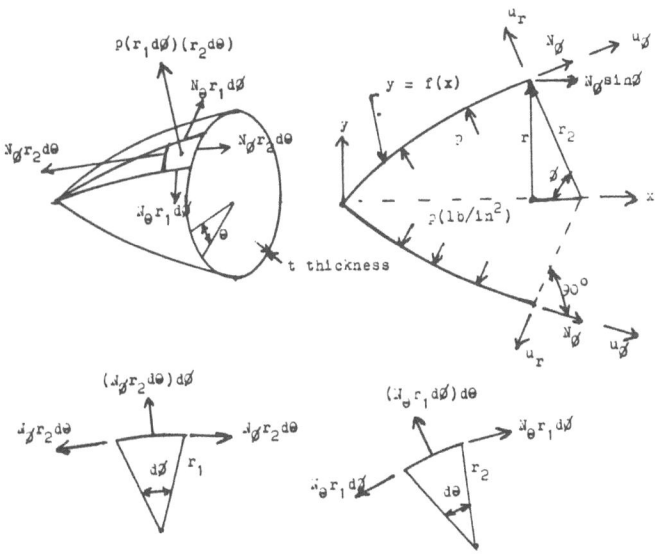

Fig. 4.1. Forces on thin shell of revolution with internal pressure.

hoop load per unit length in hoop direction (measured by θ), $N_\phi = t\sigma_\phi$ = internal load per unit length in meridional direction (measured by ϕ).

The radius of curvature r_1 for the curve $y = f(x)$ is given in calculus books as

$$r_1 = \frac{[1 + (y')^2]^{3/2}}{|y''|}, \quad y' = \frac{dy}{dx}. \tag{4.1}$$

From Figure 4.1 the r_2 radius of curvature is

$$r_2 = r/\sin\phi = y/\sin\phi. \tag{4.2}$$

Also, the relation between ϕ and the curve can be expressed as

$$y' = \frac{dy}{dx} = \cot\phi, \quad \text{or} \quad \sin\phi = [1 + (y')^2]^{-1/2}. \tag{4.3}$$

From Figure 4.1 with internal pressure $p(\text{lb/in}^2)$, equilibrium of forces on the circular cross section of radius r is

$$2\pi r N_\phi \sin\phi = (\pi r^2)p, \quad N_\phi = pr_2/2 = \frac{py}{2}[1 + (y')^2]^{1/2}. \tag{4.4}$$

Equilibrium in the normal direction on the surface element gives

$$N_\phi r_2 d\theta d\phi + N_\theta r_1 d\phi d\theta = p(r_1 d\phi)(r_2 d\theta),$$
$$(N_\phi/r_1) + (N_\theta/r_2) = p, \quad N_\theta = pr_2[1 - (r_2/2r_1)]. \tag{4.5}$$

Thus, the tension stresses in any closed shell of revolution with internal pressure $p(\text{lb/in}^2)$ are given in terms of the two principal radii of curvature r_1 and

r_2 in Equations (4.4) and (4.5). Note that in Equation (4.5) it is possible for N_θ to be a compression stress.

Since N_θ and N_ϕ are at right angles to each other they can produce a shear stress by Mohr's circle. With N_θ and N_ϕ as principal stresses the maximum shear stress $N_{\theta\phi}$ is given by Equations (1.22), (4.4), and (4.5) as

$$(N_{\theta\phi})_{\max} = (1/2)(N_\theta - N_\phi) = (pr_2/2r_1)(r_1 - r_2). \tag{4.6}$$

The deflections u_ϕ in the N_ϕ direction and u_r in the normal direction, as indicated on Figure 4.1, are given in Chapter 14 of Reference 1 in the form

$$\begin{aligned} u_\phi &= \left[\int \frac{f(\phi)}{\sin \phi} d\phi + C \right] \sin \phi, \\ f(\phi) &= (1/Et)[N_\phi(r_1 + \nu r_2) - N_\theta(r_2 + \nu r_1)], \\ u_r &= -u_\phi \cot \phi + (r_2/Et)(N_\theta - \nu N_\phi), \end{aligned} \tag{4.7}$$

where C is a constant to be determined from a reference point for u_ϕ.

For the *circular cylinder* with closed ends and radius R,

$$\begin{aligned} r_2 &= R, \quad \phi = 90°, \quad r_1 = \infty, \quad N_\phi = pR/2, \quad N_\theta = pR, \\ (N_{\theta\phi})_{\max} &= pR/4, \quad u_r = \frac{pR^2}{2Et}(2-\nu), \quad u_\phi = \frac{pR}{2Et}(1-2\nu)x, \\ u_\phi &= 0 \text{ at } x = 0. \end{aligned} \tag{4.8}$$

For the *sphere* of radius R,

$$\begin{aligned} r_1 &= r_2 = R, \quad \phi = \theta = 90°, \quad N_\phi = N_\theta = pR/2, \quad N_{\theta\phi} = 0, \\ u_r &= (pR^2/2Et)(1-\nu), \quad u_\phi = 0. \end{aligned} \tag{4.9}$$

Example 4.1. Revolve the ellipse

$$(x-2)^2 + ky^2 = 4 \tag{4.10}$$

about the x-axis to obtain a closed pressure vessel with internal pressure p. (a) Find the internal loads N_ϕ, N_θ, and $(N_{\theta\phi})_{\max}$. (b) What values of k will produce a compressive N_θ in some region of the vessel?

Solution. (a) Differentiate Equation (4.10) and use the above equations to get

$$\begin{aligned} y' &= -(x-2)/ky, \quad y'' = -4/k^2 y^3, \\ H_x &= \left[4k - (k-1)(x-2)^2\right]^{1/2}, \quad r_1 = H_x^3/4k, \quad r_2 = H_x/k, \\ N_\phi &= pH_x/2k, \quad N_\theta = p(H_x^2 - 2)/kH_x, \quad (N_{\theta\phi})_{\max} = p(4 - H_x^2)/4kH_x. \end{aligned}$$

(b) The range of x is $0 \le x \le 4$, whence

$$\begin{aligned} H_x^2 - 2 &= 4k - 2 - (k-1)(x-2)^2 \\ &= 2 \text{ for } x = 0, \\ &= 4k - 2 \text{ for } x = 2, \\ &= 2 \text{ for } x = 4. \end{aligned}$$

Thus, N_θ can have a negative value at $x = 2$ for $0 < k < 1/2$. The largest range for minus N_θ occurs for k very small, or

$$2 - 2^{1/2} < x < 2 + 2^{1/2}.$$

Note that the minus values for N_θ occur for the major axis in the y-direction, or

$$2b = 2(4/k)^{1/2} = 4/k^{1/2} = 4(2)^{1/2} \text{ for } k = 1/2$$
$$= \text{large for } k \text{ small}.$$

Thus, N_ϕ and N_θ are tension loads for rotation about the major axis, $k > 1$, and for rotation about the minor axis for $1/2 < k < 1$.

4.3. Cut-outs in thin shells with membrane stresses

A major design problem in the pressurized cabins of flight vehicles is that of transferring the tension loads from the internal pressure around the cut-outs for doors and windows. For doors it is necessary to use frames that can transfer the load by bending. However, for circular windows it is possible to use circular rings that transfer the load in tension without bending. This can be done if the principal stresses N_θ and N_ϕ are equal so that there is no shear stress and the tension stress is the same in all directions. If the cabin is a constant cross section cylinder, Equation (4.8) shows that the hoop tension N_θ is twice the N_ϕ stress. Thus, to cut N_θ in half in the shell, use flat circular hoops spaced about ten inches apart and make the area of the hoops compatible with the skin area. Besides making $N_\theta = N_\phi$, these hoops can stop fatigue cracks in the skin from reaching a critical length that would cause the cabin to explode. To have a window larger than ten inches in diameter, change the flat hoop to a small rod and let it cross the window. See Chapter 9 of Reference 2. Also, bending frames can be used around larger windows, whether circular or rectangular.

Let the radius of the circular cut-out in the thin skin be a and reinforce the edge of the hole with a flat circular ring as shown in Figure 4.2. From Figure 4.2 the tension load in the ring is

$$P_r = kcN_0 = kct\sigma_0, \tag{4.11}$$

where kN_0 represents the load transferred to the rings. Since the ring and the skin are fastened together, they must have the same tangential strain on the rivet circle.

To determine the strain in the skin, it is necessary to use polar coordinates to find the radial stress σ_r and the tangential stress σ_θ in the region of the point. For this case of axial symmetry with a hole the basic equations for the stresses are given in Chapter 4 of Reference 4 and Chapter 9 of Reference 2 as

$$\sigma_r = B - A/r^2, \quad \sigma_\theta = B + A/r^2, \tag{4.12}$$

where A and B are constants to be determined from the boundary conditions. Since the radial stress σ_r will change at the rivet circle, there will be a solution (1) in the circle from $r = a$ to $r = c$ and a different solution (2) beyond $r = c$.

Take the boundary conditions as

$$\begin{aligned}
\sigma_{r1} &= 0 \text{ at } r = a, & \sigma_{r2} &= \sigma_0 \text{ at } r = \text{large}, \\
\sigma_{r2} - \sigma_{r1} &= k\sigma_0 \text{ at } r = c, & u_{r1} &= u_{r2} \text{ at } r = c,
\end{aligned} \tag{4.13}$$

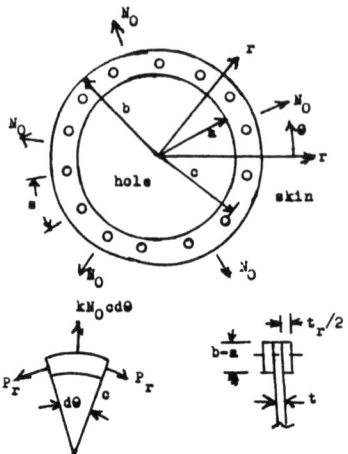

Fig. 4.2. Circular hole in thin skin.

whence the first two conditions give

$$0 = B_1 - A_1/a^2, \quad B_1 = A_1/a^2,$$
$$\sigma_{r1} = (A_1/a^2)\left(1 - \frac{a^2}{r^2}\right), \quad \sigma_{\theta 1} = (A_1/a^2)\left(1 + \frac{a^2}{r^2}\right), \quad (4.14)$$
$$\sigma_{r2} = \sigma_0 - A_2/r^2, \quad \sigma_{\theta 2} = \sigma_0 + A_2/r^2.$$

From the third condition

$$A_2 = c^2(1-k)\sigma_0 - A_1\left(\frac{c^2}{a^2} - 1\right). \quad (4.15)$$

To get the deflection u_r to use in the fourth boundary condition use Equations (4.12) and integrate the strain equation

$$e_r = u_{r,r} = \frac{1}{E}(\sigma_r - \nu\sigma_\theta) = \frac{1}{E}[B(1-\nu) - (1+\nu)(A/r^2)], \quad (4.16)$$

to get

$$u_r = \frac{1}{E}[B(1-\nu)r + (1+\nu)(A/r)], \quad (4.17)$$

whence

$$(A_1/a^2)[(1-\nu)c^2 + (1+\nu)a^2] = p(1-\nu)c^2 + (1+\nu)A_2. \quad (4.18)$$

Solve Equations (4.15) and (4.18) for A_1 and A_2, or

$$\begin{aligned}A_1 &= (\sigma_0 a^2/2)[2 - k(1+\nu)], \\ A_2 &= (\sigma_0 a^2/2)[2 - k\{(c^2/a^2)(1-\nu) + (1+\nu)\}].\end{aligned} \quad (4.19)$$

Analysis and design of pressurized structures

If $A_2 \neq 0$ in Equation (4.14) then either σ_{r2} or $\sigma_{\theta 2}$ will be greater than σ_0. Thus, an optimum value for k will make $A_2 = 0$,

$$k_{\text{opt}} = \frac{2}{(c^2/a^2)(1-\nu) + 1 + \nu}. \tag{4.20}$$

Take $\Delta k = k - k_{\text{opt}}$ so that at $r = c$

$$\sigma_{r2c} = \sigma_0\left[1 + \frac{a^2}{c^2}\frac{\Delta k}{k_{\text{opt}}}\right], \quad \sigma_{\theta 2c} = \sigma_0\left[1 - \frac{a^2}{c^2}\frac{\Delta k}{k_{\text{opt}}}\right], \tag{4.21}$$

and Δk positive makes $\sigma_{r2c} > \sigma_0$ while Δk negative makes $\sigma_{\theta 2c} > \sigma_0$. The maximum stress in region 1 is

$$\sigma_{\theta 1c} \doteq (\sigma_0/2)[2 - k(1+\nu)]\left[1 + (a^2/c^2)\right], \tag{4.22}$$

which will be less than σ_0 in realistic designs.

The tangential strain at $r = c$ is

$$\begin{aligned}
e_{\theta 1c} = e_{\theta 2c} &= \frac{1}{E}(\sigma_{\theta 1c} - \nu\sigma_{r1c}) = (A_1/Ea^2)[(1+\nu)(a^2/c^2) + 1 - \nu] \\
&= (\sigma_0/2E)[(1+\nu)(a^2/c^2) + 1 - \nu][2 - k(1+\nu)] \\
&= (\sigma_0/2E)(a^2/c^2)[2 - k(1+\nu)]/k_{\text{opt}}.
\end{aligned} \tag{4.23}$$

From Equation (4.11) the average strain in the ring is

$$e_r = P_r/E_r A_r = \frac{kct\sigma_0}{t_r(b-a)E_r}. \tag{4.24}$$

Equate the two strains to get

$$\frac{t_r}{t} = \frac{E}{E_r}\frac{c}{b-a}\frac{c^2}{a^2}\frac{kk_{\text{opt}}}{2-k(1+\nu)}. \tag{4.25}$$

The weight of the ring W_r can be compared to the weight of the skin removed from the hole as

$$\frac{W_r}{W_s} = \frac{2\pi ct_r(b-a)\rho_r}{\pi a^2 t\rho_s} = \frac{2c(b-a)}{a^2}\frac{\rho_r}{\rho_s}\frac{t_r}{t}, \tag{4.26}$$

where ρ_r and ρ_s are the weight densities of the materials.

The average tension stress in the ring is

$$\sigma_r = \frac{P_r}{t_r(b-a-D)} = \frac{kc\sigma_0}{b-a-D}\frac{t}{t_r}, \tag{4.27}$$

where D is the diameter of the rivet holes in the ring. It should be noted that the stress in the ring varies, being larger on the inner edge and smaller on the outer edge. The actual stresses in the ring can be calculated by the same procedure as used for the skin. Change the second boundary condition in Equation (4.13) to

$$\sigma_{r2} = 0 \text{ at } r = b \tag{4.28}$$

and solve for the constants in the stresses. However, the σ_r stress is less than the σ_0 stress due to the Poisson's ratio effect in the skin so that usually the increased edge stress in the ring is not critical in the design.

It should be emphasized that the design of the ring can affect Δk in Equation (4.21). If the ring is too weak, less load $k\sigma_0 t$ is transferred so that Δk is negative and $\sigma_{\theta 2c}$ increases. If the ring is too strong, more load is transferred so that Δk is positive and σ_{r2c} increases. Thus, making A_r too large is as bad as making it too small.

Example 4.2. In Figure 4.2 assume $a = 10$ in., $b = 12$ in., $c = 11$ in., $\nu = 0.3$, $D = 1/8$ in., $E_r = E$ for same materials. Use σ_0 as the reference stress, find k_{opt}, and graph t_r/t, σ_{2c}/σ_0 maximum stress in the skin, W_r/W_s, and σ_r/σ_0 for values of k above and below k_{opt}.

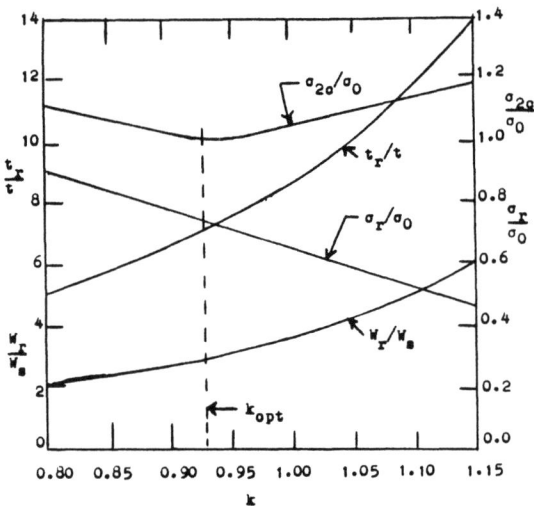

Fig. 4.3. Stresses and weights in ring reinforced skin, Fig. 4.2.

Solution. From Equations (4.20), (4.25), (4.27), and (4.21)

$k_{opt} = 0.93$, $\quad t_r/t = 6.19k/(2 - 1.3k)$,
$W_r/W_s = 2.72k/(2 - 1.3k)$, $\quad \sigma_r/\sigma_0 = (2 - 1.3k)/1.06$,
$\sigma_{2c}/\sigma_0 = 1 + 0.89 \mid k - 0.93 \mid$.

Figure 4.3 shows the results for $0.80 \leq k \leq 1.15$. With a tolerance of 3% on the skin stresses, the best design point for ring weight is $k = 0.90$.

4.4. Bending in circular cylindrical shells with axially symmetric loading

If a pressure vessel is a circular cylindrical shell with hemispherical ends, the radial deflection of the cylinder and hemisphere are different so that shear forces and bending moments must act along the juncture between the two shells to make them compatible. If a fuselage shell is restrained by a fuselage bulkhead attached on a circumference, local shear and bending stresses will be produced. See Figure 4.4. Vertical cylindrical tanks filled with a liquid have bending stresses as well as hoop stresses.

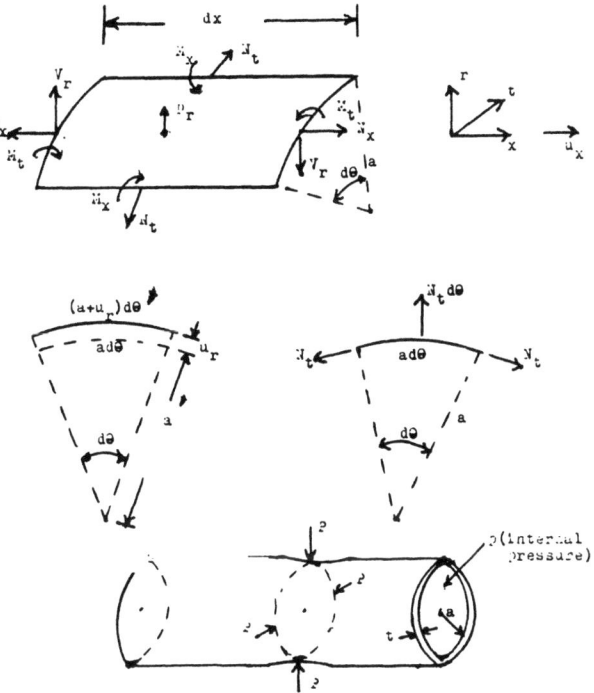

Fig. 4.4. Notation for axially symmetric cylindrical shell.

To examine these bending effects start with the tangential and axial strains e_t and e_x in Figure 4.4, or

$$(e_t + e_E)a\,d\theta = (a + u_r)d\theta - a\,d\theta,$$
$$e_t = u_r/a - e_E, \qquad e_x = u_{x,x} - e_E. \tag{4.29}$$

From the two-dimensional stress-strain Equations (1.26) in Volume 1,

$$\begin{aligned}
\sigma_x &= N_x/t = \frac{E}{1-\nu^2}(e_x + \nu e_t) - N_E/t \\
&= \frac{E}{1-\nu^2}\left(u_{x,x} + \frac{\nu u_r}{a}\right) - N_E/t, \\
\sigma_t &= N_t/t = \frac{E}{1-\nu^2}\left(\frac{u_r}{a} + \nu u_{x,x}\right) - N_E/t \\
N_E &= Ee_E t/(1-\nu).
\end{aligned} \tag{4.30}$$

Assume N_x is a known constant and eliminate $u_{x,x}$ to get

$$N_t = Etu_r/a + \nu N_x - (1-\nu)N_E, \tag{4.31}$$

whence Figure 4.4 shows that N_t produces a radial load on the circumference of

$$p_t = -N_t\,d\theta/a\,d\theta = -N_t/a$$

$$= -Etu_r/a^2 - \nu N_x/a + (1-\nu)N_E/a. \tag{4.32}$$

From Equation (4.150, Vol.1) with $-N_x$ in place of P_C, p_t in place of p_z, D in place of EI, and p_r included,

$$Du_{r,xxxx} - N_x u_{r,xx} + (Et/a^2)u_r = p_r - \nu N_x/a + (1-\nu)N_E/a,$$
$$D = \frac{Et^3}{12(1-\nu^2)}. \tag{4.33}$$

Divide by D and simplify to get

$$u_{r,xxxx} - Hu_{r,xx} + Gu_r = F(x)/D, \tag{4.34}$$
$$H = N_x/D, \quad G = Et/a^2 D = 12(1-\nu^2)/a^2 t^2,$$
$$F(x) = p_r - \nu N_x/a + (1-\nu)N_E/a.$$

If $F(x)$ is constant or linear, the particular integral u_{rp} is

$$u_{rp} = F(x)/DG = a^2 F(x)/Et. \tag{4.35}$$

The solution for the homogeneous equation has the form

$$u_r = Ce^{Bx}, \quad \text{whence} \tag{4.36}$$
$$B^4 - HB^2 + G = 0,$$
$$B^2 = (H/2) \pm [(H/2)^2 - G]^{1/2} \tag{4.37}$$
$$= B_1^2 \text{ or } B_2^2.$$

If the tension N_x is sufficiently large so that $(H/2)^2 > G$, or

$$N_x^2/4D > Et/a^2, \quad \text{or} \quad N_x^2 > (Et^2/a)^2/3(1-\nu^2), \quad \text{or} \tag{4.38}$$
$$e_x = N_x/Et > 0.61(t/a),$$

then B_1^2 and B_2^2 are positive and the four roots are $\pm B_1, \pm B_2$. Thus, the solution is

$$u_r = u_{rp} + C_1 e^{-B_1 x} + C_2 e^{-B_2 x} + C_3 e^{B_1 x} + C_4 e^{B_2 x}. \tag{4.39}$$

If the shell is very long with x positive the C_3 and C_4 constants will be zero, whence

$$u_r = u_{rp} + C_1 e^{-B_1 x} + C_2 e^{-B_2 x}. \tag{4.40}$$

If $N_x = 0$, then use

$$F = (G/4)^{1/4}, \quad B^4 = -G, \quad B^2 = \pm iG^{1/2},$$
$$B = \pm F(1 \pm i), \tag{4.41}$$

whence

$$u_r = e^{-Fx}(C_1 \cos Fx + C_2 \sin Fx) + e^{Fx}(C_3 \cos Fx + C_4 \sin Fx) +$$
$$+ u_{rp}. \tag{4.42}$$

This case is used in Chapter 15 of Reference 1.

If Equation (4.38) is not satisfied and N_x is not zero, then

$$B^2 = H/2 \pm i[G - (H/2)^2]^{1/2} = c \pm id, \qquad (4.43)$$
$$B = \pm(F_1 \pm iF_2),$$
$$2F_1^2 = (c^2 + d^2)^{1/2} + c = G^{1/2} + c, \quad 2F_2^2 = G^{1/2} - c,$$

whence

$$u_r = u_{rp} + e^{-F_1 x}(C_1 \cos F_2 x + C_2 \sin F_2 x) + \\ + e^{F_1 x}(C_3 \cos F_2 x + C_4 \sin F_2 x). \qquad (4.44)$$

For the u_r in Equation (4.40) the bending moments M_t and M_x and the shear V_r are given by

$$u_{r,x} = u_{rp,x} - [B_1 C_1 e^{-B_1 x} + B_2 C_2 e^{-B_2 x}],$$
$$M_t = D u_{r,xx} = D[B_1^2 C_1 e^{-B_1 x} + B_2^2 C_2 e^{-B_2 x}], \quad M_x = \nu M_t, \qquad (4.45)$$
$$V_r = D u_{r,xxx} = -D[B_1^3 C_1 e^{-B_1 x} + B_2^3 C_2 e^{-B_2 x}].$$

Two boundary conditions to determine C_1 and C_2 in Equations (4.40) and (4.45) must be selected from conditons on u_r, $u_{r,x}$, M_t, or V_r in the particular application.

Also, if discontinuities occur in temperature, material E, thickness t, and relative deflections at the junction of two shells, then it may be necessary to use separate solutions in Equation (4.40) on each side of the junction. In this case, take

$$u_{r1} = u_{rp1} + C_1 e^{-B_1 x} + C_2 e^{-B_2 x}, \quad x \geq 0,$$
$$u_{r2} = u_{rp2} + C_3 e^{B_3 x} + C_4 e^{B_4 x}, \quad x \leq 0. \qquad (4.46)$$

The four continuity conditions at $x = 0$ to determine the four constants $C_1 \cdots C_4$ are

$$u_{r1} = u_{r2}, \quad u_{r1,x} = u_{r2,x}, \quad M_{t1} = M_{t2}, \quad V_{r1} = V_{r2}, \text{ at } x = 0. \quad (4.47)$$

If u_{rp1} and u_{rp2} are constant, the matrix equation to determine the constants is

$$\begin{bmatrix} 1 & 1 & -1 & -1 \\ B_1 & B_2 & B_3 & B_4 \\ D_1 B_1^2 & D_1 B_2^2 & -D_2 B_3^2 & -D_2 B_4^2 \\ D_1 B_1^3 & D_1 B_2^3 & D_2 B_3^3 & D_2 B_4^3 \end{bmatrix} \begin{Bmatrix} C_1 \\ C_2 \\ C_3 \\ C_4 \end{Bmatrix} = \begin{Bmatrix} u_{rp2} - u_{rp1} \\ 0 \\ 0 \\ 0 \end{Bmatrix}. \qquad (4.48)$$

If the discontinuity occurs only in the particular integral so that $D_2 = D_1$,

$B_3 = B_1$, $B_4 = B_2$, then Equation (4.48) reduces to a simple form to give

$$C_3 = -C_1, \quad C_4 = -C_2, \quad C_1 = \frac{(u_{rp1} - u_{rp2})B_2^2}{2(B_1^2 - B_2^2)},$$

$$C_2 = -\frac{(u_{rp1} - u_{rp2})B_1^2}{2(B_1^2 - B_2^2)}, \quad M_t = 0 \text{ at } x = 0,$$

$$V_r = -\frac{(u_{rp1} - u_{rp2})B_1^2 B_2^2}{2(B_1 + B_2)} D \quad \text{at } x = 0, \quad (4.49)$$

$$x = \frac{\ln(B_1/B_2)}{B_1 - B_2} \quad \text{for } (M_t)_{\max} \text{ in Equation (4.45)}.$$

It should be noted that from Equation (4.30)

$$u_{x,x} = \frac{1-\nu^2}{Et}(N_x + N_E) - (\nu u_r/a),$$

$$u_x = \frac{1-\nu^2}{Et}(N_x + N_E)x - \frac{\nu}{a}\int u_r\, dx + C_5, \quad (4.50)$$

whence u_x is not zero although $N_x + N_E$ may be zero. Thus, the above equations for u_r apply only if the shell deflection in the x-direction is unrestrained. If restraints are present with u_x specified as a constant on the circumference, then the proper N_x to use can be determined approximately from Equation (4.50) for a long shell. Since the exponential terms in Equation (4.40) usually give a local effect over a short distance, use only the particular integral for u_r in Equation (4.50), whence

$$u_x = \frac{1-\nu^2}{Et}(N_x + N_E)x - \frac{\nu a}{Et}xF, \quad F(x) = F = \text{constant}. \quad (4.51)$$

Assume $u_x = \dot{u}_{xL}$ at $x = L$ so that

$$N_{xc} = [Etu_{xL}/L(1-\nu^2)] - N_E + \nu aF/(1-\nu^2),$$
$$F = p_r - \nu N_{xc}/a + (1-\nu)N_E/a, \quad (4.52)$$
$$N_{xc} = Etu_{xL}/L + \nu a p_r - (1-\nu)N_E.$$

Thus, for a restrained case, use this N_{xc} in place of N_x in H and F in Equation (4.34).

It should be emphasized that the $H = N_x/D$ term in Equation (4.34) is very important for pressure vessel applications with thin skins in aircraft structures. The condition in Equation (4.38) is satisfied for practically all pressurized designs for fuselages so that the deflection wave shape, which occurs for $H = 0$, does not appear. Examples 4.3 and 4.4 below show that the maximum deflection and maximum moment for $H = 0$ case may be reduced over 50% by the N_x tension. The $H = N_x/D$ term is included in Chapter 9 of Reference 2, but is not included in Chapter 15 of Reference 1, Chapter 12 of Reference 3, or Chapter 9 of Reference 5. In cases where Equation (4.38) is not satisfied but N_x is not zero, the solution for thin shells should be made using Equations (4.43) and (4.44). In cases where c is small compared to d in Equation (4.43) so that F_1 and F_2 are nearly the same, the simpler case of Equation (4.42) can be used.

Analysis and design of pressurized structures

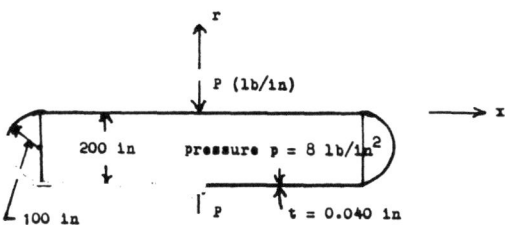

Fig. 4.5. Pressure vessel for Examples 4.3 and 4.4.

Example 4.3. Suppose a circular frame produces a load of P lb/in. on the circumference of the pressurized cylinder in Figure 4.5. (a) Assume the shell has $e_E = e_T = 0.002$ from temperature and find the deflections, moments, and shear forces in the shell with $\nu = 0.3$, $E = 10^7$ lb/in^2. (b) Use the deflection at the load and find the area of the frame.

Solution. (a) With $a = 100$ in., $p_r = 8$ lb/in^2, $t = 0.040$ in., $\nu = 0.3$, Equations (4.8), (4.34), and (4.37) give

$$N_x = 8(100)/2 = 400 \text{ lb/in}, \quad D = Et^3/12(1-\nu^2) = 58.61 \text{ lb-in},$$
$$H = N_x/D = 6.825, \quad G = 0.6825, \quad B_1^2 = 6.725, \quad B_2^2 = 0.1015,$$
$$B_1 = \pm 2.59, \quad B_2 = \pm 0.32, \quad N_E = Ete_E/(1-\nu) = 1143,$$
$$F(x) = 8.00 - 0.3(400)/100 + 0.7(1143)/100 = 14.8.$$

From Equation (4.35)

$$u_{rp} = 0.370 \text{ in. with temperature},$$
$$= 0.170 \text{ in. without temperature},$$

whence Equation (4.40) becomes

$$u_r = 0.370 + C_1 e^{-2.59x} + C_2 e^{-0.32x}. \tag{4.53}$$

With $x = 0$ at the load P, Equation (4.45) applies for x positive with symmetry for x and u_r, and the boundary conditions are

$$u_{r,x} = 0 \text{ at } x = 0,$$
$$V_r = Du_{r,xxx} = -P/2 \text{ at } x = 0^+,$$

whence Equation (4.41) gives

$$2.59 C_1 + 0.32 C_2 = 0, \quad 17.37 C_1 + 0.033 C_2 = P/2D,$$
$$C_1 = 0.029 P/D, \quad C_2 = -0.237 P/D,$$
$$u_r = 0.370 + (P/D)(-0.237 e^{-0.32x} + 0.029 e^{-2.59x}), \tag{4.54}$$
$$M_t = P(-0.0243 e^{-0.32x} + 0.1945 e^{-2.59x}),$$
$$V_r = P(0.008 e^{-0.32x} - 0.504 e^{-2.59x}).$$

It can be seen from these results that the bending effects are local, reaching short distances on both sides of the load P. The maximum values due to P are at $x = 0$,

$$u_r = -0.208 P/D \text{ at } x = 0,$$
$$M_t = 0.170 P \text{ at } x = 0,$$
$$V_r = -0.50 P \text{ at } x = 0.$$

At other locations,

$$u_r = -0.035 P/D \text{ at } x = 6 \text{ in},$$

$M_t = -0.003P$ at $z = 1$ in,
$V_r = -0.31P$ at $z = 1$ in,

which shows that the moments and shear forces are practically point values, while the deflection dies out more slowly. The tension load N_z causes the maximum deflection to be smaller and to spread out more. From Chapter 15 of Reference 1 without tension, this case gives $u_r = -0.48P/D$ at $z = 0$ and $u_r = 0$ at $z = 3.68$ in. The maximum moment is much larger at $0.39P$ but zero at $z = 1.23$ in. Thus, the tension load N_z reduces the maximum deflection by 57% and the maximum moment by 56%.

(b) Since the frame restrains the deflection of the shell, the load P puts tension in the frame so that the deflection of the frame is

$$u_{rf} = Pa^2/E_f A_f. \tag{4.55}$$

Use $E_f = E$ for the skin and $u_{rP} = -0.208P/D$ from above, or

$$u_{rf} = u_r(0) = u_{rp} + u_{rP}, \tag{4.56}$$

or for the case without temperature,

$(P/1000 A_f) + (0.208P/58.61) = 0.170$,
$(1/A_f) = (170/P) - 3.55. \tag{4.57}$

This Equation (4.57) also applies for the case of the frame temperature the same as the skin. The maximum value of P is 47.90 for which $1/A_f = 0$. However, $A_f = 0.28$ in^2 for $P = 23.95$ and $A_f = 0.13$ in^2 for $P = 15$ lb/in. Thus a frame with a small area can restrain the shell, giving considerable deflection and stresses.

For the case of $A_f = 0.13$ in^2, the deflections at $z = 0$ are

$u_{rp} = -0.208(15)/58.61 = -0.053$ in,
$u_r = u_{rf} = 0.170 - 0.053 = 0.117$ in.

The hoop load is $N_t = 4000 u_r + 120 = 588$ lb/in at $z = 0$ instead of 800 lb/in. The maximum bending moment and bending stress are

$M_t = 0.170(15) = 2.55$ in-lb,
$\sigma_b = 6M_t/t^2 = 6(2.55)/(0.040)^2 = 9600$ lb/in^2.

The hoop stress in the frame is

$\sigma_f = (15)(100)/0.13 = 11,500$ lb/in^2.

If a rigid bulkhead makes $u_r = 0$, then $M_t = 0.170(47.90) = 8.14$ in-lb, $\sigma_b = 30,500$ lb/in^2.

Example 4.4. Find the deflections, moments, and shear forces in the cylindrical shell at and near its junction with the hemispherical end caps in Figure 4.5. Use the data in Example 4.3 except assume no temperature change in the end caps.

Solution. Designate the cylindrical shell as (1) and the hemispherical shell as (2). Assume the shell (2) to act in the same manner as shell (1) for the short distance involved in this load case. Without the temperature term in (2), the particular integrals in Equation (4.35) are $u_{rp2} = 0.170$ in. and $u_{rp1} = 0.370$ in. from Example 4.3. From Equations (4.8) and (4.9) the particular integral for shell (2) must be modified, or

$u_{rp2} = (1/2)(0.170)(0.7/0.85) = 0.070$ in.

In this case Equation (4.49) applies so that

$C_1 = +(0.370 - 0.070)(0.32)^2/2[(2.59)^2 - (0.32)^2]$
$= 0.0023$,
$C_2 = -0.1523$,
$u_{r1} = 0.370 - 0.1523 e^{-0.32z} + 0.0023 e^{-2.59z}$,
$M_{t1} = 0.914(-e^{-0.32z} + e^{-2.59z})$,
$V_{r1} = 0.914(0.32 e^{-0.32z} - 2.59 e^{-2.59z})$,

$V_{r0} = -2.07$ lb/in at $z = 0.92$ in,
$(M_t)_{max} = 0.60$ in-lb/in at $z = 0.92$ in,

where Equation (4.45) has been used with $D = 58.41$.
From Chapter 15 in Reference 1 the results for the data in this example with $H = 0$, or $N_z = 0$, in Equation (4.34) are

$V_{r0} = -4.61$ lb/in,
$(M_t)_{max} = -2.32$ in-lb/in at $z = 1.23$ in.

The N_x term reduces the maximum shear force and maximum moment by 55% and 74%, respectively.

4.5. Bending in pressurized aircraft fuselages from stringers and frames

Example 4.3 above shows that a small ring frame can cause large local bending stresses in the pressurized fuselage. Most aircraft fuselages have many stringers on the skin to carry the bending from the overall fuselage air and inertia loads. These stringers have large bending stiffness in comparison to the skin bending stiffness so that they will affect the interaction between the frames and the skin. That is, the basic equations given in Section 4.4 and used in Example 4.3 must be modified to allow for the effect of the stringers.

Add the areas and stiffnesses of the stringers and divide by the fuselage circumference to get an equivalent thickness t_1 and an unit equivalent stiffness $D_1 = (EI)_1$ for the stringers. Equation (4.30) will now have the form

$$N_{zs} = \frac{Et}{1-\nu^2}(u_{x,x} + \nu u_r/a) - N_{Es}, \quad \text{skin,}$$
$$N_{z1} = Et_1 u_{x,x} - N_{E1}, \quad \text{equivalent stringers,} \quad (4.58)$$
$$N_t = \frac{Et}{1-\nu^2}(u_r/a + \nu u_{x,x}) - N_{Es},$$
$$N_{Es} = tEe_{Es}/(1-\nu), \quad N_{E1} = t_1 E_1 e_{E1}.$$

Define

$$N_x = N_{zs} + N_{z1}, \quad N = N_x + N_{Es} + N_{E1}, \quad (4.59)$$

and eliminate $u_{x,x}$ from Equation (4.58) to get

$$ap_t = -N_t = -(Etu_r/a)\frac{t+t_1}{t+(1-\nu^2)t_1} - \frac{\nu t N}{t+(1-\nu^2)t_1} + N_{Es}, \quad (4.60)$$

which corresponds to Equation (4.3). The basic Equation (4.34) becomes

$$u_{r,xxxx} - H_1 u_{r,xx} + G_1 u_r = F(x)/D_1, \quad (4.61)$$
$$H_1 = N_x/D_1, \quad G_1 = (Et/a^2 D_1)\frac{t+t_1}{t+(1-\nu^2)t_1},$$
$$aF(x) = ap_r + N_{Es} - \nu t N/[t+(1-\nu^2)t_1].$$

Since D_1 for an airplane fuselage with stringers may be between 10^4 and 10^5 lb-in, the values of H_1 and G_1 in Equation (4.61) will be small and much less than one. Although H_1 may be larger than G_1, $(H_1/2)^2$ is much smaller than G_1 so that Equations (4.43) and (4.44) apply to Equation (4.61). Also, F_1 and F_2 in Equation (4.43) for the case of Equation (4.61) will be small so that the local bending effects from one frame will spread out over a long distance and overlap the next frame. Thus for equally spaced frames of equal size, modify Equation (4.44) to put the origin of x midway between two frames and use symmetry, whence Equation (4.44) can be put into the symmetrical form

$$u_r = u_{rp} + C_1 \sinh F_1 x \sin F_2 x + C_2 \cosh F_1 x \cos F_2 x \tag{4.62}$$

as the solution for Equation (4.61).

Example 4.5. Use the data in Example 4.3 and add stringers on the fuselage giving $t_1 = 0.020$ in., $D_1 = 1000D = 56,800$ lb-in. Use a frame spacing of L in. Discuss the deflections, moments, and frame areas for several values of L. Assume the stringers have the same temperature as the skin.

Solution. In Equation (4.61)

$$H_1 = 400/58,600 = 0.006826,$$

$$G_1 = \frac{(10^7)(0.040)(0.040 + 0.020)}{(100)^2(58,600)(0.040 + 0.0182)} = 0.0007037,$$

$$N_{E_s} = Et e_B/(1-\nu) = 1143, \quad N_{E1} = Et_1 e_B = 400, \quad N = 1943,$$

$$N_t = 412 u_r - 742, \quad F(x) = 8.00 - 4.01 + 11.43 = 15.42.$$

From Equation (4.43)

$$c = H_1/2 = 0.003413, \quad d = \left[G_1 - (H_1/2)^2\right]^{1/2} = 0.01631,$$

$$F_1 = 0.1224, \quad F_2 = 0.1075.$$

Since c is small compared to d and F_1, F_2 are close together, simplify to the case in Equation (4.41), whence

$$F = (G_1/4)^{1/4} = 0.1150$$

can be used for both F_1 and F_2 in Equation (4.62).

In Equation (4.62), $u_{rp} = F(x)/G_1 D_1 = 0.374$,

$$u_r = 0.374 + C_1 \sinh Fx \sin Fx + C_2 \cosh Fx \cos Fx,$$

$$M_t = D_1 u_{r,xx} = 2D_1 F^2 (C_1 \cosh Fx \cos Fx - C_2 \sinh Fx \sin Fx), \tag{4.63}$$

$$V_r = 2D_1 F^3 [(C_1 - C_2)\sinh Fx \cos Fx - (C_1 + C_2)\cosh Fx \sin Fx].$$

Use the boundary conditions from Example 4.3 in the form

$$u_{r,x} = 0 \text{ at } x = L/2, \quad V_r = P/2 \text{ at } x = L/2,$$

and define

$$J = \sinh 0.0575L \cos 0.0575L, \quad H = \cosh 0.0575L \sin 0.0575L, \tag{4.64}$$

whence

$$C_2 = (H+J)C_1/(H-J), \quad C_1/P = -\frac{H-J}{8D_1 F^3 (H^2+J^2)}. \tag{4.65}$$

From Equations (4.56), (4.57), and (4.63) for no change in the frame temperature,

$$P/1000A_f = u_r(L/2) = 0.370 + P(u_{rP}/P),$$

$$u_{rP}/P = (C_1/P)\sinh(FL/2)\sin(FL/2) + (C_2/P)\cosh(FL/2)\cos(FL/2),$$

$$P = 374A_f/[1 - 1000A_f(u_{rP}/P)]. \tag{4.66}$$

Analysis and design of pressurized structures 141

Table 4.1. Results for example 4.5.

L	10 in.	15 in.	20 in.	30 in.
$10^3 C_1/P$	-0.2675	-0.3925	-0.4945	-0.5501
$10^3 C_2/P$	-2.419	-1.558	-1.065	-0.4100
$10^3 u_{rP}(0)/P$	-2.419	-1.558	-1.065	-0.4100
$10^3 u_{rP}(L/2)/P$	-2.463	-1.705	-1.397	-1.295
$\frac{10^3}{P}[u_{rP}(0) - u_{rP}(L/2)]$	0.044	0.147	0.332	0.885
$M_t(0)/P$	-0.4146	-0.6084	-0.7665	-0.8526
$M_t(L/2)/P$	0.8310	1.2329	1.5968	2.0829
$A_f = 0.13$ in^2				
P(lb/in)	36.83	39.80	41.15	41.61
σ_f(psi)	28,300	30,600	31,600	32,000
$u_r(0)$ (in)	0.285	0.312	0.330	0.357
$u_r(L/2)$ (in)	0.283	0.306	0.316	0.320
$u_r(0) - u_r(L/2)$ (in)	0.002	0.006	0.014	0.037
$M_t(L/2)$ (in-lb)	30.61	49.07	65.71	86.67
σ_b (psi)	5200	8370	11,200	14,700
$N_t(0)$ (lb/in)	433	546	619	730
$N_t(L/2)$ (lb/in)	425	520	561	586
$N_t(0)/800$	0.54	0.68	0.77	0.91
$A_f = 0.25$ in^2				
P(lb/in)	57.87	65.56	69.30	70.63
σ_f(psi)	23,100	26,200	27,700	28,200
$u_r(0)$ (in)	0.234	0.272	0.300	0.345
$u_r(L/2)$ (in)	0.231	0.262	0.277	0.283
$u_r(0) - u_r(L/2)$ (in)	0.003	0.010	0.023	0.062
$M_t(L/2)$ (in-lb)	48.10	80.84	110.8	147.1
σ_b(psi)	8200	13,800	18,900	24,900
$N_t(0)$ (lb/in)	223	380	495	681
$N_t(L/2)$ (lb/in)	211	338	400	433
$N_t(0)/800$	0.28	0.48	0.62	0.85
$A_f = 0.25$ in^2, No temperature effect				
P(lb/in)	26.92	30.50	32.20	32.86
σ_f(psi)	10,700	12,200	12,900	13,100
$u_r(0)$ (in)	0.109	0.127	0.140	0.161
$u_r(L/2)$ (in)	0.107	0.122	0.129	0.132
$u_r(0) - u_r(L/2)$ (in)	0.002	0.005	0.011	0.029
σ_b(psi)	3800	6400	8800	11,600
$N_t(0)$ (lb/in)	532	606	659	746
$N_t(L/2)$ (lb/in)	523	585	614	626
$N_t(0)/800$	0.66	0.76	0.82	0.93

For bending, use $c = 1$ and $I = 58,600/10^7 = 0.00586$ so that

$\sigma_b = M_t(1)/0.00586$.

Table 4.1 shows results for $L = 10$ in., 15 in., 20 in., 30 in. and $A_f = 0.13$ in^2, 0.25 in^2 for the temperature case. For the case of no temperature, change to $N_x = 400 = N$, $u_{rp} = 0.174$ in., $F(x) = 8.00 - 400(4.01)/1943 = 7.18$, $N_t = 4124 u_r + 82$. Thus, all values in Table 4.1 for the temperature case except N_t values can be reduced to the no temperature case by using the ratio $0.174/0.374 = 0.465$. Table 4.1 shows values for the no temperature case for $A_f = 0.25$ in^2. Except for the u_r deflections, these latter results apply for the temperature case with the frames at the same temperature as the skin and stringers.

Several conclusions can be drawn from the results in Table 4.1. The small values of the expression $u_r(0) - u_r(L/2)$ show that the large bending stiffness of the stringers allows very little change in the deflections u_r along the fuselage for all cases of L and A_f. However, the bending stresses σ_b in the stringers can be large. Elevated temperatures in the skin and stringers produce higher stresses in the frames and stringers but smaller N_t hoop loads in the skin. If $L = 10$ in. and $A_f = 0.44$ in^2, then $N_t(0) = 0$ for the elevated temperature case.

4.6. Bending of non-circular cross sections with internal pressure

A thin non-circular shell tends to become circular under internal pressure. To maintain the non-circular shape, it is necessary to provide bending support by some means such as shown in Figure 4.6. Obviously, the weight for the non-circular case will be much larger than for the thin circular shell. Provided internal

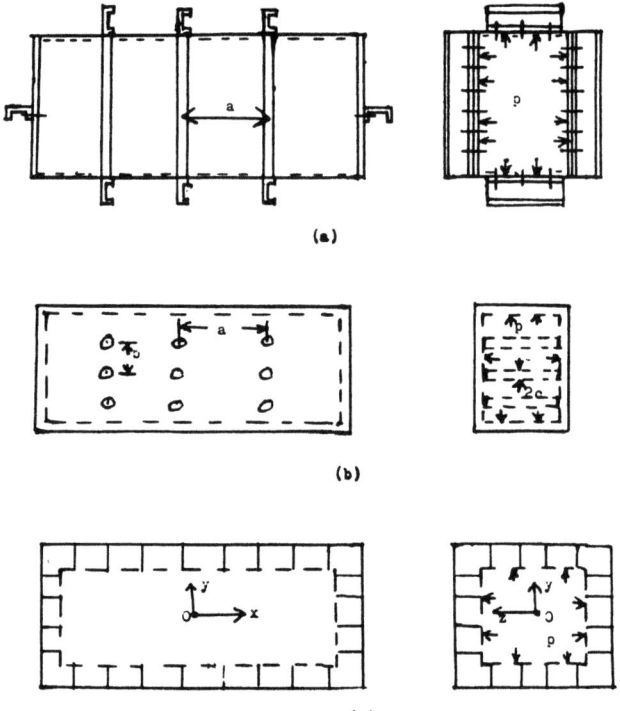

Fig. 4.6. Rectangular parallelepiped pressure vessels: (a). Bending stiffeners. (b). Cables or posts. (c). Honeycomb sandwich.

cables or posts can be used, the least weight design for the rectangular case in Figure 4.6 would be a combination of the honeycomb sandwich (c) and cables (b). If space is available and no internal structure can be used then the honeycomb sandwich (c) may be best.

The difficulty with using the internal posts is the bending moment concentration at the posts. With $b \geq a$, p = pressure (psi), c = radius of post, it is shown

Analysis and design of pressurized structures

Table 4.2. Values of A and B for moments at posts.

b/a	1.0	1.1	1.2	1.3	1.4	1.5	2.0
A	0.811	0.822	0.829	0.833	0.835	0.836	0.838
B	0.811	0.698	0.588	0.481	0.374	0.268	-0.256
A + 0.3B	1.05	1.03	1.01	0.98	0.95	0.92	0.76
B + 0.3A	1.05	0.95	0.86	0.73	0.62	0.52	0.00

on page 250 of Reference 1 that the bending moments at a post away from the edges are

$$M_x = (abp/4\pi)[(1+\nu)\ln(a/c) - (A+\nu B)], \\ M_y = (abp/4\pi)[(1+\nu)\ln(a/c) - (B+\nu A)], \tag{4.67}$$

where ν is Poisson's ratio, M_x is moment (in-lb/in) about the y-axis producing stress in the x-direction, M_y is moment about the x-axis, A and B are in Table 4.2. For the case of $b = a$ and $\nu = 0.3$,

$$\begin{aligned} M_x = M_y &= (a^2 p/4\pi)[1.3\ln(a/c) - 1.05] \\ &= 0.321 p a^2 \text{ for } a/c = 50, \\ &= 0.226 p a^2 \text{ for } a/c = 20, \\ &= 0.154 p a^2 \text{ for } a/c = 10, \\ &= 0.083 p a^2 \text{ for } a/c = 5. \end{aligned} \tag{4.68}$$

Since the maximum bending moment on a one inch wide strip of length a with fixed ends is $M_y = pa^2/12 = 0.083pa^2$, it is evident in Equation (4.68) that there are large moment concentrations at a post for the diameter $2c$ small. The bending stresses for the solid plate of thickness t in Figure 4.6a are

$$\sigma_{xx} = 6M_x/t^2, \quad \sigma_{yy} = 6M_y/t^2. \tag{4.69}$$

For a honeycomb plate of core height h and face thickness t_f,

$$\sigma_{xx} = M_x/ht_f, \quad \sigma_{yy} = M_y/ht_f. \tag{4.70}$$

If the stresses in the cases in Equations (4.69) and (4.70) are the same, then

$$M_{xh} = (6ht_f/t^2)M_{xp}, \tag{4.71}$$

and the M_{xh} honeycomb moment is much larger than the M_{xp} plate moment. If $h = 0.50$ in., $t_f = 0.010$ in., $t = 0.040$ in., then $M_{xh} = 18.8 M_{xp}$. Thus, in spite of moment concentrations of three or four for the posts, the honeycomb case with posts is more effficient than case (a) in Figure 4.6.

4.7. Bending of non-circular fuselage rings with internal pressure

Since the fuselage has a x-z plane of symmetry, consider one-half of the fuselage ring, Figure 4.7. The slopes are zero at the bottom ($z = 0$, $y = 0$, $s = 0$) and at the top ($z = 2c$, $y = 0$, $s = s_L$). Clamp the ring at the top and cut it at

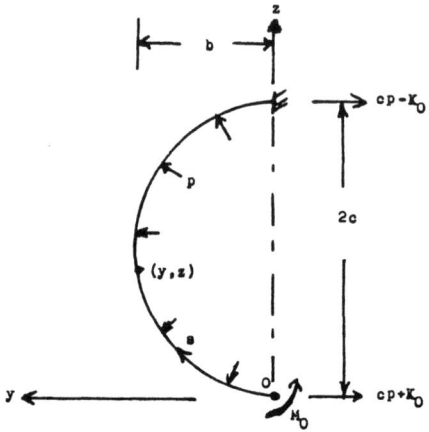

Fig. 4.7. One side of non-circular fuselage ring with internal pressure.

the bottom with the unknown moment M_0 and unknown axial load K_0. The bending moment at any point (y, z) is

$$M = M_0 + (K_0 + cp)z - (p/2)(y^2 + z^2) \qquad (4.72)$$

for a unit strip. Assume the thin skin of the fuselage shell takes negligible bending so that the total moment on any ring is Md for rings at d spacing.

Apply a unit moment and a unit load at the bottom so that the unit moment theorem in Equation (4.39, Vol.1) gives

$$\theta_0 = 0 = \int_0^{s_L} (1)HM\,ds, \quad H = (EI)_R/(EI),$$
$$u_{by}(0) = 0 = \int_0^{s_L} (z)HM\,ds. \qquad (4.73)$$

These equations can be written in the form

$$B_1 M_0 + B_2 K_0 = (C_1 - 2cB_2)(p/2),$$
$$B_2 M_0 + B_3 K_0 = (C_2 - 2cB_3)(p/2), \qquad (4.74)$$

$$B_1 = \int_0^{s_L} H\,ds, \quad B_2 = \int_0^{s_L} Hz\,ds, \quad B_3 = \int_0^{s_L} Hz^2\,ds,$$
$$C_1 = \int_0^{s_L} H(y^2 + z^2)\,ds, \quad C_2 = \int_0^{s_L} Hz(y^2 + z^2)\,ds.$$

For (EI) constant, or $H = 1$, K_0 will be zero so that

$$B_1 = s_L, \quad B_s = \int_0^{s_L} z\,ds, \quad C_1 = \int_0^{s_L} (y^2 + z^2)\,ds,$$
$$M_0 = (p/2s_L)(C_1 - 2cB_2). \qquad (4.75)$$

Analysis and design of pressurized structures 145

In general, the integrals in Equations (4.74) and (4.75) can be evaluated by summation over N elements, or

$$B_1 = \sum_{i=1}^{N} H_i(\Delta s)_i, \quad B_2 = \sum_{i=1}^{N} z_i H_i(\Delta s)_i, \quad B_3 = \sum_{i=1}^{N} z_i^2 H_i(\Delta s)_i,$$

$$C_1 = B_3 + \sum_{i=1}^{N} y_i^2 H_i(\Delta s)_i, \quad C_2 = \sum_{i=1}^{N} z_i(z_i^2 + y_i^2) H_i(\Delta s)_i,$$

$$(\Delta s)_i^2 = (\Delta y)_i^2 + (\Delta z)_i^2, \qquad (4.76)$$

where a more accurate value of $(\Delta s)_i$ may be obtained by measuring along the curve. If the curve has arcs of circles, or can be approximated by arcs of circles, then the integrations can be made for constant (EI) on the arc using $ds = R d\theta$ on each arc and expressing y and z in terms of θ.

Example 4.6. Take the curve in Figure 4.7 as the ellipse

$$\left(\frac{y}{36}\right)^2 + \left(\frac{z-48}{48}\right)^2 = 1 \qquad (4.77)$$

with

$$H = (EI)_R/(EI) = 0.25 \text{ for } 0 \leq z \leq 12$$
$$= 1.00 \text{ for } 12 \leq z \leq 96.$$

(a) Use Equation (4.76) to find M_0 and K_0 in Equation (4.74). Find the moments at the top and sides of the fuselage. (b) Give results for a constant (EI) and compare the results.

Solution. (a) Select the intervals on z and calculate the y end points of the intervals from

$$y = (3/4)(96z - z^2)^{1/2}.$$

Use y_i, z_i as the average of the end point values for each interval. Table 4.3 shows the calculations. From the sums using columns in Table 4.3

$$B_1 = 111.82, \quad B_2 = 6269.05, \quad B_\bullet = \sum_{1}^{26}(1)(8) = 445,572,$$

$$C_1 = 532,876, \quad C_2 = \sum_{1}^{26}(1)(9) = 39,304,905, \qquad (4.78)$$

whence from Equation (4.74)

$$111.82 M_0 + 6269 K_0 = [532,876 - 96(6269)](p/2) = -68,948(p/2),$$
$$6269 M_0 + 445,572 K_0 = [39,304,905 - 96(445,572)](p/2) = -3,470,007(p/2),$$
$$K_0 = 2.10 p, \quad M_0 = -426.02 p,$$
$$M/p = -426.02 + 50.10 z - (1/2)(y^2 + z^2)$$
$$= -224.42 \text{ at top } (y = 0, z = 96),$$
$$= 178.72 \text{ at sides } (y = \pm 36, z = 48).$$

(b) For constant (EI), change the 0.25 in column (6) in Table 4.3 to 1.00. From Equation (4.75) and Table 4.3,

$$B_1 = 132.16, \quad B_2 = 6357.79, \quad C_1 = 537,814, \quad M_0 = -274.42 p,$$
$$M/p = -274.42 + 48.00 z - (1/2)(y^2 + z^2),$$
$$= -274.42 \text{ at top},$$
$$= 229.58 \text{ at sides}.$$

Table 4.3. Calculations for Example 4.6.

1	2	3	4	5	6	7	8	9
z_i	$(\Delta z)_i$	(y_i)	$(\Delta y)_i$	$(\Delta s)_i$	H_i	(5)(6)	(7)z_i	(7)$(y_i^2 + z_i^2)$
0.5	1	3.65	7.31	7.38	0.25	1.84	0.92	24.97
1.5	1	8.80	2.97	3.13	0.25	0.78	1.17	62.16
3.0	2	12.28	4.11	4.57	0.25	1.14	3.42	182.17
5.0	2	15.91	3.04	3.64	0.25	0.91	4.55	253.10
7.5	3	19.21	3.56	4.66	0.25	1.16	8.70	493.31
10.5	3	22.40	2.82	4.12	0.25	1.03	10.82	630.37
14.0	4	25.32	3.02	5.01	1	5.01	70.17	4193.88
18.0	4	28.04	2.41	4.67	1	4.67	84.06	5184.83
22.5	5	30.42	2.36	5.53	1	5.53	124.40	7916.89
27.5	5	32.49	1.77	5.30	1	5.30	145.86	9602.87
33.0	6	34.11	1.49	6.18	1	6.18	204.01	13,920.40
39.0	6	35.29	0.86	6.06	1	6.06	236.39	16,764.29
45.0	6	35.86	0.28	6.01	1	6.01	270.29	19,898.75
51.0	6	35.86	0.28	6.01	1	6.01	306.51	23,360.51
57.0	6	35.29	0.86	6.06	1	6.06	345.42	27,235.99
63.0	6	34.11	1.49	6.18	1	6.18	389.34	31,718.80
68.5	5	32.49	1.77	5.30	1	5.30	363.05	30,463.61
73.5	5	30.42	2.36	5.53	1	5.53	406.46	34,991.77
78.0	4	28.04	2.41	4.67	1	4.67	364.26	32,084.03
82.0	4	25.32	3.02	5.01	1	5.01	410.82	36,899.16
85.5	3	22.40	2.82	4.12	1	4.12	352.26	32,185.48
88.5	3	19.21	3.56	4.66	1	4.66	412.41	38,217.94
91.0	2	15.91	3.04	3.64	1	3.64	331.24	31,064.23
93.0	2	12.28	4.11	4.57	1	4.57	424.01	40,215.08
94.0	1	8.80	2.97	3.13	1	3.13	294.22	27,899.07
95.5	1	3.65	7.31	7.38	1	7.38	704.79	67,405.77
						111.82	6269.05	532,876
						B_1	B_2	C_1

These moments for constant (EI) case check values for the ellipse given on page 166 in Reference 3.

The large $(EI)/(EI)_R$ across the bottom increases the moments on the bottom and decreases the moments at the top and on the sides. Since $(EI) = (EI)_R$ on the top, probably the largest bending stress will occur on the top for the increased $(EI) = 4(EI)_R$ on the bottom.

4.8. Bending of non-circular fuselage rings with point loads

Consider one half of the fuselage ring in Figure 4.8 with a point load P at the bottom. From symmetry the slopes and horizontal deflections are zero at the top and bottom. Clamp the ring at the top and cut it at the bottom with the unknown moment M_0 and unknown horizontal force T_0. The shear flows q_i produced by the load P depend upon the fuselage cross section. These shear flows can be determined by the procedures in Sections 5.3 through 5.8(Vol.1). For this discussion, the shear flows will be regarded as known.

Because of the complicated form of the moments produced by the shear flows on a non-circular ring, it is simpler to determine the moments at selected points on the ring. The selection of these points will depend upon the ring shape and

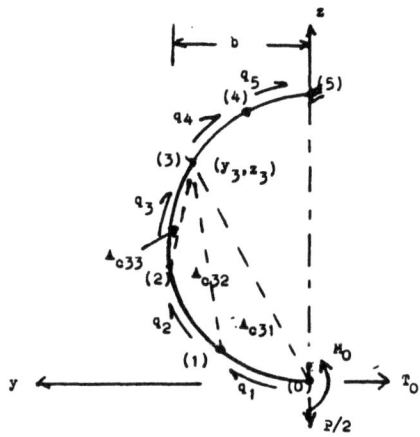

Fig. 4.8. Fuselage ring with point load P at bottom.

the shear flows. For circular rings with no fuselage stringers, moment equations for two point load cases are given in cases 24 and 25 of Table VIII in Reference 3.

For a point j at y_j, z_j on the ring in Figure 4.8 the moment M_j is

$$M_j = M_0 + T_0 z_j - (P/2)(y_j + S_j), \qquad (4.79)$$

$$q_i = w_i P/2, \qquad S_j = \sum_{i=1}^{j} 2A_{cji} w_i,$$

$$2A_{cji} = (z_j - z_i)(y_j - y_{i-1}) - (z_j - z_{i-1})(y_j - y_i),$$

where Equation (5.20), Figure 5.29, and Problem 5.6 in Volume 1 have been used. The expression for the enclosed area $2A_{cji}$ assumes a straight line between the points (y_i, z_i), (y_{i-1}, z_{i-1}). More accurate values can be obtained by using the curve. In this symmetrical case, T_0 can be determined from the shear flows as

$$T_0 = \sum_{i=1}^{N_b} q_i (\Delta y)_i = \frac{P}{2} \sum_{i=1}^{N_b} w_i (\Delta y)_i = H_b(P/2), \qquad (4.80)$$

where N_b locates the point for $y = b$ in Figure 4.8.

For a unit moment at point (0,0) in Figure 4.8, the unit load theorem in Equation (4.39, Vol.1) gives

$$\sum_{j=1}^{N} M_j (\Delta s)_j / (EI)_j = \sum_{j=1}^{N} M_j G_j = 0, \qquad (4.81)$$

$$G_j = (\Delta s)_j / (EI)_j.$$

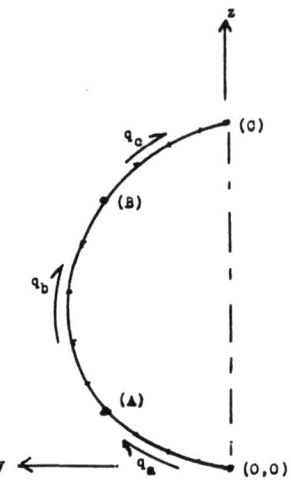

Fig. 4.9. Fuselage ring with constant shear flows on segments.

Put Equation (4.79) into Equation (4.81) to get

$$M_0 = (P/2H_1)(H_3 - H_b H_2), \qquad H_b = \sum_{i=1}^{N_b} w_i (\Delta y)_i, \qquad (4.82)$$

$$H_1 = \sum_{j=1}^{N} G_j, \qquad H_2 = \sum_{j=1}^{N} z_j G_j, \qquad H_3 = \sum_{j=1}^{N} G_j (y_j + S_j).$$

For a sequence of intervals with $w_i = w_a =$ constant,

$$S_j = \sum_{i=1}^{j} 2 A_{cji} w_i = w_a \sum_{i=1}^{j} h_i, \qquad h_i = z_i y_{i-1} - z_{i-1} y_i, \qquad (4.83)$$

where (y_i, z_i) is the end point of the interval and moments are taken about $(0,0)$. For several segments with constant shear flows as in Figure 4.9,

$$S_j = w_a \sum_{i=1}^{j} h_i, \qquad 0 < j \leq j_A,$$

$$= w_a S_A + w_b \sum_{i=j_A+1}^{j} h_i + (w_a - w_b) f_j, \qquad j_A < j \leq j_B,$$

$$= w_a S_A + w_b S_B + w_c \sum_{i=j_B+1}^{j} h_i + (w_a - w_b) f_j + (w_b - w_c) e_j, \qquad j_B < j \leq j_C,$$

$$S_A = \sum_{i=1}^{j_A} h_i, \qquad S_B = \sum_{i=j_A+1}^{j_B} h_i,$$

$$f_j = y_A z_j - z_A y_j, \qquad e_j = y_B z_j - z_B y_j. \qquad (4.84)$$

Analysis and design of pressurized structures

If the $P/2$ load in Figure 4.8 is applied at the point (y_p, z_p) instead of at $(0,0)$, then Equation (4.79) has the form

$$M_j = M_0 + T_0 z_j - (P/2) S_j, \quad 0 \le z \le z_P,$$
$$= M_0 + T_0 z_j - (P/2)(y_j - y_P + S_j), \quad z \ge z_P. \quad (4.85)$$

This modifies the sums for H_3 in Equation (4.82) by having

y_j as 0 for $1 \le j \le N_P$,

y_j as $y_j - y_P$ for $N_P < j \le N$,

$$\Delta H_3 = (H_1 - H_{1P})(-y_P) - \sum_{j=1}^{N_P} y_j G_j, \quad (4.86)$$

$$H_{1P} = \sum_{j=1}^{N_P} G_j.$$

Table 4.4. Calculations for Example 4.7.

	1	2	3	4	5	6	7	8	9
No.	z_i	y_i	h_i	$120w_i$	$\sum 120 w_i h_i$	$-0.33 f_j$	$0.33 e_j$	S_j	$G_j(y_j + S_j)_{av}$
0	0	0							
1	1	7.31	0	1	0			0	6.72
2	2	10.28	4.34	1	4.34			0.04	6.88
3	4	14.39	12.34	1	16.68			0.14	14.10
4	6	17.43	16.62	1	33.30			0.28	14.73
5	9	20.99	30.93	1	64.23			0.54	22.76
6	12	23.81	37.59	1	101.82			0.85	23.79
7	16	26.83	59.00	1.33	180.29	-19.47		1.34	132.36
8	20	29.24	68.76	1.33	271.74	-41.36		1.92	138.56
9	25	31.60	99.00	1.33	403.41	-71.30		2.77	181.18
10	30	33.37	113.75	1.33	554.70	-103.57		3.76	189.48
11	36	34.86	155.52	1.33	761.54	-144.82		5.14	238.30
12	42	35.72	178.20	1.33	998.55	-188.56		6.75	249.88
13	48	36.00	202.56	1.33	1267.95	-234.59		8.61	261.68
14	54	35.72	229.44	1.33	1573.11	-282.84		10.75	273.70
15	60	34.86	260.76	1.33	1919.92	-333.39		13.22	286.48
16	66	33.37	298.56	1.33	2317.00	-386.44		16.09	301.37
17	71	31.60	289.67	1.33	2694.28	-432.73		18.85	264.78
18	76	29.24	325.56	1.33	3127.27	-481.36		22.05	281.31
19	80	26.83	300.12	1.33	3520.43	-522.34		25.03	240.89
20	84	23.81	348.92	1.33	3990.49	-565.73		28.54	261.04
21	87	20.99	308.36	1	4298.75	-600.46	101.74	31.63	216.23
22	90	17.43	372.69	1	4671.34	-638.13	224.00	35.47	245.86
23	92	14.39	308.46	1	4979.80	-665.89	323.93	38.65	192.81
24	94	10.28	406.90	1	5388.70	-697.88	453.62	42.87	242.39
25	95	7.31	289.46	1	5678.16	-715.50	543.81	45.88	166.44
26	96	0	701.76	1	6379.92	-754.30	754.30	53.17	392.42
									4846.14
									H_3

Example 4.7. Take the curve in Figures 4.8 and 4.9 as the ellipse used in Example 4.6. Use the data in Example 4.6 with the shear flow as

$q_i = w_a P/2$ for $0 \le z \le 12$, $84 \le z \le 96$,
$\quad = w_b P/2$ for $12 \le z \le 84$,

where w_a and w_b are constant. Find M_0 and T_0 from Equation (4.82).

Solution. Use the same intervals as in Example 4.6 so that H_1 and H_2 will be the same as B_1 and B_2 given by Table 4.3 and Equation (4.78). Select $w_a = 1/120$ and $w_b = 1/90 = 1.33w_a$ so that $24w_a + 72w_b = 1$. Calculate h_i and $h_i w_i$ in Equation (4.84) with x_i and y_i at the end points of the intervals. Also,

$$H_b = \frac{23.81 + (12.19)(1.33)}{120} = 0.3335,$$

$$f_j = 23.81 x_j - 12 y_j, \quad 6 < j \le 26,$$

$$e_j = 23.81 x_j - 84 y_j, \quad 20 < j \le 26,$$

$$S_j = (1/120)\left[\sum_{i=1}^{j}(120 w_i h_i) - 0.33 f_j + 0.33 e_j\right] \quad (4.87)$$

$$= (1/120)[\text{columns } (5) + (6) + (7)] \text{ in Table 4.4},$$

G_j given in column (7) of Table 4.3.

Table 4.4 shows the calculations. From Tables 4.3 and 4.4 and Equations (4.80), (4.82), and (4.79),

$$H_1 = 111.82, \quad H_2 = 6269, \quad H_3 = 4846,$$
$$T_0 = 0.1668P, \quad M_0 = 12.42P, \quad (4.88)$$
$$M_j = [12.42 + 0.1668 x_j - \tfrac{1}{2}(y_j + S_j)]P, \quad (4.89)$$

where x_j, y_j, and S_j are given in columns (1), (2), and (8) of Table 4.4. M_0 is the maximum moment on the ring, with $M_j = 1.84P$ at the top ($z = 96$) and $M_j = -1.88P$ at the side ($z = 48$).

4.9. Effect of internal pressure upon buckling of cylindrical shells

Since internal pressure produces tension stresses in the walls of the cylinder, it tends to increase the allowable buckling loads in axial compression, bending, torsion, and transverse shear. The hoop tension stress N_t/t, Figure 4.4, due to the internal pressure p tends to increase the buckling stress, while the axial tension stress N_x/t increases the total compression load by the amount

$$P_x = 2\pi a N_x = 2\pi a(pa/2) = \pi a^2 p, \quad \sigma_x = pa/2t. \quad (4.90)$$

Thus, the total allowable compression load consists of three parts, or

$$(P_c)_{\text{all}} = P_{crc} + P_{crp} + P_x, \quad (4.91)$$

where P_{crc} is the buckling load of the cylinder without internal pressure p, P_{crp} is the increase in the buckling load due to p, and P_x is given by Equation (4.90). The actual buckling stress is

$$(F_{cr})_{\text{total}} = (P_{crc} + P_{crp})/2\pi at = F_{crc} + F_{crp}. \quad (4.92)$$

Various test data and empirical formulas for buckling of cylinders are given in Chapter 9 of Reference 5 and Chapter C8 of Reference 6. For the cylinder in *axial compression*, Equation (9.5) in Reference 5 gives

$$10^7 F_{crc}/E = 0.248(a/t) + 1.85(10^6)(t/a) - 18.1(10^{-6})(a/t)^2 - 970,$$
$$10^7 F_{crp}/E = 10^7 p_a(t/a)/(0.7 + 4.2 p_a), \quad (4.93)$$

Analysis and design of pressurized structures 151

$$p_a = (p/E)(a/t)^2, \quad (L/a)^2(a/t) > 80, \; 90\% \text{ probability,}$$

where L is the length of the cylinder.

For cylinders in *bending* empirical equations can be obtained from the 90% probability curves on Figure 9.5 in Reference 5 and Figure C8.14 in Reference 6 to get

$$\begin{aligned} 10^7 F_{crb}/E &= 1.25(a/t) + 3.47(10^6)(t/a) - 3620, \\ 10^7 F_{crp}/E &= 3.72(10^6)(p_a)^{0.22}(t/a). \end{aligned} \quad (4.94)$$

For cylinders under *torsion* Equations (9.10), (9.13), and (9.18) in Reference 5 give

$$\begin{aligned} & 10^7 F_{crT}/E = 1.27(10^7)(a/L)^{0.46}(t/a)^{1.35}, \\ & [(F_{crT})_{\text{total}}/F_{crT}]^2 = 1 \pm p_a[1.09(L/a)(a/t)^{0.5} - 0.696], \\ & (L/a)^2(a/t) > 100, \end{aligned} \quad (4.95)$$

where the + sign is for internal pressure p and the − sign is for external pressure p in $p_a = (p/E)(a/t)^2$. With p fixed

$$\sigma_T = (F_{crT})_{\text{total}} \quad \text{for M.S.} = 0.00, \quad (4.96)$$

and for a different applied σ_T,

$$\text{M.S.} = [(F_{crT})_{\text{total}}/\sigma_T] - 1. \quad (4.97)$$

For cylinders under *transverse shear* Equations (9.12) and (9.19) in Reference 5 give Equation (4.95) with F_{crT} replaced by

$$F_{crS} = 1.25 F_{crT}. \quad (4.98)$$

For various combined stress cases for buckling without pressure, see Equations (1.35)-(1.48) above.

Example 4.8. If a cylindrical shell has $a = 30$ in., $t = 0.050$ in., $p = 10$ psi, $E = 10,700,000$ psi for aluminum alloy, find the allowable axial compression buckling stresses and allowable axial compression loads with and without internal pressure p.

Solution. In this case

$$a/t = 600, \quad p_a = 10(600)^2/10,700,000 = 0.3364,$$

whence Equations (4.90) and (4.93) give

$$F_{crc} = 1.07(149 + 3083 - 6 - 970) = 2414 \text{ psi,}$$

$$F_{crp} = \frac{1.07(10)^7(0.3364)}{600[0.7 + 4.2(0.3364)]} = 2839 \text{ psi,}$$

$$P_x = \pi(30)^2(10) = 28,274 \text{ lb.}$$

Thus, the allowable buckling stresses are

$$F_{crc} = 2414, \quad \text{psi for no pressure,}$$
$$(F_{crc})_{\text{total}} = 5253, \quad \text{psi with pressure } p = 10 \text{ psi.}$$

The total allowable compression loads are

$$P_{crc} = (2414)(2\pi)(30)(0.050) = 22,751 \text{ lb for no pressure,}$$
$$(P_c)_{\text{total}} = 22,751 + 26,756 + 28,274 = 77,781 \text{ lb with pressure.}$$

In this case the internal pressure increases the allowable buckling stress by 118% and the allowable buckling load by 242%.

Example 4.9. In Example 4.8 find the allowable bending buckling stresses and allowable bending moments with and without internal pressure $p = 10$ psi.
Solution. From Equation (4.94)

$$F_{crb} = 1.07(750 + 5783 - 3620) = 3117 \text{ psi},$$

$$F_{crp} = 1.07(3.72)(10)^6(0.3364)^{0.22}/600 = 5220 \text{ psi}.$$

Thus, the allowable bending buckling stresses are

$$F_{crb} = 3117 \text{ psi}, \quad \text{for no pressure},$$

$$(F_{crb})_{\text{total}} = 8337 \text{ psi}, \quad \text{with pressure } p.$$

The allowable bending moments are

$$M_{cr} = (3117)(\pi)(30)^2(0.050) = 440,700 \text{ in-lb}, \quad \text{no pressure},$$

$$(M_{cr})_{\text{total}} = \left[8337 + \frac{(10)(30)}{2(0.050)}\right](\pi)(45) = 1,602,700 \text{ in-lb}, \quad \text{with pressure}.$$

Just as for the axial compression case in Example 4.8 the internal pressure produces large increases in the bending buckling stresses and allowable bending moments. The F_{crb} increases by 167% and the M_{cr} increases 264%.

4.10. Pressure stabilized structures

In such structures as balloons, airships, air mattresses, inflated tires, and various inflated fabric structures, the buckling stress without pressure is small or essentially zero. However, internal pressure produces tension stresses which can cancel out the applied compression stresses so that the inflated structure can take external forces. By using drop threads to form a fabric structure into a beam shape, it is possible to use internal pressure to support bending moments, torque, shear, and axial loads. The allowable stresses for bending moments and axial compression loads are given by Equation (4.90) for the circular cylinder. For any beam cross section with enclosed area A_c, Equation (4.90) becomes

$$N_y = pA_c/C_c, \tag{4.99}$$

where C_c is the length of the perimeter of the cross section.

For torsion and transverse shear the beam deflects until it can support the shear from the component of the tension load. For *transverse shear*, Figure 4.10 gives

$$\tan\theta = \theta = du_{zs}/dy = V_p/pA_c, \tag{4.100}$$

whence Equation (5.83, Vol.1) becomes

$$GA_w \frac{du_{zs}}{dy} = V_z - V_p = V_z - pA_c \frac{du_{zs}}{dy},$$

$$\frac{du_{zs}}{dy} = \frac{V_z}{GA_w + pA_c}. \tag{4.101}$$

Thus, although GA_w may be zero, the shear deflection can support transverse shear forces when p is present.

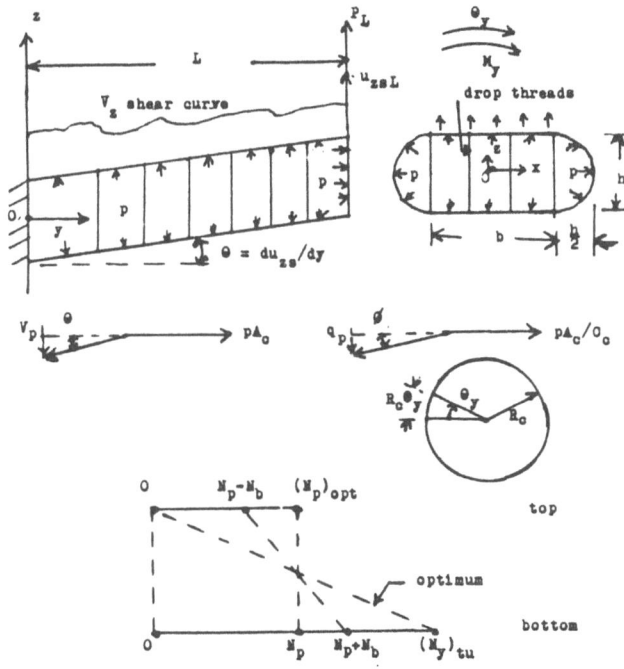

Fig. 4.10. Pressurized fabric beam.

For the *torsional moment* M_y, Figure 4.10 gives

$$q_p = M_{yp}/2A_c, \quad \tan\phi = \phi = R_c(d\theta_y/dy) = M_{yp}C_c/2pA_c^2, \quad (4.102)$$

where $\pi R_c^2 = A_c$ for the equivalent radius R_c of a circle. Equation (5.106, Vol.1) becomes

$$GJ(d\theta_y/dy) = M_y - M_{yp} = M_y - 2pA_c^2(R_c/C_c)(d\theta_y/dy),$$
$$d\theta_y/dy = \frac{M_y}{GJ + 2A_c^2 p(R_c/C_c)}. \quad (4.103)$$

Although GJ may be zero, the rotation deflection can support M_y torque when internal pressure p is present.

For *bending* Figure 4.10 shows

$$A_c = bh + (\pi h^2/4), \quad C_c = \pi h + 2b, \quad N_p = pA_c/C_c,$$
$$N_b = M_x/bh, \quad \text{no bending in side walls,}$$
$$(N_y)_t = N_p + N_b, \quad (N_y)_c = N_p - N_b, \quad (4.104)$$
$$\text{Allowables,} \quad (N_y)_c = 0, \quad (N_y)_t = (N_y)_{tu},$$
$$\text{Optimum,} \quad (N_p)_{opt} = (N_b)_{max} = (1/2)(N_y)_{tu}.$$

The maximum value of $(N_b)_{max} = (1/2)(N_y)_{tu}$ can be obtained only if the optimum value $(N_p)_{opt} = (1/2)(N_y)_{tu}$ is used. If N_p is increased or decreased from $(N_p)_{opt}$, then $(N_b)_{all} < (N_b)_{max}$.

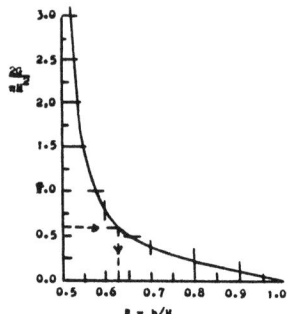

Fig. 4.11. Graph to get h approximately.

If $(N_y)_{tu}$, b, and h are specified in Figure 4.10, then

$$p_{opt} = \frac{(N_y)_{tu} C_c}{2 A_c} = \frac{(N_y)_{tu}(\pi h + 2b)}{2(bh + \pi h^2/4)},$$
$$(M_x)_{max} = bh(N_y)_{tu}/2. \qquad (4.105)$$

If the design moment is larger, then change the fabric to increase $(N_y)_{tu}$ or, if possible, increase b or h. Note that an increase in h with b fixed increases $(M_x)_{max}$ but decreases p_{opt}.

For design with $(M_x)_{max}$, p, and $(N_y)_{tu}$ specified, it is necessary to solve Equation (4.105) for b and h. Define

$$G = 2M_x/(N_y)_{tu} = bh, \quad b = G/h, \quad H = 2(N_y)_{tu}/p, \qquad (4.106)$$

whence Equation (4.105) gives a cubic equation for h, or

$$G = \pi h^2 (H - h)/(4h - 2H). \qquad (4.107)$$

Since G is positive, it is evident that h is restricted to

$$H/2 < h < H, \quad (N_y)_{tu}/p < h < 2(N_y)_{tu}/p, \qquad (4.108)$$

and the corresponding b is restricted to the range

$$2G/H > b > G/H, \quad 2pM_x/(N_y)_{tu}^2 > b > pM_x/(N_y)_{tu}^2. \qquad (4.109)$$

In order to get an approximate value for h in Equation (4.107), take

$$h = RH, \quad 0.5 < R < 1.0, \qquad (4.110)$$

and graph

$$2G/\pi H^2 = p^2 M_x/\pi(N_y)_{tu}^3 = (1 - R)R^2/(2R - 1),$$
$$R = h/H = ph/2(N_y)_{tu}, \qquad (4.111)$$

Analysis and design of pressurized structures 155

which is shown in Figure 4.11. With $2G/\pi H^2$ known read R from the graph and calculate h.

4.11. Problems

4.1. In Example 4.1, part (a), graph r_1, r_2, N_ϕ/p, N_θ/p, and $(N_{\theta\phi})_{max}/p$ against x for $k=2$.

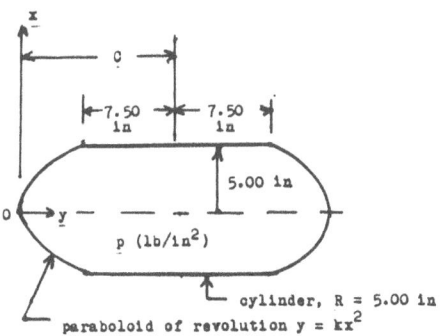

Fig. 4.12. Pressure vessel in Problem 4.2.

4.2. (a) Derive formulas for N_ϕ and N_θ in terms of y for the paraboloid region in Figure 4.12. (b) If $k=0.3$, graph N_ϕ/p and N_θ/p against y for $0 \le y \le C$.

4.3. Solve Example 4.2 for $a=5$ in., $b=7$ in., $c=6$ in., $\nu=0.3$, $E=E_R$. Compare results to Example 4.2.

4.4. Solve Example 4.3 by changing p_r to $p_r=10$ psi. Compare results to Example 4.3.

4.5. Solve Example 4.3 by changing t to $t=0.050$ in.

4.6. Graph the curves u_{r1}, M_{x1}, and V_{r1} in Example 4.4 for $0 \le x \le 4$.

4.7. In Equations (4.46)-(4.48) assume that $u_{rp1} = u_{rp2}$ but that the slope is discontinuous at $x=0$. Solve for the constants C_1-C_4 and get results corresponding to Equation (4.49).

4.8. Solve Example 4.5 for the case of $D_1 = 30,000$ lb-in.

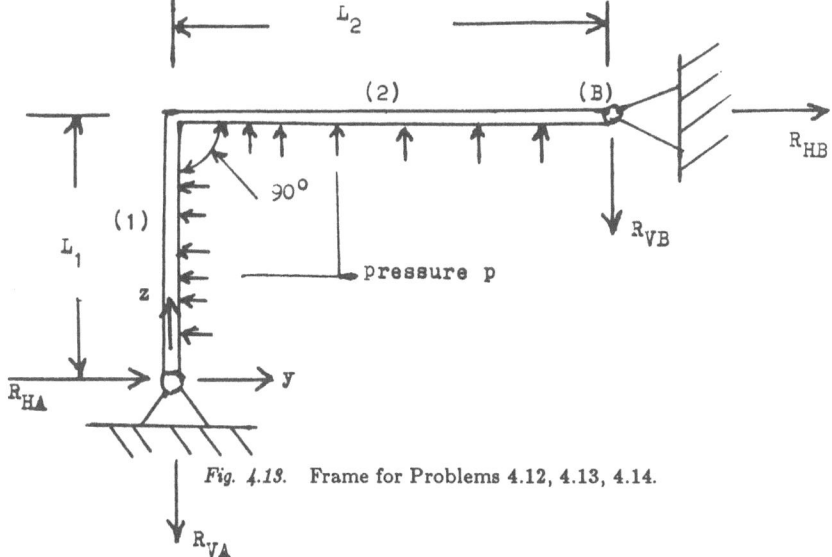

Fig. 4.13. Frame for Problems 4.12, 4.13, 4.14.

4.9. Solve Example 4.6 for the case of $EI/(EI)_R = 10$.

4.10. A fuselage ring has double symmetry with constant EI. Assume $2c = 120$ in. and $b = 50$ in. in Figure 4.7 with the ring being a rectangle except for 90° circular arcs in the corners of radius $R = 20$ in. Use Equations (4.76) and (4.75) to get M_0 for any pressure p.

4.11. Solve Problem 4.10 with $EI = 6.00(EI)_R$ across the flat 60 in. on the bottom of the ring and $EI = (EI)_R$ elsewhere.

4.12. Use the unit load theorem and find the moment equation and reactions for the pressure loaded redundant frame in Figure 4.13. Note that the reaction

$$R_{HA} = \frac{p}{8L_1(L_1 + L_2)}\left[L_1^2(3L_1 + 4L_2) - L_2^3\right]$$

can be positive, zero, or negative depending upon the values of L_1 and L_2.

4.13. In Problem 4.12 find the relation between L_1 and L_2 that makes $R_{HA} = 0$.

4.14. In Problem 4.12 draw moment curves for (a) $L_2 = L_1$, (b) $L_2 = 5L_1$. (c) Comment on the shapes of the bending deflection curves in parts (a) and (b).

4.15. Solve Example 4.7 for a constant shear flow with $w_a = w_b$. Compare results to those in Example 4.7.

4.16. Solve Example 4.7 for the case of $w_b = 2.00w_a$.

4.17. Solve Example 4.7 for constant $EI = (EI)_R$.

4.18. Solve Example 4.8 for the case of $a = 40$ in.

4.19. Solve Example 4.9 for the case of $a = 40$ in.

4.20. Use Equations (4.105) and (4.111) and Figure 4.11 to find h and b for $p = 200$ psi, $M_z = 40,000$ in-lb, $(N_y)_{tu} = 1000$ lb/in.

References

Chapter 4

1. S. Timoshenko and S. Woinowsky-Krieger: *Theory of Plates and Shells*, McGraw-Hill Book Co. (1959).
2. D. Williams: *An Introduction to the Theory of Aircraft Structures*, Edward Arnold (Publishers) LTD., London (1960).
3. R.J. Roark: *Formulas for Stress and Strain*, McGraw-Hill Book Co., Third Ed. (1954).
4. S. Timoshenko and J.N. Goodier: *Theory of Elasticity*, McGraw-Hill Book Co., 2nd Ed. (1951).
5. L.H. Abraham: *Structural Design of Missiles and Spacecraft*, McGraw-Hill Book Co. (1962).
6. E.F. Bruhn: *Analysis and Design of Flight Vehicle Structures*, Tri-State Offset Co., Cincinnati, Ohio (1965).
7. G.E. Maddux, L.A. Vorst, F.J. Giessler and T. Moritz: *Stress Analysis Manual*, AFFDL-Tr-69-42, Wright-Patterson AFB, Ohio 45433 (1970).

5

Approximate solutions using the virtual principles

5.1. Introduction

In Volume 1 and above, trusses, beams, and pressure vessel structures have been analyzed for the "exact stresses and displacements", that is, "exact" in the sense of Saint Venant's Principle and the assumptions described in Sections 1.3 through 1.9 (Vol.1). The finite element solutions for beams in Volume 1 are approximate in that average values are used for variable cross sections. In complicated beams and two and three dimensional structures it is usually necessary to solve the equations approximately. It is also necessary to make approximations in the representations for the various finite elements for two dimensional plates. This chapter is concerned with showing how the virtual principles can be used for various approximate solutions for complicated beam problems. See Chapters 6-9 for other cases of approximate solutions.

The basic procedure in obtaining approximate solutions of differential and integral equations can be outlined by using symbolic notation. Let

$$B(w) = 0, \quad w = w(x,y,z), \tag{5.1}$$

where $B(w)$ is the differential or integral equation in a dependent variable w and its derivatives. Assume that w can be represented by the sum of terms

$$w_n = \sum_{i=1}^{n} a_i F_i(x,y,z), \tag{5.2}$$

where $F_i(x,y,z)$ are a set of known functions and the a_i are unknown constants to be determined so that w_n will satisfy $B(w_n) = 0$ as close as possible.

Various methods for determining the a_i constants have been developed in the last one hundred years. Only the simpler ones will be considered here. The *Method of Least Squares* requires that the square of the error e_n in

$$e_n = B(w_n) \tag{5.3}$$

be a minimum for the entire structure, or

$$\int_V e_n^2 \, dV = \text{minimum, or} \tag{5.4}$$

$$\int_V e_n \frac{\partial e_n}{\partial a_i} \, dV = \int_V B(w_n) \frac{\partial B(w_n)}{\partial a_i} \, dV = 0, \quad i = 1, 2, \cdots, n. \tag{5.5}$$

This equation (5.5) gives n equations to determine the n a_i constants. The drawback to this method is the complicated form of the term $\frac{\partial e_n}{\partial a_i}$.

The *Galerkin Method of Approximation* is much simpler than the least squares method in that it uses the F_i functions directly, or

$$\int_V B(w_n) F_i(x, y, z) \, dV = 0, \quad i = 1, 2, \cdots, n. \tag{5.6}$$

The *Rayleigh-Ritz Method of Approximation* is usually derived from an integral expression for a physical quantity (such as strain energy in structures) by requiring the integral to have a stationary value. However, it can be derived in a more general form from the *Galerkin Method*. The procedure is to integrate Equation (5.6) by parts, reducing the order of the derivatives in $B(w)$ and producing derivatives on the known F_i functions, or for a simple case,

$$0 = \int_a^b \left(\frac{d^2 w_n}{dx^2} + p \right) F_i(x) \, dx = \left[F_i \frac{dw_n}{dx} \right]_a^b - \int_a^b \frac{dw_n}{dx} \frac{dF_i}{dx} \, dx + \int_a^b p F_i \, dx, \text{ or}$$

$$\int_a^b \left[\frac{dw_n}{dx} \frac{dF_i}{dx} - p F_i \right] dx - \theta_{bs} F_i(b) + \theta_a F_i(a) = 0, \tag{5.7}$$

where θ_{bs} is the specified slope at $x = b$. This Rayleigh-Ritz method may be simpler than the Galerkin method not only due to the lower order of the derivatives but also due to the fact that some boundary conditions are included in the approximation so that the F_i do not have to satisfy them explicitly. That is, if $\frac{dw}{dx} = \theta_{bs}$ is specified at point $x = b$ and the F_i do not satisfy this condition, then the values of the a_i given by Equation (5.7) will approximate this condition.

It should be recognized that this derivation of the Rayleigh-Ritz form of approximation is exactly the same as that used in the derivation of the virtual principles in Sections 2.2, 2.4, and 2.6 (Vol.1). If the virtual terms are expressed in terms of the same F_i functions in Equation (5.2), such as

$$w_n^V = \sum_{i=1}^n b_i^V F_i(x, y, z), \tag{5.8}$$

then the virtual principles are already in the Rayleigh-Ritz form for any desired approximations. Note that for approximations in the virtual principles the virtual terms are no longer arbitrary except for the multiplying constants, such as b_i^V in Equation (5.8). Each one of these arbitrary constants produces one of the n equations to be used to determine the a_i constants in Equation (5.2). Note that point values are included in the virtual principles.

Since the virtual principles were derived from a Galerkin form for the equations, Equations (2.1), (2.11), (2.20) (Vol.1), these equations can be used for approximations using the Galerkin method. Although the F_i satisfy all the boundary conditions in the usual application of the Galerkin method, it is possible to approximate the same boundary conditions in the Galerkin method as in the Rayleigh-Ritz method. In Equation (5.7) let $F_i(x)$ satisfy $w = 0$ at $x = a$, or $F_i(a) = 0$, but not $\frac{dw}{dx} = \theta_{b_s}$ at point b. Reverse the integration by parts in Equation (5.7) to get

$$\int_a^b \left(\frac{d^2 w_n}{dx^2} + p\right) F_i \, dx + (\theta_{b_s} - \theta_b) F_i(b) = 0, \tag{5.9}$$

where $\frac{dw}{dx} = \theta_{b_s}$ at $x = b$ is the specified value and from Equation (5.2)

$$\theta_b = \sum_{i=1}^n a_i \frac{dF_i(b)}{dx} \tag{5.10}$$

will be the approximate value for θ_{b_s}.

Both the Rayleigh-Ritz and Galerkin methods will be used in examples in this chapter.

In order to prove convergence of the Rayleigh-Ritz and Galerkin approximation procedures, it is necessary to require that the selected $F_i(x, y)$ functions in the two-dimensional case form relatively complete sets of functions. That is, the $F_i(x, y)$ in Equation (5.2) should satisfy the boundary conditions and be such that for sufficiently large n

$$|w - w_n| < \epsilon, \quad |w_{,x} - w_{n,x}| < \epsilon, \quad |w_{,y} - w_{n,y}| < \epsilon, \tag{5.11}$$

etc., for higher derivatives, for every $\epsilon > 0$ and for all (x, y) in the region R. The trigonometric functions $\sin n\pi x$ and the polynomials x^n are examples of relatively complete sets, provided they are arranged so as to satisfy the boundary conditions. Actually, in the practical use of the virtual principles, it is not necessary that the F_i satisfy all boundary conditions, provided they are such that Equation (5.11) applies for all boundary conditions for sufficiently large n. In fact, if a boundary condition is approximated in the Rayleigh-Ritz and Galerkin procedures, then Equation (5.11) does not have to be satisfied at the boundary condition points. Although the results may be poor at the boundary points, the approximation may be satisfactory elsewhere. Convergence difficulties can arise at applied concentrated loads, point reactions, at points on plate boundaries where displacement and stress boundary conditions join, and in any other situation for which Equation (5.11) does not hold. The selection of the F_i functions and the convergence problems will be considered further for each virtual principle in the following Sections.

Although the Galerkin, Rayleigh-Ritz, and least square approximation methods use only one set of F_i functions, the virtual principles do not require that the virtual functions be expressed in the F_i set. Since the virtual functions can be arbitrary when the F_i functions contain the exact solution, it is possible to use any other relatively complete set of functions G_i for the virtual functions as long as convergence occurs for the F_i set. Whether the convergence is better or worse with a different G_i set depends upon the particular problem. In the discussion of the principle of mixed virtual stresses and virtual displacements below two sets of functions will be used in some cases with one set for the stresses and one set for the displacements. In the approximations using this mixed principle, each set will be used as the virtual functions in the determination of the solution from the other set.

In this chapter the three virtual principles are used to obtain approximate solutions for various beam problems. In Sections 5.2 through 5.5 the principle of virtual displacements is used for constant cross section beams, column buckling, tapered beams with numerical integration, and a finite element procedure for column buckling. Since the principle of virtual forces gives an explicit integral for the displacements through the unit load theorem, only the procedure for numerical integration on a tapered beam is considered in Section 5.6. The mixed virtual principle with assumed functions for both displacements and stresses is considered in Section 5.7.

It should be emphasized that an approximation using the principle of virtual displacements is an approximation on the equilibrium of the structure, an approximation using the principle of virtual forces is an approximation on the compatibility of the structure, and an approximation using the mixed principle is an approximation on both equilibrium and compatibility of the structure. Comparisons among the virtual principles for different types of beam problems will be made throughout this chapter.

5.2. Approximate solutions for beams using the principle of virtual displacements

Consider the simple beam and take the principle of virtual displacements from Equation (4.5, Vol.1) for bending only, or

$$\int_0^L \int_A \sigma_{yy} e_{yy}^V \, dA \, dy - \int_0^L p_z u_z^V \, dy - \sum_{k=1}^M \Big[P_{zk} u_z^V(y_k) +$$

$$+ M_{xk} \frac{du_z^V(y_k)}{dy} \Big] = 0, \tag{5.12}$$

where P_{zk} and M_{xk} are concentrated forces and moments at any points y_k on the beam, including the ends. Take u_z and u_z^V in the form

$$u_z = \sum_{i=1}^n a_i F_i(y), \quad u_z^V = \sum_{i=1}^n b_i^V F_i(y), \tag{5.13}$$

where $F_i(y)$ are a set of known functions. From Equations (4.3, Vol.1), (4.62, Vol.1), and (5.13)

$$\sigma_{yy} = -Ez \sum_{i=1}^n a_i \frac{d^2 F_i}{dy^2} - E e_E, \quad e_{yy}^V = -z \sum_{i=1}^n b_i^V \frac{d^2 F_i}{dy^2}. \tag{5.14}$$

Put Equation (5.14) into Equation (5.12) and make the area integration to get

$$\int_0^L \Big(EI \sum_{i=1}^n a_i \frac{d^2 F_i}{dy^2} - M_{xe} \Big) \Big(\sum_{i=1}^n b_i^V \frac{d^2 F_i}{dy^2} \Big) dy - \int_0^L p_z \Big(\sum_{i=1}^n b_i^V F_i \Big) dy -$$

Approximate solutions using the virtual principles 161

$$-\sum_{k=1}^{M}\left[P_{zk}\sum_{i=1}^{n}b_i^V F_i(y_k) + M_{xk}\sum_{i=1}^{n}b_i^V \frac{dF_i(y_k)}{dy}\right] = 0. \tag{5.15}$$

Since b_i^V is arbitrary, use $F_i' = \frac{dF_i}{dy}$, $F_i'' = \frac{d^2F_i}{dy^2}$, and write the n Rayleigh-Ritz equations from Equation (5.15) for the n a_j unknowns,

$$\int_0^L \left(EI\sum_{j=1}^{n} a_j F_j''\right) F_i'' \, dy - \int_0^L p_z F_i \, dy - \int_0^L M_{xE} F_i'' \, dy - \sum_{k=1}^{M}[P_{zk} F_i(y_k) +$$

$$+ M_{xk} F_i'(y_k)] = 0, \quad i = 1, 2, \cdots, n. \tag{5.16}$$

This Equation (5.16) can be applied to determinate and redundant beams. For the determinate case, it is only necessary to select the $F_i(y)$ functions to satisfy the determinate geometric displacement restraints. For the redundant case, two procedures can be used. Either the $F_i(y)$ functions can be selected to satisfy all the geometric restraints, or the unknown redundant reactions can be included in the P_{zk} and M_{xk} terms and determined from the displacement conditions after the a_i constants have been calculated in terms of them. The latter procedure usually has simpler calculations.

In the cantilever beam of length L with $u_z(0) = 0$ and $u_z'(0) = 0$ at the support, change the y variable to the nondimensional s variable

$$s = y/L \tag{5.17}$$

in order to simplify the calculations, and in Equation (5.13) take

$$F_i(s) = \frac{s^{i+1}}{i(i+1)}, \quad i = 1, 2, \cdots, n. \tag{5.18}$$

This $F_i(s)$ satisfies the support restraints and can approach the moment and shear conditions at $s = 1$ as close as desired, whence Equation (5.16) becomes

$$\int_0^1 (EI/L^3)\left(\sum_{j=1}^{n} a_j s^{j-1}\right)(s^{i-1}) ds - \int_0^1 p_z L \frac{s^{i+1}}{i(i+1)} ds -$$

$$- \int_0^1 (M_{xE}/L) s^{i-1} ds - \sum_{k=1}^{K}\left[P_{zk}\frac{s_k^{i+1}}{i(i+1)} + M_{xk}\frac{s_k^i}{iL}\right] = 0,$$

$$i = 1, 2, \cdots, n. \tag{5.19}$$

If EI and p_z are constant, integration of Equation (5.19) gives

$$\sum_{j=1}^{n} \frac{a_j}{i+j-1} = \frac{p_z L^4}{EI}\frac{1}{i(i+1)(i+2)} + \int_0^1 \frac{M_{xE} L^2}{EI} s^{i-1} ds +$$

$$+ \sum_{k=1}^{M}\left[\frac{P_{zk} L^3}{EI}\frac{s_k^{i+1}}{i(i+1)} + \frac{M_{xk} L^2}{EI}\frac{s_k^i}{i}\right], \quad i = 1, 2, \cdots, n, \tag{5.20}$$

or in matrix form,

$$[A]\{a\} = \{Q\}, \quad \{a\} = [A]^{-1}\{Q\}, \tag{5.21}$$

$$[A] = \begin{bmatrix} 1 & 1/2 & 1/3 & \cdots & 1/n \\ 1/2 & 1/3 & 1/4 & \cdots & 1/(n+1) \\ 1/3 & 1/4 & 1/5 & \cdots & 1/(n+2) \\ \cdots & & & & \\ 1/n & 1/(n+1) & 1/(n+2) & \cdots & 1/(2n-1) \end{bmatrix}. \tag{5.22}$$

This matrix $[A]$ is the Hilbert matrix, for which the inverse is known in terms of a formula for every element (Reference 1). If a_{ij} is any element of the n by n Hilbert matrix, then the corresponding element Ia_{ij} in the inverse is

$$Ia_{ij} = \frac{(-1)^{i+j}(n+i-1)!(n+j-1)!}{(i+j-1)[(i-1)!(j-1)!]^2(n-i)!(n-j)!}, \tag{5.23}$$

where $n!$ means the product $(1)(2)(3)\cdots(n)$ with $0! = 1$. Thus, for any given n, the inverse can be written down. It should be noted that the Hilbert matrix is ill-conditioned so that the elements in Equation (5.23) become very large for n large. If $n = 20$, then the largest element is $Ia_{15,15} = 3.6(10^{27})$, with the numerator in Equation (5.23) approximately 10^{80}. Obviously, digital computers have problems with the calculations for n large.

Example 5.1. (a) Find the approximate elastic solution for the deflections, moments, and shear forces for the beam in Figure 5.1, using four constants in Equation (5.20). (b) Find the exact solution and compare the two solutions.

Solution. (a) From Figure 5.1, and Equations (5.20)-(5.23) with $s_1 = y_1/L$,

$$\begin{Bmatrix} a_1 \\ a_2 \\ a_3 \\ a_4 \end{Bmatrix} = \begin{bmatrix} 16 & -120 & 240 & -140 \\ -120 & 1200 & -2700 & 1680 \\ 240 & -2700 & 6480 & -4200 \\ -140 & 1680 & -4200 & 2800 \end{bmatrix} \left[\frac{p_xL^4}{120EI} \begin{Bmatrix} 20 \\ 5 \\ 2 \\ 1 \end{Bmatrix} + \frac{P_LL^3}{60EI} \begin{Bmatrix} 30 \\ 10 \\ 5 \\ 3 \end{Bmatrix} + \right.$$

$$\left. + \frac{P_1L^3}{60EI} \begin{Bmatrix} 30s_1^2 \\ 10s_1^3 \\ 5s_1^4 \\ 3s_1^5 \end{Bmatrix} + \frac{(M_L + M_{T0})L^2}{12EI} \begin{Bmatrix} 12 \\ 6 \\ 4 \\ 3 \end{Bmatrix} + \frac{BL^4}{60EI} \begin{Bmatrix} 20 \\ 15 \\ 12 \\ 10 \end{Bmatrix} \right],$$

$$\begin{aligned} EIa_1 &= \tfrac{1}{2}p_xL^4 + P_LL^3 + P_1L^3s_1^2(8 - 20s_1 + 20s_1^2 - 7s_1^3) + (M_L + M_{T0})L^2, \\ EIa_2 &= -p_xL^4 - P_LL^3 - P_1L^3s_1^2(60 - 200s_1 + 225s_1^2 - 84s_1^3), \\ EIa_3 &= \tfrac{1}{2}p_xL^4 + 30P_1L^3s_1^2(4 - 15s_1 + 18s_1^2 - 7s_1^3) + BL^4, \\ EIa_4 &= \phantom{\tfrac{1}{2}p_xL^4 + } -70P_1L^3s_1^2(1 - 4s_1 + 5s_1^2 - 2s_1^3). \end{aligned} \tag{5.24}$$

From Equations (5.13) and (5.18),

$$(u_z)_{\text{app}} = (a_1/2)s^2 + (a_2/6)s^3 + (a_3/12)s^4 + (a_4/20)s^5, \tag{5.25}$$

$$\begin{aligned} (M_z)_{\text{app}} &= EI(u_z'')_{\text{app}} = EI(a_1 + a_2s + a_3s^2 + a_4s^3), \\ (V_z)_{\text{app}} &= EI(u_z''')_{\text{app}} = EI(a_2 + 2a_3s + 3a_4s^2). \end{aligned} \tag{5.26}$$

(b) The exact solution for the deflection u_z of the beam in Figure 5.1 can be obtained from integration of Equations (4.6) and (4.7) (Vol.1) where M_z is the moment equation for all the

Approximate solutions using the virtual principles 163

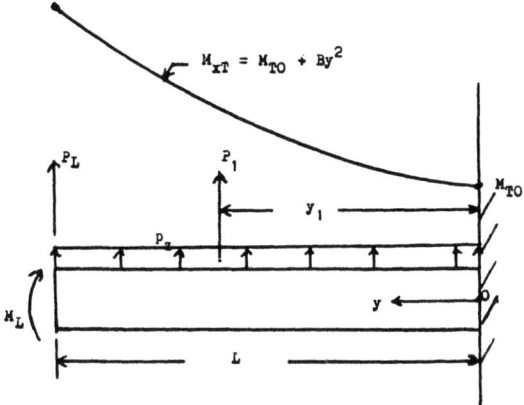

Fig. 5.1. Cantilever beam loading for Example 5.1.

loads on the beam, including M_{xT}. The result is

$$(u_z)_{ex} = \frac{p_z L^4 s^2}{24EI}(6 - 4s + s^2) + \frac{(M_L + M_{T0})L^2 s^2}{2EI} + \frac{P_L L^3 s^2}{6EI}(3 - s) + \frac{BL^4 s^4}{12EI} + K(s)P_1, \quad (5.27)$$

$$K(s) = (L^3 s^2/6EI)(3s_1 - s), \quad 0 \le s \le s_1,$$
$$K(s) = (L^3 s_1^2/6EI)(3s - s_1), \quad s_1 \le s \le 1. \quad (5.28)$$

From the a_i values in Equation (5.24), the approximate solution is exact for all loads except P_1, and agrees with the values in Equation (5.28). The comparison for the P_1 load case depends upon s_1. For $s_1 = 1/3$,

$$(u_z)_{ex} = (P_1 L^3 s^2/6EI)(1 - s), \quad 0 \le s \le 1/3,$$
$$= (P_1 L^3/162EI)(9s - 1), \quad 1/3 \le s \le 1, \quad (5.29)$$

$$(M_z)_{ex} = (P_1 L/3)(1 - 3s), \quad 0 \le s \le 1/3,$$
$$= 0, \quad 1/3 \le s \le 1, \quad (5.30)$$

$$(V_z)_{ex} = P_1, \quad 0 \le s \le 1/3, \quad (V_z)_{ex} = 0, \quad 1/3 \le s \le 1, \quad (5.31)$$

$$(u_z)_{app} = (P_1 L^3 s^2/EI)(0.1831 - 0.2819s + 0.2058s^2 - 0.0576s^3), \quad (5.32)$$

$$(M_z)_{app} = P_1 L(0.3662 - 1.6914s + 2.4696s^2 - 1.1523s^3), \quad (5.33)$$

$$(V_z)_{app} = P_1(1.6914 - 4.9392s + 3.4569s^2). \quad (5.34)$$

The comparison in this case of four constants for the P_1 load, in which the shear is discontinuous at $s = 1/3$, is shown in Figure 5.2 for the deflection u_z, moment M_z, and shear V_z. The exact and approximate deflection cannot be separated on the graph. The approximate moment follows the exact moment quite well. The approximate shear is very poor, being 69% larger than the exact shear at $y = 0$. More terms are needed to get a better approximation for the shear curve. The convergence is slow and will give $(V_z)_{app} = 0.50 P_1$ on the vertical line at $s = 1/3$. The representation at the corners of the jump never becomes exact even for a very large number of terms.

Example 5.2. Make the beam in Figure 5.1 and Example 5.1 redundant by putting a support at $y_1 = L/3$ with $u_z = 0$. (a) Assume the constant distributed p_z load is acting alone and

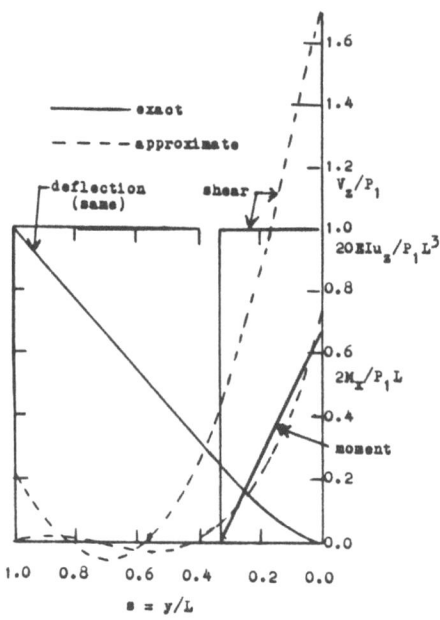

Fig. 5.2. Exact and approximate deflection, moment, and shear on beam.

determine the value of the reaction $R_1 = P_1$ from the results in Example 5.1. (b) Compare the results to the exact case.

Solution. (a) From Equations (5.24), (5.25), and (5.32),

$$24EI(u_z)_{\text{app}} = 0 = p_z L^4 (\tfrac{1}{3})^2 (6 - \tfrac{4}{3} + \tfrac{1}{9}) + 24 P_1 L^3 (\tfrac{1}{3})^2 (0.1831 -$$
$$- 0.2819/3 + 0.2058/9 - 0.0576/27), \text{ or}$$

$$(P_1)_{\text{app}} = -1.8120 p_z L = (R_1)_{\text{app}}. \tag{5.35}$$

(b) With this R_1 value, the approximate deflection for the redundant beam is

$$EI(u_z)_{\text{app}} = p_z L^4 s^2 (-0.0818 + 0.3441 s - 0.3312 s^2 + 0.1044 s^3). \tag{5.36}$$

From Equation (5.27), the exact values of $R_1 = P_1$ for $u_z(1/3) = 0$ and of u_z are

$$(R_1)_{ex} = -1.79167 p_z L, \tag{5.37}$$

$$(u_z)_{ex} = (p_z L^4 / 24 EI)(-1.1667 s^2 + 3.1667 s^3 + s^4), \quad 0 \le s \le \tfrac{1}{3},$$
$$= (p_z L^4 / 24 EI)(0.2654 - 2.3889 s + 6 s^2 - 4 s^3 + s^4), \quad 1/3 \le s \le 1. \tag{5.38}$$

The exact and approximate deflections are very close, while the moment comparison is good but the shear comparison is poor, similar to Figure 5.2. More terms give better results.

Example 5.3. (a) Find the approximate deflection u_z and stress σ_{yy} in a simply supported beam with K concentrated loads P_k at location y_k and any distributed load p_z. Use $F_i = \sin(i\pi y/L)$ in Equations (5.13) and (5.16). (b) Discuss results.

Solution. (a) The F_i functions satisfy $u_z(0) = 0$, $u_z(L) = 0$, $M_z(0) = 0$, and $M_z(L) = 0$ for the simply supported beam of length L, and Equation (5.16) becomes

$$\int_0^L \left[EI \sum_j a_j \left(\frac{j\pi}{L}\right)^2 \left(-\sin\frac{j\pi y}{L}\right) \right] \left(\frac{i\pi}{L}\right)^2 \left(-\sin\frac{i\pi y}{L}\right) dy -$$

$$-\int_0^L p_z \sin\frac{i\pi y}{L}dy - \sum_{k=1}^{K} P_k \sin\frac{i\pi y_k}{L} = 0.$$

Since the $\sin(i\pi y/L)$ functions are orthogonal, all the integrals are zero except for $j = i$, or

$$a_i = (2L^3/i^4\pi^4 EI)\left[\int_0^L p_z \sin(i\pi y/L)dy + \sum_{k=1}^{K} P_k \sin(i\pi y_k/L)\right]. \tag{5.39}$$

From Equations (5.13) and (5.14),

$$u_z = (2L^3/\pi^4 EI)\sum_{i=1}^{\infty}\left[\frac{\sin(i\pi y/L)}{i^4}\right]\left[\int_0^L p_z \sin(i\pi y/L)dy + \sum_{k=1}^{K} P_k \sin(i\pi y_k/L)\right], \tag{5.40}$$

$$\sigma_{yy} = (2Lz/\pi^2 I)\sum_{i=1}^{\infty}\left[\frac{\sin(i\pi y/L)}{i^2}\right]\left[\int_0^L p_z \sin(i\pi y/L)dy + \sum_{k=1}^{K} P_k \sin(i\pi y_k/L)\right]. \tag{5.41}$$

(b) The i^4 term in the denominator makes the u_z series converge rapidly, while the i^2 term makes the σ_{yy} series converge more slowly. However, more terms can be used for σ_{yy} to get any desired accuracy by comparing to the magnitude of the last term at points of large stress. The shear force with i in the denominator will take still more terms, particularly at points close to the P_k loads. If p_z is complicated and there are several P_k loads, the exact solution is quite involved and difficult to obtain from Equation (4.6, Vol.1). It has a different form in each interval between the point loads for both u_z and σ_{yy}. Also, four numerical integrations might be needed for u_z in Equation (4.6, Vol.1), while only one numerical integration might be needed in Equation (5.40).

If p_z is constnat and all $P_k = 0$, then one term in Equations (5.40) and (5.41) gives $(u_z)_{\text{app}} = 0.996(u_z)_{ex}$ at $y = L/2$, $(\sigma_{yy})_{\text{app}} = 1.032(\sigma_{yy})_{ex}$ at $y = L/2$. For two terms, $(\sigma_{yy})_{\text{app}} = 0.994(\sigma_{yy})_{ex}$ at $y = L/2$.

If the beam has additional redundant supports, P_k loads in Equation (5.40) can be taken as unknown reactions R_k at the supports. With $u_z = 0$ or any specified value at the supports, the R_k reactions can be obtained from Equation (5.40) by using a few terms in the sum.

Example 5.4. (a) Consider only the p_z constant distributed load in Figure 5.1 and solve for u_z using only two terms in Example 5.1, or $u_z = (a_1/2)s^2 + (a_2/6)s^3$. (b) Use the constant a_1 to make $M_z(L) = 0$ and resolve for u_z. (c) Compare the results to each other and to the exact result in Equation (5.27).

Solution. (a) From Equation (5.20) for two terms,

$$\begin{bmatrix} 1 & 1/2 \\ 1/2 & 1/3 \end{bmatrix}\begin{Bmatrix} a_1 \\ a_2 \end{Bmatrix} = \frac{p_z L^4}{24EI}\begin{Bmatrix} 4 \\ 1 \end{Bmatrix},$$

whence $a_1 = \frac{5}{12}\frac{p_z L^4}{EI}$, $a_2 = -\frac{p_z L^4}{2EI}$,

$$(u_z)_{\text{app}} = \frac{p_z L^4 s^2}{24EI}(5 - 2s). \tag{5.42}$$

(b) From Equation (5.26), $a_1 = -a_2$ in order to make $M_z(1) = 0$, whence $u_z = (a_2/6)s^2(s-3)$. Use this in Equation (5.16) to get

$$(u_z)_{\text{app}} = \frac{3p_z L^4 s^2}{48EI}(3-s). \tag{5.43}$$

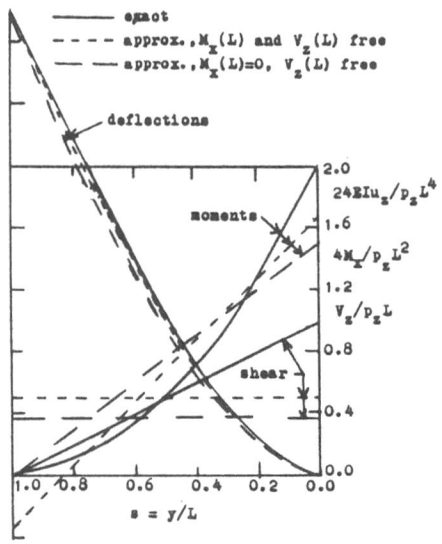

Fig. 5.3. Comparison of exact and approximate displacements, moments, and shear on cantilever beam, Example 5.4.

(c) Figure 5.3 shows the displacement, moment, and shear curves for the exact case in Equation (5.27) and the two approximate cases in Equations (5.42) and (5.43). The deflections are close for all three cases with the case of enforced $M_z(L) = 0$ having the largest deviation. The moment and shear curves do not agree very well. The regular approximate case of Equation (5.42) shows that the moment and shear curves approximate the exact curves and do not satisfy the force boundary conditions. This is to be expected since the principle of virtual displacements is the equilibrium equation for both the interior and the surface of the structure. Thus, an approximate solution obtained by using it will approximate the moment and shear equilibrium in the beam and at the ends to give the best results for the displacements. The case of Equation (5.43), which restricts the equilibrium approximation by requiring $M_z(L) = 0$, gives the worse deflection, moment and shear curves. Of course, if a sufficient number of terms is used to get accurate shear and moment curves, it makes no difference whether they are restricted in the calculations.

Equations (5.44)–(5.52) were deleted.

5.3. Approximate solutions for columns

Add the expression in Equation (4.147, Vol.1) to Equation (5.12) so that Equation (5.16) becomes

$$\int_0^L \left[EI \sum_{j=1}^n a_j F_j'' \right) F_i'' - \left(P_c \sum_{j=1}^n a_j F_j' \right) F_i' - p_z F_i - M_{xE} F_i'' \right] dy -$$
$$- \sum_{k=1}^M [P_{zk} F_i(y_k) + M_{xk} F_i'(y_k)] = 0, \quad i = 1, 2, \cdots, n. \quad (5.53)$$

Approximate solutions using the virtual principles 167

or for the buckling case alone,

$$\int_0^L \left[\left(EI\sum_{j=1}^n a_j F_j''\right) F_i'' - \left(P_c \sum_{j=1}^n a_j F_j'\right) F_i'\right] dy = 0, \quad i = 1, \cdots, n. \quad (5.54)$$

Since M_{xc} is given by Equation (4.136, Vol.1), or $M_{xc} = P_c(u_{zL} - u_z)$ for cantilever beam, and since an assumed u_z is more accurate than the corresponding u_z'', use $u_z'' = M_x/EI$ and rewrite Equation (5.54) in the form

$$\int_0^L \left[(P_c^2/EI)\sum_{j=1}^n a_j\{F_j(L) - F_j(y)\}\{F_i(L) - F_i(y)\} - \right.$$

$$\left. - \left(P_c \sum_{j=1}^n a_j F_j'\right) F_i'\right] dy = 0, \quad i = 1, 2, \cdots, n. \quad (5.55)$$

This Equation (5.55) corresponds to the classical conservation of energy equation for buckling. See Section 39 in Reference 5.

Both Equations (5.54) and (5.55) can be written in the matrix form

$$[H[A] - [B]]\{a\} = 0, \quad (5.56)$$

where $H = P_c L^2/EI$. To simplify Equation (5.56) multiply by $[A]^{-1}$ to get

$$[H[I] - [A]^{-1}[B]]\{a\} = 0. \quad (5.57)$$

The critical values of H are determined by putting the determinant of the matrix equal to zero, or

$$|H[I] - [A]^{-1}[B]| = 0. \quad (5.58)$$

The smallest value of H gives the buckling load as

$$P_{cr} = H_{\min}(EI/L^2). \quad (5.59)$$

Example 5.5. For the cantilever beam in Figure 5.4 assume the one term approximate solution as

$$u_z = a_1 s^2(3 - s), \quad (5.60)$$

which satisfies the boundary conditions $u_z(0) = u_z'(0) = u_z''(L) = 0$. Find the approximate critical buckling load P_{cr} by (a) Equation (5.54) and (b) Equation (5.55). (c) Find a_1 for a transverse load p_x in Equation (5.53).

Solution. (a) From Equation (5.60)

$$u_{z,y} = a_1 3s(2-s)/L, \quad u_{z,yy} = 6a_1(1-s)/L^2,$$

whence the a_1 terms in Equation (5.54) become

$$EIa_1\left[\int_0^1 \{36(1-s)^2/L^4 - 9s^2(2-s)^2 P_c/EI\} ds\right] = 0, \text{ or}$$

$$12/L^4 - 24P_c/5EIL^2 = 0, \quad P_{cr} = 5EI/2L^2 = 2.5000 EI/L^2. \quad (5.61)$$

This compares quite well with the exact solution in Equation (4.143, Vol.1).

(b) Equation (5.55) becomes

$$a_1\int_0^1 \left[P_c^2(2 - 3s^2 + s^3)^2/EI - 9s^2 P_c(2-s)^2/L^2\right] ds = 0, \text{ or}$$

$$68P_c/35EI - 24/5L^2 = 0, \quad P_{cr} = 42EI/17L^2 = 2.4706 EI/L^2, \quad (5.62)$$

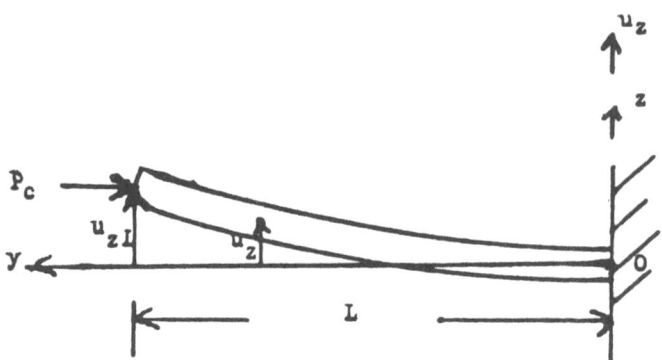

Fig. 5.4. Column loading and deflection.

which is essentially the same as Equation (4.143, Vol.1) and better than Equation (5.61).

(c) Let G_1 represent any transverse applied load terms in Equation (5.54). For a P_c load less than P_{cr}, Equation (5.61) becomes

$$EIa_1(12/L^4)[1 - (P_c/P_{cr})] = G_1, \quad \text{or} \quad a_1 = \frac{L^4 G_1}{12EI[1 - (P_c/P_{cr})]},$$

$$u_z = a_1 s^2(3 - s). \tag{5.63}$$

Thus, the deflection increases rapidly as P_c tends toward P_{cr}.

If the lateral deflection u_{zp} of the beam is known without the P_c load, then, as shown for various cases in Chapter 1 of Reference 2, the form of Equation (5.63) can be generalized to give the approximate deflection u_{zc} with P_c present in the form

$$u_{zc} = u_{zp}/[1 - (P_c/P_{cr})]. \tag{5.64}$$

Note that if $P_c = -P_t$ for a tension load, the deflection is reduced by the tension load.

Example 5.6. For the simply supported beam with P_c compression load acting, assume the one term approximate solution

$$u_z = a_1 s(1 - s), \tag{5.65}$$

which satisfies only the boundary conditons $u_z(0) = u_z(1) = 0$. Find the critical buckling load by (a) Equation (5.54) and (b) Equation (5.55).

Solution. (a) From Equation (5.65)

$$u_{z,y} = a_1(1 - 2s)/L, \quad u_{z,yy} = -2a_1/L^2,$$

whence the a_1 terms in Equation (5.54) give

$$a_1(4EI/L^2 - P_c/3L^2) = 0, \quad P_{cr} = 12EI/L^2, \tag{5.66}$$

which does not agree very well with $P_{cr} = 9.8696EI/L^2$ in Equation (4.141, Vol.1).

Approximate solutions using the virtual principles 169

(b) Equation (5.55) gives

$$a_1(P_c^2/30EI - P_c/3L^2) = 0, \qquad P_{cr} = 10EI/L^2, \tag{5.67}$$

which is much closer to P_{cr} in Equation (4.141, Vol.1) than in Equation (5.66).

Example 5.7.   Use two terms from Equation (5.18) for the cantilever beam and find the buckling load from Equation (5.54).
Solution. From Equation (5.18)

$$u_x = a_1 s^2/2 + a_2 s^3/6, \quad Lu'_x = a_1 s + a_2 s^2/2, \tag{5.68}$$
$$L^2 u''_x = a_1 + a_2 s,$$

and the two equations in Equation (5.54) involving only a_1 and a_2 are

$$\int_0^1 [(EI/L^4)(a_1 + a_2 s)(1) - (P_c/L^2)\{a_1 s + a_2(s^2/2)\}(s)]ds = 0,$$

$$\int_0^1 [(EI/L^4)(a_1 + a_2 s)(s) - (P_c/L^2)\{a_1 s + a_2(s^2/2)\}(s^2/2)]ds = 0.$$

This gives the matrices in Equation (5.56) as

$$[A] = \begin{bmatrix} 1/3 & 1/8 \\ 1/8 & 1/20 \end{bmatrix}, \quad [B] = \begin{bmatrix} 1 & 1/2 \\ 1/2 & 1/3 \end{bmatrix},$$

whence Equation (5.58) becomes

$$\left| \begin{bmatrix} H & 0 \\ 0 & H \end{bmatrix} - \begin{bmatrix} -12 & -16 \\ 40 & 140/3 \end{bmatrix} \right| = 0,$$
$$H^2 - (104/3)H + 80 = 0, \quad H = P_c L^2/EI = 2.4860, \quad 32.1807. \tag{5.69}$$

The minimum value compares very well to 2.4674 in Equation (4.143, Vol.1) but the larger value is far from the second exact value of $9\pi^2/4 = 22.21$. Although the u_x in Equation (5.68) has the same terms s^2 and s^3 as in Equation (5.60) and does not satisfy $M_{xc} = 0$ at $y = L$, the two constants give a better result than the one constant for Equation (5.61).

5.4. The tapered cantilever beam with numerical integration

Whenever EI is variable in Equation (5.16) it may be necessary to evaluate all the integrals involving EI numerically. Fix i and j and consider one EI term in Equation (5.16), or with $s = y/L$,

$$D_{ij} = (1/L^3) \int_0^1 a_j EI(s) F''_i(s) F''_j(s) ds. \tag{5.70}$$

Divide the beam into $K - 1$ subintervals with K station points, including end points. Let $g_k = (g\Delta s)_k$ be the weighting numbers, which depend upon the method of numerical integration. See Equations (A.81)-(A.88) in Volume 1. Thus, Equation (5.70) can be written as

$$D_{ij} = (a_j/L^3) \sum_{k=1}^{K} EI(s_k) F''_i(s_k) F''_j(s_k) g_k$$
$$= (1/L^3)\{F''_i\}^T [EI]_d [g]_d \{F''_i\} a_j. \tag{5.71}$$

Take the sum of n terms on j in Equations (5.16) and (5.71) to get

$$D_i = (1/L^3)\{F_i''\}^T[EI]_d[g]_d[F'']\{a\}. \tag{5.72}$$

This form applies for $i = 1, 2, \cdots, n$ so that the numerical matrix form of Equation (5.16) is

$$(1/L^3)[F'']^T[EI]_d[g]_d[F'']\{a\} = L[F]^T[g]_d\{p_z\} + [F]^T\{P_z\} +$$
$$+ (1/L)[F'']^T[g]_d\{M_{xE}\}, \tag{5.73}$$

$$[B]\{a\} = \{Q\}, \quad \{a\} = [B]^{-1}\{Q\}, \tag{5.74}$$

$$[B] = (1/L^3)[F'']^T[EI]_d[g]_d[F''], \tag{5.75}$$

$$\{Q\} = L[F]^T[g]_d\{p_z\} + (1/L)[F'']^T[g]_d\{M_{xE}\} + [F]^T\{P_z\}. \tag{5.76}$$

With the a_j constants known, the deflections, moments, and shear forces can be obtained at the same points used in the numerical integration, or at any other selected points. From Equation (5.13)

$$u_z(y_i) = \sum_{j=1}^{n} a_j F_j(y_i), \quad i = 1, 2, \cdots, K, \tag{5.77}$$

$$\{u_z\} = [F]\{a\}, \tag{5.78}$$

$$M_x = EI u_z'' = EI \sum_{j=1}^{n} a_j F_j''(y), \tag{5.79}$$

$$M_x(y_i) = EI(y_i) \sum_{j=1}^{n} a_j F_j''(y_i), \tag{5.80}$$

$$\{M_x\} = [EI]_d[F'']\{a\}, \tag{5.81}$$

$$V_z = (EI u_z'')' = EI u_z''' + (EI)' u_z'', \tag{5.82}$$

$$\{V_z\} = [[EI]_d[F'''] + [EI]_d'[F'']]\{a\}. \tag{5.83}$$

From Equation (4.28) of Reference 3 the set of functions

$$F_j(s) = (1/6)[(j+1)(j+3)s^{j+1} - 2j(j+3)s^{j+2} + j(j+1)s^{j+3}],$$
$$j = 1, 2, \cdots \tag{5.84}$$

satisfy the four boundary conditions for the cantilever beam, and can be used for the tapered cantilever beams in the above equations.

Example 5.8. Use the first two functions in Equation (5.84) to calculate the approximate deflections for the cantilever beam in Figure 5.5 by (a) direct integration and (b) numerical integration. (c) Compare to the exact values for the p_x load.

Solution. (a) The variable moment of inertia of the beam in Figure 5.5 has the form

$$EI = (EI)_0(1 - 0.8s)^3, \quad (EI)_0 = Ebh_0^3/12. \tag{5.85}$$

For the two assumed functions

$$u_z = a_1(s^2/3)(6 - 4s + s^2) + a_2(s^3/3)(10 - 10s + 3s^2),$$
$$u_z'' = 4a_1(1-s)^2 + 20a_2 s(1-s)^2. \tag{5.86}$$

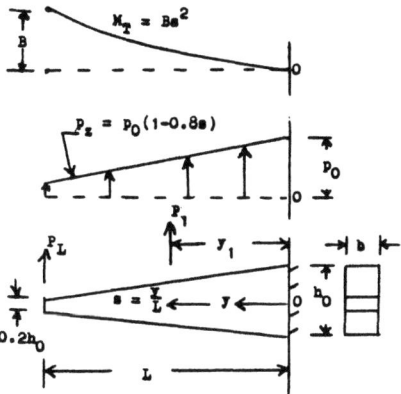

Fig. 5.5. Tapered cantilever beam.

Use Figure 5.5 and Equation (5.16) to get

$$16a_1 \int_0^1 (1-0.8s)^3(1-s)^4 ds + 80a_2 \int_0^1 (1-0.8s)^3 s(1-s)^4 ds = \frac{L^3 Q_1}{(EI)_0},$$

$$80a_1 \int_0^1 s(1-0.8s)^3(1-s)^4 ds + 400a_2 \int_0^1 s^2(1-0.8s)^3(1-s)^4 ds = \frac{L^3 Q_2}{(EI)_0},$$

$$Q_1 = (p_0 L/3) \int_0^1 (1-0.8s)(6s^2 - 4s^3 + s^4) ds + P_L(1) +$$

$$+ (4B/L) \int_0^1 s^2(1-s)^2 ds + (P_1 s_1^2/3)(6 - 4s_1 + s_1^2) \qquad (5.87)$$

$$= 0.1689 p_0 L + P_L + 2B/15L + (P_1 s_1^2/3)(6 - 4s_1 + s_1^2),$$

$$Q_2 = (p_0 L/3) \int_0^1 (1-0.8s)(10s^3 - 10s^4 + 3s^5) ds + P_L(1) +$$

$$+ (20B/L) \int_0^1 s^3(1-s)^2 ds + (P_1 s_1^3/3)(10 - 10s_1 + 3s_1^2)$$

$$= 0.1302 p_0 L + P_L + B/3L + (P_1 s_1^3/3)(10 - 10s_1 + 3s_1^2).$$

Make the integrations to get

$$2.1833 a_1 + 1.3217 a_2 = L^3 Q_1/(EI)_0,$$
$$1.3217 a_1 + 1.4325 a_2 = L^3 Q_2/(EI)_0, \qquad (5.88)$$

$$(EI)_0 L^{-3} a_1 = 1.0329 Q_1 - 0.9497 Q_2 = 0.0508 p_0 L + 0.0832 P_L -$$
$$- 0.1788 B/L + s_1^2(2.07 - 4.54 s_1 + 3.51 s_1^2 - 0.95 s_1^3) P_1,$$

$$(EI)_0 L^{-3} a_2 = -0.9497 Q_1 + 1.5688 Q_2 = 0.0439 p_0 L + 0.6191 P_L +$$
$$+ 0.3963 B/L + s_1^2(-1.90 + 6.45 s_1 - 5.55 s_1^2 + 1.57 s_1^3) P_1. \qquad (5.89)$$

(b) For numerical integration on the integrals multiplying a_1 and a_2 in Equation (5.87) divide the beam into 10 equal intervals with 11 points, including the end points. The point values for various terms are given in Table 5.1. Put the values into matrix form in Equations

Table 5.1. Point values for numerical integration, Figure 5.5, Equation (5.84).

s	$F_1(s)$	$F_2(s)$	$L^2 F_1''(s)$	$L^2 F_2''(s)$	$(EI)_d/(EI)_0$	$3g_d/0.10L$
0.0	0.0000	0.0000	4.0000	0.0000	1.0000	1
0.1	0.0187	0.0030	3.2400	1.6200	0.7787	4
0.2	0.0699	0.0217	2.5600	2.5600	0.5927	2
0.3	0.1467	0.0654	1.9600	2.9400	0.4389	4
0.4	0.2423	0.1382	1.4400	2.8800	0.3144	2
0.5	0.3542	0.2396	1.0000	2.5000	0.2160	4
0.6	0.4752	0.3658	0.6400	1.9200	0.1406	2
0.7	0.6027	0.5111	0.3600	1.2600	0.0852	4
0.8	0.7339	0.6690	0.1600	0.6400	0.0466	2
0.9	0.8667	0.8335	0.0400	0.1800	0.0219	4
1.0	1.0000	1.0000	0.0000	0.0000	0.0080	1

(5.75) and (5.74) to get

$$[B] = (EI)_0 L^{-3} \begin{bmatrix} 2.1847 & 1.3171 \\ 1.3171 & 1.4451 \end{bmatrix},$$

$$\begin{Bmatrix} a_1 \\ a_2 \end{Bmatrix} = L^3/(EI)_0 \begin{bmatrix} 1.0160 & -0.9260 \\ -0.9260 & 1.5360 \end{bmatrix} \begin{Bmatrix} Q_1 \\ Q_2 \end{Bmatrix}, \quad (5.90)$$

which compares quite well with Equations (5.88) and (5.89) in part (a). For example, for the p_x load the tip deflection is

$$\begin{aligned} u_x(1) &= 0.0947 p_0 L^4/(EI)_0 \quad \text{from Equation (5.89)}, \\ u_x(1) &= 0.0946 p_0 L^4/(EI)_0 \quad \text{from Equation (5.90)}. \end{aligned} \quad (5.91)$$

(c) For the given p_x load case and the particular EI in Equation (5.85), the differential Equation (4.6, Vol.1) can be integrated directly to get u_x, or

$$u_x = p_0 L^4 (EI)_0^{-1} \Big[0.1302 s^2 + 0.03646 s + \frac{0.003255}{1 - 0.08 s} \\ - 0.0488 \ln(1 - 0.8s) - 0.003255 \Big], \quad (5.92)$$

$$u_x(1) = 0.1011 p_0 L^4/(EI)_0. \quad (5.93)$$

This value of $u_x(1)$ for the tip is about 7% larger than the results in Equation (5.91). The two function approximate solution in parts (a) and (b) cannot represent the rapid changes in the deflection near the tip.

Example 5.9. Use

$$EI = (EI)_0 (1 - 0.5s)^3 \quad (5.94)$$

for the cantilever beam and calculate the P_{cr} buckling load using the one approximating function in Equation (5.60) and part (b) of Example 5.5.

Solution. Put the specified EI into Equation (5.62) and make the numerical integration using ten equal elements on the beam and Simpson's rule, or

$$\int_0^1 \frac{8(2 - 3s^2 + s^3)^2}{(2-s)^3} ds = 8(0.1/3)[(1)(0.5000) + 4(0.5664) + 2(0.6112) + 4(0.6289) + $$
$$+ 2(0.6226) + 4(0.5602) + 2(0.4703) + 4(0.3469) + 2(0.2028) + 4(0.0672) + 1(0.0000)]$$
$$= 3.4584,$$

whence

$$P_{cr} = 1.388 (EI)_0/L^2.$$

This compares to $P_{cr} = 1.325(EI)_0/L^2$ from page 137 of Reference 2.

5.5. Tapered beam finite element matrices for columns

From the previous Section 5.4 it is evident that to get the buckling loads and deflections of a simple tapered beam requires the selection of a sufficient number of good functions and numerical integration. In many cases it may be better to divide the beam into elements and use the matrix procedures of Sections 4.3 and 4.8 (Vol.1).

To allow for the P_c axial compression load it is necessary to add a geometric stiffness matrix $[k_G]$ to the stiffness matrix $[k]$ in Equation (4.91, Vol.1). Differentiate Equation (4.15, Vol.1) and put into Equation (4.145, Vol.1) to get

$$\sigma_{yzC} = -P_C u_{z,y}/A_s = -(P_C/LA_s)[6s(1-s)u_{zL} + s(3s-2)L\theta_{zL} \\ - 6s(1-s)u_{z0} + (1-s)(1-3s)L\theta_{z0}]. \tag{5.95}$$

Put σ_{yzC} into Equation (2.71, Vol.1) with the order $u_{zL}, \theta_{zL}, u_{z0}, \theta_{z0}$ to get

$$[k_G] = -\frac{P_C}{L} \int_0^1 \begin{Bmatrix} 6s(1-s) \\ Ls(3s-2) \\ -6s(1-s) \\ L(1-s)(1-3s) \end{Bmatrix} [6s(1-s) \quad Ls(3s-2)$$
$$- 6s(1-.2) \quad L(1-s)(1-3s)]ds$$
$$= -\frac{P_C}{30L} \begin{bmatrix} 36 & -3L & -36 & -3L \\ -3L & 4L^2 & 3L & -L^2 \\ -36 & 3L & 36 & 3L \\ -3L & -L^2 & 3L & 4L^2 \end{bmatrix}. \tag{5.96}$$

whence Equation (4.91, Vol.1) takes the form

$$\{P\}_1 = [[k]_1 + [k_G]_1]\{u\}_1. \tag{5.97}$$

This $[k_G]$ matrix checks the $[k_G]$ matrix derived in Section 15.3 of Reference 8, where the strain $e_{yy} = u_{y,y} + (1/2)u_{z,y}^2$ from Equation (2.53, Vol.1) was used in the Theorem of Minimum Strain Energy, Equations (C.27, Vol.1) and (C.9, Vol.1).

The assembled matrix for several elements is

$$\{P\} = [[k] + [k_G]]\{u\}, \tag{5.98}$$

and the reduced matrix for the boundary conditions is

$$\{P\}_R = [[k] + [k_G]]_R \{u\}_R. \tag{5.99}$$

If $\{P\}_R = 0$, then the critical buckling load load P_{cr} is the smallest value obtained by setting the determinant equal to zero, or

$$\left|[[k] + [k_G]]_R\right| = 0. \tag{5.100}$$

This procedure applies for both determinate and redundant beams with each element having a different EI and L.

Example 5.10. Find the buckling load for a two element stepped cantilever beam in Figure 5.6 with $k = (EI)_2/(EI)_1 = 3$. (b) Compare the result to the exact solution and to a two element approximation for the tapered beam in Example 5.9.

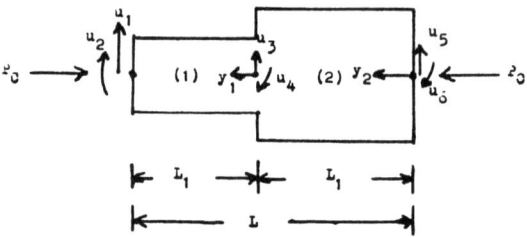

Fig. 5.6. Two element stepped beam.

Solution. (a) Since the two elements have the same length L_1, the L can be removed from Equations (4.91, Vol.1) and (5.96) by the procedure used in Equation (4.29, Vol.1). Thus, for the sequence $u_1 - u_6$ in Figure 5.6 with overlap of the two elements on u_3, u_4, the stiffness matrix has the form

$$\left(\frac{EI}{L^3}\right)_1 \begin{bmatrix} a_{11} - Cb_{11} & a_{12} - Cb_{12} & 0 \\ a_{12}^T - Cb_{12}^T & ka_{11} + a_{33} - C(b_{11} + b_{33}) & ka_{12} - Cb_{12} \\ 0 & ka_{12}^T - Cb_{12}^T & ka_{33} - Cb_{33} \end{bmatrix}^{6 \text{ by } 6},$$

$$k = (EI)_2/(EI)_1, \qquad C = P_C L_1^2/30(EI)_1,$$

$$a_{11} = \begin{bmatrix} 12 & -6 \\ -6 & 4 \end{bmatrix}, \quad a_{12} = \begin{bmatrix} -12 & -6 \\ 6 & 2 \end{bmatrix}, \quad a_{33} = \begin{bmatrix} 12 & 6 \\ 6 & 4 \end{bmatrix},$$

$$b_{11} = \begin{bmatrix} 36 & -3 \\ -3 & 4 \end{bmatrix}, \quad b_{12} = \begin{bmatrix} -36 & -3 \\ 3 & -1 \end{bmatrix}, \quad b_{33} = \begin{bmatrix} 36 & 3 \\ 3 & 4 \end{bmatrix}. \tag{5.101}$$

Various boundary conditions can be applied to get the reduced stiffness matrix and the determinant in Equation (5.100). Also, various values of k can be used.

For the cantilever beam with $u_5 = 0$ and $u_6 = 0$ in Figure 5.6, remove the last row and last column in Equation (5.101) to get Equation (5.100) as

$$\begin{vmatrix} 12(1 - 3C) & -3(2 - C) & -12(1 - 3C) & -3(2 - C) \\ -3(2 - C) & 4(1 - C) & 3(2 - C) & 2 + C \\ -12(1 - 3C) & 3(2 - C) & 12(1 + k - 6C) & 6(1 - k) \\ -3(2 - C) & 2 + C & 6(1 - k) & 4(1 + k - 2C) \end{vmatrix} = 0, \tag{5.102}$$

or for $k = 3$,

$$637.5C^4 - 2580C^3 + 2612C^2 - 656C + 24 = 0. \tag{5.103}$$

The smallest root is $C = 0.04395$ so that

$$P_{cr} = 30C(EI)_1/L_1^2 = 1.3185(EI)_1/L_1^2$$
$$= 1.7580(EI)_2/L^2, \qquad L = 2L_1. \tag{5.104}$$

(b) From Equation (b) on page 130 in Reference 2, the exact buckling load can be calculated as

$$(P_{cr})_{\text{exact}} = 1.3169(EI)_1/L_1^2 = 1.7559(EI)_2/L^2, \tag{5.105}$$

which is essentially the same as in Equation (5.104).

For the tapered beam in Example 5.9 and Equation (5.94),

$$EI = 0.1250(EI)_0 \quad \text{at } y = L,$$
$$= 0.4219(EI)_0 \quad \text{at } y = L/2,$$
$$= 1.0000(EI)_0 \quad \text{at } y = 0.$$

For two elements with $L_1 = L/2$, the average values are

$$(EI)_1 = 0.2734(EI)_0, \qquad (EI)_2 = 0.7110(EI)_0,$$

Approximate solutions using the virtual principles 175

whence

$$k = (EI)_2/(EI)_1 = 2.60.$$

Put this k into Equation (5.102) to get the smallest C as $C = 0.0405$ so that

$$P_{cr} = 1.2150(EI)_1/L_1^2 = 1.8692(EI)_2/L^2 = 1.3290(EI)_0/L^2.$$

This last value is very close to the exact value of

$$P_{cr} = 1.325(EI)_0/L^2$$

in Example 5.9. Thus, average values of EI for the elements can give good results for buckling.

5.6. The unit load theorem and numerical integration

From Equation (4.39, Vol.1) write the unit load theorem for beam bending in the form

$$u_z(y_i) = \int_0^L M_x(y_d) m_x^1(y_i, y_d) dy_d / EI(y_d), \tag{5.106}$$

which gives the deflection at point y_i of the beam. With EI variable it may be necessary to make the integration numerically. Use the same y_i points for the numerical integration, or for point y_i

$$u_z(y_i) = \sum_{k=1}^{K} M_x(y_k) m_x^1(y_i, y_k) g_k / EI(y_k)$$

$$= [m_{xi}^1]^{1 \text{ by } K} [g/EI]_d^{K \text{ by } K} \{M_x\}^{K \text{ by } 1}. \tag{5.107}$$

For all the K points

$$\{u_z\}^{K \text{ by } 1} = [m_x^1]^{K \text{ by } K} [g/EI]_d^{K \text{ by } K} \{M_x\}^{K \text{ by } 1}. \tag{5.108}$$

If the beam is redundant with R redundants as in Equation (4.42, Vol.1), separate M_x into columns for the applied loads and for the unknown redundant loads, whence Equation (5.108) has the form

$$[u_z]^{K \text{ by } (R+1)} = [m_x^1][g/EI]_d[M_x]^{K \text{ by } (R+1)}. \tag{5.109}$$

The redundants are determined from the specified u_{zi} values as by Equation (4.43, Vol.1). See Section 4.4 in Volume 1.

Influence coefficients can be calculated directly from Equation (5.108) by taking M_x as $m_x^1(y_j, y_k)$ for a unit load at y_j, whence

$$[C]^{K \text{ by } K} = [m_x^1][g/EI]_d[m_x^1]^T. \tag{5.110}$$

For applied loads P_j at some of the same points used in the numerical integration,

$$\{u_z\} = [C]\{P\}. \tag{5.111}$$

If M_x is produced by a column load as $M_x = P_C u_z$, then Equation (5.108) has the form

$$\{u_z\} = P_C[m_x^1][g/EI]_d\{u_z\} = P_C L^2 (EI)_0^{-1}[H]\{u_z\}, \tag{5.112}$$

whence the buckling load P_{cr} can be obtained from

$$\left|\frac{(EI)_0}{P_C L^2}[I] - [H]\right|_R = 0. \tag{5.113}$$

The problem with this procedure to get P_{cr} is that the determinant is K by K and difficult to evaluate except by computer. However, the smallest P_{cr} can be obtained by an iteration process. Assume a $u_{z1}(y_i)$ shape with largest value one and calculate $[H]\{u_{z1}\}$. Normalize the result to one maximum and repeat until convergence. The last normalizing factor determines the P_{cr} buckling load. See Example 5.12 below.

It is evident from the above discussion of the unit load theorem that approximate solutions for beams using the principle of virtual forces is completely different from the principle of virtual displacements used above in Sections 5.2 through 5.5. Since M_x is being used directly, no assumed functions are needed as in the principle of virtual displacements. However, in Chapter 7 on plates, it will be found that assumed functions are needed for the stress functions in the principle of virtual forces.

Example 5.11. Use Equation (5.108) and find the deflections at the eleven station points in Example 5.8 for the $p_x = p_0(1 - 0.8s)$ case. Compare the results to those given by the a_1 and a_2 constants in Equation (5.90) and by the exact Equation (5.92).
Solution. Calculate the M_x moment from

$$M_x'' = p_0(1 - 0.8s), \quad \text{or}$$
$$M_x = (p_0 L^2/6)(1.4 - 3.6s + 3s^2 - 0.8s^3). \tag{5.114}$$

Use the values in Table 5.1 in Example 5.8 to calculate the column matrix for the 11 points,

$$\{[g/EI]_d\{M_x\}\}^T = (0.10 p_0 L^3/18)(EI)_0^{-1}[1.4000 \quad 5.4922 \quad 2.6779$$
$$5.1802 \quad 2.4733 \quad 4.6296 \quad 2.0939 \quad 3.5493 \quad 1.3047 \quad 1.2420 \quad 0],$$

whence Equation (5.108) becomes

$$\{u_x\} = \frac{p_0 L^4}{180(EI)_0} \begin{bmatrix} 0 & & & & & & & & & & \\ 0.1 & 0 & & & & & & & & & \\ 0.2 & 0.1 & 0 & & & & & & & & \\ 0.3 & 0.2 & 0.1 & 0 & & \text{all zeros} & & & & & \\ 0.4 & 0.3 & 0.2 & 0.1 & 0 & & & & & & \\ 0.5 & 0.4 & 0.3 & 0.2 & 0.1 & 0 & & & & & \\ 0.6 & 0.5 & 0.4 & 0.3 & 0.2 & 0.1 & 0 & & & & \\ 0.7 & 0.6 & 0.5 & 0.4 & 0.3 & 0.2 & 0.1 & 0 & & & \\ 0.8 & 0.7 & 0.6 & 0.5 & 0.4 & 0.3 & 0.2 & 0.1 & 0 & & \\ 0.9 & 0.8 & 0.7 & 0.6 & 0.5 & 0.4 & 0.3 & 0.2 & 0.1 & 0 & \\ 1.0 & 0.9 & 0.8 & 0.7 & 0.6 & 0.5 & 0.4 & 0.3 & 0.2 & 0.1 & 0 \end{bmatrix} \begin{Bmatrix} 1.4000 \\ 5.4922 \\ 2.6779 \\ 5.1802 \\ 2.4733 \\ 4.6296 \\ 2.0939 \\ 3.5493 \\ 1.3047 \\ 1.2420 \\ 0 \end{Bmatrix},$$

or for the three cases of Equation (5.108), Equations (5.77) and (5.90), and the exact Equation (5.92),

Approximate solutions using the virtual principles

$$[u_z] = \frac{p_0 L^4}{(EI)_0} \begin{bmatrix} & \text{Equation (5.108)} & \text{Equations (5.77),} & \text{Equation (5.92),} \\ & & (5.90) & \text{exact} \\ 0.0000 & 0.0000 & 0.0000 \\ 0.0008 & 0.0011 & 0.0002 \\ 0.0046 & 0.0045 & 0.0046 \\ 0.0099 & 0.0103 & 0.0103 \\ 0.0181 & 0.0184 & 0.0181 \\ 0.0277 & 0.0285 & 0.0280 \\ 0.0398 & 0.0402 & 0.0398 \\ 0.0531 & 0.0530 & 0.0534 \\ 0.0684 & 0.0666 & 0.0684 \\ 0.0844 & 0.0805 & 0.0845 \\ 0.1011 & 0.0946 & 0.1011 \end{bmatrix} . \quad (5.115)$$

Here the points are at $s = 0.0, 0.1, 0.2, \cdots, 0.9, 1.0$. The results in Equation (5.115) show that the numerical integration in Equation (5.108) for the unit load theorem gives essentially the same deflection curve for the tapered beam as the exact solution in Equation (5.92). However, the numerical integration solution for two assumed functions in Example 5.8 deviates from the exact solution toward the tip of the beam, being smaller from $s = 0.80$ to $s = 1.00$, where the deflection has a rapid change.

Example 5.12 (a) Solve Example 5.9 by using five elements and six points on the tapered beam. Use Equation (5.112) with the trapezoid rule for the numerical integration and iterate to get the buckling load P_{cr}. (b) Comment on the convergence of the iteration.

Solution. (a) Calculate $(EI)_0/EI$ from Equation (5.94) at $s = 0, 0.2, 0.4, 0.6, 0.8, 1.0$, whence $[H]$ in Equation (5.112) is

$$[H] = 0.20 \begin{bmatrix} 0 & 0 & 0 & 0 & 0 & 0 \\ 0.2 & 0 & 0 & 0 & 0 & 0 \\ 0.4 & 0.2 & 0 & 0 & 0 & 0 \\ 0.6 & 0.4 & 0.2 & 0 & 0 & 0 \\ 0.8 & 0.6 & 0.4 & 0.2 & 0 & 0 \\ 1.0 & 0.8 & 0.6 & 0.4 & 0.2 & 0 \end{bmatrix} \begin{bmatrix} 0.5000 \\ 1.3717 \\ 1.9531 \\ 2.9155 \\ 4.6296 \\ 4.0000 \end{bmatrix}_{\text{diagonal}}$$

$$= 0.20 \begin{bmatrix} 0 & 0 & 0 & 0 & 0 & 0 \\ 0.10 & 0 & 0 & 0 & 0 & 0 \\ 0.20 & 0.2743 & 0 & 0 & 0 & 0 \\ 0.30 & 0.5486 & 0.3906 & 0 & 0 & 0 \\ 0.40 & 0.8230 & 0.7812 & 0.5831 & 0 & 0 \\ 0.50 & 1.0974 & 1.1719 & 1.1662 & 0.9259 & 0 \end{bmatrix}.$$

For the cantilever beam $M_z = P_G(u_{zL} - u_z)$ so that Equation (5.112) has the form

$$\{u\}_z = \frac{P_G L^2}{(EI)_0}[H]\{u_{zL} - u_z\}. \quad (5.116)$$

This form can be iterated but the form in Equation (5.113) does not apply in this case. Assume

$$\{u_z\}^T = [0.0 \ 0.1 \ 0.2 \ 0.4 \ 0.7 \ 1.0],$$

whence

$$\frac{(EI)_0}{P_G L^2}\{u_z\}_1 = [H] \begin{Bmatrix} 1.0 \\ 0.9 \\ 0.8 \\ 0.6 \\ 0.3 \\ 0.0 \end{Bmatrix} = 0.2 \begin{Bmatrix} 0.0 \\ 0.1 \\ 0.4469 \\ 1.1062 \\ 2.1155 \\ 3.4027 \end{Bmatrix} = 0.2(3.4027) \begin{Bmatrix} 0.0 \\ 0.0294 \\ 0.1313 \\ 0.3251 \\ 0.6217 \\ 1.0000 \end{Bmatrix},$$

$$\frac{(EI)_0}{P_C L^2}\{u_z\}_2 = [H]\begin{Bmatrix} 1.0 \\ 0.9706 \\ 0.8687 \\ 0.6749 \\ 0.3783 \\ 0.0 \end{Bmatrix} = 0.2(3.7205)\begin{Bmatrix} 0.0 \\ 0.0269 \\ 0.1253 \\ 0.3071 \\ 0.6104 \\ 1.0000 \end{Bmatrix},$$

$$\frac{(EI)_0}{P_C L^2}\{u_z\}_3 = (0.2)(3.7617)\begin{Bmatrix} 0.0000 \\ 0.0266 \\ 0.1242 \\ 0.3126 \\ 0.6085 \\ 1.0000 \end{Bmatrix},$$

which is close to convergence after three iterations. The buckling load is

$$P_{cr} = \frac{(EI)_0}{(0.2)(3.7617)L^2} = 1.3292(EI)_0/L^2, \qquad (5.117)$$

which is very close to the exact value of $1.325(EI)_0/L^2$ given in Example 5.9.

(b) The rapid convergence occurs because a deflected shape somewhat like the final shape was assumed and because the various theoretical buckling loads for the column are widely separated. From $C_n = (2n-1)^2(\pi^2/4)$, $C_1 = \pi^2/4$, $C_2 = 9\pi^2/4$, $C_3 = 25\pi^2/4$, etc. To show this separation effect, write Equation (5.112) as

$$(1/C)\{u_z\} = [H]\{u_z\}, \qquad (5.118)$$

and assume

$$\{u_z\} = a_1\{F_1\} + a_2\{F_2\} + \cdots,$$

where $\{F_i\}$ is the buckled deflected shape for the theoretical C_i buckling load. Thus,

$$[H]\{u_z\} = a_1[H]\{F_1\} + a_2[H]\{F_2\} + \cdots = \frac{a_1}{C_1}\{F_1\} + \frac{a_2}{C_2}\{F_2\} + \cdots,$$

$$[H]^m\{u_z\} = (a_1/C_1^m)\{F_1\} + (a_2/C_2^m)\{F_2\} + \cdots. \qquad (5.119)$$

Since C_1 is smaller than C_2, C_3, etc., it is evident that the convergence must be to the $\{F_1\}$ shape after m iterations when all the other terms become very small.

5.7. Approximate solutions for beams using the mixed virtual principle

Consider the beam in Figure 4.1. (Vol.1), and take the principle of mixed virtual stresses and virtual displacements from Equations (2.46, Vol.1) and and (4.44, Vol.1) for bending only, or

$$\int_0^L \int_A \left[\sigma_{yy}e_{yy}^V + (e_{yyu} - e_{yy\sigma})\sigma_{yy}^V\right] dA\, dy - \int_0^L p_z u_z^V\, dy - \sum_m \Big[P_{zm}u_{zm}^V +$$

$$+ M_{xm}\theta_{xm}^V + (u_{zm} - u_{zsm})P_{zm}^V + (\theta_{xm} - \theta_{xsm})M_{xm}^V\Big] = 0. \qquad (5.120)$$

Use Equation (4.1, Vol.1) and make the are integration to get

$$\int_0^L \left[M_x \frac{d^2 u_z^V}{dy^2} + \left(\frac{d^2 u_z}{dy^2} - \frac{M_x}{EI}\right)M_x^V\right] dy - \int_0^L p_z u_z^V\, dy - \sum_m \Big[P_{zm}u_{zm}^V +$$

$$+ M_{xm}\theta_{xm}^V + (u_{zm} - u_{zsm})P_{zm}^V + (\theta_{xm} - \theta_{xsm})M_{xm}^V\Big] = 0. \qquad (5.121)$$

In Equation (5.121) integrate the $\int_0^L M_x^V (d^2 u_z/dy^2) dy$ term by parts and separate Equation (5.121) into two equations with one involving virtual displacements and the other involving the virtual moments, or

$$\int_0^L \left(M_x \frac{d^2 u_z^V}{dy^2} - p_z u_z^V \right) dy - \sum_m \left(P_{zm} u_{zm}^V + M_{xm} \frac{du_{zm}^V}{dy} \right) = 0, \quad (5.122)$$

$$\int_0^L \left(u_z \frac{d^2 M_x^V}{dy^2} - \frac{M_x}{EI} M_x^V \right) dy - \sum_m \left(u_{zm} \frac{dM_{xm}^V}{dy} + \frac{du_{zm}}{dy} M_{xm}^V \right) = 0. \quad (5.123)$$

Note that these Equations (5.122) and (5.123) are simply the virtual form of the two differential Equations (4.56, Vol.1) and (4.60, Vol.1) with point forces and moments added, where the moment Equation (4.56, Vol.1) has been multiplied by u_z^V, and the deflection Equation (4.60, Vol.1) has been multiplied by M_x^V. Approximate solutions from Equations (5.122), (5.123) will be approximate solutions for these differential equations with any point loads and moments added.

Since only second derivatives are involved in the above equations and in the differential equations, it is often convenient to use the Galerkin form of the mixed principle for beams when no applied point forces or point moments are present. From Equations (4.2), (4.56), (4.60), (see Vol.1), (5.122), and (5.123)

$$\int_0^L \left(\frac{d^2 M_x}{dy^2} - p_z \right) u_z^V dy + [M_x(L) - M_{xL}]\theta_x^V(L) - [M_x(0) + M_{x0}]\theta_x^V(0) -$$

$$- \left[\frac{dM_x(L)}{dy} + P_{zL} \right] u_z^V(L) + \left[\frac{dM_x(0)}{dy} - P_{z0} \right] u_z^V(0) = 0, \quad (5.124)$$

$$\int_0^L \left(\frac{d^2 u_z}{dy^2} - \frac{M_x}{EI} \right) M_x^V dy + [u_z(L) - u_{zL}]\frac{dM_x^V(L)}{dy} - [u_z(0) + u_{z0}]\frac{dM_x^V(0)}{dy} -$$

$$- [\theta_x(L) + \theta_{xL}]M_x^V(L) + [\theta_x(0) - \theta_{x0}]M_x^V(0) = 0. \quad (5.125)$$

Note that if M_x and M_x^V satisfy the force boundary conditions and u_z and u_z^V satisfy the geometric boundary conditions, then all boundary terms drop out in Equations (5.124) and (5.125). There are no conditions on the higher derivatives of u_z as in the principle of virtual displacements.

Take u_z and M_x in terms of the known sets of functions $F_i(y)$ and $G_i(y)$ as

$$u_z = \sum_{i=1}^n a_i F_i(y), \quad u_z^V = \sum_{i=1}^n c_i^V F_i(y), \quad M_x = \sum_{i=1}^m b_i G_i(y),$$

$$M_x^V = \sum_{i=1}^m d_i^V G_i(y), \quad (5.126)$$

whence Equations (5.122) and (5.123) become

$$\int_0^L \left[\left(\sum_{i=1}^m b_i G_i \right) \left(\sum_{i=1}^n c_i^V \frac{d^2 F_i}{dy^2} \right) - p_z \left(\sum_{i=1}^n c_i^V F_i \right) \right] dy -$$

$$- \sum_{k=1}^K \left[P_{zk} \sum_{i=1}^n c_i^V F_i(y_k) + M_{xk} \sum_{i=1}^n c_i^V \frac{dF_i(y_k)}{dy} \right] = 0, \quad (5.127)$$

$$\int_0^L \left[\left(\sum_{i=1}^n a_i F_i\right)\left(\sum_{i=1}^m d_i^V \frac{\mathrm{d}^2 G_i}{\mathrm{d}y^2}\right) - \frac{M_x}{EI}\left(\sum_{i=1}^m d_i^V G_i\right)\right]\mathrm{d}y -$$

$$-\sum_{k=1}^K \left[u_{zk}\sum_{i=1}^m d_i^V \frac{\mathrm{d}G_i(y_k)}{\mathrm{d}y} + \frac{\mathrm{d}u_{zk}}{\mathrm{d}y}\sum_{i=1}^m d_i^V G_i(y_k)\right] = 0. \tag{5.128}$$

Since c_i^V and d_i^V are arbitrary, these two equations represent the two systems of equations

$$\int_0^L \left[\frac{\mathrm{d}^2 F_i}{\mathrm{d}y^2}\sum_{j=1}^m b_j G_j - p_z F_i\right]\mathrm{d}y - \sum_{k=1}^K \left[P_{zk}F_i(y_k) + M_{xk}\frac{\mathrm{d}F_i(y_k)}{\mathrm{d}y}\right] = 0,$$
$$i = 1, 2, \cdots, n, \tag{5.129}$$

$$\int_0^L \left[\frac{\mathrm{d}^2 G_i}{\mathrm{d}y^2}\sum_{j=1}^n a_j F_j - \frac{M_x}{EI}G_i\right]\mathrm{d}y - \sum_{k=1}^K \left[u_{zk}\frac{\mathrm{d}G_i(y_k)}{\mathrm{d}y} + \frac{\mathrm{d}u_{zk}}{\mathrm{d}y}G_i(y_k)\right] = 0,$$
$$i = 1, 2, \cdots, m. \tag{5.130}$$

Since different functions F_i and G_i are used in Equations (5.129) and (5.130) instead of one function, these equations do not have the usual Rayleigh-Ritz form. However, this form may be better because the multiplying virtual functions can be made to satisfy explicitly the two boundary conditions that may be omitted from the real functions, while in the Rayleigh-Ritz form both the real functions and the multiplying virtual functions satisfy the same two boundary conditions. Of course, both sets of functions must be capable of converging to all boundary conditions.

Note that Equation (5.129) approximates the equilibrium of the structure while Equation (5.130) approximates the compatibility of the structure. With the equations being coupled all the a_i and b_i constants must adjust to satisfy both equilibrum and compatibility as close as possible.

If $m = n$ in Equations (5.129) and (5.130) then the solution for the b_j constants for the moment can be obtained directly from Equation (5.129), where any unknown reactions for the redundant beam are included in the P_{zk} and M_{xk} term s. Once M_x is known, then u_z can be obtained from Equation (5.130), or from Equation (4.60, Vol.1), or from Equation (4.39, Vol.1). Once the u_z equation is determined, then any unknown redundant reactions included in the M_x and u_z are given by the redundant displacement restraints. If u_z satisfies the redundant restraints through the assumed functions, then it is necessary to make $m > n$ in order to couple the equations and impose the compatibility restraints on the moments.

If $m > n$, then Equations (5.129) and (5.130) are coupled so that both a_j and b_j must be solved together with M_x in Equation (5.130) expressed as in Equation (5.126). Write Equations (5.129) and (5.130) in matrix form as

$$[A]\{b\} = \{Q\}, \tag{5.131}$$
$$[B]\{b\} - [C]\{a\} = \{R\}, \tag{5.132}$$

where the elements in the matrices have the form

$$a_{ij} = \int_0^L \frac{d^2 F_i}{dy^2} G_j \, dy \quad \text{with } [A] \ n \text{ by } m,$$

$$b_{ij} = \int_0^L \frac{G_i G_j}{EI} \, dy \quad \text{with } [B] \ m \text{ by } m,$$

$$c_{ij} = \int_0^L \frac{d^2 G_i}{dy^2} F_j \, dy \quad \text{with } [C] \ m \text{ by } n,$$

$$q_i = \int_0^L p_z F_i \, dy + \sum_{k=1}^K \left[P_{zk} F_i(y_k) + M_{zk} \frac{d F_i(y_k)}{dy} \right]$$

with $\{Q\}$ n by 1,

$$r_i = -\sum_{k=1}^K \left[u_{zk} \frac{dG_i(y_k)}{dy} + \frac{du_{zk}}{dy} G_i(y_k) \right]$$

with $\{R\}$ m by 1. (5.133)

Since the $[B]$ matrix is square and non-singular, solve Equation (5.132) for $\{b\}$ in terms of $\{a\}$, or

$$\{b\} = [B]^{-1}[C]\{a\} + [B]^{-1}\{R\}. \tag{5.134}$$

Put this $\{b\}$ into Equation (5.131) to get

$$\{a\} = [G]^{-1}\{\{Q\} - [A][B]^{-1}\{R\}\}, \quad [G] = [A][B]^{-1}[C]. \tag{5.135}$$

These a_i constants give u_z in Equation (5.126), whence any unknown redundants can be determined from the redundant restraints. Put the final a_i values into Equation (5.134) to get the b_i constants and the moment from Equation (5.126). With $m = n$, the b_i are given directly by Equation (5.131), whence the a_i are given by Equation(5.132).

It should be noted that there are certain limitations, Reference 4, on the selected F_i and G_i functions in Equation (5.126). If the displacement boundary conditions are included in F_i but the moment and shear conditions are not in G_i or imposed directly, then the mixed principle reduces to the principle of virtual displacements. On the other hand, if the moment and shear boundary conditions are in G_i but the displacement conditions are not in F_i or imposed directly, then the mixed virtual principle reduces to the principle of virtual forces. Thus, in the following examples, the F_i functions will be selected to satisfy the given geometric restraints while the G_i functions will be selected to satisfy the shear and moment boundary conditions so that the mixed virtual principle will give results different from the other virtual principles.

Example 5.13. Consider only the constant distributed load case in figure 5.1 and apply the principle of mixed virtual stresses and virtual displacements with

$$M_z = b_1 \left(1 - \frac{y}{L}\right)^2, \quad u_z = a_1 \left(1 - \cos \frac{\pi y}{2L}\right). \tag{5.136}$$

(a) Find a_1 and b_1 for the approximate solutions for the displacements and moments. (b) Use the above u_z form in Equation (5.16) to find the approximate solutions using the principle of virtual displacements. (c) Compare the two results to each other and to the exact solution.

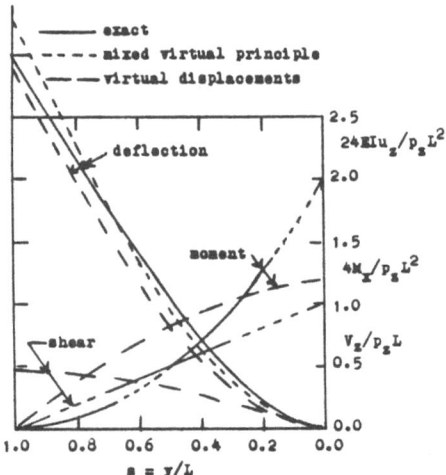

Fig. 5.7. Approximate solutions in Example 5.13.

Solution. (a) In this case the beam is determinate, the assumed u_x satisfies the support conditions, and the assumed M_x satisfies the moment and shear conditions at $y = L$. With

$$F_1 = 1 - \cos\frac{\pi y}{2L}, \qquad G_1 = \left(1 - \frac{y}{L}\right)^2, \tag{5.137}$$

Equation (5.129) gives

$$\int_0^L \left[b_1\left(1 - \frac{y}{L}\right)^2\left(\frac{\pi}{2L}\right)^2 \cos\frac{\pi y}{2L} - p_z\left(1 - \cos\frac{\pi y}{2L}\right)\right] dy = 0, \quad \text{or}$$

$$b_1 = p_z L^2/2, \qquad M_x = \left(p_z L^2/2\right)\left(1 - \frac{y}{L}\right)^2, \qquad V_x = p_z L\left(1 - \frac{y}{L}\right). \tag{5.138}$$

Since this is the exact M_x, the exact u_x can be obtained from

$$\frac{d^2 u_x}{dy^2} - \frac{M_x}{EI} = 0, \tag{5.139}$$

as given in Equation (5.27). However, Equation (5.130) gives

$$a_1 = \frac{\pi L^2 b_1}{10(\pi - 2)EI} = \frac{\pi p_z L^4}{20(\pi - 2)EI},$$

$$(u_x)_{\text{app}} = \frac{\pi p_z L^4}{20(\pi - 2)EI}\left(1 - \cos\frac{\pi y}{2L}\right). \tag{5.140}$$

(b) From Equation (5.16) for the principle of virtual displacements,

$$\int_0^L \left[\left(\frac{\pi}{2L}\right)^4 a_1 EI \cos^2\frac{\pi y}{2L} - p_z\left(1 - \cos\frac{\pi y}{2L}\right)\right] dy = 0,$$

$$a_1 = 32(\pi - 2)p_z L^4/\pi^5 EI, \qquad (u_x)_{\text{app}} = a_1\left(1 - \cos\frac{\pi y}{2L}\right),$$

$$(M_x)_{\text{app}} = \left[8(\pi - 2)p_z L^2/\pi^3\right]\cos\frac{\pi y}{2L},$$

$$(V_x)_{\text{app}} = \left[4(\pi - 2)p_z L/\pi^2\right]\sin\frac{\pi y}{2L}. \tag{5.141}$$

Approximate solutions using the virtual principles 183

(c) The exact u_z is given in Equation (5.27) and the exact M_x and V_z are in Equation (5.138). Figure 5.7 shows the comparisons for the mixed virtual principle, for which M_x and V_z are exact, the virtual displacement principle, and the exact solution. The principle of virtual displacements gives the poorest results for displacements, moments, and shear forces, with the shear curve completely reversed from the correct one. Thus, in this simple example, the principle of mixed virtual stresses and virtual displacements gives better results than the principle of virtual displacements.

Example 5.14. Repeat Example 5.13 for the redundant beam with an additional support so that $u_z = 0$ at $y/L = 1/3$. Take the functions in Equation (5.126) as

$$u_z = a_1(s - s_1)s^2, \quad s = y/L, \quad s_1 = 1/3,$$
$$M_x = (b_1 + b_2 s)(1 - s)^2. \tag{5.142}$$

Solution. (a) In Equation (5.126), $n = 1$, $m = 2$, and

$$F_1 = (s - s_1)s^2, \quad G_1 = (1 - s)^2, \quad G_2 = s(1 - s)^2. \tag{5.143}$$

From Equations (5.129) and (5.130)

$$\int_0^1 [(1-s)^2(b_1 + b_2 s)(6s - 2s_1) - p_z L^2(s - s_1)s^2]ds = 0,$$

$$\int_0^1 \left[a_1(s - s_1)s^2 \left\{ \frac{2}{-4 + 6s} \right\} - \frac{L^2}{EI}(1-s)^2(b_1 + b_2 s) \left\{ \frac{(1-s)^2}{s(1-s)^2} \right\} \right] ds = 0,$$

which can be integrated to give

$$10(3 - 4s_1)b_1 + 2(6 - 5s_1)b_2 = 5(3 - 4s_1)p_z L^2,$$
$$5(3 - 4s_1)a_1 - (6L^2/EI)b_1 - (L^2/EI)b_2 = 0,$$
$$7(6 - 5s_1)a_1 - (7L^2/EI)b_1 - (2L^2/EI)b_2 = 0. \tag{5.144}$$

For $s_1 = 1/3$, the solution is

$$a_1 = 0.04937 p_z L^4/EI, \quad b_1 = -0.13493 p_z L^2, \quad b_2 = 1.22101 p_z L^2,$$
$$(u_z)_{\text{app}} = 0.01645 s^2(3s - 1)p_z L^4/EI,$$
$$(M_x)_{\text{app}} = (1 - s)^2(-0.1349 + 1.2210s)p_z L^2,$$
$$(V_z)_{\text{app}} = (1 - s)(-1.4908 + 3.6630s)p_z L. \tag{5.145}$$

(b) From Equation (5.16) for the principle of virtual displacements

$$\int_0^1 [(4EI a_1/L^3)(3s - s_1)^2 - p_z L(s - s_1)s^2]ds = 0,$$

$$a_1 = \frac{3 - 4s_1}{48(3 - 3s_1 + s_1^2)} \frac{p_z L^4}{EI} = 0.016447 \frac{p_z L^4}{EI} \quad \text{for} \quad s_1 = 1/3,$$

$$(u_z)_{\text{app}} = 0.005482(3s - 1)s^2 p_z L^4/EI,$$
$$(M_x)_{\text{app}} = 0.010964(9s - 1)p_z L^2,$$
$$(V_z)_{\text{app}} = 0.098676 p_z L. \tag{5.146}$$

(c) The exact u_z is given in Equation (5.38) and the exact moment and shear are as follows

$$M_x = (-7 + 57s + 36s^2)p_z L^2/72, \quad 0 \le s \le 1/3,$$
$$M_x = (1 - s)^2 p_z L^2/2, \quad 1/3 \le s \le 1, \tag{5.147}$$
$$V_z = -(57 + 72s)p_z L/72, \quad 0 \le s \le 1/3$$
$$V_z = (1 - s)p_z L, \quad 1/3 \le s \le 1. \tag{5.148}$$

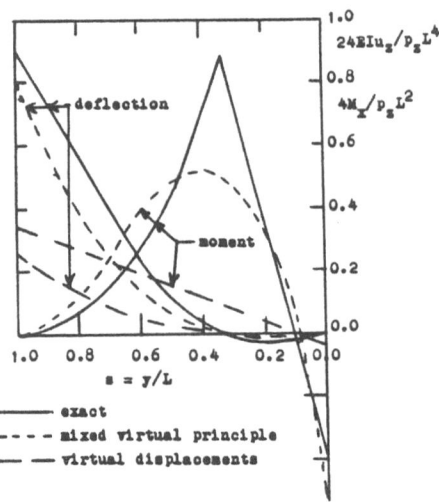

Fig. 5.8. Comparisons for redundant beam, Example 5.14.

Figure 5.8 shows the comparison of the results for the exact, the mixed virtual principle, and the principle of virtual displacements cases. For this redundant case, the one term displacement approximation in the principle of virtual displacements gives very poor results. However, the addition of the two term moment approximation to the same displacement term in the principle of mixed virtual stresses and virtual displacements gives much better results.

The large difference between the two methods in this example is due to the very poor shape of the moment curve given by $d^2 u_x/dy^2$ with the one term. Since this term occurs as a square in Equation (5.16) it can give an incorrect value for the a_1 constant. On the other hand, a better moment curve shape given by the G_1 and G_2 functions can improve the a_1 constant through Equation (5.129), where the product involves both moment expressions. Thus, the principle of mixed virtual stresses and virtual displacements can give better results than the principle of virtual displacements when a few terms are involved. Of course, if a large number of terms can be used, then both methods can approach the exact results.

5.8. Problems

5.1. Use Equation (5.23) and write out the inverse of the 5 by 5 Hilbert matrix in Equation (5.22).

5.2. Derive the exact beam deflection equation in Equation (5.27) by integration of Equations (4.65, Vol.1) and (4.66, Vol.1).

5.3. Recalculate the expressions in Equations (5.29)-(5.34) for $s_1 = 1/2$, and make a graph similar to Figure 5.2 for this case.

5.4. Calculate terms in Equation (5.24) for the case in which $p_x = p_{x0}(y/L)^2$.

5.5. Use the result in Problem 5.1 and calculate the P_1 load case in Example 5.1 using five terms in Equation (5.20).

5.6. In Example 5.1, use six terms for the P_1 load case alone and graph the approximate shear curve. Compare to the exact shear curve in Figure 5.2.

5.7. Assume $u_x = a_1(1 - \cos \frac{\pi y}{2L})$ in Figure 5.1 and find the value of a_1 for the constant p_x load acting alone. Compare to the exact case in Figure 5.2.

5.8. Repeat Example 5.2 for the P_L end load acting alone.

5.9. Repeat Example 5.2 for the M_{xT} thermal moment acting alone.

5.10. Calculate approximate values in terms of the exact values for u_x and σ_{yy} in Equations

(5.40) and (5.41) at $y = L/2$ for the case of one point load P_1 at $y_1 = L/3$.

5.11. Calculate approximate values in terms of the exact values for u_z and σ_{yy} in Equations (5.40) and (5.41) at $y = L/2$ for the case of two point loads of P_1 at $y_1 = 0.25L$ and P_2 at $y_2 = 0.60L$.

5.12. In Example 5.3, determine the series solutions for the moment M_z and shear $V_z = dM_z/dy$ for a point load P_1 alone at $y_1 = L/2$. Use sufficient terms in the series to calculate the approximate M_z and V_z at the points $y = 0.00, 0.05L, 0.25L, 0.45L, 0.50L$. Compare to the exact values.

5.13. In Example 5.3 assume the beam has additional supports at $y_1 = 0.25L$ and $y_2 = 0.60L$ with $u_z(y_1) = 0$ and $u_z(y_2) = 0$. If the p_z loading is constant, use Equation (5.40) and find the reactions P_1 and P_2 at these supports. What are the reactions at the ends of the beam?

5.14. In Example 5.3 assume the beam has additional supports at $y_1 = 0.25L$, $y_2 = 0.50L$, and $y_3 = 0.75L$ with $u_z(y_1) = 0$, $u_z(y_2) = u_{z2} = $ known, $u_z(y_3) = 0$. If the only load on the beam is a known point load P_4 at $y_4 = 0.375L$, use Equation (5.40) and find the reactions at these additional supports.

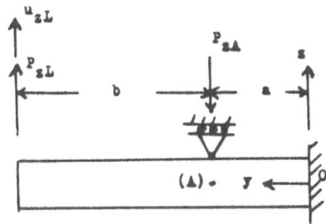

Fig. 5.9. Redundant beam deflection.

5.15. Use the redundant beam shown in Figure 5.9 and add a moment M_L at $y = a + b = L$. (a) Assume the approximate solution $u_z = y^2(y - a)(a_2 + a_3 y)$ and derive the two equations for a_2 and a_3 from Equation (5.16). (b) Take $a = L/3$, solve for a_2 and a_3, and determine the approximate u_z, M_z, and V_z equations. (c) Calculate the exact equations and compare results.

5.16. Assume $u_z = y^2(y - L)(a_2 + a_3 y)$ for the redundant beam in Figure 4.2, Vol.1. Use Equation (5.16) to determine the a_2 and a_3 constants. Compare to the exact solution for u_z and explain the results.

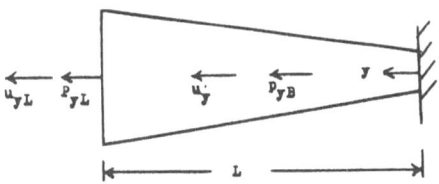

Fig. 5.10. Tapered bar for problems 5.17, 5.18.

5.17. The tapered beam shown in Figure 5.10 has a constant body force p_{yb} per unit length, $EA = (EA)_R[1 + (y/L)]^2$, and $P_{yL}/Lp_{yb} = 2.0$. (a) Use the exact axial strain $e_{yy}(y)$ and find the exact axial displacement $u_y(y)$ by the unit load theorem, Equation (2.41, Vol.1). (b) Assume $u_y = a_1 y + a_2 y^2 + a_3 y^3$ and find the approximate displacement $u_y(y)$ by using the principle of virtual displacements, Equation (2.34, Vol.1). (c) Compare the results for both u_y and σ_{yy}.

5.18. Repeat Problem 5.17 for an assumed stress in part (b) of

$$\frac{(EA)_R \sigma_{yy}}{ELp_{yb}} = (3 - \frac{y}{L})(1 - \frac{3y}{4L}) + \frac{y}{L}(1 - \frac{y}{L})(b_1 + b_2 \frac{y}{L}),$$

which satisfies the stress end conditions. Compare results to those in Problem 5.17.

5.19. In Example 5.4 derive the approximate moment and shear equations in parts (a) and (b).

5.20. Graph the cantilever beam buckled deflected shapes from Equations (5.46) and (5.60) by using

$$(u_x/u_{xL})_{\text{exact}} = 1 - \cos(\pi s/2),$$
$$(u_x/u_{xL})_{\text{approx.}} = (s^2/2)(3 - s).$$

Compare the two curves and discuss why the result in Equation (5.61) would be expected.

5.21. In Example 5.5 evaluate the G_1 in Equation (5.63) for the uniform p_x load case. Use $P_G = 0$ and compare the approximate and exact deflection curves. Discuss the relative approximations for this deflected case and the buckled case in Problem 5.20.

5.22. Repeat Problem 5.21 for $p_x = p_{x0}(1 - s)$.

5.23. In Example 5.6 use $(u_z)_{\text{exact}} = \sin \pi s$, $(u_z)_{\text{approx.}} = 4s(1 - s)$, and by comparison to the exact expressions demonstrate the separate effects of $(u_z)_{\text{approx.}}$, $(u'_z)_{\text{approx.}}$, and $(u''_z)_{\text{approx.}}$ in Equations (5.66) and (5.67).

5.24. Repeat Example 5.7 using the first three terms from Equation (5.18).

5.25. Repeat Example 5.8 using the first three terms of Equation (5.84).

5.26. Make the integration of Equation (4.6, Vol.1) to verify the result in Equation (5.92).

5.27. (a) Determine Equation (5.103) for the cases of $k = 2$ and $k = 4$. (b) Determine the resulting values of C to get the buckling loads.

5.28. Reduce Equation (5.101) to a form similar to Equation (5.102) for the simply supported beam with a column load.

5.29. Repeat the first part of Example 5.11 for the load case of $p_x = p_0(1 - 0.8s^2)$.

5.30. Solve Example 5.12 for $EI = (EI)_0(1 - 0.8s)^3$.

5.31. Repeat Example 5.13 for the case of

$$M_x = b_1(1 - s)^2, \quad u_x = a_1 s^2, \quad s = y/L.$$

Compare results to those in Figure 5.7.

5.32. In Example 5.13 take

$$M_x = c_1 + c_2 y + c_3 y^2, \quad u_x = a_1 \left(1 - \cos \frac{\pi y}{2L}\right),$$

and solve for the four constants a_1, c_1, c_2, and c_3. Show that the a_1 result is the same as for the prinicple of virtual displacements in Equation (5.141), and that the moment from c_1, c_2, c_3 values approximates the moment in Equation (5.141). This demonstrates that when M_x is not restrained, the mixed principle reduces to the principle of virtual displacements.

5.33. Repeat Example 5.14 for the case of $s_1 = 1/2$.

5.34. Repeat Example 5.14 for the case of

$$u_x = a_1(s - s_1)\left(1 - \cos \frac{\pi s}{2}\right), \quad s = y/L, \quad s_1 = 1/3,$$
$$M_x = (1 - s)^2 (b_1 + b_2 s).$$

Compare results to those in Figure 5.8.

References

Chapter 5

1. J.R. Westlake: *Handbook of Numerical Matrix Inversion and Solution of Linear Equations*, John Wiley and Sons (1968).
2. S. Timoshenko: *Theory of Elastic Stability*, McGraw-Hill Book Co., New York (1936).
3. R.L. Bisplinghoff, H. Ashley and R.L. Halfman: *Aeroelasticity*, Addison-Wesley Publishing Co., Reading, Mass. (1955).
4. O.J. Zienkiewicz and G.S. Hollister (Editors): *Stress Analysis*, Chapter 9 by B.F. de Veubeke, John Wiley and Sons, New York (1965).
5. S. Timoshenko: *Strength of Materials*, Part II, *Advanced Theory and Problems*, D. van Nostrand Co. (1936).

References for additional reading

6. C.L. Dym and I.H. Shames: *Solid Mechanics: A Variational Approach*, McGraw-Hill Book Co., New York (1973).
7. D.H. Allen and W.E. Haisler: *Introduction to Aerospace Structural Analysis*, John Wiley and Sons, New York (1985).
8. J.S. Przemieniecki, *Theory of Matrix Structural Analysis*, McGraw-Hill Book Co., New York (1968).

6

Dynamics of simple beams

6.1. Introduction

In the previous chapters is was assumed that the applied loads were independent of time. That is, the loads were assumed to be applied slowly so that no dynamic effects occurred. Actually, rapid application of loads can occur in aircraft structures due to air gusts, pilot maneuvers, and rough landings. These dynamic or time effects upon the structure of the airplane are considered in detail in References 1, 4, 5, and 6. In this chapter only the dynamic effects on simple beams will be considered. The procedures given in this chapter can be extended to more complicated aircraft structures.

In Section 6.2 the principle of virtual displacements is used to derive the bending time-dependent partial differential equations and boundary conditions for the simple beam. The homogeneous form of these equations is used to determine the bending frequencies and time varying deflected shapes of the beam for given initial conditions. In Section 6.3 the non-homogeneous form of the equations with applied load being a function of time is used to determine the forced motion of the beam. In Section 6.4 the principle of virtual displacements is used to obtain approximate solutions for the beam frequencies and deflected mode shapes.

In Section 6.5 the procedures of Sections 6.2, 6.3, and 6.4 are applied for torsional vibration of the beam. In Section 6.6 approximate solutions for beam frequencies are obtained by using finite elements.

The flutter problem for a wing segment is examined in Sections 6.7 and 6.8. Flutter is a simple harmonic diverging vibration that can occur on airplanes due to unsteady airloads out of phase with the angle of attack. It involves interaction between the deflections of the wing, tail, and aileron and the unsteady airloads, which is called the "Aeroelasticity" problem.

Since the dynamic effects change the deflections of the wing, it is evident that the stresses and strains in the wing are changed also. These dynamic stresses add to or subtract from the static stresses in the wing. These dynamic internal load effects are described in Section 6.9.

6.2. Bending vibrations of simple beams

From Equation (4.5, Vol.1) the Principle of Virtual Displacements for the simple beam has the form

$$\int_{t_0}^{t_1}\left[\int_0^L\int_A(\sigma_{yy}e_{yy}^V+\sigma_{yz}e_{yz}^V+\rho u_{z,tt}u_z^V+\rho u_{y,tt}u_y^V)dA\,dy-\right.$$
$$\left.-\int_0^L p_z u_z^V\,dy - P_{zL}u_{zL}^V - M_{xL}\theta_{xL}^V - P_{z0}u_{z0}^V - M_{x0}\theta_{x0}^V\right]dt = 0. \quad (6.1)$$

From Equation (2.30, Vol.1) with the u_{ya} axial deflection omitted,

$$u_y = -z\theta_x, \quad u_{y,tt} = -z\theta_{x,tt}, \quad u_y^V = -z\theta_x^V, \quad (6.2)$$

whence

$$\int_A(\rho u_{z,tt}u_z^V + \rho u_{y,tt}u_y^V)dA = mu_{z,tt}u_z^V + I_m\theta_{x,tt}\theta_x^V,$$
$$m = \int_A \rho\,dA, \quad I_m = \int_A \rho z^2\,dA. \quad (6.3)$$

From Equations (4.1) and (4.3) (Vol.1),

$$u_z = u_{zb} + u_{zs}, \quad u_{z,y} = \theta_x + e_{yz}, \quad e_{yz} = u_{z,y} - \theta_x,$$
$$\sigma_{yy} = -Ez\theta_{x,y}, \quad e_{yy}^V = -z\theta_{x,y}^V, \quad (6.4)$$
$$\sigma_{yz} = Ge_{yz} = G(u_{z,y} - \theta_x), \quad e_{yz}^V = u_{z,y}^V - \theta_x^V.$$

Put the terms in Equations (6.3) and (6.4) into Equation (6.1), make the area integrations, and integrate by parts on terms with $u_{z,y}^V$ and $\theta_{x,y}^V$, whence Equation (6.1) becomes

$$\int_{t_0}^{t_1}\left[\int_0^L\left[\theta_x^V\{I_m\theta_{x,tt}-(EI\theta_{x,y})_{,y}-GA_s e_{yz}\}+u_z^V\{mu_{z,tt}-(GA_s e_{yz})_{,y}-\right.\right.$$
$$\left.-p_z\}\right]dy + [EI\theta_{x,y}-M_{xL}]\theta_{xL}^V - [EI\theta_{x,y}+M_{x0}]\theta_{x0}^V +$$
$$\left.+[GA_s e_{yz}-P_{zL}]u_{zL}^V - [GA_s e_{yz}+P_{z0}]u_{z0}^V\right]dt = 0. \quad (6.5)$$

Since the virtual terms are arbitrary, the differential equations for u_z and θ_x and the equilibrium boundary conditions are

$$I_m\theta_{x,tt} - (EI\theta_{x,y})_{,y} - GA_s(u_{z,y}-\theta_x) = 0, \quad (6.6)$$
$$mu_{z,tt} - [GA_s(u_{z,y}-\theta_x)]_{,y} - p_z = 0, \quad (6.7)$$
$$EI\theta_{x,y}(L) - M_{xL} = 0, \quad EI\theta_{x,y}(0) - M_{x0} = 0, \quad (6.8)$$
$$GA_s[u_{z,y}(L) - \theta_x(L)] - P_{zL} = 0,$$
$$GA_s[u_{z,y}(0) - \theta_x(0)] - P_{z0} = 0. \quad (6.9)$$

Differentiate Equation (6.6) and put it into Equation (6.7) to get

$$mu_{z,tt} - (I_m\theta_{x,tt})_{,y} + (EI\theta_{x,y})_{,yy} - p_z = 0. \quad (6.10)$$

In many beams the rotary inertia term $(I_m\theta_{x,tt})_{,y}$ and the shear strain term e_{yz} are small so that $u_{zb,y} = \theta_x = u_{z,y}$, whence Equation (6.10) reduces to

$$\left(EIu_{z,yy}\right)_{,yy} + mu_{z,tt} - p_z = 0. \tag{6.11}$$

Note that this Equation (6.11) is Equation (4.6, Vol.1) with the inertia term $mu_{z,tt}$ added. Also, the equilibrium boundary conditions for Equation (6.11) will be those in Equation (4.7, Vol.1).

Consider the homogeneous form of Equation (6.11), or

$$\left(EIu_{z,yy}\right)_{,yy} + mu_{z,tt} = 0. \tag{6.12}$$

Separate the variables in the partial differential Equation (6.12) by taking $u_z(y,t)$ as a product, or

$$u_z(y,t) = Y(y)T(t). \tag{6.13}$$

Put into Equation (6.12) to get

$$-T_{,tt}/T = \left(EIY_{,yy}\right)_{,yy}/mY = \omega^2, \tag{6.14}$$

where ω^2 is the unknown separation constant. This gives two ordinary differential equations

$$T_{,tt} + \omega^2 T = 0, \tag{6.15}$$

$$\left(EIY_{,yy}\right)_{,yy} - m\omega^2 Y = 0. \tag{6.16}$$

If EI and m are constant for the beam the solutions of Equations (6.15) and (6.16) have the form

$$T = C_1 \sin \omega t + C_2 \cos \omega t, \tag{6.17}$$

$$Y = C_3 \sinh ky + C_4 \cosh ky + C_5 \sin ky + C_6 \cos ky, \tag{6.18}$$

$$k^4 = m\omega^2/EI, \tag{6.19}$$

$$u_z = TY, \tag{6.20}$$

where the constants $C_1, \cdots, C_6$ must be obtained from the initial conditions and the boundary conditions for the beam. For the simply supported beam of length L,

$$Y(0) = Y(L) = Y_{,yy}(0) = Y_{,yy}(L) = 0, \tag{6.21}$$

whence

$$C_4 = C_6 = C_3 = 0, \quad C_5 \sin kL = 0. \tag{6.22}$$

Thus, the only non-trivial homogeneous solutions occur for

$$kL = n\pi, \quad n \text{ an integer}, \quad \omega_n^2 = (n\pi/L)^4(EI/m), \tag{6.23}$$

whence

$$u_{zn} = (C_{1n} \sin \omega_n t + C_{2n} \cos \omega_n t)(C_{5n} \sin (n\pi y/L)) \tag{6.24}$$

Dynamics of simple beams

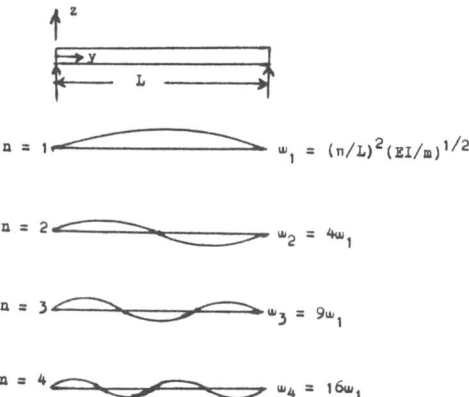

Fig. 6.1. Mode shapes and Frequencies for Simply Supported Beam.

are solutions for all n values. This gives

$$u_z = \sum_{n=1}^{\infty} u_{zn} = \sum_{n=1}^{\infty} (A_n \sin \omega_n t + B_n \cos \omega_n t) \sin(n\pi y/L), \qquad (6.25)$$

where the constants A_n and B_n can be determined from the initial conditions

$$u_z(y,0) = f(y), \qquad u_{z,t}(y,0) = g(y). \qquad (6.26)$$

Put Equation (6.26) into Equation (6.25) to get

$$\sum_{n=1}^{\infty} B_n \sin(n\pi y/L) = f(y), \qquad \sum_{n=1}^{\infty} A_n \omega_n \sin(n\pi y/L) = g(y). \qquad (6.27)$$

Since $\sin(n\pi y/L)$ are orthogonal functions with

$$\int_0^L \sin(n\pi y/L) \sin(m\pi y/L) dy = 0, \quad m \neq n, \qquad (6.28)$$
$$= L/2, \quad m = n,$$

multiply Equation (6.27) by $\sin(m\pi y/L)$ and integrate to get

$$B_n = (2/L) \int_0^L f(y) \sin(n\pi y/L) dy,$$
$$A_n = (2/L\omega_n) \int_0^L g(y) \sin(n\pi y/L) dy. \qquad (6.29)$$

The expression for T in Equation (6.17) represents simple harmonic motion with frequency ω, whence each ω_n term in Equation (6.25) represents simple harmonic motion of the beam mode shape given by $\sin(n\pi y/L)$. The case of $n = 1$ is called the fundamental frequency ω_1 with the corresponding fundamental mode shape $\sin(\pi y/L)$. Figure 6.1 shows the first four frequencies and mode shapes for the simply supported beam.

Other cases for the constant cross section beam with different boundary conditions are given in Chapter 3 of Reference 1. For the uniform cantilever beam of length L,

$$\omega_1 = (0.597)^2(\pi/L)^2(EI/m)^{1/2} = (0.597)^2\omega_r,$$
$$\omega_2 = (1.494)^2\omega_r,$$
$$\cdots\cdots\cdots\cdots$$
$$\omega_n = (\frac{2n-1}{2})\omega_r, \quad n \geq 3,$$
$$Y_n = (-1)^{n+1}[H_n(\sinh k_n y - \sin k_n y) + (\cosh k_n y - \cos k_n y)],$$
$$H_n = (\sin k_n L - \sinh k_n L)/(\cos k_n L + \cosh k_n L),$$
$$(\cosh k_n L)(\cos k_n L) = -1. \tag{6.30}$$

6.3. Forced motion of uniform beam

Use the case in Equation (6.11) with $p_z = p_z(y,t)$ a function of time t and location y. Assume that frequencies and mode shapes have been obtained from Equation (6.12) for the specified boundary conditions for Equation (6.11). From Equation (6.25) take

$$u_z(y,t) = \sum_{n=1}^{\infty} Y_n(y) T_n(t), \tag{6.31}$$

where $Y_n(y)$ are mode shapes and $T_n(t)$ are the unknown normal coordinates giving the amplitudes on $Y_n(y)$ at any specified time. In this case $T_n(t)$ will depend upon $p_z(y,t)$. Put Equation (6.31) into Equation (6.11), multiply by $Y_m(y)$, and integrate over the beam length L to get

$$\sum_{n=1}^{\infty}\left[T_n(t)\int_0^L Y_m(EIY_{n,yy})_{,yy}\,dy + T_{n,tt}\int_0^L Y_m Y_n m\,dy\right]$$
$$= \int_0^L Y_m p_z(y,t) dy. \tag{6.32}$$

To simplify Equation (6.32) use Equation (6.16) for the Y_m and Y_n mode shapes, multiply by Y_n and Y_m, respectively, and integrate to get

$$(\omega_m^2 - \omega_n^2)\int_0^L Y_m Y_n m\,dy = \int_0^L [Y_n(EIY_{m,yy})_{,yy} - Y_m(EIY_{n,yy})_{,yy}]dy$$
$$= [Y_n(EIY_{m,yy})_{,y}]_0^L - \int_0^L Y_{n,y}(EIY_{m,yy})_{,y}\,dy-$$
$$- [Y_m(EIY_{n,yy})_{,y}]_0^L + \int_0^L Y_{m,y}(EIY_{n,yy})_{,y}\,dy$$
$$= [Y_n(EIY_{m,yy})_{,y} - Y_m(EIY_{n,yy})_{,y}]_0^L - EI[Y_{n,y}Y_{m,yy} - Y_{m,y}Y_{n,yy}]_0^L +$$
$$+ \int_0^L EI(Y_{n,yy}Y_{m,yy} - Y_{m,yy}Y_{n,yy})dy. \tag{6.33}$$

Dynamics of simple beams

The last integral is zero so that

$$\int_0^L Y_m Y_n m \, dy = 0, \quad m \neq n, \tag{6.34}$$

for various combinations of zero boundary conditions on the beam. The possible pairs of boundary conditions for each end of the beam are

$$\begin{aligned} Y &= 0 \quad \text{and} \quad Y_{,y} = 0, \\ Y &= 0 \quad \text{and} \quad EIY_{,yy} = 0, \\ Y_{,y} &= 0 \quad \text{and} \quad \left(EIY_{,yy}\right)_{,y} = 0, \\ EIY_{,yy} &= 0 \quad \text{and} \quad \left(EIY_{,yy}\right)_{,y} = 0. \end{aligned} \tag{6.35}$$

Thus, Equation (6.34) shows the orthogonal properties of the natural mode shapes of the beam under the combinations in Equation (6.35). From Equations (6.33) and (6.34) it follows that Equation (6.32) reduces to

$$\begin{aligned} M_n T_{n,tt} + M_n \omega_n^2 T_n &= p_n(t), \\ M_n = \int_0^L Y_n^2 m \, dy, \quad p_n &= \int_0^L Y_n p_z(y,t) \, dy, \\ n &= 1, 2, \cdots, \infty. \end{aligned} \tag{6.36}$$

The term M_n is called the generalized mass of the nth mode and p_n is called the generalized force of the nth mode.

Example 6.1. Find the displacement $u_z(y,t)$ for a simply supported uniform beam with

$$p_z(y,t) = p(y) \sin qt, \quad u_z(y,0) = 0, \quad u_{z,t}(y,0) = 0. \tag{6.37}$$

Solution. From Equation (6.24) the Y_n mode shapes in Equation (6.31) are

$$Y_n = \sin(n\pi y/L), \tag{6.38}$$

whence Equation (6.36) gives

$$\begin{aligned} M_n &= m \int_0^L \sin^2(n\pi y/L) \, dy = mL/2, \\ p_n &= \sin qt \int_0^L p(y) \sin(n\pi y/L) \, dy = P_n \sin qt. \end{aligned} \tag{6.39}$$

From the initial conditions in Equation (6.37), $T_n(0) = 0$, $T_{n,t}(0) = 0$ are the conditions on Equation (6.36) with the form

$$T_{n,tt} + \omega_n^2 T_n = (2P_n/mL) \sin qt. \tag{6.40}$$

The particular integral for Equation (6.40) is

$$Q_n = \frac{2P_n/mL}{\omega_n^2 - q^2} \sin qt, \tag{6.41}$$

whence

$$T_n = \frac{2P_n/mL}{\omega_n^2 - q^2} [\sin qt - (q/\omega_n) \sin \omega_n t]. \tag{6.42}$$

Thus, Equation (6.31) becomes

$$u_z(y,t) = (2/mL)\sum_{n=1}^{\infty}\frac{P_n}{\omega_n^2 - q^2}[\sin(n\pi y/L)][\sin qt - \frac{q}{\omega_n}\sin\omega_n t]. \tag{6.43}$$

The motion for the particular integral Q_n in Equation (6.41) is simple harmonic and diverges when q approaches ω_n. The motion for T_n and u_z in Equations (6.42) and (6.43) is not simple harmonic and exhibits the phenomenon of *beats* when q approaches ω_n. Since some damping is always present in the structure, the terms representing the natural frequencies decay with time so that only the $\sin qt$ forcing function remains in Equation (6.42) and (6.43). See Section 3.3 of Reference 4 or books on mechanical vibrations.

6.4. Approximate solutions for frequencies and mode shapes

Omit the shear and rotary inertia effects in Equation (6.1) as well as p_z and the end forces and moments, whence the principle of virtual displacements in Equation (6.1) becomes

$$\int_{t_0}^{t_1}\left[\int_0^L\int_A(\sigma_{yy}e_{yy}^V + \rho u_{z,tt}u_z^V)dA\,dy\right]dt = 0. \tag{6.44}$$

In Equation (6.4) use

$$\sigma_{yy} = -Ezu_{z,yy}, \quad e_{yy}^V = -zu_{z,yy}^V, \tag{6.45}$$

whence Equation (6.44) becomes

$$\int_{t_0}^{t_1}\left[\int_0^L(EIu_{z,yy}u_{z,yy}^V + mu_{z,tt}u_z^V)dy\right]dt = 0. \tag{6.46}$$

From Equation (5.10) assume u_z to be approximated by

$$u_z = \sum_{i=1}^{n}a_i F_i(y)\sin(\omega t + \alpha), \tag{6.47}$$

where the natural vibrations give simple harmonic motion, and take u_z^V in the form

$$u_z^V = \sum_{i=1}^{n}b_i^V F_i(y). \tag{6.48}$$

Equation (6.46) reduces to the form

$$\int_0^L\left[EI\left(\sum_i a_i F_{i,yy}\right)\left(\sum_i b_i^V F_{i,yy}\right) - m\omega^2\left(\sum_i a_i F_i\right)\left(\sum_i b_i^V F_i\right)\right]dy = 0. \tag{6.49}$$

Since b_i^V is arbitrary, Equation (6.49) represents n equations for the n b_i^V, or

$$\int_0^L\left[EI\left(\sum_i a_i F_{i,yy}\right)F_{j,yy} - m\omega^2\left(\sum_i a_i F_i\right)F_j\right]dy = 0, \tag{6.50}$$

Dynamics of simple beams

$$\sum_{j=1}^{n} B_{ij}a_j - \omega^2 \sum_{j=1}^{n} A_{ij}a_j = 0, \quad i = 1, 2, \cdots, n, \tag{6.51}$$

$$B_{ij} = \int_0^L EIF_{i,yy}F_{j,yy}\,dy, \quad A_{ij} = \int_0^L mF_iF_j\,dy,$$

$$[B]\{a\} = \omega^2[A]\{a\}. \tag{6.52}$$

This Equation (6.52) corresponds to Equation (5.56) for columns. The $[B]$ matrix is the same for both cases but the $[A]$ matrix for vibrations has the mass m and F_iF_j functions instead of $F'_iF'_j$.

The Equation (6.52) corresponds to Equation (A.115a, Vol.1) so that the discussion in Section A.10 (Vol.1) applies, where the frequencies are the eigenvalues ω_i^2 and the mode shapes can be obtained from the eigenvectors and the $F_i(y)$ functions.

The accuracy of the approximation depends upon the selected $F_i(y)$ functions, the boundary conditions, and the number of functions used. The $F_i(y)$ should satisfy all the beam boundary conditions and should be capable of combining to represent the mode shapes. Obviously, from Figure 6.1, it is difficult to represent the mode shapes for the higher frequencies without using a large number of terms.

See References 1, 2, 3 for other approximation procedures to determine the frequencies and mode shapes for various structures. In particular, the iteration procedure described in Section 5.6 can be extended to get several vibration frequencies (Section 4.3(d) of Reference 1).

Example 6.2. Use only $F_1(y)$ in Equation (6.52) to approximate ω_1 for the uniform cantilever beam of length L for the cases

(a) $F_1(y) = 1 - 4(y/L) - \left(1 - \dfrac{y}{L}\right)^4$, (6.53)

(b) $F_1(y) = 1 - \cos(\pi y/L)$. (6.54)

Compare results to the exact ω_1 and Y_1 in Equation (6.30)

Solution. For this case Equation (6.52) becomes

$$\omega_1^2 = B_{11}/A_{11}, \quad B_{11} = EI\int_0^L F_{1,yy}^2\,dy, \quad A_{11} = m\int_0^L F_1^2\,dy. \tag{6.55}$$

(a) For the given $F_1(y)$,

$$F_{1,yy} = -(12/L^2)\left(1 - \dfrac{y}{L}\right)^2,$$

whence

$$B_{11} = 144EI/5L^3, \quad A_{11} = 104mL/45, \quad \omega_1 = (3.53/L^2)(EI/m)^{1/2}.$$

This compares quite well to the exact value in Equation (6.30), or

$$\omega_{1,\text{exact}} = (3.52/L^2)(EI/m)^{1/2}.$$

In this case the assumed $F_1(y)$ satisfies all four boundary conditions for the cantilever beam.

(b) For the given $F_1(y)$,

$$F_{1,yy} = +(\pi/L)^2\cos(\pi y/L),$$

Table 6.1. Comparison of mode shapes for example 6.2.

	$F_1(y/L)$ Equation (6.53)	$F_1(y/L)$ Equation (6.54)	Y_1 (exact) Equation (6.30)
C_1 (Equation 6.56)	3.53	3.66	3.52
y/L			
0	0.0	0.0	0.0
0.2	0.0699	0.0955	0.0639
0.4	0.2433	0.3455	0.2300
0.6	0.4752	0.6545	0.4612
0.8	0.7339	0.9045	0.7256
1.0	1.0000	1.0000	1.0000

whence

$$B_{11} = (EI/2)(\pi/L)^3, \quad A_{11} = (mL/2\pi)(2\pi + 1),$$
$$\omega_1 = (3.66/L^2)(EI/m)^{1/2},$$

which is 4% larger than the exact value. In this case, the assumed $F_1(y)$ satisfies three boundary conditions, but gives the moment

$$M(L,t) = EI(\pi/L)^2 \sin(\omega_1 t + \alpha), \quad a_1 = 1.$$

Table 6.1 shows the comparison of the mode shapes $F_1(y)$ to the exact Y_1 in Equation (6.30), where

$$\omega_1 L^2 (m/EI)^{1/2} = C_1, \quad \text{(coefficient)}. \tag{6.56}$$

In the Table 6.1, although the mode shape in Equation (6.54) is quite different from the exact one, the first vibration frequency is only 4% larger.

Example 6.3. Find the first two vibration frequencies and mode shapes of the tapered beam in Figure 5.5 and Example 5.8. Use the same assumed functions as in Example 5.8.

Solution. The $[B]$ matrix in Equation (6.52) is the same as in part (b) of Example 5.8, or

$$[B] = (EI_0/L^3) \begin{bmatrix} 2.1847 & 1.3171 \\ 1.3171 & 1.4451 \end{bmatrix}. \tag{6.57}$$

The $[A]$ matrix in Equation (6.52) has the form

$$[A] = [F][g]_d[m]_d[F]^T, \tag{6.58}$$

where $[F]$ and $[g]_d$ are given in Table 5.1 in Example 5.8 and

$$m = m_0(1 - 0.8s). \tag{6.59}$$

Make the matrix multiplications to get

$$[A] = (m_0 L/30) \begin{bmatrix} 2.7592 & 2.3635 \\ 2.3635 & 2.0751 \end{bmatrix}. \tag{6.60}$$

Use Equation (6.57) and

$$C^2 = \omega^2 L^4 m_0/30EI_0 \tag{6.61}$$

so that the determinant for Equation (6.52) is

$$\begin{vmatrix} 2.1847 - 2.7592C^2 & 1.3171 - 2.3635C^2 \\ 1.3171 - 2.3635C^2 & 1.4451 - 2.0751C^2 \end{vmatrix} = 0, \tag{6.62}$$

Dynamics of simple beams   197

or

$$0.1325C^4 - 2.2949C^2 + 1.4223 = 0,$$
$$C^2 = 0.6451, \quad 15.8058,$$
$$\omega_1 = (4.40/L^2)(EI_0/m_0)^{1/2}, \quad \omega_2 = (21.78/L^2)(EI_0/m_0)^{1/2}. \tag{6.63}$$

The corresponding coefficients given in Reference 1 for different weighting numbers in the numerical integration are 4.39 and 22.36. According to data in Reference 1, the exact coefficients are 4.29 and 15.75. Thus, the first frequency is satisfactory but the second one is no good.

To get the mode shapes use Equation (6.62) in Equation (6.52) for the calculated C^2 values, whence

$$a_1/a_2 = 0.513 \quad \text{for mode 1},$$
$$= -0.870 \quad \text{for mode 2}. \tag{6.64}$$

This gives the mode shapes

$$Y_1(s) = (1/1.513)[0.513F_1(s) + F_2(s)],$$
$$Y_2(s) = (1/0.130)[-0.870F_1(s) + F_2(s)], \tag{6.65}$$

where $F_1(s)$ and $F_2(s)$ are in Table 5.1 in Example 5.8. According to Figure 4.4 in Reference 1, $Y_1(s)$ is reasonably close to the exact curve, but $Y_2(s)$ is quite different from the exact curve.

6.5. Torsional vibrations of simple beams

From Equation (1.10, Vol.1)

$$u_z = -x\theta_y, \quad u_x = z\theta_y, \tag{6.66}$$

where θ_y is the torsional angle of rotation. From Equation (1.7, Vol.1)

$$e_{xy} = u_{x,y} = z\theta_{y,y}, \quad \sigma_{xy} = Ge_{xy},$$
$$e_{yz} = u_{z,y} = -x\theta_{y,y}, \quad \sigma_{yz} = Ge_{yz}, \tag{6.67}$$

whence the principle of virtual displacements for torsion has the form

$$\int_{t_0}^{t_1}\left[\int_0^L\int_A (\sigma_{xy}e_{xy}^V + \sigma_{yz}e_{yz}^V + \rho u_{x,tt}u_x^V + \rho u_{z,tt}u_z^V)\,dA\,dy - \right.$$
$$\left. - \int_0^L m_y\theta_y^V\,dy - M_{yL}\theta_{yL}^V - M_{y0}\theta_{y0}^V\right]dt = 0. \tag{6.68}$$

Put the expressions in Equations (6.66) and (6.67) into Equation (6.68), make the area integrations, and integrate by parts on y, or

$$\int_0^L\int_A [Gz^2\theta_{y,y}\theta_{y,y}^V + Gx^2\theta_{y,y}\theta_{y,y}^V + \rho(x^2+z^2)\theta_{y,tt}\theta_y^V]\,dA\,dy$$
$$= [(GJ\theta_{y,y})\theta_y^V]_0^L - \int_0^L (GJ\theta_{y,y})_{,y}\theta_y^V\,dy + \int_0^L I_T\theta_{t,tt}\theta_y^V\,dy, \tag{6.69}$$
$$GJ = \int_A G(x^2+z^2)\,dA, \quad I_T = \int_A \rho(x^2+z^2)\,dA.$$

Put into Equation (6.68) and collect terms for the arbitrary virtual expressions to get the differential equation and boundary conditions for torsion, or

$$\left(GJ\theta_{y,y}\right)_{,y} - I_T\theta_{y,tt} = -m_y,$$
$$\left[GJ\theta_{y,y} - M_{yL}\right]_{y=L} = 0, \quad \left[GJ\theta_{y,y} - M_{y0}\right]_{y=0} = 0. \tag{6.70}$$

Consider the homogeneous form of Equation (6.70), or

$$\left(GJ\theta_{y,y}\right)_{,y} - I_T\theta_{y,tt} = 0. \tag{6.71}$$

Assume

$$\theta_y(y,t) = Y(y)T(t), \tag{6.72}$$

whence

$$T_{,tt} + \omega^2 T = 0,$$
$$\left(GJY_{,y}\right)_{,y} + I_T\omega^2 Y = 0, \tag{6.73}$$

If GJ and I_T are constant, the solution is

$$T = C_1 \sin \omega t + C_2 \cos \omega t,$$
$$Y = C_3 \sin ky + C_4 \cos ky, \quad k^2 = \omega^2 I_T/GJ. \tag{6.74}$$

For the cantilever beam

$$Y(0) = 0, \quad Y_{,y}(L) = 0, \tag{6.75}$$

where the second condition is given by Equation (6.70). This gives

$$C_4 = 0, \quad \cos kL = 0, \quad kL = (2n-1)\pi/2,$$
$$\omega_n = \frac{(2n-1)\pi}{2L}(GJ/I_T)^{1/2}, \quad Y_n = C_n \sin \frac{(2n-1)\pi}{2L}y. \tag{6.76}$$

The initial conditions on time can be handled in the same manner as for bending in Equations (6.24)-(6.29).

The forced motion of the beam in torsion is similar to Equation (6.36) with the form

$$M_n T_{n,tt} + M_n \omega_n^2 T_n = m_n(t),$$
$$M_n = \int_0^L Y_n^2 I_T \, dy, \quad m_n = \int_0^L Y_n m_y \, dy. \tag{6.77}$$

For approximate solutions for the frequencies use Equation (6.69) so that in Equation (6.51)

$$B_{ij} = \int_0^L GJF_{i,y}F_{j,y} \, dy, \quad A_{ij} = \int_0^L I_T F_i F_j \, dy. \tag{6.78}$$

From Equations (5.106, Vol.1), and (4.39, Vol.1) the unit load theorem for torsional rotation of the beam can be written as

$$\theta_{ym} = \int_0^L (M_y m_{ym}^1/GJ) dy, \tag{6.79}$$

Dynamics of simple beams

whence the matrix forms in Equations (5.108)-(5.111) apply, or

$$\{\theta_y\} = [m_y^1][g/GJ]_d\{M_y\}, \tag{6.80}$$
$$[C_T] = [m_y^1][g/GJ]_d[m_y^1]^T, \tag{6.81}$$
$$\{\theta_y\} = [C_T]\{m_y\}. \tag{6.82}$$

For the vibration case,

$$\{m_y\} = -[I_T]\{\theta_{y,tt}\} = \omega^2[I_T]\{\theta_y\}, \tag{6.83}$$
$$\{\theta_y\} = \omega^2[C_T][I_T]\{\theta_y\}, \tag{6.84}$$

which can be solved for ω^2 by iteration as described in Section 5.6 and in References 1 and 2.

Example 6.4. Find the first torsional frequency for a cantilevered tapered thin wall frustum of a cone with wall thichness t (constant) and radius $r = r_0(1 - 0.8s)$. Assume the one approximating function to be the uniform beam solution in Equation (6.76).

Solution. From Equations (6.69), (6.76), and (6.78)

$$GJ = GJ_0(1-0.8s)^3, \quad GJ_0 = 2\pi t G r_0^3,$$
$$I_T = I_{T0}(1-0.8s)^3, \quad I_{T0} = 2\pi r_0^3 t \rho,$$
$$F_1 = a_1 \sin(\pi s/2), \quad F_{1,y} = (\pi a_1/2)\cos(\pi s/2),$$
$$B_{11} = (\pi/2)^2 GJ_0 \int_0^1 (1-0.8s)^3 \cos^2(\pi s/2) ds,$$
$$A_{11} = I_{T0} \int_0^1 (1-0.8s)^3 \sin^2(\pi s/2) ds.$$

Use Simpson's rule with 11 points so that Table 5.1 can be used for numerical integration, whence from Equation (6.51),

$$\omega_1^2 = B_{11}/A_{11} = \omega_0^2(7.429/1.832),$$
$$\omega_1 = 2.014\omega_0 = 3.163\bigl(GJ_0/L^2 I_{T0}\bigr)^{1/2}, \tag{6.85}$$

where ω_0 is the uniform beam result in Equation (6.76).

Example 6.5. Repeat Example 6.4 by iterating Equation (6.84). Use six points with the trapezoid rule.

Solution. From Example 6.4 and Table 5.1 in Example 5.8,

$$[1/GJ]_d = (1/GJ_0[1 \quad 1.6872 \quad 3.1807 \quad 7.1174 \quad 21.453 \quad 125.00],$$
$$[I_T]_{dE} = 0.2 L I_{T0}[0.5 \quad 0.5927 \quad 0.3144 \quad 0.1406 \quad 0.0466 \quad 0.0040].$$

Since the rotation angle θ_y for a point value of M_y is constant outboard of the point on a cantilever beam, Equations (6.80) and (6.81) can be simplified as

$$\{\theta_y\} = [g][1/GJ]_d\{M_y\}, \quad [C_T]_d = [g][1/GJ]_d\{1\},$$

$$[g] = 0.2L \begin{bmatrix} 0.0 & 0.0 & 0.0 & 0.0 & 0.0 & 0.0 \\ 0.5 & 0.5 & 0.0 & 0.0 & 0.0 & 0.0 \\ 0.5 & 1.0 & 0.5 & 0.0 & 0.0 & 0.0 \\ 0.5 & 1.0 & 1.0 & 0.5 & 0.0 & 0.0 \\ 0.5 & 1.0 & 1.0 & 1.0 & 0.5 & 0.0 \\ 0.5 & 1.0 & 1.0 & 1.0 & 1.0 & 0.5 \end{bmatrix}. \tag{6.86}$$

Thus, the diagonal elements of $[C_T]$ are

$$[C_T]_d = (0.2L/GJ_0)[0 \quad 1.344 \quad 3.278 \quad 8.924 \quad 23.21 \quad 96.44],$$

whence

$$[C_T] = (0.2L/GJ_0)\begin{bmatrix} 0 & 0 & 0 & 0 & 0 & 0 \\ 0 & 1.344 & 1.344 & 1.344 & 1.344 & 1.344 \\ 0 & 1.344 & 3.278 & 3.278 & 3.278 & 3.278 \\ 0 & 1.344 & 3.278 & 8.924 & 8.924 & 8.924 \\ 0 & 1.344 & 3.278 & 8.924 & 23.21 & 23.21 \\ 0 & 1.344 & 3.278 & 8.924 & 23.21 & 96.44 \end{bmatrix},$$

$$[C_T][I_T]_{dE} = \frac{0.04L^2 I_{T0}}{GJ_0}\begin{bmatrix} 0 & 0 & 0 & 0 & 0 & 0 \\ 0 & 0.796 & 0.422 & 0.189 & 0.0626 & 0.0054 \\ 0 & 0.796 & 1.030 & 0.461 & 0.1527 & 0.0131 \\ 0 & 0.796 & 1.030 & 1.255 & 0.416 & 0.0357 \\ 0 & 0.796 & 1.030 & 1.255 & 1.082 & 0.0908 \\ 0 & 0.796 & 1.030 & 1.255 & 1.082 & 0.3802 \end{bmatrix}.$$

Start with $\theta_y = \sin(\pi s/2)$ from Example 6.4 and iterate four times in Equation (6.84) to give convergence with

$$\omega_1 = (3.05/L)(GJ_0/I_{T0})^{1/2},$$

which compares to $C_1 = 3.16$ in Example 6.4. The mode shape is considerably different from $\sin(\pi s/2)$, or

$$\text{mode shape} = [0\quad 0.180\quad 0.362\quad 0.651\quad 0.892\quad 1.0],$$
$$\sin(\pi s/2) = [0\quad 0.309\quad 0.588\quad 0.809\quad 0.951\quad 1.0].$$

6.6. Finite element matrices for beam frequencies

The matrix procedures of Sections 4.3, 4.8 (Vol.1), and 5.5 can be used to obtain approximate values for beam frequencies. An equivalent mass matrix for the beam element can be obtained for the $u_{z,tt}$ term in Equation (6.1) by using Equation (4.15, Vol.1) in Equation (2.71, Vol.1), or for the order u_{zL}, θ_{xL}, u_{z0}, θ_{x0},

$$[M] = \rho AL \int_0^1 \begin{Bmatrix} 3s^2 - 2s^3 \\ L(s^3 - s^2) \\ 1 - 3s^2 + 2s^3 \\ L(s - 2s^2 + s^3) \end{Bmatrix} [3s^2 - 2s^3 \quad L(s^3 - s^2)$$

$$1 - 3s^2 + 2s^3 \quad L(s - 2s^2 + s^3)]ds$$

$$= (\rho AL/420)\begin{bmatrix} 156 & -22L & 54 & 13L \\ -22L & 4L^2 & -13L & -3L^2 \\ 54 & -13L & 156 & 22L \\ 13L & -3L^2 & 22L & 4L^2 \end{bmatrix}, \qquad (6.87)$$

whence from Equation (5.98) the assembled matrix equation for several beam elements is

$$[M]\{u_{,tt}\} + [[k] + [k_C]]\{u\} = \{P\}. \qquad (6.88)$$

The reduced matrix for the boundary conditions is

$$[M]_R\{u_{,tt}\} + [[k] + [k_C]]_R\{u\} = \{P\}_R. \qquad (6.89)$$

Dynamics of simple beams 201

If $\{P\}_R = \{0\}$ and the beam is vibrating in simple harmonic motion so that $\{u_{,tt}\} = -\omega^2\{u\}$, then ω can be obtained from

$$\left|-\omega^2[M] + [k] + [k_G]\right|_R = 0. \tag{6.90}$$

The above $[M]$ is the static mass matrix for beam translation. The rotary inertia can be added by using the matrix in Equation (5.96) with the coefficient $(\rho AL/30)(r/L)^2$, where r is the radius of gyration for the beam cross section. Also, it is shown in Chapter 10 of Reference 3 that there are dynamic correction terms which involve ω^2 and affect both $[M]$ and $[k]$. However, these terms are small for short elements, involving matrix coefficients of the form $\omega^2(\rho AL)^2(L^3/EI)10^{-3}$.

The element matrices for torsional vibration are

$$[k_\theta] = (GJ/L)\begin{bmatrix} 1 & -1 \\ -1 & 1 \end{bmatrix}, \quad [I_E] = (LI_T/6)\begin{bmatrix} 2 & 1 \\ 1 & 2 \end{bmatrix}. \tag{6.91}$$

In the assembled case

$$[I_E]\{\theta_{y,tt}\} + [k_\theta]\{\theta_y\} = \{m_y\},$$
$$\left|-\omega^2[I_E] + [k_\theta]\right|_R = 0. \tag{6.92}$$

Example 6.6. Use three elements for the uniform cantilever beam in torsion and find the first three torsion frequencies using Equations (6.91) and (6.92). Compare to the exact frequencies in Equation (6.76).

Solution. For the three element beam with each element of length $L/3$, Equations (6.91) and (6.92) give

$$\left| \frac{-\omega^2 L I_T}{18}\begin{bmatrix} 2 & 1 & 0 & 0 \\ 1 & 4 & 1 & 0 \\ 0 & 1 & 4 & 1 \\ 0 & 0 & 1 & 2 \end{bmatrix} + \frac{3GJ}{L}\begin{bmatrix} 1 & -1 & 0 & 0 \\ -1 & 2 & -1 & 0 \\ 0 & -1 & 2 & -1 \\ 0 & 0 & -1 & 1 \end{bmatrix} \right| = 0.$$

Take $g = \omega^2 L^2 I_T/54GJ$ and use $\theta_{y4} = 0$ as the boundary condition to remove the last row and last column, whence

$$\begin{vmatrix} 2g-1 & g+1 & 0 \\ g+1 & 2(2g-1) & g+1 \\ 0 & g+1 & 2(2g-1) \end{vmatrix} = 0,$$

$$(2g-1)(13g^2 - 22g + 1) = 0,$$
$$g = 0.0467, \quad 0.5000, \quad 1.6456,$$
$$\left(L^2 I_T/GJ\right)^{1/2}\omega = 1.588, \quad 5.196, \quad 9.427,$$
$$\left(L^2 I_T/GJ\right)^{1/2}\omega_{exact} = 1.571, \quad 4.712, \quad 7.854.$$

The first frequency is close, but the others are too large.

Note that the element matrices for longitudinal vibrations of a bar are

$$[k] = (EA/L)\begin{bmatrix} 1 & -1 \\ -1 & 1 \end{bmatrix}, \quad [M] = (mL/6)\begin{bmatrix} 2 & 1 \\ 1 & 2 \end{bmatrix}, \tag{6.91a}$$

which are the same as for torsion in Equation (6.91), except for the multiplying coefficients. Thus the above results apply with mL^2/EA in place of $I_T L^2/GJ$.

6.7. Flutter of wing segment with one degree of freedom

Flutter is a dynamic instability of an elastic body such as airplane wings, tails, and control surfaces moving in an airstream. The flutter velocity V_F and the associated frequency ω_F are the lowest velocity and corresponding frequency at which one or more components of an airplane under given flight conditions has simple harmonic motion. This is a neutral stability condition with damped oscillations below V_F and diverging oscillations above V_F.

The flutter problem is different from the static wing divergence and aileron reversal problems described in Sections 5.14 and 5.15 (Vol.1). In the static case the airloads are assumed to vary directly with angle of attack with no time effects. Actually, the airloads may be unsteady and out of phase with the angle of attack. At low velocities the airloads damp any wing motion but at high velocities the airloads may cause the motion to diverge.

Since flutter occurs in a simple harmonic situation, the flutter analysis assumes all dependent variables to be proportional to $e^{i\omega t}$ (ω real), whence the problem is to find V_F and ω_F for which this occurs. This gives a double eigenvalue problem where two critical numbers must be found to get V_F and ω_F. Thus, the flutter problem is more complicated than the simple vibration problems discussed in the previous sections of this chapter.

Consider the wing segment of unit width in Figure 6.2. From Equation (6.77) for $Y_n = 1$,

$$I_T \alpha_{,tt} + I_t \omega_\theta^2 \alpha = m_{yT},$$
$$m_{yT} = A(V/b\omega) + iB(V/b\omega), \tag{6.93}$$

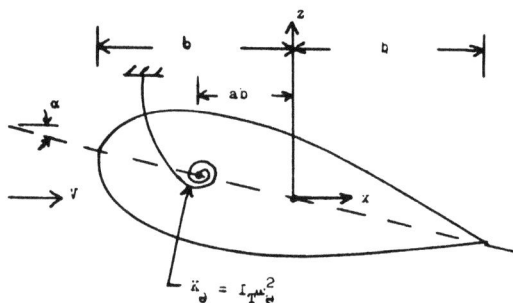

Fig. 6.2. Wing segment under torsional restraint.

where $A(V/b\omega)$ and $B(V/b\omega)$ are complicated functions of wing parameters and the unknown V_F and ω_F. They can be obtained from Equation (5.312) in Reference 1 for the torsional case in Figure 6.2, or

$$m_{yT} = \pi\rho b^2[-bV(0.5-a)\alpha_{,t} - b^2(0.125 + a^2)\alpha_{,tt} + \\ + 2\pi\rho b^2 V(0.5+a)C(k)[V\alpha + b(0.5-a)\alpha_{,t}], \tag{6.94}$$

where

$$k = b\omega/V. \tag{6.95}$$

The Theodorsen function $C(k)$ is discussed in References 1, 4, 5, 6, and 7. From

Chapter 8 of Reference 4,

$$C(k) = F(k) + iG(k),$$
$$F(k) = \frac{J_1(J_1 + Y_0) + Y_1(Y_1 - J_0)}{(J_1 + Y_0)^2 + (Y_1 - J_0)^2},$$
$$G(k) = -\frac{Y_1 Y_0 + J_1 J_0}{(J_1 + Y_0)^2 + (Y_1 - J_0)^2},$$
(6.96)

where $J_1(k)$, $J_0(k)$, $Y_1(k)$, $Y_0(k)$ are Bessel functions. Tables and curves for $F(k)$ and $G(k)$ are given in various books and reports. The values in Table 6.2 are from References 5 and 6. More detailed tables are given in Reference 7.

Table 6.2. The Theodorsen Function $C(k) = F(k) + iG(k)$.

k	F(k)	-G(k)	k	F(k)	-G(k)
∞	0.5000	0.0000	0.40	0.6250	0.1650
10.00	0.5006	0.0124	0.34	0.6469	0.1738
6.00	0.5017	0.0206	0.30	0.6650	0.1793
4.00	0.5037	0.0305	0.24	0.6989	0.1862
3.00	0.5063	0.0400	0.20	0.7276	0.1886
2.00	0.5129	0.0577	0.16	0.7628	0.1876
1.50	0.5210	0.0736	0.12	0.8063	0.1801
1.20	0.5300	0.0877	0.10	0.8320	0.1723
1.00	0.5394	0.1003	0.08	0.8604	0.1604
0.80	0.5541	0.1165	0.06	0.8920	0.1426
0.66	0.5699	0.1308	0.05	0.9090	0.1305
0.60	0.5788	0.1378	0.04	0.9267	0.1160
0.56	0.5857	0.1428	0.025	0.9545	0.0872
0.50	0.5979	0.1507	0.01	0.9824	0.0482
0.44	0.6130	0.1592	0.00	1.0000	0.0000

Assume simple harmonic motion for α, or

$$\alpha = \alpha_0 e^{i\omega t},$$
(6.97)

whence Equations (6.93) and (6.94) become

$$-I_T \omega^2 \alpha_0 + I_T \omega_\theta^2 \alpha_0 = \pi \rho b^4 \omega^2 \alpha_0 [D(k) + iE(k)],$$
$$(\omega_\theta/\omega)^2 = 1 + (\pi \rho b^4/I_T)[D(k) + iE(k)],$$
$$D(k) = k^{-2}[(a^2 + 0.125)k^2 + (2a+1)F(k) - 2k(0.25 - a^2)G(k)],$$
$$E(k) = k^{-2}[-k(0.5 - a) + 2k(0.25 - a^2)F(k) + (2a+1)G(k)]. \quad (6.98)$$

For a given value of a the unknowns k and ω can be obtained from

$$E(k) = 0, \quad (\omega_\theta/\omega)^2 = 1 + \frac{D(k)}{(I_T/\pi \rho b^4)},$$
(6.99)

by trial or by graphing $E(k)$. The value of k can be determined from $E(k) = 0$. Use this value to calculate $D(k)$, which is negative. If this value of $D(k)$ is small enough to make $(\omega_\theta/\omega)^2$ real in Equation (6.99), then this determines ω_F and hence V_F for the flutter velocity. If ω_θ/ω is imaginary, then flutter does not occur. This means that flutter is more likely for large values of $I_T/\pi \rho b^4$.

Example 6.7. Assume $a = -1$ for torsional restraint at the leading edge in Figure 6.2 and find solutions for k and ω in Equation (6.99).

Solution. For $a = -1$ Equation (6.98) gives

$$D(k) = k^{-2}[1.125k^2 - F(k) + 1.5kG(k)],$$
$$E(k) \doteq k^{-2}[-1.5k - G(k) - 1.5kF(k)]. \qquad (6.100)$$

From an approximate graph of $E(k)$, use trial and error in Table 6.2 to get k from $E(k) = 0$, whence

$$k = 0.041, \quad G = -0.1175, \quad F = 0.9349, \quad D(k) = -553,$$
$$(\omega_\theta/\omega)^2 = 1 - \frac{553}{(I_T/\pi\rho b^4)}. \qquad (6.101)$$

Thus, if $I_T/\pi\rho b^4$ is greater than 553, then ω_F will exist and flutter will occur at $V_F = 24.4b\omega_F$. Actually, practical wings have $I_T/\pi\rho b^4$ much less than 553 except at very high altitudes, where it is possible for this torsional flutter to occur in supersonic flight.

6.8. Flutter of wing segment with two degrees of freedom

The airplane wing not only rotates under load but also deflects in bending so that the two degrees of freedom are α for rotation and h for vertical translation. Coupling between the bending and torsional motions of relative large aspect

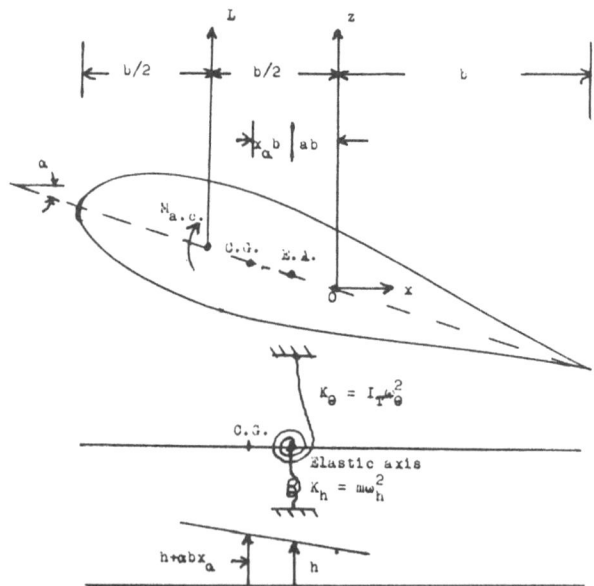

Fig. 6.3. Wing segment under torsional and bending restraints.

wings or tails can cause flutter to occur within the flight speeds of the airplane. It is possible to obtain information about the effect of various wing parameters on this type of flutter by analyzing the unit width wing segment in Figure 6.3. This segment can represent the wing approximately if it is taken about the $y/b = 0.75$ location on the semispan. Usually, it gives a lower flutter velocity than that obtained from more complicated wing analysis.

From Figure 6.3 the equations of motion are

$$mh_{,tt} + S_\alpha \alpha_{,tt} + m\omega_h^2 h = -L,$$
$$S_\alpha h_{,tt} + I_T \alpha_{,tt} + I_T \omega_\alpha^2 \alpha = m_y, \qquad (6.102)$$

where m is the mass of the unit segment, $S_\alpha = mbx_\alpha$ =static mass moment per unit segment about the elastic axis (+ when C.G. is aft), $I_T = (I_T)_{\text{C.G.}} + m(bx_\alpha)^2$ is mass moment of inertia of the unit segment about the elastic axis. From Equations (5.311) and (5.312) in Reference 1,

$$L = \pi \rho b^2 [h_{,tt} + V\alpha_{,t} - b a \alpha_{,tt}] + 2\pi \rho V b C(k)[h_{,t} + V\alpha + b(0.5-a)\alpha_{,t}],$$
$$m_y = \pi \rho b^3 a h_{,tt} + 2\pi \rho V b^2 C(k)(0.5+a)h_{,t} + m_{yT}, \qquad (6.103)$$

where m_{yT} is in Equation (6.94).

Assume simple harmonic motion

$$\alpha = \alpha_0 e^{i\omega t}, \qquad h = h_0 e^{i\omega t}, \qquad (6.104)$$

whence Equations (6.102) and (6.103) become

$$C_1(h_0/b) + C_2 \alpha_0 = 0, \qquad C_3(h_0/b) + C_4 \alpha_0 = 0, \qquad (6.105)$$
$$C_1 = (m/\pi \rho b^2)\left[1 - (\omega_\theta/\omega)^2 (\omega_h/\omega_\theta)^2\right] + 1 - (2iC(k)/k),$$
$$C_2 = (m/\pi \rho b^2)x_\alpha - a - \frac{i}{k} - \frac{(1-2a)iC(k)}{k} - \frac{2C(k)}{k^2},$$
$$C_3 = (m/\pi \rho b^2)x_\alpha - a + \frac{(1+2a)iC(k)}{k},$$
$$C_4 = (I_T/\pi \rho b^4)\left[1 - (\omega_\theta/\omega)^2\right] + D(k) + iE(k),$$

where $D(k)$ and $E(k)$ are in Equation (6.98).

The critical values of ω_F and k_F are given by

$$\begin{vmatrix} C_1 & C_2 \\ C_3 & C_4 \end{vmatrix} = 0. \qquad (6.106)$$

Because of the $C(k)$ function it is not possible to solve Equation (6.106) directly for k_F and ω_θ/ω_F. One procedure is to assume values of k_F and solve for ω_θ/ω_F and ω_h/ω_θ. This allows a graph of

$$V_F/b\omega_\theta = (\omega_F/\omega_\theta)(1/k_F) \qquad (6.107)$$

to be plotted against ω_h/ω_θ for a practical range of ω_h/ω_θ. Such graphs with $m/\pi \rho b^2$, a, and x_α as parameters are shown in Figure 9.5 of Reference 1.

For comparison purposes the divergence speed V_D can be calculated for the unit segment from

$$\theta_y = M_y/K_\theta = m_0 \theta q_D (2b)(1)(b)(0.5+a)/I_T \omega_\theta^2, \qquad (6.108)$$

or

$$V_D/b\omega_\theta = \left[mr_\theta^2/(\pi\rho b^2)(1+2a)\right]^{1/2}, \quad m_0 = 2\pi,$$
$$I_T = mb^2 r_\theta^2, \quad q_D = \rho V_D^2/2. \tag{6.109}$$

Example 6.8. Solve for

$$x = (\omega_\theta/\omega_F)^2, \quad y = (\omega_h/\omega_\theta)^2, \tag{6.110}$$

in Equations (6.105) and (6.106) for a typical wing section with

$$m/\pi\rho b^2 = 3.0, \quad I_T/\pi\rho b^4 = 0.75, \quad a = -0.2, \quad x_\alpha = 0.2. \tag{6.111}$$

Assume $k = 1.00$ and compare $V_F/b\omega_\theta$ to $V_D/b\omega_\theta$ in Equation (6.109).

Solution. In Equations (6.105) and (6.106) take

$$C_i = a_i + ib_i, \quad i = 1, 2, 3, 4, \tag{6.112}$$

whence Equation (6.106) becomes

$$a_1 a_4 - b_1 b_4 - a_2 a_3 + b_2 b_3 = 0,$$
$$a_1 b_4 + a_4 b_1 - a_2 b_3 - a_3 b_2 = 0. \tag{6.113}$$

From Equation (6.105) and the given values

$$\begin{aligned}
&a_1 = 4 - 3xy + 2G/k, \quad b_1 = -2F/k, \\
&a_2 = 0.8 + (1.4G/k) - (2F/k^2), \quad b_2 = -(1 + 1.4F)/k - (2G/k^2), \\
&a_3 = 0.8 - (0.6G/k), \quad b_3 = 0.6F/k, \\
&a_4 = 0.915 - 0.75x - 0.42G/k + 0.60F/k^2, \\
&b_4 = -0.7/k + 0.42F/k + 0.60G/k^2.
\end{aligned} \tag{6.114}$$

For $k = 1$, Table 6.2 gives

$$F = 0.5394, \quad G = -0.1003,$$

whence

$$\begin{aligned}
&a_1 = 3.8 - 3xy, \quad b_1 = -1.0789, \\
&a_2 = -0.4188, \quad b_2 = -1.5546, \\
&a_3 = 0.860, \quad b_3 = 0.3237, \\
&a_4 = 1.2807 - 0.75x, \quad b_4 = -0.534.
\end{aligned}$$

From Equation (6.113),

$$\begin{aligned}
&a_1 = -2.02a_4 + 2.776, \\
&a_4^2 - 1.3689 a_4 + 0.356 = 0, \\
&a_4 = 1.017, \quad 0.351, \\
&x_F = 0.35, \quad 1.24, \\
&y_F = 2.92, \quad 0.405, \\
&\omega_h/\omega_\theta = 1.71, \quad 0.68, \\
&V_F/b\omega_\theta = 1.69, \quad 0.90, \quad V_D/b\omega_\theta = 1.12, \\
&\omega_F/\omega_\theta = 1.69, \quad 0.90.
\end{aligned}$$

Thus, for $\omega_h/\omega_\theta = 0.68$, which is a realistic value for a wing, there exists a flutter speed V_F less than the divergence speed V_D, or for $b = 10$ ft., $\omega_\theta = 60$ cps,

$$V_F = 0.90 b\omega_\theta = 540 \text{ ft/sec} = 367 \text{ mph},$$
$$V_D = 458 \text{ mph}.$$

6.9. Dynamic loads on beams

When rapidly applied external loads act on an airplane, the structure not only deflects in translation and rotation but also may vibrate at its natural frequencies. These effects can produce load and stress variations in the structure, possibly causing large increases in the shear forces, moments, and stresses in the wing. These dynamic loads may arise on an airplane from maneuvers, gusts, turbulence, shock waves, landing, catapulting, arrested landings, and dropping large items from the airplane.

If the natural frequencies and mode shapes have been obtained by the procedures described in previous sections of this chapter, then the dynamic loads and stresses can be obtained from Equations (6.31) and (6.36). Since the external forces usually consist of a component that is a function of space and time and a component that is dependent upon the motion of the strucutre, change Equation (6.36) to

$$M_n(T_{n,tt} + \omega_n^2 T_n) = p_n^F(t) + p_n^M, \quad n = 1, 2, \cdots, N, \quad (6.115)$$

$$p_n^M = p_n^M(T_1, \cdots, T_N; T_{1,t}, \cdots, T_{N,t}; T_{1,tt}, \cdots, T_{N,tt})$$

$$= \int_0^L Y_n(\Delta p^M) dy,$$

$$p_n^F(t) = \int_0^L Y_n p_z(y,t) dy. \quad (6.116)$$

Both the $p_n^F(t)$ and p_n^M cases are considered in Chapter 10 of Reference 1, including examples.

The time varying shear forces and moments on the beam can be obtained from Equation (6.31) by differentiation, or

$$u_z(y,t) = \sum_{n=1}^{N} Y_n(y) T_n(t), \quad (6.31)$$

$$M_x(y,t) = EI u_{z,yy} = \sum_{n=1}^{N} (EI Y_{n,yy}) T_n(t)$$

$$= \sum_{n=1}^{N} M_n(y) T_n(t), \quad (6.117)$$

$$V_z(y,t) = (EI u_{z,yy})_{,y} = \sum_{n=1}^{N} (EI Y_{n,yy})_{,y} T_n(t)$$

$$= \sum_{n=1}^{N} V_n(y) T_n(t). \quad (6.118)$$

From Equation (6.16) for a cantilever beam

$$(EIY_{n,yy})_{,yy} = m\omega_n^2 Y_n,$$

$$V_n(y) = \omega_n^2 \int_y^L m(y_d) Y_n(y_d) dy_d, \qquad (6.119)$$

$$M_n(y) = \omega_n^2 \int_y^L m(y_d) Y_n(y_d)(y_d - y) dy_d.$$

Thus, at point (y, z) on the beam the bending and shear stresses can be expressed as

$$\sigma_{yy}(y, z, t) = \sum_{n=1}^{N} \sigma_n(y, z) T_n(t), \qquad (6.120)$$

where σ_n can be calculated from $M_n(y)$ and $V_n(y)$. The above procedure for calculating deflections, moments, shear forces, and stresses is the *mode displacement method*. Each item is the sum of the contributions from the mode shapes. The convergence of the sum, particularly for shear, may be quite slow.

To improve convergence, the *mode acceleration method* can be used. Take $T_n(t)$ in the form

$$T_n(t) = T_{ns}(t) + Q_n(t) \qquad (6.121)$$

so that Equation (6.31) becomes

$$u_z(y, t) = u_{zs}(y, t) + \sum_{n=1}^{N} Y_n(y) Q_n(t),$$

$$u_{zs}(y, t) = \sum_{n=1}^{N} Y_n(y) T_{ns}(t) = \sum_{n=1}^{N} Y_n(y) p_n^F(t) / M_n \omega_n^2, \qquad (6.122)$$

where $u_{zs}(y, t)$ is the *quasi-steady displacement* due to $p_z(y, t)$ with all vibrations of the beam suppressed. Also, $u_{zs}(y, t)$ can be obtained from

$$(EI u_{zs,yy})_{,yy} = p_z(y, t) \qquad (6.123)$$

by direct integrations. Equation (6.115) for no p_n^M takes the form

$$(T_{n,tt}/\omega_n^2) + T_{ns} + Q_n = p_n^F(t)/M_n \omega_n^2 = T_{ns},$$
$$Q_n = -T_{n,tt}/\omega_n^2, \qquad (6.124)$$

whence

$$u_z(y, t) = u_{zs}(y, t) - \sum_{n=1}^{N} Y_n T_{n,tt}/\omega_n^2,$$

$$M_x(y, t) = M_s(y, t) - \sum_{n=1}^{N} M_n(y) T_{n,tt}/\omega_n^2,$$

Dynamics of simple beams

$$V_x(y,t) = V_s(y,t) - \sum_{n=1}^{N} V_n(y)T_{n,tt}/\omega_n^2,$$

$$\sigma_{yy}(y,z,t) = \sigma_s(y,z,t) - \sum_{n=1}^{N} \sigma_n(y,z)T_{n,tt}/\omega_n^2,$$

$$M_s(y,t) = \int_y^L p_z(y_d,t)(y_d-y)dy_d, \quad V_s = \int_y^L p_z(y_d,t)dy_d. \qquad (6.125)$$

6.10. Problems

6.1. Solve the equation

$(\cosh k_n L)(\cos k_n L) = -1$

for values of $k_n L$ and hence check the frequencies ω_n in Equation (6.30).

6.2. Derive the solution for T_n in Equation (6.42) using zero initial conditions on

$T_n = C_1 \sin \omega_n t + C_2 \cos \omega_n t + Q_n.$

6.3. In Equation (6.42) use $p = q/\omega_n$, $z = \omega_n t$ and examine the motion of

$$U = \frac{\sin pz - p \cos z}{1 - p^2} \qquad (6.126)$$

for $p = 0.90$, $0 \le z \le 20\pi$, and discuss results. What happens for $20\pi \le z \le 40\pi$?

6.4. Find the limit for U in Problem 6.3 when p approaches 1.0. Discuss the resulting motion.

6.5. Check the results in Example 6.2 by making the integrations.

6.6. In Example 6.2 take $F_1(y)$ as

$$F_1(y) = 1 + (\pi^2/2)(y/L)^2 - \cos(\pi y/L) \qquad (6.127)$$

and calculate the frequency ω_1 and the mode shape. Compare the results to Table 6.1. Although this $F_1(y)$ satisfies all four boundary conditions, explain why the frequency is low.

6.7. In Example 6.3 make the calculations to get the $[A]$ matrix in Equation (6.60).

6.8. Use the procedure in Equations (6.24)-(6.29) and set up the general form of the torsional equation for $\theta_y(y,t)$.

6.9. Find the torsional frequencies and mode shapes for the uniform beam clamped at both ends.

6.10. Solve Example 6.6 using two elements.

6.11. Solve Example 6.6 using four elements.

6.12. Solve Example 6.7 for $a = -0.90$.

6.13. Solve Example 6.7 for $a = -0.20$.

6.14. Repeat the calculations in Example 6.8.

6.15. Solve Example 6.8 for $k = 2.0$.

6.16. Solve Example 6.8 for $k = 1.5$.

6.17. Solve Example 6.8 for $k = 3.0$.

6.18. To demonstrate the effect of the center of gravity on flutter, solve Example 6.8 with $x_\alpha = 0.1$. If a_4 is a complex number, there is no flutter.

References

Chapter 6

1. R.L. Bisplinghoff, H. Ashley, R.L. Halfman: *Aeroelasticity*, Addison-Wesley Publishing Co., Reading Mass (1955).
2. W.C. Hurty and M.F. Rubinstein: *Dynamics of Structures*, Prentice-Hall, Inc. (1965).

3. J.S. Przemieniecki: *Theory of Matrix Structural Analysis*, McGraw-Hill Book Co. (1968).
4. R.H. Scanlan and R. Rosenbaum: *Introduction to the Study of Aircraft Vibration and Flutter*, Macmillan Co. (1951).
5. Y.C. Fung: *An Introduction to the Theory of Aeroelasticity*, John Wiley and Sons (1955).
6. T. Theodorsen and I.E. Garrick: *Nonstationary Flow about a Wing, Aileron-Tab Combination including Aerodynamic Balance*, NACA R-736 (1942).
7. Y.L Luke and M.A. Dengler: Tables of the Theodorsen Circulation Function for generalized motion, *Jour. of Aero-Sciences* 18, pp. 478-483, July (1951).

References for additional reading

8. R.W. Clough: *Dynamics of Structures*, McGraw-Hill Book Co., New York (1975).
9. D.G. Fertis: *Dynamics and Vibration of Structures*, John Wiley and Sons, New York (1973).
10. R. Craig, Jr.: *Structural Dynamics: An Introduction to Computer Methods*, John Wiley and Sons, New York (1981).
11. W. Weaver Jr. and P.R. Johnson: *Structural Dynamics by Finite Elements*, Prentice-Hall, Englewood Cliffs, New Jersey (1987).

7

The plate equations

7.1. Introduction

Up to this point the analysis and design methods have been restricted to one-dimensional beams, trusses, box-beams, and simple membrane structures. As can be seen in Section 1.6 (Vol.1), the two dimensional problems for thin plates and long cylinders are more complicated than the one dimensional cases, and involve fourth order partial differential equations. Although series solutions of these equations are given in Reference 1 for various plates, there are many plate problems for which approximate methods with assumed functions must be used. Since it is necessary to use approximate methods for most two dimensional plate problems in flight vehicle structures, either for large plates or for plate finite elements, the discussion in this chapter is devoted primarily to using the virtual principles to obtain approximate solutions for plates.

There are six plate cases that need to be considered, depending upon the type of applied loads and the plate deflections, or

1. Applied forces in the plane of the plate,
2. Lateral applied forces bending the plate,
3. Combined inplane and lateral forces,
4. Combined forces with large bending deflections,
5. Buckling of the plate,
6. Plate vibrations.

The basic equations for these six cases will be examined in this chapter using the plate notation shown in Figure 7.1. This notation is different from that in Figure 1.5 (Vol.1) and in Section 1.6 (Vol.1) for plates. The notation used below is essentially the same as that used in References 1-6. Note that the M_x moment is not about the x-axis as it was in beam problems, but is defined by the σ_{xx} stress. Thus, M_x is about the y-axis and M_y is about the x-axis.

The internal loads indicated in Figure 7.1 are defined as the loads or moments per *unit width*, obtained by integration of the stresses over the thickness t of the plate, or

$$N_x = \int_t \sigma_{xx}\, dz, \quad N_y = \int_t \sigma_{yy}\, dz, \quad N_{xy} = \int_t \sigma_{xy}\, dz,$$

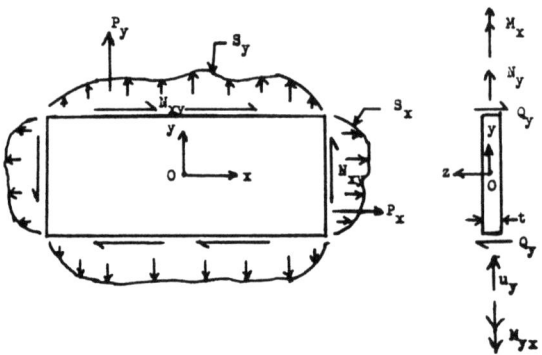

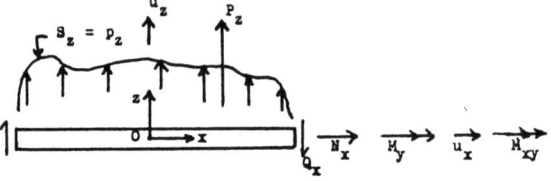

Fig. 7.1. External and internal loadings on plate element.

$$M_x = \int_t \sigma_{xx} z \, dz, \quad M_y = \int_t \sigma_{yy} z \, dz, \quad M_{yx} = -M_{xy} = \int_t \sigma_{xy} z \, dz, \quad (7.1)$$

$$Q_x = \int_t \sigma_{xz} \, dz, \quad Q_y = \int_t \sigma_{yz} \, dz, \quad \sigma_{zz} = 0.$$

Since it is desirable to use matrices in the virtual principles to obtain the differential equations and the approximate solutions, the basic plate equations will be expressed in matrix form with operator matrices for the partial derivatives using

$$d_x = \frac{\partial}{\partial x}, \quad d_{xy} = \frac{\partial^2}{\partial x \partial y}, \quad d_{xx} = \frac{\partial^2}{\partial x^2}, \quad \text{etc.} \quad (7.2)$$

The basic operator matrices and expressions for plates are taken in the form

$$[J_{pd}] = \begin{bmatrix} d_x & 0 \\ 0 & d_y \\ d_y & d_x \end{bmatrix}, \quad [J_{bd}] = \begin{bmatrix} d_y & 0 \\ 0 & d_x \\ -d_x/2 & -d_y/2 \end{bmatrix},$$

$$\{J_{cd}\} = \begin{Bmatrix} d_x \\ d_y \end{Bmatrix}, \quad \{J_{pdd}\} = \begin{Bmatrix} d_{yy} \\ d_{xx} \\ -d_{xy} \end{Bmatrix}, \quad \{J_{bdd}\} = \begin{Bmatrix} d_{xx} \\ d_{yy} \\ 2d_{xy} \end{Bmatrix}, \quad (7.3)$$

$$\{J_{bdd}\} = [J_{pd}]\{J_{cd}\}, \quad \nabla^2 = \{J_{cd}\}^T\{J_{cd}\} = d_{xx} + d_{yy},$$

$$\nabla^4 = \nabla^2 \nabla^2 = d_{xxxx} + 2d_{xxyy} + d_{yyyy}$$

$$= E\{J_{pdd}\}^T[J]\{J_{pdd}\} = \frac{1-\nu^2}{E}\{J_{pdd}\}^T[J]^{-1}\{J_{bdd}\}, \quad (7.4)$$

The plate equations

$$[J] = \frac{1}{E}\begin{bmatrix} 1 & -\nu & 0 \\ -\nu & 1 & 0 \\ 0 & 0 & 2+2\nu \end{bmatrix}, \quad [J]^{-1} = \frac{E}{1-\nu^2}\begin{bmatrix} 1 & \nu & 0 \\ \nu & 1 & 0 \\ 0 & 0 & \frac{1-\nu}{2} \end{bmatrix},$$

$$[\lambda_b]^T = \begin{bmatrix} l & 0 & m \\ 0 & m & l \end{bmatrix}, \quad l \text{ and } m \text{ are direction cosines,} \tag{7.5}$$

$$D = \frac{Et^3}{12(1-\nu^2)} = \text{plate stiffness in bending.}$$

The virtual principles are applied to the plate inplane case in Sections 7.2, 7.3, 7.4, and to the plate bending case in Sections 7.5, 7.6, 7.7. The case of combined inplane and bending loads with small bending deflections is considered in Section 7.8. The combined case with large bending deflections giving a non-linear coupled problem is examined in Section 7.9 using the principle of virtual displacements. Sections 7.10 and 7.11 are concerned with the plate buckling and plate vibration problems using the principle of virtual displacements.

It should be noted that the principle of virtual displacements bending Equation (7.100) in Section 7.5 is a key equation for approximate solutions in Sections 7.5, 7.8, 7.9, 7.10 and 7.11. If the assumed functions and bending boundary conditions are the same for all the cases of lateral loading, combined loading, buckling, and vibrations, then the left side of Equation (7.100) is the same for these cases. The lateral loading on the right side of Equation (7.100) will be different for the cases. However, in certain situations for plate buckling and plate vibration, the integrals on the right side of Equation (7.100) also will be the same so that the buckling and vibration solutions are directly related in such situations, Section 7.11.

7.2. The plate inplane case using the principle of virtual displacements

For this case the S_x, S_y, P_x, P_y external forces in Figure 7.1 act symmetrically about the midplane of the plate. The stresses are regarded as constant average values through the thickness of the plate. From Section 1.6 (Vol.1) with the notation of Figure 7.1 and the above expressions in Equations (7.2)-(7.5), the basic inplane equations relating the strains, stresses, and displacements are

$$\{e\} = \begin{Bmatrix} e_{xx} \\ e_{yy} \\ e_{xy} \end{Bmatrix} = [J_{pd}]\begin{Bmatrix} u_x \\ u_y \end{Bmatrix} = [J_{pd}]\{u\} = [J]\begin{Bmatrix} \sigma_{xx} \\ \sigma_{yy} \\ \sigma_{xy} \end{Bmatrix} + \begin{Bmatrix} e_{xxE} \\ e_{yyE} \\ e_{xyE} \end{Bmatrix}$$

$$= [J]\{\sigma\} + \{e_E\}, \tag{7.6}$$

$$\{\sigma\} = [J]^{-1}\{e - e_E\} = [J]^{-1}[J_{pd}]\{u\} - [J]^{-1}\{e_E\},$$

$$\{N\} = \begin{Bmatrix} N_x \\ N_y \\ N_{xy} \end{Bmatrix} = \int_t \{\sigma\}\,dz = t\{\sigma\}, \quad \{N_E\} = \int_t [J]^{-1}\{e_E\}\,dz, \tag{7.7}$$

$$\{e^V\} = [J_{pd}]\{u^V\}.$$

The equilibrium Equation (1.21, Vol.1) and equilibrium boundary conditions (1.22, Vol.1) are

$$[J_{pd}]^T\{\sigma\} = -\begin{Bmatrix} X_x \\ X_y \end{Bmatrix} = -\{X\} = \{J_{cd}\}V,$$
$$[\lambda_b]^T\{\sigma\} = \begin{Bmatrix} S_x \\ S_y \end{Bmatrix} = \{S\},$$
(7.8)

where V is the potential function for conservative body forces.

For this inplane case use the principle of virtual displacements in Equation (2.22, Vol.1) in the form

$$\int_{A_m}\int_t \left(\sigma_{xx}e^V_{xx} + \sigma_{yy}e^V_{yy} + \sigma_{xy}e^V_{xy}\right) dz\, dA_m - \int_{A_m} (X_x u^V_x + X_y u^V_y) t\, dA_m -$$
$$- \int_C (S_x u^V_x + S_y u^V_y) t\, ds - \sum_{m=1}^M (P_{mx} u^V_{mx} + P_{my} u^V_{my}) = 0. \quad (7.9)$$

Use the above matrix notation and rewrite Equation (7.9) as

$$\int_{A_m} \{[J_{pd}]\{u^V\}\}^T \{[J]^{-1}[J_{pd}]\{u\} - \frac{1}{t}\{N_E\}\} t\, dA_m -$$
$$- \int_{A_m} \{u^V\}^T \{X\} t\, dA_m - \int_C \{u^V\}^T \{S\} t\, ds - \sum_{m=1}^M \{u^V_m\}^T \{P_m\} = 0, \quad (7.10)$$

whence Green's identity in Equation (2.2, Vol.1) gives

$$- \int_{A_m} \{u^V\}^T \{[J_{pd}]^T [J]^{-1} [J_{pd}]\{u\} - \frac{1}{t}[J_{pd}]^T\{N_E\} - \{X\}\} t\, dA_m +$$
$$+ \int_C \{u^V\}^T [\lambda_b]\{N\} ds - \int_C \{u^V\}^T \{S\} t\, ds - \sum_{m=1}^M \{u^V_m\}^T \{P_m\} = 0. \quad (7.11)$$

Omit the point loads $\{P_m\}$ so that for arbitrary virtual displacements Equation (7.11) gives the displacement differential equations and edge conditions, or

$$[J_{pd}]^T [J]^{-1} [J_{pd}]\{u\} = -\{X\} + \frac{1}{t}[J_{pd}]^T \{N_E\},$$
$$[\lambda_b]^T [J]^{-1} [J_{pd}]\{u\} = \{S\},$$
$$[J_{pd}]^T [J]^{-1} [J_{pd}] = \frac{E}{1-\nu^2} \begin{bmatrix} d_{xx} + \frac{1-\nu}{2} d_{yy} & \frac{1+\nu}{2} d_{xy} \\ \frac{1+\nu}{2} d_{xy} & d_{yy} + \frac{1-\nu}{2} d_{xx} \end{bmatrix}.$$
(7.12)

Besides the equilibrium boundary conditions in Equation (7.12), u_x and u_y must satisfy also any displacement boundary conditons, including at least rigid body restraints.

The displacement equations in Equation (7.12) are coupled partial differential equations. They are difficult to solve for either specified boundary displacements or for specified boundary forces. There is very little on the solution of these equations in the literature on plates, which is primarily concerned with the stress

The plate equations

equations for inplane problems, Section 7.3. However, approximate solutions for Equation (7.12) can be obtained by using the principle of virtual displacements in Equation (7.10).

Assume functions for u_x and u_y in the form

$$u_x = F_x(x,y) + \sum_m \sum_n A_{mn} G_{xm}(x) H_{xn}(y),$$
$$u_y = F_y(x,y) + \sum_m \sum_n B_{mn} G_{ym}(x) H_{yn}(y), \qquad (7.13)$$

$$u_x^V = \sum_i \sum_j A_{ij}^V G_{xi}(x) H_{xj}(y),$$
$$u_y^V = \sum_i \sum_j B_{ij}^V G_{yi}(x) H_{yj}(y), \qquad (7.14)$$

where F_x, F_y represent any specified non-zero displacement boundary values and where the $G(x)H(y)$ functions satisfy the displacement support conditions for the plate and are zero where F_x, F_y are specified. When these forms of u_x and u_y are substituted into Equation (7.10), various double integrals can be identified so that these equations can be written as

$$\sum_m \sum_n [A_{mn}(I^{a1}_{mnij} + I^{a2}_{mnij}) + B_{mn}(I^{a3}_{mnij} + I^{a4}_{mnij})] = I^{a5}_{ij},$$
$$\sum_m \sum_n [A_{mn}(I^{b1}_{mnij} + I^{b2}_{mnij}) + B_{mn}(I^{b3}_{mnij} + I^{b4}_{mnij})] = I^{b5}_{ij}. \qquad (7.15)$$

Use $G' = \frac{dG}{dx}$, $H' = \frac{dH}{dy}$, etc., and assume constant material properties so that the integrals are

$$I^{a1}_{mnij} = \int_{A_m} G'_{xm} H_{xn} G'_{xi} H_{xj} \, dx \, dy,$$
$$I^{a2}_{mnij} = \int_{A_m} \frac{1-\nu}{2} G_{xm} H'_{xn} G_{xi} H'_{xj} \, dx \, dy,$$
$$I^{a3}_{mnij} = \int_{A_m} \nu G_{ym} H'_{yn} G'_{xi} H_{xj} \, dx \, dy,$$
$$I^{a4}_{mnij} = \int_{A_m} \frac{1-\nu}{2} G'_{ym} H_{yn} G_{xi} H'_{xj} \, dx \, dy, \qquad (7.16)$$

$$I^{a5}_{ij} = \int_{A_m} \Big[G'_{xi} H_{xj}(e_{xxE} + \nu e_{yyE} - F_{x,x} - \nu F_{y,y}) +$$
$$+ \frac{1-\nu}{2} G_{xi} H'_{xj}(e_{xyE} - F_{x,y} - F_{y,x}) + \frac{1-\nu^2}{E} G_{xi} H_{xj} X_x \Big] dx \, dy +$$
$$+ \frac{1-\nu^2}{Et} \Big[\int_c tS_x G_{xi} H_{xj} \, ds + \sum_k P_{xk} G_{xi}(x_k) H_{xj}(y_k) \Big], \qquad (7.17)$$

$$I^{b1}_{mnij} = \int_{A_m} \nu G'_{xm} H_{xn} G_{yi} H'_{yj} \, dx \, dy,$$

$$I^{b2}_{mnij} = \int_{A_m} \frac{1-\nu}{2} G_{xm} H'_{xn} G'_{yi} H_{yj} \, dx \, dy,$$ (7.18)

$$I^{b3}_{mnij} = \int_{A_m} G_{ym} H'_{yn} G_{yi} H'_{yj} \, dx \, dy,$$

$$I^{b4}_{mnij} = \int_{A_m} \frac{1-\nu}{2} G'_{ym} H_{yn} G'_{yi} H_{yj} \, dx \, dy,$$

$$I^{b5}_{ij} = \int_{A_m} \Big[G_{yi} H'_{yj} (\nu e_{xxE} + e_{yyE} - \nu F_{x,x} - F_{y,y}) +$$

$$+ \frac{1-\nu}{2} G'_{yi} H_{yj} (e_{xyE} - F_{x,y} - F_{y,x}) + \frac{1-\nu^2}{E} G_{yi} H_{yj} X_y \Big] dx \, dy +$$

$$+ \frac{1-\nu^2}{E} \Big[\int_C G_{yi} H_{yj} S_y t \, ds + \sum_k P_{yk} G_{yi}(x_k) H_{yj}(y_k) \Big].$$ (7.19)

Note that Equation (7.15) can be written in matrix form by using a single counter for each double counter, Equation (A.90, Vol.1). Use a procedure such as $11, 12, \cdots, 1N, 21, 22, \cdots, 2N, 31, \cdots, MN \rightarrow 1, 2, 3, \cdots, N, N+1, N+2, \cdots, 2N, 2N+1, \cdots, MN$, whence

$$\begin{bmatrix} I^{aa} & I^{ab} \\ I^{ba} & I^{bb} \end{bmatrix} \begin{Bmatrix} A \\ B \end{Bmatrix} = \begin{Bmatrix} I^{a5} \\ I^{b5} \end{Bmatrix}.$$ (7.20)

If the $G_m(x)$ and $H_n(y)$ sets of functions are orthogonal, then many of the integrals in Equations (7.16)-(7.19) will be zero and the constants may be uncoupled, which simplifies the calculations. The following example shows the solution procedure for a simple plate case with inplane loading. See References 1-6 for other examples.

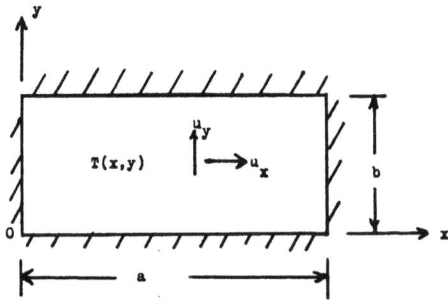

Fig. 7.2. Restrained inplane plate for Example 7.1.

Example 7.1. (a) Supose the rectangular plate in Figure 7.2 is restrained on all edges, or $u_x = 0$ on $x = 0, a$, $u_y = 0$ on $y = 0, b$, and is subjected to a temperature change $T(x,y)$ constant through the thickness. Use the same orthogonal sets of trigometric functions for both

The plate equations

u_x and u_y as $G_m(x)$ and $H_n(y)$ in Equation (7.13), or

$$G_m(x) = \sin(m\pi x/a), \quad H_n(y) = \sin(n\pi y/b), \tag{7.21}$$

and set up Equation (7.15) for the unknown constants A_{mn} and B_{mn}. Assume E, G, ν, α, t are constant. (b) Assume $T = T_0 \sin(\pi x/a) \sin(\pi y/b)$ and find the displacements and stresses in the square plate for only the A_{21} and B_{12} constants in the approximations.

Solution. (a) From the orthogonal conditions,

$$\int_0^b \int_0^a \sin(m\pi x/a) \sin(i\pi x/a) \sin(n\pi y/b) \sin(j\pi y/b) dx\, dy$$
$$= 0 \quad \text{for} \quad m \neq i \quad \text{and} \quad n \neq j,$$
$$= (a/2)(b/2) \quad \text{for} \quad i = m \quad \text{and} \quad j = n. \tag{7.22}$$

The same results apply for cosines. Also,

$$\int_0^b \int_0^a \sin(m\pi x/a) \cos(i\pi x/a) \sin(n\pi y/b) \cos(j\pi y/b) dx\, dy$$
$$= 0 \quad \text{for} \quad m \pm i \quad \text{even and} \quad n \pm j \quad \text{even},$$
$$= \frac{4mnab}{\pi^2(m^2 - i^2)(n^2 - j^2)} \quad \text{for} \quad m \pm i \quad \text{odd}, \quad n \pm j \quad \text{odd}. \tag{7.23}$$

From Equations (7.16)-(7.19) the non-zero integrals are

$$I^{a1}_{ijij} = (i\pi/a)^2(ab/4), \quad I^{a2}_{ijij} = \frac{1-\nu}{2}(j\pi/b)^2(ab/4),$$

$$I^{a3}_{mnij} = \frac{4\nu mnij}{(m^2 - i^2)(j^2 - n^2)} \quad \text{for} \quad m \pm i \quad \text{odd}, \quad n \pm j \quad \text{odd},$$

$$I^{a4}_{mnij} = \frac{2(1-\nu)mnij}{(m^2 - i^2)(j^2 - n^2)} \quad \text{for} \quad m \pm i \quad \text{odd}, \quad n \pm j \quad \text{odd}, \tag{7.24a}$$

$$I^{a5}_{ij} = (1+\nu)\int_0^b \int_0^a (i\pi/a)\alpha T \cos(i\pi x/a) \sin(j\pi y/b) dx\, dy$$
$$= (i\pi/a)(1+\nu)T_{xij},$$

$$I^{b1}_{mnij} = I^{a3}_{mnij}, \quad I^{b2}_{mnij} = I^{a4}_{mnij},$$

$$I^{b3}_{ijij} = (j\pi/b)^2(ab/4), \quad I^{b4}_{ijij} = \frac{1-\nu}{2}(i\pi/a)^2(ab/4), \tag{7.24b}$$

$$I^{b5}_{ij} = (1+\nu)\int_0^b \int_0^a (j\pi/b)\alpha T \sin(i\pi x/a) \cos(j\pi y/b) dx\, dy$$
$$= (j\pi/b)(1+\nu)T_{yij}.$$

Thus, Equation (7.15) has the form

$$\frac{\pi^2}{4} A_{ij}\left[(bi^2/a) + \frac{1-\nu}{2}(aj^2/b)\right] + \sum_m \sum_n B_{mn} \frac{2(1+\nu)mnij}{(m^2 - i^2)(j^2 - n^2)}$$
$$= (i\pi/a)(1+\nu)T_{xij},$$

$$\frac{\pi^2}{4} B_{ij}\left[(aj^2/b) + \frac{1-\nu}{2}(bi^2/a)\right] + \sum_m \sum_n A_{mn} \frac{2(1+\nu)mnij}{(m^2 - i^2)(j^2 - n^2)}$$
$$= (j\pi/b)(1+\nu)T_{yij}, \quad m \pm i \quad \text{odd}, \quad n \pm j \quad \text{odd}. \tag{7.25}$$

Although some of the integrals drop out due to the orthogonal conditions, the system is still coupled. For several constants use the procedure of Equation (7.20) with a computer program for simultaneous equations.

(b) For the approximation

$$u_x = A_{21} \sin(2\pi x/a) \sin(\pi y/a), \quad u_y = B_{12} \sin(\pi x/a) \sin(2\pi y/a), \quad (7.26)$$

in the square plate with $b = a$ and the given temperature distribution $T(x, y)$, Equation (7.24) and (7.25) give

$$T_{x21} = T_{y12} = -(a^2 \alpha T_0/3\pi),$$
$$A_{21} = B_{12} = -\frac{48(1+\nu)a\alpha T_0}{9(9-\nu)\pi^2 + 64(1+\nu)}. \quad (7.27)$$

For the case of $\nu = 1/4$,

$$u_x = -0.0700 a\alpha T_0 \sin(2\pi x/a) \sin(\pi y/a),$$
$$u_y = -0.0700 a\alpha T_0 \sin(\pi x/a) \sin(2\pi y/a), \quad (7.28)$$

$$N_{T0} = tE\alpha T_0, \quad (7.29)$$

$$N_x = t\sigma_{xx} = \frac{tE}{1-\nu^2}[u_{x,x} + \nu u_{y,y} - (1+\nu)\alpha T]$$
$$= -0.4691 N_{T0}[\cos(2\pi x/a)\sin(\pi y/a) + 0.25 \sin(\pi x/a)\cos(2\pi y/a)] -$$
$$\quad - 1.3333 N_{T0} \sin(\pi x/a)\sin(\pi y/a)$$
$$= -0.4691 N_{T0} \quad \text{at} \quad x = 0, a, \ y = a/2,$$
$$= -0.7469 N_{T0} \quad \text{at} \quad x = a/2, \ y = a/2,$$
$$N_y = -0.4691 N_{T0}[0.25 \cos(2\pi x/a)\sin(\pi y/a) + \sin(\pi x/a)\cos(2\pi y/a)] -$$
$$\quad - 1.3333 N_{T0} \sin(\pi x/a)\sin(\pi y/a)$$
$$= -0.1173 N_{T0} \quad \text{at} \quad x = 0, a, \ y = a/2,$$
$$= -0.7469 N_{T0} \quad \text{at} \quad x = a/2, \ y = a/2, \quad (7.30)$$
$$N_{xy} = t\sigma_{xy} = \frac{tE}{1-\nu^2}(u_{x,y} + u_{y,x})\left(\frac{1-\nu}{2}\right)$$
$$= -0.0880 N_{T0}[\sin(2\pi x/a)\cos(\pi y/a) + \cos(\pi x/a)\sin(2\pi y/a)]$$
$$= -0.1244 N_{T0} \quad \text{at} \quad x = a/4, \ y = a/4.$$

7.3. The plate inplane case using the principle of virtual forces

To obtain the stress form of the plate equations use the stress function ϕ as defined in Equation (1.24, Vol.1), or from Equation (7.4),

$$\{\sigma\} = \{J_{pdd}\}\phi + \{s\}V, \quad \{s\}^T = [1 \ 1 \ 0],$$
$$\{e\} = [J]\{\sigma\} + \{e_E\}, \quad \{\sigma^V\} = \{J_{pdd}\}\phi^V. \quad (7.31)$$

Note that this form of $\{\sigma\}$ satisfies Equation (7.8).

For the inplane case use the principle of virtual forces in Equation (2.23, Vol.1) in the form

$$\int_{A_m}(e_{xx}\sigma_{xx}^V + e_{yy}\sigma_{yy}^V + e_{xy}\sigma_{xy}^V)t\,dA_m - \int_{A_m}(u_x X_x^V + u_y X_y^V)t\,dA_m$$
$$- \int_C (u_x S_x^V + u_y S_y^V)t\,ds - \sum_{m=1}^{M}(u_{mx}P_{mx}^V + u_{my}P_{my}^V) = 0. \quad (7.32)$$

The plate equations

Since the virtual stresses satisfy Equation (7.8) the second integral in Equation (7.32) is zero. Use Equation (7.31) and rewrite Equation (7.32) as

$$\int_{A_m} \{\{J_{pdd}\}\phi^V\}^T \{[J]\{J_{pdd}\}\phi + [J]\{s\}V + \{e_E\}\}t\,dA_m -$$
$$- \int_C \{u\}^T \{S^V\} t\,ds - \sum_k \{u_k\}^T \{P_k^V\} = 0. \tag{7.33}$$

From Equations (7.8) and (7.31) the stress boundary conditions are

$$[\lambda_b]^T \{J_{pdd}\}\phi + [\lambda_b]^T \{s\}V = \{S\}, \tag{7.34}$$

which can be written in the form

$$\begin{Bmatrix} \phi_{,y} \\ -\phi_{,x} \end{Bmatrix} = \int_{s_0}^{s} \begin{Bmatrix} S_x - lV \\ S_y - mV \end{Bmatrix} ds. \tag{7.35}$$

Use Equation (7.35) as a virtual form and integrate the boundary integral in Equation (7.33) by parts to get

$$\int_C u_x S_x^V t\,ds = \left[t u_x \int_{s_0}^{s} s_x^V\,ds \right]_{s_0}^{s_0} - \int_C \left[u_{x,s} \int_{s_0}^{s} S_x^V\,ds \right] t\,ds$$
$$= 0 - \int_C \phi_{,y}^V u_{x,s} t\,ds, \tag{7.36}$$

$$\int_C u_y S_y^V t\,ds = \int_C \phi_{,x}^V u_{y,s} t\,ds. \tag{7.37}$$

Use the results in Equations (7.36) and (7.37), omit point values, and integrate the first integral in Equation (7.33) by using Equation (2.2, Vol.1), whence Equation (7.33) becomes

$$\int_{A_m} \phi^V [\nabla^4 \phi + E\{J_{pdd}\}^T\{e_E\} + (1-\nu)\nabla^2 V] t\,dA_m +$$
$$+ \int_C \phi_{,y}^V [m\nabla^2 \phi + (1+\nu)(\phi_{,xs}) + m\{Ee e_{Ex} + (1-\nu)V\} - \frac{El}{2} e_{Exy} + E u_{x,s}] t\,ds +$$
$$+ \int_C \phi_{,x}^V [l\nabla^2 \phi - (1+\nu)\phi_{,ys} + l\{Ee e_{Ey} + (1-\nu)V\} - \frac{Em}{2} e_{Exy} - E u_{y,s}] t\,ds -$$
$$- \int_C \phi^V [(\nabla^2 \phi)_{,n} + (1-\nu)V_{,n} + E(m e_{Ex,y} + l e_{Ey,x} - \tfrac{1}{2} e_{Exy,y} - \tfrac{m}{2} e_{Exy,x})] t\,ds$$
$$= 0. \tag{7.38}$$

Since ϕ^V, $\phi_{,x}^V$, and $\phi_{,y}^V$ are arbitrary this gives the differential equation and displacement boundary conditions for ϕ as

$$\nabla^4 \phi + \{J_{pdd}\}^T \{E e_E\} + (1-\nu)\nabla^2 V = 0, \tag{7.39}$$
$$m\nabla^2 \phi + (1+\nu)\phi_{,xs} + m(1-\nu)V + E(me_{Ex} - \tfrac{1}{2} e_{Exy}) = -E u_{x,s}, \quad \text{on } C,$$
$$l\nabla^2 \phi - (1+\nu)\phi_{,ys} + l(1-\nu)V + E(le_{Ey} - \frac{m}{2} e_{Exy} = E u_{y,s}), \quad \text{on } C, \tag{7.40}$$

$$[\nabla^2\phi + (1-\nu)V]_{,n} + E(me_{Ex,y} + le_{Ey,x} - \tfrac{1}{2}e_{Exy,y} - \tfrac{m}{2}e_{Exy,x})$$
$$= 0, \quad \text{on } C, \tag{7.41}$$

where Equation (7.41) is the condition that $\nabla^2\phi$ be single-valued.

Use Equation (7.31) in Equation (7.6) and integrate to get

$$u_x = -\frac{1+\nu}{E}\phi_{,x} + \int_{x_0}^{x}\left[e_{Ex} + \frac{1}{E}\{\nabla^2\phi + (1-\nu)V\}\right]dx + f_1(y),$$
$$u_y = -\frac{1+\nu}{E}\phi_{,y} + \int_{y_0}^{y}\left[e_{Ey} + \frac{1}{E}\{\nabla^2\phi + (1-\nu)V\}\right]dy + f_2(x), \tag{7.42}$$

$$e_{xy} = u_{x,y} + u_{y,x} = -\frac{2(1+\nu)}{E}\phi_{,xy} + f_{1,y}(y) + f_{2,x}(x) + e_{Exy} +$$
$$+ \int_{(x_0,y_0)}^{(x,y)}\{\left[e_{Ex} + \frac{1}{E}\{\nabla^2\phi + (1-\nu)V\}\right]_{,y}dx +$$
$$+ \left[e_{Ey} + \frac{1}{E}\{\nabla^2\phi + (1-\nu)V\}\right]_{,x}dy\}$$
$$= -\frac{2(1+\nu)}{E}\phi_{,xy} + e_{Exy} = (\sigma_{xy}/G) + e_{Exy}, \tag{7.43}$$

where the integral in Equation (7.43) will be zero if $e_{Ex} = e_{Ey}$, $\nabla^2\phi$, and V are all harmonic functions (see Section 57 in Reference 3). In this case

$$f_{1,y}(y) + f_{2,x}(x) = 0, \tag{7.44}$$

which adds at most a rigid body displacement to the plate. Otherwise, with V and $\nabla^2\phi$ harmonic, e_{Ex} and e_{Ey} could have the forms $e_{Ex,y}(x)$ and $e_{Ey,x}(y)$. Note that Equation (7.42) will give Equation (7.40) on the boundary C, except for the e_{Exy} term.

It is evident from Equations (7.39), (7.35), and (7.40) that a direct solution for ϕ is difficult for both the stress and displacement boundary conditions. However, closed form solutions can be obtained in many cases by using analytic functions in complex variables, Chapter 5 of Reference 8 and Chapter 9 of Reference 9. The approximation procedure using assumed functions for ϕ in Equation (7.33) will be used here. Take

$$\phi = \phi_0(x,y) + \sum_m\sum_n D_{mn}G_m(x)H_n(y),$$
$$\phi^V = \sum_i\sum_j D^V_{ij}G_i(x)H_j(y), \tag{7.45}$$

where ϕ_0 represents any specified non-zero applied edge forces including reactions for an equilibrium system, and where the $G_m(x)H_n(y)$ expressions satisfy zero stress boundary conditons.

When the forms for ϕ and ϕ^V in Equation (7.45) are put into Equation (7.33),

The plate equations

various double integrals can be identified so that Equation (7.33) has the form

$$\sum_m \sum_n D_{mn}(I^{d1}_{mnij} + I^{d2}_{mnij} + I^{d3}_{mnij} + I^{d4}_{mnij} + I^{d5}_{mnij}) = I^{d6}_{ij},$$

$$[I^d]\{D\} = \{I^{d6}\},$$
(7.46)

where

$$I^{d1}_{mnij} = \int_{A_m} G''_m H_n G''_i H_j \, dx \, dy, \qquad I^{d2}_{mnij} = -\nu \int_{A_m} G_m H''_n G''_i H_j \, dx \, dy,$$

$$I^{d3}_{mnij} = \int_{A_m} G_m H''_n G_i H''_j \, dx \, dy, \qquad I^{d4}_{mnij} = -\nu \int_{A_m} G''_m H_n G_i H''_j \, dx \, dy,$$

$$I^{d5}_{mnij} = 2(1+\nu) \int_{A_m} G'_m H'_n G'_i H'_j \, dx \, dy, \qquad (7.47)$$

$$I^{d6}_{ij} = -\int_{A_m} [G_i H''_j \{\phi_{0,yy} - \nu\phi_{0,xx} + (1-\nu)V + Ee_{Ex}\} + G''_i H_j \{\phi_{0,xx}-$$
$$-\nu\phi_{0,yy} + (1-\nu)V + Ee_{Ey}\} + G'_i H'_j \{2(1+\nu)\phi_{0,xy} + Ee_{Exy}\}] dx\,dy+$$
$$+\int_C E[u_x(lG_i H''_j - mG''_i H'_j) + u_y(mG''_i H_j - lG'_i H'_j)]ds+$$
$$+ (E\Delta A/t)\sum_k [u_{xk}G_i(x_k)H''_j(y_k) + u_{yk}G''_i(x_k)H_j(y_k)]. \qquad (7.48)$$

Once the D_{mn} constants in Equation (7.46) are obtained the stresses are given by Equation (7.31) and the displacements by Equation (7.42).

It should be noted that the $\nabla^4\phi$ term in Equation (7.39) and the force boundary conditions in Equation (7.34) do not involve material properties. Thus, the integrals involving ν in Equation (7.47) will cancel out when combined in Equation (7.46). Also, the ϕ_0 terms involving ν in Equation (7.48) will cancel out or be zero.

Because of the particular derivatives involved in the stress boundary conditions, the trigometric functions are complicated so that polynomials may be preferable for stress functions. For the rectangular plate in Figure 7.3, use

$$\phi - \phi_0 = (x^2 - a^2)^2(y^2 - b^2)^2 \sum_m \sum_n D_{mn} x^{m-1} y^{n-1},$$

$$\phi^V = (x^2 - a^2)^2(y^2 - b^2)^2 \sum_i \sum_j D^V_{ij} x^{i-1} y^{j-1},$$
(7.49)

which satisfy the zero stress boundary conditions

$$\sigma_{xy} = -\phi_{,xy} = 0, \qquad \sigma_{xx} = \phi_{,yy} = 0 \quad \text{on} \quad x = \pm a,$$
$$\sigma_{xy} = -\phi_{,xy} = 0, \qquad \sigma_{yy} = \phi_{,xx} = 0, \quad \text{on} \quad y = \pm b.$$
(7.50)

Example 7.2. (a) Find the approximate stresses in the rectangular plate with the temperature and axial loading shown in Figure 7.3. Use polynomials to satisfy the boundary conditions on the stresses. (b) Find the deflections of the plate.

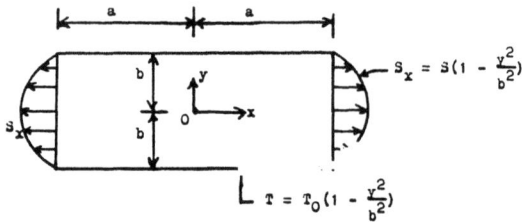

Fig. 7.3. Rectangular plate with inplane axial load and temperature.

Solution. (a) From Equations (7.31) and (7.49) and Figure 7.3 the stress boundary conditions can be written in terms of the stress function ϕ as

$$\sigma_{xy} = -\phi_{,xy} = 0, \quad \sigma_{xx} = \phi_{,yy} = S\left(1 - \frac{y^2}{b^2}\right), \quad \text{on} \quad x = \pm a, \tag{7.51}$$
$$\sigma_{xy} = -\phi_{,xy} = 0, \quad \sigma_{yy} = \phi_{,xx} = 0, \quad \text{on} \quad y = \pm b.$$

To meet these conditions, select

$$\phi_0 = \tfrac{1}{2}Sy^2\left(1 - \tfrac{1}{6}\tfrac{y^2}{b^2}\right),$$
$$\phi = \phi_0 + (x^2 - a^2)^2(y^2 - b^2)^2(D_{11} + D_{21}x^2 + D_{12}y^2), \tag{7.52}$$

where D_{11}, D_{21}, and D_{12} are the unknown constants to be determined, and in Equation (7.45)

$$G_1 = (x^2 - a^2)^2, \quad G_2 = x^2(x^2 - a^2)^2,$$
$$H_1 = (y^2 - b^2)^2, \quad H_2 = y^2(y^2 - b^2)^2. \tag{7.53}$$

Differentiate to get

$$G_1' = 4x(x^2 - a^2), \quad G_1'' = 4(3x^2 - a^2), \quad G_2' = 2x(x^2 - a^2)(3x^2 - a^2),$$
$$G_2'' = 2(15x^4 - 12a^2x^2 + a^4), \quad H_1' = 4y(y^2 - b^2), \quad H_1'' = 4(3y^2 - b^2), \tag{7.54}$$
$$H_2' = 2y(y^2 - b^2)(3y^2 - b^2), \quad H_2'' = 2(15y^4 - 12b^2y^2 + b^4),$$

whence Equations (7.47) and (7.48) give

$$I_{11,11}^{d1} = 16 \int_0^b \int_0^a (3x^2 - a^2)^2 (y^2 - b^2)^4 dx\,dy = 16\left(\frac{4a^5}{5}\right)\left(\frac{128b^9}{315}\right),$$

$$I_{11,11}^{d2} = I_{11,11}^{d4} = -\nu\left(\frac{128}{105}\right)^2 a^7 b^7, \quad I_{11,11}^{d3} = 16\left(\frac{128a^9}{315}\right)\left(\frac{4b^5}{5}\right),$$

$$I_{11,11}^{d5} = 2(1+\nu)\left(\frac{128}{105}\right)^2 a^7 b^7,$$

$$I_{11d}^{d6} = -\int_0^b \int_0^a \left[4\left\{S\left(1 - \frac{y^2}{b^2}\right) + N_{T0}\left(1 - \frac{y^2}{b^2}\right)\right\}(x^2 - a^2)^2(3y^2 - b^2) + \right. \tag{7.55}$$
$$\left. + 4(-\nu S + N_{T0})\left(1 - \frac{y^2}{b^2}\right)(3x^2 - a^2)(y^2 - b^2)^2 \right] dx\,dy$$

$$= -4\left(\frac{8a^5}{15}\right)\left(-\frac{4b^3}{15}\right)(S + N_{T0}), \quad N_{T0} = tE\alpha T,$$

with similar type expressions for the other integrals. The results from Equations (7.46) and (7.52) with $b/a = k$ are

The plate equations

$$(1 + \frac{4k^2}{7} + k^4)D_{11} + (\frac{1}{11} + \frac{k^4}{7})a^2 D_{21} + (\frac{1}{7} + \frac{k^4}{11})b^2 D_{12} = 7(S + N_{T0})/64a^4 b^2,$$

$$(\frac{1}{11} + \frac{k^4}{7})D_{11} + (\frac{3}{143} + \frac{4k^2}{77} + \frac{3k^4}{7})a^2 D_{21} + (1 + k^4)\frac{b^2 D_{12}}{77} = 7(S + N_{T0})/64a^4 b^2, \quad (7.56)$$

$$(\frac{1}{7} + \frac{k^4}{11})D_{11} + (1 + k^4)\frac{a^2 D_{21}}{77} + (\frac{3}{7} + \frac{4k^2}{77} + \frac{3k^4}{143})b^2 D_{12} = (S + N_{T0})/64a^4 b^2.$$

This system of equations can be solved for the constants D_{11}, D_{12}, D_{21} for any selected value of k. If $k = b/a = 1/2$, then

$$D_{11} = 0.07983(S + N_{T0})/a^4 b^2, \qquad D_{21} = 0.1250(S + N_{T0})/a^6 b^2,$$
$$D_{12} = 0.01826(S + N_{T0})/a^4 b^4. \tag{7.57}$$

From Equations (7.57) and (7.31), the internal loads per unit length for $k = 1/2 = b/a$ are

$$\frac{N_x}{S} = \left(1 - \frac{y^2}{b^2}\right) - \left(1 + \frac{N_{T0}}{S}\right)\left(1 - \frac{x^2}{a^2}\right)^2 \left[\left(0.3193 + \frac{x^2}{2a^2}\right)\left(1 - 3\frac{y^2}{b^2}\right) - \right.$$
$$\left. - 0.00913\left(1 - 12\frac{y^2}{b^2} + 15\frac{y^4}{b^4}\right)\right], \tag{7.58}$$

$$\frac{N_y}{S} = -\left(1 + \frac{N_{T0}}{S}\right)\left(1 - \frac{y^2}{b^2}\right)^2 \left[\left(1 - \frac{3x^2}{a^2}\right)\left(0.07983 + 0.00456\frac{y^2}{b^2}\right) - \right.$$
$$\left. - 0.06250\left(1 - 12\frac{x^2}{a^2} + 15\frac{x^4}{a^4}\right)\right], \tag{7.59}$$

$$\frac{N_{xy}}{S} = -\left(1 + \frac{N_{T0}}{S}\right)\left(\frac{xy}{ab}\right)\left(1 - \frac{x^2}{a^2}\right)\left(1 - \frac{y^2}{b^2}\right)\left[0.63704 - \right.$$
$$\left. - 0.5000\left(1 - \frac{3x^2}{a^2}\right) - 0.01824\left(1 - \frac{3y^2}{b^2}\right)\right]. \tag{7.60}$$

By assigning values to x/a, plots of the stress distributions on the cross-sections can be made by varying y/b. Figure 7.4 shows separate graphs of the stresses for the applied force and for the temperature at several cross sections in the first quadrant. The stresses N_x and N_y are symmetrical about both the x and y axes, while N_{xy} is antisymmetrical. For this $a = 2b$ plate additional terms in the Equation (7.52) have little effect so that the above solution is close to the correct solution. Since the N_x stress due to the end parabolic force is practically constant at the average stress of $0.667S$ at $x = 0$, which is at a distance $2b$ from the end and corresponds to the height of the plate, this two-dimensional solution verifies Saint Venant's principle (Section 1.7, Vol.1) for the axially loaded beam. Away from the ends at a distance at least equal to the height $2b$, the beam stresses for axial loads are $N_x = 0.667S$, $N_y = 0$, $N_{xy} = 0$. From Figure 7.4, N_y is quite small $(-0.017S)$ and N_x varies between $0.689S$ and $0.649S$. Although $N_{xy} = 0$ at $x = 0$ because of anti-symmetry in this example, it is very small near $x = 0$. Results for square plate are on page 169 of Reference 3.

The N_x thermal stresses change from zero at the ends to values at $x = 0$ nearly the same as for the long beam case, $0.675 N_{T0}$ as compared to $0.667 N_{T0}$ at $y = b$, $-0.311 N_{T0}$ as compared to $-0.333 N_{T0}$ at $y = 0$. The N_y and N_{xy} thermal stresses are the same as for the force case with T and $S_x(a, y)$ having the same distribution in Figure 7.4, and are small at $x = 0$.

(b) To get the displacements u_x and u_y, assume

$$u_x(0,0) = 0, \quad u_y(0,0) = 0, \quad u_{x,y}(0,0) = 0, \tag{7.61}$$

so that $f_1(y) = 0$, $f_2(x) = 0$ in Equation (7.42). Take $V = 0$ and use

$$e_{Ex} = e_{Ey} = \alpha T = (N_{T0}/Etb^2)(b^2 - y^2),$$
$$t\nabla^2 \phi = N_x + N_y, \quad t\phi_{,x} = \int_0^x N_y \, dx, \quad t\phi_{,y} = \int_0^y N_x \, dy, \tag{7.62}$$
$$x_r = x/a, \quad y_r = y/b$$

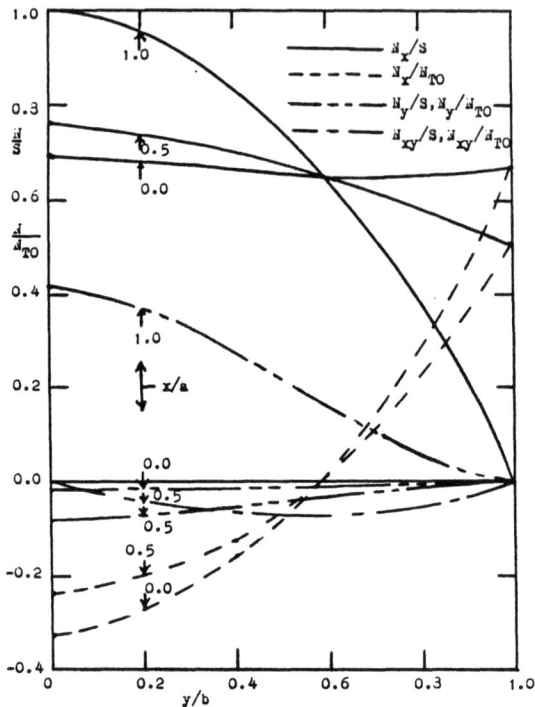

Fig. 7.4. Two-dimensional stresses in plate in Fig. 7.3 with $a = 2b$, Example 7.2.

in Equation (7.42) to get

$$Etu_x(x,y) = \int_0^x \left[N_x - \nu N_y + N_{T0}(1 - y_r^2)\right]dx,$$
$$Etu_y(x,y) = \int_0^y \left[N_y - \nu N_x + N_{T0}(1 - y_r^2)\right]dy. \tag{7.63}$$

Integration gives

$$\frac{Etu_x(x,y)}{S + N_{T0}} = x(1 - y_r^2) - x[(15 - 10x_r^2 + 3x_r^4)\{0.0213(1 - 3y_r^2) -$$
$$- 0.00061(1 - 12y_r^2 + 15y_r^4)\} + 0.00476x_r^2(35 - 42x_r^2 + 15x_r^4)(1 - 3y_r^2)] +$$
$$+ \nu x(1 - y_r^2)(1 - x_r^2)[0.07983 + 0.00456y_r^2 - 0.06250(1 - 3x_r^2)], \tag{7.64}$$

$$\frac{Etu_y(x,y)}{S + N_{T0}} = 0.3333\left(\frac{N_{T0} - \nu S}{N_{T0} + S}\right)y(3 - y_r^2) + \nu y(1 - x_r^2)[0.3193 +$$
$$+ 0.5000x_r^2(1 - y_r^2) - 0.00913(1 - y_r^2)(1 - 3y_r^2)] -$$
$$- y[(15 - 10y_r^2 + 3y_r^4)\{0.00532(1 - 3x_r^2) - 0.00417(1 - 12x_r^2 + 15x_r^4)\} +$$
$$+ 0.000043y_r^2(35 - 42y_r^2 + 15y_r^4)(1 - 3x_r^2)]. \tag{7.65}$$

The plate equations

Note that Equation (7.63) can be obtained directly from the unit load theorem in Equation (2.27, Vol.1). Take $u_z(0,0) = 0$ and use a unit strip in the x-direction. Apply $P_x^V = 1$ at a point x on the strip, whence

$$\begin{aligned} &p_x^1 = 1, \quad p_y^1 = 0, \quad t\sigma_{xx}^1 = 1, \quad t\sigma_{yy}^1 = 0, \quad t\sigma_{xy}^1 = 0, \\ &Ete_{zz} = N_x - \nu N_y + N_T, \quad X_x^1 = X_y^1 = 0, \quad S_x^1 = S_y^1 = 0. \end{aligned} \quad (7.66)$$

This gives the first equation in Equation (7.63). Use a unit strip in the y-direction for the u_y calculation.

Note that $\nabla^2 \phi$ and T are not harmonic so that the integral in Equation (7.43) is not zero and the shear strain is incompatible with u_x and u_y. Usually, the effect on u_x and u_y is small.

7.4. The plate inplane case using the mixed virtual principle

From Equation (7.7) use

$$\begin{aligned} &t\{\sigma\} = \{N\}, \quad t\{\sigma^V\} = \{N^V\}, \quad \{e\}_u = [J_{pd}]\{u\}, \\ &\{e^V\}_u = [J_{pd}]\{u^V\}, \quad t\{e\}_\sigma = [J]\{N\} + t\{e_E\} \end{aligned} \quad (7.67)$$

in the principle of mixed virtual displacements and virtual stresses in Equation (2.24, Vol.1) to get

$$\int_{A_m} \{[J_{pd}]\{u^V\}\}^T \{N\} \, dA_m - \int_{A_m} \{u^V\}^T \{X\} t \, dA_m - \int_C \{u^V\}^T \{S\} t \, ds -$$
$$- \sum_k \{u_k^V\}^T \{P_k\} + \int_{A_m} \{N^V\}^T \{[J_{pd}]\{u\} - \frac{1}{t}[J]\{N\} - \{e_E\}\} \, dA_m -$$
$$- \int_C \{[\lambda_b]^T \{N^V\}\}^T \{u - u_s\} \, ds - \sum_k \{P_k^V\}^T \{u_k - u_{sk}\} = 0. \quad (7.68)$$

Integrate the first area integral in Equation (7.68) by using Green's identity in Equation (2.2, Vol.1), or

$$\int_{A_m} \{[J_{pd}]\{u^V\}\}^T \{N\} dA_m = -\int_{A_m} \{u^V\}^T [J_{pd}]^T \{N\} dA_m +$$
$$+ \int_C \{u^V\} [\lambda_b]^T \{N\} ds. \quad (7.69)$$

For no point loads and $\{u^V\}$, $\{N^V\}$ arbitrary, Equations (7.68) and (7.69) give the differential equations and boundary conditions

$$\begin{aligned} &[J_{pd}]^T \{N\} + t\{X\} = 0, \quad [\lambda_b]^T \{N\} = t\{S\}, \\ &Et[J_{pd}]\{u\} - E[J]\{N\} - \{N_E\} = 0, \quad \{u\} = \{u_s\}. \end{aligned} \quad (7.70)$$

The set of Equations (7.70) involve five first order partial differential equations for the five unknowns u_x, u_y, N_x, N_y, and N_{xy}. The equations are coupled and must be solved together, which is a difficult problem. There is very little on the solution of these equations in the plate literature. However, approximate solutions for Equation (7.70) can be obtained using the mixed principle of virtual displacements and virtual stresses. The approximation procedure also allows the use of mixed stress and displacement boundary conditions.

Assume the approximating functions in the form

$$\{u\} = \left\{\begin{matrix} u_x \\ u_y \end{matrix}\right\} = \left\{\begin{matrix} u_{x0} \\ u_{y0} \end{matrix}\right\} + \left[\begin{matrix} F_x(x,y) & 0 \\ 0 & F_y(x,y) \end{matrix}\right] \left\{\begin{matrix} A_x \\ A_y \end{matrix}\right\}$$

$$= \{u_0\} + [F]\{A\}, \quad (7.71)$$

$$\{N\} = \left\{\begin{matrix} N_x \\ N_y \\ N_{xy} \end{matrix}\right\} = \left\{\begin{matrix} N_{x0} \\ N_{y0} \\ N_{xy0} \end{matrix}\right\} + \left[\begin{matrix} G_x(x,y) & 0 & 0 \\ 0 & G_y(x,y) & 0 \\ 0 & 0 & G_{xy}(x,y) \end{matrix}\right] \left\{\begin{matrix} B_x \\ B_y \\ B_{xy} \end{matrix}\right\}$$

$$= \{N_0\} + [G]\{B\}, \quad (7.72)$$

where the subscript zero indicates non-zero functions for non-zero specified boundary values and the F and G functions satisfy the zero boundary conditions, and

$$F_x(x,y)A_x = \sum_k A_{xk} F_{xk}(x,y), \quad F_y(x,y)A_y = \sum_k A_{yk} F_{yk}(x,y),$$
$$F_x(x,y)A_x^V = \sum_k A_{xk}^V F_{xk}(x,y), \quad F_y(x,y)A_y^V = \sum_k A_{yk}^V F_{yk}(x,y). \quad (7.73)$$

Put Equations (7.71) and (7.72) into Equation (7.68) to get

$$\int_{A_m} \{A^V\}^T \{[[J_{pd}][F]]^T \{\{N_0\} + [G]\{B\}\} - t[F]^T\{X\}\} dA_m -$$
$$- \int_C t\{A^V\}^T [F]^T \{S\} ds - \sum_m \{A^V\}^T [F_m]^T \{P_m\} = 0, \quad (7.74)$$

$$\int_{A_m} \{B^V\}^T \{[G]^T Et\{[J_{pd}]\{\{u_0\} + [F]\{A\}\} - E[J]\{\{N_0\} + [G]\{B\}\} -$$
$$- \{N_E\}\} dA_m = 0. \quad (7.75)$$

This gives the system of equations

$$\left[\begin{matrix} -[M_{11}] & [M_{12}] \\ [M_{12}]^T & [0] \end{matrix}\right] \left\{\begin{matrix} B \\ A \end{matrix}\right\} = \left\{\begin{matrix} -u_E \\ P \end{matrix}\right\}, \quad (7.76)$$

$$t[M_{11}] = \int_{A_m} [G]^T[J][G] dA_m, \quad [M_{12}] = \int_{A_m} [G]^T[J_{pd}][F] dA_m,$$

$$Et\{u_E\} = \int_{A_m} [G]^T \{Et[J_{pd}]\{u_0\} - E[J]\{N_0\} - \{N_E\}\} dA_m,$$

$$\{P\} = \int_{A_m} [F]^T \{[J_{pd}]^T\{N_0\} + t\{X\}\} dA_m + \sum_m [F_m]^T\{P_m\}.$$

If the number of B_i constants equals the number of A_i constants, then the system is determinate provided only equilibrium force boundary conditions are

The plate equations

specified. In this case

$$\{B\} = [[M_{12}]^T]^{-1}\{P\}, \quad \{A\} = [M_{12}]^{-1}\{[M_{11}]\{B\} - \{u_E\}\}. \quad (7.77)$$

This procedure could be used in Example (7.2) in Section 7.3 by using three constants for each displacement and two constants for each stress.

If any displacements are specified, any e_E strains are present, and the number of A_i and B_i constants are not equal, then the coupled Equations (7.76) must be used. Equation (7.76) can be solved directly or it can be solved in the form

$$\{A\} = [K]^{-1}\{\{P\} - [M_{12}]^T[M_{11}]^{-1}\{u_E\}\},$$
$$\{B\} = [M_{11}]^{-1}\{[M_{12}]\{A\} + \{u_E\}\}, \quad [K] = [M_{12}]^T[M_{11}]^{-1}[M_{12}]. \quad (7.78)$$

Example 7.3. Suppose a rectangular plate a by $2b$, symmetrical about the x-axis, is clamped on the edge $x = 0$ with $u_x = 0$, $u_y = 0$, Figure 7.5. If the inplane plate is subjected to a constant temperature change T_0, find the displacements and stresses in the plate.

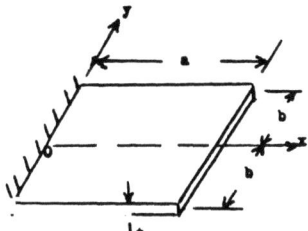

Fig. 7.5. Inplane plate for Example 7.3, clamped on $x = 0$.

Solution. The mixed boundary conditions for the plate are

$$\begin{array}{ll} u_x = 0, \quad u_y = 0, & \text{on} \quad x = 0, \\ N_x = 0, \quad N_{xy} = 0, & \text{on} \quad x = a, \\ N_y = 0, \quad N_{xy} = 0, & \text{on} \quad y = \pm b. \end{array} \quad (7.79)$$

For these conditions, the exact mathematical equations for the elastic plate show singularities at the corners $x = 0$, $y = \pm b$, with both N_x and N_y tending toward infinity. Of course, the real material plate yields at the corners which limits the stresses. Approximate solutions using the principle of virtual forces give large stresses at the corners when many terms are used, but the displacements have large deviations from zero along $x = 0$. Finite element solutions using the principle of virtual displacements show the discontinuity in N_y at the fixed corners, give realistic stresses, and satisfy the point displacements on $x = 0$.

The mixed virtual principle can satisfy all the boundary conditions in Equation (7.79) provided N_y is assumed continuous at the fixed corners. However, the slope of N_y at the two corners should be very steep. In order to reduce the number of constants in the approximations for this example, the assumed stress terms are based upon the shape of the stress curves given by a finite element solution. Otherwise, a computer solution is necessary using about six constants for each stress. Thus, the assumed functions satisfying Equation (7.79) are selected as

$$\begin{array}{ll} u_x = a\alpha T_0 x_r + A_1 x_r (1 - y_r^8), & x_r = x/a, \quad y_r = y/b, \\ u_y = b\alpha T_0 y_r x_r^{0.01} + A_2 y_r x_r^{0.01}(1 - x_r^{0.50}), & (7.80) \\ N_x = B_1(1 - x_r^{0.125})(1 - 9y_r^8), & N_y = B_2(1 - x_r^{0.50})(1 - y_r^8), \\ N_{xy} = B_3(1 - x_r^{0.50})(y_r - y_r^3). & (7.81) \end{array}$$

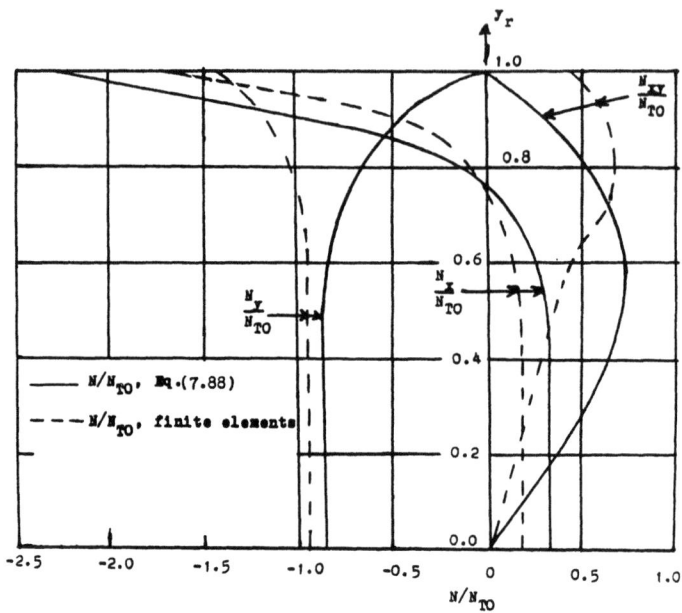

Fig. 7.6. Plate stresses on restrained edge $x_r = 0$, Fig. 7.5.

In this case the F and G matrices in Equations (7.71) and (7.72) are diagonal matrices, or

$$[F] = \begin{bmatrix} x_r(1-y_r^8) & 0 \\ 0 & y_r x_r^{0.01}(1-x_r^{0.50}) \end{bmatrix}, \tag{7.82}$$

$$[G] = \begin{bmatrix} (1-x_r^{0.125})(1-9y_r^8) & 0 & 0 \\ 0 & (1-x_r^{0.50})(1-y_r^8) & 0 \\ 0 & 0 & (1-x_r^{0.50})(y_r-y_r^3) \end{bmatrix}, \tag{7.83}$$

whence the terms in Equation (7.76) with $\nu = 1/3$ are

$$[M_{11}] = \frac{ab}{45Et} \begin{bmatrix} 3.765 & -0.375 & 0 \\ -0.375 & 6.275 & 0 \\ 0 & 0 & 1.524 \end{bmatrix}, \tag{7.84}$$

$$[M_{12}] = \begin{bmatrix} 0.4183k & 0 \\ 0 & 1.3333 \\ -0.1455 & 0.5824k \end{bmatrix}, \quad k = b/a, \tag{7.85}$$

$$\{u_E\} = \int_{A_m} [G]^T \left\{ [J_{pd}] \left\{ \begin{matrix} ax_r \\ by_r x_r^{0.01} \end{matrix} \right\} - \left\{ \begin{matrix} 1 \\ 1 \\ 0 \end{matrix} \right\} \right\} (N_{T0}/Et) dA_m$$

$$= (N_{T0}/27Et) \left\{ \begin{matrix} 0 \\ -0.1317 \\ 3.5294 \end{matrix} \right\}. \tag{7.86}$$

For the case of $k = b/a = 1.00$ in Figure 7.5 and $\nu = 1/3$, Equation (7.78) gives

$$[K] = \frac{Et}{81} \begin{bmatrix} 2.669 & -2.1071 \\ -2.1071 & 22.840 \end{bmatrix}, \quad A_1 = 0.74aN_{T0},$$

$$A_2 = -0.84aN_{T0}, \quad B_1 = 0.32N_{T0}, \quad B_2 = -0.87N_{T0}, \quad B_3 = 1.90N_{T0}, \tag{7.87}$$

whence Equations (7.80) and (7.81) become

$$u_x = \alpha T_0 x[1 + 0.74(1-y_r^8)], \quad u_y = \alpha T_0 y x_r^{0.01}[1-0.84(1-x_r^{0.5})],$$
$$N_x = 0.32 N_{T0}(1-x_r^{0.125})(1-9y_r^8), \quad N_y = -0.87 N_{T0}(1-x_r^{0.5})(1-y_r^8), \quad (7.88)$$
$$N_{xy} = 1.90 N_{T0}(1-x_r^{0.5})(y_r - y_r^3).$$

Although the assumed stress functions in Equation (7.81) are not in equilibrium and are not compatible with the assumed displacements in Equation (7.80), the results for the constants B_1, B_2, and B_3 compare reasonably well with those for the finite element solution. The A_1 constant appears to be too large. The A_2 constant is nearly the same as the B_2 constant, and appears to be reasonable. From Equation (7.86) it appears that the magnitude of these constants is very much dependent upon the restraint factor $x_r^{0.01}$, which makes $u_y = 0$ on $x = 0$. Without some x_r factor for u_y, $\{u_E\}$ in Equation (7.86) would be zero. Since N_{xy} is only a rough approximation of the finite element solution and the choice of $x_r^{0.01}$ was a guess, the results are better than might be expected. See Figure 7.6 for the comparison of the stresses on $x_r = 0$.

7.5. The plate bending case using the principle of virtual displacements

In Figure 7.1 the applied forces P_z and p_z produce bending in the plate with moments M_x, M_y, M_{xy} and shear loads Q_x and Q_y, Equation (7.1). From Equation (2.30, Vol.1) for plane bending in the plate,

$$u_x = z\theta_y, \quad u_y = -z\theta_x, \quad e_{zz} = 0, \quad \sigma_{zz} = 0, \quad (7.89)$$

where θ_y is the slope of the plate in the x-direction and θ_x is the slope in the y-direction, or

$$\theta_y = -u_{z,x}, \quad \theta_x = u_{z,y}, \quad u_x = -z u_{z,x}, \quad u_y = -z u_{z,y}. \quad (7.90)$$

Thus, by use of Equations (7.3)-(7.7), the stress-strain-displacement equations for bending are

$$\{e\} = [J_{pd}]\begin{Bmatrix} u_x \\ u_y \end{Bmatrix} = z[J_{pd}]\begin{Bmatrix} \theta_y \\ -\theta_x \end{Bmatrix} = -z\{J_{bdd}\}u_z,$$

$$\{\sigma\} = -z[J]^{-1}\{J_{bdd}\}u_z - [J]^{-1}\{e_E\} = (12z/t^3)\{M\},$$

$$\{M\} = \begin{Bmatrix} M_x \\ M_y \\ M_{yx} \end{Bmatrix} = -(t^3/12)[J]^{-1}\{J_{bdd}\}u_z - \{M_e\}, \quad (7.91)$$

$$\{M_E\} = \begin{Bmatrix} M_{Ex} \\ M_{Ey} \\ M_{Exy} \end{Bmatrix} = [J]^{-1}\int_t \{e_E\}z\,dz, \quad \{e^V\} = -z\{J_{bdd}\}u_z^V.$$

Add in the effect of the inplane loads N_x, N_y, N_{xy} by using the geometric shear strains, Equation (2.53a, Vol.1), or

$$t\sigma_{xz} = N_x u_{z,x} + N_{xy} u_{z,y},$$
$$t\sigma_{yz} = N_{xy} u_{z,x} + N_y u_{z,y}, \qquad (7.91)$$
$$e_{xz}^V = u_{z,x}^V, \quad e_{yz}^V = u_{z,y}^V. \qquad \text{(continued)}$$

Thus, from Equation (2.6, Vol.1) with $\sigma_{zz} = 0$, the principle of virtual displacements has the form

$$\int_{A_m}\int_t (\sigma_{xx}e^V_{xx} + \sigma_{yy}e^V_{yy} + \sigma_{xy}e^V_{xy} + \sigma_{xz}e^V_{xz} + \sigma_{yz}e^V_{yz})\,dt\,dA_m -$$

$$- \int_{A_m} p_z u^V_z \, dA_m - \int_C \int_t (S_x u^V_x + S_y u^V_y + S_z u^V_z)\,dt\,ds -$$

$$- \sum_k P_{ks} u^V_{zk} = 0, \tag{7.92}$$

where tS_x and tS_y include the edge values of N_x and N_y and $p_z(x,y)$ is any normal loading on the surface of the plate.

Use the above notation and expressions in Equations (7.89)-(7.91) and Equation (7.1) and rewrite Equation (7.92) in the form

$$\int_{A_m} \{\{J_{bdd}\}u^V_z\}^T \{(t^3/12)[J]^{-1}\{J_{bdd}\}u_z + \{M_E\}\}dA_m -$$

$$- \int_{A_m} (p_z + p_{zd}) u^V_z \, dA_m - \int_C [-M_n u^V_{z,n} + (Q_n - M_{ns,s})u^V_z]\,ds +$$

$$+ 2M_{ns}(P)u^V_z(P) - \sum_k (P_{zk} u^V_{zk} + M_k \theta^V_k) = 0, \tag{7.93}$$

$$p_{zd} = N_x u_{z,xx} + N_y u_{z,yy} + 2N_{xy} u_{z,xy},$$

$$M_n = \int_t \sigma_n z \, dz, \quad M_{ns} = -\int_t \sigma_{ns} z \, dz, \quad Q_n = \int_t \sigma_{nz}\, dz, \tag{7.94}$$

$$u^V_n = -z\theta^V_n, \quad \theta^V_s = -u^V_{z,n}, \quad \theta^V_n = u^V_{z,s}, \quad u^V_n = z\theta^V_s,$$

$$\int_C \int_t (S_x u^V_x + S_y u^V_y + S_z u^V_z)dz\,ds = \int_C \int_t (S_n u^V_n + S_{ns} u^V_s + S_z u^V_z)dz\,ds$$

$$= \int_C (M_n \theta^V_s + M_{ns} \theta^V_n + Q_n u^V_z)ds = \int_C (-M_n u^V_{z,n} + M_{ns} u^V_{z,s} + Q_n u^V_z)ds$$

$$= \int_C [-M_n u^V_{z,n} + (Q_n - M_{ns,s})u^V_z]ds + 2M_{ns}(P)u^V_z(P). \tag{7.95}$$

The point P is a corner point on the boundary and $2M_{ns}(P)$ is the reaction at the corner to make the plate remain flat on a simply supported boundary. See page 85 of Reference 1.

Omit point values and integrate Equation (7.93) using Equation (2.2, Vol.1), whence the differential equation and boundary conditions are

$$D\nabla^4 u_z + \{J_{bdd}\}^T \{M_E\} - p_z - p_{zd} = 0, \tag{7.96}$$

$$D = Et^3/12(1-\nu^2), \quad D\nabla^4 u_z = (t^3/12)\{J_{bdd}\}^T[J]^{-1}\{J_{bdd}\}u_z,$$

$$D(1-\nu)(l^2 u_{z,xx} + 2lm u_{z,xy} + m^2 u_{z,yy}) + \nu D\nabla^2 u_z = -M_n, \tag{7.97}$$

$$D(1-\nu)[lm(u_{z,xx} - u_{z,yy}) - (l^2 - m^2)u_{z,xy}]_{,s} - Dl(\nabla^2 u_z)_{,x} -$$

$$- Dm(\nabla^2 u_z)_{,y} = Q_n - M_{ns,s}, \quad \text{on boundary,}$$

u_z and $u_{z,n}$ may be specified on boundary.

The plate equations

For the rectangular plate on the edges $x =$ constant Equation (7.97) reduces to

$$-D(u_{z,xx} + \nu u_{z,yy}) = M_x, \quad -D\left[u_{z,xx} + (2-\nu)u_{z,yy}\right]_{,x} = Q_x - M_{xy,y}. \quad (7.98)$$

Most of the literature on plates is concerned with the solution of Equation (7.96) for p_z loading and various boundary conditions. See References 1-5 and 8. In particular, in Reference 1 about 150 pages are devoted to the rectangular plate with series solutions being used. Here, approximate solutions will be obtained using the principle of virtual displacements in Equation (7.93).

Assume functions for u_z similar to those in Equations (7.13) and (7.14), or

$$u_z(x,y) = F_z(x,y) + \sum_m \sum_n C_{mn} G_m(x) H_n(y),$$
$$u_z^V(x,y) = \sum_i \sum_j C_{ij}^V G_i(x) GJ(y). \quad (7.99)$$

When these expressions are put into Equation (7.93) various double integrals can be identified so that Equation (7.93) becomes

$$\sum_m \sum_n C_{mn}\left(I^{c1}_{mnij} + I^{c2}_{mnij} + I^{c3}_{mnij} + I^{c4}_{mnij} + I^{c5}_{mnij}\right) = I^{c6}_{ij}, \quad (7.100)$$

$$I^{c1}_{mnij} = \int_{A_m} G_m'' H_n G_i'' H_j \, dx\,dy, \quad I^{c2}_{mnij} = \nu \int_{A_m} G_m H_n'' G_i'' H_j \, dx\,dy,$$

$$I^{c3}_{mnij} = \int_{A_m} G_m H_n'' G_i H_j'' \, dx\,dy, \quad I^{c4}_{mnij} = \nu \int_{A_m} G_m'' H_n G_i H_j'' \, dx\,dy,$$

$$I^{c5}_{mnij} = 2(1-\nu) \int_{A_m} G_m' H_n' G_i' H_j' \, dx\,dy, \quad (7.101)$$

$$I^{c6}_{ij} = \frac{1}{D} \int_{A_m} \left[-G_i'' H_j (DF_{z,xx} + D\nu F_{z,yy} + M_{xE})-\right.$$
$$- G_i H_j'' (DF_{z,yy} + D\nu F_{z,xx} + M_{yE})-$$
$$- 2G_i' H_j' \{D(1-\nu)F_{z,xy} + M_{xyE}\} + G_i H_j (p_z + p_{zd})\bigg]dx\,dy+$$
$$+ \frac{1}{D}\left[\int_C \{-M_n(G_i H_j)_{,n} + (Q_n - M_{ns,s})(G_i H_j)\}ds-\right.$$
$$\left.- 2M_{ns}(P)G_i(P)H_j(P) + \sum_k \{P_{zk}(G_i H_j)_k + M_{nk}(G_i H_j)_{k,n}\}\right].$$

In the above bending equations the stresses σ_{zz}, σ_{xz}, σ_{yz} were assumed to be zero. However, with a $p_z + p_{zd}$ loading on the surface, there are shear stresses through the thickness of the plate. If these shear effects are represented by Q_x and Q_y in Equation (7.1) and Equation (7.4) is used, then equilibrium of the plate gives (Section 21 of Reference 1)

$$\begin{Bmatrix} Q_x \\ Q_y \end{Bmatrix} = [J_{pd}]^T\{M\} = -D\{J_{cd}\}\nabla^2 u_z - [J_{pd}]^T\{M_E\},$$
$$\{J_{cd}\}^T \begin{Bmatrix} Q_x \\ Q_y \end{Bmatrix} = -p_z - p_{zd}, \quad (7.102)$$

where the effect of the shear forces upon the plate bending stresses is neglected. Since the combination of these two equations gives Equation (7.96), it follows that the shear effect is a higher order effect and will be small on thin plates. The more general plate bending equations allowing for the shear stresses are derived in Section 26 of Reference 1, where it is shown that the shear effect is small for the plate thickness small when compared to the other plate dimensions.

It should be noted that the inplane stress function ϕ in Equation (7.39) and the bending deflection u_z in Equation (7.96) satisfy the same homogeneous fourth order equations $\nabla^4 \phi = 0$, $\nabla^4 u_z = 0$. Thus, there is a dual relation between the two cases so that the same general homogeneous solution applies for both cases. Also, in the approximate solutions using the virtual principles the homogeneous integrals in Equation (7.49) are the same as in Equation (7.101), provided the sign on ν is changed. Of course, the integrals I_{ij}^{d6} and I_{ij}^{c6} for the applied loads and boundary conditions are quite different in the two cases. Also, because of the boundary conditions, it may be desirable to use different approximating functions for the two cases.

Example 7.4. (a) Suppose the rectangular plate in Figure 7.2 is simply supported on all four edges with $u_z = 0$ for $x = 0, a$ and $y = 0, b$, and is loaded by a lateral load p_z distributed over the plate, a point load P_{z1} at point x_1, y_1, a thermal moment M_T, and a constant inplane load N_z. Use the trigometric functions in Equation (7.21), which satisfy the simple support conditions, and find the bending deflection u_z. (b) Determine the moments, shear forces, and corner reactions. (c) Discuss the convergence of the terms in the solution and the effect of N_z upon the u_z deflection.

Solution. (a) Use the orthogonal conditions in Equations (7.22) and (7.23) to get the non-zero integrals in Equation (7.101) as

$$I_{mnmn}^{c1} = (m\pi/a)^4 (ab/4), \quad I_{mnmn}^{c2} = \nu(mn\pi^2/ab)^2(ab/4),$$
$$I_{mnmn}^{c3} = (n\pi/b)^4 (ab/4), \quad I_{mnmn}^{c4} = I_{mnmn}^{c2},$$
$$I_{mnmn}^{c5} = 2(1-\nu)(mn\pi^2/ab)^2(ab/4),$$
$$DI_{mn}^{c6} = \int_0^b \int_0^a \left[M_T\{(m\pi/a)^2 + (n\pi/b)^2\} + p_z\right]\sin(m\pi x/a)\sin(n\pi y/b) dx\, dy -$$
$$- N_z C_{mn}(m\pi/a)^2(ab/4) + P_{z1}\sin(m\pi x_1/a)\sin(n\pi y_1/b).$$

From Equations (7.99) and (7.100)

$$C_{mn}(ab/4)\pi^4 D\left[(m/a)^2 + (n/b)^2\right]^2 = I_{mn}^{c6},$$

$$C_{mn} = \frac{4P_{ST} + 4P_{z1}\sin(m\pi x_1/a)\sin(n\pi y_1/b)}{ab\pi^4 D\left[\{(m/a)^2 + (n/b)^2\}^2 + (m^2 N_z/\pi^2 a^2 D)\right]}, \quad (7.103)$$

$$P_{ST} = \int_0^b \int_0^a \left[p_z + \pi^2 M_T\{(m/a)^2 + (n/b)^2\}\right]\sin(m\pi x/a)\sin(n\pi y/b)dx\,dy,$$

$$u_z = \sum_m \sum_n C_{mn}\sin(m\pi x/a)\sin(n\pi y/b). \quad (7.104)$$

(b) The moments are given by Equation (7.91) as

$$\{M\} = -D\begin{Bmatrix} u_{z,xx} + \nu u_{z,yy} \\ \nu u_{z,xx} + u_{z,yy} \\ (1-\nu)u_{z,xy} \end{Bmatrix} - \begin{Bmatrix} M_T \\ M_T \\ 0 \end{Bmatrix},$$

where $M_{Ex} = M_T$, $M_{Ey} = M_T$, and $M_{Exy} = 0$. For the case of constant $p_z = p_{z0}$ and

The plate equations 233

constant $M_T = M_{T0}$, C_{mn} becomes

$$C_{mn} = (16p_{x0}/\pi^6 Dmn)\left[(m/a)^2 + (n/b)^2\right]^{-2} +$$
$$+ (16M_{T0}/\pi^4 Dmn)\left[(m/a)^2 + (n/b)^2\right]^{-1}, \quad m, n, \text{ odd}. \qquad (7.105)$$

This gives

$$M_x = \pi^2 D \sum_m \sum_n C_{mn}[(m/a)^2 + \nu(n/b)^2] \sin(m\pi x/a) \sin(n\pi y/b) - M_{T0},$$

$$M_y = \pi^2 D \sum_m \sum_n C_{mn}[\nu(m/a)^2 + (n/b)^2] \sin(m\pi x/a) \sin(n\pi y/b) - M_{T0}, \qquad (7.106)$$

$$M_{yx} = -(1-\nu)\pi^2 D \sum_m \sum_n C_{mn}(mn/ab) \cos(m\pi x/a) \cos(n\pi y/b).$$

$$Q_x = +\frac{\pi^3 D}{a} \sum_m \sum_n m C_{mn}[(m/a)^2 + (n/b)^2] \cos(m\pi x/a) \sin(n\pi y/b),$$
$$\qquad (7.107)$$
$$Q_y = \frac{\pi^3 D}{b} \sum_m \sum_n n C_{mn}[(m/a)^2 + n/b)^2] \sin(m\pi x/a) \cos(n\pi y/b),$$

$$R(0,0) = M_{xy}(0,0) - M_{yx}(0,0) = 2(1-\nu)\pi^2 D \sum_m \sum_n (mn/ab)C_{mn},$$

$$R(0,b) = M_{yx}(0,b) - M_{xy}(0,b) = R(0,0), \qquad (7.108)$$
$$R(a,b) = M_{xy}(a,b) - M_{yx}(a,b) = R(0,0),$$
$$R(a,0) = M_{yx}(a,0) - M_{xy}(a,0) = R(0,0),$$

where the corner reactions R are calculated as the discontinuity at the corner while proceeding in the clockwise direction around the plate boundary. The total reaction $4R(0,0)$ balances the distributed shear reactions produced by M_{xy} in the shear expression

$$V_{xa} = \left(Q_x - M_{xy,x}\right)_{x=a} = V_{x0}, \quad V_{yb} = \left(Q_y - M_{xy,y}\right)_{y=b} = V_{y0}. \qquad (7.109)$$

(c) From C_{mn} in Equation (7.105) the u_z convergence is faster (n^5, m^5) for the p_{x0} load than for the M_{T0} thermal bending (m^3, n^3). This means that convergence for the moments in Equation (7.106) will be much slower, (m^3, n^3) for p_{x0} load and (m^1, n^1) for the M_{T0} moment. Convergence is still slower for the shear forces in Equation (7.107), $(m^3, n^2$ or $m^2, n^3)$ for p_{x0} load and $(m^1$ or $n^1)$ for the M_{T0} moment.

The N_x term for the inplane axial load in the x-direction shows that the tension load (N_x positive) decreases the deflection while a compression load (N_x negative) increases the deflection. The critical compression value for plate buckling is given by

$$\left[(m/a)^2 + (n/b)^2\right]^2 + \left(m^2 N_{xcr}/\pi^2 a^2 D\right) = 0,$$
$$N_{xcr} = -(\pi^2 a^2 D/m^2)\left[(m/a)^2 + (n/b)^2\right]^2. \qquad (7.110)$$

This result is the same as on page 328 of Reference 2.

7.6. The plate bending case using the principle of virtual forces

The stresses can be obtained by expressing the bending moments M_x, M_y, M_{xy} in terms of two stress functions W_x and W_y, which satisfy the homogeneous equilibrium equation for the plate. Take

$$\{M\} = \{M_Q\} + [J_{bd}]\{W\}, \quad \{W\} = \begin{Bmatrix} W_x \\ W_y \end{Bmatrix}, \tag{7.111}$$

where $\{M_Q\}$ is a particular equilibrium solution for any transverse loading on the surface of the plate. From Equations (7.91) and (7.96) it follows that the homogeneous equilibrium equation in terms of $\{M\}$ is

$$\{J_{bdd}\}^T\{M\} = 0, \quad \{J_{bdd}\}^T[J_{bd}]\{W\} = 0. \tag{7.112}$$

The stresses and strains can be expressed in terms of $\{M\}$ as

$$\{\sigma\} = (12z/t^3)\{M\}, \quad \{e\} = (12z/t^3)[J]\{M\} + \{e_E\},$$
$$\{\sigma^V\} = (12z/t^3)\{M^V\} = (12z/t^3)[J_{bd}]\{W^V\}. \tag{7.113}$$

Use the principle of virtual forces in Equation (2.23, Vol.1) in the form

$$\int_{A_m}\int_t \{\sigma^V\}^T\{e\}dz\,dA_m -$$
$$-\int_C \left(Q_n^V u_z - [u_{z,x}\ u_{z,y}]\int_t z[\lambda_b]^T\{\sigma^V\}dz \right) ds = 0, \tag{7.114}$$

where Equation (7.90) has been used. In this case

$$p_z^V = -(Q_{x,x}^V + Q_{y,y}^V) = -\{J_{cd}\}^T\{Q^V\} = [J_{bdd}]^T\{M^V\} = 0,$$
$$Q_n^V = lQ_x^V + mQ_y^V = [l\ m][J_{pd}]^T[J_{bd}]\{W^V\} = \tfrac{1}{2}(W_{x,x}^V - W_{y,y}^V)_{,s}. \tag{7.115}$$

Use these expressions and Equation (7.113) in Equation (7.114) and integrate the area integral by using Equation (2.2, Vol.1) and integrate the boundary integral by parts to get

$$-\int_{A_m}(12/t^3)\{W^V\}^T[J_{bd}]^T[J]\{[J_{bd}]\{W\} + \{M_Q\} + \{M_E\}\}dA_m +$$
$$+\int_C (12/t^3)\{W^V\}^T\begin{bmatrix} m & 0 & -l/2 \\ 0 & l & -m/2 \end{bmatrix}[J]\{[J_{bd}]\{W\} + \{M_Q + M_E\}\}ds -$$
$$-\int_C (W_x^V u_{z,xs} - W_y^V u_{z,ys})ds = 0, \tag{7.116}$$

where $\{M_E\}$ is defined in Equation (7.91). Since W_x^V and W_y^V are arbitrary, this gives the differential equations and displacement boundary conditions for W_x and W_y as

$$[J_{bd}]^T[J][J_{bd}]\{W\} + [J_{bd}]^T[J]\{M_Q + M_E\} = 0, \tag{7.117}$$

$$(12/t^3)\begin{bmatrix} m & 0 & -l/2 \\ 0 & l & -m/2 \end{bmatrix}[J]\{[J_{bd}]\{W\} + \{M_Q + M_E\}\} = \begin{Bmatrix} u_{z,xs} \\ -u_{z,ys} \end{Bmatrix}, \tag{7.118}$$

$$[J_{bd}]^T[J][J_{bd}] = (1/E)\begin{bmatrix} d_{yy} + \frac{1+\nu}{2}d_{xx} & \frac{1-\nu}{2}d_{xy} \\ \frac{1-\nu}{2}d_{xy} & d_{xx} + \frac{1+\nu}{2}d_{yy} \end{bmatrix}. \tag{7.119}$$

The plate equations

Note that Equation (7.119) is a dual to Equation (7.12), provided $-\nu$ replaces ν. Thus, homogeneous solutions for u_x and u_y in the inplane displacement case can be used for W_x and W_y in the bending case.

The force boundary conditions are

$$M_{Qn} + lW_{x,s} - mW_{y,s} = M_n,$$
$$-\left(M_{Qns} + mW_{x,s} + lW_{y,s}\right)_{,s} = Q_n - M_{ns,s}. \tag{7.120}$$

This method has received very little attention in the literature. However, in the approximate form it has been used for plate finite elements, and it can be used for approximate solutions in the same manner as in Section 7.2 above. Take Equation (7.114) in the form

$$(12/t^3)\int_{A_m} \{[J_{bd}]\{W^V\}\}^T [J]\{[J_{bd}]\{W\} + \{M_Q\} + \{M_E\}\}\mathrm{d}A_m +$$
$$+ \int_C (W_x^V u_{z,xs} - W_y^V u_{z,ys})\mathrm{d}s - \sum_k (P_{zk}^V u_{zk} + M_k^V \theta_k) = 0. \tag{7.121}$$

Take the approximating functions similar to Equations (7.13) and (7.14), or

$$W_y = W_{y0}(x,y) + \sum_m \sum_n A_{mn} G_{xm}(x) H_{xn}(y),$$
$$W_x = W_{x0}(x,y) + \sum_m \sum_n B_{mn} G_{ym}(x) H_{yn}(y), \tag{7.122}$$
$$W_y^V = \sum_i \sum_j A_{ij}^V G_{xi}(x) H_{xj}(y), \quad W_x^V = \sum_i \sum_j B_{ij}^V G_{yi}(x) H_{yj}(y),$$

where W_{y0}, W_{x0} represent any non-zero boundary conditions including $\{M_Q\}$. If these expressions are put into Equation (7.121), then the system of Equations (7.15) results. In this case the integrals on the left side of Equation (7.15) are the same as in Equations (7.16) and (7.18) except $-\nu$ replaces ν. The integrals I_{ij}^{a5} and I_{ij}^{b5} are different from those in Equations (7.17) and (7.19) and have the form

$$I_{ij}^{b5} = -\int_{A_m} [G_{yi} H'_{yj}\{W_{x0,y} - \nu W_{y0,x} + M_{xQ} + M_{xE} - \nu(M_{yQ} + M_{yE})\} +$$
$$+ (1+\nu)G'_{yi} H_{yj}\{\tfrac{1}{2}(W_{x0,x} + W_{y0,y}) - M_{xyQ} - M_{xyE}\}]\mathrm{d}x\,\mathrm{d}y +$$
$$+ (Et^4/12)\sum_k [\tfrac{1}{2}u_{zk}(G'_{yi} H'_{yj} - G_{xi} H''_{xj})_k + u_{zk,x}(G_{yi} H'_{yj})_k] +$$
$$+ (Et^3/12)\int_C (u_{z,xs} G_{yi} H_{yj})\mathrm{d}s, \tag{7.123}$$

$$I_{ij}^{a5} = -\int_{A_m} [G'_{xi} H_{xj}\{W_{y0,x} - \nu W_{x0,y} + M_{yQ} + M_{yE} - \nu(M_{xQ} + M_{xE})\} +$$
$$+ (1+\nu)G_{xi} H'_{xj}\{\tfrac{1}{2}(W_{x0,x} + W_{y0,y}) - M_{xyQ} - M_{xyE}\}]\mathrm{d}x\,\mathrm{d}y +$$
$$+ (Et^4/12)\sum_k [\tfrac{1}{2}u_{zk}(G'_{xi} H'_{xj} - G'''_{yi} H_{yj})_k + u_{zk,y}(G_{xi} H_{xj})_k] +$$
$$+ (Et^3/12)\int_C (u_{z,ys} G_{xi} H_{xj})\mathrm{d}s. \tag{7.124}$$

See case (16) in Section 8.8 for application of this method to finite elements.

7.7. The plate bending case using the mixed virtual principle

For this case Equation (2.20, Vol.1) reduces to

$$\int_{A_m}\int_t (\sigma_{xx}e^V_{xx} + \sigma_{yy}e^V_{yy} + \sigma_{xy}e^V_{xy}) dz\, dA_m - \int_{A_m}(p_z + p_{zd})u^V_z\, dA_m -$$

$$- \int_C \int_t (S_x u^V_x + S_y u^V_y + S_z u^V_z) dz\, ds - \sum_k (P_{kz} u^V_{zk} + M_k \theta^V_k) +$$

$$+ \int_{A_m}\int_t [(e_{xxu} - e_{xx\sigma})\sigma^V_{xx} + (e_{yyu} - e_{yy\sigma})\sigma^V_{yy} + (e_{xyu} - e_{xy\sigma})\sigma^V_{xy}] dz\, dA_m -$$

$$- \int_C \int_t [S^V_x(u_x - u_{xs}) + S^V_y(u_y - u_{ys}) + S^V_z(u_z - u_{zs})] dz\, ds -$$

$$- \sum_k [P^V_{zk}(u_{zk} - u_{zsk}) + M^V_k(\theta_k - \theta_{sk})] = 0. \qquad (7.125)$$

Use Equations (7.89)-(7.95) and rewrite Equation (7.125) in the form

$$\int_{A_m} [\{\{J_{bdd}\}u^V_z\}^T\{M\} + (p_z + p_{zd}u^V_z)] dA_m + \sum_k (P_k u^V_{zk} + M_k \theta^V_k) +$$

$$+ \int_C [-M_n u^V_{z,n} + (Q_n - M_{ns,s})u^V_z] ds + 2M_{ns}(P)u^V_z(P) = 0, \qquad (7.126)$$

$$\int_{A_m} \{M^V\}^T \{[J_{bdd}]u_z + (12/t^3)[J]\{M + M_E\}\} dA_m +$$

$$+ 2M^V_{ns}(P)[u_z(P) - u_{zs}(P)] +$$

$$+ \int_C [-M^V_n(u_{z,n} - u_{zs,n}) + (Q^V_n - M^V_{ns,s})(u_z - u_{zs})] ds +$$

$$+ \sum_k [P^V_{zk}(u_{zk} - u_{zsk}) + M^V_k(\theta_k - \theta_{sk})] = 0. \qquad (7.127)$$

Use Equation (2.2, Vol.1) twice to integrate the $\{M\}$ term in Equation (7.126) whence the differential equations and boundary conditions become

$$\{J_{bdd}\}^T\{M\} + p_z + p_{zd} = 0, \quad \{J_{bdd}\}u_z + (12/t^3)[J]\{M + M_E\} = 0,$$
$$M_n = (M_n)_s, \quad Q_n - M_{ns,s} = (Q_n - M_{ns,s})_s, \qquad (7.128)$$
$$u_{z,n} = u_{zs,n}, \quad u_z = u_{zs},$$

where point values are omitted and Equation (7.127) has been used.

Another form of Equations (7.126) and (7.127) involving only first derivatives is useful for finite elements. Use Equation (2.2, Vol.1) for both equations to get

The plate equations

$$\int_{A_m} [Q_x u_{z,x}^V + Q_y u_{z,y}^V - (p_z + p_{zd}) u_z^V] dA_m + \sum_k (P_{zk} + M_k \theta_k^V) +$$

$$+ \int_C [-\{M_n - (M_n)_s\} u_{z,n}^V + (Q_n - M_{ns,s}) u_z^V] ds + 2 M_{ns}(P) u_z^V(P)$$

$$= 0, \tag{7.129}$$

$$\int_{A_m} [Q_x^V u_{z,x} + Q_y^V u_{z,y} - (12/t^3)\{M^V\}^T [J]\{M + M_E\}] dA_m -$$

$$- \int_C [-M_n^V u_{z,n} + (Q_n^V - M_{ns,s}^V)(u_z - u_{zs})] ds -$$

$$- 2 M_{ns}^V(P) [u_z(P) - u_{zs}(P)] - \sum_k [P_{zk}^V (u_{zk} - u_{zsk}) +$$

$$+ M_k^V (\theta_k - \theta_{sk})] = 0. \tag{7.130}$$

For approximate solutions assume

$$u_z = u_{z0} + [F_z(x, y)]\{A_z\}, \quad \{M\} = \{M_0\} + [H]\{C\}, \tag{7.131}$$

$$\{M\}^T = [M_x \quad M_y \quad M_{yx}], \quad \{M_0\}^T = [M_{x0} \quad M_{y0} \quad M_{yx0}],$$

$$[H] = \begin{bmatrix} H_x(x,y) & 0 & 0 \\ 0 & H_y(x,y) & 0 \\ 0 & 0 & H_{yx}(x,y) \end{bmatrix},$$

$$\{C\}^T = [C_x \quad C_y \quad C_{yx}],$$

where subscript zero indicates non-zero boundary values and the assumed known functions $[F_z]$ and $[H]$ satisfy zero boundary conditions. The $\{A_z\}$ and $\{C\}$ are the unknown constants with one or more constants associated with each function F_z, H_x, H_y, and H_{yx}. Take the virtual functions to be the same with specified terms omitted.

Use Equations (7.3)-(7.5) and (7.102) and put the expressions in Equation (7.131) into Equations (7.129) and (7.130), whence

$$\{A_z^V\}^T \int_{A_m} [\{\{J_{cd}\}[F_z]\}^T [J_{pd}]^T \{\{M_0\} + [H]\{C\}\} - (p_z + p_{zd})[F_z]^T] dA_m +$$

$$+ \{A_z^V\}^T \sum_k P_k [F_{zk}]^T = 0, \tag{7.132}$$

$$\{C^V\}^T \int_{A_m} [[J_{pd}]^T [H]]^T \{J_{cd}\}(u_{z0} + [F_z]\{A_z\}) -$$

$$- (12/t^3)[H]^T [J]\{[H]\{C\} + \{M_0 + M_E\}\} dA_m = 0, \tag{7.133}$$

$$\begin{bmatrix} -[M_{11}] & [M_{12}] \\ [M_{12}]^T & [0] \end{bmatrix} \begin{Bmatrix} C \\ A_z \end{Bmatrix} = \begin{Bmatrix} u_E \\ P \end{Bmatrix}, \tag{7.134}$$

$$[M_{11}] = (12/t^3) \int_{A_m} [H]^T [J][H] dA_m,$$

$$[M_{12}] = \int_{A_m} [[J_{pd}]^T[H]]^T \{J_{cd}\}[F_z] dA_m,$$

$$\{u_E\} = \int_{A_m} \{(12/t^3)[H]^T[J]\{M_0 + M_E\} - [[J_{pd}]^T[H]]^T \{J_{cd}\} u_{z0}\} dA_m,$$

$$\{P\} = \int_{A_m} \{-\{\{J_{cd}\}[F_z]\}^T [J_{pd}]^T \{M_0\} + (p_z + p_{zd})[F_z]^T\} dA_m -$$

$$- \sum_k P_k [F_{zk}]^T.$$

The solution of Equation (7.134) can be expressed as [see Equations (7.76)-(7.78)]

$$[K] = [M_{12}]^T [M_{11}]^{-1} [M_{12}],$$
$$\{A_z\} = [K]^{-1} \{\{P\} + [M_{12}]^T [M_{11}]^{-1} \{u_E\}\}, \qquad (7.135)$$
$$\{C\} = [M_{11}]^{-1} \{[M_{12}]\{A_z\} - \{u_E\}\}.$$

Example 7.5. A rectangular plate in bending with mixed boundary conditions is supported on the edges as shown in Figure 7.7. Assume the plate to have a uniform constant load p_z and a constant thermal moment M_T, and take

$$u_z = A_1(1 - x_r^2)(1 - y_r^2)^2, \qquad x_r = x/a, \quad y_r = y/b,$$
$$M_z = (1 - x_r^2)[C_1 + C_2(1 - y_r^2)], \qquad M_{xy} = C_5 x_r y_r (1 - y_r^2), \qquad (7.136)$$
$$M_y = C_3(1 - x_r^2)(3y_r^2 - 1) + C_4(1 - y_r^2).$$

Determine the constants A_1, C_1-C_5, and compare the results to the series solution in Reference 1.

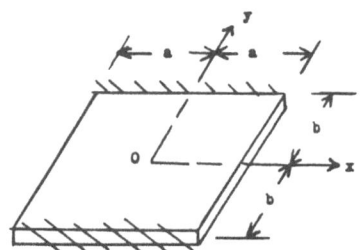

Fig. 7.7. Plate in bending for Example 7.5, clamped on $y = \pm b$, simple support on $x = \pm a$.

Solution. Note that the assumed functions satisfy the boundary conditions

$$u_z = 0, \quad u_{z,y} = 0 \quad \text{on} \quad y_r = 1, -1,$$
$$u_z = 0, \quad M_z = 0 \quad \text{on} \quad x_r = 1, -1.$$

Use Equations (7.131) and (7.134) to get

$$F_z = (1 - x_r^2)(1 - y_r^2)^2, \qquad (7.137)$$

$$[H] = \begin{bmatrix} (1-x_r^2) & (1-x_r^2)(1-y_r^2) & 0 & 0 & 0 \\ 0 & 0 & (1-x_r^2)(3y_r^2-1) & (1-y_r^2) & 0 \\ 0 & 0 & 0 & 0 & x_r y_r (1-y_r^2) \end{bmatrix},$$

$[M_{12}]^T = (32/1575)[35k \quad 30k \quad -84k^{-1} \quad 35k^{-1} \quad -20]$,

$\{u_E\} = (2abM_T/9D(1+\nu))[3 \quad 2 \quad 0 \quad 3 \quad 0]$,

$k = b/a$, $\quad P = 16abp_z/45$, $\quad D = Et^3/12(1-\nu^2)$,

$$[M_{11}] = \frac{16ab}{525Et^3} \begin{bmatrix} 210 & 140 & 0 & -175\nu & 0 \\ 140 & 112 & 56\nu & -140\nu & 0 \\ 0 & 56\nu & 168 & -70 & 0 \\ -175\nu & -140\nu & -70 & 210 & 0 \\ 0 & 0 & 0 & 0 & 20(1+\nu) \end{bmatrix}.$$

Put these values into Equation (7.135) for the case of $\nu = 1/3$ and $k = b/a = 1.00$ for the square plate, whence

$$A_1 = 0.0313(p_z a^4/D) + 0.0744(M_T a^2/D),$$

$$\{C\}^T = p_z a^2[-0.1033 \quad 0.2146 \quad -0.1297 \quad 0.0131 \quad -0.1669] + \qquad (7.138)$$

$$+ M_T[-1.4389 \quad 0.5148 \quad -0.8796 \quad -1.3288 \quad -0.3968].$$

The above results for the p_z load can be compared to results in Reference 1, page 187, for square plate, or

	$\dfrac{Du_z(0,0)}{p_z a^4}$	$\dfrac{M_z(0,0)}{p_z a^2}$	$\dfrac{M_y(0,0)}{p_z a^2}$	$\dfrac{M_y(0,\pm b)}{p_z a^2}$
Equation (7.138)	0.0313	0.1113	0.1428	-0.2594
Reference 1	0.0307	0.0976	0.1328	-0.2788

Note that $\nu = 1/3$ was used in Equation (7.138) while $\nu = 0.30$ was used in Reference 1. Nevertheless, the comparison is very good for the few constants used. Since both M_z and M_y are actually discontinuous at the corners for the thermal moment case with $M_z = -M_T =$ constant on $y = \pm b$, $M_y = -M_T =$ constant on $x = \pm a$, it is evident that the few terms in Equation (7.136) cannot give good results for the moments due to M_T. However, the deflection due to M_T appears to be reasonable.

7.8. Combined inplane and lateral forces

If the bending deflection u_z of the plate is small compared to the plate thickness, then the inplane solution is independent of the u_z deflection so that it can be obtained from Sections 7.2, 7.3, 7.4 above or it can be specified. Thus, the N_x, N_y, N_{xy} internal inplane loads are regarded as known in the following discussion. Also, the bending Equations (7.91) remain unchanged, but the p_{zd} term will be present in Equation (7.96), or

$$D\nabla^4 u_z = p_z + p_{zd} - \{J_{bdd}\}^T\{M_E\},$$
$$p_{zd} = N_x u_{z,xx} + N_y u_{z,yy} + 2N_{xy} u_{z,xy} - X_x u_{z,x} - X_y u_{z,y}. \qquad (7.139)$$

Here, $u_{z,x}$ and $u_{z,y}$ are the slopes of the plate surface and $u_{z,xx}$, $u_{z,yy}$, $u_{z,xy}$ are the changes in the slopes for an element with the forces N_x, N_y, N_{xy} per unit length acting upon it. See Chapter 12 of Reference 1.

For approximate solutions the p_{zd} term is included directly in Equations (7.93) and (7.101) for the principle of virtual displacements and in Equations (7.126), (7.129), and (7.134) for the mixed virtual principle.

The effect of a specified N_x force upon the bending deflection u_z of a rectangular plate was demonstrated in Example 7.4 above, Equation (7.103). Other examples are given in Reference 1.

Example 7.6. Add a constant N_x inplane force to Example 7.5 and recalculate the A_1 and C_i constants.

Solution. From Equations (7.139) and (7.136)

$$p_{zd} = N_x u_{z,xx} = -(2N_x A_1/a^2)(1 - y_r^2)^2, \qquad (7.140)$$

whence Equation (7.134) gives

$$P = (16abp_z/45) - (512/945)kN_x A_1.$$

This changes the A_1 and C_i in Equation (7.138) to

$$A_1 = \frac{0.0313(p_z a^4/D) + 0.0744(M_T a^2/D)}{1 + 0.0477(N_x a^2/D)}, \qquad (7.141)$$

$$\{C\}^T = \frac{0.0313 p_z a^2 + 0.0744 M_T}{1 + 0.0477(N_x a^2/D)}[-3.2992 \quad 6.8571 \quad -4.1424 \quad 0.4200 \quad -5.3333] +$$

$$+ M_T[-1.1899 \quad 0 \quad -0.5660 \quad -1.3583 \quad 0]. \qquad (7.142)$$

Thus, a tension N_x decreases the deflections and moments, while a compression N_x increases the deflections and moments. The effect on the p_z term is different from the effect on the M_T term, where part of the M_T term is not affected by N_x.

The 0.0477 value gives an approximate N_x buckling load of

$$(N_x)_{cr} = -20.96 D/a^2, \qquad (7.143)$$

which corresponds to $(N_x)_{cr} = -18.97 D/a^2$ on page 345 of Reference 2 for the series solution with $\nu = 0.25$.

7.9. Combined forces with large bending deflections

If the bending deflection u_z of the plate is not small compared to the plate thickness, then it affects the inplane loads by stretching the midplane of the plate. From Equation (1.8, Vol.1) an approximate expression for this change in length per unit length due to u_z is

$$e_{dxx} = (1/2)(u_{z,x})^2, \qquad e_{dyy} = (1/2)(u_{z,y})^2,$$
$$e_{dxy} = u_{z,x} u_{z,y}, \qquad \{e_d\}^T = [e_{dxx} \quad e_{dyy} \quad e_{dxy}], \qquad (7.144)$$

where the other non-linear terms in Equation (1.8, Vol.1) are much smaller than these. These terms can be included in the $\{e_E\}$ strains in Equation (7.39) by using the $-e_d$ in Equation (2.57, Vol.1). The bending equations are unaffected except N_x, N_y, N_{xy} in Equation (7.139) are variable, or from Equation (7.31),

$$\{N\} = t\{J_{pdd}\}\phi + t\{s\}V. \qquad (7.145)$$

Thus, Equations (7.39) and (7.139) have the coupled non-linear form

$$\nabla^4 \phi = E(u_{z,xy}^2 - u_{z,xx} u_{z,yy}) - \{J_{pdd}\}^T \{Ee_E\} - (1-\nu)\nabla^2 V,$$
$$D\nabla^4 u_z = p_z - \{J_{bdd}\}^T \{M_E\} + t\{\{J_{pdd}\}\phi\}^T \{J_{bdd}\} u_z + \qquad (7.146)$$
$$+ tV\{s\}^T \{J_{bdd}\} u_z - X_x u_{z,x} - X_y u_{z,y},$$

The plate equations

where the $\{e_d\}$ terms have been separated from the $\{e_E\}$ strains. Also, see Chapter 13 of Reference 1 for a different derivation.

Note that the $\nabla^4 \phi$ equation comes from the principle of virtual forces, while the $\nabla^4 u_z$ equation comes from the principle of virtual displacements. Thus, to solve the equations approximately by assumed functions, it is necessary to solve Equation (7.48) for the D_{mn} constants, where I_{ij}^{d6} includes the $\{e_d\}$ strains in Equation (7.144). This puts D_{mn} in terms of the C_{mn} constants for u_z in Equation (7.99) and hence $\{N\}$ and p_{zd} in Equation (7.139) in terms of C_{mn}. Since p_{zd} is in I_{ij}^{c6} in Equation (7.101), this gives Equation (7.100) as a non-linear equation for the C_{mn} constants.

Take I_{ij}^{c6} and I_{ij}^{d6} in Equations (7.100) and (7.48) in the form

$$I_{ij}^{c6} = \left(I_{ij}^{c6}\right)_0 + \left(I_{ij}^{c6}\right)_u, \quad I_{ij}^{d6} = \left(I_{ij}^{d6}\right)_0 + \left(I_{ij}^{d6}\right)_u, \tag{7.147}$$

whence equations (7.101) and (7.50) give

$$\begin{aligned}\left(I_{ij}^{c6}\right)_u &= (1/D)\int_{A_m} G_j H_j p_{zd}\, dA_m, \\ \left(I_{ij}^{d6}\right)_u &= E\int_{A_m} \left(G_i H_j'' e_{dxx} + G_i'' H_j e_{dyy} + G_i' H_j' e_{dxy}\right)_d dA_m. \end{aligned} \tag{7.148}$$

These expressions are complicated so that only a few constants C_{mn} and D_{mn} can be used in the solution.

Also, the solution can be made using the principle of virtual displacements by combining the cases in Sections 7.2 and 7.5. In this procedure the unknowns are the displacements u_x, u_y, and u_z in Equations (7.13) and (7.99). In Equations (7.15) and (7.100) take

$$\begin{aligned}I_{ij}^{a5} &= \left(I_{ij}^{a5}\right)_0 + \left(I_{ij}^{a5}\right)_u, \quad I_{ij}^{b5} = \left(I_{ij}^{b5}\right)_0 + \left(I_{ij}^{b5}\right)_u, \\ I_{ij}^{c6} &= \left(I_{ij}^{c6}\right)_0 + \left(I_{ij}^{c6}\right)_u,\end{aligned} \tag{7.149}$$

whence Equations (7.17), (7.19), and (7.101) give

$$\begin{aligned}\left(I_{ij}^{a5}\right)_u &= -\int_{A_m}\left[G_{xi}' H_{xj}(e_{dxx} + \nu e_{dyy}) + \frac{1-\nu}{2} G_{xi} H_{xj}' e_{dxy}\right] dx\,dy, \\ \left(I_{ij}^{b5}\right)_u &= -\int_{A_m}\left[G_{yi} H_{yj}'(\nu e_{dxx} + e_{dyy}) + \frac{1-\nu}{2} G_{yi}' H_{yj} e_{dxy}\right] dx\,dy, \\ \left(I_{ij}^{c6}\right)_u &= (1/D)\int_{A_m} G_i H_j p_{zd}\, dx\,dy.\end{aligned} \tag{7.150}$$

Calculate $(I_{ij}^{a5})_u$ and $(I_{ij}^{b5})_u$ in terms of the C_{mn} constants and solve Equation (7.15) for the A_{mn} and B_{mn} constants. Calculate $\{N\}$ from

$$\{N\} = t[J]^{-1}[J_{pd}]\begin{Bmatrix} u_x \\ u_y \end{Bmatrix} - \{N_E\}, \tag{7.151}$$

whence p_{zd} can be expressed in terms of C_{mn}. This gives $(I_{ij}^{c6})_u$ and the non-linear Equation (7.100) to determine C_{mn}.

Example 7.7. Suppose the plate in Figure 7.2 is square with $b = a$. The plate is restrained on the edges with $u_x = 0$ on $x = 0, a$, $u_y = 0$ on $y = 0, b$, and $u_z = 0$ on all four edges. The plate has a constant uniform p_{z0} load. Assume

$$u_x = A_{11} \sin(2\pi x/a) \sin(\pi y/a),$$
$$u_y = A_{11} \sin(\pi x/a) \sin(2\pi y/a), \qquad (7.152)$$
$$u_z = C_{11} \sin(\pi x/a) \sin(\pi y/a),$$

and find the constants A_{11} and C_{11} for the large deflection case with the plate as a membrane ($D = 0$).

Solution. For this example the terms in Equations (7.149) and (7.150) become

$$\left(I_{ij}^{a5}\right)_0 = 0, \quad \left(I_{ij}^{b5}\right)_0 = 0, \quad \left(I_{ij}^{c6}\right)_0 = (p_{z0}/D) \int_{A_m} \sin(\pi x/a)\sin(\pi y/a) dx\, dy$$
$$= 4a^2 p_{z0}/\pi^2 D, \qquad (7.153)$$

$$\left(I_{ij}^{a5}\right)_u = -\int_0^a \int_0^a \left[(\pi/a)(u_{z,x}^2 + \nu u_{z,y}^2) \cos(2\pi x/a) \sin(\pi y/a) + \right.$$
$$\left. + \{(1-\nu)\pi/2a\} u_{z,x} u_{z,y} \sin(2\pi x/a) \cos(\pi y/a)\right] dx\, dy$$
$$= -\frac{5 - 3\nu}{12a} \pi^2 C_{11}^2 = \left(I_{ij}^{b5}\right)_u, \qquad (7.154)$$

$$\left(I_{ij}^{c6}\right)_u = \frac{1}{D}\int_0^a \int_0^a (N_x u_{z,xx} + N_y u_{z,yy} + 2N_{xy} u_{z,xy}) \sin\frac{\pi x}{a} \sin\frac{\pi y}{a} dx\, dy$$
$$= \frac{C_{11}}{D}\int_0^\pi \int_0^\pi \left[-(N_x + N_y)\sin^2 t \sin^2 z + \tfrac{1}{2} N_{xy} \sin 2t \sin 2z\right] dt\, dz, \qquad (7.155)$$

$$N_x + N_y = \frac{Et(1+\nu)}{1-\nu^2}\left[u_{x,x} + u_{y,y} + \tfrac{1}{2}\left(u_{z,x}^2 + u_{z,y}^2\right)\right]$$
$$= \frac{Et}{1-\nu}\left[\frac{2\pi}{a} A_{11}(\cos 2z \sin t + \sin z \cos 2t) + \right.$$
$$\left. + (\pi^2 C_{11}^2/2a^2)(\cos^2 z \sin^2 t + \sin^2 z \cos^2 t)\right],$$
$$N_{xy} = \frac{Et}{2(1+\nu)}(u_{x,y} + u_{y,x} + u_{z,x} u_{z,y}) \qquad (7.156)$$
$$= \frac{Et}{2(1+\nu)}\left[\frac{\pi}{a} A_{11}(\sin 2z \cos t + \cos z \sin 2t) + \right.$$
$$\left. + (\pi^2 C_{11}^2/4a^2) \sin 2z \sin 2t\right].$$

Put Equation (7.156) into Equation (7.155) and integrate to get

$$\left(I_{ij}^{c6}\right)_u = \frac{\pi^2 Et C_{11}}{(1-\nu)aD}\left[A_{11}\left(\tfrac{4}{3} + \tfrac{1}{3}\tfrac{1-\nu}{1+\nu}\right) + \frac{\pi^2 C_{11}^2}{a}\left(-\tfrac{3}{64} + \tfrac{1}{64}\tfrac{1-\nu}{1+\nu}\right)\right]. \qquad (7.157)$$

From Equations (7.25)-(7.27) and (7.154) with $\nu = 1/4$

$$A_{11} = A_{21} = B_{12},$$
$$\left[\frac{9-\nu}{8}\pi^2 + \frac{8(1+\nu)}{9}\right] A_{11} = -\frac{5-3\nu}{12a}\pi^2 C_{11}^2,$$
$$A_{11} = -0.294 C_{11}^2/a. \qquad (7.158)$$

Put A_{11} and $\nu = 1/4$ into Equation (7.157) to get

$$\left(I_{ij}^{c6}\right)_u = -1.0937\pi^2 Et C_{11}^3/Da^2. \qquad (7.159)$$

The plate equations

From Equations (7.100), (7.101), (7.153), (7.159), and Example 7.4, it follows that

$$(D\pi^4/a^2)C_{11} = (4a^2/\pi^2)p_{x0} - 1.0937(\pi^2/a^2)EtC_{11}^3. \tag{7.160}$$

Take $D = 0$ for no bending stiffness of the membrane, whence

$$\begin{aligned} C_{11} &= 0.3348\left(p_{x0}a^4/Et\right)^{1/3}, \\ A_{11} &= -0.0330(1/a)\left(p_{x0}a^4/Et\right)^{2/3}. \end{aligned} \tag{7.161}$$

The results for A_{11} and C_{11} check those for c and w_0 on page 420 of Reference 1, provided c and w_0 are changed to allow for the plate here being a by a instead of $2a$ by $2a$. Also, from Equation (8.24) of Reference 9 the maximum deflection of a unit strip of length a of a long plate is

$$(u_z)_{\max} = 0.35\left(p_{x0}a^4/Et\right)^{1/3}, \tag{7.162}$$

which indicates that $(u_z)_{\max} = C_{11}$ above is probably too large for the square plate. More constants are needed to get better results.

For $D \neq 0$, it is necessary to solve the cubic Equation (7.160) for C_{11}. See Section 102 of Reference 1 for this case with eleven constants used.

7.10. Buckling of plates

If the inplane loads produce compression stresses in the plate, then the plate may buckle. In most cases the critical buckling forces can be obtained from the small deflection equations so that the procedure in Section 7.8 can be used. Take the lateral forces p_z and P_z as zero, the temperature T as constant through the thickness, and assume all inplane forces and the temperature to vary together, or

$$\begin{aligned} N_x &= cN_{x0}, \quad N_y = cN_{y0}, \quad N_{xy} = cN_{xy0}, \\ p_x &= cp_{x0}, \quad T = cT_0, \quad \text{etc.} \end{aligned} \tag{7.163}$$

Put these expressions into equation (7.139) to get

$$\begin{aligned} p_{zd} = cp_{zd0} &= -c(N_{x0}u_{z,xx} + N_{y0}u_{z,yy} + 2N_{xy0}u_{z,xy} \\ &\quad - X_{x0}u_{z,x} - X_{y0}u_{z,y}), \\ D\nabla^4 u_z + cp_{zd0} &= 0, \end{aligned} \tag{7.164}$$

where the minus sign makes cN_{x0} positive for compression.

Since Equation (7.164) is homogeneous for u_z, it has solutions only for specific values of c, whence the smallest value of $c = c_{cr}$ gives the buckling loads for the plate. Also, as demonstrated in Examples 7.4 and 7.6, the buckling load can be obtained as the critical load that makes the deflection infinite when lateral loads are present.

For approximate solutions the principle of virtual displacements in Equation (7.93) can be used with the p_{zd} term given by Equation (7.164). All other applied load terms in Equations (7.93) and (7.101) are omitted so that

$$I^{c6}_{mnij} = \int_{A_m} (G_iH_j/D)p_{zd}\,dx\,dy$$

$$= -c\int_{A_m} (G_iH_j/D)\sum_m\sum_n C_{mn}(N_{x0}G''_mH_n + N_{y0}G_mH''_n +$$

$$+ 2N_{xy0}G'_mH'_n)dx\,dy$$

$$= -(c/D)\sum_m\sum_n C_{mn}\left(N_{x0}I^{c7}_{mnij} + N_{y0}I^{c8}_{mnij} + 2N_{xy0}I^{c9}_{mnij}\right), \quad (7.165)$$

$$I^{c7}_{mnij} = \int_{A_m} G_iH_jG''_mH_n\,dx\,dy, \qquad I^{c8}_{mnij} = \int_{A_m} G_iH_jG_mH''_n\,dx\,dy,$$

$$I^{c9}_{mnij} = \int_{A_m} G_iH_jG'_mH'_n\,dx\,dy.$$

Thus, Equation (7.100) has the matrix form (see Equation (7.20))

$$[I^c]\{C\} + (c/D)[I^d]\{C\} = 0, \qquad |[I^c] + (c/D)[I^d]| = 0, \quad (7.166)$$

which determines the critical values of c_{cr}. The smallest critical c_{cr} gives the buckling load. Note that in Example 7.4 for the $N_x = cN_{x0}$ constant inplane load, the $[I^c]$ and $[I^d]$ matrices are diagonal matrices so that the c_{cr} values are uncoupled with $c_{cr}N_{x0} = N_{xcr}$ given by Equation (7.110). See Chapter 7 of Reference 2 for various buckling cases.

Example 7.8. Find the minimum values of N_{xcr} in Equation (7.110) for given a/b values, Figure 7.8, with simple supports.

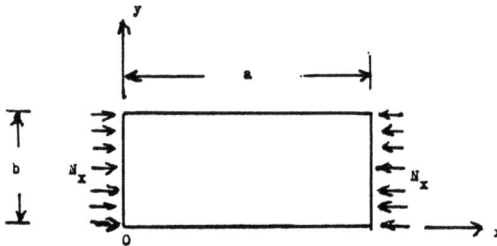

Fig. 7.8. Plate simply supported on $x = 0, a$ and various supports on $y = 0, b$.

Solution. Omit the minus sign and write Equation (7.110) in the form

$$N_{xcr} = k(\pi^2 D/b^2), \quad k = \left[m(b/a) + (n^2/m)(a/b)\right]^2. \quad (7.167)$$

With respect to n the smallest k will occur for $n = 1$. Fix m and use $dk/d(a/b) = 0$ to get the minimum k as

$$k = 4 \quad \text{for} \quad a/b = m. \quad (7.168)$$

The plate equations 245

k is slightly larger than 4 for $m < a/b < m+1$ with its largest value occuring at the change over from m half-waves to $m+1$ half-waves, or

$$k = 2 + \frac{m}{m+1} + \frac{m+1}{m} \quad \text{at} \quad a/b = \left[m(m+1)\right]^{1/2}. \tag{7.169}$$

Usually in design, $k = 4$ is used for $a/b > 1$. For $a/b < 1$,

$$k = \left[(b/a) + (a/b)\right]^2. \tag{7.170}$$

The graph of these results is shown in Figure 1.14 above, where

$$K = \pi^2 k/12(1-\nu^2) = 0.905k \quad \text{for} \quad \nu = 0.3. \tag{7.171}$$

Example 7.9. (a) Set up Equation (7.166) for a simply supported rectangular plate in Figure 7.8 subjected to a constant shear force N_{xy}. (b) Discuss the solution of the system of equations to get the buckling shear force cN_{xy0}.
Solution. (a) As in Example 7.4 assume

$$u_z = \sum_m \sum_n C_{mn} \sin(m\pi x/a)\sin(n\pi y/b), \quad G_m = \sin\frac{m\pi x}{a}, \quad H_n = \sin\frac{n\pi y}{b}, \tag{7.172}$$

whence Equations (7.165) and (7.23) give

$$I^{c9}_{mnij} = \int_0^b\int_0^a \frac{mn\pi^2}{ab} \sin\frac{i\pi x}{a}\cos\frac{m\pi x}{a}\sin\frac{j\pi y}{b}\cos\frac{n\pi y}{b}\,dx\,dy$$

$$= \frac{4mnij}{(m^2 - i^2)(n^2 - j^2)}, \quad m\pm i \text{ odd}, \quad n\pm j \text{ odd}. \tag{7.173}$$

Thus, from Example 7.4, Equation (7.100) has the form

$$Q\left(i^2 + d^2 j^2\right)^2 C_{ij} = -\sum_m\sum_n \frac{mnijC_{mn}}{(m^2-i^2)(n^2-j^2)}, \tag{7.174}$$

$$d = a/b, \quad Q = \frac{\pi^4 D}{32cN_{xy0}b^2 d^3}, \quad m\pm i \text{ odd}, \quad n\pm j \text{ odd}.$$

This shows that in Equation (7.166) $[I^c]$ is a diagonal matrix and $[I^d]$ has zeros on the diagonal and zeros for $m\pm i$ even, $n\pm j$ even. These zeros isolate the system into two groups with $m+n$ odd for one group and $m+n$ even for the other group. Since the $m=1, n=1$ term is important in representing the buckled shape of the plate and is in the $m+n$ even group, this group should give the smallest value for cN_{xy0}. See Section 67 in Reference 2.

(b) As demonstrated in Section 67 of Reference 2 it takes a large number of terms in Equation (7.174) to get accurate values of $c_{cr}N_{xy0}$ for various values of a/b. The final approximate equation in Section 67 of Reference 2 is

$$N_{xycr} = c_{cr}N_{xy0} = k(\pi^2 D/b^2), \quad k = 5.35 + 4(b/a)^2. \tag{7.175}$$

The graph of this Equation (7.175) is given in Figure 1.14, where K is given by Equation (7.171).

7.11. Plate vibrations

From Equation (6.3) add the term

$$\int_V u_z^V \rho u_{z,tt}\,dV = \int_{A_m} u_z^V t\rho u_{z,tt}\,dA_m \tag{7.176}$$

to Equation (7.93). This changes Equation (7.96) to

$$D\nabla^4 u_z + \rho t u_{z,tt} + \{J_{bdd}\}^T \{M_E\} - p_z - p_{zd} = 0. \tag{7.177}$$

Take the homogeneous case as

$$(D/t\rho)\nabla^4 u_z + u_{z,tt} = 0 \tag{7.178}$$

and separate the variables as

$$u_z(x,y,t) = X(x)Y(y)T(t),$$
$$T_{,tt} + \omega^2 T = 0, \quad \nabla^4(XY) - (t\rho\omega^2/D)(XY) = 0, \tag{7.179}$$
$$\frac{X''''}{X} + 2\frac{X''Y''}{XY} + \frac{Y''''}{Y} - \frac{t\rho\omega^2}{D} = 0.$$

From $X'' = \pm p^2 X$, $Y'' = \pm q^2 Y$, it follows that

$$T = C_1 \sin \omega t + C_2 \cos \omega t,$$
$$X_1 = C_3 \sin px + C_4 \cos px,$$
$$X_2 = C_5 \sinh px + C_6 \cosh px, \tag{7.180}$$
$$Y_1 = C_7 \sin qy + C_8 \cos qy,$$
$$Y_2 = C_9 \sinh qy + C_{10} \cosh qy,$$
$$\omega^2 = (D/\rho t)(p^2 + q^2)^2.$$

If the boundary conditions for the plate are such that X_1, Y_1 can be used, then the solution follows directly. Otherwise, a Levy-type solution, Reference 10, can be used for rectangular plates, or

$$XY = \sum_m Y_m(y) \sin(m\pi x/a), \tag{7.181}$$

whence Equation (7.179) gives

$$Y_m'''' - 2(m\pi/a)^2 Y_m'' + [(m\pi/a)^4 - (t\rho\omega^2/D)]Y_m = 0. \tag{7.182}$$

The solutions for Y_m are

$$Y_{m1} = A_m \cosh \beta_m y + B_m \sinh \beta_m y + C_m \cos \gamma_m y + D_m \sin \gamma_m y,$$
$$Y_{m2} = A_m \cosh \beta_m y + B_m \sinh \beta_m y + C_m \cosh \alpha_m y + D_m \sinh \alpha_m y, \tag{7.183}$$
$$\beta_m^2 = (t\rho/D)^{1/2}\omega + \left(\frac{m\pi}{a}\right)^2, \quad \gamma_m^2 = (t\rho/D)^{1/2}\omega - (m\pi/a)^2 \geq 0,$$
$$\alpha_m^2 = (m\pi/a)^2 - (t\rho/D)^{1/2}\omega \geq 0.$$

The boundary conditions for Equation (7.181) require simple support on $x = 0$ and $x = a$. However, simple support, clamped, free, or a mixture can be used on $y = 0$ and $y = b$. The determinant of the equations given by these homogeneous boundary conditions determine the frequencies ω for the plate. A graph of the determinant can be made to get the frequencies for each m and each b/a plate ratio. This method is used in Reference 10 for various rectangular plate boundary conditions, and tables of frequencies are given. It should be noted that data in Tables EV-3 and EV-4 of Reference 10 has been interchanged.

The plate equations

Also, an approximate solution using the principle of virtual displacements with assumed functions for XY can be used. Usually for rectangular plates, the beam mode shapes can be used for the assumed functions, which would be a combination of X_1 and X_2, Y_1 and Y_2 above.

Tables and curves for both exact and approximate solutions for frequencies and mode shapes for plates of various shapes and boundary conditions are given in Reference 12. Also, various solutions for shells are given in Reference 13. There are many references to the plate and shell literature on vibrations in References 12 and 13.

As pointed out in References 10 and 12 in many cases it is not evident which one of the two equations in Equation (7.183) gives the smallest frequency for a given a/b ratio for the plate. Some authors have used only the first equation so that some of their results may be incorrect.

For the simply supported rectangular plate, Figure 7.2, Equation (6.22) gives

$$pa = m\pi, \quad p = m\pi/a, \quad qb = n\pi, \quad q = n\pi/b,$$
$$X_m Y_n = \sin(m\pi x/a)\sin(n\pi y/b), \tag{7.184}$$

whence the $X_1 Y_1$ form in Equation (7.180) applies and the frequencies in Equation (7.180) are

$$\omega_{mn}^2 = (D\pi^4/\rho t b^4)\left[m^2\left(\frac{b}{a}\right)^2 + n^2\right]^2. \tag{7.185}$$

The procedure for initial conditions is similar to that for beams in Equations (6.24)-(6.29), or

$$u_z = \sum_m \sum_n (A_{mn}\sin\omega_{mn}t + B_{mn}\cos\omega_{mn}t)\sin\frac{m\pi x}{a}\sin\frac{n\pi y}{b},$$
$$u_z(x,y,0) = f(x,y), \quad u_{z,t}(x,y,0) = g(x,y), \tag{7.186}$$
$$B_{mn} = \frac{4}{ab}\int_0^b\int_0^a f(x,y)\sin\frac{m\pi x}{a}\sin\frac{n\pi y}{b}\,dx\,dy,$$
$$A_{mn} = \frac{4}{ab\omega_{mn}}\int_0^b\int_0^a g(x,y)\sin\frac{m\pi x}{a}\sin\frac{n\pi y}{b}\,dx\,dy.$$

For approximate solutions for plate vibrations change Equation (7.99) to

$$u_z = \sum_m \sum_n C_{mn}G_m(x)H_n(y)\sin(\omega t + \alpha), \tag{7.187}$$

whence Equations (7.100) and (7.101) apply with

$$I^{c6}_{mnij} = (t\rho\omega^2/D)\int_{A_m} G_i H_j \sum_m \sum_n C_{mn}G_m H_n\,dx\,dy$$
$$= (t\rho\omega^2/D)\sum_m \sum_n C_{mn}I^{c10}_{mnij}, \tag{7.188}$$
$$I^{c10}_{mnij} = \int_{A_m} G_i G_m H_j H_n\,dx\,dy.$$

Thus, Equations (7.100) and (7.166) give the determinant

$$\left|[I^c] - (t\rho\omega^2/D)[I^w]\right| = 0 \tag{7.189}$$

to obtain the frequencies ω_{mn}^2. Note that for the simply supported rectangular plate with $G_m H_n = \sin\frac{m\pi x}{a} \sin\frac{n\pi y}{b}$, the matrices $[I^c]$ and $[I^w]$ are diagonal and the result in Equation (7.185) follows directly from Example 7.4 and Equation (7.188).

It should be noted that the plate buckling problem and the plate vibration problem are closely related. The $[I^c]$ matrix in Equations (7.166) and (7.186) is the same for the same set of assumed functions for the same boundary conditions. If $G_m = (\text{constant})G_m''$ or $H_n = (\text{constant})H_n''$ in Equations (7.165) and (7.188), then it may be possible to get some of the vibration solutions from the buckling solutions. Also, the differential Equations (7.164) and (7.178) will be similar if $u_{z,xx} = (\text{constant})u_z$ or $u_{z,yy} = (\text{constant})u_z$. For example, if $u_z = Y(y) \sin\frac{m\pi x}{a} = XY$ so that $YX_{,xx} = -\left(\frac{m\pi}{a}\right)^2 XY$, then

$$\begin{aligned}\nabla^4(XY) - (\rho t\omega^2/D)(XY) &= 0, \\ \nabla^4(XY) - (cN_{x0}/D)(m\pi/a)^2(XY) &= 0.\end{aligned} \tag{7.190}$$

Thus, solutions of the plate buckling problems given in Section 65 of Reference 2 can be used to solve the corresponding vibration problems, Figure 7.8, for the fundamental frequencies. This analogy is noted in Equation (4.24) of Reference 12.

With k as the basic buckling coefficient in Equation (7.167), or

$$cN_{xcr} = k(\pi^2 D/b^2), \tag{7.191}$$

it follows that from Equation (7.190)

$$\omega_{m1}^2 = \frac{\pi^4 D m^2}{\rho t b^4}\left(\frac{b}{a}\right)^2 k. \tag{7.192}$$

Since k in Section 65 of Reference 2 usually applies for the smallest value for $m = 1$, only the fundamental frequency ω_{11} can be obtained directly for various a/b values. However, since only the m/a ratio appears in Equation (7.190), the same k values hold for any m provided ma is used in place of a. Thus, the ω_{m1} values in Equation (7.192) hold for any m provided ma is used in place of a and $a/b \geq 1.00$. The tables in Reference 10 demonstrate this conclusion.

Example 7.10. Use Table 32 of Reference 2 for the plate in Figure 7.8 with side $y = 0$ simply supported and side $y = b$ free, and sides $x = 0$, a simply supported. Find ω_{11} and ω_{21} from Equation (7.192) and compare the results to those in Table EV-3 of Refernce 10. Note that the correct data for Table EV-3 starts on page 50 of Reference 10.

Solution. The a/b plate ratios are different in the two tables with overlap on only part of the values. Also, $\nu = 1/4$ in Table 32 of Reference 2 and $\nu = 1/3$ in Table EV-3 of Reference 10. Table 7.1 shows the results. The small difference between the p values from Equation (7.192) and from Reference 10 is due to the difference in ν for the cases.

Example 7.11. (a) Find the first approximate vibration frequency for a rectangular plate clamped on $y = 0, b$ and simply supported on $x = 0, a$. Assume

$$Y_1 X_1 = \left(1 - \cos\frac{2\pi y}{b}\right)\sin\frac{\pi x}{a} = H_1(y)G_1(x), \tag{7.193}$$

The plate equations

Table 7.1 Comparison of vibration frequencies for simple supported plate.

$m = 1, \quad p = (\rho t b^4/\pi^4 D)^{1/2}\omega_{11}$

a/b	0.50	1.00	1.20	1.25	1.40	1.50	1.60	2.00	2.50
k, Ref. 2	4.40	1.44	1.14		0.95		0.84	0.70	0.61
p, Eq. (7.192)	4.20	1.20	0.89		0.70		0.57	0.42	0.31
p, Ref. 10	4.15	1.17		0.81		0.61		0.40	0.30

$m = 2, \quad p = (\rho t b^4/\pi^4 D)^{1/2}\omega_{21}$

a/b	1.00	2.00	2.40	2.50	2.80	3.00	3.20	4.00	5.00
k, Ref. 2	4.40	1.44	1.14		0.95		0.84	0.70	0.61
p, Eq. (7.192)	4.20	1.20	0.89		0.70		0.57	0.42	0.31
p, Ref. 10		1.17		0.81		0.61			

which satisfies the boundary conditions

$$X_{11} = X_{11}'' = 0 \quad \text{on} \quad x = 0, a; \quad Y_{11} = Y_{11}' = 0 \quad \text{on} \quad y = 0, b.$$

(b) Compare results to those given by Equation (7.192) and from Table EV-2.1 of Reference 10.

Solution. (a) From Equations (7.101) and (7.188)

$$I_{1111}^{c1} = (\pi/a)^4(3ab/4), \quad I_{1111}^{c2} = 4\nu(\pi^2/ab)^2(ab/4) = I_{1111}^{c4},$$
$$I_{1111}^{c3} = (2\pi/b)^4(ab/4), \quad I_{1111}^{c5} = 8(1-\nu)(\pi^2/ab)^2(ab/4), \quad (7.194)$$
$$I_{1111}^{c10} = 3ab/4,$$

whence Equation (7.189) gives

$$\omega_{11}^2 = (\pi^4 D/3\rho t b^4)[16 + 8(b/a)^2 + 3(b/a)^4]. \quad (7.195)$$

If the plate has b/a large, Equation (7.195) reduces to the simply supported strip, or

$$\omega_1^2 = \pi^4 D/\rho t a^4, \quad a = \text{length of strip}, \quad (7.196)$$

which agrees with the reduced form of Equation (7.185) and which corresponds to the simply supported beam frequency, Equation (6.23),

$$\omega_1^2 = \pi^4 EI/\rho A a^4. \quad (7.197)$$

If the plate in Equation (7.195) has b/a small, then the frequency reduces to

$$\omega_1^2 = 16\pi^4 D/3\rho t b^4 = 5.33\pi^4 D/\rho t b^4 \quad (7.198)$$

for the clamped strip. This corresponds to

$$\omega_1^2 = 5.20\pi^4 EI/\rho A b^4 \quad (7.199)$$

for the clamped beam.

(b) Table 7.2 shows results for the first frequency from Table 35 of Reference 2, Equations (7.192) and (7.195), and Table EV-2.1 of Reference 10. The approximate frequencies agree very well with the exact values. Other values for given a/b can be calculated from the transcendental equation on page 345 of Reference 2, or from equations on page 34 of Reference 10. Also, see Reference 12 for tables and curves.

Table 7.2 Comparison of first vibration frequencies for restrained plate.

$m = 1, \quad p = (\rho t b^4/\pi^4 D)^{1/2}\omega_{11}$

a/b	0.40	0.50	0.60	0.70	0.80	0.90	1.00
k, Ref. 2	9.44	7.69	7.05	7.00	7.29	7.83	
p, Eq. (7.192)	7.68	5.55	4.43	3.78	3.38	3.11	
p, Eq. (7.195)	7.81	5.66	4.52	3.87	3.46	3.19	3.00
p, Ref. 10	7.68	5.55			3.38		2.93

7.12. Problems

7.1. Derive Equations (7.15)-(7.19) from Equation (7.10).
7.2. Verify Equations (7.22) and (7.23).
7.3. Solve part (b) of Example 7.1 for $T = T_0(\frac{x}{a})(\frac{y}{b})$.
7.4. Make non-dimensional graphs of Equations (7.28) and (7.30) along the line $y = a/2$.
7.5. Use Equation (7.47) and Equation (7.54) to evaluate the integrals in Example 7.2 for $m = n = 2, i = j = 1$.
7.6. In Example 7.2 make the solution for the D_{11} constant alone. Graph the results and compare to Figure 7.4.
7.7. Solve the system of Equations (7.56) for $k = 1$.
7.8. Make the calculations to get Equations (7.84)-(7.86) in Example 7.3.
7.9. In Example 7.3 use $z_r^{0.001}$ instead of $z_r^{0.01}$ and repeat the calculations.
7.10. Make the integrations in Equation (7.103) for M_T constant and $p_z = p_{z0}\cos(2\pi x/a)$. For $b = a$, $M_z = 0$, $P_{z1} = 0$, $M_T = 0$, graph $Du_z/p_{z0}a^4$ along the line $y = b/2$ in Figure 7.2.
7.11. Graph $Du_z/P_{z1}a^2$ along the line $y = b/2$ for $b = a$ in Example 7.4 with only the point load P_{z1} acting at $x_1 = a/2$, $y_1 = b/2$.
7.12. In Example 7.4, calculate the expressions for the moments M_x, M_y, and M_{xy} from Equations (7.104) and (7.103) by using Equation (7.91). Discuss convergence for a constant p_z, constant M_T, and the P_{z1} load. Calculate the Q_x and Q_y shear forces from Equation (7.102) and discuss convergence.
7.13. Assume the rectangular plate is $2a$ by $2b$ with $x = 0$, $y = 0$ at the center. Let the plate be clamped on the edges $y = \pm b$ and simply supported on the edges $x = \pm a$. Assume $u_z = C_1(a^2 - x^2)(b^2 - y^2)^2$ and find C_1 from Equations (7.100) and (7.101) for a constant load p_z. Calculate the moments.
7.14. Derive the expressions in Equations (7.123) and (7.124) from Equation (7.121).
7.15. Make the calculations for $[M_{11}]$ in Example 7.5.
7.16. Solve Example 7.5 for $p_z = p_{z0}(1 - x_r^2)$.
7.17. Recalculate Equation (7.138) for $b/a = 2$.
7.18. Compare the results of Problem 7.13 and Example 7.5.
7.19. Solve Example 7.6 for $N_x = N_{x0}x_r^2$.
7.20. Make the integrations to get Equation (7.157).
7.21. Solve Example 7.7 for $p_z = p_{z0}\sin\frac{\pi x}{a}\sin\frac{\pi y}{a}$.
7.22. Solve Example 7.7 for $\nu = 0.3$.
7.23. Solve Equation (7.174) for N_{xycr} for the case of two constants C_{11} and C_{22}.
7.24. Solve Equation (7.174) for N_{xycr} for the case of three constants C_{11}, C_{22}, and C_{13}.
7.25. Put the k in Equation (7.167) into Equation (7.189) and show that Equation (7.182) results.
7.26. Repeat Example 7.11 for $m = 2$ with

$$Y_1 X_2 = \left(1 - \cos\frac{2\pi y}{b}\right)\sin\frac{2\pi x}{a}.$$

Use the procedure of Example 7.10 to get the exact results for $m = 2$.

References

Chapter 7

1. S. Timoshenko and G. Woinowsky-Krieger: *Theory of Plates and Shells*, McGraw-Hill Book Co. (1959).
2. S. Timoshenko: *Theory of Elastic Stability*, McGraw-Hill Book Co. (1936).
3. S. Timoshenko and J.N. Goodier: *Theory of Elasticity*, McGraw-Hill Book Co. (1951).
4. E.H. Mansfield: *The Bending and Stretching of Plates*, Pergamon Press (1964).
5. H.L. Cox: *The Buckling of Plates and Shells*, Pergamon Press (1963).
6. R.H. Rivello: *Theory and Analysis of Flight Structures*, McGraw-Hill Book Co. (1969).
7. O.C. Zienkiewicz and G.S. Hollister (Editors): *Stress Analysis*, John Wiley and Sons, Chapter 9 by B.F. de Veubeke (1965).
8. I.S. Sokolnikoff: *Mathematical Theory of Elasticity*, McGraw-Hill Book Co. (1956).

9. B.E. Gatewood: *Thermal Stresses*, McGraw-Hill Book Co. (1957).
10. D.J. Gorman: *Free Vibration Analysis of Rectangular Plates*, Elsevier North Holland (1982).
11. I.S. Sokolnikoff and R.M. Redheffer: *Mathematics of Physics and Modern Engineering*, McGraw-Hill Book Co. (1958).
12. A.W. Leissa: *Vibration of Plates*, NASA SP-160, 1969, U.S. Government Printing Office, Washington, D.C. 20402.
13. A.W. Leissa: *Vibration of Shells*, NASA SP-288, 1973, U.S. Government Printing Office, Washington, D.C. 20402.

References for additional reading

14. D.O. Brush and B.O. Almroth: *Buckling of Bars, Plates, and Shells*, McGraw-Hill Book Co., New York (1975).

8

Approximate matrix equations for plate finite elements

8.1. Introduction

The approximate solution procedures for plates given in Chapter 7 can be applied to plate finite elements. The procedure is to construct approximate stiffness, flexibility, and mixed matrices for the elements and then assemble the element matrices for the entire structure. This procedure was used in Chapters 3 and 4 of Volume 1 for simple bar and beam structures. This chapter will be concerned with using the matrix forms of the virtual principles given in Chapter 7 to derive the general matrix equations for plate finite elements. Examples for a simple right triangular element with constant stresses will be used to demonstrate the forms of the various matrices. The details for specific types of finite elements will be given in Chapter 9.

In Example 7.1 through 7.11 in Chapter 7 it is evident that the boundary conditions are the most important item in obtaining the plate approximate solutions. Since the boundary conditions can be very complicated for the various shaped plates that may arise in flight vehicle structures, it is often impractical to set up functions for the approximation procedures for the entire plate. Instead, it may be simpler to divide the plate into elements as in Figure 8.1, obtain approximate expressions for each element, and assemble these expressions for all the elements.

The important advantage of the finite element is that the boundary conditions do not have to be satisfied on the interior elements, except possibly for continuity between elements, Figure 8.1. The exterior boundary conditions can be satisfied at the element boundary corner points, or for large elements it may be necessary to satisfy the boundary conditions along one edge of the boundary elements. This means that the assumed functions for the boundary elements may be different from the assumed functions for the interior elements. In either case, simple polynomial expressions can be assumed for the elements.

The number of terms used in a polynomial approximation for the displacements and stresses in the element depends upon the element size. If the elements are small, then in Figure 8.1 the stresses could be taken as constant in each element. Larger elements could be used for linear stresses in each element, etc. As will be seen below there are also limitations on the number of terms that can

Approximate matrix equations for plate finite elements   253

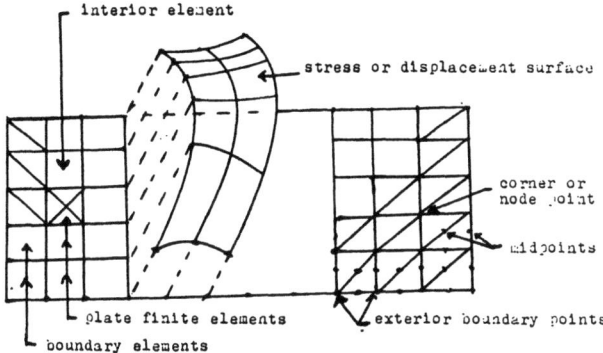

Fig. 8.1. Notation for plate finite elements.

be used. These factors of number of terms and element size are basic problems in the finite element procedures and will be considered in all the methods and discussions in Chapters 8 and 9.

In this Chapter the point unknowns for the matrices, the methods to obtain the matrix equations, and the derivation of the plate element matrices from all three virtual principles for both inplane and bending cases will be discussed. See Chapter 9 for some of the references on finite elements.

8.2. The point unknowns for the matrices

Besides the boundary conditions, another major difference between the approximation procedures in Chapter 7 and those for the plate finite elements is in the unknowns used. In Chapter 7 constants multiplying the assumed functions were the unknowns. For the plate elements it is necessary to use physical unknowns, such as displacements or stresses, that are common to adjacent elements, whence assembly for the unknowns can be made. The simplest procedure is to take the unknowns as the displacements or stresses or their derivatives at the corner or node points of the element, Figures 8.1 through 8.4. However, this may limit the number of terms that can be used in the approximations.

For example, with u_x and u_y the unknown displacements in the plate element, there are six corners (three for u_x and three for u_y) values in a triangular element, Figure 8.2. For six constants, take

$$u_x = A_1 + A_2 x + A_3 y, \qquad u_y = A_4 + A_5 x + A_6 y, \tag{8.1}$$

whence only linear displacements and hence constant stresses can be used. From Figure 8.1, this may be satisfactory if the elements are small. To use larger elements, middle points on the element edges can be added to get twelve constants and twelve displacements, but this doubles the number of unknowns and will pay off only if the elements can be twice as large. Another procedure is to use

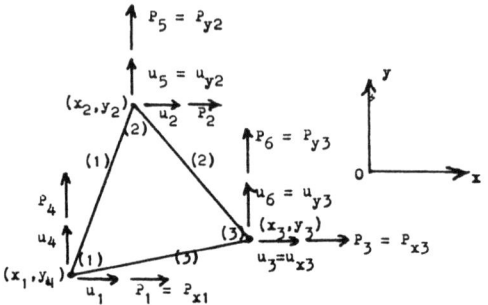

Fig. 8.2. Corner point displacements and forces in triangular plate element with inplane stresses.

additional constants which are forced to obey the virtual principle being used and which can be solved in terms of the corner values, and thus eliminated.

Suppose ten constants are used in Equation (8.1), or

$$\begin{Bmatrix} u_x \\ u_y \end{Bmatrix} = [F]\{A\}, \quad \{A\}^T = [A_1 \ A_2 \cdots A_9 \ A_{10}],$$
$$[F] = \begin{bmatrix} 1 & x & y & 0 & 0 & 0 & x^2 & y^2 & 0 & 0 \\ 0 & 0 & 0 & 1 & x & y & 0 & 0 & x^2 & y^2 \end{bmatrix}. \tag{8.2}$$

From Figure 8.2 the six corner point conditions are

$$\begin{aligned} u_x &= u_1, & u_y &= u_4 & \text{at} & \quad (x_1, y_1), \\ u_x &= u_2, & u_y &= u_5 & \text{at} & \quad (x_2, y_2), \\ u_x &= u_3, & u_y &= u_6 & \text{at} & \quad (x_3, y_3). \end{aligned} \tag{8.3}$$

Put these conditions into Equation (8.2) and write the six equations as

$$\{u\} = [H]\{A\} = [H_a \ H_b] \begin{Bmatrix} A_a \\ A_b \end{Bmatrix}, \tag{8.4}$$

$$\{u\}^T = [u_1 \ u_2 \ u_3 \ u_4 \ u_5 \ u_6],$$
$$\{A_a\}^T = [A_1 \ A_2 \cdots A_6], \quad \{A_b\}^T = [A_7 \ A_8 \ A_9 \ A_{10}],$$
$$[H_a] = \begin{bmatrix} H_1 & 0 \\ 0 & H_1 \end{bmatrix}, \quad [H_b] = \begin{bmatrix} H_2 & 0 \\ 0 & H_2 \end{bmatrix}, \tag{8.5}$$
$$[H_1] = \begin{bmatrix} 1 & x_1 & y_1 \\ 1 & x_2 & y_2 \\ 1 & x_3 & y_3 \end{bmatrix}, \quad [H_2] = \begin{bmatrix} x_1^2 & y_1^2 \\ x_2^2 & y_2^2 \\ x_3^2 & y_3^2 \end{bmatrix}.$$

Solve Equation (8.4) for $\{A_a\}$, or

$$\{A_a\} = [H_a]^{-1}\{u\} - [H_a]^{-1}[H_b]\{A_b\}, \tag{8.6}$$

Approximate matrix equations for plate finite elements

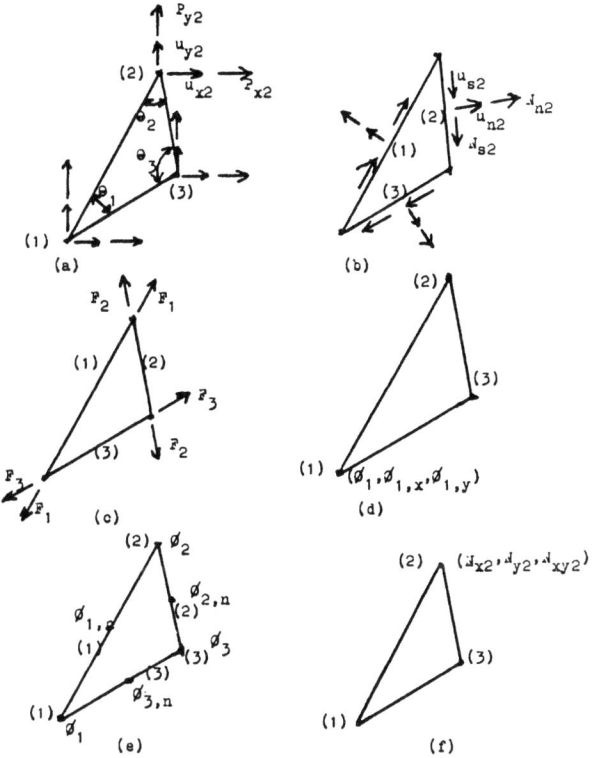

Fig. 8.3. Point unknowns for finite elements with inplane loads.

and add $\{A_b\} = \{A_b\}$ to get

$$\{A\} = \begin{Bmatrix} A_a \\ A_b \end{Bmatrix} = \begin{bmatrix} H_a^{-1} & -H_a^{-1}H_b \\ 0 & I \end{bmatrix} \begin{Bmatrix} u \\ A_b \end{Bmatrix} = [D]\{u_c\}. \tag{8.7}$$

This puts the unknown $\{A_a\}$ constants in terms of the unknown corner displacements $\{u\}$ and $\{A_b\}$.

If this $\{A\}$ is put into Equation (8.2) with $\{u_c\}$ as the unknowns and if the strains and stresses are calculated from u_x and u_y and put into the principle of virtual displacements in Equation (7.9), then the result has the form (see Section 8.4 for more details)

$$\begin{bmatrix} k_{aa} & k_{ab} \\ k_{ba} & k_{bb} \end{bmatrix} \begin{Bmatrix} u \\ A_b \end{Bmatrix} = \begin{Bmatrix} P_a \\ P_b \end{Bmatrix}. \tag{8.8}$$

Solve for $\{A_b\}$ and put it back in to get

$$\{A_b\} = [k_{bb}]^{-1}\{\{P_b\} - [k_{ba}]\{u\}\}, \tag{8.9}$$

$$[k]\{u\} = \{P\} = \{P_a\} - [k_{ab}][k_{bb}]^{-1}\{P_b\},$$
$$[k] = [k_{aa}] - [k_{ab}][k_{bb}]^{-1}[k_{ba}]. \tag{8.10}$$

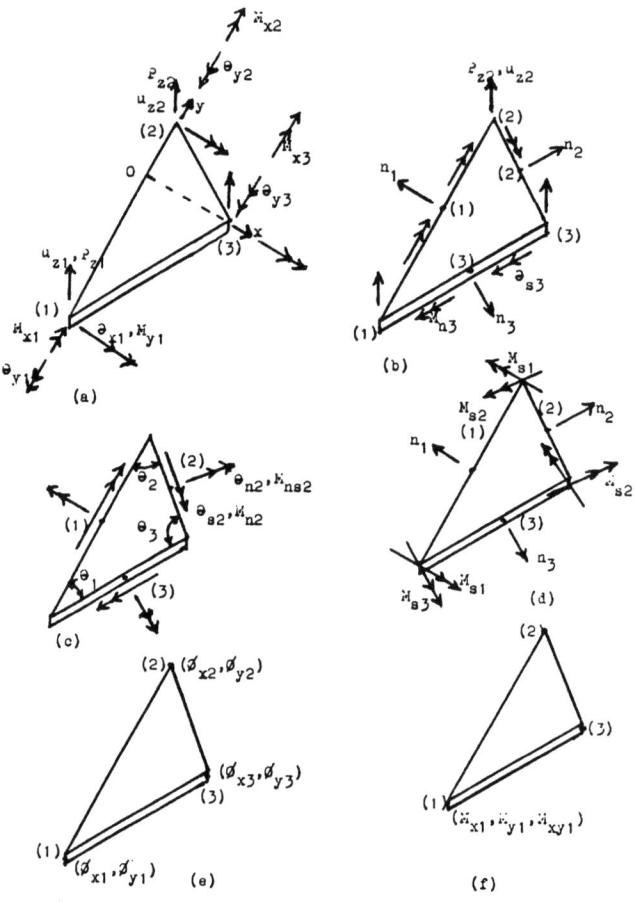

Fig. 8.4. Point unknowns for plate finite elements with bending.

This Equation (8.10) corresponds to Equation (2.69, Vol.1) with $[k]$ as the stiffness matrix for the element, $\{u\}$ as the point displacements at the corner points, and $\{P\}$ the equivalent point forces at the corner points. $\{P\}$ is zero unless it includes any applied loads and temperatures on the element. No point forces actually occur at the corner points of the element, as $\{P\}$ is an equivalent representation for the finite element. Note that by using Equation (8.10) the assembly of the elements is similar to that used in Example 3.17, Vol.1.

Although both of the above procedures for obtaining a better approximation of the stresses and displacements in the element have been used in structural analysis, the discussion here will be restricted to using corner unknowns only, midpoint unknowns only, or a combination of corner and midpoint unknowns. The various cases to be considered are shown in Figures 8.3 and 8.4.

8.3. The methods to obtain the matrix equations

As pointed out in Chapters 2-4 (Vol.1), the key matrices in the solution of structural problems with many elements are the stiffness matrix $[k_i]$, the flexibility matrix $[C_i]$, and the mixed matrix $[M_i]$ for the individual elements i. These element matrices are assembled to produce the set of matrix equations for the point displacements and point stresses in the structures. See Chapters 2-4 (Vol. 1) and 9.

In Chapters 2-4 (Vol.1), the $[k_i]$, $[C_i]$, and $[M_i]$ matrices were associated with the three virtual principles. From Equations (2.69), (2.76), and (2.86) (Vol.1) the matrix form of the equations given by the three virtual principles are

$$[k]\{u\} = \{P\} + \{P_E\}, \quad \text{PVD}, \tag{8.11}$$

$$[C_P]\{F\} = \{u\} - \{u_E\}, \quad \text{PVF}, \tag{8.12}$$

$$[M]\begin{Bmatrix} N \\ u \end{Bmatrix} = \begin{bmatrix} -M_{11} & M_{12} \\ M_{12}^T & 0 \end{bmatrix}\begin{Bmatrix} N \\ u \end{Bmatrix} = \begin{Bmatrix} [M_{11}]\{N_E\} \\ P \end{Bmatrix}, \quad \text{PMVSD}, \tag{8.13}$$

where the element subscripts are omitted.

In Equations (7.33) and (7.121) if point values of the stress functions ϕ, $\phi_x = W_x$, $\phi_y = W_y$ are used instead of equivalent loads $\{F\}$, then

$$[C_\phi]\{\phi_c\} = \{u - u_E\}_\phi \tag{8.14}$$

replaces Equation (8.12). The $[C_\phi]$ matrix is not the usual flexibility matrix and the $\{u\}_\phi$ terms are not deflections. They are matrices that can be evaluated by the principle of virtual forces and assembled to determine values of ϕ and its derivatives, whence the stresses can be calculated. Point values of ϕ and its derivatives can be specified for given values of applied forces on the exterior boundary of the plate.

To get the $[C_P]$ matrix in Equation (8.12) from the principle of virtual forces, it is necessary to select equilibrium force sets F_i for the element and assume equilibrium stresses in the element for each F_i force set. These assumed stresses are then put into Equation (7.114) to determine $[C_P]$. Rather than use force sets F_i, it may be simpler to assume equilibrium stresses in the element and express them in terms of shear and normal stresses at the midpoints of the edges of the element. This third form of Equation (8.12) is

$$[C_\sigma]\{N_m\} = \{u - u_E\}_\sigma \tag{8.15}$$

where $\{N_m\}$ are point values of the stresses and not equivalent point forces.

The various methods to obtain the $[k]$ and $[C]$ element matrices can be listed as follows:

(A) Use the virtual principles directly to get Equations (8.11)-(8.15).
(B) Use hybrid procedures with the continuous edge displacements or stresses selected independently of the assumed interior stresses or displacements.
(C) Get $[C_P]$ from $[k]$, $[k]$ from $[C_P]$, $[k]$ and $[C_\sigma]$ from $[M]$.
(D) Use the dual relations for the plate equations to get $[k]$ from $[C_\phi]$ and $[C_\phi]$ from $[k]$, Equations (7.12) and (7.119).

(E) Use an analogy between the inplane stresses and bending moments to get $[C_\sigma]_{\text{bending}}$ from $[C_\sigma]_{\text{inplane}}$ for no transverse shear cases.

A brief discussion of these methods is given below.

(A) Assume simple polynomial expressions for the displacements or stresses in terms of unknown constants. Eliminate the unknown constants by expressing them in terms of the unknown displacement or stress point values. Proceed as outlined in Equations (8.1)-(8.10) above, but omit the $\{A_b\}$ terms.

(B) The hybrid methods can be used to determine $[C_\sigma]$ from the principle of virtual displacements and $[k]$ from the principle of virtual forces. In the latter case, assume equilibrium stresses in the element in terms of unknown constants $\{A\}$. Calculate the strains in the element from the assumed stresses. Calculate the S_x, S_y edge stresses from the assumed stresses. Assume continuous displacements on the edges in terms of the unknown corner point displacements $\{u\}$ so that the displacements are continuous between elements. These edge displacements are independent of the assumed stresses. Put the expressions into the principle of virtual forces in Equation (7.114) to get

$$\{A^V\}^T[C_H]\{A\} = \{A^V\}^T[B]\{u\}, \tag{8.16}$$

where $[C_H]$ represents the area integral and $[B]$ the edge integral. See Sections 8.4 and 8.5 below for details. Since $\{A^V\}$ is arbitrary, solve Equation (8.16) for $\{A\}$ to get

$$\{A\} = [C_H]^{-1}[B]\{u\}. \tag{8.17}$$

From Equation (C.45, Vol.1) write the right side of Equation (8.16) as

$$\{P^V\}^T\{u\} = \{A^V\}^T[B]\{u\}, \quad \{P^V\}^T = \{A^V\}^T[B],$$
$$\{P^V\} = [B]^T\{A^V\}. \tag{8.18}$$

The $\{A\}$ constants can be taken in the virtual set $\{A^V\}$ so that

$$\{P\} = [B]^T\{A\} = [B]^T[C_H]^{-1}[B]\{u\} = [k]\{u\}, \tag{8.19}$$

where Equation (8.17) has been used. In a similar manner for the principle of virtual displacements where the displacements are assumed in the element and the stresses on the edges

$$\{A^V\}^T[k_H]\{A\} = \{A^V\}^T[B]\{P\}, \quad [C_\sigma] = [B]^T[k_H]^{-1}[B]. \tag{8.20}$$

(C) The $[C_P]$ matrix can be calculated from the $[k]$ matrix and the $[k]$ matrix can be calculated from $[C_P]$. To determine the flexibility $[C_P]$ matrix, select a set of equilibrium force systems for the element, such as F_1, F_2, F_3 in Figure 8.3. Since there are many possible sets of equilibrium forces, the flexibility matrix is not unique and depends upon the particular sets used. On the basis of the selected force systems, specify the rigid body restraints for the element. This reduces the $[k]$ matrix size down to the $[C_P]$ matrix size, or the number of P_i forces down to the number of F_i forces, Figure 8.5. From Equation (3.38, Vol.1)

Approximate matrix equations for plate finite elements

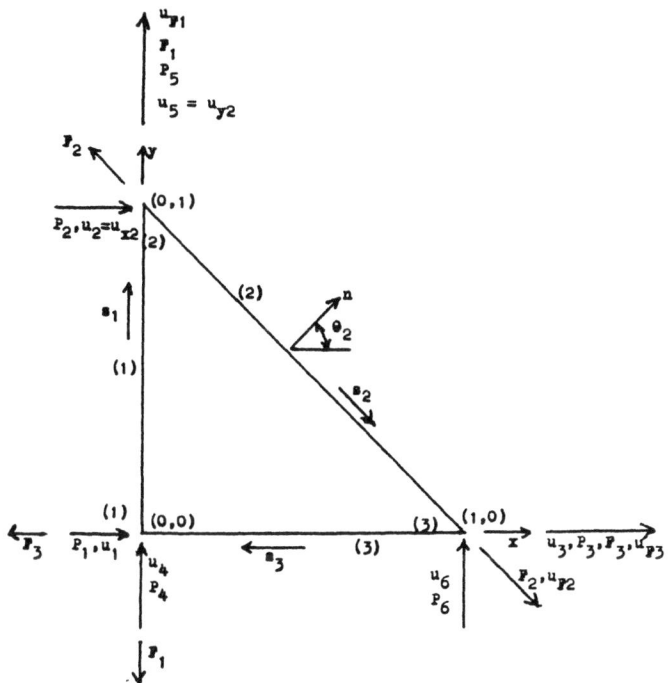

Fig. 8.5. Right triangular plate element for Examples.

reduce the $[s]^T$ equilibrium matrix to match $\{P_r\}$, whence with subscript r for the reduced matrices,

$$\{P_r\} = [k_r]\{u_r\}, \quad \{u_r\} = [k_r]^{-1}\{P_r\}, \quad \{P_r\} = [s_r]^T\{F\},$$
$$\{u_F\} = [s_r]\{u_r\}, \quad \{u_F\} = [C_P]\{F\},$$
$$[C_P] = [s_r][k_r]^{-1}[s_r]^T. \tag{8.21}$$

To get $[k]$ from $[C_P]$, use Equation (3.108, Vol.1), whence

$$[k] = [s]^T[C_P]^{-1}[s]. \tag{8.22}$$

(D) The $[k]$ and $[C_\phi]$ matrices can also be obtained from each other by the dual relations discussed in Sections 7.5 and 7.6. If $[k]$ is known for the inplane plate element, then $[C_\phi]$ for plate bending is the same with

$$\nu \to -\nu, \quad Et \to D^{-1}, \tag{8.23a}$$

provided the assumed functions for the stress functions ϕ_y, ϕ_x are the same as for the displacements u_x, u_y. If $[k]$ is known for plate bending, then $[C_\phi]$ for the inplane plate is the same, provided Equation (8.23a) is used and the approximation for ϕ is the same as for u_z.

(E) Under certain conditions there is a direct analogy between the inplane and bending $[C_\sigma]$ matrices. From Equations (7.7) and (7.91),

$$\{N\} = t[J]^{-1}[J_{pd}] \begin{Bmatrix} u_x \\ u_y \end{Bmatrix} - \{N_E\},$$

$$\{M\} = \frac{t^3}{12}[J]^{-1}[J_{pd}] \begin{Bmatrix} \theta_y \\ -\theta_x \end{Bmatrix} - \{M_E\}, \qquad (8.23\text{b})$$

$$u_x \to \theta_y, \qquad u_y \to -\theta_x, \qquad t \to \frac{t^3}{12}.$$

The analogy holds for the $[C_\sigma]$ matrices provided the transverse shear forces Q_x and Q_y can be regarded as zero.

All the above methods for determining the $[k]$, $[C]$, and $[M]$ matrices will be considered in Sections 8.4 through 8.9 and in Chapter 9. For reference purposes, the methods for both inplane and bending cases are summarized as follows:

Inplane Cases
Principle of Virtual Displacements:

1. $[k]$ in Equation (8.11),
2. Hybrid $[C_\sigma]$, Equation (8.20),   (8.24a)
3. $[k]$ from $[C_P]$, Equation (8.22),

Principle of Virtual Forces:

4. $[C_P]$ in Equation (8.12),
5. $[C_\sigma]$ in Equation (8.15),
6. $[C_\phi]$ in Equation (8.14),   (8.24b)
7. Hybrid $[k]$, Equation (8.19),
8. $[C_P]$ from $[k]$, Equation (8.21),

Principle of Mixed Virtual Stresses and Virtual Displacements

9. $[M]$ in Equation (8.13),
10. $[k]$ from $[M]$,   (8.24c)
11. $[C_\sigma]$ from $[M]$,

Bending Cases
Principle of Virtual Displacements:

12. $[k]$ in Equation (8.11) and dual to $[C_\phi]$ in case (6),
13. Hybrid $[C_\sigma]$, Equation (8.20),   (8.24d)

Principle of Virtual Forces

14. $[C_P]$ in Equation (8.12),
15. $[C_\sigma]$ in Equation (8.15),
16. $[C_\phi]$ dual to $[k]$ in case (1),   (8.24e)
17. Hybrid $[k]$, Equation (8.19),

Approximate matrix equations for plate finite elements

Principle of Mixed Virtual Stresses and Virtual Displacements:

18. $[M]$ in Equation (8.13),
19. $[M_{11}]$ by analogy from cases (9) and (5). (8.24f)

It should be noted that most of the literature on finite elements is concerned with the $[k]$ matrix in cases (1) and (12) above. Although cases (4), (7), and (17) have received considerable investigation, there are only a few papers on cases (9) and (18). Cases (2), (5), and (13) have been mentioned and cases (3), (8), and (10) have been used occasionally. Some of the literature is referenced in Chapter 9. The general matrix forms with some simple examples for all the cases are given in the following Sections 8.4 through 8.9.

8.4. Inplane plate element matrices from the principle of virtual displacements

This Section is concerned with cases (1), (2), and (3) in Equation (8.24a) on inplane plate loading with the number of constants used being the same as the number of corner or node points

Case (1) in Equation (8.24a) for [k]

Use N corner values for u_x in sequence $u_1, u_2, \cdots, u_N$, followed by N corner values for u_y as $u_{N+1}, u_{N+2}, \cdots, u_{2N}$. See Equation (8.3) and Figures 8.2 and 8.3 for the triangular plate. Use $2N$ constants A_i and take

$$\left\{ \begin{array}{c} u_x \\ u_y \end{array} \right\} = [F]\{A\} = \begin{bmatrix} F_1(x,y) & 0 \\ 0 & F_1(x,y) \end{bmatrix} \{A\}, \quad (8.25)$$

$[F_1(x,y)] = [1 \quad x \quad y \ldots N \text{ terms}]$.

Put the corner values in to get

$$\{u\} = [H]\{A\} = \begin{bmatrix} H_c & 0 \\ 0 & H_c \end{bmatrix} \{A\}, \quad [H_c] = \begin{bmatrix} F_1(x_1, y_1) \\ F_1(x_2, y_2) \\ \vdots \\ F_1(x_N, y_N) \end{bmatrix}, \quad (8.26)$$

$\{u\}^T = [u_1 \quad u_2 \ldots u_{2N}]$.

Thus,

$$\{A\} = [H]^{-1}\{u\} = \begin{bmatrix} H_c^{-1} & 0 \\ 0 & H_c^{-1} \end{bmatrix} \{u\},$$

$$\left\{ \begin{array}{c} u_x \\ u_y \end{array} \right\} = [F][H]^{-1}\{u\}, \quad (8.27)$$

and from Equations (7.2)-(7.6)

$$\{e\} = [J_{pd}][F][H]^{-1}\{u\} = [D]\{u\}, \quad [D] = [J_{pd}][F][H]^{-1}. \quad (8.28)$$

Put these expressions into the principle of virtual displacements in Equation (7.11) to get

$$\{u^V\}^T \left[\int_{A_m} t[D]^T[J]^{-1}[D]\{u\}dA_m - \int_{A_m} t[D]^T[J]^{-1}\{e_E\}dA_m - \right.$$
$$- \int_{A_m} t[H^{-1}]^T[F]^T \left\{ \begin{matrix} X_x \\ X_y \end{matrix} \right\} dA_m - \int_c t[H^{-1}]^T[F]^T \left\{ \begin{matrix} S_x \\ S_y \end{matrix} \right\} ds -$$
$$\left. - \sum_k [H^{-1}]^T [F(x_k, y_k)]^T \left\{ \begin{matrix} P_{xk} \\ P_{yk} \end{matrix} \right\} \right] = 0. \tag{8.29}$$

For arbitrary $\{u^V\}$ this equation (8.29) has the form of Equation (8.11), or

$$[k]\{u\} = \{P\} + \{P_E\}, \tag{8.30}$$

$$[k] = \int_{A_m} t[D]^T[J]^{-1}[D]dA_m, \quad \{P_E\} = \int_{A_m} t[D]^T[J]^{-1}\{e_E\}dA_m,$$

$$\{P\} = \int_{A_m} t[H^{-1}]^T[F]^T \left\{ \begin{matrix} X_x \\ X_y \end{matrix} \right\} dA_m + \int_c t[H^{-1}]^T[F]^T \left\{ \begin{matrix} S_x \\ S_y \end{matrix} \right\} ds +$$
$$+ \sum_k [H^{-1}]^T [F(x_k, y_k)]^T \left\{ \begin{matrix} P_{xk} \\ P_{yk} \end{matrix} \right\}.$$

Note that for interior elements $\{P\}$ is zero unless body forces are present and that $\{P_E\}$ includes any element temperature, initial, and inelastic strains. For elements along the exterior boundary, $\{P\}$ will include weighted values for the exterior edge forces S_x, S_y and any point forces P_k.

Example 8.1. (a) Use $N = 3$ in Equation (8.25) with a total of six constants A_i and find the $[k]$ matrix for the right triangular element in Figure 8.5. (b) Comment on the use of Equation (8.2) for a triangular element.

Solution. (a) For the triangle in Figure 8.5 the $[H_c]$ matrix in Equation (8.26) is

$$[H_c] = \begin{bmatrix} 1 & 0 & 0 \\ 1 & 0 & 1 \\ 1 & 1 & 0 \end{bmatrix}, \quad [H_c]^{-1} = \begin{bmatrix} 1 & 0 & 0 \\ -1 & 0 & 1 \\ -1 & 1 & 0 \end{bmatrix},$$

whence Equation (8.28) gives

$$[D] = \begin{bmatrix} 0 & 1 & 0 & 0 & 0 & 0 \\ 0 & 0 & 0 & 0 & 0 & 1 \\ 0 & 0 & 1 & 0 & 1 & 0 \end{bmatrix} \begin{bmatrix} H_c^{-1} & 0 \\ 0 & H_c^{-1} \end{bmatrix} = \begin{bmatrix} -1 & 0 & 1 & 0 & 0 & 0 \\ 0 & 0 & 0 & -1 & 1 & 0 \\ -1 & 1 & 0 & -1 & 0 & 1 \end{bmatrix}.$$

Since $[D]$ is constant, the integral for $[k]$ in Equation (8.30) is constant with area $= 1/2$, whence

$$[k] = \frac{Et}{4(1-\nu^2)} \begin{bmatrix} 3-\nu & \nu-1 & -2 & 1+\nu & -2\nu & \nu-1 \\ \nu-1 & 1-\nu & 0 & \nu-1 & 0 & 1-\nu \\ -2 & 0 & 2 & -2\nu & 2\nu & 0 \\ 1+\nu & \nu-1 & -2\nu & 3-\nu & -2 & \nu-1 \\ -2\nu & 0 & 2\nu & -2 & 2 & 0 \\ \nu-1 & 1-\nu & 0 & \nu-1 & 0 & 1-\nu \end{bmatrix}. \tag{8.31}$$

(b) Since the stresses are constant in the triangular element with linear displacements, they are in equilibrium in Equation (7.8) for the case of no body forces and they are compatible. Thus, this is the *only solution* for the six conditions in Equation (8.3) so that any additional constants in Equation (8.2) and (8.25) will turn out to be zero. More points on the edges must be used to get better results with more constants.

Case (2) in Equation (8.24a) for hybrid $[C_\sigma]$

Take the displacements the same as in Equation (8.25) with no constant terms, but change the strains in Equation (8.28), or

$$\begin{Bmatrix} u_x \\ u_y \end{Bmatrix} = [F]\{A\}, \quad \{e\} = [J_{pd}][F]\{A\} = [G]\{A\}. \tag{8.32}$$

Assume continuous stresses along the edges of the element and express them in terms of corner point values, or from Equation (7.8) for edge i

$$t \begin{Bmatrix} S_{xi} \\ S_{yi} \end{Bmatrix} = [\lambda_{bi}]^T \{N_i\} = [\lambda_{bi}]^T [D_i] \begin{Bmatrix} N_{ci} \\ N_{cj} \end{Bmatrix} = [R_i] \begin{Bmatrix} N_{ci} \\ N_{cj} \end{Bmatrix}, \tag{8.33a}$$

$$\{N_{ci}\}^T = [N_{xi} \quad N_{yi} \quad N_{xyi}],$$
$$[D_i] = [(1-s_i)[I] \quad s_i[I]], \quad [R_i] = [R_{ii} \quad R_{ij}],$$

where i and j are end points of edge i, $i = 1, 2, \cdots N$, and where $[D_i]$ may have more complicated terms. Combine all edges of the element into one matrix, or

$$t \begin{Bmatrix} S_x \\ S_y \end{Bmatrix}_e = t \begin{Bmatrix} S_{x1} \\ S_{y1} \\ \cdots \\ S_{xN} \\ S_{yN} \end{Bmatrix} = \begin{bmatrix} R_{11} & R_{12} & 0 & | & 0 \\ 0 & R_{22} & R_{23} & | & 0 \\ \cdots & & & & \cdots \\ R_{N1} & 0 & 0 & | & R_{NN} \end{bmatrix} \begin{Bmatrix} N_{c1} \\ N_{c2} \\ \cdots \\ N_{cN} \end{Bmatrix}$$

$$= [R]\{N_c\}. \tag{8.33b}$$

From Equation (8.32), take the displacements along the edges as

$$\begin{Bmatrix} u_x \\ u_y \end{Bmatrix}_e = \begin{Bmatrix} u_{x1e} \\ u_{y1e} \\ \cdots \\ u_{xNe} \\ u_{yNe} \end{Bmatrix} = \begin{bmatrix} F_{1e} \\ \cdots \\ F_{Ne} \end{bmatrix} \{A\} = [F_e]\{A\}. \tag{8.34}$$

Put these expressions into the principle of virtual displacements in Equation (8.11) to get

$$\{A^V\}^T \left[\int_{A_m} t[G]^T [J]^{-1}[G]\{A\}dA_m - \int_{A_m} t[G]^T [J]^{-1}\{e_E\}dA_m - \right.$$
$$- \int_{A_m} t[F]^T \begin{Bmatrix} X_x \\ X_y \end{Bmatrix} dA_m - \int_c [F_e]^T [R]\{N_c\}ds -$$
$$\left. - \sum_k [F(x_k, y_k)]^T \begin{Bmatrix} P_{xk} \\ P_{yk} \end{Bmatrix} \right] = 0. \tag{8.35}$$

For arbitrary $\{A^V\}^T$, Equation (8.35) has the form

$$[k_H]\{A\} = [B]\{N_c\} + \{P_E\} + \{P_a\}, \tag{8.36}$$

$$[k_H] = \int_{A_m} t[G]^T [J]^{-1}[G]dA_m, \quad [B] = \int_c [F_e]^T [R]ds,$$

$$\{P_e\} = \int_{A_m} t[G]^T [J]^{-1} \{e_E\} dA_m,$$

$$\{P_a\} = \int_{A_m} t[F]^T \begin{Bmatrix} X_x \\ X_y \end{Bmatrix} dA_m - \sum_k [F(x_k, y_k)]^T \begin{Bmatrix} P_{xk} \\ P_{yk} \end{Bmatrix}.$$

From the procedure discussed in Equations (8.17)-(8.20), it follows that

$$\{u\} = [B]^T \{A\} = [C_\sigma]\{N_c\} + \{u_E\} + \{u_a\}, \tag{8.37}$$

$$[C_\sigma] = [B]^T [k_H]^{-1} [B], \quad \{u_E + u_a\} = [B]^T [k_H]^{-1} \{P_E + P_a\}.$$

Since the element displacements and the edge stresses are assumed separately, the approximate result for $[C_\sigma]$ depends upon both assumptions. More terms in u_x and u_y, which do not increase the number of $\{N_c\}$ unknowns, will make $[k_H]$ more accurate and give a better value for $[F_e]$ in $[B]$. Although the assumed stresses on the edges must be in terms of the corner points to insure continuity between elements, more terms in powers of s^n can be used. It appears that for the best $[B]$ the degree of the stress terms should be consistent with the degree of the displacement terms. If x^n is in the displacements, then x^{n-1} should be in the edge stresses.

Example 8.2. (a) Find the $[C_\sigma]$ flexibility matrix for the right triangular element in Figure 8.5 for constant stresses on the edges. Assume

$$u_x = A_1 x + \tfrac{1}{2} A_3 y, \quad u_y = A_2 y + \tfrac{1}{2} A_3 x,$$

and use the edge normal stresses as the unknowns. (b) Comment on the case of linear stresses on the edges of the triangle.

Solution. (a) From Equation (1.20) the normal N_n and shear N_{ns} stresses can be expressed in terms of N_x, N_y, and N_{xy} constant stresses by

$$\begin{aligned} N_{ni} &= N_x \cos^2 \beta_i + N_y \sin^2 \beta_i + N_{xy} \sin 2\beta_i, \\ N_{nsi} &= \tfrac{1}{2}(N_y - N_x) \sin 2\beta_i + N_{xy} \cos 2\beta_i, \end{aligned} \tag{8.38}$$

where β_i is the angle from the x-axis to the outward normal n on edge i of the element, Figure 8.5. Since only three unknown independent constant stresses can be used, express the constant N_x, N_y, N_{xy} in terms of N_{ni} and hence N_{si} in terms of N_{ni}, $i = 1, 2, 3$, or in matrix form

$$\{N\} = \begin{Bmatrix} N_x \\ N_y \\ N_{xy} \end{Bmatrix} = [Z] \begin{Bmatrix} N_{n1} \\ N_{n2} \\ N_{n3} \end{Bmatrix} = [Z]\{N_n\}, \tag{8.39}$$

$$[Z] = [Z_1]^{-1}, \quad [Z_1] = \begin{bmatrix} \cos^2 \beta_1 & \sin^2 \beta_1 & \sin 2\beta_1 \\ \cos^2 \beta_2 & \sin^2 \beta_2 & \sin 2\beta_2 \\ \cos^2 \beta_3 & \sin^2 \beta_3 & \sin 2\beta_3 \end{bmatrix}, \tag{8.40}$$

$$\{N_{ns}\} = \begin{Bmatrix} N_{ns1} \\ N_{ns2} \\ N_{ns3} \end{Bmatrix} = [W]\{N\} = [W][Z]\{N_n\}, \tag{8.41}$$

$$[W] = \tfrac{1}{2} \begin{bmatrix} -\sin 2\beta_1 & \sin 2\beta_1 & 2\cos 2\beta_1 \\ -\sin 2\beta_2 & \sin 2\beta_2 & 2\cos 2\beta_2 \\ -\sin 2\beta_3 & \sin 2\beta_3 & 2\cos 2\beta_3 \end{bmatrix}. \tag{8.42}$$

For the plate element in Figure 8.5, use $\beta_1 = 180°$, $\beta_2 = 45°$, $\beta_3 = -90°$, whence

$$[Z_1] = \tfrac{1}{2} \begin{bmatrix} 2 & 0 & 0 \\ 1 & 1 & 2 \\ 0 & 2 & 0 \end{bmatrix}, \quad [Z] = [Z_1]^{-1} = \tfrac{1}{2} \begin{bmatrix} 2 & 0 & 0 \\ 0 & 0 & 2 \\ -1 & 2 & -1 \end{bmatrix}. \tag{8.43}$$

Approximate matrix equations for plate finite elements 265

Thus, Equation (8.33) becomes

$$t\begin{Bmatrix} S_{x1} \\ S_{y1} \\ S_{x2} \\ S_{y2} \\ S_{x3} \\ S_{y3} \end{Bmatrix} = \begin{Bmatrix} -N_x \\ -N_{xy} \\ b(N_x + N_{xy}) \\ b(N_y + N_{xy}) \\ -N_{xy} \\ -N_y \end{Bmatrix} = \frac{1}{2}\begin{bmatrix} -2 & 0 & 0 \\ 1 & -2 & 1 \\ b & 2b & -b \\ -b & 2b & b \\ 1 & -2 & 1 \\ 0 & 0 & -2 \end{bmatrix}\begin{Bmatrix} N_{n1} \\ N_{n2} \\ N_{n3} \end{Bmatrix}$$

$$= [R]\{N_n\}, \quad b = 2^{-1/2}, \tag{8.44}$$

where the $[Z]$ values are used directly. From Equations (8.32) and (8.34) and the given u_x, u_y,

$$[G] = [I], \quad [F_e] = \begin{bmatrix} 0 & 0 & y/2 \\ 0 & y & 0 \\ x & 0 & x/2 \\ 0 & x & x/2 \\ x & 0 & 0 \\ 0 & 0 & x/2 \end{bmatrix}, \quad \{ds\} = \begin{Bmatrix} dy \\ dy \\ dx/b \\ dx/b \\ dx \\ dx \end{Bmatrix}, \tag{8.45}$$

whence Equations (8.36) and (8.37) give

$$[k_H] = \frac{t}{2}[J]^{-1}, \quad [B] = \frac{1}{4}\begin{bmatrix} 2 & 0 & 0 \\ 0 & 0 & 2 \\ -1 & 2 & -1 \end{bmatrix},$$

$$[C_\sigma] = \frac{1}{4Et}\begin{bmatrix} 3+\nu & -2(1+\nu) & 1-\nu \\ -2(1+\nu) & 4(1+\nu) & -2(1+\nu) \\ 1-\nu & -2(1+\nu) & 3+\nu \end{bmatrix}, \tag{8.46}$$

where the $\{N_c\}$ unknowns in Equation (8.37) are N_{n1}, N_{n2}, N_{n3}.

(b) If all the three stresses are assumed linear on the edges, then nine corner point values are required for $\{N_c\}$ in Equation (8.33), and at least six A_i constants in Equation (8.32) are required. The nine unknowns require the elements to be about three times as large to compare to the three unknowns in the above constant case so that it is questionable whether the linear case is better than the constant case with respect to calculation time.

Case (3) in Equation (8.24a) to get $[k]$ from $[C_P]$

To get the equilibrium matrix $[s]^T$ needed in Equation (8.22), it is necessary to relate the P_i forces to the F_i forces by the direction cosines. For the triangular element in Figure 8.6,

$$\begin{Bmatrix} P_1 \\ P_2 \\ P_3 \\ P_4 \\ P_5 \\ P_6 \end{Bmatrix} = \begin{bmatrix} -l_{12} & 0 & l_{31} \\ l_{12} & -l_{23} & 0 \\ 0 & l_{23} & -l_{31} \\ -m_{12} & 0 & m_{31} \\ m_{12} & -m_{23} & 0 \\ 0 & m_{23} & -m_{23} \end{bmatrix}\begin{Bmatrix} F_1 \\ F_2 \\ F_3 \end{Bmatrix}, \quad \text{or} \quad \{P\} = [s]^T\{F\}, \tag{8.47}$$

where l_{ij} and m_{ij} are direction cosines for edge direction ij.

Example 8.3. Calculate the stiffness matrix $[k]$ for the right triangular plate element in Figure 8.5 using the flexibility matrix $[C_P]$ in Equation (8.58) below.

Solution. From Equations (8.47) and (8.58) for the case of Figure 8.5

$$[s] = \begin{bmatrix} 0 & 0 & 0 & -1 & 1 & 0 \\ 0 & -b & b & 0 & b & -b \\ -1 & 0 & 1 & 0 & 0 & 0 \end{bmatrix}, \quad [C_P] = \frac{2}{Et}\begin{bmatrix} 1 & d & -\nu \\ d & 2 & d \\ -\nu & d & 1 \end{bmatrix},$$

$$[C_P]^{-1} = \frac{Et}{4(1-\nu^2)}\begin{bmatrix} 3-\nu & -2d & 1+\nu \\ -2d & 2(1-\nu) & -2d \\ 1+\nu & -2d & 3-\nu \end{bmatrix}, \tag{8.48}$$

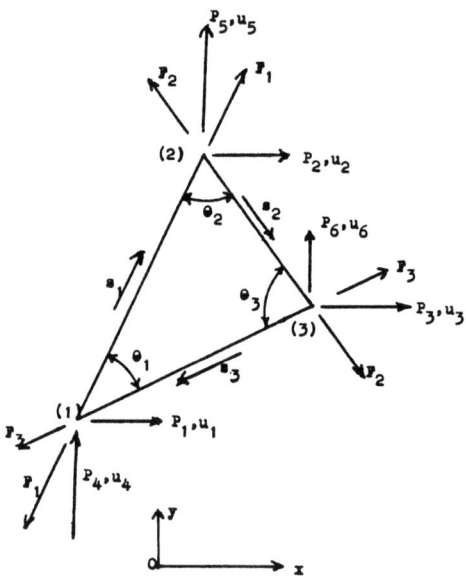

Fig. 8.6. Force systems in inplane triangular element.

where $b = 2^{-1/2}$, $d = b(1 - \nu)$. From Equation (8.22),

$$[k] = [s]^T [C_P]^{-1} [s] = [k] \quad \text{in Equation (8.31).}$$

8.5. Inplane plate element matrices from the principle of virtual forces

This section is concerned with cases (4)-(8) in Equation (8.24b) for inplane plate loading, with the number of constants the same as the number of corner or node points.

Case (4) in Equation (8.24b) for $[C_P]$

To get the $[C_P]$ flexibility matrix in Equation (8.12) it is necessary to select equilibrium force systems on the element, such as the F_i sets in Figure 8.6, and to assume stresses in the element in equilibrium with the forces, or

$$\{\sigma\} = \begin{Bmatrix} \sigma_{xx} \\ \sigma_{yy} \\ \sigma_{xy} \end{Bmatrix} = [H(x,y)]\{F\}. \tag{8.49}$$

Take

$$\{e\} = [J]\{\sigma\} + \{e_E\}, \quad \{\sigma^V\} = [H]\{F^V\}, \tag{8.50}$$

Approximate matrix equations for plate finite elements 267

whence Equations (2.76) and (2.77) in Vol.1 give

$$[C_P]\{F\} = \{u\} - \{u_E\}, \tag{8.51}$$

$$[C_P] = \int_{A_m} t[H]^T[J][H]dA_m, \tag{8.52}$$

$$\{u_E\} = \int_{A_m} t[H]^T\{e_E\}dA_m.$$

In this case, the u_i are relative displacements in the F_i directions.

Example 8.4 Assume constant stresses for the triangular element in Figure 8.6 and find the $[C_P]$ flexibility matrix for the F_i force systems in Figures 8.6 and 8.5.

Solution. Let h_i be the altitude of the triangle perpendicular to edge i in Figure 8.6. If a constant tension stress σ_{xxri} in the direction of edge (1) with $\sigma_{yyr1} = 0$ and $\sigma_{xyr1} = 0$ is acting in the plate, then the statically equivalent force is $th_1\sigma_{xxr1}$ located at $h_1/2$ midway between edge (1) and corner (3). This force is equivalent to $F_1 = \frac{1}{2}(th_1\sigma_{xxr1})$ at edge (1) and at corner (3). No effect in the plate is produced by F_1 at corner (3). Thus, three sets of statically equivalent stresses in the plate are

$$\sigma_{xxri} = 2F_i/th_i, \quad \sigma_{yyri} = 0, \quad \sigma_{xyri} = 0, \quad i = 1, 2, 3. \tag{8.53}$$

Use the direction cosines as defined in Equation (8.47) and put the above reference stresses on the right hand side in Equation (1.20) for Mohr's circle, whence

$$\sigma_{xxi} = l_{ij}^2 \sigma_{xxri} = 2l_{ij}^2 F_i/th_i,$$
$$\sigma_{yyi} = 2m_{ij}^2 F_i/th_i, \quad \sigma_{xyi} = -2m_{ij}l_{ij}F_i/th_i. \tag{8.54}$$

Put these expressions into Equation (8.49) to get

$$[H] = \frac{2}{t}\begin{bmatrix} l_{12}^2/h_1 & l_{23}^2/h_2 & l_{31}^2/h_3 \\ m_{12}^2/h_1 & m_{23}^2/h_2 & m_{31}^2/h_3 \\ -l_{12}m_{12}/h_1 & -l_{23}m_{23}/h_2 & -l_{31}m_{31}/h_3 \end{bmatrix}. \tag{8.55}$$

See Section 7.8 in Reference 1 in the Chapter 9 list of references below for a different derivation of these expressions.

Let L_i be the length of the edge i of the triangle and use the area of the triangle in the forms

$$2A_p = h_1L_1 = h_2L_2 = h_3L_3. \tag{8.56}$$

Since the interior angles θ_i in Figure 8.6 are related to the direction cosines and to the h_i and L_i values in $[H]$ and A_p, the $[C_P]$ matrix in Equation (8.52) can be expressed in terms of trigonometric functions of these three angles, or after using various trigonometric identities,

$$[C_P] = \frac{2}{Et}\begin{bmatrix} a & b_2 & b_1 \\ b_2 & c & b_3 \\ b_1 & b_3 & d \end{bmatrix}, \quad b_i = \cos\theta_i \cot\theta_i - \nu\sin\theta_i,$$
$$a = \cot\theta_1 + \cot\theta_2, \quad c = \cot\theta_2 + \cot\theta_3, \quad d = \cot\theta_1 + \cot\theta_3. \tag{8.57}$$

For the particular triangle in Figure 8.5, $\theta_1 = 90°$, $\theta_2 = 45°$, $\theta_3 = 45°$, whence

$$[C_P] = \frac{2}{Et}\begin{bmatrix} 1 & f & -\nu \\ f & 2 & f \\ -\nu & f & 1 \end{bmatrix}, \quad f = 2^{-1/2}(1-\nu). \tag{8.58}$$

See Example 8.3 above where this latter matrix is used to calculate the stiffness matrix $[k]$ for the element with constant stresses.

Case (5) in Equation (8.24b) for $[C_\sigma]$

To get the flexibility matrix $[C_\sigma]$, assume equilibrium stresses that satisfy Equation (7.8) with no body forces and express the stresses in terms of the normal and shear stresses at the midpoints of the plate edges. Take

$$\{N\} = \begin{Bmatrix} N_x \\ N_y \\ N_{xy} \end{Bmatrix} = [H(x,y)]\{A\}, \tag{8.59a}$$

$$[H] = \begin{bmatrix} 1 & 0 & 0 & y & 0 & x & 0 & xy & 0 & \cdots \\ 0 & 1 & 0 & 0 & x & 0 & y & 0 & xy & \cdots \\ 0 & 0 & 1 & 0 & 0 & -y & -x & -y^2/2 & -x^2/2 & \cdots \end{bmatrix},$$

whence at the midpoint on edge i Equation (8.38) gives

$$\begin{Bmatrix} N_{ni} \\ N_{nsi} \end{Bmatrix} = \begin{bmatrix} \cos^2\beta_i & \sin^2\beta_i & \sin 2\beta_i \\ -\tfrac{1}{2}\sin 2\beta_i & \tfrac{1}{2}\sin 2\beta_i & \cos 2\beta_i \end{bmatrix} \{N_i\}$$

$$= [b_i]\{N_i\} = [b_i][H_i]\{A\} = [q_i]\{A\}. \tag{8.59b}$$

For all the edges together

$$\{N_m\}^T = [N_{n1} \quad N_{ns1} \cdots N_{nN} \quad N_{nsN}] = \{A\}^T[q_1^T \quad q_2^T \cdots q_N^T]$$
$$= \{A\}^T[q]^T, \tag{8.59c}$$
$$\{A\} = [q]^{-1}\{N_m\}, \tag{8.59d}$$
$$\{N\} = [H][q]^{-1}\{N_m\} = [G]\{N_m\}. \tag{8.59e}$$

The procedure to get $[C_\sigma]$ by using Equation (8.59e) is the same as in Equations (8.49)-(8.52) above, whence

$$[C_\sigma]\{N_m\} = \{\{u\} - \{u_E\}\}_\sigma,$$
$$[C_\sigma] = \int_{A_m} \frac{1}{t}[G]^T[J][G]\mathrm{d}A_m,$$
$$\{u_E\}_\sigma = \int_{A_m} [G]^T\{e_E\}\mathrm{d}A_m. \tag{8.60}$$

It should be noted that the equivalent displacement terms in Equation (8.37) are directly in terms of the applied forces while here in Equation (8.60) the applied forces while here in Equation (8.60) the applied forces are absent. In cases where the $[C_\sigma]$ matrices are the same in Equations (8.37) and (8.60), use the displacement terms in Equation (8.37) for any applied forces.

Example 8.5. (a) Assume the first three columns in $[H]$ in Equation (8.59a) for constant stresses for the triangular element in Figure 8.5 and find the $[C_\sigma]$ flexbility matrix. (b) Comment on the case of linear stresses in the plate.

Solution. (a) Since only three independent stress unknowns can be used for constant stresses, take $\{N_m\}$ as the three constant normal stresses on the edges, whence Equation (8.59e) reduces to Equations (8.39) and (8.43), or

$$[G] = [Z] = \tfrac{1}{2}\begin{bmatrix} 2 & 0 & 0 \\ 0 & 0 & 2 \\ -1 & 2 & -1 \end{bmatrix}.$$

From Equation (8.60)

$$[C_\sigma] = \frac{1}{8Et}\begin{bmatrix} 2 & 0 & -1 \\ 0 & 0 & 2 \\ 0 & 2 & -1 \end{bmatrix}\begin{bmatrix} 1 & -\nu & 0 \\ -\nu & 1 & 0 \\ 0 & 0 & 2(1+\nu) \end{bmatrix}\begin{bmatrix} 2 & 0 & 0 \\ 0 & 0 & 2 \\ -1 & 2 & -1 \end{bmatrix}$$

$$= \frac{1}{4Et}\begin{bmatrix} 3+\nu & -2(1+\nu) & 1-\nu \\ -2(1+\nu) & 4(1+\nu) & -2(1+\nu) \\ 1-\nu & -2(1+\nu) & 3+\nu \end{bmatrix}, \qquad (8.61)$$

which is the same as the $[C_\sigma]$ in Equation (8.46) given by the hybrid displacement procedure.

(b) Since there are only six unknowns available from N_{ni} and N_{nsi} for the triangular element, the linear case here is different from that in the hybrid $[C_\sigma]$ case in case (2) above, Example 8.2(b). Only six columns in $[H]$ in Equation (8.59a) can be used with the six unknowns. One difficulty with this linear case is that the $[q]$ matrix in Equation (8.59c) is singular for certain types of triangles, including the one in Figure 8.5. For the general triangle in Figure 8.7 below, the fifth column in $[H]$ in Equation (8.59a) must be omitted because the z midpoints are 0, $L_3/2$, $L_3/2$, which makes $[q]$ singular. As this linear case has six unknowns as compared to three unknowns for the constant case, and the constant case applies for any triangle, the constant case may be more satisfactory for calculations.

Case (6) in Equation (8.24b) for $[C_\phi]$

Since there is only one node value for the stress function ϕ at a node point or corner, it is necessary to use additional unknowns. As the stresses are given by second derivatives of ϕ, Equation (7.31), the first derivatives of ϕ should be used as additional corner point or midpoint unknowns, Figure 8.3, so that the stresses do not become infinite at the node points. Use the corner point case and take $3N$ constants A_i,

$$\phi = [F(x,y)]\{A\}, \qquad (8.62)$$

$$\begin{Bmatrix} \phi \\ \phi_{,x} \\ \phi_{,y} \end{Bmatrix} = \begin{Bmatrix} F \\ F_{,x} \\ F_{,y} \end{Bmatrix}\{A\} = [F_d]\{A\}. \qquad (8.63)$$

Take the $3N$ unknown corner values as

$$\{\phi_c\}^T = [\phi_1 \ \phi_2 \cdots \phi_N \ \phi_{,x1} \cdots \phi_{,xN} \ \phi_{,y1} \cdots \phi_{,yN}], \qquad (8.64)$$

whence

$$\{\phi_c\} = [H]\{A\}, \quad [H] = \begin{bmatrix} H_1 \\ H_2 \\ H_3 \end{bmatrix}, \qquad (8.65)$$

$$\{A\} = [H]^{-1}\{\phi_c\} = [G]\{\phi_c\}, \quad [G] = [H]^{-1}, \quad \phi = [F][G]\{\phi_c\}. \quad (8.66)$$

The rows of $[H_1]$, $[H_2]$, $[H_3]$ are $F(x_i, y_i)$, $F_{,x}(x_i, y_i)$, $F_{,y}(x_i, y_i)$, respectively.

In the principle of virtual forces in Equation (7.33), the stresses due to ϕ expressed in terms of $N_x = t\sigma_{xx}$, $N_y = t\sigma_{yy}$, $N_{xy} = t\sigma_{xy}$, are

$$\frac{1}{t}[J_{pdd}]\phi = \frac{1}{t}\begin{Bmatrix} \phi_{,yy} \\ \phi_{,xx} \\ -\phi_{,xy} \end{Bmatrix} = \frac{1}{t}[F_{dd}][G]\{\phi_c\}, \qquad (8.67)$$

$$[F_{dd}] = \begin{Bmatrix} F_{,yy} \\ F_{,xx} \\ -F_{,xy} \end{Bmatrix}, \quad \phi^V = [F][G]\{\phi_c^V\}. \qquad (8.68)$$

whence Equation (7.33) becomes

$$[C_\phi]\{\phi_c\} = \{u - u_E - u_Q\}_\phi, \qquad (8.69)$$

$$[C_\phi] = [G]^T[D][G], \quad [D] = \int_{A_m} \frac{1}{t}[F_{dd}]^T[J][F_{dd}]\mathrm{d}A_m,$$

$$\{u_E\}_\phi = [G]^T \int_{A_m} [F_{dd}]^T\{e_E\}\mathrm{d}A_m,$$

$$\{u_Q\}_\phi = [G]^T \int_{A_m} \frac{(1+\nu)}{E}[F_{dd}]^T \begin{Bmatrix} 1 \\ 1 \\ 0 \end{Bmatrix} V\,\mathrm{d}A_m,$$

$$\{u\}_\phi = [G]^T \left[\int_c [F_{dd}]^T[\lambda_b] \begin{Bmatrix} u_{xg} + u_{xQ} \\ u_{yg} + u_{yQ} \end{Bmatrix} \mathrm{d}s + \right.$$

$$\left. + \sum_k [F_{dd}]_k^T s_k [\lambda_b]_k \begin{Bmatrix} u_{xg} + u_{xQ} \\ u_{yg} + u_{yQ} \end{Bmatrix}_k \right].$$

Example 8.6. (a) Set up the $[H]$ and $[F_{dd}]$ matrices for the triangular element in Figure 8.7 using a nine term polynomial for ϕ. (b) Discuss the difficulties of using this procedure for triangles.

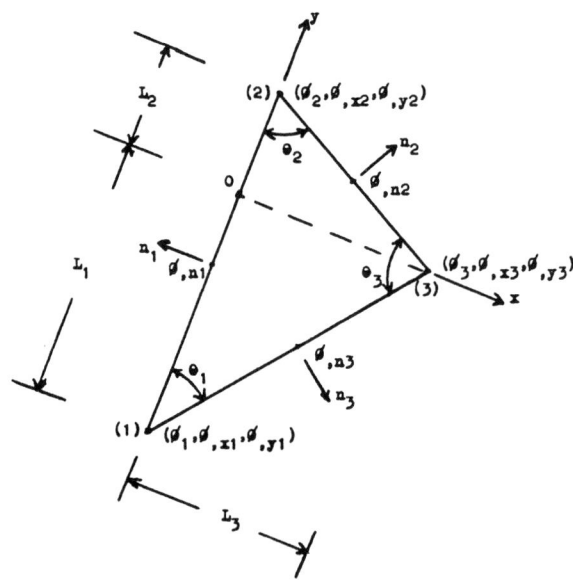

Fig. 8.7. Local coordinates for any triangular element.

Solution. (a) Take the polynomial for ϕ as

$$\phi = A_1 + A_2 x + A_3 y + A_4 x^2 + A_5 xy + A_6 y^2 + A_7 x^3 + A_8(xy^2 + x^2 y) + A_9 y^3, \quad (8.70)$$

where the A_8 term is used to have symmetry in x and y. Take the local coordinates x and y for the element as shown in Figure 8.7, where the coordinates for the three corners are $(0, y_1)$,

$(0, y_2)$, $(x_3, 0)$. From Equation (8.63), the $[F_d]$ matrix is

$$[F_d] = \begin{bmatrix} 1 & x & y & x^2 & xy & y^2 & x^3 & (x^2y+xy^2) & y^3 \\ 0 & 1 & 0 & 2x & y & 0 & 3x^2 & (2xy+y^2) & 0 \\ 0 & 0 & 1 & 0 & x & 2y & 0 & (x^2+2xy) & 3y^2 \end{bmatrix}, \quad (8.71)$$

whence $[H]$ in Equation (8.65) is

$$\begin{bmatrix} 1 & 0 & y_1 & 0 & 0 & y_1^2 & 0 & 0 & y_1^3 \\ 1 & 0 & y_2 & 0 & 0 & y_2^2 & 0 & 0 & y_2^3 \\ 1 & x_3 & 0 & x_3^2 & 0 & 0 & x_3^3 & 0 & 0 \\ 0 & 1 & 0 & 0 & y_1 & 0 & 0 & y_1^2 & 0 \\ 0 & 1 & 0 & 0 & y_2 & 0 & 0 & y_2^2 & 0 \\ 0 & 1 & 0 & 2x_3 & 0 & 0 & 3x_3^2 & 0 & 0 \\ 0 & 0 & 1 & 0 & 0 & 2y_1 & 0 & 0 & 3y_1^2 \\ 0 & 0 & 1 & 0 & 0 & 2y_2 & 0 & 0 & 3y_2^2 \\ 0 & 0 & 1 & 0 & x_3 & 0 & 0 & x_3^2 & 0 \end{bmatrix}. \quad (8.72)$$

From Equations (8.68) and (8.70)

$$[F_{dd}] = \begin{bmatrix} 0 & 0 & 0 & 0 & 0 & 2 & 0 & 2x & 6y \\ 0 & 0 & 0 & 2 & 0 & 0 & 6x & 2y & 0 \\ 0 & 0 & 0 & 0 & -1 & 0 & 0 & -2(x+y) & 0 \end{bmatrix}, \quad (8.73)$$

whence the stresses are linear in the plate. For any particular triangle calculate $[G] = [H]^{-1}$ and integrate for $[D]$ in Equation (8.69), whence $[C_\phi]$ can be calculated from Equation (8.69).

(b) One difficulty with this procedure is that for certain types of triangles, the $[H]$ matrix is singular. In fact, $[H]$ is singular for the triangle in Figure 8.5. Another difficulty is that in some arrangements of the triangles, the solution for the entire plate does not converge to the proper results as the number of elements is increased. Actually, a different procedure can be used for triangles. Instead of using the corner derivatives, satisfactory results can be obtained by using the normal derivative $\phi_{,n}$ at the midpoints of the element edges, Figure 8.7. Although this case has constant stresses in the element instead of the linear stresses in part (a) above, it has six unknowns instead of nine unknowns. This case is discussed in the following Example 8.7.

Example 8.7. Derive the $[C_\phi]$ matrix for the triangular element in Figure 8.7 using six terms in Equation (8.70) with points $\phi_1, \phi_2, \phi_3, \phi_{,n1}, \phi_{,n2}, \phi_{,n3}$.

Solution. Express $\phi_{,n}$ in terms of $\phi_{,x}$ and $\phi_{,y}$ by

$$\phi_{,n} = \phi_{,x}\frac{dx}{dn} + \phi_{,y}\frac{dy}{dn}, \quad (8.74)$$

and use the local coordinates in Figure 8.7 to write $\frac{dx}{dn}$, $\frac{dy}{dn}$ in terms of the interior angles θ_1 and θ_2. Thus, for Figure 8.7 with $\phi_{,x1}, \phi_{,x2}$, etc. evaluated at the midpoints,

$$\phi_{,n1} = -\phi_{,x1}, \quad \phi_{,n2} = \phi_{,x2}\cos\theta_2 + \phi_{,y2}\sin\theta_2,$$
$$\phi_{,n3} = \phi_{,x3}\cos\theta_1 - \phi_{,y3}\sin\theta_1. \quad (8.75)$$

With $\{\phi_c\}$ in the form

$$\{\phi_c\}^T = [\phi_1 \quad \phi_2 \quad \phi_3 \quad \phi_{,n1} \quad \phi_{,n2} \quad \phi_{,n3}], \quad (8.76)$$

the $[H]$ matrix in Equation (8.65) can be evaluated from the first six columns in Equation (8.71) by using Equation (8.75), whence

$$[H] = \begin{bmatrix} H_{11} & H_{12} \\ H_{21} & H_{22} \end{bmatrix}, \quad [H_{11}] = \begin{bmatrix} 1 & 0 & -L_1 \\ 1 & 0 & L_2 \\ 1 & L_3 & 0 \end{bmatrix}, \quad (8.77)$$

$$[H_{12}] = \begin{bmatrix} 0 & 0 & L_1^2 \\ 0 & 0 & L_2^2 \\ L_3^2 & 0 & 0 \end{bmatrix}, \quad [H_{21}] = \begin{bmatrix} 0 & -1 & 0 \\ 0 & \cos\theta_2 & \sin\theta_2 \\ 0 & \cos\theta_1 & -\sin\theta_1 \end{bmatrix},$$

$$[H_{22}] = \begin{bmatrix} 0 & \frac{1}{2}(L_1 - L_2) & 0 \\ L_3 \cos\theta_2 & \frac{1}{2}L_{23} & L_2 \sin\theta_2 \\ L_3 \cos\theta_1 & -\frac{1}{2}L_{31} & L_1 \sin\theta_1 \end{bmatrix},$$

$$L_{12} = L_1 + L_2, \quad L_{23}^2 = L_2^2 + L_3^2, \quad L_{31}^2 = L_1^2 + L_3^2, \tag{8.78}$$

$$\sin\theta_1 = L_3/L_{31}, \quad \sin\theta_2 = L_3/L_{23}.$$

From Equation (8.73)

$$[F_{dd}] = [[0] \ [F_1]], \quad [F_1] = \begin{bmatrix} 0 & 0 & 2 \\ 2 & 0 & 0 \\ 0 & -1 & 0 \end{bmatrix}, \tag{8.79}$$

whence the stresses are constant and $[D]$ in Equation (8.69) is

$$[D] = \frac{L_3 L_{12}}{Et} \begin{bmatrix} [0] & [0] \\ [0] & [D_{22}] \end{bmatrix}, \tag{8.80}$$

$$[D_{22}] = \frac{E}{2}[F_1]^T[J][F_1] = \begin{bmatrix} 2 & 0 & -2\nu \\ 0 & 1+\nu & 0 \\ -2\nu & 0 & 2 \end{bmatrix}.$$

Take $[G] = [H]^{-1}$ in the form

$$[G] = \begin{bmatrix} G_{11} & G_{12} \\ G_{21} & G_{22} \end{bmatrix}, \tag{8.81}$$

whence $[C_\phi]$ in Equation (8.69) becomes

$$[C_\phi] = \begin{bmatrix} B_{11} & B_{12} \\ B_{12}^T & B_{22} \end{bmatrix}, \quad [B_{11}] = [G_{21}]^T[D_{22}][G_{21}], \tag{8.82}$$

$$[B_{12}] = [G_{21}]^T[D_{22}][G_{22}], \quad [B_{22}] = [G_{22}]^T[D_{22}][G_{22}].$$

Thus, because of the [0] matrices in $[D]$, only $[G_{21}]$ and $[G_{22}]$ are needed in $[G]$. The result is

$$[C_\phi] = \begin{bmatrix} C_{11} & C_{12} \\ C_{12}^T & C_{22} \end{bmatrix}, \tag{8.83}$$

$$[C_{11}] = \frac{4(1+\nu)\sin^2\theta_3}{EtL_{12}^2} \begin{bmatrix} \cot\theta_2 + \cot\theta_3 & -\cot\theta_3 & -\cot\theta_2 \\ -\cot\theta_3 & \cot\theta_1 + \cot\theta_3 & -\cot\theta_1 \\ -\cot\theta_2 & -\cot\theta_1 & \cot\theta_1 + \cot\theta_2 \end{bmatrix},$$

$$[C_{22}] = \frac{2}{Et} \begin{bmatrix} \cot\theta_1 + \cot\theta_2 & b_2 & b_1 \\ b_2 & \cot\theta_2 + \cot\theta_3 & b_3 \\ b_1 & b_3 & \cot\theta_1 + \cot\theta_3 \end{bmatrix},$$

$$b_i = \cos\theta_1 \cot\theta_i - \nu \sin\theta_i, \quad i = 1, 2, 3,$$

$$[C_{12}]^T = \frac{2(1+\nu)\sin\theta_3}{EtL_{12}} \begin{bmatrix} -\frac{\cos\theta_1}{\sin\theta_2} & -\frac{\cos\theta_2}{\sin\theta_1} & \frac{\cos\theta_1}{\sin\theta_2} + \frac{\cos\theta_2}{\sin\theta_1} \\ \frac{\cos\theta_2}{\sin\theta_3} + \frac{\cos\theta_3}{\sin\theta_2} & -\frac{\cos\theta_2}{\sin\theta_3} & -\frac{\cos\theta_3}{\sin\theta_2} \\ -\frac{\cos\theta_1}{\sin\theta_3} & \frac{\cos\theta_1}{\sin\theta_3} + \frac{\cos\theta_3}{\sin\theta_1} & -\frac{\cos\theta_3}{\sin\theta_1} \end{bmatrix}.$$

Note that this form of $[C_\phi]$ is independent of the orientation of the triangle, with only the interior angles $\theta_1, \theta_2, \theta_3$ and the length L_{12} of edge (1) being used. Also, note that $[C_{22}]$ is the same as $[C_P]$ in Equation (8.57). Since $[C_{22}]$ operates on $\{\phi_{,ni}\}$ and

$$\phi_{,ni} = \int \phi_{,nni} \, dn_i = \int N_{si} \, dn_i = F_i,$$

the equality is to be expected.

Case (7) in Equation (8.24b) for Hybrid [k]

This case (7) allows the stiffness matrix $[k]$ to be calculated from the principle of virtual forces using a large number of terms for the stresses but using only

Approximate matrix equations for plate finite elements 273

the corner point displacements as the unknowns. This procedure has been designated as the *hybrid stress method* in the literature. See References in Chapter 9, particularly References 13 and 15(a).

Assume a set of stresses in the plate element that are in equilibrium, similar to Equation (8.59a), or

$$\{\sigma\} = [H]\{A\}, \quad \{e\} = [J]\{\sigma\} + \{e_E + e_Q\}, \quad \{\sigma^V\} = [H]\{A^V\}, \tag{8.84}$$

$$H = \begin{bmatrix} 1 & 0 & 0 & y & 0 & x & 0 & xy & 0 & 0 & x^2 & y^2 \\ 0 & 1 & 0 & 0 & x & 0 & y & 0 & xy & x^2 & y^2 & 0 \\ 0 & 0 & 1 & 0 & 0 & -y & -x & -y^2/2 & -x^2/2 & 0 & -2xy & 0 \end{bmatrix}, \tag{8.85}$$

where the desired number of columns can be selected. From Equation (7.8), the equilibrium edge forces can be expressed as

$$\{S\}_{\text{edge}} = [R]\{A\}. \tag{8.86}$$

Assume the u_x and u_y displacements are continuous on the edges and express them in terms of the unknown corner displacements $\{u\}$, or

$$\{u\}_{\text{edge}} = [G]\{u\}_{\text{corner}} = [G]\{u\}. \tag{8.87}$$

For this case, Equation (7.32) has the form

$$\{A^V\}^T \left[\int_{A_m} t[H]^T[J][H]\{A\}dA_m + \int_{A_m} t[H]^T\{e_E\}dA_m - \int_c t[R]^T[G]\{u\}ds \right] = 0. \tag{8.88}$$

For arbitrary $\{A^V\}^T$,

$$[C_H]\{A\} = [B]\{u\} - \{u_E\}, \tag{8.89}$$

$$[C_H] = \int_{A_m} t[H]^T[J][H]dA_m,$$

$$[B] = \int_C t[R]^T[G]ds,$$

$$\{u_E\} = \int_{A_m} t[H]^T\{e_E\}dA_m.$$

From Equations (8.19) and (8.89),

$$[k]\{u\} = \{P\} + \{P_E\}, \tag{8.90}$$

$$[k] = [B]^T[C_H]^{-1}[B],$$

$$\{P_E\} = [B]^T[C_H]^{-1}\{u_E\}.$$

Note that $\{P\}$ and $\{P_E\}$ can be taken from the principle of virtual displacements in Equation (8.30).

Example 8.8. Use the first five columns in Equation (8.85) and find the hybrid $[k]$ matrix for the triangular element in Figure 8.5. Use linear displacements on the edges.

Solution. Take $[H]$ from Equation (8.85) in the form

$$[H] = [I \quad C_2], \quad [C_2] = \begin{bmatrix} y & 0 \\ 0 & x \\ 0 & 0 \end{bmatrix}, \qquad (8.91)$$

whence Equation (8.89) gives

$$[C_H] = \int_0^1 \int_0^{1-x} t \begin{bmatrix} J & JC_2 \\ C_2^T J & C_2^T J C_2 \end{bmatrix} dy\, dx \qquad (8.92)$$

$$= \frac{t}{24E} \begin{bmatrix} 12 & -12\nu & 0 & 4 & -4\nu \\ -12\nu & 12 & 0 & -4\nu & 4 \\ 0 & 0 & 24(1+\nu) & 0 & 0 \\ 4 & -4\nu & 0 & 2 & -\nu \\ -4\nu & 4 & 0 & -\nu & 2 \end{bmatrix}, \qquad (8.93)$$

where the following integrals are used:

$$\int_0^1 \int_0^{1-x} (1, x, y, x^2, y^2, xy)\, dy\, dx = \tfrac{1}{24}(12, 4, 4, 2, 2, 1). \qquad (8.94)$$

On the three edges of the element in Figure 8.5, $S_{x1} = -\sigma_{xx1}$, $S_{y1} = -\sigma_{xy1}$, $S_{x2} = b(\sigma_{xy2} + \sigma_{xx2})$, $S_{y2} = b(\sigma_{xy2} + \sigma_{yy2})$, $S_{x3} = -\sigma_{xy3}$, $S_{y3} = -\sigma_{yy3}$, $b = 2^{-1/2}$, whence the matrix $[R]$ in Equation (8.86) is

$$[R] = \begin{bmatrix} -1 & 0 & 0 & -y & 0 \\ 0 & 0 & -1 & 0 & 0 \\ b & 0 & b & b(1-x) & 0 \\ 0 & b & b & 0 & bx \\ 0 & 0 & -1 & 0 & 0 \\ 0 & -1 & 0 & 0 & -x \end{bmatrix}. \qquad (8.95)$$

For linear displacements along the three edges, the $[G]$ matrix in Equation (8.87) is

$$[G] = \begin{bmatrix} 1-y & y & 0 & 0 & 0 & 0 \\ 0 & 0 & 0 & 1-y & y & 0 \\ 0 & 1-x & x & 0 & 0 & 0 \\ 0 & 0 & 0 & 0 & 1-x & x \\ x & 0 & 1-x & 0 & 0 & 0 \\ 0 & 0 & 0 & x & 0 & 1-x \end{bmatrix}. \qquad (8.96)$$

Use $b\, ds = dx$ on edge (2) and proceed clockwise around the element to calculate $[B]$ from Equation (8.89), integrating 0 to 1 on each term, whence

$$[B] = \frac{t}{6} \begin{bmatrix} -3 & 0 & 3 & 0 & 0 & 0 \\ 0 & 0 & 0 & -3 & 3 & 0 \\ -3 & 3 & 0 & -3 & 0 & 3 \\ -1 & 0 & 1 & 0 & 0 & 0 \\ 0 & 0 & 0 & -1 & 1 & 0 \end{bmatrix}. \qquad (8.97)$$

From Equation (8.89) the $\{A\}$ constants can be expressed as

$$\{A\} = [C_H]^{-1}[B]\{u\} = \frac{E}{2(1-\nu^2)} \begin{bmatrix} -2 & 0 & 2 & -2\nu & 2\nu & 0 \\ -2\nu & 0 & 2\nu & -2 & 2 & 0 \\ \nu-1 & 1-\nu & 0 & \nu-1 & 0 & 1-\nu \\ 0 & 0 & 0 & 0 & 0 & 0 \\ 0 & 0 & 0 & 0 & 0 & 0 \end{bmatrix} \{u\}, \qquad (8.98)$$

which gives $A_4 = A_5 = 0$. This result verifies the statement in Example 8.1 for the triangle that the only solution for six corner unknown displacements is the case of constant stresses. However, it is shown in Reference 14a of Chapter 9 that the k matrix for the rectangular element can be made more accurate by this hybrid procedure. Use the above values in Equation (8.90) to get

$$[k] = [B]^T [C_H]^{-1} [B] = [k] \quad \text{in Equation (8.31).} \qquad (8.99)$$

Case (8) in Equation (8.24b) for $[C_P]$ from $[k]$

The flexibility $[C_P]$ matrix can be obtained from the reduced stiffness $[k]$ matrix by Equation (8.21), as demonstrated in the following Example.

Example 8.9. Find the $[C_P]$ for the triangular element in figure 8.5 using the restraints $u_1 = 0$, $u_2 = 0$, and $u_4 = 0$ to reduce the $[s]$ and $[k]$ matrices.
Solution. The reduced $[s_r]$ matrix from Equations (8.47) and (8.48) is

$$[s_r] = \begin{bmatrix} 0 & 1 & 0 \\ b & b & -b \\ 1 & 0 & 0 \end{bmatrix}, \quad b = 2^{-1/2}, \quad [s_r]^T = \begin{bmatrix} 0 & b & 1 \\ 1 & b & 0 \\ 0 & -b & 0 \end{bmatrix}.$$

From Equation (8.31), the reduced $[k_r]$ matrix is

$$[k_r] = \frac{Et}{4(1-\nu^2)} \begin{bmatrix} 2 & 2\nu & 0 \\ 2\nu & 2 & 0 \\ 0 & 0 & 1-\nu \end{bmatrix},$$

$$[k_r]^{-1} = \frac{2}{Et} \begin{bmatrix} 1 & -\nu & 0 \\ -\nu & 1 & 0 \\ 0 & 0 & 2(1+\nu) \end{bmatrix}.$$

Thus, Equation (8.21) gives

$$[C_P] = [s_r][k_r]^{-1}[s_r]^T = \frac{2}{Et} \begin{bmatrix} 1 & b(1-\nu) & -\nu \\ b(1-\nu) & 2 & b(1-\nu) \\ -\nu & b(1-\nu) & 1 \end{bmatrix}, \tag{8.100}$$

which is the same as in Equation (8.58).

8.6. Inplane plate element matrices from the mixed virtual principle

As has been pointed out in the previous sections and examples, the principle of virtual displacements gives approximate solutions which approximate the equilibrium equations and the stress equilibrium boundary conditions, the principle of virtual forces gives approximate solutions which approximate the compatibility conditions and the displacement boundary conditions, and the mixed principle gives approximate solutions which approximate both the equilibrium and the compatibility equations but can satisfy both the stress and displacement boundary conditions. In order to impose both stress and displacement boundary conditions on the exterior boundary of the plate with finite elements, it is necessary to assemble the mixed $[M]$ matrix for the entire plate in Equation (7.76). This section is concerned with the calculation of the $[M]$ matrix for a plate finite element with inplane loads and covers cases (9)-(11) in Equations (8.24c).

It should be emphasized that in the mixed principle the stresses at the corner or node points are the actual approximate stresses and not equivalent point forces.

Case (9) in Equation (8.24c) for $[M]$

Assume the displacements and stresses in the plate in the form

$$\{u\} = \left\{\begin{array}{c} u_x \\ u_y \end{array}\right\} = [F(x,y)]\{A\}, \quad \{N\} = \left\{\begin{array}{c} N_x \\ N_y \\ N_{xy} \end{array}\right\} = [G(x,y)]\{B\}, \quad (8.101)$$

where F_i and G_i are the assumed functions for the element and A_i and B_i are the unknown constants. F_i and G_i should be continuous between the element edges. Refer to Equations (8.25)-(8.27) and take

$$[F] = \begin{bmatrix} F_1(x,y) & 0 \\ 0 & F_1(x,y) \end{bmatrix}, \qquad (8.102)$$

$$[G] = \begin{bmatrix} F_1(x,y) & 0 & 0 \\ 0 & F_1(x,y) & 0 \\ 0 & 0 & F_1(x,y) \end{bmatrix}. \qquad (8.103)$$

$$\{u\} = [F][H_u]^{-1}\{u_c\} = [F_u]\{u_c\},$$
$$\{N\} = [G][H_n]^{-1}\{N_c\} = [G_N]\{N_c\},$$
$$\{u_c\}^T = [u_{x1}\ u_{x2}\cdots u_{xN}\ u_{y1}\cdots u_{yN}], \qquad (8.104)$$
$$\{N\}^T = [N_{x1}\ N_{x2}\cdots N_{xN}\ N_{y1}\cdots N_{yN}\ N_{xy1}\cdots N_{xyN}],$$

$$[H_u]^{-1} = \begin{bmatrix} H_c^{-1} & 0 \\ 0 & H_c^{-1} \end{bmatrix}, \quad [H_N]^{-1} = \begin{bmatrix} H_c^{-1} & 0 & 0 \\ 0 & H_c^{-1} & 0 \\ 0 & 0 & H_c^{-1} \end{bmatrix}, \quad (8.105)$$

$$[H_c] = \begin{bmatrix} F_1(x_1,y_1) \\ F_1(x_2,y_2) \\ \cdots \\ F_1(x_N,y_N) \end{bmatrix}, \quad [G_N] = \begin{bmatrix} F_1 H_c^{-1} & 0 & 0 \\ 0 & F_1 H_c^{-1} & 0 \\ 0 & 0 & F_1 H_c^{-1} \end{bmatrix}. \quad (8.106)$$

From Equation (8.28) take the strains as

$$\{e_u\} = [D]\{u_c\}, \quad [D] = [J_{pd}][F_u],$$
$$\{e_\sigma\} = \frac{1}{t}[J]\{N\} + \{e_E\} = \frac{1}{t}[J][G_N]\{N_c\} + \{e_E\}, \qquad (8.107)$$

and use

$$\{u^V\} = [F_u]\{u_c^V\}, \quad \{e^V\} = [D]\{u_c^V\}, \quad \{N^V\} = [G_N]\{N_c^V\}, \qquad (8.108)$$

whence the mixed principle in Equations (7.68) and (7.69) becomes

$$\{u_c^V\}\left[\int_{A_m}\left\{[D]^T[G_N]\{N_c\} - t[F_y]^T\left\{\begin{array}{c}X_x\\X_y\end{array}\right\}\right\}dA_m - \int_{c\sigma}t[F_u]_e^T\left\{\begin{array}{c}S_x\\S_y\end{array}\right\}ds - \right.$$
$$-\int_c t[F_u]_e^T[\lambda_b]^T[G_N]_e\{N_c\}ds - \sum_k[F_u]_k^T\left\{\begin{array}{c}P_{xk}\\P_{yk}\end{array}\right\} +$$
$$+\{N_c^V\}^T\left[\int_{A_m}[G_N]^T\{[D]\{u_c\} - \frac{1}{t}[J][G_N]\{N_c\} - \{e_E\}\}dA_m - \right.$$
$$\left.-\int_c t[G_N]_e^T[\lambda_b][F_u]\{u_c\}ds + \int_{cu}t[G_N]_e^T[\lambda_b]\{u_s\}ds\right] = 0. \qquad (8.109)$$

Approximate matrix equations for plate finite elements 277

For arbitrary $\{u_c^V\}$ and $\{N_c^V\}$, Equation (8.109) has the form

$$[M]\left\{\begin{array}{c}N_c\\u_c\end{array}\right\}=\left[\begin{array}{cc}-[M_{11}]&[M_{12}]\\ [M_{12}]^T&[0]\end{array}\right]\left\{\begin{array}{c}N_c\\u_c\end{array}\right\}=\left\{\begin{array}{c}u_E-u_g\\P\end{array}\right\}, \qquad (8.110)$$

$$[M_{11}]=\int_{A_m}\frac{1}{t}[G_N]^T[J][G_N]dA_m,\quad \{u_g\}=\int_{cu}t[G_N]_e^T[\lambda_b]\{u_s\}ds,$$

$$[M_{12}]=\int_{A_m}[G_N]^T[D]dA_m-\int_c t[G_N]_e^T[\lambda_b][F_u]_e\,ds,$$

$$\{u_E\}=[M_{11}]\{N_e\}=\int_{A_m}[G_N]^T\{e_E\}dA_m,$$

$$\{P\}\doteq\int_{A_m}t[F_u]^T\left\{\begin{array}{c}X_x\\X_y\end{array}\right\}dA_m+\int_{c\sigma}t[F_u]^T\left\{\begin{array}{c}S_x\\S_y\end{array}\right\}ds+\sum_k[F_u]_k^T\left\{\begin{array}{c}P_{xk}\\P_{yk}\end{array}\right\}.$$

If the assumed displacements and stresses satisfy continuity between the element edges, then the edge integrals in Equation (8.110) on interior elements are either zero or they cancel out on assembly of the elements. On the exterior edges the edge integrals give weighted values for the applied forces and displacements as well as weighted corrections when the assumed displacements do not satisfy the displacement boundary conditions. These corrections are zero if displacements and stresses satisfying the boundary conditions are assumed for the exterior boundary elements. Also, if the elements are small, these corrections have little effect and can be omitted.

From the diagonal form of $[G_N]$ in Equation (8.106), it follows that for interior elements

$$[M_{11}]=\frac{1}{Et}\left[\begin{array}{ccc}B_1&-\nu B_1&0\\-\nu B_1&B_1&0\\0&0&2(1+\nu)B_1\end{array}\right],\quad [M_{12}]=\left[\begin{array}{cc}B_2&0\\0&B_3\\B_3&B_2\end{array}\right], \qquad (8.111)$$

$$[B_1]=[H_c^{-1}]^T\left[\int_{A_m}[F_1]^T[F_1]dA_m\right][H_c^{-1}],$$

$$[B_2]=[H_c^{-1}]^T\left[\int_{A_m}[F_1]^T[F_{1,x}]dA_m\right][H_c^{-1}],$$

$$[B_3]=[H_c^{-1}]^T\left[\int_{A_m}[F_1]^T[F_{1,y}]dA_m\right][H_c^{-1}].$$

Example 8.10. (a) Construct the $[M]$ matrix for the triangular element in Figure 8.5 using linear terms in $[F_1(x,y)]=[1\ x\ y]$. (b) Discuss the results for both interior and boundary elements.

Solution. (a) From Equations (8.102)-(8.106) for Figure 8.5 with the given $F_1(x,y)$,

$$[H_c]=\left[\begin{array}{ccc}1&0&0\\1&0&1\\1&1&0\end{array}\right],\quad [H_c]^{-1}=\left[\begin{array}{ccc}1&0&0\\-1&0&1\\-1&1&0\end{array}\right].$$

Use the integrals from Equation (8.94) in Equation (8.111) to get

$$[B_1]=[H_c^{-1}]^T\int_0^1\int_0^{1-x}\left\{\begin{array}{c}1\\x\\y\end{array}\right\}[1\ x\ y]dy\,dx[H_c^{-1}]$$

$$= \tfrac{1}{24}\begin{bmatrix} 1 & -1 & -1 \\ 0 & 0 & 1 \\ 0 & 1 & 0 \end{bmatrix}\begin{bmatrix} 12 & 4 & 4 \\ 4 & 2 & 1 \\ 4 & 1 & 2 \end{bmatrix}\begin{bmatrix} 1 & 0 & 0 \\ -1 & 0 & 1 \\ -1 & 1 & 0 \end{bmatrix} = \tfrac{1}{24}\begin{bmatrix} 2 & 1 & 1 \\ 1 & 2 & 1 \\ 1 & 1 & 2 \end{bmatrix},$$

$$[B_2] = \tfrac{1}{6}\begin{bmatrix} -1 & 0 & 1 \\ -1 & 0 & 1 \\ -1 & 0 & 1 \end{bmatrix}, \quad [B_3] = \tfrac{1}{6}\begin{bmatrix} -1 & 1 & 0 \\ -1 & 1 & 0 \\ -1 & 1 & 0 \end{bmatrix}. \tag{8.112}$$

Thus, with these values, the $[M]$ matrix is

$$[M] = \begin{bmatrix} -\tfrac{1}{Et}\begin{bmatrix} B_1 & -\nu B_1 & 0 \\ -\nu B_1 & B_1 & 0 \\ 0 & 0 & 2(1+\nu)B_1 \end{bmatrix} & \begin{bmatrix} B_2 & 0 \\ 0 & B_3 \\ B_3 & B_2 \end{bmatrix} \\ \begin{bmatrix} B_2^T & 0 & B_3^T \\ 0 & B_3^T & B_2^T \end{bmatrix} & \begin{bmatrix} 0 & 0 \\ 0 & 0 \end{bmatrix} \end{bmatrix} \tag{8.113}$$

(b) For interior triangular elements with all corner $\{N_c\}$ and $\{u_c\}$ values unknown, the $[M_{11}]$ matrix can be inverted to give

$$[M_{11}]^{-1} = \frac{Et}{1-\nu^2}\begin{bmatrix} B_1^{-1} & \nu B_1^{-1} & 0 \\ \nu B_1^{-1} & B_1^{-1} & 0 \\ 0 & 0 & \tfrac{1}{2}(1-\nu)B_1^{-1} \end{bmatrix}, \tag{8.114}$$

$$[B_1]^{-1} = 6\begin{bmatrix} 3 & -1 & -1 \\ -1 & 3 & -1 \\ -1 & -1 & 3 \end{bmatrix},$$

whence in Equation (8.110) the corner stresses can be expressed in terms of the corner displacements as

$$\{N_c\} = [M_{11}]^{-1}[M_{12}]\{u_c\} - \{N_E\}$$

$$= \frac{6Et}{1-\nu^2}\begin{bmatrix} B_2 & \nu B_3 \\ \nu B_2 & B_3 \\ \tfrac{1-\nu}{2}B_3 & \tfrac{1-\nu}{2}B_2 \end{bmatrix}\{u_c\} - \{N_E\}. \tag{8.115}$$

If $\{N_E\}^T$ has the form $[A\ A\ A\ B\ B\ B\ C\ C\ C]$, then the forms of $[B_2]$ and $[B_3]$ in Equation (8.112) show that the stresses in Equation (8.115) are constant, or $N_{x1} = N_{x2} = N_{x3}$, etc. This indicates that any $[k]$ and $[C_P]$ matrices determined from the $[M]$ matrix in Equation (8.113) for triangular elements will be the same as those given by the principle of virtual displacements and the principle of virtual forces. However, if Equation (8.11) is assembled for all the elements and both stress and displacement boundary conditions applied at the exterior boundary, then the resulting stresses will be linear in each element. Also, if the element has one or two corner points on the exterior boundary, then some of the stresses can be specified so that $[M]$ can be partitioned, and Equation (8.115) will no longer give constant stresses. See Example 8.12 below.

Although the above procedure gives all three stresses at the corner points and is simple to set up, it produces a very large matrix on assembly of the elements. The number of unknown stresses can be reduced by using the normal and shear stresses along the edges of the element in terms of midpoint values, as described in case (5) and Example 8.5 above. Take the displacements as in Equation (8.104) but take the stresses from Equation (8.59) as

$$\{N\} = \begin{Bmatrix} N_x \\ N_y \\ N_{xy} \end{Bmatrix} = [G_m(x,y)]\{N_m\}. \tag{8.116}$$

All the Equations (8.102)-(8.110) apply with $[G_m]$ in place of $[G_N]$ and $\{N_m\}$ in place of $\{N_c\}$. Since $[G_m]$ is not diagonal, Equation (8.111) does not apply.

Approximate matrix equations for plate finite elements

Example 8.11. Construct the $[M]$ matrix for the triangular element in Figure 8.5 using linear displacements but only three unknown normal stresses N_{ni} on the three edges of the element.

Solution. In this case $\{N_m\}$ are the three constant normal stresses on the edges of the triangle so that Equations (8.39) and (8.40) can be used, or

$$[G_m] = [Z_1]^{-1}, \quad [Z_1] \text{ in Equation (8.40).} \tag{8.117}$$

For the triangular element in Figure 8.5, Equation (8.43) gives

$$[G_m] = \tfrac{1}{2} \begin{bmatrix} 2 & 0 & 0 \\ 0 & 0 & 2 \\ -1 & 2 & -1 \end{bmatrix},$$

whence Equation (8.11) becomes

$$[M_{11}] = \int_{A_m} \tfrac{1}{t}[G_m]^T [J][G_m] dA_m$$

$$= \frac{1}{4Et} \begin{bmatrix} 3+\nu & -2(1+\nu) & 1-\nu \\ -2(1+\nu) & 4(1+\nu) & -2(1+\nu) \\ 1-\nu & -2(1+\nu) & 3+\nu \end{bmatrix}, \tag{8.118a}$$

$$[M_{12}] = \int_{A_m} [G_m]^T \begin{bmatrix} F_{1,x} & 0 \\ 0 & F_{1,y} \\ F_{1,y} & F_{1,x} \end{bmatrix} \begin{bmatrix} H_c^{-1} & 0 \\ 0 & H_c^{-1} \end{bmatrix} dA_m$$

$$= \tfrac{1}{4} \begin{bmatrix} -1 & -1 & 2 & 1 & 0 & -1 \\ -2 & 2 & 0 & -2 & 0 & 2 \\ 1 & -1 & 0 & -1 & 2 & -1 \end{bmatrix}, \tag{8.118b}$$

$$\{u_E\} = [M_{11}]\{N_e\} = \tfrac{1}{2} \int_{A_m} \begin{bmatrix} 2 & 0 & -1 \\ 0 & 0 & 2 \\ 0 & 2 & -1 \end{bmatrix} \begin{Bmatrix} e_{xxE} \\ e_{yyE} \\ e_{xyE} \end{Bmatrix} dA_m$$

$$= \int_{A_m} \begin{Bmatrix} \alpha T \\ 0 \\ \alpha T \end{Bmatrix} dA_m \quad \text{for temperature case.} \tag{8.118c}$$

Put $[M_{11}]$ and $[M_{12}]$ into Equation (8.11) to get $[M]$ for this case.

Case (10) in Equation (14.24c) for [k] from [M]

The $[k]$ matrix can be calculated from the $[M]$ matrix by using Equation (7.78), or by use of Equations (8.111) and (8.114),

$$[k] = \frac{Et}{1-\nu^2} \begin{bmatrix} B_2^T B_1^{-1} B_2 + \tfrac{1-\nu}{2} B_3^T B_1^{-1} B_3 \\ \nu B_3^T B_1^{-1} B_2 + \tfrac{1-\nu}{2} B_2^T B_1^{-1} B_3 \end{bmatrix}$$

$$\begin{matrix} \nu B_2^T B_1^{-1} B_3 + \tfrac{1-\nu}{2} B_3^T B_1^{-1} B_2 \\ B_3^T B_1^{-1} B_3 + \tfrac{1-\nu}{2} B_2^T B_1^{-1} B_2 \end{matrix}. \tag{8.119}$$

Since the same linear displacements are used in Example 8.10 as in Example 8.1, it follows that if the $[B_2]$, $[B_3]$, and $[B_1]^{-1}$ in Equations (8.112) and (8.114) are put into Equation (8.119), then the $[k]$ matrix in Equation (8.31) is obtained.

Example 8.12. Calculate the element stiffness matrix $[k]$ for a boundary element with $N_{y1} = 0$ and $N_{xy1} = 0$ in the triangular element in Example 8.10.

Solution. Partition Equation (8.110) to include specified stresses, or

$$\begin{bmatrix} -M_{11} & -M_{12} & M_{13} \\ -M_{12}^T & -M_{22} & M_{23} \\ M_{13}^T & M_{23}^T & 0 \end{bmatrix} \begin{Bmatrix} S \\ S_g \\ u \end{Bmatrix} = \begin{Bmatrix} u_E \\ u_{E_g} \\ P \end{Bmatrix}, \tag{8.120}$$

where $\{S_g\}^T = [0\ 0]$ for $N_{y1} = 0$, $N_{xy1} = 0$, and no $\{u\}$ are specified. Thus, from Example 8.10,

$$[M_{11}] = \frac{1}{Et}\begin{bmatrix} B_1 & -\nu B_5^T & 0 \\ -\nu B_5^T & B_4 & 0 \\ 0 & 0 & 2(1+\nu)B_4 \end{bmatrix}, \quad [B_5] = \frac{1}{24}\begin{bmatrix} 1 & 1 \\ 2 & 1 \\ 1 & 2 \end{bmatrix}, \quad (8.121)$$

$$[B_4] = \frac{1}{24}\begin{bmatrix} 2 & 1 \\ 1 & 2 \end{bmatrix}, \quad [M_{12}] = \frac{1}{24Et}\begin{bmatrix} -\nu B_6 \\ B_7 \\ 2(1+\nu)B_8 \end{bmatrix}, \quad [B_6] = \begin{bmatrix} 2 & 0 \\ 1 & 0 \\ 1 & 0 \end{bmatrix}, \quad (8.122)$$

$$[B_7] = \begin{bmatrix} 1 & 0 \\ 1 & 0 \end{bmatrix}, \quad [B_8] = \begin{bmatrix} 0 & 1 \\ 0 & 1 \end{bmatrix}, \quad [M_{13}] = \begin{bmatrix} B_2 & 0 \\ 0 & Q_3 \\ Q_3 & Q_4 \end{bmatrix}, \quad (8.123)$$

$$[Q_3] = \tfrac{1}{6}\begin{bmatrix} -1 & 1 & 0 \\ -1 & 1 & 0 \end{bmatrix}, \quad [Q_4] = \tfrac{1}{6}\begin{bmatrix} -1 & 0 & 1 \\ -1 & 0 & 1 \end{bmatrix},$$

$$[M_{23}] = \tfrac{1}{6}\begin{bmatrix} 0 & 0 & 0 & -1 & 1 & 0 \\ -1 & 1 & 0 & -1 & 0 & 1 \end{bmatrix}. \quad (8.124)$$

Solve Equation (8.120) for $\{S\}$ and hence get $[k_b]\{u\} = \{P\} + \{P_E\}$, where

$$[k_b] = [M_{13}]^T [M_{11}]^{-1} [M_{13}]$$

$$= \frac{2Et}{9(1-\nu^2)}\begin{bmatrix} \frac{13-4\nu+\nu^2}{4} & \nu-1 & \frac{\nu^2-9}{4} & 1+\nu & -2\nu & \nu-1 \\ \nu-1 & 1-\nu & 0 & \nu-1 & 0 & 1-\nu \\ \frac{\nu^2-9}{4} & 0 & \frac{9-\nu^2}{4} & -2\nu & 2\nu & 0 \\ 1+\nu & \nu-1 & -2\nu & 3-\nu & -2 & \nu-1 \\ -2\nu & 0 & 2\nu & -2 & 2 & 0 \\ \nu-1 & 1-\nu & 0 & \nu-1 & 0 & 1-\nu \end{bmatrix}. \quad (8.125)$$

This $[k_b]$ can be compared directly to the $[k]$ in Equation (8.31). In $[k_b]$ all elements except four are $(2/9)/(1/4) = 8/9 = 89\%$ of those in $[k]$. The other four elements are also less than the corresponding ones in $[k]$. Since the principle of virtual displacements with the approximate $[k]$ in Equation (8.31) for all elements makes the structure too stiff, the mixed principle with the boundary elements less stiff and the stresses linear in all elements may give better results.

Case (11) in Equation (8.24c) for $[C_\sigma]$

Since the $[M_{11}]$ matrix in $[M]$, Equation (8.110), is determined from the assumed stresses, it is a flexibility matrix corresponding to a $[C_\sigma]$ type matrix. In fact, the $[M_{11}]$ matrix in Equation (8.118a) is the same as the $[C_\sigma]$ matrix in Equations (8.61) and (8.46). However, $[C_\sigma]$ is based upon the number of independent equilibrium sets of stresses while $[M_{11}]$ is not. The stresses for $[M_{11}]$ need not be independent and equilibrium is approximated by the mixed virtual principle. Thus, the $[M_{11}]$ matrix in Equation (8.111) is not a $[C_\sigma]$ matrix that would be given by the principle of virtual forces. They are the same for the constant stress case, as equilibrium is satisfied exactly for both matrices.

8.7. Bending plate element matrices from the principle of virtual displacements

In the small deflection theory of plates used in Section 7.5, the transverse deflection u_z perpendicular to the plate surface is uncoupled from the inplane deflections u_x and u_y. Thus, the stiffness matrices for the u_z deflection can be calculated separately from the inplane stiffness matrices given above. Since the bending moments and shear forces in the plate with transverse loading are given by second and third derivatives of u_z, Equations (7.91) and (7.102), the bending of the finite element is more complicated than the inplane case. Not only is it desirable to have the displacements continuous between the elements but also the slopes or rotations θ_x and θ_y should be continuous between elements. Since these requirements are very difficult to meet with simple assumed polynomial terms for the displacement u_z, a large variety of other methods have been developed to obtain the stiffness matrices for the bending elements. In fact, some of the methods listed in Equation (8.24) and discussed above for the inplane case originally were developed for the plate bending case.

Since there are only three possible constants for triangular plate corner values of the displacement u_z, it is evident that additional points or additional corner variables are needed in order to use more constants. Although both procedures have been used, the case of using the first derivatives of u_z at the corners or at the midpoints of the edges as additional variables has received much study. Since u_z is the lateral deflection of the plate, the derivatives $u_{z,x}$ and $u_{z,y}$ represent the slopes of the plate as well as the rotations θ_x, θ_y of the plate, Equation (7.90). Also, from Equation (8.23b) the plate moments are related to the slopes or rotations. Thus, in the basic matrix Equation (8.11), $\{u\}$ includes point values of u_z, $u_{z,x}$, and $u_{z,y}$, $\{P\}$ includes point values of P_z forces and the moments M_x and M_y, Figure 8.4.

The primary difference between the inplane and bending plate cases is the transverse shear forces in the bending case. In spite of this physical difference, the dual relations discussed in Section 7.5 show little mathematical difference. In fact, in case (6) above on the $[C_\phi]$ matrix, the point values ϕ, $\phi_{,x}$, and $\phi_{,y}$ were used. Although the u_ϕ terms in Equation (8.69) associated with the point values of ϕ have no physical meaning, they have the form of P_{zi} corner forces associated with u_{zi}. As indicated in Example 8.7, the point values of $\phi_{,x}$ and $\phi_{,y}$ are associated with the inplane forces in the element.

The bending cases listed in Equation (8.24d) will be considered in this section. Also, the notation, procedures, and results in the previous Sections 8.4, 8.5, and 8.6 will be used.

Case (12) in Equation (8.24d) for [k]

If the displacement u_z and the slopes $u_{z,x}$ and $u_{z,y}$ are taken as unknowns at each corner point, then u_z can be taken as a polynomial with $3N$ constants for N corner points, or

$$u_z = [F(x,y)]\{A\}, \quad 3N \text{ constants } A_i, \tag{8.126}$$

$$\begin{Bmatrix} u_z \\ -\theta_y \\ \theta_x \end{Bmatrix} = \begin{Bmatrix} u_z \\ u_{z,x} \\ u_{z,y} \end{Bmatrix} = \begin{Bmatrix} F \\ F_{,x} \\ F_{,y} \end{Bmatrix}\{A\} = [F_d]\{A\}, \tag{8.127}$$

which corresponds to the dual case for the stress function ϕ in the inplane case, Equation (8.63). Take the $3N$ unknown corner values as

$$\{u_c\}^T = [u_{z1} \quad u_{z2} \cdots u_{z,x1} \quad u_{z,x2} \cdots u_{z,y1} \quad u_{z,y2} \cdots], \tag{8.128}$$

whence from Equations (8.65) and (8.66)

$$\{u_c\} = [H]\{A\}, \quad [G] = [H]^{-1}, \quad \{A\} = [G]\{u_c\},$$
$$u_z = [F][G]\{u_c\}. \tag{8.129}$$

In the principle of virtual displacements in Equation (7.93) take

$$[J_{bdd}]u_z = \begin{Bmatrix} u_{z,xx} \\ u_{z,yy} \\ 2u_{z,xy} \end{Bmatrix} = [F_{dd}][G]\{u_c\}, \tag{8.130}$$

$$[F_{dd}] = \begin{bmatrix} F_{,xx} \\ F_{,yy} \\ 2F_{,xy} \end{bmatrix}, \quad u_z^V = [F][G]\{u_c^V\},$$

whence Equation (7.93) becomes with three boundary conditions

$$[k]\{u_c\} = \{P\} + \{P_E\}, \tag{8.131}$$

$$[k] = [G]^T[D][G], \quad [D] = \int_{A_m} \frac{t^3}{12} [F_{dd}]^T [J]^{-1} [F_{dd}] dA_m,$$

$$\{P\} = [G]^T \left\{ \int_{A_m} [F]^T (p_z + p_{zd}) dA_m + \int_c [F_d]_b^T \begin{Bmatrix} Q_n \\ M_n \\ M_{ns} \end{Bmatrix} ds + \right.$$

$$\left. + \sum_k [F(x_k, y_k)]^T P_{zk} \right\},$$

$$\{P_E\} = -[G]^T \int_{A_m} [F_{dd}]^T \{M_E\} dA_m.$$

Although the procedure has been used for triangular elements in various computer programs, it can give convergence problems by converging to incorrect results for some element arrangements as the number of elements are increased. It can also produce singular matrices for some triangular elements.

Note that the dual relations in Equation (8.23a) hold between the above $[D]$ and the $[D]$ in Equation (8.69) for the $[C_\phi]$ inplane stress function. The $[F_{dd}]$ in Equation (8.130) has a factor of 2 on the $F_{,xy}$ shear term while the $[F_{dd}]$ in Equation (8.68) has a factor of -1. Now, $[J]^{-1}$ in Equation (8.131) has a factor of 1/4 on the shear term as compared to $[J]$ and in the product for $[D]$ the $[F_{dd}]$ terms give $(-1)^2 = 1$ and $(2)^2 = 4$. Thus, in the multiplication of the matrices the factors give unity, and the only difference internally in the two $[D]$ matrices is $+\nu$ in place of $-\nu$. Externally, $[D]$ in Equation (8.131) has $D = Et^3/12(1-\nu^2)$ while $[D]$ in Equation (8.69) has $1/Et$, which is the relation given in Equation (8.23a).

As discussed in case (6) above in Example 8.7 for the inplane dual case, satisfactory results can be obtained by using six constants for the triangular element. The corner values of u_z and the edge midpoint normal slopes can be used as the six unknowns. The procedure is the same as given in Example 8.7, with u_z in place of ϕ. For the triangle in Figures 8.7 and 8.8, Equations (8.74)-(8.78) are

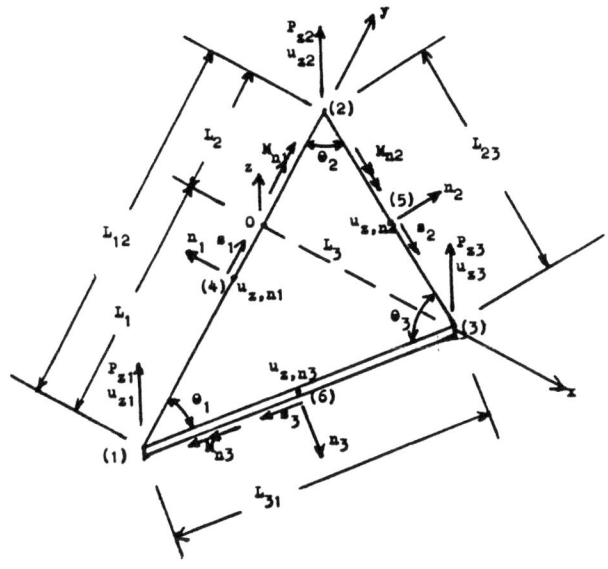

Fig. 8.8. Triangular plate element in bending.

the same for both cases. The $[F_{dd}]$ matrix in Equation (8.79) becomes

$$[F_{dd}] = [[0] \ [F_1]], \quad [F_1] = 2 \begin{bmatrix} 0 & 0 & 1 \\ 1 & 0 & 0 \\ 0 & 1 & 0 \end{bmatrix}, \tag{8.132}$$

whence $[D]$ in Equation (8.80) is

$$[D] = DL_3 L_{12} \begin{bmatrix} 0 & 0 \\ 0 & D_{22} \end{bmatrix}, \quad [D_{22}] = \begin{bmatrix} 2 & 0 & 2\nu \\ 0 & 1-\nu & 0 \\ 2\nu & 0 & 2 \end{bmatrix}. \tag{8.133}$$

This $[D]$ is dual to the $[D]$ matrix in Equation (8.80) with $-\nu$ in place of ν and $D = Et^3/12(1-\nu^2)$ in place of $1/Et$, Equation (8.23a). Since the two $[G]$ matrices, Equation (8.81), are the same, the dual relations apply for $[k]$ in place of $[C_\phi]$ in Equations (8.82) and (8.83). Thus, from Equation (8.83), the $[k]$ bending matrix for the triangular element in Figure 8.8 is

$$[k] = \begin{bmatrix} k_{11} & k_{12} \\ k_{12}^T & k_{22} \end{bmatrix}, \tag{8.134}$$

$$[k_{11}] = \frac{4D(1-\nu)\sin^2\theta_3}{L_{12}^2} \begin{bmatrix} \cot\theta_2 + \cot\theta_3 & -\cot\theta_3 & -\cot\theta_2 \\ -\cot\theta_3 & \cot\theta_1 + \cot\theta_3 & -\cot\theta_1 \\ -\cot\theta_2 & -\cot\theta_1 & \cot\theta_1 + \cot\theta_2 \end{bmatrix},$$

$$[k_{22}] = 2D \begin{bmatrix} \cot\theta_1 + \cot\theta_2 & b_2 & b_1 \\ b_2 & \cot\theta_2 + \cot\theta_3 & b_3 \\ b_1 & b_3 & \cot\theta_1 + \cot\theta_3 \end{bmatrix},$$

$$b_i = \cos\theta_i \cot\theta_i + \nu \sin\theta_i, \qquad i = 1, 2, 3,$$

$$[k_{12}]^T = \frac{2D(1-\nu)\sin\theta_3}{L_{12}} \begin{bmatrix} -\frac{\cos\theta_1}{\sin\theta_2} & -\frac{\cos\theta_2}{\sin\theta_1} & \frac{\cos\theta_1}{\sin\theta_2} + \frac{\cos\theta_2}{\sin\theta_1} \\ \frac{\cos\theta_2}{\sin\theta_3} + \frac{\cos\theta_3}{\sin\theta_2} & -\frac{\cos\theta_2}{\sin\theta_3} & -\frac{\cos\theta_3}{\sin\theta_2} \\ -\frac{\cos\theta_1}{\sin\theta_3} & \frac{\cos\theta_1}{\sin\theta_3} + \frac{\cos\theta_3}{\sin\theta_1} & -\frac{\cos\theta_3}{\sin\theta_1} \end{bmatrix}.$$

Example 8.13. (a) Calculate $[k]$ for the triangular element in Figure 8.5. (b) Discuss the equilibrium conditions given by $[k]$ for $u_{x2} = 1$ and all the other deflections and slopes zero.

Solution. (a) The interior angles and lengths of the triangle in Figure 8.5 corresponding to Figure 8.8 are $\theta_1 = 90°$, $\theta_2 = \theta_3 = 45°$, $L_{12} = 1.00$, $b = 2^{1/2}$, whence $[k]$ in Equation (8.134) becomes

$$[k] = D \begin{bmatrix} 2(1-\nu)\begin{bmatrix} 2 & -1 & -1 \\ -1 & 1 & 0 \\ -1 & 0 & 1 \end{bmatrix} & (1-\nu)\begin{bmatrix} 0 & 2b & 0 \\ -1 & -b & 1 \\ 1 & -b & -1 \end{bmatrix} \\ (1-\nu)\begin{bmatrix} 0 & -1 & 1 \\ 2b & -b & -b \\ 0 & 1 & -1 \end{bmatrix} & \begin{bmatrix} 2 & b(1+\nu) & 2\nu \\ b(1+\nu) & 4 & b(1+\nu) \\ 2\nu & b(1+\nu) & 2 \end{bmatrix} \end{bmatrix}. \qquad (8.135)$$

(b) From Equation (8.76) with u_z for ϕ and from Equation (8.131) the P_z forces and edge normal moments M_n for the $u_{x2} = 1$ case are determined by column 2 in Equation (8.135), or

$$P_{z1} = -2(1-\nu)D, \qquad P_{z2} = 2(1-\nu)D, \qquad P_{z3} = 0,$$
$$M_{n1} = -(1-\nu)D, \qquad bM_{n2} = -b(1-\nu)D, \qquad M_{n3} = (1-\nu)D. \qquad (8.136)$$

The sum of the P_{zi} forces checks zero. The moment produced by the P_{z1} and P_{z2} couple of $2(1-\nu)D(1)$ is balanced by the M_{n3} moment and the y-component of the M_{n2} moment. M_{n1} balances the x-component of M_{n2}.

Case (13) in Equation (8.24d) for Hybrid $[C_\sigma]$

Take the displacements as in Equation (8.126) but change Equation (8.130) to

$$\{J_{bdd}\}u_z = [F_{dd}]\{A\}. \qquad (8.137)$$

Assume the moments and shear forces along the edges of the element and express them in terms of the midpoint values of the normal moment M_n and shear force V_n, or for N midpoints,

$$[M_x \ M_y \ M_{xy} \ Q_x \ Q_y]_e^T = [R]\begin{Bmatrix} M_n \\ V_n \end{Bmatrix}_m = [R]\{M_m\}, \qquad (8.138)$$

$$\{M_m\}^T = [M_{n1} \ V_{n1} \ M_{n2} \ V_{n2} \ \cdots \ M_{nN} \ V_{nN}].$$

From Equation (8.127) express the displacements and slopes along the edges as

$$\begin{Bmatrix} u_z \\ u_{z,x} \\ u_{z,y} \end{Bmatrix}_e = [F_d]_e\{A\}. \qquad (8.139)$$

If these expressions are put into the principle of virtual displacements in Equation (7.93) and reference made to Equations (8.131), (8.35), and (8.36), then there results

$$[k_H]\{A\} = [B]\{M_m\} + \{P_E\} + \{P_a\}, \qquad (8.140)$$

$$[k_H] = \int_{A_m} \frac{t^3}{12}[F_{dd}]^T[J]^{-1}[F_{dd}]dA_m, \qquad [B] = \int_c [F_d]_e^T[R]ds,$$

$$\{P_E\} = -\int_{A_m} [F_{dd}]^T \{M_E\} dA_m,$$

$$\{P_a\} = \int_{A_m} [F]^T (S_z + S_{zd}) dA_m + \sum_k [F]_k^T P_{zk}.$$

Note that these equations are the same as in Equation (8.131) when $[G]$ is omitted in Equation (8.131) and the boundary integral is changed. From the hybrid procedure discussed in Equations (8.17)-(8.20) it follows that

$$\{u\} = [B]^T \{A\} = [C_\sigma]\{M_m\} + \{u_E\} + \{u_a\}, \tag{8.141}$$

$$[C_\sigma] = [B]^T [k_H]^{-1} [B], \quad \{u_E + u_a\} = [B]^T [k_H]^{-1} \{P_E + P_a\}.$$

As pointed out in the inplane case in Section 8.4, the $[C_\sigma]$ matrix must have the rigid body motion restrained so that its size is $N-3$ by $N-3$ in comparison to the corresponding N by N $[k]$ matrix. This is evident in Equation (8.133) where $[D]$ has three rows of zeros. Thus, the linear terms in u_z must be omitted when polynomials are used for the assumed approximation.

Example 8.14. (a) Find the $[C_\sigma]$ bending flexibility matrix for the triangle in Figure 8.5 by taking

$$u_z = \tfrac{1}{2}(A_1 x^2 + A_2 y^2 + A_3 xy)$$

and assuming constant moments on the edges of the plate element. (b) Discuss the analogy of the result with the inplane $[C_\sigma]$ in Equation (8.46).

Solution. (a) For constant moments in the element there are no transverse shear forces Q_x, Q_y, V_n so that these terms can be omitted in Equation (8.138). Also, in this case the corner values of u_z have no effect on the moments so that u_{zi} can be omitted in Equation (8.139), whence $[F_d]_e$ involves only values of the slopes. The procedure of Example 8.2 applies and Equation (8.44) also holds for constant moments, which determines the $[R]$ matrix in Equation (8.138). For the specified u_z, the $[F_d]_e$ matrix on the edges is the same as the $[F_e]$ matrix in Equation (8.45) so that the $[B]$ matrix from Equation (8.46) is

$$[B] = \tfrac{1}{4}\begin{bmatrix} 2 & 0 & 0 \\ 0 & 0 & 2 \\ -1 & 2 & -1 \end{bmatrix}.$$

Also, $[F_{dd}] = [I]$ for the given u_z, whence $[k_H] = (t^3/24)[J]^{-1}$. Thus, from Equation (8.46),

$$[C_\sigma] = \frac{3}{Et^3}\begin{bmatrix} 3+\nu & -2(1+\nu) & 1-\nu \\ -2(1+\nu) & 4(1+\nu) & -2(1+\nu) \\ 1-\nu & -2(1+\nu) & 3+\nu \end{bmatrix}. \tag{8.142}$$

(b) This $[C_\sigma]$ is the same as the inplane $[C_\sigma]$ in Equation (8.46) provided $t^3/12 \to t$. This demonstrates the analogy for constant inplane stresses and constant bending moments discussed in connection with Equation (8.23b).

8.8. Bending plate element matrices from the principle of virtual forces

The cases (14)-(17) in Equation (8.24e) will be considered in this section. Results for the corresponding inplane cases in Section 8.5 above will be referred to and used when applicable.

To allow for the transverse shear effects in the plate element add the terms $e_{xz}\sigma_{xz}^V$ and $e_{yz}\sigma_{yz}^V$ to the principle of virtual forces in Equation (7.114) and take

$$[M_x \ M_y \ M_{xy} \ Q_x \ Q_y]^T = \left\{ \begin{array}{c} M \\ Q \end{array} \right\}, \qquad (8.143)$$

$$Q_x = \int_t e_{xz} \, dz, \qquad Q_y = \int_t e_{yz} \, dz.$$

Use the plate shear stiffness as Gt and take $[J_s]$ as

$$[J_s] = \frac{12}{Et^3} \begin{bmatrix} 1 & -\nu & 0 & 0 & 0 \\ -\nu & 1 & 0 & 0 & 0 \\ 0 & 0 & 2(1+\nu) & 0 & 0 \\ 0 & 0 & 0 & Et^2/12G & 0 \\ 0 & 0 & 0 & 0 & Et^2/12G \end{bmatrix}. \qquad (8.144)$$

Use Equations (7.91), (7.93), and (7.114) to express the principle of virtual forces in the form

$$\int_{A_m} \left\{ \begin{array}{c} M^V \\ Q^V \end{array} \right\}^T [J_s] \left\{ \left\{ \begin{array}{c} M \\ Q \end{array} \right\} + \{M_E\} + \{M_Q\} \right\} dA_m +$$

$$+ \int_c [M_n^V(u_{z,n} + u_{zQ,n}) - V_n^V(u_z + u_{zQ})] ds - \sum_k R_{zk}^V u_{zk} = 0, \qquad (8.145)$$

where the subscript Q indicates the particular solutions for any applied $p_z + p_{zd}$ loads, $V_n^V = Q_n^V - M_{ns,s}^V$, and R_{zk} indicates the corner point reactions, Equation (7.95).

Case (14) in Equation (8.24e) for $[C_P]$

Since the P_{zi} corner forces may be involved, the equilibrium moment and force systems for the bending element are more complicated than the inplane element case. Besides pure moment equilibrium systems, there can be moments balanced by P_{zi} corner forces, and an equilibrium system for the P_{zi} forces. See Figure 8.9 for the rectangular element case and Example 8.13 above. Since these equilibrium systems are coupled and complicated for triangular elements, only the three constant moment systems, Figure 8.4(d), will be used for triangles.

Assume stresses in the element in equilibrium with each of the moment and shear equilibrium systems and express them in terms of the M_i and P_i sets, Figure 8.9, or by use of Equation (8.143),

$$\left\{ \begin{array}{c} M \\ Q \end{array} \right\} = [H(x,y)] \left\{ \begin{array}{c} M_i \\ P_i \end{array} \right\} = [H]\{M_s\}. \qquad (8.146)$$

Assume the virtual terms

$$\left\{ \begin{array}{c} M^V \\ Q^V \end{array} \right\} = [H]\{M_s^V\}, \qquad (8.147)$$

and take on the edges and corners

$$\left\{ \begin{array}{c} M_n^V \\ V_n^V \end{array} \right\}_e = [R_1] \left\{ \begin{array}{c} M^V \\ Q^V \end{array} \right\}_e = [R_1][H]_e\{M_s^V\}, \qquad (8.148)$$

Approximate matrix equations for plate finite elements 287

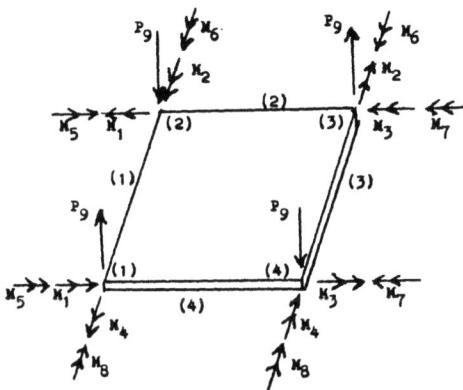

Fig. 8.9. Rectangular plate element with corner equilibrium moment and force systems.

$$\sum_k R^V_{zk} u_{zk} = \{[R_2][H]_c\{M^V_s\}\}^T \{u_c\}, \tag{8.149}$$

$$\{u_c\} = [u_{z,n1} \quad u_{z,n2} \quad \cdots \quad u_{z1} \quad u_{z2} \quad \cdots]. \tag{8.150}$$

When these expressions are put into the principle of virtual forces in Equation (8.145), the result is

$$[C_P]\{M_s\} = \{u\} - \{u_E\} - \{u_Q\}, \tag{8.151}$$

$$[C_P] = \int_{A_m} [H]^T [J_s][H] dA_m,$$

$$\{u_E + u_Q\} = \int_{A_m} [H]^T [J_s]\{M_E + M_Q\} dA_m,$$

$$\{u\} = \int_c [H]_e^T [R_1]^T \{u_c + u_{cQ}\} ds - [H]_c^T [R_2]^T \{u_c\}.$$

Example 8.15. Assume constant moments for the triangular element in Figure 8.6 and find the $[C_P]$ using the moment sets in Figure 8.4(d).
Solution. Take the reference bending moments corresponding to Equation (8.53) as

$$M_{zri} = 2M_{si}/h_i \tag{8.152}$$

and omit the shear terms, whence Equation (8.54) and $[H]$ in Equation (8.55) can be used with t omitted. Since Equation (8.56) is unchanged, it follows from Equation (8.151) and (8.57) that the $[C_P]$ for bending is

$$[C_P] = \frac{24}{Et^3} \begin{bmatrix} \cot\theta_1 + \cot\theta_2 & b_2 & b_1 \\ b_2 & \cot\theta_2 + \cot\theta_3 & b_3 \\ b_1 & b_3 & \cot\theta_1 + \cot\theta_3 \end{bmatrix}, \tag{8.153}$$

$$b_i = \cos\theta_i \cot\theta_i - \nu \sin\theta_i, \quad i = 1, 2, 3.$$

The analogy as given in Equation (8.23b) that $t \to t^3/12$ is verified here between Equations (8.57) and (8.153).

Case (15) in Equation (8.24e) for $[C_\sigma]$

Assume equilibrium stresses for the moments and shear forces in Equation (8.143) as

$$\left\{\begin{array}{c} M \\ Q \end{array}\right\} = [H(x,y)]\{A\}, \tag{8.154}$$

$$[H] = \begin{bmatrix} 1 & 0 & 0 & x & 0 & y & 0 & 0 & 0 & \cdots \\ 0 & 1 & 0 & 0 & y & 0 & x & 0 & 0 & \cdots \\ 0 & 0 & 1 & 0 & 0 & 0 & 0 & y & x & \cdots \\ 0 & 0 & 0 & 1 & 0 & 0 & 0 & 1 & 0 & \cdots \\ 0 & 0 & 0 & 0 & 1 & 0 & 0 & 0 & 1 & \cdots \end{bmatrix},$$

where Equation (7.102) with $p_z + p_{zd} = 0$ and Equation (7.112) give

$$\{J_{bdd}\}^T\{M\} = 0, \quad \left\{\begin{array}{c} Q_x \\ Q_y \end{array}\right\} = [J_{pd}]^T\{M\}, \tag{8.155}$$

as the equilibrium equations to be satisfied by each column in $[H]$.

At the midpoint of edge i use Equation (8.38) with

$$Q_n = Q_x \cos\beta_i + Q_y \sin\beta_i, \tag{8.156}$$

whence

$$\left\{\begin{array}{c} M_{ni} \\ M_{nsi} \\ Q_{ni} \end{array}\right\}_e = \begin{bmatrix} \cos^2\beta_i & \sin^2\beta_i & \sin 2\beta_i & 0 & 0 \\ -\frac{1}{2}\sin 2\beta_i & \frac{1}{2}\sin 2\beta_i & \cos 2\beta_i & 0 & 0 \\ 0 & 0 & 0 & \cos\beta_i & \sin\beta_i \end{bmatrix} \left\{\begin{array}{c} M_i \\ Q_i \end{array}\right\}_e$$

$$= [b_i]\left\{\begin{array}{c} M_1 \\ Q_i \end{array}\right\}_e = [b_i][H_i]_e\{A\} = [q_i]\{A\}. \tag{8.157}$$

Assemble all the edges to get for the element

$$\{M_m\} = \left\{\begin{array}{c} M_n \\ M_{ns} \\ Q_n \end{array}\right\} = \left\{\begin{array}{c} q_1 \\ \cdots \\ q_N \end{array}\right\}\{A\} = [q]\{A\}, \tag{8.158}$$

$$\{A\} = [q]^{-1}\{M_m\}, \quad \left\{\begin{array}{c} M \\ Q \end{array}\right\} = [H][q]^{-1}\{M_m\} = [G]\{M_m\}. \tag{8.159}$$

Put this $\left\{\begin{array}{c} M \\ Q \end{array}\right\}$ into Equation (8.145) and make use of Equations (8.146)-(8.151) to get

$$[C_\sigma]\{M_m\} = \{u - u_E - u_Q\}_\sigma, \tag{8.160}$$

$$[C_\sigma] = \int_{A_m} [G]^T [J_s][G] dA_m,$$

$$\{u_E + u_Q\} = \int_{A_m} [G]^T [J_s]\{M_E + M_Q\} dA_m,$$

$$\{u\} = \int_c [G]_e^T [R_1]^T \{u_c + u_{cQ}\} ds - [G]_c^T [R_2]^T \{u_c\}.$$

Approximate matrix equations for plate finite elements 289

With nine unknowns for the triangle, only the nine linear terms in $[H]$ in Equation (8.154) can be used. Because of the nine unknowns and because of the difficulties with $[q]$ in Equation (8.158) being singular for certain triangles, as discussed in Example 8.5(b), it is simpler to use the three unknown constant M_n moments for triangular elements. This neglects any shear effects from $[J_s]$ and assumes the transverse shear is zero. The result for $[C_\sigma]$ for the triangular element in Figure 8.5 is the same as that given in Example 8.14 for Case (13) above, Equation (8.142).

Case (16) in Equation (8.24e) for $[C_\phi]$

As discussed in Section 7.6 and Equation (7.23a), there is a dual relation between the bending stress functions ϕ_y, ϕ_x and the inplane displacements u_x, u_y. That is, if the same assumed functions given in Equation (8.25) for u_x and u_y are used for ϕ_y and ϕ_x, then the $[C_\phi]$ matrix in Equation (8.14) can be written down from the $[k]$ matrix in Equations (8.30) and (8.31). However, the right hand terms in Equation (8.14) are quite different from those in Equation (8.30), and must be derived from the principle of virtual forces in Equation (7.116).

From Equation (8.25), take

$$\begin{Bmatrix} \phi_y \\ \phi_x \end{Bmatrix} = [F]\{A\} = \begin{bmatrix} F_1(x,y) & 0 \\ 0 & F_1(x,y) \end{bmatrix} \{A\}, \qquad (8.161)$$

whence the $[H]$ and $[H_c]$ matrices in Equation (8.26) are the same, and

$$\{\phi_c\}^T = [\phi_{y1} \quad \phi_{y2} \quad \cdots \quad \phi_{yN} \quad \phi_{x1} \quad \phi_{x2} \quad \cdots \quad \phi_{xN}]. \qquad (8.162)$$

From Equations (8.27), (8.28), and (7.111) with a change in the order of M_x and M_y,

$$\{M\} = \begin{Bmatrix} M_y \\ M_x \\ M_{yx} \end{Bmatrix} = \begin{Bmatrix} M_{yQ} \\ M_{xQ} \\ M_{yxQ} \end{Bmatrix} + \begin{Bmatrix} \phi_{y,x} \\ \phi_{x,y} \\ -\tfrac{1}{2}(\phi_{x,x} + \phi_{y,y}) \end{Bmatrix}, \qquad (8.163)$$

$$\begin{Bmatrix} \phi_y \\ \phi_x \end{Bmatrix} = [F][H]^{-1}\{\phi_c\}, \qquad (8.164)$$

$$\{M\} = \{M_Q\} + [F_d][H]^{-1}\{\phi_c\} = \{M_Q\} + [D]\{\phi_c\}, \qquad (8.165)$$

$$[F_d] = \begin{bmatrix} F_{1,x} & 0 \\ 0 & F_{1,y} \\ -\tfrac{1}{2}F_{1,y} & -\tfrac{1}{2}F_{1,x} \end{bmatrix}, \qquad [D] = [F_d][H]^{-1}.$$

Put these expressions into Equation (8.145) to get

$$\{\phi_c^V\}^T \Bigg[\int_{A_m} \frac{12}{t^3}[D]^T[J]\{[D]\{\phi_c\} + \{M_E\} + \{M_Q\}\}dA_m -$$
$$- \int_c [[R][D]_c]^T \begin{Bmatrix} u_{zg,n} + u_{zQ,n} \\ u_{zg,s} + u_{zQ,s} \\ u_{zg} + u_{zQ} \end{Bmatrix} ds \Bigg] = 0, \qquad (8.166)$$

$$[R] = \begin{bmatrix} \sin^2 \beta & \cos^2 \beta & \sin 2\beta \\ \frac{1}{2}\sin 2\beta & -\frac{1}{2}\sin 2\beta & \cos 2\beta \\ \sin \beta \frac{\partial}{\partial y} & \cos \beta \frac{\partial}{\partial x} & \sin \beta \frac{\partial}{\partial x} + \cos \beta \frac{\partial}{\partial y} \end{bmatrix}_e.$$

where three boundary conditions are used. For arbitrary $\{\phi_c^V\}$

$$[C_\phi]\{\phi_c\} = \{u - u_E - u_Q\}_\phi, \tag{8.167}$$

$$[C_\phi] = \int_{A_m} \frac{12}{t^3}[D]^T[J][D] dA_m,$$

$$\{u_E + u_Q\}_\phi = \int_{A_m} \frac{12}{t^3}[D]^T[J]\{M_E + M_Q\} dA_m,$$

$$\{u\}_\phi = \int_c [[R][D]_e]^T \{u_c + u_{cQ}\} ds.$$

Note that the dual relations in Equation (8.23a) hold between the above $[C_\phi]$ and the $[k]$ in Equation (8.30). The $[D]$ matrix in $[C_\phi]$ has a factor of -1/2 on the shear term, Equation (8.163), while the factor is 1 in Equation (8.28). The $[J]$ matrix has a factor of 4 on the shear term as compared to $[J]^{-1}$. In the product of the matrices, the $(-1/2)^2 = 1/4$ cancels the 4 factor so that internally the integrals in $[C_\phi]$ and in $[k]$ are the same except for $\nu \to -\nu$. Externally, $[k]$ has $Et/(1-\nu^2)$ and $[C_\phi]$ has $D^{-1}/(1-\nu^2)$, which corresponds to the conditions in Equation (8.23a).

Example 8.16. (a) Write the $[C_\phi]$ matrix for constant moments in the triangular element in Figure 8.5 by using the dual of $[k]$ in Example 8.1. (b) Comment on the case of linear moments in the triangular elements.

Solution. (a) Apply the dual relations in Equation (8.23a) to Equation (8.31) to get

$$[C_\phi] = \frac{3}{Et^3} \begin{bmatrix} 3+\nu & -1-\nu & -2 & 1-\nu & 2\nu & -1-\nu \\ -1-\nu & 1+\nu & 0 & -1-\nu & 0 & 1+\nu \\ -2 & 0 & 2 & 2\nu & -2\nu & 0 \\ 1-\nu & -1-\nu & 2\nu & 3+\nu & -2 & -1-\nu \\ 2\nu & 0 & -2\nu & -2 & 2 & 0 \\ -1-\nu & 1+\nu & 0 & -1-\nu & 0 & 1+\nu \end{bmatrix}. \tag{8.168}$$

(b) Since the constant moments are in equilibrium and compatible, this is the *only solution* for the six corner point conditions on ϕ_y and ϕ_x. As discussed in Example 8.1(b), additional A_i constants in Equation (8.161) will be zero, unless more unknowns, such as points on the edges, are used. Thus, more unknowns are needed to get linear moments in the triangular plate element by this case (16).

Case (17) in Equation (8.24e) for hybrid [k]

This case (17) for bending corresponds to case (7) for inplane loading and allows the bending $[k]$ stiffness matrix to be calculated from the principle of virtual forces. Assume the moments and transverse shear forces in the form

$$[M \quad Q]^T = [M_x \quad M_y \quad M_{xy} \quad Q_x \quad Q_y]^T = [H]\{A\}, \tag{8.169}$$

where $[H]$ is defined in Equation (8.154) and where Equation (8.155) applies for equilibrium of the assumed moments and shear forces.

On the edges of the element put the shear forces in terms of V_n by using Equation (8.145), or

$$V_x = Q_x + M_{xy,y}, \quad V_y = Q_y + M_{xy,x}, \quad V_n = V_x \cos \beta + V_y \sin \beta, \tag{8.170}$$

where β is the angle from the x-axis to the outward normal on the edge of the plate. Thus, from Equations (8.169), (8.170), (8.145), (8.154), and (8.38) on the edges,

$$[M \quad -V]_e^T = [H_e]\{A\}, \tag{8.171}$$

$$[H_e] = \begin{bmatrix} 1 & 0 & 0 & x & 0 & y & 0 & 0 & 0 & \cdots \\ 0 & 1 & 0 & 0 & y & 0 & x & 0 & 0 & \cdots \\ 0 & 0 & 1 & 0 & 0 & 0 & 0 & y & x & \cdots \\ 0 & 0 & 0 & 1 & 0 & 0 & 0 & 2 & 0 & \cdots \\ 0 & 0 & 0 & 0 & 1 & 0 & 0 & 0 & 2 & \cdots \end{bmatrix}_e,$$

$$\begin{Bmatrix} M_{ni} \\ -V_{ni} \end{Bmatrix} = \begin{bmatrix} \cos^2\beta_i & \sin^2\beta_i & \sin 2\beta_i & 0 & 0 \\ 0 & 0 & 0 & \cos\beta_i & \sin\beta_i \end{bmatrix} \begin{Bmatrix} M_i \\ V_i \end{Bmatrix}_e$$

$$= [b_i]\begin{Bmatrix} M_i \\ -V_i \end{Bmatrix}_e = [b_i][H_{ei}]\{A\} = [R_i]\{A\}, \tag{8.172}$$

$$\{M_e\} = [M_{n1} \quad -V_{n1} \quad \cdots \quad M_{nN} \quad -V_{nN}]^T = [R_1 \cdots R_N]^T\{A\}$$
$$= [R]\{A\}. \tag{8.173a}$$

For convenience in calculations rearrange $\{M_e\}$ and $[R]$ as

$$\{M_e\}_a = [M_{n1} \quad \cdots \quad M_{nN} \quad -V_{n1} \quad \cdots \quad -V_{nN}]^T = [R_a]\{A\}. \tag{8.173b}$$

From Equations (8.38) and (7.90c), the corner forces R_{zk} in the last term of Equation (8.145) can be expressed as

$$[R_{zc}] = [R_{1c}][f_c][H_c]\{A\} = [R_{Mc}]\{A\}, \tag{8.174}$$

$$[R_{1c}] = \begin{bmatrix} 1 & 0 & 0 & \cdots & -1 \\ -1 & 1 & 0 & \cdots & 0 \\ 0 & -1 & 1 & 0 & \cdots & 0 \\ \cdots & \cdots & \cdots & \cdots & \cdots \\ 0 & 0 & 0 & \cdots & -1 & 1 \end{bmatrix}, \quad N \text{ by } N \text{ for } N \text{ corners,}$$

$$[f_c] = \tfrac{1}{2}\begin{bmatrix} -\sin 2\beta_1 & \sin 2\beta_1 & 2\cos 2\beta_1 & 0 & 0 \\ -\sin 2\beta_2 & \sin 2\beta_2 & 2\cos 2\beta_2 & 0 & 0 \\ \cdots & \cdots & \cdots & \cdots & \cdots \\ -\sin 2\beta_N & \sin 2\beta_N & 2\cos 2\beta_N & 0 & 0 \end{bmatrix}.$$

Express the u_{ze} displacement and θ_{xe}, θ_{ye} slopes on the edges for the element in terms of the unknown corner point values so that they are continuous with the adjoining elements along the edges, or

$$(\theta_x)_i = (1-s)\theta_{xi} + s\theta_{xj}, \quad (\theta_y)_i = (1-s)\theta_{yi} + s\theta_{yj},$$
$$(\theta_s)_i = (1-s)\theta_{si} + s\theta_{sj}, \quad \theta_{si} = \theta_{xi}\sin\beta_i - \theta_{yi}\cos\beta_i,$$
$$\theta_{sj} = \theta_{xj}\sin\beta_i - \theta_{yj}\cos\beta_i, \tag{8.175a}$$
$$\theta_{ni} = \theta_{xi}\cos\beta_i + \theta_{yi}\sin\beta_i, \quad \theta_{nj} = \theta_{xj}\cos\beta_i + \theta_{yj}\sin\beta_i,$$
$$(u_z)_i = (1-s)u_{zi} + su_{zj} + (L_{ij}/2)s(1-s)(\theta_{ni} - \theta_{nj}), \tag{8.175b}$$

where $s = s_i/L_{ij}$ with s_i the variable along edge i, L_{ij} the length of edge i, i and j the corner points at the ends of edge i, and β_i the angle from the local

x-axis to the outward normal on edge i. The last term in $(u_z)_i$ allows for any bending deflection along edge i. Assemble the expressions in Equation (8.175) to match Equation (8.173b), or

$$\left\{\begin{matrix} \theta_s \\ u_z \end{matrix}\right\}_e = [\,(\theta_s)_1 \quad \cdots \quad (\theta_s)_N \quad (u_z)_1 \quad \cdots \quad (u_z)_N\,]^T$$

$$= [G]\left\{\begin{matrix} \theta_y \\ \theta_x \\ u_z \end{matrix}\right\}_c = [G]\{u_c\}, \qquad (8.176)$$

$$\{u_c\}^T = [\,\theta_{y1} \quad \cdots \quad \theta_{yN} \quad \theta_{x1} \quad \cdots \quad \theta_{xN} \quad u_{z1} \quad \cdots \quad u_{zN}\,].$$

The corner values of u_{zci} can be written in matrix form in terms of $\{u_c\}$ as

$$\{u_{zc}\} = [G_c]\{u_c\}, \qquad [G_c] = [0 \quad 0 \quad I]. \qquad (8.177)$$

Put the above expressions into the principle of virtual forces in Equation (8.145) to get

$$\{A^V\}^T \left[\int_{A_m} [H]^T[J_s]\{[H]\{A\} + \{M_E\} + \{M_Q\}\}\mathrm{d}A_m - \right.$$
$$\left. - \int_c [R_a]^T[G]\{u_c + u_{cQ}\}\mathrm{d}s - [R_{Mc}]^T[G_c]\{u_c\}\right] = 0. \qquad (8.178)$$

For arbitrary $\{A^V\}$,

$$[C_H]\{A\} = [B]\{u_c\} - \{u_E + u_Q + u_b\}, \qquad (8.179)$$

$$[C_H] = \int_{A_m} [H]^T[J_s][H]\mathrm{d}A_m, \qquad [B] = \int_c [R_a]^T[G]\mathrm{d}s + [R_{Mc}]^T[G_c],$$

$$\{u_E + u_Q\} = \int_{A_m} [H]^T[J_s]\{M_E + M_Q\}\mathrm{d}A_m, \qquad \{u_b\} = \int_c [R_a]^T[G]\{u_{cQ}\}\mathrm{d}s.$$

From Equations (8.19) and (8.179),

$$[k]\{u_c\} = \{P\} + \{P_E + P_Q + P_b\}, \qquad [k] = [B]^T[C_H]^{-1}[B], \qquad (8.180)$$
$$\{P_E + P_Q + P_b\} = [B]^T[C_H]^{-1}\{u_E + u_Q + u_b\}.$$

Note that all the force terms on the right side in Equation (8.180) can be taken from the principle of virtual displacements in Equation (8.131).

This procedure for calculating the $[k]$ matrix has received considerable investigation in the literature for both triangular and rectangular elements. It does not have difficulties with singular matrices and with convergence for some types of triangular elements that occur in case (12) derivation of $[k]$. Note also that the $[k]$ for six unknowns given in case (12), Equation (8.134), can be derived by the above procedure by using constant stresses from the upper left 3 by 3 I matrix in $[H]$ and using a linear u_z with constant θ_s on the edges.

Example 8.17. Set up the $[C_H]$, $[R_a]$, $[R_{Mc}]$, and $[G]$ matrices for the general triangular element in Figures 8.4(a) and 8.8. Use the nine linear terms in Equation (8.154) for $[H]$.

Approximate matrix equations for plate finite elements 293

Solution. To simplify the calculations, rearrange the $[H]$ matrix from Equation (8.154) in the form

$$[H] = \begin{bmatrix} D_1 & D_2 & D_3 \\ D_4 & D_5 & D_6 \end{bmatrix}, \quad [D_1] = \begin{bmatrix} 1 & x & y \\ 0 & 0 & 0 \\ 0 & 0 & 0 \end{bmatrix}, \quad [D_2] = \begin{bmatrix} 0 & 0 & 0 \\ 1 & x & y \\ 0 & 0 & 0 \end{bmatrix}, \quad (8.181)$$

$$[D_3] = \begin{bmatrix} 0 & 0 & 0 \\ 0 & 0 & 0 \\ 1 & x & y \end{bmatrix}, \quad [D_4] = \begin{bmatrix} 0 & 1 & 0 \\ 0 & 0 & 0 \\ 0 & 0 & 0 \end{bmatrix}, \quad [D_5] = \begin{bmatrix} 0 & 0 & 0 \\ 0 & 0 & 1 \\ 0 & 0 & 1 \end{bmatrix}, \quad [D_6] = \begin{bmatrix} 0 & 0 & 1 \\ 0 & 1 & 0 \end{bmatrix}.$$

From Equation (8.144), take

$$[J_e] = \frac{12}{Et^3}\begin{bmatrix} J_1 & 0 \\ 0 & aI \end{bmatrix}, \quad [J_1] = \begin{bmatrix} 1 & -\nu & 0 \\ -\nu & 1 & 0 \\ 0 & 0 & 2+2\nu \end{bmatrix}, \quad a = \frac{Et^2}{12G}, \quad (8.182)$$

whence $[C_H]$ in Equation (8.179) becomes

$$[C_H] = \frac{12}{Et^3}\int_{A_m}\left[\begin{bmatrix} D_1^T J_1 D_1 & D_1^T J_1 D_2 & D_1^T J_1 D_3 \\ D_2^T J_1 D_1 & D_2^T J_1 D_2 & D_2^T J_1 D_3 \\ D_3^T J_1 D_1 & D_3^T J_1 D_2 & D_3^T J_1 D_3 \end{bmatrix}\right.$$

$$\left. + a\begin{bmatrix} D_4^T D_4 & D_4^T D_5 & D_4^T D_6 \\ D_5^T D_4 & D_5^T D_5 & D_5^T D_6 \\ D_6^T D_4 & D_6^T D_5 & D_6^T D_6 \end{bmatrix}\right]dA_m$$

$$= \frac{12}{Et^3}\int_{A_m}\left[\begin{bmatrix} D_7 & -\nu D_7 & 0 \\ -\nu D_7 & D_7 & 0 \\ 0 & 0 & 2(1+\nu)D_7 \end{bmatrix} + a[F]\right]dA_m$$

$$= \frac{12}{Et^3}\left[\begin{bmatrix} D_8 & -\nu D_8 & 0 \\ -\nu D_8 & D_8 & 0 \\ 0 & 0 & 2(1+\nu)D_8 \end{bmatrix} + (a/2)L_3 L_{12}[F]\right], \quad (8.183)$$

$$[D_7] = \begin{bmatrix} 1 & x & y \\ x & x^2 & xy \\ y & xy & y^2 \end{bmatrix}, \quad [D_8] = \frac{L_3}{24}\begin{bmatrix} 12L_{12} & 4L_3 L_{12} & 4(L_1^2 + L_2^2) \\ 4L_3 L_{12} & 2L_3^2 L_{12} & L_3(L_1^2 + L_2^2) \\ 4(L_1^2 + L_2^2) & L_3(L_1^2 + L_2^2) & 2(L_1^3 + L_2^3) \end{bmatrix},$$

$F_{22} = F_{66} = F_{88} = F_{99} = F_{29} = F_{92} = F_{68} = F_{86} = 1, \quad$ other $F_{ij} = 0$.

If the plate element is thin so that the term a in Equation (8.182) is small, then the $a[F]$ term in Equation (8.183) can be neglected and the inverse of $[C_H]$ calculated directly as

$$[C_H]^{-1} = D\begin{bmatrix} D_8^{-1} & \nu D_8^{-1} & 0 \\ \nu D_8^{-1} & D_8^{-1} & 0 \\ 0 & 0 & \frac{1-\nu}{2}D_8^{-1} \end{bmatrix}. \quad (8.184)$$

For the local coordinate system in Figure 8.8, the β_i angles for the normals are

$$\beta_1 = 180°, \quad \beta_2 = \theta_2, \quad \beta_3 = -\theta_1, \quad (8.185)$$

where θ_1 and θ_2 are interior angles, Figure 8.8. Thus, from Equations (8.172) and (8.173),

$$[R_a] = \begin{bmatrix} R_{11} & R_{12} & R_{13} \\ R_{21} & R_{22} & R_{23} \end{bmatrix}, \quad [R_{11}] = \begin{bmatrix} 1 & 0 & L_{12}s - L_1 \\ \cos^2\theta_2 & L_3 s \cos^2\theta_2 & L_2(1-s)\cos^2\theta_2 \\ \cos^2\theta_1 & -L_3 s \cos^2\theta_1 & -L_1 s \cos^2\theta_1 \end{bmatrix},$$

$$[R_{21}] = \begin{bmatrix} 0 & -1 & 0 \\ 0 & \cos\theta_2 & 0 \\ 0 & \cos\theta_1 & 0 \end{bmatrix}, \quad [R_{12}] = \begin{bmatrix} 0 & 0 & 0 \\ \sin^2\theta_2 & L_3 s \sin^2\theta_2 & L_2(1-s)\sin^2\theta_2 \\ \sin^2\theta_1 & -L_3 s \sin^2\theta_1 & -L_1 s \sin^2\theta_1 \end{bmatrix},$$

$$[R_{22}] = \begin{bmatrix} 0 & 0 & 0 \\ 0 & 0 & \sin\theta_2 \\ 0 & 0 & \sin\theta_1 \end{bmatrix}, \quad [R_{13}] = \begin{bmatrix} 0 & 0 & 0 \\ \sin 2\theta_2 & L_3 s \sin 2\theta_2 & L_2(1-s)\sin 2\theta_2 \\ \sin 2\theta_1 & -L_3 s \sin 2\theta_1 & -L_1 s \sin 2\theta_1 \end{bmatrix},$$

$$[R_{23}] = \begin{bmatrix} 0 & 0 & -1 \\ 0 & \sin\theta_2 & \cos\theta_2 \\ 0 & -\sin\theta_1 & \cos\theta_1 \end{bmatrix}, \quad (8.186)$$

where $[H_{ei}]$ on the edges is taken from Equation (8.181).

To get $[R_{Mc}]$ in Equation (8.174) with the β_i in Equation (8.185), multiply each row of the matrix $[R_{1c}][f_c]$ into the corresponding corner point value of $[H]$ in Equation (8.181). The result is

$$[R_{Mc}] = \tfrac{1}{2}[-R_{M1} \quad R_{M1} \quad R_{M2}], \tag{8.187}$$

$$[R_{M1}] = \begin{bmatrix} \sin 2\theta_1 & 0 & -L_1 \sin 2\theta_1 \\ \sin 2\theta_2 & 0 & L_2 \sin 2\theta_2 \\ -(\sin 2\theta_1 + \sin 2\theta_2) & -L_3(\sin 2\theta_1 + \sin 2\theta_2) & 0 \end{bmatrix},$$

$$[R_{M2}] = 4 \begin{bmatrix} \sin^2 \theta_1 & 0 & -L_1 \sin^2 \theta_1 \\ -\sin^2 \theta_2 & 0 & -L_2 \sin^2 \theta_2 \\ \sin^2 \theta_2 - \sin^2 \theta_1 & L_3(\sin^2 \theta_2 - \sin^2 \theta_1) & 0 \end{bmatrix},$$

From Equations (8.175) and (8.176) for the β_i in Equation (8.185), the $[G]$ matrix is

$$[G] = \begin{bmatrix} G_{11} & G_{12} & 0 \\ G_{21} & G_{22} & G_{23} \end{bmatrix}, \quad [G_{21}] = \frac{s(1-s)}{2} \begin{bmatrix} 0 & 0 & 0 \\ 0 & L_3 & -L_3 \\ L_3 & 0 & -L_3 \end{bmatrix}, \tag{8.188}$$

$$[G_{11}] = \begin{bmatrix} 1-s & s & 0 \\ 0 & -(1-s)\cos\theta_2 & -s\cos\theta_2 \\ -s\cos\theta_1 & 0 & -(1-s)\cos\theta_1 \end{bmatrix},$$

$$[G_{12}] = \begin{bmatrix} 0 & 0 & 0 \\ 0 & (1-s)\sin\theta_2 & s\sin\theta_2 \\ -s\sin\theta_1 & 0 & -(1-s)\sin\theta_1 \end{bmatrix},$$

$$[G_{22}] = \frac{s(1-s)}{2} \begin{bmatrix} -L_{12} & L_{12} & 0 \\ 0 & L_2 & -L_2 \\ -L_1 & 0 & L_1 \end{bmatrix}, \quad [G_{23}] = \begin{bmatrix} 1-s & s & 0 \\ 0 & 1-s & s \\ s & 0 & 1-s \end{bmatrix},$$

where $\sin\theta_2 = L_3/L_{23}$, $\sin\theta_1 = L_3/L_{31}$, $\cos\theta_2 = L_2/L_{23}$, $\cos\theta_1 = L_1/L_{31}$ have been used.

In terms of the above 3 by 3 matrices the $[B]$ matrix in Equation (8.179) has the form

$$[B] = \int_0^1 \begin{bmatrix} R_{11}^T & R_{21}^T \\ R_{12}^T & R_{22}^T \\ R_{13}^T & R_{23}^T \end{bmatrix} \begin{bmatrix} G_{11} & G_{12} & 0 \\ G_{21} & G_{22} & G_{23} \end{bmatrix} ds + \tfrac{1}{2} \begin{bmatrix} -R_{M1}^T \\ R_{M1}^T \\ R_{M2}^T \end{bmatrix} [0 \ 0 \ I]. \tag{8.189}$$

Thus, for any particular triangular element the $[k]$ matrix in Equation (8.180) can be calculated from this $[B]$ and the $[C_H]^{-1}$ in Equation (8.184).

8.9. Bending plate element matrices from the mixed virtual principle

The cases (18) and (19) in Equation (8.24f) will be considered in this section. Results from the corresponding inplane cases in Section 8.6 will be used when applicable.

Case (18) in Equation (8.24f) for [M]

From Equations (8.101)-(8.103) for the inplane case take

$$u_z = [F_1(x,y)]\{A\}, \quad [M] = \begin{Bmatrix} M_x \\ M_y \\ M_{xy} \end{Bmatrix} = [G(x,y)]\{B\}, \tag{8.190}$$

$$[G] = \begin{bmatrix} F_1 & 0 & 0 \\ 0 & F_1 & 0 \\ 0 & 0 & F_1 \end{bmatrix}, \quad [F_1] = [1 \quad x \quad y \cdots].$$

Follow Equations (8.104)-(8.106) to get

$$u_z = [F_1][H_c]^{-1}\{u_c\} = [F_u]\{u_c\}, \tag{8.191}$$

$$\{M\} = \begin{bmatrix} F_1 H_c^{-1} & 0 & 0 \\ 0 & F_1 H_c^{-1} & 0 \\ 0 & 0 & F_1 H_c^{-1} \end{bmatrix}\{M_c\} = [G_M]\{M_c\}, \tag{8.192}$$

$$\{u_c\}^T = [u_{z1} \ u_{z2} \cdots u_{zN}],$$
$$\{M_c\}^T = [M_{x1} \cdots M_{xN} \ M_{y1} \cdots M_{yN} \ M_{xy1} \cdots M_{xyN}],$$
$$[H_c]^T = [F_1^T(x_1,y_1) \ F_1^T(x_2,y_2) \cdots F_1^T(x_N,y_N)].$$

From Equations (7.90) and (7.102), the slopes and shear forces are

$$\begin{Bmatrix} u_{z,x} \\ u_{z,y} \end{Bmatrix} = [J_{cd}]u_z = [J_{cd}][F_u]\{u_c\} = [F_d]\{u_c\},$$

$$\begin{Bmatrix} Q_x \\ Q_y \end{Bmatrix} = [J_{pd}]^T\{M\} = [J_{pd}]^T[G_M]\{M_c\} = [G_d]\{M_c\}. \tag{8.193}$$

Take the virtual terms as

$$u_z^V = [F_u]\{u_c^V\}, \quad \begin{Bmatrix} u_{z,x}^V \\ u_{z,y}^V \end{Bmatrix} = [F_d]\{u_c^V\}, \quad \{M^V\} = [G_M]\{M_c^V\},$$

$$\begin{Bmatrix} Q_x^V \\ Q_y^V \end{Bmatrix} = [G_d]\{M_c^V\}, \tag{8.194}$$

whence the mixed virtual principle in Equations (7.129) and (7.130) is

$$\{u_c^V\}^T \left[\int_{A_m} \{[F_d]^T[G_d]\{M_c\} - [F_u]^T\{p_z + p_{zd}\}\}dA_m - \right.$$
$$- \int_c [F_{db}]^T[G_{Mb}]\{M_c\}ds - \int_{c\sigma}\left\{[F_{db}]^T\begin{Bmatrix}M_{ng}\\M_{nsg}\end{Bmatrix}_b - [F_u]_e^T\{Q_{ng}\}\right\}ds -$$
$$- \int_c [F_u]_e^T[G_{db}]\{M_c\}ds - \sum_k [F_u]_k^T P_{zk} \Bigg] +$$
$$+\{M_c^V\}^T\Bigg[\int_{A_m}\Big\{[G_d]^T[F_d]\{u_c\} -$$
$$-[G_M^T \ G_d^T][J_s]\Big\{\begin{Bmatrix}G_M\\G_d\end{Bmatrix}\{M_c\} + \{M_E\}\Big\}\Big\}dA_m -$$
$$- \int_c [G_{Mb}]^T[F_{db}]\{u_c\}ds - \int_c [G_{db}]^T[F_u]\{u_c\}ds + \int_{cu}[G_{db}]^T\{u_{zs}\}ds\Bigg]$$
$$= 0, \tag{8.195}$$

where the shear stiffness of the plate element is included through $[J_s]$ in Equation (8.144), where the subscript b indicates that the matrices are to be calculated on the edges using Equations (8.156) and (8.157), and where three boundary conditions are used. If u_z and the moments are continuous between elements on the element edges, then the edge integrals are either zero on interior elements or will cancel on the assembly of the elements. On exterior edges, they give weighted values for the applied forces and displacements as well as weighted corrections when the assumed element functions do not satisfy the boundary conditions. If a discontinuity exists on an interior edge, then it may be necessary to calculate a weighted value from the edge integral.

For arbitrary $\{u_c^V\}$ and $\{M_c^V\}$, Equation (8.195) gives

$$[M]\begin{Bmatrix} M_c \\ u_c \end{Bmatrix} = \begin{bmatrix} -M_{11} & M_{12} \\ M_{12}^T & 0 \end{bmatrix}\begin{Bmatrix} M_c \\ u_c \end{Bmatrix} = \begin{Bmatrix} u_E - u_s \\ P \end{Bmatrix}, \quad (8.196)$$

$$[M_{11}] = \int_{A_m} [G_M^T \ G_d^T][J_s]\begin{Bmatrix} G_M \\ G_d \end{Bmatrix} dA_m,$$

$$[M_{12}] = \int_{A_m} [G_d]^T[F_d]dA_m - \int_c [G_{Mb}]^T[F_{db}]ds - \int_c [G_{db}]^T[F_u]ds,$$

$$\{u_E\} = \int_{A_m} [G_M^T \ G_d^T][J_s]\{M_E\}dA_m, \quad \{u_s\} = \int_{cu} [G_{db}]^T\{u_{zg}\}ds,$$

$$\{P\} = \int_{A_m} [F_u]^T\{p_z + p_{zd}\}dA_m + \int_{c\sigma}\left\{[F_u]^T\{Q_{ng}\} - [F_{db}]^T\begin{Bmatrix} M_{ng} \\ M_{nsg} \end{Bmatrix}\right\}ds +$$
$$+ \sum_k [F_u]_k^T P_{zk}.$$

For the thin plate with $[J_s]$ in $[M_{11}]$ replaced by $(12/t^3)[J]$ and $[G_d]$ omitted, it is possible to use the diagonal form of $[G_M]$ in Equation (8.192) to write $[M_{11}]$ in the form

$$[M_{11}] = \frac{12}{Et^3}\begin{bmatrix} B_1 & -\nu B_1 & 0 \\ -\nu B_1 & B_1 & 0 \\ 0 & 0 & 2(1+\nu)B_1 \end{bmatrix}, \quad (8.197)$$

$$[B_1] = [H_c^{-1}]^T\left[\int_{A_m}[F_1]^T[F_1]dA_m\right][H_c^{-1}].$$

Also, the first integral in $[M_{12}]$ can be written as

$$[M_{12}]^T = [B_2^T \ B_3^T \ B_4^T], \quad (8.198)$$

$$[B_2] = [H_c^{-1}]^T\left[\int_{A_m}[F_{1,x}]^T[F_{1,x}]dA_m\right][H_c^{-1}],$$

$$[B_3] = [H_c^{-1}]^T\left[\int_{A_m}[F_{1,y}]^T[F_{1,y}]dA_m\right][H_c^{-1}],$$

$$[B_4] = [H_c^{-1}]^T\left[\int_{A_m}[F_{1,y}^T F_{1,x} + F_{1,x}^T F_{1,y}]dA_m\right][H_c^{-1}].$$

Example 8.18. Construct the $[M]$ matrix for the interior triangular element in Figure 8.5 using linear terms for $[F_1] = [1 \ x \ y]$ in Equation (8.190).

Solution. The $[H_c^{-1}]$ and $[B_1]$ matrices are the same as in Example 8.10, or

$$[H_c]^{-1} = \begin{bmatrix} 1 & 0 & 0 \\ -1 & 0 & 1 \\ -1 & 1 & 0 \end{bmatrix}, \quad [B_1] = \frac{1}{24}\begin{bmatrix} 2 & 1 & 1 \\ 1 & 2 & 1 \\ 1 & 1 & 2 \end{bmatrix}.$$

Approximate matrix equations for plate finite elements 297

From $[F_{1,x}] = [0\ 1\ 0]$ and $[F_{1,y}] = [0\ 0\ 1]$, the $[M_{12}]$ component matrices can be calculated from Equation (8.198) as

$$[B_2] = \tfrac{1}{2}\begin{bmatrix} 1 & 0 & -1 \\ 0 & 0 & 0 \\ -1 & 0 & 1 \end{bmatrix}, \quad [B_3] = \tfrac{1}{2}\begin{bmatrix} 1 & -1 & 0 \\ -1 & 1 & 0 \\ 0 & 0 & 0 \end{bmatrix}, \quad [B_4] = \tfrac{1}{2}\begin{bmatrix} 2 & -1 & -1 \\ -1 & 1 & 0 \\ -1 & 0 & 1 \end{bmatrix}.$$

Thus,

$$[M] = \begin{bmatrix} -\tfrac{12}{Et^3} \begin{bmatrix} B_1 & -\nu B_1 & 0 \\ -\nu B_1 & B_1 & 0 \\ 0 & 0 & 2(1+\nu)B_1 \end{bmatrix} \begin{bmatrix} B_2 \\ B_3 \\ B_4 \end{bmatrix} \\ [B_2^T\ B_3^T\ B_4^T] \qquad\qquad [0] \end{bmatrix}. \tag{8.199}$$

Just as in the inplane case (9) above, the nine moment unknowns for the triangular element can be reduced to three unknowns by using constant moments in the element. Use the same displacements u_z as in Equation (8.191) but express the constant moments $\{M\}$ in terms of the constant edge normal moments as

$$\{M\} = [Z]\{M_n\}, \tag{8.200}$$

where $[Z]$ is defined in Equation (8.40). All the Equations (8.190)-(8.196) apply with $[Z]$ in place of $[G_M]$ and $\{M_n\}$ in place of $\{M_c\}$. However, for constant moments in the elements, the $[G_d]$ transverse shear matrix is zero. Also, the $\{M_{ns}\}$ moments are discontinuous across the edges of the elements so that on the interior elements it is necessary to modify the terms in the edge integrals in Equation (8.195).

In Equation (8.195) on interior elements for constant moments, the terms in the edge integrals are $Q_n = 0$, M_n continuous across the edges so that it can be omitted, but the constant M_{ns} is discontinuous across the edges. From Equations (8.174), (8.145), and (8.41) the simplest form for the value of the edge integral due to M_{ns} is

$$[R_{1c}]\{M_{ns}\} = [R_{1c}][W][Z]\{M_n\}, \tag{8.201}$$

whence in Equations (8.195) and (8.196)

$$[M_{12}]^T = R_{1c}[W][Z], \tag{8.202}$$

where $[R_{1c}]$, $[W]$, and $[Z]$ are given in Equations (8.174), (8.42), and (8.40).

Example 8.19. Find the $[M_{11}]$, $[M_{11}]^{-1}$, and $[M_{12}]$ matrices with constant moments for the triangular element in Figure 8.8.

Solution. From Equations (8.196) and (8.200)

$$[M_{11}] = \int_{A_m} \frac{12}{t^3}[Z]^T[J][Z]\mathrm{d}A_m, \tag{8.203}$$

where $[Z_1]$ and $[Z]$ in Equation (8.40) have the following form for Figure 8.8 with the β_i angles given in Equation (8.185),

$$[Z_1] = \begin{bmatrix} 1 & 0 & 0 \\ \cos^2\theta_2 & \sin^2\theta_2 & \sin 2\theta_2 \\ \cos^2\theta_1 & \sin^2\theta_1 & -\sin 2\theta_1 \end{bmatrix}, \tag{8.204}$$

$$[Z] = [Z_1]^{-1} = \begin{bmatrix} 1 & 0 & 0 \\ -\cot\theta_1\cot\theta_2 & a_1\cot\theta_1 & a_2\cot\theta_2 \\ \tfrac{1}{2}(\cot\theta_1 - \cot\theta_2) & \tfrac{1}{2}a_1 & -\tfrac{1}{2}a_2 \end{bmatrix}, \tag{8.205}$$

$$a_1 = \cot\theta_2 + \cot\theta_3, \qquad a_2 = \cot\theta_1 + \cot\theta_3, \qquad a_3 = \cot\theta_1 + \cot\theta_2. \tag{8.206}$$

Since $[Z]$ is constant, the integration in Equation (8.203) is the area $A_p = L_3 L_{12}/2$, whence

$$[M_{11}] = \frac{6A_p}{Et^3} \begin{bmatrix} a_3^2 b_3 & c_2 \csc^2 \theta_2 & c_1 \csc^2 \theta_2 \\ c_2 \csc^2 \theta_2 & a_1^2 b_1 & c_3 \csc^2 \theta_3 \\ c_1 \csc^2 \theta_1 & c_3 \csc^2 \theta_3 & a_2^2 b_2 \end{bmatrix}, \quad (8.207)$$

$$b_i = 2 \cot^2 \theta_i + 1 + \nu, \quad c_i = 1 - \nu - 2a_i \cot \theta_i, \quad i = 1, 2, 3. \quad (8.208)$$

The inverse of $[M_{11}]$ can be calculated directly from $[Z_1]$ in Equation (8.204), or

$$[M_{11}]^{-1} = \frac{t^3}{12A_p}[Z^{-1}][J]^{-1}[Z^{-1}]^T = \frac{t^3}{12A_p}[Z_1][J]^{-1}[Z_1]^T$$

$$= \frac{D}{A_p} \begin{bmatrix} 1 & d_2 & d_1 \\ d_2 & 1 & d_3 \\ d_1 & d_3 & 1 \end{bmatrix}, \quad (8.209)$$

$$d_i = \cos^2 \theta_i + \nu \sin^2 \theta_i. \quad (8.210)$$

From Equations (8.202) and (8.205),

$$[M_{12}]^T = \frac{1}{2} \begin{bmatrix} 1 & 0 & -1 \\ -1 & 1 & 0 \\ 0 & -1 & 1 \end{bmatrix} \begin{bmatrix} 0 & 0 & 2 \\ -\sin 2\theta_2 & \sin 2\theta_2 & 2\cos 2\theta_2 \\ \sin 2\theta_1 & -\sin 2\theta_1 & 2\cos 2\theta_1 \end{bmatrix} [Z]$$

$$= \begin{bmatrix} -\cot \theta_2 & \cot \theta_2 + \cot \theta_3 & -\cot \theta_3 \\ -\cot \theta_1 & -\cot \theta_3 & \cot \theta_1 + \cot \theta_3 \\ \cot \theta_1 + \cot \theta_2 & -\cot \theta_2 & -\cot \theta_1 \end{bmatrix}. \quad (8.211)$$

Case (19) in Equation (8.24f) for $[M_{11}]$ and $[k]$

For the thin plate the $[M_{11}]$ bending matrix in Equation (8.197) is the same as the $[M_{11}]$ inplane matrix in case (9) in Equation (8.111) with $t \to t^3/12$, so that the direct analogy holds for the two cases. However, if shear stiffness effects are included for bending, then Equation (8.196) is different from Equation (8.111). Also, for the constant moment case, the analogy holds between $[M_{11}]$ and the inplane $[C_\sigma]$ in case (5). For the particular triangle in Figure 8.5, Equation (8.207) gives $[M_{11}]$ the same as $[C_\sigma]$ in Equation (8.61) with $t \to t^3/12$. This particular $[M_{11}]$ is exactly the same as the bending $[C_\sigma]$ in Equation (8.142) in case (13). Thus, the $[C_\sigma]$ for any triangular element is given by Equation (8.207).

As demonstrated in case (10), the $[k]$ matrix can be calculated from the $[M_{11}]$ and $[M_{12}]$ matrices by

$$[k] = [M_{12}]^T [M_{11}]^{-1} [M_{12}]. \quad (8.212)$$

However, for the bending case, only that part of $[k]$ for the u_{zi} corner points is obtained. That is, for the constant moment case in the triangle, Equation (8.212) will give only the $[k_{11}]$ part of $[k]$ in Equation (8.134). To get the entire $[k]$ matrix, use the following procedure.

Since the equilibrium relations between the constant edge moments for the triangular element associated with $[k]$ in Equation (8.134) and the unit constant edge moments associated with $[M_{11}]$ above are simply

$$[s_L] = \begin{bmatrix} L_{12} & 0 & 0 \\ 0 & L_{23} & 0 \\ 0 & 0 & L_{31} \end{bmatrix}, \quad (8.213)$$

Approximate matrix equations for plate finite elements 299

it is possible to calculate the entire $[k]$ in Equation (8.134) from

$$[k] = [s]^T [M_{11}]^{-1} [s] = \begin{bmatrix} M_{12}^T M_{11}^{-1} M_{12} & M_{12}^T M_{11}^{-1} s_L \\ s_L M_{11}^{-1} M_{12} & s_L M_{11}^{-1} s_L \end{bmatrix}, \tag{8.214}$$

$$[s] = [M_{12} \quad s_L]. \tag{8.215}$$

8.10. Matrices for constant stress triangular elements

In the previous sections of this chapter, various stiffness and flexibility matrices for triangular elements have been derived. Inplane constant stress matrices as well as bending constant moment matrices have been given in examples for either the particular right triangular element in Figure 8.5 or for general triangular elements in Figures 8.7 and 8.8. From the dual relations, analogies, and transformation relations described in the previous sections, it is possible to calculate all these matrices for the general triangular element from the matrices for one case. In this section the $[k]$, $[C_P]$, $[C_\sigma]$, $[C_\phi]$, and $[M_{11}]$ matrices for both the inplane and bending cases for the general triangle will be determined from the $[M_{11}]$ and $[M_{12}]$ bending matrices in Example 8.19.

Use the subscript p for inplane cases and the subscript b for bending cases. Let $[k(-\nu)]$ indicate that ν in the matrix is replaced by $-\nu$. Take the reference matrices to be used as

$[M_{11}]_b$ in Equation (8.207), $[M_{11}]_b^{-1}$ in Equation (8.209),
$[M_{12}]_b$ in Equation (8.211), $[s]_b = [M_{12} \quad s_L]$ in Equation (8.215),

$$[s]_p = \begin{bmatrix} 0 & 0 & 0 & -1 & 1 & 0 \\ 0 & -\sin\theta_2 & \sin\theta_2 & 0 & \cos\theta_2 & -\cos\theta_2 \\ -\sin\theta_1 & 0 & \sin\theta_1 & -\cos\theta_1 & 0 & \cos\theta_1 \end{bmatrix}, \tag{8.216}$$

where $[s]_p$ in Equation (8.47) has been written for the local coordinate system in Figures 8.7 and 8.8.

In terms of $[M_{11}]_b$ and the analogy in Equation (8.23b),

$$[M_{11}]_p = \frac{t^2}{12}[M_{11}]_b. \tag{8.217}$$

The case of the right triangle in Figure 8.5 is in Equation (8.118a).

From the normal moment definitions and Equation (8.23a)

$$[C_\sigma]_b = [M_{11}]_b, \tag{8.218}$$

$$[C_\sigma]_p = \frac{t^2}{12}[M_{11}]_b. \tag{8.219}$$

For Figure 8.5, see Equations (8.46), (8.61), and (8.142).
From Equation (8.214),

$$[k]_b = [s]_b^T [M_{11}]_b^{-1} [s]_b, \tag{8.220}$$

$$[k_{22}]_b = [s_L][M_{11}]_b^{-1}[s_L]. \tag{8.221}$$

See Equation (8.134) for the general triangle of Figure 8.8.

From the dual relations in Equation (8.23a) and the analogy in Equation (8.23b),

$$[C_P]_p = \frac{1}{EtD}[k_{22}(-\nu)]_b, \tag{8.222}$$

$$[C_P]_b = \frac{12}{Et^3D}[k_{22}(-\nu)]_b. \tag{8.223}$$

See Equations (8.57) and (8.153) for case in Figures 8.7 and 8.8.

From the dual relations in Equation (8.23a),

$$[C_\phi]_p = \frac{1}{EtD}[k(-\nu)]_b, \tag{8.224}$$

$$[C_{22\phi}] = [C_P]_p. \tag{8.225}$$

From Equation (8.22),

$$[k]_p = [s]_p^T [C_P]_p^{-1} [s]_p$$
$$= EtD[s]_p^T [s_L]^{-1} [M_{11}(-\nu)]_b [s_L]^{-1} [s]_p. \tag{8.226}$$

See Example 8.3 for the case of Figure 8.5.

From the dual relations in Equation (8.23a)

$$[C_\phi]_b = \frac{1}{EtD}[k(-\nu)]_p = [s]_p^T [s_L]^{-1} [M_{11}]_b [s_L]^{-1} [s]_p. \tag{8.227}$$

See Equation (8.168) for the right triangle case in Figure 8.5.

Note that the $[C_P]$ matrix can be expressed in terms of the $[C_\sigma]$ matrix by

$$[C_P]_p = \frac{1-\nu^2}{E^2 t^2}[s_L][C_\sigma(-\nu)]_p^{-1}[s_L]. \tag{8.228}$$

Thus, all the inplane and bending stiffness and flexibility matrices for the constant stress triangular element can be calculated from the five matrices in Equation (8.216).

8.11 Problems

8.1. Make the matrix multiplications in Example 8.1 to get the $[k]$ in Equation (8.31).

8.2. Repeat Example 8.1, part (a), for a triangle with corner points (0,0), (0,3), and (2,0).

8.3. If $\{e_E\}^T = [\alpha T \quad \alpha T \quad 0]$ for a constant temperature change T in the element, calculate $\{P_E\}$ in Equation (8.30) for Example 8.1.

8.4. In Figure 8.5 assume $u_1 = 0$, $u_2 = 0$, and $u_4 = 0$. Use the $[k]$ matrix in Equation (8.31) to find the displacements u_3, u_5, u_6, the reactions R_1, R_2, R_4, and the stresses in the element in terms of the applied forces P_3, P_5, P_6.

8.5. Use the procedure discussed in Section 8.2 and solve Example 8.1 with eight A_i constants by adding $A_7 xy(1-x)$ to u_x and $A_8 xy(1-x)$ to u_y.

8.6. Calculate $[C_\sigma]$ in Example 8.2 for a triangle with corner points (0,0), (0,3), (2,0).

8.7. If the temperature is constant with $\{e_E\}^T = [\alpha T \quad \alpha T \quad 0]$ in the element in Example 8.2, find $\{u_E\}$ in Equation (8.37).

Approximate matrix equations for plate finite elements 301

8.8. Calculate the $[B]$ matrix in Example 8.2 for the case of

$$u_x = A_1 x + \tfrac{1}{2}A_3 y + A_4 xy - \tfrac{1}{2}A_5 y^2,$$
$$u_y = A_2 y + \tfrac{1}{2}A_3 x - \tfrac{1}{2}A_4 x^2 + A_5 xy.$$

Use the same $[R]$ matrix as in Equation (8.44).

8.9. Calculate $[C_P]$ in Equation (8.57) for a triangle with corner points (0,0), (0,3), (2,0). Use this $[C_P]$ and repeat Example 8.3 to calculate the $[k]$ matrix. Compare the result to the $[k]$ in Problem 8.2.

8.10. Make the calculations in Example 8.4 to obtain $[C_P]$ in Equation (8.57).

8.11. Calculate $[C_P]$ in Equation (8.57) for a triangle with corner points (0,0), (1,2), (2,1).

8.12. Repeat Example 8.5, part (a), for a triangle with corner points (0,0), (0,3), (2,0).

8.13. Show that $[H]$ in Equation (8.72) is singular for a triangle with corner points (0,0), (0,1), (1,0).

8.14. Show that $[H]$ in Equation (8.72) is non-singular for a triangle with corner points (0,0), (0,3), (2,0).

8.15. Make the derivation for $[C_\phi]$ in Equation (8.83).

8.16. Evaluate $[C_\phi]$ in Equation (8.83) for a triangle with corner points (0,0), (0,3), (2,0).

8.17. Make the integrations and calculate $[B]$ in Equation (8.97).

8.18. Invert $[C_H]$ in Equation (8.93) and use it to calculate $\{A\}$ in Equation (8.98).

8.19. Repeat Example 8.8 for the case of the first three columns in Equation (8.85).

8.20. Repeat Example 8.8 for a triangle with the corner points (0,0), (0,3), (2,0).

8.21. Make the calculations to verify Equation (8.99).

8.22. In Example 8.9 use the restraints $u_2 = 0$, $u_3 = 0$, $u_6 = 0$ and show that the same $[C_P]$ matrix is obtained, Equation (8.100).

8.23. Verify the calculations in Equation (8.112).

8.24. Repeat Example 8.10 for the triangle with corner points (0,0), (0,3), (2,0).

8.25. Verify the calculations in Example 8.11.

8.26. Repeat Example 8.11 for the triangle with corner points (0,0), (0,3), (2,0).

8.27. Show that Equation (8.119) will give the $[k]$ in Equation (8.31) for the $[B_1]^{-1}$, $[B_2]$, and $[B_3]$ matrices in Example 8.10.

8.28. Make the matrix calculations in Equation (8.125).

8.29. (a) Use the $[M_{11}]$ and $[M_{13}]$ matrices in Example 8.12 to calculate the form of the corner stresses corresponding to Equation (8.120). (b) Use the results to show that in the element

$$N_x = (1 - x - y)N_{x1} + (x + y)N_{x2}, \quad N_{x2} \neq N_{x1},$$
$$N_y = (x + y)N_{y2}, \quad N_{xy} = (x + y)N_{xy2}.$$

8.30. Repeat Example 8.12 for the case of a boundary element with $N_{x1} = 0$, $N_{y1} = 0$, $N_{xy1} = 0$.

8.31. Repeat Example 8.13 for the triangle with corner points (0,0), (0,3), (2,0).

8.32. Repeat Example 8.13 for the triangle with corner points (0,0), (1,1), (2,0).

8.33. If the temperature produces a constant moment $\{M_E\}^T = \tfrac{1}{1-\nu}[M_T \quad M_T \quad 0]$ in the element in Example 8.14, find the slopes $\{u_E\}$ in Equation (8.141).

8.34. Use the $[C_\sigma]$ matrix in Equation (8.142) and calculate $[k]$ in Equation (8.135) from

$$[k] = [s]^T [C_\sigma]^{-1} [s], \quad [s] = \begin{bmatrix} -1 & 0 & 1 & 1 & 0 & 0 \\ 2 & -1 & -1 & 0 & b & 0 \\ -1 & 1 & 0 & 0 & 0 & 1 \end{bmatrix}.$$

8.35. Compare the $[C_P]$ in Equation (8.153) with the $[k_{22}]$ in Equation (8.134) by considering the duality and analogy discussion given in connection with Equation (8.23).

8.36. Calculate $[C_P]$ in Equation (8.153) for a triangle with corner points (0,0), (1,2), (2,1).

8.37. Repeat Example 8.16 for the triangle and results in Problem 8.2 above.

8.38. Make the calculations in the derivation of $[C_H]$ in Equation (8.183).

8.39. Calculate $[D_s]$ and $[D_s]^{-1}$ in Example 8.17 for the triangle in Figure 8.5.

8.40. Write out the submatrices in $[B]$ in Equation (8.189) for the triangle in Figure (8.5).

8.41. Make the calculations for the $[B_2]$, $[B_3]$, and $[B_4]$ matrices in Example 8.18.

8.42. Repeat Example 8.18 for the triangle with corner points (0,0), (0,3), (2,0).

8.43. Make the calculations to get $[Z]$ in Equation (8.205).

8.44. Make the calculations for $[M_{11}]$ and $[M_{11}]^{-1}$ in Example 8.19.

8.45. Use Equation (8.212) for the constant moment triangular element case and show that

$$[k_{11}] = \frac{12}{A_p t^3}[R_{1c}][W]\big[J\big]^{-1}\big[W\big]^T\big[R_{1c}\big]^T.$$

Use this form to calculate $[k_{11}]$ in Equation (8.134).

8.46. Show that Equation (8.214) gives Equation (8.134) for $[k]$.

8.47. Verify the sequence of matrix calculations in Equations (8.216)-(8.228) for the right triangle element in figure 8.5.

References

For references concerning this chapter see Chapter 9.

9

Matrix structural analysis using finite elements

9.1. Introduction

Although simple finite elements as truss members and beam members in frames have been used in structural analysis for many years, the general finite element methods of analysis have been developped only recently in parallel with the development of digital computers. It is the use of general matrix methods in high speed digital computers that makes the general finite element methods feasible. At the present time the finite element methods are being used extensively in the analysis and design of flight vehicle structures as well as in other types of structures. Many large general purpose computer programs as well as many specialized programs have been developed for the analysis of structures. These computer programs use various finite elements and usually calculate and assemble the element matrices by one of the three methods of analysis described in previous chapters. Most of the general programs use the displacement method of analysis; a few general programs and many specialized programs use the force method of analysis; a few programs use the mixed method, while some programs use more than one method.

As shown in Chapters 3 and 4 in Vol.1 for bar and two-dimensional beam elements, it is necessary to construct the stiffness matrices $[k_i]$ of the elements for use in the displacement method of analysis, the flexibility matrices $[C_i]$ for the force method, and the mixed matrices $[M_i]$ for the mixed method. All three of these element matrices $[k_i]$, $[C_i]$, and $[M_i]$ will be given for a few finite elements in this chapter. Other elements are described in the references for this chapter and in the documentation for various computer programs.

Many books and papers on finite elements have been published in the last twenty years. Some of these are listed in the references. See the references given in References 2 and 5 for many of the hundreds of papers on finite elements. Many different elements have been described, such as flat plate triangular, rectangular, and quadrilateral elements, curved plate elements, shell elements, solid tetrahedron elements, parametric elements with curved boundaries, etc.

It should be pointed out that most of the books and papers on finite elements are concerned with the stiffness matrices $[k_i]$ and the displacement method of analysis. The flexibility matrix $[C_i]$ does not appear to be mentioned in some of the books on finite elements. However, both $[k]$ and $[C_P]$ are derived and used in References 1, 7, and 8. Various comparisons between the two procedures are given in Reference 23. Although $[k]$ receives most of the attention in Reference 2, $[C_{Pi}]$ is mentioned and the mixed matrix $[M_i]$ receives some discussion. In books devoted to matrix methods for trusses and beams, some use both the displacement and force methods while others use only the force method. All three methods will be considered here.

The three-dimensional bar elements have been considered in Sections 3.5, 3.6, and 3.7 (Vol.1). Constant shear flow web elements and stiffener elements have been discussed in Sections 6.10 through 6.13 (Vol.1). Beam elements for the inplane case have been examined in detail in Chapter 4 (Vol.1). The equations for the general beam element with six component forces and moments will be given below in Sections 9.2 and 9.3 for the local and datum coordinate systems.

Triangular elements with inplane loading are considered in Section 9.4 for local and datum coordinates. The general plate element equations derived in Chapter 8 will be used for the triangular plate elements. All three virtual principles are used to obtain the element stiffness matrices $[k_i]$, flexibility matrices $[C_i]$, mixed matrices $[M_i]$, transformation matrices $[\lambda_i]$, and equilibrium matrices $[s_i]$. These are the element matrices needed to assemble the elements for the three methods of finite element structural analysis, as described briefly in Section 9.5. See References 1 and 5 for more details on assembly procedures.

Due to the large number of equations involved and due to the necessity of assembling the elements in matrix form, it is necessary to use matrix operations for all the elements. Many references will be made to Chapters 3, 4 (Vol.1), and 8 for the matrix forms.

As demonstrated in Chapter 8, various approximations for stresses and displacements can be made in constructing the plate element matrices. The simpler cases will be given here for the elements, with references made to more complicated approximations. It should be emphasized that the approximations used affect the accuracy and speed of convergence of the assembled elements in the structure. For this reason, it is necessary to understand completely the approximations used in constructing the element matrices so that the size and arrangement of the elements can be made properly for a given structure and a given computer program.

It should be pointed out that the derivation of the matrix expressions for plate finite elements in Chapter 8 was based upon simple polynomial expressions in rectangular coordinates in order to show the physical problems involved. This procedure requires the inversion of an $[H]$ matrix in order to express the unknown variables in terms of the unknown node point values. See Equation (8.27) as an example. It is possible to avoid this matrix inversion by using so-called natural coordinates and interpolation or shape functions for the element. Not only is the inversion of $[H]$ avoided, but also a singular $[H]$ is avoided. Also, the natural coordinates allow simpler integration of the area integrals. It is possible to use these element shape functions to establish curvilinear coordinates for an element with curved boundaries. This gives the isoparametric elements with curved edges. These procedures allow more points to be used on the edges so

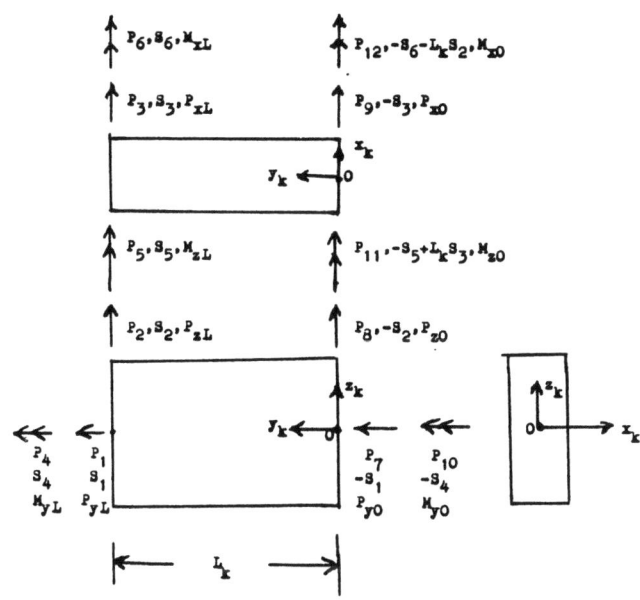

Fig. 9.1. General beam element with three forces and three moments.

that larger elements can be used. However, whether the computer time is reduced depends upon the particular application. See, for example, References 2, 5, and 9 for details on these questions of natural coordinates, shape functions, and isoparametric elements. For simplicity, the plate elements given in this chapter are based upon the matrix expressions derived in Chapter 8.

9.2. General beam elements in local coordinates

As pointed out in Chapter 4 (Vol.1) the basic matrices for finite elements are the stiffness matrix $[k]$, the flexibility matrix $[C]$, and the mixed matrix $[M]$. These matrices were derived in Chapter 4 (Vol.1) for the simple beam element with bending in one plane. See Figure 4.1 and Equation (4.91) (Vol.1) for $[k]$, Equation (4.115, Vol.1) for $[C]$, and Equation (4.135, Vol.1) for $[M]$.

The general beam element with all six generalized forces (three forces and threee moments) will be considered here. In Figure 9.1 with x_k, y_k, z_k as the local coordinates for the element, P_1 and P_7 the axial forces (P_y) in the y_k direction, P_2 and P_8 the shear forces (P_z) in the z_k direction, P_3 and P_9 the shear forces (P_x) in the x_k direction, P_4 and P_{10} the torsional moments (M_y) about the y_k axis, P_5 and P_{11} the bending moments (M_z) about the z_k axis, P_6 and P_{12} the bending moments (M_x) about the x_k axis, it follows that the six displacements u_1, u_7, u_2, u_8, u_3, u_9 correspond to the six forces and the six rotations u_4, u_{10}, u_5, u_{11}, u_6, u_{12} correspond to the six moments. The S_i are the internal point forces and the P_{x0}, P_{xL}, M_{x0}, M_{xL}, etc. are the applied point forces and moments.

The e_E strain effects will be included in the element matrix equation. In order to simplify the matrices, the x_k, z_k axes of the beam cross section will be assumed to be principle axes at the centroid. This makes the integrals S_{xE}, S_{zE}, and I_{xzE} in Equation (1.81, Vol.1) zero and makes Equation (1.91, Vol.1) the proper equation for σ_{yy}. See Equations (1.21) and (1.23) for the θ_d rotation angle to the principle axes and the values of the principle moments of inertia C_{11} and C_{22} with $C_{12}=0$, where $a_{11}=I_{xxR}$, $a_{22}=I_{zzR}$, $a_{12}=I_{xzR}$.

Since the beam deflection equations for all six generalized forces have been derived in Sections 5.10, 5.11, 5.12 (Vol.1), they can be used directly to get the flexibility $[C]$ matrix for the beam element. The stiffness matrix $[k]$ and the mixed matrix $[M]$ can then be derived from $[C]$. From Equation (2.76, Vol.1) for this determinate case for beam element (k) in Figure 9.1 with the zero end fixed,

$$\{u\}_k = [C]_k \{P\}_k + \{u_E\}_k, \tag{9.1}$$

$$\{u\}_k^T = [u_1 \ u_2 \ u_3 \ u_4 \ u_5 \ u_6]_k, \quad \{P\}_k^T = [P_1 \ P_2 \cdots P_6]_k. \tag{9.2}$$

From Equation (5.66, Vol.1) with $I_{xzE}=0$, I_{xxE} and I_{zzE} constant,

$$u_1 = (P_1 L / E_R A_E) + u_{E1}, \quad u_{E1} = \int_0^L (P_{yE}/E_R A_E) dy, \tag{9.3}$$

$$P_{yE} = \int_A E e_E \, dA,$$

$$u_{2b} = (1/E_R I_{xxE}) \int_0^L (L-y)[(L-y)P_2 + P_6 + M_{xE}]dy$$
$$= (P_2 L^3 / 3 E_R I_{xxE}) + (P_6 L^2 / 2 E_R I_{xxE}) + u_{E2b}, \tag{9.4}$$

$$u_{E2b} = \int_0^L \frac{(L-y)M_{xE}}{E_R I_{xxE}} dy, \quad M_{xE} = -\int_A E e_E z \, dA,$$

$$u_{3b} = (1/E_R I_{zzE}) \int_0^L (L-y)[(L-y)P_3 - P_5 - M_{zE}]dy$$
$$= (P_3 L^3 / 3 E_R I_{zzE}) - (P_5 L^2 / 2 E_R I_{zzE}) + u_{E3b}, \tag{9.5}$$

$$u_{E3b} = -\int_0^L \frac{(L-y)M_{zE}}{E_R I_{zzE}} dy, \quad M_{zE} = \int_A E e_E x \, dA.$$

To get a simple form for the shear deflections in Section 5.11 (Vol.1), use Equations (5.83, Vol.1) and (5.85, Vol.1) to get

$$u_{2s} = P_2 L / G A_{sz}, \quad u_{3s} = P_3 L / G A_{sx}. \tag{9.6}$$

The shear stiffnesses GA_{sz} and GA_{sx} are effective values, which include multiplying factors for the particular cross section, Equations (5.79)-(5.82) (Vol.1), as well as any inelastic and buckling effects, Equation (5.62).

For the torsional rotation use Equation (5.106, Vol.1) to get

$$u_4 = P_4 L / GJ. \tag{9.7}$$

Matrix structural analysis using finite elements

The torsional stiffness GJ is an effective value depending upon the cross section and any inelastic and buckling effects.

To get the θ_x and θ_z rotations or slopes use a unit moment instead of a unit load in Equations (9.4) and (9.5), whence by Equation (2.31, Vol.1)

$$u_5 = (-du_x/dy)_{y=L} = (1/E_R I_{zzE}) \int_0^L [-(L-y)P_3 + P_5 + M_{zE}] dy$$
$$= -(P_3 L^2/2E_R I_{zzE}) + (P_5 L/E_R I_{zzE}) + u_{E5}, \tag{9.8}$$

$$u_{E5} = \int_0^L \frac{M_{zE}\,dy}{E_R I_{zzE}}, \quad M_{zE} = \int_A E e_E x \, dA,$$

$$u_6 = (du_z/dy)_{y=L} = (1/E_R I_{xxE}) \int_0^L [(L-y)P_2 + P_6 + M_{xE}] dy$$
$$= (P_2 L^2/2E_R I_{xxE}) + (P_6 L/E_R I_{xxE}) + u_{E6}, \tag{9.9}$$

$$u_{E6} = \int_0^L (M_{xE}/E_R I_{xxE}) dy, \quad M_{xE} = -\int_A E e_E z \, dA.$$

With the above expressions for $\{u\}$ in Equations (9.3)-(9.9) it follows that in Equation (9.1)

$$[C]_k = \begin{bmatrix} C_{11} & C_{12} \\ C_{12}^T & C_{22} \end{bmatrix}_k, \tag{9.10}$$

$$[C_{11}]_k = \begin{bmatrix} \frac{L}{E_R A_E} & 0 & 0 \\ 0 & \frac{L^3(3+H_z)}{12 E_R I_{zzE}} & 0 \\ 0 & 0 & \frac{L^3(3+H_x)}{12 E_R I_{xxE}} \end{bmatrix}_k,$$

$$[C_{12}]_k = \begin{bmatrix} 0 & 0 & 0 \\ 0 & 0 & \frac{L^2}{2E_R I_{xxE}} \\ 0 & -\frac{L^2}{2E_R I_{zzE}} & 0 \end{bmatrix}_k,$$

$$[C_{22}]_k = \begin{bmatrix} \frac{L}{GJ} & 0 & 0 \\ 0 & \frac{L}{E_R I_{xxE}} & 0 \\ 0 & 0 & \frac{L}{E_R I_{zzE}} \end{bmatrix},$$

$$H_z = 1 + \frac{12 E_R I_{zzE}}{L^2 G A_{sz}}, \quad H_x = 1 + \frac{12 E_R I_{xxE}}{L^2 G A_{sx}}, \tag{9.11}$$

$$\{u_E\}_k^T = \int_0^L \left[\frac{P_{yE}}{E_R A_E} \quad \frac{(L-y)M_{zE}}{E_R I_{zzE}} \quad -\frac{(L-y)M_{xE}}{E_R I_{xxE}} \right.$$
$$\left. 0 \quad \frac{M_{xE}}{E_R I_{xxE}} \quad \frac{M_{zE}}{E_R I_{zzE}} \right]_k. \tag{9.12}$$

For element with constant cross section, constant material properties, and constant apparent loads,

$$\{u_E\}_k^T = \left[\frac{L P_{yE}}{E_R A_E} \quad \frac{L^2 M_{zE}}{2 E_R I_{zzE}} \quad -\frac{L^2 M_{xE}}{2 E_R I_{xxE}} \quad 0 \quad \frac{L M_{xE}}{E_R I_{xxE}} \quad \frac{L M_{zE}}{E_R I_{zzE}} \right]_k. \tag{9.13}$$

To get the mixed matrix $[M]$ write the equilibrium equations from Figure 9.1

in the form ($[I]$ is 3 by 3)

$$\{P\}_k = [s]_k^T \{S\}_k, \tag{9.14}$$

$$\{s\}_k = \begin{bmatrix} I & 0 & -I & -T_k \\ 0 & I & 0 & -I \end{bmatrix}, \quad [T]_k = \begin{bmatrix} 0 & 0 & 0 \\ 0 & 0 & L_k \\ 0 & -L_k & 0 \end{bmatrix}.$$

Thus, from Equations (2.86, Vol.1) and (9.10)

$$[M]_k = \begin{bmatrix} -C_{11} & -C_{12} & I & 0 & -I & -T_k \\ -C_{12}^T & -C_{22} & 0 & I & 0 & -I \\ I & 0 & 0 & 0 & 0 & 0 \\ 0 & I & 0 & 0 & 0 & 0 \\ -I & 0 & 0 & 0 & 0 & 0 \\ -T_k^T & -I & 0 & 0 & 0 & 0 \end{bmatrix}_k. \tag{9.15}$$

If the first submatrix in Equation (2.86, Vol.1) is solved for $\{S\}$ and the result put into the second row, then

$$\{P\} = [k]\{u\} - \{P_E\}, \tag{9.16}$$
$$[k] = [s]^T [C]^{-1} [s], \quad \{P_E\} = [s]^T \{S_E\},$$

which corresponds to Equation (4.90, Vol.1). The inverse of $[C]_k$ in Equation (9.10) is

$$[k_S]_k = [C]_k^{-1} = \begin{bmatrix} B_{11} & B_{12} \\ B_{12}^T & B_{22} \end{bmatrix}_k, \tag{9.17}$$

$$[B_{11}]_k = \begin{bmatrix} E_R A_E/L & 0 & 0 \\ 0 & 12 E_R I_{xxE}/L^3 H_x & 0 \\ 0 & 0 & 12 E_R I_{zzE}/L^3 H_x \end{bmatrix}_k,$$

$$[B_{12}]_k = \begin{bmatrix} 0 & 0 & 0 \\ 0 & 0 & -6 E_R I_{xxE}/L^2 H_z \\ 0 & 6 E_R I_{zzE}/L^2 H_x & 0 \end{bmatrix}_k,$$

$$[B_{22}]_k = \begin{bmatrix} GJ/L & 0 & 0 \\ 0 & (3+H_x) E_R I_{zzE}/L H_x & 0 \\ 0 & 0 & (3+H_z) E_R I_{xxE}/L H_z \end{bmatrix}_k.$$

Thus, from Equation (9.16)

$$[k]_k = \begin{bmatrix} I & 0 \\ 0 & I \\ -I & 0 \\ -T_k^T & -I \end{bmatrix} \begin{bmatrix} B_{11} & B_{12} \\ B_{12}^T & B_{22} \end{bmatrix} \begin{bmatrix} I & 0 & -I & -T_k \\ 0 & I & 0 & -I \end{bmatrix}$$

$$= \begin{bmatrix} B_{11} & B_{12} & -B_{11} & B_{12} \\ B_{12}^T & B_{22} & -B_{12}^T & -B_{24} \\ -B_{11} & -B_{12} & B_{11} & -B_{12} \\ B_{12}^T & -B_{24} & -B_{12}^T & B_{22} \end{bmatrix}_k, \tag{9.18}$$

$$[B_{24}]_k = \begin{bmatrix} GJ/L & 0 & 0 \\ 0 & (H_x-3)E_R I_{zzE}/LH_x & 0 \\ 0 & 0 & (H_z-3)E_R I_{xxE}/LH_z \end{bmatrix}_k. \quad (9.19)$$

See Reference 5 for a direct derivation of $[k]_k$ in Equation (9.18) by using the beam differential equations for the deflections. However, the sign of $[B_{12}]$ is reversed in Equation (9.18) due to a different numbering system.

Note that if there are any distributed or point loads acting on the beam element, then the equivalent end values of these loads in Equation (4.18, Vol.1) can be added to the applied load matrix $\{P\}$ in the above equations, or in general, Equation (4.91, Vol.1),

$$\{P\}_g = \{P\}_a + \{P\}_d + \{P\}_E. \quad (9.20)$$

Here $\{P\}_a$ are the applied end loads, $\{P\}_d$ the equivalent end loads for any distributed loads, and $\{P\}_E$ the equivalent end loads for temperature, inelastic, large deflection, and initial strains in $\{e_E\}$.

Example 9.1. Use the stiffness matrix $[k]$ to find the deflections and reactions for the two element beam in Figure (9.2). Assume constant thermal axial loads and moments are present in both elements.

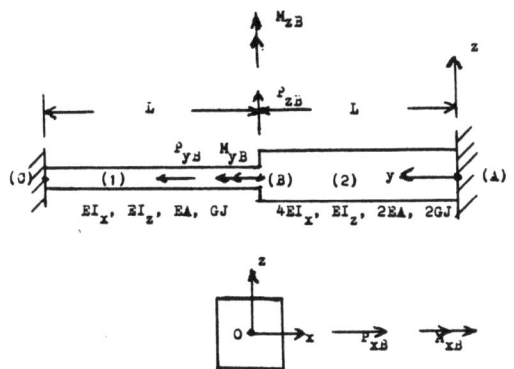

Fig. 9.2. Two element beam for Example 9.1.

Solution. Omit the shear stiffnesses so that $H_x = 1$, $H_z = 1$ in Equations (9.17)-(9.19). Use Figure (9.2) to get

$$[B_{11}]_1 = \frac{E}{L^3}\begin{bmatrix} AL^2 & 0^0 & 0 \\ 0 & 12I_z & 0 \\ 0 & 0 & 12I_x \end{bmatrix}, \quad [B_{11}]_2 = \frac{E}{L^3}\begin{bmatrix} 2AL^2 & 0 & 0 \\ 0 & 48I_z & 0 \\ 0 & 0 & 12I_x \end{bmatrix},$$

$$[B_{12}]_1 = \frac{6E}{L^2}\begin{bmatrix} 0 & 0 & 0 \\ 0 & 0 & -I_x \\ 0 & I_z & 0 \end{bmatrix}, \quad [B_{12}]_2 = \frac{6E}{L^2}\begin{bmatrix} 0 & 0 & 0 \\ 0 & 0 & -4I_x \\ 0 & I_z & 0 \end{bmatrix},$$

$$[B_{22}]_1 = \frac{1}{L}\begin{bmatrix} GJ & 0 & 0 \\ 0 & 4EI_x & 0 \\ 0 & 0 & 4EI_z \end{bmatrix}, \quad [B_{22}]_2 = \frac{1}{L}\begin{bmatrix} 2GJ & 0 & 0 \\ 0 & 4EI_x & 0 \\ 0 & 0 & 16EI_z \end{bmatrix},$$

$$[B_{24}]_1 = \frac{1}{L}\begin{bmatrix} GJ & 0 & 0 \\ 0 & -2EI_x & 0 \\ 0 & 0 & -2EI_z \end{bmatrix}, \quad [B_{24}]_2 = \frac{1}{L}\begin{bmatrix} 2GJ & 0 & 0 \\ 0 & -2EI_x & 0 \\ 0 & 0 & -8EI_z \end{bmatrix}.$$

Group the loads and displacements in threes to match the 3 by 3 B_{ij} matrices, or

$$\{P_B\}^T = [P_y \quad P_z \quad P_x]_B, \quad \{M_B\}^T = [M_y \quad M_z \quad M_x]_B, \quad \text{etc.},$$
$$\{P_T\}_1^T = [P_{yT1} \quad 0 \quad 0], \quad \{M_T\}_1^T = [0 \quad M_{zT1} \quad M_{xT1}],$$
$$\{P_T\}_2^T = [P_{yT2} \quad 0 \quad 0], \quad \{M_T\}_2^T = [0 \quad M_{zT2} \quad M_{xT2}].$$

From Equations (9.16) and (9.14),

$$\{P_E\} = \begin{Bmatrix} P_{CT} \\ M_{CT} \\ P_{BT} \\ M_{BT} \\ P_{AT} \\ M_{AT} \end{Bmatrix} = \begin{bmatrix} I & 0 & 0 & 0 \\ 0 & I & 0 & 0 \\ -I & 0 & I & 0 \\ -T_1^T & I & 0 & I \\ 0 & 0 & -I_T & 0 \\ 0 & 0 & -T_2^T & -I \end{bmatrix} \begin{Bmatrix} P_{T1} \\ M_{T1} \\ P_{T2} \\ M_{T2} \end{Bmatrix}$$

$$= \begin{Bmatrix} P_{T1} \\ M_{T1} \\ -P_{T1} + P_{T2} \\ -T_1^T P_{T1} - M_{T1} + M_{T2} \\ -P_{T2} \\ -T_2^T P_{T2} - M_{T2} \end{Bmatrix},$$

$$\begin{Bmatrix} P_C + P_{CT} \\ M_C + M_{CT} \\ P_B + P_{BT} \\ M_B + M_{BT} \\ P_A + P_{AT} \\ M_A + M_{AT} \end{Bmatrix} =$$

$$\begin{bmatrix} B_{11,1} & B_{12,1} & -B_{11,1} & B_{12,1} & 0 & 0 \\ B_{12,1}^T & B_{22,1} & -B_{12,1}^T & -B_{24,1} & 0 & 0 \\ -B_{11,1} & -B_{12,1} & B_{11,1}+B_{11,2} & B_{12,2}-B_{12,1} & -B_{11,2} & B_{12,2} \\ B_{12,1}^T & -B_{24,1} & B_{12,2}^T - B_{12,1}^T & B_{22,1}+B_{22,2} & -B_{12,2}^T & -B_{24,2} \\ 0 & 0 & -B_{11,2} & -B_{12,2} & B_{11,2} & -B_{12,2} \\ 0 & 0 & B_{12,2}^T & -B_{24,2} & -B_{12,2}^T & B_{22,2} \end{bmatrix} \begin{Bmatrix} 0 \\ 0 \\ u_B \\ \theta_B \\ 0 \\ 0 \end{Bmatrix},$$

$$\begin{Bmatrix} P_B + P_{BT} \\ M_B + M_{BT} \end{Bmatrix} = \begin{bmatrix} B_{11,1}+B_{11,2} & -B_{12,1}+B_{12,2} \\ -B_{12,1}^T+B_{12,2}^T & B_{22,1}+B_{22,2} \end{bmatrix} \begin{Bmatrix} u_B \\ \theta_B \end{Bmatrix} =$$

$$\begin{bmatrix} \dfrac{E}{L^3}\begin{bmatrix} 3AL^2 & 0 & 0 \\ 0 & 60I_z & 0 \\ 0 & 0 & 24I_x \end{bmatrix} & \dfrac{6E}{L^2}\begin{bmatrix} 0 & 0 & 0 \\ 0 & 0 & -3I_z \\ 0 & 0 & 0 \end{bmatrix} \\ \dfrac{6E}{L^2}\begin{bmatrix} 0 & 0 & 0 \\ 0 & 0 & 0 \\ 0 & -3I_z & 0 \end{bmatrix} & \dfrac{1}{L}\begin{bmatrix} 3GJ & 0 & 0 \\ 0 & 8EI_z & 0 \\ 0 & 0 & 20EI_z \end{bmatrix} \end{bmatrix}_B \begin{Bmatrix} u_y \\ u_z \\ u_x \\ \theta_y \\ \theta_z \\ \theta_x \end{Bmatrix}_B$$

Invert to get

$$u_{yB} = \frac{L}{3EA}(P_{yB} - P_{yT1} + P_{yT2}), \quad u_{zB} = \frac{L^3 P_{zB}}{24EI_z},$$

$$u_{xB} = \frac{5L^3 P_{xB}}{219EI_z} + \frac{3L^2}{146EI_z}(M_{zB} - M_{zT1} + M_{zT2}),$$

$$\theta_{yB} = \frac{LM_{yB}}{3GJ}, \quad \theta_{xB} = \frac{L}{8EI_z}(M_{xB} - M_{xT1} + M_{xT2}),$$

$$\theta_{zB} = \frac{3L^2 P_{xB}}{146EI_z} + \frac{5L}{73EI_z}(M_{zB} - M_{zT1} + M_{zT2}).$$

The reactions are

$$\left\{ \begin{array}{c} P_C + P_{CT} \\ M_C + M_{CT} \end{array} \right\} = \begin{bmatrix} -B_{11,1} & B_{12,1} \\ -B_{12,1}^T & -B_{24,1} \end{bmatrix} \left\{ \begin{array}{c} u_B \\ \theta_B \end{array} \right\},$$

$$\left\{ \begin{array}{c} P_A + P_{AT} \\ M_A + M_{AT} \end{array} \right\} = \begin{bmatrix} -B_{11,2} & B_{12,2} \\ B_{12,2}^T & -B_{24,2} \end{bmatrix} \left\{ \begin{array}{c} u_B \\ \theta_B \end{array} \right\},$$

$P_{yC} = -(P_{yB} + 2P_{yT1} + P_{yT2})/3, \quad P_{yA} = -(2P_{yB} - 2P_{yT1} - P_{yT2})/3,$
$P_{xC} = -[29LP_{xB} + 48(M_{xB} - M_{xT1} + M_{xT2})]/73L,$
$P_{xA} = -[44LP_{xB} - 48(M_{xB} - M_{xT1} + M_{xT2})]/73L,$
$P_{zC} = -[2LP_{zB} - 3(M_{zB} - M_{xT1} + M_{xT2})]/4L,$
$P_{zA} = -[2LP_{zB} + 3(M_{zB} - M_{xT1} + M_{xT2})]/4L,$
$M_{yC} = -M_{yB}/3,$
$M_{yA} = -2M_{yB}/3,$
$M_{xC} = -(LP_{zB} - M_{zB} + 5M_{xT1} - M_{xT2})/4,$
$M_{xA} = (LP_{zB} + M_{zB} - M_{xT1} + 5M_{xT2})/4,$
$M_{zC} = (13LP_{xB} + 19M_{xB} + 19M_{xT2} - 92M_{xT1})/73,$
$M_{zA} = -(28LP_{xB} - 4M_{xB} - 77M_{xT2} + 4M_{xT1})/73.$

9.3. General beam elements in datum coordinates

The equations in the previous Section 9.2 are in local coordinates (x_k, y_k, z_k). To assemble the elements for the entire structure it is necessary to convert the equations to a datum (x, y, z) coordinate system. From Figure 9.3 take the direction cosine matrix as

$$[R]_k^T = \begin{bmatrix} \cos(y, y_k) & \cos(y, z_k) & \cos(y, x_k) \\ \cos(z, y_k) & \cos(z, z_k) & \cos(z, x_k) \\ \cos(x, y_k) & \cos(z, z_k) & \cos(x, x_k) \end{bmatrix}. \tag{9.21}$$

For the twelve displacements and twelve forces in Figure 9.1, the transform matrix is

$$[\lambda]_k^T = \begin{bmatrix} R & 0 & 0 & 0 \\ 0 & R & 0 & 0 \\ 0 & 0 & R & 0 \\ 0 & 0 & 0 & R \end{bmatrix}_k^T. \tag{9.22}$$

Thus, in the datum system

$$\{P\}_d = [\lambda]_k^T \{P\}_k, \quad \{u\}_d = [\lambda]_k^T \{u\}_k, \quad \{u\}_k = [\lambda]_k \{u\}_d,$$
$$\{P\}_d = [\lambda]_k^T [k]_k [\lambda]_k \{u\}_d = [k]_d \{u\}_d, \tag{9.23}$$
$$[k]_d = [\lambda]_k^T [k]_k [\lambda]_k, \quad [s]_d^T = [\lambda]_k^T [s]_k^T.$$

The calculation of the $[R]_k$ matrix usually requires three rotations in sequence to take datum axes to the local axes. See Figure A.5 in Example A.10 in Appendix A (Vol.1) for a cylindrical element that can be any general beam element. The form of the $[R]_k$ matrix for the sequence in Figure A.5 is given in Equation

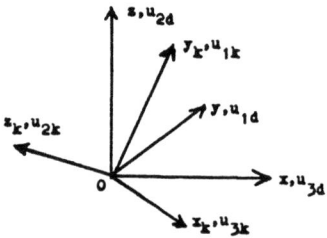

Fig. 9.3. Local (k) and datum (d) coordinate axes.

(A.114, Vol.1). Since principle axes are being used in the above matrices, let the x_k local axis be the maximum moment of inertia axis. In Figure A.5 (Vol.1) the x_k, z_k principle axes are in the same plane as the x_1, z_2 axes with the x_1 axis in the datum x, y plane. Thus, the angle θ_{yk} between the x_1 and x_k axes can be included as part of the input data along with the location of the end points in the datum system.

Since the order of the axes in Equation (A.114, Vol.1) is x, y, z while here the order is y, z, x, it is necessary to modify the rotation matrices to the form

$$[R]_k$$
$$= \begin{bmatrix} 1 & 0 & 0 \\ 0 & \cos\theta_{yk} & \sin\theta_{yk} \\ 0 & -\sin\theta_{yk} & \cos\theta_{yk} \end{bmatrix} \begin{bmatrix} \cos\theta_{x1} & \sin\theta_{x1} & 0 \\ -\sin\theta_{x1} & \cos\theta_{x1} & 0 \\ 0 & 0 & 1 \end{bmatrix} \begin{bmatrix} \cos\theta_z & 0 & \sin\theta_z \\ 0 & 1 & 0 \\ -\sin\theta_z & 0 & \cos\theta_z \end{bmatrix}$$

$$= \begin{bmatrix} 1 & 0 & 0 \\ 0 & \cos\theta_{yk} & \sin\theta_{yk} \\ 0 & -\sin\theta_{yk} & \cos\theta_{yk} \end{bmatrix} \begin{bmatrix} b/e & c/e & a/e \\ -cb/ed & d/e & -ca/ed \\ -a/d & 0 & b/d \end{bmatrix}, \qquad (9.24)$$

$a = x_j - x_i, \quad b = y_j - y_i, \quad c = z_j - z_i,$
$d^2 = a^2 + b^2, \quad e^2 = a^2 + b^2 + c^2 = L_k^2, \quad$ Figure A.5, Volume 1.

From Equations (9.14), (9.22), and (9.23), for element (k),

$$[s]_{dk}^T = [\lambda]_k^T = \begin{bmatrix} R_k^T & 0 \\ 0 & R_k^T \\ -R_k^T & 0 \\ -R_k^T T_k^T & -R_k^T \end{bmatrix} = \begin{bmatrix} B_{kk} \\ -S_{k+1,k} \end{bmatrix}. \qquad (9.25)$$

If $[s]_k^T$ is assembled for several elements, with only two elements joining at node points, then it has a block lower bi-diagonal form, or for $M-1$ elements,

$$[s]_d^T = \begin{bmatrix} B_{11} & 0 & 0 & 0 & 0 & 0 \\ -S_{21} & B_{22} & 0 & 0 & 0 & 0 \\ 0 & -S_{32} & B_{33} & 0 & 0 & 0 \\ \hline 0 & 0 & 0 & 0 & -S_{M,M-1} & B_{MM} \end{bmatrix}. \qquad (9.26)$$

Since the inverse of $[R]_k^T$ is

$$[[R]_k^T]^{-1} = [R]_k, \tag{9.27}$$

it follows that

$$[B_{kk}]^{-1} = \begin{bmatrix} R_k & 0 \\ 0 & R_k \end{bmatrix}. \tag{9.28}$$

Thus, only matrix multiplications are needed to get the inverse of $[s]_d^T$. Define

$$[C_{k,k-1}] = [B_{kk}]^{-1}[S_{k,k-1}], \tag{9.29}$$

whence by use of Equation (A.54b, Vol.1),

$$[p]_d = [[s]_d^T]^{-1} = \begin{bmatrix} B_{11}^{-1} & 0 & 0 & --- \\ C_{21}B_{11}^{-1} & B_{22}^{-1} & 0 & --- \\ C_{32}C_{21}B_{11}^{-1} & C_{32}B_{22}^{-1} & B_{33}^{-1} & --- \\ -- & -- & -- & -- \end{bmatrix}. \tag{9.30}$$

Note that $[C_{k,k-1}]$ times row $(k-1)$ gives row (k) with $[B_{kk}]^{-1}$ added. See Equation (4.109, Vol.1) for the plane beam case.

All the beam supports are represented by columns in Equation (9.30) with the corresponding P_i forces being reactions. Thus, the matrix $[p]_d$ can be partioned as in Equation (4.111, Vol.1) and the necessary $[p]$ and $[q]$ obtained from Equation (4.112, Vol.1) for the force method of solution. The discussion following Equation (4.112) in Section 4.9 (Vol.1) also applies to the general beam case under consideration here.

Example 9.2. (a) Find the point deflections and internal forces for the two element beam in Figure 9.4 by using the flexibility $[C]$ matrix. Assume the two elements have the same properties. (b) Comment on the redundant beam with additional supports.

Solution. (a) Use the x, y, z axes with element (2) along the y-axis as the datum axes. For element (1) assume the θ_{y1} rotation angle to be zero so that from Figure 9.4 and Equation (9.24),

$$[R]_1 = \tfrac{1}{3}\begin{bmatrix} 2 & 2 & 1 \\ -4h & 5h & -2h \\ -3h & 0 & 6h \end{bmatrix}, \quad h^2 = 1/5. \tag{9.31}$$

From Equations (9.25)–(9.30),

$$[s]_d^T = \begin{bmatrix} B_{11} & 0 \\ -S_{21} & B_{22} \end{bmatrix}, \quad [B_{11}] = \begin{bmatrix} R_1^T & 0 \\ 0 & R_1^T \end{bmatrix},$$

$$[B_{11}]^{-1} = \begin{bmatrix} R_1 & 0 \\ 0 & R_1 \end{bmatrix}, \quad [S_{21}] = \begin{bmatrix} R_1^T & 0 \\ R_1^T T_1^T & R_1^T \end{bmatrix},$$

$$[B_{22}] = [I],$$

$$[C_{21}] = [I][S_{21}] = [S_{21}], \quad [C_{21}][B_{11}]^{-1} = \begin{bmatrix} I & 0 \\ K & I \end{bmatrix},$$

$$[K] = [R_1^T T_1^T R_1] = (L_1/3)\begin{bmatrix} 0 & -1 & 2 \\ 1 & 0 & -2 \\ -2 & 2 & 0 \end{bmatrix},$$

$$[p]_d = \begin{bmatrix} R_1 & 0 & 0 & 0 \\ 0 & R_1 & 0 & 0 \\ I & 0 & I & 0 \\ K & I & 0 & I \end{bmatrix}. \tag{9.32}$$

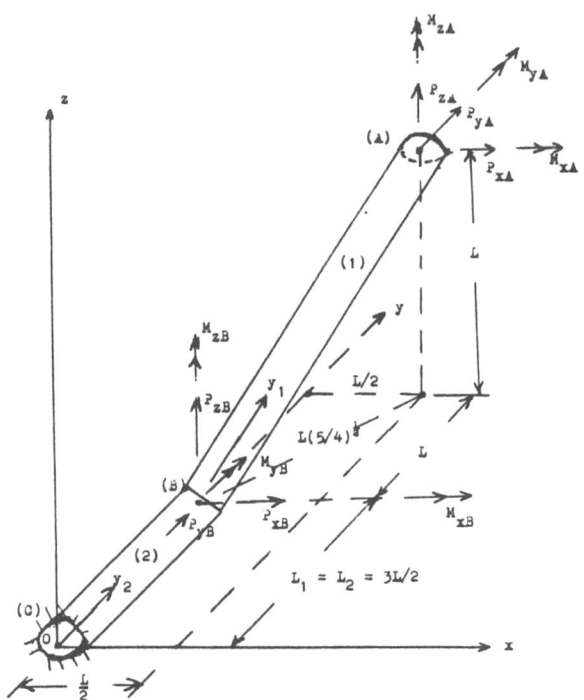

Fig. 9.4. Two element beam with different directions

From Equations (9.10) and (4.115, Vol.1),

$$[C]_d = [p]_d^T [C_F][p]_d,$$

$$[C_F] = \begin{bmatrix} C_{11,1} & C_{12,1} & 0 & 0 \\ C_{12,1}^T & C_{22,1} & 0 & 0 \\ 0 & 0 & C_{11,2} & C_{12,2} \\ 0 & 0 & C_{12,2}^T & C_{22,2} \end{bmatrix}.$$

The form of $[C]_d$ to calculate the deflections at points (A) and (B) relative to the fixed origin at (C) is

$$\begin{Bmatrix} u_A \\ \theta_A \\ u_B \\ \theta_B \end{Bmatrix} = \begin{bmatrix} D_{11} & D_{12} & D_{13} & D_{14} \\ D_{12}^T & D_{22} & C_{12,2}^T & C_{22,2} \\ D_{13}^T & C_{12,2} & C_{11,2} & C_{12,2} \\ D_{14}^T & C_{22,2} & C_{12,2}^T & C_{22,2} \end{bmatrix} \begin{Bmatrix} P_A \\ M_A \\ P_B \\ M_B \end{Bmatrix}, \qquad (9.33)$$

$$\{u_A\}^T = [u_{yA} \quad u_{xA} \quad u_{zA}], \qquad \{P_A\}^T = [P_{yA} \quad P_{xA} \quad P_{zA}], \quad \text{etc.}$$

Omit the shear stiffness so that $H_z = H_x = 1$, whence

$$[D_{14}]^T = [C_{12,2}]^T + [C_{22,2}][K]$$

$$= (L_1^2/3E) \begin{bmatrix} 0 & -E/GJ & 2E/GJ \\ 1/I_z & 0 & -7/2I_z \\ -2/I_z & 7/2I_z & 0 \end{bmatrix}, \quad L_1 = 3L/2,$$

$$[D_{13}]^T = [C_{11,2}] + [C_{12,2}][K]$$

$$= (L_1^3/3E) \begin{bmatrix} 3/AL_1^2 & 0 & 0 \\ -1/I_z & 2/I_z & 0 \\ -1/2I_x & 0 & 2/I_x \end{bmatrix},$$

$$[D_{12}]^T = [R_1]^T[C_{12,1}]^T[R_1] + [D_{14}]^T$$

$$= (L_1^2/3E) \begin{bmatrix} \frac{1}{5}\left(\frac{1}{I_x} - \frac{1}{I_z}\right) & -\frac{E}{GJ} - \frac{1}{2I_z} & \frac{2E}{GJ} + \frac{4}{5I_z} + \frac{1}{5I_x} \\ \frac{3}{2I_z} & 0 & -\frac{9}{2I_z} \\ -\frac{1}{5I_x} - \frac{14}{5I_z} & \frac{9}{2I_z} & \frac{2}{5}\left(\frac{1}{I_z} - \frac{1}{I_x}\right) \end{bmatrix},$$

$$[D_{11}] = [R_1]^T[C_{11,1}][R_1] + [D_{13}]^T + [K]^T[D_{14}]^T$$

$$= \frac{L_1^3}{27E} \begin{bmatrix} \frac{39}{AL_1^2} + \frac{76}{5I_z} + \frac{24}{5I_x} & \frac{12}{AL_1^2} - \frac{25}{I_z} & \frac{6}{AL_1^2} + \frac{8}{5I_x} - \frac{141}{10I_z} \\ \frac{12}{AL_1^2} - \frac{25}{I_z} & \frac{12}{AL_1^2} + \frac{44}{I_z} + \frac{3E}{GJ} & \frac{6}{AL_1^2} - \frac{2}{I_z} - \frac{6E}{GJ} \\ \frac{6}{AL_1^2} + \frac{8}{5I_x} - \frac{141}{10I_z} & \frac{6}{AL_1^2} - \frac{2}{I_z} - \frac{6E}{GJ} & \frac{3}{AL_1^2} + \frac{4}{5I_x} + \frac{221}{5I_z} + \frac{12E}{GJ} \end{bmatrix},$$

$$[D_{22}] = [R_1]^T[C_{22,1}][R_1] + [C_{22,2}]$$

$$= \frac{L_1}{9E} \begin{bmatrix} \frac{13E}{GJ} + \frac{9}{5I_z} + \frac{16}{5I_x} & \frac{4E}{GJ} - \frac{4}{I_z} & \frac{2E}{GJ} - \frac{18}{5I_z} + \frac{8}{5I_x} \\ \frac{4E}{GJ} - \frac{4}{I_z} & \frac{4E}{GJ} + \frac{14}{I_z} & \frac{2E}{GJ} - \frac{2}{I_z} \\ \frac{2E}{GJ} - \frac{18}{5I_z} + \frac{8}{5I_x} & \frac{2E}{GJ} - \frac{2}{I_z} & \frac{E}{GJ} + \frac{36}{5I_z} + \frac{4}{5I_x} \end{bmatrix}.$$

The internal shear forces and moments at points (A), (B), and (C) are

$$\begin{aligned} &\{S\} = [p]\{P\}, \quad \{S_{PA}\} = [R_1]\{P_A\}, \quad \{S_{MA}\} = [R_1]\{M_A\}, \\ &\{S_{PB}\} = \{P_A\} + \{P_B\} = -\{S_{PC}\}, \\ &\{S_{MB}\} = \{M_A\} + \{M_B\} + [K]\{P_A\}, \\ &\{S_{MC}\} = -\{S_{MB}\} - L_1\{P_A + P_B\}. \end{aligned} \tag{9.34}$$

(b) Any additional supports with zero displacements in Figure 9.4 can be put into Equation (9.33) to determine the corresponding reactions. If the $\{u_A\} = 0$, then the $\{P_A\}$ become reactions, or

$$\{P_{AR}\} = -[D_{11}]^{-1}\{D_{12}M_A + D_{13}P_B + D_{14}M_B\}. \tag{9.35}$$

This $\{P_{AR}\}$ can be used directly in Equation (9.33) to get $\{\theta_A\}$, $\{u_B\}$, $\{\theta_B\}$. It is evident that numerical values are necessary to invert $[D_{11}]$. Equation (9.34) still applies for the internal forces provided the $\{P_{AR}\}$ loads in Equation (9.35) are used as the applied forces for $\{P_A\}$.

9.4. Triangular plate elements with inplane forces

The general matrix equations for plate elements with inplane forces have been derived in Sections 8.4, 8.5, and 8.6. Also, in those sections examples for special triangles were given, in particular for the right triangle in Figure 8.5. Here, the $[k]$, $[C_P]$, $[M]$, $[s]$, and $[\lambda_k]$ matrices for the general triangular element in Figure 9.5 will be determined. Only the case of linear displacements and constant stresses will be considered for $[k]$ and $[C_P]$, while both linear displacements and linear stresses will be user for $[M]$.

The [k] and $[\lambda_k]$ matrices

Although the inplane $[k]$ matrix for constant stresses can be calculated from Equation (8.226), it is simpler to calculate it directly from Equation (8.30) for

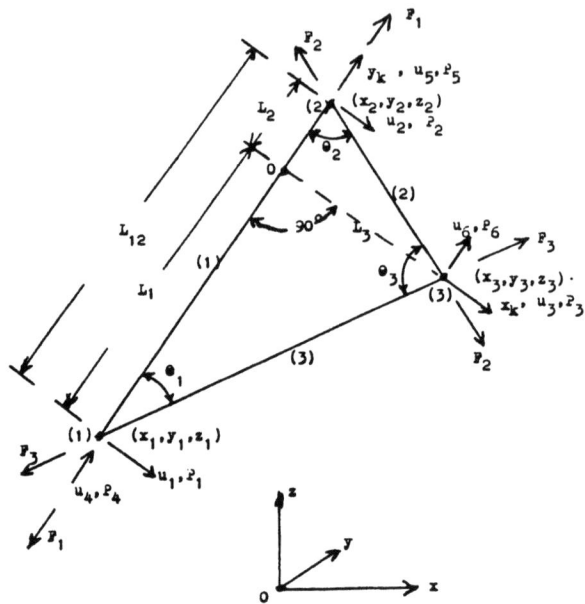

Fig. 9.5. Local and datum coordinates for triangular plate elements.

the local coordinates in Figure 9.5. The $[H_c]$ matrix in Equation (8.26) is

$$[H_c] = \begin{bmatrix} 1 & 0 & -L_1 \\ 1 & 0 & L_2 \\ 1 & L_3 & 0 \end{bmatrix}, \tag{9.36}$$

whence $[H_c]^{-1}$ and $[D]$ in Equation (8.28) are

$$[H_c]^{-1} = \frac{1}{L_3 L_{12}} \begin{bmatrix} L_2 L_3 & L_1 L_3 & 0 \\ -L_2 & -L_1 & L_{12} \\ -L_3 & L_3 & 0 \end{bmatrix}, \tag{9.37}$$

$$[D] = \frac{1}{L_3 L_{12}} \begin{bmatrix} -L_2 & -L_1 & L_{12} & 0 & 0 & 0 \\ 0 & 0 & 0 & -L_3 & L_3 & 0 \\ -L_3 & L_3 & 0 & -L_2 & -L_1 & L_{12} \end{bmatrix}. \tag{9.38}$$

Put this $[D]$ into Equation (8.30) to get

$$[k_k] = \frac{Et}{2(1-\nu^2)L_3 L_{12}} \begin{bmatrix} k_{11} & k_{12} \\ k_{12}^T & k_{22} \end{bmatrix}, \tag{9.39}$$

$$[k_{11}] = \begin{bmatrix} L_2^2 + aL_3^2 & L_1 L_2 - aL_3^2 & -L_2 L_{12} \\ L_1 L_2 - aL_3^2 & L_1^2 + aL_3^2 & -L_1 L_{12} \\ -L_2 L_{12} & -L_1 L_{12} & L_{12}^2 \end{bmatrix}, \quad a = \frac{1-\nu}{2},$$

$$[k_{12}] = \begin{bmatrix} \frac{1+\nu}{2}L_3L_2 & -\nu L_2 L_3 + aL_1 L_3 & -aL_3 L_{12} \\ \nu L_1 L_3 - aL_2 L_3 & -\frac{1+\nu}{2}L_1 L_3 & aL_3 L_{12} \\ -\nu L_3 L_{12} & \nu L_3 L_{12} & 0 \end{bmatrix},$$

$$[k_{22}] = \begin{bmatrix} L_3^2 + aL_2^2 & -L_3^2 + aL_1 L_2 & -aL_2 L_{12} \\ -L_3^2 + aL_1 L_2 & L_3^2 + aL_1^2 & -aL_1 L_{12} \\ -aL_2 L_{12} & -aL_1 L_{12} & aL_{12}^2 \end{bmatrix}.$$

Note that by factoring out a term such as L_3^2, these matrices can be expressed in terms of trigonometric functions of the interior angles θ_1, θ_2, θ_3 in Figure 9.5.

In general applications, it is necessary to determine a $[\lambda_k]$ matrix to put the plate element into a three-dimensional datum system. The simplest way to do this is to select the local coordinates x_k, y_k as shown in Figure 9.5. In this case the direction cosines for the y_k-axis on edge (1) are

$$\begin{aligned} l_{yk} &= \cos(y_k, x) = (x_2 - x_1)/L_{12} = x_{21}/L_{12}, \\ m_{yk} &= \cos(y_k, y) = (y_2 - y_1)/L_{12} = y_{21}/L_{12}, \\ n_{yk} &= \cos(y_k, z) = (z_2 - z_1)/L_{12} = z_{21}/L_{12}, \\ L_{12}^2 &= x_{21}^2 + y_{21}^2 + z_{21}^2. \end{aligned} \quad (9.40)$$

The datum coordinates of the local origin 0 in Figure 9.5 are

$$\begin{aligned} & x_1 + l_{yk}L_1, \quad y_1 + m_{yk}L_1, \quad z_1 + n_{yk}L_1, \\ & L_1 = l_{yk}x_{31} + m_{yk}y_{31} + n_{yk}z_{31}, \end{aligned} \quad (9.41)$$

where L_1 is the distance from point (1) to the local origin of x_k, y_k in Figure 9.5. Thus, the direction cosines for the x_k-axis are

$$\begin{aligned} l_{xk} &= \cos(x_k, x) = \frac{x_3 - x_1 - l_{yk}L_1}{L_3} = \frac{x_{31} - l_{yk}L_1}{L_3}, \\ m_{xk} &= \cos(x_k, y) = \frac{y_{31} - m_{yk}L_1}{L_3}, \\ n_{xk} &= \cos(x_k, z) = \frac{z_{31} - n_{yk}L_1}{L_3}, \\ L_3^2 &= x_{31}^2 + y_{31}^2 + z_{31}^2 - L_1^2 = L_{31}^2 - L_1^2. \end{aligned} \quad (9.42)$$

At the corner (1) in Figure 9.5,

$$\left\{\begin{matrix} u_1 \\ u_4 \end{matrix}\right\}_k = \begin{bmatrix} R_{xk} \\ R_{yk} \end{bmatrix} \left\{\begin{matrix} u_{1x} \\ u_{1y} \\ u_{1z} \end{matrix}\right\}, \quad \begin{bmatrix} R_{xk} \\ R_{yk} \end{bmatrix} = \begin{bmatrix} l_{xk} & m_{xk} & n_{xk} \\ l_{yk} & m_{yk} & n_{yk} \end{bmatrix}, \quad (9.43)$$

and for all three corners in the order $u_1, u_2, \cdots, u_6$,

$$\{u\}_k = [\lambda_k]\{u\}_d, \quad (9.44)$$

$$[\lambda_k] = \begin{bmatrix} R_{xk} & 0 & 0 \\ 0 & R_{xk} & 0 \\ 0 & 0 & R_{xk} \\ R_{yk} & 0 & 0 \\ 0 & R_{yk} & 0 \\ 0 & 0 & R_{yk} \end{bmatrix} = \begin{bmatrix} R_x \\ R_y \end{bmatrix}_k,$$

$$\{u\}_d^T = [u_{1x} \quad u_{1y} \quad u_{1z} \quad u_{2x} \quad u_{2y} \quad u_{2z} \quad u_{3x} \quad u_{3y} \quad u_{3z}]_d, \tag{9.45}$$

where subscript d indicates datum system. From Equation (9.23), $[k]$ in datum coordinates is

$$[k] = [\lambda_k]^T [k_k][\lambda_k], \tag{9.46}$$

where $[k_k]$ is in Equation (9.39). Also,

$$\{P + P_E\}_d = [\lambda_k]\{P + P_E\}_k, \tag{9.47}$$

where $\{P + P_E\}_k$ is given in Equation (8.30).

If all the elements are in the same plane then a simple rotation θ_k exists between the x_k, y_k and x, y coordinates, and $[\lambda_k]$ is much simpler, or

$$[R_{xk}] = [\cos\theta_k \quad \sin\theta_k], \quad [R_{yk}] = [-\sin\theta_k \quad \cos\theta_k], \tag{9.48}$$

$$\{u\}_d^T = [u_{1x} \quad u_{1y} \quad u_{2x} \quad u_{2y} \quad u_{3x} \quad u_{3y}]. \tag{9.49}$$

In this case $[\lambda_k]$ in Equation (9.44) is 6 by 6 instead of 6 by 9 and $[k]$ in Equation (9.46) is 6 by 6 instead of 9 by 9.

The $[s]$ and $[C_P]$ Matrices

The equilibrium matrix $[s_k]^T$ between the P_i and F_i forces in the local coordinates of Figure 9.5 is given by $[s]_p$ in Equation (8.216), or

$$[s_k] = \begin{bmatrix} 0 & 0 & 0 & | & -1 & 1 & 0 \\ 0 & -\sin\theta_2 & \sin\theta_2 & | & 0 & \cos\theta_2 & -\cos\theta_2 \\ -\sin\theta_1 & 0 & \sin\theta_1 & | & -\cos\theta_1 & 0 & \cos\theta_1 \end{bmatrix}$$

$$= [S_x \quad S_y]_k. \tag{9.50}$$

In the datum system

$$[s] = [s_k][\lambda_k] = [S_x R_x + S_y R_y]_k, \tag{9.51}$$

whence the datum equilibrium equations are

$$\{P\}_d = \{P_g\}_d + [\lambda_k]^T \{P_a\}_k = [s]^T \{F\}, \tag{9.52}$$

where $\{P_g\}_d$ are any specified point component forces in the datum system and $\{P_a\}_k$ are element statically equivalent point applied forces and body forces in the local system. $\{P_a\}_k$ can be taken as $\{P\}$ in Equation (8.30).

As described in Section 4.9 (Vol.1), the assembly using the principle of virtual forces is made on Equation (9.52) to determine auxiliary matrices, such as $[p]$ and $[q]$ in Equations (4.106)-(4.112) (Vol.1), which are then used directly with $[C_P]_k$

and $\{u_E\}_k$ to determine $\{F\}$. If the system is determinate, then the assembled $[s]$ is square and $\{F\}$ is given by the assembled Equation (9.52), or

$$\{F\} = [[s]^T]^{-1}. \tag{9.53}$$

The $[C_P]_k$ matrix for the triangular element in Figure 9.5 with constant stresses is given in Equation (8.57). If the $\{u_E\}_k$ strains are constant, then for Figure 9.5 with $[H]_k$ in Equation (8.55),

$$\{u_E\}_k = \frac{1}{tA_k} \int_{A_m} t \begin{bmatrix} 0 & L_{12} & 0 \\ L_{23}\sin^2\theta_2 & L_{23}\cos^2\theta_2 & L_{23}\sin\theta_2\cos\theta_2 \\ L_{31}\sin^2\theta_1 & L_{31}\cos^2\theta_1 & -L_{31}\sin\theta_1\cos\theta_1 \end{bmatrix} \begin{Bmatrix} e_{xxE} \\ e_{yyE} \\ e_{xyE} \end{Bmatrix}_k dA_m$$

$$= \begin{bmatrix} L_{12}(e_{yyE})_k & L_{23}(e_E)_{23} & L_{31}(e_E)_{31} \end{bmatrix}_k^T, \tag{9.54}$$

$$(e_E)_{23k} = \left(e_{xxE}\sin^2\theta_2 + e_{yyE}\cos^2\theta_2 + e_{xyE}\sin\theta_2\cos\theta_2\right)_k,$$

$$(e_E)_{31k} = \left(e_{xxE}\sin^2\theta_1 + e_{yyE}\cos^2\theta_1 - e_{xyE}\sin\theta_1\cos\theta_1\right)_k.$$

The [M] matrix

For the exact elements in Sections 9.2 and 9.3, the $[M]$ matrix was given directly by the $[C_P]$ and $[s]$ matrices. However, for the approximate elements the $[M]$ matrix depends upon the separate approximations made for the stresses and displacements so that the $[M_{11}]$ and $[M_{12}]$ may be different from $[C_P]$ and $[s]$. Both the displacements and stresses will be assumed to be linear and the coordinate systems of Figure 9.5 will be used. This linear case was considered in Example (8.10) for the particular element in Figure 8.5 so that the equations given there will be modified to fit the general case of Figure 9.5.

The following integrals can be used to make the integrations for the triangle in Figure 9.5,

$$\int_{A_p} dy\,dx = (1/2)L_3 L_{12}, \qquad \int_{A_p} x\,dy\,dx = (1/6)L_3^2 L_{12},$$

$$\int_{A_p} y\,dx\,dy = (1/6)L_3(L_1^2 + L_2^2),$$

$$\int_{A_p} x^2 dy\,dx = (1/12)L_3^3 L_{12}, \tag{9.55a}$$

$$\int_{A_p} y^2 dy\,dx = (1/12)L_3(L_1^3 + L_2^3),$$

$$\int_{A_p} xy\,dy\,dx = (1/24)L_3^2(L_1^2 + L_2^2).$$

Use the integrals in Equation (9.55a) and the $[H_c]^{-1}$ matrix in Equation (9.37) so that expressions in the $[M_{11}]$ and $[M_{12}]$ matrices in Equation (8.111) become

$$[B_1] = \frac{L_3}{24L_{12}^2} \begin{bmatrix} 2(L_{12}^3 - 6L_1^2 L_2) & L_{12}^3 - 6L_1^3 + 6L_1^2 L_2 & L_{12}(L_{12}^2 - 2L_1^2) \\ L_{12}^3 - 6L_1^3 + 6L_1^2 L_2 & 2(L_{12}^3 + 6L_1^3) & L_{12}(L_{12}^2 + 2L_1^2) \\ L_{12}(L_{12}^2 - 2L_1^2) & L_{12}(L_{12}^2 + 2L_1^2) & 2L_{12}^3 \end{bmatrix},$$

$$[B_2] = \frac{1}{6L_{12}^2} \begin{bmatrix} L_2(2L_1^2 - L_{12}^2) & L_1(2L_1^2 - L_{12}^2) & -L_{12}(2L_1^2 - L_{12}^2) \\ -L_2(2L_1^2 + L_{12}^2) & -L_1(2L_1^2 + L_{12}^2) & L_{12}(2L_1^2 + L_{12}^2) \\ L_2 L_{12}^2 & -L_1 L_{12}^2 & L_{12}^3 \end{bmatrix}, \quad (9.55b)$$

$$[B_3] = \frac{L_3}{6L_{12}^2} \begin{bmatrix} 2L_1^2 - L_{12}^2 & -2L_1^2 + L_{12}^2 & 0 \\ -2L_1^2 - L_{12}^2 & 2L_1^2 + L_{12}^2 & 0 \\ -L_{12}^2 & L_{12}^2 & 0 \end{bmatrix}.$$

With these values, the $[M]$ matrix has the form in Equation (8.113) with local coordinates for the triangular element in Figure (9.5). See Reference 15(b) for application of this mixed matrix case with linear stresses to some plate problems.

The $[M_{11}]$ and $[M_{12}]$ matrices for the case of linear displacements and constant stresses, as described in connection with Equation (8.116), can be obtained from Equations (8.217), (8.207), (8.118b), (9.37), and 8.205) with $[G_m] = [Z]$. The results in local coordinates for the general triangle of Figure 9.5 are

$$[M_{11}] = \frac{L_3 L_{12}}{4Et} \begin{bmatrix} a_3^2 b_3 & c_2 \csc^2 \theta_2 & c_1 \csc^2 \theta_1 \\ c_2 \csc^2 \theta_2 & a_1^2 b_1 & c_3 \csc^2 \theta_3 \\ c_1 \csc^2 \theta_1 & c_3 \csc^2 \theta_3 & a_2^2 b_2 \end{bmatrix},$$

$$[M_{12}] = \frac{L_3}{4} \begin{bmatrix} -a_3 & -a_3 & 2a_3 & a_3 \cot \theta_2 & -a_3 \cot \theta_1 & a_3(\cot \theta_1 - \cot \theta_2) \\ -a_1 & a_1 & 0 & -a_1(a_3 + \cot \theta_1) & a_1 \cot \theta_1 & a_1 a_3 \\ a_2 & -a_2 & 0 & -a_2 \cot \theta_2 & a_2(a_3 + \cot \theta_2) & -a_2 a_3 \end{bmatrix}, \quad (9.56)$$

where the a_i, b_i, c_i constants are defined in Equations (8.206), (8.208).

In datum coordinates assume $z = z_k$ and consider rotation in the x-y plane from local axes x_k, y_k in Figure 9.5 to the datum x, y axes. The direction cosine matrices for both the displacement and stress rotations are

$$[\lambda_k] = \begin{bmatrix} r_1 & r_2 \\ -r_2 & r_1 \end{bmatrix}, \quad [R_k] = \begin{bmatrix} R_{11} & R_{12} & 2R_{13} \\ R_{12} & R_{11} & -2R_{13} \\ -R_{13} & R_{13} & R_{11} - R_{12} \end{bmatrix}, \quad (9.57)$$

with r_i and R_{ij} 3 by 3 diagonal matrices,

$$[r_1] = [l_{xk} \quad l_{xk} \quad l_{xk}]_d, \quad [r_2] = [m_{xk} \quad m_{xk} \quad m_{xk}]_d,$$
$$[R_{11}] = [l_{xk}^2 \quad l_{xk}^2 \quad l_{xk}^2]_d, \quad [R_{12}] = [m_{xk}^2 \quad m_{xk}^2 \quad m_{xk}^2]_d,$$
$$[R_{13}] = [l_{xk} m_{xk} \quad l_{xk} m_{xk} \quad l_{xk} m_{xk}]_d,$$
$$\{u\}^T = [u_{1x} \quad u_{2x} \quad u_{3x} \quad u_{1y} \quad u_{2y} \quad u_{3y}],$$
$$\{N\}^T = [N_{x1} \quad N_{x2} \quad N_{x3} \quad N_{y1} \quad N_{y2} \quad N_{y3} \quad N_{xy1} \quad N_{xy2} \quad N_{xy3}],$$

Matrix structural analysis using finite elements 321

where Equations (A.105) and (A.107) (Vol.1) have been used. Thus, the $[M]$ matrix in Equation (8.113) can be expressed in datum coordinates as

$$[M]_d = \begin{bmatrix} R_k^T & 0 \\ 0 & \lambda_k^T \end{bmatrix} \begin{bmatrix} -M_{11} & M_{12} \\ M_{12}^T & 0 \end{bmatrix} \begin{bmatrix} R_k & 0 \\ 0 & \lambda_k \end{bmatrix}$$
$$= \begin{bmatrix} -R_k^T M_{11k} R_k & R_K^T M_{12k} \lambda_k \\ \lambda_k^T M_{12k}^T R_k & 0 \end{bmatrix}, \tag{9.58}$$

where the $[B_1]$, $[B_2]$, and $[B_3]$ matrices in $[M]_k$ are given in Equation (9.55b).

For the constant stress case with the three normal stresses on the edges, the $[\lambda_k]$ is the same as in Equation (9.57) but no transformation is needed on the normal stresses, which must be continuous between the two edges and at 180° to each other. Thus,

$$[M]_d = \begin{bmatrix} -M_{11k} & M_{12k} \lambda_k \\ \lambda_k^T M_{12k}^T & 0 \end{bmatrix}, \tag{9.59}$$

where $[M_{11k}]$ and $[M_{12k}]$ are given by Equation (9.56).

Example 9.3. A triangular plate element has corners in datum coordinates at $(x_i, y_i, z_i) = (1,1,1), (2,3,3), (3,1,2)$. Calculate the data needed to evaluate the various matrices in this Section 9.4.

Solution. From Equations (9.40)-(9.42)

$x_{21} = 2 - 1 = 1,$ $y_{21} = 3 - 1 = 2,$ $z_{21} = 3 - 1 = 2,$
$x_{31} = 3 - 1 = 2,$ $y_{31} = 1 - 1 = 0,$ $z_{31} = 2 - 1 = 1,$
$L_{12}^2 = (1)^2 + (2)^2 + (2)^2 = 9,$ $L_{12} = 3,$
$L_{23}^2 = (1)^2 + (2)^2 + (1)^2 = 6,$ $L_{23} = 6^{1/2},$
$L_{31}^2 = (2)^2 + (0)^2 + (1)^2 = 5,$ $L_{31} = 5^{1/2},$
$l_{yk} = 1/3,$ $m_{yk} = 2/3,$ $n_{yk} = 2/3,$
$L_1 = [(2)(1) + (0)(2) + (1)(2)]/3 = 4/3,$
$L_3^2 = 5 - (4/3)^2 = 29/9,$ $L_3 = (1/3)(29)^{1/2},$
$L_2 = L_{12} - L_1 = 5/3,$
$l_{xk} = (196/261)^{1/2},$ $m_{xk} = -(64/261)^{1/2},$ $n_{xk} = (1/261)^{1/2}.$

Put these values into Equations (9.43), (9.44), (9.39), and (9.46) to get the $[\lambda_k]$, $[k_k]$, and $[k]$ matrices.

From Figure 9.5,

$\sin\theta_1 = L_3/L_{31} = (29/45)^{1/2},$ $\cos\theta_1 = (16/45)^{1/2},$
$\sin\theta_2 = L_3/L_{23} = (29/54)^{1/2},$ $\cos\theta_2 = (25/54)^{1/2},$
$\sin\theta_3 = \sin(\theta_1 + \theta_2) = (29/30)^{1/2},$ $\cos\theta_3 = (1/30)^{1/2}.$

Put these values into Equations (9.50), (9.51), and (9.54) to get $[s_k]$, $[s]$, and $\{u_E\}_k$.

The above values can be used to evaluate $[B_1], [B_2], [B_3], [\lambda_k],$ and $[R_k]$ in Equations (9.55) and (9.57). In Equation (9.56) for the $[M]$ matrix

$\cot\theta_1 = (16/29)^{1/2},$ $\cot\theta_2 = (25/29)^{1/2},$ $\cot\theta_3 = (1/29)^{1/2},$
$a_1 = (36/29)^{1/2},$ $a_2 = (25/29)^{1/2},$ $a_3 = (81/29)^{1/2},$
$b_1 = \nu + (61/29),$ $b_2 = \nu + (79/29),$ $b_3 = \nu + (30/29),$
$c_1 = -\nu - (19/29),$ $c_2 = -\nu - (21/29),$ $c_3 = -\nu + (9/29).$

9.5. Assembly of finite elements by the virtual principles

In Chapter 8 and the previous sections of this chapter, the virtual principles were used to obtain equations for the individual plate elements. Since the virtual principles apply to the entire structure, they can be also used directly to assemble the components of the structure. To make the assembly it is necesssary to put all the elements into a datum coordinate system. The conversion of the elements from local coordinates to datum coordinates has been given for several elements in the previous sections of this chapter. Also, the assembly procedures were demonstrated for simple cases in Chapters 3,4, and 6 (Vol.1).

The conversion equations for the real expressions also apply to the virtual expressions. For example,

$$\{u\}_k = [\lambda]_k \{u\}_d, \quad \{u^V\}_k = [\lambda]_k \{u^V\}_d,$$
$$\{u^V\}_k^T = \{u^V\}_d^T [\lambda]_k^T. \tag{9.60}$$

Since the $[\lambda]_k^T$ term has been included in the transformation of the real expressions, such as in Equation (9.22), the $\{u^V\}_d^T$ term applies directly, such as in Equation (8.29) for assembly of the elements.

Assembly by the Principle of Virtual Displacements

Assume that Equation (8.30) has been changed to datum coordinates, Equations (9.41)-(9.47), then from Equation (8.29),

$$\sum_{k=1}^{N} (\{u^V\}_d^T \{[k]\{u\} - \{P + P_E\}\}_d)_k = 0, \tag{9.61}$$

where the sum is over all the N elements. Each u_{id}^V may occur several times in the sum so that terms on each arbitrary u_{id}^V can be collected into an equation. Thus, the total number of separate equations for the structure is equal to the number of datum point deflections.

For example, in Figure 9.6 there are twelve node points with two deflection components or 24 datum deflections, or 24 equations for the structure. However, each of the twelve triangular elements has three node points, or six datum deflections, or six equations for the elements alone. That is, the six equations for each element do not add to $(6)(12) = 72$ equations for the entire plate. In Figure 9.6 the assembled matrix has 24 equations, or

$$([k]\{u\} - \{P + P_E\})_a = 0, \tag{9.62}$$

with $[k]_a$ 24 by 24.

Take $\{u\}_a$ in Equation (9.62) for Figure 9.6 in the order

$$\{u\}_a^T = \begin{bmatrix} \overset{1}{u_{1x}} & \overset{2}{u_{1y}} & \overset{3}{u_{2x}} & \overset{4}{u_{2y}} & \cdots & \overset{23}{u_{12x}} & \overset{24}{u_{12y}} \end{bmatrix}, \tag{9.63}$$

whence the $[k]_a$ and $\{P + P_E\}_a$ matrices can be assembled step by step by adding the elements in sequence in the proper locations. The 6 by 6 $[k]_{d1}$ datum

Matrix structural analysis using finite elements 323

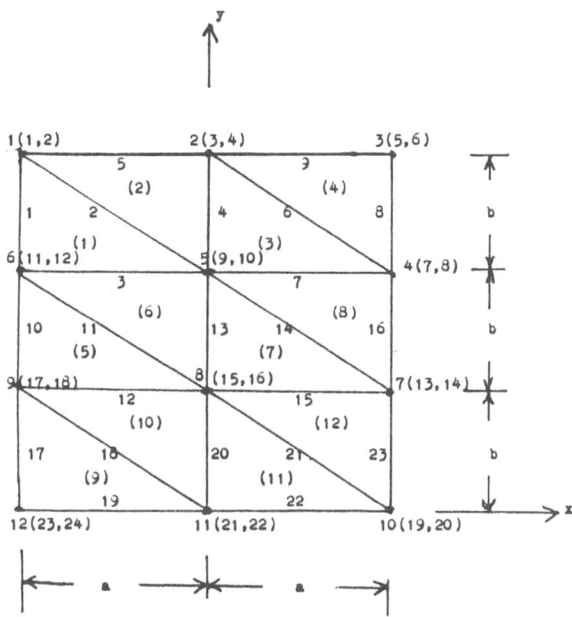

Fig. 9.6. Plate divided into triangular elements (inplane forces) with constant thickness t and one material.

matrix for element (1) in Figure 9.6 puts values into $[k]_a$ at locations (1,1), (1,2), (1,9), (1,10), (1,11), (1,12), (2,1), (2,2), (2,9), (2,10), (2,11), (2,12), (9,1), (9,2), (9,9), (9,10), (9,11), (9,12), (10,1), (10,2), (10,9), (10,10), (10,11), (10,12), (11,1), (11,2), (11,9), (11,10), (11,11), (11,12), (12,1), (12,2), (12,9), (12,10), (12,11), (12,12). Values from $[k]_{d2}$ for element (2) will add to the 16 element locations involving points 1,2,9,10 and also put values at locations (1,3), (1,4), (2,3), (2,4), (3,9), (3,10), (4,1), (4,2), (4,3), (4,4), (4,9), (4,10), (9,3), (9,4), (10,3), (10,4), (3,1), (3,2), (3,3), (3,4).

Once the equations have been assembled for the entire structure, it is necessary to impose the displacement boundary conditions. As demonstrated in Chapters 3 and 4 (Vol.1), this can be done by partitioning the assembled set of Equations (9.62) as

$$\begin{Bmatrix} P_u\; P_{Eu} \\ P_g\; P_{Eg} \end{Bmatrix} = \begin{bmatrix} k_{uu} & k_{ug} \\ k_{gu} & k_{gg} \end{bmatrix} \begin{Bmatrix} u_u \\ u_g \end{Bmatrix}, \tag{9.64}$$

where $\{u_g\}$ are given displacements and $\{P_g\}$ are reactions. Thus,

$$\begin{aligned}\{u_u\} &= [k_{uu}]^{-1}\{\{P_u + P_{Eu}\} - [k_{ug}]\{u_g\}\}, \\ \{P_g\} &= [k_{gu}]\{u_u\} + [k_{gg}]\{u_g\} - \{P_{Eg}\}.\end{aligned} \tag{9.65}$$

Once $\{u_u\}$ is known, the stresses can be obtained from the applicable stress-displacement equations.

Example 9.4. Construct row 1 or column 1 for the $[k]_a$ matrix for Figure 9.6 with $a = b$. Use the element $[k_k]$ in Equation (9.39) and a two-dimensional $[\lambda_k]$ from Equation (9.44).

Solution. Column 1 of $[k]_a$ operating on u_{1x} involves only elements (1) and (2) in Figure 9.6. Take local coordinates for element (1) at node point 6 with $x_k = x$, $y_k = y$. From Figure 9.5, $L_2 = b$, $L_3 = b$, $L_{12} = b$, $L_1 = 0$. Also, $[\lambda_1] = [I]$. Thus,

$$[k_1]_d = [k_1] = (Et/2b^2(1-\nu^2))\begin{bmatrix} k_{11} & k_{12} \\ k_{12}^T & k_{22} \end{bmatrix}, \tag{9.66}$$

$$[k_{11}] = \frac{b^2}{2} \begin{array}{c} \\ \\ \end{array} \begin{bmatrix} \overset{11}{3-\nu} & \overset{1}{-(1-\nu)} & \overset{9}{-2} \\ -(1-\nu) & 1-\nu & 0 \\ -2 & 0 & 2 \end{bmatrix} \begin{array}{c} 11 \\ 1, \\ 9 \end{array}$$

$$[k_{12}] = \frac{b^2}{2} \begin{bmatrix} \overset{12}{1+\nu} & \overset{2}{-2\nu} & \overset{10}{-(1-\nu)} \\ -(1-\nu) & 0 & 1-\nu \\ -2\nu & 2\nu & 0 \end{bmatrix} \begin{array}{c} 11 \\ 1, \\ 9 \end{array}$$

$$[k_{22}] = \frac{b^2}{2} \begin{bmatrix} \overset{12}{3-\nu} & \overset{2}{-2} & \overset{10}{-(1-\nu)} \\ -2 & 2 & 0 \\ -(1-\nu) & 0 & 1-\nu \end{bmatrix} \begin{array}{c} 12 \\ 2, \\ 10 \end{array}$$

whence row 1 in $[k]_a$ due to element (1) is

$$[k_{1j}]_{a1} = \frac{Et}{4(1-\nu^2)} \begin{bmatrix} \overset{1}{(1-\nu)} & \overset{2}{0} & \cdots & \overset{9}{0} & \overset{10}{1-\nu} & \overset{11}{-(1-\nu)} & \overset{12}{-(1-\nu)} & \overset{13}{0} & \cdots & \overset{24}{0} \end{bmatrix}. \tag{9.67}$$

Take local coordinates for element (2) at node 2 with $x_2 = -y$, $y_2 = x$, whence from Figure 9.5, $L_1 = L_{12} = b$, $L_3 = b$, $L_2 = 0$. In this case

$$[\lambda_2] = \begin{bmatrix} 0 & -1 & 0 & 0 & 0 & 0 \\ 0 & 0 & 0 & -1 & 0 & 0 \\ 0 & 0 & 0 & 0 & 0 & -1 \\ 1 & 0 & 0 & 0 & 0 & 0 \\ 0 & 0 & 1 & 0 & 0 & 0 \\ 0 & 0 & 0 & 0 & 1 & 0 \end{bmatrix}, \tag{9.68}$$

whence Equations (9.39) and (9.46) give

$$[k_2]_d = [\lambda_2]^T [k_2][\lambda_2]$$

$$= \frac{Et}{4(1-\nu^2)} \begin{bmatrix} \overset{1}{2} & \overset{2}{0} & \overset{3}{-2} & \overset{4}{-2\nu} & \overset{9}{0} & \overset{10}{2\nu} \\ 0 & 1-\nu & -(1-\nu) & -(1-\nu) & 1-\nu & 0 \\ -2 & -(1-\nu) & 2 & 1+\nu & -(1-\nu) & -2\nu \\ -2\nu & -(1-\nu) & 1+\nu & 3-\nu & -(1-\nu) & -2 \\ 0 & 1-\nu & -(1-\nu) & -(1-\nu) & 1-\nu & 0 \\ 2\nu & 0 & -2\nu & -2 & 0 & 2 \end{bmatrix}. \tag{9.69}$$

Thus, from Equation (9.67) and the first row of Equation (9.69)

$$[k_{1j}]_a = \frac{Et}{4(1-\nu^2)} \begin{bmatrix} \overset{1}{3-\nu} & \overset{2}{0} & \overset{3}{-2} & \overset{4}{-2\nu} & \overset{5}{0} & \cdots & \overset{9}{0} & \overset{10}{1+\nu} & \overset{11}{-(1-\nu)} \\ & & & & & & \overset{12}{-(1-\nu)} & \overset{13}{0} & \cdots & \overset{24}{0} \end{bmatrix}. \tag{9.70}$$

Assembly by the Principle of Virtual Forces

Write the element equilibrium Equation (9.52) in the virtual displacement form

$$(\{u^V\}_d^T\{\{P\} - [s]^T\{F\}_d\})_k = 0, \tag{9.71}$$

and use the principle of virtual displacements to sum over all elements in the structure as

$$\sum_k (\{u^V\}_d^T\{\{P\} - [s]^T\{F\}\}_d)_k = 0. \tag{9.72}$$

This Equation (9.72) corresponds to Equation (9.61) so that the virtual terms can be used to collect and assemble the matrices in the same manner as described for the principle of virtual displacements above. The assembled matrix equation for the structure is

$$\{P\}_a = [s]_a^T\{F\}. \tag{9.73}$$

For the structure in Figure 9.6, the F_i unknown forces act along the triangular element edges, Figure 9.5, and each internal element edge will have different F_i for the two adjacent elements. Thus, in Figure 9.6, there are 36 F_i and $(24 - N_R)u_i^V$ terms, where N_R is the number of supports. This gives $(12 + N_R)$ redundants.

The Jordan successive transformation procedure in Equation (3.82, Vol.1) can be applied to Equation (9.73) to select the redundants and to generate the $[p]$ and $[q]$ auxiliary matrices in Equation (3.81, Vol.1) for the entire structure. With $[p]$ and $[q]$ known and with $\{Q\}$ as the unknown redundants, Equation (3.66, Vol.1) is

$$\{F\} = [p]_a\{P\}_a + [q]\{Q\}. \tag{9.74}$$

To evaluate $\{Q\}$ use the principle of virtual forces in Equation (8.51). Since there is no overlap between elements, the assembled matrix can be taken as

$$\sum_k \{F^V\}_k^T\{[C_P]\{F\} - \{u - u_E\}\}_k = 0, \tag{9.75}$$

where $[C_P]$ is a block diagonal matrix. Since $\{F\}$ in Equation (9.74) is expressed in the datum system, use

$$\{F^V\}_d^T = \{P^V\}_d^T[p]^T + \{Q^V\}^T[q]^T, \tag{9.76}$$

whence Equation (9.75) becomes

$$\begin{aligned}\{P^V\}^T[p]^T\{[C_P]\{[p]\{P\} + [q]\{Q\}\} - \{u - u_E\}_r\} &= 0, \\ \{Q^V\}^T[q]^T\{[C_P]\{[p]\{P\} + [q]\{Q\}\} - \{u - u_E\}_r\} &= 0, \end{aligned} \tag{9.77}$$

where $\{u - u_E\}$ are the assembled element relative displacements. For arbitrary $\{P^V\}$ and $\{Q^V\}$, Equation (9.77) becomes

$$\begin{Bmatrix} u \\ 0 \end{Bmatrix} = \begin{Bmatrix} [p]^T\{u\}_r \\ [q]^T\{u\}_r \end{Bmatrix} = \begin{bmatrix} C_{PP} & C_{PQ} \\ C_{QP} & C_{QQ} \end{bmatrix} \begin{Bmatrix} P \\ Q \end{Bmatrix} + \begin{Bmatrix} u_{EP} \\ u_{EQ} \end{Bmatrix}, \tag{9.78}$$

$$[C_{PP}] = [p]^T[C_P][p], \quad [C_{PQ}] = [C_{QP}]^T = [p]^T[C_P][q],$$
$$[C_{QQ}] = [q]^T[C_P][q], \quad \{u_{EP}\} = [p]^T\{u_E\}_r, \quad \{u_{EQ}\} = [q]^T\{u_E\}_r,$$

where $\{u_Q\} = \{0\}$ for compatibility at the redundant forces and $\{u\}$ represents the datum node point displacements. Solve for $\{Q\}$ to get

$$\{Q\} = -[C_{QQ}]^{-1}\{[C_{QP}]\{P\} + \{u_{EQ}\}\}. \tag{9.79}$$

With $\{Q\}$ known the F_i forces are given by Equation (9.74) and the displacements by Equation (9.78). The reactions at the supports are

$$\{R\} = [s]_R^T\{F\}. \tag{9.80}$$

For the case in Figure 9.6, the 3 by 3 $[C_P]_k$ matrices are given by Equation (9.57) so that the assembled $[C_P]$ matrix is 36 by 36. Assume four supports and six applied loads P_i so that there are 16 Q_i redundants. Thus, in Equation (9.78), $[p]$ is 36 by 6, $[q]$ is 36 by 16, $[C_{PP}]$ is 6 by 6, $[C_{PQ}]$ is 6 by 16, $[C_{QQ}]$ is 16 by 16, $\{u\}$ and $\{u_{EP}\}$ are 6 by 1, $\{u_{EQ}\}$ is 16 by 1.

Assembly by the Mixed Virtual Principle

Use Equations (8.109) and (8.110) to get the element virtual form

$$\begin{aligned}\{N_c^V\}_k^T\{-[M_{11}]\{N_c\} + [M_{12}]\{u_c\} - \{u_E\}\}_k &= 0, \\ \{u_c^V\}_k^T\{[M_{12}]^T\{N_c\} - \{P\}\}_k &= 0\end{aligned} \tag{9.81}$$

in local coordinates. Use Equations (9.57) and (9.58) for the datum coordinates, or

$$\begin{aligned}\{N_c\}_k &= [R_k]\{N_c\}_d, \quad \{u_c\}_k = [\lambda_k]\{u_c\}_d, \\ [M_{11}]_d &= [R_k]^T[M_{11}]_k[R_k], \quad [M_{12}]_d = [R_k]^T[M_{12}]_k[R_k], \\ \{u_E\}_d &= [R_k]^T\{u_E\}_k, \quad \{P\}_d = [\lambda_k]^T\{P\}_k,\end{aligned} \tag{9.82}$$

$$\begin{aligned}\{N_c^V\}_d^T\{-[M_{11}]\{N_c\} + [M_{12}]\{u_c\} - \{u_E\}\}_d &= 0, \\ \{u_c^V\}_d^T\{[M_{12}]^T\{N_c\} - \{P\}\}_d &= 0\end{aligned} \tag{9.83}$$

Thus, the assembly is similar to that for the principle of virtual displacements in Equation (9.62), or

$$\begin{bmatrix} -[M_{11}] & [M_{12}] \\ [M_{12}]^T & [0] \end{bmatrix}_a \begin{Bmatrix} N_c \\ u_c \end{Bmatrix} = \begin{Bmatrix} -u_E \\ P \end{Bmatrix}_a. \tag{9.84}$$

If the system of Equations (9.84) is reduced for boundary conditions on stresses and displacements, then it can be solved directly, or by the procedure in Equation (7.78).

For the inplane case in Figure 9.6, $\{N_c\}$ is 36 by 1 and $\{u_c\}$ is 24 by 1. In Figure (9.6) all 14 normal node point stresses on the boundary are zero, except at node points where either N_{un} normal displacements or N_{Pn} normal load components are specified. Thus,

$$\begin{aligned}\{N_c\}_{\text{reduced}} \text{ is } (36 - 14 + N_{un} + N_{Pn}) \text{ by } 1, \\ \{u_c\}_{\text{reduced}} \text{ is } (24 - N_{un} - N_{ut}) \text{ by } 1,\end{aligned} \tag{9.85}$$

where N_{ut} is the number of specified tangential displacements.

References

Chapter 9

1. J.S. Przemieniecki: *Theory of Matrix Structural Analysis*, McGraw-Hill Book Co. (1968).
2. R.H. Gallagher: *Finite Element Analysis, Fundamentals*, Prentice-Hall, Inc. (1975).
3. P. Tong and J.N. Rossettos: *Finite Element Method, Basic Technique and Implementation*, MIT Press (1977).
4. H.C. Martin and G.F. Carey: *Introduction to Finite Element Analysis*, McGraw-Hill Book Co. (1973).
5. O.C. Zienkiewicz: *The Finite Element Method*, Third Edition, McGraw-Hill Book Co. (1977).
6. T.J. Chung: *Solid Mechanics and Finite Elements*, McGraw-Hill Book Co. (1979).
7. J.H. Argyris: *Energy Theorem and Structural Analysis*, Butterworth Scientific Publications, London (1960).
8. J.H. Argyris: *Recent Advances in Matrix Methods of Structural Analysis*, Macmillan Co., New York (1964).
9. C.S. Desai and J.F. Abel: *Introduction to the Finite Element Method*, Van Nostrand Reinhold Co. (1972).
10. D.H. Norrie and G. Devries: *An Introduction to Finite Element Analysis*, Academic Press (1978).
11. G. Strang and G.J. Fix: *An Analysis of the Finite Element Method*, Prentice-Hall, Inc. (1973).
12. T.H.H. Pian: Derivation of element stiffness matrices, *AIAA Journal* **2**, p. 576, March (1964).
13. T.H.H. Pian: Derivation of element stiffness matrices by assumed stress distribution, *AIAA Journal* **2**, p. 1333, July (1964).
14. J.S. Przemieniecki, R.M. Bader, W.F. Bozich, J.R. Johnson and W.J. Mykytow (Editors): *Proceedings of Conference on Matrix Methods in Structural Mechanics*, AFFDL-TR-66-80, 1966. Various papers on finite elements, including:
    (a) T.H.H. Pian: Element stiffness matrices for boundary compatibility and for prescribed boundary stresses, pp. 457-478.
    (b) R.W. Clough and J.L. Tocher: Finite element stiffness matrices for analysis of plate bending, pp. 515-546.
    (c) L.R. Herrmann: A bending analysis of plates, pp. 577-602.
    (d) G.P. Bezeley, Y.K. Chenng, B.M. Irons and O.C. Zienkiewicz: Triangular elements in plate bending: conforming and nonconforming solutions, pp. 547-576.
    (e) F.K. Bogner, R.L. Fox, and L.A. Schmit, Jr.: The generation of inter-element compatible stiffness and mass matrices by the use of interpolation formulas, pp. 397-443.
15. L. Berke, R.M. Bader, W.J. Mykytow, J.S. Przemieniecki and M.H. Shirk (Editors): *Proceedings of the Second Conference on Matrix Methods in Structural Mechanics*, AFFDL-TR-68-150 (1968). Various papers on finite elements, including:
    (a) T.H.H. Pian and Pin Tong: Rationalization of deriving element stiffness matrices by assumed stress approach, pp. 448-469.
    (b) R.S. Dunham and K.S. Pister: Finite element application of the hellinger-reissner variational theorem, pp. 471-487.
    (c) J.E. Walz, R.E. Fulton, and Nancy J. Cyrus: Accuracy and convergence of finite element approximations, pp. 995-1027.
16. R.H. Gallagher, Y. Yamada, and J.J. Oden (Editors): *Recent Advances in Matrix Methods of Structural Analysis and Design*, The University of Alabama Press, University, Alabama (1971). Various papers on finite elements.
17. L.R. Herrmann: Finite element boundary analysis for plates, *Journal of the Engineering Mechanics Division*, ASCE, **93**, EM5, pp. 13-26 (1967).
18. R.D. Cook: Some elements for analysis of plate bending, *Proc. ASCE, Journal of the Engineering Mechanics Division* **98**, EM6, pp. 1457-1470 (1972).
19. W. Visser: A Refined mixed-type plate bending element, *AIAA Journal* **7**, pp. 1801-1802 (1969).
20. L.S.D. Morley: The constant-moment plate bending element, *Journal of Strain Analysis* **6**,

pp. 20-24 (1971).
21. J. Bron and G. Khatt: Mixed quadrilateral elements for bending, *AIAA Journal* **10**, pp. 1359-1361, Oct. (1972).
22. Z.M. Elias: Duality in finite element methods, *Proc. of ASCE, Journal of Engineering Mechanics Division* **94**, pp. 931-946 (1968).
23. J. Robinson: *Integrated Theory of Finite Element Methods*, John Wiley and Sons (1973).

References for additional reading

24. C.A. Brebbia and J.J. Connor: *Fundamentals of Finite Elements Techniques for Structural Engineers*. John Wiley and Sons, New York (1974).
25. K.V. Rockey: *The Finite Element Method: A Basic Introduction*, John Wiley and Sons, New York (1975).
26. K.J. Bathe and E.L. Wilson: *Numerical Methods in Finite Element Analysis*, Prentice-Hall, Englewood Cliffs, New Jersey (1976).
27. R.D. Cook: *Concepts and Applications of Finite Element Analysis*, 2nd Ed., John Wiley and Sons, New York (1981).
28. J.N. Reddy: *An Introduction to the Finite Element Method*, McGraw-Hill Book Co., New York (1984).
29. L.J. Segerlind: *Applied Finite Element Analysis*, 2nd Ed., John Wiley and Sons, New York (1985).
30. I.H. Shames and C.L. Dym: *Energy and Finite Element Methods in Structural Mechanics*, McGraw-Hill Book Co., New York (1985).
31. H. Gradin Jr.: *Fundamentals of the Finite Element Method*, Macmillan, New York (1986).

10

Composite materials

10.1. Introduction

Composite materials may be of three different types:
1. Fibrous composites with fibers in a matrix material,
2. Laminated composites with layers of various materials,
3. Particulate composites with particles in a matrix material.

1. In long fibers of a material the crystals tend to be aligned along the axis of the fiber so that the strength of the fiber may be much larger than the strength of the bulk material. A long steel fiber may have $F_t = 600,000$ psi but a steel bar may have $F_t = 100,000$ psi.

The long fibers can be bound together by a matrix material to form a useful structural element. Usually the matrix material has lower density, stiffness, and strength than the fibers. Matrix materials may be metal or may be organic low density epoxy materials. The combination of the high strength fibers in the low density matrix can produce a material of high strength and stiffness with low density.

2. Two or more different materials can be bonded together to produce a laminated composite. The layers of the materials may be oriented in different directions to produce a more useful material. Fibrous composites can be bonded together to make laminated fiber-reinforced composites, such as fiberglass automobile bodies and boat hulls, graphite-epoxy aircraft components, tennis rackets, and golf clubs

3. Particles of one or more materials can be supported in a matrix of another material to produce a particulate composite. The particles and the matrix can be either metallic or non-metallic. Some examples of particulate composites are concrete, solid rocket propellants, lead in copper alloys for bearings, cermets.

In the previous chapters of this book the materials have been assumed to be *homogeneous* and *isotropic*. That is, the material properties are not functions of position and direction in the structural body. However, composite materials are usually *nonhomogeneous* and *nonisotropic* so that the material properties are functions of both position and direction in the body.

Because of the nonhomogeneous nature of composite materials, they can be examined from two viewpoints: micromechanics and macromechanics.

Micromechanics is the study of composite material behavior of the component materials on a microscopic scale.

Macromechanics is the study of composite material behavior where the materials is assumed to be homogeneous, and the effects of the component materials are detected only as averaged apparent properties of the composite.

Only the macromechanics viewpoint will be considered here for laminated fiber-reinforced composite plates. See Reference 1 for details on micromechanics analysis and design.

The construction of a laminated fiber-reinforced composite involves *laminae* and *laminates*:

A *lamina* is a flat (curved for shell) arrangement of unidirectional fibers or woven fibers in a matrix material.

A *laminate* is a stack of laminae with various orientations in the laminae. The layers of the laminate are usually bonded together by the same matrix material as used in the laminae. Lamination is used to tailor the directional dependence of strength and stiffness to match the loading on the structural element.

Shearing stresses between layers may be a problem in the construction of the laminates. The shearing stresses occur because each layer tends to deform independently. These shearing stresses, which are in the matrix material, may cause delamination at the edges of the laminate.

It should be pointed out that the basic equilibrium equations, strain-displacement equations, and virtual principles in Chapters 1 and 2 of Volume 1 apply to composite materials. The nonisotropic composite materials affect the *stress-strain* equations so that Equations (1.11) and (1.12) (Vol.1) will be changed and the $[J]$ matrix in Equation (2.68, Vol.1) will be changed. The stress-strain equations for composite materials are given in Section 10.2 and for plane stress orthotropic materials in Section 10.3.

The discussion in this chapter will be restricted to orthotropic plate laminates. Primarily, the plane section analysis of classical lamination plate theory will be used. In Section 10.4 the forces and moments on laminated plates will be determined from the properties of the laminae. The inplane forces and the bending moments are coupled in laminated plates, except for cases of symmetrical laminae arrangements. The coupled equations will be derived, but the examples will be restricted to inplane uncoupled cases. Thermal forces and moments will be included.

In Section 10.5 the procedure to obtain the stresses in each lamina from the forces and moments on the laminate is described. This is a complicated procedure and requires a computer for a coupled laminate with a large number of laminae. Examples of the three layer inplane symmetrical laminate will be given.

In Section 10.6 allowable stresses and strains for the laminae, and combined stress theories of failure for the laminae are examined.

In Section 10.7 the problem of interlamina stresses in laminated plates is described. This can be a serious problem in the manufacturing and use of composite structures.

Joining of composite materials may be a difficult problem. Bolted joints have large stress concentrations and possible delamination problems. Bonded joints

Composite materials

have interlamina stress effects as well as the effects considered above in bonded metal joints, Section 2.10. A brief introduction to the joint problem together with some references is given in Section 10.8.

The problems of the laminated plate deflections, buckling, and vibration are described in Sections 10.9, 10.10 and 10.11. These problems are related to the corresponding isotropic plate cases covered in Chapter 7.

Most composite materials have low ductility and act more as brittle materials. Because of low strain limits, many composites cannot redistribute loads by becoming inelastic in local regions. This means that great care must be used in designing, analyzing, fabrication, and assembly of composite materials. Many tests are needed to verify design and fabrication.

Because high strength low density composites are lighter than metals for the same loading, they are being used to reduce structural weight in a number of airplanes. Approximately, the F-15 is 1% by weight composite materials, the F/A-18 is 10%, and the AV-8B is 30%.

10.2. Stress-strain equations for nonisotropic materials

From References 1 and 2 for an anistropic material lamina the most general form of the $[J]$ material property matrix is symmetric with 21 independent constants, or

$$\{e\} = [J]\{\sigma\} + \{e_E\} \tag{10.1}$$

$$\{e\}^T = [e_{xx}\ e_{yy}\ e_{zz}\ e_{xy}\ e_{xz}\ e_{yz}],$$
$$\{\sigma\}^T = [\sigma_{xx}\ \sigma_{yy}\ \sigma_{zz}\ \sigma_{xy}\ \sigma_{xz}\ \sigma_{yz}],$$
$$[J] = [J_{ij}], \quad \text{6 by 6}, \quad J_{ij} = J_{ji}. \tag{10.2}$$

$\{e_E\}$ includes any temperature effects.

If the body has material property symmetry about the z-axis, then

$$[J] = \begin{bmatrix} J_{11} & J_{12} & J_{13} & J_{14} & 0 & 0 \\ J_{12} & J_{22} & J_{23} & J_{24} & 0 & 0 \\ J_{13} & J_{23} & J_{33} & J_{34} & 0 & 0 \\ J_{14} & J_{24} & J_{34} & J_{44} & 0 & 0 \\ 0 & 0 & 0 & 0 & J_{55} & J_{56} \\ 0 & 0 & 0 & 0 & J_{56} & J_{66} \end{bmatrix}, \tag{10.3}$$

which has 13 independent constants.

If a body has three orthogonal planes of material symmetry, then the stress-strain relations in coordinates aligned with the principal material directions has

9 independent constants, or

$$[J] = \begin{bmatrix} J_{11} & J_{12} & J_{13} & 0 & 0 & 0 \\ J_{12} & J_{22} & J_{23} & 0 & 0 & 0 \\ J_{13} & J_{23} & J_{33} & 0 & 0 & 0 \\ 0 & 0 & 0 & J_{44} & 0 & 0 \\ 0 & 0 & 0 & 0 & J_{55} & 0 \\ 0 & 0 & 0 & 0 & 0 & J_{66} \end{bmatrix}. \quad (10.4)$$

This is an *orthotropic* material in which there is no interaction between normal stresses and shearing strains, or between normal strains and shearing stresses, or between shearing stresses and strains in different planes.

If a body has one plane in which the material properties are the same in all directions, the material is *transversely isotropic* with 5 independent constants. If the xy-plane is the special plane of isotropy, then

$$[J] = \begin{bmatrix} J_{11} & J_{12} & J_{13} & 0 & 0 & 0 \\ J_{12} & J_{11} & J_{13} & 0 & 0 & 0 \\ J_{13} & J_{23} & J_{33} & 0 & 0 & 0 \\ 0 & 0 & 0 & (J_{11}-J_{12})/2 & 0 & 0 \\ 0 & 0 & 0 & 0 & J_{44} & 0 \\ 0 & 0 & 0 & 0 & 0 & J_{44} \end{bmatrix}. \quad (10.5)$$

If the body has an infinite number of planes of material symmetry, then the body is *isotropic* with 2 independent constants. $[J]$ is in Equation (2.68, Vol.1) for this case.

In terms of the material properties E_i, G_i, ν_{ij} the orthotropic $[J]$ in Equation (10.4) has the form

$$[J] = \begin{bmatrix} 1/E_x & -\nu_{xy}/E_x & -\nu_{xz}/E_x & 0 & 0 & 0 \\ -\nu_{xy}/E_x & 1/E_y & -\nu_{yz}/E_y & 0 & 0 & 0 \\ -\nu_{xz}/E_x & -\nu_{yz}/E_y & 1/E_z & 0 & 0 & 0 \\ 0 & 0 & 0 & 1/G_{xy} & 0 & 0 \\ 0 & 0 & 0 & 0 & 1/G_{xz} & 0 \\ 0 & 0 & 0 & 0 & 0 & 1/G_{yz} \end{bmatrix}, (10.6)$$

$$\nu_{xy}/E_x = \nu_{yx}/E_y, \quad \nu_{xz}/E_x = \nu_{zx}/E_z,$$
$$\nu_{yz}/E_y = \nu_{zy}/E_z. \quad (10.7)$$

The relations in Equation (10.7) as well as various restrictions on the orthotropic material properties given in Section 2.4.2 of Reference 1 can be used in checking test procedures used to obtain the properties of the materials.

10.3. Stress-strain equations for plane stress in an orthotropic material

From Equations (1.26, Vol.1) and (10.4) the plane stress case for the orthotropic material lamina with the z-axis perpendicular to the lamina plane gives

$$\sigma_{zz} = 0, \quad \sigma_{xz} = 0, \quad \sigma_{yz} = 0, \quad e_{xz} = 0, e_{yz} = 0,$$
$$e_{zz} = J_{13}\sigma_{xx} + J_{23}\sigma_{yy},$$

$$\left\{\begin{array}{c} e_{xx} \\ e_{yy} \\ e_{xy} \end{array}\right\}_k = \left[\begin{array}{ccc} J_{11} & J_{12} & 0 \\ J_{12} & J_{22} & 0 \\ 0 & 0 & J_{33} \end{array}\right]_k \left\{\begin{array}{c} \sigma_{xx} \\ \sigma_{yy} \\ \sigma_{xy} \end{array}\right\}_k + \{e_E\}_k, \tag{10.8}$$

$$\{e_k\}_k = [J]_k\{\sigma\}_k + \{e_E\}_k \quad \text{for lamina} \quad (k).$$

From Equations (10.6) and (10.7) for lamina (k)

$$\begin{aligned} J_{11k} &= 1/E_{xk}, & J_{12k} &= -(\nu_{xy}/E_x)_k = -(\nu_{yx}/E_y)_k, \\ J_{22k} &= 1/E_{yk}, & J_{33k} &= 1/G_{xyk}, \\ J_{13k} &= -(\nu_{xz}/E_x)_k = -(\nu_{zx}/E_z)_k, \\ J_{23k} &= -(\nu_{yz}/E_y)_k = -\nu_{zy}/E_z)_k. \end{aligned} \tag{10.9}$$

In these equations the x-direction is in the direction of the fibers in the lamina while the y-direction is perpendicular to the fibers in the plane of the lamina.

If a laminated plate is assembled from laminae with different orientations, it is necessary to refer the material properties of the laminae to the laminate coordinates. Let x_d, y_d be the datum coordinates for the laminate and x_k, y_k the principal material coordinates for the lamina (k) so that Equations (10.8) and (10.9) apply to each lamina. Use Equations (1.20) and (A.107, Vol.1) to get

$$\left\{\begin{array}{c} \sigma_{xx} \\ \sigma_{yy} \\ \sigma_{xy} \end{array}\right\}_{dk} = [T_s]_k^{-1} \left\{\begin{array}{c} \sigma_{xx} \\ \sigma_{yy} \\ \sigma_{xy} \end{array}\right\}, \tag{10.10}$$

$$[T_s]_k^{-1} = \left[\begin{array}{ccc} \cos^2\theta & \sin^2\theta & -2\sin\theta\cos\theta \\ \sin^2\theta & \cos^2\theta & 2\sin\theta\cos\theta \\ \sin\theta\cos\theta & -\sin\theta\cos\theta & \cos^2\theta - \sin^2\theta \end{array}\right]_k,$$

$$[T_s(\theta)]_k = [T_s(-\theta)]_k^{-1},$$

where θ_k is measured from the datum axes to lamina (k) axes and subscript (dk) indicates stresses in datum system due to lamina (k).

The rotation of axes for the strains gives

$$\left\{\begin{array}{c} e_{xx} \\ e_{yy} \\ e_{xy} \end{array}\right\}_{dk} = [T_s]_k^T \left\{\begin{array}{c} e_{xx} \\ e_{yy} \\ e_{xy} \end{array}\right\}_k. \tag{10.11}$$

Put Equations (10.8) and (10.10) into Equation (10.11) to get

$$\begin{aligned} \{e\}_{dk} &= [T_s]_k^T [J]_k \{\sigma_k\} + [T_s]_k^T \{e_E\}_k \\ &= [J]_{dk}\{\sigma\}_{dk} + \{e_E\}_{dk}, \end{aligned} \tag{10.12}$$

$$[J]_{dk} = [T_s]_k^T [J]_k [T_s]_k = \left[\begin{array}{ccc} J_{11} & J_{12} & J_{13} \\ J_{12} & J_{22} & J_{23} \\ J_{13} & J_{23} & J_{33} \end{array}\right]_{dk}, \tag{10.13}$$

$$J_{11d} = J_{11} \cos^4 \theta + (2J_{12} + J_{33}) \sin^2 \theta \cos^2 \theta + J_{22} \sin^4 \theta,$$
$$J_{12d} = J_{12}(\sin^4 \theta + \cos^4 \theta) + (J_{11} + J_{22} - J_{33}) \sin^2 \theta \cos^2 \theta,$$
$$J_{22d} = J_{11} \sin^4 \theta + (2J_{12} + J_{33}) \sin^2 \theta \cos^2 \theta + J_{22} \cos^4 \theta,$$
$$J_{33d} = 2(2J_{11} + 2J_{22} - 4J_{12} - J_{33}) \sin^2 \theta \cos^2 \theta + J_{33}(\sin^4 \theta + \cos^4 \theta),$$
$$J_{13d} = (2J_{11} - 2J_{12} - J_{33}) \sin \theta \cos^3 \theta - (2J_{22} - 2J_{12} - J_{33}) \sin^3 \theta \cos \theta,$$
$$J_{23d} = (2J_{11} - 2J_{12} - J_{33}) \sin^3 \theta \cos \theta - (2J_{22} - 2J_{12} - J_{33}) \sin \theta \cos^3 \theta,$$

where subscript k is omitted.

The $[J]_{dk}$ material property matrix in Equation (10.13) defines apparent material properties for the lamina loaded at angle θ_k to its material axes, or

$$\begin{aligned}
(1/E_x)_{dk} &= J_{11dk}, & (\nu_{xy}/E_x)_{dk} &= J_{12dk}, \\
(1/E_y)_{dk} &= J_{22dk}, & (1/G_{xy})_{dk} &= J_{33dk}, \\
(n_{xyx}/E_x)_{dk} &= J_{13dk}, & (n_{xyy}/E_y)_{dk} &= J_{23dk},
\end{aligned} \quad (10.14)$$

where n_{xyxdk} and n_{xyydk} are mutual influence coefficients representing shear in the xy-plane caused by a normal stress in the x or y direction. It is apparent from Equation (10.9) and (10.13) that these apparent material properties depend upon the lamina properties and the angle θ_k.

Equation (10.12) gives the strains in terms of the stresses. To calculate loads and moments, it is necessary to have equations giving the stresses in terms of the strains. To derive these expressions invert Equation (10.8) to get

$$\{\sigma\}_k = [J]_k^{-1} \{e - e_E\}_k, \quad (10.15)$$

$$[J]_k^{-1} = \begin{bmatrix} F_{11} & F_{12} & 0 \\ F_{12} & F_{22} & 0 \\ 0 & 0 & F_{33} \end{bmatrix}_k,$$

$$F_{11k} = E_x/(1 - \nu_{xy}\nu_{yx}), \quad F_{22k} = E_y/(1 - \nu_{xy}\nu_{yx}),$$
$$F_{12k} = \nu_{xy} E_y/(1 - \nu_{xy}\nu_{yx}) = \nu_{yx} E_x/(1 - \nu_{xy}\nu_{yx}),$$
$$F_{33k} = 1/G_{xy}, \quad \nu_{xy}/E_x = \nu_{yx}/E_y, \quad \text{subscript } k \text{ omitted.}$$

To put Equation (10.15) into datum coordinates, use Equations (10.10), (10.11), and (10.15) to get

$$\{\sigma\}_{dk} = [T_\sigma]_k^{-1} [J]_k^{-1} \{e - e_E\}_k$$
$$= [J]_{dk}^{-1} \{e - e_E\}_{dk}, \quad (10.16)$$

$$[J]_{dk}^{-1} = [T_\sigma]_k^{-1} [J]_k^{-1} [T_\sigma]_k^T = \begin{bmatrix} F_{11} & F_{12} & F_{13} \\ F_{12} & F_{22} & F_{23} \\ F_{13} & F_{23} & F_{33} \end{bmatrix}_{dk}, \quad (10.17)$$

$$F_{11d} = F_{11} \cos^4 \theta + 2(F_{12} + 2F_{33}) \sin^2 \theta \cos^2 \theta + F_{22} \sin^4 \theta,$$
$$F_{12d} = F_{12}(\sin^4 \theta + \cos^4 \theta) + (F_{11} + F_{22} - 4F_{33}) \sin^2 \theta \cos^2 \theta,$$
$$F_{22d} = F_{11} \sin^4 \theta + 2(F_{12} + 2F_{33}) \sin^2 \theta \cos^2 \theta + F_{22} \cos^4 \theta,$$
$$F_{33d} = (F_{11} + F_{22} - 2F_{12} - 2F_{33}) \sin^2 \theta \cos^2 \theta + F_{33}(\sin^4 \theta + \cos^4 \theta),$$

Composite materials

$$F_{13d} = (F_{11} - F_{12} - 2F_{33})\sin\theta\cos^3\theta + (F_{12} - F_{22} + F_{33})\sin^3\theta\cos\theta,$$
$$F_{23d} = (F_{11} - F_{12} - 2F_{33})\sin^3\theta\cos\theta + (F_{12} - F_{22} + F_{33})\sin\theta\cos^3\theta.$$

where subscript k is omitted.

Example 10.1.   Consider a boron-epoxy lamina (k) with

$$(E_x/E_y)_k = 10, \quad (G_{xy}/E_y)_k = 1/3, \quad (\nu_{xy})_k = 0.3, \tag{10.18}$$

and (a) Calculate (E_{xd}/E_{yk}), (G_{xyd}/G_{xyk}), (ν_{xyd}), n_{xyzd} as functions of θ from Equations (10.13) and (10.14). (b) Give values at $\theta = 0°$, $45°$ and $90°$.

Solution. (a) From Equations (10.13), (10.14), and (10.9),

$$\begin{aligned}
E_{yk}/E_{xd} &= 0.1\cos^4\theta + 2.94\sin^2\theta\cos^2\theta + \sin^4\theta, \\
\nu_{xyd} &= (E_{xd}/E_{yk})[0.03(\sin^4\theta + \cos^4\theta) + 1.9\sin^2\theta\cos^2\theta], \\
G_{xyd}/G_{xyk} &= -0.5333\sin^2\theta\cos^2\theta + \sin^4\theta + \cos^4\theta, \\
n_{xyzd} &= (E_{xd}/E_{yk})[-2.74\sin\theta\cos^3\theta + 0.94\sin^3\theta\cos\theta].
\end{aligned} \tag{10.19}$$

(b) For $\theta = 0°$, $45°$, $90°$, respectively,

$$\begin{aligned}
E_{xd}/E_{yk} &= 10.00, \quad 1.02, \quad 1.00, \\
\nu_{xyd} &= 0.30, \quad 0.50, \quad 0.03, \\
G_{xyd}/G_{xyk} &= 1.00, \quad 2.73, \quad 1.00, \\
n_{xyd} &= 0.00, \quad -0.45, \quad 0.00.
\end{aligned} \tag{10.20}$$

It is evident that there are large variations in the material properties with the angle θ between the datum directions and the lamina material property directions.

10.4. Forces and moments for laminated plates

Two or more flat laminae can be bonded together to make an integral plate. The laminae principal material directions can be oriented to produce a plate that can take loads in several directions. The effective plate material properties of the laminated plate must be obtained from the properties of the laminae in the plate. Assume the laminate to have perfectly bonded laminae and assume the bonds to have zero thickness with no shear deformation. Assume the plate to be unrestrained on the edges. Thus, away from the edges, the plate has plane sections so that the strain-displacement relations in Equations (7.6), (7.7), and (7.91) apply.

However, in Chapter 7 the plate was isotropic so that the inplane loads and bending moments were uncoupled for small deflections. Here the orthotropic properties for each lamina may be different so that coupling between inplane and bending loads may be present.

From Equations (7.6) and (7.91),

$$\{e\} = [J_{pd}]\{u\} - z[J_{bdd}]u_z = \{e_0\} + z\{e_b\}, \tag{10.21}$$

whence Equation (10.16) for lamina (k) becomes (see Figure 10.1)

$$\{\sigma\}_{dk} = [J]_{dk}^{-1}\{\{e_0\} - \{e_E\} + z\{e_b\}\}_{dk}, \quad z_{k-1} \le z \le z_k. \tag{10.22}$$

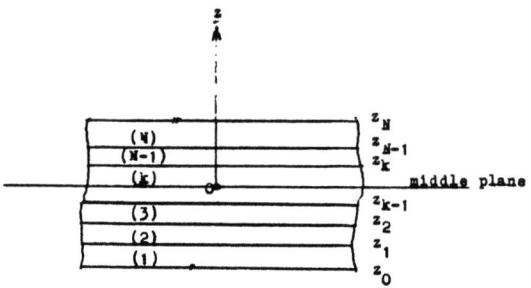

Fig. 10.1.   Laminate with N layers.

From Equation (7.1) and Figure 16.1,

$$\{N\} = \begin{Bmatrix} N_x \\ N_y \\ N_{xy} \end{Bmatrix} = \int_{z_0}^{z_N} \{\sigma\}_{dk}\, dz, \quad \{M\} = \begin{Bmatrix} M_x \\ M_y \\ M_{yx} \end{Bmatrix} = \int_{z_0}^{z_N} \{\sigma\}_{dk} z\, dz, \quad (10.23)$$

or

$$\{N\} = \sum_{k=1}^{N} [J]_{dk}^{-1}\{(z_k - z_{k-1})\{e_0\} + \tfrac{1}{2}(z_k^2 - z_{k-1}^2)\{e_b\}\} - \{N_E\},$$

$$\{M\} = \sum_{k=1}^{N} [J]_{dk}^{-1}\{\tfrac{1}{2}(z_k^2 - z_{k-1}^2)\{e_0\} + \tfrac{1}{3}(z_k^3 - z_{k-1}^3)\{e_b\}\} - \{M_E\},$$

$$\{N_E\} = \int_{z_0}^{z_k} [J]_{dk}^{-1}\{e_E\}_{dk}\, dz, \quad \{M_E\} = \int_{z_0}^{z_k} [J]_{dk}^{-1}\{e_E\}_{dk} z\, dz.$$

(10.24)

The Equation (10.24) can be expressed as

$$\{N\} = [A]\{e_0\} + [B]\{e_b\} - \{N_E\},$$
$$\{M\} = [B]\{e_0\} + [C]\{e_b\} - \{M_E\},$$

$$A_{ij} = \sum_{k=1}^{N} (F_{ij})_{dk}(z_k - z_{k-1}), \quad B_{ij} = \sum_{k=1}^{N} (F_{ij})_{dk}(z_k^2 - z_{k-1}^2)/2,$$

$$C_{ij} = \sum_{k=1}^{N} (F_{ij})_{dk}(z_k^3 - z_{k-1}^3)/3,$$

(10.25)

where the $(F_{ij})_{dk}$ are in Equation (10.17) for each lamina (k) of thickness $t_k = z_k - z_{k-1}$.

Symmetric Laminates

If laminates are symmetric in both geometry and material properties about the middle surface $z = 0$, then from

$$(1/2)(z_k^2 - z_{k-1}^2) = t_k(z_k + z_{k-1})/2 \quad (10.26)$$

Composite materials

it follows that $B_{ij} = 0$, $[B] = 0$, in Equation (10.25) so that the inplane loads and bending moments can be determined separately. Symmetric laminates are used in many cases not only because of the simpler analysis but also because of simpler manufacturing procedures. Unsymmetrical laminates tend to twist during cooling following the curing process in fabrication.

For this symmetric case, Equation (10.25) with $B_{ij} = 0$ can be compared to the basic isotropic plate Equations (7.7), (7.92), and (7.4). For the inplane case

$$\{N\} = t[J]^{-1}\{e_0\} - \int_t [J]^{-1}\{e_E\}dz,$$

$$[J]^{-1} = \frac{E}{1-\nu^2}\begin{bmatrix} 1 & \nu & 0 \\ \nu & 1 & 0 \\ 0 & 0 & \frac{1-\nu}{2} \end{bmatrix}, \quad \text{isotropic,}$$

$$\{N\} = [A]\{e_0\} - \int_{z_0}^{z_N} [J]_{dk}^{-1}\{e_E\}_{dk}dz,$$

$$[A] = \begin{bmatrix} A_{11} & A_{12} & A_{13} \\ A_{12} & A_{22} & A_{13} \\ A_{13} & A_{23} & A_{33} \end{bmatrix}, \quad \text{symmetric laminate,}$$

$$A_{ij} = \sum_{k=1}^{N}(F_{ij})_{dk}(z_k - z_{k-1}),$$

(10.27)

$[J]_{dk}^{-1}$, $(F_{ij})_{dk}$ in Equation (10.17) for lamina (k).

If the 3 by 3 $t[J]^{-1}$ matrix is replaced by the 3 by 3 $[A]$ matrix in Sections 7.2 through 7.4, then all the plate equations can be derived for symmetric laminated plates. Necessarily, the equations will be more complicated than for the isotropic case, particularly as non-zero A_{13} and A_{23} will add more terms to the equations. Note that in Equation (10.17) F_{13d}, F_{23d} and hence A_{13}, A_{23} will be zero for $\theta = 0°$ and $\theta = 90°$. Thus, symmetric isotropic laminates and cross-ply symmetric orthotropic laminates at $0°$ and $90°$ will have $A_{13} = 0$, $A_{23} = 0$.

Example 10.2. Suppose the three layer symmetric laminate in Figure 10.2 has two different isotropic materials with E_1, ν_1, α_1, t_1 for the center lamina and E_2, ν_2, α_2, t_2 for the outer laminae. Calculate the $[A]$ and $\{N_E\}$ matrices for the laminate.

Solution. For the isotropic laminae

$$(F_{11})_1 = (F_{22})_1 = E_1/(1-\nu_1^2), \quad (F_{12})_1 = \nu_1 E_1/(1-\nu_1^2),$$
$$(F_{33})_1 = E_1/2(1+\nu_1), \quad (F_{11})_2 = (F_{22})_2 = E_2/(1-\nu_2^2),$$
$$(F_{12})_2 = \nu_2 E_2/(1-\nu_2^2), \quad (F_{33})_2 = E_2/2(1+\nu_2).$$

(10.28)

For the $(F_{ij})_{dk}$ with $\theta_k = 0$

$$(F_{ij})_{dk} = (F_{ij})_k, \quad (F_{13})_{dk} = (F_{23})_{dk} = 0.$$

Thus, from Equation (10.25),

$$A_{11} = A_{22} = E_1 t_1(1+f)/(1-\nu_1^2), \quad f = 2E_2 t_2(1-\nu_1^2)/E_1 t_1(1-\nu_2^2),$$
$$A_{12} = E_1 t_1(\nu_1 + f\nu_2)/(1-\nu_1^2),$$
$$A_{33} = [E_1 t_1/(1-\nu_1^2)][(1-\nu_1) + f(1-\nu_2)]/2,$$

(10.29)

$$[A] = \frac{E_1 t_1}{1-\nu_1^2}\begin{bmatrix} 1+f & \nu_1 + f\nu_2 & 0 \\ \nu_1 + f\nu_2 & 1+f & 0 \\ 0 & 0 & \frac{1-\nu_1}{2} + f(\frac{1-\nu_2}{2}) \end{bmatrix}.$$

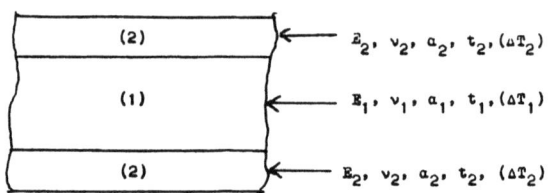

Fig. 10.2. Three layer symmetric laminate with two isotropic materials.

If $\nu_1 = \nu_2$, then

$$[A] = t_1(1+f)\left[J_1\right]^{-1}, \quad \left[J_1\right]^{-1} = \frac{E_1}{(1-\nu_1^2)}\begin{bmatrix} 1 & \nu_1 & 0 \\ \nu_1 & 1 & 0 \\ 0 & 0 & (1-\nu_1)/2 \end{bmatrix}.$$

From Equation (10.24),

$$\{N_E\} = \frac{E_1 t_1}{1-\nu_1^2}\begin{bmatrix} 1 & \nu_1 & 0 \\ \nu_1 & 1 & 0 \\ 0 & 0 & \frac{1}{2}(1-\nu_1) \end{bmatrix}\begin{Bmatrix} \alpha_1(\Delta T)_1 \\ \alpha_1(\Delta T)_1 \\ 0 \end{Bmatrix} +$$

$$+ \frac{2E_2 t_2}{1-\nu_2^2}\begin{bmatrix} 1 & \nu_2 & 0 \\ \nu_2 & 1 & 0 \\ 0 & 0 & \frac{1}{2}(1-\nu_2) \end{bmatrix}\begin{Bmatrix} \alpha_2(\Delta T)_2 \\ \alpha_2(\Delta T)_2 \\ 0 \end{Bmatrix}$$

$$= \frac{E_1 t_1}{1-\nu_1^2}\begin{Bmatrix} (1+\nu_1)\alpha_1(\Delta T)_1 + f(1+\nu_2)\alpha_2(\Delta T)_2 \\ (1+\nu_1)\alpha_1(\Delta T)_1 + f(1+\nu_2)\alpha_2(\Delta T)_2 \\ 0 \end{Bmatrix} \quad (10.31)$$

$$= \frac{E_1 t_1(\Delta T)_1}{1-\nu_1}\begin{Bmatrix} \alpha_1 + f\alpha_2 \\ \alpha_1 + f\alpha_2 \\ 0 \end{Bmatrix}, \quad \nu_2 = \nu_1, \quad (\Delta T)_2 = (\Delta T)_1. \quad (10.32)$$

Thus, if $\nu_2 = \nu_1$, $(\Delta T)_2 = (\Delta T)_1$, the available isotropic plate solutions for material lamina (1) can be used provided the constant multiplying factor $(1+f)$ is used on the $[J_1]^{-1}$ elements. Only a weighted $\alpha = \alpha_1 + f\alpha_2$ is needed in the thermal load term.

Example 10.3. Solve Example 7.1, part (b), for the laminated plate in Example 10.2 with $\nu = \nu_1 = \nu_2$, $T = (\Delta T)_1 = (\Delta T)_2$.

Solution. Use the equivalent $\alpha = \alpha_1 + f\alpha_2$ and take $N_{T0} = t_1 E_1(\alpha_1 + f\alpha_2)T_0$, $f = 2E_2 t_2/E_1 t_1$, whence Equation (7.28) becomes

$$u_x = -\frac{0.0700a}{1+f}(\alpha_1 + f\alpha_2)T_0 \sin(2\pi x/a) \sin(\pi y/a),$$

$$u_y = -\frac{0.0700a}{1+f}(\alpha_1 + f\alpha_2)T_0 \sin(\pi x/a) \sin(2\pi y/a), \quad (10.33)$$

$$\{N\} = t_1(1+f)[J_1]^{-1}\begin{Bmatrix} u_{x,x} \\ u_{y,y} \\ u_{x,y} + u_{y,x} \end{Bmatrix} - \frac{N_{T0}T}{(1-\nu_1)T_0}\begin{Bmatrix} 1 \\ 1 \\ 0 \end{Bmatrix}$$

= same as in Equation (7.30) with above N_{T0}.

Thus, the equivalent $\alpha = \alpha_1 + f\alpha_2$ determines the loads while α and f determine the deflections.

Example 10.4. Suppose a three layer cross-ply symmetric laminate in Figure 10.3 has one orthotropic material with E_x, E_y, ν_{xy}, G_{xy}, α_x, α_y but with t_2 for cross-ply (90°) middle layer and t_1 for outside (0°) layers. (a) Determine the $[A]$ and $\{N_E\}$ matrices for the laminate. (b) Calculate typical numerical values for graphite-epoxy laminate.

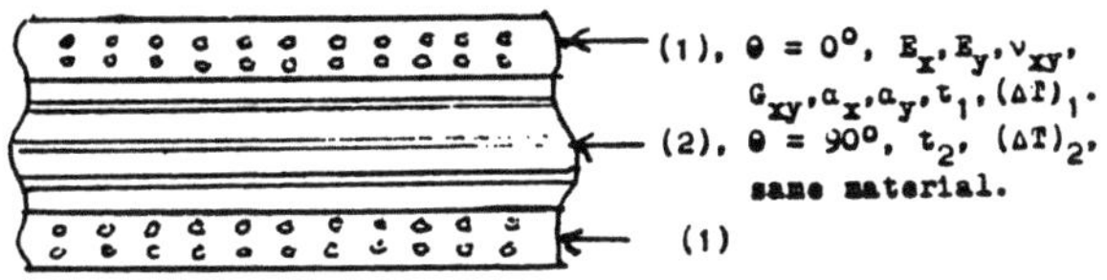

Composite materials 339

```
(1), θ = 0°, E_x, E_y, ν_xy,
     G_xy, α_x, α_y, t_1, (ΔT)_1.
(2), θ = 90°, t_2, (ΔT)_2,
     same material.
(1)
```

Fig. 10.3. Three layer cross-ply symmetric laminate with orthotropic material.

Solution. (a) From Equations (10.15) and (10.17),

$\theta = 0°$ for lamina (1):

$$(F_{11})_{d1} = F_{11} = E_x/(1 - \nu_{xy}\nu_{yx}),$$
$$(F_{12})_{d1} = F_{12} = \nu_{xy}E_y/(1 - \nu_{xy}\nu_{yx}),$$
$$(F_{22})_{d1} = F_{22} = E_y/(1 - \nu_{xy}\nu_{yx}),$$ (10.34)
$$(F_{33})_{d1} = F_{33} = G_{xy}, \quad (F_{13})_{d1} = 0, \quad (F_{23})_{d1} = 0.$$

$\theta = 90°$ for laminae (2):

$$(F_{11})_{d2} = F_{22}, \quad (F_{12})_{d2} = F_{12}, \quad (F_{22})_{d2} = F_{11},$$
$$(F_{33})_{d2} = F_{33}, \quad (F_{13})_{d2} = 0, \quad (F_{23})_{d2} = 0.$$ (10.35)

From Equation (10.25),

$$A_{11} = \frac{2t_1 E_x + t_2 E_y}{1 - \nu_{xy}\nu_{yx}}, \quad \nu_{yx} = (E_y/E_x)\nu_{xy},$$
$$A_{22} = \frac{2t_1 E_y + t_2 E_x}{1 - \nu_{xy}\nu_{yx}}, \quad A_{12} = (2t_1 + t_2)\frac{\nu_{xy}E_y}{1 - \nu_{xy}\nu_{yx}},$$ (10.36)
$$A_{33} = (2t_1 + t_2)G_{xy}, \quad A_{13} = 0, \quad A_{23} = 0.$$

From Equations (10.24), (10.34), and (10.35),

$$(1 - \nu_{xy}\nu_{yx})N_{Tx} = 2t_1(\Delta T)_1(E_x\alpha_x + \nu_{xy}E_y\alpha_y) + t_2(\Delta T)_2(E_y\alpha_y + \nu_{xy}E_x\alpha_x),$$
$$(1 - \nu_{xy}\nu_{yx})N_{Ty} = 2t_1(\Delta T)_1(\nu_{xy}E_y\alpha_x + E_y\alpha_y) + t_2(\Delta T)_2(\nu_{xy}E_y\alpha_y + E_x\alpha_x).$$ (10.37)

(b) For the graphite-epoxy lamina use the values

$E_x = 20(10^6)$ psi, $E_y = 1.5(10^6)$ psi, $G_{xy} = 1.0(10^6)$ psi,
$\nu_{xy} = 0.30$, $\alpha_x = 0.0$, $\alpha_y = 4.0(10^{-6})/°F$,
$\nu_{yx} = \nu_{xy}(E_y/E_x) = 0.0225$,
$1 - \nu_{xy}\nu_{yx} = 0.99$.

Take $t = 2t_1 + t_2 = 6t_1$ for $t_2 = 4t_1$, $(\Delta T) = (\Delta T)_1 = (\Delta T)_2$. From Equations (10.36) and (10.37),

$$A_{11}/10^6 t = [2(20) + 4(1.5)]/6(0.99) = 7.74,$$
$$A_{22}/10^6 t = [2(1.5) + 4(20)]/6(0.99) = 13.97,$$
$$A_{12}/10^6 t = 0.3(1.5)/0.99 = 0.45,$$ (10.38)
$$A_{33}/10^6 t = 1.00,$$
$$N_{Tx}/t(\Delta T) = [2(0 + 1.80) + 4(6.0 + 0)]/6(0.99) = 4.65,$$
$$N_{Ty}/t(\Delta T) = [2(0 + 6.0) + 4(1.8 + 0)]/6(0.99) = 3.23.$$ (10.39)

Example 10.5. Suppose a three layer symmetric laminate in Figure 10.4 has an isotropic middle layer and orthotropic outer layers. (a) Determine the $[A]$ and $\{N_E\}$ matrices for the laminate. (b) Calculate typical numerical values for aluminum alloy middle layer (2) and graphite-epoxy outer layers (1).

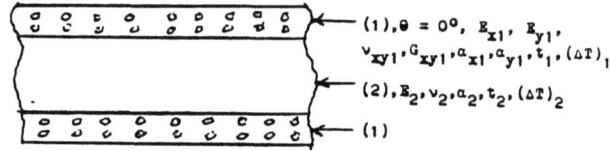

Fig. 10.4. Three layer symmetric laminate with orthotropic outer layers (1) and isotropic middle layer (2).

Solution. (a) From Figure 10.4 and Equations (10.28), (10.34), and (10.25),

$$A_{11} = \frac{2t_1 E_{x1}}{1 - \nu_{xy1}\nu_{yx1}} + \frac{t_2 E_2}{1 - \nu_2^2},$$

$$A_{22} = \frac{2t_1 E_{y1}}{1 - \nu_{xy1}\nu_{yx1}} + \frac{t_2 E_2}{1 - \nu_2^2},$$

$$A_{12} = \frac{2t_1 \nu_{xy1} E_{y1}}{1 - \nu_{xy1}\nu_{yx1}} + \frac{t_2 \nu_2 E_2}{1 - \nu_2^2}, \tag{10.40}$$

$$A_{33} = 2t_1 G_{xy1} + \frac{t_2 E_2}{2(1 + \nu_2)},$$

$$\left\{\begin{matrix} N_{Tx} \\ N_{Ty} \end{matrix}\right\} = \frac{2t_1(\Delta T)_1}{1 - \nu_{xy1}\nu_{yx1}} \left\{\begin{matrix} E_{x1}\alpha_{x1} + \nu_{xy1} E_{y1}\alpha_{y1} \\ \nu_{xy1} E_{y1}\alpha_{x1} + E_{y1}\alpha_{y1} \end{matrix}\right\} + \frac{E_2 t_2 \alpha_2 (\Delta T)_2}{1 - \nu_2} \left\{\begin{matrix} 1 \\ 1 \end{matrix}\right\}. \tag{10.41}$$

(b) For graphite-epoxy with the same material properties as in Example 10.4, part (b), use

$$E_{x1} = 20(10^6) \text{ psi}, \quad E_{y1} = 1.5(10^6), \quad \nu_{xy1} = 0.30,$$

$$G_{xy1} = 1.0(10^6) \text{ psi}, \quad \alpha_{x1} = 0.0, \quad \alpha_{y1} = 4(10^{-6})/°F.$$

For aluminum alloy, use

$$E_2 = 10^7 \text{ psi}, \quad \nu_2 = 0.3, \quad \alpha_2 = 13.0(10^{-6})/°F.$$

Take $t_2 = 2t_1$, $t = 4t_1$, $(\Delta T) = (\Delta T)_1 = (\Delta T)_2$, whence

$$A_{11}/10^6 t = 2(20)/4(0.99) + 2(10)/4(0.91) = 15.60,$$
$$A_{22}/10^6 t = 2(1.5)/4(0.99) + 2(10)/4(0.91) = 6.25,$$
$$A_{12}/10^6 t = 2(0.3)(1.5)/4(0.99) + 2(0.3)(10)/4(0.91) = 1.88, \tag{10.42}$$
$$A_{33}/10^6 t = 2(1.0)/4 + 2(10)/4(2)(1.3) = 2.42,$$
$$N_{Tx}/t(\Delta T) = 2[(20)(0) + 0.3(1.5)(4.0)]/4(0.99) + 2(13)(10)/4(0.70) = 93.77,$$
$$N_{Ty}/t(\Delta T) = 2[0 + 1.5(4.0)]/4(0.99) + 2(13)(10)/4(0.70) = 95.89. \tag{10.43}$$

Note that the thermal loads N_{Tx} and N_{Ty} are much larger for this case than for the case in Example 10.4. This comparison will be examined further in the following Sections 10.5 and 10.6 for the lamina stresses, and the load carrying ability of the laminate.

10.5. Stresses in laminated plates

When the plate inplane loads $\{N\}$ and moments $\{M\}$ are known, the stresses for the isotropic plate are simply

$$\{\sigma\}_a = \{N\}/t, \quad \{\sigma\}_b = (12z/t^3)\{M\}. \tag{10.45}$$

However, for the laminated plate, the calculations of the stresses in each lamina can be complicated. It is necessary to get $\{e_0\}$ and $\{e_b\}$ from Equation (10.25),

Composite materials

get $\{\sigma\}_{dk}$ from Equation (10.22), and get $\{\sigma\}_k$ from Equation (10.10). Take Equation (10.25) in the form

$$\begin{Bmatrix} N + N_E \\ M + M_E \end{Bmatrix} = \begin{bmatrix} A & B \\ B & C \end{bmatrix} \begin{Bmatrix} e_0 \\ e_b \end{Bmatrix}, \tag{10.46}$$

which can be inverted to give

$$\begin{Bmatrix} e_0 \\ e_b \end{Bmatrix} = \begin{bmatrix} H_1 & H_2 \\ H_2 & H_3 \end{bmatrix} \begin{Bmatrix} N + N_E \\ M + M_E \end{Bmatrix}, \tag{10.47}$$

$$[Q] = [C] - [B][A]^{-1}[B], \quad [H_3] = [Q]^{-1},$$
$$[H_1] = [A]^{-1}[I + BQ^{-1}BA^{-1}], \quad [H_2] = -[A]^{-1}[B][Q]^{-1}.$$

From Equations (10.10) and (10.22),

$$\{\sigma\}_k = [T_s]_k \{\sigma\}_{dk} = [T_s]_k [J]_{dk}^{-1} \{\{e_0\} + z\{e_b\} - [T_s]_k^T \{e_E\}_{dk}\}. \tag{10.48}$$

For the symmetrical laminate $[H_2] = 0$ so that Equation (10.47) simplifies to

$$\{e_0\} = [A]^{-1}\{N + N_E\}, \quad \{e_b\} = [C]^{-1}\{M + M_E\}. \tag{10.49}$$

It should be noted that the elastic stresses given by these equations are for the unrestrained plate so that the thermal stresses, which must be zero at the edges, apply away from the edges. There are also shear lag effects from variable applied loads on the edges. These shear effects will be considered in Section 10.7 below.

The following examples demonstrate the stress calculations for the cases considered in Examples 10.2, 10.4, and 10.5 above.

Example 10.6. Solve for the stresses in each lamina in Example 10.2 for constant temperature change (ΔT). See Figure 10.2.

Solution. For the $\nu = \nu_1 = \nu_2$, $\Delta T = (\Delta T)_1 = (\Delta T)_2$ case, use $[A]$ from Equation (10.30) in Equation (10.49) to get

$$\{e_0\} = [A]^{-1}\{N + N_E\} = \frac{1}{t_1(1+f)}[J_1]\{N + N_e\}$$

$$= \frac{1}{E_1 t_1(1+f)} \begin{Bmatrix} N_x + N_{Tx} - \nu N_y - \nu N_{Ty} \\ N_y + N_{Ty} - \nu N_x - \nu N_{Tx} \\ 2(1+\nu)N_{xy} \end{Bmatrix}, \tag{10.50}$$

Use Equation (10.48) for Example 10.2 and Equation (10.32) to get

$$\{\sigma\}_k = \{\sigma\}_{dk} = [J]_k^{-1}\{e_0 - e_E\},$$

$$\{\sigma\}_1 = \frac{1}{t_1(1+f)}[J_1]^{-1}[J_1]\{N + N_E\} - [J_1]^{-1}\{e_E\}_1$$

$$= \frac{1}{t_1(1+f)}\{N + N_E\} - \frac{E_1 \alpha_1 (\Delta T)_1}{1-\nu} \begin{Bmatrix} 1 \\ 1 \\ 0 \end{Bmatrix}$$

$$= \frac{\{N\}}{t_1(1+f)} + \frac{E_1 f (\Delta T)(\alpha_2 - \alpha_1)}{(1-\nu)(1+f)} \begin{Bmatrix} 1 \\ 1 \\ 0 \end{Bmatrix}, \quad f = 2E_2 t_2 / E_1 t_1,$$

$$\{\sigma\}_2 = \frac{f}{2t_2(1+f)}[J_1]^{-1}[J_1]\{N + N_E\} - (E_2/E_1)[J_1]^{-1}\{e_E\}_2$$

$$= \frac{f\{N\}}{2t_2(1+f)} - \frac{E_2(\Delta T)(\alpha_2 - \alpha_1)}{(1-\nu)(1+f)}\begin{Bmatrix}1\\1\\0\end{Bmatrix} = \frac{f\{N\}}{2t_2(1+f)} - \frac{t_1}{2t_2}\{\sigma\}_{1T}. \qquad (10.51)$$

Note that these stresses check the equilibrium and strain compatibility equations

$$\begin{aligned}2t_2\{\sigma\}_2 + t_1\{\sigma\}_1 &= \{N+0\},\\ \{e\}_{d1} &= \{e\}_{d2} \text{ from Equation (10.12)},\\ &= \{e_0\} \text{ from Equation (10.50)}.\end{aligned} \qquad (10.52)$$

Example 10.7. Find the stresses in each lamina in Example 10.4 for constant temperature change (ΔT). See Figure 10.3.

Solution. From Equations (10.38), (10.39), (10.49), and (10.48),

$$[A] = 10^6 t \begin{bmatrix} 7.74 & 0.45 & 0 \\ 0.45 & 13.97 & 0 \\ 0 & 0 & 1.00 \end{bmatrix},$$

$$[A]^{-1} = (1/10^6 t)\begin{bmatrix} 0.1294 & -0.0042 & 0 \\ -0.0042 & 0.0717 & 0 \\ 0 & 0 & 1.00 \end{bmatrix},$$

$$\{e_0\} = [A]^{-1}\begin{Bmatrix} N_x + 4.65t(\Delta T) \\ N_y + 3.23t(\Delta T) \\ N_{xy} \end{Bmatrix}, \qquad (10.53)$$

$$\{\sigma\}_1 = (10^6/0.99)\begin{bmatrix} 20 & 0.45 & 0 \\ 0.45 & 1.5 & 0 \\ 0 & 0 & 0.99 \end{bmatrix}\{e_0 - e_{E1}\}$$

$$= \begin{Bmatrix} 2.6122(N_x/t) - 0.0524(N_y/t) + 10.16(\Delta T) \\ 0.0524(N_x/t) + 0.1067(N_y/t) - 5.46(\Delta T) \\ 1.0000(N_{xy}/t) \end{Bmatrix},$$

$$\{\sigma\}_2 = (10^6/0.99)\begin{bmatrix} 1.5 & 0.45 & 0 \\ 0.45 & 20 & 0 \\ 0 & 0 & 0.99 \end{bmatrix}\{e_0 - e_{E2}\} \qquad (10.54)$$

$$= \begin{Bmatrix} 0.1922(N_x/t) + 0.0260(N_y/t) - 5.08(\Delta T) \\ -0.0257(N_x/t) + 1.4468(N_y/t) + 2.73(\Delta T) \\ 1.00(N_{xy}/t) \end{Bmatrix},$$

where

$$\{e_E\}_1 = \begin{Bmatrix} 0 \\ 4(10^{-6})(\Delta T) \\ 0 \end{Bmatrix}, \quad \{e_E\}_2 = \begin{Bmatrix} 4(10^{-6})(\Delta T) \\ 0 \\ 0 \end{Bmatrix}.$$

Note that if (ΔT) is the reduction in temperature from the curing temperature, then (ΔT) is negative. Also, in this case the temperature terms in the stresses will be residual stresses in the laminated plate.

Example 10.8. Find the stresses in both laminae in Example 10.5 for constant temperature change (ΔT). See Figure 10.4.

Solution. From Equations (10.42), (10.43), (10.49), and (10.48),

$$[A] = 10^6 t \begin{bmatrix} 15.60 & 1.88 & 0 \\ 1.88 & 6.25 & 0 \\ 0 & 0 & 2.42 \end{bmatrix},$$

$$[A]^{-1} = (1/10^6 t)\begin{bmatrix} 0.0665 & -0.0200 & 0 \\ -0.0200 & 0.1660 & 0 \\ 0 & 0 & 0.4132 \end{bmatrix},$$

$$\{e_0\} = [A]^{-1}\begin{Bmatrix} N_x + 93.77t(\Delta T) \\ N_y + 95.89t(\Delta T) \\ N_{xy} \end{Bmatrix}, \qquad (10.55)$$

Composite materials 343

$$\{\sigma\}_1 = \frac{10^6}{0.99}\begin{bmatrix} 20.00 & 0.45 & 0 \\ 0.45 & 1.50 & 0 \\ 0 & 0 & 0.99 \end{bmatrix}\{e_0 - e_{E1}\},$$

$$= \left\{ \begin{array}{l} 1.3343(N_x/t) - 0.3286(N_y/t) + 91.79(\Delta T) \\ -0.0001(N_x/t) + 0.2424(N_y/t) + 17.17(\Delta T) \\ 0.4134(N_{xy}/t) \end{array} \right\},$$

$$\{\sigma\}_2 = \frac{10^6}{0.91}\begin{bmatrix} 10 & 3 & 0 \\ 3 & 10 & 0 \\ 0 & 0 & 0.91(3.85) \end{bmatrix}\{e_0 - e_{E2}\}$$

$$= \left\{ \begin{array}{l} 0.6648(N_x/t) + 0.3279(N_y/t) - 91.97(\Delta T) \\ -0.0005(N_x/t) + 1.7502(N_y/t) - 17.16(\Delta T) \\ 1.59(N_{xy}/t) \end{array} \right\},$$

(10.56)

where

$$\{e_E\}_1 = \left\{ \begin{array}{c} 0 \\ 4(10^{-6})(\Delta T) \\ 0 \end{array} \right\}, \quad \{e_E\}_2 = \left\{ \begin{array}{c} 13(10^{-6})(\Delta T) \\ 13(10^{-6})(\Delta T) \\ 0 \end{array} \right\}.$$

10.6. Allowable stresses in laminated plates

Allowable stresses for isotropic materials have been given in Chapter 1. The combined isotropic stress case with various modes of failure are given in Section 1.5. For laminated plates the allowable stresses needed for each lamina are

$$F_{xt} \text{ (tension)}, \quad F_{yt} \text{ (tension)}, \quad F_{xy} \text{ (shear)},$$
$$F_{xc} \text{ (compression)}, \quad F_{yc} \text{ (compression)}.$$

(10.57)

Also, if a maximum strain theory of failure is used, then maximum strains are needed for the laminae, or

$$S_{xt} \text{ (tension)}, \quad S_{yt} \text{ (tension)}, \quad S_{xy} \text{ (shear)},$$
$$S_{xc} \text{ (compression)}, \quad S_{yc} \text{ (compression)}.$$

(10.58)

Biaxial strength theories for orthotropic laminae described in Section 2.9 of Reference 1 are maximum stress theory, maximum strain theory, Tsai-Hill theory, and Tsai-Wu tensor theory. The Tsai-Hill theory is a modification of the distortion energy theory or octohedral shear stress theory given in Equation (1.34) above. It has the form

$$(\sigma_x/F_x)^2 - (\sigma_x\sigma_y/F_x^2) + (\sigma_y/F_y)^2 = 1 \tag{10.59}$$

for a $\theta = 0°$ lamina, which is different in the product term in Equation (1.34). Also,

$$\frac{\cos^4\theta}{F_x^2} + \left(\frac{1}{F_{xy}^2} - \frac{1}{F_x^2}\right)\sin^2\theta\cos^2\theta + \frac{\sin^4\theta}{F_y^2} = \frac{1}{\sigma_x^2} \tag{10.60}$$

for $0° \leq \theta \leq 90°$ with an applied stress σ_x on a single lamina. This theory checks test data for E-glass-epoxy laminae very closely (Reference 1).

The maximum strain theory, the maximum stress theory, and the Tsai-Hill theory will be used in the following examples.

It should be noted that the allowables in Equations (10.57) and (10.58) are ultimate values so that the actual temperature changes (ΔT) in the stress and strain equations in the previous Section 10.5 should be multiplied by an ultimate factor k_u. Probably, this factor k_u can be $k_u = 1.00$ for the case of residual laminate stresses caused by cooling from the laminate curing temperature.

Some allowable ultimate stress values for glass-epoxy, boron-epoxy, and graphite-epoxy laminae are given in Table 2.3 in Reference 1. The following values for uni-directional graphite-epoxy SP-286T300 are from Reference 3:

$$\begin{aligned}
&\text{Ply thickness } = 0.0051 \text{ in.,} \\
&E_x = 21.9(10^6) \text{ psi}, \quad E_y = 1.53(10^6) \text{ psi}, \\
&G_{xy} = 0.96(10^6) \text{ psi}, \quad \nu_{xy} = 0.31, \quad \nu_{yx} = 0.014, \\
&F_{xt} = 214{,}000 \text{ psi}, \quad S_{xt} = 0.00972, \\
&F_{xc} = 164{,}000 \text{ psi}, \quad S_{xc} = 0.01040, \\
&F_{yt} = 7800 \text{ psi}, \quad S_{yt} = 0.00541, \\
&F_{yc} = 30{,}000 \text{ psi}, \quad S_{yc} = 0.02565, \\
&F_{xy} = 10{,}500 \text{ psi}, \quad S_{xy} = 0.0110.
\end{aligned} \tag{10.61}$$

Example 10.9. Find the allowable ultimate applied load N_x for the graphite-epoxy cross-ply laminate in Examples 10.4 and 10.7. Assume $k_u = 1.0$ for the thermal stresses. (a) Use the maximum strain theory. (b) Use the Tsai-Hill theory. (c) Use the maximum stress theory. (d) Comment on thermal stress effects and the results for the three methods.

Solution (a) From Equation (10.53) the strains in the laminate are

$$10^6 \{e_0\} = \left\{ \begin{array}{c} 0.1294(N_x/t) + 0.5881(\Delta T) \\ -0.0042(N_x/t) + 0.2121(\Delta T) \\ 0 \end{array} \right\}. \tag{10.62}$$

From Equation (10.61) the allowable strains are

$$\begin{aligned}
\{S_0\}_t^T &= [0.00972 \quad 0.00541 \quad 0.0110], \\
\{S_0\}_c^T &= [0.01040 \quad 0.02565 \quad 0.0110],
\end{aligned} \tag{10.63}$$

whence from maximum strain theory,

$$\begin{aligned}
(N_x/t)_t &= [9720 - 0.5881(\Delta T)]/0.1294 \\
&= 75{,}100 - 4.5448(\Delta T) \\
&= 76{,}460 \text{ psi for } (\Delta T) = -300°\text{F}, \\
(N_x/t)_c &= [-10{,}400 - 0.5881(\Delta T)]/0.1294 \\
&= -80{,}400 - 4.5448(\Delta T) \\
&= -79{,}040 \text{ psi for } (\Delta T) = -300°\text{F}.
\end{aligned} \tag{10.64}$$

Note that

$$e_{0y} = [-0.0042(76{,}460) - 64] = -0.000385,$$

which is less than the allowable $S_{0t} = 0.00541$.

(b) From Equation (10.54) the stresses in lamina (1) are

$$\{\sigma\}_1 = \left\{ \begin{array}{c} \sigma_x \\ \sigma_y \\ \sigma_{xy} \end{array} \right\}_1 = \left\{ \begin{array}{c} 2.6122(N_x/t) + 10.16(\Delta T) \\ 0.0524(N_x/t) - 5.46(\Delta T) \\ 0 \end{array} \right\}. \tag{10.65}$$

Put Equation (10.59) in the form $(\sigma_{xy} = 0)$

$$(\sigma_x)^2 - \sigma_x\sigma_y + (F_x/F_y)^2(\sigma_y)^2 = (F_x)^2 \tag{10.66}$$

and substitute in σ_{x1}, σ_{y1}, F_x, F_y from Equations (10.61) and (10.65), whence

$$(N_x/t)_t = 19.85(\Delta T) + \left[4950(10^6) - 2070(\Delta T)^2\right]^{1/2}$$
$$= -5955 + 69{,}020 = 63{,}065 \text{ psi for } (\Delta T) = -300°\text{F},$$
$$= 70{,}356 \text{ psi for } (\Delta T) = 0°, \tag{10.67}$$
$$(N_x/t)_c = -3.797(\Delta T) - \left[3973(10^6) - 135.6(\Delta T)^2\right]^{1/2}$$
$$= 1159 - 62{,}935 = -61{,}796 \text{ psi for } (\Delta T) = -300°\text{F},$$
$$= -63{,}032 \text{ psi for } (\Delta T) = 0°.$$

(c) For the maximum stress theory of failure for both σ_{x1} and σ_{y1}, Equations (10.61) and (10.65) give

$$(N_x/t)_t = [214{,}000 - 10.16(\Delta T)]/2.6122$$
$$= 81{,}923 - 3.89(\Delta T)$$
$$= 83{,}090 \text{ psi for } (\Delta T) = -300°\text{F},$$
$$(N_x/t)_c = [-164{,}000 - 10.16(\Delta T)]/2.6122$$
$$= -62{,}782 - 3.89(\Delta T)$$
$$= -61{,}615 \text{ psi for } (\Delta T) = -300°\text{F}, \tag{10.68}$$
$$(N_x/t)_t = [7800 + 5.08(\Delta T)]/0.0524$$
$$= 148{,}555 + 196.9(\Delta T)$$
$$= 119{,}470 \text{ psi for } (\Delta T) = -300°\text{F},$$
$$= 90{,}385 \text{ psi for } (\Delta T) = -500°\text{F}.$$

(d) With $\alpha_x = 0$ and $\alpha_y = 4.0(10^{-6})$ for the graphite-epoxy, the thermal stress effects are small for all the failure theories considered, except for $(N_x/t)_t$ for the Tsai-Hill theory. The 70,356 psi is 12% larger than the 63,065 psi for $(\Delta T) = -300°\text{F}$. This reduction in $(N_x/t)_t$ is due to the large effect of the thermal stresses upon the σ_{y1} stress. The above values of $(N_x/T)_t$ in part (c) demonstrate this effect when $(N_x/t)_t$ is calculated from the $F_{yt} = 7800$ psi allowable tension strength σ_{y1}. The term

$$(F_x/F_y)^2(\sigma_y)^2 = F_x^2(\sigma_y/F_y)^2$$

in Equation (10.66) reduces the $(N_x/t)_t$ value to account for the relative large stress ratio for (σ_y/F_y).

It should be noted that the product term in Equation (10.59) is small for graphite-epoxy laminates in tension. In the case here,

$$(2.6122)(63{,}065)(0.0524)(63{,}065)/(214{,}000)^2 = 0.0171,$$

and for the material properties in Equation (10.61) the maximum possible value is

$$(214{,}000)(7800)/(214{,}000)^2 = 0.0364.$$

Thus, Equations (10.59) and (10.66) could be reduced to Equation (1.33) for graphite-epoxy laminae in tension.

For compression the Tsai-Hill theory and the maximum stress theory give approximately the same results, with the temperature change having little effect.

Example 10.10. For the three layer laminate in Examples 10.5 and 10.8 and Figure 10.4, (a) Find the ultimate applied load N_{xu} by an inelastic strain analysis. (b) Find the limit applied load N_{xL}. (c) Find the residual stresses after one load cycle. (d) Find the tension N_x load that gives permanent set but no residual stresses after one cycle. (e) Comment on the three theories of failure described above for this example, and discuss the temperature effects.

Solution. (a) From Equations (10.55) and (10.56)

$$10^6\{e_0\} = \left\{ \begin{array}{c} 0.0665(N_x/t) + 4.3189(\Delta T) \\ -0.0200(N_x/t) + 14.0437(\Delta T) \\ 0 \end{array} \right\},$$

$$10^6\{e_{0T}\} = \left\{ \begin{array}{c} -1296 \\ -4213 \\ 0 \end{array} \right\} \text{ for } \Delta T = -300°F,$$

$$\{\sigma_{1T}\} = \left\{ \begin{array}{c} -27,537 \\ -5151 \\ 0 \end{array} \right\} \text{ for } \Delta T = -300°F,$$

$$\{\sigma_{2T}\} = -\{\sigma_{1T}\}.$$

(10.69)

For the aluminum alloy (2) use $F_{ty} = 40,000$ psi, $F_{tu} = 60,000$ psi, $F_{cy} = -40,000$ psi, $F_{cu} = -60,000$ psi. The aluminum alloy is ductile with the stress-strain curve defined in Table 1.1 (Vol.1) with $n = 10$. From Table 1.1 (Vol.1) the elastic proportional limit is approximately $F_p = 0.60(40,000) = 24,000$ psi, which is smaller than the thermal stress $\sigma_{2Tx} = 27,537$ psi. Thus, for a completely elastic design no tension load N_x can be applied. However, a compression N_x can be applied.

Since this laminate is approximately uniaxial in the x-direction, an inelastic analysis can be made from the stress-strain curves. See Section 1.15 and Figure 1.15 (Vol. 1). In this case

$$(\alpha T)_{eq.} = 27,537/10^7 + 27,537/2(10^7)$$
$$= 0.00276 + 0.00138 = 0.00414,$$

which locates the stress-strain curves on Figure 10.5 to give $N_x = 0$ at $e_x = 0$. The graphite-epoxy curve (1) has origin at $e_x = 0.00138$ to give $\sigma_{1Tx} = -27,537$ psi at $e_x = 0$, while the aluminum alloy curve (2) has origin at $e_x = -0.00276$.

Use the maximum tension strain of 0.00972 for the graphite-epoxy, whence

$$e_{1t} = 0.00972, \quad e_{xtu} = 0.00972 + 0.00138 = 0.01110,$$
$$e_{2t} = 0.01110 + 0.00276 = 0.0136,$$
$$(10^7)(0.01386)/40,000 = 3.465,$$
$$\sigma_{2t} = 40,000(1.18), \quad \text{from Table 1.1(Vol.1)},$$
$$= 47,200 \text{ psi},$$
$$\sigma_{1t} = (0.00972)(20,000,000) = 194,400 \text{ psi},$$
$$(N_x/t)_t = [2t_1(194,400) + 2t_1(47,200)]/4t_1$$
$$= 120,800 \text{ psi for } \Delta T = -300°F,$$
$$= 119,200 \text{ psi for } \Delta T = 0°F.$$

(10.70)

For compression

$$e_{1c} = -164,000/20,000 = -0.00820,$$
$$e_{xcu} = -0.00820 + 0.00138 = -0.00682,$$
$$e_{2c} = -0.00682 + 0.00276 = -0.00406,$$
$$(10^7)(0.00406)/40,000 = 1.018, \quad \text{Table 1.1(Vol.1)},$$
$$\sigma_{2c} = -(40,000)(0.890) = -35,600 \text{ psi},$$
$$(N_x/t)_c = (-164,000 - 35,600)/2$$
$$= -99,500 \text{ psi for } \Delta T = -300°F,$$
$$= -103,500 \text{ psi for } \Delta T = 0°F.$$

(10.71)

(b) For tension limit loads, use Equation (10.70) to get

$$(N_x/t)_{tL} = (2/3)(120,800) = 80,500 \text{ psi}.$$

From Figure 10.5 the σ_{2t} will change very little for this limit load. Thus, assume $\sigma_{2tL} = 46,500$ psi, whence

Composite materials

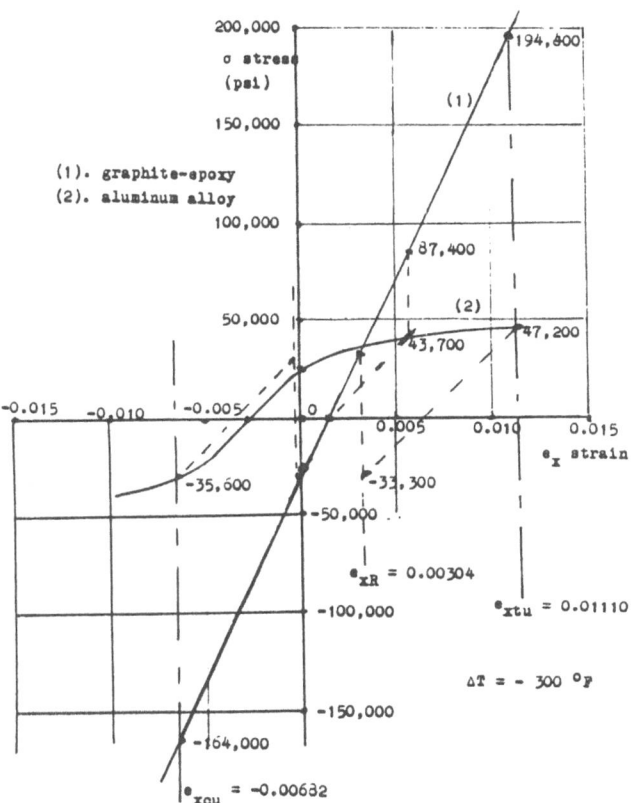

Fig. 10.5. Inelastic strain analysis for Example 10.10.

$$(\sigma_{1tL} + 46,500)/2 = 80,500,$$
$$\sigma_{1tL} = 114,500 \text{ psi},$$
$$e_{xtL} = 0.00138 + [114,500/2(10^7)] = 0.00715.$$
(10.72)

For compression limit load assume laminae (1) and (2) are elastic, whence

$$(N_x/t)_{cL} = (2/3)(-99,500) = -66,300$$
$$= [2(10^7)e_{1cL} + 10^7(e_{1cL} + 0.00414)]/2,$$
$$e_{1cL} = -0.00580,$$
$$e_{xcL} = -0.00580 + 0.00138 = -0.00442,$$
$$\sigma_{1cL} = -116,000 \text{ psi},$$
$$\sigma_{2cL} = -17,600 \text{ psi},$$
(10.73)

which are elastic.

(c) If the aluminum alloy unloads elastically, the removal of the N_x load will give residual stresses, as shown in Figure 10.5. For the ultimate tension $(N_x/t)_t$,

$$194,400 - \sigma_R = 2(47,200 + \sigma_R),$$
$$\sigma_{1tR} = 33,300 \text{ psi}, \quad \sigma_{2tR} = -33,300 \text{ psi},$$
$$e_{xR} = 0.00304.$$
(10.74)

Not only are these residual stresses larger than the original residual thermal stresses but also they have reversed signs. For the ultimate compression case,

$$\sigma_{1cR} = -30,000 \text{ psi}, \quad \sigma_{2cr} = 30,300 \text{ psi},$$
$$e_{xR} = -0.00014.$$

These stresses are slightly larger than the original thermal residual stresses.
For the limit load case,

$$\sigma_{1tR} = 7200 \text{ psi}, \quad \sigma_{2tR} = -7200 \text{ psi},$$
$$\sigma_{1cR} = -27,500 \text{ psi}, \quad \sigma_{2cR} = 27,500 \text{ psi}.$$

Note that in these cases cycling of the N_x loads is elastic after the first application and removal of N_x.

(d) From Figure 10.5 it is evident that there is a tension load N_x that will give no residual stresses when it is removed. This load can be read approximately from Figure 10.5 or calculated from Equation (1.94, Vol.1) where

$$10^7 e_2/40,000 = (\sigma_2/40.000) + (3/7)(\sigma_2/40,000)^{10}.$$

Use $e_2 = (\sigma_2/10^7) + 0.00414$, whence

$$\left(\sigma_2/40,000\right)^{10} = (7/3)(41,400/40,000),$$
$$\sigma_{2t} = 43,700 \text{ psi}, \quad \sigma_{1t} = 87,400 \text{ psi}, \quad (10.75)$$
$$(N_x/t)_t = 65,500 \text{ psi}.$$

Note that this load is 54% of the N_{xu} ultimate load and 81% of the N_{xL} limit load obtained above. If most of the tension operating loads on the laminate are less than $(N_x/t)_t = 65,000$ psi, then consideration could be given to loading the laminate to this load to remove the residual thermal stresses and allow the laminate to operate in the elastic range. Of course, there would be a small permanent set of $e_{ps} = 0.00138$.

(e) In this case of high strength elastic graphite-epoxy laminae bonded to a ductile aluminum alloy lamina, where the metal must go into the inelastic range to develop the strength of the graphite-epoxy laminae, an inelastic strain analysis must be used. The maximum strain theory of failure described above applies to laminae in an elastic laminate so that it does not apply to the case here. The maximum stress theory of failure and the Tsai-Hill theory give values too low for tension but satisfactory for compression. The comparison is shown in Equation (10.76).

Allowable Ultimate Loads

	Inelastic		Maximum Stress		Tsai-Hill	
	$-300°$	$0°$	$-300°$	$0°$	$-300°$	$0°$
$(N_x/t)_t$	120,800	119,200	49,000	90,200	52,500	90,200
$(N_x/t)_c$	$-99,500$	$-103,500$	$-103,300$	$-90,200$	$-99,100$	$-90,200$

(10.76)

The inelastic analysis shows that the temperature has little effect upon the ultimate loads, but the maximum stress and Tsai-Hill theories show large temperature effects for tension.

10.7. Interlamina stresses

The plane section plate theory used for the laminated plates in the above sections applies away from the free edges of the plate. Since the stresses are zero on the free edges but not zero in the interior, a shear redistribution of the loads must occur near the free edges. This problem is somewhat like the shear lag problem discussed in Chapter 6 (Vol.1). However, it is more complicated in that the

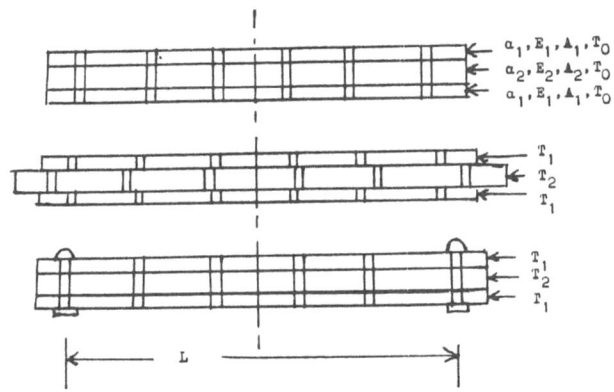

Fig. 10.6. Behaviour of bolted joints under thermal loads.

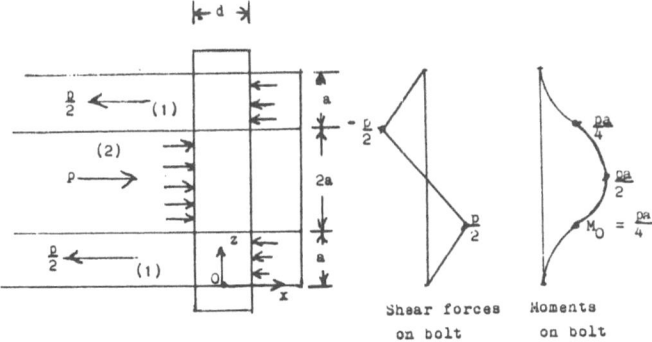

Fig. 10.7. Thermal forces on end bolts in long plate strip.

three dimensional stresses through the thickness of the laminated plate must be considered.

The case of residual thermal stresses in an unrestrained symmetric laminated plate will be examined here. From Example 10.10 above the residual thermal stresses away from the edges can be very large, but they must be zero at the free edges.

First, consider the case of two bars in Figure 10.6. If the two end bolts or rivets have no deflections (bending, shear, bearing, etc.), then they take all the thermal load, while the bolts between take no load. As shown in Section 1.3 of Reference 4, there are deflections so that bolts near the ends will take some of the load. However, if the length L is large so that $\Delta L = L(\alpha_2 - \alpha_1)(\Delta T)$ is about 0.10 inch or less, then the end bolts take practically all the load. If the end bolts or rivets fail, then the adjacent bolts will pick up the load. Failure of bolts or rivets will continue until the length is short enough so that the deflections reduce the loads enough to prevent failure. See Example 2.10 above.

For long laminated plates, it follows that the bonds between the laminae will have large stresses near the free edges so that delamination may occur.

In order to examine the stresses near the edges, consider the bolt case in Figure 10.7. On the bolt in bottom lamina (1)

$$M = M_0(z/a)^2, \qquad \sigma_{zz} = Mx/I = (M_0x/I)(z/a)^2, \tag{10.77}$$

where $-d/2 \leq x \leq d/2$, $0 \leq z \leq a$. From the equilibrium Equations (1.4, Vol.1),

$$\begin{aligned}
&\sigma_{xz,x} + \sigma_{yz,y} + \sigma_{zz,z} = 0, \\
&\sigma_{xx,x} + \sigma_{xy,y} + \sigma_{xz,z} = 0, \\
&\sigma_{yz,y} = 0, \qquad \sigma_{xy,y} = 0, \\
&\sigma_{xz,x} = -\sigma_{zz,z} = -2M_0 x^2/Ia^2 \\
&\sigma_{xz} = -(M_0 z x^2/Ia^2) + C_1 z + C_2, \\
&\sigma_{xx,x} = -\sigma_{xz,z} = (M_0 x^2/Ia^2) - C_1, \\
&\sigma_{xx} = (M_0 x^3/3Ia^2) - C_1 x + C_3,
\end{aligned} \tag{10.78}$$

where for simplicity it has been assumed that the bolt is a square pin.

To get the constants of integration use

$$\begin{aligned}
&\sigma_{xz} = 0 \text{ at } x = \pm d/2, \\
&\sigma_{xx} = 0 \text{ at } x = d/2,
\end{aligned} \tag{10.79}$$

whence

$$\begin{aligned}
&C_1 = M_0 d^2/4Ia^2, \qquad C_2 = 0, \qquad I = d^4/12, \qquad M_0 = pa/4, \\
&\sigma_{xz} = (M_0/Ia^2)[x^2 - (d/2)^2] = (3p/4d^2)[4(x/d)^2 - 1], \\
&C_3 = M_0 d^3/12Ia^2, \\
&\sigma_{xx} = (p/4ad)[4(x/d)^3 - 3(x/d) + 1], \\
&\sigma_{zz} = (3pa/d^3)(x/d)(z/a)^2.
\end{aligned} \tag{10.80}$$

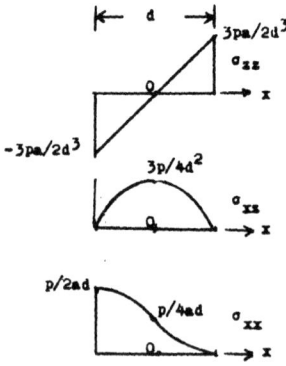

Fig. 10.8. Stress distributions on square pin at z=a, Fig. 10.7.

Composite materials

The stress distributions on the bolt at $z = a$ are shown in Figure 10.8. There are two stresses σ_{zz} and σ_{xz} due to the σ_{xx} stress acting on the shear plane where the outer lamina (1) and the middle lamina (2) join, at $z = a$ in Figure 10.8. If the layers are bonded instead of bolted, then the bond must transmit these two stresses σ_{zz} and σ_{xz} between the layers. Figure 10.9 shows a sketch of the stress distributions that act on the bond near the free edge of the laminated plate.

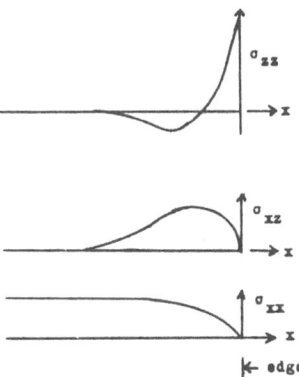

Fig. 10.9. Interlamina stresses on bond line at free edge of laminated plate, due to σ_{xx}.

These stresses require a more complicated analysis than the bolt case, but they have a somewhat similar form and may be considerably larger in magnitude. Not only must the bond take large shear stresses σ_{xz} near the edge, it must take the tension stress σ_{zz}, which tends to peel the lamina apart at the edge.

This discussion has been restricted to thermal forces in a simple three layer laminate. See Section 4.6 of Reference 1 for similar type interlamina stresses for various applied loading situations on angle-ply laminates.

If a shear stress σ_{xy} also acts on the laminate, then it must be zero on the free edge. From Equation (1.4, Vol.1) with $\sigma_{yy,y} = 0$,

$$\sigma_{xy,x} + \sigma_{yz,z} = 0, \tag{10.81}$$

whence

$$\sigma_{yz} = -\int_z \sigma_{xy,x}\, dz + C_4. \tag{10.82}$$

Figure 10.10 shows a sketch of these stresses near the free edge of the laminated plate (Reference 1).

The interlamina stresses not only occur on free edges but also at holes, broken fibers, and splices.

Example 10.11. (a) Calculate the maximum bolt stresses for the thermal stresses in Equation (10.69), Example 10.10, using Figure 10.8. (b) Discuss the results as applied to interlamina

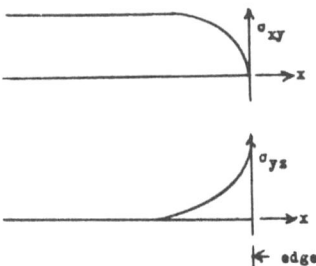

Fig. 10.10. Interlamina stresses on bond line at free edge of laminated plate, due to σ_{xy}.

stresses in Figure 10.9. (c) Comment on the thermal stress results for the cross-ply graphite-epoxy laminate in Example 10.7.

Solution. (a) Assume $d = 1/4$ inch square pins at 1.00 inch spacing on the edge in Figure 10.7. For $a = t_1$, Figure 10.8 and Equation (10.69) give

$$p = 2t_1(1)(27,537) = 55,000t_1,$$
$$(\sigma_{zz})_{max} = 3pa/2d^3 = 5,280,000t_1^2$$
$$= 52,800 \text{ psi for } t_1 = 0.10 \text{ in.,}$$
$$(\sigma_{xz})_{max} = 3p/4d^2 = 660,000t_1$$
$$= 66,000 \text{ psi for } t_1 = 0.10 \text{ in.}$$
(10.83)

The bearing stresses on the bolt and laminate are

$$\sigma_{br} = p/2t_1 d = 55,000t_1/(2t_1/0.25)$$
$$= 110,000 \text{ psi independent of } t_1.$$
(10.84)

(b) Since the allowable epoxy bond stresses in Equation (10.61) are about $F_{xz} = 7800$ psi and $F_{zz} = 10,500$ psi, the stresses in part (a) would cause delamination. To reduce the stresses to the allowables would require that the thin bond be able to spread the stresses over a distance of about twice the laminate thickness of $4t_1 = 0.40$ in. However, in Section 4.6 of Reference 1 it is shown that three dimensional solutions and test data indicate the interlamina stresses occur in a distance approximately equal to the laminate thickness. Thus, the aluminum alloy and graphite-epoxy laminate considered in Examples 10.5, 10.8, 10.10, and 10.11(a) would probably delaminate on the free edges from the thermal stresses induced by cooling about 300°F from the curing temperature of the laminate.

(c) The maximum thermal stress in Equation (10.54), Example 10.7, is

$$\sigma_{1xT} = 10.16(\Delta T) = -3048 \text{ psi for } \Delta T = -300°F.$$

Since this stress is about 11% of that in Equation (10.83) above, the interlamina stresses will be small so that no delamination will occur. Of course, the applied loads may produce interlamina stresses that could be a problem.

10.8. Joints in laminated plates

Bolted and bonded joints or a combination of the two are used in laminated plate splices. Failure of the bolted joint may occur as bolt shear, plate tension, bearing, tear-out, and cleavage (tension in laminates perpendicular to load direction). Bearing, tear-out, and cleavage may involve the matrix material only, which has low allowable stresses. Also, most composite materials have low elongation so that there are large stress concentrations at the bolt holes, and it is difficult to redistribute the loads near the holes.

Bonded joints for metals have been dicussed in Section 2.10 above. Most of the results there apply to laminates. However, interlamina stresses may add to the splice effects on shear and tension stresses in the bond.

See Section 4.7.4 in Reference 1 for a discussion of laminated joints. Also, see References 6-8 and 11.

As demonstrated in Chapter 2 above the design of efficient joints in ductile metals requires complicated analyses. The design problem for joints in brittle composite materials is much more complicated and difficult than for metals. In many cases it is necessary to increase the weight and space for the joint in order to take the full design load. Titanium laminae can be added to the laminate in the splice region to increase bearing and cleavage strength for a bolted joint.

10.9. Bending deflections of laminated plates

Isotropic plate bending equations have been given above in Sections 7.5, 7.6, and 7.7 for small deflections. These isotropic equations can be modified in a simple manner for the particular orthotropic case in which $[B] = 0$ in Equation (10.25). Also, the equations will be simpler if the laminated plates are made in such a way that the C_{13} and C_{23} terms are zero in the $[C]$ matrix in Equation (10.25), or

$$[C] = \begin{bmatrix} C_{11} & C_{12} & 0 \\ C_{12} & C_{22} & 0 \\ 0 & 0 & C_{33} \end{bmatrix}. \tag{10.85}$$

Under these conditions the isotropic $[J]^{-1}$ matrix in Equation (7.96) becomes the $[C]$ matrix, or

$$(t^3/12)[J]^{-1} \to [C]. \tag{10.86}$$

Thus, the differential Equation (7.96) for the bending deflection u_z of the laminated plate becomes

$$[d_{xx} \quad d_{yy} \quad 2d_{xy}] \begin{bmatrix} C_{11} & C_{12} & 0 \\ C_{12} & C_{22} & 0 \\ 0 & 0 & C_{33} \end{bmatrix} \begin{Bmatrix} d_{xx} \\ d_{yy} \\ 2d_{xy} \end{Bmatrix} u_z$$
$$- \{J_{bdd}\}^T \{M_E\} - p_z - p_{zd} = 0, \tag{10.87}$$

or

$$C_{11}u_{z,xxxx} + 2(C_{12} + 2C_{33})u_{z,xxyy} + C_{22}u_{z,yyyy} -$$
$$- \{J_{bdd}\}^T \{M_E\} - p_z - p_{zd} = 0. \tag{10.88}$$

Note that the force boundary conditions in Equations (7.97) and (7.98) will be changed. Also, the approximate solution form in Equation (7.99) can be used in Equation (10.93) with the change in Equation (10.86). This will change the multiplying factors in the integrals in Equation (7.101) and remove the terms D from I_{ij}^{c6}. For the integrals in Equation (7.101),

$$I^{c1}_{mnij} = C_{11} \int_{A_m} G''_m H_n G''_i H_j dx\, dy,$$

$$I^{c2}_{mnij} = C_{12} \int_{A_m} G_m H''_n G''_i H_j dx\, dy,$$

$$I^{c3}_{mnij} = C_{22} \int_{A_m} G_m H''_n G_i H''_j dx\, dy, \qquad (10.89)$$

$$I^{c4}_{mnij} = C_{12} \int_{A_m} G''_m H_n G_i H''_j dx\, dy,$$

$$I^{c5}_{mnij} = 4C_{33} \int_{A_m} G'_m H'_n G'_i H'_j dx\, dy,$$

Omit D from I_{ij}^{c6}.

Example 10.12. Solve Example 7.4, part (a), for the bending deflections of a simply supported rectangular laminated plate.
Solution. In Example 7.4 modify the results for the integrals to

$$\begin{aligned} I^{c1}_{mnmn} &= C_{11}(m\pi/a)^4(ab/4), \quad I^{c3}_{mnmn} = C_{22}(n\pi/b)^4(ab/4), \\ I^{c2}_{mnmn} &= C_{12}(mn\pi^2/ab)^2(ab/4) = I^{c4}_{mnmn}, \\ I^{c5}_{mnmn} &= 4C_{33}(mn\pi^2/ab)^2(ab/4), \end{aligned} \qquad (10.90)$$

whence

$$(ab\pi^4/4)[(m/a)^4 C_{11} + 2(mn/ab)^2(C_{12} + 2C_{33}) + (n/b)^4 C_{22}]C_{mn} = F_{mn} C_{mn}$$
$$= P_{ST} - (m\pi/a)^2(ab/4)N_x C_{mn} + P_{z1}\sin(m\pi x_1/a)\sin(n\pi y_1/b),$$

$$C_{mn} = \frac{P_{ST} + P_{z1}\sin(m\pi x_1/a)\sin(n\pi y_1/b)}{F_{mn} + (m\pi/a)^2(ab/4)N_x}, \qquad (10.91)$$

$$u_z = \sum_m \sum_n C_{mn} \sin(m\pi x/a)\sin(n\pi y/b).$$

If $p_z = p_{z0} = $ constant and all other loads are zero, then

$$\begin{aligned} C_{mn} &= 16 p_{z0}/\pi^6 mn G_{mn}, \\ G_{mn} &= (m/a)^4 C_{11} + 2(mn/ab)^2(C_{12} + 2C_{33}) + (n/b)^4 C_{22}, \end{aligned} \qquad (10.92)$$

for m, n odd. This result checks Section 88 in Reference 5.

Other results for bending deflections of orthotropic plates are given in Section 5.3 of Reference 1 and Chapter 11 of Reference 5.

Note that a negative N_x, or compression axial load, can make the denominator zero for C_{mn} in Equation (10.91). This defines a N_{xcrmn} buckling load for the plate, or

$$N_{xcrmn} = -\left(\frac{a\pi}{m}\right)^2 G_{mn}, \quad G_{mn} \text{ in Equation (10.92)}, \qquad (10.93)$$

which is the same as Equation (5.65) in Reference 1.

10.10. Buckling loads for laminated plates

The buckling of isotropic plates has been considered in Section 7.10 above. Use Equation (10.88) to change the buckling Equation (7.164) to

$$C_{11} u_{z,xxxx} + 2(C_{12} + 2C_{33}) u_{z,xxyy} + C_{22} u_{z,yyyy} + c p_{zd0} = 0, \qquad (10.94)$$

where cp_{zd0} is defined in Equation (7.164). Equations (7.165) and (7.166) still apply provided D is removed and the integrals in $[I^c]$ have the form in Equation (10.89).

For the simply supported case with $N_x = cN_{x0}$, $N_y = cN_{y0}$,

$$u_z = \sum_m \sum_n C_{mn} \sin(m\pi x/a) \sin(n\pi y/b), \tag{10.95}$$

it follows from Equation (10.94) that

$$c_{crmn} = \frac{\pi^2 G_{mn}}{(m/a)^2 N_{x0} + (n/b)^2 N_{y0}}, \tag{10.96}$$

where G_{mn} is in Equation (10.92).

Example 10.13. Find the minimum values of N_{xcrmn} in Equation (10.93) or (10.96) for given a/b values with $N_y = 0$.

Solution. Omit the minus sign and write Equation (10.93) in the form

$$N_{xcrmn} = k_{mn}\pi^2 C_{11}/b^2, \tag{10.97}$$
$$k_{mn} = m^2(b/a)^2 + 2(C_{12} + 2C_{33})/C_{11} n^2 + (n^4/m^2)(C_{22}/C_{11})(a/b)^2.$$

Follow the procedure for the isotropic plate in Example 7.8. With respect to n, the smallest k_{mn} will occur at $n = 1$. Fix m and use $dk_m/d(a/b) = 0$ to get the minimum $k_{\min}$ as

$$k_{\min} = 2(C_{22}/C_{11})^{1/2} + (2C_{12} + 4C_{33})/C_{11} \tag{10.98}$$
$$\text{at } a/b = m(C_{11}/C_{22})^{1/4}.$$

Since m is a positive integer, this determines values of a/b for given (C_{22}/C_{11}) and m. For a/b values between m and $m+1$, k_m will be larger with the largest values occurring at the change-over from m half-waves to $m+1$ half-waves, or

$$k_m = \frac{2C_{12} + 4C_{33}}{C_{11}} + \left(\frac{m}{m+1} + \frac{m+1}{m}\right)\left(\frac{C_{22}}{C_{11}}\right)^{1/2} \tag{10.99}$$
$$\text{at } a/b = [m(m+1)]^{1/2}(C_{11}/C_{22})^{1/4}.$$

The graph of k_m against a/b for $n = 1$ in Equation (10.97) is similar to the isotropic case in Figure 1.14 above.

See Section 5.4 of Reference 1 for other cases of laminated plate buckling.

10.11. Vibrations of laminated plates

Put the form in Equation (10.94) into Equation (7.178) to get

$$C_{11}u_{z,xxxx} + 2(C_{12} + 2C_{33})u_{z,xxyy} + C_{22}u_{z,yyyy} +$$
$$+ (t\rho)_{eq} u_{z,tt} = 0, \quad (t\rho)_{eq} = \sum_{i=1}^N t_i \rho_i. \tag{10.100}$$

Equation (7.179) becomes

$$T_{,tt} + \omega^2 T = 0, \tag{10.101}$$
$$C_{11}\left(\frac{X''''}{X}\right) + 2(C_{12} + 2C_{33})\left(\frac{X''Y''}{XY}\right) + C_{22}\left(\frac{Y''''}{Y}\right) - (t\rho)_{eq}\omega^2 = 0.$$

Follow the procedure in Section 7.11 above and use Equation (7.180) with

$$(t\rho)_{eq}\omega^2 = C_{11}p^4 + 2(C_{12} + 2C_{33})p^2q^2 + C_{22}q^4. \tag{10.102}$$

From Equation (7.184) for the simply supported plate

$$p = m\pi/a, \quad q = n\pi/b, \tag{10.103}$$
$$X_m Y_n = \sin(m\pi x/a)\sin(n\pi y/b),$$
$$\omega_{mn}^2 = [\pi^4/b^4(t\rho)_{eq}][C_{11}(b/a)^4 + 2(C_{12} + 2C_{33})m^2n^2(b/a)^2 +$$
$$+ C_{22}n^4]. \tag{10.104}$$

This result checks formulas given in Section 5.5 of Reference 1 and in Chapter 9 of Reference 9. Results for other cases are given in both of these references.

Equations (7.188) and (7.189) still apply provided that D is removed and the integrals in $[I^c]$ have the form in Equation (10.69).

10.12. Problems

10.1. Check the expressions in Equation (10.19).

10.2. (a) Draw curves for the datum material properties in Equation (10.19) for $0° \leq \theta \leq 90°$. (b) Find the maximum and minimum values on the curves.

10.3. Repeat Example 10.1 for a fiber glass-epoxy lamina with

$$(E_x/E_y)_k = 3.0, \quad (G_{xy}/E_y)_k = 0.5, \quad \nu_{xyk} = 0.25.$$

10.4. Find the stresses in each lamina in Example 10.3 by using expressions from Equation (7.30).

10.5. Solve Example 10.4, part (b), for glass-epoxy laminae with $E_x = 7.8(10^6)$ psi, $E_y = 2.6(10^6)$ psi, $\nu_{xy} = 0.25$, $G_{xy} = 1.3(10^6)$ psi, $\alpha_x = 3.5(10^{-6})/°F$, $\alpha_y = 11.4(10^{-6})/°F$.

10.6. Solve Example 10.5, part (b), for glass-epoxy outer laminae with properties as in Problem 10.5, and a aluminum alloy middle lamina with $E_2 = 10^7$ psi, $\nu_{12} = 0.3$, $\alpha_2 = 13(10^{-6})$.

10.7. Check the calculations in Example 10.6.

10.8. Calculate numerical values in Example 10.6 for steel alloy lamina (1) with $E_1 = 3(10^6)$ psi, $\nu = 0.3$, $\alpha_1 = 7(10^{-6})$, and aluminum alloy (2) with $E_2 = 10^7$ psi, $\nu_{12} = 0.3$, $\alpha_2 = 13(10^{-6})$. Use $t_2/t_1 = 1.0$, $t = 3t_1$, and express the stresses in terms of N_x/t, N_y/t, N_{xy}/t, and ΔT.

10.9. Check the calculations in Example 10.7.

10.10. Show that the results in Example 10.7 satisfy the equilibrium and compatibility equations for the laminate.

10.11. Use the results from Problem 10.4 above and repeat Example 10.7.

10.12. Check the calculations in Example 10.8.

10.13. Show that the results in Example 10.8 satisfy the equilibrium and strain compatibility equations for the laminate.

10.14. Solve Example 10.5, part (b), for glass-epoxy instead of graphite-epoxy, using properties in Problem 10.5 above.

10.15. Use the results from Problem 10.14 and repeat Example 10.8.

10.16. Check the calculations in Example 10.9.

10.17. Solve Example 10.9 for $N_y = 0.80N_x$ instead of 0.

10.18. Check the calculations in Example 10.10

10.19. Solve Example 10.10 for $\Delta T = -200°F$.

10.20. Solve Example 10.10 for $\Delta T = +200°F$.

10.21. Solve Example 10.10 for $N_y = 0.5N_x$.

10.22. Solve Example 10.11 for $D = \frac{3}{8}$ in. at 1.5 in. spacing.

10.23. (a) Evaluate the C_{ij} in Equation (10.25) for the cross-ply symmetric laminate in Example 10.4 for $t_1 = 0.100$. (b) Put the results into Equations (10.92) and (10.91) with

$m = 1, n = 1$, to get the center deflections for $a = b$. (c) Compare the results to isotropic plate Equation (7.105) with $E_{av} = 7.67(10^6)$ psi, $\nu = 0.30$, and $t = 6t_1 = 0.60$ in. (d) Calculate buckling loads from Equations (10.93) and (7.110).

10.24. Use the C_{ij} from part (a) of Problem 10.23 and graph Equation (10.97) for k_m against a/b with $n = 1$. Use $m = 1, 2$, and 3 and get minimum and maximum applicable values from Equations (10.98) and (10.99).

10.25. Use the C_{ij} from part (a) of Problem 10.23 and compare frequencies given by Equations (10.104) and (7.185).

References

Chapter 10

1. R.M. Jones : *Mechanics of Composite Materials*, McGraw-Hill Book Co. (1975).
2. I.S. Sokolnikoff: *Mathematical Theory of Elasticity*, McGraw-Hill Book Co. (1956).
3. I.M. Daniel: *Strain amd Failure Analysis of Graphite-epoxy Plates with Cracks*. Experimental Mechanics, Vol. 18, July (1978).
4. B.E. Gatewood: *Thermal Stresses*, McGraw-Hill Book Co. (1957).
5. S. Timoshenko and S. Woinowsky-Krieger: *Theory of Plates and Shells*, 2nd Ed., McGraw-Hill Book Co. (1959).
6. M. Goland and E. Reissner: The stresses in cemented joints, *J. of Applied Mechanics*, Pages A-17–A-27, March (1944).
7. *Plastics for Aerospace Vehicles, Part I, Reinforced Plastics*, Military Handbook 17A, latest revised edition.
8. *Structural Design Guide for Advanced Composite Applications*, Air Force Materials Laboratory, Advanced Composite Divison, latest revised editon.
9. A.W. Leissa: *Vibration of Plates*, NASA SP-160 (1969).
10. S.W. Tsai and H.T. Hahn: *Introduction to Composite Materials*, Technomic Publishing Co., Westport, CT 06880 (1980).
11. J.R. Vinson and R.L. Sierakowski: *The Behavior of Structures Composed of Composite Materials*. Martinus Nijhoff Publishers, Dordrecht/Boston/Lancaster (1986).

References for additional reading

12. B.B. Hoskin and A.A. Baker: *Composite Materials for Aircraft Structures*, AIAA Education Series (1986).
13. T.J. Reinhart, (Editor): *Engineered Materials Handbook, Volume 1, Composites*, ASM International (1987).

Index

aerodynamic loads 207-209
aeroelasticity 188
allowable combined stresses 22
allowable compression stresses 11
  columns 19
  combined buckling 28
  local buckling 11
  local crippling 18
  plate buckling 243
  yield 3-5, 8, 9
allowable stresses
  bearing 3-9
  combined 22
  compression (*see* allowable compression stresses)
  crack effects 40
  creep effects 31
  fatigue effects 35
  laminated plates 343
  shear 3-9
  temperature effects 6-7, 38
  tension 3-9
aluminum alloys
  design curves 89-111
  elevated temperature 7-9
  room temperature 3-5
approximate solutions by
  assumed displacement functions 160, 215, 229, 239, 243-246
  assumed mixed displacement and stress functions 179, 225, 236
  assumed stress functions 218, 234
  finite elements 252-302
approximate solutions for
  beams 160-172, 178-184
  columns 173
  frequencies and mode shapes 194
  plates 211-250
assembly of finite elements 321

beam-columns 173
beam finite elements 305-310

beams
  approximate solutions 162-172, 177-183
  diagonal tension 121-123
  fabric 152-154
  tapered 189
  thermal stresses 309
  two elements 173
  vibration 188-201
bending frequency 191-192
bending stresses in thin shells 132-150
Bessel functions 203
biaxial tension stresses 24
bonded joints 81-85
box beams 114-121
buckling coefficients
  angles 13
  angle stiffened panels 15
  channels 14
  flat plates 12
  Z-sections 14
  Z-stiffened panels 16
buckling stresses (*see* allowable compression stresses)
buckling of
  composite plates 354
  plates 243
  thin shells 150
bulkheads 132
butt joints 49

characteristic values (*see* eigenvalues)
$[C]$ matrix (*see* flexibility matric $[C]$)
columns
  buckling 19, 167-175
  tapered 169-175
combined stresses 23
comparisons of solutions by virtual principles
  beams 176, 184
  finite element derivations 257-260
  finite elements 321-326
composite materials 329-356

358

Index

coordinate axes
  plates 212
  pressure vessels 126-127
crack effects 40
creep effects 31
crippling stresses 18
critical values (*see* eigenvalues)
cut-out in shell 129
cylindrical structures
  axially-symmetric 132
  circular 126-142
  non-circular fuselage 144-150

datum coordinates
  beam elements 311
  triangular plate elements 315, 316
deflections from the principle of virtual displacements
  beam approximate solutions 160-174
  laminated plates 353
  plate approximate solutions 215, 229
  plate differential equations 215, 229
  plate finite elements 260
deflections from the mixed virtual principle
  beam approximate solutions 179
  beam differential equations 179
  plate approximate solutions 225, 236
  plate differential equations 225, 236
  plate finite elements 260
deflections from the principle of virtual forces
  beam approximate solutions 175-178
  plate approximate solutions 220, 236
  plate differential equations 218, 234
  plate finite elements 260
deflection of thin shells 127
design curves
  buckling and crippling stresses 96-100
  columns 91-96
  columns with local buckling 100-104
  fabric beam 153-155
  shear web attachments 66
  stiffened panels 105-114
design procedures
  box beams 117-121
  compression of thin web structures 121-123
  diagonal tension beams 121-123
  load intensity on wings 114
  pressure stabilized structure 152-155
  splices 65-69
determinants for characteristic equations 167, 190, 201, 244, 248
diagonal tension beams 121-123
dual and analogy relations for plate solutions
  bending stress functions and inplane displacements 235
  inplane stress functions and bending displacements 232
  inplane and bending $[C_\sigma]$ finite element matrices 258, 259

$[k]$ and $[C_g]$ finite element matrices 259
plate buckling and vibrations 248
dynamics of beams 188-210
dynamic loads on beams 207-209
dynamic magnification of loads 208

eccentric loads on joints 54
effect of load intensity on wing design 114
effective areas for stiffened panels 112, 114, 119
effective skin width 114, 119
efficiencies of aircraft materials for
  columns 89-96
  columns with local buckling 100-104
  local buckling and crippling 96-100
  local buckling and crippling of stiffened panels 105-109
  stiffened panel columns 109-112
  temperature effects 96, 101
eigenvalues for
  columns 167
  plates 244, 249
  vibrations 190, 201
elastic constants 3-5, 8, 9, 332
equilibrium equations for
  approximations 161
  plates 215, 229-231
equilibrium matrix $[s]^T$ for beams 312-313
Euler's column formula 19

fabric beams 152-154
failure stresses (*see* ultimate stresses)
failure theories for combined stresses 25, 343
fatigue effects on allowable stresses 35
fatigue life 35-40
  crack effects on 40
  temperature effects on 38
finite elements for
  general beam 305-315
  plates (*see* plate finite elements)
flat plates (*see* plates)
flexibility matrix $[C]$
  general beams 306
  plates 260, 266-275, 285-294
flutter 201-202
  one degree of freedom 202
  two degrees of freedom 204
forced vibration of beams 192-193
frame structures 137, 155
frequencies
  approximate solution for 194, 200
  bending 191
  bending with finite elements 200
  torsion 197
fuselage
  bulkheads 132
  circular rings 132
  frames 137
  loading 142-147
  non-circular rings 142-150

pressurized 138, 142-150
stringers 139

Galerkin method 158
geometric shear stresses 173
geometric stiffness matrix 173
graphite-epoxy laminate 339, 344
harmonic motion 191, 194
Hilbert matrix 162
honeycomb panels 116-118, 143
hybrid flexibility matrix $[C_\sigma]$ 251-260
hybrid stiffness matrix $[k]$ 257-260, 272
hybrid stress method 275

Inconel X-750 89-112
indeterminate structures (*see* redundant structures)
inelastic effects
  columns 15-21
  composite plates 343-348
  splices 74-75
influence coefficients, flexibility 175
interlamina stresses 348
internal pressure 126-156
  bending stresses 132-156
  membrane stresses 126-141
  redundant beam 155
  redundant rings 143-152
  stabilized beams 152

$[J]$ matrix 213
joints
  bonded 80-81
  butt 49
  eccentric loading 54
  laminated plates 352
  lap 45
  welded 77
  (*see* splices)
Jordan matrix transformation $[T_i]$ 325

$[k]$ stiffness matrix (*see* stiffness matrix $[k]$ and hybrid stiffness matrix $[k]$)

$[\lambda]$ rotation matrix (*see* rotation matrices)
lamina 329
laminate 329
laminated plates
  allowable stresses 343
  buckling 354
  deflections 353
  forces 335
  joints 352
  moments 335
  stresses 340
  vibrations 355
Larson-Miller parameter 32
life-fraction law for
  creep 34
  fatigue 37
list of selected equations in Volume 1, xv

load factor 114
load intensity design for wings 114
local coordinates 270, 305, 316

margin of safety (M.S.) 45-49
material property matrix $[J]$ 213, 331, 332
matrices
  apparent load $\{P_E\}$ 262, 282
  apparent moment $\{M_E\}$ 229
  applied load $\{P\}$ 262, 282
  displacement $\{u\}$, $\{u_E\}$ 267
  equilibrium $[s]^T$ 312, 318
  flexibility $[C]$ (*see* flexibility matrix $[C]$)
  hybrid flexibility $[C_\sigma]$ 252-260
  hybrid stiffness $[k]$ 257-260, 272
  Jordan transformation $[T_i]$ 325
  material property $[J]$ 213, 331, 332
  mixed $[M]$ (*see* mixed matrix $[M]$)
  rotation $[\lambda]$ 311, 318, 320
  rotation $[R]$ 311, 312-320
  stiffness $[k]$ (*see* stiffness matrix $[k]$)
membrane stresses 126-142
Miner's law 37
minimum weight design 51, 89-112
mixed matrix $[M]$
  general beam 308
  plates 260, 294, 296, 319-321
mode acceleration method 208
mode displacement method 208
mode shapes 191
modulus of elasticity 8, 10
Mohr's circle 23
moments
  plate bending axes 212
  plate M_x, M_y, M_{xy} 212, 229-339

natural vibrations 194
non-homogeneous materials 329
non-isotropic materials 329
numerical integration 169, 175

optimum design 118
orthogonal functions 165, 192-193
orthogonal mode shapes 192-193
orthotropic materials 331

$\{P\}$ applied load matrix 262, 282
$\{P_E\}$ apparent load matrix 262, 282
plate finite elements
  rectangular with bending 287
  triangular with bending 280-300
  triangular with inplane loads 260-280, 299

plates 211
  approximate solutions 213-242
  bending stresses 229-239
  buckling loads 243-245
  inplane stresses 213-229
  laminated (*see* laminated plates)
  large deflections 240-243

Index

vibration 245-249
potential functions 219
pressure loads
  distributed on plates 232, 233
  internal 126-156
pressurized fuselage 137, 143-150
  (*see* pressurized shells)
pressurized shells
  bending stresses 132, 137
  circular frames or rings 133-141
  circular holes 129
  deflections 127
  frames and stringers 137
  membrane stresses 126-141
  non-circular rings 143-150
  stabilized 152
principle of mixed virtual stresses and virtual displacements
  approximate solutions 178-184, 225, 236
  beams 308
  finite elements 260, 270-280, 294-300
  plates 225, 236
principle of virtual displacements
  approximate solutions 160-174, 200, 215, 231
  beams 308
  columns 166
  dynamics of beams 188-200
  finite elements 200, 260-265, 280-285
  plates 213, 229, 243, 247
principle of virtual forces
  approximate solutions 175, 220, 235
  beams 307
  finite elements 260, 265-275, 285-294
  inelastic joints 75-77
  plates 218 233

quasi-steady displacements 209

Ramberg-Osgood equation 20
Rayleigh-Ritz method 158
redundant structures
  frame 155
  rings 144, 147
rotation of axes 23
rotation matrices $[R]$ and $[\lambda]$
  beams 311-313
  triangular plate elements 315

Saint Venant's principle 157
shear modulus of elasticity G 3-5
shear stresses
  buckling 12
  diagonal tension beams 121-123
shear web splices 66
shells (*see* pressurized shells)
Simpson's rule 172
S-N fatigue curve 35-39
splices
  deflection effect 69
  I-beam 65

inelastic effects 74, 75
multi-row tension 49
rectangular beam cross section 57
shear web 65
temperature effects 74, 75
tension 44-45
T-section 53
  (*see* joints)
$[s]^T$ equilibrium matrix for beams 312-313
stiffness matrix $[k]$
  general beam 308
  geometric 173
  plates 260, 316
strains
  additional e_E 75, 213, 276
  geometric shear 173
  inelastic 76
  non-linear displacement 75, 240
  temperature or thermal 75
strain e_E effects
  displacements $\{u_E\}$ 267
  internal plate loads $\{N_E\}$ 213-225
  internal plate moments $\{M_E\}$ 229, 232, 237
stress-density ratio 89, 90
stresses
  axial 44
  bending 229, 234, 236
  inplane 213, 218, 224, 225, 228
  interlamina 348
  laminated plate 340
  membrane 126-142
  plates 212-245
stress functions
  finite elements 257, 260
  inplane loading 218
  plate bending 233, 234
stress-strain equations
  non-isotropic 331-335
  three dimensional 331
  two dimensional 213, 229, 333
surface forces 214

tapered beams 169
temperature effects
  beams 310
  columns 94-96, 100-104
  composite plates 340, 348
  creep 31-35
  displacements 162, 163, 171, 172, 218
  fatigue 35-40
  inelastic 74
  joints 75-77
  material properties 6-10, 89, 90, 95
  plates 214-244, 262
  two-dimensional 214-244
Theodorssen function 202-203
thin shells 126-137
  axially symmetric 132
  bending stresses 132
  cut-outs 129

membrane stresses 126-142
titanium alloy Ti-4Al-3Mo-1V 89-112
trapezoid rule 177
two dimensional equations 214, 229
two element beams 173

ultimate stresses
   bearing, tension, shear 3-11, 45
   temperature effects 6-11, 38
   (*see* allowable stresses and allowable compression stresses)
unit load theorem
   fuselage rings 144-150
   frames 155
   joints 155
   numerical integration 169, 175

vibrations
   beam in bending 188-189
   beam in torsion 197
   laminated plates in bending 355, 356
   plates in bending 246
virtual displacements (*see* principle of virtual displacements)
virtual forces (*see* principle of virtual forces)

weighting numbers 169
welded joints 77-80
wing loading 114, 115
yield allowable stresses
   bearing 3-11
   combined loads 25
   compression 3-11
   tension 3-11

MIX
Papier aus verantwortungsvollen Quellen
Paper from responsible sources
FSC® C105338

If you have any concerns about our products,
you can contact us on
ProductSafety@springernature.com

In case Publisher is established outside the EU,
the EU authorized representative is:
**Springer Nature Customer Service Center GmbH
Europaplatz 3, 69115 Heidelberg, Germany**

Printed by Libri Plureos GmbH
in Hamburg, Germany

VIRTUAL PRINCIPLES IN AIRCRAFT STRUCTURES
Volume 1: Analysis

MECHANICS OF STRUCTURAL SYSTEMS
Editors: J.S. Przemieniecki and G.Æ. Oravas

L. Fryba, Vibration of solids and structures under moving loads. 1973
ISBN 90-01-32420-7
K. Marguerre and H. Wölfel, Mechanics of vibrations. 1979
ISBN 90-286-0086-6
E.B. Magrab, Vibrations of elastic structural members. 1979
ISBN 90-286-0207-0
R.T. Haftka and M.P. Kamat, Elements of structural optimization. 1985
ISBN 90-247-2950-5(hardbound) ISBN 90-247-3062-7(paperback)
J.R. Vinson and R.L. Sierakowski, The behavior of structures composed of composite materials. 1986
ISBN 90-247-3125-9(hardbound) ISBN 90-247-3578-5(paperback)
B.E. Gatewood, Virtual Principles in Aircraft Structures Volume 1. 1989.
ISBN 90-247-3754-0
B.E. Gatewood, Virtual Principles in Aircraft Structures Volume 2. 1989.
ISBN 90-247-3755-9
ISBN 90-247-3753-2 (set).

Volume 6

B.E. GATEWOOD

*Professor Emeritus, Dept. of Aeronautical and Astronautical Engineering,
The Ohio State University, Columbus, Ohio, U.S.A.*

Virtual Principles in Aircraft Structures

Volume 1: Analysis

Springer-Science+Business Media, B.V.

Library of Congress Cataloging-in-Publication Data

Gatewood, Burford Echols, 1913-
   Virtual principles of aircraft structures.

   (Mechanics of structural systems)
   Includes bibliographies and indexes.
   Contents: v. 1. Analysis -- v. 2. Design, plates,
finite elements.
   1. Airframes. 2. Structures, Theory of.
3. Strength of materials. I. Title. II. Series.
TL671.6.G37 1989   629.134'1   88-13303

ISBN 978-94-010-7018-8   ISBN 978-94-009-1165-9 (eBook)
DOI 10.1007/978-94-009-1165-9

printed on acid free paper

All Rights Reserved
©**1989 by** Springer Science+Business Media Dordrecht
Originally published by Kluwer Academic Publishers in 1989
Softcover reprint of the hardcover 1st edition 1989

No part of the material protected by this copyright notice may be reproduced or
utilised in any form or by any means, electronic or mechanical,
including photocopying, recording or by any information storage and
retrieval system, without written permission from the copyright owner

Contents
Volume 1: Analysis

Preface	xiii
Chapter 1 / The basic three, two, and one dimensional equations in structural analysis	1
1.1 Introduction	1
1.2 Three dimensional equations	2
1.3 The displacement method of solution	6
1.4 The stress method of solution	7
1.5 The combined method of solution	8
1.6 Two dimensional equations	8
1.7 Saint Venant's principle	10
1.8 One dimensional beam equations	11
1.9 No shear stresses in the beam	13
1.10 Beam cross section of a thin plate with one shear stress	13
1.11 Thin web beams with large flange areas and one shear stress	17
1.12 Torsion of circular cross section and thin wall closed box	18
1.13 Thin web box beam with general loading	20
1.14 Inelastic effects in beams with temperature	21
1.15 Example of inelastic axial stresses and strains with temperature	25
1.16 Sequence loading and thermal cycling in beams	27
1.17 Load-strain design curves for beams	31
1.18 Problems	33
References	35
Chapter 2 / Virtual displacement and virtual force methods in structural analysis	37
2.1 Introduction	37
2.2 The principle of virtual displacements	39
2.3 The unit displacement theorem	41
2.4 The principle of virtual forces	43

2.5	The unit load theorem	45
2.6	The principle of mixed virtual stresses and virtual displacements	46
2.7	The mixed unit displacement and unit load theorem	47
2.8	Two dimensional form of the virtual principles	48
2.9	One dimensional forms of the virtual principles	49
2.10	The one dimensional virtual principles with temperature, inelastic and large displacement effects	54
2.11	Matrix forms of the virtual principles	56
2.12	Problems	59
	References	61

Chapter 3 / The virtual principles for pin-jointed trusses — 62

3.1	Introduction	62
3.2	The unit displacement theorem for trusses	64
3.3	The unit load theorem for trusses	70
3.4	Inelastic effects with temperature changes in trusses	74
3.5	Matrix equations for trusses from the unit displacement theorem	81
3.6	Matrix equations for trusses from the unit load theorem	87
3.7	Matrix equations for trusses from the mixed unit displacement and unit load theorem	96
3.8	Problems	98
	References	102

Chapter 4 / The virtual principles for simple beams — 103

4.1	Introduction	103
4.2	Principle of virtual displacements for beams	104
4.3	Point values for beam elements by the principle of virtual displacements	108
4.4	Principle of virtual forces for beams	114
4.5	Point values for beam elements by the unit load theorem	119
4.6	Principle of mixed virtual stresses and virtual displacements for beams	122
4.7	Inelastic and temperature effects in simple beams	123
4.8	Matrix equations for beams from the unit displacement theorem	130
4.9	Matrix equations for beams from the unit load theorem	136
4.10	Matrix equations for beams from the mixed unit displacement and unit load theorem	145
4.11	The beam column equations	147
4.12	Problems	149
	References	152

Chapter 5 / Box beam shear stresses and deflections — 153

5.1	Introduction	153
5.2	Shear stresses in beams	154

Contents of Volume 1: Analysis vii

5.3	Torsional shear stresses in beams	158
5.4	Shear flows in open box beams	160
5.5	Shear flows in single cell box beams	163
5.6	Shear flows in multi-cell box beams	165
5.7	Shear center for closed box beams	168
5.8	Shear flows in tapered box beams	170
5.9	Inelastic and buckling shear stresses in beams	173
5.10	Axial and bending deflections of box beams with inelastic and temperature effects	174
5.11	Shear deflections of beams	177
5.12	Torsional rotation of beams	183
5.13	Rotation of swept wings	189
5.14	Spanwise airload distribution and static wing divergence under rotation	190
5.15	Static aileron effectiveness and reversal speed under wing rotation	195
5.16	Problems	199
	References	201

Chapter 6 / Shear lag in thin web structures — 202

6.1	Introduction	202
	Part 1. Solutions for determinate cases	203
6.2	Shear flows due to concentrated loads into thin webs	203
6.3	Shear flows around cut-outs in thin web beams	206
6.4	Cut-outs in box beams	208
6.5	Shear flows in ribs and bulkheads	210
6.6	Forces on ribs due to airloads and taper effects	214
	Part 2. Solutions for redundant cases	216
6.7	Restraint effects in thin web structures	216
6.8	Shear flows in redundant beams in one plane	216
6.9	Deflections of thin web structures	226
6.10	Flexibility matrices for shear web elements and stiffener elements	231
6.11	Matrix solutions for thin web beams in one plane	233
6.12	Matrix solutions for box beams	241
6.13	Load redistribution in swept back wings	248
6.14	Problems	250
	References	252

Appendix A / Notes on matrix algebra — 254

A.1	Definition of matrices	254
A.2	Addition, subtraction, multiplication of matrices	255
A.3	Determinants	256
A.4	Matrix inversion	257
A.5	Solution of systems of simultaneous equations by matrices	260
A.6	Solution of systems of simultaneous equations by tri-diagonal matrices	261

A.7	Solution of systems of equations by Jordan successive transformations	265
A.8	Matrix representations	267
A.9	Orthogonal matrices	270
A.10	Eigenvalues and eigenvectors of matrices	275
A.11	Note on matrix notation	276
	References	277

Appendix B / External forces on flight vehicles 279

B.1	Introduction	279
B.2	Inertial forces for rigid body translation and rotation in a vertical plane	281
B.3	Air forces on airplane wing	285
B.4	Airplane equilibrium equations in flight. Load factors	287
B.5	Velocity-load factor $(V\text{-}n)$ diagram for design	289
B.6	Wing spanwise lift coefficient distribution	292
B.7	Spanwise lift coefficient distribution on twisted wings	295
B.8	Spanwise airload, shear, and moment distributions on wing	297
B.9	Distribution of inertia forces on wing and fuselage	301
B.10	Forces and moments on landing gear structures	304
B.11	Thermal forces	308
B.12	Miscellaneous forces	310
B.13	Deflection effects on the external forces	310
B.14	Criteria for the structure to support the external forces	312
B.15	Problems	314
	References	316

Appendix C / Derivation of the strain energy theorems from the virtual principles 317

C.1	Work and strain energy	317
C.2	Maximum and minimum strain energy and total potential energy	319
C.3	Theorem of minimum total potential energy	321
C.4	Theorem of minimum strain energy	322
C.5	Castigliano's theorem (Part I)	323
C.6	Hamilton's principle	323
C.7	Theorem of minimum total complementary potential theory	324
C.8	Theorem of minimum complementary strain energy	325
C.9	Castigliano's theorem (Part II)	326
C.10	Reissner's variational principle	326
C.11	Comparison of the virtual principles and the strain energy theorems	327
	References	328

Index 329

Contents
Volume 2: Design, Plates, Finite Elements

Preface

List of selected equations in Volume 1 referred to in Volume 2

Chapter 1 / Allowable stresses of flight vehicle materials

1.1 Introduction
1.2 Tension, shear, and bearing allowable stresses
1.3 Temperature effects on allowable stresses
1.4 Allowable compression stresses
1.5 Allowable combined stresses
1.6 Creep effects on allowable stresses
1.7 Room temperature fatigue effects upon allowable stresses
1.8 Temperature effects upon allowable fatigue stresses
1.9 Crack effects upon allowable fatigue stresses
1.10 Problems
     References

Chapter 2 / Analysis and design of joints and splices

2.1 Introduction
2.2 Analysis of plate splices with axial tension forces
2.3 Multi-row tension splices
2.4 Joints with eccentric loading
2.5 Minimum weight design of splice for beam with rectangular cross section
2.6 Design of splices for I-beams and thin shear webs
2.7 Deflection effects on load distribution in splices
2.8 Temperature and inelastic effects on load distribution in splices
2.9 Welded joints
2.10 Bonded joints
2.11 Problems
     References

Chapter 3 / Structural design of aircraft components

3.1 Introduction
3.2 Design of minimum weight columns without local buckling
3.3 Design of minimum weight sections with local buckling and crippling
3.4 Design of minimum weight columns with local buckling
3.5 Minimum weight design for stiffened panels in compression
3.6 Effective areas for stiffened panels
3.7 Effect of load intensity on wing design
3.8 Design of box beam cross sections with four spar caps
3.9 Analysis of diagonal tension beams
3.10 Problems
     References

Chapter 4 / Analysis and design of pressurized structures

4.1 Introduction
4.2 Membrane stresses in thin shells
4.3 Cut-outs in thin shells with membrane stresses
4.4 Bending in circular cylindrical shells with axially symmetric loading
4.5 Bending in pressurized aircraft fuselages from stringers and frames
4.6 Bending of non-circular cross sections with internal pressure
4.7 Bending of non-circular fuselage rings with internal pressure
4.8 Bending of non-circular fuselage rings with point loads
4.9 Effect of internal pressure on buckling of cylindrical shells
4.10 Pressure stabilized structures
4.11 Problems
     References

Chapter 5 / Approximate solutions using the virtual principles

5.1 Introduction
5.2 Approximate solutions for beams using the principle of virtual displacements
5.3 Approximate solutions for columns
5.4 The tapered cantilever beam with numerical integration
5.5 Tapered beam finite element matrices for columns
5.6 The unit load theorem and numerical integration
5.7 Approximate solutions for beams using the mixed virtual principle
5.8 Problems
     References

Chapter 6 / Dynamics of simple beams

6.1 Introduction
6.2 Bending vibrations of simple beams

Contents of Volume 2: Design, Plates, Finite Elements   xi

6.3   Forced motion of uniform beam
6.4   Approximate solutions for frequencies and mode shapes
6.5   Torsional vibrations of simple beams
6.6   Finite element matrices for beam frequencies
6.7   Flutter of wing segment with one degree of freedom
6.8   Flutter of wing segment with two degrees of freedom
6.9   Dynamic loads on beams
6.10 Problems
      References

Chapter 7 / The plate equations

7.1   Introduction
7.2   The plate inplane case using the principle of virtual displacements
7.3   The plate inplane case using the principle of virtual forces
7.4   The plate inplane case using the mixed virtual principle
7.5   The plate bending case using the principle of virtual displacements
7.6   The plate bending case using the principle of virtual forces
7.7   The plate bending case using the mixed virtual principle
7.8   Combined inplane and lateral forces
7.9   Combined forces with large bending deflections
7.10 Buckling of plates
7.11 Plate vibrations
7.12 Problems
      References

Chapter 8 / Approximate matrix equations for plate finite elements

8.1   Introduction
8.2   The point unknowns for the matrices
8.3   The methods to obtain the matrix equations
8.4   Inplane plate element matrices from the principle of virtual displacements
8.5   Inplane plate element matrices from the principle of virtual forces
8.6   Inplane plate element matrices from the mixed virtual principle
8.7   Bending plate element matrices from the principle of virtual displacements
8.8   Bending plate element matrices from the principle of virtual forces
8.9   Bending plate element matrices from the mixed virtual principle
8.10 Matrices for constant stress triangular elements
8.11 Problems

Chapter 9 / Matrix structural analysis using finite elements

9.1   Introduction
9.2   General beam elements in local coordinates
9.3   General beam elements in datum coordinates
9.4   Triangular plate elements with inplane forces

9.5 Assembly of finite elements by the virtual principles
References

Chapter 10 / Composite Materials

10.1 Introduction
10.2 Stress-strain equations for nonisotropic materials
10.3 Stress-strain equations for plane stress in an orthotropic material
10.4 Forces and moments in laminated plates
10.5 Stresses in laminated plates
10.6 Allowable stresses for laminated plates
10.7 Interlamina stresses
10.8 Joints in laminated plates
10.9 Bending deflections of laminated plates
10.10 Buckling loads for laminated plates
10.11 Vibrations of laminated plates
10.12 Problems
References

Preface

The basic partial differential equations for the stresses and displacements in classical three dimensional elasticity theory can be set up in three ways: (1) to solve for the displacements first and then the stresses; (2) to solve for the stresses first and then the displacements; and (3) to solve for both stresses and displacements simultaneously. These three methods are identified in the literature as (1) the displacement method, (2) the stress or force method, and (3) the combined or mixed method. Closed form solutions of the partial differential equations with their complicated boundary conditions for any of these three methods have been obtained only in special cases.

In order to obtain solutions, various special methods have been developed to determine the stresses and displacements in structures. The equations have been reduced to two and one dimensional forms for plates, beams, and trusses. By neglecting the local effects at the edges and ends, satisfactory solutions can be obtained for many cases. The procedures for reducing the three dimensional equations to two and one dimensional equations are described in Chapter 1, Volume 1, where the various approximations are pointed out.

Integral transform methods, energy methods, Rayleigh-Ritz and Galerkin approximation methods, virtual principles, and finite element methods have been developed to aid in solving more complicated structural problems. In recent years (see Chapter 2, Volume 1, References 1 and 2 and Introduction) it has been shown that three virtual principles, which correspond to the three methods of solution described above, give a rational basis for the energy methods, Rayleigh-Ritz methods, and finite element methods.

By using integral transform methods, it is possible to convert the three sets of three dimensional elasticity equations described above to integral forms involving stresses, strains, and/or displacements directly. These integral forms can be identified as (1) the principle of virtual displacements, (2) the principle of virtual forces, and (3) the principle of mixed virtual stresses and virtual displacements. The principles can be used directly to obtain exact solutions for simple structures, or they can be put into matrix forms and used directly for finite elements. Assumed functions can be put into the virtual principles to obtain a system of simultaneous equations, which correspond exactly to the Rayleigh-Ritz system of simultaneous equations.

The three virtual principles are derived in Chapter 2, Volume 1, and used throughout the text. Temperature and inelastic effects are included in the one dimensional forms. Comparisons of the three methods are made to show that for a given structure one method may be much simpler to use than the others.

The virtual principles are used to obtain differential equations and boundary conditions for beams and beam columns in Chapter 4, Volume 1, for vibration of beams in Chapter 6, Volume 2, and for plates in Chapter 7, Volume 2. They are used to obtain stresses and displacements in determinate and redundant trusses and beams in Chapter 3 and 4, Volume 1, respectively; to get deflections of box beam aircraft type structures in Chapter 5, Volume 1; to solve redundant shear lag problems in Chapter 6, Volume 1, and Chapter 2, Volume 2; to obtain approximate solutions using assumed functions in Chapters 5, 6, 7, and 8, Volume 2; to solve complex structures with finite elements and matrix equations in Chapters 3, 4, and 6, in Volume 1, and Chapters 5, 6, 8, and 9 in Volume 2; to assemble finite element matrices in Chapters 3, 4, and 6 in Volume 1 and Chapter 9 in Volume 2.

All the energy theorems are derived from the virtual principles in Appendix C, Volume 1.

Although the external forces on airplanes are described in Appendix B, Volume 1, and aircraft type structures are emphasized in the text, the virtual principles and the procedures given apply to all types of structures.

There are 230 solved examples in the two volumes so that they can be used not only as textbooks but also as reference books by practicing engineers in structural analysis. A pocket calculator was used to solve the examples and only a pocket calculator is needed to solve the 540 problems in the two volumes. Of course, computers can be used to solve the matrix equations for more complex structures with assumed functions or finite elements.

The material in Volume 1 together with Chapters 1 to 6 in Volume 2 originated from lecture notes used by the author in aircraft structures courses at the Ohio State University. The more advanced Chapters 7-10 (Volume 2) on plates, finite elements, and composite materials have been added. It is assumed that students have a knowledge of differential equations and mechanics of materials. Although not necessary, some knowledge of integral transforms (such as Laplace Transforms or Fourier Transforms) would help in understanding the derivation of the virtual principles in Chapter 2, Volume 1. The necessary matrix algebra is given in the Appendix A in the same volume.

As may be gathered from the above, the book is published in two volumes. Volume 1 contains subject matter usually covered in the two undergraduate courses in analysis of aircraft structures. Volume 2 contains the subject matter for the usual undergraduate design course as well as chapters on plates, dynamics, finite elements, and composite materials for a more advanced course.

B.E. GATEWOOD

1

The basic three, two, and one dimensional equations in structural analysis

1.1. Introduction

Structural materials, when viewed with a microscope, are seen to consist of crystals of various kinds and various orientations. However, these crystals are very small and usually have a random distribution so that the measured elastic properties of most materials represent averages of the properties of the crystals. Since the geometrical dimensions of materials used in physical structures are large in comparison to the dimensions of a single crystal, most structural materials can be assumed to be homogeneous and isotropic. That is, in the mathematical representation of the material, the properties are continuous and the same in all directions.

The assumptions of homogeneity and isotropy will be used in the derivations of the equations in this chapter 1. Also, it will be assumed that the material is elastic (deformation proportional to force) in the derivation of the general three dimensional equations. Temperature variations will be included in the equations. The three dimensional equations are given in Section 1.2 and the forms of the equations for the displacement method of solution, the stress method of solution, and the combined method of solution are given in Sections 1.3, 1.4, and 1.5, respectively.

Since the general three dimensional equations will be used primarily to obtain the two dimensional and one dimensional equations and to set up the virtual principles, only an outline of their derivation will be given here. See Reference 1 for details of the derivations.

Since in general the three dimensional equations cannot be solved explicitly either by theoretical analysis or by the largest computers, it is necessary to simplify the equations by considering particular types of structures. The equations can be reduced to a two dimensional form for plates and for long cylinders, Section 1.6. Only the elastic case will be considered for the two dimensional problems. Although many solutions have been obtained for special two dimensional cases, there are still difficulties in obtaining solutions for plates with a general boundary.

A further simplification in the equations to a one dimensional form can be made for rod and beam type structures, provided certain assumptions can be made about the distribution of the applied loads, Section 1.7. The details of the derivation of the one dimensional equations are given in Section 1.8. To obtain a better understanding of the equation simplifications and approximations, various beam shapes and beam loadings are considered in Sections 1.9 through 1.13. The beam inelastic and temperature effects are considered in Sections 1.14 through 1.17.

1.2. Three dimensional equations

When external component surface forces S_x, S_y, S_z, body component forces X_x, X_y, X_z, concentrated surface forces P_m, and a temperature distribution T in the body act on a homogeneous and isotropic deformable structure, the effects are transmitted to all points of the continuous structure and internal forces are produced between the particles in the structure. Also, every point in the structure will be displaced to a new position, except for any support points to prevent rigid body motion.

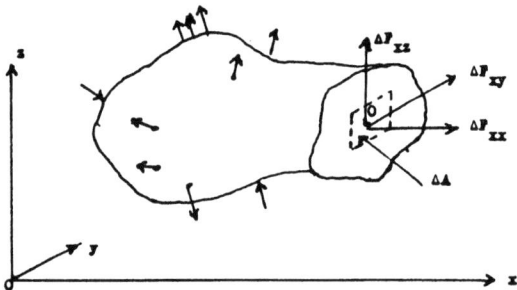

Fig. 1.1. Component internal forces on plane.

The internal forces can be identified by cutting the structure with a plane to form a free body. In Figure 1.1, the internal forces on the plane must maintain the body in equilibrium. The *stress* at the point Q is defined as the intensity of the internal force ΔF per unit area at the point, or

$$\text{stress} = \sigma = \lim_{\Delta A \to 0} (\Delta F / \Delta A) = \frac{dF}{dA}. \tag{1.1}$$

It is evident that the internal force ΔF can be resolved into components normal and parallel to the plane. In Figure 1.1, let the subscript x locate the plane and a second subscript x, y, or z give the direction of the internal force component. Thus

$$\sigma_{xx} = \lim_{\Delta A \to 0} (\Delta F_{xx}/\Delta A) = \frac{dF_{xx}}{dA}, \quad \sigma_{xy} = \frac{dF_{xy}}{dA}, \quad \sigma_{xz} = \frac{dF_{xz}}{dA}, \tag{1.2}$$

where σ_{xx} is a normal stress perpendicular to the plane and σ_{xy}, σ_{xz} are shear stresses parallel to the plane. Since planes can be drawn perpendicular to the y and z axes, there are nine possible component stresses in the structure associated

The basic three, two, and one dimensional equations

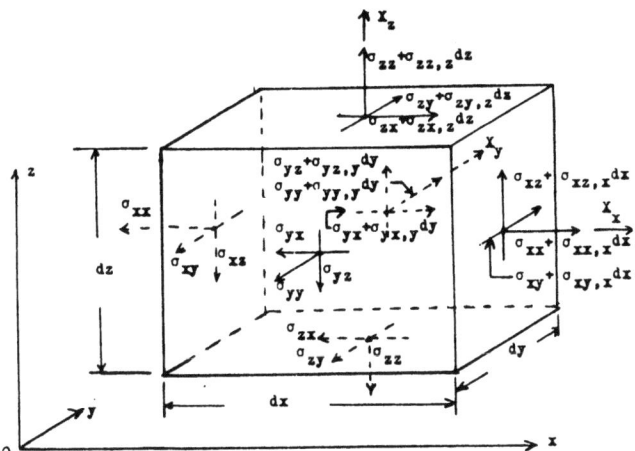

Fig. 1.2. Stress equilibrium on infinitesimal element.

with the coordinate axes: three normal stresses σ_{xx}, σ_{yy}, σ_{zz}, and six shear stresses σ_{xy}, σ_{yx}, σ_{xz}, σ_{zx}, σ_{yz}, σ_{zy}. The stresses for all other planes can be calculated from these nine stresses.

From Figure 1.2 for a small element in the structure, moment equilibrium about the center of the element gives

$$\sigma_{xy} = \sigma_{yx}, \qquad \sigma_{xz} = \sigma_{zx}, \qquad \sigma_{yz} = \sigma_{zy} \tag{1.3}$$

so that there are only three shear stresses to go with the three normal stresses. Summation of loads on the element gives the point body equilibrium equations

$$\begin{aligned} \sigma_{xx,x} + \sigma_{xy,y} + \sigma_{xz,z} + X_x &= 0, \\ \sigma_{xy,x} + \sigma_{yy,y} + \sigma_{yz,z} + X_y &= 0, \\ \sigma_{xz,x} + \sigma_{yz,y} + \sigma_{zz,z} + X_z &= 0, \\ \sigma_{xx,x} &= \frac{\partial \sigma_{xx}}{\partial x}, \quad \text{etc. for notation,} \end{aligned} \tag{1.4}$$

where Equation (1.3) has been used and where the X_x, X_y, X_z applied body forces are forces per unit volume.

Equilibrium of a small element on the surface of the structure gives

$$\begin{aligned} \sigma_{xx}l + \sigma_{xy}m + \sigma_{xz}n &= S_x, \\ \sigma_{xy}l + \sigma_{yy}m + \sigma_{yz}n &= S_y, \\ \sigma_{xz}l + \sigma_{yz}m + \sigma_{zz}n &= S_z, \end{aligned} \tag{1.5}$$

where l, m, n are the direction cosines for an outward normal, and where S_x, S_y, S_z are the applied surface forces or reactions per unit area.

The structure must also be in equilibrium as a rigid body under all applied forces and reactions at the support points, whence the three force equations and the three moment equations are

$$\int_V X_x \, dV + \int_S S_x \, dS + \sum_{m=1}^{M} P_{mx} = 0,$$

$$\int_V X_y \, dV + \int_S S_y \, dS + \sum_{m=1}^{M} P_{my} = 0,$$

$$\int_V X_z \, dV + \int_S S_z \, dS + \sum_{m=1}^{M} P_{mz} = 0,$$

$$\int_V (yX_z - zX_y) \, dV + \int_S (yS_z - zS_y) \, dS +$$
$$+ \sum_{m=1}^{M} (y_m P_{mz} - z_m P_{my}) + \sum_{n=1}^{N} M_{nx} = 0, \quad (1.6)$$

$$\int_V (zX_x - xX_z) \, dV + \int_S (zS_x - xS_z) \, dS +$$
$$+ \sum_{m=1}^{M} (z_m P_{mx} - x_m P_{mz}) + \sum_{n=1}^{N} M_{ny} = 0,$$

$$\int_V (xX_y - yX_x) \, dV + \int_S (xS_y - yS_x) \, dS +$$
$$+ \sum_{m=1}^{M} (x_m P_{my} - y_m P_{mx}) + \sum_{n=1}^{N} M_{nz} = 0.$$

Besides producing stresses in the structure the applied forces and temperature changes also produce displacements of all points in the structure. The displacement u of a point is described by three displacement components u_x, u_y, u_z, measured from a reference point. To examine the infinitesimal changes in the displacements at a point, consider the three lines QQ_x, QQ_y, QQ_z in Figure 1.3. Under deformation the three lines move to the positions $Q'Q'_x$, $Q'Q'_y$, $Q'Q'_z$. Not only have the lengths of the lines changed, but also the ninety degree angles between the lines have changed. The changes in length can be described by *longitudinal strains* defined as a displacement change per unit length and the angle changes can be described by *shearing strains* defined from Figure 1.3, or

$$e_{xx} = \lim_{\Delta x \to 0} \left(\frac{\Delta u_x}{\Delta x} \right) = \frac{\partial u_x}{\partial x}, \quad e_{yy} = \frac{\partial u_y}{\partial y}, \quad e_{zz} = \frac{\partial u_z}{\partial z},$$

$$e_{xy} = \lim_{\Delta x \to 0} \left(\frac{\Delta u_y}{\Delta x} \right) + \lim_{\Delta y \to 0} \left(\frac{\Delta u_x}{\Delta y} \right) = \frac{\partial u_y}{\partial x} + \frac{\partial u_x}{\partial y}, \quad (1.7)$$

$$e_{xz} = \frac{\partial u_z}{\partial x} + \frac{\partial u_x}{\partial z}, \quad e_{yz} = \frac{\partial u_z}{\partial y} + \frac{\partial u_y}{\partial z}.$$

The Equations (1.7) are the small displacement linear strain displacement equations in three dimensions. Higher order non-linear large displacement terms can be included in the equations by using the final lengths of the lines $Q'Q'_x$,

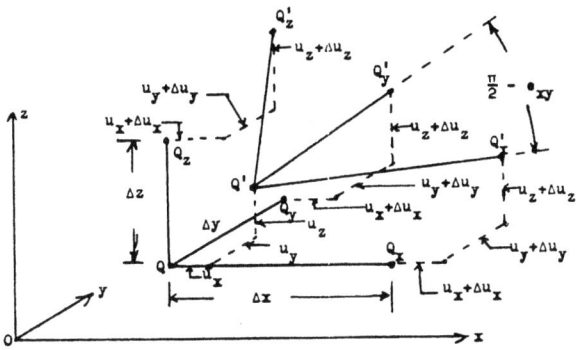

Fig. 1.3. Displacements of three orthogonal lines.

$Q'Q'_y$, $Q'Q'_z$, in Figure 1.3 for the longitudinal strains and by using the cosines of the angles in the final position for the shearing strains. The results with second order terms are

$$\begin{aligned}
e_{xx} &= u_{x,x} + (1/2)(u_{x,x}^2 + u_{y,x}^2 + u_{z,x}^2), \\
e_{yy} &= u_{y,y} + (1/2)(u_{x,y}^2 + u_{y,y}^2 + u_{z,y}^2), \\
e_{zz} &= u_{z,z} + (1/2)(u_{x,z}^2 + u_{y,z}^2 + u_{z,z}^2), \\
e_{xy} &= u_{x,y} + u_{y,x} + u_{x,x}u_{x,y} + u_{y,x}u_{y,y} + u_{z,x}u_{z,y}, \\
e_{xz} &= u_{x,z} + u_{z,x} + u_{x,x}u_{x,z} + u_{y,x}u_{y,z} + u_{z,x}u_{z,z}, \\
e_{yz} &= u_{y,z} + u_{z,y} + u_{x,y}u_{x,z} + u_{y,y}u_{y,z} + u_{z,y}u_{z,z},
\end{aligned} \quad (1.8)$$

$$u_{x,x} = \frac{\partial u_x}{\partial x}, \quad u_{y,x} = \frac{\partial u_y}{\partial x}, \quad \text{etc. for notation.}$$

In Equation (1.7) the six strains are expressed in terms of three displacements so that the strains must be related and cannot be specified arbitrarily. If the displacements u_x, u_y, and u_z are eliminated from Equation (1.7) by differentiation and combination of equations, there result six *compatibility* equations relating the strains to each other, or

$$\begin{aligned}
e_{xy,xy} &= e_{xx,yy} + e_{yy,xx}, & e_{xz,xz} &= e_{xx,zz} + e_{zz,xx}, \\
e_{yz,yz} &= e_{yy,zz} + e_{zz,yy}, & 2e_{xx,yz} &= -e_{yz,xx} + e_{xz,xy} + e_{xy,xz}, \\
2e_{yy,xz} &= e_{yz,xy} - e_{xz,yy} + e_{xx,yz}, & 2e_{zz,xy} &= e_{yz,xz} + e_{xz,yz} - e_{xy,zz},
\end{aligned} \quad (1.9)$$

$$e_{xy,xy} = \frac{\partial^2 e_{xy}}{\partial x \partial y}, \quad e_{xx,yy} = \frac{\partial^2 e_{xx}}{\partial y^2}, \quad \text{etc. for notation.}$$

In addition to the compatibility conditions (1.9), which must be satisfied at all interior points and all surface points where no displacements are specified, the displacements must satisfy all specified surface displacement values. These are the displacement boundary conditions, which may involve small rotation angles as well as displacements. For small rigid body rotation angles θ_x, θ_y, θ_z, the rigid body displacements can be expressed as

$$\begin{aligned}
u_x &= u_{x0} - y\theta_z + z\theta_y, & u_y &= u_{y0} - z\theta_x + x\theta_z, \\
u_z &= u_{z0} - x\theta_y + y\theta_x.
\end{aligned} \quad (1.10)$$

To prevent rigid body motion the translation (u_{x0}, u_{y0}, u_{z0}) and the rotation $(\theta_x, \theta_y, \theta_z)$ must be restrained.

So far, six stresses, three displacements, and six strains have been defined, all of which are produced by the external forces and temperatures. The stresses have been related to the external forces by the equilibrium equations (1.4). The strains and displacements have been related to each other, Equations (1.7) and (1.9). It is evident that there must be relations between the stresses and the strains. These relations depend upon the properties of the materials used in the structure. Since many of the materials used in structures are approximately homogeneous and isotropic and since in many cases the materials can be restricted to the elastic range in design, only the linear elastic stress-strain relations will be given for the three dimensional case, Ref. 1, or

$$
\begin{aligned}
e_{xx} &= (1/E)[\sigma_{xx} - \nu(\sigma_{yy} + \sigma_{zz})] + \alpha T, \\
e_{yy} &= (1/E)[\sigma_{yy} - \nu(\sigma_{xx} + \sigma_{zz})] + \alpha T, \\
e_{zz} &= (1/E)[\sigma_{zz} - \nu(\sigma_{xx} + \sigma_{yy})] + \alpha T, \\
e_{xy} &= \sigma_{xy}/G, \quad e_{xz} = \sigma_{xz}/G, \quad e_{yz} = \sigma_{yz}/G.
\end{aligned}
\tag{1.11}
$$

Here E, ν, and G are material properties to be obtained from test data and

$$G = \frac{E}{2(1+\nu)}. \tag{1.12}$$

E is the modulus of elasticity for elongation, G is the shear modulus, ν is Poisson's ratio, T is temperature change, and α is the coefficient of thermal expansion.

The Equation (1.11) can be solved for the stresses to give

$$
\begin{aligned}
(1-2\nu)\sigma_{xx} &= 2G[(1-\nu)e_{xx} + \nu(e_{yy} + e_{zz})] - E\alpha T, \\
(1-2\nu)\sigma_{yy} &= 2G[(1-\nu)e_{yy} + \nu(e_{xx} + e_{zz})] - E\alpha T, \\
(1-2\nu)\sigma_{zz} &= 2G[(1-\nu)e_{zz} + \nu(e_{xx} + e_{yy})] - E\alpha T, \\
\sigma_{xy} &= Ge_{xy}, \quad \sigma_{xz} = Ge_{xz}, \quad \sigma_{yz} = Ge_{yz}.
\end{aligned}
\tag{1.13}
$$

Theoretically, the fifteen unknowns (three displacements, six strains, and six stresses) can be obtained from the fifteen equations, (1.4), (1.7), and (1.11). The stresses and displacements are the primary unknowns and there are three procedures for formally solving for them: the *displacement method* in which the displacements are obtained first, the *stress method* in which the stresses are obtained first, and the *combined method* in which the stresses and displacements are obtained together. The three methods are described below.

1.3. The displacement method of solution

In solving for the displacements the strain-displacement Equations (1.7) are the compatibility equations, giving compatible strains for all possible continuous displacements. The equilibrium equations are the Equations (1.4) and (1.5), which can be used to select those displacements which not only give compatible strains and stresses but also give equilibrium stresses. Put Equation (1.7) into Equation (1.13) to get

$$(1-2\nu)\sigma_{xx} = 2G[(1-\nu)u_{x,x} + \nu(u_{y,y} + u_{z,z})] - E\alpha T$$
$$(1-2\nu)\sigma_{yy} = 2G[(1-\nu)u_{y,y} + \nu(u_{x,x} + u_{z,z})] - E\alpha T$$
$$(1-2\nu)\sigma_{zz} = 2G[(1-\nu)u_{z,z} + \nu(u_{x,x} + u_{y,y})] - E\alpha T \qquad (1.14)$$
$$\sigma_{xy} = G(u_{x,y} + u_{y,x}), \qquad \sigma_{xz} = G(u_{x,z} + u_{z,x}),$$
$$\sigma_{yz} = G(u_{y,z} + u_{z,y}),$$

which can be put into Equation (1.4) to get three equations for the three displacements,

$$(\lambda + G)(e)_{,x} + G\nabla^2 u_x - (3\lambda + 2G)(\alpha T)_{,x} + X_x = 0,$$
$$(\lambda + G)(e)_{,y} + G\nabla^2 u_y - (3\lambda + 2G)(\alpha T)_{,y} + X_y = 0,$$
$$(\lambda + G)(e)_{,z} + G\nabla^2 u_z - (3\lambda + 2G)(\alpha T)_{,z} + X_z = 0,$$
$$e = u_{x,x} + u_{y,y} + u_{z,z}, \qquad \lambda = \frac{\nu E}{(1+\nu)(1-2\nu)}, \qquad (1.15)$$
$$\nabla^2 = \frac{\partial^2}{\partial x^2} + \frac{\partial^2}{\partial y^2} + \frac{\partial^2}{\partial z^2}, \qquad (e)_{,x} = \frac{\partial e}{\partial x}, \quad \text{etc. for notation.}$$

If Equation (1.15) can be solved for the displacements subject to the surface equilibrium conditions (1.5) and the displacement restraints, then the strains are given by Equation (1.7) and the stresses by Equation (1.14).

1.4. The stress method of solution

In solving for the stresses first the strain compatibility equations are given by Equation (1.9), which can be written in terms of the stresses by putting Equation (1.11) into Equation (1.9). If Equation (1.4) is used to simplify the equations, then the stress compatibility equations are

$$\nabla^2 \sigma_{xx} + \mu F_{1,xx} = F_2 - 2X_{x,x} - \mu E(\alpha T)_{,xx},$$
$$\nabla^2 \sigma_{yy} + \mu F_{1,yy} = F_2 - 2X_{y,y} - \mu E(\alpha T)_{,yy},$$
$$\nabla^2 \sigma_{zz} + \mu F_{1,zz} = F_2 - 2X_{z,z} - \mu E(\alpha T)_{,zz},$$
$$\nabla^2 \sigma_{xy} + \mu F_{1,xy} = -\mu(X_{x,y} + X_{y,x}) - \mu E(\alpha T)_{,xy},$$
$$\nabla^2 \sigma_{xz} + \mu F_{1,xz} = -\mu(X_{x,z} + X_{z,x}) - \mu E(\alpha T)_{,xz}, \qquad (1.16)$$
$$\nabla^2 \sigma_{yz} + \mu F_{1,yz} = -\mu(X_{y,z} + X_{z,y}) - \mu E(\alpha T)_{,yz},$$
$$F_1 = \sigma_{xx} + \sigma_{yy} + \sigma_{zz}, \qquad \mu = \frac{1}{1+\nu},$$
$$F_2 = -\mu E \left(\frac{1+\nu}{1-\nu}\right) \nabla^2(\alpha T) - \frac{\nu}{1-\nu}(X_x + X_y + X_z).$$

Theoretically, the solution for the six stresses can be obtained from Equations (1.4) and (1.16) by using stress functions. If $-V$ is the potential of the body forces such that

$$X_x = -V_{,x}, \quad X_y = -V_{,y}, \quad X_z = -V_{,z}, \tag{1.17}$$

then the stresses can be expressed in terms of V and six stress functions ϕ_1, ϕ_2, ϕ_3, ϕ_4, ϕ_5, ϕ_6, which identically satisfy the equilibrium Equations (1.4), or

$$\begin{aligned}
\sigma_{xx} &= V + \phi_{1,yy} + \phi_{2,zz} + 2\phi_{4,yz}, \\
\sigma_{yy} &= V + \phi_{1,xx} + \phi_{3,zz} + 2\phi_{5,xz}, \\
\sigma_{zz} &= V + \phi_{2,xx} + \phi_{3,yy} + 2\phi_{6,xy}, \\
\sigma_{xy} &= -\phi_{1,xy} - \phi_{4,xz} - \phi_{5,yz} + \phi_{6,zz}, \\
\sigma_{xz} &= -\phi_{2,xz} - \phi_{4,xy} + \phi_{5,yy} - \phi_{6,yz}, \\
\sigma_{yz} &= -\phi_{3,yz} + \phi_{4,xx} - \phi_{5,xy} - \phi_{6,xz}.
\end{aligned} \tag{1.18}$$

Put Equation (1.18) into Equation (1.16) to get six fourth order partial differential equations for the six stress functions. Although only three of the stress functions are needed to get a solution, the equations have been solved only for special cases.

1.5. The combined method of solution

Put the strain-displacement Equations (1.7) into the stress-strain Equations (1.11) to get six equations involving the three displacements and the six stresses, or

$$\begin{aligned}
E u_{x,x} &= \sigma_{xx} - \nu(\sigma_{yy} + \sigma_{zz}) + E\alpha T, \\
E u_{y,y} &= \sigma_{yy} - \nu(\sigma_{xx} + \sigma_{zz}) + E\alpha T, \\
E u_{z,z} &= \sigma_{zz} - \nu(\sigma_{xx} + \sigma_{yy}) + E\alpha T, \\
G(u_{x,y} + u_{y,x}) &= \sigma_{xy}, \\
G(u_{x,z} + u_{z,x}) &= \sigma_{xz}, \\
G(u_{y,z} + u_{z,y}) &= \sigma_{yz}, \\
G &= \frac{E}{2(1+\nu)}.
\end{aligned} \tag{1.19}$$

These Equations (1.19) and the equilibrium Equations (1.4) give nine partial differential equations involving only first order derivatives for the displacements and stresses. However, the surface conditions in Equations (1.5) and the displacement restraints make it difficult to obtain solutions.

1.6. Two dimensional equations

In reality all structures are three dimensional, but in certain structures some of the displacements, stresses, and strains may be small compared to the others so that they can be taken as zero. Two such structures will be considered, (1) the long cylinder which is loaded so that the cross section has zero axial strain, or

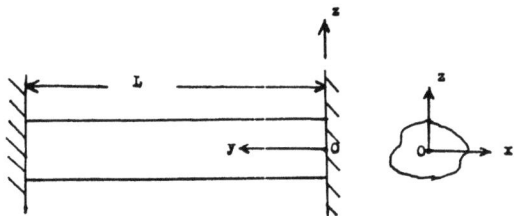

Fig. 1.4. Long cylinder with plane strain.

the *plane strain problem*, and (2) the thin plate with no normal stresses in the thickness direction, or the *plane stress problem*.

In Figure 1.4 for the plane strain case, the conditions are:
1. The ends of the cylinder are restrained so that $u_y = 0$.
2. On the cylindrical surface, $S_y = 0$ and S_x, S_z are independent of y.
3. Within the cylinder, $X_y = 0$, and X_x, X_z, and T are independent of y.

Under these conditions the strain-displacement and stress-strain relations are simplified to

$$\begin{aligned} u_x &= u_x(x,z), \quad u_z = u_z(x,z), \quad u_y = 0, \\ e_{yy} &= 0, \quad e_{xy} = 0, \quad e_{yz} = 0, \quad \sigma_{xy} = 0, \quad \sigma_{yz} = 0, \\ \sigma_{yy} &= \nu(\sigma_{xx} + \sigma_{zz}) - E\alpha T, \\ 2Ge_{xx} &= 2Gu_{x,x} = \sigma_{xx} - \nu(\sigma_{xx} + \sigma_{zz}) + E\alpha T, \\ 2Ge_{zz} &= 2Gu_{z,z} = \sigma_{zz} - \nu(\sigma_{zz} + \sigma_{xx}) + E\alpha T, \\ Ge_{xz} &= G(u_{x,z} + u_{z,x}) = \sigma_{xz}. \end{aligned} \quad (1.20)$$

In this case there are eight unknowns: two displacements, three strains, and three stresses.

The equilibrium Equations (1.4) and (1.5) become

$$\sigma_{xx,x} + \sigma_{xz,z} + X_x = 0, \quad \sigma_{xz,x} + \sigma_{zz,z} + X_z = 0, \quad (1.21)$$
$$\sigma_{xx}l + \sigma_{xz}n = S_x, \quad \sigma_{xz}l + \sigma_{zz}n = S_z. \quad (1.22)$$

The six compatibility Equations (1.9) reduce to one,

$$e_{xz,xz} = e_{xx,zz} + e_{zz,xx}, \quad (1.23)$$

and the six stress functions in Equation (1.18) reduce to one,

$$\sigma_{xx} = V + \phi_{,zz}, \quad \sigma_{zz} = V + \phi_{,xx}, \quad \sigma_{xz} = -\phi_{,xz}. \quad (1.24)$$

For the stress method of solution in Section 1.4, put Equation (1.24) into Equation (1.20) and the result into Equation (1.23), whence the equation for the stress function is

$$\nabla^4 \phi + \frac{E}{1-\nu}\nabla^2(\alpha T) + \frac{1-2\nu}{1-\nu}\nabla^2 V = 0, \quad \nabla^2 = \frac{\partial^2}{\partial x^2} + \frac{\partial^2}{\partial z^2}. \quad (1.25)$$

This Equation (1.25) has received extensive study in the literature, particularly for simple cross sections of the cylinder such as rectangle, circle, ellipse. See Reference 1 for examples. The boundary surface conditions (1.22) cause the main difficulty. To handle various boundaries, it is possible to use complex variable conformal mapping of the boundary into a unit circle and solve Equation (1.25) with two analytic complex functions (References 2, 3).

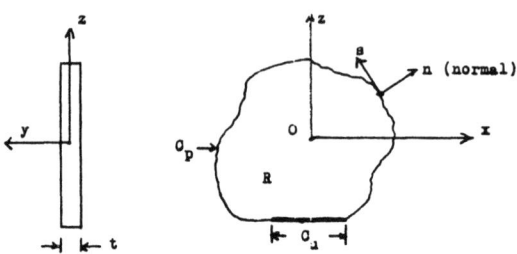

Fig. 1.5. Thin plate with plane stress.

The plane stress two dimensional case involves the stresses in the plane of a thin plate, Figure 1.5. The conditions are:
1. No surface forces, $S_x = 0$, $S_y = 0$, $S_z = 0$, on the surface planes of the plate at $y = \pm t/2$.
2. On the edge surface C_p, $S_y = 0$, and S_x, S_z are independent of y.
3. Within the plate region R, $X_y = 0$, and X_x, X_z, and T are independent of y.

Under these conditions,

$$\sigma_{yy} = 0, \quad \sigma_{yz} = 0, \quad \sigma_{xy} = 0, \quad e_{xy} = 0, \quad e_{yz} = 0,$$
$$Ee_{yy} = -\nu(\sigma_{xx} + \sigma_{zz}) + E\alpha T,$$
$$Ee_{xx} = Eu_{x,x} = \sigma_{xx} - \nu\sigma_{zz} + E\alpha T, \qquad (1.26)$$
$$Ee_{zz} = Eu_{z,z} = \sigma_{zz} - \nu\sigma_{xx} + E\alpha T,$$
$$Ge_{xz} = G(u_{x,z} + u_{z,x}) = \sigma_{xz}.$$

The Equations (1.21) - (1.24) also apply to the plane stress case, but Equation (1.25) becomes

$$\nabla^4 \phi + E\nabla^2(\alpha T) + (1 - \nu)\nabla^2 V = 0. \qquad (1.27)$$

See Chapter 7, Volume 2, for more details on the plate equations, including plate bending.

1.7. Saint Venant's principle

Before presentation of the one dimensional beam equations in the next section, it is necessary to examine the approximations that must be made in the region where the surface forces are applied. To have one axial stress in a long beam, it is evident that the stress at the ends of the beam must be distributed in exactly the same way as in the beam. Usually, it is not possible to apply such a distribution of stress on the ends of the beam. The effects of using a different stress distribution on the ends is explained by a principle stated by St. Venant in 1855:

If the distribution of forces on a small region of the surface of an elastic structure is replaced by another distribution which has the same resultant force and moment as the original distribution, then the two distributions produce essentially the same stresses in the structure except in a region near the surface where the force distribution is changed.

The region affected by the redistribution of the surface forces will usually extend into the structure for a distance approximately equal to the largest linear dimension of the portion of the surface on which the forces are changed. In other words, the end effects on beams will extend into the beam a distance of the order of the height of the beam. Test data and two dimensional analysis of the beam verify St. Venant's principle for beams. See Reference 1.

The conclusion is that the one dimensional analysis of the cross section does not apply near the ends, whence the beam must be relative long compared to its height. The same local effects occur in the region of the point of application of a concentrated load on the beam. These local effects must be considered separately in the detail design of the beam. If the material is ductile, then local yielding may occur to change the stress distribution and prevent possible failure. However, these local effects can cause failure in brittle materials and thus produce complicated local design problems.

1.8. One dimensional beam equations

Modify the plane strain cylinder in Figure 1.4 to the beam in Figure 1.6 and take the conditions as:
1. The surface forces on the lateral surface are zero, $S_x = 0$, $S_y = 0$, $S_z = 0$.
2. The body forces and temperature are zero, $X_x = 0$, $X_y = 0$, $X_z = 0$, $T = 0$.
3. Only surface forces S_x, S_y, S_z are applied on the end surfaces $y = 0$ and $y = L$. Under these conditions away from the ends by St. Venant's principle

$$\sigma_{xx} = 0, \quad \sigma_{zz} = 0, \quad \sigma_{xz} = 0, \quad e_{zz} = 0, \quad e_{xx} = -\frac{\nu}{E}\sigma_{yy},$$

$$e_{zz} = -\frac{\nu}{E}\sigma_{yy}, \quad e_{yy} = \sigma_{yy}/E = u_{y,y}, \tag{1.28}$$

$$e_{xy} = u_{x,y} + u_{y,x} = \sigma_{xy}/G, \quad e_{yz} = u_{y,z} + u_{z,y} = \sigma_{y,z}/G.$$

On the end $y = L$, by St. Venant's principle, the applied distributed forces can be replaced by total forces and moments, Equation (1.6),

$$P_{yL} = \int_A S_y \, dA, \quad P_{xL} = \int_A S_x \, dA, \quad P_{zL} = \int_A S_z \, dA,$$

$$M_{xL} = -\int_A z S_y \, dA, \quad M_{zL} = \int_A x S_y \, dA, \tag{1.29}$$

$$M_{yL} = \int_A (z S_x - x S_z) \, dA.$$

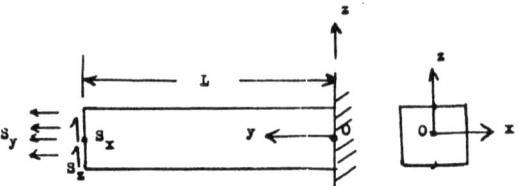

Fig. 1.6. Cantilever beam with end forces S_x, S_y, S_z.

Use the stated conditions and Equation (1.28) in the compatibility Equations (1.16) to get

$$\sigma_{yy,xx} = 0, \quad \sigma_{yy,yy} = 0, \quad \sigma_{yy,zz} = 0, \quad \sigma_{yy,xz} = 0,$$
$$\nabla^2 \sigma_{xy} + \mu \sigma_{yy,xy} = 0, \quad \nabla^2 \sigma_{yz} + \mu \sigma_{yy,yz} = 0, \quad \mu = \frac{1}{1+\nu}. \tag{1.30}$$

The equilibrium Equations (1.4) become

$$\sigma_{xy,y} = 0, \quad \sigma_{yz,y} = 0, \quad \sigma_{xy,x} + \sigma_{yy,y} + \sigma_{yz,z} = 0, \tag{1.31}$$

while the surface Equations (1.5) are

$$\sigma_{xy} = S_x, \quad \sigma_{yy} = S_y, \quad \sigma_{yz} = S_z, \quad \text{on } y = L, \tag{1.32}$$

which are replaced by Equations (1.29).

The first four equations in (1.30) can be solved to give

$$\sigma_{yy} = C_0 + C_1 x + C_2 y + C_3 z + C_4 xy + C_5 yz, \tag{1.33}$$

where $C_0, \cdots, C_5$ are unknown constants. Thus, Equations (1.30) and (1.31) become

$$\nabla^2 \sigma_{xy} + \mu C_4 = 0, \quad \nabla^2 \sigma_{yz} + \mu C_5 = 0, \quad \sigma_{xy} = \sigma_{xy}(x,z),$$
$$\sigma_{yz} = \sigma_{yz}(x,z), \quad \sigma_{xy,x} + (C_2 + C_4 x + C_5 z) + \sigma_{yz,z} = 0. \tag{1.34}$$

Although the axial or normal stress σ_{yy} in Equation (1.33) is a linear function of the cross section variables x and z for any cross section away from the ends so that the strain

$$e_{yy} = \sigma_{yy}/E = \frac{1}{E}[(C_0 + C_1 x + C_3 z) + y(C_2 + C_4 x + C_5 z)] \tag{1.35}$$

is a plane strain, the shear stresses σ_{xy} and σ_{yz} in Equation (1.34) are complicated two dimensional functions.

For this plane strain cross section of the beam, the displacements u_x and u_z of the beam can be determined for the neutral strain plane, which is normal to the cross section. From Equation (1.10)

$$u_y = u_{y0} - z\theta_x + x\theta_z, \quad \theta_x = \frac{du_z}{dy}, \quad \theta_z = -\frac{du_x}{dy}, \tag{1.36}$$

The basic three, two, and one dimensional equations 13

whence
$$e_{yy} = u_{y,y} = \frac{du_{y0}}{dy} - x\frac{d^2u_x}{dy^2} - z\frac{d^2u_z}{dy^2}. \tag{1.37}$$

From Equation (1.35) for corresponding terms
$$E\frac{du_{y0}}{dy} = C_0 + C_2 y, \quad E\frac{d^2u_x}{dy^2} = -(C_1 + C_4 y), \quad E\frac{d^2u_z}{dy^2} = -(C_3 + C_5 y), \tag{1.38}$$

are the differential equations for the deflections of the beam due to the e_{yy} strain.

Because of the complicated equations for the shear stresses in the beam, Equations (1.34), further approximations and simplifications must be made to get simple shear stresses. To get an idea of these approximations for one dimensional beam theory, several simple cases will be investigated in the following sections:
1. No shear stresses in the beam.
2. Beam cross section a thin plate, with one shear stress.
3. Thin web beam with large area flanges and one shear stress.
4. Torsion of circular cross section and thin wall closed box.
5. Thin web box beam with general loading.

1.9. No shear stresses in the beam

If $S_x = 0$, $S_z = 0$ on the end $y = L$, Figure 1.6, then in Equation (1.29),
$$P_{yL} = 0, \quad P_{zL} = 0, \quad M_{yL} = 0, \tag{1.39}$$

whence the shear stresses $\sigma_{xy} = 0$, $\sigma_{yz} = 0$ in Equation (1.34) give $C_2 = 0$, $C_4 = 0$, $C_5 = 0$, and
$$\sigma_{yy} = C_0 + C_1 x + C_3 z \tag{1.40}$$

is constant along the beam. The three constants C_0, C_1, C_3, can be obtained from the cross section equilibrium equations corresponding to Equation (1.29) at $y = L$, or
$$\int_A \sigma_{yy}\, dA = P_y, \quad \int_A z\sigma_{yy}\, dA = -M_x, \quad \int_A x\sigma_{yy}\, dA = M_z, \tag{1.41}$$

which are constant along the beam. See Section 1.14 for evaluation of the constants. Thus, the solution σ_{yy} in Equation (1.40) with all other stresses zero satisfies the equilibrium and compatibility equations, and the boundary equations in the sense of the St. Venant approximation.

1.10. Beam cross section of a thin plate with one shear stress

Take the cross section in Figure 1.6 of constant width t in the x-direction. Let t be small compared to the height $2h$ in the z-direction. If $S_x = 0$ on $y = L$, then σ_{xy} is small and can be taken as zero in the sense of the plane stress case in

Section 1.6. Assume S_z acts through the centroid so that $M_{yL} = 0$ in Equation (1.29), as well as $P_{xL} = 0$. Thus, Equations (1.33) and (1.34) become

$$\sigma_{yy} = C_0 + C_3 z + y(C_2 + C_5 z), \quad \nabla^2 \sigma_{yz} + C_5 = 0,$$
$$\sigma_{yz,z} + C_2 + C_5 z = 0, \quad \sigma_{yz} = \sigma_{yz}(z), \tag{1.42}$$

where the $(1+\nu)$ expression does not appear in the thin plate approximation. Use

$$P_z = \int_A \sigma_{yz}\, dA = t \int_{-h}^{h} \sigma_{yz}\, dz, \quad \sigma_{yz} = 0 \text{ on } z = \pm h, \tag{1.43}$$

and integrate the equilibrium equation in Equation (1.42) to get

$$\sigma_{yz} = \frac{P_z}{2I_{xx}}(h^2 - z^2), \quad I_{xx} = t(2h)^3/12,$$
$$\sigma_{yy} = C_0 + C_3 z + (P_z/I_{xx})yz - E\alpha T(z), \tag{1.44}$$

where the temperature $T(z)$ has been included in σ_{yy} on the basis of Equation (1.27) with $V = 0$. With $C_5 = P_z/I_{xx}$, this solution for σ_{yz} satisfies the compatibility equation in Equation (1.42). Note that if the beam is thin in the z-direction, similar results are obtained with

$$\sigma_{xy} = P_x(b^2 - x^2)/2I_{zz}, \quad I_{zz} = t(2b)^3/12,$$
$$\sigma_{yy} = C_0 + C_1 x + (P_x/I_{zz})xy - E\alpha T(x). \tag{1.45}$$

Thus, for thin plate bending in the plane of the beam from end forces, the two stresses σ_{yy}, σ_{yz}, or σ_{xy}, with the other stresses zero, satisfy the three dimensional equilibrium and compatibility equations in the sense of thin plate average values and St. Venant's principle.

Since the e_{yy} total strain is a plane strain in these constant shear force cases, the bending deflection along the neutral axis can be obtained from Equation (1.38). From Equations (1.7), (1.11), and (1.45) the e_{yz} shear strain is

$$e_{yz} = u_{y,z} + u_{z,y} = (P_z/2GI_{xx})(h^2 - z^2), \tag{1.46}$$

and the cross section is not a plane for shear, due to the z^2 term. In Chapter 2 of Reference 1 a two-dimensional solution for this thin plate beam with a constant shear force shows that the cross section warps from the plane and may produce additional deflections of the neutral axis, depending upon the support conditions. If the warping is unrestrained then the stresses in Equations (1.44) and (1.45) are unaffected.

From Chapter 2 of Reference 1 with no temperature, the deflections u_y and u_z have the form

$$u_y = \frac{P_z(L-y)^2 z}{2EI_{xx}} + \frac{\nu P_z z^3}{6EI_{xx}} - \frac{P_z z^3}{6GI_{xx}} + \left(\frac{P_z h^2}{2GI_{xx}} - D_3\right)z + D_2,$$
$$u_z = \frac{\nu P_z(L-y)z^2}{2EI_{xx}} + \frac{P_z(L-y)^3}{6EI_{xx}} + D_3 y + D_4. \tag{1.47}$$

The basic three, two, and one dimensional equations

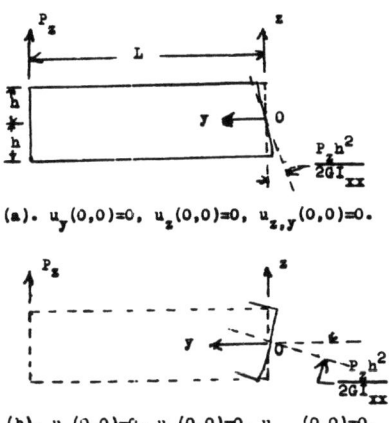

Fig. 1.7. Warping of thin plate beam cross section under constant shear force.

If the support conditions are taken as

$$u_y(0,0) = 0, \quad u_z(0,0) = 0, \quad u_{z,y}(0,0) = 0, \tag{1.48}$$

then in Equation (1.47)

$$D_2 = 0, \quad D_4 = -P_z L^3/6EI_{xx}, \quad D_3 = P_z L^2/2EI_{xx}, \tag{1.49}$$

and Figure 1.7(a) shows the warping of the cross section. In this case the u_z deflection of the neutral plane $z = 0$ is the same as the u_z deflection given by Equation (1.38). However, if the support conditions are taken as

$$u_y(0,0) = 0, \quad u_z(0,0) = 0, \quad u_{y,z}(0,0) = 0, \tag{1.50}$$

then in Equation (1.47)

$$D_2 = 0, \quad D_4 = -P_z L^3/6EI_{xx}, \quad D_3 = \frac{P_z L^2}{2EI_{xx}} + \frac{P_z h^2}{2GI_{xx}}, \tag{1.51}$$

and Figure 1.7(b) shows the warping of the cross section. In this case the u_z deflection of the neutral plane $z = 0$ is

$$u_z = \frac{P_z(L-y)^3}{6EI_{xx}} - \frac{P_z L^3}{6EI_{xx}} + \frac{P_z L^2 y}{2EI_{xx}} + \frac{P_z h^2 y}{2GI_{xx}}, \tag{1.52}$$

where the first three terms are the same as given by Equation (1.38), but the last term represents the shear deflection for the particular supports in Equation (1.50) and Figure 1.7(b). Note that this shear deflection term can be obtained from Equation (1.46) with $z = 0$, or

$$\frac{du_{zs}}{dy} = (e_{yz})_{z=0} = \frac{P_z h^2}{2GI_{xx}}. \tag{1.53}$$

The practical support of the beam cannot be either of the cases in Equations (1.48) or (1.50). However, the practical support usually maintains the cross section in a vertical plane. This can be achieved in Figure 1.7(b) by applying a different statically equivalent stress distribution on the beam at the support. Then by St. Venant's principle the case of Figure 1.7(b) and Equations (1.50)-(1.53) will apply at a sufficient distance from the support. Because of the restraint at the end the u_{zs} in Equation (1.53) will be slightly larger than the actual deflection and the stresses near the support will be different from those in Equation (1.44).

If a force $p(y)$ per unit length is distributed along the edge $z = -h$ of the thin plate beam, then the solution is two dimensional, involving $\sigma_{yy}, \sigma_{zz}, \sigma_{yz}$. If M_x is due to M_{xL}, P_{zL}, and $p(y)$ and if a temperature distribution $T(y)$ is applied to the beam, then from Reference 4 the two dimensional stresses can be expressed in the following form:

$$\sigma_{yy} = \frac{M_x z}{I_{xx}} + \frac{z(3h^2 - 5z^2)}{15 I_{xx}} \frac{d^2 M_x}{dy^2} + \frac{\alpha E(h^2 - 3z^2)}{6} \frac{d^2 T}{dy^2}$$
$$- \left[\frac{z(87h^4 - 70h^2 z^2 - 105z^4)}{4200 I_{xx}} - \frac{h^3(h^2 - 3z^2)}{18 I_{xx}} \right] \frac{d^4 M_x}{dy^4} +$$
$$+ \frac{\alpha E(7h^4 - 30h^2 z^2 + 15z^4)}{180} \frac{d^4 T}{dy^4} + \cdots, \tag{1.54}$$

$$\sigma_{zz} = \frac{(h-z)^2(2h+z)}{6 I_{xx}} \frac{d^2 M_x}{dy^2} - \frac{z(h^2 - z^2)}{60 I_{xx}} \frac{d^4 M_x}{dy^4}$$
$$- \frac{\alpha E(h^2 - z^2)}{24} \frac{d^4 T}{dy^4} + \cdots, \tag{1.55}$$

$$\sigma_{yz} = \frac{(h^2 - z^2)}{2 I_{xx}} \left[\frac{dM_x}{dy} + \frac{h^2 - 5z^2}{30} \frac{d^3 M_x}{dy^3} \right.$$
$$\left. - \frac{17h^4 - 70h^2 z^2 - 35z^4 + (h^3 z/9)}{4200} \frac{d^5 M_x}{dy^5} \right] -$$
$$- \frac{\alpha E z(h^2 - z^2)}{180} \left[30 \frac{d^3 T}{dy^3} + (7h^2 - 3z^2) \frac{d^5 T}{dy^5} \right] + \cdots. \tag{1.56}$$

If the $T(y)$ and $M_x(y)$ in Equations (1.54)-(1.56) are linear in the y variable, then the stresses reduce to those in Equation (1.44) for the end load case. If the distributed load per unit length $p(y) = p_0 =$ constant, then in Equations (1.54) and (1.55) $r d^2 M_x / dy^2 = -p_0$ and the higher derivatives are zero. In this case, M_x will have terms of order $-p_0 y^2 / 2$ so that Equation (1.54) has the form

$$\sigma_{yy} = -\frac{p_0 z L^2}{2 I_{xx}} \left[\left(\frac{y}{L} \right)^2 - \frac{1}{30} \left(\frac{2h}{L} \right)^2 \left\{ 3 - 5 \left(\frac{z}{h} \right)^2 \right\} \right].$$

If the beam is long (L) compared to its depth ($2h$), then the second term is negligible as compared to the first term for the plate design conditions at $y = L$, $z = \pm h$. The first term in σ_{zz} is simply $-p_0/t$ on $z = -h$, where the up load p_0 is applied. The second term in σ_{yz} is zero. It is evident that if $p(y)$ and $T(y)$

The basic three, two, and one dimensional equations 17

vary slowly with small higher derivatives, then the effects on the stresses will be small so that the simple beam theory can be used on long beams.

1.11. Thin web beams with large flange areas and one shear stress

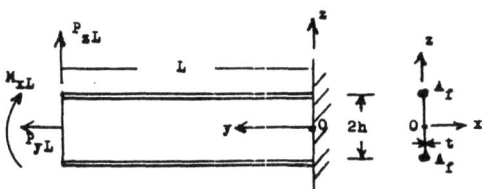

Fig.1.8. Thin Web I-beam with large flange areas.

In Figure 1.8 regard the flange area A_f as a point area with σ_{yy} constant over the area A_f. In Equation (1.42) take

$$\sigma_{yy} = C_3 z = \frac{zM_x}{I_{xx}} - \frac{z(1)P_z}{I_{xx}}, \quad I_{xx} = 2h^2 A_f \left(1 + \frac{A_w}{6A_f}\right), \quad (1.57)$$

where $A_w = 2ht$ and M_x is the moment produced by P_z for a unit length. Equilibrium for a unit length at the juncture of the flange and web gives

$$(1)(t)\sigma_{yz} = A_f \sigma_{yy}, \quad z = \pm h, \quad \text{or} \quad \sigma_{yz} = \frac{hA_f P_z}{tI_{xx}}, \quad z = \pm h. \quad (1.58)$$

Use these boundary conditions in Equation (1.43) and integrate the shear equilibrium equation in Equation (1.42) to get

$$\begin{aligned}
\sigma_{yz} &= (\sigma_{yz})_{\pm h} + \frac{P_z - A_f(\sigma_{yz})_{\pm h}}{(4/3)th^3}(h^2 - z^2) \\
&= \frac{P_z}{tI_{xx}}[hA_f + \frac{t}{2}(h^2 - z^2)] \\
&= \frac{3P_z}{4ht}\frac{4 + (A_w/A_f)[1 - (z/h)^2]}{6 + (A_w/A_f)}.
\end{aligned} \quad (1.59)$$

If A_w/A_f is small (large flange area and thin web), the shear stress is nearly constant, or

$$\sigma_{yz} = P_z/2ht = P_z/A_w = \text{constant}. \quad (1.60)$$

Figure 1.9 shows a small range from 93% to 104% of the average for $A_w/A_f = 1/2$. Thus, the constant shear stress approximation in thin web beams with large flange areas is quite good.

To allow for variable thickness of the web, it is more convenient to use the constant shear flow q with

$$q = t\sigma_{yz} = P_z/2h = \text{constant} \quad (1.61)$$

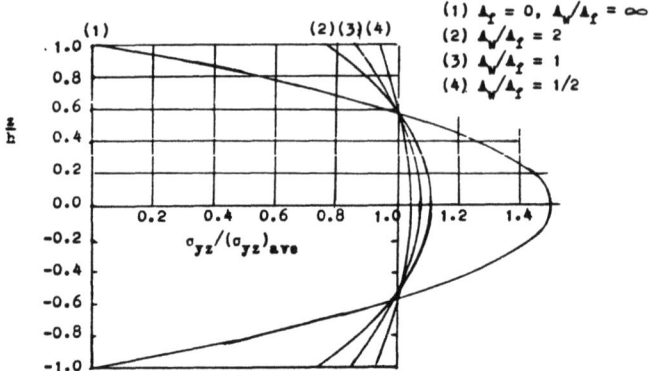

Fig. 1.9. Shear stress distribution in I-beam web.

instead of the shear stress in shear calculations for thin webs. Note that q is in the plane of the web in the direction of P_z.

The shear deflection u_{zs} is given by Equation (1.53) in the form

$$(e_{yz})_{z=0} = du_{zs}/dy = \sigma_{yz}/G = P_z/2htG, \tag{1.62}$$

and the bending deflection is given by Equation (1.38).

It should be pointed out that for A_w/A_f small nearly all the moment M_x is taken by the flange areas with the flange load P_f and the flange stress given by

$$P_f = \pm M_x/2h, \quad \sigma_{yy} = P_f/A_f = \pm M_x/2hA_f. \tag{1.63}$$

This result also follows from Equations (1.44) and (1.58) and Figure 1.8, or

$$\begin{aligned} I_{xx} &= 2h^2 A_f, \quad \sigma_{yy} = M_x z/I_{xx} = M_x z/2h^2 A_f, \\ (\sigma_{yy})_{\max} &= M_x(\pm h)/2h^2 A_f = \pm M_x/2hA_f. \end{aligned} \tag{1.64}$$

Although the moment taken by the thin web is small and can be included in the moment couple taken by the flanges, the tension and compression stresses are still present in the web and must be combined with the shear stress in the design of the web. This means that the constant shear stress is not compatible in the web or with the flange members. However, the effect on the stresses and deflections is small in ductile materials.

1.12. Torsion of circular cross section and thin wall closed box

In Equation (1.29) assume $S_y = 0$ and S_x and S_z are such that $P_{xL} = 0$, $P_{zL} = 0$, but M_{yL} is not zero. From Equations (1.5), (1.29), and (1.34),

$$\begin{aligned} &\nabla^2 \sigma_{xy} = 0, \quad \nabla^2 \sigma_{yz} = 0, \quad \sigma_{xy} = \sigma_{xy}(x,z), \quad \sigma_{yz} = \sigma_{yz}(x,z), \\ &\sigma_{xy,x} + \sigma_{yz,z} = 0, \quad M_y = M_{yL} = \int_A (z\sigma_{xy} - x\sigma_{yz})\,dA, \\ &\sigma_{xy} = S_x, \quad \sigma_{yz} = S_z \quad \text{on } y = L, \\ &\sigma_{xy}l + \sigma_{yz}n = 0 \quad \text{on the lateral surface.} \end{aligned} \tag{1.65}$$

The basic three, two, and one dimensional equations

Although all other stresses are zero, it is evident that σ_{xy} and σ_{yz} are in general two dimensional. However, for the solid circular cross section in Figure 1.10 a plane section remains plane under rotation θ_y about the y-axis so that for the cantilever bar

$$\theta_y = Ky, \quad K = \text{constant}. \tag{1.66}$$

For this rigid body rotation of a plane section, Equation (1.10) gives

$$u_x = z\theta_y = Kyz, \quad u_z = -x\theta_y = -Kxy, \tag{1.67}$$

and Equations (1.7) and (1.13) give

$$\sigma_{xy} = +GKz, \quad \sigma_{yz} = -GKx. \tag{1.68}$$

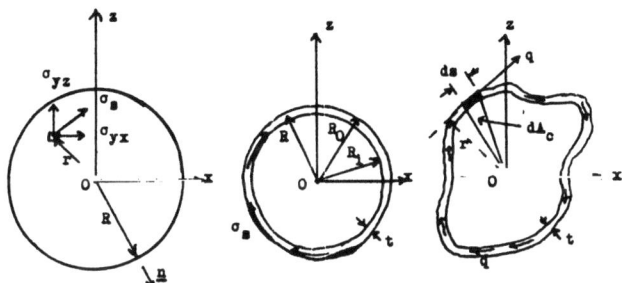

Fig. 1.10. Torsion shear stresses.

From Figure 1.10 the surface equation becomes

$$\sigma_{xy}{}^t + \sigma_{yz}{}^n = +GKz(x/R) - GKx(z/R) = 0$$

so that the stresses in Equation (1.68) satisfy the equilibrium, compatibility, and surface equations in Equation (1.65). Also, from Equation (1.65)

$$M_{yL} = \int_A (+GKz^2 + GKx^2)\, dA = +GKI_p, \quad K = +M_{yL}/GI_p, \tag{1.69}$$

where I_p is the polar moment of inertia of the solid circular cross section, or

$$I_p = \int_A (x^2 + z^2)\, dA = \pi R^4/2.$$

The shear stresses in Equation (1.68) become

$$\sigma_{xy} = M_{yL}z/I_p, \quad \sigma_{yz} = -M_{yL}x/I_p. \tag{1.70}$$

For the circular tube in Figure 1.10,

$$I_p = (\pi/2)(R_0^4 - R_i^4) = (\pi/2)(R_0 - R_i)(R_0 + R_i)(R_0^2 + R_i^2),$$

or approximately

$$I_p = (\pi/2)(t)(2R)(2R^2) = 2\pi R^3 t = 2A_c Rt, \tag{1.71}$$

where A_c is the enclosed area of the tube. Since

$$\sigma_s = (\sigma_{xy}^2 + \sigma_{yz}^2)^{\frac{1}{2}} = M_{yL} R/I_p = M_{yL}/2A_c t, \tag{1.72}$$

it follows that the constant shear flow q around the tube is

$$q = t\sigma_s = M_{yL}/2A_c. \tag{1.73}$$

Although the shear stresses in any solid cross section other than the circle are two dimensional and complicated to calculate because of the surface condition in Equation (1.65), the shear flow in a thin wall tube of any cross section is simple to calculate, provided its cross section is free to warp and the tube shape is held by ribs. In Figure 1.10, moment equilibrium about any point 0 gives

$$M_y = \oint rq\,ds = \oint q(2\mathrm{d}A_c) = 2A_c q, \quad q = M_y/2A_c, \tag{1.74}$$

where A_c is the enclosed area of the tube, or thin wall closed box. With the shear stress assumed constant through the wall thickness, and all other stresses zero, the compatibility, equilibrium, and boundary equations are satisfied by the solution in Equation (1.74).

1.13. Thin web box beam with general loading

In Sections 1.9 through 1.12 above it has been found that the stresses for each of the possible end loads P_{xL}, P_{yL}, P_{zL}, M_{xL}, M_{yL}, M_{zL}, on a beam can be obtained by simple formulas, provided the beam cross section is properly restricted. The simplest cross section that can take all six loads simultaneously and still meet the cross section restraints is the thin web box beam with corner flange areas, Figure 1.11, the cross section being constant along the length.

In Figure 1.11 the four flanges take the axial stresses σ_{yy} produced by P_y, M_x, and M_z, or

$$\begin{aligned}
\sigma_{yy1} &= \frac{P_y}{4A} + \frac{M_x}{2Ah} - \frac{M_z}{2Ab}, & \sigma_{yy2} &= \frac{P_y}{4A} - \frac{M_x}{2Ah} - \frac{M_z}{2Ab}, \\
\sigma_{yy3} &= \frac{P_y}{4A} - \frac{M_x}{2Ah} + \frac{M_z}{2Ab}, & \sigma_{yy4} &= \frac{P_y}{4A} + \frac{M_x}{2Ah} - \frac{M_z}{2Ab}.
\end{aligned} \tag{1.75}$$

These axial stresses are approximate in that the webs can take tension stresses produced by P_y, M_x, M_z, but can take very small compression stresses due to buckling. In Figure 1.11 webs (1) and (3) take P_z, webs (2) and (4) take P_x, and all four webs take the constant shear flow produced by M_y, or

$$\begin{aligned}
q_1 &= \frac{P_z}{2h} + \frac{M_y}{2bh}, & q_2 &= \frac{P_x}{2b} + \frac{M_y}{2bh}, & q_3 &= -\frac{P_z}{2h} + \frac{M_y}{2bh}, \\
q_4 &= -\frac{P_x}{2b} + \frac{M_y}{2bh}.
\end{aligned} \tag{1.76}$$

It should be noted that not only do these simple one dimensional solutions for the stresses satisfy the three dimensional compatibility and equilibrium equations in the sense of the St. Venant's principle, but also this simple thin web box beam cross section is very efficient. Within fixed geometry for h and b the flange areas

The basic three, two, and one dimensional equations 21

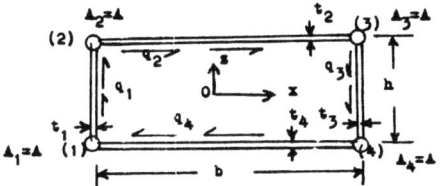

Fig. 1.11. Simple box beam cross section.

are as far away as possible, the enclosed area as large as possible, and with constant shear flows, it follows that the maximum applied axial stresses σ_{yy} and shear stresses σ_s are as small as possible for the total cross section area.

The above results are based on applied end forces only. If concentrated and distributed forces are applied on the lateral surface of the beam, then Equation (1.28) does not hold and the stresses $\sigma_{xx}, \sigma_{zz}, \sigma_{xz}$ are not necessarily zero. However, if these lateral surface forces are put into the box beam by means of ribs, which distribute the forces to the shear webs of the box approximately as constant shear flows, then the effects on the flanges are very local in the sense of St. Venant's principle. Also, if these lateral surface forces vary slowly with y so that their higher derivatives are small, then from Equation (1.54) and the discussion in Section 1.10 above the effects of these forces and of temperature variations in the y-direction are small compared to the one dimensional stresses, such as in Equations (1.75) and (1.76).

It should be noted that the typical shear and moment curves for all types of structural beams in bridges, buildings, airplanes, etc. vary slowly so that the above one dimensional equations can be used in many applications.

In the above discussion the body forces were taken as zero. The inertia forces and the weight of the structure are body forces, but in most cases they can be represented as surface forces and combined with the other forces on the surface without appreciable error in the stresses.

The more general beam equations with unsymmetrical bending are given in the following Section 1.14.

1.14. Inelastic effects in beams with temperature

The Equations (1.75) and (1.76) for Figure 1.11 represent an elastic solution for one material with a constant or linear temperature variation on the beam cross section. For mixed elastic materials, variable elastic material properties, inelastic materials, and the temperature $T(x, z)$ non-linear in x and z, the solutions will be different. To determine these effects consider Figure 1.12, where the restrained cross section will have plane strain with each element i maintaining the same length. However, the variation in the stress due to these various effects will produce end forces S_y. On the basis of St. Venant's principle, these end forces can be replaced by end moments M_{xE} and M_{zE} and an end force P_{yE}. To obtain

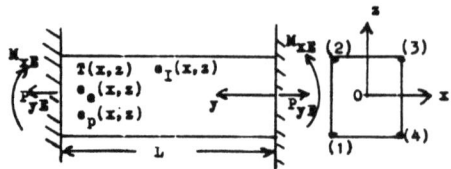

Fig. 1.12. End loads and moments due to temperature and inelastic effects.

equations for these effects, use Figure 1.13 and write the total strain on the plane section as

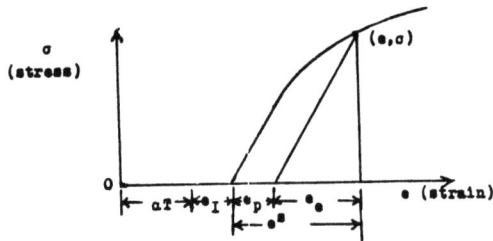

Fig. 1.13. Total strains and stress-strain curve.

$$e = e_e + e_E = D_0 + D_1 x + D_2 z,$$
$$e_E = e_p + e_I + e_T, \qquad e_T = \alpha T, \tag{1.77}$$

where e_e is the elastic strain for the stress

$$e_e = \sigma_{yy}/E , \tag{1.78}$$

e_p is the inelastic strain, e_I is any initial strain, e_T is the strain due to any temperature change, and D_0, D_1, D_2 are constants representing the plane section. The constants D_0, D_1, D_2 can be obtained from the three equilibrium equations for the cross section

$$\int_A \sigma_{yy}\,dA = P_y, \qquad \int_A z\sigma_{yy}\,dA = -M_x, \qquad \int_A x\sigma_{yy}\,dA = M_z. \tag{1.79}$$

Combine Equations (1.77)-(1.79) and divide by a reference E_R to get

$$\begin{aligned}
D_0 A_E + D_1 S_{xE} + D_2 S_{zE} &= (P_y + P_{yE})/E_R, \\
D_0 S_{xE} + D_1 I_{xxE} + D_2 I_{xzE} &= (M_z + M_{zE})/E_R, \\
D_0 S_{zE} + D_1 I_{xzE} + D_2 I_{zzE} &= -(M_x + M_{xE})/E_R,
\end{aligned} \tag{1.80}$$

$$A_E = \int_A (E/E_R)\,dA, \qquad S_{xE} = \int_A (E/E_R)x\,dA,$$

The basic three, two, and one dimensional equations

$$S_{zE} = \int_A (E/E_R)z \, dA, \quad I_{zzE} = \int_A \left(\frac{E}{E_R}\right) x^2 \, dA \tag{1.81}$$

$$I_{xzE} = \int_A \left(\frac{E}{E_R}\right) xz \, dA, \quad I_{zzE} = \int_A \left(\frac{E}{E_R}\right) z^2 \, dA,$$

$$P_{yE} = P_{yp} + P_{yI} + P_{yT} = \int_A E(e_p + e_I + e_T) \, dA, \tag{1.82}$$

$$M_{zE} = M_{zp} + M_{zI} + M_{zT} = \int_A E(e_p + e_I + e_T)x \, dA, \tag{1.83}$$

$$M_{xE} = M_{xp} + M_{xI} + M_{xT} = -\int_A E(e_p + e_I + e_T)z \, dA. \tag{1.84}$$

Define the centroid such that

$$S_{xE} = 0, \quad S_{zE} = 0, \tag{1.85}$$

or from

$$x = x_R - \bar{x}, \quad z = z_R - \bar{z}$$

$$A_E \bar{x} = \int_A \left(\frac{E}{E_R}\right) x_R \, dA, \quad A_E \bar{z} = \int_A \left(\frac{E}{E_R}\right) z_R \, dA. \tag{1.86}$$

With this centroid simplification in Equation (1.80) it follows that

$$E_R A_E D_0 = P_y + P_{yE}, \quad H = I_{xxE} I_{zzE} - I_{xzE}^2, \tag{1.87}$$

$$E_R H D_1 = I_{zzE}(M_x + M_{xE}) + I_{xzE}(M_z + M_{zE}), \tag{1.88}$$

$$E_R H D_2 = -I_{xxE}(M_x + M_{xE}) - I_{xzE}(M_z + M_{zE}). \tag{1.89}$$

From Equations (1.77) and (1.78) the stress now becomes

$$\sigma_{yy} = (E/E_R)[-E_R e_E + (P_y + P_{yE})/A_E - (zI_{zzE} - xI_{xzE}) \times \\ \times (M_x + M_{xE})/H + (xI_{xxE} - zI_{xzE})(M_z + M_{zE})/H]. \tag{1.90}$$

For the case when $I_{xzE} = 0$ (both area A and E symmetrical about an axis), Equation (1.90) reduces to

$$\sigma_{yy} = (E/E_R)[-E_R e_E + (P_y + P_{yE})/A_E - (M_x + M_{xE})(z/I_{zzE}) + \\ + (M_z + M_{zE})(x/I_{xxE})]. \tag{1.91}$$

For the case of bending of the simple beam about the x-axis with $I_{xzE} = 0$, $M_z = 0$, and $M_{zE} = 0$,

$$\sigma_{yy} = (E/E_R)[-E_R e_E + (P_y + P_{yE})/A_E - (M_x + M_{xE})(z/I_{zzE})]. \tag{1.92}$$

For the simplest case of all with no temperature change, $P_y = 0$, constant E, no inelastic or initial effects, and bending about the x-axis only, Equation (1.92) gives the bending stress formula from "Mechanics of Materials" textbooks,

$$\sigma_{yy} = -M_x(z/I_{xx}). \tag{1.93}$$

In Equation (1.77) the initial strains e_I and the thermal strains e_T are usually regarded as known in any given case, while the inelastic strains e_p are unknown in the sense that they depend upon the material stress-strain curve in Figure 1.13 and upon the applied loads.

Since the Ramberg-Osgood non-dimensional stress-strain equation, Reference 5, applies to many structural materials, it can be used to determine e_p. The form of the equation is

$$\begin{aligned} E_{yy} e_{yy}^* / F_y &= (\sigma_{yy}/F_y) + (3/7)(\sigma_{yy}/F_y)^n \\ &= (\sigma_{yy}/F_y) + (E_{yy} e_p/F_y), \\ e_p &= (3F_y/7E_{yy})(\sigma_{yy}/F_y)^n, \end{aligned} \qquad (1.94)$$

where F_y is the yield stress of the material, n is an integer representing the shape of the stress-strain curve, and e_p is the inelastic strain in Figure 1.13.

Table 1.1. Ramberg-Osgood Equation (1.94).

$\dfrac{\sigma_{yy}}{F_y}$	$n=10$ $\dfrac{E_{yy} e_{yy}^*}{F_y}$	$n=20$ $\dfrac{E_{yy} e_{yy}^*}{F_y}$	$n=50$ $\dfrac{E_{yy} e_{yy}^*}{F_y}$
0.000	0.000	0.000	0.000
0.400	0.400	0.400	0.400
0.600	0.603	0.600	0.600
0.700	0.712	0.700	0.700
0.800	0.846	0.805	0.800
0.900	1.049	0.952	0.902
0.950	1.207	1.104	0.983
0.975	1.308	1.233	1.096
1.000	1.429	1.429	1.429
1.025	1.574	1.727	2.498
1.050	1.748	2.187	5.965
1.075	1.958	2.896	17.013
1.100	2.212	3.983	51.41
1.125	2.517	5.644	–
1.150	2.884	8.164	–
1.175	3.325	11.959	–
1.200	3.854	17.630	–
1.225	4.486	26.04	–
1.250	5.241	38.42	–
1.300	7.208	–	–
1.400	13.797	–	–
1.500	26.21	–	–

Table 1.1 shows values of Equation (1.94) for a range of n values for steel and aluminium alloys, which can be used for straight line interpolation to get the stress when the strain is known. Failure of the material occurs for Ee^*/F_y about 50. Table 1.1 and Equation (1.94) apply for both tension and compression. Similar tables can be calculated for other values of n. From Figure 1.13 and Equations (1.77), (1.78), (1.90),

$$\begin{aligned} e_{yy}^* &= (\sigma_{yy}/E_{yy}) + e_p = -e_T - e_I + (P_y + P_{yE})/E_R A_E - \\ &\quad - (zI_{zzE} - xI_{xzE})(M_x + M_{xE})/E_R H + \\ &\quad + (xI_{xxE} - zI_{xzE})(M_z + M_{zE})/E_R H, \end{aligned} \qquad (1.95)$$

where by Equations (1.82)-(1.84)

$$\begin{aligned} P_{yE} &= P_{yI} + P_{yT} + P_{yp}, \quad M_{zE} = M_{zI} + M_{zT} + M_{zp}, \\ M_{xE} &= M_{xI} + M_{xT} + M_{xp}. \end{aligned} \qquad (1.96)$$

Since the equilibrium Equation (1.79) must be satisfied whether the material is elastic or inelastic, the unknown values P_{yp}, M_{zp}, M_{xp} in Equations (1.82)-(1.84) can be obtained by iteration from Equation (1.79). The procedure is as follows:

1. Calculate all the constants in Equations (1.95) and (1.96) except P_{yp}, M_{xp}, M_{zp} from Equations (1.81)-(1.87). These constants and the centroid do not change during the iteration.
2. Calculate $(e_{yy}^{s})_0$ from Equation (1.95) for each element on the cross section with P_{yp}, M_{xp}, M_{zp} all zero.
3. Get $(\sigma_{yy})_0$ from Table 1.1 for each element, using E_{yy} and F_y for the element.
4. Use the $(\sigma_{yy})_0$ values and calculate $(P_y)_0$, $(M_x)_0$, $(M_z)_0$ from Equation (1.79).
5. Calculate $(P_{yp})_1 = P_y - (P_y)_0$, $(M_{xp})_1 = M_x - (M_x)_0$, $(M_{zp})_1 = M_z - (M_z)_0$. If all are zero, then there are no inelastic effects and no need to go further.
6. Calculate $(e_{yy}^{s})_1$ from Equation (1.95) by using the values of $(P_{yp})_1$, $(M_{xp})_1$, $(M_{zp})_1$.
7. Use the $(e_{yy}^{s})_1$ values and get $(\sigma_{yy})_1$ from Table 1.1.
8. Calculate $(P_y)_1$, $(M_x)_1$, $(M_z)_1$ from Equation (1.79).
9. Calculate

$$(P_{yp})_2 = (P_{yp})_1 + P_y - (P_y)_1, \quad (M_{xp})_2 = (M_{xp})_1 + M_x - (M_x)_1,$$
$$(M_{zp})_2 = (M_{zp})_1 + M_z - (M_z)_1. \tag{1.97}$$

10. Repeat steps 6, 7, 8, 9 for R times to convergence with Equation (1.97) as

$$(P_{yp})_r = (P_{yp})_{r-1} + P_y - (P_y)_{r-1},$$
$$(M_{xp})_r = (M_{xp})_{r-1} + M_x - (M_x)_{r-1}, \tag{1.98}$$
$$(M_{zp})_r = (M_{zp})_{r-1} + M_z - (M_z)_{r-1},$$

for $r = 3, 4, \cdots, R$.

11. Determine convergence as a tolerance of

$$|(e_{yy})_R - (e_{yy})_{R-1}| \le 0.00001 \tag{1.99}$$

on all elements at step R.

1.15 Example of inelastic axial stresses and strains with temperature

Find the inelastic stresses in the two element symmetrical beam in Figure 1.14 with no bending and no buckling. Element (1) is split in two equal parts and put on each side of element (2) to avoid any bending on the cross section. The data are $A_1 = 4.00$ in^2, $E_1 = 10^7$ psi, $F_{y1} = 30,000$ psi, $e_{T1} = \alpha_1 T_1 = 0.004$, $A_2 = 0.20$ in^2, $E_2 = 3(10)^7$ psi, $F_{y2} = 60,000$ psi, $e_{T2} = 0.0005$, $P_y = 50,000$ lb.

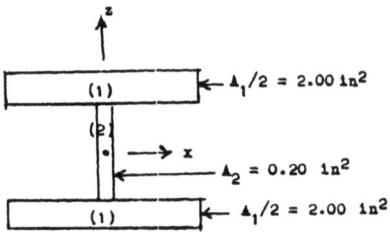

Fig. 1.14. Two element symmetrical bar for Sec. 1.15

Solution. Take $E_R = 10^7$ psi, and follow the steps as outlined above in Section 1.14:

(1) $A_E = 4.00 + (3)(0.20) = 4.6$ in^2,
$P_{yT} = (0.004)(10^7)(4.00) + (0.0005)(3)(10^7)(0.20) = 163,000$ lb
$(P_y + P_{yT})/E_R A_E = (50,000 + 163,000)/(4.6)(10^7) = 0.004630$.

(2) $(e^s_{yy1})_0 = -0.004 + 0.004630 = 0.000630$,
$(e^s_{yy2})_0 = -0.0005 + 0.00463 = 0.004130$.

(3) $(E_{yy} e^s_{yy}/F_y)_{1,0} = (10^7)(0.00063)/30,000 = 0.210$,
$(E_{yy} e^s_{yy}/F_y)_{2,0} = 3(10^7)(0.004130)/60,000 = 2.065$,
$(\sigma_{yy1})_0 = (0.210)(30,000) = 6300$ psi,
$(\sigma_{yy2})_0 = (1.085)(60,000) = 65,100$ psi,

(4) $(P_y)_0 = (4.00)(6300) + (0.20)(65,100) = 38,220$ lb,

(5) $(P_{yp})_1 = 50,000 - 38,220 = 11,780$ lb,

(6) $(e^s_{yy1})_1 = 0.000630 + (11,780/4.6(10^7)) = 0.00088861$,
$(e^s_{yy2})_1 = 0.004130 + 0.0002561 = 0.0043861$.

(7) $(E_{yy} e^s_{yy}/F_y)_{1,1} = 0.295$, $(\sigma_{yy1})_1 = 8850$ psi,
$(E_{yy} e^s_{yy}/F_y)_{2,1} = 2.193$, $(\sigma_{yy2})_1 = 65,880$ psi,

(8) $(P_y)_1 = 48,576$ lb,

(9) $(P_{yp})_2 = 11,780 + 50,000 - 48,576 = 13,204$ lb,
$13,204/4.6(10^7) = 0.0002870$.

(10) $(e^s_{yy1})_2 = 0.0009170$, $(\sigma_{yy1})_2 = 9170$ psi,
$(e^s_{yy2})_2 = 0.0044170$, $(\sigma_{yy2})_2 = 66,000$ psi,
$(P_y)_2 = 49,880$ lb,
$(P_{yp})_3 = 13,204 + 50,000 - 49,880 = 13,324$ lb,
$13,324/4.6(10^7) = 0.0002896$.

Change in strain from step 2 to step 3 is 0.0002896-0.0002870=0.0000026, which is within the specified tolerance in Equation (1.99). Thus, the stresses are $\sigma_{yy1} = 9170$ psi, $\sigma_{yy2} = 66,000$ psi, which check 120 lb low for $P_y = 50,000$ lb force.

These final stresses show that element (1) is elastic while element (2) is in-

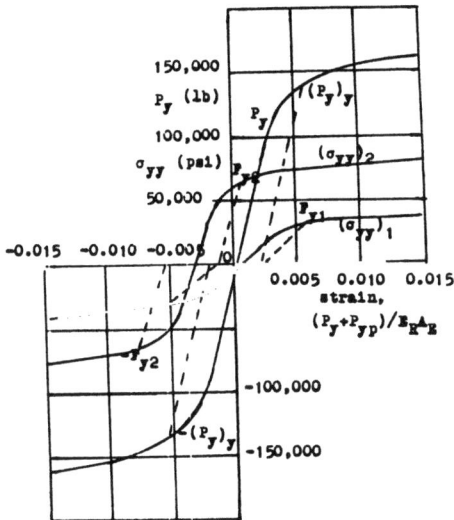

Fig. 1.15. Load-strain design curve for example in Sec. 1.15.

elastic, which means that the beam cross section can take more axial force P_y without failing. In fact, a design curve can be constructed for P_y showing the load carrying ability of the cross section in the presence of the thermal stresses and the inelastic effects. Plot P_y against the strain $(P_y + P_{yp})/E_R A_E$ to obtain a load-strain design curve for the cross section. In this example, this curve can be constructed without iteration by assuming values of the strain $(P_y + P_{yp})/E_R A_E$, calculating e^*_{yy1} and e^*_{yy2}, getting σ_{yy1} and σ_{yy2} from Table 1.1, and calculating $P_y = 4.00\sigma_{yy1} + 0.20\sigma_{yy2}$. Figure 1.15 shows the resulting design curve. Several conclusions can be made from Figure 1.15 about thermal stress effects. Although element (2) is inelastic for the thermal stresses alone $(P_y = 0)$, the entire cross section (two elements in this case) can carry large tension or compression P_y forces without inelastic deformation. From the 0.002 strain offset definition for yield stress and yield load, Figure 1.15 shows that the cross section yields at approximately the same strain as element (1) does giving $(P_y)_y = \pm 140,000$ lb. The conclusion is that the cross section as a unit can be elastic with some of the elements inelastic, provided the materials are ductile. Also, with ductile materials, failure cannot occur until all elements are beyond the yield point (with no buckling). Although there are large thermal stresses at small P_y, they practically disappear beyond $(P_y)_y$ and have little effect on the failing P_y force. The 0.0035 thermal strain offset between the two stress strain curves has little effect on the stresses at large inelastic strains. Since thermal strains have a definite limit, they cannot cause failure in ductile materials (with no buckling).

1.16. Sequence loading and thermal cycling in beams

See References 6 - 9. Since the load may already be acting when the temperature is applied to the structure, or the structure may be hot when the load is applied, or the temperature may cycle with a steady load acting, it is necessary to write Equation (1.95) to account for sequence application and removal of load and temperature. Under such conditions unloading of some elements from an inelastic position on the stress-strain curve may occur. It is assumed that for unloading the element n strain will follow a straight line of slope E_n (elastic) to zero stress and then follow a stress-strain curve with a new origin when loading in the opposite direction. The change in origin for any element n is

$$e_{n0} = e_n^s - (\sigma_n/E_n). \tag{1.100}$$

If reloading takes place from any point on the straight line or from an elastic position on the new stress-strain curve, then the straight line is followed up to the original stress-strain curve (for strain-hardening materials) and the original stress-strain curve followed for larger strains. If reloading takes place from an inelastic point on the new stress-strain curve, then the element may not return to the old curve but instead may follow a new stress-strain curve from a new origin, Figure 1.16. Which effect actually occurs depends upon how much yielding has occurred in the reverse direction in the material. It is assumed that a new stress-strain curve can arise whenever unloading occurs from a point on the stress-strain curve for which

$$\sigma_n/F_{yn} \geq 0.80, \tag{1.101}$$

where F_{yn} is the yield stress of element n on the cross section. This corresponds to

$$e_{n0} \geq 0.046(F_{yn}/E_n) \tag{1.102}$$

if the stress-strain curve for element n is represented by the Ramberg-Osgood Equation (1.94) with $n = 10$, Figure 1.16. From the above discussion it is evident that at any step in the loading and temperature sequence the strains and stresses must depend upon the previous steps as to whether some elements have a new stress-strain curve with a new origin. If at each step the e_n^s is assumed to start from a new origin as determined by Equation (1.100), then the strain, Equation (1.95), at step j can be written in terms of the previous step $j-1$ and the load or temperature application or removal at step j as

$$e_{nj}^s = (\sigma_{n,j-1})/E_{n,j-1}) + \Delta e_{nj}^s, \tag{1.103}$$

where E_{nj} is the elastic value at step j and Δe_{nj}^s is given by Equation (1.95) at step j.

For the definition of e_{nj}^s as used in Equation (1.103) the stress σ_{nj} is directly associated with e_{nj}^s either as an elastic value or as a value on the stress-strain curve of the element, except for the case in which the element returns to the old stress-strain curve on reloading. In this latter case the strain associated with the stress can be expressed as

$$e_{nj}^s + q_{nj} \quad \text{with} \quad q_{nj} = \sum_{i=k}^{j-1} e_{n0i}, \tag{1.104}$$

The basic three, two, and one dimensional equations

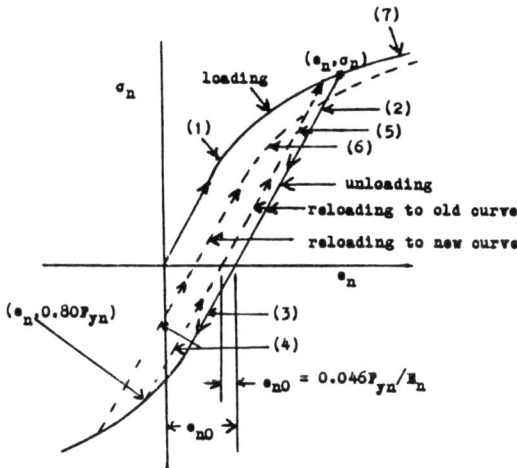

Fig. 1.16. Shift in origin of stress-strain curves.

where e_{n0i} is determined by Equation (1.100) but is not included in the sum unless it satisfies Equation (1.102), and k is the step at which the last sign reversal of e_{n0i} occurred.

With the strains known, the stresses can be obtained from the Ramberg-Osgood Equation (1.94) of Table 1.1. To determine the stresses at step j in the sequence, modify Equation (1.94) to

$$\frac{E_{nj}}{F_{ynj}}(e_{nj}^* + q_{nj}) = \frac{\sigma_{nj}}{F_{ynj}} + \frac{3}{7} r_{nj} \left(\frac{\sigma_{nj}}{F_{ynj}}\right)^n, \tag{1.105}$$

where

$$q_{nj} = 0 \quad \text{or} \quad \sum_{i=k}^{j-1} e_{n0i}, \quad r_{nj} = 0 \text{ or } 1, \tag{1.106}$$

and E_{nj} is the elastic modulus and F_{ynj} the yield stress of element n at step j. The values of q_{nj} and r_{nj} to be used at any step depend upon the previous steps and the signs of the terms in Equation (1.102). For the various points numbered in Figure 1.16 the values of q_{nj} and r_{nj} are as follows: $q_{nj} = 0$ for all points except (7), and $r_{nj} = 0$ for points (2), (4), and (5).

In the case of *thermal cycling* with an applied load on the beam it is possible for strain accumulation to occur when an inelastic strain occurs in one element on temperature application and in a different element on temperature removal. Consider the case of two bars with different stress-strain curves that are fastened together, Figure 1.17. The two bars have the same areas with a temperature change T applied to bar (2). The elastic thermal stresses and strains are

$$P_{T1}/A = 1, \quad P_{T2}/A = -1, \quad e_{T1} = 0.25, \quad e_{T2} = -1.00. \tag{1.107}$$

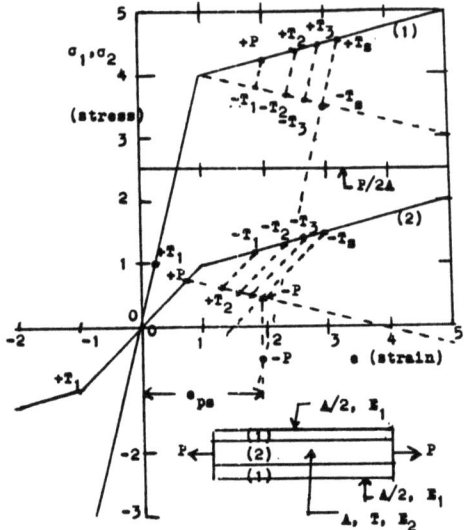

Fig. 1.17. Thermal cycling to shakedown (P=5A).

In Figure 1.17 apply the temperature to get the two points marked $+T_1$. Apply a tension load $P = 5A$ with stresses

$$P_1/A = 4.25, \quad P_2/A = 0.75, \quad P/2A = 2.5, \qquad (1.108)$$

which produces the same strain change in members (1) and (2) to give the points $+P$. Remove the temperature so that inelastic member (1) unloads on an elastic line to $-T_1$ as shown, while member (2) loads to the point $-T_1$, where the strains are equal and the load $P/2A$ is maintained. In this two element case the points can be located directly on Figure 1.17 without calculating the inelastic strains and using the above procedure in detail. The dashed lines are the reflections of the inelastic portions of the stress-strain curves placed so as to maintain $P/2A$. When the temperature is applied for the second time, points $+T_2$ are obtained, followed by points $-T_2$ on removal of T. The cycling continues until the dashed lines marked $+T_s$, $-T_s$ give elastic thermal strain of ± 1. This is the elastic shakedown cycling point in the Figure 1.17 load case. Repeat of the thermal cycle simply runs up and down the two lines $+T_s$, $-T_s$.

If the load P is removed from the shakedown position, then points $-P$ are obtained with a permanent set $e_{ps} = 1.93$ and residual stresses ± 0.45.

In Figure 1.18 for the same bars as in Figure 1.17 except the stress-strain curves have maximum failing stresses, a larger load $P = 5.90A$ is used. However, there is no shakedown in this case, because the dashed lines $+T_d$, $-T_d$ give a stress less than the elastic thermal stress ± 1. Thus, divergence occurs and the bars will fail at the material elongation strain, Reference 9. See References 6 through 9 for other examples of strain accumulation effects. The following Section 1.17 shows how load-strain design curves can be constructed for inelastic cycling cases.

The basic three, two, and one dimensional equations   31

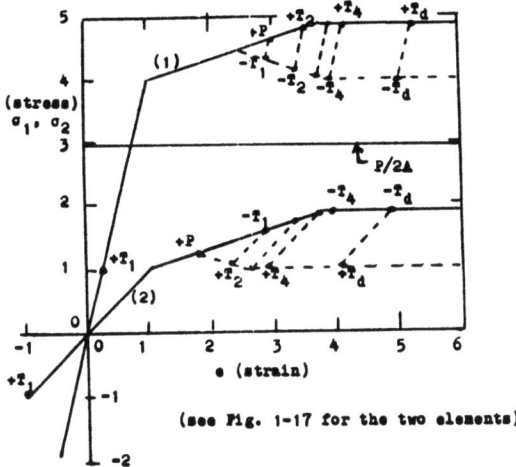

Fig. 1.18. Thermal cycling-divergence (P=5.9A).

1.17 Load-strain design curves for beams

It is evident in Figures 1.17 and 1.18 that the load P can be removed after each temperature cycle to give permanent sets for one cycle, two cycles, etc., to shakedown. For example, for the load case of $P = 5A$ in Equation (1.108) and Figure 1.17, the permanent set e_{ps} for the cross section is $e_{ps} = 0.90$ for one cycle, $e_{ps} = 1.35$ for two cycles, $e_{ps} = 1.93$ for shakedown. These calculations can be repeated for various values of the load P to get permanent sets as a function of P. With these values it is possible to calculate an equivalent stress-strain or load-strain curve for the cross section. See Figure 1.15 for a simple case without cycling.

By using the procedure of the previous Section 1.16, convergence for the inelastic effects can be calculated by Equation (1.98) for step j so that the inelastic strain on element n for step j is given by Equation (1.95) as

$$e^s_{pnj} = \frac{P_{ypj}}{E_R A_E} - (z_n I_{zzE} - x_n I_{xzE})\frac{M_{xpj}}{E_R H} + (x_n I_{xxE} - z_n I_{xzE})\frac{M_{zpj}}{E_R H}. \quad (1.109)$$

After a finite number of steps ending with the removal of all load and temperature, the expression

$$e^s_{pns} = \sum_j e^s_{pnj} \quad (1.110)$$

represents the total inelastic effect occuring for all steps, and hence is related to the permanent set on each element n of the cross section with reference to the origin of the original stress-strain curve. Since residual stresses will usually be present after removal of load and temperature, the strain e^s_{pns} in Equation (1.110) represents the permanent set at zero stress for element n plus the strain σ_{rn}/E_n for the residual stress σ_{rn}. However, unless the residual stresses change by possible creeping at room temperature, the strain e^{o}_{pns} may be regarded as the permanent strain of element n as far as permanent deformation of the structure is concerned.

If these calculations are carried out under a given temperature distribution, given materials, a given cross section, and a given load-temperature cycle for various values of the applied load and moments, then an allowable load-strain curve can be constructed by graphing σ_{apm} against e^{o}_{m} for fixed ratios among P_y, M_x, and M_z, where

$$\sigma_{apm} = \frac{E_m}{E_R}\left[\frac{P_y}{A_E} - (z_m I_{xxE} - x_m I_{xzE})\frac{M_x}{H} + (x_m I_{zzE} - z_m I_{xzE})\frac{M_z}{H}\right], \quad (1.111)$$

$$e^{o}_{m} = (\sigma_{apm}/E_m) + e^{o}_{pms}.$$

Here m designates the element in Equation (1.110) with the maximum permanent strain. The graph of Equation (1.111) is equivalent to an applied stress-strain curve for element m (do not confuse with the true stress-strain curve for element m either at room or elevated temperature), including temperature effects on material properties, thermal stress effects, and inelastic effects on the entire cross section. It is in terms of the element with the largest permanent strain and may be used as an applied stress-strain curve for design purposes, giving an equivalent applied yield stress for the 0.2 per cent offset strain and an ultimate applied stress for the maximum strain. The maximum strain corresponds to the strain in element m when some element of the cross section reaches the permissible elongation of the material or reaches a cutoff strain produced by a stability type of failure.

These load-strain curves give the load-carrying capacity of the entire cross section and not that of the highest stressed element. Even at room temperature the design based upon an allowable stress for the highest stressed element is not always optimum, while at elevated temperatures the allowable stress for the highest stressed element may have little relationship to the actual load-carrying capacity of the cross section. The element with the largest permanent strain is important as it is used to determine the shape of the design load-strain curve.

Note that the above method applies at room temperature and to mixed materials to give the total load-carrying capacity of the cross section.

The above procedure can be used to construct interaction load curves for the combination of applied axial load and applied bending moment. If the load-strain curves are calculated for various ratios between the axial load and the bending moment, then a cross-plot of the axial load against the bending moment can be made for selected values of the total strain or of the permanent strains. This can be done for the room temperature case, or for any given temperature distribution with specified materials on a given cross section.

As an example of the construction of a load-strain curve, consider the case of

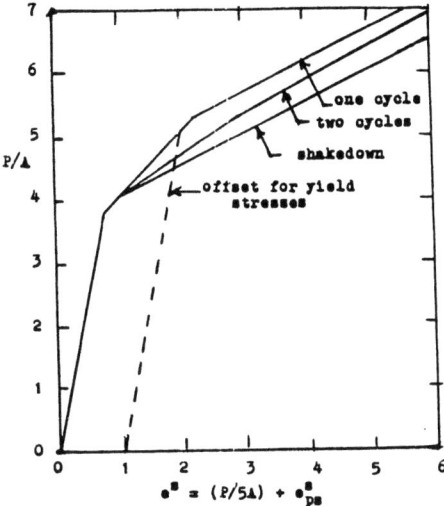

Fig. 1.19. Allowable load-strain curves with thermal cycling for case in Fig. 1.17.

the thermal cycling example in Figure 1.17. Equation (1.111) gives

$$\sigma_{ap1} = 4P/5A, \quad \sigma_{ap2} = P/5A,$$
$$e^s = e_1^s = e_2^s = (P/5A) + e_{ps}^s. \quad (1.112)$$

Figure 1.19 shows the one cycle, two cycle, and shakedown design curves with the applied stress as $P/A = \sigma_{ap1} + \sigma_{ap2}$ rather than σ_{ap1} or σ_{ap2}. The one cycle case is $+T+P-T-P$ and the two cycle case is $+T+P-T+T-T-P$, with the + sign indicating temperature or load application and the minus sign indicating removal. See Reference 6 for other cases of allowable load-strain curves for axial loads, bending moments, combined axial loads and moments, and interaction curves for combined axial loads and moments.

1.18. Problems.

1.1. If the body forces are zero, show that the stresses

$$\sigma_{xx} = C_1[y^2 + C_2(x^2 - y^2)], \quad \sigma_{yy} = C_1[x^2 + C_2(y^2 - x^2)],$$
$$\sigma_{zz} = C_1 C_2(x^2 + y^2), \quad \sigma_{xy} = -2C_1 C_2 xy, \quad \sigma_{xz} = 0, \quad \sigma_{yz} = 0,$$

satisfy the equilibrium Equations (1.4).

1.2. If the body forces and shear stresses are zero, determine the most general forms of σ_{xx}, σ_{yy}, σ_{zz} that will be in equilibrium by Equations (1.4).

1.3. For the stresses in Problem 1.1 find the surface forces on the faces $z = a$ and $z = c$ of the structure shown in Figure 1.20.

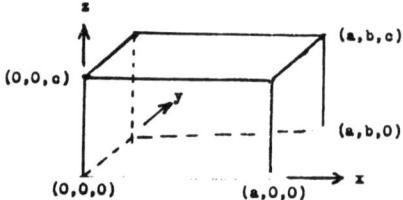

Fig. 1.20. Problem 1.3.

1.4. Derive Equations (1.9).

1.5. Show that

$$e_{xx} = K_1(x^2 + y^2), \quad e_{yy} = K_1(y^2 + z^2), \quad e_{xy} = K_2 xyz,$$
$$e_{zz} = 0, \quad e_{xz} = 0, \quad e_{yz} = 0,$$

is not a compatible state of strain.

1.6. If the thermal strains in the unrestrained body are

$$e_{xx} = e_{yy} = e_{zz} = \alpha T(x, y, z), \quad e_{xy} = e_{xz} = e_{yz} = 0,$$

integrate Equations (1.9) to get the most general permissible expression for $T(x, y, z)$.

1.7. Derive Equations (1.15).

1.8. If the body forces and temperature are zero in Equations (1.15), show that the solution

$$u_x = D_1 - aH,_x, \quad u_y = D_2 - aH,_y, \quad u_z = D_3 - aH,_z,$$
$$a = \frac{1}{4(1-\nu)}, \quad H = D_0 + xD_1 + yD_2 + zD_3,$$
$$\nabla^2 D_0 = 0, \quad \nabla^2 D_1 = 0, \quad \nabla^2 D_2 = 0, \quad \nabla^2 D_3 = 0,$$

satisfies the Equation (1.15).

1.9. Derive Equations (1.16).

1.10. If the body forces and temperature are zero, are the stresses in Problem 1.1 compatible? Show by using Equations (1.16).

1.11. Suppose V and T are zero in Equations (1.24) and (1.27), or in the plane stress case for plates

$$\nabla^4 \phi = 0, \quad \sigma_{xx} = \phi,_{yy}, \quad \sigma_{yy} = \phi,_{xx}, \quad \sigma_{xy} = -\phi,_{xy}.$$

Assume polynomial forms for ϕ that satisfy $\nabla^4 \phi = 0$, calculate the stresses and specify what stresses must be applied on the boundaries of a rectangular plate with sides parallel to the coordinate axes to produce the internal stresses. Use

(a) $\phi = c_1 x^2 + c_2 xz + c_3 z^2$,

(b) $\phi = a_1 x^3 + a_2 x^2 z + a_3 x z^2 + a_4 z^3$,

(c) $\phi = b_1 x^4 + b_2 x^3 z + b_3 x^2 z^2 + b_4 x z^3 + b_5 z^4$.

1.12. Add the curve for $A_w/A_f = 1/4$ to Figure 1.9.

1.13. Use Equation (1.90) for the elastic case with temperature and find the axial stresses $\sigma_{yy1}, \sigma_{yy2}, \sigma_{yy3}, \sigma_{yy4}$ in a four stringer box beam cross section similar to Figure 1.11. Use $P_y = 60,000$ lb, $M_x = 10^7$ in-lb, $M_z = -2(10^6)$ in-lb. The data for the stringers in the order $i=1, 2, 3, 4$ is

$A_i = 12.00, 18.00, 12.00, 25.00$ in^2,

$x_{Ri} = 0.00, 0.00, 30.00, 30.00$ in,

$z_{Ri} = 0.00, 25.00, 25.00, 0.00$ in,

$E_i/10^7 = 0.70, 0.50, 0.70, 0.60$ psi,

The basic three, two, and one dimensional equations 35

$T_i = 100, 200, 50, 150 \, °C$,
$\alpha_i = \text{constant} = 24(10^6) \, /°C$.

Neglect axial stresses in the webs and use only the stringer areas in the calculations. Assume the temperature variations in the webs are such that the four point areas give a satisfactory representation. Check the results by the equilibrium Equations (1.79).

1.14. In the simple symmetrical beam cross section in Figure 1.8 use the two flanges as point areas but use the continuous web and make the integrations to get the equation for the σ_{yy} stresses in Equation (1.92) with temperature effects. Take $A_f = 15 \, in^2$, $h = 20 \, in$, $t = 0.100 \, in$, $P_y = 0$, $M_z = -8(10^6)$ in-lb, $\alpha = 20(10^{-6})/°C$, $E_f = (0.5)10^7$ psi, $T = 200(\frac{x}{20})^2 \, °C$, $E_{web} = 2(10^7)[1 - (\frac{x}{40})^2]$ in webs and flanges. Graph the distribution of the σ_{yy} stresses in the beam cross section.

1.15. Repeat Problem 1.14 without temperature effects and compare the results to Problem 1.14.

1.16. Repeat Problem 1.14 with $E_{web} = 0.5(10^7)[1 - (\frac{x}{40})^2]$ and compare the results to Problem 1.14.

1.17. In Problem 1.13 above use the calculated constants and calculate the thermal stresses and the applied stresses separately. Check the results by using Equation (1.79).

1.18. Solve the Example in Section 1.15 for $P_y = 0$ and compare the resulting stresses to those in Figure 1.15.

1.19. Solve the Example in Section 1.15 for $P_y = 100,000$ lb and compare the results to Figure 1.15.

1.20. Construct Figure 1.17 for the case of $P = 5.31 \, A$.

1.21. Construct Figure 1.17 for the case of $P_{T1}/A = 1.25$, $P_{T2}/A = -1.25$, $e_{T1} = 0.31$, $e_{T2} = -2.00$. Use the procedure in Figure 1.16 to get the new stress-strain curve for element 2.

1.22. Determine the approximate equation for the shakedown curve in Figure 1.19 for $e^* > 1$ by using the stress-strain curves and the shakedown conditions in Figure 1.17.

References

Chapter 1

1. S. Timoshenko and J.N. Goodier: *Theory of Elasticity*, Mcgraw-Hill Book Co., New York, (1951).
2. N.I. Muskhelishvili: *Some Basic Problems of the Mathematical Theory of Elasticity*, translated from the Russian by J.R.M. Radok, P. Noordhoff, N.V., Groningen, Netherlands (1953).
3. B.E. Gatewood: *Thermal Stresses*, McGraw-Hill Book Co., New York (1957).
4. B.E. Gatewood and R. Dale: Note on two-dimensional stresses in long beams with spanwise variation of load and temperature, *J. of Applied Mechanics* **29**, No. 4, December (1962), pp 747-749.
5. W. Ramberg and W.R. Osgood: Description of stress-strain curves by three parameters, *NACA TN 902*, (1943).
6. B.E. Gatewood and R.W. Gehring: Allowable axial loads and bending moments for inelastic structures under nonuniform temperature distribution, *J. of Aerospace Sciences* **29**, no. 5, May (1962).
7. B.E. Gatewood and R.W. Gehring: Inelastic analysis with sequence application of load and temperature and changing material properties, *Proc. of the Ninth Midwestern Mechanics Conference* (1965) pp 315-322.
8. B.E. Gatewood: The problem of strain accumulation under thermal cycling, *J. of Aerospace Sciences*, **27**, June (1960).
9. B.E. Gatewood, A.P. Grothouse and W.W. Von Hausen: Experimental data on strain accumulation under equivalent thermal cycling, *J. of Aerospace Sciences*, **28**, June (1961).

References for additional reading

10. H. Reismann and P.S. Pawlik: *Elasticity Theory and Applications*, John Wiley and Sons, New

York (1980).
11. D.J. Peery and J.J. Azar: *Aircraft Structures*, 2nd Ed., McGraw-Hill Book Co., New York (1982).
12. D.H. Allen and W.E. Haisler: *Introduction to Aerospace Structural Analysis*, John Wiley and Sons, New York (1985).

2

Virtual displacement and virtual force methods in structural analysis

2.1. Introduction

It was demonstrated in Chapter 1 that a satisfactory one-dimensional analysis, except for local effects, can be made for simple beams. However, the stress analysis was made only for the beam cross section so that only the equilibrium equations were needed to get the solution. If the simple beams are assembled into a truss type structure, or into a beam with many supports, then the analysis becomes complicated again because deflection equations as well as equilibrium equations may be needed to get the stresses in the structure. Because of these complicated redundant problems, various special methods have been developed to determine stresses and deflections in redundant beams, trusses, and plates.

In 1855 Saint Venant pointed out that the redundant problem in trusses could be avoided by using the displacements in the equilibrium equations and solving for the displacements first. Of course, the drawback to this approach was the large system of equations to be solved. The digital computers have changed the picture so that this method now is used in many problems. In 1873 A. Castigliano introduced his strain energy methods, which were used to calculate displacements and solve redundant problems for beams and trusses. Since then, many papers and books have been published on strain energy methods, variational energy methods, and many other special methods. However, little progress was made in giving the strain energy methods and other special methods a rational basis until recent years when it was shown that all the classical strain energy theorems can be derived from the virtual displacement and virtual force methods of analysis, References 1 and 2.

Also, there is a third virtual method involving a combination of virtual stresses and virtual displacements. Although it has received little attention in the literature, it is very useful in certain cases, particularly for approximate solutions and for finite element representations.

Since the three virtual methods are more general than the strain energy methods with fewer restrictions, apply directly to finite elements, have advantages over the energy methods in many applications, and can be used directly for approximate solutions in three, two, and one dimensional cases, they will be derived in this chapter and used throughout the later chapters. The classical strain energy theorems are derived from the virtual methods in Appendix C. Also, further discussion of the advantages of the virtual methods is given in Appendix C.

The virtual displacement, virtual force, and mixed virtual methods are parallel methods based upon the three fundamental approaches to finding the stresses and displacements in a structure. These three approaches, the displacement method, the stress method, and the combined method, are described in Sections 1.3, 1.4, and 1.5 for the general three dimensional elasticity equations. As will be seen in Sections 2.2, 2.4, and 2.6, the three virtual methods are derived directly from the basic equations in the three approaches of Sections. 1.3, 1.4, 1.5, and expressed in integral forms. These integral forms are described and identified by three principles, or

(1) The principle of virtual displacements.
(2) The principle of virtual forces.
(3) The principle of mixed virtual stresses and virtual displacements.

Although any one of the principles can be used on most any static structural problem, each one has definite advantages over the others in certain types of structure so that all three will be developed and applied to various types of structures. The principles of virtual displacements and virtual forces are used extensively in flight vehicle analysis and design, as well as in all types of structures such as buildings, automobiles, bridges, and machines.

If a beam structure is determinate, then the stresses can be obtained from the equilibrium equations as in Sections 1.8 through 1.15 and the virtual methods may not be needed. Although the deflections for the determinate structure can be obtained from the determinate strains, it is usually much simpler to use the virtual force method to get the deflections. All three methods can be used to obtain stresses and displacements in a redundant structure. The displacement method gives the displacements and then the stresses in the structure without any regard to whether the structure is determinate or redundant. The virtual force method requires identification of redundant elements and supports for the structure in order to solve for the redundant loads first and then the stresses and displacements. The mixed method gives simultaneous coupled equations for both redundant stresses and displacements.

The three virtual principles will be derived in Sections 2.2, 2.4, and 2.6 by using integral transform methods. These transform methods have been used by the mathematicians to obtain exact and approximate solutions of complicated partial differential equations, References 3, 4, and 5. Basically, the procedure is to multiply the differential equations by some kernel function, or weighting function, or virtual function or functions, which may be specified or may be arbitrary, and then integrate over some finite or infinite region to produce an integral form. The specified kernel functions used may contain an unknown parameter, which becomes the new variable in the new and much simpler equation after the integration (such as the Laplace transform). Sneddon, in Reference 3, solves for the stresses in various two dimensional plate problems by using the Fourier transform procedure. The virtual functions may have no parameter but may be used directly in the transformations of the Green identities as known or arbitrary functions. When the virtual function is properly selected, the Green identities can be used to solve various partial differential equations, Reference 5. Also, the Green identities can be used to reduce the order of the derivatives of the real functions while increasing the order of the derivatives of the virtual functions, thus simplifying the integral form. In addition, the virtual function can be selected so as to give a Green's function for the problem, Reference 5.

The principles of virtual displacements and virtual forces will be derived using Green's first identity, where very general virtual functions that have definite physical meanings are used. The procedure will be applied to the three dimensional equilibrium Equations (1.4) to get the principle of virtual displacements and to the three dimensional strain displacement compatibility Equations (1.7) to get the principle of virtual forces. The principle of mixed virtual stresses and virtual displacements can then be obtained from the stress-strain Equations (1.19) and the principle of virtual displacements. Once the principles are obtained for the three dimensional case, they can be reduced to the two dimensional and one dimensional cases.

The three principles are used to derive the unit displacement theorem, the unit load theorem, and the mixed unit displacement and unit load theorem in Sections 2.3, 2.5, and 2.7. The two dimensional forms are given in Section 2.8, while the one dimensional forms are given in Section 2.9. Temperature and inelastic effects for the one dimensional case are considered in Section 2.10. Matrix forms of the virtual principles for point values are given in Section 2.11.

2.2. The principle of virtual displacements

This principle is an integral form of the differential equations for the displacement method of solution described in Section 1.3. Multiply each equation in Equations (1.4) in sequence by the arbitrary virtual functions $u_x^V(x,y,z)$, $u_y^V(x,y,z)$, $u_z^V(x,y,z)$, integrate over the volume V of the structure, and combine the three equations into one equation. The result is

$$\int_V \left[(\sigma_{xx,x} + \sigma_{xy,y} + \sigma_{xz,z})u_x^V + (\sigma_{xy,x} + \sigma_{yy,y} + \sigma_{yz,z})u_y^V + \right.$$
$$\left. + (\sigma_{xz,x} + \sigma_{yz,y} + \sigma_{zz,z})u_z^V \right] dV + \int_V (X_x u_x^V + X_y u_y^V + X_z u_z^V) dV \quad (2.1)$$
$$= 0, \quad \text{where } \sigma_{xx,x} = \frac{\partial \sigma_{xx}}{\partial x}, \text{ etc.}$$

This Equation (2.1) can be integrated by parts to remove the derivatives on the stresses by using Green's first identity

$$\int_V (R_{,x} S_{,x} + R_{,y} S_{,y} + R_{,z} S_{,z}) dV = \int_S R(lS_{,x} + mS_{,y} + nS_{,z}) dS -$$
$$- \int_V R(S_{,xx} + S_{,yy} + S_{,zz}) dV, \quad R_{,x} = \frac{\partial R}{\partial x}, \text{ etc.,} \quad (2.2)$$

where l, m, n are the direction cosines for an outward normal on the surface S of the structure.

Let R represent the stresses and let

$$S_{,x} = u_x^V, \quad S_{,y} = u_y^V, \quad S_{,z} = u_z^V, \tag{2.3}$$

so that the term $\int_V \sigma_{xx,x} u_x^V \, dV$ in Equation (2.1) becomes

$$\int_V \sigma_{xx,x} u_x^V \, dV = \int_S \sigma_{xx} u_x^V \, dS - \int_V \sigma_{xx} u_{x,x}^V \, dV \tag{2.4}$$

by Equation (2.2). Use the same procedure for each stress term in Equation (2.1), use the surface equilibrium Equation (1.5), and define the virtual terms

$$\begin{aligned} e_{xx}^V &= u_{x,x}^V, & e_{yy}^V &= u_{y,y}^V, & e_{zz}^V &= u_{z,z}^V, & e_{xy}^V &= u_{x,y}^V + u_{y,x}^V, \\ e_{xz}^V &= u_{x,z}^V + u_{z,x}^V, & e_{yz}^V &= u_{y,z}^V + u_{z,y}^V, \end{aligned} \tag{2.5}$$

whence Equation (2.1) takes the form

$$\int_V (\sigma_{xx} e_{xx}^V + \sigma_{yy} e_{yy}^V + \sigma_{zz} e_{zz}^V + \sigma_{xy} e_{xy}^V + \sigma_{xz} e_{xz}^V + \sigma_{yz} e_{yz}^V) \, dV -$$

$$- \int_V (X_x u_x^V + X_y u_y^V + X_z u_z^V) \, dV - \int_S (S_x u_x^V + S_y u_y^V + S_z u_z^V) \, dS - \tag{2.6}$$

$$- \sum_{m=1}^M (P_{mx} u_{mx}^V + P_{my} u_{my}^V + P_{mz} u_{mz}^V) = 0.$$

Any concentrated point forces on the surface have been explicitly included by defining a point force component as

$$P_{mx} = \int_{\Delta S_m} S_x \, dS,$$

where ΔS_m is a small area around the point m.

Since the relations among the virtual terms in Equation (2.5) are exactly the same as the strain-displacement Equations (1.7), it follows that a physical interpretation may be given to the virtual expressions, with u_x^V, u_y^V, u_z^V regarded as virtual displacements and e_{xx}^V, e_{yy}^V, e_{zz}^V, e_{xy}^V, e_{xz}^V, e_{yz}^V regarded as virtual strains. For this reason, Equation (2.6) is called the *principle of virtual displacements*. Some authors express Equation (2.6) in a strain energy and virtual work form and call it the principle of virtual work. The physical interpretation of the virtual displacements gives units of work or energy to the terms in Equation (2.6) and is convenient in selecting the *arbitrary* virtual displacements, but it is not necessary. In particular, since compatibility Equation (2.5) gives the only conditions on the virtual terms, *no virtual stresses* need be associated with the virtual strains.

On the other hand, the real stresses in the principle of virtual displacements, Equation (2.6), must either be compatible or be expressed in terms of the real displacements by means of the real stress-strain and real strain-displacement relations such as Equation (1.14). In this latter case, Equation (2.6) is an integral form of the differential Equations (1.15). However, any stress-strain and strain-displacement relations can be used in the principle. Note that Equation (2.6) includes the equilibrium surface or boundary conditions in Equations (1.5) so that only the displacement surface or boundary conditions have to be specified separately. The minimum number of displacement conditions are those necessary to prevent rigid body motion of the static structure. Note that at points where the displacements are specified the forces at these points in Equation (2.6) may be regarded as the unknown reactions. The virtual displacements at these points will give separate equilibrium equations for the reactions.

If the forces in Equation (2.6) are functions of time t, then for a time interval t_0 to t_1, the principle of virtual displacements is

$$\int_{t_0}^{t_1}\left[\int_V (\sigma_{xx}e^V_{xx} + \sigma_{yy}e^V_{yy} + \sigma_{zz}e^V_{zz} + \sigma_{xy}e^V_{xy} + \sigma_{xz}e^V_{xz} + \sigma_{yz}e^V_{yz})\,dV\right.$$
$$-\int_V \{(X_x - \rho u_{x,tt})u^V_x + (X_y - \rho u_{y,tt})u^V_y + (X_z - \rho u_{z,tt})u^V_z\}\,dV$$
$$\left. -\int_S (S_x u^V_x + S_y u^V_y + S_z u^V_z)\,dS - \sum_{m=1}^{M}(P_{mx}u^V_{mx} + P_{my}u^V_{my} + P_{mz}u^V_{mz})\right]dt$$
$$= 0, \quad u_{x,tt} = \frac{\partial^2 u_x}{\partial t^2}, \text{ etc.}, \qquad (2.7)$$

where ρ is mass density and $u_{x,tt}$, $u_{y,tt}$, $u_{z,tt}$ are the accelerations.

The integral form represented by the principle of virtual displacements in Equation (2.6) has many advantages over the equilibrium Equations (1.4) and (1.5) and over the Equations (1.15) in the displacement method of solution.

(1) Six equations are combined into one equation.

(2) Point forces and point displacements can be used.

(3) Deflection differential equations can be determined directly for either determinate or redundant structures, including the equilibrium boundary conditions, Section 4.2.

(4) The virtual displacements can be selected to simplify the calculations, such as using the point forces separately, Sections 2.3, 2.5, and 2.7.

(5) Approximate solutions can be obtained directly, Chapters 5 and 7, Volume 2.

(6) Finite elements can be used directly, Chapters 4, 6, and Chapters 8 and 9, Volume 2.

(7) Thermal, initial, large displacement, and inelastic strains can be used, Section 2.10.

(8) The equilibrium boundary conditions are included directly so that only displacement boundary conditions must be separately specified.

(9) Matrix forms of the equations can be used directly for point forces and point displacements, Sections 2.11, 3.5, 4.8, etc.

2.3. The unit displacement theorem

Since the virtual relations in Equation (2.5) are linear, superposition of virtual terms can be used in the principle of virtual displacements, Equation (2.6). This superposition may simplify the calculations for point forces and point displacements. Apply a point virtual displacement u_m^V at point m in the direction of a point force P_m at point m. P_m may be zero or it may be the resultant force over a small area. Assume the point virtual displacement u_m^V to produce continuous virtual displacements and compatible virtual strains, Equation (2.5), in the structure or in an element of the structure, with the elements being joined at points. These virtual displacements and virtual strains will be proportional to u_m^V, or

$$\begin{aligned}
&u_x^V = u_{xm}^1(x,y,z)u_m^V, \quad u_y^V = u_{ym}^1 u_m^V, \quad u_z^V = u_{zm}^1 u_m^V, \\
&e_{xx}^V = u_{x,x}^V = u_{xm,x}^1 u_m^V = e_{xxm}^1 u_m^V, \quad e_{yy}^V = e_{yym}^1 u_m^V, \\
&e_{zz}^V = e_{zzm}^1 u_m^V, \quad e_{xy}^V = e_{xym}^1 u_m^V, \quad e_{xz}^V = e_{xzm}^1 u_m^V, \\
&e_{yz}^V = e_{yzm}^1 u_m^V,
\end{aligned} \qquad (2.8)$$

where u_{xm}^1, u_{ym}^1, u_{zm}^1 are virtual displacements per unit displacement and $e_{xxm}^1 \cdots e_{yzm}^1$ are virtual strains per unit displacement due to $u_m^V = 1$ at the point m in the direction of the force P_m.

Use only the one point virtual term $P_m u_m^V$ and Equation (2.8) in the principle of virtual displacements, Equation (2.6), and divide out the arbitrary u_m^V term, whence the *unit displacement theorem* is obtained,

$$\begin{aligned}
P_m = &\int_V (\sigma_{xx} e_{xxm}^1 + \sigma_{yy} e_{yym}^1 + \sigma_{zz} e_{zzm}^1 + \sigma_{xy} e_{xym}^1 + \sigma_{xz} e_{xzm}^1 + \sigma_{yz} e_{yzm}^1)\, dV \\
&- \int_V (u_{xm}^1 X_x + u_{ym}^1 X_y + u_{zm}^1 X_z)\, dV - \\
&- \int_S (u_{xm}^1 S_x + u_{ym}^1 S_y + u_{zm}^1 S_z)\, dS.
\end{aligned} \qquad (2.9)$$

In the case of beams these virtual displacements and virtual strains may be per unit rotation due to a unit virtual rotation $\theta_m^V = 1$.

For point moments in beams due either to a unit displacement or a unit rotation at point n,

$$\begin{aligned}
M_n = &\int_V (\sigma_{xx} e_{xxn}^1 + \cdots + \sigma_{yz} e_{yzn}^1)\, dV - \int_V (u_{xn}^1 X_x + u_{yn}^1 X_y + u_{zn}^1 X_z)\, dV - \\
&- \int_S (u_{xn}^1 S_x + u_{yn}^1 S_y + u_{zn}^1 S_z)\, dS.
\end{aligned} \qquad (2.10)$$

Note that if the true stresses in Equations (2.9) and (2.10) are due only to P_m or M_n, then the integrals involving the body forces and surface forces do not appear. Usually, equivalent point values for any body and surface forces are used so that these integrals can be omitted.

There are two major applications of the unit displacement theorem (2.9) or (2.10),

(A) To determine point forces and moments, including reactions and zero values, when the real stresses or real displacements are known.

(B) To determine unknown point displacements and point rotations when the point forces and point moments, including zero values, are specified. In this case

Virtual displacement and virtual force methods

the stresses in Equations (2.9) and (2.10) must be expressed in terms of the point displacements. Both of these applications are used in later Chapters.

Since the unit displacement theorem in Equations (2.9) and (2.10) can be used for each point force and point moment, the results can be assembled into a matrix form for the unit displacement theorem. This matrix form of the principle of virtual displacements is very useful for large structures that are divided into finite elements. The matrix methods will be set up and used in the following chapters.

2.4. The principle of virtual forces

This principle is an integral form of the differential equations for the stress method of solution described in Section 1.4. Multiply the six compatible Equations (1.7) by the arbitrary virtual functions $\sigma_{xx}^V, \sigma_{yy}^V, \sigma_{xy}^V, \sigma_{xz}^V, \sigma_{yz}^V$ in sequence, integrate over the volume V, and combine the equations into one equation. The result is

$$\int_V [(e_{xx} - u_{x,x})\sigma_{xx}^V + (e_{yy} - u_{y,y})\sigma_{yy}^V + (e_{zz} - u_{z,z})\sigma_{zz}^V +$$
$$+ (e_{xy} - u_{y,x} - u_{x,y})\sigma_{xy}^V + (e_{xz} - u_{z,x} - u_{x,z})\sigma_{xz}^V +$$
$$+ (e_{yz} - u_{z,y} - u_{y,z})\sigma_{yz}^V]\, dV = 0. \qquad (2.11)$$

This Equation (2.11) can be integrated by parts to remove the derivatives on the displacements by using Green's first identity, Equation (2.2). Follow the same procedure as in Section 2.2 above and define the virtual terms

$$-X_x^V = \sigma_{xx,x}^V + \sigma_{xy,y}^V + \sigma_{xz,z}^V,$$
$$-X_y^V = \sigma_{xy,x}^V + \sigma_{yy,y}^V + \sigma_{yz,z}^V, \qquad (2.12)$$
$$-X_z^V = \sigma_{xz,x}^V + \sigma_{yz,y}^V + \sigma_{zz,z}^V,$$

$$S_y^V = \sigma_{xy}^V l + \sigma_{yy}^V m + \sigma_{yz}^V n,$$
$$S_z^V = \sigma_{xz}^V l + \sigma_{yz}^V m + \sigma_{zz}^V n, \qquad (2.13)$$
$$S_x^V = \sigma_{xx}^V l + \sigma_{xy}^V m + \sigma_{xz}^V n,$$

whence Equation (2.11) becomes

$$\int_V (e_{xx}\sigma_{xx}^V + e_{yy}\sigma_{yy}^V + e_{zz}\sigma_{zz}^V + e_{xy}\sigma_{xy}^V + e_{xz}\sigma_{xz}^V + e_{yz}\sigma_{yz}^V)\, dV -$$
$$- \int_V (u_x X_x^V + u_y X_y^V + u_z X_z^V)\, dV - \int_S (u_x S_x^V + u_y S_y^V + u_z S_z^V)\, dS -$$
$$- \sum_{m=1}^{M} (u_{mx} P_{mx}^V + u_{my} P_{my}^V + u_{mz} P_{mz}^V) = 0, \qquad (2.14)$$

where point displacements on the surface have been explicitly included.

Since the relations among the virtual terms in Equations (2.12) and (2.13) are the same as the equilibrium Equations (1.4) and (1.5), it follows that a physical interpretation can be given to the virtual expressions, with σ_{xx}^V, σ_{yy}^V, σ_{zz}^V, σ_{xy}^V, σ_{xz}^V, σ_{yz}^V regarded as virtual stresses, X_x^V, S_y^V, S_z^V regarded as virtual body forces and S_x^V, S_y^V, S_z^V regarded as virtual surface forces. For this reason, Equation (2.14) is called the *principle of virtual forces*. Some authors express Equation (2.14) in terms of complementary strain energy and complementary virtual work and call it the principle of complementary virtual work. The physical interpretation of the virtual terms gives units of work and energy to the terms in Equation (2.14) and is convenient in selecting the arbitrary virtual forces of virtual stresses but is not necessary. In particular, since equilibrium Equations (2.12) and (2.13) give the only conditions on the virtual terms, *no virtual strains* need be associated with these virtual stresses.

On the other hand, the real strains in the principle of virtual forces, Equations (2.14), must be expressed in terms of the real stresses by a stress-strain law and these real stresses must either be in equilibrium with the external forces or must be expressed in terms of stress functions by Equation (1.18). In this latter case Equation (2.14) is an integral form of the differential Equations (1.16). However, any stress-strain relation can be used in the principle. Note that Equation (2.14) includes the displacement surface or boundary conditions and restraints so that only the equilibrium body and surface forces have to be specified separately. Also, at points where the forces are specified, the displacements are unknown so that the virtual forces at these points will give separate equations for these displacements from Equation (2.14)

It should be noted that the strains in Equation (2.14) are total strains so that thermal strains, initial strains, and inelastic strains can be included by Equation (1.77) and Figure 1.13.

In determinate structures the stresses can be obtained directly from equilibrium so that the principle of virtual forces may not be needed, but it can be used directly to get point displacements in determinate structures. In redundant structures, the principle can be used to get the redundant stresses and then used again on a determinate substructure to get point deflections. In the redundant structure, the procedure is to remove selected supports and cut selected members to make the structure determinate. Then a virtual force is applied at each of the selected supports and at each cut, separately, to prodcue the necessary compatibility equations from Equation (2.14) to determine the unknown reactions and cut member loads. The real point displacements used in Equation (2.14) are zero in the cut members and any specified value (usually zero) at the selected redundant supports.

The integral form represented by the principle of virtual forces in Equation (2.14) has the same advantages over the compatibility Equations (1.9) or (1.16) as those listed for the principle of virtual displacements in Section 2.2. It is only necessary to use reciprocal relations between stresses and displacements and between equilibrium boundary conditons and displacement boundary conditions. In this connection, it should be noted that a reciprocal relation exists between the principle of virtual displacements, Equation (2.6), and the principle of virtual forces, Equation (2.14). Interchange the real and virtual terms in either principle and the other principle is obtained. However, real displacements and virtual displacements are the primary functions in Equation (2.6), while real stresses and virtual stresses are the primary functions in Equation (2.14).

2.5. The unit load theorem

Since the virtual relations in Equations (2.12) and (2.13) are linear, superposition of the virtual terms can be used in the principle of virtual forces, Equation (2.14). Thus, the equilibrium virtual stresses can be calculated with all the other point virtual forces taken as zero. In Equation (2.14) take all the point virtual forces zero at the point displacements except for P_m^V at the point m where a point displacement u_m may be present. This virtual force produces virtual stresses in the structure which are proportional to P_m^V and which are in equilibrium with the supports, Equation (2.13), or

$$\sigma_{xx}^V = \sigma_{xxm}^1(x,y,z)P_m^V, \quad \sigma_{yy}^V = \sigma_{yym}^1 P_m^V, \quad \sigma_{zz}^V = \sigma_{zzm}^1 P_m^V,$$
$$\sigma_{xy}^V = \sigma_{xym}^1 P_m^V, \quad \sigma_{xz}^V = \sigma_{xzm}^1 P_m^V, \quad \sigma_{yz}^V = \sigma_{yzm}^1 P_m^V. \tag{2.15}$$

With the virtual term $u_m P_m^V$ and Equations (2.15), (2.12), and (2.13) in Equation (2.14), the arbitrary term P_m^V cancels out and the *unit load theorem* results

$$u_m = \int_V (e_{xx}\sigma_{xxm}^1 + e_{yy}\sigma_{yym}^1 + e_{zz}\sigma_{zzm}^1 + e_{xy}\sigma_{xym}^1 + e_{xz}\sigma_{xzm}^1 + e_{yz}\sigma_{yzm}^1)\,dV$$
$$- \int_V (u_x X_{xm}^1 + u_y X_{ym}^1 + u_z X_{zm}^1)\,dV -$$
$$- \int_S (u_x S_{xm}^1 + u_y S_{ym}^1 + u_z S_{zm}^1)\,dS, \tag{2.16}$$

where $\sigma_{xxm}^1, \cdots, \sigma_{yzm}^1$ are the virtual stresses per unit force or per unit moment due to $P_m^V = 1$, or $M_m^V = 1$. For point rotations

$$\theta_n = \int_V (e_{xx}\sigma_{xxn}^1 + \cdots + e_{yz}\sigma_{yzn}^1)\,dV - \int_V (u_x X_{xn}^1 + u_y X_{yn}^1 + u_z X_{zn}^1)\,dV -$$
$$- \int_S (u_x S_{xn}^1 + u_y S_{yn}^1 + u_z S_{zn}^1)\,dS. \tag{2.17}$$

All other virtual forces at the point displacements are zero for the calculations of the virtual stresses in Equations (2.16) and (2.17).

There are two major applications of the unit load theorem,

(A) To determine point deflections when the real strains, elastic or inelastic, are known.

(B) To determine unknown redundant loads when the point deflections are specified (usually zero).

Application (A) can be used for any substructure in a determinate or redundant structure to obtain any desired deflections.

Since the unit load theorem in Equations (2.16) and (2.17) can be used for each point displacement and each point rotation, the results can be assembled into a matrix form. This matrix form can be used in large structures with many elements, and will be used in Section 2.11 and later Chapters in the setup of the matrix methods of structural analysis.

2.6. The principle of mixed virtual stresses and virtual displacements

In the combined method of solution discussed in Section 1.5, Equations (1.19) are the compatibility equations and Equations (1.4) and (1.5) are the equilibrium equations. Rewrite Equation (1.19) in the form

$$e_{xxu} - e_{xx\sigma} = 0, \quad e_{yyu} - e_{yy\sigma} = 0, \quad e_{zzu} - e_{zz\sigma} = 0,$$
$$e_{xyu} - e_{xy\sigma} = 0, \quad e_{xzu} - e_{xz\sigma} = 0, \quad e_{yzu} - e_{yz\sigma} = 0, \qquad (2.18)$$

where $e_{xxu} \cdots e_{yzu}$ are the strains expressed in terms of the displacements u_x, u_y, u_z by Equation (1.7) and $e_{xx\sigma} \cdots e_{yz\sigma}$ are the strains expressed in terms of the stresses $\sigma_{xx} \cdots \sigma_{yz}$ by Equations (1.11). The rigid body restraints and any other displacement conditions in the body and on the surface can be expressed as

$$u_x - u_{xs} = 0, \quad u_y - u_{ys} = 0, \quad u_z - u_{zs} = 0, \qquad (2.19)$$

where u_{xs}, u_{ys}, u_{zs} are the specified displacements.

Multiply the equations in Equation (2.18) by the virtual stresses $\sigma^V_{xx}, \cdots, \sigma^V_{yz}$ in order and multiply the equations in Equation (2.19) by X^V_x, X^V_y, X^V_z in order, by S^V_x, S^V_y, S^V_z in order, and by P^V_{mx}, P^V_{my}, P^V_{mz} in order. Integrate the resulting equations over the volume or surface as required, and add them to the principle of virtual displacements in Equation (2.6) to get

$$\int_V \left[\sigma_{xx} e^V_{xx} + (e_{xxu} - e_{xx\sigma})\sigma^V_{xx} + \cdots + \sigma_{yz} e^V_{yz} + (e_{yzu} - e_{yz\sigma})\sigma^V_{yz} \right] dV -$$

$$- \int_V \left[X_x u^V_x + (u_x - u_{xs}) X^V_x + X_y u^V_y + (u_y - u_{ys}) X^V_y \right.$$

$$\left. + X_z u^V_z + (u_z - u_{zs}) X^V_z \right] dV -$$

$$- \int_S \left[S_x u^V_x + (u_x - u_{xs}) S^V_x + S_y u^V_y + (u_y - u_{ys}) S^V_y \right.$$

$$\left. + S_z u^V_z + (u_z - u_{zs}) S^V_z \right] dS -$$

$$- \sum_{m=1}^M (P_{mx} u^V_{mx} + P_{my} u^V_{my} + P_{mz} u^V_{mz}) - \sum_{m=1}^N [(u_{mx} - u_{mxs}) P^V_{mx} +$$

$$+ (u_{my} - u_{mys}) P^V_{my} + (u_{mz} - u_{mzs}) P^V_{mz}] = 0, \qquad (2.20)$$

which is the *principle of mixed virtual stresses and virtual displacements*. This principle includes all the basic equations of three dimensional elasticity theory: the three body equilibrium Equations (1.4), the three surface equilibrium

Virtual displacement and virtual force methods 47

Equations (1.5), the six strain-displacement Equations (1.7), the six stress-strain Equations (1.11), and the three displacement conditions in Equation (2.19). In cases where the displacement conditions are satisfied exactly, these terms drop out.

The conditions on the virtual strains and virtual displacements, Equations (2.5), are the same as in the principle of virtual displacements, while the conditions on the virtual stresses and virtual forces, Equations (2.12) and (2.13), which are independent of the virtual strains, are the same as in the principle of virtual forces. On the other hand, with $e_{xxu}, \cdots, e_{yzu}$ in terms of displacements and $e_{xx\sigma}, \cdots, e_{yz\sigma}$ in terms of stresses, any strain-displacement and any stress-strain relations can be used in Equation (2.20) for the real stresses and strains.

The integral form represented by the principle of mixed virtual stresses and virtual displacements in Equation (2.20) has the same types of advantages over the differential equations as those listed for the principle of virtual displacements in Section 2.2.

It should be noted that in the principle of virtual displacements, Equation (2.6), the stresses are expressed in terms of the displacements, but in the mixed principle both the stresses and displacements are unknowns. The displacements occur directly and in the e_u terms, while the stresses appear directly and in the e_σ terms.

2.7. The mixed unit displacement and unit load theorem

Apply a virtual displacement u_m^V and a virtual force P_m^V at point m and follow the procedures in Sections 2.3 and 2.5. For arbitrary u_m^V and P_m^V, Equation (2.20) gives two equations

$$P_m = \int_V (\sigma_{xx} e_{xxm}^1 + \cdots + \sigma_{yz} e_{yzm}^1) \, dV -$$
$$- \int_V (X_x u_{xm}^1 + X_y u_{ym}^1 + X_z u_{zm}^1) \, dV -$$
$$- \int_S (S_x u_{xm}^1 + S_y u_{ym}^1 + S_z u_{zm}^1) \, dS,$$
$$0 = \int_V \left[(e_{xxu} - e_{xx\sigma}) \sigma_{xxm}^1 + \cdots + (e_{yzu} - e_{yz\sigma}) \sigma_{yzm}^1 \right] dV,$$
(2.21)

where the displacement boundary conditions are assumed to be satisfied.

The first equation in (2.21) is the same as the unit displacement theorem, Equation (2.9), except that the stresses are the unknowns here while the displacements are the unknowns in Equation (2.9). Since the stresses also occur in the second equation in (2.21), the two equations are usually coupled. In determinate cases it is possible to obtain the stresses from the first equation.

Since Equation (2.21) can be used for each point, the results can be assembled into a matrix form, Section 2.11.

2.8. Two dimensional forms of the virtual principles

Although all structures are three dimensional, it is possible to represent certain structures in a two or one dimensional form under certain assumptions, as described in Sections 1.6 and 1.8. Consider the *two dimensional thin plate* with thickness t in Figure 1.5 and with the conditions specified for Figure 1.5 and in Equation (1.26). For this case of inplane loading without bending, the principle of virtual displacements in Equation (2.6) becomes

$$\int_{A_m} (\sigma_{xx} e_{xx}^V + \sigma_{zz} e_{zz}^V + \sigma_{xz} e_{xz}^V) t\, dA - \int_{A_m} (X_x u_x^V + X_z u_z^V) t\, dA -$$
$$- \int_C (S_x u_x^V + S_z u_z^V) t\, ds - \sum_{m=1}^{M} (P_{mx} u_{mx}^V + P_{mz} u_{mz}^V) = 0, \qquad (2.22)$$

where A_m is the area of the midplane of the plate, C is the boundary edge of the plate, $dA = dx dz$, s is arclength variable on edge of the plate, and point forces have been explicitly included.

The principle of virtual forces in Equation (2.14) becomes

$$\int_{A_m} (e_{xx} \sigma_{xx}^V + e_{zz} \sigma_{zz}^V + e_{xz} \sigma_{xz}^V) t\, dA - \sum_{m=1}^{M} (u_{mx} P_{mx}^V + u_{mz} P_{mz}^V) -$$
$$- \int_{A_m} (u_x X_x^V + u_z X_z^V) t\, dA - \int_C (u_x S_x^V + u_z S_z^V) t\, ds = 0. \qquad (2.23)$$

The principle of mixed virtual stresses and displacements in Equation (2.20) becomes

$$\int_{A_m} [\sigma_{xx} e_{xx}^V + (e_{xxu} - e_{xx\sigma})\sigma_{xx}^V + \sigma_{zz} e_{zz}^V + (e_{zzu} - e_{zz\sigma})\sigma_{zz}^V + \sigma_{xz} e_{xz}^V +$$
$$+ (e_{xzu} - e_{xz\sigma})\sigma_{xz}^V] t\, dA - \sum_{m=1}^{M} (P_{mx} u_{mx}^V + P_{mz} u_{mz}^V) -$$
$$- \int_{A_m} [X_x u_x^V + (u_x - u_{xs}) X_x^V + X_z u_z^V + (u_z - u_{zs}) X_z^V] t\, dA -$$
$$- \int_C [S_x u_x^V + (u_x - u_{xs}) S_x^V + S_z u_z^V + (u_z - u_{zs}) S_z^V] t\, ds -$$
$$- \sum_{m=1}^{N} [(u_{mx} - u_{mxs}) P_{mx}^V + (u_{mz} - u_{mzs}) P_{mz}^V] = 0. \qquad (2.24)$$

The plate bending equations are considered in Chapters 7, 8, 9, Volume 2.

The unit displacement theorem in Equations (2.9) and (2.10) reduces to

$$P_m = \int_{A_m} (\sigma_{xx} e_{xxm}^1 + \sigma_{zz} e_{zzm}^1 + \sigma_{xz} e_{xzm}^1) t\, dA -$$
$$- \int_{A_m} (X_x u_{xm}^1 + X_z u_{zm}^1) t\, dA - \int_C (S_x u_{xm}^1 + S_z u_{zm}^1) t\, ds, \qquad (2.25)$$

Virtual displacement and virtual force methods 49

$$M_n = \int_{A_m} (\sigma_{xx} e^1_{xxn} + \sigma_{zz} e^1_{zzn} + \sigma_{xz} e^1_{xzn}) t\, dA -$$
$$- \int_{A_m} (X_x u^1_{xn} + X_z u^1_{zn}) t\, dA - \int_C (S_x u^1_{xn} + S_z u^1_{zn}) t\, ds, \qquad (2.26)$$

The unit load theorem in Equations (2.16) and (2.17) becomes

$$u_m = \int_{A_m} (e_{xx} \sigma^1_{xxm} + e_{zz} \sigma^1_{zzm} + e_{xz} \sigma^1_{xzm}) t\, dA -$$
$$- \int_{A_m} (u_x X^1_{xm} + u_z X^1_{zm}) t\, dA - \int_C (u_x S^1_{xm} + u_z S^1_{zm}) t\, ds, \qquad (2.27)$$

$$\theta_n = \int_{A_m} (e_{xx} \sigma^1_{xxn} + e_{zz} \sigma^1_{zzn} + e_{xz} \sigma^1_{xzn}) t\, dA -$$
$$- \int_{A_m} (u_x X^1_{xn} + u_z X^1_{zn}) t\, dA - \int_C (u_x S^1_{xn} + u_z S^1_{zn}) t\, ds, \qquad (2.28)$$

The mixed unit theorem in Equation (2.21) becomes

$$P_m = \int_{A_m} (\sigma_{xx} e^1_{xxm} + \sigma_{zz} e^1_{zzm} + \sigma_{xz} e^1_{xzm}) t\, dA -$$
$$- \int_{A_m} (X_x u^1_{xm} + X_z u^1_{zm}) t\, dA - \int_C (S_x u^1_{xm} + S_z u^1_{zm}) t\, ds, \qquad (2.29)$$

$$0 = \int_{A_m} \left[(e_{xxu} - e_{xx\sigma}) \sigma^1_{xxm} + (e_{zzu} - e_{zz\sigma}) \sigma^1_{zzm} + (e_{xzu} - e_{xz\sigma}) \sigma^1_{xzm} \right] t\, dA.$$

2.9. One dimensional forms of the virtual principles

For the one dimensional beam structures discussed in Sections 1.8 through 1.15 with general loading as shown in Figure 2.1, it is desirable to include moments and rotations as well as forces and deflections in the equations. At any cross section of the plane section beam, from Equation (1.10),

$$u_x = u_{xa} + z\theta_y, \qquad u_y = u_{ya} - z\theta_x + x\theta_z, \qquad u_z = u_{za} - x\theta_y, \qquad (2.30)$$

where u_{xa}, u_{ya}, u_{za} are the deflection components of the elastic axis of the beam, θ_y is the rotation about the elastic axis, θ_x and θ_z are the slopes of the elastic axis with

$$\theta_x = \frac{du_{za}}{dy} = -\frac{du_y}{dz}, \qquad \theta_z = -\frac{du_{xa}}{dy} = \frac{du_y}{dx}. \qquad (2.31)$$

Note that u_{xa}, u_{ya}, u_{za}, θ_x, θ_y, θ_z are functions of y only for the plane section beam.

The other relations among the stresses, strains, and displacements for this one

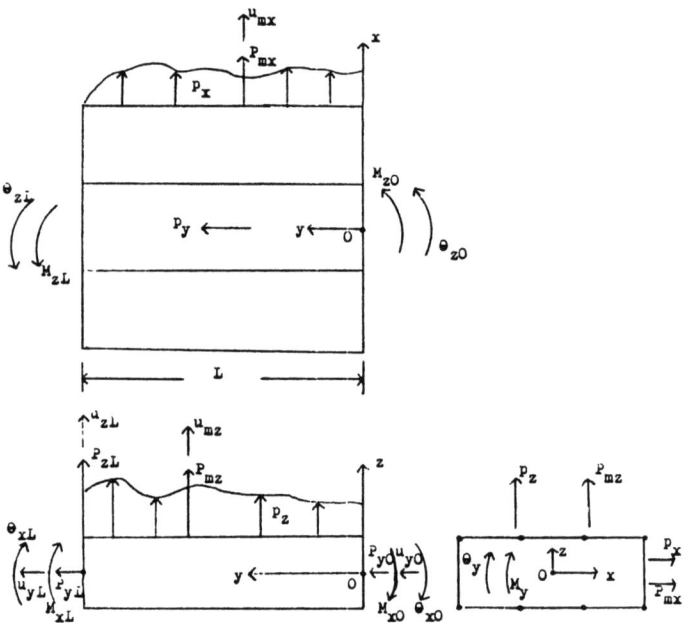

Fig. 2.1. Beam with general loading

dimensional beam case are

$$\sigma_{yy} = E e_{yy}, \quad \sigma_{yz} = G e_{yz} = q_{iz}/t_i, \quad \sigma_{xy} = G e_{xy} = q_{ix}/t_i,$$

$$e_{yy} = u_{y,y} = \frac{du_{ya}}{dy} - z\frac{d^2 u_{za}}{dy^2} - x\frac{d^2 u_{xa}}{dy^2}, \qquad (2.32)$$

$$e_{xy} = \frac{du_{xs}}{dy}, \quad e_{yz} = \frac{du_{zs}}{dy},$$

where u_{xa}, u_{za} are bending deflections and u_{xs}, u_{zs} are shear deflections and where Equations (1.37) and (1.62) have been used. The σ_{yy} stress is given in terms of axial forces P_y and bending moments M_x, M_z in Sections 1.13, 1.14, 1.15. The shear stresses σ_{xy}, σ_{yz} are given in terms of shear flows q_i, shear forces V_x and V_z, and the torsional moment M_y in various cases in Sections 1.11 and 1.12.

Since the relations in Equations (2.30)-(2.32) can be used as virtual relations for the virtual displacements and virtual rotations, it follows that on the end surfaces of the beam with forces S_x, S_y, S_z,

$$\int_A (S_x u_x^V + S_y u_y^V + S_z u_z^V)\, dA = \int_A [S_x(u_{xa}^V + z\theta_y^V) +$$

$$+ S_y(u_{ya}^V - z\theta_x^V + x\theta_z^V) + S_z(u_{za}^V - x\theta_y^V)]\, dA$$

$$= P_x u_{xa}^V + P_y u_{ya}^V + P_z u_{za}^V + M_x \theta_x^V + M_y \theta_y^V + M_z \theta_z^V, \qquad (2.33)$$

where Equation (1.29) has been used. Also, this form in Equation (2.33) can be

Virtual displacement and virtual force methods

used at any cross section of the beam.

Use Equation (2.33) and write the expression for the *principle of virtual displacements*, Equation (2.6), for the one dimensional beam in Figure 2.1 as

$$\int_0^L \int_A (\sigma_{yy} e_{yy}^V + \sigma_{xy} e_{xy}^V + \sigma_{yz} e_{yz}^V)\, dA\, dy -$$

$$- \int_0^L (p_x u_x^V + p_y u_y^V + p_z u_z^V)\, dy -$$

$$- \sum_{m=1}^M (P_{xm} u_{xam}^V + P_{ym} u_{yam}^V + P_{zm} u_{zam}^V +$$

$$+ M_{xm} \theta_{xm}^V + M_{ym} \theta_{ym}^V + M_{zm} \theta_{zm}^V) = 0, \tag{2.34}$$

where the applied distributed forces p_x, p_y, p_z are forces per unit length of the beam and include any inertia forces and the weight. The point applied forces P_{xm}, P_{ym}, P_{zm} are regarded as applied on the elastic axis with the moments M_{xm}, M_{ym}, M_{zm} including any couples for the translation from the actual points of application, Equation (1.29). Also, the end point values are included in the sums in Equation (2.34). Note that rigid body restraints must be imposed on the beam to give origin points for the deflections and rotations.

From Equations (2.9) and (2.10) the *unit displacement theorem* for the one dimensional case in Equation (2.34) becomes

$$P_m = \int_0^L \int_A (\sigma_{yy} e_{yym}^1 + \sigma_{xy} e_{xym}^1 + \sigma_{yz} e_{yzm}^1)\, dA\, dy -$$

$$- \int_0^L (p_x u_{xm}^1 + p_y u_{ym}^1 + p_z u_{zm}^1)\, dy, \tag{2.35}$$

$$M_n = \int_0^L \int_A (\sigma_{yy} e_{yyn}^1 + \sigma_{xy} e_{xyn}^1 + \sigma_{yz} e_{yzn}^1)\, dA\, dy -$$

$$- \int_0^L (p_x u_{xn}^1 + p_y u_{yn}^1 + p_z u_{zn}^1)\, dy. \tag{2.36}$$

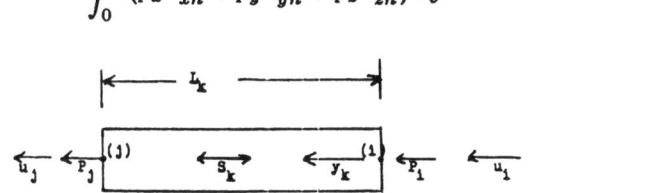

Fig. 2.2. Bar member k of truss (A_k and E_k are constant).

For the axially loaded bar k in Figure 2.2, the stresses are

$$\sigma_{yyk} = E_k e_{yyk} = E_k (u_j - u_i)/L_k = S_k/A_k,$$
$$\sigma_{xyk} = 0, \quad \sigma_{yzk} = 0. \tag{2.37}$$

Apply a unit virtual displacement at point j with point i fixed, whence

$$e^1_{yyj} = 1/L_k, \quad e^1_{xyj} = 0, \quad e^1_{yzj} = 0. \tag{2.38}$$

Equation (2.35) gives

$$\begin{aligned} P_j &= \int_0^{L_k} \int_{A_k} (E/L^2)_k (u_j - u_i) \, dA_k dy_k \\ &= (EA/L)_k (u_j - u_i) = S_k. \end{aligned} \tag{2.39}$$

For the beam in Figure 2.1 the *principle of virtual forces* in Equation (2.14) becomes

$$\int_0^L \int_A (e_{yy}\sigma^V_{yy} + e_{xy}\sigma^V_{xy} + e_{yz}\sigma^V_{yz}) \, dAdy -$$

$$- \int_0^L (u_x p^V_x + u_y p^V_y + u_z p^V_z) \, dy -$$

$$- \sum_{m=1}^M (u_{xam} P^V_{xm} + u_{yam} P^V_{ym} + u_{zam} P^V_{zm} +$$

$$+ \theta_{xm} M^V_{xm} + \theta_{ym} M^V_{ym} + \theta_{zm} M^V_{zm}) = 0, \tag{2.40}$$

$$p^V_x = \int_C S^V_x \, ds + \int_A X^V_x \, dA, \quad p^V_y = \int_C S^V_y \, ds + \int_A X^V_y \, dA,$$

$$p^V_z = \int_C S^V_z \, ds + \int_A X^V_z \, dA,$$

where p^V_x, p^V_y, p^V_z are distributed virtual forces per unit length and S^V_x, S^V_y, S^V_z, X^V_x, X^V_y, X^V_z are given by Equations (2.12) and (2.13). When the virtual terms in Equation (2.40) are produced by a point virtual force on the beam, it can be shown that p^V_x, p^V_y, p^V_z are zero.

From Equations (2.16) and (2.17) the *unit load theorem* for the one dimensional case in Equation (2.40) becomes

$$u_m = \int_0^L \int_A (e_{yy}\sigma^1_{yym} + e_{xy}\sigma^1_{xym} + e_{yz}\sigma_{yzm}) \, dAdy, \tag{2.41}$$

$$\theta_n = \int_0^L \int_A (e_{yy}\sigma^1_{yyn} + e_{xy}\sigma^1_{xyn} + e_{yz}\sigma_{yzn}) \, dAdy, \tag{2.42}$$

For the axially loaded bar in Figure 2.2

$$S_k = P_j = -P_i, \quad e_{yyk} = \sigma_{yyk}/E_k = (S/EA)_k,$$
$$e_{xyk} = 0, \quad e_{yzk} = 0. \tag{2.43}$$

Apply a unit load at j in the direction of P_j with $u_i = 0$, whence

$$\sigma^1_{yyk} = 1/A_k, \quad \sigma^1_{xyk} = 0, \quad \sigma^1_{yzk} = 0. \tag{2.44}$$

Thus, Equation (2.41) gives

$$u_j = \int_0^{L_k} \int_{A_k} (S/EA^2)_k \, dA_k \, dy_k = (SL/EA)_k, \quad u_i = 0. \tag{2.45}$$

The u_i can be included as $u_j - u_i$ in Equation (2.45) by use of Equation (2.39).

For the beam in Figure 2.1, the *principle of mixed virtual stresses and virtual displacements* in Equation (2.20) becomes

$$\int_0^L \int_A [\sigma_{yy} e_{yy}^V + (e_{yy u} - e_{yy \sigma})\sigma_{yy}^V + \sigma_{xy} e_{xy}^V + (e_{xy u} - e_{xy \sigma})\sigma_{xy}^V +$$

$$+ \sigma_{yz} e_{yz}^V + (e_{yz u} - e_{yz \sigma})\sigma_{yz}^V] \, dA dy - \int_0^L (p_x u_x^V + p_y u_y^V + p_z u_z^V) \, dy -$$

$$- \sum_{m=1}^M (P_{xm} u_{xam}^V + P_{ym} u_{yam}^V + P_{zm} u_{zam}^V + M_{xm} \theta_{xm}^V + M_{ym} \theta_{ym}^V + M_{zm} \theta_{zm}^V) -$$

$$- \sum_{m=1}^N [(u_{xam} - u_{xsm})P_{xm}^V + \cdots + (\theta_{zm} - \theta_{zsm})M_{zm}^V] = 0. \tag{2.46}$$

From Equation (2.21) the *mixed unit displacement and unit load theorem* for the one dimensional case in Equation (2.46) becomes

$$P_m = \int_0^L \int_A (\sigma_{yy} e_{yym}^1 + \sigma_{xy} e_{xym}^1 + \sigma_{yz} e_{yzm}^1) \, dA dy -$$

$$- \int_0^L (p_x u_{xm}^1 + p_y u_{ym}^1 + p_z u_{zm}^1) \, dy,$$

$$0 = \int_0^L \int_A [(e_{yy u} - e_{yy \sigma})\sigma_{yym}^1 + (e_{xy u} - e_{xy \sigma})\sigma_{xym}^1 +$$

$$+ (e_{yz u} - e_{yz \sigma})\sigma_{yzm}^1] \, dA dy. \tag{2.47}$$

For the axially loaded bar k in Figure 2.2, use Equations (2.37) and (2.43) to get

$$\sigma_{yyk} = S_k/A_k, \quad e_{yyuk} = \frac{u_j - u_i}{L_k}, \quad e_{yy\sigma k} = \frac{S_k}{E_k A_k}, \tag{2.48}$$

with other terms zero. Apply a unit virtual displacement and a unit virtual load at j with the virtual displacement zero at point i and the equilibrium virtual load of -1 at i, whence

$$e_{yyk}^1 = 1/L_k, \quad \sigma_{yyk}^1 = 1/A_k, \tag{2.49}$$

with other terms zero. Equation (2.47) gives

$$P_j = \int_0^{L_k} \int_{A_k} (S/AL)_k \, dA_k \, dy_k = S_k,$$

$$0 = \int_0^{L_k} \int_{A_k} \left(\frac{u_j - u_i}{L_k} - \frac{S_k}{E_k A_k}\right) dA_k \, dy_k = (u_j - u_i) A_k - \frac{S_k L_k}{E_k}, \tag{2.50}$$

which agrees with the results in Equations (2.39) and (2.45).

The above forms of the virtual principles and the unit theorems for the one dimensional case apply not only to the beam in Figure 2.1 and the bar in Figure 2.2 but also to any structure with a combination of bars and beams, to pin-jointed trusses, to trusses with beam members, to the wing and box beam structures in airplanes, to bridges and buildings, and to any structure in which the elements have plane sections.

Applications and examples of these one dimensional forms will be given in Chapters 3, 4, and 5.

2.10. The one dimensional virtual principles with temperature, inelastic and large displacement effects

In the derivation of the virtual principles in Sections 2.2, 2.4 and 2.6, no stress-strain relations were used and no restrictions other than compatibility were placed on the strains so that the principles can be applied to any material, whether isotropic or anisotropic, elastic or inelastic, linear or non-linear deflections, with or without temperature variations.

In the derivation of the principle of virtual displacements in Section 2.2 only the equilibrium equations (in the final position if large displacements are present) were used with operations on the stresses only. In most cases in using the principle of virtual displacements the stresses must be expressed in terms of the displacements, but any relation between stresses and displacements can be used.

In the derivation of the principle of virtual forces in Section 2.4, the strain-displacement equations (1.7) were used, but operations were performed only on the linear displacement terms and no restrictions were imposed on the strain terms. Thus any relations between the strains and stresses can be used, including large displacement strains.

In the derivation of the principle of mixed virtual stresses and virtual displacements in Section 2.6, only operations on the stresses as in the principle of virtual displacements were made so that there are no restrictions on relations between the stresses and displacements.

Since the inelastic effects in two and three dimensions are complicated and require the use of some type of failure law, the discussion here will be limited to the *one dimensional case*, involving the virtual principles as given in Section 2.9.

From Figure 1.13 and Equation (1.77) the total uniaxial strain in the y-direction is

$$e_{yy} = e_e + e_p + e_I + e_T, \tag{2.51}$$

where e_e is the elastic strain

$$e_e = \sigma_{yy}/E_{yy}, \tag{2.52}$$

e_p is the inelastic strain, $e^s = e_e + e_p$ is the strain associated with the stress on the stress-strain curve, e_I is the initial strain, and e_T is the thermal strain. From

Equation (1.8) for the uniaxial case the total strain can be expressed as

$$e_{yy} = u_{y,y} \pm e_d,$$
$$e_d = \frac{1}{2}(u_{x,y}^2 + u_{y,y}^2 + u_{z,y}^2), \tag{2.53}$$

where u_x, u_y, u_z are the deflections of the beam in Figures 1.11 and 2.1 due to the axial loads and bending moments. The plus sign in Equation (2.53) applies when the displacements are measured from the initial unloaded beam coordinate axes. The minus sign applies for coordinate axes located in the beam in the deflected position.

It should be noted that $u_{x,y}$ and $u_{z,y}$ in Equation (2.53) are geometric shear strain terms so that from Equation (1.8)

$$(e_{xy})_g = u_{x,y}, \qquad (e_{yz})_g = u_{z,y}. \tag{2.53a}$$

If $\frac{1}{2}u_{x,y}^2$ and $\frac{1}{2}u_{z,y}^2$ are large enough to affect e_{yy}, then the much larger geometric shear strains $(e_{xy})_g$ and $(e_{yz})_g$ should be included in the virtual principles. As demonstrated later in Section 4.11, and in Sections 5.3, 5.5, Volume 2, if these linear geometric shear strains are used in the principle of virtual displacements with the linear e_{yy} the column equations are obtained directly.

An approximate calculation can be made for e_d by assuming the shape of the deflection curves for the beam. If

$$u_z = (u_z)_{\max} \sin(\pi y / L), \tag{2.54}$$

then

$$u_{z,y} = (\pi/L)(u_z)_{\max} \cos(\pi y / L), \tag{2.55}$$

$$(e_d)_{av} = \frac{1}{L}\int_0^L \frac{1}{2}(u_{z,y})^2 \, dy = \left(\frac{\pi}{2L}\right)^2 (u_z)_{\max}^2. \tag{2.56}$$

Define the strain expression

$$e_E = e_I + e_T + e_p \mp e_d, \tag{2.57}$$

which represents all strain effects beyond the linear elastic strain for the applied stresses. From Equations (2.51)-(2.53),

$$u_{y,y} = (\sigma_{yy}/E_{yy}) + e_E, \tag{2.58}$$
$$(\sigma_{yy})_u = E_{yy}(u_{y,y} - e_E), \tag{2.59}$$
$$e_{yyu} = u_{y,y} \pm e_d, \tag{2.60}$$
$$e_{yy\sigma} = (\sigma_{yy}/E_{yy}) + e_E \pm e_d. \tag{2.61}$$

These expressions can be used directly in the virtual principles and unit theorems in Section 2.9 above.

For the principle of virtual displacements replace σ_{yy} in Equations (2.34)-(2.37) by $(\sigma_{yy})_u$ in Equation (2.59). In Equations (2.37) and (2.39)

$$\begin{aligned}
\sigma_{yyk} &= (E_k/L_k)(u_j - u_i) - E_k e_{Ek}, \\
P_j = S_k &= (EA/L)_k(u_j - u_i) - S_{Ek}, \\
S_{Ek} &= \int_0^{L_k} \int_{A_k} E_k e_{Ek}\left(\frac{1}{L_k}\right) dA_k\, dy_k = \frac{1}{L_k}\int_0^{L_k} (AEe_E)_k\, dy_k \\
&= E_k A_k e_{Ek}
\end{aligned} \quad (2.62)$$

for e_{Ek} constant along the bar. S_k is the force in the bar and S_{Ek} is the apparent force in the bar due to the e_{Ek} strains.

Since the principle of virtual forces in Equation (2.14) was derived from the linear Equations (1.7), it is necessary to modify the e_{yy} terms in Equations (2.40)-(2.42) to $e_{yy} \mp e_d$ for the non-linear case. Replace this latter e_{yy} by $e_{yy\sigma}$ in Equation (2.61) to get $(\sigma_{yy}/E_{yy}) + e_E$. In Equations (2.43) and (2.45)

$$\begin{aligned}
e_{yyk} &= (\sigma_{yyk}/E_k) + e_{Ek} = (S/EA)_k + e_{Ek}, \\
u_j &= (P_j L_k/E_k A_k) + u_{Ek}, \\
u_{Ek} &= \int_0^{L_k}\int_{A_k} e_{Ek}\left(\frac{1}{A_k}\right) dA_k\, dy_k = \int_0^{L_k} e_{Ek}\, dy_k \\
&= L_k e_{Ek} = (S_E L/EA)_k
\end{aligned} \quad (2.63)$$

for e_{Ek} constant.

For the principle of mixed virtual displacements and virtual stresses, replace $e_{yyu} - e_{yy\sigma}$ in Equations (2.46) and (2.47) by $e_{yyu} - e_{yy\sigma}$ from Equations (2.60) and (2.61), or

$$e_{yyu} - e_{yy\sigma} = u_{y,y} - \left(\frac{\sigma_{yy}}{E_{yy}} + e_E\right). \quad (2.64)$$

In Equation (2.50) this expression gives the same u_{Ek} as in Equation (2.63).

The effects of the e_E strains in Equation (2.57) will be considered in various examples in Chapters 3-5.

2.11. Matrix forms of the virtual principles

The unit displacement theorem in Equation (2.6) can be written in the form

$$\begin{aligned}
P_m &= \int_V [e_m^1]\{\sigma\}\, dV, \\
[e_m^1] &= [e_{xxm}^1 \quad e_{yym}^1 \quad e_{zzm}^1 \quad e_{xym}^1 \quad e_{xzm}^1 \quad e_{yzm}^1], \\
\{\sigma\}^T &= [\sigma_{xx} \quad \sigma_{yy} \quad \sigma_{zz} \quad \sigma_{xy} \quad \sigma_{xz} \quad \sigma_{yz}].
\end{aligned} \quad (2.65)$$

Thus, for all the point forces in Equation (2.9) and point moments in Equation (2.10), the matrix form for all the points can be assembled into the matrix

equation
$$\{P\} = \int_V [e^1]\{\sigma\}\,dV, \qquad (2.66)$$

where $\{P\}$ is M by 1 for M points, $[e^1]$ is M by 6, and $\{\sigma\}$ is 6 by 1. The $[e^1]$ is the matrix of virtual strains produced by the unit virtual displacements and unit virtual rotations.

If the real stresses $\{\sigma\}$ can be expressed in terms of the point displacements u_m at the point loads, then

$$\{\sigma\} = [\sigma_1]\{u\} - [J]^{-1}\{e_E\}, \qquad (2.67)$$

where the 6 by $M[\sigma_1]$ matrix are the stresses produced by the unit point displacements and where Equation (1.11) has been used in the matrix form

$$\{e\} = [J]\{\sigma\} + \{e_E\}, \quad [J] = \frac{1}{E}\begin{bmatrix} 1 & -\nu & -\nu & 0 & 0 & 0 \\ -\nu & 1 & -\nu & 0 & 0 & 0 \\ -\nu & -\nu & 1 & 0 & 0 & 0 \\ 0 & 0 & 0 & \frac{E}{G} & 0 & 0 \\ 0 & 0 & 0 & 0 & \frac{E}{G} & 0 \\ 0 & 0 & 0 & 0 & 0 & \frac{E}{G} \end{bmatrix}. \qquad (2.68)$$

Put Equation (2.67) into Equation (2.66) to get

$$\{P\} = \int_V \{[e^1][\sigma_1]\{u\} - [e^1][J]^{-1}\{e_E\}\}\,dV$$
$$= [k]\{u\} - \{P_E\}, \qquad (2.69)$$
$$[k] = \int_V [e^1][\sigma_1]\,dV, \quad \{P_E\} = \int_V [e^1][J]^{-1}\{e_E\}\,dV. \qquad (2.70)$$

The matrix $[k]$ is the *stiffness influence coefficient* matrix for the structure. Once it is known, it can be used to get external forces and displacements for any given loads. The Equation (2.69) is the general matrix form of the unit displacement theorem for point forces and point displacements. It is the basis for the *matrix displacement method* of analysis and is directly applicable to finite elements. It will be used in the one dimensional and two dimensional forms in the following Chapters.

Since the virtual strains $[e^1]$ can be selected as the real elastic strains, the $[k]$ matrix in Equation (2.70) can be written as

$$[k] = \int_V [\sigma_1]^T [J][\sigma_1]\,dV. \qquad (2.71)$$

It may be difficult to get the $[\sigma_1]$ matrix so that it may be possible to approximate $[k]$ by taking the stresses $[\sigma_1]$ as virtual stresses from the simpler virtual strains $[e^1]$ that may be selected. In this case

$$[k]_{\text{approx.}} = \int_V [e^1][J]^{-1}[e^1]^T\,dV. \qquad (2.72)$$

The unit load theorem in Equation (2.16) can be written in the matrix form

$$u_m = \int_V [\sigma_m^1]\{e\}\,dV,$$
$$[\sigma_m^1] = [\sigma_{xxm}^1 \ \sigma_{yym}^1 \ \sigma_{zzm}^1 \ \sigma_{xym}^1 \ \sigma_{xzm}^1 \ \sigma_{yzm}^1], \tag{2.73}$$
$$\{e\}^T = [e_{xx} \ e_{yy} \ e_{zz} \ e_{xy} \ e_{xz} \ e_{yz}].$$

Thus, for the point displacements in Equation (2.16) and the point rotations in Equations (2.17), all the point equations can be assembled into the matrix form

$$\{u\} = \int_V [\sigma^1]\{e\}\,dV, \tag{2.74}$$

where $[\sigma^1]$ is the matrix of virtual stresses produced by the unit virtual forces and unit virtual moments.

If the real strains can be expressed in terms of point forces P_m at the point displacements, then

$$\{e\} = [e_1]\{P\} + \{e_E\}, \tag{2.75}$$

where the 6 by $M[E_1]$ matrix are the strains produced by the unit loads. Put Equation (2.75) into Equation (2.74) to get

$$\{u\} = \int_V \{[\sigma^1][e_1]\{P\} + [\sigma^1]\{e_E\}\}\,dV$$
$$= [C]\{P\} + \{u_E\}, \tag{2.76}$$
$$[C] = \int_V [\sigma^1][e_1]\,dV, \quad \{u_E\} = \int_V [\sigma^1]\{e_E\}\,dV. \tag{2.77}$$

The matrix $[C]$ is the *flexibility influence coefficient* matrix for the structure. This Equation (2.76) is the general matrix form for the unit load theorem for point values. It is directly applicable to finite elements and is the basis for the *matrix force method* of analysis. It will be used in one dimensional and two dimensional forms in the following Chapters.

Since the virtual stresses $[\sigma^1]$ can be selected as the real elastic stresses, the $[C]$ matrix can be written as

$$[C] = \int_V [e_1]^T [J]^{-1} [e_1]\,dV. \tag{2.78}$$

Also, an approximate $[C]$ can be obtained by using the virtual stresses, or

$$[C]_{\text{approx}} = \int_V [\sigma^1][J][\sigma^1]^T\,dV. \tag{2.79}$$

The mixed unit displacement and unit load theorem in Equation (2.21) can be written in the form

$$P_m = \int_V [e_m^1]\{\sigma\}\,dV,$$
$$0 = \int_V [\sigma_m^1]\{e_u - e_\sigma\}\,dV, \tag{2.80}$$

where the matrices are defined in Equations (2.65) and (2.73). All the point equations can be assembled to give

$$\{P\} = \int_V [e^1]\{\sigma\}\, dV,$$
$$\{0\} = \int_V [\sigma^1]\{e_u - e_\sigma\}\, dV. \tag{2.81}$$

In this case the $\{\sigma\}$ and $\{e_\sigma\}$ matrices must be expressed in terms of the internal loads S_i and the $\{e_u\}$ matrix in terms of the displacements u_i. Take

$$\{\sigma\} = [D_1]\{S\}, \quad \{e_u\} = [G_1]\{u\},$$
$$\{e_\sigma\} = [J]\{\sigma\} + \{e_E\} = [J][D_1]\{S\} + \{e_E\}, \tag{2.82}$$

where $[D_1]$ are the stresses due to unit values of the internal element forces. Put Equation (2.82) into Equation (2.81) to get

$$\{P\} = \int_V [e^1][D_1]\{S\}\, dV = [s]^T\{S\}, \tag{2.83}$$

$$\{0\} = \int_V [\sigma^1]\{[G_1]\{u\} - [J][D_1]\{S\} - \{e_E\}\}\, dV \tag{2.84}$$

$$= [s]\{u\} - [C_F]\{S\} - \{v_E\}, \quad \text{or} \tag{2.85}$$

$$[M]\begin{Bmatrix} S \\ u \end{Bmatrix} = \begin{bmatrix} -[C_F] & [s] \\ [s]^T & [0] \end{bmatrix}\begin{Bmatrix} S \\ u \end{Bmatrix} = \begin{Bmatrix} v_E \\ P \end{Bmatrix}, \tag{2.86}$$

$$[s]^T = \int_V [e^1][D_1]\, dV, \quad [s] = \int_V [\sigma^1][G_1]\, dV,$$
$$[C_F] = \int_V [\sigma^1][J][D_1]\, dV, \quad \{v_E\} = \int_V [\sigma^1]\{e_E\}\, dV. \tag{2.87}$$

The fact that $[s]^T$ is the transpose of $[s]$ and that $[C_F]$ is the flexibility matrix for the elements will be demonstrated in one dimensional and two dimensional applications below.

The $[M]$ matrix is the mixed matrix for the structure. The Equation (2.86) is the general matrix form for the mixed unit displacement and unit load theorem. It is the basis for the *mixed method* of analysis and is directly applicable to finite elements. It will be used in one dimensional and two dimensional forms in the following Chapters.

2.12. Problems

2.1. Derive Equation (2.6) by calculating the other terms by the procedure in Equation (2.4).

2.2. With body forces zero, the stresses

$$\sigma_{xx} = C_1 x^2, \quad \sigma_{yy} = C_1 y^2 \quad \sigma_{xy} = -2C_1 xy, \quad \sigma_{zz} = \sigma_{xz} = \sigma_{yz} = 0,$$

are in equilibrium. Use these stresses in the principle of virtual displacements in Equation (2.6), assume

$$u_x^V = C_2 xy, \quad u_y^V = 0, \quad u_z^V = 0,$$

and show that Equation (2.6) integrates to zero for the rectangular block in Figure 1.20.

2.3. Repeat Problem 2.2. with

$$u_x^V = 0, \quad u_y^V = x^2 y^2, \quad u_z^V = 0.$$

2.4. Show that the equilibrium stresses used in Problems 2.2 and 2.3 are not compatible and explain why Equation (2.6) is still satisfied.

2.5. In the principle of virtual forces in Equation (2.14) take

$$u_x = C_2 xy, \quad u_y = 0, \quad u_z = 0,$$

and assume the equilibrium virtual stresses

$$\sigma_{xx}^V = C_1 x^2, \quad \sigma_{yy}^V = C_1 y^2, \quad \sigma_{xy}^V = -2C_1 xy, \quad \sigma_{zz}^V = \sigma_{xz}^V = \sigma_{yz}^V = 0.$$

Show that the integrations in Equations (2.14) are identical to those in Problem 2.2 and hence that Equation (2.14) is satisfied.

2.6. Although a reciprocal relation exists between Problems 2.2 and 2.5, the physical problems are completely different. Determine the real stresses in Problem 2.5 by using Equations (1.7) and (1.15) and show that the resulting stresses are not in equilibrium. Thus, Problem 2.2 involves an incompatible system while Problem 2.5 involves a non-equilibrium system. The reciprocal relation exists for permissible stress systems as well as for non-permissible stress systems of the type in Problems 2.2 and 2.5.

2.7. To obtain permissible stress systems from Equation (1.14) with $T = 0$, select the displacements by the procedure in Problem 1.8. Repeat Problem 2.2 using the functions

$$D_0 = 0, \quad D_1 = C_1 xy, \quad D_2 = 0, \quad D_3 = 0,$$

in Problem 1.8 to get the stresses and by using

$$D_0 = C_2 x, \quad D_1 = 0, \quad D_2 = C_3 xy, \quad D_3 = 0,$$

to get the virtual displacements.

2.8. Derive Equation (2.32).

2.9. Assume the bar in Figure 2.2 has a variable area $A_k(y_k)$, a variable axial load $p_y = 200 y_k$, and the load $P_j = 1000$ lb at $y_k = L_k$ with $u_i = 0$ at $y_k = 0$. Calculate σ_{yy} in the bar and apply the unit displacement theorem in Equation (2.35) at the point j to demonstrate that the theorem gives $P_j = 1000$ lb.

2.10. In Problem 2.9 assume $A_k = a + by_k$ and E_k is constant. Find the displacement u_j at $y = L_k$ by using the unit load theorem in Equation (2.41).

2.11. Use Equation (2.70) and show that the $[k]$ matrix for the bar in Figure 2.2 is

$$[k] = \left(\frac{EA}{L}\right)_k \begin{bmatrix} 1 & -1 \\ -1 & 1 \end{bmatrix}. \tag{2.88}$$

2.12. (a) Use Equation (2.77) and show that the $[C]$ matrix for the bar in Figure 2.2 with point i fixed is

$$C = (L/EA)_k. \tag{2.89}$$

(b) If $e_E = e_T =$ constant, calculate u_E in Equation (2.77).

2.13. If $e_E = e_T =$ constant for the bar in Figure 2.2, calculate $\{P_E\}$ in Equation (2.70).

2.14. Show that the $[M]$ matrix in Equation (2.86) for the bar in Figure 2.2 is

$$[M] = \begin{bmatrix} -(L/EA)_k & 1 & -1 \\ 1 & 0 & 0 \\ -1 & 0 & 0 \end{bmatrix}. \tag{2.90}$$

References

Chapter 2

1. J.H. Argyris: *Energy Theorems and Structural Analysis*, Butterworth Scientific Publications, London (1960).
2. J.S. Przemieniecki: *Theory of Matrix Structural Analysis*, McGraw-Hill Book Co., New York (1968).
3. I.N. Sneddon: *Fourier Transforms*, McGraw-Hill Book Co., New York (1951).
4. W. Nowacki: *Thermoelasticity*, Addison Wesley Publishing Co., Reading, Mass. (1962).
5. I.S. Sokolnikoff and R.M. Redheffer: *Mathematics of Physics and Modern Engineering*, McGraw-Hill book Co., New York (1958).

Reference for additional reading

6. D.H. Allen and W.E. Haisler: *Introduction to Aerospace Structural Analysis*, Chapter 5, John Wiley and Sons, New York (1985).

3

The virtual principles for pin-jointed trusses

3.1. Introduction

The truss member, Figure 3.1, is assumed to have constant area A_k, constant modulus of elasticity E_k, and take only constant axial forces. Let S_k be the force

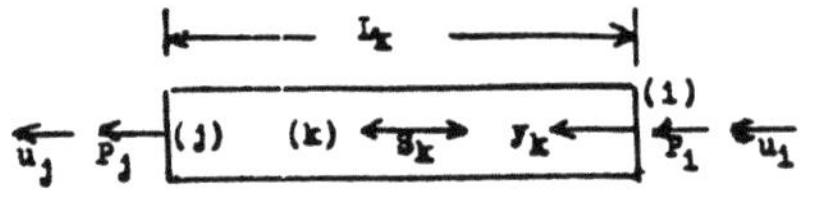

Fig. 3.1. Bar member of truss.

in the member k in Figure 3.1, whence for equilibrium

$$P_{kj} = -P_{ki} = S_k. \tag{3.1}$$

The relation between the force S_k and the end displacements u_j and u_i in the member k has been obtained from the virtual principles in Equations (2.39), (2.45), and (2.50), or

$$\begin{aligned} u_{kj} - u_{ki} &= (SL/EA)_k, & S_k &= (EA/L)_k(u_{kj} - u_{ki}), \\ \sigma_{yyk} &= S_k/A_k, & e_{yyk} &= (S/EA)_k = (u_{kj} - u_{ki})/L_k. \end{aligned} \tag{3.2}$$

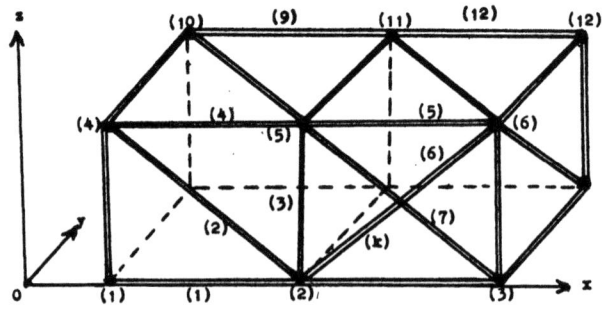

Fig. 3.2. Space redundant truss.

The virtual principles for pin-jointed trusses

When the bar members k are combined into a truss, Figure 3.2, it is necessary to transform the coordinate system for each bar to a base coordinate system for the entire truss. Figure 3.3 shows the deflection and force components of the bar k in the base x, y, z system with $u_{kj} - u_{ki}$ the elongation of bar k. Since the

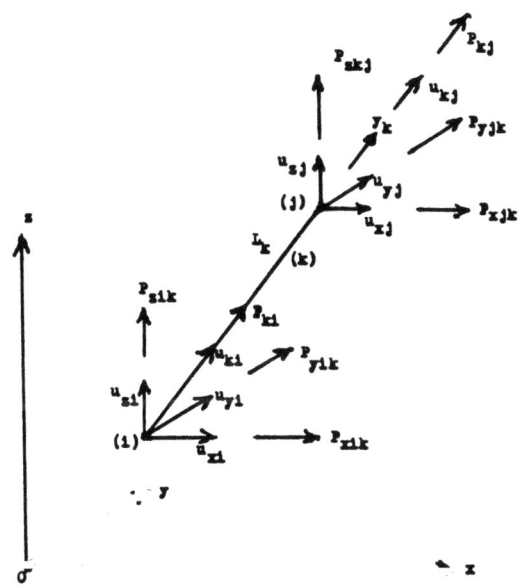

Fig. 3.3. Truss member in three dimensions.

deflections and forces are vectors, their components are given by the direction cosines so that

$$u_{kj} - u_{ki} = (u_{xj} - u_{xi}) \cos(y_k, x) + (u_{yj} - u_{yi}) \cos(y_k, y) + \\ + (u_{zj} - u_{zi}) \cos(y_k, z), \quad (3.3)$$
$$P_{kj} = P_{xj} \cos(y_k, x) + P_{yj} \cos(y_k, y) + P_{zj} \cos(y_k, z).$$

Thus, the expressions in Equation (3.2) must be modified by Equation (3.3) in a truss structure. This is done in the following sections.

Since the loads are applied at the joints of the truss and the displacements can be calculated at the joints, only point values are needed in the virtual principles so that the unit theorems can be used directly. The integrations can be made for each truss member and the results summed for all members in the truss.

The three virtual principles give three different methods for calculation of the member loads S_k and the component joint displacements u_{xj}, u_{yj}, u_{zj} in the truss. The displacement method or stiffness method is based upon the unit displacement theorem from the principle of virtual displacements. This method is described in Section 3.2. The force method or flexibility method is based upon the unit load theorem from the principle of virtual forces. It is described in

Section 3.3. The mixed method is based upon the mixed unit displacement and unit load theorem from the principle of mixed virtual displacements and virtual stresses. For simple trusses this mixed method reduces to the displacement method. However, its matrix form is different and will be discussed in Section 3.7.

The inelastic and temperature effects in trusses are considered in Section 3.4.

In large trusses many members and many joints are involved so that it is necessary to express the equations in matrix form, which can be solved by high speed digital computers. The matrix equations are given in Sections 3.5, 3.6, and 3.7 for the displacement, force, and mixed methods, respectively.

3.2. The unit displacement theorem for trusses

Consider the joint j, Figures 3.2 and 3.3, and apply a unit virtual displacement at joint j in the x-*direction*. Let the virtual displacements at the other ends of all K_j members coming to joint j be zero. Thus, for each of the K_j members, Equation (3.3), the virtual strain is

$$e^1_{kx} = (1/L_k) \cos(y_k, x). \tag{3.4}$$

Put Equations (3.2) and (3.4) into Equation (2.35) to get

$$\begin{aligned}P_{xjk} &= \int_0^{L_k} \int_{A_k} (S_k/A_k)(1/L_k) \cos(y_k, x) \, dA_k \, dy_k \\ &= S_k \cos(y_k, x).\end{aligned} \tag{3.5}$$

For all K_j members coming to joint j with similar expressions for the y and z directions,

$$\begin{aligned}P_{xj} &= \sum_{k=1}^{K_j} P_{xjk} = \sum_{k=1}^{K_j} S_k \cos(y_k, x). \\ P_{yj} &= \sum_{k=1}^{K_j} S_k \cos(y_k, y), \qquad P_{zj} = \sum_{k=1}^{K_j} S_k \cos(y_k, z).\end{aligned} \tag{3.6}$$

These are the component equilibrium equations for joint j. The y_k direction to determine the direction cosines is toward point j in all members coming to point j.

Put Equation (3.3) into Equation (3.2) to get

$$\begin{aligned}S_k &= (EA/L)_k(u_{kj} - u_{ki}) = (EA/L)_k[(u_{xj} - u_{xi}) \cos(y_k, x) + \\ &\quad + (u_{yj} - u_{yi}) \cos(y_k, y) + (u_{zj} - u_{zi}) \cos(y_k, z)] \\ &= (EA/L)_k \sum_{m=1}^{3} (u_{mj} - u_{mi}) \cos(k, m),\end{aligned} \tag{3.7}$$

where k is the direction toward point j with m as the datum directions x, y, z

The virtual principles for pin-jointed trusses

in sequence. Put the S_k in Equation (3.7) into Equation (3.6) to get

$$P_{xj} = \sum_{k=1}^{K_j}(EA/L)_k \cos(k,x) \sum_{m=1}^{3}(u_{mj} - u_{mi}) \cos(k,m),$$

$$P_{yj} = \sum_{k=1}^{K_j}(EA/L)_k \cos(k,y) \sum_{m=1}^{3}(u_{mj} - u_{mi}) \cos(k,m), \qquad (3.8)$$

$$P_{zj} = \sum_{k=1}^{K_j}(EA/L)_k \cos(k,z) \sum_{m=1}^{3}(u_{mj} - u_{mi}) \cos(k,m).$$

Although these Equations (3.6)-(3.8) can be used to get the displacements and member loads in a simple truss (see examples below), it is simpler to put the equations into a matrix form for more complicated trusses, Section 3.5. The Equations (3.6)-(3.8) are the basic equations for the displacement method of analysis and apply for both determinate and redundant trusses. However, if the truss is *determinate*, the member loads can be determined directly from Equation (3.6) as in the usual equilibrium case. Also, as will be seen below, for this determinate case it is simpler to get the joint displacements by the principle of virtual forces, rather than from Equation (3.7).

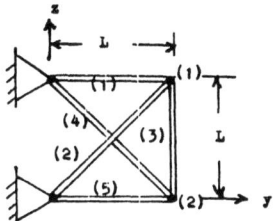

Fig. 3.4. Redundant truss with known compatible internal loads, S_1=-7000 lb, S_2=9900 lb, S_3=4000 lb, S_4=-5657 lb, S_5=15,000 lb.

Example 3.1. In Figure 3.4 with $E_i = E=$ constant, $A_1 = A_3 = A_5 = A =$ constant, $A_2 = A_4 = 0.707A$, find the external forces P_{y1}, P_{z1} and the deflections u_{y1}, u_{z1} by the above unit displacement equations.

Solution. From Equation (3.6) with members (1), (2), and (3) coming to joint (1),

$$P_{y1} = S_1 \cos 0° + S_2 \cos 45° + S_3 \cos 90°$$
$$= -7000 + 0.707(9900) = 0 \text{ lb},$$
$$P_{z1} = S_1 \cos 90° + S_2 \cos 45° + S_3 \cos 0°$$
$$= 0.707(9900) + 4000 = 11,000 \text{ lb}.$$

From Equation (3.7) with member (1)

$$-7000 = (EA/L)(u_{y1} - 0), \quad u_{y1} = -7000(L/EA).$$

From member (2),

$$9900 = \frac{E(0.707A)}{1.414L}(u_{y1} \cos 45° + u_{z1} \cos 45° - 0),$$
$$u_{z1} = 35,000(L/EA).$$

Example 3.2. In Figure 3.5 with $E_i A_i = EA = 10^7$ lb, find the internal loads S_k in the members.

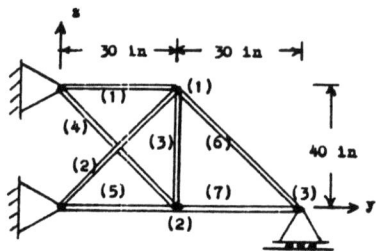

Fig. 3.5. Redundant truss with known displacements, $u_{y1} = 0.05$ in., $u_{z1} = -0.04$ in., $u_{y2} = 0.03$ in., $u_{z2} = -0.01$ in., $u_{y3} = 0.005$ in., $u_{z3} = 0.00$ in.

Solution. Use the stress-displacement Equation (3.7) to get

$S_k = (EA/L)_k (u_{kj} - u_{ki}) = (10^7/L_k)(u_{kj} - u_{ki})$,

$S_1 = (10^7/30)(u_{y1} - 0) = 16,667$ lb,

$S_2 = (10^7/50)(\frac{3}{5} u_{y1} + \frac{4}{5} u_{z1} - 0) = -400$ lb,

$S_3 = (10^7/40)(u_{z1} - u_{z2}) = -7500$ lb,

$S_4 = (10^7/50)(\frac{3}{5} u_{y2} - \frac{4}{5} u_{z2} - 0) = 5200$ lb,

$S_5 = (10^7/30)(u_{y2} - 0) = 10,000$ lb,

$S_6 = (10^7/50)(\frac{3}{5} u_{y3} - \frac{4}{5} u_{z3} - \frac{3}{5} u_{y1} + \frac{4}{5} u_{z1}) = -11,800$ lb,

$S_7 = (10^7/30)(u_{y3} - u_{y2}) = -8333$ lb.

It should be noted that if measured displacements are used to calculate member forces and external loads, then an accuracy problem exists. Because of the large multiplying factor $(EA/L)_k$ small errors in the displacements cause large errors in the forces. A tolerance of ± 0.01 in. in displacements would mean that the S_1 force could be off by ± 3333 lb.

Example 3.3. In the determinate truss in Figure 3.6(b), assume $P_{z1} = 4000$ lb, $P_{z1} = 6000$ lb, $Q_{z2} = 10,000$ lb, $E_i A_i = 10^7$ lb. Find the deflection u_{z1}.

Solution. Since the truss is determinate, the member loads can be obtained from the equilibrium equations. Since only the determinate subtruss of members (1) and (2) is needed to get u_{z1}, it is necessary to find only S_1 and S_2. Take moments about joint (3) to get

$100[-R_4 + 10^3\{-6 + 4 + 2(10)\}] = 0$, $R_4 = 18,000$ lb, $S_1 = 18,000$ lb.

The total shear force carried by member (2) is $6000 - 10,000 = -4000$ lb down. Thus, $S_2 = 1.414(-4000) = -5656$ lb. With the internal loads known, Equation (3.7) can be used to get u_{z1}. From member (2)

$-5656 = (10^7/141.4)(u_{y1} \cos 45° + u_{z1} \cos 45° - 0)$, or

$u_{z1} = -u_{y1} - 0.113$.

Get u_{y1} from member (1),

$u_{y1} \cos 0° - 0 = (100/10^7)(18,000) = 0.180$, or

$u_{y1} = 0.180$ in., $u_{z1} = -0.293$ in.

Example 3.4. Solve for the loads S_k in the redundant truss in Figure 3.6 for $u_{z2} = 0.00$, $P_{z1} = 0$ lb, $P_{z1} = 60,000$ lb, $E_i A_i =$ constant $= 10^7$ lb by the displacement method.

The virtual principles for pin-jointed trusses

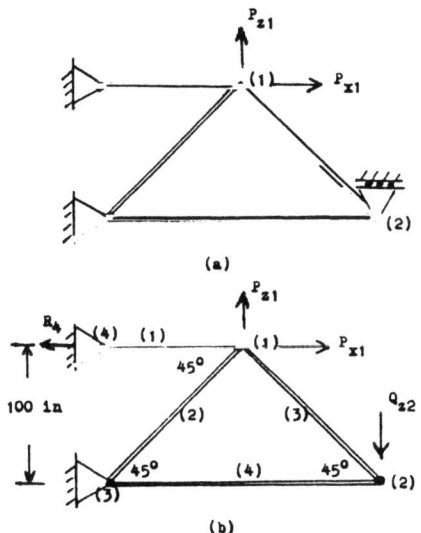

Fig. 3.6. (a). Truss with redundant supports.
(b). Redundant support replaced by unknown reaction.

Solution. From Equation (3.7)

$$10^{-3}S_1 = (10^4/100)(u_{x1} - 0) = 100u_{x1},$$
$$10^{-3}S_2 = (10^4/141.4)(u_{x1}\cos 45° + u_{z1}\cos 45° - 0) = 50(u_{x1} + u_{z1}),$$
$$10^{-3}S_3 = 50(-u_{x1} + u_{x1} + u_{x2}), \quad 10^{-3}S_4 = 50u_{z2}.$$

whence Equation (3.6) or Equation (3.8) becomes

$$0 = 100u_{x1}(1) + 50(u_{x1} + u_{z1})(0.707) + 50(-u_{x1} + u_{x1} + u_{x2})(-0.707),$$
$$60 = 0 + 50(u_{x1} + u_{z1})(0.707) + 50(-u_{x1} + u_{x1} + u_{x2})(0.707),$$
$$0 = 50u_{z2} + 50(-u_{x1} + u_{x1} + u_{x2})(0.707).$$

The solution of the three simultaneous equations gives

$$u_{x1} = 1.098 \text{ in.} \quad u_{z1} = -0.103 \text{ in.}, \quad u_{z2} = -0.498 \text{ in.},$$

whence the above equations for S_k give

$$S_1 = -10,300 \text{ lb}, \quad S_2 = 49,750 \text{ lb}, \quad S_3 = 35,140 \text{ lb}, \quad S_4 = -24.900 \text{ lb}.$$

Example 3.5. In Figure 3.7 find the internal member loads S_1, S_2, S_3 and the deflection components u_{yA}, u_{zA} of point A by using the displacement method. Assume E_1, E_2, E_3, A_1, A_2, A_3 are constant. Figure 3.8 is used for this case in Example 3.10.

Solution. From Equation (3.3)

$$u_{1A} = u_{yA}, \quad u_{2A} = u_{yA}\cos\theta + u_{zA}\sin\theta,$$
$$u_{3A} = u_{yA}\cos\theta - u_{zA}\sin\theta,$$

and from Equation (3.8)

$$P_{yA} = (EA/L)_1 u_{yA} + (EA/L)_2(u_{yA}\cos\theta + u_{zA}\sin\theta)\cos\theta +$$
$$+ (EA/L)_3(u_{yA}\cos\theta - u_{zA}\sin\theta)\cos\theta,$$
$$P_{zA} = (EA/L)_2(u_{yA}\cos\theta + u_{zA}\sin\theta)\sin\theta +$$
$$+ (EA/L)_3(u_{yA}\cos\theta - u_{zA}\sin\theta)(-\sin\theta).$$

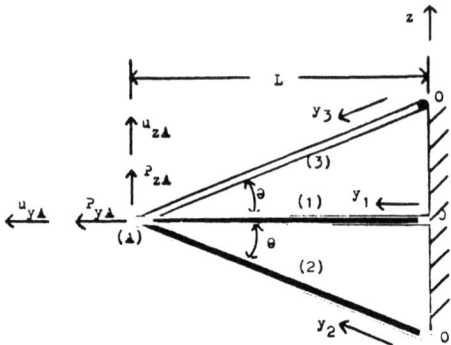

Fig. 3.7. Redundant pin-jointed truss.

Use $L_1 = L$, $L_2 = L \sec \theta$, $L_3 = L \sec \theta$, whence

$$\left[1 + \left(\frac{E_2 A_2}{E_1 A_1} + \frac{E_3 A_3}{E_1 A_1}\right) \cos^3 \theta\right] u_{yA} + \left[\left(\frac{E_2 A_2}{E_1 A_1} - \frac{E_3 A_3}{E_1 A_1}\right) \sin \theta \cos^2 \theta\right] u_{zA}$$
$$= (L/E_1 A_1) P_{yA},$$
$$\left[\left(\frac{E_2 A_2}{E_1 A_1} - \frac{E_3 A_3}{E_1 A_1}\right) \sin \theta \cos^2 \theta\right] u_{yA} + \left[\left(\frac{E_2 A_2}{E_1 A_1} + \frac{E_3 A_3}{E_1 A_1}\right) \sin^2 \theta \cos \theta\right] u_{zA}$$
$$= (L/E_1 A_1) P_{zA}.$$

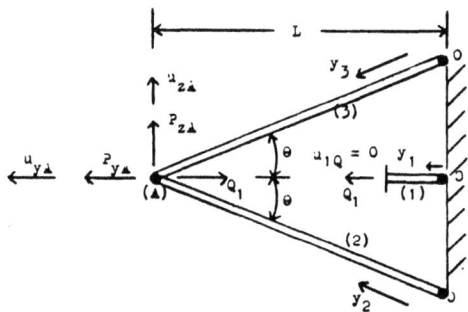

Fig. 3.8. Virtual load Q_1 on redundant truss.

Define the bracket terms as a_{11}, a_{12}, a_{12}, a_{22} in order and rewrite as

$$a_{11} u_{yA} + a_{12} u_{zA} = \frac{P_{yA} L}{E_1 A_1}, \quad a_{12} u_{yA} + a_{22} u_{zA} = \frac{P_{zA} L}{E_1 A_1}, \quad \text{or}$$

$$u_{yA} = \frac{L}{E_1 A_1} \frac{a_{22} P_{yA} - a_{12} P_{zA}}{a_{11} a_{22} - a_{12}^2}, \quad u_{zA} = \frac{L}{E_1 A_1} \frac{a_{11} P_{zA} - a_{12} P_{yA}}{a_{11} a_{22} - a_{12}^2}. \tag{3.9a}$$

From Equation (3.7)

$$S_1 = (EA/L)_1 u_{1A} = (E_1 A_1/L) u_{yA},$$
$$S_2 = (EA/L)_2 u_{2A} = (E_2 A_2/L)(u_{yA} \cos \theta + u_{zA} \sin \theta) \cos \theta, \tag{3.9b}$$
$$S_3 = (EA/L)_3 u_{3A} = (E_3 A_3/L)(u_{yA} \cos \theta - u_{zA} \sin \theta) \cos \theta.$$

These formulas can be used for any angle θ and any $E_i A_i$ values.

Example 3.6. Suppose the bar in Figure 3.1 has a variable area A_k, variable E_k and a variable body force P_y per unit length. See Figure 3.9. Find equations for the variable deflection $u_y(y)$ of the bar by the principle of virtual displacements.

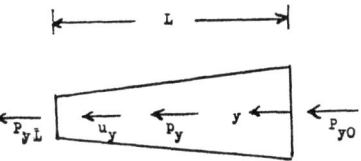

Fig. 3.9. Bar with E and A variable.

Solution. From the principle of virtual displacements in Equation (2.34)

$$\int_0^L \int_A \sigma_{yy} e_{yy}^V \, dA \, dy - \int_0^L p_y u_y^V \, dy - P_{yL} u_{yL}^V - P_{y0} u_{y0}^V = 0, \qquad (3.10)$$

where

$$\sigma_{yy} = E e_{yy} = E \frac{du_y}{dy}, \qquad e_{yy}^V = \frac{du_y^V}{dy}, \qquad (3.11)$$

$$u_y = u_{y0} \text{ at } y = 0, \qquad u_y = u_{yL} \text{ at } y = L.$$

Integrate the first term in Equation (3.10) by parts to get

$$\int_0^L \int_A E \frac{du_y}{dy} \frac{du_y^V}{dy} \, dA \, dy = \int_0^L (EA \frac{du_y}{dy}) \frac{du_y^V}{dy} \, dy$$

$$= \left[\left(EA \frac{du_y}{dy} \right) u_y^V \right]_0^L - \int_0^L \frac{d}{dy} \left(EA \frac{du_y}{dy} \right) u_y^V \, dy,$$

whence Equation (3.10) becomes

$$\int_0^L \left[\frac{d}{dy} \left(EA \frac{du_y}{dy} \right) + p_y \right] u_y^V \, dy + \left[P_{yL} - \left(EA \frac{du_y}{dy} \right)_L \right] u_{yL}^V +$$

$$+ \left[P_{y0} + \left(EA \frac{du_y}{dy} \right)_0 \right] u_{y0}^V = 0.$$

Since u_y^V, u_{yL}^V, u_{y0}^V are arbitrary, each bracket expression in the last equation above must be zero, or

$$\frac{d}{dy} \left(EA \frac{du_y}{dy} \right) + p_y = 0, \qquad \left(EA \frac{du_y}{dy} \right)_0 = -P_{y0}, \qquad \left(EA \frac{du_y}{dy} \right)_L = P_{yL}, \qquad (3.12)$$

which are the differential equation and the equilibrium boundary conditions for the deflection u_y of the bar. Note that only two of the four boundary conditions in Equations (3.11) and (3.12) can be specified. To prevent rigid body translation for the static bar, at least one of the displacement conditions in Equation (3.11) must be specified. Furthermore, if u_{y0} is specified, then P_{y0} is a reaction given by Equation (3.12) or by the force equilibrium equation and cannot be specified. Thus, both a displacement condition and a corresponding force condition cannot be specified at the same point. To get the equation for u_y, integrate Equation (3.12) twice using two specified permissible boundary conditions.

3.3. The unit load theorem for trusses

Assume loads S_k are known in the truss members, whether the truss is determinate or redundant, and apply a unit virtual force at joint j in the x-direction. Select a determinate subtruss of H_j members consistent with the supports. In Figure 3.2 with joints (1), (4), and (10) as fixed supports and joint (6) as $j = 6$, determinate subtrusses could be members (5) and (4) for the x-direction, members (4), (5), (8), (10), and (11), for the y-direction, and members (1) through (6) for the z-direction. Note that if the truss is determinate the loads in the members can always be calculated from Equation (3.6).

For each member k in the subtruss, the virtual stress for a unit virtual force is

$$\sigma^1_{kj} = p^1_{xkj}/A_k, \tag{3.13a}$$

where p^1_{xkj} is the force in member k due to the unit force in the direction of x at point j. The p^1_{xkj} values can be determined from the equilibrium equations (3.6). Since $e_k = \sigma_k/E_k = S_k/E_k A_k$, the unit load theorem in Equation (2.41) gives

$$u_{xjk} = \int_0^{L_k} \int_{A_k} \frac{p^1_{xkj}}{A_k} \left(\frac{S}{EA}\right)_k dA_k\, dy_k = p^1_{xkj}(SL/EA)_k. \tag{3.13b}$$

For all H_{xj}, H_{yj}, H_{zj} members in the subtrusses for the x, y, z directions, the deflections at joint j are

$$u_{xj} = \sum_{k=1}^{H_{xj}} p^1_{xkj}(SL/EA)_k, \qquad u_{yj} = \sum_{k=1}^{H_{yj}} p^1_{ykj}(SL/EA)_k,$$

$$u_{zj} = \sum_{k=1}^{H_{zj}} p^1_{zkj}(SL/EA)_k. \tag{3.14}$$

If the truss has *redundant supports*, Figure 3.6, so that the S_k loads in the members cannot be obtained directly from equilibrium, then selected supports can be imagined as temporarily removed to make the truss determinate, and unknown reaction forces applied at the redundant supports. For G applied component forces P_m and R redundant component reactions Q_m the loads S_k in the members can be written as

$$S_k = \sum_{m=1}^{G} p^1_{km} P_m + \sum_{m=1}^{R} q^1_{km} Q_m, \tag{3.15}$$

where p^1_{km} is the load in member k due to a unit value of P_m and q^1_{km} is the load in member k due to a unit value of Q_m. Here each component is numbered and counted in G and R. Equation (3.14) now applies at each redundant support with a specified deflection u_{Q_m} (usually zero) at the supports with S_k given by

The virtual principles for pin-jointed trusses

Equation (3.15), or for compatibility,

$$u_{Qm} = \sum_{k=1}^{H} \frac{q_{km}^1 L_k}{E_k A_k} \left(\sum_{j=1}^{G} p_{kj}^1 P_j + \sum_{i=1}^{R} q_{ki}^1 Q_i \right), \quad m = 1, \cdots, R, \qquad (3.16)$$

where H includes all members of the truss. This system of R equations can be solved for the R reactions Q_m, and the results put into Equation (3.15) for the loads in the members of the redundant truss. If desired, Equation (3.14) can then be used on selected determinate subtrusses to get the joint deflections.

If the truss has *redundant members*, Figure 3.10, then selected members can be imagined as cut temporarily and replaced by unknown loads to make the truss determinate. At the cut, Figure 3-10(c), the relative deflection must be zero so that

$$u_{mi} + u_{mj} = 0, \text{ at cut,} \qquad (3.17)$$

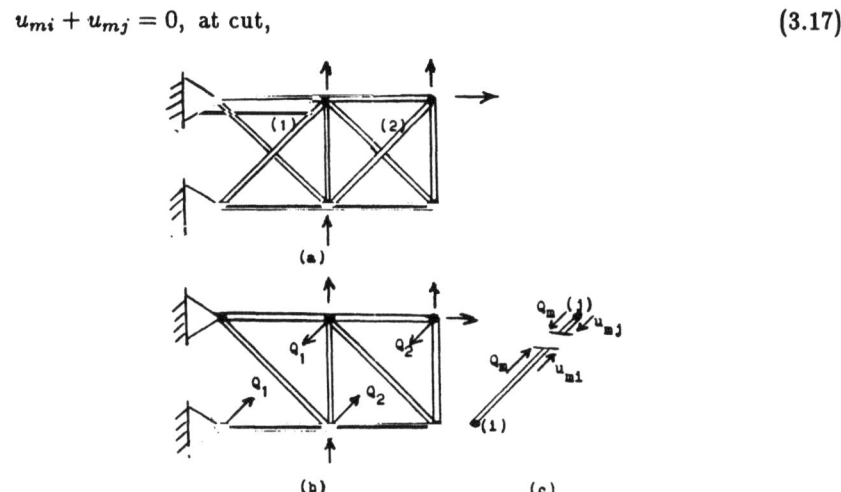

Fig. 3.10. (a). Truss with redundant members. (b). Selected members replaced by unknown forces to make truss determinate. (c). Deflections at cut in member.

where u_{mi} and u_{mj} are given by Equation (3.14) with the Q_m forces included in S_k. Define

$$q_{km}^1 = q_{kmi}^1 + q_{kmj}^1, \qquad (3.18)$$

which represents the self-equilibrium effect of unit values of the opposite Q_m forces at the cut. Note that q_{km}^1 involves only a few members and can be calculated directly using both of the opposite Q_m together. It is not necessary to calculate the more complicated q_{kmi}^1 and q_{kmj}^1, which would cancel in most of the members anyway. See examples below. With the q_{km}^1 in Equation (3.18) and Q_m the forces in the selected redundant members, it follows that Equations (3.15) and (3.16) apply to this case with $u_{Qm} = 0$. Thus, the Q_m can be solved directly from Equation (3.16).

The above procedure for obtaining the p_{km}^1 and q_{km}^1 unit values in Equation (3.15) by removing supports and cutting truss members to get a determinate truss is the classical method in the redundant force analysis. A different method that uses the equilibrium equations directly without requiring any direct unit load calculations for a determinate truss will be described in Section 3.6. It is very useful in the matrix force procedure for large structures.

Example 3.7. Solve Example 3.1 by using the above unit load equations and comment on the differences in the unit displacement and unit load procedures.

Solution. In Figure 3.4 cut member (3) and combine its 4000 lb force with P_{x1} as $P_{x1} - 4000$. Use the determinate structure of members (1) and (2) and Equation (3.15). Use unit loads at joint (1) to get

$$P_{y11}^1 = 1.00, \quad p_{y21}^1 = 0, \quad p_{x11}^1 = -1.00, \quad p_{x21}^1 = 1.414,$$
$$-7000 = (1.00)P_{y1} + (-1.00)(P_{x1} - 4000),$$
$$9900 = (0)P_{y1} + (1.414)(P_{x1} - 4000),$$
$$P_{y1} = 0 \text{ lb}, \quad P_{x1} = 11{,}000 \text{ lb}.$$

For the u_{y1} deflection, select the determinate truss with member (1) so that in Equation (3.14) $p_{y11}^1 = 1.00$ and

$$u_{y1} = (1.00)(L/EA)(-7000) = -7000(L/EA).$$

For the u_{x1} deflection, select the determinate truss with members (1) and (2) so that $p_{x11}^1 = -1.00$, $p_{x21}^1 = 1.414$, whence

$$u_{x1} = (-1.00)(L/EA)(-7000) + \frac{(1.414)(1.414L)}{0.707EA}(9900)$$
$$= 35{,}000(L/EA).$$

Examples 3.1 and 3.7 show the reciprocal relation between the unit displacement theorem and the unit load theorem. When the internal stresses or internal loads are known, then the unit displacement theorem gives the external forces directly from the equilibrium Equation (3.6), while the unit load theorem gives the deflections directly from Equation (3.14).

Example 3.8. Solve Example 3.3 by using the unit load equations.

Solution. The internal loads are known from Example 3.3 as $S_1 = 18{,}000$ lb and $S_2 = -5656$ lb. Apply a unit load up at joint (1) in Figure 3.6(b) to get $p_{x11}^1 = -1.00$, $p_{x21}^1 = 1.414$, whence Equation (3.14) gives

$$u_{x1} = 10^{-5}[-1.00(180)(100) + 1.414(-56.56)(141.4)]$$
$$= -0.293 \text{ in.}$$

Example 3.9. Solve Example 3.4 by the unit load equations and comment on the two methods.

Solution. In Figure 3.6 support (2) is removed and the unknown load Q_{x2} used as the reaction. In Equation (3.15), $G = 1$ for the one given load $P_{x1} = 60{,}000$ lb and $R = 1$ for the one redundant load Q_{x2}. In Figure 3.6(b) use a unit load up at joint (1) and a unit load down at joint (2) to get

$$p_{x11}^1 = -1.00, \quad p_{x21}^1 = 1.414, \quad p_{x31}^1 = 0, \quad p_{x41}^1 = 0,$$
$$q_{x12}^1 = 2.00, \quad q_{x22}^1 = -1.414, \quad q_{x32}^1 = 1.414, \quad q_{x42}^1 = -1.00.$$

Thus, Equation (3.15) gives

$$S_1 = -60{,}000 + 2Q_{x2}, \quad S_2 = 1.414(60{,}000 - Q_{x2}), \quad S_3 = 1.414Q_{x2}, \quad S_4 = -Q_{x2}.$$

Get Q_{x2} from Equation (3.16) by

$$0 = 2.00(-60{,}000 + 2Q_{x2}) - (1.414)^3(60{,}000 - Q_{x2}) + (1.414)^3 Q_{x2} + 2.00 Q_{x2},$$
$$Q_{x2} = 24{,}850 \text{ lb}, \quad \text{whence}$$
$$S_1 = -10{,}300 \text{ lb}, \quad S_2 = 49{,}710 \text{ lb}, \quad S_3 = 35{,}140 \text{ lb}, \quad S_4 = -24{,}850 \text{ lb}.$$

The virtual principles for pin-jointed trusses

In this simple truss with one redundant support, the unit load theorem gives the internal member forces more directly and with simpler calculations than the unit displacement theorem, which requires the solving of three simultaneous equations for the three displacements before the member forces S_k can be calculated. In small trusses with a few redundants, the unit load theorem usually is the simpler method to get the member forces and the joint deflections. However, in large trusses with many redundants in which matrices and computers must be used, the unit displacement theorem may be the better method. See Section 3.7 for further discussion of the two methods.

Example 3.10. Solve Example 3.5 by the unit load equations.

Solution. In Figure 3.8. cut member (1) to make the truss determinate. From Equation (3.15) for the S_k loads $G = 2$ and $R = 1$, and

$$p^1_{y1A} = 0, \quad p^1_{y2A} = p^1_{y3A} = 1/(2\cos\theta), \quad p^1_{z1A} = 0,$$
$$p^1_{z2A} = -p^1_{z3A} = 1/(2\sin\theta), \quad q^1_{y1} = 1, \quad q^1_{y2} = q^1_{y3} = -1/(2\cos\theta),$$

it follows that

$$S_1 = Q_1, \quad S_2 = (1/2)(P_{yA} - Q_1)\sec\theta + (1/2)P_{zA}\csc\theta,$$
$$S_3 = (1/2)(P_{yA} - Q_1)\sec\theta - (1/2)P_{zA}\csc\theta. \tag{3.19}$$

With $u_{1Q} = 0$, Equation (3.16) becomes

$$0 = Q_1 L_1/E_1 A_1 - S_2 L_2 \sec\theta/4 E_2 A_2 - S_3 L_3 \sec\theta/4 E_3 A_3, \quad \text{or}$$
$$b_{11} Q_1 = b_{22} P_{yA} + b_{12} P_{zA}, \quad b_{11} = 4 + b_{22}, \tag{3.20}$$
$$b_{22} = \left(\frac{E_1 A_1}{E_2 A_2} + \frac{E_1 A_1}{E_3 A_3}\right)\sec^3\theta, \quad b_{12} = \left(\frac{E_1 A_1}{E_2 A_2} - \frac{E_1 A_1}{E_3 A_3}\right)\sec^2\theta\csc\theta.$$

Put this Q_1 into Equation (3.19) to get the S_k forces. From Equation (3.14) with selected determinate members the deflections are

$$u_{yA} = S_1 L/E_1 A_1, \quad \text{(from member 1)},$$
$$u_{zA} = \frac{L\sec\theta}{2\sin\theta}[(S_2/E_2 A_2) - (S_3/E_3 A_3)], \quad \text{(members 2, 3)}. \tag{3.21}$$

These formulas are quite different and somewhat simpler than those in Example 3.5.

Example 3.11. Solve Example 3.6 by the principle of virtual forces and compare the two methods.

Solution. In Equation (2.40) apply a virtual force P^V_y at point y with the point $y = 0$ as a support with deflection u_{y0}, whence

$$\int_0^L \int_A e_{yy} \sigma^V_{yy} \, dy \, dA - u_y P^V_y + u_{y0} P^V_y = 0. \tag{3.22}$$

Here,

$$e_{yy} = \sigma_{yy}/E = P_y/EA = \left(P_{yL} + \int_y^L p_y \, dy\right)/EA,$$
$$\sigma^V_{yy} = P^V_y/A, \quad 0 \leq y_d \leq y,$$
$$= 0, \quad y \leq y_d \leq L,$$

whence for arbitrary p^V_y, Equation (3.22) becomes

$$u_y - u_{y0} = \int_0^y \left(P_{yL} + \int_{y_d}^L p_y \, dy_d\right) dy_d/EA. \tag{3.23}$$

Note that this expression in Equation (3.22) will be different with $y = L$ as the support.

The principle of virtual forces is more direct than the principle of virtual displacements in that it gives an integrated form of the differential equation given by the principle of virtual displacements. However, this form requires that the boundary conditions be specified in the principle of virtual forces. This is not necessary for the virtual displacements.

3.4. Inelastic effects with temperature changes in trusses

Since the stress-strain curves of materials change with temperature, it is possible for some members of the truss with high temperature to have low yield stresses. Thus, the temperature and inelastic problems tend to occur together in a structure. In a redundant structure the temperatures may cause some members to become inelastic, thus shifting the loads to other elastic members. Since the thermal strain is always a limited value, the temperature effects cannot cause a redundant ductile structure to fail until all members are inelastic and the applied loads cause the failure (see Section 1.15).

If the strain e_E in Equation (2.57) is constant along the axially loaded bar in Figure 3.1, then Equations (2.62) and (2.63) apply, or

$$\begin{aligned} u_j - u_i &= (L/EA)_k(S_k + S_{Ek}) = (SL/EA)_k + u_{Ek}, \\ S_{Ek} &= E_k A_k e_{Ek} = E_k A_k (\mp e_d + e_p + e_I + e_T)_k \\ &= S_{dk} + S_{pk} + S_{Ik} + S_{Tk}, \\ u_{Ek} &= (S_E L/EA)_k = e_{Ek} L_k. \end{aligned} \tag{3.24}$$

The basic Equations (3.6)-(3.8) for the principle of virtual displacements in the truss can be rewritten for the S_{Ek} term present as follows:

$$\begin{aligned} P_{xj} &= \sum_{k=1}^{K} S_k \cos(y_k, x), \quad S_k = \left(\frac{EA}{L}\right)_k (u_{kj} - u_{ki}) - S_{Ek}, \\ P_{xj} &= \sum_{k=1}^{K} \left[\left(\frac{EA}{L}\right)_k (u_{kj} - u_{ki}) - S_{Ek}\right] \cos(y_k, x), \end{aligned} \tag{3.25}$$

with similar terms for P_{yj} and P_{zj}.

The basic Equations (3.13)-(3.16) for the principle of virtual forces in the truss can be rewritten for the S_{Ek} term present, or

$$\begin{aligned} u_{xj} &= \sum_{k=1}^{H} p_{xkj}^1 (L/EA)_k (S_k + S_{Ek}), \\ S_k &= \sum_{m=1}^{G} p_{km}^1 P_m + \sum_{m=1}^{R} q_{km}^1 Q_m, \\ u_{Qm} &= \sum_{k=1}^{H} q_{km}^1 (L/EA)_k \left(\sum_{j=1}^{G} p_{kj}^1 P_j + \sum_{i=1}^{R} q_{ki}^1 Q_i + S_{Ek}\right), \\ m &= 1, 2, \cdots, R. \end{aligned} \tag{3.26}$$

The solution for the Q_m will include the S_{Ek} terms, which will then affect the S_k member forces. If the truss is determinate the S_k forces are unaffected by S_{Ek}, but the deflections are affected.

Note that if large deflections are present the geometry of the truss may change so that the direction cosines in Equation (3.25) and in the p^1_{kj} and q^1_{kj} terms in Equation (3.26) may change. This effect is in addition to the S_{dk} term in S_{Ek}, which could be zero in the members, with the geometry change coming from large values of S_{pk}.

If the stress-strain curve is represented in the nondimensional Ramberg-Osgood form in Equation (1.94) and in Table 1.1, then

$$S_{pk} = E_k A_k e_{pk} = (3/7) S_{yk} (S_k/S_{yk})^{10}, \quad S_{yk} = A_k F_{yk}. \tag{3.27}$$

Note that if S_k is known, such as in a determinate truss, then S_{pk} can be calculated directly.

If S_k is not known, as in a redundant truss, rewrite Equation (1.94) as

$$S_k^a/S_{yk} = (S_k/S_{yk}) + (3/7)(S_k/S_{yk})^{10}, \tag{3.28}$$

which has the same values as in Table 1.1, and where

$$S_k^a = S_k + S_{pk} = (EA/L)_k (u_j - u_i) - S_{dk} - S_{Ik} - S_{Tk}. \tag{3.29}$$

For the iteration, start with $(S_{pk})_0 = 0$ and use $(S_k^a)_0$ to get $(S_k)_0$ from Table 1.1, then

$$(S_{pk})_1 = (S_k^a)_0 - (S_k)_0, \tag{3.30}$$

which can be used to get a new $(S_k^a)_{01}$. Then

$$(S_k^a)_1 = (S_k^a)_{01} + (S_{pk})_1 \quad \text{gives } (S_k)_1. \tag{3.31}$$

Thus $(S_{pk})_2 = (S_k^a)_1 - (S_k)_1$, etc., and at the r iteration

$$(S_{pk})_r = (S_k^a)_{r-1} - (S_k)_{r-1}, \quad (S_k^a)_r = (S_k^a)_{r-1,r} + (S_{pk})_r. \tag{3.32}$$

Convergence occurs when $(S_k^a)_{r-1,r} = (S_k)_{r-1}$ within a specified tolerance.

Since the iteration process outlined above may converge slowly, it is possible to use a speed-up procedure on the $(S_{pk})_r$ feedback term, References 1,2. Use the corrected $(S_{pk})_r$ as

$$(S_{pk})_{r,c} = \frac{(S_{pk})_r (S_{pk})_{r-2,c} - (S_{pk})_{r-1}(S_{pk})_{r-1,c}}{(S_{pk})_r - (S_{pk})_{r-1} - (S_{pk})_{r-1,c} + (S_{pk})_{r-2,c}}. \tag{3.33}$$

See Example 3.14 below.

To allow for sequence loading and thermal cycling in Equation (3.26), use the procedure of Section 1.16 and write Equation (3.26) in the form

$$u_{Qm} = \sum_{k=1}^{H} q^1_{km} \left(\frac{L}{EA}\right)_k \left[\sum_{n=1}^{G} p^1_{kn}(P_n^{j-1} + P_n^j) + \sum_{n=1}^{R} q^1_{kn} Q_n^j + \right.$$
$$\left. + (S_{Ek}^{j-1} + S_{Ek}^j) \right]. \tag{3.34}$$

The P_n^{j-1} and S_{Ek}^{j-1} terms are the known results at step $j - 1$, while the P_n^j and S_{Ek}^j terms are changes in step j. If S_{pk}^{j-1} terms are present then Equations (1.105) and (1.106) must be used in getting the form of S_{pk}^j.

Example 3.12. Solve Examples 3.3 and 3.8 for the case of a thermal strain of $e_T = 0.006$ for member (1) with yield loads of $S_{y1} = 14,000$ lb, $S_{y2} = -8000$ lb.

Solution. Since the truss is determinate, the applied loads $S_1 = 18,000$ lb and $S_2 = -5656$ lb remain unchanged. The apparent thermal load in member (1) is

$$S_{T1} = (EA)_1 \, e_T = (10^7)(0.006) = 60,000 \text{ lb.}$$

From Equation (3.27) the apparent inelastic loads are

$$S_{p1} = (3/7)(14,000)(18/14)^{10} = 74,100 \text{ lb,}$$
$$S_{p2} = (3/7)(-8000)(5656/8000)^{10} = -107 \text{ lb,}$$

whence

$$S_{E1} = 60,000 + 74,100 = 134,100 \text{ lb,} \qquad S_{E2} = -107 \text{ lb.}$$

With these values and the values in Example 3.3, Equation (3.26) gives

$$u_{x1} = (1/10^7)[(-1.00)(100)(18,000 + 134,100) + \\ + (1.414)(141.4)(-5656 - 107)] = -1.63 \text{ in.,}$$

which is much larger than the value of -0.293 in Example 3.8.

The displacement method in Equation (3.25) gives for members (2) and (1) in order

$$-5656 = (10^7/141.4)(u_{y1} \cos 45° + u_{x1} \cos 45°) - (-107),$$
$$18,000 = (10^7/100)(u_{y1} \cos 0° - 0) - 134,100,$$

whence

$$u_{y1} = 1.521 \text{ in.,} \qquad u_{x1} = -1.521 - 0.111 = -1.632 \text{ in.}$$

Example 3.13. Solve Examples 3.4 and 3.9 for the elastic case with thermal strains of $e_{T1} = -0.006$, $e_{T2} = 0.004$ in members (1) and (2). Compare results to Example 3.9.

Solution. The apparent thermal loads are

$$S_{T1} = (EA)_1 \, e_{T1} = (10^7)(-0.006) = -60,000 \text{ lb } = S_{E1},$$
$$S_{T2} = (EA)_2 \, e_{T2} = (10^7)(0.004) = 40,000 \text{ lb } = S_{E2}.$$

From Equation (3.26) with the expressions in Example 3.9.

$$0 = 2.00(-60,000 + 2Q_2 - 60,000) - (1.414)^3[60,000 - Q_2 + \\ + (40,000)(0.707)] + (1.414)^3 Q_2 + 2.00 Q_2$$
$$= 11.656 Q_2 - 489,000,$$
$$Q_2 = 42,010 \text{ lb, and}$$
$$S_1 = 24,020 \text{ lb,} \qquad S_2 = 25,440 \text{ lb,} \qquad S_3 = 59,410 \text{ lb,}$$
$$S_4 = -42,010 \text{ lb,} \qquad \text{with temperature.}$$

The corresponding loads from Example 3.9 without temperature are

$$S_1 = -10,300 \text{ lb,} \qquad S_2 = 49,710 \text{ lb,} \qquad S_3 = 35,140 \text{ lb,}$$
$$S_4 = -24,850 \text{ lb,} \qquad \text{without temperature.}$$

It is evident that in a redundant truss temperature changes in some members can cause large changes in the loads in all members, increasing some loads and decreasing others. The loads in some members may change signs.

The solution could have been made using the principle of virtual displacements, Equation (3.25), but it would be considerably longer.

Example 3.14. Suppose member (3) in the redundant truss in Example 3.13 has a yield load of $S_{y3} = 45,000$ lb. (a). Recalculate the member loads in Example 3.13 for the inelastic case. (b). Comment on using the principle of virtual displacements for this case.

Solution (a) Since $S_3 = 1.414 Q_2$, Equation (3.17) gives

$$S_{p3} = (3/7)(45,000)(1.414 Q_2/45,000)^{10} = 617,000(Q_2/45,000)^{10},$$

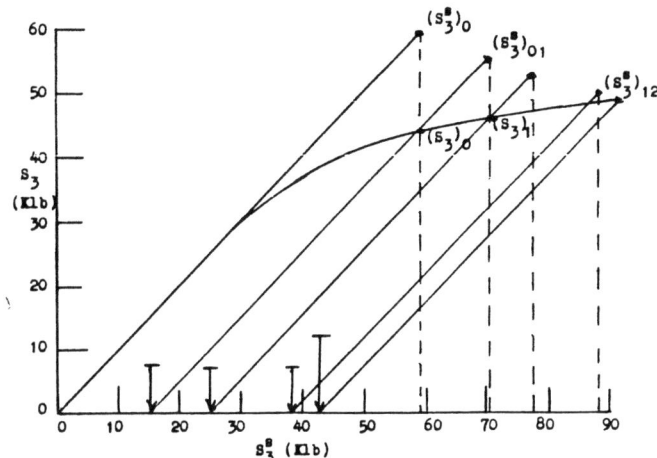

Fig. 3.11. Convergence of iteration procedure.

whence the Q_2 equation in Example 3.13 becomes

$$11.656 Q_2 = 489,000 - 2 S_{p3} = 489,000 - 1,234,000 (Q_2/45,000)^{10}.$$

In this simple case of one redundant, the solution for the redundant Q_2 can be made by trial and error, by graphing, or by solving the tenth degree equation. By trial,

$Q_2 = 34,510$ lb, whence
$S_1 = 9020$ lb, $\quad S_2 = 36,050$ lb, $\quad S_3 = 48,800$ lb, $\quad S_4 = -34,510$ lb.

In order to demonstrate the method, the solution will be made also by the iteration procedure with the speed-up method in Equations (3.28)-(3.35). To simplify the iteration calculations use the units Klb = 1000 lb for all loads. Thus, the results will correspond to the above results divided by 1000.

From Example 3.13, $(S_3^s)_0 = 59.41$, whence in Table 1.1

$(S_3^s)_0 / S_{y3} = 59.41/45.00 = 1.32, \quad (S_3)_0 / S_{y3} = 0.98,$
$(S_3)_0 = 44.10, \quad (S_{p3})_1 = 59.41 - 44.10 = 15.31,$
$11.656 (Q_2)_1 = 489 - 2(15.31) = 458, \quad (Q_2)_1 = 39.33,$
$(S_3^s)_{01} = 1.414 (Q_2)_1 = 55.61,$
$(S_3^s)_1 / S_{y3} = (55.61 + 15.31)/45 = 1.58, \quad (S_3)_1 / S_{y3} = 1.03,$
$(S_3)_1 = 46.35, \quad (S_{p3})_2 = 55.61 + 15.31 - 46.35 = 24.57.$

Equation (3.33) can now be used to change $(S_{p3})_2$, or

$(S_{p3})_{2,c} = \dfrac{-(15.31)(15.31)}{24.57 - 15.31 - 15.31} = 38.74,$
$11.656 (Q_2)_2 = 489 - 2(38.74), \quad (Q_2)_2 = 35.31,$
$(S_3^s)_{12} = 1.414 (Q_2)_2 = 49.94,$
$(S_3^s)_2 / S_{y3} = (49.94 + 38.74)/45 = 88.68/45 = 1.97,$
$(S_3)_2 / S_{y3} = 1.076, \quad (S_3)_2 = 45(1.076) = 48.42,$
$(S_{p3})_3 = 88.68 - 48.42 = 40.26,$
$(S_{p3})_{3,c} = \dfrac{(40.26)(15.31) - (24.57)(38.74)}{40.26 - 24.57 - 38.74 + 15.31} = 43.34,$
$11.656 (Q_2)_3 = 489 - 2(43.34), \quad (Q_2)_3 = 34.52,$
$(S_3^s)_{23} = 1.414 (Q_2)_3 = 48.82$ Klb,

which is within 0.40 Klb of $(S_3)_2$ for convergence, and which agrees with the results given by the trial solution. Figure 3.11 shows the steps in in the iteration. It should be noted that it takes about ten iterations to converge without using the speed-up procedure in Equation (3.33).

(b) If the Equations (3.25) are solved for the displacements (see Example 3.4) in terms of S_{p3}, then the results can be used to get the member loads in terms of S_{p3}. In the case here $S_3 = 59.41 - 0.248 S_{p3}$, whence Equation (3.27) becomes

$$S_{p3} = (3/7)(45)\left(\frac{59.41 - 0.248 S_{p3}}{45}\right)^{10},$$

and $S_{p3} = 43.0$, which agrees with the result obtained in part (a). Also, the iteration would proceed as in part (a) using

$$(S_3^\circ)_r = (59.41 - 0.248 S_{p3})_r + (S_{p3})_r = 59.41 + 0.752(S_{p3})_r,$$
$$(S_3^\circ)_0 = 59.41, \quad (S_3^\circ)_1 = 59.41 + 0.752(15.31) = 70.92, \quad \text{etc.}$$

Thus, both the principle of virtual forces and the principle of virtual displacements iterate on the member loads in the inelastic case, forcing the compatible truss to the inelastic equilibrium position. However, in simple trusses with a few redundants, the equations for the principle of virtual forces are easier to set up because they involve the forces directly while the principle of virtual displacements procedure requires going through the displacements first.

Example 3.15 (a). Use the unit load theorem to find the internal loads in the members of the truss in Figure 3.12 with $E_i A_i = 10^4$ Klb for all members. Member (2) has a thermal strain of $e_{T2} = 0.004$. Member (3) is inelastic with $S_{y3} = -30$ Klb. Member (4) is shaped as a sine wave with an initial maximum deflection of $u_{z,\max} = 10.000$ in. and has an Euler buckling load of $S_{b4} = 20$ Klb. The support at (A) is $u_A = 2.00$ in. too high so that there are initial loads in the truss. (b). Compare the results to the applied load case with none of the e_E effects.

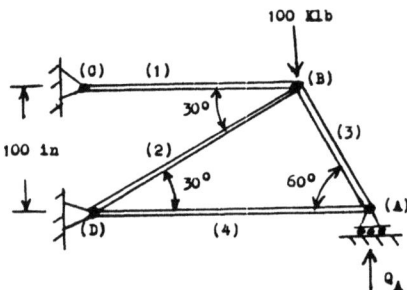

Fig. 3.12. Redundant truss with thermal, inelastic, initial and large deflection strains.

Solution. (a) Remove the support at (A) to make the truss determinate and let Q_A be the unknown reaction at (A). The internal loads due to the applied force at joint (B) are

$$S_{1a} = 173.21 \text{ Klb}, \quad S_{2a} = -200 \text{ Klb}, \quad S_{3a} = 0, \quad S_{4a} = 0.$$

Apply a unit load up at support (A) to get

$$q_{A1}^1 = -2.309, \quad q_{A2}^1 = 2.000, \quad q_{A3}^1 = -1.155, \quad q_{A4}^1 = 0.577.$$

Thus, the loads in the members due to Q_A are

$$S_{Q1} = -2.309 Q_A, \quad S_{Q2} = 2.000 Q_A, \quad S_{Q3} = -1.155 Q_A, \quad S_{Q4} = 0.577 Q_A$$

The virtual principles for pin-jointed trusses

The apparent thermal loads are zero except for member (2) with
$$S_{T2} = 10^4(0.004) = 40 \text{ Klb}.$$
The apparent inelastic loads are zero except for member (3) for which Equation (3.27) gives
$$S_{p3} = (3/7)(-30)(S_3/30)^{10} = -54.3206(Q_A/30)^{10}.$$
For the large deflection of member (4), the plus sign in Equation (3.24) applies, and the approximate deflection due to the S_4 load is
$$u_{zd} = u_{z,\max}/\left(1 + \frac{S_4}{S_{b4}}\right) = 10\left(1 + \frac{S_4}{20}\right)^{-1} = 10\left(1 + \frac{Q_A}{34.662}\right)^{-1}.$$
Equation (2.56) now gives
$$S_{d4} = 10^4 e_d = 10^4 \left(\frac{\pi}{2L_4}\right)^2 (u_{z,\max}^2 - u_{zd}^2)$$
$$= 46.2638\left[1 - \left(1 + \frac{Q_A}{34.662}\right)^{-2}\right].$$
The initial deflection at point (A) is $u_{QA} = 2.00$ in. in Equation (3.26).
Combine the above terms and use Equation (3.26) to get the equation for Q_A as

$$\frac{10^4}{100}(2.00) = (1.732)(-2.309)(173.21 - 2.309Q_A) +$$
$$+ (2.000)(2.000)(-200 + 2.000Q_A + 40) +$$
$$+ (1.155)(-1.155)\left[0 - 1.55Q_A - 54.321\left(\frac{Q_A}{30}\right)^{10}\right] +$$
$$+ (2.309)(0.577)\left[0 + 0.577Q_A + 46.264\left\{1 - \left(1 + \frac{Q_A}{34.662}\right)^{-2}\right\}\right],$$

or

$$200 = -692.70 + 9.234Q_A - 640.00 + 8.000Q_A + 1.541Q_A +$$
$$+ 72.466(Q_A/30)^{10} + 0.769Q_A + 61.637(1 - [1 + (Q_A/34.662)]^{-2}),$$

or

$$1532.70 - 19.544Q_A = 72.466(Q_A/30)^{10} +$$
$$+ 61.637(1 - [1 + (Q_A/34.662)]^{-2}).$$

This equation can be solved by trial and error for Q_A, or $Q_A = 37.87$ Klb and the loads in the members are

$S_1 = 173.21 - 2.309Q_A = 85.76$ Klb, $S_2 = -200 + 2Q_A = -124.26$ Klb,
$S_3 = -1.155Q_A = -43.74$ Klb, $S_4 = 0.577Q_A = 21.85$ Klb.

(b). If none of the temperature, inelastic, large deflection, and initial effects are present, the basic equation above becomes

$0 = -692.70 - 800.000 + 19.544Q_A$,   or $Q_A = 76.38$ Klb,
$S_1 = -3.14$ Klb, $S_2 = -47.25$ Klb, $S_3 = -88.21$ Klb, $S_4 = 44.07$ Klb.

For this elastic case, support (A), which is more directly under the applied load in Figure 3.12 takes most of the load and member (1) does practically nothing. However, for the inelastic case with the large effects in member (3), most of the load must go to support (D) with large loads in members (1) and (2). The initial deflection of support (A) and the curved member (4) have small effects on the loads in this case.

Example 3.16. In Figure 3.12 use 50 Klb instead of 100 Klb for the applied load and take $E_i A_i = 10^4$ Klb for all members. Member (1) has a thermal strain of $E_{T1} = 0.005$; members

(2) and (3) are inelastic with $S_{y2} = -35$ Klb and $S_{y3} = -30$ Klb. The support at (A) is $u_A = 2.00$ in. too high so that there are initial loads in the truss. (a). Make use of Example 3.15 to find the internal loads. (b). Cycle the temperature in member (1)

Solution (a) From Example 3.15

$$S_{1a} = 86.6 \text{ Klb}, \quad S_{2a} = -100 \text{ Klb}, \quad S_{3a} = 0, \quad S_{4a} = 0,$$
$$S_{p3} = -54.321(Q_A/30)^{10}, \quad S_{p2} = -15.00[(Q_A - 50)/17.5]^{10},$$
$$S_{T1} = (10^4)(0.005) = 50 \text{ Klb},$$

whence

$$200 = -4.00(86.6 - 2.309Q_A + 50) + 4.00\left[-100 + 2Q_A - 15.00\left(\frac{Q_A - 50}{17.5}\right)^{10}\right] -$$
$$- 1.333\left[0 - 1.155Q_A - 54.321\left(\frac{Q_A}{30}\right)^{10}\right] + 1.333(0 + 0.577Q_A), \quad (3.35)$$

$$1146.4 = 19.544Q_A - 60.000\left(\frac{Q_A - 50}{17.5}\right)^{10} + 72.466\left(\frac{Q_A}{30}\right)^{10}, \quad (3.36)$$

$$Q_A = 36.0 \text{ Klb}.$$

The internal loads are

$$S_1 = 3.5 \text{ Klb}, \quad S_2 = -28.0 \text{ Klb}, \quad S_3 = -41.6 \text{ Klb}, \quad S_4 = 20.8 \text{ Klb}.$$

The elastic solution with the temperature and initial values present is

$$Q_A = 58.7 \text{ Klb}, \quad S_1 = -48.8 \text{ Klb}, \quad S_2 = 17.4 \text{ Klb},$$
$$S_3 = -67.8 \text{ Klb}, \quad S_4 = 33.9 \text{ Klb}.$$

In this case the inelastic effects reduce the loads in all members and change member (2) load from tension to compression.

(b) In the above inelastic solution it is assumed that the load and temperature are applied simultaneously. If the load stays on and the temperature is removed, then the inelastic member (3) will unload on an elastic line and inelastic member (2) will load on its stress-strain curve. Thus, from Equations (3.34) and (3.35), $P_n^j = 0$ for all members,

$$S_{E1}^{j-1} = 50 \text{ Klb}, \quad S_{E1}^{j} = -50 \text{ Klb}, \quad S_{E2}^{j-1} = 0, \quad S_{E2}^{j} = -15\left(\frac{Q_A^j - 50}{17.5}\right)^{10},$$
$$S_{E3}^{j-1} = 336.6 \text{ Klb}, \quad S_{E3}^{j} = 0.$$

With these values, Equation (3.35) becomes

$$946.4 = 19.544Q_A^j - 60\left(\frac{Q_A^j - 50}{17.5}\right)^{10} + 448.7, \quad (3.37)$$

$$Q_A^j = 31.5 \text{ Klb}.$$

Thus, with the temperature removed the internal loads are

$$S_1 = 13.9 \text{ Klb}, \quad S_2 = -37.0 \text{ Klb}, \quad S_3 = -36.4 \text{ Klb}, \quad S_4 = 18.2 \text{ Klb}.$$

If the temperature cycle is repeated, the results for Q_A for several cycles are

	$+P+T$	$-T$	$+T$	$-T$	$+T$	$-T$
Q_A(Klb)	36.0	31.5	36.6	30.5	37.0	30.0

Convergence to shakedown occurs at $Q_A = 38.6$ Klb for T on and $Q_A = 28.4$ Klb for T off. The internal loads for these cases are

(Klb)	Q_A	S_1	S_2	S_3	S_4
T on	38.6	-2.5	-22.8	-44.6	22.3
T off	28.4	21.0	-43.2	-32.8	16.4

3.5. Matrix equations for trusses from the unit displacement theorem

For joint j in a truss the equilibrium Equations (3.6) can be written in matrix form as

$$\begin{Bmatrix} P_{xj} \\ P_{yj} \\ P_{zj} \end{Bmatrix} = \begin{bmatrix} \cos(x,y_1) & \cos(x,y_2) & \cdots & \cos(x,y_K) \\ \cos(y,y_1) & \cos(y,y_2) & \cdots & \cos(y,y_K) \\ \cos(z,y_1) & \cos(z,y_2) & \cdots & \cos(z,y_K) \end{bmatrix}_j \begin{Bmatrix} S_1 \\ S_2 \\ \vdots \\ S_K \end{Bmatrix}_j,$$

$$\{P\}_j = [s]_j^T \{S\}_j. \tag{3.38a}$$

These equations can be assembled for all the joints including support points into the form

$$\{P\} = [s]^T \{S\}. \tag{3.38b}$$

Some of the P_j will be zero and the P_j at the supports will be unknown reactions. Note that $[s]^T$ will have many zeros since only nonzero terms arise from those truss members coming to the joint.

Write Equation (3.38b) as

$$\begin{Bmatrix} P_a \\ P_R \end{Bmatrix} = \begin{bmatrix} s_a^T \\ s_R^T \end{bmatrix} \{S\}, \tag{3.39}$$

where $\{P_R\}$ are the reactions at the supports. If the truss is determinate, then $[s]_a^T$ is a square non-singular matrix, whence the member loads can be obtained directly as

$$\{S\} = [s_a^T]^{-1}\{P_a\}. \tag{3.40}$$

If the truss is redundant, then $[s]_a^T$ is rectangular and the member loads must be obtained through the displacements $\{u\}$.

Use Equation (3.7) for all members coming to joint j to get

$$\{S\}_j = [k_S]_j [s]_j \{u_j - u_i\} - \{S_E\}_j, \tag{3.41a}$$
$$\{S\}_j^T = [S_1 \; S_2 \cdots S_K]_j, \quad [s]_j = [s_j^T]^T,$$

$$[k_S]_j = \begin{bmatrix} (EA/L)_1 & 0 & \cdots & 0 \\ 0 & (EA/L)_2 & \cdots & 0 \\ \cdots & \cdots & \cdots & \cdots \\ 0 & 0 & \cdots & (EA/L)_K \end{bmatrix},$$

$$\{u\}_j^T = [u_{xj} \; u_{yj} \; u_{zj}],$$
$$\{u\}_i^T = [u_{xi} \; u_{yi} \; u_{zi}]_k \quad \text{for each member } k,$$

where Equation (3.25) has been used to add the S_{Ek} effects.

Because of repeats in S_k at the adjacent joints, the assembly for the S_k member forces should be made from Equation (3.7) rather than Equation (3.41a). For all members

$$\{S\} = [k_S][s]\{u\} - \{S_E\}, \tag{3.41b}$$

where $[s]$ is the transpose of $[s]^T$ in Equation (3.38b).

Put the $\{S\}$ from Equation (3.41b) into Equation (3.38b) to get

$$\{P\} = [k]\{u\} - \{P_E\}, \tag{3.42}$$

$$[k] = [s]^T[k_S][s], \quad \{P_E\} = [s]^T\{S_E\},$$

where $[k]$ is the *stiffness influence coefficient* matrix for the truss. Each element in $[k]$ is the force in a component load direction m due to a unit component deflection n. For example, in Equation (3.42)

$$P_1 = \sum_n k_{1n} u_n = k_{11}u_1 + k_{12}u_2 + \cdots, \tag{3.43}$$

where $P_1 = k_{12}$ for $u_2 = 1$ and other $u_n = 0$. Note that Equation (3.42) applies for both determinate and redundant trusses, and is the matrix form of Equation (3.8).

To allow for the supports with specified displacements $\{u_R\}$ in Equation (3.42), partition the matrix equation into

$$\begin{Bmatrix} P_a \\ P_R \end{Bmatrix} = \begin{bmatrix} k_{aa} & k_{aR} \\ k_{Ra} & k_{RR} \end{bmatrix} \begin{Bmatrix} u \\ u_R \end{Bmatrix} - \begin{Bmatrix} P_{Ea} \\ P_{ER} \end{Bmatrix}, \tag{3.44}$$

$$[k_{aa}] = [s_a]^T[k_S][s_a].$$

The deflections are

$$\{u\} = [k_{aa}]^{-1}\{P_a + P_{Ea} - P_{ua}\}, \quad \{P_{ua}\} = [k_{aR}]\{u_R\}, \tag{3.45}$$

and the reactions are

$$\{P_R\} = [k_{Ra}]\{u\} + [k_{RR}]\{u_R\} - \{P_{ER}\}. \tag{3.46}$$

Usually it is simpler to get the reactions from the member forces in Equations (3.39), or

$$\{P_R\} = [s]_R^T\{S\}, \tag{3.47}$$

where only the $\{P_R\}$ equations are needed.

Once the deflections $\{u\}$ are obtained from Equation (3.45), the member forces $\{S\}$ are given by either Equation (3.7) or Equation (3.41b). In this case Equation (3.41b) has the form

$$\{S\} = [k_S][s]\begin{Bmatrix} u \\ u_R \end{Bmatrix} - \{S_E\}. \tag{3.48}$$

The $[k]$ matrix also can be constructed element by element. From Equations (3.5) and (3.7)

$$P_{mk} = S_k \cos(k,m), \quad S_{kn} = (EA/L)_k u_n \cos(k,n), \tag{3.49}$$

whence for all members coming to point n,

$$P_m = k_{mn} u_n, \quad k_{mn} = \sum_k (EA/L)_k \cos(m,k) \cos(k,n). \tag{3.50}$$

Here, points m will be at n and at the other end of the members coming to point n. Otherwise, $k_{mn} = 0$ for the particular n.

Another procedure to calculate the $[k]$ matrix is to use Equation (2.71) to calculate the element $[k_k]$ matrix, which can be assembled to give $[k]$. Take Equation (3.7) in the form

$$S_k = [S_k]_1 \{u\}_k,$$

whence the $[\sigma_1]$ matrix in Equation (2.71) is

$$[\sigma_1]_k = [S_k]_1/A_k = (E/L)_k [l \quad m \quad n \quad -l \quad -m \quad -n]_k, \qquad (3.51a)$$

where l, m, n are the direction cosines for member k. The Equation (2.71) becomes

$$[k_k] = \int_0^{L_k} \int_{A_k} \left(\frac{E}{L}\right)_k^2 \begin{Bmatrix} l \\ m \\ n \\ -l \\ -m \\ -n \end{Bmatrix}_k \left(\frac{1}{E}\right)_k [l \quad m \quad n \quad -l \quad -m \quad -n] dA_k\, dy_k$$

$$= \left(\frac{EA}{L}\right)_k \begin{bmatrix} k_0 & -k_0 \\ -k_0 & k_0 \end{bmatrix}_k, \quad [k_0]_k = \begin{bmatrix} l^2 & lm & ln \\ lm & m^2 & mn \\ ln & mn & n^2 \end{bmatrix}_k. \qquad (3.51b)$$

There is overlap in the $[k_k]$ element matrices on assembly with the $[k_0]_k$ matrices adding at each joint for all K members coming to joint j, or

$$[k_0]_j = (L/EA)_{\text{Ref.}} \sum_{k=1}^{K} (EA/L)_k [k_0]_{kj}. \qquad (3.51c)$$

This $[k_0]_j$ matrix is the 3 by 3 joint matrix at joints on the diagonal of the $[k]$ matrix. Note that it can be assembled from Equation (3.50) for m, n at the joint. There is no addition on the blocks off the diagonal. The blocks are 2 by 2 for the truss in a plane.

The methods of calculating $[k]$ by both Equations (3.42) and (3.51) will be used in Example 3.17 below. See References 3-5 for examples and problems for trusses using the displacement method.

Example 3.17 (a). In the redundant truss in Figure 3.13 with $A_1 = A_3 = A_5 = A$, $A_2 = A_4 = 2^{-1/2}A$, construct the $[s]^T$ matrix for all four joints, disregarding the supports, and calculate the $[k]$ matrix from Equation (3.42). (b). Construct the $[k]$ matrix from elements $[k_k]$ in Equation (3.51).

Solution (a) For the given areas and lengths in Figure 3.13 the element stiffness matrix is

$$[k_S] = \frac{EA}{L} \begin{bmatrix} 1 & 0 & 0 & 0 & 0 \\ 0 & 0.5 & 0 & 0 & 0 \\ 0 & 0 & 1 & 0 & 0 \\ 0 & 0 & 0 & 0.5 & 0 \\ 0 & 0 & 0 & 0 & 1 \end{bmatrix}, \qquad (3.52)$$

where the order of the truss elements is according to the numbering of the elements in Figure 3.13. From Equation (3.38) and Figure 3.13 with $b = 2^{-1/2}$ and $b^2 = 0.5$, the equilibrium

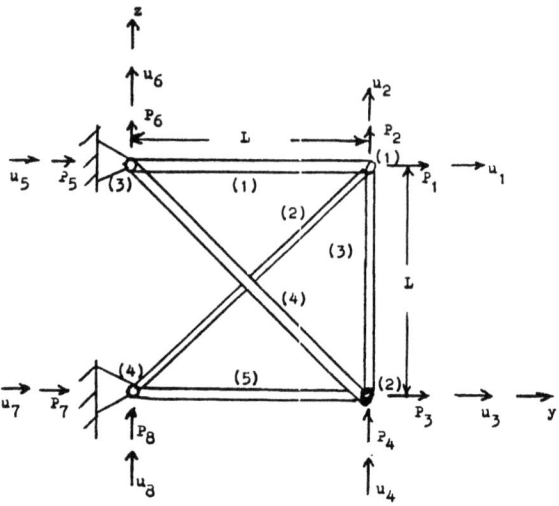

Fig. 3.13. Stiffness coefficients for redundant truss.

equations at the joints give

$$\begin{Bmatrix} P_1 \\ P_2 \\ P_3 \\ P_4 \\ P_5 \\ P_6 \\ P_7 \\ P_8 \end{Bmatrix} = \begin{bmatrix} 1 & b & 0 & 0 & 0 \\ 0 & b & 1 & 0 & 0 \\ 0 & 0 & 0 & b & 1 \\ 0 & 0 & -1 & -b & 0 \\ -1 & 0 & 0 & -b & 0 \\ 0 & 0 & 0 & b & 0 \\ 0 & -b & 0 & 0 & -1 \\ 0 & -b & 0 & 0 & 0 \end{bmatrix} \begin{Bmatrix} S_1 \\ S_2 \\ S_3 \\ S_4 \\ S_5 \end{Bmatrix}, \qquad (3.53)$$

which defines $[s]^T$. Note that all columns add to zero, showing equilibrium check on each member. Multiply the diagonal matrix $[k_S]$ into $[s]$ so that from Equation (3.42)

$$[k] = \frac{EA}{L} \begin{bmatrix} 1 & b & 0 & 0 & 0 \\ 0 & b & 1 & 0 & 0 \\ 0 & 0 & 0 & b & 1 \\ 0 & 0 & -1 & -b & 0 \\ -1 & 0 & 0 & -b & 0 \\ 0 & 0 & 0 & b & 0 \\ 0 & -b & 0 & 0 & -1 \\ 0 & -b & 0 & 0 & 0 \end{bmatrix} \begin{bmatrix} 1 & 0 & 0 & 0 & -1 & 0 & 0 & 0 \\ .5b & .5b & 0 & 0 & 0 & 0 & -.5b & -.5b \\ 0 & 1 & 0 & -1 & 0 & 0 & 0 & 0 \\ 0 & 0 & .5b & -.5b & -.5b & .5b & 0 & 0 \\ 0 & 0 & 1 & 0 & 0 & 0 & -1 & 0 \end{bmatrix}$$

$$= \frac{EA}{4L} \begin{bmatrix} 1 & 2 & 3 & 4 & 5 & 6 & 7 & 8 \\ 5 & 1 & 0 & 0 & -4 & 0 & -1 & -1 \\ 1 & 5 & 0 & -4 & 0 & 0 & -1 & -1 \\ 0 & 0 & 5 & -1 & -1 & 1 & -4 & 0 \\ 0 & -4 & -1 & 5 & 1 & -1 & 0 & 0 \\ -4 & 0 & -1 & 1 & 5 & -1 & 0 & 0 \\ 0 & 0 & 1 & -1 & -1 & 1 & 0 & 0 \\ -1 & -1 & -4 & 0 & 0 & 0 & 5 & 1 \\ -1 & -1 & 0 & 0 & 0 & 0 & 1 & 1 \end{bmatrix}. \qquad (3.54)$$

(b) From Equation (3.51b) and Figure 3.13

$$[k_0]_1 = \begin{bmatrix} 1 & 0 \\ 0 & 0 \end{bmatrix}, \quad [k_0]_2 = \tfrac{1}{2}\begin{bmatrix} 1 & 1 \\ 1 & 1 \end{bmatrix}, \quad [k_0]_3 = \begin{bmatrix} 0 & 0 \\ 0 & 1 \end{bmatrix},$$

The virtual principles for pin-jointed trusses

$$[k_0]_4 = \tfrac{1}{2}\begin{bmatrix} 1 & -1 \\ -1 & 1 \end{bmatrix}, \quad [k_0]_5 = \begin{bmatrix} 1 & 0 \\ 0 & 0 \end{bmatrix}.$$

From Equation (3.51c)

$$[k_0]_{j=1} = [k_0]_1 + \tfrac{1}{2}[k_0]_2 + [k_0]_3 = \tfrac{1}{4}\begin{bmatrix} 5 & 1 \\ 1 & 5 \end{bmatrix},$$

$$[k_0]_{j=2} = [k_0]_3 + \tfrac{1}{2}[k_0]_4 + [k_0]_5 = \tfrac{1}{4}\begin{bmatrix} 5 & -1 \\ -1 & 5 \end{bmatrix},$$

$$[k_0]_{j=3} = [k_0]_1 + \tfrac{1}{2}[k_0]_4 = \tfrac{1}{4}\begin{bmatrix} 5 & -1 \\ -1 & 1 \end{bmatrix},$$

$$[k_0]_{j=4} = \tfrac{1}{2}[k_0]_2 + [k_0]_5 = \tfrac{1}{4}\begin{bmatrix} 5 & 1 \\ 1 & 1 \end{bmatrix}.$$

In Figure 3.13 members (1), (2), (3) come to joint (1), etc., so that with $EA/4L$ as a common factor the matrix $[k]$ is

$$[k] = \frac{EA}{4L}\begin{bmatrix} 4[k_0]_{j=1} & -4[k_0]_3 & -4[k_0]_1 & -2[k_0]_2 \\ -4[k_0]_3 & 4[k_0]_{j=2} & -2[k_0]_4 & -4[k_0]_5 \\ -4[k_0]_1 & -2[k_0]_4 & 4[k_0]_{j=3} & [0] \\ -2[k_0]_2 & -4[k_0]_5 & [0] & 4[k_0]_{j=4} \end{bmatrix}$$

$= [k]$ in Equation (3.54).

It is evident that the two procedures (a) and (b) are quite different. Procedure (a) requires matrix multiplication while procedure (b) requires matrix addition and proper identification of the element matrices.

Example 3.18. In Figure 3.13 and Example 3.17 take $u_5 = u_6 = u_7 = u_8 = 0$ and find the displacements u_1, u_2, u_3, u_4, the forces S_k in the five truss members, and the reactions $P_5 - P_8$.
Solution. From Equations (3.42) and (3.54)

$$\begin{Bmatrix} P_1 \\ P_2 \\ P_3 \\ P_4 \end{Bmatrix} = \frac{EA}{4L}\begin{bmatrix} 5 & 1 & 0 & 0 \\ 1 & 5 & 0 & -4 \\ 0 & 0 & 5 & -1 \\ 0 & -4 & -1 & 5 \end{bmatrix}\begin{Bmatrix} u_1 \\ u_2 \\ u_3 \\ u_4 \end{Bmatrix}, \quad (3.55)$$

which can be inverted to give

$$\begin{Bmatrix} u_1 \\ u_2 \\ u_3 \\ u_4 \end{Bmatrix} = \frac{L}{11EA}\begin{bmatrix} 10 & -6 & -1 & -5 \\ -6 & 30 & 5 & 25 \\ -1 & 5 & 10 & 6 \\ -5 & 25 & 6 & 30 \end{bmatrix}\begin{Bmatrix} P_1 \\ P_2 \\ P_3 \\ P_4 \end{Bmatrix}. \quad (3.56)$$

The S_k member loads can be obtained from Equation (3.56), (3.54), and (3.48), or from Equation (3.7). The reactions are given by Equation (3.47) and Figure 3.13. The results are ($b = 2^{-1/2}$)

$$\begin{Bmatrix} S_1 \\ S_2 \\ S_3 \\ S_4 \\ S_5 \\ -- \\ P_5 \\ P_6 \\ P_7 \\ P_8 \end{Bmatrix} = \frac{1}{11}\begin{bmatrix} 10 & -6 & -1 & -5 \\ 2b & 12b & 2b & 10b \\ -1 & 5 & -1 & -5 \\ 2b & -10b & 2b & -12b \\ -1 & 5 & 10 & 6 \\ -- & -- & -- & -- \\ -11 & 11 & 0 & 11 \\ 1 & -5 & 1 & -6 \\ 0 & -11 & -11 & -11 \\ -1 & -6 & -1 & -5 \end{bmatrix}\begin{Bmatrix} P_1 \\ P_2 \\ P_3 \\ P_4 \end{Bmatrix}. \quad (3.57)$$

Example 3.19. Suppose there is a vertical support at joint (2) in Figure 3.13 so that $u_4 = 0$. Find the displacements and the member loads.
Solution. Delete the last row and last column in Equation (3.55) so that

$$\begin{Bmatrix} P_1 \\ P_2 \\ P_3 \end{Bmatrix} = \frac{EA}{4L}\begin{bmatrix} 5 & 1 & 0 \\ 1 & 5 & 0 \\ 0 & 0 & 5 \end{bmatrix}\begin{Bmatrix} u_1 \\ u_2 \\ u_3 \end{Bmatrix}. \quad (3.58)$$

Inversion gives

$$\left\{\begin{array}{c}u_1\\u_2\\u_3\end{array}\right\} = \frac{L}{6EA}\begin{bmatrix}5 & -1 & 0\\-1 & 5 & 0\\0 & 0 & 24/5\end{bmatrix}\left\{\begin{array}{c}P_1\\P_2\\P_3\end{array}\right\}. \tag{3.59}$$

From Equation (3.7) and Equation (3.59)

$$S_1 = \frac{EA}{L}u_1 = \tfrac{1}{6}(5P_1 - P_2), \qquad S_2 = \frac{EAb}{2L}(u_1 + u_2) = \tfrac{b}{3}(P_1 + P_2),$$
$$S_3 = \frac{EA}{L}u_2 = \tfrac{1}{6}(-P_1 + 5P_2), \qquad S_4 = \frac{EAb}{2L}u_3 = \tfrac{2b}{5}P_3, \tag{3.60}$$
$$S_5 = \frac{EA}{L}u_3 = \tfrac{4}{5}P_3, \qquad b = 2^{-1/2}.$$

Example 3.20 (a) Assume all five truss members in Figure 3.13 and Example 3.18 have a known strain e_{Ek} due to temperature change, inelastic effects, initial strains, and large deflections. Assume u_{R5} in Equation (3.44) and Figure 3.13 is not zero. Add these effects to the results in Example 3.18. (b) Discuss the results, including the procedure for inelastic effects.

Solution (a) For member k, $S_{Ek} = E_k A_k e_{Ek}$, and from Equation (3.42)

$$\{P_E\} = [s]^T\{S_E\}.$$

From the $[s]^T$ matrix in Equation (3.53)

$$P_{E1} = S_{E1} + bS_{E2}, \qquad P_{E2} = bS_{E2} + S_{E3}, \qquad P_{E3} = bS_{E4} + S_{E5},$$
$$P_{E4} = -S_{E3} - bS_{E4}.$$

From Equations (3.45) and (3.54)

$$P_{ua1} = -\frac{EA}{L}u_{R5}, \qquad P_{ua2} = 0, \qquad P_{ua3} = -\frac{EA}{4L}u_{R5}, \qquad P_{ua4} = \frac{EA}{4L}u_{R5}.$$

Thus, in Equations (3.55) and (3.56) replace the four P_i terms by $P_i + P_{Ei} - P_{uai}$, whence the additional deflections are

$$u_{1a} = u_{R5} + \frac{L}{EA}(S_{E1} - \tfrac{1}{11}S^E),$$
$$u_{2a} = -u_{R5} + \frac{L}{EA}(-S_{E1} + 4bS_{E2} + \tfrac{5}{11}S^E),$$
$$u_{3a} = \frac{L}{EA}(S_{E5} - \tfrac{1}{11}S^E), \tag{3.61}$$
$$u_{4a} = -u_{R5} + \frac{L}{EA}(S_{E5} - 4bS_{E4} - \tfrac{5}{11}S^E),$$

$$S^E = S_{E1} - 4bS_{E2} + S_{E3} - 4bS_{E4} + S_{E5}, \qquad b = 2^{-1/2}. \tag{3.62}$$

From Equations (3.25), (3.56), (3.61), and from Figure 3.13,

$$S_3 = \frac{EA}{L}(u_2 - u_4) - S_{E3} = \tfrac{1}{11}(-P_1 + 5P_2 - P_3 - 5P_4 - S^E). \tag{3.63}$$

With S_3 known, the other member forces and the reactions can be obtained from the determinate equilibrium equations, or

$$S_1 = S_3 + P_1 - P_2, \qquad S_2 = -2b(S_3 - P_2), \qquad S_4 = -2b(S_3 + P_4),$$
$$S_5 = S_3 + P_3 + P_4, \qquad P_{5R} = -P_1 + P_2 + P_4, \qquad P_{6R} = -S_3 - P_4,$$
$$P_{7R} = -P_2 - P_3 - P_4, \qquad P_{8R} = S_3 - P_2, \tag{3.64}$$
$$S_1^E = S_3^E = S_5^E = -\tfrac{1}{11}S^E, \qquad S_2^E = S_4^E = \tfrac{2b}{11}S^E,$$
$$P_{5R}^E = P_{7R}^E = 0, \qquad P_{6R}^E = -P_{8R}^E = \tfrac{1}{11}S^E.$$

(b) In this example, the support displacement u_{R5} affects the displacements but does not affect the member loads. This is to be expected since a small u_{R5} produces a rigid body rotation about support (4). Each of the additional strain effects e_{Ek} produces a self-equilibrating load redistribution in the redundant truss so that the S_{Ek} loads combine into the S^E expression in Equation (3.62). Because of interactions among the S_{Ek} in S^E, it is possible for S^E to be positive, zero, or negative. Suppose all members are subjected to the same increase in temperature to produce $S_{Ek} = S_{Tk} = E_k A_k e_T$, then $S^E = -EAe_T$ gives tension thermal stresses in members (1), (3), (5), and compression thermal stresses in the diagonal members (2), (4). Suppose only member (5) is inelastic and that the inelastic load S_{p5} is given by Equation (3.27), then

$$S_{p5} = (3/7)S_{y5}(S_5/S_{y5})^{10}. \tag{3.65}$$

The unknown S_5 and S_{p5} can be obtained by trial for any specific case, and S_{p5} added to S^E in Equations (3.62)-(3.64). If member (5) is inelastic in tension, then from Equation (3.64) the loads in members (1), (3), (5) will be reduced while the loads in members (2), (4) will increase in tension or decrease in compression. The deflections in Equations (3.61) will increase, except for u_{1a}. If member (1) becomes inelastic in compression due to the new S^E, then it will be necessary to iterate between members (1) and (5).

3.6. Matrix equations for trusses from the unit load theorem

The Equation (3.15) can be written for all truss members and assembled into the matrix equation

$$\{S\} = [p]\{P\} + [q]\{Q\}, \tag{3.66}$$

where $[p]$ and $[q]$ are arrays of forces in the truss members due to unit values of the loads $\{P\}$ and $\{Q\}$, respectively.

The flexibility $[C]$ matrix in Equation (2.76) can be obtained from Equations (3.66) and (2.78). From Equation (3.66)

$$e_k = (S/EA)_k = (1/EA)_k [p \quad q]_k \begin{Bmatrix} P \\ Q \end{Bmatrix} = [e_1]_k \begin{Bmatrix} P \\ Q \end{Bmatrix},$$
$$[e_1]_k = (1/EA)_k [p \quad q]_k. \tag{3.67}$$

From Equation (2.78)

$$[C]_k = \int_0^{L_k} \int_{A_k} \begin{Bmatrix} p \\ q \end{Bmatrix}_k E_k (1/EA)_k^2 [p \quad q]_k \, dA_k \, dy_k$$
$$= \begin{Bmatrix} p \\ q \end{Bmatrix}_k (L/EA)_k [p \quad q]_k, \tag{3.68}$$

where $[C]_k$ is not an element matrix but is the contribution of element (k) to $[C]$. The $[C]_k$ matrix is the same size as $[C]$. The assembly can be made by stacking the rows of $[e_1]_k$ for all elements, whence

$$[C] = \begin{bmatrix} p^T \\ q^T \end{bmatrix} [C_F][p \quad q] = \begin{bmatrix} C_{PP} & C_{PQ} \\ C_{QP} & C_{QQ} \end{bmatrix},$$
$$[C_{PP}] = [p]^T [C_F][p], \quad [C_{PQ}] = [p]^T [C_F][q], \tag{3.69}$$
$$[C_{QP}] = [q]^T [C_F][p], \quad [C_{QQ}] = [q]^T [C_F][q].$$

The $[C_F]$ matrix is a diagonal matrix with elements $(L/EA)_k$.

Partition Equation (2.76) to match the $[C]$ matrix in Equation (3.69),

$$\begin{Bmatrix} u \\ u_Q \end{Bmatrix} = \begin{bmatrix} C_{PP} & C_{PQ} \\ C_{QP} & C_{QQ} \end{bmatrix} \begin{Bmatrix} P \\ Q \end{Bmatrix} + \begin{Bmatrix} u_{EP} \\ u_{EQ} \end{Bmatrix}, \tag{3.70}$$

$$\{u_{EP}\} = [p]^T[C_F]\{S_E\}, \quad \{u_{EQ}\} = [q]^T[C_F]\{S_E\}, \tag{3.71}$$

where $S_{Ek} = (EAe_E)_k$ and $\{u_Q\} = 0$ for truss members with loads in $\{Q\}$. Solve for $\{Q\}$ to get

$$\{Q\} = -[C_{QQ}]^{-1}[C_{QP}]\{P\} + [C_{QQ}]^{-1}\{u_Q - u_{EQ}\}, \tag{3.72}$$

which can be put into Equation (3.70) to get

$$\{u\} = [C]\{P\} + \{u_E\}, \quad [C] = [C_{PP}] - [C_{QP}]^T[C_{QQ}]^{-1}[C_{QP}], \\ \{u_E\} = \{u_{EP}\} + [C_{QP}]^T[C_{QQ}]^{-1}\{u_Q - u_{EQ}\}. \tag{3.73}$$

Also, put $\{Q\}$ into Equation (3.66) to get $\{S\}$, which can be expressed as

$$\{S\} = [p_P]\{P\} - \{S_{EQ}\}, \tag{3.73a}$$
$$[p_P] = [p] - [q][C_{QQ}]^{-1}[C_{QP}], \\ \{S_{EQ}\} = [q][C_{QQ}]^{-1}\{u_{EQ} - u_Q\}.$$

The displacements $\{u\}$ can be expressed in terms of $\{S\}$ by assembling all the component deflections in Equations (3.14) and (3.26) into the matrix form

$$\{u\} = [p]^T[C_F]\{S + S_E\} = [p]^T[C_F]\{S\} + \{u_{EP}\}. \tag{3.74}$$

In the above equations it is evident that the $[p]$ and $[q]$ matrices are the important matrices that must be determined before the member loads and joint deflections can be calculated. In Section 3.3 $[p]$ and $[q]$ were obtained by replacing redundant supports with applied loads and by cutting redundant members to produce a determinate truss. For large trusses this method is complicated not only for the calculations but also for the selection of the proper redundants to get a determinate truss.

It is possible to avoid some of these complications by getting $[p]$ and $[q]$ from the equilibrium Equations (3.38b), which are simple to calculate and assemble for any truss. Partition Equation (3.38b) as follows

$$\begin{Bmatrix} P \\ R \end{Bmatrix} = [s]^T\{S\} = \begin{bmatrix} [s]_P^T \\ [s]_R^T \end{bmatrix} \{S\}, \tag{3.75}$$

$$\{R\} = [s]_R^T\{S\}, \tag{3.76}$$
$$\{P\} = [s]_P^T\{S\}. \tag{3.77}$$

If Equation (3.77) has M equations and $M+N$ truss members, then it is possible to express the forces in almost any M members $\{S\}_0$ in terms of $\{P\}$ and the remaining N members with loads $\{Q\}$. Partition Equation (3.77) in the form

$$\{P\} = [[s]_{P0}^T \quad [s]_{PQ}^T] \begin{Bmatrix} S_0 \\ Q \end{Bmatrix}, \tag{3.78}$$

where $[s]_{P0}^T$ is a square matrix M by M, and arrange the equilibrium P_i rows and the S_j columns so that $[s]_{P0}^T$ has no rows with all zeros. This can be accomplished

The virtual principles for pin-jointed trusses 89

by arranging the columns so that there are no zero elements on the main diagonal in $[s]_{P0}^T$.

After $[s]_{P0}^T$ is arranged so that it is non-singular, solve Equation (3.78) for $\{S_0\}$ to get

$$\{S_0\} = [[s]_{P0}^T]^{-1}\{\{P\} - [s]_{PQ}^T\{Q\}\}, \tag{3.79}$$

$$\{S\} = \begin{Bmatrix} S_0 \\ Q \end{Bmatrix} = \begin{bmatrix} [[s]_{P0}^T]^{-1} & -[[s]_{P0}^T]^{-1}[s]_{PQ}^T \\ [0] & [I] \end{bmatrix} \begin{Bmatrix} P \\ Q \end{Bmatrix}. \tag{3.80}$$

Compare this $\{S\}$ to Equation (3.66) to get

$$[p] = \begin{bmatrix} [[s]_{P0}^T]^{-1} \\ [0] \end{bmatrix}, \quad [q] = \begin{bmatrix} -[[s]_{P0}^T]^{-1}[s]_{PQ}^T \\ [I] \end{bmatrix}. \tag{3.81}$$

Note that the N truss members in $\{Q\}$ represent the difference between the number M of equilibrium equations at joints with free deflections and the number $M + N$ of actual truss members. In the classical procedure of Section 3.3 this N would represent the total of redundant supports and redundant members.

If the $[p]$ and $[q]$ matrices are obtained from Equation (3.81), it is necessary to invert $[s]_{P0}^T$. However, this matrix has many zeros so that a computer can invert it for large structures. In many cases a sequence solution can be made directly. Also, the inversion can be made by the Jordan successive transformation procedure, Section A.7 and Reference 3. This latter procedure can also be used to select the columns that go into $[s]_{P0}^T$ by interchanging the columns to put the largest row element on the diagonal.

Modify the Jordan procedure in Equation (A.57) to apply to Equations (3.77) - (3.79) in the form

$$[T_M][T_{M-1}]\cdots[T_1][[s]_P^T \quad [I]\{P\}]$$
$$= [[I] \quad [[s]_{P0}^T]^{-1}[s]_{PQ}^T\{Q\} \quad [[s]_{P0}^T]^{-1}\{P\}]. \tag{3.82}$$

Thus, the procedure not only identifies the $[s]_{P0}^T$ and $[s]_{PQ}^T$ matrices but also gives the matrices needed for $[p]$ and $[q]$ in Equation (3.81). The procedure can also use a selected partitioning by interchanging columns only in a specified $[s]_{P0}^T$. See Examples below.

Since truss member forces are used in $\{Q\}$ in Equations (3.78)-(3.80), the $\{u_Q\}$ in Equation (3.70) will be zero. If one of the redundant truss supports has a non-zero deflection, then it is necessary to remove the corresponding equilibrium equation from Equation (3.76) and add it to Equation (3.78), plus an additional column in Equation (3.78) for Q_R unknown reactions. The $\{u_Q\}$ will then include the specified support deflection.

See References 3, 4, 5 for examples and problems using the unit load method of analysis for trusses.

Example 3.21. Find the member loads $S_1 - S_4$ and the deflections $u_1 - u_4$ in Figure 3.14 in terms of the applied loads $P_1 - P_4$. Take $E_k A_k = EA$.

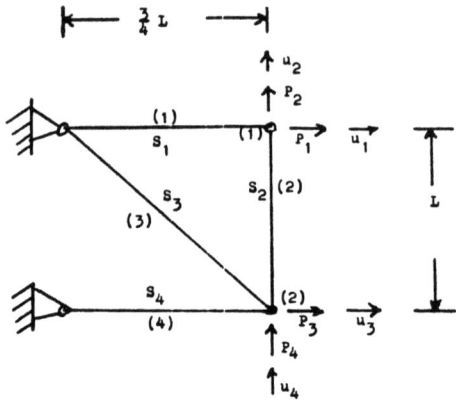

Fig. 3.14. Simple determinate truss.

Solution. The equilibrium equations in Figure 3.14 are

$$P_1 = S_1, \quad P_2 = S_2, \quad P_3 = S_4 + \tfrac{3}{5}S_3, \quad P_4 = -S_2 - \tfrac{4}{5}S_3,$$

which give

$$S_1 = P_1, \quad S_2 = P_2, \quad S_3 = -\tfrac{5}{4}(P_2 + P_4), \quad S_4 = P_3 + \tfrac{3}{4}(P_2 + P_4),$$
$$\{S\} = [p]\{P\},$$

where

$$[p] = (1/4) \begin{bmatrix} 4 & 0 & 0 & 0 \\ 0 & 4 & 0 & 0 \\ 0 & -5 & 0 & -5 \\ 0 & 3 & 4 & 3 \end{bmatrix}. \tag{3.83}$$

From Equation (3.74)

$$\{u\} = [p]^T[C_F]\{S\} = [p]^T[C_F][p]\{P\} = [C_{PP}]\{P\}.$$

The diagonal elements of $[C_F]$ are $(L/4EA)(3\ \ 4\ \ 5\ \ 3)$, whence

$$\begin{Bmatrix} u_1 \\ u_2 \\ u_3 \\ u_4 \end{Bmatrix} = \frac{L}{64EA} \begin{bmatrix} 4 & 0 & 0 & 0 \\ 0 & 4 & -5 & 3 \\ 0 & 0 & 0 & 4 \\ 0 & 0 & -5 & 3 \end{bmatrix} \begin{bmatrix} 3 & 0 & 0 & 0 \\ 0 & 4 & 0 & 0 \\ 0 & 0 & 5 & 0 \\ 0 & 0 & 0 & 3 \end{bmatrix} \begin{bmatrix} 4 & 0 & 0 & 0 \\ 0 & 4 & 0 & 0 \\ 0 & -5 & 0 & -5 \\ 0 & 3 & 4 & 3 \end{bmatrix} \begin{Bmatrix} P_1 \\ P_2 \\ P_3 \\ P_4 \end{Bmatrix}$$

$$= \frac{L}{16EA} \begin{bmatrix} 12 & 0 & 0 & 0 \\ 0 & 54 & 9 & 38 \\ 0 & 9 & 12 & 9 \\ 0 & 38 & 9 & 38 \end{bmatrix} \begin{Bmatrix} P_1 \\ P_2 \\ P_3 \\ P_4 \end{Bmatrix}. \tag{3.84}$$

Note that if P_2 is the only applied load, then $[p]$ reduces to the second column, while Equation (3.84) reduces to one element, or $u_2 = 54LP_2/16EA$. If $P_1 = P_3 = P_4 = 0$ in Equation (3.84), then $u_1 - u_4$ due to P_2 are obtained.

Example 3.22. Solve for the member loads S_k, deflections u_i, and reactions R_j for the truss in Figure 3.15. Use $E_k A_k = EA$ and include the e_{Ek} strains for temperature and inelastic effects.

Solution. The four equilibrium equations are

$$P_1 = S_1 + \tfrac{3}{5}S_5, \quad P_2 = S_2 + \tfrac{4}{5}S_5, \quad P_3 = S_4 + \tfrac{3}{5}S_3, \quad P_4 = -S_2 - \tfrac{4}{5}S_3.$$

The virtual principles for pin-jointed trusses 91

Solve for the S_k in terms of P_i and S_2, or

$$S_3 = -\tfrac{5}{4}(P_4 + S_2), \quad S_5 = \tfrac{5}{4}(P_2 - S_2), \quad S_4 = P_3 + \tfrac{3}{4}(P_4 + S_2),$$
$$S_1 = P_1 - \tfrac{3}{4}(P_2 - S_2), \quad S_2 = Q_2,$$

whence Equation (3.66) becomes

$$\begin{Bmatrix} S_1 \\ S_2 \\ S_3 \\ S_4 \\ S_5 \end{Bmatrix} = \tfrac{1}{4}\begin{bmatrix} 4 & -3 & 0 & 0 \\ 0 & 0 & 0 & 0 \\ 0 & 0 & 0 & -5 \\ 0 & 0 & 4 & 3 \\ 0 & 5 & 0 & 0 \end{bmatrix}\begin{Bmatrix} P_1 \\ P_2 \\ P_3 \\ P_4 \end{Bmatrix} + \tfrac{Q_2}{4}\begin{Bmatrix} 3 \\ 4 \\ -5 \\ 3 \\ -5 \end{Bmatrix}.$$

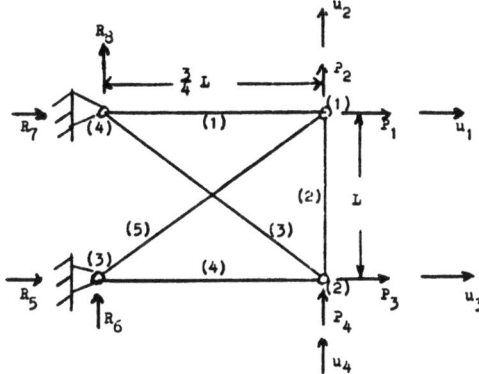

Fig. 3.15. Redundant truss for Examples 3.22 and 3.25.

The diagonal elements of the flexibility $[C_F]$ matrix for the five truss members are $(L/4EA)(3\ 4\ 5\ 3\ 5)$. From Equations (3.69)-(3.73),

$$C_{QQ} = [q]^T[C_F][q] = \frac{L}{64EA}(3^3 + 4^3 + 5^3 + 3^3 + 5^3) = \frac{23L}{4EA},$$

$$[C_{QP}] = [q]^T[C_F][p] = \frac{L}{16EA}[9\ \ -38\ \ 9\ \ 38],$$

$$u_{EQ} = [q]^T[C_F]\{S_E\} = LS^E/16EA,$$

$$S^E = 9S_{E1} + 16S_{E2} - 25S_{E3} + 9S_{E4} - 25S_{E5}, \tag{3.85}$$

$$\{u_{EP}\} = [p]^T[C_F]\{S_E\} = \frac{L}{16EA}\begin{Bmatrix} 12S_{E1} \\ -9S_{E1} + 25S_{E5} \\ 12S_{E4} \\ -25S_{E3} + 9S_{E4} \end{Bmatrix},$$

$$Q_2 = S_2 = -[C_{QQ}]^{-1}\{[C_{QP}]\{P\} + \{u_{EQ}\}\}$$
$$= \tfrac{1}{92}(-9P_1 + 38P_2 - 9P_3 - 38P_4 - S^E).$$

With $Q_2 = S_2$ known the forces in the members become

$$\begin{Bmatrix} S_1 \\ S_2 \\ S_3 \\ S_4 \\ S_5 \end{Bmatrix} = \tfrac{1}{368}\begin{bmatrix} 341 & -162 & -27 & -114 & -3 \\ -36 & 152 & -36 & -152 & -4 \\ 45 & -190 & 45 & -270 & 5 \\ -27 & 114 & 341 & 162 & -3 \\ 45 & 270 & 45 & 190 & 5 \end{bmatrix}\begin{Bmatrix} P_1 \\ P_2 \\ P_3 \\ P_4 \\ S^E \end{Bmatrix}. \tag{3.86}$$

Use Equation (3.74) for the deflections, or

$$\begin{Bmatrix} u_1 \\ u_2 \\ u_3 \\ u_4 \end{Bmatrix} = \frac{L}{1472EA} \begin{bmatrix} 1023 & -486 & -81 & -342 & -9 \\ -486 & 2052 & 342 & 1444 & 38 \\ -81 & 342 & 1023 & 486 & -9 \\ -342 & 1444 & 486 & 2052 & -38 \end{bmatrix} \begin{Bmatrix} P_1 \\ P_2 \\ P_3 \\ P_4 \\ S^E \end{Bmatrix} + \{u_{EP}\}, \quad (3.87)$$

where $\{u_{EP}\}$ is given above. Use Figure 3.15 and Equation (3.86) to get the reactions as

$$R_5 = -S_4 - \tfrac{3}{5}S_5, \quad R_6 = -\tfrac{4}{5}S_5, \quad R_7 = -S_1 - \tfrac{3}{5}S_3, \quad R_8 = \tfrac{4}{5}S_3,$$

$$\begin{Bmatrix} R_5 \\ R_6 \\ R_7 \\ R_8 \end{Bmatrix} = \frac{1}{92} \begin{bmatrix} 0 & -69 & -92 & -69 & 0 \\ -9 & -54 & -9 & -38 & -1 \\ -92 & 69 & 0 & 69 & 0 \\ 9 & -38 & 9 & -54 & 1 \end{bmatrix} \begin{Bmatrix} P_1 \\ P_2 \\ P_3 \\ P_4 \\ S^E \end{Bmatrix}. \qquad (3.88)$$

Example 3.23. Find the loads S_k and the deflections u_3 and u_5 in the truss in Figure 3.16. Assume that member (3) has a temperature change giving $e_{E3} = e_{T3}$, and member (4) is inelastic with $e_{E4} = e_{p4}$. Take $E_k A_k = EA$.

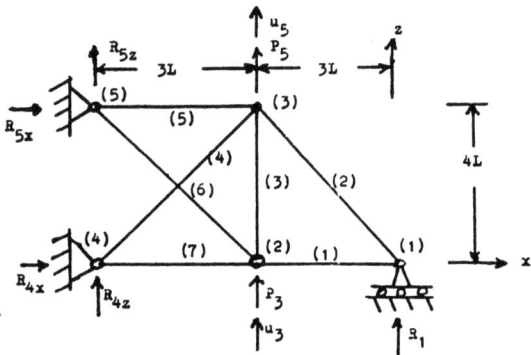

Fig. 3.16. Redundant truss for Examples 3.23, 3.24, 3.25

Solution. Use the redundants $4Q_1 = R_1$ at support (1) and $5Q_4 = S_4$ in member (4). Construct the $[p]$ and $[q]$ matrices by using the unit loads as $P_3 = 4$, $P_5 = 4$, $Q_1 = 4$, $Q_4 = 5$ in Figure 3.16, whence

$$\begin{aligned}
[p]^T &= \begin{bmatrix} 0 & 0 & 0 & 0 & 0 & -5 & 3 \\ 0 & 0 & 4 & 0 & 0 & -5 & 3 \end{bmatrix} \begin{matrix} P_3 \\ P_5 \end{matrix}, \\
[q]^T &= \begin{bmatrix} 3 & -5 & 4 & 0 & -3 & -5 & 6 \\ 0 & 0 & -4 & 5 & -3 & 5 & -3 \end{bmatrix} \begin{matrix} Q_1 \\ Q_4 \end{matrix}.
\end{aligned} \qquad (3.89)$$

The elements on the diagonal of the $[C_F]$ matrix are $(L/EA)(3\ 5\ 4\ 5\ 3\ 5\ 3)$. From Equations (3.69)–(3.73)

$$[C_{QQ}] = [q]^T[C_F][q] = \frac{4L}{EA} \begin{bmatrix} 119 & -54 \\ -54 & 92 \end{bmatrix},$$

$$[C_{QP}] = [q]^T[C_F][p] = \frac{L}{EA} \begin{bmatrix} 179 & 243 \\ -152 & -216 \end{bmatrix},$$

$$\{S_E\}^T = [0\ 0\ S_{T3}\ S_{p4}\ 0\ 0\ 0], \quad S_{T3} = EAe_{T3}, \quad S_{p4} = EAe_{p4},$$

$$\begin{Bmatrix} Q_1 \\ Q_4 \end{Bmatrix} = -[C_{QQ}]^{-1}[C_{QP}] \begin{Bmatrix} P_3/4 \\ P_5/4 \end{Bmatrix} - [C_{QQ}]^{-1}[q]^T[C_F]\{S_E\}$$

The virtual principles for pin-jointed trusses

$$= \begin{bmatrix} -0.0643 & -0.0832 & -0.0189 & -0.0420 \\ 0.0655 & 0.0979 & 0.0324 & -0.0926 \end{bmatrix} \begin{Bmatrix} P_3 \\ P_5 \\ S_{T3} \\ S_{p4} \end{Bmatrix}. \qquad (3.90)$$

From Equations (3.66), (3.89), and (3.90),

$$\begin{Bmatrix} S_1 \\ S_2 \\ S_3 \\ S_4 \\ S_5 \\ S_6 \\ S_7 \end{Bmatrix} = \begin{bmatrix} -0.1929 & -0.2496 & -0.0567 & -0.1260 \\ 0.3215 & 0.4160 & 0.0945 & 0.2100 \\ -0.5192 & 0.2756 & -0.2052 & 0.2024 \\ 0.3277 & 0.4895 & 0.1620 & -0.4630 \\ -0.0036 & -0.0441 & -0.0405 & 0.4038 \\ -0.6010 & -0.3445 & 0.2565 & -0.2530 \\ 0.1677 & -0.0429 & -0.2106 & 0.0258 \end{bmatrix} \begin{Bmatrix} P_3 \\ P_5 \\ S_{T3} \\ S_{p4} \end{Bmatrix}. \qquad (3.91)$$

From Equation (3.74)

$$\begin{Bmatrix} u_3 \\ u_5 \end{Bmatrix} = \frac{L}{EA} \begin{bmatrix} 3.9755 & 2.0571 & -2.0770 & 1.3268 \\ 1.8506 & 3.1595 & 1.1022 & 2.1436 \end{bmatrix} \begin{Bmatrix} P_3 \\ P_4 \\ S_{T3} \\ S_{p4} \end{Bmatrix}. \qquad (3.92)$$

In Equations (3.27) and (3.89) take S_{p4} as

$$S_{p4} = (3/7)P_{y4}(5Q_4/P_{y4})^{10}, \qquad (3.93)$$

whence Equation (3.90) gives

$$Q_4 + 0.0926(3/7)P_{y4}(5Q_4/P_{y4})^{10} = 0.0655P_3 + 0.0979P_5 + 0.0324S_{T3}. \qquad (3.94)$$

For any given values for P_{y4} yield load, P_3, P_5, and S_{T3}, this Equation (3.94) can be solved for Q_4 by trial. This gives the value for S_{p4} in Equation (3.93), which can be used directly in Equations (3.91) and (3.92).

Note that the deflections can be calculated directly by Equation (3.14) for selected determinate subtrusses. In this case, calculate u_5 from members (4) and (5), or

$$u_5 = \frac{L}{EA}[(-\tfrac{3}{4})(3)S_5 + \tfrac{5}{4}(5)S_4] = \frac{L}{EA}(1.85P_3 + 3.16P_5 + 1.10S_{T3} + 2.14S_{p4}), \qquad (3.95)$$

which checks the result for u_5 in Equation (3.92).

Example 3.24. (a). Solve Example 3.23 by using the equilibrium equations. (b). Comment on the two procedures.

Solution (a) From Figure 3.16 the equilibrium equations are

$$P_{1x} = 0 = S_1 + \tfrac{3}{5}S_2, \qquad P_{2z} = 0 = -S_1 + S_7 + \tfrac{3}{5}S_6,$$
$$P_{2x} = P_3 = -S_3 - \tfrac{4}{5}S_6, \qquad P_{3z} = 0 = S_5 - \tfrac{3}{5}(S_2 - S_4), \qquad (3.96)$$
$$P_{3x} = P_5 = S_3 + \tfrac{4}{5}(S_2 + S_4).$$

Express the S_i in terms of S_4, S_6, P_3, and P_5, or

$$S_4 = Q_4, \quad S_6 = Q_6, \quad S_3 = -P_3 - \tfrac{4}{5}Q_6, \quad S_2 = \tfrac{5}{4}(P_3 + P_5) + Q_6 - Q_4,$$
$$S_1 = -\tfrac{3}{4}(P_3 + P_5) - \tfrac{3}{5}(Q_6 - Q_4), \quad S_7 = -\tfrac{3}{4}(P_3 + P_5) - \tfrac{3}{5}(2Q_6 - Q_4),$$
$$S_5 = \tfrac{3}{4}(P_3 + P_5) + \tfrac{3}{5}(Q_6 - 2Q_4).$$

Thus, the [p] and [q] matrices in Equation (3.66) are

$$[p]^T = \tfrac{1}{4}\begin{bmatrix} -3 & 5 & -4 & 0 & 3 & 0 & -3 \\ -3 & 5 & 0 & 0 & 3 & 0 & -3 \end{bmatrix} \begin{matrix} P_3 \\ P_5 \end{matrix},$$
$$[q]^T = \tfrac{1}{5}\begin{bmatrix} 3 & -5 & 0 & 5 & -6 & 0 & 3 \\ -3 & 5 & -4 & 0 & 3 & 5 & -6 \end{bmatrix} \begin{matrix} Q_4 \\ Q_6 \end{matrix}. \qquad (3.97)$$

Follow the steps in Example 3.23 using these matrices to get

$$[C_{QQ}] = \frac{4L}{25EA}\begin{bmatrix} 103 & -65 \\ -65 & 119 \end{bmatrix}, \quad [C_{QP}] = \frac{L}{20EA}\begin{bmatrix} -233 & -233 \\ 297 & 233 \end{bmatrix},$$

$$\begin{Bmatrix} Q_4 \\ Q_6 \end{Bmatrix} = \begin{bmatrix} 0.3277 & 0.4895 & 0.1620 & -0.4630 \\ -0.6010 & -0.3445 & 0.2565 & -0.2530 \end{bmatrix} \begin{Bmatrix} P_3 \\ P_5 \\ S_{T3} \\ S_{p4} \end{Bmatrix}, \quad (3.98)$$

which checks the S_4 and S_6 rows in Equation (3.91) in Example 3.23. Put the above $[p]$ and $[q]$ into Equation (3.66) to verify the other member loads in Equation (3.91).

From Equations (3.76) and (3.91), and Figure 3.16, the reactions are

$$R_1 = -\tfrac{4}{5}S_2, \quad R_{4z} = -S_7 - \tfrac{3}{5}S_4, \quad R_{4x} = -\tfrac{4}{5}S_4,$$
$$R_{5z} = -S_5 - \tfrac{3}{5}S_6, \quad R_{5x} = \tfrac{4}{5}S_6,$$

$$\begin{Bmatrix} R_1 \\ R_{4z} \\ R_{4x} \\ R_{5z} \\ R_{5x} \end{Bmatrix} = \begin{bmatrix} -0.2572 & -0.3328 & -0.0756 & -0.1680 \\ -0.3643 & -0.2508 & 0.1134 & 0.2580 \\ -0.2622 & -0.3916 & -0.1296 & 0.3704 \\ 0.3642 & 0.2508 & -0.1134 & -0.2580 \\ -0.4808 & -0.2756 & 0.2052 & -0.2024 \end{bmatrix} \begin{Bmatrix} P_3 \\ P_5 \\ S_{T3} \\ S_{p4} \end{Bmatrix}.$$

(b) It appears that any of the twenty one possible combinations of two S_i loads in Equation (3.96) can be selected, except S_1, S_2 and S_3, S_6, to determine the $[p]$ and $[q]$ matrices. Also, a sequence solution can be made for many of the combinations. This procedure is simpler than the unit load procedure in Example 3.23.

Example 3.25. Use the Jordan successive transformation procedure to find $[p]$ and $[q]$ matrices in Example 3.22.

Solution. From the equilibrium equations in Example 3.22 and Equation (3.82),

$$[[s]_P^T \ [I]\{P\}] = \begin{bmatrix} \begin{array}{cccc|cccc} S_1 & S_2 & S_3 & S_4 & S_5 & P_1 & P_2 & P_3 & P_4 \\ 1 & 0 & 0 & 0 & \tfrac{3}{5} & 1 & 0 & 0 & 0 \\ 0 & 1 & 0 & 0 & \tfrac{4}{5} & 0 & 1 & 0 & 0 \\ 0 & 0 & \tfrac{3}{5} & 1 & 0 & 0 & 0 & 1 & 0 \\ 0 & -1 & -\tfrac{4}{5} & 0 & 0 & 0 & 0 & 0 & 1 \end{array} \end{bmatrix}.$$

From the Jordan procedure in Section A.7, take $[T_1] = I$ and

$$[T_2] = \begin{bmatrix} 1 & 0 & 0 & 0 \\ 0 & 1 & 0 & 0 \\ 0 & 0 & 1 & 0 \\ 0 & 1 & 0 & 1 \end{bmatrix}, \text{ whence}$$

$$[T_2][T_1][[s]_P^T \ I\{P\}] = \begin{bmatrix} \begin{array}{cccc|cccc} 1 & 0 & 0 & 0 & \tfrac{3}{5} & 1 & 0 & 0 & 0 \\ 0 & 1 & 0 & 0 & \tfrac{4}{5} & 0 & 1 & 0 & 0 \\ 0 & 0 & \tfrac{3}{5} & 1 & 0 & 0 & 0 & 1 & 0 \\ 0 & 0 & -\tfrac{4}{5} & 0 & \tfrac{4}{5} & 0 & 1 & 0 & 1 \end{array} \end{bmatrix}.$$

Interchange columns 3 and 4 to put one on the diagonal. This makes $[T_3] = I$ and

$$[T_4] = \begin{bmatrix} 1 & 0 & 0 & 0 \\ 0 & 1 & 0 & 0 \\ 0 & 0 & 1 & 3/4 \\ 0 & 0 & 0 & -5/4 \end{bmatrix}, \text{ whence}$$

$$[T_4][T_2][[s]_P^T \ I\{P\}] = \begin{matrix} 1 \\ 2 \\ 4 \\ 3 \end{matrix} \begin{bmatrix} [I] & Q_5 \begin{bmatrix} \tfrac{3}{5} \\ \tfrac{4}{5} \\ \tfrac{5}{3} \\ -1 \end{bmatrix} \begin{bmatrix} 1 & 0 & 0 & 0 \\ 0 & 1 & 0 & 0 \\ 0 & \tfrac{3}{4} & 1 & \tfrac{3}{4} \\ 0 & -\tfrac{5}{4} & 0 & -\tfrac{5}{4} \end{bmatrix} \begin{Bmatrix} P_1 \\ P_2 \\ P_3 \\ P_4 \end{Bmatrix} \end{bmatrix}. \quad (3.99)$$

Interchange rows 3 and 4 to put the member loads in sequence, whence Equations (3.81) and (3.82) give

$$[p] = \tfrac{1}{4} \begin{array}{c} \begin{array}{cccc} P_1 & P_2 & P_3 & P_4 \end{array} \\ \begin{bmatrix} 4 & 0 & 0 & 0 \\ 0 & 4 & 0 & 0 \\ 0 & -5 & 0 & -5 \\ 0 & 3 & 4 & 3 \\ 0 & 0 & 0 & 0 \end{bmatrix} \end{array}, \quad [q] = \tfrac{1}{5} \begin{array}{c} Q_5 = S_5 \\ \begin{bmatrix} -3 \\ -4 \\ 5 \\ -3 \\ 5 \end{bmatrix} \end{array}. \qquad (3.100)$$

Thus, the Jordan procedure selects member (5) as the redundant so that $[p]$ and $[q]$ are different from Example 3.22. However, if S_5 is used as the unknown in Example 3.22, the $[p]$ and $[q]$ in Equation (3.100) are obtained.

Example 3.26. Solve Example 3.24 by applying the Jordan procedure in Equation (3.82) to a $[s]_P^T$ matrix which is arranged so that the same members (4) and (6) as in Example 3.24 are obtained for Q_k.

Solution. From Figure 3.16 write the five equilibrium equations in the following arrangement, including all five possible loads,

$$\begin{Bmatrix} P_{1x} \\ P_{3x} = P_5 \\ P_{2x} \\ P_{3x} \\ P_{2x} = P_3 \end{Bmatrix} = \begin{bmatrix} 1 & 0 & 0 & 0 & 3/5 & 0 & 0 \\ 0 & 1 & 0 & 0 & 4/5 & 4/5 & 0 \\ -1 & 0 & 1 & 0 & 0 & 0 & 3/5 \\ 0 & 0 & 0 & 1 & -3/5 & 3/5 & 0 \\ 0 & -1 & 0 & 0 & 0 & 0 & -4/5 \end{bmatrix} \begin{Bmatrix} S_1 \\ S_3 \\ S_7 \\ S_5 \\ S_2 \\ S_4 \\ S_6 \end{Bmatrix}. \qquad (3.101)$$

This arrangement of the equations and the S_k columns is made to simplify the Jordan transformations on the first five columns for $\{S_0\}$ so that no column interchanges are needed. In this case

$$[T_1] = \begin{bmatrix} 1 & 0 & 0 & 0 & 0 \\ 0 & 1 & 0 & 0 & 0 \\ 1 & 0 & 1 & 0 & 0 \\ 0 & 0 & 0 & 1 & 0 \\ 0 & 0 & 0 & 0 & 1 \end{bmatrix}, \quad [T_2] = \begin{bmatrix} 1 & 0 & 0 & 0 & 0 \\ 0 & 1 & 0 & 0 & 0 \\ 0 & 0 & 1 & 0 & 0 \\ 0 & 0 & 0 & 1 & 0 \\ 0 & 1 & 0 & 0 & 1 \end{bmatrix}.$$

$$[T_3] = [T_4] = I, \quad [T_5] = \begin{bmatrix} 1 & 0 & 0 & 0 & -3/4 \\ 0 & 1 & 0 & 0 & -1 \\ 0 & 0 & 1 & 0 & -3/4 \\ 0 & 0 & 0 & 1 & 3/4 \\ 0 & 0 & 0 & 0 & 5/4 \end{bmatrix},$$

where any two of the three diagonal members (2), (4), and (6) can be selected. $[T_5]$ selects (4) and (6).

Use Equation (3.101) in Equation (3.82) and put the results of these three transformations on Equation (3.82) into Equation (3.81) to get

$$[p] = \tfrac{1}{4} \begin{array}{c} \begin{array}{ccccc} P_{1x} & P_5 & P_{2x} & P_{3x} & P_3 \end{array} \\ \begin{bmatrix} 4 & -3 & 0 & 0 & -3 \\ 0 & 0 & 0 & 0 & -4 \\ 4 & -3 & 4 & 0 & -3 \\ 0 & 3 & 0 & 4 & 3 \\ 0 & 5 & 0 & 0 & 5 \\ 0 & 0 & 0 & 0 & 0 \\ 0 & 0 & 0 & 0 & 0 \end{bmatrix} \end{array}, \quad [q] = \tfrac{1}{5} \begin{array}{c} \begin{array}{cc} Q_4 & Q_6 \end{array} \\ \begin{bmatrix} 3 & -3 \\ 0 & -4 \\ 3 & -6 \\ -6 & 3 \\ -5 & 5 \\ 5 & 0 \\ 0 & 5 \end{bmatrix} \end{array}. \qquad (3.102)$$

If the rows in $[p]$ and $[q]$ are rearranged to the $S_1, \cdots, S_7$ sequence from the column sequence in Equation (3.101), then the P_3, P_5, Q_4, Q_6 columns in Equation (3.97) are obtained.

3.7. Matrix equations for trusses from the mixed unit displacement and unit load theorem

By using Equation (3.41a) for truss member k, the expressions in Equation (2.82) can be written as

$$\sigma_k = S_k/A_k, \quad D_1 = 1/A_k, \quad e_{\sigma k} = (S/EA)_k + e_{Ek},$$

$$e_{uk} = \frac{1}{L_k}[1 \quad m \quad n]_k \begin{Bmatrix} u_{xj} - u_{xi} \\ u_{yj} - u_{yi} \\ u_{zj} - u_{zi} \end{Bmatrix}. \tag{3.103}$$

Take σ_k^1 and e_k^1 in Equations (2.83) and (2.84) as

$$\sigma_k^1 = 1/A_k, \quad \{e_k^1\} = \frac{1}{L_k}[l \quad m \quad n \quad -l \quad -m \quad -n]_k^T. \tag{3.104}$$

Thus, the integrals in Equation (2.87) for element (k) are

$$[s_k] = [l \quad m \quad n \quad -l \quad -m \quad -n]_k, \quad \{s\}_k = [s]_k^T,$$
$$C_{Fk} = (L/EA)_k, \quad v_{Ek} = L_k e_{Ek} = C_{Fk} S_{Ek}. \tag{3.105}$$

These element expressions can be assembled for all members of the truss to give the forms in Equations (2.85) and (2.86) as

$$\begin{Bmatrix} [C_F]\{S_E\} \\ \{P\} \end{Bmatrix} = [M]\begin{Bmatrix} S \\ u \end{Bmatrix}, \quad [M] = \begin{bmatrix} -[C_F] & [s]_P \\ [s]_P^T & [0] \end{bmatrix}, \tag{3.106}$$

where the support reactions with $\{u_R\} \neq 0$ have been omitted. Since $[C_F]$ is the diagonal matrix of element flexibilities and since the equilibrium matrix $[s]_P^T$ has many zeros, the $[M]$ matrix can be inverted by computers for relatively large structures. Thus, both the member loads and the deflections can be obtained directly by

$$\begin{Bmatrix} S \\ u \end{Bmatrix} = [M]^{-1}\begin{Bmatrix} [C_F]\{S_E\} \\ \{P\} \end{Bmatrix}. \tag{3.107}$$

It is only necessary to construct the $[s]_P^T$ equilibrium matrix in Equation (3.77) and the flexibility $[C_F]$ matrix to set up the $[M]$ matrix.

By using the procedure in Equations (A.25) - (A.35) the inverse of $[M]$ can be written as

$$[M]^{-1} = \begin{bmatrix} -k_S + k_S s_P k^{-1} s_P^T k_S & k_S s_P k^{-1} \\ k^{-1} s_P^T k_S & k^{-1} \end{bmatrix}, \tag{3.108}$$

$$[k] = [s]_P^T [k_S][s]_P, \quad [k_S] = [C_F]^{-1},$$

whence Equation (3.107) becomes

$$\begin{Bmatrix} S \\ u \end{Bmatrix} = \begin{bmatrix} -I + k_S s_P k^{-1} s_P^T & k_S s_P k^{-1} \\ k^{-1} s_P^T & k^{-1} \end{bmatrix} \begin{Bmatrix} S_E \\ P \end{Bmatrix}. \tag{3.109}$$

This form can be reduced to the displacement method of solution in Equations (3.45) and (3.41b).

The virtual principles for pin-jointed trusses 97

On the other hand, if $[s]_P^T$ in Equation (3.106) is partitioned as in Equation (3.78), then the second row of Equation (3.106) gives Equations (3.79)-(3.81) so that $\{S\}$ has the form in Equation (3.66). Put this $\{S\}$ into the first row in Equation (3.106) to get

$$-[C_F][p \ q]\begin{Bmatrix}P\\Q\end{Bmatrix} + \begin{bmatrix}s_{P0}\\s_{PQ}\end{bmatrix}\{u\} = [C_F]\{S_E\}. \tag{3.110}$$

Multiply this Equation (3.110) by $\begin{bmatrix}p^T\\q_T\end{bmatrix}$ and use Equation (3.81) to get

$$\begin{Bmatrix}u\\0\end{Bmatrix} = \begin{bmatrix}p^T C_F p & p^T C_F q\\q^T C_F p & q^T C_F q\end{bmatrix}\begin{Bmatrix}P\\Q\end{Bmatrix} + \begin{Bmatrix}u_{EP}\\u_{EQ}\end{Bmatrix}, \tag{3.111}$$

which is Equation (3.70) with $\{Q\}$ being given by Equation (3.72). This is the force method of solution.

Thus, all three methods of solution can be made from the three basic matrices $[s]^T$, $[C_F]$, and $\{S_E\}$. In the case of M members and N displacements, the mixed method requires the inversion of the $(M+N)$ by $(M+N)$ matrix $[M]$ in Equation (3.106). The displacement method requires the inversion of the N by N matrix $[k]$ in Equation (3.108), plus matrix operations to set up $[k]$ and to get $\{S\}$ from $\{u\}$. The force method requires the inversion of the $(M-N)$ by $(M-N)$ matrix C_{QQ} and may require the inversion of the N by N matrix $[s]_{P0}^T$, plus matrix operations to get both $\{S\}$ and $\{u\}$.

For a large structure the best method may depend upon the available computers and computer programs. For a small structure with the M members much larger than the N displacements, the displacement method is probably the best method. For a small structure with $(M-N)$ small the force method has simpler calculations.

Example 3.27. Construct the $[s]_P^T$, $[C_F]$ and $[s]_R^T$ matrices for the truss in Figure 3.17 with $E_k = E$.

Solution. The equilibrium equations for the applied loads with $b = (34)^{-1/2}$ are

$$P_1 = S_1 + \tfrac{4}{5}(S_3 + S_7), \quad P_2 = -S_2 - \tfrac{3}{5}S_7, \quad P_3 = -\tfrac{3}{5}S_3,$$
$$P_4 = \tfrac{4}{5}(S_4 + S_5 + 5bS_8) + S_6, \quad P_5 = S_2 + \tfrac{4}{5}S_5 + 3bS_8, \quad P_6 = 3bS_8 - \tfrac{3}{5}S_4,$$

whence in $\{P\} = [s]_P^T\{S\}$,

$$[s]_P^T = \tfrac{1}{5}\begin{bmatrix}5 & 0 & 4 & 0 & 0 & 0 & 4 & 0\\0 & -5 & 0 & 0 & 0 & 0 & -4 & 0\\0 & 0 & -3 & 0 & 0 & 0 & 0 & 0\\0 & 0 & 0 & 4 & 4 & 5 & 0 & 20b\\0 & 5 & 0 & 0 & 4 & 0 & 0 & 15b\\0 & 0 & 0 & -3 & 0 & 0 & 0 & -15b\end{bmatrix}. \tag{3.112}$$

The elements on the diagonal of $[C_F]$ are

$$\frac{L}{4E}\left(\frac{4}{A_1} \quad \frac{3}{A_2} \quad \frac{5}{A_3} \quad \frac{5}{A_4} \quad \frac{5}{A_5} \quad \frac{4}{A_6} \quad \frac{5}{A_7} \quad \frac{1}{bA_8}\right). \tag{3.113}$$

Take the reactions at the supports (3), (4), (5), (6) in the order R_{3x}, R_{3y}, $\cdots$, R_{6y}, R_{6z},

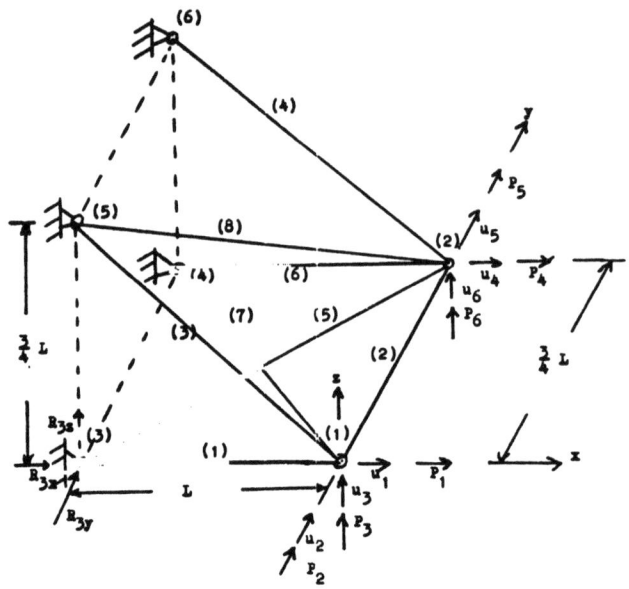

Fig. 3.17. Three dimensional redundant truss.

whence in $\{R\} = [s]_R^T \{S\}$,

$$[s]_R^T = \tfrac{1}{5}\begin{bmatrix} -5 & 0 & 0 & 0 & -4 & 0 & 0 & 0 \\ 0 & 0 & 0 & 0 & -3 & 0 & 0 & 0 \\ 0 & 0 & 0 & 0 & 0 & 0 & 0 & 0 \\ 0 & 0 & 0 & 0 & 0 & -5 & -4 & 0 \\ 0 & 0 & 0 & 0 & 0 & 0 & 3 & 0 \\ 0 & 0 & 0 & 0 & 0 & 0 & 0 & 0 \\ 0 & 0 & -4 & 0 & 0 & 0 & 0 & -20b \\ 0 & 0 & 0 & 0 & 0 & 0 & 0 & -15b \\ 0 & 0 & 3 & 0 & 0 & 0 & 0 & 15b \\ 0 & 0 & 0 & -4 & 0 & 0 & 0 & 0 \\ 0 & 0 & 0 & 0 & 0 & 0 & 0 & 0 \\ 0 & 0 & 0 & 3 & 0 & 0 & 0 & 0 \end{bmatrix} \qquad (3.114)$$

3.8. Problems

3.1. Repeat Example 3.1 for the external forces P_{y2} and P_{x2} and the deflections u_{y2} and u_{x2} at joint (2).

3.2. In order to check the results in Problem 3.1 and Example 3.1, calculate the elongation of each member k from $(SL/EA)_k$ and from the deflections and compare the results.

3.3. In Example 3.3 find the deflections u_{x2} and u_{z2} at joint (2).

3.4. In Example 3.3 take $E_1 A_1 = 5(10^6)$ lb, $E_2 A_2 = 12(10^6)$ lb, and find u_{x1} and u_{z1}.

3.5. Solve Example 3.4 for the case of $u_{x2} = 0.100$ in., $P_{x1} = 80,000$ lb, $P_{z1} = 60,000$ lb.

3.6. Solve for the deflections and internal loads in Example 3.5 for the case of $E_2 A_2 = 2 E_1 A_1$, $E_3 A_3 = 0.50 E_1 A_1$, $\theta = 30°$.

3.7. If $E_2 A_2 = E_1 A_1 = E_3 A_3$ and $S_3 = 0$ in Example 3.5, graph the load ratio P_{xA}/P_{yA} against the angle θ for $0 \leq \theta \leq 90°$. Explain why there are two values of θ for each load ratio within the permissible range $0 \leq P_{xA}/P_{yA} \leq 0.41$.

3.8. Repeat Example 3.7 for the external forces P_{y2} and P_{x2} and the deflections u_{y2} and u_{x2} at joint (2).

3.9. Solve Example 3.2 by using Equation (3.14).

3.10. In Example 3.8 find the deflections u_{x2} and u_{z2} at joint (2).

3.11. Solve Problem 3.4 by using the unit load equations.

3.12. Solve Problem 3.5 by using the unit load equations.

3.13. Solve Example 3.9 for the case of $E_1A_1 = 5(10^6)$ lb, $E_2A_2 = E_3A_3 = 10^7$ lb, and $E_4A_4 = 2(10^6)$ lb.

3.14. In Example 3.9 let $E_1A_1 = E_2A_2 = E_3A_3 = 10^7$ lb and let E_4A_4 be variable. Graph S_1, S_2, S_3, and S_4 against E_4A_4 for the range $10^5 \leq E_4A_4 \leq 10^9$. Comment on the results.

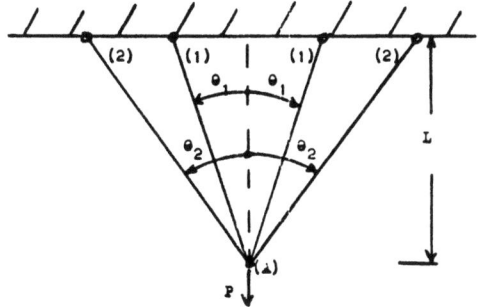

Fig. *3.18.* Problem 3.15.

3.15. Assume $E_k A_k = EA$ in the redundant symmetrical planar truss in Figure 3.18 and solve for the internal loads S_1 and S_2. Find the vertical deflection at point (A). Use both the unit load and the unit displacement theorems.

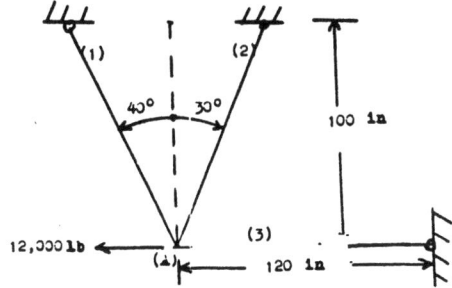

Fig. *3.19.* Problem 3.16.

3.16. Assume $E_k A_k = EA$ in the redundant planar truss in Figure 3.19 and solve for the internal loads S_1, S_2, and S_3. Find the vertical and horizontal deflections at joint (A).

3.17. In Example 3.6 and Figure 3.9 take p_y = constant, $u_y = 0$ at $y = 0$, $P_{yL} = 2Lp_y$ at $y = L$, and $EA = (EA)_R(1 + \frac{y}{L})^2$. Integrate the differential Equation (3.12) for the deflection u_y of the bar as a function of y.

3.18. Solve Problem 3.16 above for the case in which $E_1A_1 = E_2A_2 = (EA)_R$, $E_3A_3 = (EA)_R(1 + \frac{y}{120})^2$. Use the result of Problem 3.17 for the deflection of member (3).

3.19 Solve Problem 3.15 for the case $(EA)_2 = 2(EA)_1$.

3.20. In Example 3.12 assume that member (2) in the truss also has a thermal strain $e_{T2} = 0.004$ with $S_{y2} = -5000$ lb. Find the deflection u_{x1} by the unit load equations.

3.21. In Figure 3.6(b), find the deflections u_{x2} and u_{z2} at joint (2) for $P_{x1} = 40$ Klb, $P_{z1} = 60$ Klb, $Q_{x2} = 100$ Klb, $e_{T1} = 0.006$, $e_{T2} = 0.004$, $S_{y1} = S_{y2} = S_{y3} = 120$ Klb, $S_{y4} = -50$ Klb. Use the unit load equations.

3.22. Solve Example 3.13 with member (3) also having a thermal strain $e_{T3} = -0.003$.

3.23. In Example 3.13 let S_{E1} and S_{E2} be variables and solve for Q_2 and the S_k loads in terms of S_{E1} and S_{E2}. With $S_{E2} = 0$ graph the S_k loads against S_{E1} in the range $-10^5 \leq S_{E1} \leq +10^5$ and discuss the effects of the S_{E1} variation.

3.24. Solve Problem 3.16 with $e_{T1} = 0.005$ and the 12,000 lb load omitted. Assume $E_k A_k = 10^7$ lb.

3.25. Solve Example 3.14 with $S_{y2} = 20$ Klb and the other members elastic.

3.26. Solve Example 3.14 with both members (2) and (3) inelastic with $S_{y2} = 20$ Klb, $S_{y3} = 45$ Klb.

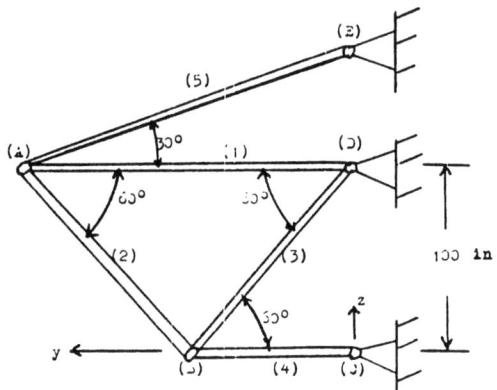

Fig. 3.20. Problems 3.27, 3.28, 3.29, 3.30

3.27. Use the unit displacement equations to find the external component forces $P_{yA}, P_{zA}, P_{yB}, P_{zB}$ at joints (A) and (B) in Figure 3.20, where $u_{yA} = 1.00$ in., $u_{zA} = 0.85$ in. $u_{yB} = 0.20$ in., $u_{zB} = -0.30$ in., $(EA)_k = 10^7$ lb for all members, $e_{T2} = 0.004$, $e_{T4} = 0.005$. Also, calculate the internal member loads S_k from Equation (3.25).

3.28. In Problem 3.27 find the component reactions at the truss supports by using the unit displacement equations.

3.29. Use the unit displacement equations to find the displacement components $u_{yA}, u_{zA}, u_{yB}, u_{zB}$ in Figure 3.20, where $P_{yA} = 100$ Klb, $P_{zA} = 80$ Klb, $P_{yB} = 0$, $P_{zB} = 40$ Klb, $e_{T1} = 0.006$, $e_{T4} = 0.004$, $(EA)_k = 10^4$ Klb.

3.30. Solve Problem 3.29 by using the unit load equations.

3.31. In Example 3.15 use the unit load equations to find the component deflections u_{Ah}, u_{Bh}, and u_{Bv} for (a) and (b).

3.32. In Example 3.15 omit the applied load in Figure 3.12 and find the internal loads due to (a) the $e_{T2} = 0.004$ alone, (b) the $u_{Av} = 2.00$ in. initial deflection alone.

3.33. In Example 3.15 reverse the direction of the applied load in Figure 3.12 and determine (a) the internal loads for part (a), (b) the component deflections u_{Ah}, u_{Bh}, u_{Bv} for part (a). Note that member (4) is in compression and approaching its buckling load so that the magnitude of Q_A is less than 34.66 Klb, whence member (4) is the critical member instead of the inelastic member (3).

3.34. Solve Example 3.16 with only the initial displacement of point (A) and the thermal strain in member (1) acting on the truss. In this case of tension load in member (2), use $S_{y2} = 35$ Klb.

3.35. Solve Example 3.16 with the applied load as 60 Klb instead of 50 Klb. If the cut-off ultimate loads are $S_{u2} = 1.50 S_{y2} = 52.5$ Klb and $S_{u3} = 1.50, S_{y3} = 45$ Klb, show that strain divergence occurs for thermal cycling.

3.36. Find the maximum applied load for ultimate static failure in Problem 3.35.

3.37. Repeat Examples 3.17 and 3.18 for $A_1 = A_5 = A$, $A_3 = 3A$, $A_2 = 2A$, $A_4 = 0.5A$, $E_k = E$.

3.38. Suppose joints (1) and (2) in Figure 3.13 are fixed with joints (3) and (4) free. Use $[k]$ from Equation (3.54) to find the deflections u_5, u_6, u_7, u_8 in terms of P_5, P_6, P_7, P_8 in matrix form.

3.39. Assume $(EA)_k = EA$ for all six members in the truss in Figure 3.21 and find the

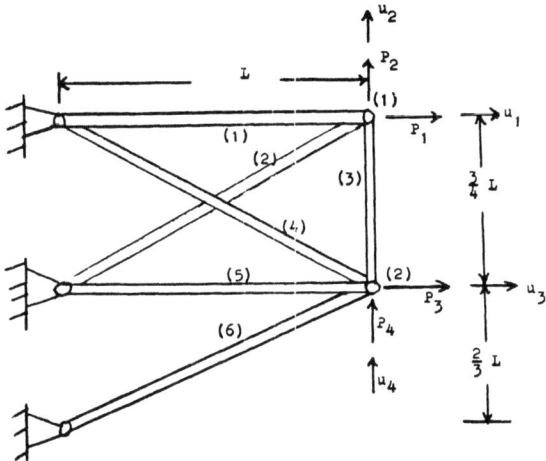

Fig. 3.21. Problems 3.39, 3.40, 3.41.

stiffness influence coefficients for joints (1) and (2).

3.40. Use the results of Problem 3.39 and find the displacements u_1, u_2, u_3, u_4 and the loads S_k in terms of the external forces P_1, P_2, P_3, P_4. Compare the results to Equations (3.56) and (3.57).

3.41. Omit member (5) and repeat Problems 3.39 and 3.40.

3.42. In Example 3.20 assume the applied loads are $P_1 = P_3 = P_4 = 0$, $P_2 = 100$ Klb. If $S_{E1} = S_{T1} + S_{p1}$ with $S_{T1} = 200$ Klb and there are no temperature or inelastic effects in the other members, find the loads in the truss members from Equations (3.62)-(3.65) using (a) $S_{y1} = -50$ Klb, (b) $S_{y1} = -40$ Klb. (c) Compare the deflections for the part (b) case to those in Example 3.20 without the temperature and inelastic effects.

3.43. Repeat Problem 3.42 with the temperature omitted and only the inelastic effect in member (1) present.

3.44. Repeat Problem 3.42 with member (3) also inelastic with $S_{y3} = 30$ Klb. Note that with both S_{p1} and S_{p3} present, it is necessary to iterate to convergence. Start with the S_{p1} result in Problem 3.42 to get S_{p3}, etc.

3.45. In Example 3.21 take $(EA)_3 = 2EA$ and recalculate the displacements. Compare results to Equation (3.84).

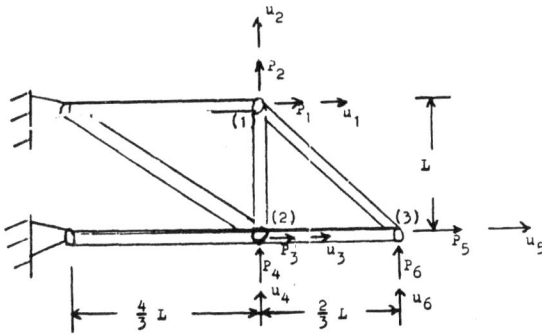

Fig. 3.22. Problems 3.46, 3.47.

3.46. If $(EA)_k = EA$ in Figure 3.22, find the flexibility influence coefficients for joints (1), (2), and (3).

3.47. Use the results in Problem 3.46 and find the u_6 deflection for $P_6 = P_4 = 40$ Klb, $P_1 = P_2 = P_3 = P_5 = 0$.

3.48. In Problem 3.39 find the flexibility matrix $[C]$.

3.49. Repeat Problem 3.40 using results of Problem 3.48.

3.50. Repeat Problem 3.41 using the flexibility method.

3.51. Solve Example 3.22 using $Q_3 = S_3$ as the unknown load.

3.52. Repeat Example 3.22 with $(EA)_1 = 3EA$.

3.53. Assume $P_3 = 132$ Klb is the only applied load acting in Example 3.20 and Figure 3.13. If $S_{y5} = 60$ Klb, use Equation (3.65) to get S_5 and $S^E = S_{E5} = S_{p5}$. Compare the member loads and the joint deflections with and without S_{p5}.

3.54. Repeat Problem 3.53 for the truss in Example 3.22 and Figure 3.15 but with member (4) having $S_{y4} = 60$ Klb.

3.55. Solve Example 3.25 with $S_6 = Q_6$, $S_7 = Q_7$ as the unknowns.

3.56. In Example 3.23 let $P_3 = 100$ Klb, $P_5 = 100$ Klb, $S_{T3} = 150$ Klb, $P_{y4} = 80$ Klb. (a) Find S_{p4} and the member loads and displacements in Equations (3.91) and (3.92). (b) Compare the elastic case with $S_{p4} = 0$ to the results. (c) Compare the results to the elastic case with $S_{T3} = 0$.

3.57. Solve Example 3.23 with member (2) as the inelastic member with S_{p2} instead of member (4).

3.58. In Example 3.26 apply the Jordan procedure to the arrangement with the S_k in the sequence $S_1, \cdots, S_7$ and the equations are in the same order as in Equation (3.96).

3.59. In Example 3.27 use the two redundants as $Q_7 = S_7$, $Q_8 = S_8$ and make a sequence solution for the $[p]$ and $[q]$ matrices and calculate the S_k forces in the truss by the force method.

References

Chapter 3

1. B.E. Gatewood and R.W. Gehring: Allowable axial loads and bending moments for inelastic structures under nonuniform temperature distribution, *J. of Aerospace Sciences* 29, No. 5, May (1962).
2. J.H. Wegstein: Accelerating convergence of iteration procedures, *Commun. of Assoc. of Computing Machinery*, June (1958).
3. J.S. Przemieniecki: *Theory of Matrix Structural Analysis*, McGraw-Hill Book Co. (1968).
4. H.C. Martin: *Introduction to Matrix Methods of Structural Analysis*, McGraw-Hill Book Co. (1966).
5. M.D. Vanderbilt: *Matrix Structural Analysis*, Quantum Publishers, New York (1974).

References for additional reading

6. J.L. Meek: *Matrix Structural Analysis*, McGraw-Hill Book Co., New York (1971).
7. R.K. Livesley: *Matrix Methods of Structural Analysis*, 2nd Ed., Pergamon, New York (1976).
8. J.V. Meyers: *Matrix Analysis of Structures*, Harper and Row, New York (1985).
9. W. Weaver Jr. and J.W. Gere: *Matrix Analysis of Framed Structures*, van Nostrand, New York, 2nd Ed. (1980).
10. C.K. Wang and C.G. Salmon: *Introductory Structural Analysis*, Prentice Hall, Englewood Cliffs, N.J. (1984).

4

The virtual principles for simple beams

4.1. Introduction

The beam in one plane is regarded as a simple beam. Take the particular case in Figure 2.1 with forces in the $y-z$ plane only and bending moments about the x-axis only. See Figure 4.1 for the notation and loading for the simple beam. From Equations (1.93) and (2.32) the stress, strain, and displacement relations for the beam in Figure 4.1 are

$$\sigma_{yy} = Ee_{yy} = -Ez\frac{d^2 u_{zb}}{dy^2} = -\frac{M_x z}{I_{xx}}, \quad u_z = u_{zb} + u_{zs}$$

$$\theta_x = \frac{du_{zb}}{dy}, \quad \sigma_{yz} = Ge_{yz} = G\frac{du_{zs}}{dy} = \frac{V_z}{A_s},$$

(4.1)

$$u_z = u_{zL} \text{ at } y = L, \quad u_z = u_{z0} \text{ at } y = 0,$$
$$\theta_x = \theta_{xL} \text{ at } y = L, \quad \theta_x = \theta_{x0} \text{ at } y = 0.$$

(4.2)

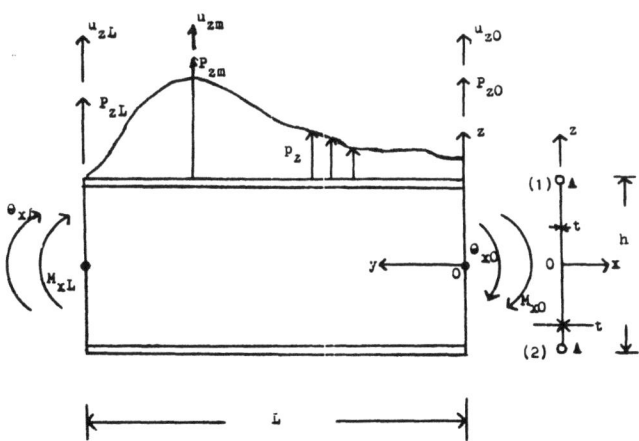

Fig. 4.1. Symmetrical beam with bending about the x-axis.

The relations for the virtual terms are

$$e_{yy}^V = -z\frac{d^2 u_{zb}^V}{dy^2}, \qquad \theta_x^V = \frac{du_{zb}^V}{dy}, \qquad e_{yz}^V = \frac{du_{zs}^V}{dy}, \qquad (4.3)$$
$$u_{zb}^V = u_{zm}^1 u_{zm}^V, \qquad u_{zs}^V = u_{zsm}^1 u_{zm}^V,$$

$$\sigma_{yy}^V = -\frac{M_x^V z}{I_{xx}}, \qquad \sigma_{yz}^V = \frac{V_z^V}{A_s}, \qquad M_x^V = m_{xm}^1 P_{zm}^V, \qquad V_z^V = p_{zm}^1 P_{zm}^V. \, (4.4)$$

From Equation (4.1) the stresses σ_{yy} and σ_{yz} can be obtained as a function of y if the equations for u_{zb} and u_{zs} are known, or if the moment M_x and shear V_z equations are known. In the beam analysis for the stresses, the moments and shear forces are the primary unknowns to be determined. They are independent of the beam cross section and can be used to get the stresses for any specified cross section. If the beam has a number of point loads or a number of point supports, then the u_{zb}, u_{zs}, M_x, and V_z equations as a function of y are complicated so that it may be simpler to divide the beam into elements and get point values for u_{zb}, u_{zs}, M_x, V_z. The equations for the point values can be put into matrices so that computers can be used to obtain the solutions.

The node point locations to divide the beam into elements can be taken at the point loads and point supports. If the beam has a variable cross section, then more elements can be used and the cross section approximated as constant for each element. If a distributed load p_z, Figure 4.1, is acting on the beam, then it is necessary to represent p_z by equivalent point forces and moments at the ends of the elements. This gives the correct shear and moment values at the node points but approximates the shear as constant and the moment as linear in each element. More elements can be used if p_z varies rapidly, or after node point values are obtained, a detailed analysis of the element using the actual p_z can be made.

In the following sections both the continuous and the finite element cases will be obtained from the three virtual principles. The displacement method from the principle of virtual displacements is described in Section 4.2 for the continuous beam and in Section 4.3 for the point or finite beam element case. The force method from the principle of virtual forces is described in Section 4.4 for the continuous beam and in Section 4.5 for the finite element case. The mixed displacement and force method is described in Section 4.6 for the continuous beam case. Inelastic and temperature effects in the beam are discussed in Section 4.7. The matrix equations for the displacement, force, and mixed methods are given in Sections 4.8, 4.9, 4.10, respectively. These matrix equations are finite element equations using only simple beam elements.

Examples demonstrating the use of the equations are given at the end of each section.

4.2. Principle of virtual displacements for beams.

For the simple beam in Figure 4.1 the principle of virtual displacements in Equation (2.34) becomes

$$\int_0^L \int_A (\sigma_{yy} e^V_{yy} + \sigma_{yz} e^V_{yz}) \, dA \, dy - \int_0^L p_z u^V_z \, dy - P_{zL} u^V_{zL} -$$
$$- M_{xL} \theta^V_{xL} - P_{z0} u^V_{z0} - M_{x0} \theta^V_{x0} - \sum_m P_{zm} u^V_{zm} = 0. \quad (4.5)$$

The first integral can be integrated by parts, or with use of Equations (4.1) and (4.3),

$$\int_0^L \int_A (\sigma_{yy} e^V_{yy} + \sigma_{yz} e^V_{yz}) \, dA \, dy$$
$$= \int_0^L \int_A \left[\left(-Ez \frac{d^2 u_{zb}}{dy^2} \right) \left(-z \frac{d^2 u^V_{zb}}{dy^2} \right) + \left(G \frac{du_{zs}}{dy} \right) \left(\frac{du^V_{zs}}{dy} \right) \right] dA \, dy$$
$$= \left[\left(EI \frac{d^2 u_{zb}}{dy^2} \right) \theta^V_x \right]_0^L - \left[\frac{d}{dy} \left(EI \frac{d^2 u_{zb}}{dy^2} \right) u^V_{zb} \right]_0^L + \int_0^L \frac{d^2}{dy^2} \left(EI \frac{d^2 u_{zb}}{dy^2} \right) u^V_{zb} \, dy +$$
$$+ \left[\left(GA_s \frac{du_{zs}}{dy} \right) u^V_{zs} \right]_0^L - \int_0^L \frac{d}{dy} \left(GA_s \frac{du_{zs}}{dy} \right) u^V_{zs} \, dy,$$

whence Equation (4.5) becomes

$$\int_0^L \left[\frac{d^2}{dy^2} \left(EI \frac{d^2 u_{zb}}{dy^2} \right) - p_z \right] u^V_{zb} \, dy - \int_0^L \left[\frac{d}{dy} \left(GA_s \frac{du_{zs}}{dy} \right) + p_z \right] u^V_{zs} \, dy +$$
$$+ \left[EI \frac{d^2 u_{zb}}{dy^2} - M_{xL} \right] \theta^V_{xL} - \left[EI \frac{d^2 u_{zb}}{dy^2} + M_{x0} \right] \theta^V_{x0} -$$
$$- \left[\frac{d}{dy} \left(EI \frac{d^2 u_{zb}}{dy^2} \right) + P_{zL} \right] u^V_{zbL} + \left[\frac{d}{dy} \left(EI \frac{d^2 u_{zb}}{dy^2} \right) - P_{z0} \right] u^V_{zb0} +$$
$$+ \left[GA_s \frac{du_{zs}}{dy} - P_{zL} \right] u^V_{zsL} - \left[GA_s \frac{du_{zs}}{dy} + P_{z0} \right] u^V_{zs0} = 0,$$

where the P_{zm} terms have been omitted. Since the virtual terms are arbitrary, each bracket term must be zero, whence the differential equations and equilibrium boundary conditions for u_{zb} and u_{zs} are

$$\frac{d^2}{dy^2} \left(EI_{xx} \frac{d^2 u_{zb}}{dy^2} \right) - p_z = 0, \quad (4.6)$$

$$\left(EI \frac{d^2 u_{zb}}{dy^2} - M_{xL} \right)_L = 0, \quad \left(EI \frac{d^2 u_{zb}}{dy^2} + M_{x0} \right)_0 = 0,$$
$$\left[\frac{d}{dy} \left(EI \frac{d^2 u_{zb}}{dy^2} \right) + P_{zL} \right]_L = 0, \quad \left[\frac{d}{dy} \left(EI \frac{d^2 u_{zb}}{dy^2} \right) - P_{z0} \right]_0 = 0, \quad (4.7)$$

$$\frac{d}{dy} \left(GA_s \frac{du_{zs}}{dy} \right) + p_z = 0,$$
$$\left(GA_s \frac{du_{zs}}{dy} - P_{zL} \right)_L = 0, \quad \left(GA_s \frac{du_{zs}}{dy} + P_{z0} \right)_0 = 0. \quad (4.8)$$

For bending, four permissible boundary conditions must be selected from the four displacement conditions (select at least two of these to prevent rigid body motion) and the four equilibrium conditons, Equations (4.2) and (4.7). For shear, at least one displacement condition must be selected from the two needed, Equations (4.2) and (4.8). No two related conditions can be selected at the same point. If θ_{x0} is specified, then M_{x0} is a reaction to be determined. If P_{zl} is specified, then u_{zL} must be free. Note that if more than two displacement conditions for bending are specified, then the beam is redundant but the solution for the displacements follows directly.

If the beam is redundant and the u_{zb} solution has been obtained, then the moment distribution M_x and shear distribution V_z are given by

$$M_x = EI_{xx}\frac{d^2u_{zb}}{dy^2}, \quad V_z = \frac{dM_x}{dy}. \tag{4.9}$$

The support reactions are given by Equation (4.7).

If the beam is determinate, then the reactions can be determined directly from the equilibrium equations

$$\sum M_x = 0, \quad \sum P_z = 0. \tag{4.10}$$

For this case the moment, shear and displacement distributions are given by

$$\frac{dV_z}{dy} = p_z, \quad \frac{dM_x}{dy} = V_z, \quad \frac{d^2u_{zb}}{dy^2} = \frac{M_x}{EI}, \quad \frac{d}{dy}\left(GA_s\frac{du_{zs}}{dy}\right) = -p_z. \tag{4.11}$$

If graphs are made for V_z and M_x, then point loads and point moments can be included in Equation (4.10) and on the graphs.

If point loads P_{zm} act on the beam at $y = y_m$ either as applied loads or as additional supports, or if p_z, GA_s, or EI_{xx} are discontinuous at $y = y_m$, then the integrations in the principle of virtual displacements in Equation (4.5) must be split into intervals. The force boundary conditions in Equations (4.7) and (4.8) may change at the ends of the intervals but u_z and $\theta_x = u_{z,y}$ must be continuous over the entire beam. Also, the arbitrary virtual expressions u_z^V and $\theta_x^V = \theta_{z,y}^V$ must be continuous over the entire beam. Let u_{zk} be the displacement in the interval k. Then, for any interval k and for the point y_m between intervals k and $k+1$, Equation (4.5) gives

$$\frac{d^2}{dy^2}\left(EI\frac{d^2u_{zb}}{dy^2}\right)_k - p_{zk} = 0, \quad \frac{d}{dy}\left(GA_s\frac{du_{zs}}{dy}\right)_k + p_{zk} = 0,$$

in the interval k, and at point $y = y_m$,

$$u_{zbk} = u_{zb,k+1}, \quad u_{zsk} = u_{zs,k+1}, \quad \theta_{xk} = \theta_{x,k+1},$$

$$\left(EI\frac{d^2u_{zb}}{dy^2}\right)_k - \left(EI\frac{d^2u_{zb}}{dy^2}\right)_{k+1} = M_{xam}, \tag{4.12}$$

$$\frac{d}{dy}\left(EI\frac{d^2u_{zb}}{dy^2}\right)_k - \frac{d}{dy}\left(EI\frac{d^2u_{zb}}{dy^2}\right)_{k+1} = -P_{zm},$$

$$\left(GA_s\frac{du_{zs}}{dy}\right)_k - \left(GA_s\frac{du_{zs}}{dy}\right)_{k+1} = P_{zm},$$

The virtual principles for simple beams

where M_{xam} is any point applied moment at y_m.

It is evident that if there are several point loads and point discontinuities on the beam, then the solution of Equation (4.12) is complicated. It may be simpler to use the beam intervals as finite elements and find only point values at the y_m points for the displacements, shear forces, and moments. This latter procedure is described in Section 4.3 below.

Example 4.1. Take $p_z = p$ =constant, EI_{xx} =constant=elastic, $u_{x0} = 0$, $u_{xL} = 0$, $\theta_{x0} = 0$, in Figure 4.2 and solve for the bending deflection equation for u_{xb} and the reactions P_{zL}, P_{z0}, M_{x0} by using the differential equation. Give the shear and moment equations.

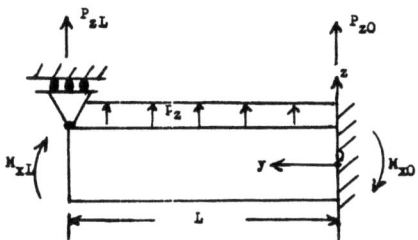

Fig. 4.2. Simple redundant beam.

Solution. From Equations (4.6) and (4.7), and the specified boundary conditons

$$\frac{d^4 u_{xb}}{dy^4} = \frac{p}{EI_{xx}}, \quad u_{x0} = 0, \quad u_{xL} = 0, \quad \theta_{x0} = 0, \quad \left[\frac{d^2 u_{xb}}{dy^2}\right]_{y=L} = \frac{M_{xL}}{EI_{xx}},$$

whence

$$EI_{xx} \frac{d^2 u_{xb}}{dy^2} = (1/2)py^2 + C_1 y + C_2,$$

$$EI_{xx} \frac{du_{xb}}{dy} = EI_{xx}\theta_x = (1/6)py^3 + (1/2)C_1 y^2 + C_2 y + C_3,$$

$$EI_{xx} u_{xb} = (1/24)py^4 + (1/6)C_1 y^3 + (1/2)C_2 y^2 + C_3 y + C_4.$$

The four boundary conditions give

$$C_3 = 0, \quad C_4 = 0, \quad C_1 = (3/2)(M_{xL}/L) - (5/8)pL,$$
$$C_2 = -(1/2)M_{xL} + (1/8)pL^2,$$

so that

$$EI_{xx} u_{xb} = (py^2/48)(L-y)(3L-2y) - (M_{xL} y^2/4L)(L-y). \tag{4.13a}$$

From Equation (4.7) the reactions are

$$P_{z0} = -(5/8)pL + (3/2)(M_{xL}/L), \quad P_{zL} = -(3/8)pL - (3/2)(M_{xL}/L),$$
$$M_{x0} = -(1/8)pL^2 + (1/2)M_{xL}. \tag{4.13b}$$

The shear and moment equations are

$$V_z = py + (3/2L)M_{xL} - (5/8)pL,$$
$$M_x = \tfrac{1}{2}py^2 + (\tfrac{3}{2L}M_{xL} - \tfrac{5}{8}pL)y - \tfrac{1}{2}M_{xL} + \tfrac{1}{8}pL^2. \tag{4.14}$$

Example 4.2. Find the equations and draw the shear and moment curves for the determinate beam in Figure 4.3. The distributed p load is constant.

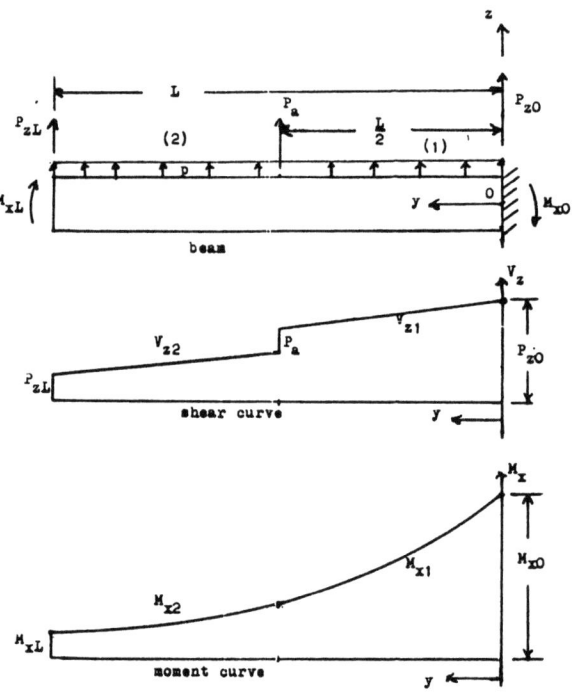

Fig. 4.3. Determinate beam with shear and moment curves.

Solution. The reactions are

$$P_{z0} = -P_{zL} - P_a - pL, \quad M_{x0} = -M_{xL} - LP_{zL} - \tfrac{L}{2}P_a - \tfrac{1}{2}pL^2.$$

In the two regions of the shear and moment curves in Figure 4.3

$$V_{z2} = P_{zL} + p(L-y), \quad (L/2) \leq y \leq L,$$
$$V_{z1} = P_{zL} + P_a + p(L-y), \quad 0 \leq y \leq (L/2),$$
$$M_{x2} = M_{xL} + P_{zL}(L-y) + \tfrac{p}{2}(L-y)^2, \quad (L/2) \leq y \leq L,$$
$$M_{x1} = M_{xL} + P_{zL}(L-y) + \tfrac{p}{2}(L-y)^2 + \tfrac{P_a}{2}(L-2y), \quad 0 \leq y \leq (L/2).$$

The graphs are shown in Figure 4.3.

4.3. Point values for beam elements by the principle of virtual displacements

Consider the simple beam in Figure 4.1 as an element in bending with EI_{xx} constant. Integrate Equation (4.6) with $p_z = 0$ and $P_{zm} = 0$ and determine the four constants of integration in terms of the end deflections u_{z0}, u_{zL} and end slopes θ_{x0}, θ_{xL}, or

$$u_z = (3u_{zL} - L\theta_{xL})(\tfrac{y}{L})^2 + (-2u_{zL} + L\theta_{xL})(\tfrac{y}{L})^3 +$$
$$+ (3u_{z0} + L\theta_{x0})(1 - \tfrac{y}{L})^2 + (-2u_{z0} - L\theta_{x0})(1 - \tfrac{y}{L})^3. \tag{4.15}$$

Calculate the stresses from Equation (4.1) to get

$$\sigma_{yy} = -\frac{Ez}{L^3}[-6(L-2y)u_{z0} - 2L(2L-3y)\theta_{x0} + 6(L-2y)u_{zL} -$$
$$- 2L(L-3y)\theta_{xL}] = -\frac{Ez}{L^3}\sum_{i=1}^{4} b_i(y)u_i, \tag{4.16}$$

where $u_1 = u_{z0}$, $u_2 = \theta_{x0}$, $u_3 = u_{zL}$, $u_4 = \theta_{xL}$. The moment and shear equations are

$$M_x = \frac{EI_{xx}}{L^3}\sum_{i=1}^{4} b_i(y)u_i, \quad b_i(y) = \text{linear},$$
$$V_z = 6\frac{EI_{xx}}{L^3}(-2u_1 - Lu_2 + 2u_3 - Lu_4), \quad \text{constant}. \tag{4.17}$$

Usually the node points joining the beam elements are taken at the point loads P_{zm}, the point moments M_{xm}, at any supports not at the ends, and at any discontinuities in E, A, I, and the distributed load p_z. To include the $p_z(y)$ load between the node points and any P_{zm} and M_{xm} point loads not at the node points, it is necessary to use equivalent end point values to represent p_z, P_{zm}, and M_{xm}. Take u_z^V as the u_z in Equation (4.15) and use it for the p_z and P_{zm} terms in the principle of virtual displacements in Equation (4.5). Use $\theta_x^V = u_{z,y}^V$ for the point moments. Thus, from Equation (4.5),

$$\int_0^L p_z u_z^V \, dy + \sum_m P_{zm} u_{zm}^V + \sum_n M_{xn}\theta_{xn}^V$$
$$= P_{zLd}u_{zL}^V + M_{xLd}\theta_{xL}^V + P_{z0d}u_{z0}^V + M_{x0d}\theta_{x0}^V, \tag{4.18}$$

$$P_{zLd} = \int_0^L p_z\left(\tfrac{y}{L}\right)^2\left(3 - 2\tfrac{y}{L}\right) dy + \sum_m P_{zm}\left(\tfrac{y_m}{L}\right)^2\left(3 - 2\tfrac{y_m}{L}\right) +$$
$$+ \sum_n \frac{6M_{xn}}{L}\left(\tfrac{y_n}{L}\right)\left(1 - \tfrac{y_n}{L}\right),$$

$$M_{xLd} = L\int_0^L p_z\left(\tfrac{y}{L}\right)^2\left(-1 + \tfrac{y}{L}\right) dy + \sum_m LP_{zm}\left(\tfrac{y_m}{L}\right)^2\left(-1 + \tfrac{y_m}{L}\right) +$$
$$+ \sum_n M_{xn}\left(\tfrac{y_n}{L}\right)\left(-2 + 3\tfrac{y_n}{L}\right),$$

$$P_{z0d} = \int_0^L p_z\left(1 - \tfrac{y}{L}\right)^2\left(1 + 2\tfrac{y}{L}\right) dy + \sum_m P_{zm}\left(1 - \tfrac{y_m}{L}\right)^2\left(1 + 2\tfrac{y_m}{L}\right) -$$
$$- \sum_n \frac{6M_{xn}}{L}\left(\tfrac{y_n}{L}\right)\left(1 - \tfrac{y_n}{L}\right),$$

$$M_{x0d} = L\int_0^L p_x\left(\frac{y}{L}\right)\left(1-\frac{y}{L}\right)^2 dy + \sum_m LP_{xm}\left(\frac{y_m}{L}\right)\left(1-\frac{y_m}{L}\right)^2 +$$
$$+ \sum_n M_{xn}\left(1-\frac{y_n}{L}\right)\left(1-3\frac{y_n}{L}\right).$$

These terms in Equation (4.18) can be added directly to the $P_{xL}u_{xL}^V$ etc. terms in Equation (4.5).

Although the same u_x^V from Equation (4.15) can be used for the e_{yy}^V term in Equation (4.5), it is simpler to use e_{yy}^V from Equation (4.3) and integrate by parts. With σ_{yy} from Equation (4.16), Equation (4.5) now becomes

$$\frac{EI}{L^3}\int_0^L \sum_{m=1}^4 b_m(y)u_m \frac{d^2 u_x^V}{dy^2} dy - (P_{x0} + P_{x0d})u_{x0}^V - (P_{xL} + P_{xLd})u_{xL}^V -$$
$$-(M_{x0} + M_{x0d})\theta_{x0}^V - (M_{xL} + M_{xLd})\theta_{xL}^V = 0. \tag{4.19}$$

Since $b_m(y)$ is linear, integration by parts gives

$$\left[\frac{EI}{L^3}\sum_{m=1}^4 b_{m,y}(0)u_m - (P_{x0} + P_{x0d})\right]u_{x0}^V -$$
$$-\left[\frac{EI}{L^3}\sum_{m=1}^4 b_m(0)u_m + (M_{x0} + M_{x0d})\right]\theta_{x0}^V -$$
$$-\left[\frac{EI}{L^3}\sum_{m=1}^4 b_{m,y}(L)u_m + (P_{xL} + P_{xLd})\right]u_{xL}^V +$$
$$+\left[\frac{EI}{L^3}\sum_{m=1}^4 b_m(L)u_m - (M_{xL} + M_{xLd})\right]\theta_{xL}^V = 0, \tag{4.20}$$

whence

$$P_{x0} + P_{x0d} = \frac{EI}{L^3}(12u_{x0} + 6L\theta_{x0} - 12u_{xL} + 6L\theta_{xL}),$$
$$M_{x0} + M_{x0d} = \frac{EI}{L^3}(6Lu_{x0} + 4L^2\theta_{x0} - 6Lu_{xL} + 2L^2\theta_{xL}),$$
$$P_{xL} + P_{xLd} = \frac{EI}{L^3}(-12u_{x0} - 6L\theta_{x0} + 12u_{xL} - 6L\theta_{xL}),$$
$$M_{xL} + M_{xLd} = \frac{EI}{L^3}(6Lu_{x0} + 2L^2\theta_{x0} - 6Lu_{xL} + 4L^2\theta_{xL}).$$
$$\tag{4.21}$$

This Equation (4.21) relates the end forces and end moments to the end deflections and end slopes for any simple beam element. It can be written directly in matrix form so that any number of beam elements in a beam structure can be assembled together in a matrix form. If the point moments and point loads are given, then the point deflections and point slopes can be solved for specified boundary conditions by using a computer program for simultaneous equations. See Section 4.8 on matrix solutions.

The virtual principles for simple beams

Since the shear V_z is constant and the moment M_x is linear in the beam elements in Equation (4.17), only the element end point values for V_z and M_x are actually used to calculate the stresses in Equation (4.1). In Figure 4.4 let

$$S_1 = V_{z1}, \quad S_2 = M_{x2}, \quad P_1 = P_{z1} \text{ (applied)},$$
$$P_2 = M_{x2a} \text{ (applied, usually zero)}, \quad u_1 = u_{zb1}, \quad u_2 = \theta_{x2}, \text{ etc.} \quad (4.22)$$

From Equation (4.17) for the point (1) end of member (1) in Figure 4.4,

$$S_1 = \left(\frac{6EI}{L^3}\right)_1 (2u_1 - L_1 u_2 - 2u_3 - L_1 u_4),$$
$$S_2 = \left(\frac{2EI}{L^2}\right)_1 (-3u_1 + 2L_1 u_2 + 3u_3 + L_1 u_4), \quad (4.23)$$

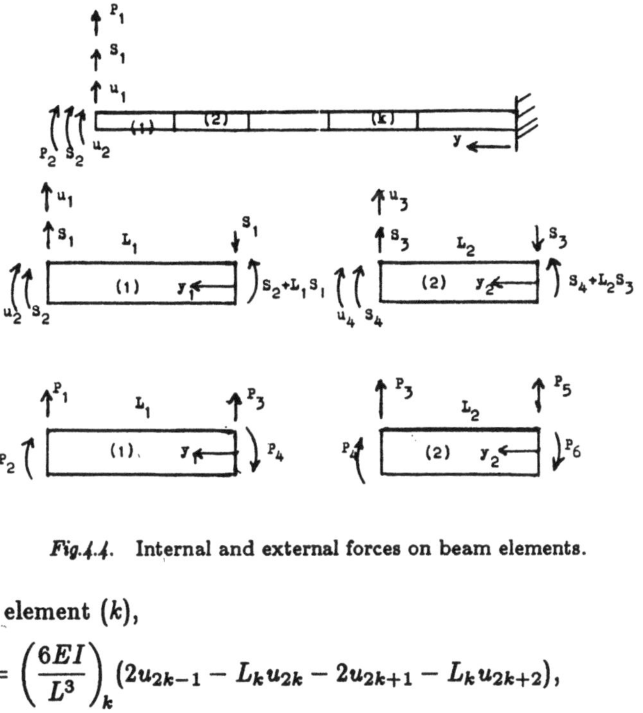

Fig.4.4. Internal and external forces on beam elements.

and for any element (k),

$$S_{2k-1} = \left(\frac{6EI}{L^3}\right)_k (2u_{2k-1} - L_k u_{2k} - 2u_{2k+1} - L_k u_{2k+2}),$$
$$S_{2k} = \left(\frac{2EI}{L^2}\right)_k (-3u_{2k-1} + 2L_k u_{2k} + 3u_{2k+1} + L_k u_{2k+2}). \quad (4.24)$$

From Figure 4.4 the equilibrium equations for the elements are

$$P_1 = S_1, \quad P_2 = S_2, \quad P_3 = -S_1, \quad P_4 = -S_2 - L_1 S_1,$$
$$P_3 = S_3, \quad P_4 = S_4, \quad P_5 = -S_3, \quad P_6 = -S_4 - L_2 S_3,$$
$$P_{2k-1} = S_{2k-1}, \quad P_{2k} = S_{2k}, \quad P_{2k+1} = -S_{2k+1}, \quad (4.25)$$
$$P_{2k+2} = -S_{2k} - L_k S_{2k-1}, \quad \text{etc.}$$

These equilibrium equations can be assembled in a matrix form for beams similar to that for the truss in Equation (3.38). If the forces at any redundant supports are regarded as known, then Equation (4.25) can be solved in sequence for the S_k point shear and moment values. From Figure 4.4, the form for S_k for two elements is

$$S_1 = P_1, \quad S_2 = P_2, \quad S_3 = P_1 + P_3, \quad S_4 = P_2 + P_4 + L_1 P_1,$$
$$S_5 = P_1 + P_3 + P_5, \quad S_6 = P_2 + P_4 + P_6 + (L_1 + L_2)P_1 + L_2 P_3. \quad (4.26)$$

Equation (4.21) can be written in the notation of Figure 4.4 for element (k) as

$$(P + P_d)_{2k-1} = \left(\frac{EI}{L^3}\right)_k (12 u_{2k-1} - 6 L_k u_{2k} - 12 u_{2k+1} - 6 L_k u_{2k+2}),$$

$$(P + P_d)_{2k} = \left(\frac{EI}{L^2}\right)_k (-6 u_{2k-1} + 4 L_k u_{2k} + 6 u_{2k+1} + 2 L_k u_{2k+2}), \quad (4.27)$$

$$(P + P_d)_{2k+1} = \left(\frac{EI}{L^3}\right)_k (-12 u_{2k-1} + 6 L_k u_{2k} + 12 u_{2k+1} + 6 L_k u_{2k+2}),$$

$$(P + P_d)_{2k+2} = \left(\frac{EI}{L^2}\right)_k (-6 u_{2k-1} + 2 L_k u_{2k} + 6 u_{2k+1} + 4 L_k u_{2k+2}).$$

Example 4.3. Use Equation (4.18) and Equation (4.21) to calculate the reactions in Equation (4.13b) in Example 4.1.

Solution. With u_{x0}, u_{xL}, and θ_{x0} zero solve for θ_{xL} from the last equation in Equation (4.21), or

$$\theta_{xL} = (L/4EI)(M_{xL} + M_{xLd}).$$

From Equation (4.18) for the constant distributed load p_x,

$$M_{xLd} = -\tfrac{1}{12} p_x L^2 = -M_{x0d}, \quad P_{xLd} = P_{x0d} = \tfrac{1}{2} p_x L.$$

Put these values into the other three equations in Equation (4.21) to get

$$P_{x0} = (3/2L)M_{xL} - (5/8)p_x L, \quad P_{xL} = -(3/2L)M_{xL} - (3/8)p_x L,$$
$$M_{x0} = (1/2)M_{xL} - (1/8)p_x L^2, \quad (4.28)$$

which check the reactions in Equation (4.13b).

Example 4.4. (a) use Equation (4.21) to find deflections θ_4, u_5, θ_6 and reactions P_1, M_2, P_3 in the two element beam in Figure 4.5 with $(EI)_1 = (EI)_2 = EI$. (b) Use Equation (4.15) and Equation (4.16) to get the equations for deflections u_x and stresses σ_{yy} in each element. (c) Use Equation (4.26) to get the point shear and moment values.

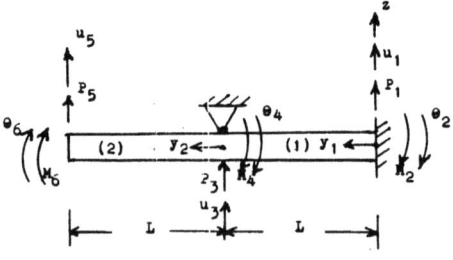

Fig. 4.5. Redundant two element beam

The virtual principles for simple beams

Solution. (a) The Equation (4.21) applies to both elements in Figure 4.5 with overlap on P_3, u_3 and M_4, θ_4. Thus

$$\begin{Bmatrix} LP_1 \\ M_2 \\ LP_3 \\ M_4 \\ LP_5 \\ M_6 \end{Bmatrix} = \frac{EI}{L^2} \begin{bmatrix} 12 & 6 & -12 & 6 & 0 & 0 \\ 6 & 4 & -6 & 2 & 0 & 0 \\ -12 & -6 & 24 & 0 & -12 & 6 \\ 6 & 2 & 0 & 8 & -6 & 2 \\ 0 & 0 & -12 & -6 & 12 & -6 \\ 0 & 0 & 6 & 2 & -6 & 4 \end{bmatrix} \begin{Bmatrix} u_1 \\ \theta_2 L \\ u_3 \\ \theta_4 L \\ u_5 \\ \theta_6 L \end{Bmatrix}. \tag{4.29}$$

From Figure 4.5, $u_1 = 0$, $\theta_2 = 0$, and $u_3 = 0$, whence θ_4, u_5, θ_6 can be obtained from the matrix equation

$$\begin{Bmatrix} M_4 \\ LP_5 \\ M_6 \end{Bmatrix} = \frac{2EI}{L^2} \begin{bmatrix} 4 & -3 & 1 \\ -3 & 6 & -3 \\ 1 & -3 & 2 \end{bmatrix} \begin{Bmatrix} \theta_4 L \\ u_5 \\ \theta_6 L \end{Bmatrix}. \tag{4.30}$$

The solution is

$$\begin{Bmatrix} \theta_4 L \\ u_5 \\ \theta_6 L \end{Bmatrix} = \frac{L^2}{12EI} \begin{bmatrix} 3 & 3 & 3 \\ 3 & 7 & 9 \\ 3 & 9 & 15 \end{bmatrix} \begin{Bmatrix} M_4 \\ P_5 L \\ M_6 \end{Bmatrix}, \tag{4.31}$$

$$\begin{Bmatrix} LP_1 \\ M_2 \\ LP_3 \end{Bmatrix} = \tfrac{1}{2} \begin{bmatrix} 3 & 3 & 3 \\ 1 & 1 & 1 \\ -3 & -5 & -3 \end{bmatrix} \begin{Bmatrix} M_4 \\ P_5 L \\ M_6 \end{Bmatrix}. \tag{4.32}$$

(b) From Equations (4.15) and (4.31) for elements (1) and (2)

$$u_{x1} = -L\theta_4 \left(\frac{y_1}{L}\right)^2 + L\theta_4 \left(\frac{y_1}{L}\right)^3 = -\frac{L^2}{4EI}(M_4 + LP_5 + M_6)\left(\frac{y_1}{L}\right)^2 \left(1 - \frac{y_1}{L}\right), \tag{4.33}$$

$$u_{x2} = L\theta_4 \left(\frac{y_2}{L}\right)\left(1 - \frac{y_2}{L}\right)^2 + (3u_5 - L\theta_6)\left(\frac{y_2}{L}\right)^2 + (-2u_5 + L\theta_6)\left(\frac{y_2}{L}\right)^3$$

$$= \frac{L^2}{4EI}\left(\frac{y_2}{L}\right)\left[M_4 + LP_5\left\{1 + 2\left(\frac{y_2}{L}\right) - \tfrac{2}{3}\left(\frac{y_2}{L}\right)^2\right\} + M_6\left(1 + \frac{2y_2}{L}\right)\right]. \tag{4.34}$$

From Equations (4.16) and (4.31)

$$\sigma_{yy1} = \frac{z}{2I}(M_4 + LP_5 + M_6)\left(1 - \frac{3y_1}{L}\right), \quad \sigma_{yy2} = -\frac{z}{I}[M_6 + P_5(L - y_2)]. \tag{4.35}$$

(c) From Figure 4.5 and Equation (4.26) the internal point shear forces and moments can be expressed as

$$S_6 = M_6, \quad S_5 = P_5, \quad S_4 = M_6 + LP_5 + M_4, \quad S_3 = -\frac{3}{2L}(M_6 + LP_5 + M_4),$$
$$S_2 = M_2 = \tfrac{1}{2}(M_6 + LP_5 + M_4), \quad S_1 = P_1 = \frac{3}{2L}(M_6 + LP_5 + M_4). \tag{4.36}$$

The moment values check those in Equation (4.35) at the corresponding points, where $\sigma_{yy} = -M_x z/I$.

Example 4.5. In Example 4.4 and Figure 4.5 assume there is no support at point (3) but P_3 and M_4 are both applied loads. Find the deflections u_3, θ_4, u_5, θ_6 and use Equation (4.15) to get the deflection equations for u_x in each element.

Solution. From Example 4.4, Equation (4.30) becomes

$$\begin{Bmatrix} LP_3 \\ M_4 \\ LP_5 \\ M_6 \end{Bmatrix} = \frac{2EI}{L^2} \begin{bmatrix} 12 & 0 & -6 & 3 \\ 0 & 4 & -3 & 1 \\ -6 & -3 & 6 & -3 \\ 3 & 1 & -3 & 2 \end{bmatrix} \begin{Bmatrix} u_3 \\ L\theta_4 \\ u_5 \\ L\theta_6 \end{Bmatrix},$$

which gives

$$\begin{Bmatrix} u_3 \\ L\theta_4 \\ u_5 \\ L\theta_6 \end{Bmatrix} = \frac{L^2}{6EI} \begin{bmatrix} 2 & 3 & 5 & 3 \\ 3 & 6 & 9 & 6 \\ 5 & 9 & 16 & 12 \\ 3 & 6 & 12 & 12 \end{bmatrix} \begin{Bmatrix} LP_3 \\ M_4 \\ LP_5 \\ M_6 \end{Bmatrix}.$$

Use these values in Equation (4.15) to get

$$u_{x1} = (3u_3 - L\theta_4)(y_1/L)^2 + (-2u_3 + L\theta_4)(y_1/L)^3$$
$$= (y_1^2/6EI)[(3L - y_1)P_3 + (6L - y_1)P_5 + 3M_4 + 3M_6],$$
$$u_{x2} = (3u_5 - L\theta_6)(y_2/L)^2 + (-2u_5 + L\theta_6)(y_2/L)^3 +$$
$$+ (3u_3 + L\theta_4)\left(1 - \frac{y_2}{L}\right)^2 - (2u_3 + L\theta_4)\left(1 - \frac{y_2}{L}\right)^3$$
$$= u_{x1}(L) + (Ly_2/6EI)[3LP_3 + \{9 + 3(y_2/L) - (y_2/L)^2\}LP_5 +$$
$$+ 6M_4 + 3\{2 + (y_2/L)\}M_6].$$

4.4. Principle of virtual forces for beams

For the simple beam in Figure 4.1 the unit load theorem in Equations (2.41) and (2.42) becomes

$$u_{zm} = \int_0^L \int_A (e_{yy}\sigma^1_{yym} + e_{yz}\sigma^1_{yzm}) dA\, dy,$$
$$\theta_{xn} = \int_0^L \int_A (e_{yy}\sigma^1_{yyn} + e_{yz}\sigma^1_{yzn}) dA\, dy,$$
(4.37)

where σ^1_{yym} and σ^1_{yzm} are the stresses due to a unit virtual load at point m on the beam and σ^1_{yyn} and σ^1_{yzn} are due to a unit virtual moment at point n on the beam. From Equation (4.4), these unit terms can be expressed as

$$\sigma^1_{yym} = -m^1_{xm}z/I_{xx}, \quad \sigma^1_{yzm} = p^1_{zm}/A_s,$$
$$\sigma^1_{yyn} = -m^1_{xn}z/I_{xx}, \quad \sigma^1_{yzn} = p^1_{zn}/A_s.$$
(4.38)

Put e_{yy} and e_{yz} from Equation (4.1) and the terms from Equation (4.38) into Equation (4.37), and make the area integrations to get

$$u_{zm} = \int_0^L \left(\frac{M_x m^1_{xm}}{EI_{xx}} + \frac{V_z p^1_{zm}}{GA_s}\right) dy,$$
$$\theta_{xn} = \int_0^L \left(\frac{M_x m^1_{xn}}{EI_{xx}} + \frac{V_z p^1_{zn}}{GA_s}\right) dy.$$
(4.39)

The area integration is over the effective bending cross section for I_{xx} and over the effective shear area for A_s.

The expressions for m^1_{xm}, p^1_{zm}, m^1_{xn}, and p^1_{zn} depend upon the particular determinate beam used. For the cantilever beam in Figure 4.1 with $u_{z0} = 0$, $\theta_{x0} = 0$, and a unit load and a unit moment applied at any point y on the beam

$$p^1_{zm} = p^1_z(y) = 1, \quad m^1_{xm} = m^1_x(y) = y - y_d, \quad 0 \leq y_d \leq y,$$
$$p^1_{zm} = p^1_z(y) = 0, \quad m^1_{xm} = m^1_x(y) = 0, \quad y \leq y_d \leq L,$$
$$p^1_{zn} = p^1_z(y) = 0, \quad m^1_{xn} = m^1_x(y) = 1, \quad 0 \leq y_d \leq y,$$
$$p^1_{zn} = p^1_z(y) = 0, \quad m^1_{xn} = m^1_x(y) = 0, \quad y \leq y_d \leq L,$$
(4.40)

The virtual principles for simple beams 115

where y_d is the integration variable. Thus, for the cantilever beam, Equation (4.39) gives

$$u_z(y) = \int_0^y \left[\frac{M_x(y - y_d)}{EI_{xx}} + \frac{V_z}{GA_s} \right] dy_d, \quad \theta_x(y) = \int_0^y \frac{M_x}{EI_{xx}} dy_d, \quad (4.41)$$

where M_x, E, I_{xx}, V_z, G, A_s may be functions of y_d. Also, $u_z = u_{zb} + u_{zs}$ for the combined bending and shear deflections.

If the beam has applied point forces and point moments P_{zm} and has redundant supports with unknown reaction forces or moments Q_{zm}, then by assuming the supports temporarily removed to make the beam determinate, it follows that M_x and V_z in Equation (4.39) can be written as

$$M_x = M_{xa} + \sum_{m=1}^G m_{xm}^{1P} P_{zm} + \sum_{m=1}^R m_{xm}^{1Q} Q_{zm},$$

$$V_z = V_{za} + \sum_{m=1}^G p_{zm}^{1P} P_{zm} + \sum_{m=1}^R p_{zm}^{1Q} Q_{zm}. \quad (4.42)$$

M_{xa} and V_{za} are due to the applied distributed forces acting on the selected determinate beam, G is the number of applied point forces, R is the number of redundants, m_{xm}^{1P} and m_{xm}^{1Q} are the moments on the determinate beam produced by unit values of P_{zm} and Q_{zm}, p_{zm}^{1P} and p_{zm}^{1Q} are the shear curves produced by unit values of P_{zm} and Q_{zm}. Apply Equation (4.39) at each redundant support with a specified deflection u_{zmQ} (usually zero) with M_x, V_z given by Equation (4.42), or for compatibility,

$$u_{zmQ} = \int_0^L \left(\frac{M_{xa} m_{xm}^{1Q}}{EI_{xx}} + \frac{V_{za} p_{zm}^{1Q}}{GA_s} \right) dy + \sum_{k=1}^G P_{zk} \int_0^L \left(\frac{m_{xm}^{1Q} m_{xk}^{1P}}{EI_{xx}} + \frac{p_{zm}^{1Q} p_{zk}^{1P}}{GA_s} \right) dy +$$

$$+ \sum_{i=1}^R Q_{zi} \int_0^L \left(\frac{m_{xm}^{1Q} m_{xi}^{1Q}}{EI_{xx}} + \frac{p_{zm}^{1Q} p_{zi}^{1Q}}{GA_s} \right) dy, \quad m = 1, 2, \cdots, R. \quad (4.43)$$

Note that the range of integration may be different for each m, i, and k. This system of equations can be solved for the R reactions Q_{zm} and the results put into Equation (4.42) for the final shear and moment values. This Equation (4.43) corresponds to Equation (3.16) for trusses.

It should be pointed out that the beam can be made determinate by making cuts in the beam at the redundant supports with unknown moments being used to make the rotations $\theta_{xmi} + \theta_{xmj} = 0$. Equation (4.43) can be used for this case.

The beam can be divided into elements so that point values can be used. This case for the force method is considered in the next Section 4.5.

Example 4.6. Find the bending deflection u_{zA} at point A in Figure 4.6 for the given moment curve on the redundant beam by using the unit load theorem. Take $EI_{xx} = 10^5$ Klb-in.

Solution. Select a determinate beam by removing support B and apply a unit load at point A. Thus, Equation (4.41) applies with $y = 200$ in. and

$$M_x = -60 + 1.80 y_d, \quad 0 \leq y_d \leq 100,$$

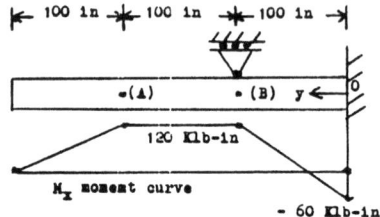

Fig. 4.6. Redundant beam with given bending moment.

$M_z = 120$, $100 \leq y_d \leq 200$,
$m_{zA}^1 = 200 - y_d$, $0 \leq y_d \leq 200$,

whence

$$10^5 u_{zA} = \int_0^{100} (-60 + 1.80 y_d)(200 - y_d) dy_d$$
$$+ \int_{100}^{200} (120)(200 - y_d) dy_d$$
$$= \left[-12,000 y_d + 210 y_d^2 - 0.60 y_d^3 \right]_0^{100}$$
$$+ \left[24,000 y_d - 60 y_d^2 \right]_{100}^{200}, \quad \text{or}$$

$u_{zA} = 9.00$ in.

If the determinate beam is selected by removing the moment at $y = 0$, then

$m_{zA}^1 = y_d$, $0 \leq y_d \leq 100$,
$m_{zA}^1 = 200 - y_d$, $100 \leq y_d \leq 200$,

and the integration over the two intervals using this m_{zA}^1 gives the same value for u_{zA}.

Example 4.7. Solve Example 4.1 by using the unit load theorem.
  Solution. Remove the support at $y = L$ in Figure 4.2 and apply the unknown force $Q_{zL} = P_{zL}$, whence Equation (4.42) gives

$$M_z = M_{zL} + (p/2)(L - y)^2 + Q_{zL}(L - y), \quad m_z^{1Q} = L - y,$$

and Equation (4.43) becomes

$$0 = (1/EI_{zz}) \int_0^L \left[M_{zL} + (p/2)(L - y_d)^2 + Q_{zL}(L - y_d) \right](L - y_d) dy_d.$$

Integration gives

$$(1/2) M_{zL} L^2 + (1/8) p L^4 + (1/3) Q_{zL} L^3 = 0,$$
$$Q_{zL} = P_{zL} = -(3/8) p L - (3/2)(M_{zL}/L),$$

which is the same as P_{zL} in Equation (4.13). Equilibrium of the beam gives P_{z0} and M_{z0} directly. Put the known Q_{zL} into M_z above and use Equation (4.41) for u_{zb}, or

$$EI_{zz} u_{zb} = \int_0^y \left[M_{zL} + (p/2)(L - y_d)^2 + Q_{zL}(L - y_d) \right](y - y_d) dy_d$$
$$= \left(M_{zL} y^2/2 \right) + (py^2/24)(6L^2 - 4Ly + y^2) -$$
$$- (y^2/6)(3L - y)\left[(3pL/8) + (3M_{zL}/2L) \right]$$

$$= (py^2/48)(L-y)(3L-2y) - (M_{zL}y^2/4L)(L-y),$$

which is the same as given by the principle of virtual displacements. Note that three of the boundary conditions are used in writing the equation for M_z while the fourth condition is used in writing Equation (4.43). This demonstrates a major difference in using the two principles.

Example 4.8. Find the deflection u_{zA} at point (A) of the beam shown in Figure 4.7 by using the unit load theorem with EI constant.

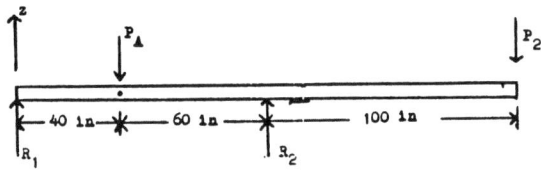
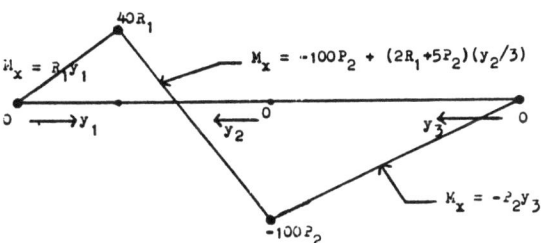
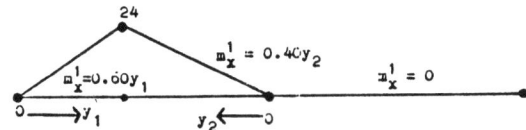

Fig. 4.7. Deflection of simple supported beam.

Solution. Equation (4.39) can be used for this beam as

$$EI u_{zA} = \int_0^L M_z m_z^1 \, dy. \tag{4.44}$$

From Figure 4.7 the reactions are

$$R_1 = 0.60P_A - P_2, \quad R_2 = 0.40P_A + 2P_2,$$

and the resulting moment curves for M_z are shown in Figure 4.7. Apply a unit force at point (A) in the direction P_A to obtain the m_z^1 moment curves shown in Figure 4.7. Note that different origins have been used for the equations in order to simplify the integrations. This is permissible because the integral simply represents an area under a curve with dummy integration variables. Equation (4.44) gives

$$EI u_{zA} = \int_0^{40} (R_1 y_1)(0.60y_1) dy_1 + \int_0^{60} \left[-100P_2 + (2R_1 + 5P_2)\frac{y_2}{3}\right](0.40y_2) dy_2$$

$$= 12{,}800 R_1 + 2880(4P_A - 15P_2)$$

$$= 19{,}200 P_A - 56{,}000 P_2,$$

where u_{zA} is positive down. Note that it is not necessary that the load P_A be present at point (A). The unit load can be applied at any point whether a load is present or not.

Example 4.9. Use the unit load theorem to find (a) the reactions and bending moments in the redundant beam frame in Figure 4.8 and (b) the horizontal bending deflection u_D at point (D). Assume $(EI)_1 = EI$, $(EI)_2 = 2EI$, $(EI)_3 = 3EI$ and neglect shear and axial deflections.

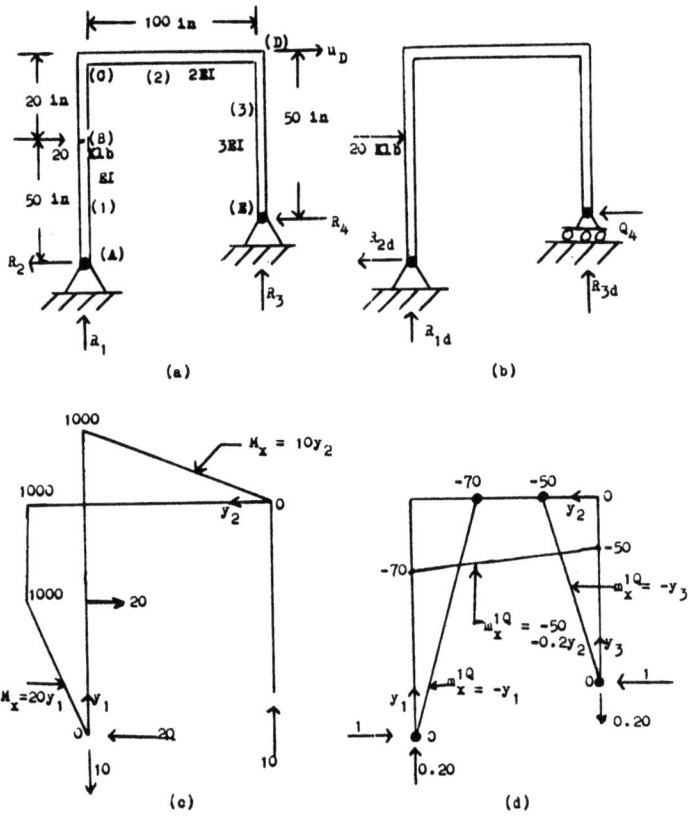

Fig. 4.8. Redundant beam frame.

Solution (a) Make the frame determinate by removing the horizontal restraint at support (E) and replacing it with the unknown redundant force $Q_4 = R_4$, Figure 4.8(b). The moment curves on the determinate frame due to the applied force at point (B) are shown in Figure 4.8(c), while the moment curves due to a unit value of Q_4 are shown in Figure 4.8(d). From Equation (4.42) the moment equations on sections (1) and (2) of the frame are

$$M_{z1} = 20y_1 - Q_4 y_1, \quad 0 \leq y_1 \leq 50,$$
$$M_{z1} = 1000 - Q_4 y_1, \quad 50 \leq y_1 \leq 70,$$
$$M_{z2} = 10y_2 - Q_4(50 + 0.2y_2), \quad 0 \leq y_2 \leq 100,$$
$$M_{z3} = -Q_4 y_3, \quad 0 \leq y_3 \leq 50,$$

whence Equation (4.39) becomes

$$0 = \int_0^{50} \frac{(20y_1 - Q_4 y_1)(-y_1)dy_1}{EI} + \int_{50}^{70} \frac{(1000 - Q_4 y_1)(-y_1)dy_1}{EI} +$$
$$+ \int_0^{100} \frac{[10y_2 - Q_4(50 + 0.2y_2)](-50 - 0.2y_2)dy_2}{2EI} + \int_0^{50} \frac{Q_4(-y_3)^2 dy_3}{3EI},$$

or

$$0 = \left[(-20 + Q_4)(y_1^3/3)\right]_0^{50} + \left[(-500 + \frac{Q_4 y_1}{3})y_1^2\right]_{50}^{70} + \left[(1/3)(Q_4 y_3^3/3)\right]_0^{50} +$$
$$+ (1/2)\left[-y_2^2(250 + \tfrac{2}{3}y_2) + Q_4 y_2(2500 + 10y_2 + \tfrac{0.04}{3}y_2^2)\right]_0^{100}.$$

The results are $Q_4 = 11.67$ Klb $= R_4$, $R_2 = 8.33$ Klb, $R_3 = 7.67$ Klb, $R_1 = -7.67$ Klb,

$M_{x1} = 8.33y_1, \quad 0 \leq y_1 \leq 50,$
$M_{x1} = 1000 - 11.67y_1, \quad 50 \leq y_1 \leq 70,$
$M_{x2} = -584 + 7.67y_2, \quad 0 \leq y_2 \leq 100,$
$M_{x3} = -11.67y_3, \quad 0 \leq y_3 \leq 50.$

(b) To find the deflection u_D, select the same determinate frame as in Figure 4.8(b) and apply a unit load at (D) in the direction of u_D. The resulting m_z^1 moments for Equation (4.39) are

$$m_{x1}^1 = y_1, \quad m_{x2}^1 = 0.70y_2, \quad m_{x3}^1 = 0,$$

whence

$$EIu_{xD} = \int_0^{50} (8.33y_1)y_1 \, dy_1 + \int_{50}^{70} (1000 - 11.67y_1)y_1 \, dy_1 +$$
$$+ (1/2)\int_0^{100} (-584 + 7.67y_2)(0.70y_2)dy_2$$
$$= 57.19(10)^4 \text{ Klb} - \text{in}^3.$$

4.5. Point values for beam elements by the unit load theorem

In figure 4.1 take $u_{x0} = 0$ and $\theta_{x0} = 0$ for the cantilever beam case and let the end values P_{xL} and M_{xL} be the only applied forces, then

$$V_x = P_{xL}, \qquad M_x = M_{xL} + (L-y)P_{xL}. \tag{4.45}$$

Put these expressions into Equation (4.41) to get u_{xL} and θ_{xL} for the beam element or

$$u_{xL} = \frac{M_{xL}L^2}{2EI} + \frac{P_{xL}L^3}{3EI} + \frac{P_{xL}L}{GA_s}, \qquad \theta_{xL} = \frac{M_{xL}L}{EI} + \frac{P_{xL}L^2}{2EI}. \tag{4.46}$$

The reactions at the wall $y = 0$ are

$$P_{x0} = -P_{xL}, \qquad M_{x0} = -M_{xL} - P_{xL}L. \tag{4.47}$$

The Equation (4.46) relates to Equation (4.21) for the beam element, except in Equation (4.46) $u_{x0} = 0$ and $\theta_{x0} = 0$, and the shear deflection is included.

If Equation (4.46) is solved for P_{zL} and M_{xL}, the results check Equation (4.21) without the shear term, or

$$P_{zL} = \frac{EI}{L^3 H_z}(12u_{zL} - 6L\theta_{xL}), \quad H_z = 1 + \frac{12EI}{L^2 GA_s},$$
$$M_{xL} = \frac{EI}{L^3 H_z}[-6Lu_{zL} + (3 + H_z)L^2\theta_{xL}]. \tag{4.48}$$

From Equations (4.26), (4.42), and (3.15) the point shear force and point moment for the left end of element (k), Figure 4.4, can be expressed in the form

$$S_{2k-1} = \sum_{i=1}^{2k-1} c_i P_i + \sum_{i=n}^{2k-1} d_i Q_i, \quad S_{2k} = \sum_{i=1}^{2k-1} a_i P_i + \sum_{i=n}^{2k-1} b_i Q_i. \tag{4.49}$$

The Q_i are the unknown values of some of the P_i, either forces or moments, for the redundant beam.

From Equation (4.45) for element (k), the shear and moment equations are

$$V_{2k} = S_{2k-1}, \quad M_{xk} = S_{2k} + S_{2k-1}(L_k - y_k), \tag{4.50}$$

whence the element deflections are given by Equation (4.46). From Equation (4.41) the deflections at the left end of element (m) for a beam with N elements can be expressed in the form

$$u_{2m-1} = \sum_{k=m}^{N}(\Delta u)_{zk} + \sum_{k=m}^{N-1} L_k u_{2k+2}, \tag{4.51}$$
$$(\Delta u)_{zk} = (L^3/3EI)_k S_{2k-1} + (L^2/2EI)_k S_{2k} + (L/GA_s)_k S_{2k-1},$$
$$u_{2m} = \sum_{k=m}^{N}(\Delta u)_{xk}, \quad (\Delta u)_{xk} = \left(\frac{L^2}{2EI}\right)_k S_{2k-1} + \left(\frac{L}{EI}\right)_k S_{2k}. \tag{4.52}$$

If Q_i in Equation (4.49) is an unknown point force at point $2n - 1$, then use Equation (4.51) with $u_{2n-1,Q}$ as a specified value, which is usually zero.

From Equations (4.41) and (4.50) the relative deflections along the beam element in terms of the S_{2k-1} and S_{2k} forces are

$$u_{2k-1} = \frac{S_{2k}y_k^2}{2E_k I_k} + \frac{S_{2k-1}y_k^2}{6E_k I_k}(3L_k - y_k) + \frac{S_{2k-1}y_k}{G_k A_{sk}},$$
$$u_{2k} = \frac{S_{2k}y_k}{E_k I_k} + \frac{S_{2k-1}y_k}{2E_k I_k}(2L_k - y_k). \tag{4.53}$$

Example 4.10. Use the unit load procedure to solve Example 4.4 for the beam in Figure 4.5.

Solution (a) Remove the support at point (3) in Figure 4.5 and apply the unknown load $Q_3 = P_3$. Use the notation of Figure 4.4 and Equations (4.49), (4.51), and (4.52) to get

$$S_1 = P_5, \quad S_2 = M_6, \quad S_3 = Q_3 + P_5, \quad S_4 = M_6 + M_4 + LP_5,$$
$$6EI(\Delta u)_{z1} = 2L^3 P_5 + 3L^2 M_6, \quad 6EI(\Delta u)_{x1} = 3L^2 P_5 + 6LM_6,$$
$$6EI(\Delta u)_{z2} = 2L^3(P_5 + Q_3) + 3L^2(M_6 + M_4 + LP_5),$$
$$6EI(\Delta u)_{x2} = 3L^2(P_5 + Q_3) + 6L(M_6 + M_4 + LP_5). \tag{4.54a}$$

The virtual principles for simple beams

Set $u_3 = 0$, whence

$$0 = (\Delta u)_{x2} = (L^2/6EI)[2L(P_5 + Q_3) + 3(M_6 + M_4 + LP_5)],$$
$$LQ_3 = LP_3 = -(1/2)(3M_4 + 3M_6 + 5LP_5). \tag{4.54b}$$

The reactions at the supports are

$$P_1 = -S_3 = -Q_3 - P_5 = (3/2L)(M_4 + M_6 + LP_5),$$
$$M_2 = -S_4 - LS_3 = (1/2)(M_4 + M_6 + LP_5).$$

These values agree with Equation (4.32) in Example 4.4.
Use Equations (4.51)-(4.54b) to get the point deflections

$$u_4 = \theta_4 = (\Delta u)_{x2} = (L/4EI)(M_4 + M_6 + LP_5),$$
$$u_2 = \theta_6 = (\Delta u)_{x1} + (\Delta u)_{x2} = (L/4EI)(M_4 + 3LP_5 + 5M_6),$$
$$u_1 = u_5 = (\Delta u)_{x1} + (\Delta u)_{x2} + Lu_4 = (\Delta u)_{x1} + 0 + L\theta_4$$
$$= (L^2/12EI)(3M_4 + 9M_6 + 7LP_5),$$

which agree with Equation (4.31).
(b) From Equation (4.53) the deflections in each beam element are (shear omitted)

$$u_{x1} = \frac{S_4 y_1^2}{2EI} + \frac{S_3 y_1^2}{6EI}(3L - y_1) = -\frac{L^2}{4EI}(M_4 + M_6 + LP_5)\left(\frac{y_1}{L}\right)^2\left(1 - \frac{y_1}{L}\right),$$

$$u_{x2} = \theta_4 y_2 + \frac{M_6 y_2^2}{2EI} + \frac{P_5 y_2^2}{6EI}(3L - y_2)$$
$$= \frac{L^2}{12EI}\left(\frac{y_2}{L}\right)\left[3M_4 + LP_5\left\{3 + 6\left(\frac{y_2}{L}\right) - 2\left(\frac{y_2}{L}\right)^2\right\} + M_6\left\{3 + 6\left(\frac{y_2}{L}\right)\right\}\right],$$

which agree with Equations (4.33) and (4.34).
This Example 4.10 using the principle of virtual forces and Example 4.4 using the principle of virtual displacements demonstrate the completely different sequence of calculations of the two methods on the same beam.

Example 4.11. (a) Find the point shear forces and bending moments for the redundant beam in Figure 4.9. (b) Find the deflection u_1 and slope u_6. Omit the shear term, use $(EI)_k = EI$, and take the unknown redundant forces as Q_3 and Q_5.

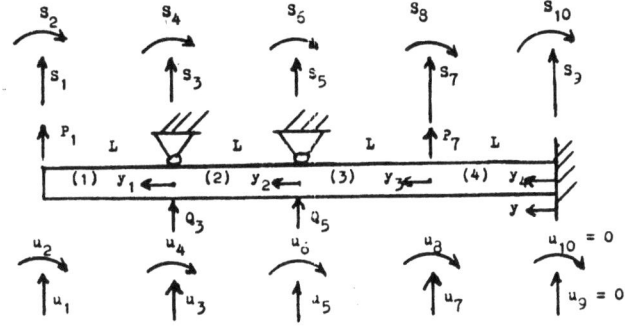

Fig. 4.9. Beam with two redundants (EI constant).

Solution (a) From Figure 4.9 and Equations (4.49) and (4.50)

$$S_1 = P_1, \quad S_2 = 0, \quad S_3 = P_1 + Q_3, \quad S_4 = LP_1, \quad S_5 = P_1 + Q_3 + Q_5,$$
$$S_6 = 2LP_1 + LQ_3, \quad S_7 = P_1 + Q_3 + Q_5 + P_7, \quad S_8 = 3LP_1 + 2LQ_3 + LQ_5,$$
$$S_9 = S_7, \quad S_{10} = 4LP_1 + 3LQ_3 + 2LQ_5 + LP_7.$$

From Equations (4.51) and (4.52) with $R = 6EI/L^2$,

$R(\Delta u)_{z1} = 2LP_1, \quad R(\Delta u)_{z1} = 3P_1, \quad R(\Delta u)_{z2} = 2L(P_1 + Q_3) + 3LP_1,$
$R(\Delta u)_{z2} = 3(P_1 + Q_3) + 6P_1, \quad R(\Delta u)_{z3} = 2L(P_1 + Q_3 + Q_5) + 3L(2P_1 + Q_3),$
$R(\Delta u)_{z3} = 3(P_1 + Q_3 + Q_5) + 6(2P_1 + Q_3),$
$R(\Delta u)_{z4} = 2L(P_1 + Q_3 + Q_5 + P_7) + 3L(3P_1 + 2Q_3 + Q_5),$
$R(\Delta u)_{z4} = 3(P_1 + Q_3 + Q_5 + P_7) + 6(3P_1 + 2Q_3 + Q_5),$
$0 = u_3 = (\Delta u)_{z2} + (\Delta u)_{z3} + (\Delta u)_{z4} + L(\Delta u)_{z3} + 2L(\Delta u)_{z4},$
$0 = u_5 = (\Delta u)_{z3} + (\Delta u)_{z4} + L(\Delta u)_{z4},$
$0 = 81P_1 + 54Q_3 + 28Q_5 + 8P_7, \quad 0 = 40P_1 + 28Q_3 + 16Q_5 + 5P_7,$
$Q_3 = -(11/5)P_1 + (3/20)P_7, \quad Q_5 = (27/20)P_1 - (23/40)P_7.$

The point shear forces and moments become

$S_1 = P_1, \quad 20S_3 = -24P_1 + 3P_7, \quad 40S_5 = 6P_1 - 17P_7, \quad 40S_7 = 6P_1 + 23P_7,$
$40S_9 = 6P_1 + 23P_7, \quad S_2 = 0, \quad S_4 = LP_1, \quad 20S_6 = -4LP_1 + 3LP_7,$
$40S_8 = -2LP_1 - 11LP_7, \quad 10S_{10} = LP_1 + 3LP_7.$

(b) From Equations (4.51) and (4.52) the deflection u_1 and slope u_6 are

$u_1 = (\Delta u)_{z1} + Lu_4 = (L^3/6EI)(47P_1 + 27Q_3 + 12Q_5 + 3P_7)$
$= (L^3/120EI)(76P_1 + 3P_7),$
$u_6 = (\Delta u)_{z3} + (\Delta u)_{z4} = (L^2/6EI)(36P_1 + 24Q_3 + 12Q_5 + 3P_7)$
$= -(L^2/20EI)(2P_1 + P_7).$

4.6. Principle of mixed virtual stresses and virtual displacements for beams

For the simple beam case in Figure 4.1 the principle of mixed virtual stresses and displacements in Equation (2.46) has the form

$$\int_0^L \int_A [\sigma_{yy} e_{yy}^V + (e_{yy}u - e_{yy}\sigma)\sigma_{yy}^V + \sigma_{yz} e_{yz}^V + (e_{yzu} - e_{yz\sigma})\sigma_{yz}^V] dA\, dy -$$
$$- \int_0^L p_z u_z^V\, dy - P_{zL} u_{zL}^V - P_{z0} u_{z0}^V - M_{zL} \theta_{zL}^V - M_{z0} \theta_{z0}^V = 0, \qquad (4.55)$$

where the P_{zm} terms have been omitted in order to simplify the equations. Use Equations (4.1) and (4.3) and integrate the first integral by parts, or

$$\int_0^L \int_A \left[\left(-\frac{M_x z}{I_{xx}}\right)\left(-z\frac{d^2 u_{zb}^V}{dy^2}\right) + \left(-z\frac{d^2 u_{zb}}{dy^2} + \frac{M_x z}{EI_{xx}}\right)\left(-\frac{M_x^V z}{I_{xx}}\right) + \right.$$
$$\left. + \left(\frac{V_z}{A_s}\right)\left(\frac{du_{zs}^V}{dy}\right) + \left(\frac{du_{zs}}{dy} - \frac{V_z}{GA_s}\right)\left(\frac{V_z^V}{A_s}\right)\right] dA\, dy$$
$$= \int_0^L \left[M_x \frac{d^2 u_{zb}^V}{dy^2} + \left(\frac{d^2 u_{zb}}{dy^2} - \frac{M_x}{EI_{xx}}\right)M_x^V + V_z \frac{du_{zs}^V}{dy} + \left(\frac{du_{zs}}{dy} - \frac{V_z}{GA_s}\right)V_z^V\right] dy$$
$$= \left[M_x \theta_x^V\right]_0^L - \left[\frac{dM_x}{dy} u_z^V\right]_0^L + \int_0^L u_{zb}^V \frac{d^2 M_x}{dy^2} dy + \left[u_{zs}^V V_z\right]_0^L -$$

The virtual principles for simple beams 123

$$-\int_0^L u_{zs}^V \frac{dV_z}{dy} dy + \int_0^L \left[\left(\frac{d^2 u_{zb}}{dy^2} - \frac{M_x}{EI_{xx}}\right) M_x^V + \left(\frac{du_{zs}}{dy} - \frac{V_z}{GA_s}\right) V_z^V\right] dy.$$

Put this result into Equation (4.55) and use the arbitrary $M_x^V, V_z^V, \theta_x^V, u_z^V = u_{zb}^V + u_{zs}^V$ to get the differential equations and equilibrium boundary conditions,

$$\frac{d^2 M_x}{dy^2} - p_z = 0, \tag{4.56}$$

$$\left(\frac{dM_x}{dy}\right)_L + P_{zL} = 0, \quad \left(\frac{dM_x}{dy}\right)_0 - P_{z0} = 0, \tag{4.57a}$$

$$M_x(L) - M_{xL} = 0, \quad M_x(0) + M_{x0} = 0, \tag{4.57b}$$

$$\frac{dV_z}{dy} + p_z = 0, \tag{4.58}$$

$$V_z(L) - P_{zL} = 0, \quad V_z(0) + P_{z0} = 0, \tag{4.59}$$

$$\frac{d^2 u_{zb}}{dy^2} - \frac{M_x}{EI_{xx}} = 0, \tag{4.60}$$

$$\frac{du_{zs}}{dy} - \frac{V_z}{GA_s} = 0, \tag{4.61}$$

where the boundary conditions for the displacements are given in Equation (4.2). See the discussion on the permissible combinations of boundary conditions given above for the principle of virtual displacements equations in Section 4.2. The primary difference between the results given by the two principles in Equations (4.6)-(4.8) and in Equations (4.56)-(4.61) is that the mixed principle has split the differential equations into two sets of lower order as compared to the higher order equations given by the principle of virtual displacements. There is little difference in making the integrations for this simple beam case. However, the matrix forms for combined beam elements with points and for approximate solutions are quite different for the two principles. See Section 4.10 below for the matrix forms.

4.7. Inelastic and temperature effects in simple beams

If the temperature varies on the beam cross section, it may cause thermal stresses, material property changes, and inelastic effects. The strain term e_E in Equation (2.57), which represents the various additional effects from temperature strain, initial strain, deflection strain, and inelastic strain, will be a function of the cross section variable z and possibly a function of the length variable y in the beam, or

$$e_E(y, z) = e_T(y, z) + e_I(y, z) \mp e_d(y) + e_p(y, z). \tag{4.62a}$$

Assume that e_E affects only the beam bending, whence Equations (4.1), (1.77),

(1.78) and (1.92) give

$$e_{total} = (\sigma_{yy}/E_{yy}) + e_E = -z\frac{d^2 u_{zb}}{dy^2},$$

$$\sigma_{yy} = -E_{yy}\left(z\frac{d^2 u_{zb}}{dy^2} + e_E\right) = \frac{E_{yy}}{E_R}[-E_R e_E - (M_x + M_{xE})\frac{z}{I_{xx}}]. \tag{4.62b}$$

Put the σ_{yy} expression into the first term in the principle of virtual displacements in Equation (4.5) to get

$$\int_0^L \int_A \sigma_{yy} e_{yy}^V \, dA \, dy = \int_0^L \int_A \left(-E_{yy} z \frac{d^2 u_{zb}}{dy^2} - E_{yy} e_E\right)\left(-z\frac{d^2 u_{zb}^V}{dy^2}\right) dA \, dy$$

$$= \int_0^L \left(E_{yy} I_{xx} \frac{d^2 u_{zb}}{dy^2} - M_{xE}\right)\frac{d^2 u_{zb}^V}{dy^2} \, dy, \tag{4.63}$$

where

$$M_{xE} = -\int_A E_{yy} e_E z \, dA = M_{xT} + M_{xI} + M_{xd} + M_{xp},$$

$$M_{xT} = -\int_A E_{yy} e_T z \, dA, \quad M_{xI} = -\int_A E_{yy} e_I z \, dA, \text{ etc.,} \tag{4.64}$$

are the apparent moments due to the additional strain terms. Thus, the Equations (4.6) and (4.7) from the principle of virtual displacements take the form

$$\frac{d^2}{dy^2}\left(EI_{xx}\frac{d^2 u_{zb}}{dy^2} - M_{xE}\right) - p_z = 0, \tag{4.65}$$

$$\left[EI_{xx}\frac{d^2 u_{zb}}{dy^2} - M_{xE} - M_{xL}\right]_{y=L} = 0,$$

$$\left[EI_{xx}\frac{d^2 u_{zb}}{dy^2} - M_{xE} + M_{x0}\right]_{y=0} = 0,$$

$$\left[\frac{d}{dy}\left(EI_{xx}\frac{d^2 u_{zb}}{dy^2} - M_{xE}\right) + P_{zL}\right]_{y=L} = 0,$$

$$\left[\frac{d}{dy}\left(EI_{xx}\frac{d^2 u_{zb}}{dy^2} - M_{xE}\right) - P_{z0}\right]_{y=0} = 0. \tag{4.66}$$

For the principle of virtual forces put the strain e_{total} from Equation (4.62b) into Equation (4.37). This gives the following form for Equation (4.39),

$$u_{zm} = \int_0^L \left[\frac{(M_x + M_{xE})m_{xm}^1}{EI_{xx}} + \frac{V_z p_{zm}^1}{GA_s}\right] dy,$$

$$\theta_{xn} = \int_0^L \left[\frac{(M_x + M_{xE})m_{xn}^1}{EI_{xx}} + \frac{V_z p_{zn}^1}{GA_s}\right] dy. \tag{4.67}$$

Also,

Replace M_{xa} with $M_{xa} + M_{xE}$ in Equations (4.41) – (4.43). \hfill (4.68)

For the principle of mixed virtual stresses and virtual displacements

Replace M_x with $M_x + M_{xE}$ in Equations (4.57), (4.56), (4.60). (4.69)

In the Ramberg-Osgood Equation (1.94) for the inelastic stress-strain curve, the strain e_{yy}^s on any cross section of the simple beam considered here is given by Equations (1.95) and (4.62), as

$$e_{yy}^s = e_e + e_p = (\sigma_{yy}/E_{yy}) + e_p$$
$$= -e_T - e_I - e_d - (M_x + M_{xE})(z/E_R I_{xxE}), \quad (4.70)$$

where e_I, e_T, e_d may be functions of z. When this expression is determined at selected points on the cross section, or for selected elements on the cross section, then the stress σ_{yy} at the corresponding points is given by Table 1.1. The iteration procedure for the inelastic case on the cross section is the same as described in Section 1.14. If the moment M_x and the apparent moment M_{xE} vary along the beam length, then the inelastic analysis must be made at several cross sections, where some cross sections may be elastic. In this case M_{xE} will be known only at certain points so that the integration for deflections in Equation (4.67) could be a summation process.

If the beam cross section has a thin web that takes essentially no bending moment so that the simple beam may be considered as a two element cross section for bending, Figure 4.1, and if the temperature variation on the cross section is approximately linear so that M_{xT} may exist without thermal stresses on the cross section, then the solution for the inelastic effects is much simpler than that described above. In this case for spar caps (1) and (2) in Figure 4.1, the expressions in the principle of virtual forces, Equation (4.67), are

$$e_{yy1} = -\frac{M_x}{hA_1E_1}, \quad e_{yy2} = \frac{M_x}{hA_2E_2}, \quad \sigma_{yy1}^1 = -\frac{m_x^1}{hA_1},$$
$$\sigma_{yy2}^1 = \frac{m_x^1}{hA_2}, \quad e_{E1} = e_{T1} + e_{p1}, \quad e_{E2} = e_{T2} + e_{p2}, \quad (4.71)$$
$$e_{p1} = \frac{3F_{y1}}{7E_1}\left[\frac{-M_x}{hA_1F_{y1}}\right]^{10}, \quad e_{p2} = \frac{3F_{y2}}{7E_2}\left[\frac{M_x}{hA_2F_{y2}}\right]^{10}.$$

For this case, Equation (4.67) has the form

$$u_z = \int_0^L \left[\left(\frac{M_x}{h}\right)\left(\frac{1}{A_1E_1} + \frac{1}{A_2E_2}\right) + e_{T2} - e_{T1} + e_{p2} - e_{p1}\right] \frac{m_x^1}{h} dy. \quad (4.72)$$

Thus, with M_x known on the determinate beam, e_{p1} and e_{p2} can be calculated at various cross sections and the integrations in Equation (4.72) made by summations. In some cases, it may be possible to make the integrations directly. Note that h, A_1, A_2, E_1, E_2, e_{T1}, e_{T2}, F_{y1}, F_{y2} as well as M_x may be functions of y along the beam.

If the beam is redundant, then M_x will change due to the e_E or M_{xE} effects. That is, when M_{xE} or e_E is added in Equation (4.43), the Q_{zi} will have different values and M_x in Equation (4.42) will change. For this case iteration is necessary starting with the elastic redundant solution with e_{p1} and e_{p2} zero, and proceeding to change the e_{p1}, e_{p2}, and Q_{xi} values until convergence on the Q_{xi} is obtained. The procedure at each cross section is similar to that outlined for trusses in Equations (3.28)-(3.32). For a single redundant a trial and error solution can be used.

For the beam element case with points it is necessary to use equivalent point values for the e_E strain effects. From Equation (4.62b) put the e_E term from

$$\sigma_{yy} = E_{yy}(-zu_{xb,yy} - e_E) \tag{4.73}$$

into Equation (4.5) and use e_{yy}^V from Equation (4.16) as

$$e_{yy}^V = \sigma_{yy}^V/E_{yy} = (z/L^3)[6(L-2y)u_{x0}^V + 2L(2L-3y)\theta_{x0}^V - \\ - 6(L-2y)u_{xL}^V + 2L(L-3y)\theta_{xL}^V]. \tag{4.74}$$

This gives the end point terms

$$-P_{x0E}u_{x0}^V - M_{x0E}\theta_{x0}^V + P_{xLE}u_{xL}^V - M_{xLE}\theta_{xL}^V, \tag{4.75}$$

$$P_{x0E} = -(6/L^3)\int_0^L M_{xE}(L-2y)\,dy,$$

$$M_{x0E} = -(2/L^2)\int_0^L M_{xE}(2L-3y)\,dy,$$

$$P_{xLE} = (6/L^3)\int_0^L M_{xE}(L-2y)\,dy, \tag{4.76}$$

$$M_{xLE} = -(2/L^2)\int_0^L M_{xE}(L-3y)\,dy,$$

$$M_{xE} = -\int_A E_{yy}e_E z\,dA,$$

which can be added to Equation (4.19). Thus, the end point loads and moments in Equations (4.19) and (4.21) consist of three parts,

$$\begin{aligned} P_0 &= P_{x0} + P_{x0d} + P_{x0E}, & M_0 &= M_{x0} + M_{x0d} + M_{x0E}, \\ P_L &= P_{xL} + P_{xLd} + P_{xLE}, & M_L &= M_{xL} + M_{xLd} + M_{xLE}. \end{aligned} \tag{4.77}$$

In the notation of Equation (4.22) and Figure 4.4 it follows that Equation (4.33) becomes

$$\begin{aligned} S_1 + S_{E1} &= (6EI/L^3)_1(2u_1 - L_1u_2 - 2u_3 - L_1u_4), \\ S_2 + S_{E2} &= (2EI/L^2)_1(-3u_1 + 2L_1u_2 + 3u_3 + L_1u_4), \end{aligned} \tag{4.78}$$

$$S_{E1} = (6/L_1^3)\int_0^{L_1} M_{xE1}(L_1 - 2y_1)\,dy_1, \quad M_{xE1} = \int_{A_1} E_1 e_{E1} z\,dA_1,$$

$$S_{E2} = -(2/L_1^2)\int_0^{L_1} M_{xE1}(L_1 - 3y_1)\,dy_1. \tag{4.79}$$

The following examples show temperature and inelastic effects on simple beams. See References 1 and 2 for other examples.

The virtual principles for simple beams

Example 4.12. Add a thermal moment distribution $M_{zT} = (y_1/100)(M_{zT})_2$, $0 \le y_1 \le 200$, to the beam in Figure 4.7 and Example 4.8. Find the deflection u_{zA} at point (A).

Solution. From Equations (4.67) and (4.44),

$$EIu_z = \int_0^L (M_z + M_{zT})m_z^1\, dy. \tag{4.80}$$

Rewrite M_{zT} to fit the m_z^1 equations in Figure 4.7 so that the M_{zT} term in Equation (4.80) gives

$$(M_{zT})_2 \int_0^{40} (y_1/100)(0.60y_1)dy_1 + (M_{zT})_2 \int_0^{60}\left(1 - \frac{y_2}{100}\right)(0.40y_2)dy_2$$

$$= (M_{zT})_2 (128 + 432) = 560(M_{zT})_2.$$

From Example 4.8, the final solution is

$$EIu_{zA} = 19,200P_A - 56,000P_2 + 560(M_{zT})_2.$$

Example 4.13. Suppose the redundant beam in Figure 4.2 is subjected to a thermal moment $M_{zT} = (M_{zT})_0 + (M_{zT})_1 (y/L)^2$. Find the bending deflection equation for u_{zb} and the reactions P_{zL}, P_{z0}, and M_{z0} by using the unit load theorem.

Solution. Follow the solution of Example 4.7 and Equation (4.68), whence $M_z = (L - y)Q_{zL}$ and

$$0 = \int_0^L \left[(M_{zT})_0 + (M_{zT})_1 (y_d/L)^2 + (L - y_d)Q_{zL}\right](L - y_d)dy_d,$$

$$0 = (M_{zT})_0 (L^2/2) + (M_{zT})_1 (L^2/12) + Q_{zL}(L^3/3),$$

$$Q_{zL} = P_{zL} = -(3/2)(M_{zT})_0/L - (M_{zT})_1/4L,$$

$$P_{z0} = (3/2)(M_{zT})_0/L + (M_{zT})_1/4L, \tag{4.81}$$

$$M_{z0} = (3/2)(M_{zT})_0 + (1/4)(M_{zT})_1.$$

The deflection equation is

$$EIu_{zb} = \int_0^y \left[(M_{zT})_0 + (M_{zT})_1\left(\frac{y_d}{L}\right)^2 - (L - y_d)\left\{\frac{3(M_{zT})_0}{2L} + \frac{(M_{zT})_1}{4L}\right\}\right](y - y_d)dy_d$$

$$= -(M_{zT})_0 \frac{y^2(L-y)}{4L} - (M_{zT})_1 \left(\frac{y}{L}\right)^2 \frac{(L-y)(3L+2y)}{24}. \tag{4.82}$$

Example 4.14. For the beam in Figure 4.10, let $A_1E_1 = 225,000$ Klb, $A_2E_2 = 250,000$ Klb, $A_1F_{y1} = -750$ Klb, $A_2F_{y2} = 1000$ Klb, $(\alpha T)_1 = 0.001$, $(\alpha T)_2 = 0.004$. Find the deflection u_{zA} at point (A) allowing for temperature and inelastic effects.

Solution. From the M_z equations in Figure 4.10 and the specified data, Equation (4.71) gives

$$e_{p1} = \frac{3}{7}\left(\frac{-750}{225,000}\right)\left(\frac{-600y_1}{-15,000}\right)^{10} = -\frac{1}{700}(y_1/25)^{10}, \quad 0 \le y_1 \le 40,$$

$$e_{p2} = \frac{3}{7}\left(\frac{1000}{250,000}\right)\left(\frac{600y_1}{20,000}\right)^{10} = \frac{3}{1750}(3y_1/100)^{10}, \quad 0 \le y_1 \le 40,$$

$$e_{p1} = -\frac{1}{700}(2y_2/75)^{10}, \quad e_{p2} = \frac{3}{1750}(y_2/50)^{10}, \quad 0 \le y_2 \le 60.$$

In this particular beam with the selected origins for each part of the moment curve, the Equa-

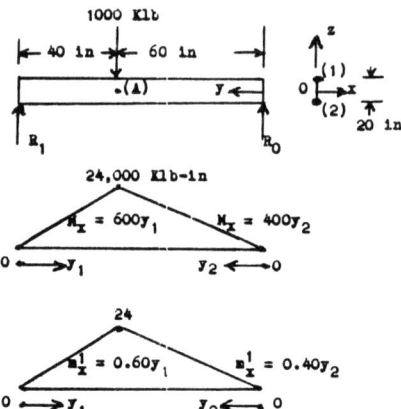

Fig. 4.10. Deflection of inelastic beam with temperature.

tion (4.72) can be integrated directly to give the deflection U_{zA}, or with $(1/20)[(1/225,000) + (1/250000)] = 0.42222(10)^{-6}$,

$$u_{zA} = \int_0^{40} \left[(0.42222)(10)^{-6} 600 y_1 + 0.004 - 0.001 + (3/1750)(3y_1/100)^{10} + (1/700)(y_1/25)^{10}\right](0.03 y_1) dy_1 +$$

$$+ \int_0^{60} \left[(0.42222)(10)^{-6} 400 y_2 + 0.003 + (3/1750)(y_2/50)^{10} + (1/700)(2y_2/75)^{10}\right](0.02 y_2) dy_2$$

$$= 0.90 + 1.36 = 2.26 \text{ in.}$$

$u_{zA,\text{elastic}} = 0.16 + 0.24 = 0.40$ in.,

$u_{zA,\text{temp.}} = 0.07 + 0.11 = 0.18$ in.

From the magnitude of the terms raised to the tenth power, it is evident that the yield region occurs around point (A), extending 10 to 15 in. on either side.

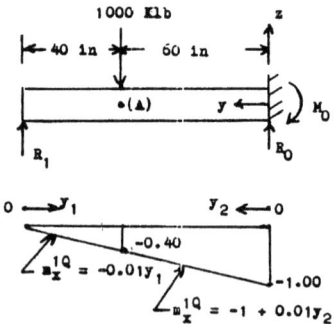

Fig. 4.11. Redundant inelastic beam.

Example 4.15. Suppose the inelastic beam in Figure 4.10 and Example 4.14 is made redundant by requiring the beam to have zero slope at the right end, Figure 4.11. Find the redundant end

The virtual principles for simple beams

moment M_0 for the beam for (a) elastic beam with applied force at point (A) alone, (b) elastic beam with both the applied force and temperature from Example 4.14, (c) inelastic beam with applied force alone and with both the applied force and temperature, and (d) comment on the results. Use the data given in Example 4.14.

Solution (a) The moment curve for the applied force is given in Figure 4.10 and the moment curve for the unit value of the moment M_0 is given in Figure 4.11. From Equation (4.43) and data in Example 4.14,

$$0 = 0.42222(10^{-6})\left[\int_0^{40}(600y_1)(-0.01y_1)dy_1 + \right.$$

$$+ \int_0^{60}(400y_2)(-1 + 0.01y_2)dy_2 + M_0\left\{\int_0^{40}(-0.01y_1)^2 dy_1 + \right.$$

$$\left.\left. + \int_0^{60}(-1 + 0.01y_2)^2 dy_2\right\}\right]$$

$$= 0.42222(10^{-6})(-560,000 + \frac{100}{3}M_0),$$

$M_0 = 16,800$ Klb $-$ in.

(b) Use Equation (4.72) and add the thermal strain term $(\alpha T)_2 - (\alpha T)_1 = 0.004 - 0.001 = 0.003$ to the equation in part (a), or

$$0 = 0.42222(10^{-6})(\frac{100}{3}M_0 - 560,000) + 0.003\left[\int_0^{40}(-0.01y_1)dy_1 + \right.$$

$$\left.+ \int_0^{60}(-1 + 0.01y_2)dy_2\right],$$

whence $M_0 = 27,460$ Klb-in.

(c) The moment curve from Equation (4.42) and Figures 4.10 and 4.11 is

$$M_x = 600y_1 - 0.001M_0 y_1, \quad 0 \le y_1 \le 40,$$
$$M_x = 400y_2 + (-1 + 0.01y_2)M_0, \quad 0 \le y_2 \le 60.$$

From Equation (4.71) and values in Example 4.14, the inelastic strains can be expressed as

$$e_{p1} = -\frac{1}{700}\left(1 - \frac{M_0}{60,000}\right)^{10}\left(\frac{y_1}{25}\right)^{10}, \quad 0 \le y_1 \le 40,$$

$$e_{p2} = \frac{3}{1750}\left(1 - \frac{M_0}{60,000}\right)^{10}\left(\frac{3y_1}{100}\right)^{10}, \quad 0 \le y_1 \le 40,$$

$$e_{p1} = -\frac{1}{700}\left[\left(1 + \frac{M_0}{40,000}\right)\left(\frac{2y_2}{75}\right) - \frac{M_2}{15,000}\right]^{10}, \quad 0 \le y_2 \le 60,$$

$$e_{p2} = \frac{3}{1750}\left[\left(1 + \frac{M_0}{40,000}\right)\left(\frac{y_2}{50}\right) - \frac{M_2}{20,000}\right]^{10}, \quad 0 \le y_2 \le 60.$$

Since the stresses in spar caps (1) and (2) change from tension to compression in the y_2 range, assume the spar caps to have the same stress-strain curves in tension and compression to avoid having to split the y_2 interval in the integrations. Also, note that the brackets in the e_{p1} and e_{p2} expressions in the y_2 range must be positive whether in tension or compression, and that the signs change on the strains.

Use Equation (4.72) and add the inelastic terms to the equation in part (a), or

$$0 = 0.42222(10^{-6})(\frac{100}{3}M_0 - 560,000) +$$

$$+ \left(1 - \frac{M_0}{60,000}\right)^{10}\int_0^{40}\left[\frac{3}{1750}\left(\frac{3y_1}{100}\right)^{10} + \frac{1}{700}\left(\frac{y_1}{25}\right)^{10}\right](-0.01y_1)dy_1 +$$

$$+ \int_0^{60} \left\{ \frac{3}{1750} \left[\left(1 + \frac{M_0}{40,000}\right)\left(\frac{y_2}{50}\right) - \frac{M_0}{20,000} \right]^{10} + \right.$$

$$\left. + \frac{1}{700} \left[\left(1 + \frac{M_0}{40,000}\right)\left(\frac{2y_2}{75}\right) - \frac{M_0}{15,000} \right]^{10} \right\} (-1 + 0.01 y_2) dy_2.$$

Thus, for both the load case and the combined load and temperature case,

$$M_0 = \left\{ \begin{matrix} 16,800 \\ 27,460 \end{matrix} \right\} + 15,890 \left(1 - \frac{M_0}{60,000}\right)^{10} + \left(1 + \frac{M_0}{40,000}\right)^{-2} \left[\left\{ 554\left(1 + \frac{M_0}{40,000}\right) - \right. \right.$$

$$\left. - 254\left(1.2 + \frac{M_0}{28,950}\right) \right\} \left(1.2 - \frac{M_0}{50,000}\right)^{11} + \left\{ -554\left(1 + \frac{M_0}{40,000}\right) - 23 \right\} \left(\frac{M_0}{20,000}\right)^{11} +$$

$$+ \left\{ 346\left(1 + \frac{M_0}{40,000}\right) - 119\left(1.6 + \frac{M_0}{21,710}\right) \right\} \left(1.6 - \frac{M_0}{37,500}\right)^{11}$$

$$\left. + \left\{ -346\left(1 + \frac{M_0}{40,000}\right) - 11 \right\} \left(\frac{M_0}{15,000}\right)^{11} \right].$$

Since the terms $[1.6 - (M_0/37,500)]^{11}$ and $(M_0/15,000)^{11}$ are the largest and most sensitive terms in the equation, it is a simple procedure to get a trial solution for M_0 as

$M_0 = 16,900$ Klb − in for applied force alone,

$M_0 = 20,200$ Klb − in for applied force and temperature.

(d) In the applied force case (a), the elastic moment M_0 has approximately the same magnitude as the moment at point (A). With the material properties the same in tension and compression, the inelastic effects would cause minor changes in M_0. However, the apparent thermal moment is maximum at the wall and tends to increase M_0, as seen in part (b). Thus, large inelastic effects will occur near the wall reducing M_0 from its elastic value. Since the temperature gives a limited strain, it cannot force an inelastic failure in a stable ductile material. If the thermal strain in the above case in part (c) is doubled, M_0 increases only to about 22,000 Klb-in.

4.8. Matrix equations for beams from the unit displacement theorem

From Equations (4.1) and (4.50) for beam element (1) in Figure 4.4,

$$\sigma_{yy} = -\frac{M_x z}{I_{xx}} = -\frac{z}{I_{xx}}[S_2 + S_1(L_1 - y_1)]. \tag{4.83a}$$

From Equation (4.16) for unit values of the displacements, take

$$\left\{ \begin{matrix} e^1_{yy1} \\ e^1_{yy2} \\ e^1_{yy3} \\ e^1_{yy4} \end{matrix} \right\} = -\frac{z}{L_1^3} \left\{ \begin{matrix} 6(L_1 - 2y_1) \\ -2L_1(L_1 - 3y_1) \\ -6(L_1 - 2y_1) \\ -2L_1(2L_1 - 3y_1) \end{matrix} \right\}, \tag{4.83b}$$

whence the unit displacement theorem in Equations (2.35) and (2.36) gives

$$\left\{ \begin{matrix} P_1 \\ M_2 \\ P_3 \\ M_4 \end{matrix} \right\} = \int_0^{L_1} \int_A \left(-\frac{z}{L^3}\right)\left(-\frac{z}{I_{xx}}\right) \left\{ \begin{matrix} 6(L_1 - 2y_1) \\ -2L_1(L_1 - 3y_1) \\ -6(L_1 - 2y_1) \\ -2L_1(2L_1 - 3y_1) \end{matrix} \right\} [L_1 - y_1 \ 1] \left\{ \begin{matrix} S_1 \\ S_2 \end{matrix} \right\} dA\, dy_1$$

$$= \begin{bmatrix} 1 & 0 \\ 0 & 1 \\ -1 & 0 \\ -L_1 & -1 \end{bmatrix} \begin{Bmatrix} S_1 \\ S_2 \end{Bmatrix}, \text{ or} \tag{4.83c}$$

$$\{P\}_1 = [s]_1^T \{S\}_1. \tag{4.84}$$

Note that Equation (4.25) also gives this result. All the elements can be assembled into the equilibrium matrix form

$$\{P\} = [s]^T \{S\}, \tag{4.85}$$

where $[s]^T$ is a lower block bi-diagonal triangular matrix for the cantilever beam in Figure 4.4. For the *determinate* case write Equation (4.85) as

$$\begin{Bmatrix} P \\ P_R \end{Bmatrix} = \begin{bmatrix} s_d \\ s_R \end{bmatrix}^T \{S\}, \tag{4.86}$$

where $\{P_R\}$ are the support reactions. In this case the point shear forces and point moments $\{S\}$ can be obtained directly as

$$\{S\} = [s_d^T]^{-1} \{P\}. \tag{4.87}$$

See the corresponding truss Equations (3.38)-(3.40).

Although the $\{S\}$ values can be obtained by the procedure in Equation (4.26), the procedure becomes complicated for a large number of elements. On the other hand, the inversion of the lower triangular matrix $[s_d]^T$ is simple to do by the procedure in Section 4.9 below.

The compatibility Equation (4.78) can be written in matrix form as

$$\begin{Bmatrix} S_1 + S_{1E} \\ S_2 + S_{2E} \end{Bmatrix} = \left(\frac{EI}{L^3}\right) \begin{bmatrix} 12 & -6L_1 \\ -6L_1 & 4L_1^2 \end{bmatrix} \begin{bmatrix} 1 & 0 & -1 & -L_1 \\ 0 & 1 & 0 & -1 \end{bmatrix} \begin{Bmatrix} u_1 \\ u_2 \\ u_3 \\ u_4 \end{Bmatrix}$$

$$= [k_S]_1 [s]_1 \{u\}_1, \tag{4.88}$$

where $[k_S]_1$ is the stiffness matrix for element (1) and $[s]_1$ is the transpose of $[s]_1^T$ in Equation (4.84). All the elements can be assembled to get

$$\{S\} = [k_S][s]\{u\} - \{S_E\}, \tag{4.89}$$

where $[k_S]$ is a square diagonal matrix with 2 by 2 blocks of the element stiffnesses.

Put the $\{S\}$ from Equation (4.89) into Equation (4.85) to get

$$\{P\} = [k]\{u\} - \{P_E\}, \tag{4.90}$$
$$[k] = [s]^T [k_S][s], \quad \{P_E\} = [s]^T \{S_E\},$$

where $[k]$ is the *stiffness influence coefficient* matrix for the beam. Equation (4.90) has the same form as Equation (3.42) for trusses so that with similar notation all the truss Equations (3.43)-(3.48) can be used for the beam case to calculate displacements, support reactions, and the internal point shear forces and point bending moments.

The $[k]$ matrix can also be constructed by assembling the $[k]_k$ element matrices. In fact, Equation (4.27) gives the $[k]_k$ matrix for element (k), or for element (1)

$$\begin{Bmatrix} P_1 + P_{1d} + P_{1E} \\ P_2 + P_{2d} + P_{2E} \\ P_3 + P_{3d} + P_{3E} \\ P_4 + P_{4d} + P_{4E} \end{Bmatrix} = \left(\frac{EI}{L^3}\right)_1 \begin{bmatrix} 12 & -6L & -12 & -6L \\ -6L & 4L^2 & 6L & 2L^2 \\ -12 & 6L & 12 & 6L \\ -6L & 2L^2 & 6L & 4L^2 \end{bmatrix}_1 \begin{Bmatrix} u_1 \\ u_2 \\ u_3 \\ u_4 \end{Bmatrix}_1, \quad (4.91)$$

$$\{P\}_1 = [k]_1\{u\}_1.$$

This element $[k]_1$ matrix can be calculated from Equation (2.70) with the $\{e^1\}$ in Equation (4.83b) and with $[\sigma_1]$ from Equation (4.16), or

$$[\sigma_1] = -\left(\frac{Ez}{L^3}\right)_1 [6(L-2y) \; -2L(L-3y) \; -6(L-2y) \; -2L(2L-3y)]_1. \quad (4.92)$$

If the beam elements are at an angle to each other in the same plane, then the direction cosine matrix can be used, as in the truss case in Section 3.5. To allow for the u_y as well as the u_z deflection components and rotation angle θ_x, use Equation (A.105) to rotate the beam element through the angle ϕ_x about the x-axis, or

$$[R]_k = \begin{bmatrix} \cos\phi_x & \sin\phi_x & 0 \\ -\sin\phi_x & \cos\phi_x & 0 \\ 0 & 0 & 1 \end{bmatrix}_k,$$

$$\begin{Bmatrix} u_y \\ u_z \\ \theta_x \end{Bmatrix}_k = [R]_k \begin{Bmatrix} u_y \\ u_z \\ \theta_x \end{Bmatrix}_d, \quad \begin{Bmatrix} P_y \\ P_z \\ M_x \end{Bmatrix}_k = [R]_k \begin{Bmatrix} P_y \\ P_z \\ M_x \end{Bmatrix}_d, \quad (4.93)$$

where (k) indicates the beam element (k) and (d) indicates reference datum values. Add $(EA/L)_k$ for the axial stiffness of the member (k) to the element matrix in Equation (4.91), whence

$$[k]_k = \left(\frac{EI}{L^3}\right)_k \begin{bmatrix} AL^2/I & 0 & 0 & -AL^2/I & 0 & 0 \\ 0 & 12 & -6L & 0 & -12 & -6L \\ 0 & -6L & 4L^2 & 0 & 6L & 2L^2 \\ -AL^2/I & 0 & 0 & AL^2/I & 0 & 0 \\ 0 & -12 & 6L & 0 & 12 & 6L \\ 0 & -6L & 2L^2 & 0 & 6L & 4L^2 \end{bmatrix}_k. \quad (4.94)$$

From Equations (4.91), (4.93), and (4.94)

$$[k]_d = \begin{bmatrix} R_k^T & 0 \\ 0 & R_k^T \end{bmatrix} [k]_k \begin{bmatrix} R_k & 0 \\ 0 & R_k \end{bmatrix},$$

$$\{P\}_d = [k]_d\{u\}_d, \quad (4.95)$$

which represents the element (k) in the datum system. See Example 4.17 below.

See References 3-5 for examples and problems for simple beams using the displacement or stiffness method of analysis.

The virtual principles for simple beams

Example 4.16. Solve Example 4.11 using the matrix displacement method. Add constant thermal moments M_{T1} and M_{T2} to elements (1) and (2).

Solution. From Equation (4.79)

$$S_{1E} = \frac{6}{L^3} \int_0^L M_{T1}(L-2y)dy = 0, \quad S_{2E} = -\frac{2}{L^2} \int_0^L M_{T1}(L-3y)dy = M_{T1},$$

$$S_{3E} = 0, \quad S_{4E} = M_{T2}. \tag{4.96}$$

From Equation (4.90),

$$\begin{Bmatrix} P_{1E} \\ P_{2E} \\ P_{3E} \\ P_{4E} \\ P_{5E} \\ P_{6E} \end{Bmatrix} = \begin{bmatrix} 1 & 0 & 0 & 0 \\ 0 & 1 & 0 & 0 \\ -1 & 0 & 1 & 0 \\ -L & -1 & 0 & 1 \\ 0 & 0 & -1 & 0 \\ 0 & 0 & -L & -1 \end{bmatrix} \begin{Bmatrix} 0 \\ M_{T1} \\ 0 \\ M_{T2} \end{Bmatrix} = \begin{Bmatrix} 0 \\ M_{T1} \\ 0 \\ -M_{T1}+M_{T2} \\ 0 \\ -M_{T2} \end{Bmatrix}. \tag{4.97}$$

Assemble the four elements, which have the same $[k]_k$, from Equations (4.91), (4.97), and Figure 4.9 to get

$$\begin{Bmatrix} LP_1 \\ M_{T1} \\ --- \\ LQ_3 \\ -M_{T1}+M_{T2} \\ --- \\ LQ_5 \\ -M_{T2} \\ --- \\ LP_7 \\ 0 \\ --- \\ LP_9 \\ P_{10} \end{Bmatrix} = \frac{EI}{L^2} \begin{bmatrix} B_1 & C_1 & 0 & 0 & 0 \\ C_1^T & B_2 & C_1 & 0 & 0 \\ 0 & C_1^T & B_2 & C_1 & 0 \\ 0 & 0 & C_1^T & B_2 & C_1 \\ 0 & 0 & 0 & C_1^T & B_3 \end{bmatrix} \begin{Bmatrix} u_1 \\ Lu_2 \\ --- \\ 0 \\ Lu_4 \\ --- \\ 0 \\ Lu_6 \\ --- \\ u_7 \\ Lu_8 \\ --- \\ 0 \\ 0 \end{Bmatrix}, \tag{4.98}$$

$$[B_1] = \begin{bmatrix} 12 & -6 \\ -6 & 4 \end{bmatrix}, \quad [C_1] = \begin{bmatrix} -12 & -6 \\ 6 & 2 \end{bmatrix}, \quad [B_2] = \begin{bmatrix} 24 & 0 \\ 0 & 8 \end{bmatrix}, \quad [B_3] = \begin{bmatrix} 12 & 6 \\ 6 & 4 \end{bmatrix}.$$

Partition the $[k]$ matrix in Equation (4.98) using the form in Equation (3.44) to get

$$[k_{aa}] = \frac{EI}{L^2} \begin{bmatrix} 12 & -6 & -6 & 0 & 0 & 0 \\ -6 & 4 & 2 & 0 & 0 & 0 \\ -6 & 2 & 8 & 2 & 0 & 0 \\ 0 & 0 & 2 & 8 & 6 & 2 \\ 0 & 0 & 0 & 6 & 24 & 0 \\ 0 & 0 & 0 & 2 & 0 & 8 \end{bmatrix}. \tag{4.99}$$

Invert $[k_{aa}]$ by using the procedure in Equations (A.25)-(A.35), whence

$$\begin{Bmatrix} u_1 \\ Lu_2 \\ Lu_4 \\ Lu_6 \\ u_7 \\ Lu_8 \end{Bmatrix} = \frac{L^2}{240EI} \begin{bmatrix} 152 & 192 & 72 & -24 & 6 & 6 \\ 192 & 312 & 72 & -24 & 6 & 6 \\ 72 & 72 & 72 & -24 & 6 & 6 \\ -24 & -24 & -24 & 48 & -12 & -12 \\ 6 & 6 & 6 & -12 & 13 & 3 \\ 6 & 6 & 6 & -12 & 3 & 33 \end{bmatrix} \begin{Bmatrix} LP_1 \\ M_{T1} \\ -M_{T1}+M_{T2} \\ -M_{T2} \\ LP_7 \\ 0 \end{Bmatrix}. \tag{4.100}$$

From Equations (4.98) and (4.100),

$$\begin{Bmatrix} LQ_3 \\ LQ_5 \\ LP_9 \\ P_{10} \end{Bmatrix} = \frac{EI}{L^2} \begin{bmatrix} -12 & 6 & 0 & -6 & 0 & 0 \\ 0 & 0 & 6 & 0 & -12 & -6 \\ 0 & 0 & 0 & 0 & -12 & 6 \\ 0 & 0 & 0 & 0 & -6 & 2 \end{bmatrix} \begin{Bmatrix} u_1 \\ Lu_2 \\ Lu_4 \\ Lu_6 \\ u_7 \\ Lu_8 \end{Bmatrix}$$

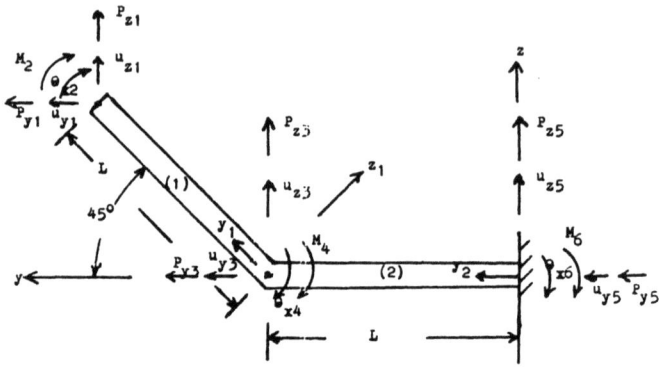

Fig. 4.12. Two element beam with different directions.

$$= \frac{1}{40} \begin{bmatrix} -88 & -48 & -48 & -24 & 6 & 6 \\ 54 & 54 & 54 & 12 & -23 & -33 \\ -6 & -6 & -6 & 12 & -23 & -27 \\ -4 & -4 & -4 & 8 & -12 & 8 \end{bmatrix} \begin{Bmatrix} LP_1 \\ M_{T1} \\ -M_{T1} + M_{T2} \\ -M_{T2} \\ LP_7 \\ 0 \end{Bmatrix}. \quad (4.101)$$

From Equations (4.88), (4.89), and (4.100) with columns for the supports omitted.

$$\begin{Bmatrix} LS_1 \\ S_2 \\ LS_3 \\ S_4 \\ LS_5 \\ S_6 \\ LS_7 \\ S_8 \end{Bmatrix} = \frac{EI}{L^2} \begin{bmatrix} 12 & -6 & -6 & 0 & 0 & 0 \\ -6 & 4 & 2 & 0 & 0 & 0 \\ 0 & 0 & -6 & -6 & 0 & 0 \\ 0 & 0 & 4 & 2 & 0 & 0 \\ 0 & 0 & 0 & -6 & -12 & -6 \\ 0 & 0 & 0 & 4 & 6 & 2 \\ 0 & 0 & 0 & 0 & 12 & -6 \\ 0 & 0 & 0 & 0 & -6 & 4 \end{bmatrix} \begin{Bmatrix} u_1 \\ Lu_2 \\ Lu_4 \\ Lu_6 \\ u_7 \\ Lu_8 \end{Bmatrix} - \begin{Bmatrix} 0 \\ M_{T1} \\ 0 \\ M_{T2} \\ 0 \\ 0 \\ 0 \\ 0 \end{Bmatrix}$$

$$= \frac{1}{40} \begin{bmatrix} 40 & 0 & 0 & 0 & 0 & 0 \\ 0 & 0 & 0 & 0 & 0 & 0 \\ -48 & -48 & -48 & -24 & 6 & 6 \\ 40 & 40 & 40 & 40 & 0 & 0 \\ 6 & 6 & 6 & -12 & -17 & -27 \\ -8 & -8 & -8 & 16 & 6 & 6 \\ 6 & 6 & 6 & -12 & 23 & -27 \\ -2 & -2 & -2 & 4 & -11 & 19 \end{bmatrix} \begin{Bmatrix} LP_1 \\ M_{T1} \\ -M_{T1} + M_{T2} \\ -M_{T2} \\ LP_7 \\ 0 \end{Bmatrix}. \quad (4.102)$$

The results not only check those in Example 4.11 but also give the slopes and u_7 deflection in Equation (4.100). Also, the effects of the thermal moments M_{T1} and M_{T2} in elements (1) and (2) are demonstrated. The M_{T1} on element (1) affects only the displacement u_1 and slope u_2 on element (1), and produces no shear forces or moments in the beam. However, the M_{T2} in element (2) affects all the displacements and produces shear forces and moments in the beam, Equations (4.100)-(4.102). This difference between the M_{T1} and M_{T2} effects is evident in Figure 4.9, where element (1) is determinate but the element (2) deflection is restrained by the supports.

Example 4.17. Find the deflections of the two element beam in Figure 4.12.
Solution From Equation (4.93) the rotation matrix for element (1) is

$$[R] = \frac{b}{2} \begin{bmatrix} 1 & 1 & 0 \\ -1 & 1 & 0 \\ 0 & 0 & b \end{bmatrix}, \quad b = 2^{1/2},$$

The virtual principles for simple beams

whence Equation (4.95) with $H = AL^2/2I$ gives

$$[k]_1 = \frac{EI}{L^2} \begin{bmatrix} H+6 & H-6 & 3b & -H-6 & -H+6 & 3b \\ H-6 & H+6 & -3b & -H+6 & -H-6 & -3b \\ 3b & -3b & 4 & -3b & 3b & 2 \\ -H-6 & -H+6 & -3b & H+6 & H-6 & -3b \\ -H+6 & -H-6 & 3b & H-6 & H+6 & 3b \\ 3b & -3b & 2 & -3b & 3b & 4 \end{bmatrix}.$$

Use Equation (4.94) for $[k]_2$, whence the assembled matrix is

$$\begin{Bmatrix} LP_{y1} \\ LP_{z1} \\ M_2 \\ LP_{y3} \\ LP_{z3} \\ M_4 \\ LP_{y5} \\ LP_{z5} \\ M_6 \end{Bmatrix} = \frac{EI}{L^2} [B] \begin{Bmatrix} u_{y1} \\ u_{z1} \\ L\theta_{z2} \\ u_{y3} \\ u_{z3} \\ L\theta_{z4} \\ u_{y5} \\ u_{z5} \\ L\theta_{z6} \end{Bmatrix}, \qquad (4.103)$$

$$[B] = \begin{bmatrix} H+6 & H-6 & 3b & -H-6 & -H+6 & 3b & 0 & 0 & 0 \\ H-6 & H+6 & -3b & -H+6 & -H-6 & -3b & 0 & 0 & 0 \\ 3b & -3b & 4 & -3b & 3b & 2 & 0 & 0 & 0 \\ -H-6 & -H+6 & -3b & 3H+6 & H-6 & -3b & -2H & 0 & 0 \\ -H+6 & -H-6 & 3b & H-6 & H+18 & 3b-6 & 0 & -12 & -6 \\ 3b & -3b & 2 & -3b & 3b-6 & 8 & 0 & 6 & 2 \\ 0 & 0 & 0 & -2H & 0 & 0 & 2H & 0 & 0 \\ 0 & 0 & 0 & 0 & -12 & 6 & 0 & 12 & 6 \\ 0 & 0 & 0 & 0 & -6 & 2 & 0 & 6 & 4 \end{bmatrix}.$$

To get the reduced $[k_{aa}]$ matrix take out the reactions with $u_{y5} = 0$, $u_{z5} = 0$, $L\theta_{z6} = 0$. Also, to simplify the the inversion of $[k_{aa}]$, assume

$$H = 6 = AL^2/2I, \quad \text{or} \quad L/EA = L^3/12EI, \qquad (4.104)$$

whence

$$[k_{aa}] = \frac{EI}{L^2} \begin{bmatrix} 12 & 0 & 3b & -12 & 0 & 3b \\ 0 & 12 & -3b & 0 & -12 & -3b \\ 3b & -3b & 4 & -3b & 3b & 2 \\ -12 & 0 & -3b & 24 & 0 & -3b \\ 0 & -12 & 3b & 0 & 24 & 3b-6 \\ 3b & -3b & 2 & -3b & 3b-6 & 8 \end{bmatrix}.$$

Invert $[k_{aa}]$ to get

$$\begin{Bmatrix} u_{y1} \\ u_{z1} \\ L\theta_{z2} \\ u_{y3} \\ u_{z3} \\ L\theta_{z4} \end{Bmatrix} = \frac{L^2}{24EI} [C] \begin{Bmatrix} LP_{y1} \\ LP_{z1} \\ M_2 \\ LP_{y3} \\ LP_{z3} \\ M_4 \end{Bmatrix}, \qquad (4.105)$$

$$[C] = \begin{bmatrix} 19 & -15-6b & -18b & 2 & -6b & -12b \\ -15-6b & 25+12b & 12+18b & 0 & 8+6b & 12+12b \\ -18b & 12+18b & 48 & 0 & 12 & 24 \\ 2 & 0 & 0 & 2 & 0 & 0 \\ -6b & 8+6b & 12 & 0 & 8 & 12 \\ -12b & 12+12b & 24 & 0 & 12 & 24 \end{bmatrix},$$

where $L^2/24EI = L/2EA$ from Equation (4.104). The reactions can be obtained from Equation (4.103) or in this case from equilibrium for the determinate beam. From equilibrium, the internal point forces are

$$S_1(\text{axial}) = (b/2)(P_{y1} + P_{z1}), \qquad S_1(\text{shear}) = (b/2)(-P_{y1} + P_{z1}),$$

S_2(moment) $= M_2$, $\quad S_3$(axial) $= P_{y1} + P_{y3}$, $\quad S_3$(shear) $= P_{z1} + P_{z3}$,
S_4(moment) $= M_2 + M_4 + (bL/2)(P_{z1} - P_{y1})$.

4.9. Matrix equations for beams from the unit load theorem

From Equation (4.26) and Figure 4.4 for the two element beam

$$\begin{Bmatrix} S_1 \\ S_2 \\ S_3 \\ S_4 \\ 0 \\ 0 \end{Bmatrix} = \begin{bmatrix} 1 & 0 & 0 & 0 & 0 & 0 \\ 0 & 1 & 0 & 0 & 0 & 0 \\ 1 & 0 & 1 & 0 & 0 & 0 \\ L_1 & 1 & 0 & 1 & 0 & 0 \\ 0 & 1 & 0 & 1 & 0 & 1 & 0 \\ L_1+L_2 & 1 & L_2 & 1 & 0 & 1 \end{bmatrix} \begin{Bmatrix} P_1 \\ P_2 \\ P_3 \\ P_4 \\ P_5 \\ P_6 \end{Bmatrix}, \text{ or}$$

$$\{S\} = [p]\{P\}, \tag{4.106}$$

where the last two rows are the external equilibrium equations. For the redundant case, Equation (4.49) can be written as

$$\{S\} = [p]\{P\} + [q]\{Q\}. \tag{4.107}$$

These unit load $[p]$ and $[q]$ matrices cannot be assembled directly from the element matrices, but must be constructed for the particular beam. For this reason, in large structures it may be desirable to get $[p]$ and $[q]$ from the equilibrium Equation (4.85).

If the equilibrium Equation (4.83c) is assembled for $M-1$ beam elements in a straight line, then $[s]^T$ in Equation (4.85) has the form

$$[s]^T = \begin{bmatrix} I & 0 & 0 & 0 & \cdots & 0 & 0 \\ -S_{21} & I & 0 & 0 & \cdots & 0 & 0 \\ 0 & -S_{32} & I & 0 & \cdots & 0 & 0 \\ 0 & 0 & -S_{43} & I & \cdots & 0 & 0 \\ \cdots & \cdots & \cdots & \cdots & \cdots & \cdots & \cdots \\ 0 & 0 & 0 & 0 & \cdots & -S_{M,M-1} & I \end{bmatrix}, \tag{4.108}$$

$$[I] = \begin{bmatrix} 1 & 0 \\ 0 & 1 \end{bmatrix}, \quad [S_{k+1,k}] = \begin{bmatrix} 1 & 0 \\ L_k & 1 \end{bmatrix}.$$

Thus, $[s]^T$ is a lower triangular bi-diagonal matrix in two by two blocks and has a simple inverse from Equations (A.53)-(A.54b), or in this case

$$[s^T]^{-1} = \begin{bmatrix} I & 0 & 0 & 0 & \cdots & \cdots & 0 \\ S_{21} & I & 0 & 0 & \cdots & \cdots & 0 \\ S_{32}S_{21} & S_{32} & I & 0 & \cdots & \cdots & 0 \\ S_{43}S_{32}S_{21} & S_{43}S_{32} & S_{43} & I & \cdots & \cdots & 0 \\ \cdots & \cdots & \cdots & \cdots & \cdots & \cdots & \cdots \\ S_{M,M-1} \cdots S_{21} & \cdots & \cdots & \cdots & \cdots & S_{M,M-1} & I \end{bmatrix} \tag{4.109}$$

$$= \begin{bmatrix} I & 0 & 0 & 0 & \cdots & \cdots & 0 \\ a_{21} & I & 0 & 0 & \cdots & \cdots & 0 \\ a_{31} & a_{32} & I & 0 & \cdots & \cdots & 0 \\ \cdots & \cdots & \cdots & \cdots & \cdots & \cdots & \cdots \\ a_{M1} & a_{M2} & a_{M3} & \cdots & \cdots & a_{M,M-1} & I \end{bmatrix}, \quad (4.110)$$

$$[a_{ij}] = \begin{bmatrix} 1 & 0 \\ \sum_{k=j}^{i-1} L_k & 1 \end{bmatrix}, \quad j < i, \quad i = 2, 3, \cdots, M.$$

Thus, all the a_{ij} block elements in the inverse can be calculated directly. Also, this inverse can be obtained by a sequence solution of Equation (4.85) using Equation (4.108).

All the beam supports are represented by columns in Equation (4.110) with the corresponding P_i forces being reactions. Let $\{R_d\}$ be the reactions at the two selected determinate supports and let $\{Q\}$ be the unknown reactions at the additional supports. The Equations (4.106) and (4.110) can be partitioned to give

$$\{S\} = [s^T]^{-1}\{P\} = [A]\{P\} + [B]\{R_d\} + [D]\{Q\}, \qquad (4.111)$$

where $[A]$, $[B]$, and $[D]$ are columns in Equation (4.110). Since the last row of two by two matrices in Equation (4.110) contains the two equilibrium equations for the entire beam, it can be used to get $\{R_d\}$, or

$$\begin{aligned} \{R_d\} &= [F]\{P\} + [G]\{Q\}, \quad \{S\} = [p]\{P\} + [q]\{Q\}, \\ [p] &= [A] + [B][F], \quad [q] = [D] + [B][G]. \end{aligned} \qquad (4.112)$$

For the determinate case, $[q] = 0$. For the case in which $\{R_d\}$ are the cantilever beam reactions at the end of element $M - 1$ in the last row of Equation (4.110), the matrix $[B]$ can be taken as $[B] = 0$. This can be done because the last column in Equation (4.110) is zero except for I in the last row.

Because of the simple inverse of the $[s]^T$ matrix, the partitioning to get the Q_i unknowns is made after the inversion. In the method for trusses in Section 3.6 the partitioning was made before the inversion and Q_i represented loads in some of the truss elements. That method can be used for the above beam starting with Equation (3.78) and getting the $[p]$ and $[q]$ matrices in Equation (3.81). However, when rows are removed from Equation (4.108) for the additional supports and when the Q_i are selected as some of the S_i to get a square matrix, the inverse is no longer as simple as in Equation (4.110). The sequence solution can still be made in terms of P_i and the selected $Q_i = S_i$. In more complicated beam structures with internal redundants where three or more beam elements join, the above procedure does not apply, but the procedure in Equations (3.78)-(3.81) can still be used.

The flexibility $[C]$ matrix in Equation (2.76) can be obtained from Equations (4.107) and (2.78). The bending strain in the beam element (k) is

$$e_k = -\left(\frac{Mz}{EI}\right)_k = -\left(\frac{z}{EI}\right)_k [L_k - y_k \quad 1] \begin{Bmatrix} S_{2k-1} \\ S_{2k} \end{Bmatrix} = [e_1]_k \begin{Bmatrix} P \\ Q \end{Bmatrix}, \quad (4.113)$$

$$[e_1]_k = -\left(\frac{z}{EI}\right)_k [L_k - y_k \quad 1] \begin{bmatrix} p_{2k-1} & q_{2k-1} \\ p_{2k} & q_{2k} \end{bmatrix}.$$

From Equation (2.78)

$$[C]_k = \int_0^{L_k} \int_{A_k} E_k \left(\frac{z}{EI}\right)_k^2 \cdot$$

$$\cdot \begin{bmatrix} p_{2k-1} & p_{2k} \\ q_{2k-1} & q_{2k} \end{bmatrix} \begin{Bmatrix} L_k - y_k \\ 1 \end{Bmatrix} [L_k - y_k \quad 1] \begin{bmatrix} p_{2k-1} & q_{2k-1} \\ p_{2k} & q_{2k} \end{bmatrix} dA_k \, dy_k$$

$$= \begin{bmatrix} p^T \\ q^T \end{bmatrix}_k \left(\frac{L}{6EI}\right)_k \begin{bmatrix} 2L^2 & 3L \\ 3L & 6 \end{bmatrix}_k [p \quad q]_k, \qquad (4.114)$$

which is the contribution of element (k) to the flexibility matrix $[C]$. Assemble all elements to get

$$[C] = \begin{bmatrix} p^T \\ q^T \end{bmatrix} [C_F] [p \quad q] = \begin{bmatrix} p^T C_F p & p^T C_F q \\ q^T C_F p & q^T C_F q \end{bmatrix} = \begin{bmatrix} C_{PP} & C_{PQ} \\ C_{QP} & C_{QQ} \end{bmatrix}, \qquad (4.115)$$

where $[C_F]$ is a diagonal matrix with two by two blocks. This Equation (4.115) corresponds to Equation (3.69) for trusses. In fact, the truss Equations (3.69)-(3.74) apply for beams with the proper interpretation of the symbols.

From Equation (4.114) the element flexibility matrix is

$$[C_F]_k = (L/6EI)_k \begin{bmatrix} 2L^2 H_f & 3L \\ 3L & 6 \end{bmatrix}_k, \qquad (4.116)$$

where the shear flexibility term in Equation (4.46) has been added as

$$H_f = 1 + (3EI/GA_s L^2). \qquad (4.117)$$

With the matrices $[p]$, $[q]$, and $[C]$ known, the displacements are given by Equation (2.76) in the form

$$\begin{Bmatrix} u \\ u_Q \end{Bmatrix} = \begin{bmatrix} C_{PP} & C_{PQ} \\ C_{QP} & C_{QQ} \end{bmatrix} \begin{Bmatrix} P \\ Q \end{Bmatrix} + \begin{Bmatrix} u_{EP} \\ u_{EQ} \end{Bmatrix}, \qquad (4.118)$$

$$\{u_{EP}\} = [p]^T [C_F] \{S_E\}, \qquad \{u_{EQ}\} = [q]^T [C_F] \{S_E\}, \qquad (4.119)$$

with S_{Ek} in Equation (4.79). Solve for $\{Q\}$ to get

$$\{Q\} = -[C_{QQ}]^{-1}[C_{QP}]\{P\} + [C_{QQ}]^{-1}\{u_Q - u_{EQ}\}. \qquad (4.120)$$

Put $\{Q\}$ into Equation (4.107) to get the internal point shear forces and moments, whence the shear and moment curves for the beam can be drawn. Put $\{Q\}$ into Equation (4.118) to get the displacements $\{u\}$. Also, $\{u\}$ can be expressed in terms of $\{S\}$ as

$$\{u\} = [p]^T [C_F] \{S + S_E\} = [p]^T [C_F] \{S\} + \{u_{EP}\}. \qquad (4.121)$$

If the beam elements are at an angle to each other in the same plane, then the rotation matrix in Equation (4.93) can be used. From Equations (4.84) and (4.93) for element (k),

$$\begin{aligned} \{P\}_k &= [R]_k \{P\}_d = [s]_k^T \{S\}_k, \\ \{P\}_d &= [R]_k^T [s]_k^T \{S\}_k = [s]_d^T \{S\}_k. \end{aligned} \qquad (4.122)$$

The virtual principles for simple beams 139

These element matrices can be assembled in equation (4.85) or Equation (4.106), whence the $[p]_d$ matrix can be calculated from a modified Equation (4.108). See Section 9.3, Volume 2, for general case.

To allow for axial loads and deflections, use the $(L/EA)_k$ flexibility expression and modify $[C_F]_k$ in Equation (4.116) to

$$[C_F]_k = (L/6EI)_k \begin{bmatrix} 6I/A & 0 & 0 \\ 0 & 2L^2 H_f & 3L \\ 0 & 3L & 6 \end{bmatrix}_k. \tag{4.123}$$

For this case, the $[s]_k^T$ in Equations (4.83c) and (4.84) becomes

$$[s]_k^T = \begin{bmatrix} I \\ T_k \end{bmatrix}, \quad [T_k] = -\begin{bmatrix} 1 & 0 & 0 \\ 0 & 1 & 0 \\ 0 & L & 1 \end{bmatrix}_k, \quad I \text{ is 3 by 3,} \tag{4.124}$$

$$[s]_d^T = \begin{bmatrix} [R]_k^T & 0 \\ 0 & [R]_k^T \end{bmatrix} [s]_k^T = \begin{bmatrix} [R]_k^T \\ [R]_k^T [T_k] \end{bmatrix},$$

where $[R]_k$ is given by Equation (4.93).

Example 4.18. Use the unit load matrices to solve Example 4.11.

Solution (a) In Figure 4.9 take the determinate reactions at the wall so that $[B] = 0$ in Equation (4.111) and so that there are two applied forces P_1 and P_7 and two unknown forces Q_3 and Q_5. Put columns 1, 7, 3, and 5 from Equation (4.110) into Equation (4.107), whence

$$\{S\} = [p] \begin{Bmatrix} P_1 \\ P_7 \end{Bmatrix} + [q] \begin{Bmatrix} Q_3 \\ Q_5 \end{Bmatrix}, \tag{4.125}$$

$$\{S\}^T = [S_1 \ S_2 \ S_3 \ S_4 \ S_5 \ S_6 \ S_7 \ S_8 \ R_9 \ R_{10} \ R_3 \ R_5],$$

$$[p]^T = \begin{bmatrix} 1 & 0 & 1 & L & 1 & 2L & 1 & 3L & -1 & -4L & 0 & 0 \\ 0 & 0 & 0 & 0 & 0 & 0 & 1 & 0 & -1 & -L & 0 & 0 \end{bmatrix},$$

$$[q]^T = \begin{bmatrix} 0 & 0 & 1 & 0 & 1 & L & 1 & 2L & -1 & -3L & 1 & 0 \\ 0 & 0 & 0 & 0 & 1 & 0 & 1 & L & -1 & -2L & 0 & 1 \end{bmatrix}.$$

Use Equation (4.116) with $H_f = 1$ for each of the four elements and use zeros for the reactions in the 12 by 12 $[C_F]$ matrix. From Equations (4.115) and (4.120),

$$[C_{QQ}] = [q]^T [C_F][q] = (L^3/3EI) \begin{bmatrix} 27 & 14 \\ 14 & 8 \end{bmatrix},$$

$$[C_{QP}] = [q]^T [C_F][p] = (L^3/6EI) \begin{bmatrix} 81 & 8 \\ 40 & 5 \end{bmatrix},$$

$$\begin{Bmatrix} Q_3 \\ Q_5 \end{Bmatrix} = -[C_{QQ}]^{-1}[C_{QP}]\{P\} = -(1/40) \begin{bmatrix} 88 & -6 \\ -54 & 23 \end{bmatrix} \begin{Bmatrix} P_1 \\ P_7 \end{Bmatrix},$$

which agrees with Q_3 and Q_5 in Example 4.11. Put the Q_3 and Q_5 values into Equation (4.125) and collect terms to get the S_i values and reactions given in Example 4.11.

(b) Since only the displacements u_1 and u_7 will be given by the above $[p]$ matrix, add the P_6 column to $[p]$ to get the slope u_6, or

$$\{p\}_6^T = [0 \ 0 \ 0 \ 0 \ 0 \ 1 \ 0 \ 1 \ 0 \ -1 \ 0 \ 0].$$

The $[C_{PP}]$ and $[C_{PQ}]$ matrices in Equation (4.118) are

$$[C_{PP}] = [p]^T [C_F][p] = \frac{L^2}{6EI} \begin{bmatrix} 128L & 36 & 11L \\ 36 & 12 & 3 \\ 11L & 3 & 2L \end{bmatrix},$$

$$[C_{PQ}] = [C_{QP}]^T = \frac{L^2}{6EI} \begin{bmatrix} 81L & 40L \\ 24 & 12 \\ 8L & 5L \end{bmatrix}.$$

With $P_6 = 0$, Equation (4.118) gives

$$\begin{Bmatrix} u_1 \\ u_6 \\ u_7 \end{Bmatrix} = [C_{PP}]\{P\} + [C_{PQ}]\{Q\} = \frac{L^2}{240EI} \begin{bmatrix} 152L & 6L \\ -24 & -12 \\ 6L & 13L \end{bmatrix} \begin{Bmatrix} P_1 \\ P_7 \end{Bmatrix}, \quad (4.126)$$

which checks Example 4.11 for u_1 and u_+.

Example 4.19 Solve Examples 4.11 and 4.18 by using the internal moments $Q_6 = M_6 = P_6$ and $Q_{10} = -M_{10} = R_{10}$ as the unknown redundants. Let the constant moment M_{T2} act on element (2) and let element (4) be inelastic with $M_{p4} = [1 - (y_4/L)]^3 M_{p10}$.

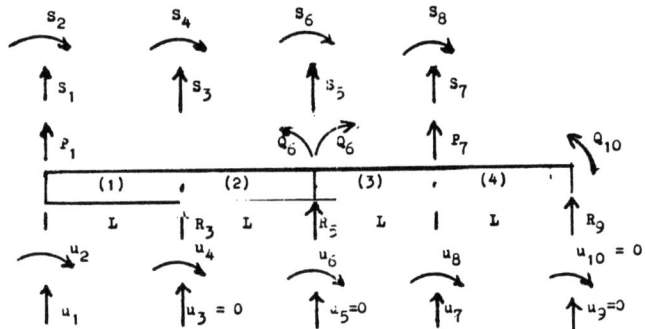

Fig. 4.13. The beam in Fig. 4.9 with Q_6 and Q_{10} as unknown moments

Solution. Calculate the reactions and the $[p]$ and $[q]$ matrices by using Figure 4.13, or by making a sequence solution using Equation (4.108). Thus,

$$\{S\} = [p]\begin{Bmatrix} P_1 \\ P_7 \end{Bmatrix} + [q]\begin{Bmatrix} Q_6 \\ Q_{10} \end{Bmatrix}, \quad (4.127)$$

$$\{S\}^T = [S_1\ S_2\ S_3\ S_4\ S_5\ S_6\ S_7\ S_8\ R_9\ R_{10}\ R_3\ R_5],$$

$$[p]^T = \frac{1}{2}\begin{bmatrix} 2 & 0 & -2 & 2L & 0 & 0 & 0 & 0 & 0 & 0 & -4 & 2 \\ 0 & 0 & 0 & 0 & -1 & 0 & 1 & -L & -1 & 0 & 0 & -1 \end{bmatrix},$$

$$[q]^T = \frac{1}{2L}\begin{bmatrix} 0 & 0 & 2 & 0 & -1 & 2L & -1 & L & 1 & 0 & 2 & -3 \\ 0 & 0 & 0 & 0 & 1 & 0 & 1 & L & -1 & -2L & 0 & 1 \end{bmatrix}.$$

From Equation (4.79)

$$S_{p8} = -(2/L^2)\int_0^L M_{p10}\left(1 - \frac{y_4}{L}\right)^3 (L - 3y_4)dy_4 = -M_{p10}/5,$$

$$LS_{p7} = (6/L^2)\int_0^L M_{p10}\left(1 - \frac{y_4}{L}\right)^3 (L - 2y_4)dy_4 = (9/10)M_{p10},$$

whence

$$\{S_E\}^T = [0\ 0\ 0\ M_{T2}\ 0\ 0\ 0.9M_{p10}\ -0.2M_{p10}].$$

From Equations (4.115) and (4.120)

$$[C_{QQ}] = [q]^T[C_F][q] = \frac{L}{3EI}\begin{bmatrix} 3 & 1 \\ 1 & 2 \end{bmatrix}, \quad [C_{QP}] = \frac{L^2}{12EI}\begin{bmatrix} 2 & -3 \\ 0 & -3 \end{bmatrix},$$

$$\begin{Bmatrix} Q_6 \\ Q_{10} \end{Bmatrix} = -[C_{QQ}]^{-1}[C_{QP}]\{P\} - [C_{QQ}]^{-1}[q]^T[C_F]\{S_E\}$$

$$= \frac{1}{200}\begin{bmatrix} -40 & 30 & -120 & 21 \\ 20 & 60 & 60 & -78 \end{bmatrix}\begin{Bmatrix} LP_1 \\ LP_7 \\ M_{T2} \\ M_{p10} \end{Bmatrix}.$$

The virtual principles for simple beams

Put these expressions into Equation (4.127) to get

$$\begin{Bmatrix} LS_1 \\ S_2 \\ LS_3 \\ S_4 \\ LS_5 \\ S_6 \\ LS_7 \\ S_8 \\ LR_9 \\ R_{10} \\ LR_3 \\ LR_5 \end{Bmatrix} = \frac{1}{400} \begin{bmatrix} 400 & 0 & 0 & 0 \\ 0 & 0 & 0 & 0 \\ -480 & 60 & -240 & 42 \\ 400 & 0 & 0 & 0 \\ 60 & -170 & 180 & -99 \\ -80 & 60 & -240 & 42 \\ 60 & 230 & 180 & -99 \\ -20 & -110 & -60 & -57 \\ -60 & -230 & -180 & 99 \\ -40 & -120 & -120 & 156 \\ -880 & 60 & -240 & 42 \\ 540 & -230 & 420 & -141 \end{bmatrix} \begin{Bmatrix} LP_1 \\ LP_7 \\ M_{T2} \\ M_{p10} \end{Bmatrix}.$$

From Equation (4.121),

$$\begin{Bmatrix} u_1 \\ u_7 \end{Bmatrix} = [p]^T [C_F]\{S + S_E\} = \frac{L^2}{2400EI} \begin{bmatrix} 1520 & 60 & 960 & 42 \\ 60 & 130 & 180 & 111 \end{bmatrix} \begin{Bmatrix} LP_1 \\ LP_7 \\ M_{T2} \\ M_{p10} \end{Bmatrix}.$$

Use the maximum bending stress at $z = h$ on the beam and define $S_{y4} = (I/h)\sigma_{y4}$, $S_{10} = (I/h)\sigma_{10}$, $M_{p10} = (I/h)Ee_{p10}$, whence from Equation (3.27) and from above with $S_{10} = Q_{10}$,

$$M_{p10} = (\tfrac{3}{7})S_{y4}\left(S_{10}/S_{y4}\right)^{10},$$

$$S_{10} + \tfrac{78}{200}(\tfrac{3}{7})S_{y4}\left(S_{10}/S_{y4}\right)^{10} = \tfrac{1}{10}LP_1 + \tfrac{3}{10}LP_7 + \tfrac{3}{10}M_{T2}.$$

For given values of LP_1, LP_7, M_{T2}, and S_{y4}, S_{10} can be obtained by trial. This gives M_{p10} which can be used with the given values to get the internal loads, reactions, and deflections in the above equations.

Example 4.20. Solve Example 4.17 using the force method.

Solution. For the two element beam in figure 4.12, Equations (4.124) and (4.122) give

$$\begin{Bmatrix} P_{y1} \\ P_{x1} \\ M_{x2} \\ P_{y3} \\ P_{x3} \\ M_{x4} \end{Bmatrix} = \begin{bmatrix} b/2 & -b/2 & 0 & 0 & 0 & 0 \\ b/2 & b/2 & 0 & 0 & 0 & 0 \\ 0 & 0 & 1 & 0 & 0 & 0 \\ -b/2 & b/2 & 0 & 1 & 0 & 0 \\ -b/2 & -b/2 & 0 & 0 & 1 & 0 \\ 0 & -L & -1 & 0 & 0 & 1 \end{bmatrix}_d \begin{Bmatrix} S_{y1} \\ S_{x1} \\ S_{x2} \\ S_{y3} \\ S_{x3} \\ S_{x4} \end{Bmatrix}_k,$$

$$[p]_d = [s_d^T]^{-1} = \begin{bmatrix} b/2 & b/2 & 0 & 0 & 0 & 0 \\ -b/2 & b/2 & 0 & 0 & 0 & 0 \\ 0 & 0 & 1 & 0 & 0 & 0 \\ 1 & 0 & 0 & 1 & 0 & 0 \\ 0 & 1 & 0 & 0 & 1 & 0 \\ -bL/2 & bL/2 & 1 & 0 & 0 & 1 \end{bmatrix}.$$

Use Equation (4.123) with $H_f = 1$ for each element and Equation (4.118) to calculate $[C_{PP}]_d$. Also, use $H = AL^2/2I$, whence

$$\begin{Bmatrix} u_{y1} \\ u_{x1} \\ L\theta_{z2} \\ u_{y3} \\ u_{x3} \\ L\theta_{z4} \end{Bmatrix} = \frac{L^2}{24EI}[B] \begin{Bmatrix} LP_{y1} \\ LP_{x1} \\ M_{x2} \\ LP_{y3} \\ LP_{x3} \\ M_{x4} \end{Bmatrix}, \qquad (4.128)$$

$$[B] = \begin{bmatrix} a_{11} & a_{12} & -18b & 12/H & -6b & -12b \\ a_{21} & a_{22} & 12 + 18b & 0 & 8 + 6b & 12 + 12b \\ -18b & 12 + 18b & 48 & 0 & 12 & 24 \\ 12/H & 0 & 0 & 12/H & 0 & 0 \\ -6b & 8 + 6b & 12 & 0 & 8 & 12 \\ -12b & 12 + 12b & 24 & 0 & 12 & 24 \end{bmatrix},$$

$$a_{11} = \frac{18}{H} + 16, \quad a_{21} = a_{12} = \frac{6}{H} - 16 - 6b, \quad a_{22} = \frac{6}{H} + 24 + 12b, \quad b = 2^{1/2}.$$

The H term was used as $H = 6$ in Example 4.17 in order to simplify the inversion of the 6 by 6 $[k_{aa}]$ matrix. Here the H term does not occur in the inversion of the simple 6 by 6 $[s]_d^T$ matrix. If $H = 6$ is used in Equation (4.128) the result is the same as in Equation (4.105).

Example 4.21 (a). Find the internal point forces S_i and the four reactions for the beam in Figure 4.14. Assume all elements have the same EI, except $(EI)_3 = EI/6$. (b). Find the displacements at the applied loads. (c). Find $\{S\}$ and $\{u\}$ for a constant thermal moment M_T acting on the entire beam and for a support (5) deflection u_{Q5}.

Solution (a) From Equations (4.110)-(4.112),

$$[A]^T = \begin{bmatrix} 1 & 0 & 1 & L & 1 & 2L & 1 & 3L & 1 & 4L & 1 & 5L \\ 0 & 1 & 0 & 1 & 0 & 1 & 0 & 1 & 0 & 1 & 0 & 1 \\ 0 & 0 & 0 & 0 & 0 & 0 & 0 & 0 & 0 & 1 & 0 & 0 \\ 0 & 0 & 0 & 0 & 0 & 0 & 0 & 0 & 0 & 0 & 0 & 1 \end{bmatrix},$$

$$[B]^T = \begin{bmatrix} 0 & 0 & 0 & 0 & 0 & 1 & 0 & 1 & L & 1 & 2L \\ 0 & 0 & 0 & 0 & 0 & 0 & 0 & 1 & 0 & 1 & L \end{bmatrix},$$

$$[D]^T = \begin{bmatrix} 0 & 0 & 1 & 0 & 1 & L & 1 & 2L & 1 & 3L & 1 & 4L \\ 0 & 0 & 0 & 0 & 1 & 0 & 1 & L & 1 & 2L & 1 & 3L \end{bmatrix},$$

$$\begin{Bmatrix} R_7 \\ R_9 \end{Bmatrix} = \begin{bmatrix} -4 & -1/L & 1 & -1/L \\ 3 & 1/L & -2 & 1/L \end{bmatrix} \begin{Bmatrix} P_1 \\ P_2 \\ P_{11} \\ P_{12} \end{Bmatrix} + \begin{bmatrix} -3 & -2 \\ 2 & 1 \end{bmatrix} \begin{Bmatrix} Q_3 \\ Q_5 \end{Bmatrix},$$

$$[p]^T = \begin{bmatrix} 1 & 0 & 1 & L & 1 & 2L & -3 & 3L & 0 & 0 & 0 & 0 \\ 0 & 1 & 0 & 1 & 0 & 1 & -1/L & 1 & 0 & 0 & 0 & 0 \\ 0 & 0 & 0 & 0 & 0 & 0 & 1 & 0 & -1 & L & 0 & 0 \\ 0 & 0 & 0 & 0 & 0 & 0 & -1/L & 0 & 0 & -1 & 0 & 0 \end{bmatrix},$$

$$[q]^T = \begin{bmatrix} 0 & 0 & 1 & 0 & 1 & L & -2 & 2L & 0 & 0 & 0 & 0 \\ 0 & 0 & 0 & 0 & 1 & 0 & -1 & L & 0 & 0 & 0 & 0 \end{bmatrix}.$$

From Equations (4.115), (4.116), (4.120), and (4.112) with $H_f = 1$, and all EI the same except element (3) has $EI/6$,

$$[C_{QQ}] = [q]^T[C_F][q] = \frac{L^3}{3EI} \begin{bmatrix} 47 & 17 \\ 17 & 7 \end{bmatrix},$$

$$[C_{QP}] = \frac{L^3}{6EI} \begin{bmatrix} 155 & 61/L & 2 & -2/L \\ 54 & 20/L & 1 & -1/L \end{bmatrix},$$

$$\begin{Bmatrix} Q_3 \\ Q_5 \\ R_7 \\ R_9 \\ S_1 \\ S_2 \\ S_3 \\ S_4 \\ S_5 \\ S_6 \\ S_7 \\ S_8 \\ S_9 \\ S_{10} \\ S_{11} \\ S_{12} \end{Bmatrix} = \frac{1}{80} \begin{bmatrix} -167 & -87/L & 3 & -3/L \\ 97 & 97/L & -13 & 13/L \\ -13 & -13/L & 97 & -97/L \\ 3 & 3/L & -167 & 87/L \\ 80 & 0 & 0 & 0 \\ 0 & 80 & 0 & 0 \\ -87 & -87/L & 3 & -3/L \\ 80L & 80 & 0 & 0 \\ 10 & 10/L & -10 & 10/L \\ -7L & -7 & 3L & -3 \\ -3 & -3/L & 87 & -87/L \\ 3L & 3 & -7L & 7 \\ 0 & 0 & -80 & 0 \\ 0 & 0 & 80L & -80 \\ 0 & 0 & 0 & 0 \\ 0 & 0 & 0 & 0 \end{bmatrix} \begin{Bmatrix} P_1 \\ P_2 \\ P_{11} \\ P_{12} \end{Bmatrix}$$

(b) Use Equation (4.121) to get the deflections, or

The virtual principles for simple beams

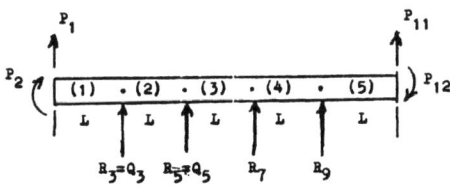

Fig. 4.14. Redundant beam for Example 4.21.

$$\{u\} = [p]^T[C_F]\{S\},$$

$$\begin{Bmatrix} u_1 \\ Lu_2 \\ u_{11} \\ Lu_{12} \end{Bmatrix} = \frac{L^2}{480EI} \begin{bmatrix} 313 & 393 & 3 & -3 \\ 393 & 633 & 3 & -3 \\ 3 & 3 & 313 & -393 \\ -3 & -3 & -393 & 633 \end{bmatrix} \begin{Bmatrix} LP_1 \\ P_2 \\ LP_{11} \\ P_{12} \end{Bmatrix}.$$

(c) From Equation (4.79),

$$S_{Ek}(k \text{ odd}) = 0, \quad S_{Ek}(k \text{ even}) = M_{zEk} = M_T, \text{ whence}$$
$$\{S_E\}^T = M_T[0\ 1\ 0\ 1\ 0\ 1\ 0\ 1\ 0\ 1\ 0\ 0].$$

From Equations (4.119) and (4.120)

$$\{u_{EQ}\} = [q]^T[C_F]\{S_E\} = \frac{7L^2 M_T}{2EI} \begin{Bmatrix} 3 \\ 1 \end{Bmatrix},$$

$$\begin{Bmatrix} Q_3 \\ Q_5 \\ R_7 \\ R_9 \\ \{S\} \end{Bmatrix} = \begin{bmatrix} -51 & -1 \\ 141 & 1 \\ -129 & 1 \\ 39 & -1 \\ [q_1] \end{bmatrix} \begin{Bmatrix} \frac{EI}{40L^3} u_{Q5} \\ \frac{21}{20L} M_T \end{Bmatrix},$$

$$[q_1]^T = \begin{bmatrix} 0 & 0 & -51 & 0 & 90 & -51L & -39 & 39L & 0 & 0 & 0 & 0 \\ 0 & 0 & -1 & 0 & 0 & -L & 1 & -L & 0 & 0 & 0 & 0 \end{bmatrix}.$$

The deflections are

$$\{u\} = [p]^T[C_F]\{S + S_E\},$$

$$\begin{Bmatrix} u_1 \\ Lu_2 \\ u_3 \\ Lu_4 \end{Bmatrix} = \frac{1}{80} \begin{bmatrix} -97 & 66 \\ -97 & 106 \\ 13 & 66 \\ -13 & -106 \end{bmatrix} \begin{Bmatrix} u_{Q5} \\ \frac{M_T L^2}{EI} \end{Bmatrix}.$$

Example 4.22. Solve Example 4.21 (a) by the procedure in Equations (3.75)-(3.81).
Solution. From Figure 4.14 and Equations (4.108), (3.75)-(3.77),

$$R_3 = -S_1 + S_3, \quad R_5 = -S_3 + S_5, \quad R_7 = -S_5 + S_7, \quad R_9 = -S_7 + S_9,$$
$$S_{11} = 0, \quad S_{12} = 0, \quad S_1 = P_1, \quad S_2 = P_2, \quad S_4 = LP_1 + P_2, \quad S_9 = -P_{11},$$
$$S_{10} = LP_{11} - P_{12}, \quad P_6 = 0 = -LS_3 - S_4 + S_6, \quad P_8 = 0 = -LS_5 - S_6 + S_8,$$
$$P_{10} = 0 = -LS_7 - S_8 + S_{10}.$$

Any two unknowns can be selected from S_3, S_5, S_6, S_7, S_8 except the combinations S_3, S_6 and S_7, S_8. Select $S_5 = Q_5$ and $S_6 = Q_6$ and solve for $S_3, S_8,$ and S_7, whence $[p]$ and $[q]$ become

$$[p]^T = \begin{bmatrix} 1 & 0 & -1 & L & 0 & 0 & 0 & 0 & 0 & 0 \\ 0 & 1 & -1/L & 1 & 0 & 0 & 0 & 0 & 0 & 0 \\ 0 & 0 & 0 & 0 & 0 & 1 & 0 & -1 & L & 0 \\ 0 & 0 & 0 & 0 & 0 & -1/L & 0 & 0 & -1 & 0 \end{bmatrix},$$

$$[q]^T = \begin{bmatrix} 0 & 0 & 0 & 0 & 1 & 0 & -1 & L & 0 & 0 \\ 0 & 0 & 1/L & 0 & 0 & 1 & -1/L & 1 & 0 & 0 \end{bmatrix},$$

$$[C_{QQ}] = \frac{L}{3EI} \begin{bmatrix} 7L^2 & 10L \\ 10L & 20 \end{bmatrix}, \quad [C_{QP}] = \frac{L}{6EI} \begin{bmatrix} 0 & 0 & L^2 & -L \\ L & 1 & L & -1 \end{bmatrix},$$

$$\begin{Bmatrix} Q_5 \\ Q_6 \end{Bmatrix} = \frac{1}{80} \begin{bmatrix} 10 & 10/L & -10 & 10/L \\ 7L & -7 & 3L & -3 \end{bmatrix} \begin{Bmatrix} P_1 \\ P_2 \\ P_{11} \\ P_{12} \end{Bmatrix}.$$

This $S_5 = Q_5$, $S_6 = Q_6$ agree with the result in Example 4.21 (a). The rest of the S_i are given by $\{S\} = [p]\{P\} + [q]\{Q\}$ and the reactions are given by the equations above.

Example 4.23. Solve Example 4.9 by using finite elements.

Solution. Divide the frame in Figure 4.8 into four elements as shown in Figure 4.15. Assume the A_i areas are large so that the term $6I/A = 0$ in Equation (4.123). Also, take $H_f = 1$. Start with point (A) with S_1 = axial load, S_2 = shear force, S_3 = moment, etc. for all five points (A)-(E). From Equation (4.124)

$$\begin{Bmatrix} P_A \\ P_B \\ P_C \\ P_D \\ P_E \end{Bmatrix} = \begin{bmatrix} R_A^T & 0 & 0 & 0 \\ R_A^T T_1 & R_B^T & 0 & 0 \\ 0 & R_B^T T_2 & R_C^T & 0 \\ 0 & 0 & R_C^T T_3 & R_D^T \\ 0 & 0 & 0 & R_D^T T_4 \end{bmatrix} \begin{Bmatrix} S_A \\ S_B \\ S_C \\ S_D \end{Bmatrix},$$

$$[T_k] = \begin{bmatrix} -1 & 0 & 0 \\ 0 & -1 & 0 \\ 0 & -L_k & -1 \end{bmatrix}, \quad [R]_A^T = [R]_B^T = \begin{bmatrix} -1 & 0 & 0 \\ 0 & -1 & 0 \\ 0 & 0 & 1 \end{bmatrix},$$

$$[R]_C^T = \begin{bmatrix} 0 & 1 & 0 \\ -1 & 0 & 0 \\ 0 & 0 & 1 \end{bmatrix}, \quad [R]_D^T = I,$$

$$\{P_A\} = \begin{Bmatrix} R_1 \\ R_2 \\ P_3 = 0 \end{Bmatrix}, \quad \{P_E\} = \begin{Bmatrix} R_{13} \\ R_{14} \\ P_{15} = 0 \end{Bmatrix}.$$

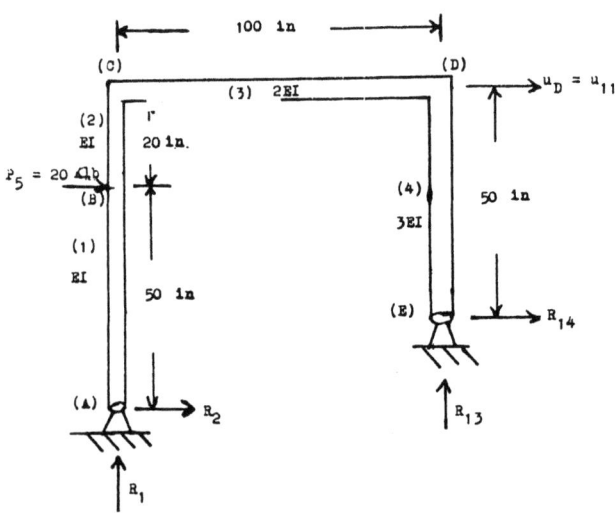

Fig. 4.15. Finite beam elements in frame.

Delete the four support equations to give eleven equations for twelve S_i forces. Select $S_2 = Q_2$

as the unknown and solve the equations in sequence to get

$$\{S\} = \{p_5\}P_5 + \{q_2\}Q_2,$$
$$\{p_5\}^T = (1/10)[3\ 0\ 0\ 3\ -10\ 0\ -10\ -3\ -200\ -3\ 10\ -500],$$
$$\{q_2\}^T = (1/5)[1\ 5\ 0\ 1\ 5\ 250\ 5\ -1\ 350\ -1\ -5\ 250].$$

Use Equation (4.123) with the proper EI and L for each element in Figure 4.15 to get

$$C_{QQ} = \{q_2\}^T[C_F]\{q_2\} = \frac{309,889}{EI}, \quad C_{QP} = -\frac{129,056}{EI},$$
$$Q_2 = S_2 = 0.4167 P_5 = 8.33 \text{ Klb}.$$

The reactions and S_i forces are

$R_1 = -7.67$ Klb, $\quad R_2 = -8.33$ Klb, $\quad R_{13} = 7.67$ Klb, $\quad R_{14} = -11.67$ Klb,
$S_1 = 7.67$ Klb, $\quad S_2 = 8.33$ Klb, $\quad S_3 = 0$, $\quad S_4 = 7.67$ Klb, $\quad S_5 = -11.67$ Klb,
$S_6 = 416$ Klb–in, $\quad S_7 = -11.67$ Klb, $\quad S_8 = -7.67$ Klb, $\quad S_9 = 183$ Klb–in,
$S_{10} = -7.67$ Klb, $\quad S_{11} = 11.67$ Klb, $\quad S_{12} = -584$ Klb–in.

To get the $u_D = u_{11}$ deflection apply a unit load for the P_{11} force in $\{P_D\}$ in the above equilibrium equations, whence

$$\{p\}_{11}^T = (1/2)[1\ 0\ 0\ 1\ 0\ 0\ 0\ -1\ 0\ -1\ 2\ -100],$$
$$u_D = u_{11} = \{p\}_{11}^T[C_F]\{P_5\{p\}_5 + Q_2\{q\}_2\}$$
$$= \frac{1}{EI}(63,889 P_5 - 84,722 Q_2) = 57.20(10^4)/EI.$$

These results are the same as in Example 4.9.

4.10. Matrix equations for beams from the mixed unit displacement and unit load theorem

The σ_{yy}, e_{yym}^1, e_{yyu}, $e_{yy\sigma}$, and σ_{yym}^1 terms in Equation (2.47) can be obtained from Equations (4.1), (4.50), and (4.16) for the simple beam element in bending, or for element (1),

$$\sigma_{yy} = -\frac{M_x z}{I_{xx}} = -\frac{z}{I_{xx}}[S_2 + S_1(L_1 - y_1)],$$

$$e_{yy\sigma} = \frac{\sigma_{yy}}{E} + e_E = -\frac{z}{EI_{xx}}[S_2 + S_1(L_1 - y_1)] + e_E,$$

$$e_{yyu} = -\frac{2z}{L^3}[3(L_1 - 2y_1)u_1 - L_1(L_1 - 3y_1)u_2 - 3(L_1 - 2y_1)u_3 - L_1(2L_1 - 3y_1)u_4], \quad (4.129)$$

$$\left\{\begin{array}{c}\sigma_{yy1}^1 \\ \sigma_{yy2}^1\end{array}\right\} = -\frac{z}{I_{xx}}\left\{\begin{array}{c}L_1 - y_1 \\ 1\end{array}\right\}, \quad \{e_{yy}^1\} \text{ in Equation (4.83b)}.$$

Thus for the beam element, the P_m equation in Equation (2.80) or Equation (2.47) is given by Equation (4.83c). For the second equation in (2.47) or (2.80),

$$0 = \int_0^{L_1} \int_A \left(-\frac{z}{I}\right) \begin{Bmatrix} L_1 - y_1 \\ 1 \end{Bmatrix} \left[\left(-\frac{2z}{L^3}\right)[3(L_1 - 2y_1) \quad -L_1(L_1 - 3y_1) - \right.$$

$$\left. -3(L_1 - 2y_1) \quad -L_1(2L_1 - 3y_1)] \begin{Bmatrix} u_1 \\ u_2 \\ u_3 \\ u_4 \end{Bmatrix} - \right.$$

$$\left. -\left(-\frac{z}{EI}\right)[L_1 - y_1 \quad 1] \begin{Bmatrix} S_1 \\ S_2 \end{Bmatrix} - e_E \right] dA\, dy$$

$$= \begin{bmatrix} 1 & 0 & -1 & -L_1 \\ 0 & 1 & 0 & -1 \end{bmatrix} \{u\}_1 - [C_F]_1 \begin{Bmatrix} S_1 + S_{E1} \\ S_2 + S_{E2} \end{Bmatrix}$$

$$= [s]_1 \{u\}_1 - [C_F]_1 \{S + S_E\}_1. \tag{4.130}$$

Assemble Equation (4.130) for all elements and use Equation (4.85) so that Equation (2.47) gives

$$\begin{bmatrix} -[C_F] & [s] \\ [s]^T & [0] \end{bmatrix} \begin{Bmatrix} S \\ u \end{Bmatrix} = \begin{Bmatrix} [C_F]\{S_E\} \\ \{P\} \end{Bmatrix}. \tag{4.131}$$

Since for the beam, Equations (4.88) and (4.116),

$$[C_F] = [k_S]^{-1}, \tag{4.132}$$

Equation (4.131) can be written in the reduced form

$$\begin{Bmatrix} S_E \\ P_a \end{Bmatrix} = [M] \begin{Bmatrix} S \\ u_a \end{Bmatrix}, \quad [M] = \begin{bmatrix} -[I] & [k_S][s]_a \\ [s]_a^T & [0] \end{bmatrix}. \tag{4.133}$$

Both the point forces S_i and the displacements u_{ai} can be obtained directly as

$$\begin{Bmatrix} S \\ u_a \end{Bmatrix} = [M]^{-1} \begin{Bmatrix} S_E \\ P_a \end{Bmatrix}. \tag{4.134}$$

It is only necessary to construct the $[s]^T$ and $[k_S]$ matrices for the beam and use a computer to make the inversion.

The $[M]$ matrix for the simple beam element (k) is

$$[M]_k = \begin{bmatrix} -\begin{bmatrix} 1 & 0 \\ 0 & 1 \end{bmatrix} & \frac{2EI}{L^3}\begin{bmatrix} 6 & -3L & -6 & -3L \\ -3L & 2L^2 & 3L & L^2 \end{bmatrix} \\ \begin{bmatrix} 1 & 0 \\ 0 & 1 \\ -1 & 0 \\ -L & -1 \end{bmatrix} & \begin{bmatrix} 0 & 0 & 0 & 0 \\ 0 & 0 & 0 & 0 \\ 0 & 0 & 0 & 0 \\ 0 & 0 & 0 & 0 \end{bmatrix} \end{bmatrix}_k. \tag{4.135}$$

Note that the discussion of the three methods of analysis in Section 3.7 also applies to this beam case.

The virtual principles for simple beams 147

4.11 The beam column equations

When compressive axial loads act on beams and truss members, as in Figure 4.16, the members tend to deflect laterally and may fail by column buckling. The effect of lateral deflections on a compression truss member in a redundant truss was considered in Example 3.15 above. These lateral deflections may be present due to initial deflections or due to deflections from applied lateral loads. In either case the P_c compression load tends to increase the deflection by producing a moment in the beam, or from Figure 4.16

$$M_{xc} = -P_c u_{xb}. \tag{4.136}$$

If a distributed lateral load p_x acts on the beam, then the applied moment on the beam due to p_x can be calculated from Equation (4.56), whence

$$M_x = M_{xc} + M_{xa}. \tag{4.137}$$

From Equation (4.60),

$$EI_{xx} u_{xb,yy} + P_c u_{xb} = M_{xa}. \tag{4.138}$$

The solution is

$$u_{xb} = u_{xbp}(y_r) + C_1 \sin K y_r + C_2 \cos K y_r, \tag{4.139}$$
$$y_r = y/L, \quad K^2 = P_c L^2 / EI_{xx},$$

where $u_{xbp}(y_r)$ is the particular solution for M_{xa}.

To determine the buckling load for the simply supported beam with $u_{xb} = 0$ at $y_r = 1$ and $y_r = 0$, assume $M_{xa} = 0$ so that $u_{xbp} = 0$, whence

$$0 = C_2, \quad 0 = C_1 \sin K,$$
$$\sin K = 0 \text{ for non} - \text{trivial solution,}$$
$$K = \pi, 2\pi, \cdots, n\pi. \tag{4.140}$$

Thus, the smallest Euler column failing or critical load is

$$K_{cr}^2 = P_c L^2 / EI_{xx} = \pi^2,$$
$$P_{c,cr} = \pi^2 EI_{xx}/L^2 = \pi^2 EA/(L/\rho)^2, \tag{4.141}$$
$$u_{xc} = \sin \pi y_r \text{ for deflected shape.}$$

If $P_c < P_{c,cr}$ and $u_{xbp}(y_r)$ is present, then the solution in Equation (4.139) is

$$u_{xb} = u_{xbp}(y_r) - u_{xbp}(0) \cos K y_r +$$
$$+ [u_{xbp}(0) \cos K - u_{xbp}(1)](\sin K y_r / \sin K). \tag{4.142}$$

If P_c is close to $P_{c,cr}$, then $\sin K$ is small and the deflection is large.

For some other cases

$$P_{c,cr} = \pi^2 EI_{xx}/4L^2, \quad \text{cantilever beam,}$$
$$P_{c,cr} = 4\pi^2 EI_{xx}/L^2, \quad \text{ends clamped.} \tag{4.143}$$

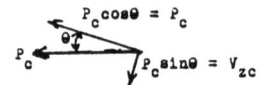
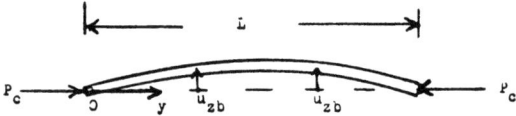

Fig. 4.16. Column buckling of long beam.

A more general beam column equation can be derived from the principle of virtual displacements. In Figure 4.16 the primary effect of deflection is the shear force

$$V_{zc} = -P_c \sin\theta = -P_c \tan\theta = -P_c u_{zb,y} \text{ for } \theta \text{ small.} \quad (4.144)$$

Also, from Equation (4.136),

$$V_{zc} = M_{xc,y} = -P_c u_{zb,y},$$
$$\sigma_{yzc} = -P_c u_{zb,y}/A_s, \quad (4.145)$$
$$e_{yzc}^V = (e_{yzc}^V)_r + (e_{yzc}^V)_g = -(P_c/GA_s)u_{zb,y}^V + u_{zb,y}^V, \quad (4.146)$$

where $(e_{yzc}^V)_r$ is the virtual form of the real shear strain and $(e_{yzc}^V)_g$ is the virtual form of the geometric shear strain produced by the u_{zb} deflection. See equation (2.53a). Also, the term

$$\int_{A_s} \sigma_{yzc}(e_{yzc}^V)_g \, dA_s = -\int_{A_s} (P_c/A_s) u_{zb,y} u_{zb,y}^V \, dA_s$$
$$= -P_c u_{zb,y} u_{zb,y}^V \quad (4.147)$$

is the virtual form of the large deflection term

$$P_c e_d = (P_c/2)(u_{z,y})^2 \quad (4.148)$$

in Equation (2.53). Usually, the term $(e_{yzc})_r$ is small compared to $(e_{yzc})_g$ so that it can be omitted.

Add the term $\sigma_{yzc} e_{yzcg}^V$ to the first integral in Equation (4.5), whence

$$\int_0^L \int_{A_s} \sigma_{yzc}(e_{yzc}^V)_g \, dA_s \, dy = -\int_0^L P_c u_{zb,y} u_{zb,y}^V \, dy$$
$$= -P_c [u_{zb,y} u_{zb}^V]_0^L + \int_0^L P_c u_{zb,yy} u_{zb}^V \, dy. \quad (4.149)$$

Thus, Equations (4.6) and (4.7) become

$$(EI u_{zb,yy})_{,yy} + P_c u_{zb,yy} - p_z = 0, \quad (4.150)$$
$$(EI u_{zb,yy} - M_{xL})_L = 0, \quad (EI u_{zb,yy} + M_{x0})_0 = 0,$$

The virtual principles for simple beams 149

$$[(EIu_{zb,yy})_{,y} + P_c u_{zb,y} + P_{zL}]_L = 0, \tag{4.151}$$
$$[(EIu_{zb,yy})_{,y} + P_c u_{zb,y} - P_{z0}]_0 = 0.$$

The fourth order Equation (4.150) allows more displacement and force boundary conditions to be used than the second order Equation (4.138).

See Section 1.4, Volume 2, for details on allowable compression stresses, including local buckling and inelastic effects.

Open section columns that are weak in torsional stiffness can fail by torsional buckling at loads below the bending buckling loads given by Equations (4.138)-(4.151). See Chapter 19 of Reference 6 and Chapter 3 of Reference 7 for equations on torsional column failure.

Also, beams under the usual bending loads that have small torsional stiffness can fail laterally at loads below the yield loads. See Chapter 4 in Reference 7 and Chapter 5 in Reference 8 on this buckling problem.

4.12 Problems

4.1. Solve Example 4.1 for the cantilever beam with P_{zL} an applied load at $y = L$ and no support at $y = L$.

4.2. Solve Example 4.1 with p_z variable, or

(a) $p_z = p_{zL} + \left(1 - \frac{y}{L}\right) p_{z0},$

(b) $p_z = \left(1 - \frac{y}{L}\right)^2 p_{z0}.$

4.3. Solve Example 4.2 for (a) $p = (1 - \frac{y}{L})^2 p_{z0}$, (b) $p = (y/L)^2 p_{zL}$.

4.4. Solve Example 4.1 for the simply supported beam. Use M_{z0}, M_{zL}, and constant p_z as applied loads.

4.5. In Equation (4.18) calculate P_{zLd}, M_{zLd}, P_{z0d}, and M_{z0d} for the distributed loading,
(a) $p_z = p_{zL} + (1 - \frac{y}{L}) p_{z0}$, (b) $p_z = (1 - \frac{y}{L})^2 p_{z0}$.

4.6. Solve Example 4.4 with $(EI)_1 = 2EI$, $(EI)_2 = EI$.

4.7. Solve Example 4.4 with $L_1 = 2L$, $L_2 = L$.

4.8. Solve Example 4.5 with $L_1 = 2L$, $L_2 = L$.

4.9. Solve Example 4.5 for a uniform load p_z on element (1).

4.10. In the fixed end beam in Figure 4.17, P_3, M_4, and constant p_z are applied loads. With EI constant find P_1, M_2, P_5, M_6, u_3, θ_4 by the displacement method.

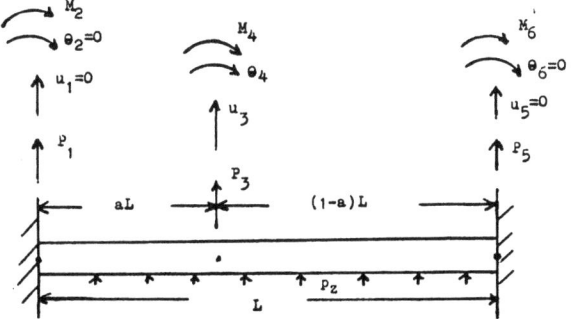

Fig. 4.17. Problems 4.10, 4.11.

4.11. Solve Problem 4.10 with the displacement u_1 free and P_1 an applied load. Find M_2, P_5, M_6, u_1, u_3, θ_4.

4.12. In Example 4.6 find the deflection at the tip of the beam.

4.13. In Example 4.6 verify that the deflection at point (B) is zero for the given moment curve.

4.14. Draw the shear curve for the beam in Figure 4.6.

4.15. Use Example 4.7 and solve Problem 4.2 by using the unit load theorem.

4.16. Solve Problem 4.4 by using the unit load theorem.

4.17. In Example 4.8 and Figure 4.7 find the deflection at the load P_2.

4.18. In Example 4.8 and Figure 4.7 find the deflection at any point $y_1 = a$ between the points R_1 and R_2.

4.19. Use the unit load theorem to find the deflection at point (A) in Figure 4.7 for a distributed loading p_z=constant on the entire length of the beam. Omit P_A and P_2.

4.20. Solve Example 4.8 for $EI = 3(EI)_R$ between R_1 and P_A and $EI = (EI)_R$ between P_A and R_2.

4.21. Use the unit load theorem to find the bending deflection u_{z2} at the load P_2 in Figure 4.7. Take EI as in Problem 4.20.

4.22. Use the unit load theorem to find (a) the slope θ_{z2} at the load P_2 with EI constant in Figure 4.7. (b) Find the slopes at the reactions R_1 and R_2.

4.23. Repeat Example 4.9 for $(EI)_1 = (EI)_2 = (EI)_3 = EI$.

4.24. Solve Example 4.9 for the 20 Klb applied force in Figure 4-8 located at the top end of member (1) in line with member (2).

4.25. Solve Example 4.9 with the 20 Klb applied force in Figure 4.8 omitted and a uniformly distributed load of 200 lb/in applied up on member (2).

4.26. Solve Example 4.9 for an applied vertical force of -40 Klb at the midpoint of element (2).

4.27. Use the unit load theorem to find (a) the reactions and bending moments in the redundant beam frame in Figure 4.18. (b) the vertical bending deflection at point (A), and (c), the slope at point (D).

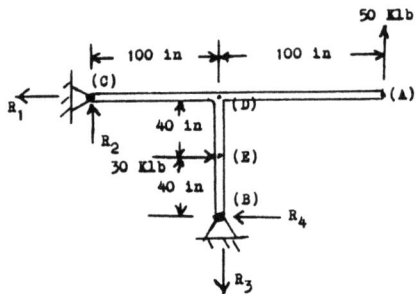

Fig. 4.18. Problems 4.27, 4.28, 4.42, 4.43.

4.28. Solve Problem 4.27 for the case of a zero slope restraint at support (C) with the additional moment reaction M_C.

4.29. In Example 4.11 calculate the deflection u_7 and slope u_8. Check the results with Equation (4.100).

4.30. Solve Example 4.11 for the case of $(EI)_1 = (EI)_2 = EI$, $(EI)_3 = (EI)_4 = 3EI$.

4.31. In Example 4.11 draw the final shear and moment curves for $L = 100$ in., $P_1 = 80$ Klb, $P_7 = 200$ Klb.

4.32. Solve Example 4.11 for slope u_{10} at the wall free and $S_{10} = P_{10}$ as an applied moment. Use Q_3 as the unknown redundant with Q_5 a reaction.

4.33. Use the temperature distribution in Example 4.12 and find the deflection at the load P_2 in Figure 4.7.

4.34. Solve Problem 4.33 with

$$M_{zT} = M_{zT1} + \left(1 - \frac{y_1}{100}\right)M_{zT2}, \quad 0 \leq y_1 \leq 200.$$

4.35. Solve Example 4.13 for the thermal moment

$$M_{xT} = \left(1 - \frac{y}{L}\right)M_{xT2} + \left(\frac{y}{L}\right)^3 M_{xT3}.$$

4.36. Repeat Problem 4.22 for the thermal moment distribution in Example 4.13.
4.37. Verify the integrations for u_{xA}, $u_{xA,\text{elastic}}$, and $u_{xA,\text{temp}}$ in Example 4.14.
4.38. Solve Example 4.14 for $A_1 F_{y1}$ =-600 Klb instead of -750 Klb.
4.39. Repeat Example 4.14 for the deflections at a point (B) located at $y = 40$ in. in Figure 4.10.
4.40. Verify the integrations to get the M_0 equation in Example 4.15, part (c).
4.41. In Example 4.15 change $A_1 F_{y1}$ to -600 Klb as in Problem 4.38, and solve for M_0 in part (c).
4.42. Solve Problem 4.27 for a thermal moment M_{xT0} constant in the (C)-(D) section of the frame.
4.43. Solve Problem 4.27 for $M_{xT} = M_{xT0}(\frac{y}{100})$ in the (C)-(D) section of the frame.
4.44. Use the unit load theorem to find the component deflections u_{xA} and u_{yA} at point (A) for the beam in Figure 4.19, where $(EI_{zz})_k = 4(10^7)$ Klb-in^2, $(GA_s)_k = 16,000$ Klb for all beam elements, and $(M_{xT})_3 = 6000$ Klb-in. Neglect axial deflections but include shear deflections.

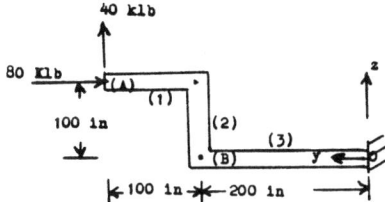

Fig. 4.19. Problems 4.44, 4.45, 4.46.

4.45. Assume $u_{xB} = 0$ at point (B) in Figure 4.19 and calculate (a) the Q_{xB} reaction and (b) the deflection u_{xA} by using the unit load theorem. Use data in Problem 4.44.
4.46. Calculate the slope at point (A) for both Problems 4.44 and 4.45 by using a unit moment at point (A).
4.47. Make the integrations in Equation (4.83c).
4.48. Verify that Equation (4.88) gives Equation (4.78).
4.49. Make the inversion for the $[k_{\alpha\alpha}]$ matrix in Equation (4.99).
4.50. Solve Example 4.16 with $(EI)_1 = (EI)_2 = EI$, $(EI)_3 = (EI)_4 = 3EI$.
4.51. In Equation (4.98) in Example 4.16 change the beam supports to $u_1 = u_3 = u_5 = u_7 = u_9 = 0$ and use applied moments M_2 and M_{10} as well as the given thermal moments. Solve for slopes, reactions, and the point S_1 forces.
4.52. Solve Example 4.17 with element (1) at 90° to element (2).
4.53. Construct the $[k]_1$ matrix in Equation (4.91) from $[s]_1^T[k_S]_1[s]_1$.
4.54. Construct $[k]$ in Equation (4.29) from $[s]^T[k_S][s]$ for two elements.
4.55. Solve Example 4.18 for the case of applied M_2 and M_8 moments.
4.56. In Example 4.18 and Figure 4.9 add a support with $u_1 = 0$. Use Q_1, Q_3, Q_5 as the unknown redundants with applied loads as M_2 and P_7. Find the S_i internal forces, reactions, slope u_2, and deflection u_7.
4.57. Repeat Problem 4.56 using unknown moments M_4, M_6, and M_{10} as the redundants. See Example 4.19.
4.58. Add constant thermal moments M_{T1} and M_{T2} to elements (1) and (2) in Example 4.18. See Equation (4.96), (3.70), and (3.71).
4.59. In Example 4.19 let $LP_1 = LP_7 = M_{T2} = 400$ Klb-in, $S_{y4} = 100$ Klb-in and find M_{p10}. Discuss the effect of M_{p10} upon the S_i forces, reactions, and deflections by comparing the results to the elastic solution.
4.60. Solve Problem 4.59 without the thermal moment, or $M_{T2} = 0$.
4.61. Solve Problem 4.59 with $LP_1 = LP_7 = 1000$ Klb-in, $M_{T2} = 400$ Klb-in, $S_{y4} = 100$ Klb-in.

4.62. Verify that the 6 by 6 matrix in Equation (4.128) is the inverse of the 6 by 6 matrix in Equation (4.103) in the upper left corner.

4.63. Rearrange Equation (4.103) so that H occurs in the load terms and the 9 by 9 matrix is independent of H.

4.64. Make use of the results in Problem 4.63 and rearrange Equation (4.128) so that the 6 by 6 matrix is independent of H.

4.65. Solve Example 4.21 with the same EI for all elements.

4.66. Solve example 4.21 for the case in which $L_4 = 2.0L$.

4.67. Solve Example 4.23 with the same EI for all elements.

4.68. In Figure 4.15 put the point (B) and the P_5 load at 35 in. from point (A) and repeat Example 4.23.

4.69. Use Equation (4.150) with $M_{xa} = 0$ and find the buckling load for the beam with clamped ends to check Equation (4.143).

References

Chapter 4

1. B.E. Gatewood and R.W. Gehring: Inelastic redundant analysis and test-data comparison for a heated ring frame, *Journal of Aerospace Sciences*, **29**, March (1962).
2. B.E. Gatewood and R.W. Gehring: Allowable axial loads and bending moments for inelastic structures under nonuniform temperature distribution, *Journal of Aerospace Sciences* **29**, May (1962).
3. J.S. Przemieniecki: *Theory of Matrix Structural Analysis*, McGraw-Hill Book Co. (1968).
4. H.C. Martin: *Introduction to Matrix Structural Analysis*, McGraw-Hill Book Co. (1966).
5. M.D. Vanderbilt: *Matrix Structural Analysis*, Quantum Publishers, New York (1974).
6. A.S. Niles and J.S. Newell: *Airplane Structures*, 3rd Ed., Vol. II, John Wiley and Sons (1943).
7. F. Bleich: *Buckling Strength of Metal Structures*, McGraw-Hill Book Co. (1952).
8. S. Timoshenko: *Theory of Elastic Stability*, McGraw-Hill Book Co. (1936).

References for additional reading

9. D.J. Peery and J.J. Azar: *Aircraft Structures*, 2nd Ed., McGraw-Hill Book Co., New York (1982).
10. D.H. Allen and W.E. Haisler: *Introduction to Aerospace Structural Analysis*, John Wiley and Sons, New York (1985).
11. J.L. Meek: *Matrix Structural Analysis*, McGraw-Hill Book Co., New York (1971).
12. R.K. Livesley: *Matrix Methods of Structural Analysis*, 2nd Ed., Pergamon, New York (1976).
13. J.V. Meyers: *Matrix Analysis of Structures*, Harper and Row, New York (1985).
14. W. Weaver, Jr. and J.W. Gere: *Matrix Analysis of Framed Structures*, 2nd Ed., van Nostrand, New York (1980).
15. C.K. Wang and C.G. Salmon: *Introductory Structural Analysis*, Prentice Hall, Englewood Cliffs, N.J. (1984).

5

Box beam shear stresses and deflections

5.1. Introduction

The simple beam with loading in one plane was considered in Chapter 4. This type of beam has many applications, such as floor beams for buildings and bridges. However, there are many applications such as airplane wings and fuselages where the beam must take bending in two directions as well as torsional moments. This type of beam which can take all six generalized loads (three forces and three moments) will be considered in this Chapter 5.

Although any beam with a solid cross section such as a solid circular drive shaft can take the six generalized forces, the box beam with thin webs and stringer members is usually the most efficient type of structure for general loading. This is particularly true for aircraft structures where some of the webs provide skin covering for the airplane. In other applications it is possible to replace the thin webs by diagonal truss members, provided the joints at the corners are not too heavy and complicated. Beams with solid cross sections as well as box beams will be considered. The bending stress and bending deflection formulas are the same for both types of beams, but the shear stresses and shear deflection formulas are completely different.

Since the general axial stress and bending stress formulas for the axial load P_y and the bending moments M_x and M_z are given in Section 1.14, Equation (1.90), the shear stresses due to the shear forces P_x and P_z and torsional moment M_y will be the primary concern in Sections 5.2 through 5.9. Bending and axial deflections are given in Section 5.10. The shear deflections and torsional rotation of beams, including the warping effects in the box beam cross section, are given in Sections 5.11 and 5.12. The warping of the cross section can be caused by the thin webs between stringers having different shear strains.

If the webs in one plane do not have the same shear deflection, they are incompatible unless they can warp to the plane strain cross section. See Figure 5.19. That is, the real strain angle plus the warping strain angle must be the same for each web in order to have compatibility. Also, the webs may warp out of the plane strain cross section for the box beam under torsional rotation. If this shear warping of the cross section is not restrained by the beam supports, then the bending stresses, shear flows, and bending deflections are not affected.

However, all stresses and deflections are changed if the warping is restrained. This restrained warping case will be considered in Chapter 6. If the beam has cut-outs in the webs, then large warping effects occur and the beam cross section is not plane for either shear or bending. In this case, the bending moment on the beam cannot be used directly to get the bending stresses and deflections.

Since the unit load theorem is a compatibility equation, it can be used to impose compatibility on the beam with warping present. If the beam has a plane section with no warping present or if the known warping is included in the e_E strain, then the virtual stresses produced by the unit load can be arbitrary on the beam or sub-beams, except for external equilibrium requirements of the beam. With unknown warping present and the real strains being used, the virtual stresses can no longer be arbitrary as unit loads on the sub-beams or individual webs will give different deflections. In this case the unit load must be applied to the entire cross section with the virtual stresses being determined in the same manner as the real elastic stresses. This insures that the virtual stresses are in equilibrium throughout the cross section as a plane section and provide the proper weighting factors for each sub-beam or web so that the unit load theorem will give the deflections including all unrestrained warping for a plane section.

As pointed out in Section B.13 and Figure B.24, the bending deflection of the swept wing elastic axis produces changes in the wing angle of attack. This changes the airload distribution on the wing and hence changes the wing shear forces and bending moments. Some of these deflection effects are considered in Sections 5.13, 5.14 and 5.15.

5.2. Shear stresses in beams

In the derivation of the beam equations in Chapter 1 from the three dimensional elasticity equations it was found that the shear stresses on the beam cross section had to satisfy complicated second order partial differential equations (1.30) and (1.34). As shown in Sections 1.10, 1.11, and 1.12, it is possible to simplify the equations for beams with thin web cross sections. In this Section 5.2 procedures to obtain solutions for the maximum shear stresses in beams with various cross sections are described.

On the beam cross section in Figure 5.1 assume $V_x = 0$ so that V_z is the only shear force on the cross section. Assume $\sigma_{xy} = 0$ so that the equilibrium Equation (1.31) becomes

$$\sigma_{yz,z} + \sigma_{yy,y} = 0. \tag{5.1}$$

Since σ_{yz} and σ_{yy} are independent of x, take the variable width of the cross section in the x-direction as $2t_x(z)$ and rewrite the equilibrium Equation (5.1) as

$$\frac{d}{dz}(t_x \sigma_{yz}) + \frac{d}{dy}(t_x \sigma_{yy}) = 0. \tag{5.2}$$

Box beam shear stresses and deflections

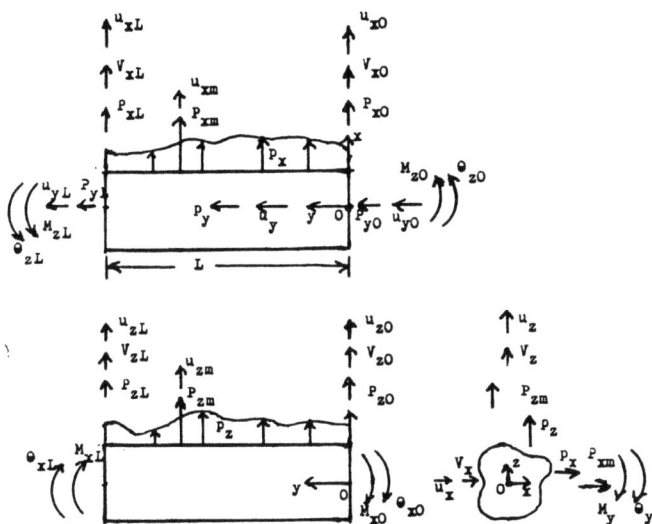

Fig. 5.1. Beam with general loading.

This gives equilibrium on any line across the beam width.

For simple bending of the beam produced by the V_z shear,

$$\sigma_{yy} = -\frac{zM_x}{I_{xx}}, \quad V_z = \frac{dM_x}{dy},$$

$$\frac{d}{dy}(t_x\sigma_{yy}) = -\frac{zt_x}{I_{xx}}\frac{dM_x}{dy} = -\frac{zt_xV_z}{I_{xx}}, \tag{5.3}$$

$$t_x\sigma_{yz} = \frac{V_z}{I_{xx}}\int_z^h zt_x\,dz = \frac{V_z}{2I_{xx}}\int_z^h z\,dA, \tag{5.4}$$

$$\sigma_{yz} = \frac{V_z}{2t_xI_{xx}}\int_z^h z\,dA, \tag{5.5}$$

where $2h$ is the height of the beam in the z-direction.

Although Equation (5.5) applies only for t_x small and $\sigma_{xy} = 0$, it actually gives a good approximation for $(\sigma_{yz})_{\max}$ for many solid cross sections with t_x large and $\sigma_{xy} \neq 0$, when compared to two dimensional cross section solutions for σ_{yz} and σ_{xy} as given in Chapter 12 of Reference 1.

For the solid circular cross section of radius R, it is shown in Reference 1 that

$$\sigma_{yz} = \frac{(3+2\nu)V_z}{8(1+\nu)I_{xx}}\left(R^2 - z^2 - \frac{1-2\nu}{3+2\nu}x^2\right), \quad \sigma_{xy} = -\frac{1+2\nu}{4(1+\nu)}\frac{xzV_z}{I_{xx}}, \tag{5.6}$$

$$(\sigma_{yz})_{\max} = \frac{(3+2\nu)R^2V_z}{8(1+\nu)I_{xx}} \tag{5.7}$$

$$= 0.346(R^2V_z/I_{xx}) = 1.384V_z/A \text{ for } \nu = 0.30. \tag{5.8}$$

In this case Equation (5.5) gives

$$(\sigma_{yz})_{\max} = \frac{V_z}{2RI_{xx}} \int_0^R 2z(R^2 - z^2)^{\frac{1}{2}} \, dz = \frac{R^2 V_z}{3I_{xx}} = \frac{4V_z}{3A}, \qquad (5.9)$$

which is about 4% smaller than in Equation (5.8).

For the solid rectangular cross section $2t_x$ by $2h$, it is shown in Reference 1 that the $(\sigma_{yz})_{\max} = 3V_z/2A$ given by Equation (5.5) is nearly exact for $h/t_x \geq 2$. For the square cross section ($h/t_x = 1$), the maximum shear stress in Equation (5.5) is about 11% smaller for $\nu = 0.25$ and about 20% smaller for $\nu = 0.30$. See Chapter 12 of Reference 1 for other cases of solid beam cross sections.

For the thin web *constant* cross section beam, such as in Figure 5.2, Equation (5.4) can be used to get the change in shear flow at the stringers, or at stringer

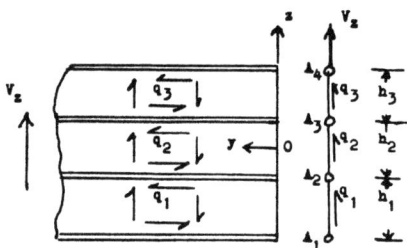

Fig. 5.2. Simple beam with several webs.

i with point area A_i and t_i = thickness of web i,

$$q_i = t_i \sigma_{yzi} = q_{i-1} - \frac{A_i z_i V_z}{I_{xx}}. \qquad (5.10)$$

Thus, the web shear flows can be obtained directly from the stringers.

If the stringer areas, temperatures and material properties vary, take Equation (5.2) in the form

$$q_z = t_x \sigma_{yz} = -\frac{d}{dy} \int_z^h t_x \sigma_{yy} \, dz = -\frac{d}{dy} \int_z^h \sigma_{yy} \, dA. \qquad (5.11)$$

For point stringer areas with A_i varying with y, Equation (5.11) is

$$q_i = q_{i-1} - \frac{d}{dy}(A\sigma_{yy})_i = q_{i-1} - \frac{dP_i}{dy}$$

$$= q_{i-1} - \frac{\Delta P_i}{\Delta y}, \quad P_i = A_i \sigma_{yyi}, \qquad (5.12)$$

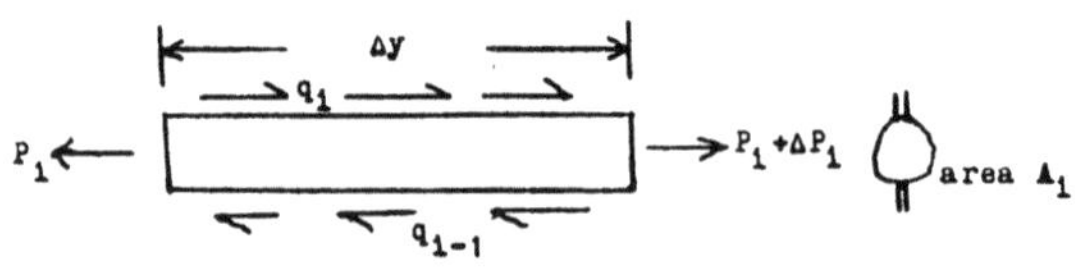

Fig. 5.3. Equilibrium of stringer element.

Box beam shear stresses and deflections

which agrees with the stringer element equilibrium in Figure 5.3. If the P_i stringer loads are calculated on the cross section at intervals at distance b apart, then the average web shear flows in the interval are

$$q_i = q_{i-1} - \frac{(P_i)_y - (P_i)_{y-b}}{b}. \tag{5.13}$$

This Equation (5.13) can be used for any beam with constant width and constant height but with variable E_i, T_i, A_i, e_{Ei}, where $(\sigma_{yy})_y$ is given by Equation (1.90) at the cross sections. Also, inelastic effects in the stringers can be included through Equation (1.90).

Example 5.1. In Figure 5.2 take $h_1 = h_2 = h_3 = 10$ in., $A_1 = A_2 = 4.00$ in^2., $A_3 = A_4 = 8.00$ in^2., $V_z = 30,000$ lb, and find the shear flows q_1, q_2, q_3.

Solution. From stringer (1), the centroid is

$$\bar{z} = [8(30+20) + 4(10)]/24 = 18.33 \text{ in.},$$

whence

$$z_1 = -18.33, \quad z_2 = -8.33, \quad z_3 = 1.67, \quad z_4 = 11.67,$$
$$I_{zz} = 8(30^2 + 20^2) + 4(10^2) - 24(18.33)^2 = 2733.33 \text{ in}^4.$$

From Equation (5.10),

$$q_i = q_{i-1} - (30,000/2733.33)A_i z_i = q_{i-1} - 10.9756 A_i z_i,$$
$$q_1 = 0 - 10.9756(4)(-18.33) = 805 \text{ lb/in.},$$
$$q_2 = 805 - 10.9756(4)(-8.33) = 1171 \text{ lb/in.},$$
$$q_3 = 1171 - 10.9756(8)(1.67) = 1024 \text{ lb/in.},$$
$$q_4 = 1024 - 10.9756(8)(11.67) = -1 \text{ lb/in.},$$

which checks zero shear flow where there is no web.

Example 5.2. In Figure 5.2 find the average shear flows between cross sections at $y = 200$ in. and $y = 160$ in. using Equations (1.92) and (5.13) with the following data:

	$y = 200$ in.				$y = 160$ in.			
	(1)	(2)	(3)	(4)	(1)	(2)	(3)	(4)
A_i (in^2)	4.00	4.00	8.00	8.00	6.00	6.00	8.00	8.00
h_i (in)		10.00	10.00	10.00		10.00	10.00	10.00
$10^{-7} E_i$ (psi)	0.7	0.7	0.7	0.7	0.6	0.6	0.7	0.7
T_i (°C)	0	0	0	0	200	200	0	0
$10^{-7} M_x$ (in – lb)		1.00				1.12		

$P_y = 0$, $\quad \alpha_i = $ constant $= 20(10^{-6})/°$C.

Solution. Neglect any axial thermal stresses in the webs and take $E_R = 0.7(10^7)$ psi. Make the calculations in parallel for the two stations, Equation (1.92).

	$y = 200$ in.	$y = 160$ in.
A_E	24.00 in^2	26.29 in^2
$\bar{z}$	18.33 in.	17.17 in.
z_i	-18.33, -8.33, 1.67, 11.67 in.	-17.17, -7.17, 2.83, 12.83 in.
I_{zzE}	2733.33 in^4	3163.76 in^4
P_{yT}	0.00 lb	288,000 lb
M_{xT}	0.00 in-lb	3,505,000 in-lb
σ_{yy1}	67,061 psi	53,795 psi
σ_{yy2}	30,476 psi	13,954 psi
σ_{yy3}	-6,110 psi	-2,200 psi
σ_{yy4}	-42,695 psi	-48,678 psi
P_i	268,200, 121,900, -48,900, -341,600 lb	322,767, 83,736, -17,600, -389,426 lb

From Equation (5.13),

$q_1 = 0 - (268,200 - 322,767)/40 = 1364$ lb/in,
$q_2 = 1364 - (121,900 - 83,736)/40 = 410$ lb/in,
$q_3 = 410 - (-48,900 + 17,600)/40 = 1192$ lb/in,
$q_4 = 1192 - (-341,600 + 389,426)/40 = -4$ lb/in, checks.

Although the height of the beam and the shear force V_z are the same in the two Examples 5.1 and 5.2, the results in the two examples show the large effects on the shear flows and stringer stresses caused by the temperature and area changes in Example 5.2.

5.3. Torsional shear stresses in beams

As indicated in Section 1.12 the shear stresses produced by the torsional moment M_y are two dimensional on the beam cross section. However, as shown in Section 1.12, there are two types of cross sections for which the shear stresses are simple to calculate: the solid circular rod and the thin wall closed tube. Two dimensional solutions for some other solid cross sections are given in References 1, 2, and 3.

From Chapter 11 of Reference 1, the maximum shear stress for a solid elliptical cross section with major axis $2h$ and minor axis $2t_x$ is

$$(\sigma_{yz})_{max} = -2M_y/\pi h t_x^2, \tag{5.14}$$

located at the ends of the minor axis. This reduces to the solid circular case in Equation (1.70) for $h = t_x = R$ and $x = R$. The twist θ_y of the constant cross section ellipse can be determined from

$$\frac{d\theta_y}{dy} = M_y(h^2 + t_x^2)/\pi h^3 t_x^3 G. \tag{5.15}$$

For the solid rectangular cross section $2t_x$ by $2h$ it is shown in Reference 1 that

$$(\sigma_{yz})_{max} = -M_y/C_1 h t_x^2, \tag{5.16}$$

where C_1 varies from 1.664 for the square cross section ($h = t_x$) to 2.667 for the limit of $h/t_x \to \infty$. Tables of $C_1/8$ for various ratios of h/t_x are given in References 1 and 2, and a curve for $C_1/8$ is given in Reference 3. Note that in Equation (5.14) C_1 for the ellipse is $C_1 = \pi/2 = 1.57$ for all h/t_x ratios, whence the ellipse has larger shear stresses than the rectangle. The twist θ_y can be determined from

$$\frac{d\theta_y}{dy} = M_y/C_2 h t_x^3 G, \tag{5.17}$$

where C_2 varies from 2.243 for the square cross section to 5.333 for $h/t_x \to \infty$. Values of $C_2/16$ are given in References 1, 2, and 3 for various h/t_x. See References 1, 2, and 3 for other cases of solid beam cross sections.

Box beam shear stresses and deflections

For the thin web closed tube or box beam the shear stress or shear flow is given by Equation (1.74). However, if the box beam has stringers then the shear flows will be different between stringers so that Equation (1.74) must be modified. Also, if the box beam is open and not capable of taking torque, it is necessary to apply the shear forces V_x and V_z so that no torsional moment M_y is produced.

Consider the two stringer beam cross section in Figure 5.4 and find the distance e_x so that V_z produces no M_y. In Figure 5.4 the torsional moment about point 0 produced by V_z must equal the moment produced by the constant shear flow q in the curved web, or

$$e_x V_z = \int_{s_0}^{s_1} rq\, ds = \int_{s_0}^{s_1} q(2\, dA_c) = 2A_c q = 2A_c V_z/h, \qquad (5.18)$$

$$e_x = 2A_c/h, \qquad (5.19)$$

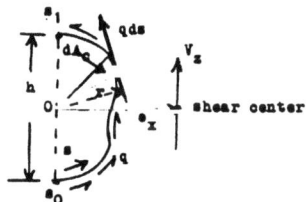

Fig. 5.4. Shear flow in curved web.

where A_c is the area enclosed by the web and the lines from 0 to s_0 and s_1. The point s.c. at the distance e_x from 0 is the *shear center* for no twist of the beam.

Figure 5.5 shows a closed box beam with six stringers and six shear flows in the thin webs between the stringers. If torsional moments are taken about stringer (1), then

$$M_y(\text{about } 1) = \sum_{i=1}^{6} 2A_{ci} q_i, \qquad (5.20)$$

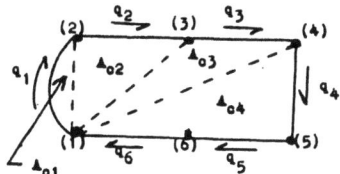

Fig. 5.5. Six stringer thin web box beam.

where in Figure 5.5, $A_{c5} = A_{c6} = 0$. In general for N stringers,

$$M_y(\text{Ref. point}) = \sum_{i=1}^{N} 2A_{ci} q_i. \qquad (5.21)$$

To find the shear flows q_i, this Equation (5.21) must be combined with the V_z and V_x shear flow Equations (5.10) or (5.13).

The following Sections 5.4 through 5.8 show the procedures for calculating the q_i shear flows in various types of box beams with any number of stringers.

5.4. Shear flows in open box beams

If the shear forces are applied at the shear center of the open box beam cross section, then the shear flows are given by Equation (5.10) or (5.13). For the general unsymmetrical open box beam with constant cross section, constant area, constant P_y, E, and T in the y-direction, σ_{yyi} is given by Equation (1.90) and q_i by Equation (5.12). Thus,

$$q_i = q_{i-1} - \frac{E_i A_i}{E_R H}\left[-(z_i I_{xxE} - x_i I_{xzE})\left(\frac{dM_x}{dy}\right) + \right.$$
$$\left. + (x_i I_{xxE} - z_i I_{xzE})\left(\frac{dM_z}{dy}\right)\right] \quad (5.22)$$

$$= q_{i-1} - \frac{E_i A_i}{E_R H}[(z_i I_{xxE} - x_i I_{xzE})V_z + (x_i I_{xxE} - z_i I_{xzE})V_x], \quad (5.23)$$

$$= q_{i-1} - \frac{E_i A_i}{E_R H}[(V_z I_{xxE} - V_x I_{xzE})z_i - (V_z I_{xzE} - V_x I_{xxE})x_i], \quad (5.24)$$

where $(dM_x/dy) = -V_z$ and $(dM_z/dy) = V_x$ for applied shear flows. For the case of $I_{xzE} = 0$, Equation (5.24) reduces to

$$q_i = q_{i-1} - \frac{E_i A_i}{E_R}\left(\frac{V_z z_i}{I_{xxE}} + \frac{V_x x_i}{I_{zzE}}\right). \quad (5.25)$$

For the general case with variations in the y-direction, Equation (5.13) must be used.

Take Equation (5.23) in the form

$$q_i = q_{i-1} + H_{i0}V_z + J_{i0}V_x, \quad (5.26)$$

$$H_{i0} = -(E_i A_i / E_R H)(z_i I_{xxE} - x_i I_{xzE}),$$
$$J_{i0} = (E_i A_i / E_R H)(z_i I_{xzE} - x_i I_{zzE}), \quad (5.27)$$

$$H = I_{xxE} I_{zzE} - I_{xzE}^2.$$

In the open box beam take $q_0 = 0$ in the open region, whence Equation (5.26) gives

$$q_1 = H_{10}V_z + J_{10}V_x,$$
$$q_2 = q_1 + H_{20}V_z + J_{20}V_x = (H_{10} + H_{20})V_z + (J_{10} + J_{20})V_x,$$
$$------------$$
$$q_i = V_z \sum_{j=1}^{i} H_{j0} + V_x \sum_{j=1}^{i} J_{j0} \quad (5.28)$$

Box beam shear stresses and deflections

$$= V_z H_i + V_x J_i, \tag{5.29}$$

$$H_i = \sum_{j=1}^{i} H_{j0}, \quad J_i = \sum_{j=1}^{i} J_{j0}. \tag{5.30}$$

Note that the H_i are the shear flows per unit value of V_z and the J_i are the shear flows per unit value of V_x.

To get the shear center of the open box use Figures 5.4 and 5.6 and take moments about an axis in y-direction at some point 0. Then, from Figure 5.6 with moment M_y and the shear flows positive clockwise

$$-e_x V_z + e_z V_x = \sum_{i=1}^{N} r_i q_i s_i, \tag{5.31}$$

$$= \sum_{i=1}^{N} 2 A_{ci} q_i, \tag{5.32}$$

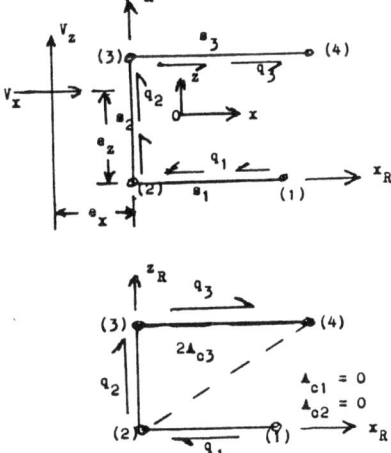

Fig. 5.6. Open box beam cross section.

where s_i is the length of web i, r_i is the perpendicular distance from 0 to web i (must integrate as in Equation (5.18) for a curved web), A_{ci} is the area enclosed by web i and the straight lines from 0 to stringers i and $i+1$, e_x and e_z have the same signs as x and z. Put Equation (5.29) into Equation (5.32) to get

$$e_x = -\sum_{i=1}^{N} 2 A_{ci} H_i, \quad e_z = \sum_{i=1}^{N} 2 A_{ci} J_i, \tag{5.33}$$

which locates the shear center of the open box independent of the shear forces. H_i and J_i are given in Equation (5.30).

Example 5.3. Find the shear center of the open box in Figure 5.6 with $A_1 = 3.00$ in² $= A_2$, $A_3 = 4.00$ in², $A_4 = 2.00$ in², $s_1 = s_3 = 10.00$ in, $s_2 = 20.00$ in. Also find the shear flows for

$V_z = 30,000$ lb and $V_x = 20,000$ lb, E and T are constant on the cross section.
Solution. From the origin of the reference axes x_R, z_R,

$$\bar{x} = (3+2)(10)/(12) = 4.17 \text{ in.}, \quad \bar{z} = (6)(20)/(12) = 10.00 \text{ in.},$$

whence the (x_i, z_i) location of the stringers from the centroid is

$$(x_1, z_1) = (5.83, -10.00), \quad (x_2, z_2) = (-4.17, -10.00),$$
$$(x_3, z_3) = (-4.17, 10.00), \quad (x_4, z_4) = (5.83, 10.00).$$

Thus, the moments of inertia are

$$I_{xx} = (3+3+4+2)(10)^2 = 1200 \text{ in}^4,$$
$$I_{zz} = (2+3)(5.83)^2 + (3+4)(4.17)^2 = 292 \text{ in}^4,$$
$$I_{xz} = (2)(10)(20) - (12)(4.17)(10) = -100 \text{ in}^4,$$
$$H = I_{xx}I_{zz} - I_{xz}^2 = 340,400 \text{ in}^8,$$

and Equation (5.27) becomes

$$H_{i0} = -(10^{-3})A_i(0.294x_i + 0.858z_i),$$
$$J_{i0} = -(10^{-3})A_i(3.525x_i + 0.294z_i).$$

Use a table form to calculate the sums in Equations (5.30) and (5.33), taking M_y moments about stringer (2). From Equation (5.33) and the sums in Table 5.1

$$e_x = -4.12 \text{ in.}, \quad e_z = 9.43 \text{ in.}$$

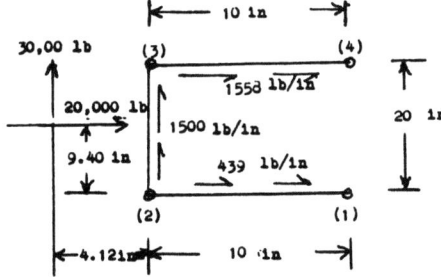

Fig. 5.7. Shear center and shear flows for Example 5.3

as measured from stringer (2). Since H_i and J_i are the unit shear flows, the table values of H_i and J_i can be used for any values of V_x and V_z. The last row in Table 5.1 and Figure 5.7 show the shear flows due to the specified shear forces.

Table 5.1. Table for Example 5.3.

	$S(1)$	$W(1)$	$S(2)$	$W(2)$	$S(3)$	$W(3)$	$S(4)$	$W(4)$	Σ
A_i (in^2)	3.00		3.00		4.00		2.00		
x_i (in)	5.83		-4.17		-4.17		5.83		
z_i (in)	-10.00		-10.00		10.00		10.00		
$10^3 H_{i0}$	20.59		29.42		-29.42		-20.59		0
$10^3 H_i$		20.59		50.01		20.59		0	
$10^3 J_{i0}$	-52.86		52.86		47.14		-47.14		0
$10^3 J_i$		-52.86		0		47.14		0	
$2A_{ci}$		0		0		200			
$2A_{ci}H_i$		0		0		4.12			4.12
$2A_{ci}J_i$		0		0		9.43			9.43
q_i (lb/in.)		-439		1500		1558		0	

5.5. Shear flows in single cell box beams

In the previous section on the open box beam it was assumed that the shear forces were at the shear center so that the box would have no twisting moment M_y on it. Actually, the open box may be able to take some torque, but usually the torque will produce large twisting deformations and large stresses in the stringers. If two separate single beams essentially parallel to each other can be found in an open box, then it is evident that the torque can be represented by a force couple putting a shear force in each beam, Figure 5.8. Since the two beams

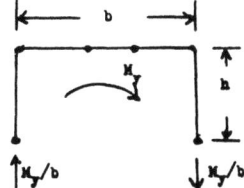

Fig. 5.8. Open box beam with torque moment M_y.

bend in opposite directions, a large rotation can occur. Also, large shear flows occur in the two beams with no shear flows in the other webs. The open box is very inefficient for torsion.

Since the box beams in wings, fuselages, and control surfaces of flight vehicles must take large torsional moments, it is essential that an efficient box for torsion be used and that open boxes be avoided in so far as possible. As pointed out in Section 1.13, the closed thin wall box beam is very efficient for torsion so that this section is devoted to the determinate single cell box beam with shear forces V_z and V_x and the torsional moment M_y acting.

From Figure 1.10 and Equation (1.74) the shear flow due to M_y acting on a thin web single cell box is

$$q_0 = M_y/2A_c, \quad \text{constant}, \tag{5.34}$$

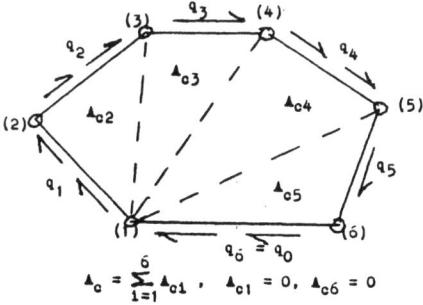

Fig. 5.9. Single cell box beam cross section.

where A_c is the enclosed area of the cell. In Figure 5.9 assume web 6 or web

0 to be temporarily removed so that the shear flows q_{is} in the open box can be calculated by the procedure of the previous Section 5.4. Then by Equation (5.34) the shear flows in the closed box will be

$$q_i = q_0 + q_{is}. \tag{5.35}$$

The torque equilibrium equation about some reference point such as stringer (1) can be expressed as, Equation (5.21),

$$M_y = \sum_{i=1}^{N}(2A_{ci})q_i = q_0\sum_{i=1}^{N}(2A_{ci}) + \sum_{i=1}^{N}(2A_{ci})q_{is}$$
$$= 2A_c q_0 + M_{ys}. \tag{5.36}$$

Thus,

$$q_0 = (M_y - M_{ys})/2A_c, \tag{5.37}$$

where M_y is the given torque about the reference point and

$$M_{ys} = \sum_{i=1}^{N}(2A_{ci})q_{is}. \tag{5.38}$$

From Equation (5.29)

$$q_{is} = V_z H_i + V_x J_i, \tag{5.39}$$

whence

$$M_{ys} = V_z \sum_{i=1}^{N} 2A_{ci}H_i + V_x \sum_{i=1}^{N} 2A_{ci}J_i. \tag{5.40}$$

Note that these sums for M_{ys} are the same as those in Equation (5.33) for the shear center calculation of the open box. Thus, the analysis for the closed single cell box proceeds exactly as for the open box with Equations (5.40) and (5.37) added with q_0 added to the open box shear flows.

Example 5.4. In Example 5.3 assume the box is closed with a web (4) from stringer (4) to stringer (1). Assume M_y about stringer (2) is $M_y = 100,000$ in-lb and $V_z = 30,000$ lb, $V_x = 20,000$ lb as in Example 5.3. Find the shear flows in the closed box.

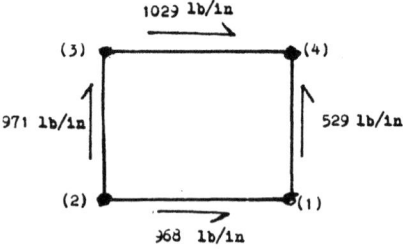

Fig. 5.10. Shear flows for Example 5.4.

Solution. From Equation (5.40) and Table 5.1,

$$M_{ys} = 30,000(4.12) + 20,000(9.43) = 311,600 \text{ in} - \text{lb}.$$

Box beam shear stresses and deflections

From Equation (5.37)

$$q_0 = (100{,}000 - 311{,}600)/400 = -529 \text{ lb/in},$$

whence Equation (5.35) and the last row in the table of Example 5.3 give

$$q_1 = -968 \text{ lb/in.}, \quad q_2 = 971 \text{ lb/in.}, \quad q_3 = 1029 \text{ lb/in.}$$

The shear flows on the cross section are shown in Figure 5.10.

Example 5.5. Find the shear flow equations for a three stringer closed box beam cross section, Figure 5.11.

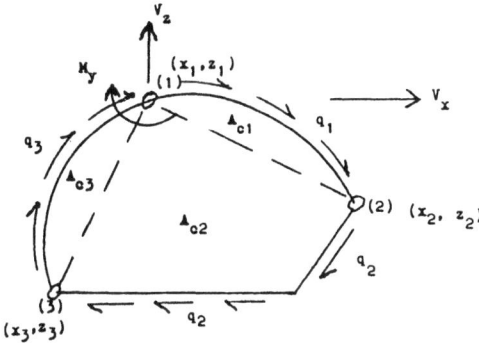

Fig. 5.11. Three stringer cell box beam cross section.

Solution. Since three equilibrium equations are available to find three shear flows, it is not necessary to use the above procedure. The equilibrium equations are, Figure 5.11,

$$(z_2 - z_1)q_1 + (z_3 - z_2)q_2 + (z_1 - z_3)q_3 = V_x,$$
$$(x_2 - x_1)q_1 + (x_3 - x_2)q_2 + (x_1 - x_3)q_3 = V_z,$$
$$2A_{c1}q_1 + 2A_{c2}q_2 + 2A_{c3}q_3 = M_y, \text{ about (1)},$$

which can be solved for q_1, q_2, q_3 for any particular case.

5.6. Shear flows in multi-cell box beams

The multi-cell box beam is redundant so that rotation or twist equations must be added to the equilibrium equations in order to obtain the shear flows. From Equation (5.103) below, the rotation of any single cell in a cantilever box beam at station y is

$$\theta_y = \int_0^y (1/2A_c) \sum_{k=1}^N (s/Gt)_k q_k \, dy \tag{5.41}$$

for N webs in the cell and where s_k is the length of web k, t_k is the thickness of web k, and G_k is the shear modulus of elasticity. In wing analysis the airfoil must maintain its shape so that rotation of all the cells should be the same, or per unit length,

$$(d\theta/dy)_{\text{cell } a} = (d\theta/dy)_{\text{cell } b} = (d\theta/dy)_{\text{cell } c}, \quad \text{etc.}, \tag{5.42}$$

$$(d\theta/dy)_{\text{cell } i} = \left[(1/2A_c)\sum_{k=1}^{N}(s/Gt)_k q_k\right]_{\text{cell } i}. \quad (5.43)$$

The procedure to determine the shear flows is to cut certain webs temporarily to make the box determinate, or an open box as used in the previous Sections 5.4 and 5.5. Cut one web in each cell or cut a web in cell a and then cut the common webs in the cells. The following equations are set up for cutting webs (a), (ab), (bc) in the box beam cross section in Figure 5.12. With these webs cut, the q_{si} shear flows for the open box are calculated as in Sections 5.4 and 5.5. The equations are written for a three cell box, but they can be extended to any number of cells. To simplify the calculations, take

$$q_{ab} = q_a - q_b, \quad q_{bc} = q_b - q_c, \quad (5.44)$$

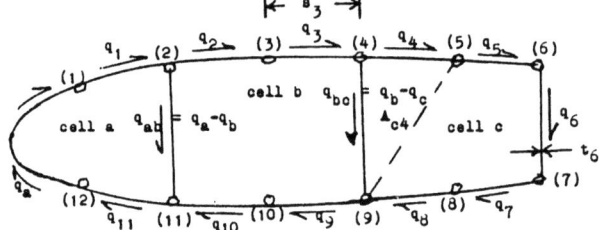

Fig. 5.12. Multicell box beam.

so that q_a is added to the q_{si} shear flows in cell a, q_b is added in cell b, etc. Thus,

$$q_{ai} = q_a + q_{si}, \quad q_{bi} = q_b + q_{si}, \quad q_{ci} = q_c + q_{si}, \quad (5.45)$$

and Equation (5.42) becomes

$$(1/2A_c)_a \left[q_a \sum_a (s/Gt)_i + \sum_a q_{si}(s/Gt)_i - q_b(s/Gt)_{ab}\right]$$
$$= (1/2A_c)_b \left[q_b \sum_b (s/Gt)_i + \sum_b q_{si}(s/Gt)_i - q_a(s/Gt)_{ab} - q_c(s/Gt)_{bc}\right]$$
$$= (1/2A_c)_c \left[q_c \sum_c (s/Gt)_i + \sum_c q_{si}(s/Gt)_i - q_b(s/Gt)_{bc}\right]. \quad (5.46)$$

The torque equation about a selected reference point with M_y given is Equation (5.36), or

$$(2A_c)_a q_a + (2A_c)_b q_b + (2A_c)_c q_c = M_y - M_{ys}, \quad (5.47)$$

$$M_{ys} = \sum_{i=1}^{N} (2A_{ci}) q_{si}, \quad \text{all webs.} \quad (5.48)$$

The two independent equations in (5.46) and the Equation (5.47) give three equations to determine q_a, q_b, q_c.

Box beam shear stresses and deflections

Once the shear flows have been obtained, any one of the expressions in Equation (5.46) will give the rotation of the box per unit length at the cross section. See Sections 5.12 through 5.15 for examples of the calculations of the rotation θ_y.

Example 5.6. Find the shear flows in the symmetrical two cell box beam cross section in Figure 5.13 with no taper and with G and E constant.

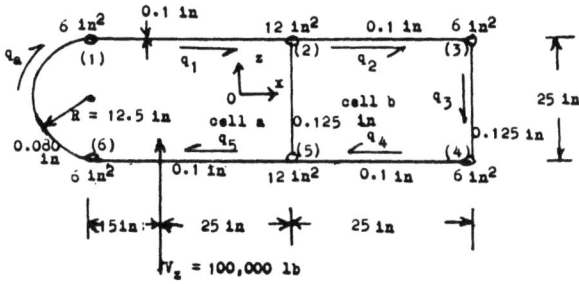

Fig. 5.13. Two cell box beam, Example 5.6.

Solution. From the procedure in Section 5.4

$$I_{zz} = 2(6 + 12 + 6)(25/2)^2 = 7500 \text{ in}^4,$$
$$z_i = \pm 12.5 \text{ in},$$

$$H_{i0} = -A_i z_i / 7500, \quad H_i = \sum_{j=1}^{i} H_{j0}, \quad q_{i\bullet} = 10^5 H_i.$$

Calculate torsion moments about stringer (5). Use Table 5.2 to calculate the terms and the sums for each cell in Equations (5.46) and (5.47) separately. Divide G out in Equation (5.46).

Table 5.2. Table for Example 5.6.

	H_{i0}	H_i	$q_{\bullet i}$	$2A_{ci}$	$\frac{2A_{ci}q_{\bullet i}}{10^6}$	$(s/t)_i$	$\frac{q_{\bullet i}(s/t)_i}{1000}$	q_i (lb/in)
$S(1)$	-0.01							
$W(1)$		-0.01	-1000	1000	-1.000	400	-400	496
$S(2)$	-0.02							
$W(2)$		-0.03	-3000	625	-1.875	250	-750	317
$S(3)$	-0.01							
$W(3)$		-0.04	-4000	625	-2.500	200	-800	-683
$S(4)$	0.01							
$W(4)$		-0.03	-3000	0	0	250	-750	317
$S(5)$	0.02							
$W(5)$		-0.01	-1000	0	0	400	-400	496
$S(6)$	0.01							
$W(a)$		0	0	1491	0	491	0	1496
$W(ab)$		0	0	0	0	200	0	-1821
Sum					-5.375			

From Table 5.2 and Figure 5.13,

$$(2A_c)_a = 1000 + 1491 = 2491 \text{ in}^2,$$
$$(2A_c)_b = 625 + 625 = 1250 \text{ in}^2,$$

$$\sum_A (s/t)_i = 400 + 200 + 400 + 491 = 1491,$$

$$\sum_b (s/t)_i = 250 + 200 + 250 + 200 = 900,$$

$$\sum_a q_{si}(s/t)_i = -400,000 - 400,000 = -800,000,$$

$$\sum_b q_{si}(s/t)_i = -750,000 - 800,000 - 750,000 = -2,300,000.$$

Equations (5.46) and (5.47) give

$$(1/2491)(1491q_a - 800,000 - 200q_b) = (900q_b - 2,300,000 - 200q_a)/1250,$$
$$2491q_a + 1250q_b = 25(100,000) - (-5,375,000) = 7,875,000.$$

The solution is

$$q_a = 1496 \text{ lb/in}, \qquad q_b = 3317 \text{ lb/in},$$

and the shear flows are shown in the last column of the table. Note that a negative sign on the shear flow means that it goes in the reverse direction from that in Figure 5.13.

5.7. Shear center for closed box beams

In Section 5.4 the shear center for the open box was obtained from the torque equilibrium equation and is given in Equation (5.33). However, to get the shear center for a single cell or multi-cell box beam, it is necessary to use the rotation equations of the cells as well as the torque equilibrium equation of the box. For no twist the rotation per unit length is zero, or from Equation (5.43),

$$\left[\sum_{k=1}^{N} q_k (s/Gt)_k \right]_{\text{cell } i} = 0 \tag{5.49}$$

for each cell. For the three cell case, this Equation (5.49) corresponds to setting the expressions in brackets in Equation (5.46) equal to zero, or

$$q_a \sum_a R_i + \sum_a R_i q_{si} - q_b R_{ab} = 0,$$

$$q_b \sum_b R_i + \sum_b R_i q_{si} - q_a R_{ab} - q_c R_{bc} = 0, \tag{5.50}$$

$$q_c \sum_c R_i + \sum_c R_i q_{si} - q_b R_{bc} = 0,$$

$$R_i = (s/Gt)_i \text{ for web } i.$$

Use q_{si} from Equation (5.39) in Equation (5.50) to get

$$q_a \sum_a R_i - q_b R_{ab} = -V_z \sum_a R_i H_i - V_x \sum_a R_i J_i,$$

$$q_b \sum_b R_i - q_a R_{ab} - q_c R_{bc} = -V_z \sum_b R_i H_i - V_x \sum_b R_i J_i, \tag{5.51}$$

Box beam shear stresses and deflections

$$q_c \sum_c R_i - q_b R_{bc} = -V_z \sum_c R_i H_i - V_x \sum_c R_i J_i,$$

which can be solved for q_a, q_b, q_c in terms of V_z and V_x as

$$q_a = -V_z Q_a - V_x S_a, \quad q_b = -V_z Q_b - V_x S_b, \quad q_c = -V_z Q_c - V_x S_c. \quad (5.52)$$

The Q_a, Q_b, Q_c, S_a, S_b, S_c have complicated formulas so that it is simpler to evaluate them numerically in the particular application. In terms of these constants the equilibrium Equation (5.47) become

$$-(2A_c)_a (V_z Q_a + V_x S_a) - (2A_c)_b (V_z Q_b + V_x S_b) - (2A_c)_c (V_z Q_c + V_x S_c)$$
$$= -e_x V_z + e_z V_x - V_z \sum_{i=1}^{N} 2A_{ci} H_i - V_x \sum_{i=1}^{N} 2A_{ci} J_i, \quad (5.53)$$

where Equations (5.39) and (5.48) have been used. M_y in Equation (5.47) represents the moment about the reference point due to the shear forces V_z and V_x at the shear center. Equate corresponding terms for V_z and V_x in Equation (5.53) to get

$$e_x = (2A_c)_a Q_a + (2A_c)_b Q_b + (2A_c)_c Q_c - \sum_{i=1}^{N} 2A_{ci} H_i,$$
$$e_z = -(2A_c)_a S_a - (2A_c)_b S_b - (2A_c)_c S_c + \sum_{i=1}^{N} 2A_{ci} J_i, \quad (5.54)$$

which applies for both single cell and multi-cell beams. For the single cell box, Equation (5.54) simplifies to

$$e_x = \left(2A_c / \sum_{i=1}^{N} H_i\right)\left(\sum_{i=1}^{N} R_i H_i\right) - \sum_{i=1}^{N} 2A_{ci} H_i,$$
$$e_z = -\left(2A_c / \sum_{i=1}^{N} R_i\right)\left(\sum_{i=1}^{N} R_i J_i\right) + \sum_{i=1}^{N} 2A_{ci} J_i. \quad (5.55)$$

If the shear center is calculated at various cross sections of the wing box beam, then these points give the *elastic axis* of the wing about which it twists under the moments M_y. The deflections of this axis gives the bending and shear deflections of the wing.

Example 5.7. Find the shear center of the two cell box beam cross section in Example 5.6.

Solution. The cross section is symmetrical about the x-axis so that $e_z = 12.50$ in. from stringer (5), the reference axis in Example 5.6. From the table and calculations in Example 5.6, the terms in Equation (5.51) involving V_z are

$$\sum_a R_i = 1491, \quad \sum_b R_i = 900, \quad R_{ab} = 200,$$

$$\sum_a R_i H_i = -\frac{800,000}{100,000} = -8.00, \quad \sum_b R_i H_i = -\frac{2,300,000}{100,000} = -23.00,$$

whence

$$1491q_a - 200q_b = 8.00V_z,$$
$$-200q_a + 900q_b = 23.00V_z.$$

This gives

$$q_a = -0.0090637V_z, \qquad q_b = -0.0275697V_z$$

for the form in Equation (5.52), whence Equation (5.54) gives

$$e_z = (2491)(-0.0090637) + (1250)(-0.0275697) - (-53.75) = -3.29 \text{ in.}$$

Thus, from stringer (5) as origin

$$x_{s.c.} = -3.29 \text{ in.}, \qquad z_{s.c.} = 12.50 \text{ in.},$$

and the centroid is at

$$\bar{x} = -3.75 \text{ in.}, \qquad \bar{z} = 12.50 \text{ in.}$$

5.8. Shear flows in tapered box beams

For most wings the box beam tapers in depth and width, has variable stringer areas, and may have variable temperature and material properties in the y-direction. In such cases, Equation (5.13) must be used to get the average shear flows between two cross sections. The taper of the box also produces component forces in the x and z-directions, which must be included in the shear flow calculations. In Figure 5.14 the axial force in the stringer must act along the stringer so that the horizontal load $P_i = A_i \sigma_{yyi}$ will have components along the stringer and perpendicular to P_i. Since the taper angle is usually small so that its cosine is nearly one, take the stringer load as P_i. To get the components of the x and z-directions, take the average load

$$P_i = [(P_i)_y + (P_i)_{y-b}]/2 \tag{5.56}$$

and calculate the average $(P_i)_x$ and $(P_i)_z$ components by

$$\begin{aligned}(P_i)_x &= [(x_{Ri})_{y-b} - (x_{Ri})_y](P_i/b), \\ (P_i)_z &= [(z_{Ri})_{y-b} - (z_{Ri})_y](P_i/b).\end{aligned} \tag{5.57}$$

These component forces will produce torsional moments $(M_{yi})_{\text{taper}}$ about the reference y-axis. With average values

$$z_{Ri} = [(z_{Ri})_y + (z_{Ri})_{y-b}]/2, \qquad x_{Ri} = [(x_{Ri})_y + (x_{Ri})_{y-b}]/2, \tag{5.58}$$

it follows that

$$(M_{yi})_{\text{taper}} = -(P_i)_z x_{Ri} + (P_i)_x z_{Ri} \tag{5.59}$$

for clockwise moments and the right hand rule. For this tapered case, the single cell box q_0 shear flow in Equation (5.37) becomes

$$q_0 = \left[M_y - M_{ys} + \sum_{i=1}^{N}(M_{yi})_{\text{taper}}\right]/(2A_c), \tag{5.60}$$

Box beam shear stresses and deflections

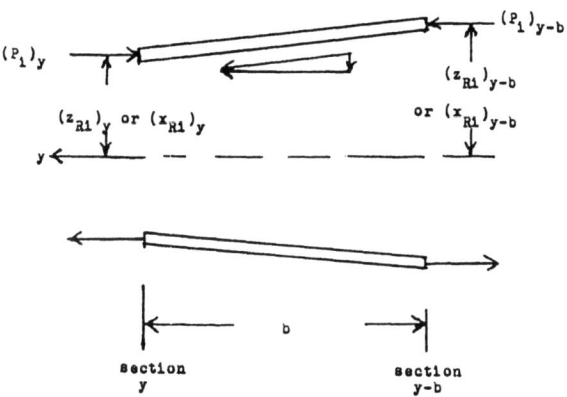

Fig. 5.14. Stringers in tapered box beam.

where $2A_c$, M_y, M_{ys} are calculated at the average cross section. Also, for the multi-cell box, add $\sum(M_{yi})_{\text{taper}}$ on the right side of Equation (5.47)

The net average shear forces on the cross section are

$$(V_z)_w = V_z + \sum_{i=1}^{N}(P_i)_z, \quad (V_x)_w = V_x + \sum_{i=1}^{N}(P_i)_x, \tag{5.61}$$

where $(V_z)_w$ and $(V_x)_w$ are the shear forces supported by the webs.

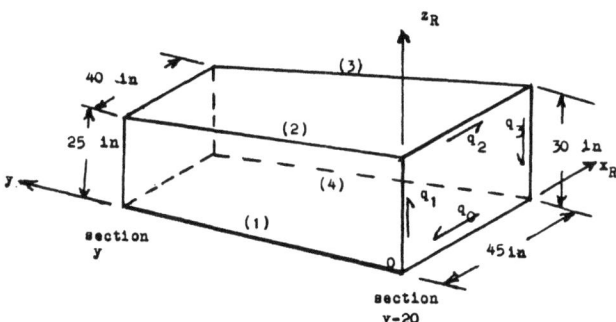

Fig. 5.15. Tapered box beam, Example 5.8.

Example 5.8. Find the average shear flows in the symmetrical tapered box beam with rectangular cross section in Figure 5.15. E and G are constant and the data is as follows: At y, $A_1 = 6.00$ in^2, $A_2 = 6.00$ in^2, $A_3 = A_4 = 12.00$ in^2, $M_z = 10^7$ in-lb, $M_x = 0$; at y-20, $A_1 = A_2 = 9.00$ in^2, $A_3 = A_4 = 12.00$ in^2, $M_z = 1.2(10^7)$ in-lb, $M_x = 0$; $V_z = 10^5$ lb, $V_x = 0$, $(M_y)_1 = -10^6$ in-lb about stringer (1) as reference axis.

Solution. At y, $I_{xx} = (36)(12.5)^2 = 5625$ in^4, at y-20, $I_{xx} = (42)(15)^2 = 9450$ in^4, whence

$$(P_i)_y = (A_i \sigma_{yyi})_y = -(A_i M_x z_i / I_{xx})_y = -1777.78(A_i z_i)_y,$$

$$(P_i)_{y-20} = -1269.84(A_i z_i)_{y-20}.$$

Use Equation (5.13) for q_{si} calculations with 0 in web 0 and put calculations in a table.

From Table 5.3, Equation (5.60), and the given $(M_y)_1$,

$$q_0 = q_4 = (-1,000,000 - 2,233,000 + 928,500)/2338 = -986 \text{ lb/in.}$$

The shear flows are shown in the last row of Table 5.3. From Equation (5.61),

$$\left(V_z\right)_w = 100,000 - (38,200 + 61,900) = 0,$$
$$\left(V_z\right)_w = 0 - (-61,900 + 61,900) = 0,$$

which check the shear flows in Table 5.3 for the webs.

Table 5.3. Table for Example 5.8.

	S(1)	W(1)	S(2)	W(2)	S(3)	W(3)	S(4)	W(4)	Σ
$(P_i)_y/10^3$	133.3		-133.3		-266.7		266.7		
$(P_i)_{y-20}/10^3$	171.4		-171.4		-228.6		228.6		
$(P_i)_{av}/10^3$	152.4		-152.4		-247.6		247.6		
$(P_i)_z/10^3$	0		0		-61.9		61.9		0
$(P_i)_z/10^3$	0		-38.2		-61.9		0		-100
$(M_{yi})_t/10^3$	0		0		928.5		0		928.5
$-\Delta P_i/20$	1910		-1910		1910		-1910		
q_{oi}		1910		0		1910		0	
$(2A_{ci})_{av}$		0		1169		1169		0	2338
$2A_{ci}q_{oi}/10^3$		0		0		2233		0	2233
q_i (lb/in)		924		-986		924		-986	

Note in Example 5.8 that all the V_z shear force is carried as vertical components in the stringers. This is due to the relation between the shear force V_z and the moment M_x in this example. Figure 5.16 shows that if a moment at a cross section is divided by the shear force at the cross section, then a reference distance d_i is obtained which gives an apparent location point of the force producing the

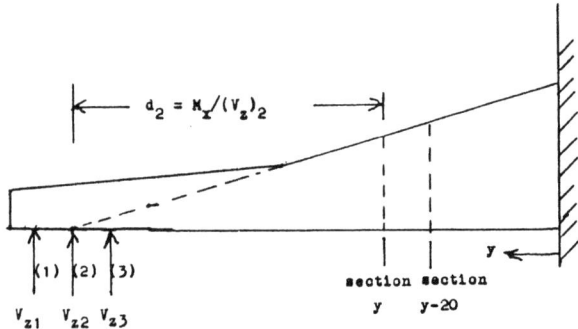

Fig. 5.16. Effect of taper on shear force in web.

moment. If this point (2) in Figure 5.16 is at the projected intersection of the stringers, then the stringers act as truss members with no shear in the webs. For point (3), part of the shear force is in the webs and part in the stringer components. For point (1), the shear force in the web is reversed, with the stringer components being larger than $(V_z)_1$. By chance, d_i in Example 5.8 gives the point (2). If the taper converges toward the wall with the intersection to the right of the wall, then the stringer components add to V_z to produce large shear flows.

Because of the shear components in the stringers, it is evident that there is little relation between the shear flows in a tapered beam and in a constant cross section beam. Figure 5.17 shows a comparison of the shear flows in Example 5.8 to those calculated at the end of the tapered section as if the cross section at

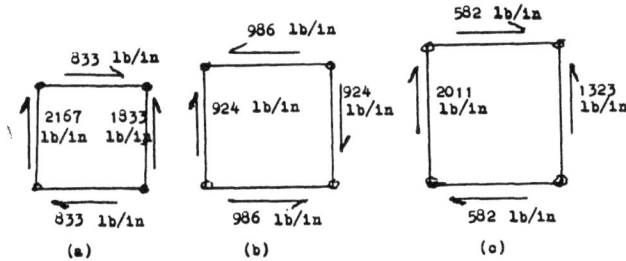

Fig. 5.17. Shear flows (a) constant section at y, (b) tapered section in Example 5.8, (c) constant section at y-20.

each end was constant. Not only is the magnitude of the shear flows different but also the directions in three of the four webs are reversed.

5.9. Inelastic and buckling shear stresses in beams

Since the shear stresses σ_{yz} are related to the axial stresses σ_{yy} through the equilibrium Equations (5.1) and (5.2), the inelastic effects in σ_{yz} and σ_{yy} will interact with each other. The Equation (1.90) includes the inelastic effects in σ_{yy} and hence in P_i in Equation (5.13). Thus, any inelastic effects in σ_{yy} can change the shear flows in Equation (5.13). On the other hand, if the thin webs are inelastic or buckle under the shear flows, they can cause a redistribution of the loads in the stringers. That is, if there are several webs and one yields or buckles, the shear forces may shift to other webs. Whether any redistribution occurs depends upon the beam restraints. Away from the ends and away from any applied point shear forces, considerable web inelastic and web buckling effects can occur without causing load redistribution.

Although the stringer loads and shear flows are unchanged in the unrestrained beam, the shear deflections are increased when some of the webs are inelastic or buckle. Since the shear stresses are known the inelastic effect for each web can be represented by an effective shear modulus G_{kE} from the shear stress-strain curve. Also, the change in deflection of the elastic buckled web can be represented by an effective shear modulus G_{kB}. From Reference 4 an approximate value for G_{kB} is $0.25E$. Also, some curves for G_{kB}/G_k are given in Figure 3.10a of Reference

4. The combination of the two effects gives

$$(G_k)_{\text{eff}} = (G_{kB}/G_k)(G_{kE}/G_k)G_k. \tag{5.62}$$

See sections 1.4 and 1.5, Volume 2.

When a thin web buckles the web takes part of the shear force in shear and part in tension at an angle of about 45°. The tension in the web affects the stringer or flange loads in the beam. This problem of diagonal tension is examined in Section 3.9, Volume 2.

The calculation of the shear flows in Sections 5.2 through 5.5 is a determinate problem. However, any redistribution of shear flows due to beam restraints is a redundant problem. The redistribution can change the beam cross section so that it is no longer a plane for the bending stresses. This redistribution problem will be considered in Chapter 6 below.

5.10. Axial and bending deflections of box beams with inelastic and temperature effects

From the unit load theorem in Equation (4.37), the axial and bending deflections are

$$u_{ya} = \int_0^L \int_A e_{yy} \sigma^1_{yya} \, dA \, dy, \qquad u_{xab} = \int_0^L \int_A e_{yy} \sigma^1_{yyxab} \, dA \, dy,$$

$$u_{zab} = \int_0^L \int_A e_{yy} \sigma^1_{yyzab} \, dA \, dy, \tag{5.63}$$

where, from Equations (1.77), (2.57), (1.87)-(1.89),

$$e_{yy} = (\sigma_{yy}/E_{yy}) + e_E = D_0 + D_1 x + D_2 z,$$
$$e_E = e_p + e_T + e_I \mp e_d,$$
$$D_0 = (P_y + P_{yE})/E_R A_E, \qquad H = I_{xxE} I_{zzE} - (I_{xzE})^2, \tag{5.64}$$
$$D_1 = (E_R H)^{-1}[I_{xxE}(M_z + M_{zE}) + I_{xzE}(M_x + M_{xE})],$$
$$D_2 = -(E_R H)^{-1}[I_{zzE}(M_x + M_{xE}) + I_{xzE}(M_z + M_{zE})].$$

Here, the various terms are defined in Equations (1.81)-(1.84).

Assume the centroids of the beam cross sections along the tapered box beam, with stringer and spar cap areas varying, to be approximately in a straight line, which is taken as the y-axis. Apply unit loads separately at point y on the y-axis in the x, y, z directions. The unit loads produce a unit axial load and the bending moments m_x^1, m_z^1, respectively, on the beam. Assume the beam to have no cut-outs or large area variations so that all bending sub-beams will be in the planes of the box cross sections. With these conditions, the e_{yy} strain in Equation (5.64) applies at each cross section with $D_0(y)$, $D_1(y)$, and $D_2(y)$ as functions of y. Also, for this plane strain cross section, the deflections u_{ya}, u_{xab}, and u_{zab} depend only upon the corresponding term D_0, $D_1 x$, and $D_2 z$, respectively. Thus, the virtual stresses produced by the unit loads can be selected to have the simple uncoupled forms

$$\sigma^1_{yya} = 1/A_y, \quad \sigma^1_{yyxab} = m^1_z x/\left(\int_{A_x} x^2\, dA_x\right),$$

$$\sigma^1_{yyzab} = -m^1_x z/\left(\int_{A_z} z^2\, dA_z\right),$$
(5.65)

where A_y, A_x, A_z are arbitrary except that A_x and A_z must have the same centroids as the beam. Put corresponding terms from Equations (5.64) and (5.65) into Equation (5.63) to get

$$u_{ya} = \int_0^L \int_{A_y} (D_0/A_y)\,dA_y\, dy = \int_0^L D_0\, dy = \int_0^L (P_y + P_{yE})(dy/E_R A_E),$$

$$u_{xab} = \int_0^L \int_{A_x} D_1 x m^1_z x\, dA_x\, dy/\left(\int_{A_x} x^2\, dA_x\right) = \int_0^L D_1 m^1_z\, dy$$

$$= \int_0^L m^1_z(E_R H)^{-1}[I_{xxE}(M_z + M_{zE}) + I_{xzE}(M_x + M_{xE})]\,dy, \quad (5.66)$$

$$u_{zab} = \int_0^L \int_{A_z} -D_2 z m^1_x z\, dA_z\, dy/\left(\int_{A_z} z^2\, dA_z\right) = \int_0^L -D_2 m^1_x\, dy$$

$$= \int_0^L m^1_x(E_R H)^{-1}[I_{zzE}(M_x + M_{xE}) + I_{xzE}(M_z + M_{zE})]\,dy.$$

For the cantilever wing the m^1_x and m^1_z unit moments are

$$m^1_x = y - y_d, \quad 0 \leq y_d \leq y,$$
$$m^1_z = -(y - y_d), \quad 0 \leq y_d \leq y,$$
(5.67)

whence the bending deflections in Equation (5.66) become

$$u_{xab}(y) = -\int_0^y [I_{xxE}(M_z + M_{zE}) + I_{xzE}(M_x + M_{xE})]\frac{(y - y_d)}{E_R H}\,dy_d,$$

$$u_{zab}(y) = \int_0^y [I_{zzE}(M_x + M_{xE}) + I_{xzE}(M_z + M_{zE})]\frac{(y - y_d)}{E_R H}\,dy_d.$$
(5.68)

Note that for the case of $I_{xzE} = 0$, u_{zab} reduces to the bending term in Equation (4.39) for the simple beam. Note also that mixed materials and variable elastic E's are included in Equation (5.68)

Since the unit load theorem can be applied to any subbeam in the box beam, such as a spar or two stringers with one directly above the other in the z-direction for u_{zab}, or directly in line with each other in the x-direction for u_{xab}, Equation (5.68) can be expressed in the form, Figure 5.18,

$$u_{zab} = \int_0^y \left[\left(\frac{\sigma_{yy}}{E_{yy}} + e_E\right)_L - \left(\frac{\sigma_{yy}}{E_{yy}} + e_E\right)_U\right]\frac{y - y_d}{z_U - z_L}\,dy_d,$$

$$u_{xab} = \int_0^y \left[\left(\frac{\sigma_{yy}}{E_{yy}} + e_E\right)_F - \left(\frac{\sigma_{yy}}{E_{yy}} + e_E\right)_A\right]\frac{y - y_d}{x_A - x_F}\,dy_d.$$
(5.69)

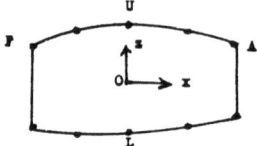

Fig. 5.18. Deflections from point strains on box beam cross section.

Here L refers to the lower cap or stringer and U refers to the upper spar cap or stringer directly above, where $z_U - z_L$ is the distance between them. Also, F refers to the forward stringer and A to the aft stringer in line with F in the x-direction, where $x_A - x_F$ is the x-distance between them. Thus, if the stresses σ_{yy} and strains e_E have been calculated for a given flight condition, it may be simpler to use Equation (5.69) instead of Equation (5.68).

As the wing is usually tapered, all the terms in both Equations (5.68) and (5.69) are functions of y_d so that in most cases numerical integrations are necessary to get the deflections for various values of y. Any of the cases in Equations (5.68) and (5.69) has the form

$$u_{ab} = y \int_0^y G(y_d) dy_d - \int_0^y y_d G(y_d) dy_d, \tag{5.70}$$

whence with N equal Δy_d intervals on one wing and L the wing structural semi-span, take

$$\Delta y_d = L/N, \quad y_1 = 0, \quad y_2 = \Delta y_d = L/N, \quad y_3 = 2(L/N),$$
$$y_4 = 3(L/N), \cdots, \quad y_i = (i-1)(L/N), \cdots, y_{N+1} = L. \tag{5.71}$$

Use the trapezoid rule of numerical integration and Equation (5.71) in Equation (5.70) to get

$$u_{ab}(y_i) = (1/2)(L/N)^2 \Big[(i-1) \sum_{k=1}^{i-1}(G_k + G_{k+1}) -$$
$$- \sum_{k=1}^{i-1}\{(k-1)G_k + kG_{k+1}\}\Big], \quad G_k = G(y_k). \tag{5.72}$$

If the wing has an efficient design with the strains in the upper and lower spar caps or stringers approximately constant, then the bracket terms in Equation (5.69) are approximately constant $= C_{zab}$, and it may be possible to integrate to get the deflections. If the wing has a straight line taper, then $z_U - z_L = (h_0/d)(d - y_d)$, where h_0 is the depth of the wing at $y_d = 0$ and $d > L$ is the taper parameter. Thus, u_{zab} in Equation (5.69) becomes approximately

$$u_{zab} = (d/h_0) C_{zab} \int_0^y \frac{y - y_d}{d - y_d} dy_d$$
$$= (d/h_0) C_{zab} \big[y + (d-y)\ln(\frac{d-y}{d})\big]. \tag{5.73}$$

For the case of a beam with constant depth $z_U - z_L = h_0$, Equation (5.72) has the form

$$u_{zab} = (C_{zab}/2h_0)y^2. \tag{5.74}$$

The inelastic strain e_p has been included in the above Equations (5.64)-(5.74) although the deflections are usually calculated for the limit load conditions. This has been done because some inelastic strain is present at limit load conditions. For materials meeting the design criterion (A) in Equation (B.74) the inelastic strain will correspond to that for the yield stress F_y, or

$e_p = 0.002$, offset yield stress,

$e_p = (3/7)(F_y/E_{yy})$, yield stress in Equation (1.94).

This e_p strain may increase the deflections by as much as fifty per cent.

Example 5.9. Compare the deflections in Equations (5.73) and (5.74) at the points $y/L = 0.25$, 0.50, and 1.00 for the parameter d/L having values $d/L = 1, 2, 5, 10$. The wings have the same C_{zab}, h_0, and L.

Solution. Divide Equation (5.74) into Equation (5.73) to get

$$R_z = \frac{(u_{zab})_{taper}}{(u_{zab})_{constant}} = \frac{2d}{y}\left[1 + (\frac{d}{y} - 1)\ln(1 - \frac{y}{d})\right]$$

$$= 2(\frac{d}{L})(\frac{L}{y})\left[1 + (\frac{d}{L}\frac{L}{y} - 1)\ln(1 - \frac{L}{d}\frac{y}{L})\right]. \tag{5.75}$$

The values of R_z for the specified values of y/L and d/L can be calculated directly as

$d/L =$	1,	2,	5,	10
$R_z, y/L=$ 0.25, =	1.10,	1.04,	1.02,	1.01
$R_z, y/L=$ 0.50, =	1.23,	1.10,	1.04,	1.02
$R_z, y/L=$ 1.00, =	2.00,	1.23,	1.07,	1.04,

which shows that the deflection of the tapered wing is not much larger than the constant wing, except for $d/L < 2.00$. Most wings have $d/L > 2.00$ so that the simple parabola in Equation (5.74) can be used to get approximate values for the wing deflections.

Example 5.10. Suppose a well-designed wing with constant cross section has $h_0 = 50$ in., $L = 1500$ in., $(\sigma_{yy}/E_{yy})_L = -(\sigma_{yy}/E_{yy})_U = 0.004$, $(e_E)_L = (e_p)_L = 0.002$, and $(e_E)_U = (e_p)_U = -0.001$ along the wing. Calculate the u_{zab} tip deflection of the wing from Equations (5.69) and (5.74).

Solution. From Equation (5.69),

$$C_{zab} = (0.004 + 0.002)_L - (-0.004 - 0.001)_U = 0.011,$$

whence Equation (5.74), at $y = L = 1500$ in. gives

$$(u_{zab})_{tip} = \frac{(0.011)(1500)^2}{2(50)} = 248 \text{ in.}$$

This tip deflection is about 16% of L, which is a typical deflection of airplane wings at limit load.

5.11. Shear deflections of beams

From Equation (1.46) the shear strain e_{yz} in a thin rectangular beam t by $2h$ is

$$e_{yz} = u_{zs,y} + u_{ys,z} = P_z(h^2 - z^2)/2GI_{xx}. \tag{5.76}$$

From Equation (1.47) with the constants in Equation (1.51), the shear terms give

$$u_{zs,y} = P_z h^2/2GI_{xx}, \qquad u_{ys,z} = -P_z z^2/2GI_{xx}, \tag{5.77}$$

which checks Equation (5.76). Thus,

$$u_{zs} = (P_z h^2/2GI_{xx})y = y(e_{yz})_{z=0}, \tag{5.78}$$

which corresponds to Equation (1.53), and which gives the shear deflection away from the fixed end of the cantilever beam.

Since u_{zs} can be regarded as the rigid body shear deflection in the z-direction, $u_{zs,y}$ is a constant for the beam cross section. Also, u_{ys} is an odd function in z with the value 0 at $z = 0$ at the centroid. Thus, for solid cross sections, the u_{zs} shear deflection, or the u_{xs} shear deflection in the x-direction, can be calculated directly from the shear strain at $z = 0$ or $x = 0$.

For the cases of maximum shear stresses at $z = 0$ given in Section 5.2, take

$$u_{zas,y} = (e_{yz})_{z=0} = (\sigma_{yz}/G)_{z=0} = C_3 V_z/GA, \tag{5.79}$$
$$u_{xas,y} = (e_{xy})_{z=0} = (\sigma_{xy}/G)_{z=0} = C_4 V_x/GA. \tag{5.80}$$

From Equation (5.8) for the solid circular cross section

$$C_3 = C_4 = 1.384 \text{ for } \nu = 0.30. \tag{5.81}$$

For the solid rectangular cross section $2t_x$ by $2h$, or $A = 4ht_x$, Equation (1.53) gives $C_3 = 1.50$. However, it is shown in Reference 1 that this C_3 is satisfactory for $h/t_x \geq 2$, but for $h/t_x = 1$,

$$\begin{aligned} C_3 &= 1.50, \quad h/t_x \geq 2, \\ C_3 &= C_4 = 1.69 \text{ for } \nu = 0.25, \quad h/t_x = 1, \\ C_3 &= C_4 = 1.88 \text{ for } \nu = 0.30, \quad h/t_x = 1. \end{aligned} \tag{5.82}$$

From Equation (1.62) for the thin web with constant shear flow between large flange areas,

$$u_{zas,y} = V_z/GA_w. \tag{5.83}$$

It should be noted that the area A in Equations (5.79) and (5.80) is the total area but the area A_w in Equation (5.83) is only the area of the web and does not include the flange areas or stringer areas. This means that the shear deflections of thin web beams may be much larger than for beams with solid wide cross sections. It can be seen from Equation (1.52) that if h is small compared to L the shear deflection term is small compared to the bending deflection term. Thus, the shear deflection for beams of the type in Equations (5.79) and (5.80) is usually neglected, but it should be included for thin web beams.

For the simple thin web between two flanges

$$e_s = \sigma_s/G = q/Gt, \tag{5.84}$$

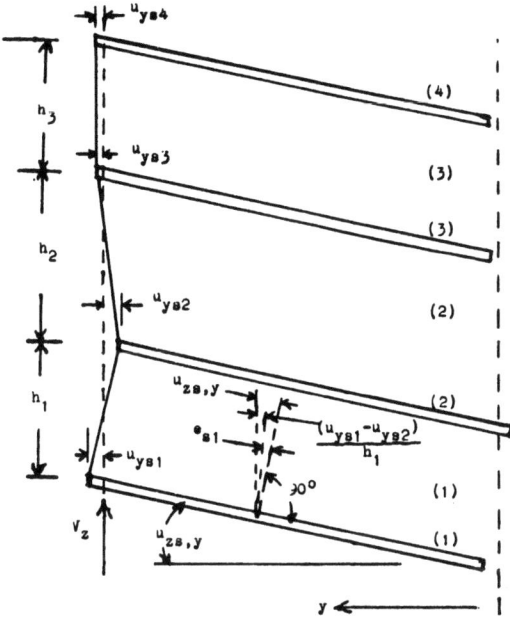

Fig. 5.19. Warping of three web beam cross section from shear forces.

and the unit load theorem gives

$$u_{xas} = \int_0^y \int_{A_w} e_s \sigma_{xas}^1 \, dA_w \, dy = \int_0^y (q_b/G_b t_b)(t_b s_b)(1/t_b s_b) dy$$
$$= \int_0^y (q_b/G_b t_b) dy = V_z y/G_b t_b s_b, \tag{5.85}$$

$$u_{zas} = \int_0^y (q_h/G_h t_h) dy = V_z y/G_h t_h s_h, \tag{5.8}$$

where $\sigma_{xas}^1 = 1/t_b s_b$, $\sigma_{zas}^1 = 1/t_h s_h$ at point y. Note that Equation (5. agrees with Equation (5.83).

For more than one web such as in Figure 5.2 and Figure 5.19, warpin occur with $u_{ys,z} \neq 0$. For instance, in Example 5.1 with $G_k t_k = Gt$ for the webs, Equation (5.85) gives

$$(u_{zas})_1 = 805(y/Gt), \quad (u_{zas})_2 = 1171(y/Gt),$$
$$(u_{zas})_3 = 1024(y/Gt), \quad (u_{zas})_{ave.} = 1000(y/Gt).$$

Thus, the webs are incompatible with each other so that the u_{ys} war be included in the strain.

From Equations (5.76) and (5.84) for web k

$$e_{sk} = (q/Gt)_k = u_{zs,y} + (u_{ys,k+1} - u_{ysk})/s_k,$$
$$\perp (u_{usk} - u_{ys,k+1})/s_k,$$

where u_{ysk} is the warping deflection of stringer k in Figure 5.19. For the three web case in Figure 5.19, Equation (5.87) gives three equations with five unknowns, $u_{zs,y}$, u_{ys1}, u_{ys2}, u_{ys3}, and u_{ys4}. The deflection u_{ysk} can produce an axial load in stringer k when it is restrained. Since this is a local effect with self-equilibrating loads and moments near a restraint, there are two additional conditions to add to Equation (5.87). As the stringer k load will be proportional to u_{ysk}, take

$$\sum_{k=1}^{N} E_k A_k u_{ysk} = 0, \tag{5.88}$$

$$\sum_{k=1}^{N} E_k A_k z_k u_{ysk} = 0 \tag{5.89}$$

at any cross section of the beam away from restraints. Here, z_k is measured from the centroid. By adding the equations in Equations (5.87), the u_{ysk} can be expressed in terms of $u_{zs,y}$ and u_{ys1}, whence Equations (5.88) and (5.89) will give $u_{zs,y}$ and u_{ys1}. Actually, with z_k measured from the centroid of the cross section, u_{ys1} drops out in Equation (5.89) so that Equation (5.89) gives $u_{zs,y}$ in terms of all the e_{sk}.

The fact that the shear deflection $u_{zs,y}$ per unit length at a cross section depends upon all the webs at the cross section means that the unit load theorem must be applied to the entire cross section. That is, the virtual shear strains produced by the unit load must be real shear strains in order to be not only compatible but also in equilibrium.

To use the unit load theorem on the entire cross section, apply unit virtual forces, separately, in the x and z directions at the shear center of the open box beam to get

$$\sigma^1_{xas} = (1)q^1_{xk}/t_k, \quad \sigma^1_{zas} = (1)q^1_{zk}/t_k, \tag{5.90}$$

where t_k, q^1_{xk}, and q^1_{zk} are constant in each web. Let the box cross section have N webs with s_k the length of web k so that the web area is $t_k s_k$. Put Equations (5.84) and (5.90) into Equation (5.85), or

$$u_{xas} = \int_0^y \sum_{k=1}^{N} (s/Gt)_k q_k q^1_{xk} \, dy, \quad u_{zas} = \int_0^y \sum_{k=1}^{N} (s/Gt)_k q_k q^1_{zk} \, dy. \tag{5.91}$$

The equations for the calculations of q_k are given in Equation (5.13) for the general open box beam cross section, and in Equations (5.10), (5.22)-(5.30) for various special cases. The formulas for the shear center of the open box are given in Equation (5.33). For the unsymmetrical constant cross section Equation (5.29) gives

$$q_k = J_k V_x + H_k V_z, \quad q^1_{xk} = J_k, \quad q^1_{zk} = H_k, \tag{5.92}$$

where J_k and H_k are defined in Equations (5.30) and (5.27), and where the unit virtual load shear flows in the total box cross section are taken directly for $V_x^V = 1$ and $V_z^V = 1$ in the real q_k.

Box beam shear stresses and deflections

For the constant open box cross section put Equation (5.92) into Equation (5.91) to get

$$u_{xa_s} = \int_0^y \sum_{k=1}^N (s/Gt)_k J_k (J_k V_x + H_k V_z) dy, \qquad (5.93)$$

$$u_{za_s} = \int_0^y \sum_{k=1}^N (s/Gt)_k H_k (J_k V_x + H_k V_z) dy. \qquad (5.94)$$

Note that in these Equations (5.93) and (5.94) only V_x, V_z, and $(Gt)_k$ may be slowly varying functions of y.

For the general open box beam with changes in area, temperature, and material properties in the y-direction, Equation (5.13) must be used to calculate the average shear flows q_k between two cross sections at distance b apart. In this case the warping Equations (5.87)-(5.89) apply only if average values $(A_k)_{av}$ and $(z_k)_{av}$ are used. Also, the unit load theorem in Equation (5.91) can be used by taking

$$q_{xk}^1 = (J_k)_{av}, \qquad q_{zk}^1 = (H_k)_{av}. \qquad (5.95)$$

If $(G_k t_k)_{av}$ along with the above average values is used in Equation (5.91), then integration on y can be made as a sum over the intervals with the average values, or for M intervals with average values indicated by subscript i,

$$u_{xa_s} = \sum_{i=1}^M \left[\sum_{k=1}^N (s/Gt)_k J_k q_k \right]_i b_i,$$

$$u_{za_s} = \sum_{i=1}^M \left[\sum_{k=1}^N (s/Gt)_k H_k q_k \right]_i b_i. \qquad (5.96)$$

As long as the cross section is free to warp, the warping of the plane section under the constant shear forces does not change the shear flows, the stresses in the stringers, and the bending deflections. If the cross section is restrained from warping by the supports, then the shear flows, stringer stresses, and deflections will be affected near the restraint. This case of restrained warping of the beam is considered in Chapter 6 below.

Example 5.11. Find the deflection u_{xa_s} of the three web beam in Example 5.1 by using (a) Equations (5.87)-(5.89) and (b) Equation (5.94). (c) Calculate the $u_{y_s k}$ warping deflections.

Solution (a) From Equation (5.87) and Example 5.1,

$$u_{y_s 1}/10 = u_{y_s 1}/10, \qquad u_{y_s 2}/10 = 805/Gt - u_{x_s,y} + u_{y_s 1}/10,$$
$$u_{y_s 3}/10 = 1976/Gt - 2u_{x_s,y} + u_{y_s 1}/10,$$
$$u_{y_s 4}/10 = 3000/Gt - 3u_{x_s,y} + u_{y_s 1}/10,$$

whence with stringer areas A_k and locations z_k from Example 5.1, Equation (5.89) gives

$$(0)(u_{y_s 1}/10) - 273.48 u_{x_s,y} + 279{,}657/Gt = 0,$$
$$u_{x_s,y} = 1022.6/Gt, \qquad u_{xa_s} = (1022.6)y/Gt.$$

(b) For the data in Example 5.1 and Figure 5.2,

$$H_1 = 805/30{,}000, \qquad H_2 = 1171/30{,}000, \qquad H_3 = 1024/30{,}000,$$

whence Equation (5.94) gives

$$u_{z\bullet} = \frac{10y}{30,000t}\left[(805)^2 + (1171)^2 + (1024)^2\right]$$
$$= 1022.6y/Gt = 1.0226V_zy/GA_w, \qquad (5.97)$$

where $A_w = 30t$, $V_z = 30,000$ lb.

(c) From the expressions in part (a) and Equation (5.88),

$$24u_{y\bullet 1}/10 - 44u_{z\bullet,y} + 43,028/Gt = 0,$$

$$u_{y\bullet 1}/10 = 81.9/Gt, \quad u_{y\bullet 2}/10 = -135.7/Gt,$$
$$u_{y\bullet 3}/10 = 12.7/Gt, \quad u_{y\bullet 4}/10 = 14.1/Gt. \qquad (5.98)$$

The approximate form of this warping is shown in Figure 5.19.

Example 5.12. Solve Example 5.11, parts (b) and (c), for the case of $Gt = (Gt)_1 = (Gt)_3 = 2(Gt)_2$, where web (2) buckles with $(G_2)_{\text{eff}} = 0.5G$, and comment on the changes in the warping deflections.

Solution (b) In Equation (5.97)

$$u_{z\bullet} = \frac{10y}{30,000t}\left[(805)^2 + 2(1171)^2 + (1024)^2\right]$$
$$= 1480(y/Gt) = 1.48(V_zy/GA_w),$$

which shows a 45% increase in the shear deflection as compared to Equation (5.97).

(c) From Equation (5.87) and Example 5.11

$$u_{y\bullet 2} = \frac{8050}{Gt} - \frac{14800}{Gt} + u_{y\bullet 1}, \quad u_{y\bullet 3} = \frac{31470}{Gt} - \frac{29600}{Gt} + u_{y\bullet 1},$$
$$u_{y\bullet 4} = \frac{41710}{Gt} - \frac{44400}{Gt} + u_{y\bullet 1},$$

whence Equation (5.88) gives

$$u_{y\bullet 1}/10 = 106/Gt, \ u_{y\bullet 2}/10 = -569/Gt, \ u_{y\bullet 3}/10 = 393/Gt, \ u_{y\bullet 4}/10 = -163/Gt.$$

The $u_{y\bullet k}$ warping deflections are much larger in this case as compared to Equation (5.98). Thus, if the beam is restrained, there would be a much larger load redistribution in this case. Also, the web buckling will produce changes in the shear flows.

Example 5.13 (a) Find the deflection $u_{z\bullet}$ for the three web beam in Example 5.11 for the conditions and data given in Example 5.2. Use $(Gt)_k = Gt =$ constant. (b) Calculate the warping deflections.

Solution (a) In Example 5.2 apply a unit load at station 160 in. and use the data in Example 5.2 in Equation (5.10) to get

$$q_1^1 = 0 - (1)(0.6/0.7)(6)(-17.17)/3163.76 = 0.027911 = H_1,$$
$$q_2^1 = 0.027911 - (1)(0.6/0.7)(6)(-7.17)/3163.76 = 0.039566 = H_2,$$
$$q_3^1 = 0.039566 - (1)(0.7/0.7)(8)(2.83)/3163.76 = 0.032410 = H_3,$$
$$q_4^1 = 0.032410 - (1)(0.7/0.7)(8)(12.83)/3163.76 = 0.$$

Average these H_k values with the H_k values in Example 5.11 to get $(H_1)_{av} = 0.02737$, $(H_2)_{av} = 0.03930$, $(H_3)_{av} = 0.03326$. Use the shear flows in Example 5.2 to get

$$(u_{z\bullet})_{\text{interval}} = 40(10/Gt)[(1364)(0.02737) + (410)(0.03930) + (1192)(0.03326)]$$
$$= 40(931/Gt) = 37240/Gt. \qquad (5.99)$$

(b) From Equation (5.87)

$$u_{y\bullet 2} = 10(1364 - 931)/Gt + u_{y\bullet 1}, \ u_{y\bullet 3} \doteq 10(1774 - 1862)/Gt + u_{y\bullet 1},$$
$$u_{y\bullet 4} = 10(2966 - 2793)/Gt + u_{y\bullet 1},$$

whence with average areas 4.57, 4.57, 8, 8 and average z_k values -17.47, -7.74, 2.24, 12.24, Equation (5.88) gives

$$0 = 25.14 u_{y \bullet 1} + 4.57(4330/Gt) - 8(880/Gt) + 8(1730/Gt),$$
$$u_{y \bullet 1} = -1060/Gt, \quad u_{y \bullet 2} = 3270/Gt, \quad u_{y \bullet 3} = -1940/Gt, \quad u_{y \bullet 4} = 670/Gt.$$

These values check Equation (5.89) within the round-off error. Thus, the temperature and area changes produce more warping deflections than in Example 5.11.

Example 5.14. With $(Gt)_i = GT = $ constant, find the $u_{z a \bullet}$ and $u_{x a \bullet}$ shear deflections of the open box beam in Figures 5.6 and 5.7 and Example 5.3. Discuss the rotation and warping of the cross section.

Solution. In this case, Equations (5.93) and (5.94) apply with $H_1 = H_3 = 0.02059$, $H_2 = 0.05000$, $J_1 = -0.05286$, $J_2 = 0$, $J_3 = 0.04714$ from Example 5.3. Thus,

$$u_{z a \bullet} = (y/Gt)(0.05016 V_z - 0.00118 V_x),$$
$$u_{x a \bullet} = (y/Gt)(-0.00118 V_z + 0.05848 V_x).$$

For the shear forces of $V_z = 20,000$ lb and $V_x = 30,000$ lb in Example 5.3,

$$u_{z a \bullet} = 968(y/Gt), \quad u_{x a \bullet} = 1731(y/Gt). \tag{5.100}$$

The shear flows in the three webs are

$$q_1 = -0.05286 V_z + 0.02059 V_x = -439 \text{ lb/in},$$
$$q_2 = 0.05000 V_z = 1500 \text{ lb/in}, \quad q_3 = 0.04714 V_z + 0.02059 V_x = 1558 \text{ lb/in},$$

whence Equation (5.85) gives

$$(u_{x a \bullet})_1 = 439(y/Gt), \quad (u_{x a \bullet})_3 = 1558(y/Gt), \quad (u_{x a \bullet})_2 = 1500(y/Gt).$$

Although the shear forces are applied at the shear center, Figure 5.7, so that the beam has no twisting moment, the beam cross section still has a small rotation θ from the shear strains. Apply a unit moment to give a force couple of 1/20 on webs (1) and (3), whence

$$\theta = (1558 - 439)(y/20 Gt) = 56(y/Gt). \tag{5.101}$$

This rotation occurs about the shear center so that

$$(u_{x a \bullet})_2 = 1731 y/Gt - 4.12\theta = 1500 y/Gt,$$
$$(u_{x a \bullet})_1 = 968 y/Gt - 9.40\theta = 441 y/Gt,$$
$$(u_{x a \bullet})_3 = 968 y/Gt + 10.60\theta = 1562 y/Gt,$$

which agree with the individual web shear deflections.

Use Equations (5.87) and (5.88) with the areas from Example 5.3 to get the warping deflections,

$$0 = 12 u_{y \bullet 1} + (10)(1590 + 268 + 1314)/Gt,$$
$$u_{y \bullet 1} = -2640/Gt, \quad u_{y \bullet 2} = 2650/Gt, \quad u_{y \bullet 3} = -1970/Gt, \quad u_{y \bullet 4} = 3930/Gt.$$

5.12. Torsional rotation of beams

The shear stresses in beams due to the torsional moment M_y are discussed in Sections 5.3 and 5.5. The θ_y rotation for some beams with solid cross sections is given in Equations (5.15) and (5.17).

If a unit moment $M_y = 1$ is applied to a single cell cantilever box beam at the point y, then the constant shear flow in all webs of the box is given by Equation (5.34) as

$$q_0^1 = 1/2 A_c, \tag{5.102}$$

where A_c is the enclosed area of the box at the cross section y. If this equation (5.102) is put into Equation (5.91), then the θ_y rotation of the box is

$$\theta_y(y) = \int_0^y \left[\sum_{k=1}^N (s/Gt)_k (q_0 + q_{ks})\right] dy/2A_c, \qquad (5.103)$$

where q_{ks} are the shear flows in the open box with one web cut and q_0 is the torque shear flow to close the box, Section 5.5.

The shear deflections u_{xas} and u_{zas} can be calculated at the reference point used in the shear flow calculations by using Equations (5.91) with q_{xk}^1 and q_{zk}^1 being determined from unit loads at the reference axis applied to the entire box beam cross section and calculated in the same manner as the real shear flows $q_k = q_0 + q_{ks}$. By this procedure the u_{xas} and u_{zas} deflections at the reference axis will include the shear center deflections and the rotation deflection about the shear center. These u_{xas} and u_{zas} values and θ_y from Equation (5.103) can be put into Equation (2.30) to get u_{xs} and u_{zs} at any point on the cross section without knowing the shear center location.

For the constant cross section case, Equations (5.93) and (5.94) can be used at the reference axis for u_{xas} and u_{zas} and Equation (5.103) for θ_y has the form

$$\theta_y(y) = (1/2A_c) \int_0^y \sum_{k=1}^N (s/Gt)_k \left[V_x\left(J_k - \frac{1}{2A_c}\sum_{i=1}^N 2A_{ci}J_i\right) + \right.$$
$$\left. + V_z\left(H_k - \frac{1}{2A_c}\sum_{i=1}^N 2A_{ci}H_i\right) + \frac{M_{yR}}{2A_c}\right] dy, \qquad (5.104)$$

where only V_x, V_z, and M_{yR} may vary with y, and where Equations (5.37), (5.39), and (5.40) have been used.

For the M_y moment acting alone about the shear center of the closed single cell box beam,

$$q_0 = M_y/2A_c, \qquad (5.105)$$

and in equation (5.103),

$$\theta_y(y) = \int_0^y \sum_{k=1}^N (s/Gt)_k (M_y/4A_c^2) dy,$$

$$\frac{d\theta_y}{dy} = M_y/GJ, \quad 1/GJ = (1/4A_c^2)\sum_{k=1}^N (s/Gt)_k. \qquad (5.106)$$

In Equation (5.104), M_{yR} about the reference axis can be written as

$$M_{yR} = M_y - e_x V_z + e_z V_x, \qquad (5.107)$$

where e_x and e_z are the distances from the reference axis to the shear center. See Figure 5.6 and Equation (5.33) for the open box. Put M_{yR} from Equation

(5.107) into Equation (5.104) to get

$$\frac{d\theta_y}{dy} = V_x \left[\frac{1}{2A_c} \sum_{k=1}^{N} \left(\frac{s}{Gt}\right)_k J_k - \frac{1}{GJ} \sum_{i=1}^{N} 2A_{ci} J_i + \frac{e_z}{GJ} \right] +$$
$$+ V_z \left[\frac{1}{2A_c} \sum_{k=1}^{N} \left(\frac{s}{Gt}\right)_k H_k - \frac{1}{GJ} \sum_{i=1}^{N} 2A_{ci} H_i - \frac{e_x}{GJ} \right] + \frac{M_y}{GJ}, \quad (5.108)$$

where GJ is defined in Equation (5.106). Since Equations (5.108) and (5.106) must be the same, the coefficients of V_x and V_z must be zero, whence the location of the shear center is given by

$$e_x = \frac{GJ}{2A_c} \sum_{k=1}^{N} \left(\frac{s}{Gt}\right)_k H_k - \sum_{i=1}^{N} 2A_{ci} H_i,$$
$$e_z = -\frac{GJ}{2A_c} \sum_{k=1}^{N} \left(\frac{s}{Gt}\right)_k J_k + \sum_{i=1}^{N} 2A_{ci} J_i. \quad (5.109)$$

Note that the last terms in e_x and e_z are the same as for the shear center of the open box in Equation (5.33).

To determine the warping of the single cell box beam of any shape it is necessary to write Equation (5.87) in a more general form. Take u_{zs} and u_{xs} as the

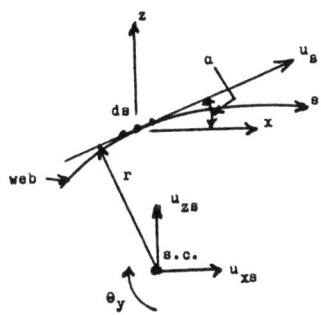

Fig. 5.20. Notation for shear web element.

deflections of the shear center, whence the shear deflection in the plane of of any web is

$$u_s = r\theta_y + u_{zs} \sin \alpha + u_{xs} \cos \alpha, \quad (5.110)$$

where the symbols are defined in Figure 5.20. Rewrite Equation (5.87) in differential form as

$$du_{ys} = \frac{q}{Gt} ds - ds \left(r\theta_y + u_{zs} \sin \alpha + u_{xs} \cos \alpha \right)_{,y}, \quad (5.111)$$

and integrate by summing over the webs between stringers, or at stringer i

$$u_{ysi} = u_{ys1} + \sum_{k=1}^{i} q_k \left(\frac{s}{Gt}\right)_k - \theta_{y,y} \sum_{k=1}^{i} 2A_{ck} -$$

$$- \sum_{k=1}^{i} s_k (u_{zs} \sin \alpha_k + u_{xs} \cos \alpha_k)_{,y}$$

$$= u_{ys,i-1} + q_i \left(\frac{s}{Gt}\right)_i - 2A_{ci}\theta_{y,y} - s_i (u_{zs} \cos \alpha_i + u_{xs} \sin \alpha_i)_{,y}. \quad (5.112)$$

Equation (5.88) can be used to get the u_{ys1} constant. It is shown in Section 7.6 of Reference 4 that the moments M_x and M_z for restrained warping are zero and satisfy Equation (5.89) provided the shear center is used for the calculations in Equation (5.112).

The case of the restrained box beam is considered in Chapter 6.

Example 5.15 (a) Find the rotation θ_y of the constant cross section symmetrical box beam in Figure 5.21 with V_z, E, and G constant along the beam. (b) Find u_{zas} and u_{xas} at stringer (2). (c) Find the u_{ysi} warping deflections of the beam.

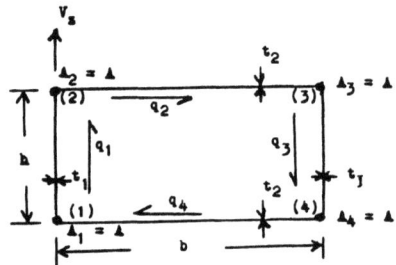

Fig. 5.21. Constant box beam cross section for Example 5.15.

Solution (a) Since the webs and stringer areas are symmetrical in Figure 5.21 the shear center is at the centroid of the box. Thus

$$M_y = V_z(b/2), \quad A_c = bh, \quad \frac{1}{GJ} = \frac{1}{4b^2h^2G}\left(\frac{2h}{t_1} + \frac{2b}{t_2}\right),$$

whence Equation (5.106) gives

$$\theta_y = \frac{yV_z}{4bh^2G}\left(\frac{b}{t_2} + \frac{h}{t_1}\right). \quad (5.113)$$

(b) Apply unit loads at the shear center to get

$$q_{1z}^1 = 1/2h, \quad q_{2z}^1 = 0, \quad q_{3z}^1 = -1/2h, \quad q_{4z}^1 = 0,$$
$$q_{1x}^1 = 0, \quad q_{2x}^1 = 1/2b, \quad q_{3x}^1 = 0, \quad q_{4x}^1 = -1/2b.$$

The shear flows due to V_z are

$$q_1 = V_z/2h + (bV_z/2)(1/2bh) = 3V_z/4h, \quad q_2 = 0 + V_z/4h = V_z/4h,$$
$$q_3 = -V_z/2h + V_z/4h = -V_z/4h, \quad q_4 = 0 + V_z/4h = V_z/4h,$$

whence Equation (5.91) gives at the shear center,

$$u_{zas} = 0, \quad u_{xas} = yV_z/2Ght_1.$$

Box beam shear stresses and deflections

At stringer (2) use θ_y from Equation (5.113) to get the rotation, or

$$(u_{xa\bullet})_2 = 0 + \theta_y h/2 = \frac{V_z y}{8bh^2 G}\left(\frac{b}{t_2} + \frac{h}{t_1}\right),$$

$$(u_{xa\bullet})_2 = \frac{V_z y}{2ht_1 G} + \theta_y b/2 = \frac{3V_z y}{4Ght_1}\left[1 + \frac{t_1}{6h}\left(\frac{b}{t_2} - \frac{h}{t_1}\right)\right]. \tag{5.114}$$

(c) Because of symmetry in Figure 5.21, the warping will be zero on the axes through the shear center. Thus, the calculations in Equation (5.112) will be started at the middle of web 1 with $u_{y\bullet 0} = 0$. Use the above data to get

$$u_{y\bullet 2} = \frac{3V_z}{4h}\frac{h}{2Gt_1} - \frac{V_z}{4Gbh^2}\left(\frac{b}{t_2} + \frac{h}{t_1}\right)\frac{bh}{4} - \frac{h}{2}\left[0 + \frac{V_z}{2Ght_1}(1)\right]$$

$$= -\frac{V_z}{16Gh}\left(\frac{b}{t_2} - \frac{h}{t_1}\right),$$

$$u_{y\bullet 3} = u_{y\bullet 2} + \frac{V_z b}{4hGt_2} - \frac{V_z}{4Gbh^2}\left(\frac{b}{t_2} + \frac{h}{t_1}\right)\frac{bh}{2} + b(0+0) = -u_{y\bullet 2}.$$

$$u_{y\bullet 4} = u_{y\bullet 3} - \frac{V_z h}{4hGt_1} - \frac{V_z}{4Gbh^2}\left(\frac{b}{t_2} + \frac{h}{t_1}\right)\frac{bh}{2} - h\left[0 + \frac{V_z}{2Ght_1}(-1)\right] = u_{y\bullet 2},$$

$$u_{y\bullet 1} = u_{y\bullet 4} + \frac{V_z b}{4hGt_2} - \frac{V_z}{4Gbh^2}\left(\frac{b}{t_2} + \frac{h}{t_1}\right)\frac{bh}{2} + 0 = -u_{y\bullet 2}.$$

As might be expected the warping at the four corners of the box has the same magnitude with alternating signs. Note that there is no warping for $b/t_2 = h/t_1$.

Example 5.16. Find (a) shear center, (b) shear flows, (c) deflections, and (d) warping deflections of the single cell box beam in Figure 5.22.

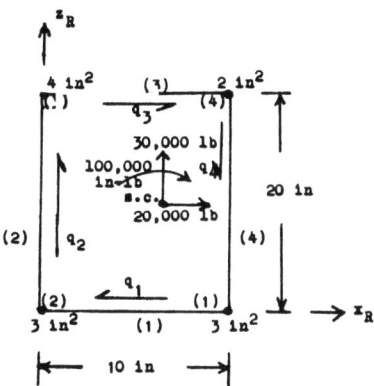

Fig. 5.22. Box beam for Example 5.16 with cross section and Gt constant.

Solution (a) The cross section in Figure 5.22 is the same as in Figure 5.6 with web (4) added and the data calculated in Example 5.3 can be used here. In Equations (5.106) and (5.109)

$$\frac{1}{GJ} = \frac{1}{(400)^2}\left(\frac{60}{Gt}\right),$$

$$\frac{1}{Gt}\sum_{k=1}^{4} s_k H_k = \frac{1}{10^3 Gt}[10(20.59) + 20(50.00) + 10(20.59) + 0] = \frac{1.4118}{Gt},$$

$$\frac{1}{Gt}\sum_{k=1}^{4} s_k J_k = \frac{1}{10^3 Gt}[10(-52.86) + 0 + 10(47.14) + 0] = -\frac{0.0572}{Gt},$$

$$e_z = \frac{(400)^2 Gt}{60(400)} \frac{1.4118}{Gt} - 4.12 = 5.29 \text{ in.},$$

$$e_z = -\frac{20}{3}(-0.0572) + 9.43 = 9.81 \text{ in.}$$

(b) From Equation (5.38) and the shear flows in Example 5.3, M_{y_s} about the shear center is

$$M_{y_s} = (-439)(9.81)(10) + (1500)(5.29)(20) + (1558)10.19)(10) + 0$$
$$= 274,394 \text{ in} - \text{lb},$$

whence in Equation (5.37)

$$q_0 = (100,000 - 274,394)/400 = -436 \text{ lb/in}.$$

The shear flows are

$$q_1 = -439 - 436 = -875 \text{ lb/in}, \qquad q_2 = 1500 - 436 = 1064 \text{ lb/in},$$
$$q_3 = 1558 - 436 = 1122 \text{ lb/in}, \qquad q_4 = -436 \text{ lb/in}.$$

(c) From Equation (5.106)

$$\theta_{y,y} = \frac{100,000(60)}{(400)^2 Gt} = \frac{37.50}{Gt}.$$

To calculate $u_{z_s,y}$ and $u_{z_s,y}$ apply unit loads at the shear center and calculate the shear flows in the webs. In this case use the results for H_k and J_k in Example 5.3 and translate from the open box shear center to the closed box shear center. Thus,

$$q_{zk}^1 = H_k - \frac{9.41}{400} = H_k - 0.02353,$$

$$q_{z1}^1 = -0.00294, \quad q_{z2}^1 = 0.02647, \quad q_{z3}^1 = -0.00294, \quad q_{z4}^1 = -0.02353,$$

$$q_{zk}^1 = J_k + \frac{0.38}{400} = J_k + 0.00095,$$

$$q_{z1}^1 = -0.05191, \quad q_{z2}^1 = 0.00095 = q_{z4}^1, \quad q_{z3}^1 = 0.04809.$$

From Equation (5.91),

$$Gtu_{z_s,y} = 10(-0.00294)(-875) + 20(0.02647)(1064) + 10(-0.00294)(1122) +$$
$$+ 20(-0.02353)(-436) = 761.2,$$
$$u_{z_s,y} = 761.2/Gt,$$
$$u_{z_s,y} = 1005.7/Gt.$$

(d) Use the warping Equation (5.112) to get

$$Gtu_{y,2} = Gtu_{y,1} + 10(-875) - 10(9.81)(37.50) - 10(1005.7)(-1)$$
$$= Gtu_{y,1} - 2372,$$
$$Gtu_{y,3} = Gtu_{y,1} - 283,$$
$$Gtu_{y,4} = Gtu_{y,1} - 2942.$$

Use Equation (5.88) to get $u_{y,1}$, or

$$0 = 12Gtu_{y,1} + 3(-2372) + 4(-283) + 2(-2942) + 0,$$
$$u_{y,1} = 1178/Gt, \quad u_{y,2} = -1194/Gt, \quad u_{y,3} = 895/Gt, \quad u_{y,4} = -1764/Gt.$$

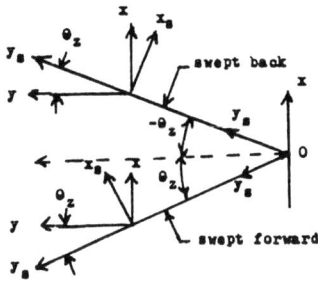

Fig. 5.23. Axes on swept wings.

5.13. Rotation of swept wings

As pointed out in Section B.13 and Figure B.24, the u_z deflection of the wing elastic axis on a sweptback wing produces a decrease in the angle of attack of the wing for the air flow in the x-direction. In Figures B.24 and 5.23 the deflections $u_{zab} + u_{zas}$ and θ_{ys} of the swept wing can be calculated for the x_s, y_s axes of the wing by the formulas in the previous sections of this Chapter 5, except for warping effects near the wing root which will be considered in Chapter 6. Let θ_{xs}, θ_{ys}, and θ_z be the angles of rotation about the x_s, y_s, and z axes and let θ_x, θ_y, and θ_z be the angles about the x, y, z axes. From Equation (2.31) take θ_{xs} as

$$\theta_{xs} = \frac{d}{dy_s}(u_{zab} + u_{zas}). \tag{5.115}$$

With the angles as vectors along their axes, it can be seen from Figure 5.23 that

$$\theta_y = \theta_{ys} \cos\theta_z + \theta_{xs} \sin\theta_z$$
$$= \theta_{ys} \cos\theta_z + \frac{d}{dy_s}(u_{zab} + u_{zas}) \sin\theta_z, \tag{5.116}$$

where θ_z is the swept forward angle of the wing.

Since only the component $V \cos\theta_z$ of the velocity in the x_s-direction produces lift on the wing, it is necessary to find the effective angle of attack α_{es} for the x_s-direction due to the rotation and deflection. From Figure 5.23 the component of α_{es} about the y-axis is

$$\alpha_{es} \cos\theta_z = \theta_y, \text{ or } \alpha_{es} = \theta_{ys} + \frac{d}{dy_s}(u_{zab} + u_{zas}) \tan\theta_z, \tag{5.117}$$

where Equation (5.116) has been used. Physically, the second term in α_{es} is the induced angle from the spanwise velocity component $V \sin\theta_z$ with the velocity $(V \sin\theta_z)\frac{d}{dy_s}(u_{zab} + u_{zas})$ perpendicular to the wing surface.

It is evident that $\tan\theta_z$ is positive for swept forward wings and negative for swept back wings. Thus, the bending deflections tend to decrease the twist of the swept back wing, but to increase the twist of the swept forward wing. To obtain an idea of the relative magnitude of these θ_{y_s} and deflection effects, use Equation (5.74) to get

$$\frac{du_{zab}}{dy_s} = \frac{C_{zab}}{h_0} y_s, \qquad (5.118)$$

which can be larger than θ_{y_s}. For example, if $C_{zab} = 0.01$, $h_0 = 70$ in., $(y_s)_{\text{tip}} = 1500$ in., then $0.01(1500/70)(180/\pi) = 12.28°$ at the wing tip. Thus, if $\theta_z = \mp 45°$ and $\theta_{y_s} = 6°$ at the tip of the wing, then in Equation (5.117)

$$\begin{aligned}\alpha_{e_s} &= 6.00° - 12.28° = -6.28°, \text{ swept back case,} \\ \alpha_{e_s} &= 6.00° + 12.28° = 18.28°, \text{ swept forward case.}\end{aligned} \qquad (5.119)$$

Since the large positive twist tends to shift the lift loads toward the wing tip, Equation (5.119) demonstrates a serious problem in the design of swept forward wings. Also, this large twist causes another difficulty with the swept forward wing, which will be discussed in the next Section 5.14.

5.14. Spanwise airload distribution and static wing divergence under rotation

In the above Sections 5.12 and 5.13 the wing rotation θ_y was calculated from a known M_y torsional moment on the wing. However, it is evident that the θ_y will change the lift load distribution on the wing and hence will change V_x, V_z, M_x, M_z, and M_y on the wing. Thus, to get the final θ_y on an actual wing under flight loads, it is necessary to iterate using the above formulas. That is, calculate θ_y for the flight condition using the M_y for no twist, recalculate the load distribution and M_y, calculate a new θ_y, etc. Satisfactory convergence usually occurs in about two iterations.

The θ_y rotation can be obtained without iteration by setting up the differential equation for θ_y. From Equations (5.106), (B.57), and (B.60) the differential equation for θ_y on the wing can be written as

$$\frac{d}{dy}\left(GJ\frac{d\theta_y}{dy}\right) = p_z x_a = x_a c c_l q. \qquad (5.120)$$

From $c_l = m_0 \alpha$, where m_0 is the slope of the wing lift curve and α is the section angle of attack,

$$c_l = m_0(\alpha_R + \theta_y) \qquad (5.121)$$

when wing twist is present. Thus Equation (5.120) becomes

$$\frac{d}{dy}\left(GJ\frac{d\theta_y}{dy}\right) - m_0 x_a c q \theta_y = m_0 x_a c q \alpha_R - n(mg)x_d, \qquad (5.122)$$

where the moment produced by the $w = mg$ weight per unit span of the wing has been added, and where n is the load factor and x_d is the distance from the elastic axis to the center of mass of the cross section. x_a and x_d have the signs of x on the wing.

Since GJ on a tapered wing is usually known only at points, it is evident that Equation (5.122) would have to be solved by numerical integration for tapered wings. Only the special case of the uniform constant wing is given here. For this case, with x_a negative for the case of interest, Equation (5.122) becomes

$$\frac{d^2\theta_y}{dy^2} + k^2\theta_y = -f, \quad k^2 = -\frac{m_0 x_a c q}{GJ}, \quad f = k^2\alpha_R + \frac{n w x_d}{GJ}, \quad (5.123)$$

whence

$$\theta_y = A\cos ky + B\sin ky - (f/k^2). \quad (5.124)$$

The boundary conditions are

$$\theta_y(0) = 0, \quad \frac{d\theta_y(L)}{dy} = 0, \quad (5.125)$$

where $M_y = 0$ at $y = L$ gives the second condition. Thus

$$\theta_y = (f/k^2)\left[\frac{\cos k(L-y)}{\cos kL} - 1\right]. \quad (5.126)$$

Since $q = (1/2)\rho V^2$, k in Equations (5.123)-(5.126) is a function of the velocity V. Hence, in Equation (5.126), θ_y becomes infinite and the wing fails at $\cos kL = 0$, or

$$kL = \pi/2, \quad \text{or} \quad -\frac{m_0 x_a c \rho V^2 L^2}{2GJ} = \frac{\pi^2}{4}, \quad \text{or} \quad V_D = \frac{\pi}{2L}\left(\frac{-2GJ}{m_0 x_a c \rho}\right)^{1/2}, \quad (5.127)$$

where V_D is the *divergence velocity* for wing failure. Obviously, the wing must be designed so that V_D is larger than $V_{\max}$ on the V-n diagram. V_D can be increased by increasing GJ or by decreasing the magnitude of x_a. Since the air density ρ is smaller at altitude, V_D is larger at higher altitudes.

Note that if $x_d = 0$ in Equation (5.123), then $f/k^2 = \alpha_R$ and Equation (5.126) becomes

$$\theta_y = \alpha_R\left[\frac{\cos k(L-y)}{\cos kL} - 1\right], \quad (5.128)$$

whence

$$\alpha = \alpha_R + \theta_y = \alpha_R\frac{\cos kL(1-y_d)}{\cos kL}, \quad y_d = y/L,$$
$$\frac{\alpha}{\alpha_R} = 1 + \frac{\theta_y}{\alpha_R} = \frac{\cos kL(1-y_d)}{\cos kL} = \frac{c_l}{c_{lR}}. \quad (5.129)$$

Thus, there is a large increase in the section lift coefficient outboard unless kL is small for all points on the V-n diagram. If $kL = \pi/3$, the c_l is doubled at the wing tip, $y = L$. This increase in the angle of attack outboard puts more load outboard and increases the M_x bending moment on the wing.

For the case of $C_L = 0.00$ on the twisted wing, calculate α_{w0} from Equation (B.51) with θ_y in Equation (5.126) in place of α_{aR}, or

$$\alpha_{w0} = \frac{2c}{2Lc}\int_0^L \theta_y\, dy = (f/k^2)\int_0^1\left[\frac{\cos kL(1-y_d)}{\cos kL} - 1\right]dy_d$$

$$= (f/k^2)\left(\frac{\tan kL}{kL} - 1\right). \tag{5.130}$$

Thus, for the case of $x_d = 0$, Equation (5.129) has the form

$$\frac{c_l}{c_{lR}} = \frac{\alpha_R + \theta_y - \alpha_{w0}}{\alpha_R} = 1 - \frac{\tan kL}{kL} + \frac{\cos kL(1-y_d)}{\cos kL} \tag{5.131}$$

for the spanwise section lift coefficient distribution with the same total load as given by c_{lR}. If the c_d drag term is omitted in Equation (B.56), then the spanwise airload distribution is

$$\frac{p_z}{p_{zR}} = \frac{c_z}{c_{zR}} = \frac{c_l}{c_{1R}}, \tag{5.132}$$

whence from Equations (B.58)-(B.60),

$$\frac{V_z}{V_{zR}} = \left[p_{zR}\int_{y_d}^1 \frac{p_z}{p_{zR}}dy_d\right] / \left[p_{zR}\int_{y_d}^1 dy_d\right]$$

$$= 1 - \frac{\tan kL}{kL} + \frac{\sin kL(1-y_d)}{(1-y_d)kL\cos kL}, \tag{5.133}$$

$$\frac{M_x}{M_{xR}} = 1 - \frac{\tan kL}{kL} + \frac{2[1-\cos kL(1-y_d)]}{(1-y_d)^2(kL)^2\cos kL}, \tag{5.134}$$

$$\frac{M_y}{M_{yR}} = 1 - \frac{\tan kL}{kL} + \frac{\sin kL(1-y_d)}{(1-y_d)kL\cos kL} = \frac{V_z}{V_{zR}}. \tag{5.135}$$

Suppose a flight condition has

$$kL = \left[-\frac{m_0 x_a c\rho V^2 L^2}{2GJ}\right]^{1/2} = \pi/3, \tag{5.136}$$

which corresponds to $V = (2/3)V_D$ in Equation (5.127). For this case on a uniform constant wing, Figure 5.24 shows the magnification of the shear V_z, bending moment M_x, and torsional moment M_y due to the wing rotation and redistribution of the spanwise air loading. Although the total lift remains the same, the large effects on the shear and moments are evident in the Figure 5.24.

The above discussion of the load redistribution and the divergence velocity V_D applies for the straight wing. For the swept wing it is necessary to bring in the wing elastic axis deflection u_{za}, as discussed in Section 5.13, in order to get the proper θ_y rotation. For the coordinates of the structural wing with shear deflections omitted, Equations (4.6), (5.117), and (5.122) give

$$\frac{d^2}{dy_s^2}\left(EI\frac{d^2u_{zab}}{dy_s^2}\right) - c_s m_0 \alpha_{es} q \cos^2\theta_z$$

$$= c_s m_0 \alpha_{Rs} q \cos^2\theta_z - nw, \tag{5.137a}$$

$$\frac{d}{dy_s}\left(GJ\frac{d\theta_{ys}}{dy_s}\right) - m_0 x_{as} c_s \alpha_{es} q \cos^2\theta_z$$

$$= m_0 x_{as} c_s \alpha_{Rs} q \cos^2\theta_z - nw x_{as}. \tag{5.137b}$$

Here $\alpha_s = \alpha_{Rs} + \alpha_{es}$ and the $\cos^2\theta_z$ term arises from the $V\cos\theta_z$ component of V in the x_s-direction in q, which remains $q = (1/2)\rho V^2$.

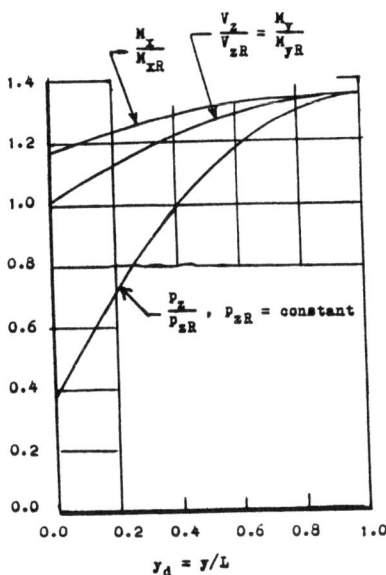

Fig. 5.24. Effect of wing rotation on load, shear, and moment distribution for $V = (2/3)V_D$.

From Equation (5.117), α_{es} involves θ_{ys} and du_{zab}/dy_s so that the two equations in Equation (5.137) for θ_{ys} and u_{zab} are coupled. Solutions of equations (5.137) for θ_{ys}, u_{zab}, and the divergence velocity V_D have been made in Reference 6 for the cases of the uniform constant wing and for the tapered wing with EI and GJ varying as the fourth power of the wing chord. Also, see Reference 8 for discussion of swept wing divergence. Although the exact solution in Reference 6 involves reading curves to get the divergence velocity, an approximate solution, which is within the tolerances on the actual values of EI and GJ, is given also. This approximate solution for both the uniform and the tapered wing can be written in the form

$$V_D^2 = \frac{2(GJ)_R}{\rho m_0 c_R L^3 \cos^2\theta_z}\left(\frac{L}{-x_a}\right)\left(\frac{K_1}{K_D}\right),$$

$$K_D = 1 + K_2 \frac{(GJ)_R}{(EI)_R}\left(\frac{L}{-x_a}\right)\tan\theta_z,$$

(5.138)

where $(GJ)_R$, $(EI)_R$, and c_R are root values for the wing, $-x_a$ is the distance from the y_s elastic axis to the aerodynamic center (see Figure B.9), and K_1, K_2 have the following values for taper ratio c_T/c_R (see Equation (B.47)),

$$\begin{aligned}
c_T/c_R &= 0.20, \quad 0.50, \quad 1.00, \quad 1.50, \\
K_1 &= 2.81, \quad 2.74, \quad 2.47, \quad 2.22, \\
K_2 &= 0.614, \quad 0.497, \quad 0.390, \quad 0.326.
\end{aligned}$$

(5.139)

For the straight wing, $\theta_z = 0°$, and Equation (5.138) gives

$$V_{D0} = \frac{(K_1)^{1/2}}{L} \left[\frac{-2(GJ)_R}{\rho m_0 x_a c_R} \right]^{1/2}, \tag{5.140}$$

which checks Equation (5.127) for $c_T/c_R = 1.00$. Use V_{D0} as a reference divergence velocity and divide it into Equation (5.138) to get V_{D_s} for swept wings as

$$V_{D_s}/V_{D0} = (\cos\theta_z)^{-1}(1 + K_2 H \tan\theta_z)^{-1/2},$$
$$H = \frac{(GJ)_R}{(EI)_R} \left(\frac{L}{-x_a} \right). \tag{5.141}$$

This expression becomes infinite at

$$\theta_z = -\arctan(1/K_2 H), \tag{5.142}$$

whence no divergence occurs on swept back wings for larger values of $-\theta_z$. This is due to the bending deflection, Equation (5.119), which rotates the swept back wing to move the airload inboard and reduce the angle of attack. On the other hand, on the swept forward wing the bending deflection increases the rotation

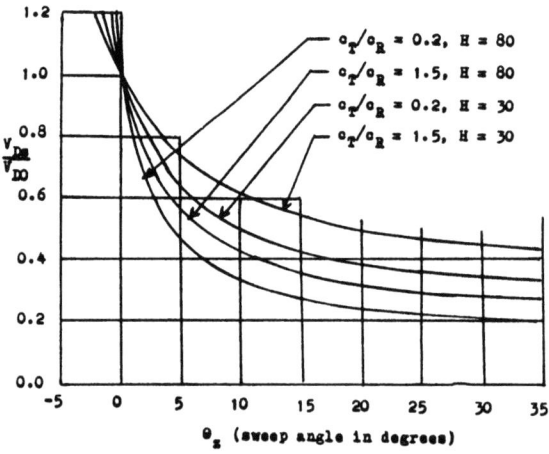

Fig. 5.25. Divergence speed ratio for swept wings.

to produce a lower divergence velocity. Figure 5.25 shows graphs of Equation (5.141) for two values of the H parameter and the taper ratios 0.2 and 1.5. On Figure 5.25 values for other taper ratios can be approximated by linear interpolation between the curves. The Figure 5.25 demonstrates the large decrease in divergence speed on the swept forward wing at small angles of sweep. The curves reach a minimum value at

$$\theta_z = (1/2)\arctan K_2 H, \tag{5.143}$$

which ordinarily is between 40° and 45°.

It should be noted that x_a is in both the parameters H and V_{D0} in Equation (5.141) and Figure 5.25. Also, x_a will be different for different taper ratios. Thus, each curve on Figure 5.25 gives V_{D_s} against θ_x for some V_{D0}, but the different curves do not necessarily give a comparison of V_{D_s} at some other taper ratio or x_a value.

In References 7 and 8 calculations are made for the spanwise load distribution on both swept back and swept forward uniform wings. As might be expected from Equation (5.119), the load on the outer portion of the wing is decreased on swept back wings and increased on swept forward wings. For a case of $\theta_x = 30°$ and $V = 0.5V_{D_s}$ for the swept forward wing, or $V = 0.2V_{D0}$ approximately for $H = 32$, the load distribution given in Reference 8 is similar to that in Figure 5.24. For the same wing swept back 30° at the same speed, the p_x/p_{xR} is about 0.85 at the tip and about 1.20 at the root. Thus, swept back wings have practically no divergence problems with the load reduced toward the tip from the deformations. However, due to the low divergence speeds and high loads toward the tip, swept forward wings are not very practical. For these reasons, very few swept forward wings have been built.

Example 5.17. (a) Find the divergence speed of the swept wing for bending deflection alone, with $x_a = 0$ and $x_d = 0$ so that the airload and inertia load are on the elastic axis. (b) What happens when x_a is positive, or the lift is behind the elastic axis?

Solution (a) Take the limit in Equation (5.138) at $x_a = 0$, or

$$V_D = \frac{1}{L}\left(\frac{K_1}{K_2}\right)^{1/2}\left[\frac{4(EI)_R}{\rho m_0 c_R L \sin 2\theta_x}\right]^{1/2}. \tag{5.144}$$

The swept back and straight wings do not diverge, but the swept forward wing diverges at lower speeds as θ_x increases. The minimum V_D occurs at $\theta_x = 45°$.

(b) If x_a is positive in Equations (5.138) and (5.141), then Equation (5.141) takes the form

$$V_{D_s}/V_{\text{Ref}} = (\cos\theta_x)^{-1}(K_2 H \tan\theta_x - 1)^{1/2},$$

$$V_{\text{Ref}} = \frac{K_1^{1/2}}{L}\left[\frac{2(GJ)_R}{\rho m_0 x_a c_R}\right]^{1/2}, \quad H = \left(\frac{L}{x_a}\right)\frac{(GJ)_R}{(EI)_R}, \tag{5.145}$$

where V_{Ref} is a reference velocity. No divergence occurs for

$$\theta_s \leq \arctan(1/K_2 H). \tag{5.146}$$

The graph of Equation (5.145) is similar to Figure 5.25, except shifted to the right.

5.15. Static aileron effectiveness and reversal speed under wing rotation

When the aileron on the trailing edge of the wing deflects through an angle β, it changes the lift on the wing in the region of the aileron, Figure 5.26. This lift force, which acts up on one wing and down on the other, produces a moment about the x axis causing the airplane to roll, giving θ_x roll angle, $\theta_{x,t}$ angular velocity, and $\theta_{x,tt}$ angular acceleration. The angular acceleration produces an inertia force and moment per unit span at any point y on the wing of

$$F_m = -my\theta_{x,tt}, \quad m_{ym} = myx_d\theta_{x,tt}, \tag{5.147}$$

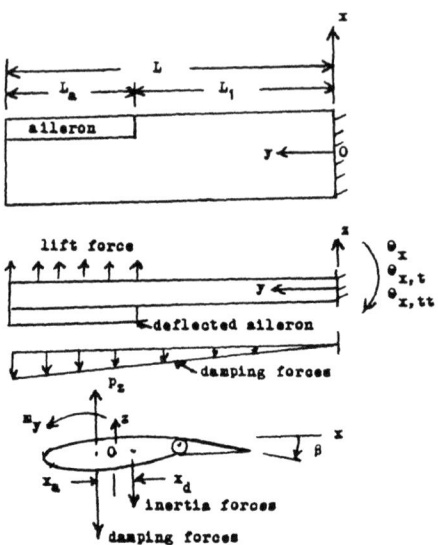

Fig. 5.26. Forces on wing due to aileron deflection.

where m is the mass per unit span at the point y. Since the angular velocity component $y\theta_{x,t}$ is perpendicular to the air velocity V, it produces a damping angle of attack $y\theta_{x,t}/V$ or the damping force and moment

$$F_D = -m_0 cq\theta_{x,t} y/V, \qquad m_{yD} = m_0 cq\theta_{x,t} x_a y/V. \tag{5.148}$$

For the uniform constant wing with $\alpha_R = 0$, with the above moments in Equations (5.147) and (5.148) and with the loads in Figure (5.26), $p_z = c_{lR,\beta} cq\beta$, $m_y = c_{mac,\beta} c^2 q\beta$, it follows that Equation (5.123) for θ_y becomes

$$\theta_{y,yy} + k^2 \theta_y = Hk^2(y/L) - (k^2 \beta \delta_a/m_0)(c_{lR,\beta} - \frac{c}{x_a} c_{mac,\beta}), \tag{5.149}$$

$$H = (\theta_{x,t} L/V) - (m x_d L \theta_{x,tt}/k^2 GJ),$$

where δ_a has the value of unity over the aileron span and the value of zero elsewhere. The solution of Equation (5.149) is

$$\theta_y = HG_1(y) + \beta G_2(y), \qquad G_1(y) = \frac{y}{L} - \frac{\sin ky}{kL \cos kL}, \tag{5.150}$$

$$G_2(y) = -(1/m_0)(c_{lR,\beta} - \frac{c}{x_a} c_{mac,\beta})\Big[\delta_a\{1 - \cos k(y - L_1)\} - \frac{\sin k(L - L_1)}{\cos kL} \sin ky\Big].$$

The angle of attack at any point y on the rising wing is

$$\alpha(y) = (c_{lR,\beta} \beta/m_0) - (\theta_{x,t} L/V)(y/L) + \theta_y(y), \tag{5.151}$$

and the equilibrium equation for the rolling wing is given by Equation (B.19) as

$$M_x = I_{xx} \theta_{x,tt}. \tag{5.152}$$

The rolling moment about the x-axis is

$$M_x = 2q \int_0^L cm_0 \alpha(y) y \, dy$$

$$= 2m_0 cq \int_0^L \left[\{c_{lR,\beta}/m_0 + G_2(y)\}\beta - (\theta_{x,t}L/V)\{(y/L) - G_1(y)\} - \right.$$
$$\left. - (mLx_d/k^2 GJ)G_1(y)\theta_{x,tt} \right] y \, dy, \qquad (5.153)$$

whence Equation (5.152) becomes

$$A\theta_{x,tt} + B(\theta_{x,t}L/V) - C\beta = 0.$$

$$A = (I_{xx}/2qcm_0) + (mx_d L/k^2 GJ) \int_0^L G_1(y) y \, dy,$$

$$B = \int_0^L [(y/L) - G_1(y)] y \, dy, \qquad (5.154)$$

$$C = \int_0^L [(c_{lR,\beta}/m_0) + G_2(y)] y \, dy.$$

Put $G_1(y)$ and $G_2(y)$ from Equation (5.150) into Equation (5.154) to get

$$A = (I_{xx}/2qcm_0) + (mx_d/k^2 GJ)\left[\frac{L^3}{3} + \frac{L}{k^2} - \frac{\tan kL}{k^3}\right],$$

$$B = (1/k^2)\left(\frac{\tan kL}{kL} - 1\right),$$

$$C = (c_{lR,\beta}/m_0 k^2)\left(\frac{\cos kL_1}{\cos kL} - 1\right) - \frac{cc_{mac,\beta}}{m_0 x_a k^2}\left[\frac{\cos kL_1}{\cos kL} - 1 - \right.$$
$$\left. - (k^2/2)(L^2 - L_1^2)\right]. \qquad (5.155)$$

Although Equation (5.154) can be solved for θ_x as a function of time, it is not necessary since the most important cases for the design of the wing and aileron occur at the start of the motion and in the steady roll. At the start, $\theta_{x,t} = 0$ and $\theta_{x,tt}$ is maximum for instantaneous aileron deflection. This maximum $\theta_{x,tt}$ gives the load factor $n_z = (y\theta_{x,tt}/g)$ along the wing for local design conditions. With $\theta_{x,t} = 0$, Equation (5.154) gives $\theta_{x,tt}$ directly in terms of the aileron angle β, or

$$\theta_{x,tt} = (C/A)\beta, \qquad (5.156)$$

with C and A in Equation (5.155). For the steady roll, $\theta_{x,tt} = 0$, and equation (5.154) gives

$$\theta_{x,t}L/V = (C/B)\beta, \qquad (5.157)$$

with C and B in Equation (5.155).

Since k depends upon the velocity V, the terms A, B, C in Equation (5.155)-(5.157) depend upon V. The effectiveness of the aileron in producing $\theta_{x,t}$, $\theta_{x,tt}$ in Equations (5.156) and (5.157) depends upon C/A and C/B as functions of V. If C becomes zero at a velocity V_R, the aileron is ineffective and V_R is called the aileron reversal speed. At higher velocities, the airplane rolls in the other direction, which is somewhat embarrassing to the pilot. Obviously, the aileron and wing should be designed to have V_R greater than $V_{\max}$ on the V-n diagram. Note that if $V_R > V_D$, both A and B in Equation (5.155) become infinite at $kL = \pi/2$ in the $\tan kL$ term and the aileron is ineffective because of wing failure. See Equation (5.127).

The effectiveness of the aileron for steady roll is shown by plotting C/B in Equation (5.157) against kL, which is proportional to V. To make the graph, the terms $c_{lR,\beta}/m_0$, $cc_{mac,\beta}/m_0 x_a$, and L_1/L must be known for the given aileron and wing. The form of the equation is

$$\frac{(\theta_{x,t} L/V)}{\beta} = \frac{C_1}{B_1}, \qquad B_1 = \frac{\tan kL}{kL} - 1,$$

$$C_1 = (c_{lR,\beta}/m_0)\left[\frac{\cos kL(L_1/L)}{\cos kL} - 1\right] - \qquad (5.158)$$

$$- (cc_{mac,\beta}/m_0 x_a)\left[\frac{\cos kL(L_1/L)}{\cos kL} - 1 - \frac{(kL)^2}{2}\left(1 - \frac{L_1^2}{L^2}\right)\right],$$

$$kL = (-\rho m_0 x_a c V^2 L^2 / 2GJ)^{1/2}.$$

Figure 5.27 shows the graph of Equation (5.158) for the case of

$$c_{lR,\beta}/m_0 = 0.45, \qquad cc_{mac,\beta}/m_0 x_a = 0.60, \qquad L_1/L = 0.70, \qquad (5.159)$$

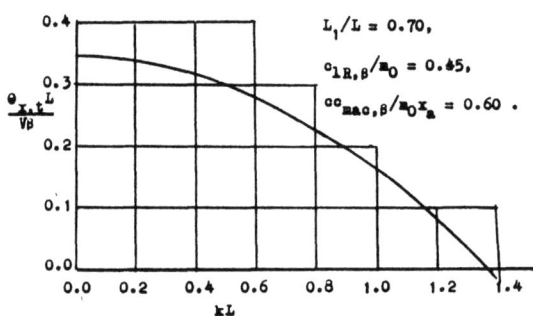

Fig. 5.27. Aileron effectiveness of uniform straight wing.

where the reversal speed is given by $kL = 1.37$. Thus, from Equations (5.127) and (5.158),

$$V_R/V_D = 1.37/1.57 = 0.87, \qquad (5.160)$$

and the reversal speed is less than the divergence speed.

The aileron deflection at a given flight condition changes the load distribution on the wing, putting more load outboard for the down aileron and less load outboard for the up aileron. See Chapter 8 of Reference 8 for some examples of load distribution.

For swept back wings with the bending deflection reducing the load outboard, the aileron is less effective and the reversal speed is lower than for the straight wing. On some airplanes with swept back wings, it is necessary to add spoilers on the wing to get more roll effectiveness at higher speeds, where the normal ailerons have little effect. See Reference 8 for examples.

5.16. Problems

5.1. In Figure 5.2 take $h_1 = h_3 = 12$ in., $h_2 = 8$ in., $A_1 = A_2 = A_3 = 6.00$ in^2, $A_4 = 4.00$ in^2, $V_z = 50,000$ lb, and find the shear flows q_1, q_2, q_3.

5.2. Solve Example 5.2 for no area change, using the areas at station $y = 200$ in., as constants in the beam. Compare results to those in Examples 5.1 and 5.2.

5.3. Find the shear center for the open box in Figure 5.6 with $s_1 = s_3 = 6$ in., $s_2 = 10$ in., $A_1 = A_4 = 2.00$ in^2, $A_2 = 1.00$ in^2, $A_3 = 3.00$ in^2.

5.4. Solve Example 5.4 with $V_x = 50,000$ lb, $V_z = 20,000$ lb, and $M_y = 300,000$ in-lb.

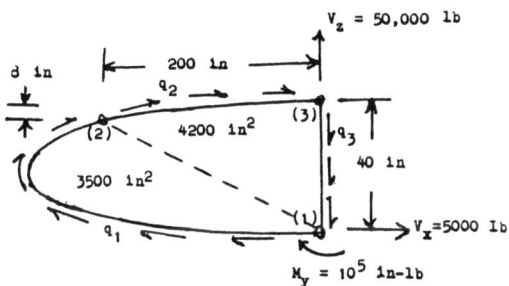

Fig. 5.28. Problem 5.5.

5.5. Find the shear flows in the three stringer single cell wing box beam cross section in Figure 5.28.

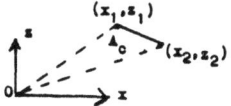

Fig. 5.29. Problem 5.6.

5.6. For a straight web between points (x_1, z_1) and (x_2, z_2) in Figure 5.29 show that the enclosed area $2A_c$ for moments about point $(0,0)$ is $2A_c = x_2 z_1 - x_1 z_2$.

5.7. Solve Example 5.6 for $V_z = 0$ and $M_y = 1.5(10^6)$ in-lb.

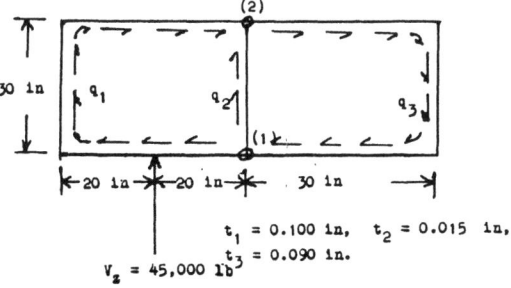

Fig. 5.30. Problem 5.8.

5.8. Find the shear flows in the two-stringer two-cell box beam cross section in Figure 5.30.

5.9. Find the shear center in Problem 5.8.

5.10. Solve Example 5.8 for the case in which only the cross section dimensions change, with the stringer areas remaining constant at the y-section values.

5.11. Solve Example 5.8 for the case in which only the stringer areas change, with the cross section remaining constant at the y-section values.

5.12. Compare the shear flows in Problems 5.10 and 5.11 to those shown in Figure 5.17 and discuss the differences.

5.13. Add the thermal strains $(e_T)_L = 0.002$, and $(e_T)_U = 0.003$ to Example 5.10 and calculate the deflections u_{xab} at the tip of the wing.

5.14. Verify that the approximate numerical solution in Equation (5.72) is exact at points y_i for the case in Equation (5.74).

5.15. In Example 5.10 let $E_{yy} = E_R$, $P_Y = 0$, $(E_R A_E)_L = (E_R A_E)_U$, $(e_T)_L = 0.001$, and $(e_T)_U = 0.003$. Find the u_{y*} deflection at the tip of the wing.

5.16. Use $N = 10$ in Equation (5.72) and calculate the values of $u_{xb}(y_i)$ in terms of G_0 and L for the case in which $G(y_d) = G_0(1 + \frac{y_d}{2L})$.

5.17. Solve Example 5.11 for case of $(Gt) = (Gt)_1 = (Gt)_3 = 0.5(Gt)_2$.

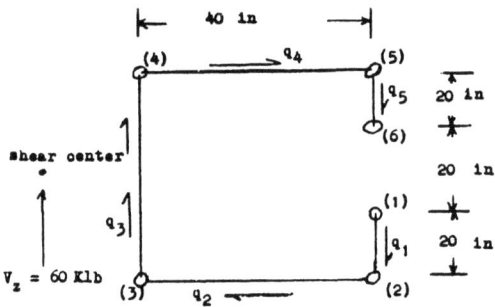

Fig. 5.31. Problems 5.18, 5.19, and 5.20.

5.18. A cantilevered open box beam with the constant cross section in Figure 5.31 has length L with $A_1 = A$ for the stringers, $(Gt)_i = GT$ for the webs. Calculate the shear flows for a unit value of V_z and determine the shear deflection u_{za*} at the tip of the beam.

5.19. In Problem 5.18 examine the warping of the beam cross section by following the procedure in Example 5.11.

5.20. Apply $V_z = 60$ Klb to the beam of Problem 5.18 and find the shear deflection u_{za*} at the tip.

5.21. Apply unit loads at stringer (3) in Figure 5.21 and find the shear deflections u_{xa*} and u_{za*} using Example 5.15.

5.22. Solve Example 5.15 for the case in which stringers (1) and (2) have area $2A$ each.

5.23. Calculate the shear center in Problem 5.22 and use it to get θ_y from Equation (5.106). Compare to the θ_y in Problem 5.22.

5.24. Let the webs in Figure 5.21 have the thicknesses t_1, t_2, t_3, and t_4 and let a moment M_y be applied with no shear forces V_x and V_z. Find the warping deflections at the four stringers.

5.25. Let the webs in Figure 5.21 have thicknesses t_1, t_2, t_3, and t_4. Calculate the shear center and get θ_y for the box cross section with V_z as shown in Figure 5.21.

5.26. If $M_y/G = 0.60$ and $G_i = G =$ constant, find (a) the rotation $d\theta_y/dy$ and (b) the warping deflections at the stringers in the box beam cross section in Figure 5.32.

5.27. Let the box beam cross section in Figure 5.21 vary along the wing with $G_k t_k = Gt =$ constant, $h = h_0(1 - Ry)$, $b = b_0(1 - Ry)$, $RL < 1$, and

$$M_y = M_{y0}\left[\frac{(1-Ry)^3 - (1-RL)^3}{1 - (1-RL)^3}\right].$$

(a) Find the equation for the angle of twist $\theta_y(y)$ of the cantilever beam. (b) Graph the equation $\theta_y/\theta_{y0} =$ function of y/L for $RL = 1/2$, $0 \leq y/L \leq 1$, where θ_{y0} is in terms of the constants t, M_{y0}, L, h_0, and b_0.

Box beam shear stresses and deflections

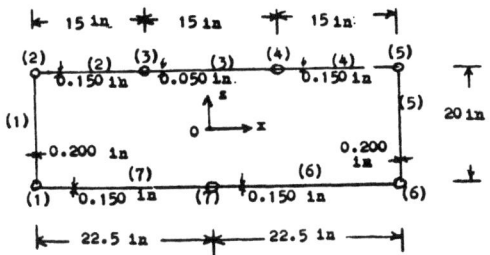

Fig. 5.32. Problem 5.26.

5.28. Solve Equations (5.123) and (5.125) to get θ_y in Equation (5.126).

5.29. Make the integrations to get the shear and moment Equations (5.133)-(5.135).

5.30. Draw curves similar to Figure 5.24 for the cases of (a) $kL = 0.45\pi$ with $V = 0.90V_D$; (b) $kL = 0.40\pi$ with $V = 0.80V_D$; (c) $kL = 0.25\pi$ with $V = 0.50V_D$.

5.31. (a) Construct curves similar to Figure 5.25 for the $C_T/C_R = 1.00$ wing with $H = 60$ and $H = 30$. (b) Assume the change in H from 60 to 30 is due to a change in x_a. Replot the $H = 60$ curve using $2^{1/2}V_{Ds}/V_{D0}$ on the ordinate and compare to the $H = 30$ curve.

5.32. Suppose a straight uniform wing has $GJ = 2.0(10)^{11}$ lb-in^2, $L = 1200$ in., $m_0 = 5.0/\text{rad}$, $c = 250$ in., $x_a = -40$ in. (a) What is the divergence speed of the wing at sea level? (b) If $EI = GJ$, what is the divergence speed of the wing for 30° forward sweep? (c) Will divergence occur at a sweep back angle of $\theta_z = -5°$?

5.33. For the wing in Problem 5.32, (a) find the divergence speed for $x_a = 0$ and $\theta_z = 45°$; (b) find the divergence speed for $x_a = 20$ in. and $\theta_z = 45°$.

5.34. Integrate Equation (5.149) to get Equation (5.150).

5.35. Derive Equations (5.154) and (5.155) from Equations (5.152) and (5.153).

5.36. Integrate Equation (5.154) on time and discuss the motion of the airplane in roll for a fixed β and V. Does the motion become steady after a short time?

References

Chapter 5

1. S. Timoshenko and J.N. Goodier: *Theory of Elasticity*, McGraw-Hill Book Co., New York (1951).
2. D.J. Peery and J.J. Azar: *Aircraft Structures*, McGraw-Hill Book Co., New York, 2nd ed. (1982).
3. R.M. Rivello: *Theory and Analysis of Flight Structures*. McGraw-Hill Book Co., New York (1969).
4. P. Kuhn: *Stresses in Aircraft and Shell Structures*, McGraw-Hill Book Co., New York (1956).
5. J.S. Przemieniecki: *Theory of Matrix Structural Analysis*, McGraw-Hill Book Co. (1968).
6. F.W. Diederich and B. Budianksy: *Divergence of Swept Wings*, NACA(NASA) TN-1680, August (1948).
7. F.W. Diederich and K.A. Foss: *Charts and Approximate Formulas for the Estimation of Aeroelastic Effects on the Loading of Swept and Unswept Wings*, NACA(NASA) TN 2608, February (1952).
8. R.L. Bisplinghoff, H. Ashley and R.L. Halfman: *Aeroelasticity*, Addison-Wesley Publishing Co., Reading, Mass. (1957).

Reference for additional reading

9. D.H. Allen and W.E. Haisler: *Introduction to Aerospace Structural Analysis*, John Wiley and Sons, New York (1985).

6

Shear lag in thin web structures

6.1. Introduction

In Chapter 5 the stringer loads and the web shear flows were determined at cross sections of thin web beams. This Chapter 6 is concerned with the calculation of shear flows and stringer or stiffener loads in box beams and simple thin web structures under conditions in which the cross section analysis may not be applicable. The procedures in this Chapter also apply to local end problems for thin web beams. When concentrated loads are applied to a thin web or when cut-outs of webs and stringers are made in a structure, then the plane strain cross section analysis used in the previous Chapters does not apply so that an analysis of the shear flows along the length or spanwise direction of the beam must be made. The term *beam*, which usually implies a long structure with a plane strain cross section, will be used here for any long structure with or without plane sections. The beam may be very short in the following analysis. Although this spanwise problem is redundant requiring deflection equations, in some cases sufficient assumptions can be made to keep the solution determinate so that only equilibrium equations are needed.

When a concentrated force is applied to a stiffener and the stiffener then distributes the force into a thin web, the resulting shear flows are variable. These variable shear flows depend upon the materials of the stiffeners and webs, the area of the stiffener and the thickness of the web so that their calculation requires the use of deflections. This is the *shear lag* redundant problem in structures. In this Chapter 6 the shear flows will be approximated as constant average values in each thin web panel, bounded by four stiffeners. If the webs and stiffeners are designed on the basis of this constant shear flow and if the materials are sufficiently ductile, then the structure tends to work as designed before failure occurs.

In Part 1 of this Chapter 6 the number of shear panels and the number of supports will be restricted to keep the structure determinate. In Part 2 redundant cases will be considered. Sections 6.2 through 6.5 will consider determinate problems for the cases of shear flows due to concentrated loads into thin webs, shear flows around cut-outs in thin webs, cut-outs in box beams, and shear flows in ribs and bulkheads. The unit load theorem is used to find the shear flows

Shear lag in thin web structures

and deflections in plane redundant beams directly in Sections 6.8 and 6.9, and by flexibility matrices in Sections 6.10 and 6.11. In Section 6.12 restrained box beam shear flows are obtained by using flexibility matrices. In Section 6.13 the load redistribution in swept back wings is examined by using matrix analysis.

The bending, shear, and rotation deflection equations in Chapter 5 do not apply to the structures in this Chapter 6. However, as discussed in Section 6.9, point deflections can be obtained by using the unit load theorem directly. Also, point solutions are given by the matrix solutions in Sections 6.11, 6.12, and 6.13. See Section 6.7 for further discussion of the redundant problems.

PART 1

SOLUTIONS FOR DETERMINATE CASES

6.2. Shear flows due to concentrated loads into thin webs

When a concentrated load is applied in the plane of a thin web, a stiffening member is required to distribute the load into the web. This member should be in the direction of the load, or the load should be applied at the intersection of two stiffeners so that each stiffener resists the load component in its direction, Figure 6.1. In fact, no stiffener should start or stop in a web unless there is a transverse stiffener for it to join. Otherwise, there will be stress concentrations and local buckling in the thin web.

In order to obtain the shear flows from equilibrium of the stiffeners or from the equilibrium shear diagram for the simple beam, it is necessary to restrict

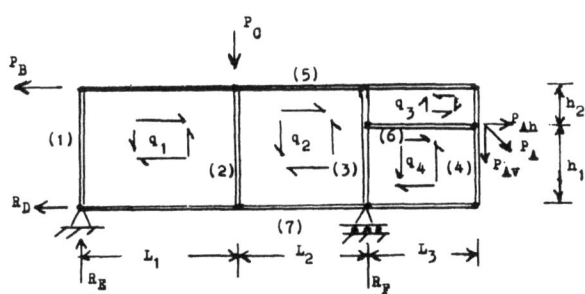

Fig. 6.1. Concentrated force into thin shear webs.

the number of separate shear webs in relation to the number of stiffeners and supports. If stiffener (6) in Figure 6.1 is extended across web (2) to stiffener (2), making an additional web, then there are more unknown shear flows (five) and reactions (three) than possible stiffener equilibrium equations (seven) so that the beam becomes redundant. Sufficient restrictions will be used in this Part 1 of Chapter 6 to keep the beam determinate. That is, N_w webs plus N_{su} supports equals N_{st} stiffeners. Also, it is assumed that the load in stringer (6) in Figure 6.1 is zero at its end joining stringer (3). This is an approximation since web (2) can apply a small tension load to the end of stringer (6).

Once the shear flows are obtained, load diagrams can be drawn for the stiffeners. Since it is necessary to know whether the stiffener is in tension or in

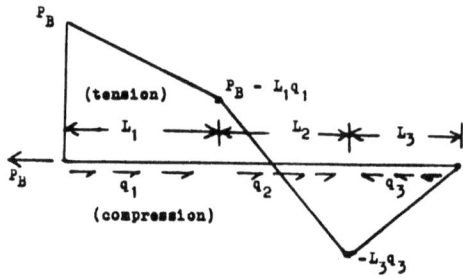

Fig. 6.2. Load diagram for stringer.

compression or both, the shear flows are usually calculated as reactions so that they can be used directly to draw the proper load diagrams for the stiffeners. Figure 6.2 shows a typical load diagram for a stringer or stiffener.

To calculate the shear flows in Figure 6.1, determine the reactions R_D, R_F, and R_E as usual from external equilibrium equations for the beam, summation of forces in the y and z-directions, and summation of moments about the x-axis. Assume the directions of the shear flows as reactions. If a shear flow calculates as plus, its direction is as assumed. If a shear flow calculates as negative, reverse its direction before using it to draw load diagrams. From equilibrium of stiffener (1) in Figure 6.1,

$$R_E - (h_1 + h_2)q_1 = 0, \qquad q_1 = R_E/(h_1 + h_2). \tag{6.1}$$

From equilibrium of stiffener (2),

$$(h_1 + h_2)q_1 - (h_1 + h_2)q_2 - P_C = 0,$$
$$q_2 = q_1 - (P_C)/(h_1 + h_2) = (R_E - P_C)/(h_1 + h_2). \tag{6.2}$$

From equilibrium of stringer (7),

$$-R_D - L_1 q_1 - L_2 q_2 - L_3 q_4 = 0,$$
$$q_4 = -(L_1/L_3)q_1 - (L_2/L_3)q_2 - (R_D/L_3). \tag{6.3}$$

From equilibrium of stiffener (6),

$$P_{Ah} + L_3 q_4 + L_3 q_3 = 0, \qquad q_3 = -q_4 - (P_{Ah}/L_3). \tag{6.4}$$

Note that one of the other stiffeners or stringers not used in the calculations can be used to check the results. Also, note that the usual shear and moment diagrams can be drawn for the beam. However, in most cases they are not needed

Shear lag in thin web structures

except the shear diagram may be useful for shear flow calculations. Also, both diagrams can be used for checking the shear flows and stringer loads at any cross section.

Example 6.1. In Figure 6.1, take $L_1 = L_3 = L_2 = 10$ in., $h_1 = h_2 = 5$ in., $P_B = 15,000$ lb, $P_C = 6000$ lb, $P_{Ah} = 20$ Klb, $P_{Av} = 4$ Klb, and find the shear flows in the beam webs. Draw load diagrams for all the stiffeners.

Solution. From summation of horizontal forces,

$$-R_D - 15 + 20 = 0, \quad R_D = 5 \text{ Klb} = 5000 \text{ lb}.$$

Summation of moments about support F gives

$$20R_E - (10)(15) - (10)(6) + (10)(4) + (5)(20) = 0, \quad R_E = 3.5 \text{ Klb}.$$

Summation of vertical forces gives

$$3.5 - 6 + R_F - 4 = 0, \quad R_F = 6.5 \text{ Klb}.$$

From Equations (6.1)-(6.4),

$$q_1 = 3500/10 = 350 \text{ lb/in}, \quad q_2 = 350 - (6000/10) = -250 \text{ lb/in},$$
$$q_4 = -350 + 250 - (5000/10) = -600 \text{ lb/in}, \quad q_3 = -600 + (20,000/10) = 1400 \text{ lb/in}.$$

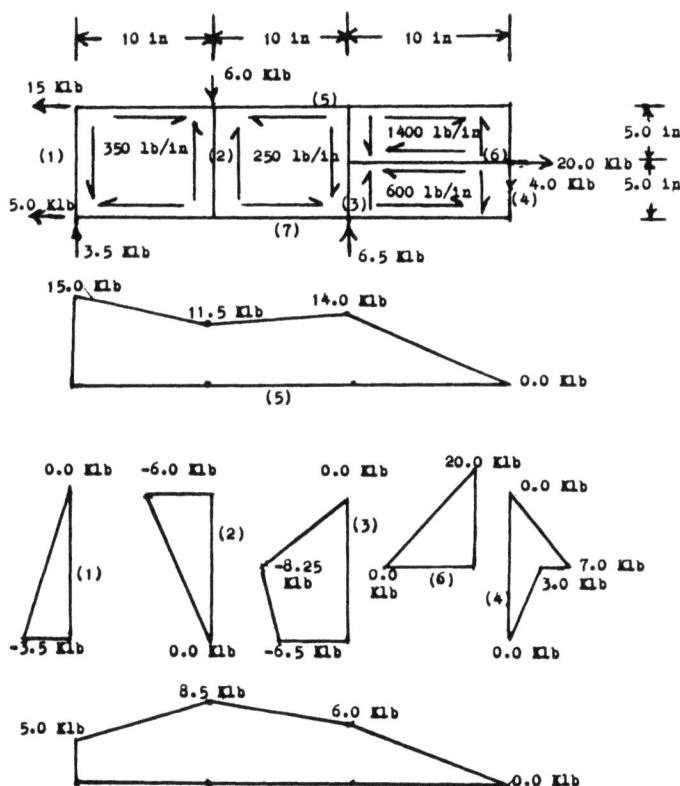

Fig. 6.3. Shear flows and load diagrams for Example 6.1.

The calculated shear flows with signs as reactions and the load diagrams are shown in Figure 6.3. The known end loads on all the stiffeners check when the load diagrams are drawn starting at one end and using the shear flows. Tension loads (positive sign) are up and to the right and compression loads (negative sign) are plotted down and to the left, Figure 6.3.

6.3. Shear flows around cut-outs in thin web beams

In order to have access to equipment and to run fuel, control, and electrical lines throughout the vehicle, it is necessary to cut holes in the thin webs of the box beam structure. These cut-outs must be framed by stiffeners and the shear forces redistributed to the adjacent webs. In adding the stiffeners to redistribute the shear forces, it is desirable to keep the beam determinate by using only the shear webs adjacent to the cut-out and assuming the rest of the structure to be unaffected. This is an approximation, but with ductile materials and a proper design of the stiffeners and adjacent shear webs, the structure tends to redistribute the loads as assumed in the analysis.

Consider the simple beam shown in Figure 6.4 with a cut-out for the fuel lines, where the reactions R_1, R_2, R_3 are the support points for the cantilever beam. The procedure for obtaining the shear flows and drawing the load diagrams is

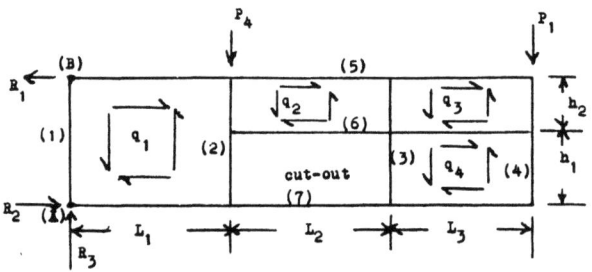

Fig. 6.4. Web cut-out in simple beam

the same as in the previous Section 6.2. Get the reactions and use equilibrium of selected stiffeners to get the shear flows.

Consider the case of a cut-out in the wing or fuselage skin which has stringers carrying the axial loads. The webs also have shear flows from torque and shear forces, Figure 6.5. By assuming that the redistribution of the loads around the cut-out is restricted to one shear bay on each of the four sides, it is necessary to add only the four stiffeners (1)-(4) and to make use of two stringers on each side with stringers (6) and (8) and stiffeners (2) and (3) framing the cut-out. In order to determine the shear flows around the cut-out in Figure 6.5, make a separate analysis for the cases of the P_i loads, the q_0 shear flow, and the Q shear force. For the P_i axial loads in the stringers assume that $R_i = P_i$ and that the P_3 load divides equally to the other four stringers (actually stringers (6) and (8) will take more of the load than stringers (5) and (9), depending upon areas, web

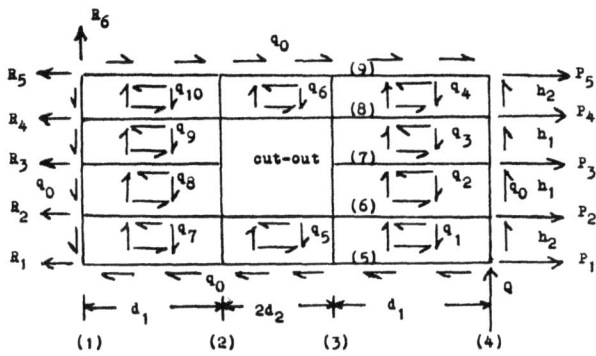

Fig. 6.5. Cut-out in skin-stringer panel.

thickness, etc.). With these assumptions, the shear flows due to the P_i loads alone are

$$q_5 = q_6 = 0, \quad q_2 = P_3/2d_1 = -q_3, \quad q_1 = P_3/4d_1 = -q_4,$$
$$q_7 = -q_1, \quad q_8 = -q_2, \quad q_9 = -q_3, \quad q_{10} = -q_4. \tag{6.5}$$

For the q_0 shear flow alone in Figure 6.5, assume symmetry so that $q_6 = q_5$, $q_3 = q_2 = q_8 = q_9$, $q_4 = q_1 = q_7 = q_{10}$. From shear equilibrium at the cut-out,

$$(2h_1 + 2h_2)q_0 - h_2(q_5 + q_6) = 0, \quad q_5 = q_6 = q_0(h_1 + h_2)/h_2. \tag{6.6}$$

From equilibrium of stringer (5),

$$-2(d_1 + d_2)q_0 + 2d_2 q_5 + 2d_1 q_1 = 0,$$
$$q_1 = -d_2 q_0 (h_1 + h_2)/d_1 h_2 + q_0(d_1 + d_2)/d_1 = [1 - (h_1 d_2/h_2 d_1)]q_0. \tag{6.7}$$

From equilibrium of stiffener (4),

$$2(h_1 + h_2)q_0 - 2h_1 q_2 - 2h_2 q_1 = 0, \quad q_2 = q_0(d_1 + d_2)/d_1. \tag{6.8}$$

The shear flows due to the shear force Q alone in Figure 6.5 depend upon how the moment produced by Q is taken by the reactions R_1 to R_5. If $R_2 = R_3 = R_4 = 0$ so that the moment is taken as a couple by R_1 and R_5, then the shear flows correspond to those for q_0 with

$$[Q/2(h_1 + h_2)] \text{ in place of } q_0. \tag{6.9}$$

The load diagrams for some of the stiffeners will be different from those for q_0.

The combination of all the applied loads in Figure 6.5 gives the reaction shear flows

$$q_1 = q_{10} = \frac{P_3}{4d_1} + \left(1 - \frac{h_1 d_2}{h_2 d_1}\right)\left[q_0 + \frac{Q}{2(h_1 + h_2)}\right],$$
$$q_2 = q_9 = \frac{P_3}{2d_1} + \left(1 + \frac{d_2}{d_1}\right)\left[q_0 + \frac{Q}{2(h_1 + h_2)}\right], \tag{6.10}$$
$$q_3 = q_8 = q_2 - (P_7/d_1), \quad q_4 = q_7 = q_1 - (P_7/2d_1),$$
$$q_5 = q_6 = \left(1 + \frac{h_1}{h_2}\right)q_0 + \frac{Q}{2h_2}.$$

Example 6.2. Find the shear flows in Figure 6.4 for $L_1 = L_2 = L_3 = 40$ in., $h_1 = 20$ in., $h_2 = 10$ in., $P_1 = 20$ Klb, $P_2 = 10$ Klb, $P_3 = 0$, $P_4 = 15$Klb.

Solution. Calculate the reaction R_1 by taking moments about support (A),

$$-30R_1 + 15(40) + 20(120) + 10(20) = 0, \quad R_1 = 106.7 \text{ Klb}.$$

From summation of horizontal and vertical forces,

$$R_2 = 96.7 \text{ Klb}, \quad R_3 = 35 \text{ Klb}.$$

From equilibrium of stiffeners (1), (2), (7), and (3),

$$35,000 - 30q_1 = 0, \quad q_1 = 1167 \text{ lb/in},$$
$$30(1167) - 15,000 - 10q_2 = 0, \quad q_2 = 2000 \text{ lb/in},$$
$$96,667 - 40(1167) - 0 - 40q_4 = 0, \quad q_4 = 1250 \text{ lb/in},$$
$$10(2000) - 10q_3 - 20(1250) = 0, \quad q_3 = -500 \text{ lb/in}.$$

6.4. Cut-outs in box beams

When any web of a box beam is cut out or omitted, the shear flows and stringer or spar cap loads are completely changed not only in the region of the cut-out but also on both sides of the cut-out. If the cut-out occurs between two ribs or bulkheads, then the effects extend at least to the next rib or bulkhead in both directions. The ribs or bulkheads must redistribute the shear flows around the cut-out in the same manner as the stringers and stiffeners do in the simple beam discussed above in Section 6.3.

Consider the case of one web out for the entire length of the box beam. This corresponds to the open box problem already considered in Section 5.4, provided the shear forces are applied at the shear center. However, if the forces are not at the shear center a torque M_y is produced. This torque can be taken only if there are sufficient stringers or spar caps present to provide two separate simple beams. These two beams can take the torque M_y as a couple by bending in opposite directions. From Figure 6.6

$$P_z = M_y/b, \tag{6.11}$$
$$q_1 = q_3 = M_y/bh, \tag{6.12}$$

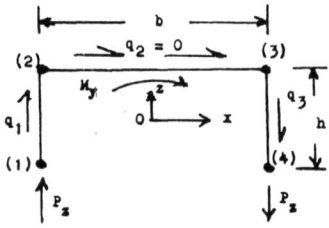

Fig. 6.6. Torque couple in open box.

Shear lag in thin web structures

and the moment in the beams is $M_x = P_z y$ from the free end of each beam. The additional cap loads are

$$P_1 = -P_2 = P_z y/h = P_3 = -P_4. \tag{6.13}$$

Thus, the shear flows in Equation (6.12) and the cap loads in Equation (6.13) combine with the shear flows and cap loads calculated for the open box in the usual manner.

Consider the case of one web cut between two ribs and assume that the effects of the cut-out extend equally to adjacent ribs on both sides, Figure 6.7. In bays $n-2$ and $n+2$ take the torque shear flow as

$$q_t = M_y/2bh, \tag{6.14}$$

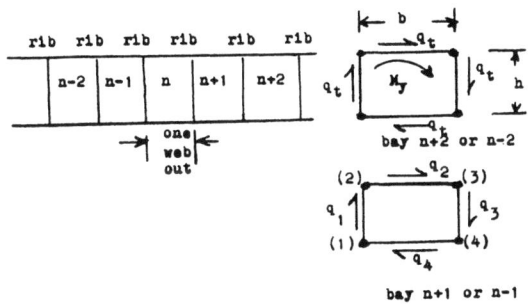

Fig. 6.7. Cut-out in constant wing box.

whence in bay n, with the cut-out of the bottom web, Equation (6.12) and Figure 6.6 give

$$q_1 = q_3 = 2q_t, \qquad q_2 = 0, \qquad \text{bay } n. \tag{6.15}$$

This $2q_t$ shear flow in the two simple beams (1)-(2) and (3)-(4) produces moments

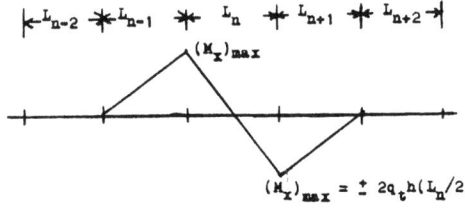

Fig. 6.8. Moment curve M_x in beam (1)-(2), or beam (3)-(4) with reversed signs, Fig. 6.7.

and flange loads, with the moment distribution in each beam taken as in Figure

6.8. In bay $n+1$, the shear flows can be determined from Figures 6.7 and 6.8 as

$$q_1 = q_3, \quad q_2 = q_4, \quad L_{n+1}(q_1 - q_2) = -L_n q_t,$$
$$q_1 bh + q_2 bh = M_y, \quad q_1 + q_2 = 2q_t, \quad \text{or}$$
$$q_1 = q_3 = \left(1 - \frac{L_n}{2L_{n+1}}\right) q_t, \quad q_2 = q_4 = \left(1 + \frac{L_n}{2L_{n+1}}\right) q_t. \tag{6.16}$$

Similarly, for bay $n - 1$,

$$q_1 = q_3 = \left(1 - \frac{L_n}{2L_{n-1}}\right) q_t, \quad q_2 = q_4 = \left(1 + \frac{L_n}{2L_{n-1}}\right) q_t. \tag{6.17}$$

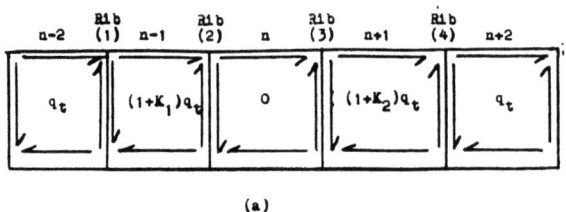

(a)

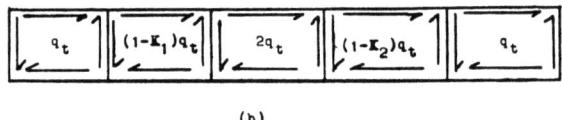

(b)

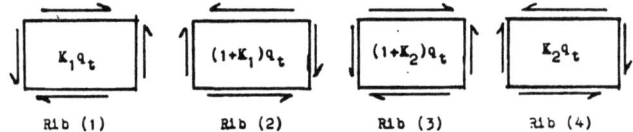

Fig. 6.9. Shear flows in box beam and ribs for cut-out in top or bottom web, (a) top and bottom, (b) front and back, $K_1 = L_n/2L_{n-1}$, $K_2 = L_n/2L_{n+1}$.

It is evident that the ribs must change the shear flows between the bays. All the shear flows in the webs and in the ribs are obtained from Figure 6.8 and Equations (6.15)-(6.17) and shown in Figure 6.9.

6.5. Shear flows in ribs and bulkheads

Ribs and bulkheads may have large shear flows when they must transmit concentrated forces (landing gear loads, fuel loads, thrust loads, etc.) into the wing or fuselage structure, or must redistribute shear forces and torque forces due to cut-outs in the structure. The rib or bulkhead is considered as a free body with the loads applied to it by the external forces and by the wing or fuselage skins as reactions. The rib or bulkhead should be regarded as a beam which is supported by the skins and webs of the main structure, which are attached to it. This means that the distribution of the shear flows around the edges of the rib or bulkhead is determined by the primary structure and not by the rib or bulkhead. In Figure 6.9 of Section 6.4, the shear flows in the ribs were determined directly from the shear flows in the skins and webs attached to the ribs.

Consider the rib in the single cell box beam in Figure 6.10. The shear flows $q_1, q_2, \cdots, q_8$ in the box are determined from the given forces P_x and P_z by the procedures given in Section 5.5. These shear flows are reversed in direction and

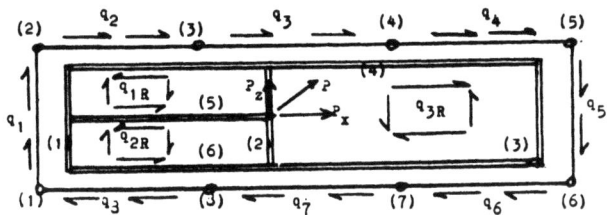

Fig. 6.10. Rib in single cell box beam (attachments of rib edge stiffeners not shown).

applied to the rib stiffeners (1), (3), (4), (6). The rib beam shear flows q_{1R}, q_{2R}, q_{3R} are then obtained from equilibrium of the rib stiffeners, as in Section 6.2. In Figure 6.10 stiffener (3) gives $q_{3R} = -q_5$, and stiffeners (4) and (1) can be used to get q_{1R} and q_{2R}.

Example 6.3. Find the rib shear flows in Figure 6.11, which is the rib that puts the shear forces into the single cell box in Example 5.4. Use the box shear flows of Figure 5.10 as shown in Figure 6.11. The forces have been located on the rib to give the proper moment M_y in Example 5.4.

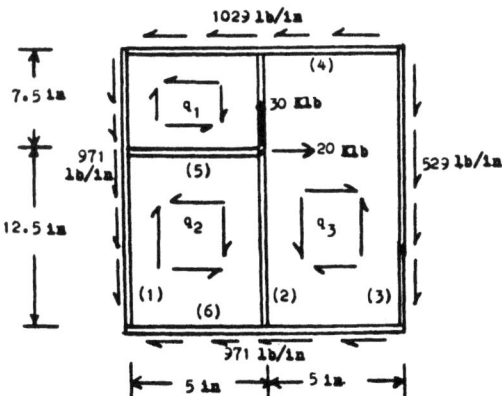

Fig. 6.11. Rib shear flows in single cell box beam.

Solution. From equilibrium of stiffeners (3), (4), and (1) in Figure 6.11

$$20q_3 - 20(529) = 0, \quad q_3 = 529 \text{ lb/in},$$
$$-10(1029) - 5q_1 + 5(529) = 0, \quad q_1 = -1529 \text{ lb/in},$$
$$-20(971) + 7.5(-1529) + 12.5q_2 = 0, \quad q_2 = 2471 \text{ lb/in}.$$

Check stringer (6), $-10(971)+5(2471)-5(529) = 0$.

Example 6.4. Find the shear flows in the bulkhead in Figure 6.12, which is transmitting the landing gear forces into the fuselage symmetrical box beam. Draw load diagram for stiffener (8).

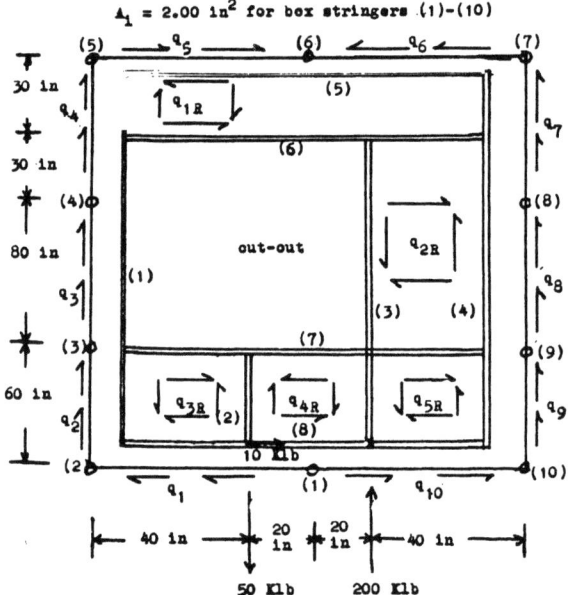

Fig. 6.12. Fuselage bulkhead shear flows (attachments of bulkhead stiffeners to box webs not shown)

Solution. First, the shear flows q_i in the box beam must be obtained for the given forces. Since the box has double symmetry, Equations (5.25), (5.35), and (5.37) can be used. Equation (5.25) gives

$$q_{i,\epsilon} = q_{i-1,\epsilon} - 2.00 \left(\frac{150,000 z_i}{132,800} + \frac{10,000 x_i}{57,600} \right)$$
$$= q_{i-1,\epsilon} - 2.2590 z_i - 0.3472 x_i.$$

From the symmetry, webs (3) and (8) have zero shear flows for the $V_x = 10,000$ lb force at the centroid, while the load per unit length in stringers (1) and (6) can be split, with one half going to each side, for the $V_z = 150,000$ lb force at the centroid. With the superposition of these shear flows in the closed box and with torsional moments about the centroid, where $M_{y,\epsilon}$ in Equation (5.37) is zero, it follows that

$$q_0 = \frac{-(100)(10,000) - (20)(50,000 + 200,000)}{(2)(200)(120)} = -125 \text{ lb/in}.$$

Table 6.1 shows the calculations for the box shear flows using the combination of the q_{ix}, q_{iz}, and q_0 shear flows as described above and using the direction of the shear flows shown in

Shear lag in thin web structures

Table 6.1. Calculations for Example 6.4.

	$-2.2590z_i$	q_{iz}	$-0.3472x_i$	q_{iz}	q_0	q_i (lb/in)
$S(1)$	226		0			
$W(1)$		113		-42	-125	-54
$S(2)$	226		21			
$W(2)$		339		-21	-125	193
$S(3)$	90		21			
$W(3)$		429		0	-125	304
$S(4)$	-90		21			
$W(4)$		339		21	-125	235
$S(5)$	-226		21			
$W(5)$		113		42	-125	30
$S(6)$	-226		0			
$W(6)$		113		-42	125	196
$S(7)$	-226		-21			
$W(7)$		339		-21	125	443
$S(8)$	-90		-21			
$W(8)$		429		0	125	554
$S(9)$	90		-21			
$W(9)$		339		21	125	485
$S(10)$	226		-21			
$W(10)$		113		42	125	280

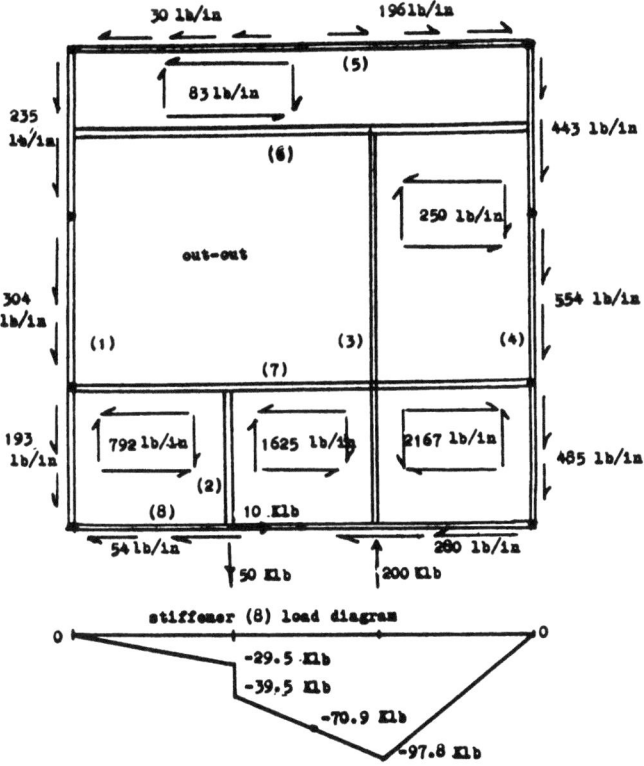

Fig. 6.13. Results for Example 6.4.

Figure 6.12.
The reversed shear flows from Table 6.1 for the box are shown on the free body of the

bulkhead in Figure 6.13. From equilibrium of bulkhead stiffener (5),

$$(60)(-30) + (60)(196) - 120q_{1R} = 0, \quad q_{1R} = 83 \text{ lb/in.}$$

From equilibrium of stiffener (6),

$$(120)(83) + (40)(q_{2R}) = 0, \quad q_{2R} = -250 \text{ lb/in.}$$

From equilibrium of stiffener (1),

$$-235(60) - 304(80) - 193(60) + 83(30) - 60q_{3R} = 0, \quad q_{3R} = -792 \text{ lb/in.}$$

From equilibrium of stiffener (2),

$$-60(792) - 50,000 + 60q_{4R} = 0, \quad q_{4R} = 1625 \text{ lb/in.}$$

From equilibrium of stiffener (3),

$$200,000 - 60(1625) + 110(250) - 60q_{5R} = 0, \quad q_{5R} = 2167 \text{ lb/in.}$$

The final bulkhead shear flows and the load diagram for stiffener (8) are shown in Figure 6.13.

6.6. Forces on ribs due to airloads and taper effects

Besides transmitting the concentrated forces discussed in the previous Section 6.5 into the wing box beam structure, the ribs must also support and transmit to the box beam all the forces applied to them by the stringers. Figure 6.14 shows component forces applied to the rib due to airloads, change in direction of the stringers due to change in taper, change in centroid of the stringer due to a splice or attachment, and change in direction of the stringer due to wing bending deflection.

The force applied to the rib by the airload on the stringer, Figure 6.14(a), depends upon the flight condition, the chordwise airload distribution on the upper and lower surfaces of the wing, and the chordwise location of the stringer. If p_u and p_L are the pressure differences on the upper and lower surfaces, d the rib spacing, and s the stringer spacing, Figure 6.14, then the forces R_u and R_L applied to the rib by one upper surface stringer and one lower surface stringer are

$$R_u = p_u(s)(d), \quad R_L = p_L(s)(d). \tag{6.18}$$

The component forces on the rib due to taper changes and centroid changes in the stringer, Figures 6.14(b) and 6.14(c), are

$$R_u = P_s \tan \alpha, \quad R_L = P_s \tan \beta, \tag{6.19}$$

where α and β are the angle changes in the direction of the stringer centroid, or the box taper. Note that such changes in angle in the surface planes of the wing will produce tangential force components in the rib, Figure 6.14(e).

Under the applied forces the wing has a bending deflection due to the rotation of the plane strain sections of the box beam. This rotation angle α, Figure 6.14(d), per unit length, is

$$\tan \alpha = \frac{e_{st}}{h/2}, \tag{6.20}$$

Shear lag in thin web structures

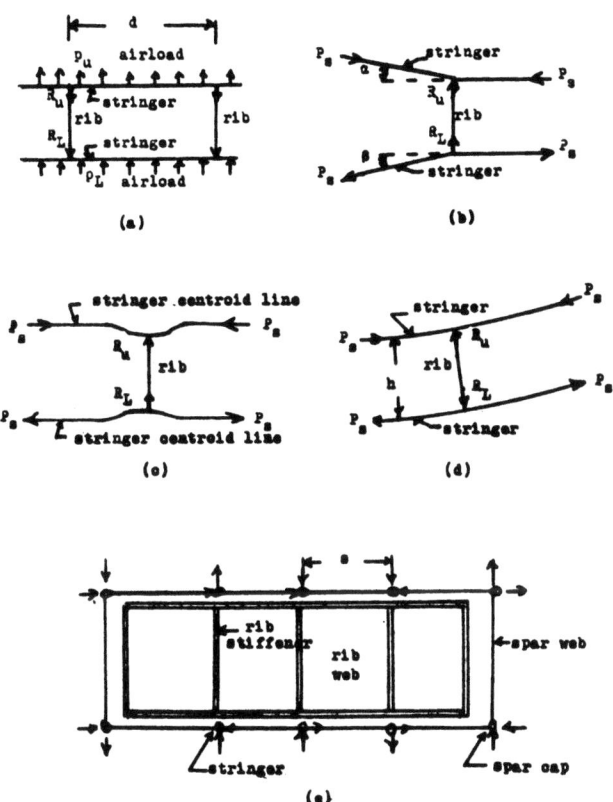

Fig. 6.14. Stringer component forces on ribs due to: (a) airloads, (b) change in taper, (c) change in stringer centroid line, (d) wing deflection, (e) rib with stiffeners to distribute loads to box beam.

where e_{st} is the stringer strain. The component force for the rib spacing d is

$$R_U = P_s d \tan \alpha = 2P_s e_{st} d/h = R_L, \tag{6.21}$$

for the symmetrical case. Note that these forces produce compression in the rib, tending to crush it. In some cases it may be necessary to use stiffeners on the rib to take the compression loads. With ductile materials, the strain is quite large as the cross section approaches failure. If $P_s = 20$ Klb, $e_{st} = 0.08$, $d = 80$ Klb, $h = 20$ in., then $R_u = 12.8$ Klb crushing force from one stringer. In a few cases, due to improper rib design, these forces have produced wing failures in static tests. Also, note that the crushing forces produced by the spar cap loads must be taken by the stiffeners on the spar webs.

The above described forces must be considered in the design of ribs for the control surfaces and for bulkheads in the fuselage.

The analysis of the rib for the forces described in this section is made in the same manner as in Section 6.5.

PART 2

SOLUTIONS FOR REDUNDANT CASES

6.7. Restraint effects in thin web structures

In the analysis in Chapter 5 and Part 1 of Chapter 6 sufficient approximations and assumptions have been used to make the structure determinate so that only the equilibrium equations were needed to solve for the stresses. As long as the assumptions are realistic and the materials are sufficiently ductile, these procedures give satisfactory results. That is, if the structure is properly designed by these analysis methods, it tends to work the way it is designed by yielding and making itself compatible with the stresses before failure. However, such a structural design based upon a determinate analysis may not be a least weight design. Also, such a design procedure usually cannot be used for materials with low ductility, such as composite materials. Furthermore, there are many redundant situations in the structure of an airplane that cannot be approximated by a determinate structure, such as complicated cut-outs, torsional restraints on the wing supports at the fuselage, more than two supports for the wing and control surfaces, swept back wing warped cross sections near the fuselage, buckling effects, closed ring bulkheads and ribs, shear lag and inelastic effects, joints and attachments in the structure, low aspect ratio wings and delta wings, short fuselages with many cut-outs.

As shown in Chapter 5, warping of thin web beams occurs from the shear stresses and this warping must be unrestrained for the determinate solution to apply. Obviously, the warping must be restrained at the beam supports, at the ends of the beam, and at the fuselage center line for a wing box beam.

In all these redundant problems the distribution of the stringer or stiffener loads into the thin shear webs is usually called the *shear lag problem*. In many cases this problem has been attacked by using differential equations for a continuously varying shear flow along the beam. However, as pointed out previously, the shear flow in thin webs tends to be constant between stiffeners and discontinuous at stiffeners. In the following sections this redundant shear lag problem will be attacked using the discontinuous shear flows together with the unit load theorem. Comparison of some of the results will be made to continous solutions given in References 1 and 2.

In previous chapters the temperature and inelastic effects on the beam plane section were calculated away from the ends of the beam. At the free end of the beam the thermal stresses are zero so that there is a shear lag problem for the thermal stresses. Also, there is a shear lag problem in the redistribution of the stringer loads in regions of inelastic strains in the stringers. Both of these problems will be considered in examples in the following sections.

6.8. Shear flows in redundant beams in one plane

See Figures 6.1, 6.3, 6.4, and 6.5 for simple determinate beams in one plane,

Shear lag in thin web structures

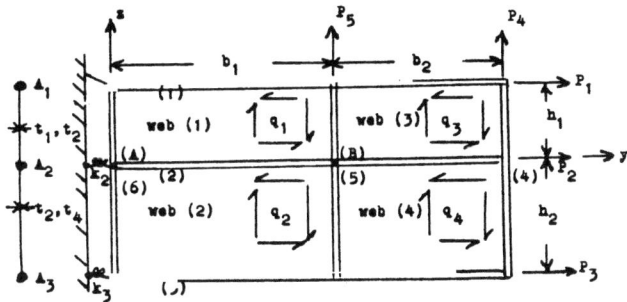

Fig. 6.15. Shear flows in redundant planar beam.

where the shear flows in the panels were calculated from the equilibrium equations. However, if the sum of the number of shear webs and number of supports exceeds the number of stiffeners, the structure is redundant. The equilibrium equations of the stiffeners can be used for the support reactions and to express all the shear flows in terms of unknown selected redundant shear flows, such as q_1 and q_3 in Figure 6.15. Then the load diagrams for all the stiffeners can be drawn in terms of the unknown shear flows and the applied loads. Thus, the strain distributions in the axially loaded members and in the shear webs can be expressed in terms of the unknown shear flows, whence the unit load theorem can be used. Cut a sufficient number of stiffeners to make the beam determinate, and apply unit loads at the cuts. Now the shear flows and the load diagrams can be calculated for the unit loads at each cut, which gives the necessary unit virtual stresses to use in the unit load theorem.

For the above described procedure on the planar beam, the unit load theorem in Equation (2.41) has the form (see Equation 2.63)

$$u_m = 0 = \int_0^L \int_A \{(e_{yy} + e_E)\sigma_{yym}^1 + (e_s + e_{Es})\sigma_{sm}^1\}dA\,dy$$

$$= \sum_{j=1}^{J} \int_0^{L_j} \int A_j \left(\frac{P_y}{EA} + e_E\right)_j \left(\frac{p_y^1}{A}\right)_{mj} dA_j\,dy_j$$

$$+ \sum_{k=1}^{N} \int_0^{L_k} \int_{A_k} \left(\frac{q}{Gt} + e_{Es}\right)_k \left(\frac{q^1}{t}\right)_{mk} dA_k\,dy_k$$

$$= \sum_{j=1}^{J} \int_0^{L_j} \left(\frac{P_y}{EA} + e_E\right)_j p_{ymj}^1\,dy_j + \sum_{k=1}^{N} \left(\frac{q}{Gt} + e_{Es}\right)_k q_{mk}^1 h_k b_k, \quad (6.22)$$

where $m = 1, 2, \cdots, M$ for M redundants, J is the number of stringers and stiffeners, and N is the number of shear webs. In Figure 6.15, $M = 2$, $J = 6$, $N = 4$.

To demonstrate the procedure, the shear flows in Figure 6.15 will be determined for the case in which each of the six stiffeners have a constant $(EA)_j$ and

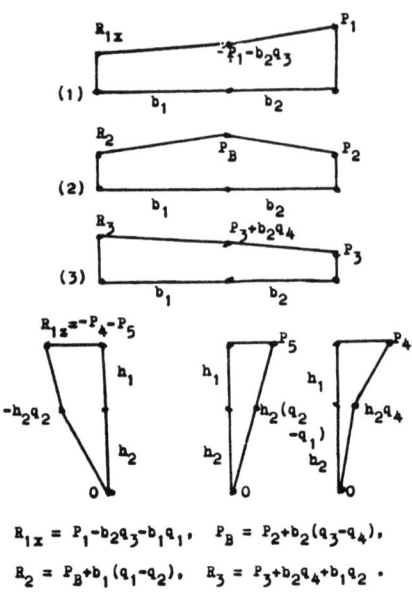

Fig. 6.16. Load diagrams for stiffeners in Fig. 6.15.

each web has a constant $(Gt)_k$. Also, the supports (2) and (3) will be assumed to be flexible with spring constants k_2 and k_3, respectively. From the shear force diagrams produced by the forces P_4 and P_5, equilibrium gives

$$P_4 = h_1 q_3 + h_2 q_4, \qquad P_4 + P_5 = h_1 q_1 + h_2 q_2,$$
$$q_4 = (P_4 - h_1 q_3)/h_2, \qquad q_2 = (P_4 + P_5 - h_1 q_1)/h_2, \quad (6.23)$$

where q_1 and q_3 are selected as the unknown shear flows to be determined from Equation (6.22). The next step is to draw the load diagrams for the six stiffeners, Figure 6.16. Because of the constant shear flow, all the diagrams have a linear variation of the load P_{yj} in stiffener j.

Cut stringer (2) at points (A) and (B) on Figure 6.15 and apply unit tension loads. The unit loads at (A) give the deflection $1/k_2$ for the spring at support (2), and the shear flows

$$q_{1A}^1 = (h_2/b_1)/(h_1 + h_2), \qquad q_{2A}^1 = -(h_1/b_1)/(h_1 + h_2),$$
$$q_{3A}^1 = 0, \qquad q_{4A}^1 = 0. \quad (6.24)$$

Figure 6.17 shows the load diagrams for this case of the unit loads at point (A).

For the unit loads at point (B) in stringer (2), the shear flows are

$$q_{1B}^1 = -(h_2/b_1)/(h_1 + h_2), \qquad q_{2B}^1 = (h_1/b_1)/(h_1 + h_2),$$
$$q_{3B}^1 = (h_2/b_2)/(h_1 + h_2), \qquad q_{4B}^1 = -(h_1/b_2)/(h_1 + h_2). \quad (6.25)$$

Figure 6.18 shows the load diagrams for this case of the unit loads at point (B).

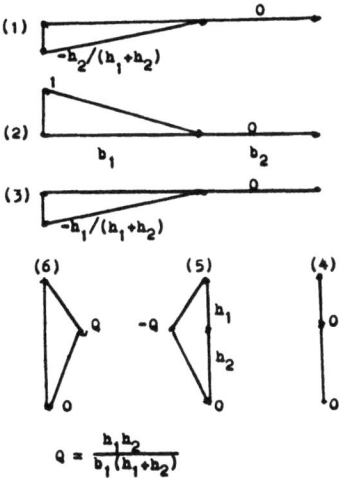

Fig. 6.17. Load diagrams for unit load at point (A) in Fig. 6.15.

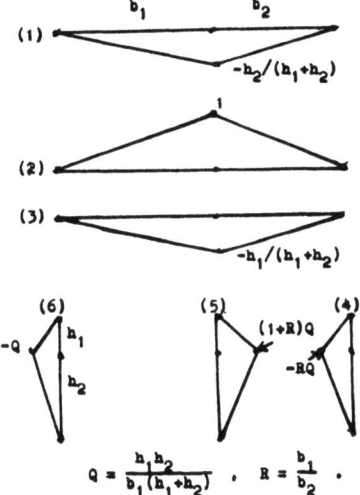

Fig. 6.18. Load diagrams for unit load at point (B) in Fig. 6.15.

The terms in Equation (6.22) can be calculated separately for each stiffener and each web for the cases of unit loads at (A) and (B) by using Equations (6.23)-(6.25) and Figures 6.16, 6.17, and 6.18. Since one end of the linear unit load diagrams is zero, the integrations can be made starting at the zero end, as indicated in Figure 6.19. From Equation (6.22) for a typical case with the

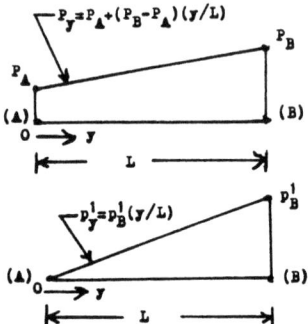

Fig. 6.19. Typical case for stiffener load diagrams.

notation in Figure 6.19.

$$\int_0^L (\frac{P_y}{EA} + e_E) p_y^1 \, dy = (1/EA) \int_0^L \{P_A + (P_B - P_A)(y/L) + e_E EA\} p_B^1 (\frac{y}{L}) dy$$

$$= \frac{p_B^1 L}{6EA}(P_A + 2P_B + 6S_E), \quad S_E = \frac{EA}{L^2} \int_0^L e_E y \, dy. \quad (6.26)$$

For example, take the case of stringer (3) in Figures 6.16 and 6.18 where in interval b_1, $P_A = R_3$, $P_B = P_3 + b_2 q_4$, $p_B^1 = -h_1/(h_1 + h_2)$, and in interval b_2, $P_A = P_3$, $P_B = P_3 + b_2 q_4$, $p_B^1 = -h_1/(h_1 + h_2)$, or from Equation (6.26)

$$\int_0^{b_1} (\frac{P_y}{EA} + e_E)_3 \, p_{yB3}^1 \, dy + \int_0^{b_2} (\frac{P_y}{EA} + e_E)_3 \, p_{yB3}^1 \, dy$$

$$= -\frac{h_1}{6E_3 A_3 (h_1 + h_2)} \{b_1 R_3 + 2b_1(P_3 + b_2 q_4) + 6b_1 S_{E3,1} + b_2 P_3 +$$

$$+ 2b_2(P_3 + b_2 q_4) + 6b_2 S_{E3,2}\}$$

$$= \frac{h_1 b_2}{6(h_1 + h_2) E_3 A_3} \{-3(b_1 + b_2)(P_3/b_1) - (b_1 + b_2)(b_1 + 2b_2)(P_4/b_1 h_2) -$$

$$- (b_1 P_5/h_2) + (b_2 h_1/h_2) q_1 + \frac{b_2 h_1}{b_1 h_2}(3b_1 + 2b_2) q_3 - 6S_{E3,1} - 6S_{E3,2}(\frac{b_2}{b_1})\},$$

where Equation (6.23) has been used.

All the stiffeners can be calculated in the same manner for the two cases and the results put into Equation (6.22) to get the two equations

$$A_{11} q_1 + A_{12} q_3 = B_{11} P_1 + B_{12} P_2 + B_{13} P_3 + B_{14} P_4 + B_{15} P_5 + u_{1E},$$
$$A_{21} q_1 + A_{22} q_3 = B_{21} P_1 + B_{22} P_2 + B_{23} P_3 + B_{24} P_4 + B_{25} P_5 + u_{2E}. \quad (6.27)$$

where

$$A_{11} = \frac{h_2 b_1^2}{3(h_1 + h_2) E_1 A_1} + \frac{(h_1 + h_2) b_1^2}{3 h_2 E_2 A_2} + \frac{(h_1 + h_2) b_1}{h_2 k_2} + \frac{h_1^2 b_1^2}{3 h_2 (h_1 + h_2) E_3 A_3} +$$

$$+ \frac{b_1 h_1^2}{h_2(h_1+h_2)k_3} + \frac{h_1^2 h_2}{3b_1 E_5 A_5} + \frac{h_1^2 h_2}{3b_1 E_6 A_6} + \frac{h_1 h_2}{(h_1+h_2)G_1 t_1} +$$
$$+ \frac{h_1^2}{(h_1+h_2)G_2 t_2}, \tag{6.28}$$

$$A_{12} = \frac{h_2 b_1 b_2}{2(h_1+h_2)E_1 A_1} + \frac{(h_1+h_2)b_1^2}{2h_2 E_2 A_2} + \frac{(h_1+h_2)b_2}{h_2 k_2} + \frac{h_1^2 b_1 b_2}{2h_2(h_1+h_2)E_3 A_3} +$$
$$+ \frac{b_2 h_1^2}{h_2(h_1+h_2)k_3} - \frac{h_1^2 h_2}{3b_1 E_5 A_5}, \tag{6.29}$$

$$A_{21} = \frac{h_2 b_1^2}{6(h_1+h_2)E_1 A_1} + \frac{(h_1+h_2)b_1^2}{6h_2 E_2 A_2} + \frac{h_1^2 b_1^2}{6h_2(h_1+h_2)E_3 A_3} -$$
$$- \frac{(b_1+b_2)h_1^2 h_2}{3b_1 b_2 E_5 A_5} - \frac{h_1^2 h_2}{3b_1 E_6 A_6} - \frac{h_1}{h_1+h_2}\left(\frac{h_2}{G_1 t_1} + \frac{h_1}{G_2 t_2}\right), \tag{6.30}$$

$$A_{22} = \frac{b_2(3b_1+2b_2)}{6}\left\{\frac{h_2}{(h_1+h_2)E_1 A_1} + \frac{h_1+h_2}{h_2 E_2 A_2} + \frac{h_1^2}{h_2(h_1+h_2)E_3 A_3}\right\} +$$
$$+ \frac{(b_1+b_2)h_1^2 h_2}{3b_2 b_1 E_5 A_5} + \frac{h_1^2 h_2}{3b_2 E_4 A_4} + \frac{h_1}{h_1+h_2}\left(\frac{h_2}{G_3 t_3} + \frac{h_1}{G_4 t_4}\right), \tag{6.31}$$

$$B_{11} = \frac{h_2 b_1}{2(h_1+h_2)E_1 A_1}, \quad B_{21} = \frac{h_2(b_1+b_2)}{2(h_1+h_2)E_1 A_1},$$
$$B_{12} = -\frac{b_1}{2E_2 A_2} - \frac{1}{k_2}, \quad B_{22} = -\frac{b_1+b_2}{2E_2 A_2}, \tag{6.32}$$
$$B_{13} = \frac{h_1}{h_1+h_2}\left(\frac{b_1}{2E_3 A_3} + \frac{1}{k_3}\right), \quad B_{23} = \frac{h_1(b_1+b_2)}{2(h_1+h_2)E_3 A_3},$$

$$B_{14} = \frac{b_1(3b_2+2b_1)}{6h_2}\left\{\frac{1}{E_2 A_2} + \frac{h_1}{(h_1+h_2)E_3 A_3}\right\} + \frac{b_2+b_1}{h_2}\left\{\frac{1}{k_2} + \frac{h_1}{(h_1+h_2)k_3}\right\} +$$
$$+ \frac{h_1 h_2(3h_1+2h_2)}{6b_1(h_1+h_2)E_6 A_6} + \frac{h_1}{(h_1+h_2)G_2 t_2}, \tag{6.33}$$

$$B_{24} = \frac{(b_1+b_2)(b_1+2b_2)}{6h_2}\left\{\frac{1}{E_2 A_2} + \frac{h_1}{(h_1+h_2)E_3 A_3}\right\} - \frac{h_1}{h_1+h_2}\left(\frac{1}{G_2 t_2} - \frac{1}{G_4 t_4}\right) +$$
$$+ \frac{h_1 h_2(3h_1+2h_2)}{6b_2(h_1+h_2)}\left(\frac{1}{E_4 A_4} - \frac{1}{E_6 A_6}\right), \tag{6.34}$$

$$B_{15} = \frac{b_1^2}{3h_2 E_2 A_2} + \frac{b_1}{h_2 k_2} + \frac{h_1 b_1}{h_2(h_1+h_2)}\left(\frac{b_1}{3E_3 A_3} + \frac{1}{k_3}\right) +$$
$$+ \frac{h_1 h_2(3h_1+2h_2)}{6b_1(h_1+h_2)}\left(\frac{1}{E_5 A_5} + \frac{1}{E_6 A_6}\right) + \frac{h_1}{(h_1+h_2)G_2 t_2}, \tag{6.35}$$

$$B_{25} = \frac{b_1^2}{6h_2 E_2 A_2} + \frac{h_1 b_1^2}{6h_2(h_1+h_2)E_3 A_3} - \frac{h_1}{(h_1+h_2)G_2 t_2} -$$
$$- \frac{h_1 h_2(3h_1+2h_2)}{6b_1 b_2(h_1+h_2)}\left(\frac{b_1+b_2}{E_5 A_5} + \frac{b_2}{E_6 A_6}\right), \tag{6.36}$$

$$u_{1E} = \frac{h_2 b_1}{h_1+h_2}\left[\frac{(S_{E1,1})_1}{E_1 A_1} - \frac{h_1+h_2}{h_1}\frac{(S_{E2,1})_1}{E_2 A_2} + \frac{h_1}{h_2}\frac{(S_{E3,1})_1}{E_3 A_3} +\right.$$

$$+ \frac{h_1^2 S_{E5,1} + h_1 h_2 S_{E5,2}}{b_1^2 E_5 A_5} - \frac{h_1^2 S_{E6,1} + h_1 h_2 S_{E6,2}}{b_1^2 E_6 A_6} -$$

$$- \frac{h_1}{b_1}\left(\frac{q_{E1}}{G_1 t_1} - \frac{q_{E2}}{G_2 t_2}\right)\bigg], \tag{6.37}$$

$$u_{2E} = \frac{h_2 b_1}{h_1 + h_2}\bigg[\frac{b_1(S_{E1,1})_2 + b_2(S_{E1,2})_2}{b_1 E_1 A_1} - \frac{h_1 + h_2}{h_2}\frac{b_1(S_{E2,1})_2 + b_2(S_{E2,2})_2}{b_1 E_2 A_2} +$$

$$+ \frac{h_1}{h_2}\frac{b_1(S_{E3,1})_2 + b_2(S_{E3,2})_2}{b_1 E_3 A_3} + \frac{h_1^2 S_{E4,1} + h_1 h_2 S_{E4,2}}{b_1 b_2 E_4 A_4} -$$

$$- \frac{b_1 + b_2}{b_2}\frac{h_1^2 S_{E5,1} + h_1 h_2 S_{E5,2}}{b_1 b_1 E_5 A_5} + \frac{h_1^2 S_{E6,1} + h_1 h_2 S_{E6,2}}{b_1 b_1 E_6 A_6} -$$

$$- \frac{h_1}{b_1}\left(-\frac{q_{E1}}{G_1 t_1} + \frac{q_{E2}}{G_2 t_2} + \frac{q_{E3}}{G_3 t_3} - \frac{q_{E4}}{G_4 t_4}\right)\bigg]. \tag{6.38}$$

Although Equation (6.27) can be solved for q_1 and q_3 in terms of the complicated constants, it is simpler to start from Equation (6.27) for various special cases. From the Equations (6.27)-(6.38) it is evident that the stiffness $E_j A_j$ of every stiffener and the stiffness $G_k t_k h_k$ of every web affects the final shear flows in the redundant beam. Also, the stiffness spring constants k_2 and k_3 for the supports, the temperature, and the inelastic effects affect the shear flows. Some of these effects are considered in the following examples for various special cases of the beam in Figure 6.15.

Example 6.5. In Figure 6.15 assume web (3) is cut out and find the shear flows and discuss the results for (a) applied load P_1 acting alone, (b) applied load P_4 acting alone.

Solution (a) To remove web (3) take $G_3 t_3 = 0$ in Equations (6.27)-(6.38), whence $q_3 = 0$ and the second equation of Equation (6.27) is not present. Thus,

$$q_1 = B_{11} P_1 / A_{11} = 1.50(P_1/b_1)/D_{11},$$
$$q_2 = -(h_1/h_2)q_1, \quad q_3 = 0, \quad q_4 = 0, \tag{6.39}$$

$$D_{11} = 1 + \frac{(h_1 + h_2)^2}{h_2^2}\left(\frac{E_1 A_1}{E_2 A_2} + \frac{3 E_1 A_1}{b_1 k_2}\right) + \frac{h_1^2}{h_2^2}\left(\frac{E_1 A_1}{E_3 A_3} + \frac{3 E_1 A_1}{b_1 k_3}\right) +$$

$$+ \frac{h_1^2(h_1 + h_2)}{b_1^3}\left(\frac{E_1 A_1}{E_5 A_5} + \frac{E_1 A_1}{E_6 A_6}\right) + \frac{3 h_1^2}{b_1^2}\left(\frac{E_1 A_1}{G_1 t_1 h_1} + \frac{E_1 A_1}{G_2 t_2 h_2}\right). \tag{6.40}$$

The results for this simple axial load case demonstrates the shear lag problem directly. Although the load P_1 is in stringer (1), which goes directly to the support (1), the load may transfer to the other supports by shear depending upon the $E_j A_j$ and $G_k t_k h_k$ stiffnesses. If $E_1 A_1$ is small as compared to $E_2 A_2$, $E_3 A_3$, $E_5 A_5$, $E_6 A_6$, $G_1 t_1 h_1$, $G_2 t_2 h_2$, k_2, and k_3, then q_1 can be large. However, if $E_1 A_1$ is large as compared to any one of the element stiffnesses, q_1 will be small. If all the stiffnesses are approximately the same, q_1 is small and very little load transfers to the other supports. If $h_1 = h_2$ is small compared to b_1, then the transverse stiffeners (5) and (6), the webs, and the spring supports will have less effect than stringers (2) and (3). On the other hand, if b_1 is small as compared to $h_1 = h_2$, q_1 will be very small with the P_1 load going to support (1), as might be expected in a short beam.

(b) For the shear force P_4 acting alone on the beam,

$$q_3 = 0, \quad q_4 = P_4/h_2, \quad q_2 = (P_4/h_2) - (h_1/h_2)q_1,$$
$$q_1 = B_{14} P_4 / A_{11} = (P_4/2h_2)(C_{14}/D_{11}), \tag{6.41}$$

$$C_{14} = \frac{(3b_2 + 2b_1)(h_1 + h_2)}{h_2 b_1}\left(\frac{E_1 A_1}{E_2 A_2} + \frac{h_1}{h_1 + h_2}\frac{E_1 A_1}{E_3 A_3}\right) +$$

$$+ \frac{6(b_1+b_2)(h_1+h_2)}{h_2 b_1}\left(\frac{E_1 A_1}{b_1 k_2} + \frac{h_1}{h_1+h_2}\frac{E_1 A_1}{b_1 k_3}\right)+$$

$$+ \frac{h_1 h_2 (3h_1+2h_2)}{b_1^3}\frac{E_1 A_1}{E_6 A_6} + \frac{6 h_1 h_2}{b_1^2}\frac{E_1 A_1}{G_2 t_2 h_2}, \tag{6.42}$$

and D_{11} is in Equation (6.40). In this case q_1 is small if $E_1 A_1$, or $G_1 t_1 h_1$, or $E_5 A_5$ is small compared to the other stiffnesses, and web (2) takes the load P_4. However, since $E_2 A_2$, $E_3 A_3$, $b_1 k_2$, $b_1 k_3$, $E_6 A_6$, and $G_2 t_2 h_2$ appear in both C_{14} and D_{11}, the division of the load P_4 to q_1 and q_2 must be examined for each of these elements individually. If $G_2 t_2$ is very small, then in the limit, $q_1 = P_4/h_1$, $q_2 = 0$. If support (2) is removed by making $k_2 = 0$, then $q_1 = (P_4/b_1)(b_1+b_2)/(h_1+h_2)$ and q_2 may be positive, zero, or negative depending upon the values of b_1, b_2, h_1, and h_2.

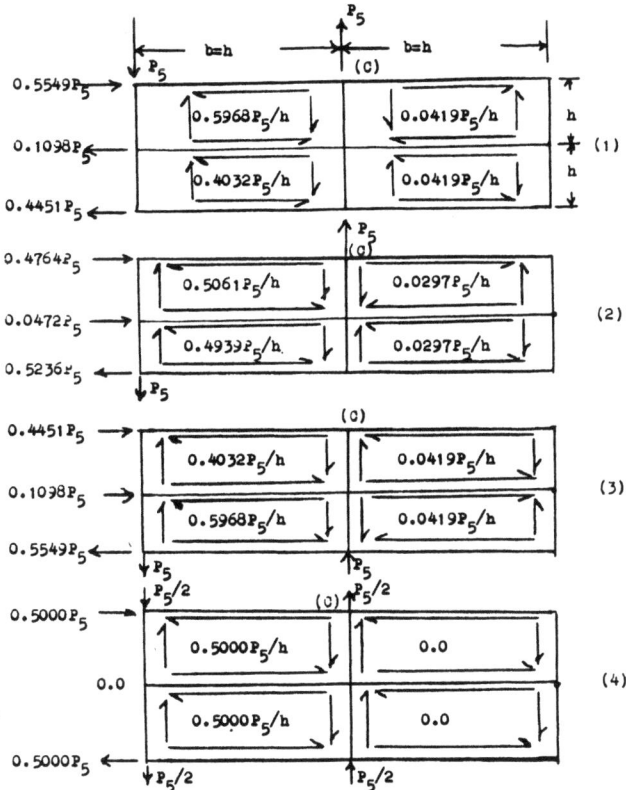

Fig. 6.20. Shear flows in redundant beam with loads applied at different locations on transverse stiffeners.

Example 6.6. (a) In Figure 6.15, assume the supports (2) and (3) are fixed by taking $k_2 = k_3 = $ large, all stiffeners have $E_j A_j = EA$, all webs have $G_k t_k h_k = Gth$, $h_1 = h_2 = h$, $b_1 = b_2 = b$, and find the shear flows for the load P_5 acting alone. (b) Discuss the effects of applying the load P_5 in different ways to stiffener (5) and of taking the load out at different support points.

Solution (a) From Equations (6.28)-(6.36), the required constants are

$$\begin{aligned}
EAA_{11}/b^2 &= H + 1 + \tfrac{2}{3}d^3, \quad H = EAh/Gtb^2, \quad d = h/b,\\
EAA_{12}/b^2 &= \tfrac{3}{2} - \tfrac{1}{3}d^3, \quad EAA_{21}/b^2 = -H + \tfrac{1}{2} - d^3,\\
EAA_{22}/b^2 &= H + \tfrac{5}{2} + d^3, \quad EAB_{15}/b^2 = (H + \tfrac{1}{2} + \tfrac{5}{3}d^3)/2h,\\
EAB_{25}/b^2 &= (-H + \tfrac{1}{2} - \tfrac{5}{3}d^3)/2h.
\end{aligned} \tag{6.43}$$

Put these terms into Equation (6.27) and solve for q_1 and q_3 to get

$$\frac{2hq_1}{P_5} = \frac{12H^2 + 60H + 21 + 28Hd^3 + 109d^3 + 10d^6}{12H^2 + 60H + 21 + 16Hd^3 + 52d^3 + 4d^6},$$

$$\frac{2hq_3}{P_5} = 2\left[\frac{3h + 3 + 5d^3 - (3H + 3 + 2d^3)(2hq_1/P_5)}{9 - 2d^3}\right], \quad (6.44)$$

$$q_4 = -q_3, \quad q_2 = (P_5/h) - q_1,$$

where Equation (6.23) has been used. To get an idea of the distribution of the shear flows in the webs, take $d = 1$ for square panels and $H = 4$, whence

$$q_1 = 0.5968(P_5/h), \quad q_2 = 0.4032(P_5/h),$$
$$q_3 = -0.0419(P_5/h), \quad q_4 = 0.0419(P_5/h).$$

Although this is a cantilever beam, the shear lag gives non-zero values for q_3 and q_4 outboard of the load P_5.

(b) Figure 6.20 shows the shear flows and support reactions for several cases of applying the P_5 load at different locations on stiffener (5) and taking the load out at different supports. The numbers used are for square panels, $b = h$, and $H = EAh/Gtb^2 = 4$. It is evident in Figure 6.20 that the transverse stiffeners have considerable effect on the shear flows, so that in the redundant beam the results depend upon where the load is applied on the stiffener and how the supports take the shear load. In the case of the stiffnesses in this example the shear flow is larger when the load is supported on the same side of the beam where it is applied. Note that if the transverse stiffeners are assumed to have infinite stiffness, the d^3 terms in Equations (4.43) and (6.44) are zero and all the cases in Figure 6.20 are the same with $q_1 = q_2 = 0.5000(P_5/h)$, $q_3 = q_4 = 0$.

Example 6.7. Consider the particular case of Figure 6.15 with symmetrical axial loads as shown in Figure 6.21. Take $P_4 = 0$, $P_5 = 0$, $P_2 = 0$, $P_3 = P_1$, $k_2 = k_3 =$large, $E_4A_4 = E_5A_5 = E_6A_6 =$ large, $E_3A_3 = E_1A_1$, $E_3 = E_2 = E_1 = E$, $h_1 = h_2 = h$, $L = b_1 + b_2$, $G_k t_k = Gt$ for all webs, which corresponds to the shear lag problem discussed in Chapter 4 of Reference 1. Find the shear flows q_1 and q_3 in Figure 6.21 and compare them as a two step approximation to the continuous shear flow q given in Reference 1.

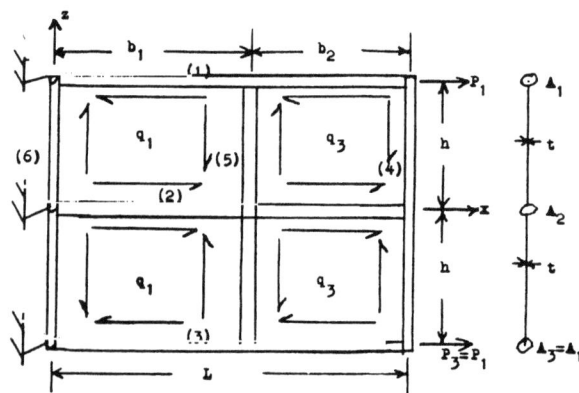

Fig. 6.21. Two step shear lag problem in symmetrical panel.

Solution. In Reference 1, the continuous solution for the shear flow q is obtained by assuming the web to have infinite transverse stiffness with stiffeners (4), (5) and (6) omitted in Figure 6.21 so that q is constant in the z-direction but varies continuously in the x-direction to $q = 0$ at the supports. The solution for q in Reference 1 is

$$q = \frac{FP_1}{L}\left(\frac{A_2}{2A_1 + A_2}\right)\frac{\sinh F(x/L)}{\cosh F}, \quad F^2 = \frac{Gt}{Eh}\left(\frac{L^2}{A_1}\right)\left(\frac{2A_1 + A_2}{A_2}\right), \quad (6.45)$$

Shear lag in thin web structures

where $L = b_1 + b_2$. For convenience in the comparisons, the constants in Equation (6.27) will be expressed in terms of the parameter F. From Equations (6.28)-(6.32),

$$\frac{Gt}{h}A_{11} = 1 + \left(\frac{b_1}{L}\right)^2 \frac{F^2}{3}, \quad \frac{Gt}{h}A_{12} = \left(\frac{b_1}{L}\right)\left(\frac{b_2}{L}\right)\frac{F^2}{2}, \tag{6.46}$$

$$\frac{Gt}{h}A_{21} = -1 + \left(\frac{b_1}{L}\right)^2 \frac{F^2}{6}, \quad \frac{Gt}{h}A_{22} = 1 + \left(2 + 3\frac{b_1}{b_2}\right)\left(\frac{b_2}{L}\right)^2 \frac{F^2}{6}, \tag{6.47}$$

$$\frac{Gt}{h}(B_{11} + B_{13}) = \frac{b_1}{2L^2}\left(\frac{A_2}{2A_1 + A_2}\right)F^2, \tag{6.48}$$

$$\frac{Gt}{h}(B_{21} + B_{23}) = \frac{1}{2L}\left(\frac{A_2}{2A_1 + A_2}\right)F^2, \tag{6.49}$$

whence Equation (6.27) can be solved for q_1 and q_2. Use a non-dimensional form for q in Equation (6.45) and for the q_1 and q_3 results, or

$$\begin{aligned}
Q &= \frac{L}{FP_1}\left(\frac{2A_1+A_2}{A_2}\right)q = \frac{\sinh(Fx/L)}{\cosh F}, \\
Q_1 &= \frac{L}{FP_1}\left(\frac{2A_1+A_2}{A_2}\right)q_1 = \frac{3F}{D}\left(\frac{b_1}{L}\right)\left\{6 - \left(\frac{b_2}{L}\right)^2 F^2\right\}, \\
Q_3 &= \frac{L}{FP_1}\left(\frac{2A_1+A_2}{A_2}\right)q_3 = \frac{3F}{D}\left\{6\left(1 + \frac{b_1}{L}\right) + \left(1 + \frac{b_2}{L}\right)\left(\frac{b_1}{L}\right)^2 F^2\right\}, \\
D &= 36 + 12\left(1 + \frac{b_1}{L}\frac{b_2}{L}\right)F^2 + \left(4 - \frac{b_1}{L}\right)\left(\frac{b_1}{L}\right)^2\left(\frac{b_2}{L}\right)F^4.
\end{aligned} \tag{6.50}$$

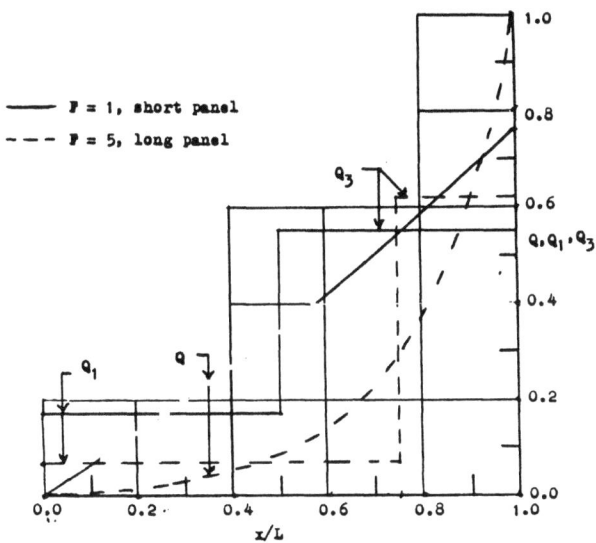

Fig. 6.22. Shear lag results for Example 6.7.

Figure 6.22 shows the graphs of Equation (6.50) for the two cases

$$\begin{aligned}
F &= 1, \quad b_1/L = 0.50, \quad b_2/L = 0.50, \quad Q_1 = 0.17, \quad Q_3 = 0.55, \\
F &= 5, \quad b_1/L = 0.75, \quad b_2/L = 0.25, \quad Q_1 = 0.07, \quad Q_3 = 0.62.
\end{aligned} \tag{6.51}$$

The two step values of Q_1 and Q_3 give a good average of the continuous curve for Q in each interval. In the actual case of a thin web with buckling, these constant values between the transverse stiffeners represent the shear flows better than the continuous values. Also, the transverse stiffeners with finite stiffness and the webs with variable web thickness can be included in the calculation of Q_1 and Q_3 to give more realistic results for the shear flows.

This Section 6.8 has been restricted to derivation of the shear flow equations for the general case of Figure 6.15 and to examples of particular cases of Figure 6.15. The procedure can be extended to larger beams with more webs and stiffeners. However, the calculations become more complicated so that it is simpler to set up the procedure in terms of matrices. The webs and stiffeners can be treated as finite elements with each stiffener in Figure 6.15 having two elements so that Figure 6.15 would include twelve axially loaded elements and four constant shear flow web elements. This matrix procedure using finite elements will be used in Sections 6.10 through 6.13 below.

6.9. Deflections of thin web structures

In the previous Section 6.8 it was found that all the stiffeners and all the webs affected the shear stresses and stiffener stresses. Thus, the deflections will vary throughout the beam structure. Since the beam cross sections are no longer plane for either bending or shear, no separation of bending and shear deflections can be made. However, the total deflections at any junction point of the stiffeners can be calculated by using the unit load theorem.

Since all the stresses obtained by the procedures in Section 6.8 are compatible and in equilibrium, and include all warping effects of the supports and web redundants, the unit load theorem can be used for any determinate substructure in equilibrium with the supports to get the deflections. This substructure will usually include one or more shear webs and the stiffeners framing the webs. In some cases only one stringer is involved in the deflection. For example, to get the vertical u_z deflection at point (B) in Figure 6.15, use web (1) and the four stiffener sections framing web (1). To get the horizontal u_y deflection at point (B), use stringer (2) between points (A) and (B) and add the support deflections due to the spring k_2.

Modify Equation (6.26) and Figure 6.19 to include p_A^1 in the unit load equation, or

$$p_y^1 = p_A^1 + (p_B^1 - p_A^1)(\frac{y}{L}). \tag{6.52}$$

Thus, Equation (6.26) becomes

$$\int_0^L (\frac{P_y}{EA} + e_E)p_y^1 \, dy = (p_A^1 L/6EA)(P_B + 2P_A + 6S_{EA} + $$
$$+ (p_B^1 L/6EA)(P_A + 2P_B + 6S_{EB}), \tag{6.53}$$

$$S_{EA} = (EA/L^2)\int_0^L e_E(L-y)dy, \quad S_{EB} = (EA/L^2)\int_0^L e_E y \, dy. \tag{6.54}$$

From Equations (6.22) and (6.53) the deflection at a junction point m in a

Shear lag in thin web structures 227

direction m is

$$u_m = \sum_{i=1}^{J}(L/6EA)_i\left[p^1_{Am}(P_B+2P_A+6S_{EA})+p^1_{Bm}(P_A+2P_B+6S_{EB})\right]_i+$$

$$+\sum_{k=1}^{K}\left[(q/Gt)+e_{Es}\right]_k b_k h_k q^1_{mk}, \qquad (6.55)$$

where the sums are over the selected determinate substructure and p^1_{Am}, p^1_{Bm} are produced by the unit load at m.

Since Equation (6.55) gives point deflections which include the effects of both the stiffeners and shear webs, it applies to the determinate thin web beams in Sections 6.2 and 6.3. The procedure also applies to point deflections on box beams.

To determine the inelastic terms in e_E in Equation (6.54), use Equation (3.27) for S_p, or

$$EAe_p = S_p = \tfrac{3}{7}S_y\left(\tfrac{P_A+(P_B-P_A)y_r}{S_y}\right)^{10}, \quad y_r = \tfrac{y}{L}. \qquad (6.56a)$$

Put into Equation (6.54) to get

$$S_{pB} = \frac{3S_y}{847}\left(\frac{S_y}{P_B-P_A}\right)^2\left[11\left\{\left(\frac{P_B}{S_y}\right)^{12}-\left(\frac{P_A}{S_y}\right)^{12}\right\}-\right.$$
$$\left.-12\left(\frac{P_A}{S_y}\right)\left\{\left(\frac{P_B}{S_y}\right)^{11}-\left(\frac{P_A}{S_y}\right)^{11}\right\}\right],$$

$$S_{pA} = \frac{3S_y}{847}\left(\frac{S_y}{P_B-P_A}\right)^2\left[-11\left\{\left(\frac{P_B}{S_y}\right)^{12}-\left(\frac{P_A}{S_y}\right)^{12}\right\}+\right.$$
$$\left.+12\left(\frac{P_B}{S_y}\right)\left\{\left(\frac{P_B}{S_y}\right)^{11}-\left(\frac{P_A}{S_y}\right)^{11}\right\}\right]. \qquad (6.56b)$$

Example 6.8. Take $E_iA_i = EA$, $G_it_i = Gt$ for all the stiffeners and webs in Example 6.1 and Figures 6.1, 6.3. Assume stiffener (2) in Figure 6.3 to have a yield load $S_{y2} = -4.0$ Klb and stiffener (7) to have an apparent thermal load of $S_{T7} = 6.0$ Klb. Find the u_{x2} deflection at the upper end of stiffener (2).

Solution. Use the subbeam shown in Figure 6.23 with the unit load applied. From Equations (6.54) and (6.56) $S_{T7A} = S_{T2}/2 = 3$ Klb, $S_{T7B} = 3.0$ Klb, $S_{p2A} = -0.817$ Klb, $S_{p2B} = -8.987$ Klb. Put these values and the values from Figures 6.3 and 6.23 into Equation (6.55) to get

$$u_m = (10/6EA)[(-0.5)(2)(-3.5)+(-1)\{2(-6.0)+6(-0.817)+6(2)(-8.987)\}+$$
$$+(5/10)\{(-0.25)(2)(-8.25)+(-0.25)(-6.5-16.5)+$$
$$+(-0.5)(-8.25-13.00)\}+(-0.5)\{(14.0+23.0)+(15.0+23.0)\}+$$
$$+(0.5)\{(6.0+17.0)+(5.0+17.0)+(2)(6)(3+6)\}]+$$
$$+(10/Gt)[(0.050)(0.350+0.250)]$$
$$= \frac{295.83}{EA}+\frac{0.300}{Gt} \text{ (in)},$$

where EA has units of Klb and Gt of Klb/in. The largest term in the $295.83/EA$ deflection is the inelastic effect of $187.91/EA$ in stiffener (2). The temperature in stiffener (7) produces $90/EA$ deflection. On the other hand, the tension load in stiffener (5) reduces the deflection by $-62.5/EA$.

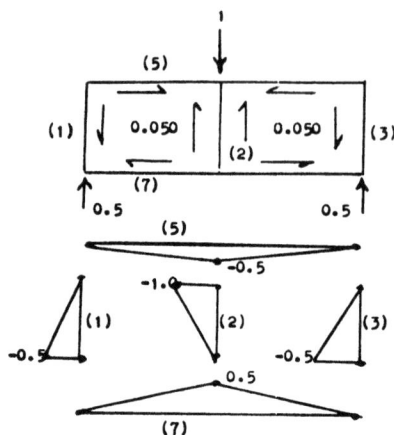

Fig. 6.23. Unit loads in subbeam for Example 6.8.

Example 6.9. Compare the deflections of the cantilever beam in Figure 6.24 at the load P_1 for the cases with and without the cut-out. Use $E_i A_i = EA$, $G_i t_i = Gt$.

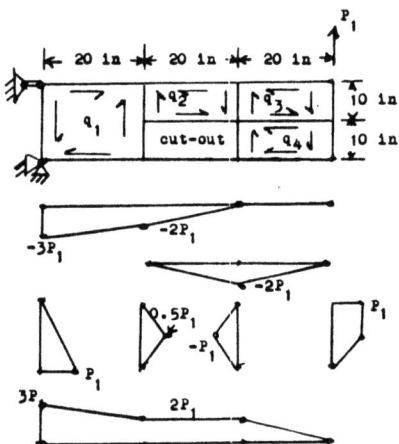

Fig. 6.24. Beam and load diagrams for Example 6.9.

Solution. For no cut-out, the middle stringer has no load and the beam has a plane section with $I = 2A(10)^2$, whence

$$u_{P1} = \frac{P_1 L^3}{3EI} + \frac{Lq}{Gt} = \frac{360 P_1}{EA} + \frac{3P_1}{Gt}.$$

Actually, the vertical stiffeners would add $13.33 P_1/EA$ to this deflection.

With the cut-out, the shear flows are $q_1 = P_1/20$, $q_2 = P_1/10$, $q_3 = 0$, $q_4 = P_1/10$. The stiffener load diagrams are shown in Figure 6.24. The unit load diagrams are the same with $P_1 = 1$. From Equation (6.55)

$$u_{P1} = (P_1/6EA)[(20)(-2)(2)(-2) + 20(-2)(-3-4) + 20(-3)(-2-6)+$$

Shear lag in thin web structures

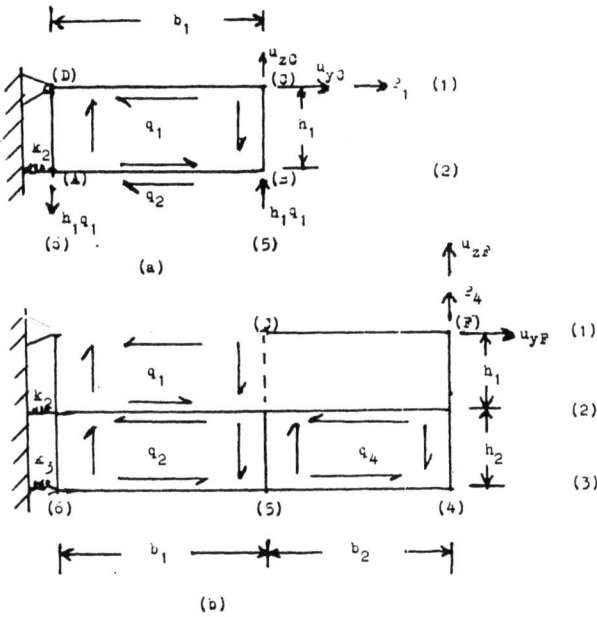

Fig. 6.25. Determinate substructures for Example 6.10.

$$+ 2(20)(-2)(2)(-2) + 20(2)(2)(2) + 2(20)(2)(2 + 4) +$$
$$+ 20(2)(3 + 4) + 20(3)(2 + 6) + 10(1)(2)(1) + 2(10)(-1)(2)(-1) +$$
$$+ 2(10)(\tfrac{1}{2})(2)(\tfrac{1}{2}) + 20(1)(2)(1) + 2(10)(1)(1 + 2)] +$$
$$+ (P_1/Gt)[(\tfrac{1}{20})(20)(20)(\tfrac{1}{20}) + (\tfrac{1}{10})(10)(20)(\tfrac{1}{10}) + 0 + (\tfrac{1}{10})(10)(20)(\tfrac{1}{10})]$$
$$= \frac{468.33 P_1}{EA} + \frac{5 P_1}{Gt}.$$

Thus, the cut-out causes a 30% increase in the deflection at P_1 from the stiffeners and a 67% increase from the shear webs. The per cent changes will be different at other points on the beam. Note that the four short vertical stiffeners contribute only $28.33 P_1/EA$ to the deflection.

Example 6.10. (a) Make use of Example 6.5 and find the deflection components at point (C) in Figure 6.25a for the P_1 load. (b) Find the deflection components at point (F) in Figure 6.25b for the P_4 load. (c) Discuss results.

Solution. (a) Use stringer (1) in Figure 6.25a to get the u_{yC} deflection. From Figure 6.25a and Equations (6.39), (6.52), (6.55),

$$P_A = P_1, \quad P_B = P_1 - b_1 q_1 = P_1 - (1.50 P_1/D_{11}),$$
$$p^1_{AyC} = 1 = p^1_{ByC},$$
$$u_{yC} = (b/6EA)_1[(1)(P_1 + 2P_1 - 2b_1 q_1) + (1)(P_1 - b_1 q_1 + 2P_1)]$$
$$= (b/EA)_1 P_1 [1 - (0.75/D_{11})],$$

where D_{11} is given by Equation 6.40.

To get the deflection u_{xC}, use web (1) and the four stiffeners on the edges of web (1) in Figure 6.25a. Use the stiffener order (1), (5), (2), (6) with the $(A)_i$ ends in Figure 6.19 having the order (C), (C), (B), (D). Thus,

$$q^1_{1xC} = 1/h_1, \quad p^1_{AxCi} = 0, \ 1, \ 0, \ -1, \quad p^1_{BxCi} = -b_1/h_1, \ 0, \ b_1/h_1, \ 0,$$
$$q_1 = 1.50 P_1/b_1 D_{11}, \quad P_{Ai} = P_1, \ 0, \ 0, \ 0,$$
$$P_{Bi} = P_1 - b_1 q_1, \ -h_1 q_1, \ b_1(h_1+h_2)q_1/h_2, \ +h_1 q_1.$$

Thus, from Equation (6.55),

$$u_{xC} = -\frac{P_1 b_1^2}{2E_1 A_1 h_1} + \frac{P_1 b_1^2}{4E_1 A_1 h_1 D_{11}}\left[2 + 2(h_1+h_2)\left(\frac{E_1 A_1}{E_2 A_2 h_2}\right) - \frac{E_1 A_1}{E_5 A_5}\left(\frac{h}{b}\right)^3_1 \right.$$
$$\left. - \frac{E_1 A_1}{E_6 A_6}\left(\frac{h}{b}\right)^3_1 + 6(h_1+h_2)\left(\frac{E_1 A_1}{k_2 b_1 h_2}\right) + 6\left(\frac{E_1 A_1 h_1}{G_1 t_1 b_1^2}\right)\right]. \quad (6.57)$$

(b) The shear flows are given in Equation (6.41), whence Figure 6.25b and the u_{yC} above show that

$$u_{yF} = -\frac{b_1^2 q_1}{2E_1 A_1} = -\frac{b_1^2 P_4}{4E_1 A_1 h_2}\frac{C_{14}}{D_{11}}, \quad (6.58)$$

where C_{14} is given by Equation (6.42).

To get u_{xF} in Figure 6.25b use the determinate substructure including webs (2) and (4), stiffeners (2), (3), (4), (5), (6) and springs k_2 and k_3. Thus, with two parts on stiffeners (2), (3), (4), (6) in order and zero virtual load in stiffener (5),

$$q^1_{4F} = 1/h_2 = q^1_{2F}, \quad p^1_{AxFi} = 0, \ -b_2/h_2, \ 0, \ b_2/h_2, \ 1, \ 1, \ 0, \ -1,$$
$$p^1_{BxFi} = -b_2/h_2, \ -(b_1+b_2)/h_2, \ b_2/h_2, \ (b_1+b_2)/h_2, \ 1, \ 0, \ -1, \ -1,$$
$$q_4 = P_4/h_2, \quad q_2 = (P_4/h_2) - (h_1 q_1/h_2), \quad q_1 = C_{14} P_4/2h_2 D_{11},$$
$$P_{Ai} = 0, \ -\frac{b_2}{h_2}P_4, \ 0, \ \frac{b_2}{h_2}P_4, \ P_4, \ P_4, \ 0, \ -h_2 q_2,$$
$$P_{Bi} = -\frac{b_2}{h_2}P_4, \ \left[-\frac{b_1+b_2}{h_2}P_4 + \frac{b_1}{h_2}(h_1+h_2)q_1\right], \ \frac{b_2}{h_2}P_4, \ \left[\frac{b_1+b_2}{h_2}P_4 - \frac{b_1 h_1}{h_2}q_1\right],$$
$$P_4, \ 0, \ -h_2 q_2, \ -P_4.$$

Put these expressions into Equation (6.55) to get

$$u_{xF} = H_1 P_4 - H_2 q_1 = \left(H_1 - \frac{H_2 C_{14}}{2h_2 D_{11}}\right) P_4, \quad (6.59)$$

$$H_1 = \frac{1}{3h_2^2}\left(\frac{1}{E_2 A_2} + \frac{1}{E_3 A_3}\right)(b_1+b_2)^3 + \frac{3h_1+h_2}{3}\left(\frac{1}{E_4 A_4} + \frac{1}{E_6 A_6}\right) +$$
$$+ \frac{(b_1+b_2)^2}{h_2^2}\left(\frac{1}{k_2} + \frac{1}{k_3}\right) + \frac{b_2}{h_2 G_4 t_4} + \frac{b_1}{h_2 G_2 t_2},$$

$$H_2 = \frac{b_1^2(2b_1+3b_2)}{6h_2^2}\left(\frac{h_1+h_2}{E_2 A_2} + \frac{h_1}{E_3 A_3}\right) + \frac{h_1(3h_1+2h_2)}{6E_6 A_6} + \frac{b_1 h_1}{h_2 G_2 t_2} +$$
$$+ \frac{b_1(b_1+b_2)}{h_2^2}\left(\frac{h_1+h_2}{k_2} + \frac{h_1}{k_3}\right).$$

(c) The relative contributions of the longitudinal stiffeners, the transverse stiffeners, the shear webs, and the springs to the u_{xF} deflection can be seen in H_1 in Equation (6.59). In symbolic form,

$$H_1 = \frac{h}{(EA)_L}\left[\left(\frac{b}{h}\right)^3 + \frac{(EA)_L}{bk}\left(\frac{b}{h}\right)^3 + \frac{(EA)_L}{Gth}\left(\frac{b}{h}\right) + \frac{(EA)_L}{(EA)_T}\right]. \quad (6.60)$$

Thus, if the beam is long with b/h large, the order from maximum to minimum is longitudinal stiffeners, springs, shear webs, and transverse stiffeners. On the other hand for short beams, the order may reverse with b/h less than one.

Shear lag in thin web structures 231

Example 6.11. Calculate the u_{xC} deflection at the point (C) for the four cases in Figure 6.20. Discuss results.

Solution. In Example 6.6 and Figure 6.20 the solution is based upon EA and Gt the same for all stringers and webs. Also, $b = h$ with $b_i = b$, $h_i = h$, and

$$Gt = EA/hH = EA/4h, \quad H = 4. \tag{6.61}$$

In Figure 6.20 use the web in the upper left corner and its edge stiffeners as the determinate substructure. Thus, $q_C^1 = 1/h$ and from Equation (6.55) u_{xC} for the four cases is

$$
\begin{align*}
(1) \quad & u_{xC} = 0.9948 P_5 b/EA + 0.5968 P_5/Gt = 3.3820 P_5 b/EA, \\
(2) \quad & u_{xC} = 1.4608 P_5 b/EA + 0.5061 P_5/Gt = 3.4852 P_5 b/EA, \\
(3) \quad & u_{xC} = 0.9085 P_5 b/EA + 0.4032 P_5/Gt = 2.5213 P_5 b/EA, \\
(4) \quad & u_{xC} = 0.5000 P_5 b/EA + 0.5000 P_5/Gt = 2.5000 P_5 b/EA.
\end{align*}
\tag{6.62}
$$

These results show that the supports and location of the shear forces can have large effects upon the deflections of the structure. In this short beam, the shear terms and the transverse stiffeners produce the major part of the deflection. As pointed out in Example 6.6, if the transverse stiffeners are assumed to be infinitely stiff, then the shear flows of case (4) apply to all four cases and thus for all four cases

$$u_{xC} = 0.1667 P_5 b/EA + 0.5000 P_5/Gt = 2.1667 P_5 b/EA. \tag{6.63}$$

This case also occurs with normal transverse stiffeners for a long beam with $d = h/b$ small. See part (c) of Example 6.10 and Equation (6.44). With H in Equations (6.43) and (6.61) small, the shear term in Equation (6.63) becomes small.

6.10. Flexibility matrices for shear web elements and stiffener elements

As demonstrated in the above sections of this chapter 6, it is necessary to use the stiffener elements and the shear web elements in the calculations for deflections and for redundant beam type structures. To use these finite elements in a matrix form it is convenient to set up the flexibility matrices for the elements.

For the constant shear flow thin rectangular web in Figure 6.26,

$$e_{yzk} = q_k/G_k t_k, \quad \sigma_{yzk}^1 = 1/t_k, \tag{6.64}$$

where a *unit shear flow* is used. Thus, Equation (2.77) gives

$$[C_F]_k = \int_0^{t_k} \int_0^{h_k} \int_0^{b_k} \left(\frac{1}{t_k}\right)\left(\frac{1}{G_k t_k}\right) dV_k = \left(\frac{bh}{Gt}\right)_k, \tag{6.65}$$

$$B_k = h_k u_{zsk} = b_k u_{ysk} = C_{Fk} q_k,$$

where B_k is proportional to the web deflection.

For the stiffener element in Figure 6.26 in which the constant shear flows produce a linear axial loading, use

$$
\begin{align*}
\left(EAe_{yy}\right)_i &= S_{iL} y_r + S_{iR}(1 - y_r) + \left(EAe_E\right)_i \\
&= [y_r \ \ 1 - y_r] \begin{Bmatrix} S_{iL} \\ S_{iR} \end{Bmatrix} + \left(EAe_E\right)_i, \tag{6.66} \\
y_r &= y_i/b_i, \quad \text{or} \quad y_r = z_i/h_i.
\end{align*}
$$

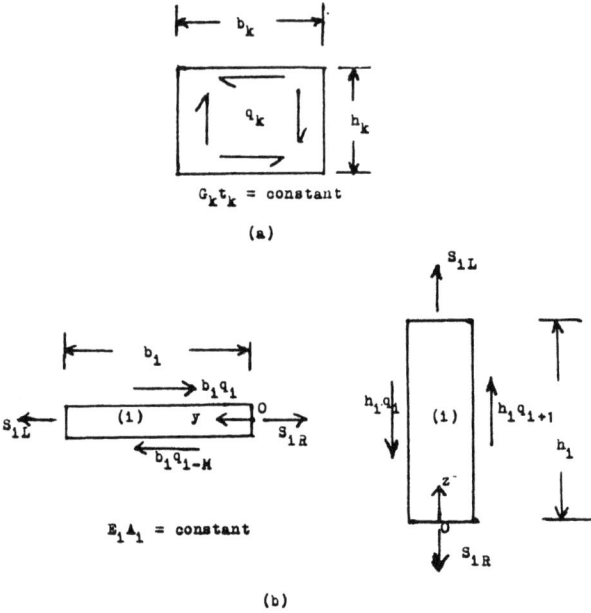

Fig. 6.26. (a). Shear web element. (b). Stiffener elements.

The S_{iL} and S_{iR} internal loads correspond to the P_B and P_A terms in Equations (6.26) and (6.53). From Equation (2.78)

$$[C_F]_i = b_i \int_0^i \int_{A_i} \begin{Bmatrix} y_r \\ 1 - y_r \end{Bmatrix} \frac{1}{E_i A_i^2} [y_r \quad 1 - y_r] \, dA_i \, dy_r$$

$$= \left(\frac{b}{6EA}\right)_i \begin{bmatrix} 2 & 1 \\ 1 & 2 \end{bmatrix},$$

$$\begin{Bmatrix} u_L \\ u_R \end{Bmatrix}_i = [C_F]_i \begin{Bmatrix} S_L \\ S_R \end{Bmatrix}_i + \begin{Bmatrix} u_{EL} \\ u_{ER} \end{Bmatrix}_i, \qquad (6.67)$$

$$\begin{Bmatrix} u_{EL} \\ u_{ER} \end{Bmatrix}_i = b_i \int_0^1 \begin{Bmatrix} y_r \\ 1 - y_r \end{Bmatrix} e_{Ei} \, dy_r,$$

where the expressions for u_{ELi} and u_{ERi} correspond to the S_{EB} and S_{EA} terms in Equations (6.54) and (6.55).

The above element flexibility matrices and element deflections due to the additional thermal and inelastic strains will be used in the following sections in the assembly of the matrices for the entire structure.

The stiffness matrices for these elements are given in Chapter 6 of Reference 3. The $[k_S]$ matrix is 4 by 4 for the shear web element and 3 by 3 for the stiffener element. The assembled $[k]$ matrix for the thin web structure is quite large and requires computers for the inversion. For example, the structure in Figure 6.28 and Example 6.12 below gives a 26 by 26 reduced $[k]$ matrix. On the other hand, there are only two redundants in the flexibility solution given in Example 6.12.

6.11. Matrix solutions for thin web beams in one plane

In Figure 6.27 include the support reactions in the P_i external forces and write the equilibrium equations in terms of the internal S_i forces, the q_i shear flows,

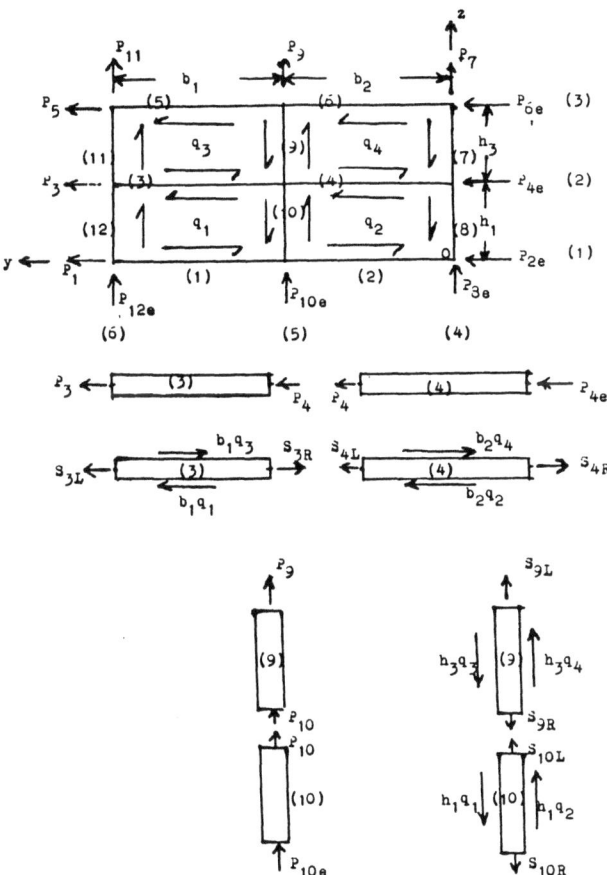

Fig. 6.27. Equilibrium stiffener elements.

and the P_i applied forces. For stiffener (2) with elements (3) and (4) and for stiffener (5) with elements (9) and (10), the equations are

$$
\begin{aligned}
S_{3L} &= P_3, \\
S_{3R} &= P_3 + b_1 q_1 - b_1 q_3, \\
S_{4L} &= P_3 + P_4 + b_1 q_1 - b_1 q_3, \\
S_{4R} &= P_3 + P_4 + b_1 q_1 + b_2 q_2 - b_1 q_3 - b_2 q_4, \\
0 &= P_3 + P_4 + P_{4e} + b_1 q_1 + b_2 q_2 - b_1 q_3 - b_2 q_4,
\end{aligned}
\qquad (6.68)
$$

$$S_{9L} = P_9,$$
$$S_{9R} = P_9 + h_3 q_4 - h_3 q_3,$$
$$S_{10L} = P_9 + P_{10} + h_3 q_4 - h_3 q_3, \tag{6.69}$$
$$S_{10R} = P_9 + P_{10} - h_1 q_1 + h_1 q_2 - h_3 q_3 + h_3 q_4,$$
$$0 = P_9 + P_{10} + P_{10e} - h_1 q_1 + h_1 q_2 - h_3 q_3 + h_3 q_4.$$

Similar equations can be written for the other stiffeners. Also, similar equations apply for any number of elements in the stiffeners, or for stiffener (1) with M elements,

$$S_{iL} = \sum_{k=1}^{i} P_k + \sum_{k=1}^{i-1} b_k q_k,$$
$$S_{iR} = \sum_{k=1}^{i} P_k + \sum_{k=1}^{i} b_k q_k, \quad i = 1, 2, \cdots, M, \tag{6.70}$$
$$0 = P_{Me} + \sum_{k=1}^{M} P_k + \sum_{k=1}^{M} b_k q_k.$$

The last equation in each of Equations (6.68)-(6.70) is the equilibrium equation for the entire stiffener. These equilibrium equations can be used to determine the support reactions and some of the shear flows. The remaining shear flows can be taken as the unknown redundants. For example, in Figure 6.15, the equilibrium equations for the six stiffeners would give the four reactions and two shear flows. Two shear flows will be unknown, as found in Section 6.8.

Write Equation (6.68) for stiffener (2) in the matrix form

$$\begin{Bmatrix} S_{3L} \\ S_{3R} \\ S_{4L} \\ S_{4R} \\ 0 \end{Bmatrix} = \begin{bmatrix} 1 & 0 & 0 & 0 & 0 & 0 & 0 \\ 1 & 0 & 0 & b_1 & 0 & -b_1 & 0 \\ 0 & 0 & -1 & 0 & -b_2 & 0 & b_2 \\ 0 & 0 & -1 & 0 & 0 & 0 & 0 \\ 1 & 1 & 1 & b_1 & b_2 & -b_1 & -b_2 \end{bmatrix} \begin{Bmatrix} P_3 \\ P_4 \\ P_{4e} \\ q_1 \\ q_2 \\ q_3 \\ q_4 \end{Bmatrix}, \tag{6.71}$$

where the last row 5 has been used to simplify rows 3 and 4. In compact form,

$$\begin{Bmatrix} S_2 \\ 0 \end{Bmatrix} = \begin{bmatrix} p_{2P} & p_{2q} \\ r_{2P} & r_{2q} \end{bmatrix} \begin{Bmatrix} P_{2P} \\ q \end{Bmatrix}. \tag{6.72}$$

Assemble the $\{S_i\}$ matrices for all the stiffeners in the beam in Figure 6.27 to

Shear lag in thin web structures

get

$$\begin{Bmatrix} S_1 \\ S_2 \\ S_3 \\ S_4 \\ S_5 \\ S_6 \end{Bmatrix} = \begin{bmatrix} p_{1P} & 0 & 0 & 0 & 0 & 0 & p_{1q} \\ 0 & p_{2P} & 0 & 0 & 0 & 0 & p_{2q} \\ 0 & 0 & p_{3P} & 0 & 0 & 0 & p_{3q} \\ 0 & 0 & 0 & p_{4P} & 0 & 0 & p_{4q} \\ 0 & 0 & 0 & 0 & p_{5P} & 0 & p_{5q} \\ 0 & 0 & 0 & 0 & 0 & p_{6P} & p_{6q} \end{bmatrix} \begin{Bmatrix} P_{1P} \\ P_{2P} \\ P_{3P} \\ P_{4P} \\ P_{5P} \\ P_{6P} \\ q \end{Bmatrix}. \quad (6.73)$$

The equilibrium equations can be assembled in a similar manner, or in compact form

$$\begin{Bmatrix} S \\ 0 \end{Bmatrix} = \begin{bmatrix} p_P & p_q \\ r_P & r_q \end{bmatrix} \begin{bmatrix} P \\ q \end{bmatrix} \quad (6.74)$$

This latter general matrix form applies for any number of stiffeners and shear webs in a thin web structure.

The reactions $\{R\}$ are included in $\{P\}$ and the determinate shear flows $\{q_D\}$ are included in $\{q\}$. Thus, for given supports and selected unknown redundant shear flows $\{q_R\}$, Equation (6.74) can be partitioned and the columns interchanged to give

$$\begin{Bmatrix} S \\ 0 \end{Bmatrix} = \begin{bmatrix} p_P & p_S & p_D & p_R \\ r_P & r_S & r_D & r_R \end{bmatrix} \begin{Bmatrix} P \\ R \\ q_D \\ q_R \end{Bmatrix}. \quad (6.75)$$

Since the equilibrium equations to determine the $\{q_D\}$ shear flows do not involve any reactions, write the equilibrium equations in the form

$$\begin{Bmatrix} 0 \\ 0 \end{Bmatrix} = \begin{bmatrix} r_{PD} & 0 & r_{DD} & r_{RD} \\ r_{PS} & r_{SS} & r_{DS} & r_{RS} \end{bmatrix} \begin{Bmatrix} P \\ R \\ q_D \\ q_R \end{Bmatrix}. \quad (6.76)$$

Solve for $\{q_D\}$ to get

$$\{q_D\} = -[r_{DD}]^{-1}\{[r_{PD}]\{P\} + [r_{RD}]\{q_R\}\}$$
$$= [B_P]\{P\} + [B_R]\{q_R\}. \quad (6.77)$$

This gives $\{R\}$ as

$$\{R\} = -[r_{SS}]^{-1}\{[r_{PS} + r_{DS}B_P]\{P\} + [r_{RS} + r_{DS}B_R]\{q_R\}\}$$
$$= [F_P]\{P\} + [F_R]\{q_R\}, \quad (6.78)$$

where $[r_{SS}]$ is a diagonal matrix.

Put the above expressions into Equation (6.75) to get

$$\begin{Bmatrix} S \\ R \\ q_D \\ q_R \end{Bmatrix} = \begin{bmatrix} p & p_Q \\ F_P & F_R \\ B_P & B_R \\ 0 & I \end{bmatrix} \begin{Bmatrix} P \\ q_R \end{Bmatrix}, \quad (6.79)$$

$$[p] = [p_P] + [p_S][F_P] + [p_D][B_P],$$
$$[p_Q] = [p_R] + [p_S][F_R] + [p_D][B_R]. \tag{6.80}$$

The flexibility matrices can be obtained from Equation (6.65) and (6.67). For stiffener (2)

$$[C_{F2}] = \begin{bmatrix} (b_1/6E_3A_3)\begin{bmatrix} 2 & 1 \\ 1 & 2 \end{bmatrix} & \begin{bmatrix} 0 & 0 \\ 0 & 0 \end{bmatrix} \\ \begin{bmatrix} 0 & 0 \\ 0 & 0 \end{bmatrix} & (b_2/6E_4A_4)\begin{bmatrix} 2 & 1 \\ 1 & 2 \end{bmatrix} \end{bmatrix}. \tag{6.81}$$

For the shear webs use Equation (6.65) directly. For supports with springs use $C_S = 1/k_S$. For fixed supports, $C_S = 0$. The assembled flexibility matrix $[C_F]$ is

$$[C_F]_{\text{diagonal}} = [C_{F1} \quad C_{F2} \quad C_{F3} \quad C_{F4} \quad C_{F5} \quad C_{F6} \quad C_{FS} \quad C_{FD} \quad C_{FR}]$$
$$= [C_{FP} \quad C_{FS} \quad C_{FD} \quad C_{FR}]. \tag{6.82}$$

From Equations (4.115), (6.79), and (6.82), the total flexibility matrix for the structure is

$$[C] = \begin{bmatrix} C_{PP} & C_{PQ} \\ C_{QP} & C_{QQ} \end{bmatrix}, \tag{6.83}$$
$$[C_{PP}] = [p^T C_{FP} p + F_P^T C_{FS} F_P + B_P^T C_{FD} B_P],$$
$$[C_{PQ}] = [p^T C_{FP} p_Q + F_P^T C_{FS} F_R + B_P^T C_{FD} B_R], \quad [C_{QP}] = [C_{PQ}]^T,$$
$$[C_{QQ}] = [p_Q^T C_{FP} p_Q + F_R^T C_{FS} F_R + B_R^T C_{FD} B_R + C_{FR}].$$

Take $Q = q_R$ so that the Equations (4.118)-(4.121) apply directly. Thus,

$$\{q_R\} = \{Q\} = -[C_{QQ}]^{-1}[[C_{QP}]\{P\} + \{u_{EQ}\}], \tag{6.84}$$
$$\{u_{EQ}\} = [p_Q]^T \begin{Bmatrix} u_{EL} \\ u_{ER} \end{Bmatrix},$$

where $\{u_R\} = \{u_Q\} = 0$ for continuity of the selected web redundants. Once $\{q_R\}$ is obtained, the $\{q_D\}$ shear flows, the reactions $\{R\}$, and the point stiffener loads $\{S\}$ are given by Equation (6.79). Also, stiffener load diagrams can be drawn directly from the shear flows and applied loads. The deflections at the applied loads are given by

$$\{u\} = [C_{PP}]\{P\} + [C_{PQ}]\{q_R\} + \{u_{EP}\}, \quad \{u_{EP}\} = [p]^T \begin{Bmatrix} u_{EL} \\ u_{ER} \end{Bmatrix}. \tag{6.85}$$

To determine deflections at selected points, unit loads can be used on determinate substructures consistent with the supports.

Note that the number of redundants N_R can be obtained from

$$N_R = N_w + N_{su} - N_{st}, \tag{6.86}$$

where N_w is the number of webs, N_{su} the number of supports, and N_{st} the number of stiffeners.

Shear lag in thin web structures

Example 6.12. (a) Find the four shear flows in the beam in Figure 6.28. Assume $(EA)_i = EA$, $(Gt)_i = Gt$, $h_i = h = b_i$ for square webs. (b) Find the displacements at the applied load points. (c) Assume additional strains e_{E1} act in stiffener element (1) and find the shear flows.

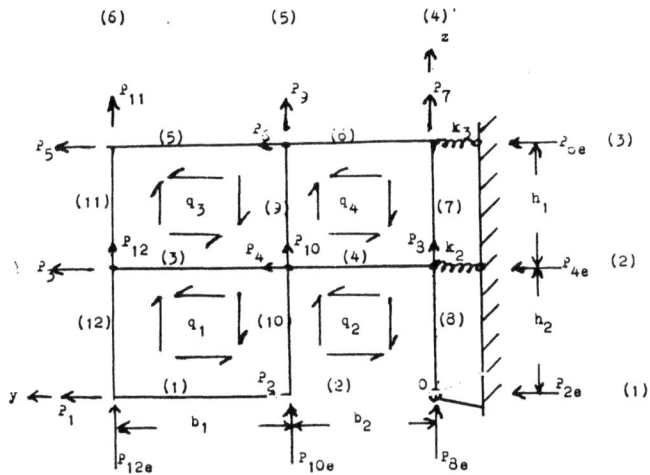

Fig. 6.28. Structure for Example 6.12.

Solution (a) In Figure 6.28, $N_w = 4$ webs, $N_{su} = 4$ supports, $N_{st} = 6$ stiffeners, whence Equation (6.86) gives $N_R = 2$ redundant webs. Select webs (3) and (4) as redundant so that

$$\{q_D\} = \begin{Bmatrix} q_1 \\ q_2 \end{Bmatrix}, \quad \{q_R\} = \begin{Bmatrix} q_3 \\ q_4 \end{Bmatrix}. \tag{6.87}$$

From Figure 6.28 there are 14 applied loads and 4 reactions, or

$$\{P\}^T = [P_1\ P_2\ P_3\ P_4\ P_5\ P_6\ P_7\ P_8\ P_9\ P_{10}\ P_{10e}\ P_{11}\ P_{12}\ P_{12e}] \\ \{R\}^T = [P_{2e}\ P_{4e}\ P_{6e}\ P_{8e}]. \tag{6.88}$$

Use Equation (6.71) in the basic Equation (6.74) to get

$$\begin{Bmatrix} S_1 \\ S_2 \\ S_3 \\ S_4 \\ S_5 \\ S_6 \\ 0 \\ 0 \\ 0 \\ 0 \\ 0 \\ 0 \end{Bmatrix} = \begin{bmatrix} a & 0 & 0 & 0 & 0 & 0 & -c_1 & c_2 & 0 & 0 \\ 0 & a & 0 & 0 & 0 & 0 & c_1 & -c_2 & -c_1 & c_2 \\ 0 & 0 & a & 0 & 0 & 0 & 0 & 0 & c_1 & -c_2 \\ 0 & 0 & 0 & a & 0 & 0 & 0 & c_2 & 0 & -c_1 \\ 0 & 0 & 0 & 0 & a & 0 & c_2 & -c_2 & -c_1 & c_1 \\ 0 & 0 & 0 & 0 & 0 & a & -c_2 & 0 & c_1 & 0 \\ d & 0 & 0 & 0 & 0 & 0 & -h & -h & 0 & 0 \\ 0 & d & 0 & 0 & 0 & 0 & h & h & -h & -h \\ 0 & 0 & d & 0 & 0 & 0 & 0 & 0 & h & h \\ 0 & 0 & 0 & d & 0 & 0 & 0 & -h & 0 & -h \\ 0 & 0 & 0 & 0 & d & 0 & -h & h & -h & h \\ 0 & 0 & 0 & 0 & 0 & d & h & 0 & h & 0 \end{bmatrix} \begin{Bmatrix} P_{1P} \\ P_{2P} \\ P_{3P} \\ P_{4P} \\ P_{5P} \\ P_{6P} \\ q_1 \\ q_2 \\ q_3 \\ q_4 \end{Bmatrix}, \tag{6.89}$$

$$[a] = \begin{bmatrix} 1 & 0 & 0 \\ 1 & 0 & 0 \\ 0 & 0 & -1 \\ 0 & 0 & -1 \end{bmatrix}, \quad \{c_1\} = h \begin{Bmatrix} 0 \\ 1 \\ 0 \\ 0 \end{Bmatrix}, \quad \{c_2\} = h \begin{Bmatrix} 0 \\ 0 \\ 1 \\ 0 \end{Bmatrix}, \quad [d] = [1\ 1\ 1],$$

where S_i represents the four internal loads in stiffener i and P_{iP} represents the three applied forces on stiffener i.

Solve the last two rows in Equation (6.89) for $\{q_D\}$, whence in Equation (6.77)

$$[B_P] = -\frac{1}{h}\begin{bmatrix} 0 & 0 & 0 & 0 & 0 & 0 & 0 & 0 & 0 & 1 & 1 & 1 \\ 0 & 0 & 0 & 0 & 0 & 0 & 0 & 1 & 1 & 1 & 1 & 1 \end{bmatrix}, \quad [B_R] = -\begin{bmatrix} 1 & 0 \\ 0 & 1 \end{bmatrix}.$$

Use the other four equilibrium equations in Equation (6.89) to get $\{R\}$ in Equation (6.78), or

$$[F_P] = -\begin{bmatrix} 1 & 1 & 0 & 0 & 0 & 0 & 0 & 0 & 1 & 1 & 1 & 2 & 2 & 2 \\ 0 & 0 & 1 & 1 & 0 & 0 & 0 & 0 & -1 & -1 & -1 & -2 & -2 & -2 \\ 0 & 0 & 0 & 0 & 1 & 1 & 0 & 0 & 0 & 0 & 0 & 0 & 0 & 0 \\ 0 & 0 & 0 & 0 & 0 & 0 & 1 & 1 & 1 & 1 & 1 & 1 & 1 & 1 \end{bmatrix},$$

$$[F_R] = -h\begin{bmatrix} 1 & 1 \\ -2 & -2 \\ 1 & 1 \\ 0 & 0 \end{bmatrix}.$$

To get the $[p_P]$ and $[p_S]$ matrices in Equation (6.80), partition the first four $[a]$ matrices in Equation (6.89) as

$$[a_P] = \begin{bmatrix} 1 & 0 \\ 1 & 0 \\ 0 & 0 \\ 0 & 0 \end{bmatrix}, \quad [a_S] = \begin{bmatrix} 0 \\ 0 \\ -1 \\ -1 \end{bmatrix},$$

whence

$$[p_P] = \begin{bmatrix} a_P & 0 & 0 & 0 & 0 & 0 \\ 0 & a_P & 0 & 0 & 0 & 0 \\ 0 & 0 & a_P & 0 & 0 & 0 \\ 0 & 0 & 0 & a_P & 0 & 0 \\ 0 & 0 & 0 & 0 & a & 0 \\ 0 & 0 & 0 & 0 & 0 & a \end{bmatrix}, \quad [p_S] = \begin{bmatrix} a_S & 0 & 0 & 0 \\ 0 & a_S & 0 & 0 \\ 0 & 0 & a_S & 0 \\ 0 & 0 & 0 & a_S \\ 0 & 0 & 0 & 0 \\ 0 & 0 & 0 & 0 \end{bmatrix}.$$

With these values and with $[p_D]$ and $[p_R]$ as in Equation (6.89), calculate $[p]$ and $[p_Q]$ from Equation (6.80) as

$$[p] = \begin{bmatrix} f & 0 & 0 & 0 & g_1 & g_2 \\ 0 & f & 0 & 0 & -g_1 & -g_2 \\ 0 & 0 & f & 0 & 0 & 0 \\ 0 & 0 & 0 & f & g_1 & g_1 \\ 0 & 0 & 0 & 0 & g_3 & 0 \\ 0 & 0 & 0 & 0 & 0 & g_3 \end{bmatrix}, \quad [p_Q] = h\begin{bmatrix} f \\ -2f \\ f \\ f_1 \\ f_2 \\ f_3 \end{bmatrix}, \quad (6.90)$$

$$[f] = \begin{bmatrix} 1 & 0 \\ 1 & 0 \\ 1 & 1 \\ 1 & 1 \end{bmatrix}, \quad [g_1] = \begin{bmatrix} 0 & 0 & 0 \\ 0 & 0 & 0 \\ 0 & 0 & 0 \\ 1 & 1 & 1 \end{bmatrix}, \quad [g_2] = \begin{bmatrix} 0 & 0 & 0 \\ 1 & 1 & 1 \\ 1 & 1 & 1 \\ 2 & 2 & 2 \end{bmatrix}, \quad [g_3] = \begin{bmatrix} 1 & 0 & 0 \\ 1 & 0 & 0 \\ 1 & 1 & 0 \\ 0 & 0 & -1 \end{bmatrix},$$

$$[f_1] = \begin{bmatrix} 0 & 0 \\ 0 & -1 \\ 0 & -1 \\ 0 & 0 \end{bmatrix}, \quad [f_2] = \begin{bmatrix} 0 & 0 \\ -1 & 1 \\ -1 & 1 \\ 0 & 0 \end{bmatrix}, \quad [f_3] = \begin{bmatrix} 0 & 0 \\ 1 & 0 \\ 1 & 0 \\ 0 & 0 \end{bmatrix}.$$

From Equations (6.81) and (6.82) the flexibility matrices are

$$[C_{FP}]_i = \frac{h}{6EA}\begin{bmatrix} 2 & 1 & 0 & 0 \\ 1 & 2 & 0 & 0 \\ 0 & 0 & 2 & 1 \\ 0 & 0 & 1 & 2 \end{bmatrix}, \quad [C_{FD}] = [C_{FR}] = \frac{h^2}{Gt}\begin{bmatrix} 1 & 0 \\ 0 & 1 \end{bmatrix},$$

$[C_{FS}] = 0$ for large k_2 and k_3.

Use the expression

$$H = EA/Gth \qquad (6.91)$$

Shear lag in thin web structures

and calculate $[C_{QQ}]$, $[C_{QP}]$ in Equation (6.83). The results are

$$[C_{QQ}] = \frac{h^3}{3EA}\begin{bmatrix} 28+6H & 7 \\ 7 & 10+6H \end{bmatrix},$$

$$[C_{QP}] = \frac{h^2}{6EA}\begin{bmatrix} 9 & 6 & -18 & -12 & 9 & 6 & 0 & 0 & 4 & 8 & 10 \\ 3 & 3 & -6 & -6 & 3 & 3 & -6 & -3 & 10+6H & 7+6H & 4+6H \end{bmatrix}$$
$$\begin{bmatrix} 38+6H & 35+6H & 32+6H \\ 14+6H & 14+6H & 14+6H \end{bmatrix}.$$

In Equation (6.84) put the P_i applied forces above the corresponding column rather than in a column to the right, whence

$$\{q_R\} = \begin{Bmatrix} q_3 \\ q_4 \end{Bmatrix} = -\frac{1}{2h(231+228H+36H^2)}$$

$$\begin{array}{cccc} & P_1 & P_2 & P_3 \\ & \begin{bmatrix} 69+54H & 39+36H & -138-108H \\ 21+18H & 42+18H & -42-36H \end{bmatrix} \end{array}$$

$$\begin{array}{ccc} P_4 & P_5 & P_6 & P_7 & P_8 \\ -78-72H & 69+54H & 39+36H & 42 & 21 \\ -84-36H & 21+18H & 42+18H & -168-36H & -84-18H \end{array}$$

$$\begin{array}{ccc} P_9 & P_{10} & P_{10e} \\ -30-18H & 31+6H & 72+18H \\ 252+228H+36H^2 & 140+210H+36H^2 & 42+192H+36H^2 \end{array}$$

$$\begin{array}{ccc} P_{11} & P_{12} & P_{12e} \\ 282+246H+36H^2 & 252+228H+36H^2 & 222+210H+36H^2 \\ 126+210H+36H^2 & 147+210H+36H^2 & 168+210H+36H^2 \end{array}\Bigg]. \quad (6.92)$$

The q_1 and q_2 shear flows are given by Equation (6.77), whence for the case of $H=4$,

$$\begin{Bmatrix} q_1 \\ q_2 \\ q_3 \\ q_4 \end{Bmatrix} = \frac{1}{h}\begin{bmatrix} P_1 & P_2 & P_3 & P_4 & P_5 & P_6 & P_7 \\ 0.0829 & 0.0532 & -0.1658 & -0.1065 & 0.0829 & 0.0532 & 0.0122 \\ 0.0271 & 0.0332 & -0.0541 & -0.0663 & 0.0271 & 0.0332 & -0.0908 \\ -0.0829 & -0.0532 & 0.1658 & 0.1065 & -0.0829 & -0.0532 & -0.0122 \\ -0.0271 & -0.0332 & 0.0541 & 0.0663 & -0.0271 & -0.0332 & 0.0908 \end{bmatrix}$$

$$\begin{array}{ccccccc} P_8 & P_9 & P_{10} & P_{10e} & P_{11} & P_{12} & P_{12e} \\ 0.0061 & -0.0297 & 0.0131 & 0.0419 & -0.4638 & -0.4935 & -0.5232 \\ -0.0454 & -0.4939 & -0.5474 & -0.5969 & -0.5515 & -0.5454 & -0.5393 \\ -0.0061 & 0.0297 & -0.0131 & -0.0419 & -0.5362 & -0.5065 & -0.4768 \\ 0.0454 & -0.5061 & -0.4526 & -0.4031 & -0.4484 & -0.4546 & -0.4607 \end{array}\Bigg]. \quad (6.93)$$

Note that the results for the P_{10e} and P_9 forces correspond to cases (1) and (2) in Figure 6.20.

(b) Use the $[p]$ and $[B_p]$ matrices from part (a) to calculate $[C_{PP}]$ in Equation (6.83). From

Equation (6.85) the point deflections are

$$\frac{6EA}{h}\{u\} = \begin{bmatrix}
12 & 6 & 0 & 0 & 0 & 0 & 0 & 0 & 3 & 3 & 3 \\
6 & 12 & 0 & 0 & 0 & 0 & 0 & 0 & 3 & 3 & 3 \\
0 & 0 & 12 & 6 & 0 & 0 & 0 & 0 & -3 & -3 & -3 \\
0 & 0 & 6 & 12 & 0 & 0 & 0 & 0 & -3 & -3 & -3 \\
0 & 0 & 0 & 0 & 12 & 6 & 0 & 0 & 0 & 0 & 0 \\
0 & 0 & 0 & 0 & 6 & 12 & 0 & 0 & 0 & 0 & 0 \\
0 & 0 & 0 & 0 & 0 & 0 & 12 & 6 & 3 & 3 & 3 \\
0 & 0 & 0 & 0 & 0 & 0 & 6 & 12 & 3 & 3 & 3 \\
3 & 3 & -3 & -3 & 0 & 0 & 3 & 3 & 14+6H & 8+6H & 5+6H \\
3 & 3 & -3 & -3 & 0 & 0 & 3 & 3 & 8+6H & 8+6H & 5+6H \\
3 & 3 & -3 & -3 & 0 & 0 & 3 & 3 & 5+6H & 5+6H & 8+6H \\
12 & 9 & -12 & -9 & 0 & 0 & 3 & 3 & 12+6H & 12+6H & 12+6H \\
12 & 9 & -12 & -9 & 0 & 0 & 3 & 3 & 12+6H & 12+6H & 12+6H \\
12 & 9 & -12 & -9 & 0 & 0 & 3 & 3 & 12+6H & 12+6H & 12+6H
\end{bmatrix}$$

(column headings above the matrix: P_1 P_2 P_3 P_4 P_5 P_6 P_7 P_8 P_9 P_{10} P_{10e})

$$\begin{bmatrix}
12 & 12 & 12 & 9 & 3 \\
9 & 9 & 9 & 6 & 3 \\
-12 & -12 & -12 & -18 & -6 \\
-9 & -9 & -9 & -12 & -6 \\
0 & 0 & 0 & 0 & 3 \\
0 & 0 & 0 & 6 & 3 \\
3 & 3 & 3 & 0 & -6 \\
3 & 3 & 3 & 0 & -3 \\
12+6H & 12+6H & 12+6H & 4 & 10+6H \\
12+6H & 12+6H & 12+6H & 8 & 7+6H \\
12+6H & 12+6H & 12+6H & 10 & 4+6H \\
42+12H & 36+12H & 33+12H & 38+6H & 14+6H \\
36+12H & 36+12H & 33+12H & 35+6H & 14+6H \\
33+12H & 33+12H & 36+12H & 32+6H & 14+6H
\end{bmatrix} . \quad (6.94)$$

(column headings above second matrix: P_{11} P_{12} P_{12e} hq_3 hq_4)

With $H = 4$, the u_9 and u_{10e} deflections for P_9 and P_{10e} in Equation (6.94) check the deflections in Equation (6.62). If $H = EA/Gth$ is large then the deflections in the z-direction in Figure 6.28 may be very large.

(c) From Equations (6.67), (6.84), and (6.90) the additional strains e_{E1} in element (1) produce the following additional shear flows,

$$\left\{ \begin{matrix} u_{EL} \\ u_{ER} \end{matrix} \right\}_1 = h \int_0^1 \left\{ \begin{matrix} y_r e_{E1} \\ (1-y_r)e_{E1} \end{matrix} \right\} dy_r,$$

$$\left\{ \begin{matrix} u_{EL} \\ u_{ER} \end{matrix} \right\}^T = [u_{EL1} \quad u_{ER1} \quad 0 \quad 0 \quad \cdots \quad 0],$$

$$\{u_{EQ}\} = [p_Q]^T \left\{ \begin{matrix} u_{EL} \\ u_{ER} \end{matrix} \right\} = h \left\{ \begin{matrix} u_{EL1} + u_{ER1} \\ 0 \end{matrix} \right\}$$

$$\{q_{ER}\} = \left\{ \begin{matrix} q_{E3} \\ q_{E4} \end{matrix} \right\} = -[C_{QQ}]^{-1}\{u_{EQ}\} = -\frac{3EA}{h^2 D} \left\{ \begin{matrix} (10+6H)(u_{EL1} + u_{ER1}) \\ -7(u_{EL1} + u_{ER1}) \end{matrix} \right\},$$

$$D = 231 + 228H + 36H^2, \quad H = EA/Gth, \quad q_{E1} = -q_{E3}, \quad q_{E2} = -q_{E4}.$$

Shear lag in thin web structures

Assume a uniform temperature change with e_{T1} = constant and an inelastic e_{p1} in stiffener element (1) so that (Equation 6.56)

$$e_{E1} = e_{T1} + e_{p1}, \quad e_{p1} = \tfrac{3}{7}\tfrac{S_y}{EA}\left(\tfrac{S_1}{S_y}\right)^{10}, \quad S_{T1} = EAe_{T1}.$$

Use Equation (6.66) for S_1, whence by Equation (6.56)

$$u_{EL1} + u_{ER1} = h\int_0^1 e_{E1}\, dy_r = \frac{hS_{T1}}{EA} + \frac{3hS_y^2}{77EA(S_L - S_R)}\left[\left(\frac{S_L}{S_y}\right)^{11} - \left(\frac{S_R}{S_y}\right)^{11}\right].$$

In this case, $S_L = P_1$, $S_R = P_1 - hq_1$, whence from Equation (6.92) and the above results

$$\frac{hq_1}{S_y} = \frac{P_1}{2DS_y}(69 + 54H) + \frac{3(10 + 6H)}{D}\left[\frac{S_{T1}}{S_y} + \tfrac{3}{77}\left(\frac{S_y}{hq_1}\right)\left\{\left(\frac{P_1}{S_y}\right)^{11} - \left(\frac{P_1 - hq_1}{S_y}\right)^{11}\right\}\right].$$

Take $H = 2$, $P_1/S_y = 1.3$, $S_{T1}/S_y = 2.0$, and solve for hq_1/S_y by trial, or

$$\frac{hq_1}{S_y} = 0.1384 + 0.1588 + 0.003094\left(\frac{S_y}{hq_1}\right)\left\{17.92 - \left(1.30 - \frac{hq_1}{S_y}\right)^{11}\right\},$$

$hq_1/S_y = 0.43, \quad q_1 = 0.33P_1/h,$

$hq_2/S_y = 0.0446 - 0.0505 - 0.0405 = -0.0464, \quad q_2 = -0.04P_1/h.$

These results show that both the temperature and the inelastic effect in element (1) cause a large increase in the q_1 shear flow, which transfers more load from element (1) to the other stiffeners. In this case the q_2 shear flow changes sign, which transfers some of the load back into stiffener (1) through element (2). That is, the tension load in stiffener (1) decreases from P_1 to $0.67P_1$ and then increases to $0.71P_1$ at the support. Without these effects, it decreases from P_1 to $0.89P_1$ and then decreases to $0.86P_1$ at the support. The above results can be added to Equation (6.85) to obtain the effects upon the deflections in Equation (6.94).

6.12. Matrix solutions for box beams

Consider the box beam in Figure 6.29 to be the local section near the fuselage

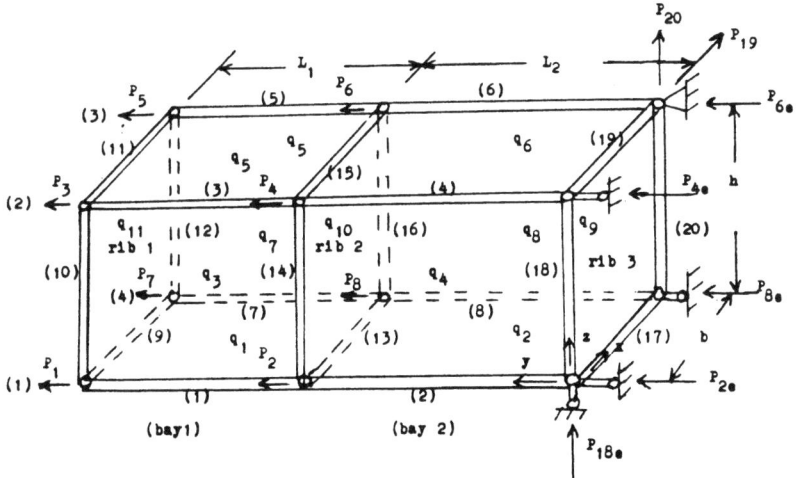

Fig. 6.29. Redundant box beam and restraints, (see Fig. 6.31 for other forces).

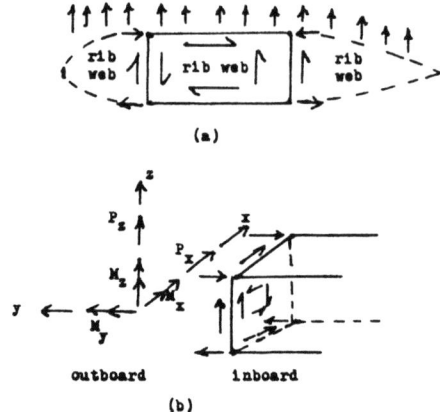

Fig. 6.30. (a). Wing cross section at rib. (b) Loading from outboard forces.

supports of the primary load carrying box of a wing. As shown in Figure 6.30, the rest of the wing structure applies loads to the box beam stiffeners either as point loads or as shear flows. In many cases these applied loads can be obtained by determinate analysis using simple beams and the single cell box beam analysis in Section 5.5. However, the redundant supports in Figure 6.29 restrain the warping of the webs so that the shear flows will change from those for the determinate single cell beam. Some of the change will occur at rib (1) and more change will occur at rib (2).

The matrix equations to obtain the shear flows and stiffener loads can be set up in the same manner as in the previous Section 6.11. For the case in Figure 6.29 there are eleven shear flows, including the rib shear flows ($N_w = 11$), seven reactions ($N_{su} = 7$), and sixteen stiffeners $N_{st} = 16$), whence Equation (6.86) gives two redundant shear flows. In Figure 6.29 and 6.31 select the shear flows q_{11} in rib (1) and q_{10} in rib (2) as redundants.

The internal force S_i equations for stringers (1)-(4) with two elements each are similar to Equations (6.68) and (6.71). From Figure 6.31 the other stiffeners have only one element with the following equations for stiffeners 13 and 14

$$S_{13L} = P_{13},$$
$$S_{13R} = P_{13} + b(q_1 - q_2 - q_{10}) = -P_{13e}, \qquad (6.95a)$$
$$0 = P_{13} + P_{13e} + b(q_1 - q_2 - q_{10}),$$
$$S_{14L} = P_{14},$$
$$S_{14R} = P_{14} + h(q_{2b} + q_3 - q_4 - q_{10}) = -P_{14e}, \qquad (6.95b)$$
$$0 = P_{14} + P_{14e} + h(q_{2b} + q_3 - q_4 - q_{10}).$$

From Equation (6.95) the end point stresses for one element stiffeners depend only upon the end loads and not upon the shear flows. Thus, the flexibilities of these one element stiffeners do not affect the shear flow calculations so that they

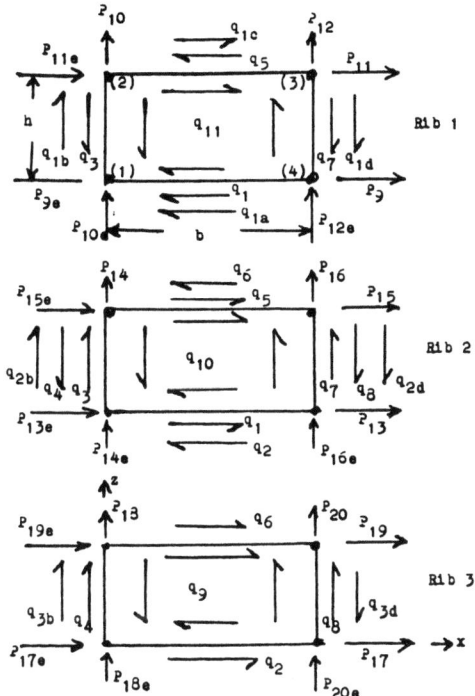

Fig. 6.31. Box beam ribs in Fig. 6.29 (view looking outboard in y-direction).

can be omitted in the $[C_F]$ expressions. They may have small effects upon the deflections, which can be added separately for the deflections of the end points. Also, note that in Equation (6.95) for single element stiffeners the applied loads can be combined into one load, such as

$$Q_{14} = P_{14} + P_{14e} + hq_{2b}. \tag{6.96}$$

These conclusions apply when the end loads are known. They do not apply if one end load is an unknown redundant support reaction. In this case the flexibility of the one element stiffener can affect the reaction and the shear flows.

The above simplifications for the box beam case can be used in the general Equations (6.74)-(6.85) of Section 6.11 to solve for shear flows in single cell and multicell box beams. The following examples demonstrate the procedure for several load cases for Figures 6.29 and 6.31.

Example 6.13. In the box beam in Figure 6.29 take $E_i A_i = EA$, $G_i = G$, rib webs = t_r, top and bottom webs = t_2, side webs = t_1, $L_1 = L_2 = L$, and use the applied loads P_1, P_2, Q_9, Q_{10}, Q_{14}. (a) Find the shear flows and the stiffener loads at the supports. (b) Compare the results for the Q_{10} load case to those for the plane section unrestrained box beam analysis.

Solution (a) From Figures 6.29 and 6.31, and Equation (6.96), take

$$Q_1 = P_1, \quad Q_2 = P_2, \quad Q_9 = P_9 + P_{9e} - bq_{1a}, \quad Q_{10} = P_{10} + P_{10e} + hq_{1b},$$
$$Q_{14} = P_{14} + P_{14e} + hq_{2b},$$
$$\{P\}^T = [Q_1 \quad Q_2 \quad Q_9 \quad Q_{10} \quad Q_{14}],$$

$$\{q_D\}^T = [q_1 \quad q_2 \quad q_3 \quad q_4 \quad q_5 \quad q_6 \quad q_7 \quad q_8 \quad q_9],$$
$$\{q_R\}^T = [q_{10} \quad q_{11}],$$
$$\{R\}^T = [P_{18e} \quad P_{19} \quad P_{20} \quad P_{2e} \quad P_{4e} \quad P_{6e} \quad P_{8e}].$$

Note that in Figure 6.29 there are only three determinate supports in the plane of rib 3 so that the flexibilities of all one element stiffeners can be omitted in the calculations.

From Equation (6.64) the four two element stiffeners or stringers in Figure 6.29 give sixteen $\{S\}$ equations in the basic Equation (6.75). Equilibrium of the sixteen stiffeners in Figure 6.29 gives sixteen $\{0\}$ equations in Equation (6.75). Use Figures 6.29 and 6.31 and Equation (6.68) to obtain these 32 equations. Use the S_i equations in order on the first eight elements and take the equilibrium equations in the stiffener order (9)-(20), (1), (2), (3), (4). The matrices in Equation (6.75) become

$[p_S] = 0;\ [p_R] = 0;$

$[p_P]$ is 16 by 5 with all zeros except 1 in locations $(1,1), (2,1), (3,1), (4,1), (3,2), (4,2);$

$[r_P]$ is 16 by 5 with all zeros except 1 in locations $(1,3), (2,4), (6,5), (13,1), (13,2);$

$$[r_S] = \begin{bmatrix} [0]^{9\,by\,7} \\ [I]^{7\,by\,7} \end{bmatrix};$$

$$[r_R]^T = \begin{bmatrix} 0 & B & 0 & 0 \\ B & 0 & 0 & 0 \end{bmatrix}^{2\,by\,16}, \quad [B] = [-1 \quad -1 \quad 1 \quad 1];$$

$$[p_D] = L \begin{bmatrix} a & a & 0 & 0 & 0 \\ 0 & -a & a & 0 & 0 \\ 0 & 0 & -a & -a & 0 \\ -a & 0 & 0 & a & 0 \end{bmatrix}^{16\,by\,9}, \quad [a] = \begin{bmatrix} 0 & 0 \\ 1 & 0 \\ 1 & 0 \\ 1 & 1 \end{bmatrix};$$

$$[r_D] = \begin{bmatrix}
-1 & 0 & 0 & 0 & 0 & 0 & 0 & 0 \\
0 & 0 & -1 & 0 & 0 & 0 & 0 & 0 \\
0 & 0 & 0 & 0 & -1 & 0 & 0 & 0 \\
0 & 0 & 0 & 0 & 0 & 0 & -1 & 0 \\
1 & -1 & 0 & 0 & 0 & 0 & 0 & 0 \\
0 & 0 & 1 & -1 & 0 & 0 & 0 & 0 \\
0 & 0 & 0 & 0 & 1 & -1 & 0 & 0 \\
0 & 0 & 0 & 0 & 0 & 1 & -1 & 0 \\
0 & 1 & 0 & 0 & 0 & 0 & 0 & -1 \\
0 & 0 & 0 & 1 & 0 & 0 & 0 & -1 \\
0 & 0 & 0 & 0 & 0 & 1 & 0 & 0 & 1 \\
0 & 0 & 0 & 0 & 0 & 0 & 1 & 1 \\
1 & 1 & 1 & 1 & 0 & 0 & 0 & 0 \\
0 & 0 & -1 & -1 & 1 & 1 & 0 & 0 \\
0 & 0 & 0 & 0 & -1 & -1 & -1 & -1 & 0 \\
-1 & -1 & 0 & 0 & 0 & 0 & 1 & 1 & 0
\end{bmatrix}.$$

From the above equilibrium equations, Equation (6.77) can be obtained directly as

$$[B_P]^T = \begin{bmatrix}
0 & 0 & 0 & 0 & 0 & 0 & 0 & 0 \\
0 & 0 & 0 & 0 & 0 & 0 & 0 & 0 \\
\frac{1}{b} & \frac{1}{b} & 0 & 0 & 0 & 0 & 0 & \frac{1}{b} \\
0 & 0 & \frac{1}{h} & \frac{1}{h} & 0 & 0 & 0 & 0 \\
0 & 0 & 0 & \frac{1}{h} & 0 & 0 & 0 & 0
\end{bmatrix}, \quad (6.97)$$

$$[B_R]^T = \begin{bmatrix} 0 & -1 & 0 & -1 & 0 & 1 & 0 & 1 & -1 \\ -1 & -1 & -1 & -1 & 1 & 1 & 1 & 1 & -1 \end{bmatrix}.$$

From Equation (6.80) with $v = L/b,\ w = L/h$,

$$[p]^T = \begin{bmatrix}
1 & 1 & 1 & 1 & 0 & 0 & 0 & 0 & 0 & 0 & 0 & 0 & 0 & 0 & 0 \\
0 & 0 & 1 & 1 & 0 & 0 & 0 & 0 & 0 & 0 & 0 & 0 & 0 & 0 & 0 \\
0 & v & v & 2v & 0 & 0 & 0 & 0 & 0 & 0 & 0 & 0 & -v & -v & -2v \\
0 & w & w & 2w & 0 & -w & -w & -2w & 0 & 0 & 0 & 0 & 0 & 0 & 0 \\
0 & 0 & 0 & w & 0 & 0 & 0 & -w & 0 & 0 & 0 & 0 & 0 & 0 & 0
\end{bmatrix},$$

Shear lag in thin web structures

$$[p_Q]^T = L \begin{bmatrix} 0 & 0 & 0 & -2 & 0 & 0 & 0 & 2 & 0 & 0 & 0 & -2 & 0 & 0 & 0 & 2 \\ 0 & -2 & -2 & -4 & 0 & 2 & 2 & 4 & 0 & -2 & -2 & -4 & 0 & 2 & 2 & 4 \end{bmatrix}.$$

The flexibility matrices in Equation (6.83) are

$$[C_{FF}]_i = \frac{L}{6EA} \begin{bmatrix} 2 & 1 & 0 & 0 \\ 1 & 2 & 0 & 0 \\ 0 & 0 & 2 & 1 \\ 0 & 0 & 1 & 2 \end{bmatrix}, \quad [C_{FR}] = \frac{hb}{Gt_r} \begin{bmatrix} 1 & 0 \\ 0 & 1 \end{bmatrix}, \quad [C_{FS}] = 0,$$

$$[C_{FD}]_{\text{diagonal}} = \frac{L}{G} \begin{bmatrix} \frac{b}{t_2} & \frac{b}{t_2} & \frac{h}{t_1} & \frac{h}{t_1} & \frac{b}{t_2} & \frac{b}{t_2} & \frac{h}{t_1} & \frac{h}{t_1} & \frac{bh}{Lt_r} \end{bmatrix}.$$

From the above expressions it follows from Equation (6.83) that

$$[C_{QQ}] = \frac{8L^3}{3EA} \begin{bmatrix} 2 + H_1 + 2H_2 & 5 + H_1 + H_2 \\ 5 + H_1 + H_2 & 16 + 2H_1 + 2H_2 \end{bmatrix}, \tag{6.98}$$

$$H_1 = H\left(\frac{b}{t_2} + \frac{h}{t_1}\right), \quad H_2 = H\left(\frac{bh}{Lt_r}\right), \quad H = \frac{3EA}{4GL^2},$$

where H_1 represents the box shear webs and H_2 represents the rib webs. Also,

$$[C_{PQ}]^T = [C_{QP}] = -\frac{2L^2}{3EA} \begin{bmatrix} \frac{3}{2} & \frac{3}{2} & \frac{L}{b}\{5 + 2H(\frac{b}{t_2} + \frac{bh}{Lt_r})\} \\ 6 & \frac{9}{2} & \frac{L}{b}\{16 + 2H(\frac{2b}{t_2} + \frac{bh}{Lt_r})\} \\ \frac{L}{h}\{5 + 2H\frac{h}{t_1}\} & \frac{L}{h}\{2 + 2H\frac{h}{t_1}\} \\ \frac{L}{h}\{16 + 4H\frac{h}{t_1}\} & \frac{L}{h}\{5 + 2H\frac{h}{t_1}\} \end{bmatrix}. \tag{6.99}$$

From Equation (6.84)

$$4LDq_{10} = -3(2 + H_1 + H_2)Q_1 + \tfrac{3}{2}(1 - H_1 - H_2)Q_2 +$$
$$+ \frac{L}{b}\{6H_1J_w + 2H_2(8 + H_1 + H_2)\}Q_9 - \frac{6L}{h}(H_1J_w + H_2)Q_{10} +$$
$$+ \frac{L}{h}\{D - H_1J_w(11 - H_1 + H_2) - 3H_2(9 + H_1 + H_2)\}Q_{14}, \tag{6.100}$$

$$4LDq_{11} = \tfrac{3}{2}(3 + 3H_1 + 7H_2)Q_1 + \tfrac{3}{2}(1 + 2H_1 + 5H_2)Q_2 +$$
$$+ \frac{L}{b}\{D + H_1J_w(H_1 + 3H_2 - 1) - H_2(5 + H_1 + H_2)\}Q_9 +$$
$$+ \frac{L}{h}\{D - (H_1J_w + H_2)(H_1 + 3H_2 - 1)\}Q_{10} +$$
$$+ \frac{L}{h}\{H_1J_w(H_2 + 3) + H_2(H_1 + 8)\}Q_{14},$$

$$D = 7 + 10H_1 + H_1^2 + H_2(26 + 4H_1 + 3H_2),$$

$$J_w = \left(\frac{b}{t_2} - \frac{h}{t_1}\right) / \left(\frac{b}{t_2} + \frac{h}{t_1}\right).$$

From Equations (6.77) and (6.97) the other shear flows are

$$bq_1 = Q_9 - bq_{11}, \quad bq_2 = Q_9 - bq_{10} - bq_{11}, \quad hq_3 = Q_{10} - hq_{11},$$
$$hq_4 = Q_{10} + Q_{14} - hq_{10} - hq_{11}, \quad q_5 = q_{11}, \quad q_6 = q_{10} + q_{11}, \quad q_7 = q_{11}, \tag{6.101}$$
$$q_8 = q_{10} + q_{11}, \quad bq_9 = Q_9 - bq_{10} - bq_{11}.$$

Table 6.2. Table for Example 6.13.

Shear flows and stringer loads for several values of the parameters H_1 and J_w with $H_2 = 0$.

	plane section	$J_w = 0$	$J_w = 0.4$ $H_1 = 12$	$J_w = 0.4$ $H_1 = 1$	$J_w = 0.8$ $H_1 = 12$	$J_w = 0.8$ $H_1 = 1$
$4hq_1/Q_{10}$	-1	-1	-0.805	-1	-0.610	-1
$4hq_2/Q_{10}$	-1	-1	-0.699	-0.867	-0.398	-0.733
$4hq_3/Q_{10}$	3	3	3.195	3	3.390	3
$4hq_4/Q_{10}$	3	3	3.301	3.133	3.602	3.267
$4hq_5/Q_{10}$	1	1	0.805	1	0.610	1
$4hq_6/Q_{10}$	1	1	0.699	0.867	0.398	0.733
$4hq_7/Q_{10}$	1	1	0.805	1	0.610	1
$4hq_8/Q_{10}$	1	1	0.699	0.867	0.398	0.733
Stringer loads at rib (2)						
hS_1/LQ_{10}	0.5	0.5	0.597	0.5	0.685	0.5
hS_2/LQ_{10}	-0.5	-0.5	-0.597	-0.5	-0.685	-0.5
hS_3/LQ_{10}	-0.5	-0.5	-0.403	-0.5	-0.315	-0.5
hS_4/LQ_{10}	0.5	0.5	0.403	0.5	0.315	0.5
Stringer loads at supports						
hR_1/LQ_{10}	1	1	1.248	1.067	1.496	1.133
hR_2/LQ_{10}	-1	-1	-1.248	-1.067	-1.496	-1.133
hR_3/LQ_{10}	-1	-1	-0.752	-0.933	-0.504	-0.867
hR_4/LQ_{10}	1	1	0.752	0.933	0.504	0.867

From Figures 6.29 and 6.31 the reactions are

$$P_{18c} = \frac{h}{b}Q_9 - Q_{10} - Q_{14}, \quad P_{19} = -Q_9, \quad P_{20} = -\frac{h}{b}Q_9$$

$$P_{2c} = -Q_1 - Q_2 - \frac{2L}{b}Q_9 - \frac{2L}{h}Q_{10} - \frac{L}{h}Q_{14} + 2L(q_{10} + 2q_{11}),$$

$$P_{4c} = \frac{2L}{h}Q_{10} + \frac{L}{h}Q_{14} - 2L(q_{10} + 2q_{11}), \quad P_{6c} = 2L(q_{10} + 2q_{11}),$$

$$P_{8c} = \frac{2L}{b}Q_9 - 2L(q_{10} + 2q_{11}).$$

(6.102)

(b). From Equations (6.100)-(6.102) for the Q_{10} load alone

$q_{10} = -(6Q_{10}/4hD)(H_1 J_w + H_2),$

$q_{11} = (Q_{10}/4hD)\{D - (H_1 J_w + H_2)(H_1 + 3H_2 - 1)\},$ (6.103)

$q_1 = -q_{11}, \quad q_2 = -q_{10} - q_{11}, \quad hq_3 = Q_{10} - hq_{11}, \quad hq_4 = Q_{10} - hq_{10} - hq_{11},$

$q_5 = q_{11}, \quad q_6 = q_{10} + q_{11}, \quad q_7 = q_{11}, \quad q_8 = q_{10} + q_{11}, \quad q_9 = -q_{10} - q_{11},$

$P_{18c} = -Q_{10}, \quad P_{19} = P_{20} = 0, \quad P_{2c} = -(2Q_{10}L/h) + L(2q_{10} + 4q_{11}),$

$P_{4c} = -P_{2c}, \quad P_{6c} = 2L(q_{10} + 2q_{11}) = -P_{8c}.$

From Example 5.15 and Figure 5.21 the shear flows in the unrestrained beam are

$$q_3 = q_4 = 3Q_{10}/4h, \quad q_1 = q_2 = -Q_{10}/4h, \quad q_5 = q_6 = q_7 = q_8 = \frac{Q_{10}}{4h},$$

$$q_{10} = 0, \quad q_{11} = Q_{10}/4h,$$

(6.104)

which correspond to Equation (6.103) with the warping term $J_w = 0$ and the rib web term $H_2 = 0$.

From the definitions of H_1, H_2, D, J_w in Equations (6.98) and (6.100), it is evident in Equation (6.103) that these parameters can have a large effect upon the shear flows and stiffener loads, particularly for short beams with L small. The J_w parameter is proportional

Shear lag in thin web structures

to the box beam warping deflections given in part (c) of Example 5.15. If the rib webs have a small t_r so that the parameter H_2 is large, then both q_{10} and q_{11} are small due to H_2 in D so that there is very little transfer of the Q_{10} load from the beam webs q_3 and q_4. If J_w is not zero, then a large H_1 parameter makes q_{10} and q_{11} small so that the Q_{10} does not transfer to the other beam. Table 6.2 for this Example 6.13 shows some effects of these parameters.

Example 6.14. In part (b) of Example 6.13 use the load $Q_{12} = -Q_{10}$ in addition to the Q_{10} load to produce a constant torsional moment $M_y = bQ_{10}$ on the box beam. (a) Find the shear flows and reactions. (b) Compare this two-step solution to the continuous solution given in Chapter 6 of Reference 1 for the torsion case.

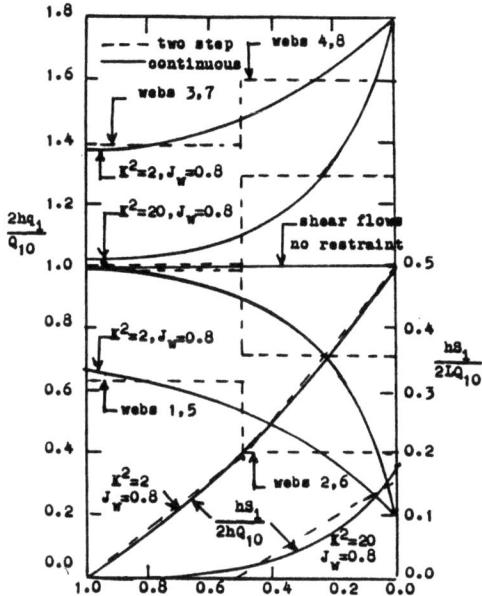

Fig. 6.32. Shear flows and stringer loads in restrained box beam.

Solution (a) Since a Q_{12} load will reverse the signs on q_{10} and q_{11}, the $Q_{12} = -Q_{10}$ load will give the same q_{10} and q_{11} as in Equation (6.103). Thus for this torque case, Equation (6.103) becomes

$$q_{10} = -(6Q_{10}/2hD)(H_1 J_w + H_2),$$
$$q_{11} = (Q_{10}/2hD)\{D - (H_1 J_w + H_2)(H_1 + 3H_2 - 1)\}, \qquad (6.105)$$
$$q_1 = -q_{11}, \quad q_2 = -q_{10} - q_{11}, \quad hq_3 = Q_{10} - hq_{11},$$
$$hq_4 = Q_{10} - hq_{10} - hq_{11}, \quad q_5 = q_{11}, \quad q_6 = q_{10} + q_{11}, \quad hq_7 = -Q_{10} + hq_{11},$$
$$hq_8 = -Q_{10} + hq_{10} + hq_{11}, \quad q_9 = -q_{10} - q_{11},$$
$$P_{18c} = -Q_{10}, \quad P_{19} = 0, \quad P_{20} = Q_{10},$$
$$P_{4c} = P_{8c} = -P_{2c} = -P_{6c} = (2L/h)Q_{10} - 2L(q_{10} + 2q_{11}).$$

In this case at the supports for the four stringers

$$S_i = P_{4c} = P_{8c} = -P_{2c} = -P_{6c}$$
$$= (2LQ_{10}/hD)(H_1 J_w + H_2)(H_1 + 3H_2 + 2). \qquad (6.106)$$

For the unrestrained beam the S_i reactions in Equation (6.106) are zero and the shear flows are

$$q_{11} = q_3 = q_4 = -q_1 = -q_2 = q_5 = q_6 = q_7 = q_8 = Q_{10}/2h. \qquad (6.107)$$

(b) From Chapter 6 in Reference 1, the continuous solution for the shear flows q_b for the case of ribs with infinite stiffness spaced infinitesimally close together is

$$\frac{2hq_b}{Q_{10}} = 1 \pm \frac{J_w \cosh K(1-y_r)}{\cosh K}, \quad y_r = \frac{y}{2L}, \quad K^2 = \frac{24}{H_1}, \quad H_2 = 0. \qquad (6.108)$$

The stringer loads are

$$\frac{hS_i}{2LQ_{10}} = \pm \frac{J_w \sinh K(1-y_r)}{K \cosh K}. \qquad (6.109)$$

The cases of $K^2 = 2$ ($H_1 = 12$), $J_w = 0.8$ and $K^2 = 20$ ($H_1 = 1.2$), $J_w = 0.8$ are shown in Figure 6.32 to compare the two-step solution to the continuous solution. The two-step shear flows are an approximate average of the continuous shear flow in each interval while the stringer loads compare very well. It is evident from Figure 6.32 that the restraints at the end of the box beam can have a large effect upon the shear flows and stringer loads. The shear flows in the top and bottom webs are reduced from the constant torque shear flows while the shear flows in the spar or side webs are increased. The stringer loads, which are zero under unrestrained torsion, are quite large for the small K^2 and large J_w case. Note that these effects are reduced for finite stiffness of the rib webs with $H_2 \neq 0$.

6.13. Load redistribution in swept back wings

In Figure 6.33 the additional length L_F of the front spar will cause a local redistribution of the shear flows and stiffener loads in the wing. Part of loads in the

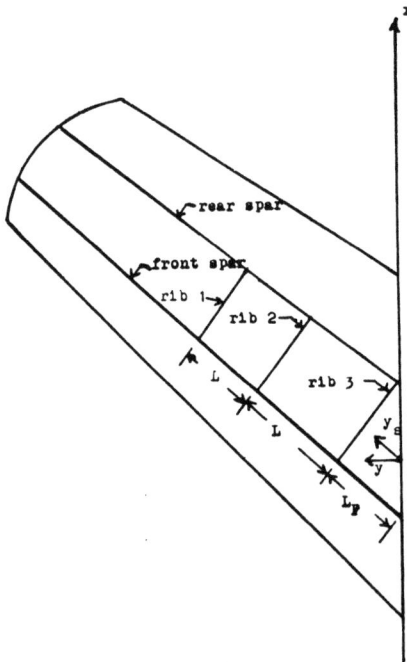

Fig. 6.33. Two spar swept back wing.

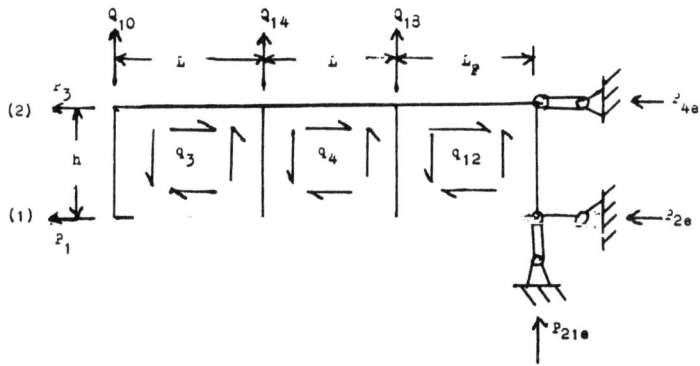

Fig. 6.34. Front spar for wing box beam in Fig. 6.33.

front spar will tend to transfer to the rear spar with the shorter length. This redistribution of the shear flows is a shear lag problem and can be solved by the same procedure as used in Sections 6.11 and 6.12.

Use the box beam in Figure 6.29 and add a simple section of length L_F to the front spar as shown in Figure 6.34. With the same rear spar supports as in Figure 6.29 the additional shear flow q_{12} is determinate so that web 12 will have no effect on the redundant shear flows. However, stringers (1) and (2) now have three elements so that there are two additional stress equations in Equation (6.68). This adds four rows to the $\{S\}$ matrix in the basic Equation (6.75). The following example demonstrates the procedure by modifying Example 6.13.

Example 6.15. In Figures 6.33 and 6.34 take $L_F = L$ and modify the box beam in Example 6.13 and Figure 6.29 to include the additional front spar elements. (a) Find the shear flows and stiffener loads at the supports. (b) Discuss the effects of the flexibility parameters for the Q_{10} load case.

Solution (a) Add four rows to the $[p_P]$ and $[p_D]$ matrices in Example 6.13 by repeating rows 4 and 8 twice each giving additional rows 4a, 4b, 8a, 8b. This gives two repeats of rows 4 and 8 in the $[p]$ and $[p_Q]$ matrices. Take EA for the additional elements to be the same as for the other elements, whence the new Equations (6.98) and (6.99) become

$$[C_{QQ}] = \frac{8L^3}{3EA} \begin{bmatrix} 5 + H_1 + 2H_2 & 11 + H_1 + H_2 \\ 11 + H_1 + H_2 & 28 + 2H_1 + 2H_2 \end{bmatrix}, \tag{6.110}$$

$$[C_{QP}] = -\frac{2L^2}{3EA}\begin{bmatrix} \frac{9}{2} & \frac{9}{2} & \frac{L}{b}\{11 + 2H(\frac{b}{t_2} + \frac{bh}{Lt_r})\} \\ 12 & \frac{21}{2} & \frac{L}{b}\{28 + 2H(2\frac{b}{t_2} + \frac{bh}{Lt_r})\} \\ \frac{L}{h}\{17 + 2H\frac{h}{t_1}\} & \frac{L}{h}\{8 + 2H\frac{h}{t_1}\} \\ \frac{L}{h}\{40 + 4H\frac{h}{t_1}\} & \frac{L}{h}\{17 + 2H\frac{h}{t_1}\} \end{bmatrix}. \tag{6.111}$$

From Equation (6.84) with $D = 19 + 16H_1 + H_1^2 + H_2(44 + 4H_1 + 3H_2)$,

$$4LDq_{10} = -3(2 + H_1 + H_2)Q_1 + \tfrac{3}{2}(7 - H_1 - H_2)Q_2 +$$

$$+ \frac{L}{b}\{6H_1 J_w + 2H_2(14 + H_1 + H_2)\}Q_9 + \frac{L}{h}\{36 - 6(H_1 J_w + H_2)\}Q_{10} +$$

$$+ \frac{L}{h}\{D + 18 - H_1 J_w(17 - H_1 - H_2) - 3H_2(15 + H_1 + H_2)\}Q_{14},$$

$$4LDq_{11} = \tfrac{3}{2}(7 + 5H_1 + 13H_2)Q_1 + \tfrac{3}{2}(2 + 4H_1 + 11H_2)Q_2 + \qquad (6.112)$$

$$+ \frac{L}{b}\{D + H_1J_w(H_1 + 3H_2 - 1) - H_2(H_1 + H_2 + 11)\}Q_9 +$$

$$+ \frac{L}{h}\{D + 6(H_1 - 1) - H_1J_w(H_1 + 3H_2 - 1) - H_2(H_1 + 3H_2 - 19)\}Q_{10} +$$

$$+ \frac{L}{h}\{6(H_1 - 1) + H_1J_w(H_2 + 3) + H_2(H_1 + 26)\}Q_{14}.$$

The other shear flows and reactions are given in Equations (6.101) and (6.102), except that

$$q_{12} = Q_{10} + Q_{14} - Q_9,$$
$$P_{2e} = -Q_1 - Q_2 - (3L/b)Q_9 - (3L/h)Q_{10} - (2L/h)Q_{14} + 2L(q_{10} + 2q_{11}), \qquad (6.113)$$
$$P_{4e} = (3L/h)Q_{10} + (2L/h)Q_{14} - 2L(q_{10} + 2q_{11}).$$

(b). For the Q_{10} load alone in Figure 6.34, Equation (6.112) gives

$$q_{10} = (Q_{10}/4hD)\{36 - 6(H_1J_w + H_2)\},$$
$$q_{11} = (Q_{10}/4hD)\{D + 6(H_1 - 1) - H_1J_w(H_1 + 3H_2 - 1) - H_2(H_1 + 3H_2 - 19)\},$$
$$q_7 = q_5 = -q_1 = q_{11}, \qquad q_6 = -q_9 = -q_2 = q_8 = q_{10} + q_{11},$$
$$q_3 = (Q_{10}/h) - q_{11}, \qquad q_4 = (Q_{10}/h) - q_{10} - q_{11}, \qquad q_{12} = Q_{10}/h, \qquad (6.114)$$
$$P_{19} = P_{20} = 0, \qquad P_{21e} = -Q_{10},$$
$$P_{4e} = -P_{2e} = (3L/h)Q_{10} - 2L(q_{10} + 2q_{11}),$$
$$P_{6e} = -P_{8e} = 2L(q_{10} + 2q_{11}).$$

Consider the case in which the three parameters J_w, H_1, and H_2 are all zero, or only the flexibility of the four stringers is included. The shear flows and reactions are

$$q_{10} = \frac{9}{19h}Q_{10}, \qquad q_{11} = \frac{13}{76h}Q_{10} = q_5 = q_7 = -q_1, \qquad q_{12} = Q_{10}/h,$$

$$q_6 = q_8 = -q_2 = -q_9 = \frac{49}{76h}Q_{10}, \qquad q_3 = \frac{63}{76h}Q_{10}, \qquad q_4 = \frac{27}{76h}Q_{10}, \qquad (6.115)$$

$$R = \pm\frac{26L}{19h}Q_{10}, \quad \text{front spar}; \quad R = \pm\frac{31L}{19h}Q_{10}.$$

These results show the large transfer of shear force from the front spar to the rear spar for the swept back wing, as compared to the straight wing results in Equation (6.104).

In Equation (6.114) with D in Equation (6.112), it is evident that non-zero values of the flexibility parameters J_w, H_1, and H_2 will reduce q_{10} and q_{11} and thus reduce the shear transfer to the rear spar from that in Equation (6.115).

6.14. Problems

6.1. In Figure 6.1 take $L_1 = L_2 = 30$ in., $L_3 = 50$ in., $h_1 = 10$ in., $h_2 = 5$ in., $P_B = 0$, $P_C = 15.0$ Klb, $P_{Ah} = 15.0$ Klb, $P_{Av} = 10.0$ Klb, and find the shear flows. Draw load diagrams for all stiffeners.

6.2. Use the results in Example 6.2 and draw load diagrams for all the stiffeners in Figure 6.4.

6.3. In Figure 6.4 assume that there is a partial web in the cut-out panel that can carry an equivalent shear flow of $q_5 = Kq_2$. Solve for the shear flows in Example 6.2 in terms of K and graph the shear flows against K for $0 \leq K \leq 1$.

6.4. In Figure 6.5 and Equation (6.10) take $P_1 = P_2 = P_3 = P_4 = P_5 = 8.0$ Klb, $q_0 = 900$ lb/in, $Q = 40.0$ Klb, $h_1 = h_2 = 15$ in., $d_1 = 60$ in., $d_2 = 15$ in., and find the reactions and the shear flows and draw load diagrams for stiffeners (1) through (6).

6.5. In Figure 6.8, assume that the moment divides 60% to the right side and 40% to the left side and recalculate the shear flows. Show the new values on a sketch similar to Figure 6.9.

6.6. In Example 6.3 draw load diagrams for stiffeners (2), (4), and (6).

6.7. In Example 6.4 assume there is no side force so that the applied forces in Figure 6.12 are 75.0 Klb up on each of stiffeners (2) and (3). Recalculate the shear flows in Table 6.1 and in the bulkhead and show the results as in Figure 6.13, including load diagrams for all stiffeners.

6.8. In Figure 6.14 take $p_u = 10.0$ lb/in^2, $P_s = 30.0$ Klb, $s = 12$ in., $d = 50$ in., $e_{st} = 0.10$, $\alpha = 10°$, $h = 20$ in., and calculate the net force R_u on the rib from one upper surface stringer.

6.9. Repeat Example 6.5, part (a), for the applied load P_2 acting alone. Consider the cases of $k_2 = \infty$ and $k_2 = 0$.

6.10. Repeat Example 6.6 for the load P_4 acting alone, with P_4 being applied in different ways to stiffener (4).

6.11. In Figure 6.20, reverse the direction of the P_5 load in the top beam and combine it with the third beam from the top, and hence solve for the shear flows with a balancing P_5 force acting at the ends of stiffener (5).

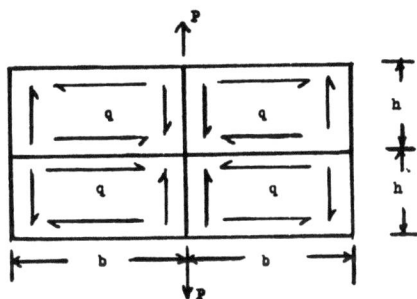

Fig. 6.35. Problem 6.12.

6.12. Let $E_j A_j = EA$, $G_k t_k h_k = Gth$, and find the shear flow q for the no support case in Figure 6.35.

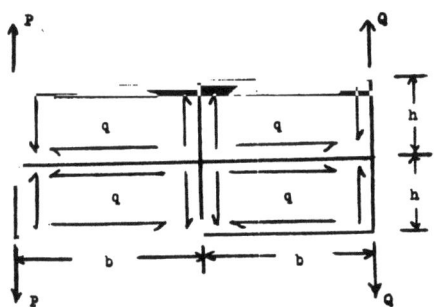

Fig. 6.36. Problem 6.13.

6.13. In Figure 6.36 let $E_j A_j = EA$, $G_k t_k h_k = Gth$, and find the shear flow q.

6.14. In Equation (6.44), graph $2hq_1/P_5$ and $2hq_3/P_5$ against H for $d = h/b = 1.0$, and for $d = 0.5$.

6.15. What happens in Equation (6.44) when $d^3 = 9/2$? Find the new formula for $2hq/P_5$ for this case.

6.16. Draw graphs similar to Figure 6.22 for the cases of (a) $F = 2.5$, $b_1/L = 0.50$, $b_2/L = 0.50$. (b) $F = 2.5$, $b_1/L = 0.70$, $b_2/L = 0.30$.

6.17. Use Equations (6.46)-(6.49) and solve Equation (6.27) to get the results for q_1 and q_3 in Equation (6.50).

6.18. In Example 6.7, assume k_2 is finite instead of infinite and solve for the q_1 and q_3 shear flows.

6.19. In Example 6.7, omit the P_1 load and put in a P_2 tension load. Solve for the q_1 and q_3 shear flows.

6.20. Use the data in Example 6.8 and Figure 6.3 to find the deflection at the left end of stringer (5) in the direction of the 15.0 Klb load.

6.21. Use the data in Example 6.8 and Figure 6.3 to find the horizontal and vertical deflections at the right end of stringer (7). Why does the inelastic effect in stiffener (2) have no effect upon these deflections?

6.22. Omit the inelastic effect in Example 6.8 and assume $EA/Gt = 40.0$. (a) Use the resulting u_m deflection and find the up force P_m at m to reduce u_m to zero. (b) Combine the given shear flows and the shear flows from P_m to get the results for a redundant support at m.

6.23. In Example 6.9 compare the vertical deflections at the upper end of stiffener (5) for the two cases.

6.24. In Example 6.9 compare the horizontal end deflections for stringers (1), (2), (3) and thus demonstrate that the end cross section is not plane for the cut-out case. What value of EA/Gt would make the end cross section plane?

6.25. In Equation (6.59) take $E_1A_1 = E_2A_2 = E_3A_3 = (EA)_L$, $E_4A_4 = E_5A_5 = E_6A_6 = (EA)_T$, $h_1 = h_2 = h$, $b_1 = b_2 = b$, $k_2 = k_3 = k$, $G_it_i = Gt$, $d = b/h$, $K_1 = (EA)_L/bh$, $K_2 = (EA)_L/Gth$, $K_3 = (EA)_L/(EA)_T$, and express U_{xF} in the form

$$u_{xF} = [2hP_4/3(EA)_L]f(d, K_1, K_2, K_3).$$

6.26. Solve Example 6.11 for the case of $d = 1.2$ and $H = 1$.

6.27. Verify that if $P_5 = 1$ is used as the unit load for the real stresses in all elements, then the same deflection is obtained as in case (1) in Example 6.11.

6.28. Calculate Equation (6.93) for the cases of (a) $H = 1$, (b) $H = 10$.

6.29. (a) Use Equation (6.92) for the single load P_{11} and graph hq_3/P_{11} and Hq_4/P_{11} against H for $0 \leq H \leq 10$. (b) Use the results in part (a) in Equation (6.94) and graph Eau_{11}/hP_{11} against H for the range in part (a).

6.30. Solve part (c) of Example 6.12 for the case of $P_1/S_y = 1.5$ instead of 1.3.

6.31. (a) In part (c) of Example 6.12 graph hq_1/P_1 against H for $0 \leq H \leq 10$. (b) Repeat part (a) without temperature.

6.32. Use the results in part (c) of Example 6.12 in Equations (6.94) and (6.85) to get the deflection EAu_1/hP_1 with the e_E effects. Compare to the deflection without e_E.

6.33. Repeat the table in Example 6.13 for $H_2 = 4$.

6.34. (a) Calculate the $[C_{PP}]$ matrix in Example 6.13. (b). Use the result in Equation (6.85) to calculate the u_{10} deflection for the Q_{10} load.

6.35. Calculate the shear flows in Equation (6.100) for case of $H_1 = 4$, $H_2 = 2$, $J_w = 0.4$.

6.36. Add a constant thermal load S_{T1} to stiffener element (1) in Example 6.13 and calculate the term to add to the shear flows in Equation (6.100).

6.37. In Example 6.14 draw curves similar to those in Figure 6.32 for the case of $K^2 = 8$ and $J_w = 0.8$.

6.38. Use Equation (6.100) for the Q_{14} applied load and repeat part (a) of Example 6.14 with $Q_{16} = -Q_{14}$. Graph this two-step case as in Figure 6.32.

6.39. In Example 6.14 apply a unit load at the Q_{10} load to the subbeam with stringers (1) and (2) and webs (3), (4), and find the u_{10} deflection for the torsion case. This gives the end rotation angle $\theta_y = 2u_{10}/b$.

6.40. Repeat Problem 6.39 for a unit load at the Q_{14} load location, but with only the end torque bQ_{10} acting.

6.41. Repeat part (b) of Example 6.15 for the Q_9 load acting alone.

6.42. (a) Add expressions for a $Q_{12}L/h$ applied load to Equations (6.111) and (6.112). (b) Combine the Q_{10} solution with the $Q_{12} = -Q_{10}$ solution to get the torsion shear flows in the swept back wing.

References

Chapter 6.

1. P. Kuhn: *Stresses in Aircraft and Shell Structures,* McGraw-Hill Book Co., New York (1956).
2. B.E. Gatewood: Shear distributions in beams with variable webs, *Journal of Aeronautical Sciences* 16, No. 12, pages 749-753, Dec. (1949).

3. J.S. Przemieniecki: *Theory of Matrix Structural Analysis*, McGraw-Hill Book Co., New York (1968).

Reference for additional reading

4. D.J. Peery and J.J. Azar: *Aircraft Structures*, McGraw-Hill Book Co., New York, 2nd Ed. (1982).

A

Notes on matrix algebra

A.1. Definition of matrices

Consider the set of equations

$$y_1 = a_{11}x_1 + a_{12}x_2 + a_{13}x_3 + a_{14}x_4$$
$$y_2 = a_{21}x_1 + a_{22}x_2 + a_{23}x_3 + a_{24}x_4 \quad \text{(A.1)}$$
$$y_3 = a_{31}x_1 + a_{32}x_2 + a_{33}x_3 + a_{34}x_4$$

or

$$y_i = \sum_{k=1}^{4} a_{ik}x_k, \quad i = 1, 2, 3 \quad \text{(A.2)}$$

If the equations are arranged in the form

$$\begin{bmatrix} y_1 \\ y_2 \\ y_3 \end{bmatrix} = \begin{bmatrix} a_{11} & a_{12} & a_{13} & a_{14} \\ a_{21} & a_{22} & a_{23} & a_{24} \\ a_{31} & a_{32} & a_{33} & a_{34} \end{bmatrix} \begin{bmatrix} x_1 \\ x_2 \\ x_3 \\ x_4 \end{bmatrix} \quad \text{(A.3)}$$

or

$$[y_i] = [a_{ik}][x_k], \quad \text{or} \quad Y = AX \quad \text{(A.4)}$$

then the array of numbers in each bracket is called a *matrix*. The numbers in the matrix are the *elements* of the matrix. The elements are arranged in rows and columns, with all the rows in a matrix having the same length and all the columns having the same length. The first subscript indicates the row number and the second subscript indicates the column number. A matrix may be designated by a capital letter and written without brackets, as in Equation (A.4). The size of a matrix is m by n if it has m rows and n columns. A one by one matrix is a scalar number. A matrix with one row is a *row vector*; a matrix with one column is a *column vector*, such as Y and X in Equation (A.3). A *square matrix* has the same number of rows as columns. A *diagonal matrix* is a square matrix with all elements zero except on the main diagonal. A *unit matrix* I is a diagonal matrix with elements on the main diagonal all unity. A *zero matrix* has all elements zero. The *transpose* of a matrix is obtained by interchanging rows and columns.

A *symmetrical matrix* is a square matrix which is equal to its own transpose, or $a_{ij} = a_{ji}$.

A.2. Addition, subtraction, multiplication of matrices

Two matrices are equal only if they are the same size and all corresponding elements are equal, $A = B$, if $a_{ij} = b_{ij}$. Matrices can be added or subtracted only if they are the same size and corresponding elements are added or subtracted. Thus

$$A = B \pm C, \quad \text{if} \quad a_{ij} = b_{ij} \pm c_{ij} \tag{A.5}$$

If each equation in Equation (A.1) is multiplied by a constant k, then in Equation (A.4)

$$kY = [kA]X = A[kX] \tag{A.6}$$

so that in *scalar* multiplication by k, every term in the matrix is multiplied by k.

Since Equations (A.3) and (A.4) represent Equation (A.1), *matrix multiplication* must be defined so that AX in Equation (A.4) is the same as the right side of Equation (A.1). In general

$$A = BC, \quad \text{if} \quad a_{ij} = b_{i1}c_{1j} + b_{i2}c_{2j} + \cdots + b_{ir}c_{rj} \tag{A.7}$$

or row i in B times column j in C, corresponding elements in the row and column being multiplied and the products summed. It is evident that multiplication can be performed only if the number of elements in the row equals the number of elements in the column that it multiplies, or the number of columns in B must equal the number of rows in C. Thus

$$A^{mn} = B^{pq}C^{rs} \quad \text{only if} \quad q = r, \; m = p, \; n = s \tag{A.8}$$

or in general

$$A^{mn} = B^{mr}C^{rn}, \quad a_{ij} = \sum_{k=1}^{r} b_{ik}c_{kj} \tag{A.9}$$

allows matrix multiplication.

It is evident that

$$A(B + C) = AB + AC, \quad ABC = A(BC) = (AB)C \tag{A.10}$$

However, in general

$$BC \neq CB \tag{A.11}$$

For example, from Equation (A.3) XA, (4 by 1)(3 by 4), disobeys Equation (A.9). Even though both products in Equation (A.11) obey Equation (A.9), generally the products are not equal. For example

$$\begin{bmatrix} 2 & 1 \\ 0 & 1 \end{bmatrix} \begin{bmatrix} 1 & 5 \\ -3 & 1 \end{bmatrix} = \begin{bmatrix} -1 & 11 \\ -3 & 1 \end{bmatrix}, \quad \begin{bmatrix} 1 & 5 \\ -3 & 1 \end{bmatrix} \begin{bmatrix} 2 & 1 \\ 0 & 1 \end{bmatrix} = \begin{bmatrix} 2 & 6 \\ -6 & -2 \end{bmatrix}$$

$$[1 \ 4] \begin{bmatrix} 3 \\ -2 \end{bmatrix} = [-5], \quad \begin{bmatrix} 3 \\ -2 \end{bmatrix} [1 \ 4] = \begin{bmatrix} 3 & 12 \\ -2 & -8 \end{bmatrix}$$

which indicates that the size of the products may be unequal.

A.3. Determinants

Although a matrix is not one number but an array of numbers it is convenient to define a numerical value for *square matrices*. The *determinant* of a square matrix is the sum of all possible products, taking one element from each row and one from each column without repetition in any product, and using the proper sign on each product. The determinant of matrix A is

1 by 1, $\quad |A| = |a_{11}| = a_{11}$

2 by 2, $\quad |A| = \begin{vmatrix} a_{11} & a_{12} \\ a_{21} & a_{22} \end{vmatrix} = a_{11}a_{22} - a_{12}a_{21}$

3 by 3, $\quad |A| = \begin{vmatrix} a_{11} & a_{12} & a_{13} \\ a_{21} & a_{22} & a_{23} \\ a_{31} & a_{32} & a_{33} \end{vmatrix}$

$$= a_{11} \begin{vmatrix} a_{22} & a_{23} \\ a_{32} & a_{33} \end{vmatrix} - a_{12} \begin{vmatrix} a_{21} & a_{23} \\ a_{31} & a_{33} \end{vmatrix} + a_{13} \begin{vmatrix} a_{21} & a_{22} \\ a_{31} & a_{32} \end{vmatrix}$$

$$= a_{11}(a_{22}a_{33} - a_{23}a_{32}) - a_{12}(a_{21}a_{33} - a_{23}a_{31}) +$$
$$+ a_{13}(a_{12}a_{32} - a_{22}a_{31}), \quad \text{row 1 expansion,} \quad (A.12)$$

m by m, $\quad |A| = \sum_{j=1}^{m}(-1)^{i+j}a_{ij}|M_{ij}|, \quad$ row i expansion, $\quad$ (A.13)

where $|M_{ij}|$ is the determinant of the $m-1$ by $m-1$ matrix obtained by omitting row i and column j in matrix A.

The above definition of the determinant is based on the solution of systems of simultaneous equations. If

$$\begin{matrix} a_{11}x_1 + a_{12}x_2 = b_1 \\ a_{21}x_1 + a_{22}x_2 = b_2 \end{matrix}, \quad \text{or} \quad AX = B,$$

then

$$x_1 = \frac{\begin{vmatrix} b_1 & a_{12} \\ b_2 & a_{22} \end{vmatrix}}{\begin{vmatrix} a_{11} & a_{12} \\ a_{21} & a_{22} \end{vmatrix}}, \quad x_2 = \frac{\begin{vmatrix} a_{11} & b_1 \\ a_{21} & b_2 \end{vmatrix}}{\begin{vmatrix} a_{11} & a_{12} \\ a_{21} & a_{22} \end{vmatrix}} \quad (A.14)$$

In general

$$x_i = |C_i|/|A| \tag{A.15}$$

where matrix C_i is matrix A with column i replaced by the column matrix B. If $|C_i| \neq 0$, then a solution for x_i exists only if $|A| \neq 0$. On the other hand, if $B = 0$, then $|C_i| = 0$ and a solution for x_i exists only if $|A| = 0$. Thus the condition to solve homogeneous equations is that

$$|A| = 0, \quad \text{homogeneous equations,} \tag{A.16}$$

Every square matrix has a value for its determinant, and since the value is a number it follows that

$$|A||B| = |AB| = |B||A| = |BA| \tag{A.17}$$

A.4. Matrix inversion

If A is a square matrix such that $|A| \neq 0$, then there exits a matrix A^{-1} such that

$$A^{-1}A = AA^{-1} = I \tag{A.18}$$

A^{-1} is called the *inverse* of A. It is square, of the same order as A, and unique. In the matrix equation

$$AX = B \tag{A.19}$$

multiply by A^{-1} to get

$$A^{-1}AX = A^{-1}B, \quad \text{or} \quad X = A^{-1}B \tag{A.20}$$

Thus to solve the system of equations in Equation (A.19) for X it is only necessary to find A^{-1} and multiply it into B.

Of the many methods of calculating A^{-1} the most obvious one is based on the solution of Equation (A.19) by determinants. From Equations (A.15) and (A.13)

$$x_i = \sum_{j=1}^{m} b_j (-1)^{i+j} |M_{ji}|/|A|, \quad \text{column } i \text{ expansion,}$$

$$= \sum_{j=1}^{m} c_{ij} b_j, \quad c_{ij} = (-1)^{i+j} |M_{ji}|/|A|, \tag{A.21}$$

Thus $X = CB$ and the c_{ij} must be the elements of A^{-1} by Equation (A.20), whence

$$A^{-1} = \left[(-1)^{i+j} |M_{ij}|/|A| \right]^T \tag{A.22}$$

where capital T indicates the transpose of the matrix, or the interchange of the rows and columns.

Example A.1. Use Equation (A.22) to find the inverse of

$$A = \begin{bmatrix} 2 & -3 & 1 \\ 1 & -1 & 2 \\ 3 & -2 & 1 \end{bmatrix}. \tag{A.23}$$

Solution. From Equation (A.12)

$$|A| = 2(3) - (-3)(-5) + (1)(1) = -8$$

$$|M_{11}| = \begin{bmatrix} -1 & 2 \\ -2 & 1 \end{bmatrix} = 3, \quad |M_{12}| = \begin{bmatrix} 1 & 2 \\ 3 & 1 \end{bmatrix} = -5,$$

$$|M_{13}| = 1, \quad |M_{21}| = -1, \quad |M_{22}| = -1, \quad |M_{23}| = 5,$$
$$|M_{31}| = -5, \quad |M_{32}| = 3, \quad |M_{33}| = 1$$

$$A^{-1} = (1/8) \begin{bmatrix} -3 & -1 & 5 \\ -5 & 1 & 3 \\ -1 & 5 & -1 \end{bmatrix} \tag{A.24}$$

A more convenient method for inversion of large matrices than that of Equation (A.22) is given by using *submatrices*. Any matrix can be partitioned into submatrices and matrix operations can be performed using the submatrices as long as the matrix rules are obeyed. For example

$$\begin{bmatrix} \begin{bmatrix} 1 & 0 \\ 0 & 1 \end{bmatrix} & \begin{bmatrix} 2 \\ -2 \end{bmatrix} \end{bmatrix} \begin{bmatrix} \begin{bmatrix} 1 & 0 \\ 0 & 1 \end{bmatrix} \\ \begin{bmatrix} 3 & -1 \end{bmatrix} \end{bmatrix} = \begin{bmatrix} \begin{bmatrix} 1 & 0 \\ 0 & 1 \end{bmatrix} \begin{bmatrix} 1 & 0 \\ 0 & 1 \end{bmatrix} + \begin{bmatrix} 2 \\ -2 \end{bmatrix} [3 \quad -1] \end{bmatrix}$$

$$= \begin{bmatrix} \begin{bmatrix} 1 & 0 \\ 0 & 1 \end{bmatrix} + \begin{bmatrix} 6 & -2 \\ -6 & 2 \end{bmatrix} \end{bmatrix} = \begin{bmatrix} 7 & -2 \\ -6 & 3 \end{bmatrix}$$

which is the same result as that obtained by direct multiplication without partitioning.

Partition the given matrix G into four submatrices

$$G = \begin{bmatrix} G_{11} & | & G_{12} \\ \hline G_{21} & | & G_{22} \end{bmatrix} \tag{A.25}$$

where G_{11} and G_{22} are square matrices but not necessarily the same order, and $|G_{11}| \neq 0$. Let the inverse of G be

$$G^{-1} = \begin{bmatrix} H_{11} & | & H_{12} \\ \hline H_{21} & | & H_{22} \end{bmatrix} \tag{A.26}$$

where the H_{ij} matrices are the same size as the corresponding G_{ij} matrices. Now GG^{-1} gives the four equations

$$G_{11}H_{11} + G_{12}H_{21} = I \tag{A.27}$$
$$G_{11}H_{12} + G_{12}H_{22} = 0 \tag{A.28}$$
$$G_{21}H_{11} + G_{22}H_{21} = 0 \tag{A.29}$$
$$G_{21}H_{12} + G_{22}H_{22} = I \tag{A.30}$$

Notes on matrix algebra

Multiply Equation (A.28) by G_{11}^{-1} and then by G_{21}, both on left, to get

$$H_{12} + G_{11}^{-1}G_{12}H_{22} = 0 \qquad (A.31)$$
$$G_{21}H_{12} + G_{21}G_{11}^{-1}G_{12}H_{22} = 0 \qquad (A.32)$$

Put $G_{21}H_{12}$ from Equation (A.32) into Equation (A.30) to get

$$(G_{22} - G_{21}G_{11}^{-1}G_{12})H_{22} = I$$

or

$$H_{22} = Q^{-1}, \qquad Q = G_{22} - G_{21}G_{11}^{-1}G_{12} \qquad (A.33)$$

Put H_{22} into Equation (A.31) to get

$$H_{12} = -G_{11}^{-1}G_{12}Q^{-1} \qquad (A.34)$$

From Equations (A.27) and (A.29) by similar procedure

$$H_{21} = -Q^{-1}G_{21}G_{11}^{-1}, \qquad H_{11} = G_{11}^{-1}(I + G_{12}Q^{-1}G_{21}G_{11}^{-1}) \qquad (A.35)$$

Thus the four submatrices in the inverse G^{-1} in Equation (A.26) are given by Equations (A.33)-(A.35). In the calculations the submatrices G_{11} and Q can be inverted by using exactly the same procedure of partitioning them into four smaller submatrices.

Example A.2. Use submatrices to invert the matrix

$$G = \begin{bmatrix} 2 & -3 & | & 1 \\ 1 & -1 & | & 2 \\ \hline 3 & -2 & | & 1 \end{bmatrix}$$

Solution. Partition the 2 by 2 G_{11} into 1 by 1 elements

$$G_{11} = \begin{bmatrix} 2 & | & -3 \\ \hline 1 & | & -1 \end{bmatrix}, \qquad g_{11} = 2, \quad g_{12} = -3, \quad g_{21} = 1, \quad g_{22} = -1$$

whence $g_{11}^{-1} = 1/2$, $q = -1 - (1)(1/2)(-3) = 1/2$, $h_{22} = q^{-1} = 2$, $h_{12} = -(1/2)(-3)(2) = 3$, $h_{21} = -(2)(1)(1/2) = -1$, $h_{11} = (1/2)[1 + (-3)(2)(1)(1/2)] = -1$, and

$$G_{11}^{-1} = \begin{bmatrix} -1 & 3 \\ -1 & 2 \end{bmatrix}$$

For the matrix G

$$Q = 1 - [3 \quad -2]\begin{bmatrix} -1 & 3 \\ -1 & 2 \end{bmatrix}\begin{bmatrix} 1 \\ 2 \end{bmatrix} = 1 - [3 \quad -2]\begin{bmatrix} 5 \\ 3 \end{bmatrix} = -8$$

$$H_{22} = Q_{-1} = -(1/8)$$

$$H_{21} = (1/8)[3 \quad -2]\begin{bmatrix} -1 & 3 \\ -1 & 2 \end{bmatrix} = (1/8)[-1 \quad 5]$$

$$H_{12} = -\begin{bmatrix} -1 & 3 \\ -1 & 2 \end{bmatrix}\begin{bmatrix} 1 \\ 2 \end{bmatrix}(-1/8) = (1/8)\begin{bmatrix} 5 \\ 3 \end{bmatrix}$$

$$H_{11} = \begin{bmatrix} -1 & 3 \\ -1 & 2 \end{bmatrix}\left[\begin{bmatrix} 1 & 0 \\ 0 & 1 \end{bmatrix} - (1/8)\begin{bmatrix} 1 \\ 2 \end{bmatrix}[-1 \quad 5]\right] = (1/8)\begin{bmatrix} -3 & -1 \\ -5 & 1 \end{bmatrix}$$

$$G^{-1} = (1/8)\begin{bmatrix} -3 & -1 & 5 \\ -5 & 1 & 3 \\ -1 & 5 & -1 \end{bmatrix}$$

A.5. Solution of systems of simultaneous equations by matrices

Consider the matrix equation

$$SU = G \qquad (A.36)$$

where S is a p by p square matrix of known constants with $|S| \neq 0$, G is a p by q matrix of known constants, and U is a p by q unknown matrix. If $q = 1$, Equation (A.36) represents a system of simultaneous equations. In fact for each column in G the corresponding column in U gives a solution of a system of simultaneous equations. If S^{-1} were known, then by Equation (A.20) the solution for U would be $U = S^{-1}G$. However, it is possible to obtain U without using S^{-1}.

Partition S into M by M submatrices S_{ij} with the properties

$$\begin{aligned} S_{ii} \text{ square}, & \quad |S_{ii}| \neq 0, \quad i = 1, 2, \cdots, M, \\ S_{i,i\pm j} &= 0 \text{ for } j > N, \quad N < M, \quad i = 1, 2, \cdots, M \end{aligned} \qquad (A.37)$$

That is, all submatrices in S are zero outside of $2N+1$ diagonals with N diagonals above the main diagonal and N diagonals below the main diagonal. If S has no zero submatrices then $N = M - 1$. Partition U and G into a column of M submatrices U_i and G_i, $i = 1, 2, \cdots, M$, with the number of rows in U_i and G_i the same as in the corresponding S_{ii}. Note that the S_{ii} do not have to be the same size, although in many practical problems they will be taken the same size. Now S can be expressed in the form

$$S = \begin{bmatrix} S_{i,i-N} \cdots S_{i,i-1} \; S_{ii} \; S_{i,i+1} \cdots S_{i,i+N} \end{bmatrix}_1^M \qquad (A.38)$$

The matrix S in Equation (A.38) can be expressed as the product of a lower triangular matrix B times an upper triangular matrix C with B and C partitioned in the same way as S, or

$$\begin{aligned} S = BC, \quad B &= \begin{bmatrix} B_{i,i-N} \cdots B_{i,i-1} \; B_{ii} \end{bmatrix}_1^M \\ C &= \begin{bmatrix} I_{ii} \; C_{i,i+1} \cdots C_{i,i+N} \end{bmatrix}_1^M \end{aligned} \qquad (A.39)$$

Multiply BC and equate to the corresponding submatrices in S to get recursion formulas for the calculation of $B_{i,i-j}$ and $C_{i,i+j}$:

$$\begin{aligned} B_{i,i-j} &= S_{i,i-j} - \sum_{k=1}^{N-j} B_{i,i-j-k} C_{i-j-k,i-j}, \\ & \quad j = N, N-1, \cdots 1, 0; \; j < i; \; k < (i-j) \\ C_{i,i+j} &= B_{ii}^{-1}\left(S_{i,i+j} - \sum_{k=1}^{N-j} B_{i,i-k} C_{i-k,i+j} \right), \\ & \quad j = 1, 2, \cdots, N; \; j \leq (M-i); \; k < i \\ i &= 1, 2, 3, \cdots, M \end{aligned} \qquad (A.40)$$

If P is defined by

$$CU = P \qquad (A.41)$$

where P is partitioned in a column of M submatrices P_i the same as U, then from Equations (A.36) and (A.39)

$$BP = G; \quad P_i = B_{ii}^{-1}\left(G_i - \sum_{k=1}^{N} B_{i,i-k} P_{i-k}\right), \quad (A.42)$$
$$k < i; \quad i = 1, 2, 3, \cdots, M$$

Note that this recursion for P_i can be carried out simultaneously with the recursion for $B_{i,i-j}$ and $C_{i,i+j}$ in Equation (A.40).

From Equations (A.39) and (A.41) the backsweep recursion formula for U_i is

$$U_i = P_i - \sum_{k=1}^{N} C_{i,i+k} U_{i+k}, \quad k \le (M-i); \quad i = M, M-1, \cdots, 1 \quad (A.43)$$

Thus for any number of columns in G the solution for the matrix U can be obtained by a forward recursion calculating the submatrices $B_{i,i-j}$, $C_{i,i+j}$, P_i and a backward recursion calculating the submatrices U_i of U. Note that the submatrices $C_{i,i+j}$ and P_i must be kept for use on the backward recursion. Note also that after the U_i are obtained for $i = M, M-1, \cdots, M-N+1$, and if $S_{i,i-N}$ in Equation (A.38) is such that $|S_{i,i-N}| \neq 0$, then the backward recursion may be continued by using Equations (A.36) and (A.38), or

$$U_i = S_{i+N,i}^{-1}\left(G_{i+N} - \sum_{k=1}^{N} S_{i+N,i+k} U_{i+k}\right), \quad i = M-N, \cdots, 1 \quad (A.44)$$

The determinant of S can be obtained directly from the submatrices B_{ii}, as

$$|S| = |B||C| = |B_{11}||B_{22}|\cdots|B_{MM}||I| \quad (A.45)$$

The inverse of S can be obtained by the above procedure by taking $G = I$, whence $P = B^{-1}$ and $U = S^{-1}$. Although this method is one of the shortest for getting S^{-1} when S is large, it is evident that a system of simultaneous equations with one column for G can be solved much quicker than S^{-1} can be calculated.

A.6. Solution of systems of simultaneous equations by tri-diagonal matrices

A particular case of Equation (A.38) that arises in many structural problems and in finite difference equations is that of $N = 1$ giving S three diagonals in submatrices. The equations for calculating the unknown U are considerably simplified for this case from those in Section A.5. Equations (A.38)-(A.44) become

$$SU = G, \quad S = \left[S_{i,i-1}\ S_{ii}\ S_{i,i+1}\right]_1^M \quad (A.46)$$
$$S = BC, \quad B = \left[S_{i,i-1}\ B_{ii}\right]_1^M, \quad C = \left[I_{ii}\ C_{i,i+1}\right]_1^M \quad (A.47)$$

$$B_{11} = S_{11}, \quad B_{ii} = S_{ii} - S_{i,i-1}C_{i-1,i},$$
$$C_{i,i+1} = B_{ii}^{-1}S_{i,i+1}, \quad i = 1, 2, \cdots, M \tag{A.48}$$

$$P_1 = B_{11}^{-1}G_1, \quad P_i = B_{ii}^{-1}(G_i - S_{i,i-1}P_{i-1}), \quad i = 2, \cdots, M \tag{A.49}$$
$$U_M = P_M, \quad U_i = P_i - C_{i,i+1}U_{i+1}, \quad i = M-1, \cdots, 1 \tag{A.50}$$
$$U_M = P_M, \quad U_{i-1} = S_{i-1,i}^{-1}(G_i - S_{ii}U_i - S_{i,i+1}U_{i+1}),$$
$$i = M, M-1, \cdots, 2; \quad |S_{i-1,1}| \neq 0 \tag{A.51}$$

Note that although $|S_{i-1,i}| = 0$ in Equation (A.51) the backward recursion of Equation (A.51) can still be accomplished by rearranging the equations in the original system to give a partitioning of S that will have $|S_{i-1,i}| \neq 0$.

Example A.3. Solve the following system of simultaneous equations by using several tri-diagonal partitionings. Also find the inverse of the 5 by 5 matrix by using $G = I$:

$$\begin{bmatrix} 1 & 1 & 0 & 0 & 0 \\ 1 & 2 & -1 & 0 & 0 \\ 0 & -1 & -2 & 3 & 0 \\ 0 & 0 & 3 & 1 & 4 \\ 0 & 0 & 0 & 4 & 2 \end{bmatrix} \begin{bmatrix} U_1 \\ U_2 \\ U_3 \\ U_4 \\ U_5 \end{bmatrix} = \begin{bmatrix} 1 \\ -2 \\ 0 \\ 3 \\ 2 \end{bmatrix} \tag{A.52}$$

Solution. Partition into 1 by 1 submatrices so that by Equation (A.48) $B_{11} = 1$, $C_{12} = (1)(1) = 1$, $B_{22} = 2 - (1)(1) = 1$, $C_{23} = (1)(-1) = -1$, $B_{33} = -2 - (-1)(-1) = -3$, $C_{34} = (-1/3)(3) = -1$, $B_{44} = 1 - (3)(-1) = 4$, $C_{45} = (1/4)(4) = 1$, $B_{55} = 2 - (4)(1) = -2$. From Equation (A.49) $P_1 = (1)(1) = 1$, $P_2 = (1)[-2 - (1)(1)] = -3$, $P_3 = (-1/3)[0 - (-1)(-3)] = 1$, $P_4 = (1/4)[3 - (3)(1)] = 0$, $P_5 = (-1/2)[2 - (4)(0)] = -1$, and from Equation (A.50) $U_5 = -1$, $U_4 = 0 - (1)(-1) = 1$, $U_3 = 1 - (-1)(1) = 2$, $U_2 = -3 - (-1)(2) = -1$, $U_1 = 1 - (1)(-1) = 2$. If $G = I$, the values of the B_{ii} and $C_{i,i+1}$ are unchanged and from Equation (A.49)

$$P_1 = (1)[1 \ 0 \ 0 \ 0 \ 0] = [1 \ 0 \ 0 \ 0 \ 0],$$
$$P_2 = (1)\big[[0 \ 1 \ 0 \ 0 \ 0] - (1)[1 \ 0 \ 0 \ 0 \ 0]\big] = [-1 \ 1 \ 0 \ 0 \ 0]$$
$$P_3 = (-1/3)\big[[0 \ 0 \ 1 \ 0 \ 0] - (-1)[-1 \ 1 \ 0 \ 0 \ 0]\big]$$
$$= (-1/3)[-1 \ 1 \ 1 \ 0 \ 0],$$
$$P_4 = (1/4)\big[[0 \ 0 \ 0 \ 1 \ 0] - (3)(-1/3)[-1 \ 1 \ 1 \ 0 \ 0]\big]$$
$$= (1/4)[-1 \ 1 \ 1 \ 1 \ 0],$$
$$P_5 = (-1/2)\big[[0 \ 0 \ 0 \ 0 \ 1] - (4)(1/4)[-1 \ 1 \ 1 \ 1 \ 0]\big]$$
$$= (1/2)[-1 \ 1 \ 1 \ 1 \ -1]$$

From Equation (A.50) the U_i give

$$U = S^{-1} = (1/12) \begin{bmatrix} 17 & -5 & 7 & 3 & -6 \\ -5 & 5 & -7 & -3 & 6 \\ 7 & -7 & -7 & -3 & 6 \\ 3 & -3 & -3 & -3 & 6 \\ -6 & 6 & 6 & 6 & -6 \end{bmatrix}$$

Now suppose the S matrix in Equation (A.52) is partitioned to give

$$S_{11} = \begin{bmatrix} 1 & 1 \\ 1 & 2 \end{bmatrix} \quad S_{12} = \begin{bmatrix} 0 & 0 \\ -1 & 0 \end{bmatrix} = S_{21}^T, \quad S_{22} = \begin{bmatrix} -2 & 3 \\ 3 & 1 \end{bmatrix},$$
$$S_{31} = [0 \ 4] = D_{13}^T, \quad S_{33} = 2,$$

whence

$$B_{11} = \begin{bmatrix} 1 & 1 \\ 1 & 2 \end{bmatrix}, \quad B_{11}^{-1} = \begin{bmatrix} 2 & -1 \\ -1 & 1 \end{bmatrix}, \quad C_{12} = \begin{bmatrix} 2 & -1 \\ -1 & 1 \end{bmatrix} \begin{bmatrix} 0 & 0 \\ -1 & 0 \end{bmatrix} = \begin{bmatrix} 1 & 0 \\ -1 & 0 \end{bmatrix},$$

$$B_{22} = \begin{bmatrix} -2 & 3 \\ 3 & 1 \end{bmatrix} - \begin{bmatrix} 0 & -1 \\ 0 & 0 \end{bmatrix} \begin{bmatrix} 1 & 0 \\ -1 & 0 \end{bmatrix} = \begin{bmatrix} -3 & 3 \\ 3 & 1 \end{bmatrix},$$

$$B_{22}^{-1} = (1/12) \begin{bmatrix} -1 & 3 \\ 3 & 3 \end{bmatrix}, \quad C_{23} = (1/12) \begin{bmatrix} -1 & 3 \\ 3 & 3 \end{bmatrix} \begin{bmatrix} 0 \\ 4 \end{bmatrix} = \begin{bmatrix} 1 \\ 1 \end{bmatrix},$$

$$B_{33} = 2 - \begin{bmatrix} 0 & 4 \end{bmatrix} \begin{bmatrix} 1 \\ 1 \end{bmatrix} = -2, \quad B_{33}^{-1} = -1/2$$

$$P_1 = \begin{bmatrix} 2 & -1 \\ -1 & 1 \end{bmatrix} \begin{bmatrix} 1 \\ -2 \end{bmatrix} = \begin{bmatrix} 4 \\ -3 \end{bmatrix},$$

$$P_2 = (1/12) \begin{bmatrix} -1 & 3 \\ 3 & 3 \end{bmatrix} \left[\begin{bmatrix} 0 \\ 3 \end{bmatrix} - \begin{bmatrix} 0 & -1 \\ 0 & 0 \end{bmatrix} \begin{bmatrix} 4 \\ -3 \end{bmatrix} \right] = \begin{bmatrix} 1 \\ 0 \end{bmatrix},$$

$$P_3 = (-1/2) \left[2 - \begin{bmatrix} 0 & 4 \end{bmatrix} \begin{bmatrix} 1 \\ 0 \end{bmatrix} \right] = -1,$$

$$U_3 = -1, \quad U_2 = \begin{bmatrix} 1 \\ 0 \end{bmatrix} - \begin{bmatrix} 1 \\ 1 \end{bmatrix}(-1) = \begin{bmatrix} 2 \\ 1 \end{bmatrix},$$

$$U_1 = \begin{bmatrix} 4 \\ -3 \end{bmatrix} - \begin{bmatrix} 1 & 0 \\ -1 & 0 \end{bmatrix} \begin{bmatrix} 2 \\ 1 \end{bmatrix} = \begin{bmatrix} 2 \\ -1 \end{bmatrix},$$

which is the same solution for U_i as before.

It should be pointed out that the submatrices B_{ii} and $C_{i,i+1}$ can be used to give the inverses of B, C, and S directly in submatrix form. Take S as symmetrical and change the sign on $C_{i,i+1}$ in Equation (A.47) so that in Equations (A.47) and (A.48)

$$\begin{aligned} C &= [I_{ii} \quad -C_{i,i+1}], \quad C_{i,i+1} = -B_{ii}^{-1} S_{i,i+1}, \\ B_{ii} &= S_{ii} + S_{i,i-1} C_{i-1,i}. \end{aligned} \qquad (A.53)$$

Apply the submatrix inversion procedure in Equations (A.25)-(A.35) to C in sequence as

$$C_1 = I_{11}, \quad C_1^{-1} = I_{11},$$

$$C_2 = \begin{bmatrix} C_1 & -C_{12} \\ 0 & I_{22} \end{bmatrix}, \quad C_2^{-1} = \begin{bmatrix} C_1^{-1} & C_{12} \\ 0 & I_{22} \end{bmatrix},$$

$$C_3 = \left[\begin{array}{c|c} C_2 & \begin{matrix} 0 \\ -C_{23} \end{matrix} \\ \hline 0 \quad 0 & I_{33} \end{array} \right], \quad C_3^{-1} = \left[\begin{array}{c|c} C_2^{-1} & \begin{matrix} C_{12}C_{23} \\ C_{23} \end{matrix} \\ \hline 0 \quad 0 & I_{33} \end{array} \right],$$

etc., with the $i+1$ row and $i+1$ column being added to C_i to get C_{i+1}. When $i = M$, then $C_M = C$, and $C_M^{-1} = C^{-1}$, or

$$C^{-1} = \begin{bmatrix} I_{11} & C_{12} & (C_{12}C_{23}) & (C_{12}C_{23}C_{34}) & \cdots & (C_{12}\cdots C_{M-1,M}) \\ 0 & I_{22} & C_{23} & (C_{23}C_{34}) & \cdots & (C_{23}\cdots C_{M-1,M}) \\ 0 & 0 & I_{33} & C_{34} & \cdots & (C_{34}\cdots C_{M-1,M}) \\ \multicolumn{6}{c}{\dotfill} \\ 0 & 0 & 0 & 0 & \cdots & I_{MM} \end{bmatrix} \quad (A.54a)$$

In a similar manner,

$$B^{-1} =$$

$$\begin{bmatrix} B_{11}^{-1} & 0 & 0 & \cdots & 0 & 0 \\ B_{22}^{-1}C_{12}^T & B_{22}^{-1} & 0 & \cdots & 0 & 0 \\ B_{33}^{-1}C_{23}^TC_{12}^T & B_{33}^{-1}C_{23}^T & B_{33}^{-1} & \cdots & 0 & 0 \\ \cdots & \cdots & \cdots & \cdots & \cdots & \cdots \\ B_{MM}^{-1}C_{M-1,M}^T \cdots C_{12}^T & B_{MM}^{-1} \cdots C_{23}^T & \cdots & \cdots & B_{MM}^{-1}C_{M-1,M}^T & B_{MM}^{-1} \end{bmatrix}.$$

(A.54b)

The inverse of the matrix S is

$$S^{-1} = C^{-1}B^{-1} = \begin{bmatrix} P_{i1} & P_{i2} & P_{i3} \cdots P_{ii} \cdots P_{iM} \end{bmatrix}_1^M, \quad (A.55)$$

$$\begin{aligned} P_{MM} &= B_{MM}^{-1}, \\ P_{ii} &= B_{ii}^{-1} + C_{i,i+1}P_{i+1,i+1}C_{i,i+1}^T, & i &= M-1, M-2, \cdots, 1, \\ P_{ij} &= P_{i,j+1}C_{j,j+1}^T, & j < i, & i &= M, M-1, \cdots, 2, \\ P_{ij} &= C_{i,i+1}P_{i+1,j}, & j > i, & i &= M-1, M-2, \cdots, 1. \end{aligned} \quad (A.56)$$

This gives the inverse of S from the P_{MM} submatrix in reverse order as P_{MM}, $P_{M-1,M-1}, \cdots, P_{11}$, and $P_{M,M-1} = P_{MM}C_{M-1,M}^T$, or $P_{M-1,M} = C_{M-1,M}P_{MM}$, or $P_{M,M-1} = P_{M-1,M}^T$, etc.

Example A.4. Calculate the inverse of the upper bi-diagonal matrix

$$C = \begin{bmatrix} 1 & -2 & 0 & 0 & 0 \\ 0 & 1 & -5 & 0 & 0 \\ 0 & 0 & 1 & -1 & 0 \\ 0 & 0 & 0 & 1 & -3 \\ 0 & 0 & 0 & 0 & 1 \end{bmatrix}.$$

Solution. In Equation (A.54) with 1 by 1 submatrices the C_i values are known, or $C_{12} = 2$, $C_{23} = 5$, $C_{34} = 1$, $C_{45} = 3$, so that

$$C^{-1} = \begin{bmatrix} 1 & 2 & 10 & 10 & 30 \\ 0 & 1 & 5 & 5 & 15 \\ 0 & 0 & 1 & 1 & 3 \\ 0 & 0 & 0 & 1 & 3 \\ 0 & 0 & 0 & 0 & 1 \end{bmatrix},$$

where it is evident that $CC^{-1} = I$.

Example A.5. Calculate the inverse of the lower bi-diagonal matrix

$$B = \begin{bmatrix} 1/2 & 0 & 0 & 0 & 0 \\ 3 & 1/3 & 0 & 0 & 0 \\ 0 & 2 & 1 & 0 & 0 \\ 0 & 0 & 5 & 1/2 & 0 \\ 0 & 0 & 0 & 1 & 1/4 \end{bmatrix}.$$

Solution. In Equation (A.47) with 1 by 1 submatrices, the B_{ii} and $S_{i,i-1}$ values are known so that in Equation (A.53) with $S_{i,i+1} = S_{i+1,i}$, $C_{12} = -B_{11}^{-1}S_{12} = -(2)(3) = -6$, $C_{23} = -6$, $C_{34} = -5$, $C_{45} = -2$, and Equation (A.54b) gives

$$B^{-1} = \begin{bmatrix} 2 & 0 & 0 & 0 & 0 \\ -18 & 3 & 0 & 0 & 0 \\ 36 & -6 & 1 & 0 & 0 \\ -360 & 60 & -10 & 2 & 0 \\ 1440 & -240 & 40 & -8 & 4 \end{bmatrix}.$$

Example A.6. Use the 1 by 1 B_{ii} and $C_{i,i+1}$ values in Example A.3 to calculate the inverse of S in Equation (A.52) by using Equation (A.56).

Solution. From Equation (A.56), $P_{55} = B_{55}^{-1} = -1/2$, $P_{44} = 1/4 + (-1)(-1/2)(-1) = -1/4$. $P_{33} = -1/3 + (1)(-1/4)(1) = -7/12$, $P_{22} = 1 + (1)(-7/12)(1) = 5/12$, $P_{11} = 1 + 5/12 = 17/12$, $P_{54} = (-1/2)(-1) = 1/2$, $P_{53} = (1/2)(1) = 1/2$, $P_{52} = (1/2)(1) = 1/2$, $P_{51} = (1/2)(-1) = -1/2$, $P_{43} = -1/4$, $P_{42} = -1/4$, $P_{41} = -(1/4)(-1) = 1/4$, $P_{32} = (-7/12)(1) = -7/12$, $P_{31} = (-7/12)(-1) = 7/12$, $P_{21} = (5/12)(-1) = -5/12$, which checks the inverse given in Example A.3.

A.7. Solution of systems of equations by Jordan successive transformations

Arrange the system of equations in Equation (A.36), $SU = G$, in the form

$$T_M T_{M-1} \cdots T_2 T_1 [S \mid G] = [I \mid U], \qquad (A.57)$$

where $[S \mid G]$ and $[I \mid U]$ are rectangular matrices and G and U can be one column to M columns. G can be I so that $U = S^{-1}$. The $T_1, T_2, \cdots, T_M$ matrices operate on S to transform it into I one column at the time. Take S in the general form

$$S = \begin{bmatrix} a_{11} & a_{12} & \cdots & a_{1M} \\ a_{21} & a_{22} & \cdots & a_{2M} \\ \cdots & \cdots & \cdots & \cdots \\ a_{M1} & a_{M2} & \cdots & a_{MM} \end{bmatrix}, \qquad (A.58)$$

and then take T_1 as

$$T_1 = \left[\begin{array}{c|c} 1/a_{11} & 0 \\ \hline -a_{21}/a_{11} & \\ -a_{31}/a_{11} & I_{M-1} \\ \cdots & \\ -a_{M1}/a_{11} & \end{array} \right]. \qquad (A.59)$$

Thus, multiply T_1 into S and G to get

$$S_1 = T_1 S = \left[\begin{array}{c|cccc} 1 & b_{12} & b_{13} & \cdots & b_{1M} \\ \hline & b_{22} & b_{23} & \cdots & b_{2M} \\ 0 & \cdots & & & \\ & b_{M2} & b_{M3} & \cdots & b_{MM} \end{array} \right], \quad G_1 = T_1 G, \qquad (A.60)$$

$$b_{1j} = a_{1j}/a_{11}, \quad b_{ij} = a_{ij} - (a_{i1} a_{1j}/a_{11}) \text{ for } i \neq 1. \qquad (A.61)$$

To get the second column of the inverse, take

$$T_2 = \begin{bmatrix} 1 & | & -b_{12}/b_{22} & | & 0 \\ \hline 0 & | & 1/b_{22} & | & 0 \\ \hline & | & -b_{32}/b_{22} & | & \\ 0 & | & \ldots\ldots\ldots & | & I_{M-2} \\ & | & -b_{M2}/b_{22} & | & \end{bmatrix}, \quad (A.62)$$

whence

$$S_2 = T_2 S_1 = T_2 T_1 S = \begin{bmatrix} & | & c_{13} & c_{14} & \cdots & c_{1M} \\ I_2 & | & & & & \\ & | & c_{23} & c_{24} & \cdots & c_{2M} \\ \hline & | & c_{33} & c_{34} & \cdots & c_{3M} \\ 0 & | & \ldots\ldots\ldots\ldots\ldots & & & \\ & | & c_{M3} & c_{M4} & \cdots & c_{MM} \end{bmatrix}, \quad (A.63)$$

$$G_2 = T_2 G_1 = T_2 T_1 G, \quad c_{2j} = b_{2j}/b_{22},$$
$$c_{ij} = b_{ij} - (b_{i2} b_{2j}/b_{22}) \text{ for } i \neq 2. \quad (A.64)$$

Continue the procedure until T_M produces I and U in Equation (A.57).

In the calculation of the $T_1, T_2, \cdots, T_M$ matrices, one of the diagonal elements may be zero so that it is necessary to interchange the affected column with one of the following columns. After the calculations are completed, all the interchanged columns can be changed back to their original positions. In fact, it is evident that column interchange can be made at each step in order to use the largest element in each row on the diagonal and thus improve the accuracy of the calculations. This procedure is used in Chapter 3 to automatically select the redundant members in the matrix force method of analysis.

Example A.7. Use the Jordan successive transformation procedure to solve the system of equations

$$\begin{bmatrix} 4 & 1 & 2 \\ 2 & 1 & 3 \\ 1 & -2 & 3 \end{bmatrix} \begin{bmatrix} U_1 \\ U_2 \\ U_3 \end{bmatrix} = \begin{bmatrix} -3 \\ 2 \\ 1 \end{bmatrix}.$$

Solution. From Equations (A.59)-(A.64),

$$T_1 = \begin{bmatrix} 1/4 & 0 & 0 \\ -1/2 & 1 & 0 \\ -1/4 & 0 & 1 \end{bmatrix}, \quad S_1 = T_1 S = T_1 \begin{bmatrix} 4 & 1 & 2 \\ 2 & 1 & 3 \\ 1 & -2 & 3 \end{bmatrix} = \begin{bmatrix} 1 & 1/4 & 1/2 \\ 0 & 1/2 & 2 \\ 0 & -9/4 & 5/2 \end{bmatrix},$$

$$G_1 = T_1 G = \begin{bmatrix} 1/4 & 0 & 0 \\ -1/2 & 1 & 0 \\ -1/4 & 0 & 1 \end{bmatrix} \begin{bmatrix} -3 \\ 2 \\ 1 \end{bmatrix} = \begin{bmatrix} -3/4 \\ 7/2 \\ 7/4 \end{bmatrix}, \quad T_2 = \begin{bmatrix} 1 & -1/2 & 0 \\ 0 & 2 & 0 \\ 0 & 9/2 & 1 \end{bmatrix},$$

$$S_2 = T_2 S_1 = \begin{bmatrix} 1 & 0 & -1/2 \\ 0 & 1 & 4 \\ 0 & 0 & 23/2 \end{bmatrix}, \quad G_2 = T_2 G_1 = \begin{bmatrix} -5/2 \\ 7 \\ 35/2 \end{bmatrix},$$

$$T_3 = \begin{bmatrix} 1 & 0 & 1/23 \\ 0 & 1 & -8/23 \\ 0 & 0 & 2/23 \end{bmatrix}, \quad S_3 = T_3 S_2 = I, \quad G_3 = T_3 G_2 = \tfrac{1}{23}\begin{bmatrix} -40 \\ 21 \\ 35 \end{bmatrix} = U.$$

This result for U checks the original system of equations.

Example A.8. Find the inverse of the S matrix in Example (A.7).

Solution. Change G to I in Example (A.7) so that

$$S^{-1} = T_3 T_2 T_1 I = T_3 \begin{bmatrix} 1/2 & -1/2 & 0 \\ -1 & 2 & 0 \\ -5/2 & 9/2 & 1 \end{bmatrix} = \tfrac{1}{23}\begin{bmatrix} 9 & -7 & 1 \\ -3 & 10 & -8 \\ -5 & 9 & 2 \end{bmatrix},$$

which checks $SS^{-1} = I$.

A.8. Matrix representations

From the definition of matrix multiplication, any sum of products can be written in matrix form, Equation (A.9), as

$$a_{ij} = \sum_{k=1}^{r} b_{ik} c_{kj} = [b_i]^T [c_j] = b_{ic}^T c_{jc}, \tag{A.65}$$

where b_{ic}^T is a 1 by r row with b_{ic} and c_{jc} as r by 1 columns. Here i and j are regarded as fixed so that a_{ij} is a scalar. If j is fixed but i takes values $i = 1, 2, \cdots, M$, then

$$[a_j] = a_{jc} = [B][c_j] = B c_{jc}, \tag{A.66}$$

where a_{jc} is a M by 1 column and B is a M by r matrix.

For a sum of triple products with fixed i and j, the form is

$$a_{ij} = \sum_{k=1}^{r} b_{ik} d_k c_{kj} = [b_i]^T [d][c_j] = b_{ic}^T D c_{jc}, \tag{A.67}$$

where D is a r by r diagonal matrix with the d_k elements on the diagonal. If i and j both take values $i = 1, 2, \cdots, M$, $j = 1, 2, \cdots, M$, then a_{ij} becomes a square matrix A, or

$$A = BDC, \tag{A.68}$$

where A is M by M, B is M by r, D is r by r, and C is r by M. Note that if the $b_{jc} = c_{jc}$, then Equation (A.68) is

$$A = C^T DC. \tag{A.69}$$

Consider the differential equation

$$\frac{d^2 U}{dx^2} + p(x) \frac{dU}{dx} + q(x) U = r(x) \tag{A.70}$$

over the interval $0 \leq x \leq L$ with $U = 0$ at $x = 0, L$. Divide the interval into $M + 1$ equal elements of length Δx and take

$$\begin{aligned}\frac{dU}{dx} &= \frac{U(x + \Delta x) - U(x - \Delta x)}{2\Delta x} = \frac{U_{i+1} - U_{i-1}}{2\Delta x}, \\ \frac{d^2 U}{dx^2} &= \frac{U(x + \Delta x) - 2U(x) + U(x - \Delta x)}{(\Delta x)^2} = \frac{U_{i+1} - 2U_i + U_{i-1}}{(\Delta x)^2}.\end{aligned} \tag{A.71}$$

Take $p(x_i) = p_i$, $q(x_i) = q_i$, $r(x_i) = r_i$, and substitute Equation (A.71) into Equation (A.70) to get

$$(2 - p_i \Delta x)U_{i-1} + 2[q_i(\Delta x)^2 - 2]U_i + (2 + p_i \Delta x)U_{i+1} = 2r_i(\Delta x)^2,$$
$$i = 1, 2, \cdots, M; \quad U_0 = 0, \quad U_{M+1} = 0, \quad (A.72)$$

or

$$S_{i,i-1}U_{i-1} + S_{ii}U_i + S_{i,i+1}U_{i+1} = G_i, \quad i = 1, \cdots, M, \quad (A.73)$$

which is the tri-diagonal form of Equation (A.46).

The vector r can be written in matrix form as follows:

$$\mathbf{r} = \mathbf{i}r_x + \mathbf{j}r_y + \mathbf{k}r_z = [r_x \ r_y \ r_z] \begin{bmatrix} \mathbf{i} \\ \mathbf{j} \\ \mathbf{k} \end{bmatrix} = R_c^T \mathbf{I}_c = \mathbf{I}_c^T R_c, \quad (A.74)$$

where R_c is the column of vector components and $\mathbf{I}_c$ is the column of unit vectors. The transpose R_c^T of R_c is the row of vector components. If $\mathbf{q} = Q_c^T \mathbf{I}_c$, then the matrix form of the dot product is

$$\mathbf{r} \cdot \mathbf{q} = r_x q_x + r_y q_y + r_z q_z = R_c^T Q_c = Q_c^T R_c. \quad (A.75)$$

The cross product is

$$\mathbf{p} = \mathbf{r} \times \mathbf{q} = \mathbf{i}(r_y q_z - r_z q_y) + \mathbf{j}(r_z q_x - r_x q_z) + \mathbf{k}(r_x q_y - r_y q_x)$$
$$= \mathbf{I}_c^T R_s Q_c = \mathbf{I}_c^T Q_s^T R_c = Q_c^T R_s^T \mathbf{I}_c = R_c^T Q_s \mathbf{I}_c, \quad (A.76)$$

$$R_s = \begin{bmatrix} 0 & -r_z & r_y \\ r_z & 0 & -r_x \\ -r_y & r_x & 0 \end{bmatrix}, \quad Q_s = \begin{bmatrix} 0 & -q_z & q_y \\ q_z & 0 & -q_x \\ -q_y & q_x & 0 \end{bmatrix}, \quad (A.77)$$

where R_s is a square matrix involving the components of the vector r. Note that

$$R_s^T = -R_s. \quad (A.78)$$

If the vector p is taken as $\mathbf{p} = \mathbf{I}_c P_c$, then from Equation (A.76) the components in the vector product are

$$P_c = R_s Q_c = Q_s^T R_c = -Q_s R_c \quad (A.79)$$

so that R_s is a transformation matrix which transforms the components of the vector q into the components of the vector p in the same manner as the vector operation $\mathbf{r} \times \mathbf{q}$ does. Note from Equation (A.75) that

$$\mathbf{r} \cdot \mathbf{r} = r^2 = r_x^2 + r_y^2 + r_z^2 = R_c^T R_c. \quad (A.80)$$

Consider the definite integral

$$w = \int_0^L f(x)g(x)\,dx \quad (A.81)$$

Notes on matrix algebra

and divide the interval 0 to L into K subintervals, equal or unequal in length. The integral can be expressed as a sum

$$w = \sum_{k=0}^{K} f(x_k)g(x_k)(p\Delta x)_k = \sum_{k=0}^{K} f_k g_k q_k$$

$$= [f_0 \quad f_1 \quad \cdots \quad f_K] \begin{bmatrix} q_0 & 0 & \cdots & 0 \\ 0 & q_1 & \cdots & 0 \\ \vdots & & & \vdots \\ 0 & 0 & \cdots & q_K \end{bmatrix} \begin{bmatrix} g_0 \\ g_1 \\ \vdots \\ g_K \end{bmatrix} = F_c^T Q_d G_c, \quad (A.82)$$

where F_c^T is a 1 by $K+1$ row matrix, Q_d is a $K+1$ by $K+1$ square diagonal matrix, and G_c is a $K+1$ by 1 column matrix, the product being the 1 by 1 number w. The $q_k = (p\Delta x)_k$ elements in Q_d are weighting numbers which depend upon the method of numerical integration employed. For the *trapezoid rule*,

$$q_0, q_1, \cdots, q_K = (1/2, 1, 1, \cdots, 1, 1/2)\Delta x. \qquad (A.83)$$

while for *Simpson's rule* (K even)

$$q_0, q_1, \cdots, q_K = (1, 4, 2, 4, 2, \cdots, 2, 4, 1)(\Delta x/3). \qquad (A.84)$$

Note that

$$w = \int_0^L [f(x)]^2 dx = F_c^T Q_d F_c. \qquad (A.85)$$

Consider the definite integral

$$w(y) = \int_0^L C(y, x) P(x) dx, \qquad (A.86)$$

which is to be evaluated numerically for N values of the parameter y. Thus

$$w(y_i) = w_i = \sum_{k=1}^{K} C(y_i, x_k) P(x_k)(p\Delta x)_k$$

$$= \sum_{k=1}^{K} C_{ik} P_k q_k, \quad i = 1, 2, \cdots, N, \qquad (A.87)$$

$$W_c = CQ_d P_c, \qquad (A.88)$$

where W_c is N by 1 column matrix, C is N by K matrix, Q_d is K by K diagonal matrix, and P_c is K by 1 column matrix. In many practical problems $N = K$ and the matrix C is square and symmetrical.

The double integral with two parameters

$$w(x, y) = \int_0^L \int_0^c C(x, y; u, v) P(u, v) du\, dv \qquad (A.89)$$

can be written in the form, ($i = 1, 2, \cdots, K$; $j = 1, 2, \cdots, R$),

$$w(x_i, y_j) = \sum_{r=1}^{R} \sum_{k=1}^{K} C(x_i, y_j; u_k, v_r) P(u_k, v_r)(p\Delta u)_r (p\Delta v)_r,$$

$$w_n = \sum_{m=1}^{M} C_{nm} P_m q_m, \quad n = 1, 2, \cdots, M; \quad M = (K)(R), \quad \text{(A.90)}$$

where n counts over all values given by i and j together and m counts over k and r together. Hence the matrix equation is the same as Equation (A.88) with C a square matrix M by M.

Matrices can be used to solve *integral equations* in the same manner as differential equations (see Equations A.70 and A.73 above). The integral equation

$$w(y) = \int_0^L C(y, x) P(x) [w(x) + r(x)] dx \quad \text{(A.91)}$$

becomes

$$w_i = \sum_{k=1}^{K} C_{ik} P_k w_k q_k + \sum_{k=1}^{K} C_{ik} P_k r_k q_k, \quad i = 1, 2, \cdots, K,$$

$$W_c = C P_d Q_d W_c + C P_d Q_d R_c, \quad \text{(A.92)}$$

$$[I - C P_d Q_d] W_c = C P_d Q_d R_c, \quad \text{(A.93)}$$

which can be solved for W_c by the procedure of Section A.5. If $r(x) = 0$ the the integral equation is homogeneous and by Equation (A.16) a solution for W_c exists only if the determinant

$$|I - C P_d Q_d| = 0. \quad \text{(A.94)}$$

A.9. Orthogonal matrices

Consider the matrix equation

$$Q_c = A P_c \quad \text{(A.95)}$$

in which the N by N square matrix A transforms the N components of the column vector P_c into the column vector Q_c. If certain restrictions are placed on the vectors P_c and Q_c then the matrix A must have certain properties. For example suppose the length of the vector P_c is unchanged when transformed into the vector Q_c, or by Equation (A.80),

$$Q_c^T Q_c = P_c^T P_c \quad \text{(A.96)}$$

Take $B = A$ as the special form of A in Equation (A.95) to make Equation (A.96) hold, whence

$$P_c^T P_c = P_c^T B^T B P_c, \quad B^T B = I = B B^T, \quad \text{(A.97)}$$

$$B^T = B^{-1}, \quad |B^T B| = 1, \quad |B| = \pm 1. \quad \text{(A.98)}$$

From these conditions on B there exists certain relations among the elements of B. If B_{ir} is the row matrix for the elements in row i of B, then B_{ir}^T is the column matrix for the elements in column i of B^T, and from $BB^T = I$,

$$B_{ir} B_{ir}^T = b_{i1}^2 + b_{i2}^2 + \cdots + b_{iN}^2 = 1, \quad i = 1, 2, \cdots, N$$
$$B_{ir} B_{jr}^T = b_{i1} b_{j1} + b_{i2} b_{j2} + \cdots + b_{iN} b_{jN} = 0, \quad i \neq j, \quad \text{(A.99)}$$
$$i, j = 1, 2, \cdots, N.$$

Thus every row or column of B is a vector of unit length and the vectors in different rows or columns are orthogonal to each other, and B is called an *orthogonal transformation matrix*.

If P_c and Q_c are transformed simultaneously by B into P_{1c} and Q_{1c} then

$$P_{1c} = BP_c, \quad Q_{1c} = BQ_c \quad Q_{1c}^T P_{1c} = Q_c^T B^T B P_c = Q_c^T P_c. \quad \text{(A.100)}$$

Thus the scalar product of two vectors remains unchanged under an orthogonal transformation by the B matrix. Since the lengths of the vectors are unchanged the absolute angle between the vectors must be unchanged. In particular, orthogonal vectors transform into orthogonal vectors.

If p_1, p_2, p_3 are the components of a vector **p** in the x_1, x_2, x_3 coordinate system and if q_1, q_2, q_3 are the components of the same vector in the y_1, y_2, y_3 coordinate system, then in matrix form

$$Q_c = \begin{bmatrix} q_1 \\ q_2 \\ q_3 \end{bmatrix} = \begin{bmatrix} \cos(y_1, x_1) & \cos(y_1, x_2) & \cos(y_1, x_3) \\ \cos(y_2, x_1) & \cos(y_2, x_2) & \cos(y_2, x_3) \\ \cos(y_3, x_1) & \cos(y_3, x_2) & \cos(y_3, x_3) \end{bmatrix} \begin{bmatrix} p_1 \\ p_2 \\ p_3 \end{bmatrix}$$
$$= \begin{bmatrix} L_{11} & L_{12} & L_{13} \\ L_{21} & L_{22} & L_{23} \\ L_{31} & L_{32} & L_{33} \end{bmatrix} = RP_c. \quad \text{(A.101)}$$

Since Q_c and P_c are components of the same vector, R must be an orthogonal matrix with the properties of B above, $R^{-1} = R^T$, $|R| = 1$. Further

$$Y_c = RX_c \quad \text{(A.102)}$$

rotates the orthogonal coordinate system x_1, x_2, x_3 into the system y_1, y_2, y_3 so that R is the *rotation matrix* of transformation. See Figure A.1.

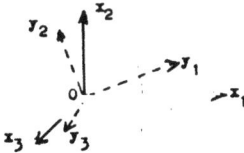

Fig. A.1. Rectangular coordinate axes.

The *reflection matrix* of transformation which reverses the direction of one or

272   Appendix A

more of the coordinate axes has the properties of B and the form

$$R_d = \begin{bmatrix} \pm 1 & 0 & 0 \\ 0 & \pm 1 & 0 \\ 0 & 0 & \pm 1 \end{bmatrix}, \tag{A.103}$$

the minus sign being used if the direction of the axis is reversed.

If the coordinate system for Q_c and P_c in Equation (A.95) is rotated by the matrix R, then

$$Q_{1c} = RQ_c, \quad Q_c = R^{-1}Q_{1c}; \quad P_{1c} = RP_c, \quad P_c = R^{-1}P_{1c};$$
$$Q_c = AP_c, \quad R^{-1}Q_{1c} = AR^{-1}P_{1c};$$
$$Q_{1c} = (RAR^{-1})P_{1c} = (RAR^T)P_{1c} = A_1 P_{1c};$$
$$A_1 = RAR^{-1}, \tag{A.104}$$

where Q_c and P_c have only three components and the A_1 transformation is *similar* to the A transformation because both transformations produce the same effect relative to two different coordinate systems.

Example A.9. Determine the rotation matrix R for rotation of the orthogonal coordinate system x_1, x_2, x_3 about axis x_3 through the angle θ into the y_1, y_2, y_3 coordinate system. Use the resulting R to calculate the transformation matrix A_1 in Equation (A.104) for the new coordinate system.

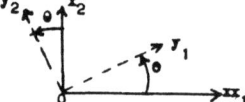

Fig. A.2. Rotation about x_3 axis, Fig. A.1.

Solution. From Figure A.2 and Equation (A.104) with $y_3 = x_3$, $L_{11} = \cos(y_1, x_1) = \cos\theta$, $L_{12} = \cos(y_1, x_2) = \cos(90° - \theta) = \sin\theta$, $L_{13} = \cos(y_1, x_3) = \cos 90° = 0$, $L_{21} = \cos(y_2, x_1) = \cos(90° + \theta) = -\sin\theta$, $L_{22} = \cos(y_2, x_2) = \cos\theta$, $L_{23} = \cos(y_2, x_3) = \cos 90° = 0$, $L_{31} = \cos(y_3, x_1) = \cos 90° = 0$, $L_{32} = \cos(y_3, x_2) = \cos 90° = 0$, $L_{33} = \cos(y_3, x_3) = \cos 0° = 1$, whence

$$R = \begin{bmatrix} \cos\theta & \sin\theta & 0 \\ -\sin\theta & \cos\theta & 0 \\ 0 & 0 & 1 \end{bmatrix}. \tag{A.105}$$

In Equation (A.104) use $R^{-1} = R^T$ and take

$$A = \begin{bmatrix} a_{11} & a_{12} & a_{13} \\ a_{21} & a_{22} & a_{23} \\ a_{31} & a_{32} & a_{33} \end{bmatrix}, \quad A_1 = \begin{bmatrix} c_{11} & c_{12} & c_{13} \\ c_{21} & c_{22} & c_{23} \\ c_{31} & c_{32} & c_{33} \end{bmatrix}. \tag{A.106}$$

Multiply RAR^T to get the elements of A_1 as follows:

$$c_{11} = a_{11}\cos^2\theta + a_{21}\sin\theta\cos\theta + a_{12}\sin\theta\cos\theta + a_{22}\sin^2\theta,$$
$$c_{12} = a_{12}\cos^2\theta + (a_{22} - a_{11})\sin\theta\cos\theta - a_{21}\sin^2\theta,$$
$$c_{21} = a_{21}\cos^2\theta + (a_{22} - a_{11})\sin\theta\cos\theta - a_{12}\sin^2\theta,$$
$$c_{22} = a_{22}\cos^2\theta - (a_{12} + a_{21})\sin\theta\cos\theta + a_{11}\sin^2\theta, \tag{A.107}$$
$$c_{13} = a_{13}\cos\theta + a_{23}\sin\theta, \quad c_{31} = a_{31}\cos\theta + a_{32}\sin\theta,$$
$$c_{23} = a_{23}\cos\theta - a_{13}\sin\theta, \quad c_{32} = a_{32}\cos\theta - a_{31}\sin\theta,$$
$$c_{33} = a_{33}.$$

Notes on matrix algebra

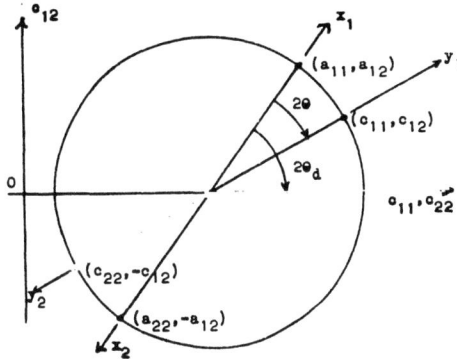

Fig. A.3. Matrix components for rotation of coordinate axes.

In particular, if A is symmetric and $a_{13} = a_{23} = a_{31} = a_{32} = 0$, then by using the relations

$$2\sin^2\theta = 1 - \cos 2\theta, \quad 2\cos^2\theta = 1 + \cos 2\theta, \tag{A.108}$$

Equation (A.107) becomes

$$A_1 = \begin{bmatrix} c_{11} & c_{12} & 0 \\ c_{12} & c_{22} & 0 \\ 0 & 0 & a_{33} \end{bmatrix}, \tag{A.109}$$

$$\begin{aligned} c_{11} &= (1/2)(a_{11} + a_{22}) + (1/2)(a_{11} - a_{22})\cos 2\theta + a_{12}\sin 2\theta, \\ c_{12} &= a_{12}\cos 2\theta - (1/2)(a_{11} - a_{22})\sin 2\theta, \\ c_{22} &= (1/2)(a_{11} + a_{22}) - (1/2)(a_{11} - a_{22})\cos 2\theta - a_{12}\sin 2\theta, \end{aligned} \tag{A.110}$$

With θ as a variable parameter a graph of Equation (A.110) can be made by putting c_{11} and c_{22} on the abscissa and c_{12} on the ordinate. In fact, by elimination of θ,

$$\begin{aligned}[] [c_{11} - (1/2)(a_{11} + a_{22})]^2 + c_{12}^2 &= (1/4)(a_{11} - a_{22})^2 + a_{12}^2 \\ &= a^2, \\ [c_{22} - (1/2)(a_{11} + a_{22})]^2 + c_{12}^2 &= a^2, \end{aligned} \tag{A.111}$$

and the graph is a circle with center at $[(1/2)(a_{11} + a_{22}), 0]$ and radius a. The graph of the circle in Figure A.3 shows the location of the original axes x_1, x_2, the new axes y_1, y_2, and the rotation angle θ in terms of the coordinate variables c_{12} and c_{11} (or c_{22}). If the original constants a_{11}, a_{12}, a_{22} are plotted to determine the $\theta = 0$ diameter line of the circle, then the new values c_{11}, c_{12}, c_{22} can be read directly from the circle at the diameter line given by rotating 2θ clockwise. This is Mohr's circle for stresses and strains.

From Figure A.3 it is evident that there exists a value of $\theta = \theta_d$ which will make $c_{12} = 0$, c_{11} a maximum, and c_{22} a minimum. It is given by

$$\tan 2\theta_d = 2a_{12}/(a_{11} - a_{22}), \tag{A.112}$$

or as indicated in Figure A.3. Note that this value of θ_d makes the matrix A_1 in Equation (A.109) a diagonal matrix, whence the components of Q_{1c} in Equation (A.104) are proportional to the components of P_{1c}.

Example A.10. Calculate the $[R]$ transformation matrix in Equation (A.101) to give the unit vectors r_1, θ_1, ϕ_1 in spherical coordinates in terms of the unit vectors $\mathbf{i}$, $\mathbf{j}$, $\mathbf{k}$ in rectangular coordinates, Figure A.4.

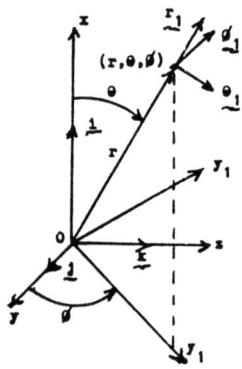

Fig. A.4. Spherical coordinates.

Solution. Use the $[R]$ matrix in Equation (A.105) twice, rotating first about the x-axis in Figure A.4 through the angle ϕ and then about the new x_1-axis through the angle θ, or

$$\begin{Bmatrix} r_1 \\ \theta_1 \\ \phi_1 \end{Bmatrix} = \begin{bmatrix} \cos\theta & \sin\theta & 0 \\ -\sin\theta & \cos\theta & 0 \\ 0 & 0 & 1 \end{bmatrix} \begin{bmatrix} 1 & 0 & 0 \\ 0 & \cos\phi & \sin\phi \\ 0 & -\sin\phi & \cos\phi \end{bmatrix} \begin{Bmatrix} i \\ j \\ k \end{Bmatrix}$$

$$= \begin{bmatrix} \cos\theta & \sin\theta\cos\phi & \sin\theta\sin\phi \\ -\sin\theta & \cos\theta\cos\phi & \cos\theta\sin\phi \\ 0 & -\sin\phi & \cos\phi \end{bmatrix} \begin{Bmatrix} i \\ j \\ k \end{Bmatrix}. \tag{A.113}$$

Example A.11. Determine the form of the three rotation matrices to get the $[R]$ matrix in Equation (A.101) for the general case of a cylinder with the ends located in the x, y, z system, Figure A.5.

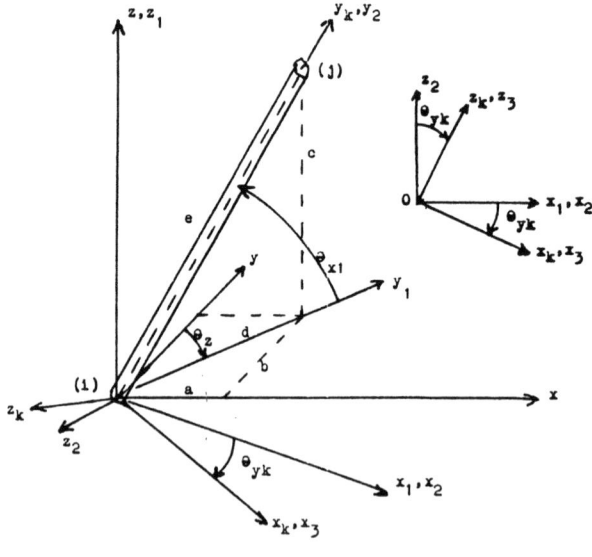

Fig. A.5. Rotation of axes for a cylindrical structure.

Solution. Let p_1, p_2, p_3 in Equation (A.101) be the x, y, z system in Figure A.5. There are several ways to make three rotations to carry the x, y, z axes into the x_k, y_k, z_k axes. In Figure A.5 the sequence is as follows:

(1) Rotate $-\theta_z$ about the z-axis taking the y-axis to the y_1-axis, which is along the projection of the cylinder on the xy-plane. This angle θ_z is known from the given values of a and b.

(2) Rotate θ_{x1} about the new x_1-axis taking the y_1-axis into the y_k-axis of the cylinder. This angle θ_{x1} is known from the given values of a, b, c.

(3) Rotate $\theta_{yk} = \theta_{y2}$ about the y_k-axis taking the x_1 or x_2 axis to the x_k-axis in the cylinder. This angle θ_{yk} will depend upon the desired axes on the cross section for the particular case.

The form of the $[R]$ matrix in Equation (A.101) is

$$[R] = \begin{bmatrix} \cos\theta_{yk} & 0 & \sin\theta_{yk} \\ 0 & 1 & 0 \\ -\sin\theta_{yk} & 0 & \cos\theta_{yk} \end{bmatrix} \begin{bmatrix} 1 & 0 & 0 \\ 0 & \cos\theta_{x1} & \sin\theta_{x1} \\ 0 & -\sin\theta_{x1} & \cos\theta_{x1} \end{bmatrix} \begin{bmatrix} \cos\theta_z & -\sin\theta_z & 0 \\ \sin\theta_z & \cos\theta_z & 0 \\ 0 & 0 & 1 \end{bmatrix},$$
(A.114)

$\cos\theta_z = b/d, \quad \sin\theta_z = a/d, \quad d^2 = a^2 + b^2,$

$\cos\theta_{x1} = d/e, \quad \sin\theta_{x1} = c/e, \quad e^2 = d^2 + c^2.$

A.10. Eigenvalues and eigenvectors of matrices

Consider the homogeneous matrix equation

$$BX_c = \lambda AX_c,$$
(A.115a)

where X_c is an unknown column matrix, λ is an unknown constant scalar parameter, A and B are real square symmetric matrices, A has an inverse. Multiply by A^{-1} to get

$$A^{-1}BX_c = \lambda A^{-1}AX_c = \lambda IX_c, \quad \text{or}$$
(A.115b)

$$[F - \lambda I]X_c = 0, \quad F = A^{-1}B.$$
(A.115c)

This system of simultaneous homogeneous equations has a solution for X_c only if

$$|F - \lambda I| = 0.$$
(A.116)

With F known, the expansion of Equation (A.116) gives a polynomial in λ, or for F a M by M matrix,

$$\lambda^M + a_1\lambda^{M-1} + \cdots + a_{M-1}\lambda + a_M = 0.$$
(A.117)

This is the *characteristic equation* of the matrix F. The roots $\lambda_1, \lambda_2, \cdots, \lambda_M$ of the Equation (A.117) are the *characteristic values* or *eigenvalues* of the matrix F.

These eigenvalues must obey

$$\sum_{i=1}^{M} \lambda_i = \sum_{i=1}^{M} f_{ii}, \quad \lambda_1\lambda_2\cdots\lambda_{M-1}\lambda_M = |F|,$$
(A.118)

which can be used to check the calculated eigenvalues. The diagonal terms of F are the f_{ii} values.

If one of the known eigenvalues is put into Equation (A.115c) and one of the values in the X_c vector is specified, then the other values of the X_c vector can be calculated from Equation (A.115c) to give the eigenvector for that particular eigenvalue. This eigenvector can be normalized to 1.0 as its maximum component or it can be restricted to have a unit length, Equation (A.80). There is a eigenvector for each eigenvalue, and it can be shown that these eigenvectors are orthogonal to each other. The matrix of these eigenvectors is called the *modal matrix* in structural vibrations problems, with the eigenvectors being the mode shapes for the λ_i vibration frequencies.

If the matrix is large, it may be difficult to calculate all the roots of Equation (A.117). If the roots are all real, such as in static buckling problems, then the roots can be located by graphing the determinant for assumed values of λ.

Example A.12. If $|D| = \lambda^3 - 10\lambda^2 + 24\lambda - 16$, find the roots of $|D| = 0$ by constructing a table for $|D|$.

Solution. Assume values of λ and construct a table for $|D|$ as follows:

λ	0	1	1.1	1.2	1.17	1.5	1.8	2.0	2.5
$|D|$	-16	-1	-0.37	0.13	-0.01	0.88	0.63	0.00	-2.88
λ	3.0	4.0	5.0	6.0	7.0	6.8	6.82	6.83	
$|D|$	-7.00	-16.0	-21.0	-16.0	5.0	-0.77	-0.23	0.04	

Thus, the three roots are approximately 1.17, 2.00, 6.83. The exact roots are $4 - 2(2)^{1/2} = 1.171573$, 2.000000, $4 + 2(2)^{1/2} = 6.828427$.

Example A.13. Find the characteristic values and the modal matrix for the characteristic equation

$$\left| \begin{bmatrix} 2 & -3^{1/2} & 0 \\ -3^{1/2} & 2 & -1 \\ 0 & -1 & 2 \end{bmatrix} - \lambda I \right| = 0.$$

Solution. The characteristic equation becomes

$$(2 - \lambda)^3 - 4(2 - \lambda) = 0,$$

whence

$$\lambda_1 = 0, \quad \lambda_2 = 2, \quad \lambda_3 = 4.$$

The equations

$$(2 - \lambda_i)x_1 - 3^{1/2}x_2 = 0,$$
$$-3^{1/2}x_1 + (2 - \lambda_i)x_2 - x_3 = 0,$$
$$-x_2 + (2 - \lambda_i)x_3 = 0,$$

determine the columns of the modal matrix. For $\lambda_1 = 0$, $x_2 = (2/3^{1/2})x_1$, $x_3 = (1/3^{1/2})x_1$, $x_1^2 + x_2^2 + x_3^2 = 1$, whence $x_1 = (3/8)^{1/2}$, $x_2 = 2^{1/2}/2$, $x_3 = (1/8)^{1/2}$. For $\lambda_2 = 2$, $x_2 = 0$, $x_3 = -3^{1/2}x_1$, $x_1^2 + x_2^2 + x_3^2 = 1$, whence $x_1 = 1/2$, $x_2 = 0$, $x_3 = -3^{1/2}/2$. For $\lambda_3 = 4$, $x_2 = -(2/3^{1/2})x_1$, $x_3 = (1/3^{1/2})x_1$, $x_1^2 + x_2^2 + x_3^2 = 1$, whence $x_1 = (3/8)^{1/2}$, $x_2 = -(1/2)^{1/2}$, $x_3 = (1/8)^{1/2}$. Thus, $B^{-1} = B^T$ in Equation (A.98), where B is

$$B = \begin{bmatrix} (3/8)^{1/2} & 1/2 & (3/8)^{1/2} \\ (1/2)^{1/2} & 0 & -(1/2)^{1/2} \\ (1/8)^{1/2} & -(3/4)^{1/2} & (1/8)^{1/2} \end{bmatrix}.$$

A.11. Note on matrix notation

In this Appendix A capital letters have been used to represent matrices of all sizes. This is convenient for discussing matrix theory, but it is too limited for actual working matrix operations where capital letters, lower case letters, Greek letters, and some standard symbols are needed in matrix form. In this text braces, such as

$$\{A\}, \quad \{b\}, \quad \{\sigma\} = \begin{Bmatrix} \sigma_{xx} \\ \sigma_{yy} \\ \sigma_{zz} \end{Bmatrix},$$

are used for column matrices and

$$\{\sigma\}^{\mathrm{T}} = [\, \sigma_{xx} \quad \sigma_{yy} \quad \sigma_{xy} \,]$$

is a row matrix. Brackets are used for rectangular and square matrices, such as

$$[H], \quad [k], \quad [\lambda].$$

The braces { } and brackets [ ] make it easier to follow the matrix operations on the various equations.

References

Appendix A: Brief bibliography on matrices

S.O. Asplund: *Inversion of band matrices*, ASCE 2nd Conference on Electronic Computation, Pittsburgh, Pa., Sept. No. 1960-42 (1960).

E. Bodewig: *Matrix Calculus*, North-Holland Publishing Co., Amsterdam (1959).

A.F. Cornock: *The numerical solution of Poisson's and the bi-harmonic equations by matrices*, Proc. of Cambridge Philosophical Society 50, pp. 524-535 (1954).

V.N. Faddeeva: *Computational methods of linear algebra*, Dover Publications, Inc., New York (1959).

R.A. Frazer, W.J. Duncan and A.R. Collar: *Elementary Matrices*, The Macmillan Co., New York (1946).

B.E. Gatewood and N. Ohanian: Note on solution of a system of three moment equations, *AIAA Journal* 1, p. 1965 (1963).

Olle Karlqvist: Numerical solution of elliptic differential equations, *Tellus* 4, pp. 374-384 (1952).

Eric Kosko: Matrix Inversion by Partitioning, *Aeronautical Quarterly* VIII, pp. 157-184, May (1957).

R.K. Livesley: *Matrix Methods of Structural Analysis* 2nd Ed., Pergamon, New York (1976).

C.C. MacDuffee: *The Theory of Matrices*, Chelsea Publishing Co., New York (1946).

H.C. Martin: *Introduction to Matrix Methods of Structural Analysis*, McGraw-Hill Book Co., New York (1966).

S.J. McMinn: *Matrices for Structural Analysis*, John Wiley & Sons, New York (1962).

J.L. Meek: *Matrix Structural Analysis*, McGraw-Hill Book Co., New York (1971).

J.V. Meyers: *Matrix Analysis of Structures*, Harper and Row, New York (1985).

T.A. Oliphant: An extrapolation procedure for solving linear systems, *Quar. of Applied Math.* 20, pp. 257-265, October (1962).

E. Pestel and F.A. Leckie: *Matrix Methods in Elastomechanics*, McGraw-Hill Book Co., New York (1963).

J.S. Przemieniecki: *Theory of Matrix Structural Analysis*, McGraw-Hill Book Co., New York (1968).

S. Schechter: Quasi-tri-diagonal matrices and type-insensitive difference equations, *Quar. of Applied Math.* 18, pp. 285-295 (1960).

Hans Schwerdtfeger: *Introduction to Linear Algebra and the Theory of Matrices*, P. Noordhoff N.V., Groningen, Holland (1950).

I.S. Sokolnikoff: *Tensor Analysis*, John Wiley & Sons, New York, Chap. 1 (1951).

R.R. Stoll: *Linear Algebra and Matrix Theory*, McGraw-Hill Book Co., New York (1952).

M.D. Vanderbilt: *Matrix Structural Analysis*, Quantum Publishers, New York (1974).
R.S. Varga: *Matrix Iterative Analysis*, Prentice-Hall, Englewood Cliffs, New Jersey (1962).

B

External forces on flight vehicles

B.1. Introduction

Since flight vehicles may have motion on the ground, in the atmosphere, and in space, they are subjected to various external forces. The primary forces are the thrust or booster forces from the engines, the lift and drag forces from the air, and the inertia forces from gravity and acceleration. These forces can be classified in various ways. In the atmosphere there are air maneuver forces produced by the pilot in his operation of the controls, and air gust forces produced by the air itself. On the ground there are forces due to taxiing, towing, hoisting, and jacking. Large forces may occur during take-off and landing, particularly for catapults and boosters in take-off, and for arrested landings. Also, there are internal pressurization forces, control forces, and internal thermal forces due to variable temperatures.

Since the flight vehicle is a moving body the equations of dynamics apply to it so that the external forces must be in equilibrium with the internal gravity and inertia forces. Although the flight vehicle can usually be considered as a moving rigid body, it is necessary to find the distribution of the air forces over the actual vehicle structure and to find the distribution of the inertia forces to obtain the design loads for the vehicle structure.

The major difference between flight vehicles and other vehicles is *weight*. All flight vehicles must have the lowest possible weight in order to get off the ground with a reasonable thrust force. Since the structure of the flight vehicle must be lifted and carried on every flight during the life of the vehicle, it is evident the structure must be an absolute minimum weigth design. One essential factor in obtaining this minimum weight design is the specification of realistic flight and ground performance conditions for the vehicle so that the external forces and the corresponding inertia forces are as small as possible. Not only must the external forces be as small as possible but also the distribution of the air and inertia forces must be obtained as accurately as possible in order to have a safe and minimum weight design for the structure.

In order to design a flight vehicle with a realistic structural weight and with a realistic aerodynamic performance, it is evident that the vehicle must be restricted to do a specific mission. A commerical transport is designed to transport

280   *Appendix B*

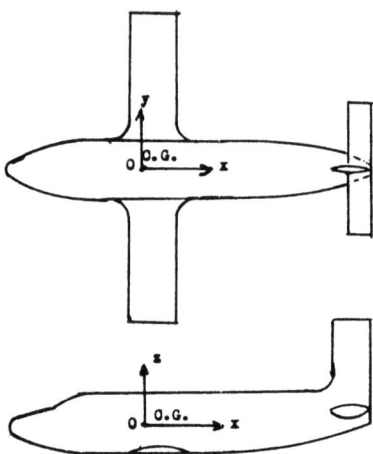

Fig. B.1. Coordinate axes for airplane.

a certain number of passengers safely and efficiently over various distances between airports. A military fighter is designed to maneuver at high speeds at various altitudes in order to contact enemy aircraft. The design conditions are entirely different for the two airplanes. Thus, there are many categories of aircraft for both commercial, military, and civil requirements. The requirements for various aircraft types are given in specification manuals and reports such as in References 1 and 2.

Since the procedures to determine the external forces and their distributions are essentially the same for the various types of flight vehicles, only values for a typical airplane will be given here. Also, various assumptions will be made to simplify the calculations for the air forces and the inertia forces. The assumptions to be made correspond to those normally used in a preliminary design of the vehicle. Basically, the ideas, procedures, and simple solutions are covered without getting into the final design with more refined calculations used with test data and corrections.

Some of the assumptions and simplifications used are steady state aerodynamic forces, motion in a vertical plane for take-off, flight, and landing, constant acceleration cases for motion and inertia forces, and instantaneous gust conditions. A set of right-hand rectangular coordinate axes, Figure B.1, with origin at the center of mass, or center of gravity (C.G.), of the airplane will be used.

The items covered here are inertia forces for rigid body translation and rotation in a vertical plane, air forces on airplane wing, airplane equilibrium equations in steady flight in a vertical plane, load factors, velocity-load factor diagram for design, wing spanwise lift coefficient distribution on untwisted and on twisted wings, spanwise airload, shear, and moment distributions, distributions of inertia forces on wing and fuselage, forces and moments on landing gear structures, thermal forces, miscellaneous forces, deflection effects on external forces, and

External forces on flight vehicles

criteria for the structure to support the external forces.

B.2. Inertial forces for rigid body translation and rotation in a vertical plane

For rectilinear motion of a particle, or a rigid body considered as a particle at its center of mass, let s = distance, v = velocitiy, a = acceleration, t = time. In general by definition

$$v_{average} = \frac{\Delta s}{\Delta t}, \quad v = \lim_{\Delta t \to 0} \left(\frac{\Delta s}{\Delta t}\right) = \frac{ds}{dt} = s_{,t}, \tag{B.1}$$

$$a_{average} = \frac{\Delta v}{\Delta t}, \quad a = \lim_{\Delta t \to 0} \left(\frac{\Delta v}{\Delta t}\right) = \frac{dv}{dt} = v_{,t} = s_{,tt}. \tag{B.2}$$

If time is eliminated from Equations (B.1) and (B.2),

$$a = \frac{dv}{dt} = \frac{dv}{dt}\frac{ds}{ds} = \frac{dv}{ds}\frac{ds}{dt} = v\frac{dv}{ds}, \quad \text{or}$$

$$a\,ds = v\,dv, \quad s_{,tt}\,ds = s_{,t}\,d(s_{,t}). \tag{B.3}$$

If the acceleration a is *constant* and if $s = s_0$, $v = v_0$, $t = 0$, are the initial conditions, then Equations (B.1)-(B.3) can be integrated directly to give

$$v = v_0 + at, \tag{B.4}$$
$$s = s_0 + v_0 t + \tfrac{1}{2}at^2, \tag{B.5}$$
$$2a(s - s_0) = v^2 - v_0^2. \tag{B.6}$$

The force F for rectilinear motion is given by Newton's law

$$F = ma = \frac{W}{g}a, \tag{B.7}$$

where m is mass, W is weight, and g is the acceleration of gravity. The components of F satisfy the equilibrium equations

$$\sum F_x = ma_x, \quad \sum F_y = ma_y, \quad \sum F_z = ma_z, \tag{B.8}$$

where a_x, a_y, a_z are the components of the acceleration a.

Example B.1. The airplane in Figure B.2 weights 16 Klb and has a constant braking force F of 6 Klb. (a) Find the wheel reactions R_1 and R_2. (b) Find the landing run $s - s_0$ if the airplane lands at 180 ft/sec.

Solution (a) From summation of forces in x and z directions and summation of moments about the main wheels contact point R_2,

$$6 - Ma_x = 0, \quad Ma_x = 6 \text{ Klb},$$
$$140 R_1 - 16(20) - 6(48) = 0, \quad R_1 = 3.31 \text{ Klb},$$
$$3.31 + R_2 - 16 = 0, \quad R_2 = 12.69 \text{ Klb}.$$

(b) From Equations (B.7) and (B.6)

$$a_x = Fg/W = -6(32.174)/16 = -12.07 \text{ ft/sec}^2,$$
$$s - s_0 = \frac{(0)^2 - (180)^2}{-2(12.07)} = 1342 \text{ ft}.$$

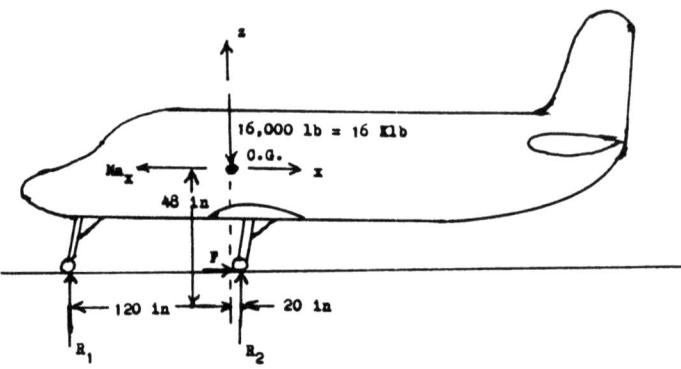

Fig. B.2. Airplane with constant braking force F.

For angular motion of a line AB, Figure B.3, about a point 0, let θ = angle (radians), ω = angular velocity (rad/sec), and α = angular acceleration (rad/s²). By direct analogy to Equations (B.1)-(B.3),

$$\omega = \frac{d\theta}{dt} = \theta_{,t}, \quad \alpha = \frac{d\omega}{dt} = \omega_{,t} = \theta_{,tt},$$
$$\alpha\, d\theta = \omega\, d\omega, \quad \theta_{,tt}\, d\theta = \theta_{,t}\, d(\theta_{,t}). \tag{B.9}$$

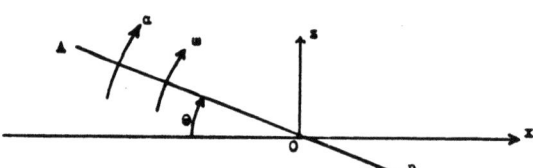

Fig. B.3. Angular motion of a line AB.

For the special case of α = constant, by analogy to Equations (B.4)-(B.6), Equation (B.9) gives

$$\omega = \omega_0 + \alpha t, \quad \theta = \theta_0 + \omega_0 t + \tfrac{1}{2}\alpha t^2,$$
$$2\alpha(\theta - \theta_0) = \omega^2 - \omega_0^2. \tag{B.10}$$

Example B.2. In Figure B.2, suppose the nose wheel is 24 in. above the runway when the main wheels contact the runway and suppose the pitching acceleration of the airplane about its center of mass is $\alpha_y = -4.00$ rad/s² = constant until the nose wheel contacts the runway. Find the angular velocity at nose wheel contact and the time of rotation to contact. Disregard the compression of the main gear shock absorbers and assume the main wheel reactions produce no moment about the center of mass.

Solution. From Figure B.2

$$\sin(\theta - \theta_0) = 24/140, \quad \theta - \theta_0 = -0.172 \text{ rad}.$$

From Equation (B.10) with $\omega_0 = 0$

$$2\alpha(\theta - \theta_0) = 2(-4.00)(-0.172) = \omega^2, \quad \omega \doteq 1.17 \text{ rad/s},$$
$$t^2 = 2(\theta - \theta_0)/\alpha = 2(-0.172)/(-4.00), \quad t = 0.29 \text{ s}.$$

External forces on flight vehicles

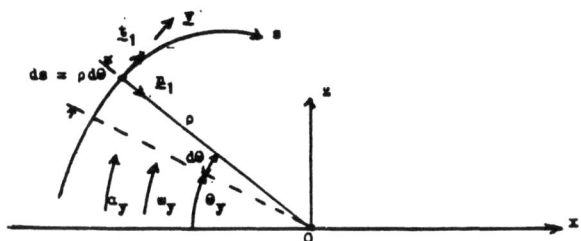

Fig. B.4. Vector components in curvilinear motion in a plane.

For curvilinear motion in plane with normal and tangential coordinates, Figure B.4, take t_1 and n_1 as unit vectors in the tangential and normal directions. From Figure B.4, the magnitude of the velocity $\mathbf{v}$ is

$$v = s_{,t} = \rho \theta_{,t}, \tag{B.11}$$

and the vector change in the velocity $\mathbf{v}$ is

$$d\mathbf{v} = d\mathbf{v}_n + d\mathbf{v}_t, \tag{B.12}$$

$$|d\mathbf{v}_n| = v\,d\theta, \quad |d\mathbf{v}_t| = d(s_{,t}) = d(\rho\theta_{,t}), \tag{B.13}$$

whence the vector acceleration is

$$\mathbf{a} = \mathbf{v}_{,t} = \mathbf{v}_{n,t} + \mathbf{v}_{t,t} = \mathbf{a}_n + \mathbf{a}_t. \tag{B.14}$$

The magnitudes of $\mathbf{a}_n$ and $\mathbf{a}_t$ are

$$a_n = v\theta_{,t} = \rho(\theta_{,t})^2 = v^2/\rho, \quad a_t = v_{,t} = s_{,tt}. \tag{B.15}$$

For circular motion with $\rho = r = $ constant, $\theta_{,t} = \omega$, $\theta_{,tt} = \alpha$, then

$$v = r\omega, \quad a_n = r\omega^2 = v^2/r, \quad a_t = r\alpha. \tag{B.16}$$

In terms of components in rectangular coordinates Equation (B.16) and Figure B.4 give

$$\begin{aligned}
v_x &= x_{,t} + r\omega_y \sin\theta_y = x_{,t} + z\omega_y, \\
v_z &= z_{,t} + r\omega_y \cos\theta_y = z_{,t} - x\omega_y, \\
a_x &= x_{,tt} + r\omega_y^2 \cos\theta_y + r\alpha_y \sin\theta_y = x_{,tt} - x\omega_y^2 + z\alpha_y, \\
a_z &= z_{,tt} - r\omega_y^2 \sin\theta_y + r\alpha_y \cos\theta_y = z_{,tt} - z\omega_y^2 - x\alpha_y.
\end{aligned} \tag{B.17}$$

Note that this Equation (B.17) shows that the acceleration and hence the inertia forces can vary throughout the flight vehicle.

For the rigid body rotating in a plane about a fixed axis y_R through 0 in Figure B.5 the moment is

$$\begin{aligned}
M_0 &= \int_V r\,dF_t = \int_V \alpha r^2\,dM = \alpha I_0, \\
I_0 &= \int_V r^2\,dM = \int_V (x_R^2 + z_R^2)\,dM = \text{mass moment of inertia}.
\end{aligned} \tag{B.18}$$

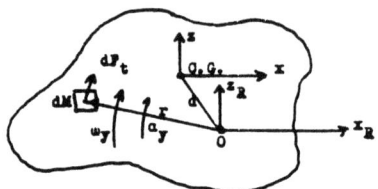

Fig. B.5. Rotation of rigid body in a plane.

For the flight vehicle, Newton's equilibrium equations for the three axes with origin at the center of mass are

$$\sum M_x = \alpha_x I_{xx}, \quad \text{Roll;} \quad \sum M_y = \alpha_y I_{yy}, \quad \text{Pitch;}$$
$$\sum M_z = \alpha_z I_{zz}, \quad \text{Yaw.} \tag{B.19}$$

It should be noted that these Equations (B.19) apply only for principle axes of the structure and for zero angular velocities. The general equations of motion of the vehicle with the conventional center of mass axes fixed in the vehicle are given in Reference 14. Note that if d is the distance from 0 to the C.G., Figure B.5, then the translation formulas can be used, or

$$I_{yy} = I_0 - Md^2. \tag{B.20}$$

Example B.3. Suppose the airplane in Figure B.2 makes a hard landing in soft ground on the two main wheels so that $k_2 = 64$ Klb and $F = 24$ Klb. The pitching mass moment of inertia is $I_{yy} = 320$ Klb-s²-in. (a) Find the accelerations $x_{,tt}$, $z_{,tt}$, and α_y. (b) After 0.20 sec, find the velocities v_x and v_z and the accelerations a_x and a_z acting at a point on the tail with $x = 160$ in., $z = 20$ in., using $x_{,t} = -150$ ft/s, $z_{,t} = -15$ ft/s. (c) What is the resultant force acting on a 40 lb black box at the location and conditions in part (b)?

Solution. (a) From summation of forces in the x and z directions, Equation (B.8), and moments about the C.G., Equation (B.19),

$$24 - (16/g)x_{,tt} = 0, \quad x_{,tt} = 1.50\,g,$$
$$64 - 16 - (16/g)z_{,tt} = 0, \quad z_{,tt} = 3.00\,g,$$
$$-(48)(924) - (20)(64) = 320\alpha_y, \quad \alpha_y = -7.60 \text{ rad/s}^2.$$

(b) With $\omega_0 = 0$ and $t = 0.20$ s, Equation (B.10) gives

$$\omega_y = (-7.60)(0.20) = -1.52 \text{ rad/s}.$$

From Equation (B.17),

$$v_x = -150 + (20/12)(-1.52) = -152.33 \text{ ft/s},$$
$$v_z = -15 - (160/12)(-1.52) = 5.27 \text{ ft/s},$$
$$a_x = 1.50\,g - (-1.52)^2(160/12)(g/32.174) + (-7.60)(\tfrac{20}{12})(\tfrac{g}{32.174})$$
$$= 0.15\,g,$$
$$a_z = 3.00\,g - (-1.52)^2(20/12)(g/32.174) - (-7.60)(\tfrac{160}{12})(\tfrac{g}{32.174})$$
$$= 6.03\,g.$$

(c) The resultant force on the box is

$$F_R = 40\,g[(0.15)^2 + (1+6.03)^2]^{1/2}/g = 281 \text{ lb},$$

B.3. Air forces on airplane wing

The air forces on the wing provide the lift so that the airplane can fly. These forces are distributed over the entire wing surface with the magnitude and distribution dependent upon angle of attack, altitude, airplane velocity, type of airfoil, flaps, wing taper, etc. Figure B.6 shows some typical chordwise distributions, where the resultant forces p_D and p_L per unit span are parallel (drag)

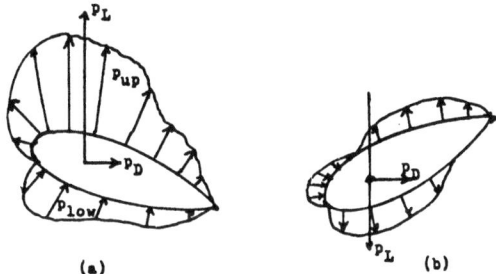

Fig. B.6. Chordwise air force distribution for (a) positive angle of attack (b) negative angle of attack.

and perpendicular (lift) to the flight path, and located at the center of pressure. Typical non-dimensional pressure distributions for p_{up}/q and p_{low}/q, where q is the dynamic pressure

$$q = \tfrac{1}{2}\rho V^2 \qquad (\text{B.21})$$

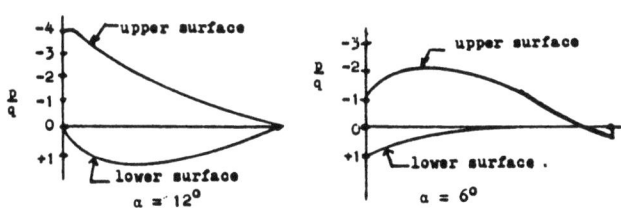

Fig. B.7. Typical airfoil non-dimensional pressure distributions.

with ρ the air mass density and V the free stream velocity, are shown in Figure B.7. See Reference 9. These pressure distributions p_{up} and p_{low} in force per unit area can be calculated for specific airfoils at specific angles of attack α and can be measured in wind tunnels. Note in Figure B.7 that the plus sign indicates pressure differences above atmospheric pressure and the minus sign indicates pressure differences below atmospheric pressure.

The resultant lift force $p_L = cc_l q$ per unit span can be obtained by integration of the chordwise distribution in Figure B.7, where c is the chord and c_l is the section lift coefficient. For a given angle of attack and a two dimensional wing with constant chord, this lift force is constant along the span so that the total lift force is $L = bP_L$ with span b. However, on a finite wing, p_L or c_l varies due to downwash effects, which also produces a varying p_D drag force per unit span. Since these finite wing downwash effects depend upon the planform of the wing, it is evident that the calculation of the p_L or $cc_l q$ and p_D or $cc_d q$ spanwise distributions can best be done from the wing as a unit rather than from the chordwise distributions in Figure B.7, which are affected by the downwash angles. The procedures to calculate the spanwise distribution of c_l are given below in Section B.6.

In Figure B.7 the chordwise location of the center of pressure can be calculated. This is the point of zero moment for the chordwise distribution, and obviously varies with angle of attack. However, for many airfoils in subsonic flight, the moment about the point at $0.25c$ is nearly constant for any permissible angle of attack. Thus, the forces p_L and p_D per unit span can be located at this airfoil

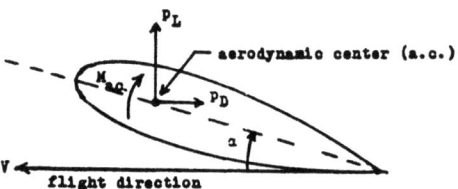

Fig. B.8. Lift and drag forces per unit span at aerodynamic center.

aerodynamic center, Figure B.8, with a moment $m_{a.c.}$ about this point for the particular airfoil. Once the p_L and p_D spanwise distributions are obtained, the aerodynamic center for the wing can be obtained, Figure B.9. If $p_L = cc_l q$ is the lift distribution along the quarter chord line in Figure B.9, then the total wing lift L is

$$L = 2q \int_0^{b/2} cc_l \, dy, \tag{B.22}$$

and from moments about the y and x axes

$$L\bar{x}/2 = q \int_0^{b/2} cc_l (0.25c + y \tan \beta) dy,$$

$$\bar{x} = \left[\int_0^{b/2} cc_l (0.25c + y \tan \beta) dy \right] \Big/ \left[\int_0^{b/2} cc_l \, dy \right],$$

$$\bar{y} = \left[\int_0^{b/2} cc_l y \, dy \right] \Big/ \left[\int_0^{b/2} cc_l \, dy \right]. \tag{B.23}$$

External forces on flight vehicles

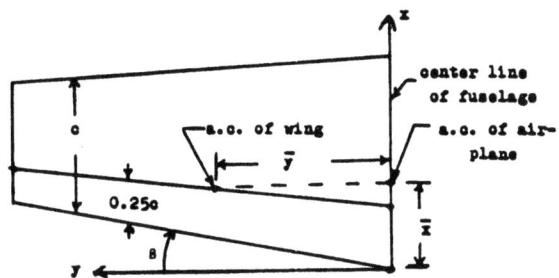

Fig. B.9. Wing aerodynamic center.

Thus, for the airplane as a rigid body, the distributed air forces on the wings can be replaced by a lift force L and a drag force D at the airplane aerodynamic center and a moment $M_{a.c.}$ about a reference y-axis through the aerodynamic center, Figure B.10. The air forces on the tail surfaces can be represented in a similar way.

In the next section the air forces are combined with the airplane weight and inertia forces to give the equilibrium equations of motion of the airplane in flight in a vertical plane.

B.4. Airplane equilibrium equations in flight. Load factors

Since the major external air forces and inertia forces on the airplane occur for motion in the x-z plane, the equilibrium equations given in this section apply

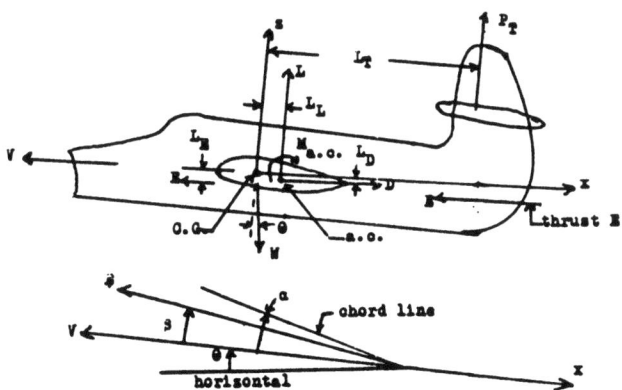

Fig. B.10. External forces on airplane in flight in fixed x-z plane.

only for the x-z plane with zero angular velocities about the x and z axes. In

Figure B.10, take summation of forces in the x and z directions and moments about the y-axis through point 0, or from Equations (B.9) and (B.19)

$$D + W \sin \theta - E \cos \beta = \frac{W}{g} a_x,$$
$$L - W \cos \theta + E \sin \beta + P_T = \frac{W}{g} a_z, \qquad (B.24)$$
$$M_{a.c.} - LL_L - P_T L_T - DL_D + EL_E = I_{yy} \alpha_y.$$

where E is thrust force, P_T is tail force, L is lift force, D is drag force, W is weight, $M_{a.c.}$ is moment about aerodynamic center due to air forces, L_L, L_T, L_D, L_E are lengths defined in Figure B.10, θ is angle of flight path from horizontal plane, β is angle of thrust line from flight path, α is angle of attack from flight path, and a_x, a_z, α_y are the accelerations. For steady flight $a_x = 0$, $a_z = 0$, $\alpha_y = 0$. Unsteady flight occurs due to pilot maneuvers (changing α and θ) and due to gusts in the air (changing α).

In working with the forces acting on the flight vehicle it is very convenient to use non-dimensional forces in certain cases. The term *load factor* n is defined as

$$n = \frac{F}{W}, \qquad (B.25)$$

where F is any force and W is the weight of any item being acted on by the force F. The weight W is taken as a scalar number so that the load factor n has the direction of the force F. Thus, n is simply a non-dimensional measure of F in terms of that important gravity force W that the earth imposes on everything.

From Equation (B.24) the load factors at the center of mass for the external applied forces are

$$n_z = (L + P_T + E \sin \beta)/W = \cos \theta + (a_z/g),$$
$$n_x = (D - E \cos \beta)/W = -\sin \theta + (a_x/g), \qquad (B.26)$$

where the non-dimensional accelerations a_z/g and a_x/g may produce large effects on the airplane load factors at the center of mass. Note that $n_z = 1$ and $n_x = 0$ in steady level flight. Also small n_y load factors can occur on the airplane due to forces on the vertical tail and due to a side drift landing. If the accelerations in Equation (B.17) are put into Equation (B.26), then the load factors at any point on the airplane can be calculated for translational accelerations $x_{,tt}, z_{,tt}$ and for angular acceleration α_y and angular velocity ω_y. These load factors give a direct measure of the forces acting on any item (black box, fuel tank, engine, person, etc.) at any point in the airplane. The equations are

$$n_z = \cos \theta + (z_{,tt}/g) - (\omega_y^2 z/g) - (\alpha_y x/g),$$
$$n_x = -\sin \theta + (x_{,tt}/g) - (\omega_y^2 x/g) + (\alpha_y z/g). \qquad (B.27)$$

Example B.4. Suppose the airplane in Figure B.10 weights 12 Klb and is flying horizontally when the pilot pulls up into a curved path of 2400 ft radius. At the position $\theta = 30°$ on the circle, assume $V = 750$ ft/s, $\beta = 0°$, $E - D = 400$ lb, $L_T = 200$ in., $L_L = 8$ in., $L_E = L_D = 4$ in. above the x-axis, $M_{a.c.} = 0$. Find accelerations a_x, a_z, and α_c, forces L and P_T, load factors n_z and n_x at the C.G. for the airplane at the instant specified. Here α_c is the angular acceleration about the center of the circle.

External forces on flight vehicles

Solution. Since the airplane is moving as a particle at its C.G. on a circle, Equation (B.16) applies with $a_x = a_n$, $a_z = a_t = r\alpha_c$ or

$$a_x = V^2/r = (750)^2/2400 = 234 \text{ ft/s}^2, \qquad a_z = 2400\alpha_c.$$

From Equation (B.26),

$$n_x = -400/12,000 = -0.03 = -0.50 + (a_x/g),$$
$$a_x = 0.47g = 15.12 \text{ ft/s}^2, \qquad \alpha_c = a_z/2400 = 0.0063 \text{ rad/s}^2,$$
$$n_z = 0.866 + \frac{234}{32.174} = 8.14 = (L + P_T + 0)/W,$$
$$L + P_T = 97,670 \text{ lb.}$$

Take moments about the C.G., Equation (B.24),

$$0 - 8L - 200 P_T - 4(400) = 0, \qquad -L - 25 P_T = 200,$$
$$P_T = -4136 \text{ lb}, \qquad L = 101,800 \text{ lb}.$$

Example B.5. Suppose the airplane in Example B.4 at the instant used there is further maneuvered by the pilot to give instantaneously a pitching acceleration of $\alpha_y = -4$ rad/s². Take $I_{yy} = 180,000$ lb-s²-in. If the tail load is the only external force to change, (a) Find the new tail load P_T, the acceleration a_z, and the load factor n_z. (b) If the engine is located at $x = 120$ in., $z = 0$ in., find the load factors n_x and n_z on the engine.

Solution (a) From the moment equation in Equation (B.24)

$$-8L - 200 P_T - 4(500) = (-4)(180,000), \qquad P_T = -482 \text{ lb}.$$

From Equation (B.26),

$$n_z = (101,800 - 482)/12,000 = 8.44,$$
$$a_z = (-0.866 + 8.44)g = 7.57g = 243.79 \text{ ft/s}^2.$$

(b) From Equation (B.17) with $\omega_y = 0$,

$$a_x = 15.12 - 0 - 0 = 15.12 \text{ ft/s}^2, \qquad \text{as in Example B.4,}$$
$$a_z = 243.79 - 0 - (-4)(120/12) = 283.79 \text{ ft/s}^2,$$
$$n_x = -0.03, \quad \text{as in Example B.4,} \qquad n_z = \frac{283.79}{32.174} + 0.866 = 9.69.$$

It is evident that the maneuvers of the airplane in Examples B.4 and B.5 put very large forces on the airplane. Since the pilot would black out above $n_z = 6.0$, and the engine supports would break about $n_z = 9.0$, the pilot had best use a larger radius for his maneuvers. The point is that limitations must be used in the design and operation of flight vehicles. In the next section the design conditions for the load factors and velocities, as specified by regulations, will be considered.

B.5. Velocity-load factor (V-n) diagram for design

The lift and drag forces, L and D, are related to the angle of attack α through the lift and drag coefficients C_{za} and C_{xa} for the airplane,

$$C_{za} = C_L + C_T, \qquad C_L = m_0 \alpha, \qquad \alpha = \alpha_0 + \frac{(1+k_2)C_L}{\pi(\text{A.R.})},$$
$$C_{xa} = C_{D0} + \frac{(1+k_1)C_L^2}{\pi(\text{A.R.})}, \tag{B.28}$$

where C_L is the wing lift coefficient, m_0 is slope of lift curve for wing airfoil, C_T is tail lift coefficient, α_0 is angle of attack on infinite wing, C_{D0} is profile drag coefficient for infinite wing, (A.R.) is aspect ratio (b^2/S), and k_1, k_2 are small

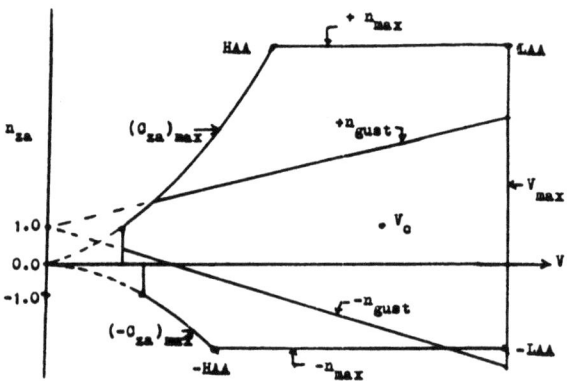

Fig. B.11. V-n diagram with gust lines.

correction factors for wing taper ratio. The lift and drag forces for the airplane are

$$L_a = C_{za}qS = \tfrac{1}{2}C_{za}\rho V^2 S, \qquad D_a = C_{xa}qS, \tag{B.29}$$

where Equation (B.21) is used.

From Equation (B.25)

$$n_{za} = L_a/W = \frac{C_{za}\rho V^2}{2(W/S)}, \qquad n_{za} = \frac{C_{za}\rho V^2}{2(W/S)}, \tag{B.30}$$

where (W/S) is wing loading in force per unit area. For a given wing loading, altitude, and lift coefficient C_{za}, the load factor n_{za} varies with velocity squared. Thus, the load factor n_{za} can be plotted against velocity V, Figure B.11, where $(C_{za})_{max}$ is used for the stall curves. It is evident that to have reasonable design forces for the airplane, n_{za} must have limits, and V must have a maximum value, as shown in Figure B.11. These limits are specified for the type of airplane in government regulations, References 1 and 2. The airplane must be designed to have the proper aerodynamic parameters, proper stability, and proper structural strength to fly at any point inside the V-n diagram. Naturally, much of the flying is done around the point V_c, the cruise velocity, and $n_{za} = 1$, Figure B.11.

To design the structure, only a few points with the largest forces are selected, such as HAA (high angle of attack) with maximum $+n_{za}$, LAA (low angle of attack) with maximum $+n_{za}$ and different tail load from HAA, - HAA, and - LAA. Other points may be checked, depending upon the maneuver requirements of the airplane.

The V-n diagram discussed above is the maneuver diagram, which the pilot is supposed to fly inside of at all times. There are also forces produced by air gusts, which are not directly under pilot control. In Figure B.12, a vertical upward gust of velocity U changes the direction of the resultant air motion so that the angle of attack on the wing is changed, or

$$\Delta\alpha_g = U/V, \quad \text{small angles}, \tag{B.31}$$

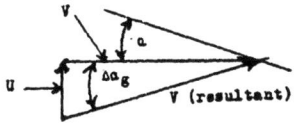

Fig. B.12. Change in angle of attack due to up gust velocity U.

where U is perpendicular to the airplane velocity vector V. Approximately (if there is no tail load change), from Equations (B.28), (B.30), and (B.31),

$$\Delta C_{za} = m_0 \Delta \alpha_g = m_0 U/V, \quad \Delta n_g = \frac{(\Delta C_{za})\rho V^2}{2(W/S)} = \frac{m_0 \rho U V}{2(W/S)}. \quad \text{(B.32)}$$

Although a gust can occur at any point in the V-n diagram, the most probable points are on the line $n_{za} = 1$. Thus, since the gust may be up or down,

$$n_g = 1 \mp \Delta n_g = 1 \mp \frac{m_0 \rho U V}{2(W/S)}. \quad \text{(B.33)}$$

Note that the gust may be in any direction, but only the vertical components affect the angle of attack. The horizontal components produce a small change in V and usually are not serious, except in low speed flight near the ground.

The gust load factor in Equation (B.33) is a linear function of V so that for fixed U and altitude n_g versus V graphs as a straight line, Figure B.11. The gust velocity U is specified as some $U_{\max}$ based upon recorded flight data over many years on all types of airplanes (References 1 and 2). On some airplanes the gust conditions may produce larger load factors than the maneuver diagram (see -LAA corner in Figure B.11), which must be allowed for in the design. Obviously, from Figure B.11, the gust effects can be reduced by flying at lower velocities (when the air gets rough, slow down).

The V-n diagrams are usually drawn for various altitudes and these diagrams become the basis for all the air and inertia forces, except landing and take-off, acting on the airplane.

Example B.6. Draw a sea level V-n diagram for an airplane with $n_{\max} = +6, -3, V_{\max} = 500$ mph, $(C_{za})_{\max} = +1.8, -1.2, m_0 = 4.0/\text{rad}, U = 45$ ft/s, $W/S = 50$ lb/ft^2.

Solution. The sea level density of air is

$$\rho_0 = 0.002378 \text{ lb-s}^2/\text{ft}^4$$

so that Equation (B.30) gives

$$n_{za} = \frac{1.80(0.002378)}{2(50)} \left(\frac{5280}{3600}\right)^2 V^2 = 0.00009208 V^2, \quad V \text{ in mph,}$$

$$n_{za} = -0,00006137 V^2 \text{ for negative load factors,}$$

whence $(n_{za}, V) = (6, 255), (1, 104), (-3, 221), (4, 208), (2, 148), (-1, 128), (-2, 181)$ for V in mph.
From Equation (B.33),

$$n_g = 1 \mp \frac{(4)(0.002378)}{2(50)} \left(\frac{5280}{3600}\right)(45) V = 1 \mp 0.006278 V$$

$$= -2.14, \ 4.14 \quad \text{at } V = 500 \text{ mph.}$$

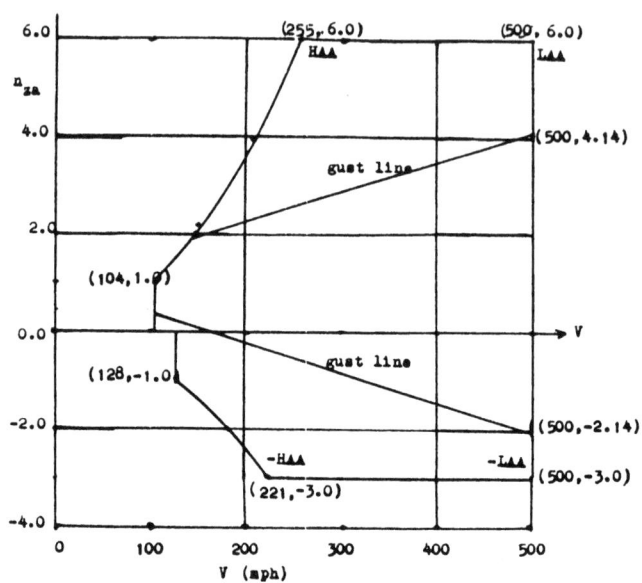

Fig. B.13. V-n diagram for Example B.6.

Figure B.13 shows the graph.

B.6. Wing spanwise lift coefficient distribution

In order to get the distribution of the external forces acting on the airplane wing for a given flight condition on the V-n diagram, it is necessary first to obtain the wing section lift coefficient distribution, $c_l = m_0 \alpha$. On a finite wing the angle of attack α varies along the span due to the flow of the air around the wing tips. The angle of attack can be expressed as

$$\alpha = \alpha_0 - \alpha_i = \alpha_0 - \frac{w}{V}, \tag{B.34}$$

where α_0 is the angle of attack for the infinite wing, α_i is the induced angle of attack, w is the downwash velocity tending to reduce the angle of attack, and Equation (B.31) has been used for $\alpha_i = w/V$. The downwash angle α_i varies along the span for all wings except for those with an elliptical planform. In this case

$$\alpha_i = w/V = C_L/\pi(\text{A.R.}) = SC_L/\pi b^2 = \text{constant, (ellipse)}, \tag{B.35}$$

whence the spanwise lift distribution per unit span

$$p_L = c_e c_l q = c_e m_0 \alpha q, \quad \text{(ellipse)}, \tag{B.36}$$

is proportional to the chord c_e of the ellipse.

External forces on flight vehicles 293

For the ellipse, the chord c_e and area S_e are

$$c_e = c_s[1-(2y/b)^2]^{\frac{1}{2}}, \qquad S_e = \pi b c_s/4, \tag{B.37}$$

where c_s is the root chord at $y = 0$. Since $c_l = C_L$ in this case, Equations (B.34), (B.35), and (B.37) give

$$\begin{aligned}c_l &= m_0\alpha = m_0\alpha_0 - (m_0 w/V) = m_0\alpha_0 - (m_0 S_e c_l/\pi b^2)\\ &= m_0\alpha_0 - (m_0 c_s c_l/4b).\end{aligned} \tag{B.38}$$

Thus,

$$c_l = \frac{m_0\alpha_0}{1+(m_0/\pi(\text{A.R.}))} = \frac{m_0\alpha_0}{1+(m_0 c_s/4b)}, \quad \text{ellipse.} \tag{B.39}$$

For other wing planforms, the downwash w is variable along the span. From the Prandtl wing theory using a wing vortex system, Reference 3, it can be shown that at a point y on the wing

$$w(y) = (1/4\pi)\int_{-b/2}^{b/2} \frac{dC_i(y_p)}{y-y_p}, \tag{B.40}$$

where $dC_i(y_p)$ is the infinitesimal circulation for a trailing vortex at point y_p and y_p is the dummy integration variable. Since the lift per unit span is $L = cc_l q = \rho V C_i$ by the Kutta-Joukowsky law, it follows that

$$C_i = cc_l V/2. \tag{B.41}$$

Put Equation (B.41) into Equation (B.40) and the result into Equation (B.38) to get the basic integral equation for c_l,

$$c_l(y) = m_0\alpha_0 - (1/8\pi)\int_{-b/2}^{b/2} \frac{d[c(y_p)c_l(y_p)]}{y-y_p}. \tag{B.42}$$

If $m_0\alpha_0$ is constant, it can be verified that this Equation reduces to Equation (B.38) for the ellipse. For other cases it is necessary to use Fourier series or singular integral equation theory in complex variables to get a solution.

Due to the complicated calculations in solving Equation (B.42) and due to the fact that Equation (B.42) is still approximate around the wing tips, a simple approximation procedure based on the elliptical wing is usually used to get the spanwise lift distribution. The Schrenk method (References 3, 4) averages the spanwise lift of the elliptical wing of area S and span b with a planform distribution of the actual wing of area S and span b. That is,

$$cc_l = C_L(c+c_e)/2, \tag{B.43}$$

where C_L is the wing lift coefficient for the flight condition considered, c is the chord of the actual wing, c_l is section lift coefficient of actual wing, and c_e is chord of the ellipse. From Equation (B.37),

$$c_s = 4S/\pi b, \qquad c_e = (4S/\pi b)[1-(2y/b)^2]^{\frac{1}{2}}, \tag{B.44}$$

where S and b are for the actual wing.

Since the wing lift coefficient C_L has different values for various flight conditions, or from Equations (B.28) and (B.30)

$$C_L = C_{za} - C_T = (n_{za}W - P_T)/qS, \tag{B.45}$$

the spanwise lift distribution is usually calculated for $C_L = 1.00$,

$$cc_{l1} = (c + c_e)/2, \quad C_L = 1.00. \tag{B.46}$$

The values are then multiplied by the proper C_L for any flight point on the V-n diagram.

For a wing planform with straight taper and no rounding of the tip, the equation for the chord is

$$c = c_R[1 - \{1 - (c_T/c_R)\}(2y/b)], \tag{B.47}$$

where c_R is the root chord and c_T is the tip chord. In such a case, Equation (B.46) gives

$$c_{l1} = (1/2) + (c_e/2c) = 0.50 + \frac{(2S/\pi bc_R)[1 - (2y/b)^2]^{\frac{1}{2}}}{1 - \{1 - (c_T/c_R)\}(2y/b)}, \tag{B.48}$$

which can be calculated at several points and a c_{l1} curve drawn.

Since the wing tips are usually rounded and the taper ratio may change, it is simpler to use Equation (B.46) directly and calculate at various points on the wing by using a table.

Example B.7. Find the spanwise distribution of the section lift coefficient c_{l1} for the wing in Figure B.14. Assume the flaps are in a neutral position, $C_L = 1.00$ and m_0 is constant. The

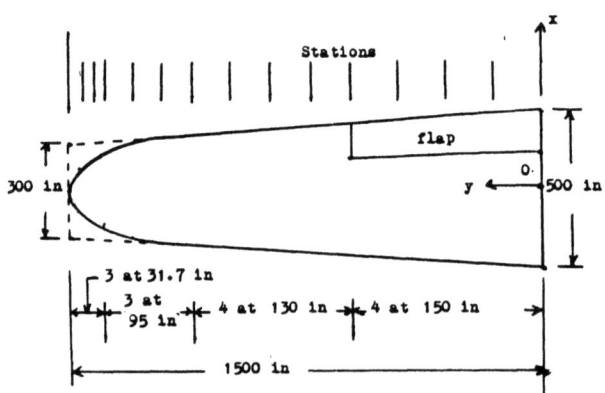

Fig. B.14. Wing planform for c_l distribution.

values of the chord in the tip region are shown in the table of calculations and $S = 7986$ ft^2. Show calculations and results in Table B.1.

Solution. In Table B.1, column (1) is the station location shown in Figure B.14, column (2) is $(2y/b) = y/1500$, column (3) gives the chord $c = 500[1 - 0.4(2y/b)]$ except in the tip region, column (4) gives the variable term in the chord of the equivalent ellipse, Equation (B.44),

External forces on flight vehicles

column (5) is the chord c_e of the equivalent ellipse with $c_e = 4S/\pi b = 488.08$ in., column (6) is cc_{l1} from Equation (B.46), column (7) is $c_{l1} = (cc_{l1})/c$.

Table B.1.   Spanwise section lift coefficient distribution in example B.7.

(1)	(2)	(3)	(4)	(5)	(6)	(7)
y, in.	$2y/b$	c, in.	$\left[1-\left(\frac{2y}{b}\right)^2\right]^{\frac{1}{2}}$	c_e, in. (B.44)	cc_{l1} (B.46)	c_{l1}
0	0.000	500.0	1.000	488.1	494.0	0.988
150	0.100	480.0	0.995	485.6	482.8	1.006
300	0.200	460.0	0.980	478.2	469.1	1.020
450	0.300	440.0	0.954	465.6	452.8	1.029
600	0.400	420.0	0.916	447.3	433.7	1.033
730	0.487	402.7	0.873	426.3	414.5	1.029
860	0.573	385.3	0.820	400.0	392.7	1.019
990	0.660	368.0	0.751	366.7	367.3	0.998
1120	0.747	350.7	0.665	324.5	337.6	0.963
1215	0.810	338.0	0.586	286.2	312.1	0.923
1310	0.873	300.0	0.488	238.0	269.0	0.897
1405	0.937	225.0	0.349	170.5	197.8	0.879
1437	0.958	200.0	0.287	139.9	170.0	0.850
1469	0.979	160.0	0.202	98.7	129.3	0.808
1500	1.000	0.0	0.0	0.0	0.0	0.0

B.7. Spanwise lift coefficient distribution on twisted wings

If the angle of attack for zero lift of the airfoil sections, as measured from a reference plane, varies along the wing, then the wing is regarded as twisted. This may be due to construction of the wing, the use of different airfoils inboard and outboard, deflections of flaps or ailerons, rotation of the wing due to the airloads, or any effect that produces a variation in the angle of attack along the wing.

The airload distribution on the twisted wing is calculated in two parts, and the two parts combined to get the final distribution for the flight condition. One part for twist alone is calculated for $C_L = 0$ on the entire wing, or for no wing lift. The other part for lift alone is calculated as in the previous Section B.6. For the part due to twist alone some sections of the wing will have positive lift while other sections will have negative lift with the total lift being zero. To find the reference angle of attack plane for zero lift of the wing, take

$$cc_l = cm_0(\alpha_{aR} - \alpha_{w0}), \tag{B.49}$$

where α_{aR} is the angle of attack of the section from a reference plane for the wing and α_{w0} is the angle of attack plane for wing zero lift. To evaluate α_{w0} integrate the lift over the entire wing and equal to zero, or

$$2q \int_0^{b/2} m_0(\alpha_{aR} - \alpha_{w0})c\,dy = 0, \quad \text{or}$$

$$\alpha_{w0} = \int_0^{b/2} m_0 \alpha_{aR} c\,dy \bigg/ \int_0^{b/2} m_0 c\,dy. \tag{B.50}$$

If m_0 is constant along the wing (same airfoils with same thickness ratio), then

$$\alpha_{w0} = (2/S) \int_0^{b/2} \alpha_{aR} c \, dy. \tag{B.51}$$

Once α_{w0} is obtained the angle of attack of each section is simply $\alpha_{aR} - \alpha_{w0}$ so that the spanwise airload distribution for the twisted wing at zero lift is

$$(cc_l)_0 = \tfrac{1}{2} c m_0 (\alpha_{aR} - \alpha_{w0}), \qquad (c_l)_0 = (m_0/2)(\alpha_{aR} - \alpha_{w0}). \tag{B.52}$$

Here, the Schrenk approximation (References 3, 4, 5), which takes the induced angle of attack α_i due to the downwash as one half the angle for the infinite wing, has been used.

The combination of the zero lift part and the lift part due to any wing C_L gives the section lift coefficient on the twisted wing as

$$c_l = (c_l)_0 + c_{l1} C_L. \tag{B.53}$$

Example B.8. The flap in Figure B.14 is deflected 30° to produce an effective change in wing angle of attack of 7.00° in the region of the flap. Find the spanwise section lift coefficient distribution for $C_L = 0$ (no lift), $C_L = 1.00$, $C_L = 1.90$. Let $m_0 = 0.100$ per degree for all sections of the wing.

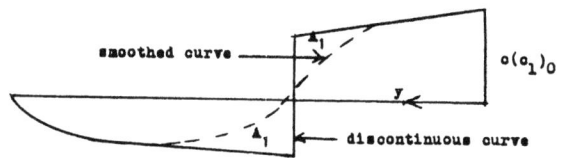

Fig. B.15. Smoothing $c(c_l)_0$ distribution for zero lift.

Solution. Take the reference plane for α_{aR} in Equation (B.51) as the zero-lift chords of the outboard section of the wing so that $\alpha_{aR} = 7.00°$ in the flap region, $\alpha_{aR} = 0°$ elsewhere. Since $c = 500 - 0.1333y$ in the region of the flap, Equation (B.51) gives

$$\alpha_{w0} = \frac{2(7.00)}{115(10)^4} \int_0^{600} (500 - 0.1333y) dy = 3.36°.$$

From Equation (B.52), $(c_l)_0 = (0.100/2)(7.00 - 3.36) = 0.182$ in the flap region and $(c_l)_0 = (0.100/2)(0 - 3.36) = -0.168$ elsewhere. Since the pressure of the air on the wing cannot be discontinous, the discontinuity in $(c_l)_0$ does not actually occur, so that it is necessary to modify the $(c_l)_0$ variation near the end of the flap. Since the total lift from $c(c_l)_0$ must remain zero, the simplest procedure to modify $(c_l)_0$ is to graph $c(c_l)_0$ and then smooth the curve as shown in Figure B.15. The numerical values for $(c_l)_0$ discontinuous, $c(c_l)_0$ smoothed, and $(c_l)_0$ smoothed are shown in Table B.2. By using $c_{l,1}$ from Table B.1, Equation (B.53) gives the c_l for any flight condition. Table B.2 shows the results for $C_L = 1.00$ and $C_L = 1.90$.

The above procedures for calculation of the airload distributions on the wing assume a steady state situation, a simple wing planform, and no interference effects from engines and stores on the wing or from the fuselage. To account for these effects, it is necessary to use more refined procedures to get the airloads. With the availability of large computers it is now possible in some cases to use a finite element approach to get the airload distribution over the wing and fuselage with allowance for interference effects. For the quasi-steady state case, the vortex lattice method, References 10 and 11, can be used. For the unsteady case with the elastic wing vibrating due to transcient disturbances, the doublet lattice method, References 12 and 13, can be used.

Table B.2. Spanwise lift coefficient distribution for twisted wing in example B.8.

(1) y, in	(2) c, in	(3) $(c_l)_0$ (B.52)	(4) $c(c_l)_0$	(5) $c(c_l)_0$ smooth	$C_L = 0$ (6) $(c_l)_0$ smooth	(7) c_{l1} Table B.1	$C_L = 1.00$ (8) (c_l) (B.53)	$C_L = 1.90$ (9) c_l (B.53)
0	500.0	0.182	91.00	91.00	0.182	0.988	1.170	2.059
150	480.0	0.182	87.36	87.36	0.182	1.006	1.188	2.093
300	460.0	0.182	83.72	80.00	0.174	1.020	1.194	2.112
450	440.0	0.182	80.08	45.00	0.102	1.029	1.131	2.057
600	420.0	0.182	76.44	3.00	0.007	1.033	1.040	1.970
600	420.0	-0.168	-70.56	3.00	0.007	1.033	1.040	1.970
730	402.7	-0.168	-67.65	-20.00	-0.050	1.029	0.979	1.905
860	385.3	-0.168	-64.73	-45.00	-0.117	1.019	0.902	1.819
990	368.0	-0.168	-61.82	-60.00	-0.163	0.998	0.835	1.733
1120	350.7	-0.168	-58.92	-58.92	-0.168	0.963	0.795	1.662
1215	338.0	-0.168	-56.78	-56.78	-0.168	0.923	0.755	1.586
1310	300.0	-0.168	-50.40	-50.40	-0.168	0.897	0.729	1.536
1405	225.0	-0.168	-37.80	-37.80	-0.168	0.879	0.711	1.502
1437	200.0	-0.168	-33.60	-33.60	-0.168	0.850	0.682	1.447
1469	160.0	-0.168	-26.88	-26.88	-0.168	0.808	0.640	1.367
1500	0.0	-0.168	0.0	0.0	-0.168	0.0	0.0	0.0

B.8. Spanwise airload, shear, and moment distributions on wing

To calculate the airload, shear, and moment distributions on the wing it is necessary to know both the drag coefficient and the lift coefficient distributions. The drag coefficient distribution will be approximated by taking the constant drag for the elliptical wing of the same area and span as the actual wing. Since the local flow direction is changed by the downwash angle $\alpha_i = w/V$ on the wing section, the drag on the wing consists of the profile drag D_0 and the induced drag $L \tan \alpha_i$ from the lift component, or

$$D = D_0 + \alpha_i L, \quad \text{(small angles)}. \tag{B.54}$$

Rewrite Equation (B.54) in coefficient form and use Equation (B.35) for the ellipse planform to get

$$c_d = d_{d0} + [C_L^2/\pi(\text{A.R.})], \quad \text{(constant)}. \tag{B.55}$$

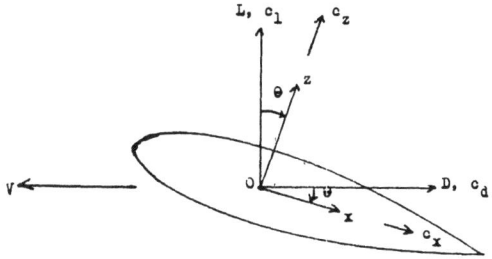

Fig. B.16. Wing cross section axes x and z.

Since the structural axes in the wing cross section are fixed in the wing, it is evident that the direction of the lift and drag forces on the wing will be different from the wing axes for each flight condition. From Figure B.16,

$$c_z = c_l \cos\theta + c_d \sin\theta, \qquad c_x = -c_l \sin\theta + c_d \cos\theta, \tag{B.56}$$

where c_l is given by Equation (B.53) and c_d by Equation (B.55) for any specified flight condition, Equation (B.45).

The airload distributions p_x and p_z per unit span can be obtained from Equation (B.56) for a specified condition as

$$p_z = cc_z q(1), \qquad p_x = cc_x q(1). \tag{B.57}$$

The shear force distributions V_z and V_x, starting from zero at the wing tip, are

$$V_{zi} = \int_{y_i}^{b/2} p_z \, dy = \sum_{j=1}^{y_i} (p_z)_{av,j} (\Delta y)_j,$$

$$V_{xi} = \int_{y_i}^{b/2} p_x \, dy = \sum_{j=1}^{y_i} (p_x)_{av,j} (\Delta y)_j. \tag{B.58}$$

The bending moment distributions M_x and M_z are

$$M_{xi} = \int_{y_i}^{b/2} V_z \, dy = \sum_{j=1}^{y_i} (V_z)_{av,j} (\Delta y)_j,$$

$$M_{zi} = -\int_{y_i}^{b/2} V_x \, dy = -\sum_{j=1}^{y_i} (V_x)_{av,j} (\Delta y)_j. \tag{B.59}$$

The forces p_x and p_z will also produce M_y moments about the y-axis for the wing cross section axes. Since the M_y moment from the p_x force is small, it is usually neglected, whence

$$M_{yi} = -\int_{y_i}^{b/2} x_a p_z \, dy = -\sum_{j=1}^{y_i} (x_a)_{av,j} (p_z)_{av,j} (\Delta y)_j. \tag{B.60}$$

Here x_a is the distance chordwise from the location of the section airload to the reference y-axis for the structural wing.

Example B.9. Find the airload, shear, and moment distributions on the wing in Figure B.14 and Example B.7 for the high angle of attack (HAA) flight condition on the airplane in Example B.6 and Figure B.13. Take $c_{d0} = 0.01$, $\theta = 21°$, $P_T = -50,000$ lb, section center of lift at the 30% chord, reference y-axis for M_y torque at the 50% chord, and use the c_{l1} distribution in Example B.7.

Solution. From the HAA flight condition in Figure B.13,

$$n_{za} = 6.00, \qquad V = 225 \text{ mph}, \qquad c_{za} = 1.80.$$

From data in Examples B.6 and B.7,

$$W = (W/S)S = 50(7986) = 399,300 \text{ lb} = \text{weight of plane}.$$

External forces on flight vehicles

From Equation (B.29),

$$L_a = n_{za}W = C_{za}qS, \quad q = n_{za}W/C_{za}S = \frac{6(399,900)}{1.80(7986)}$$

$$= 166.67 lb/ft^2$$
$$= 1.1574 lb/in^2.$$

From Equation (B.45),

$$C_L = C_{za} - P_T/qS = 1.80 - \frac{(-50,000)}{(166.67)(7986)}$$
$$= 1.80 + 0.04 = 1.84.$$

From Equations (B.35) and (B.55),

A.R. $= b^2/S = (250)^2/7986 = 7.83,$

$c_d = 0.01 + (1.84)^2/2(7.83) = 0.15.$

From Equation (B.53) with no twist of the wing, $c_l = 1.84 c_{l1}$ with c_{l1} in Table B.1. Thus, Equation (B.56) becomes

$c_x = 1.84 c_{l1} \cos 21° + 0.15 \sin 21° = 1.718 c_{l1} + 0.054,$
$c_z = -1.84 c_{l1} \sin 21° + 0.15 \cos 21° = -0.659 c_{l1} + 0.140.$

The calculations for the spanwise distribution of c_x and c_z are shown in Table B.3, using data from Table B.1.

Equation (B.57) becomes

$p_x = qcc_x = 1.1574 cc_x,$
$p_z = qcc_z = 1.1574 cc_z,$

which are shown in columns (6) and (7) in Table B.3.

From Equations (B.58)-(B.60), the calculations for the shear forces and moments in Table B.3 are made as

$$(\Delta V_x)_j = (p_x)_{av,j}(\Delta y)_j, \quad V_{xi} = \sum_{j=1}^{y_i}(\Delta V_x)_j, \text{ from tip,}$$

$$(\Delta V_z)_j = (p_z)_{av,j}(\Delta y)_j, \quad V_{zi} = \sum_{j=1}^{y_i}(\Delta V_z)_j, \text{ from tip,}$$

$$(\Delta M_x)_j = (V_z)_{av,j}(\Delta y)_j, \quad M_{xi} = \sum_{j=1}^{y_i}(\Delta M_x)_j, \text{ from tip,}$$

$$(\Delta M_z)_j = -(V_x)_{av,j}(\Delta y)_j, \quad M_{zi} = \sum_{j=1}^{y_i}(\Delta M_z)_j, \text{ from tip,}$$

$$(\Delta M_y)_j = -(z_a)_{av,j}(\Delta V_x)_j, \quad M_{yi} = \sum_{j=1}^{y_i}(\Delta M_y)_j, \text{ from tip.}$$

Approximate checks for the V_x and V_z results at $y = 0$ in Table B.3 can be made from

$$L = 6W - P_T = 2,445,800 \text{ lb}, \quad D = 0.15qS = 199,650 lb.$$

Put Equation (B.56) into force form to get

$V_x = \frac{1}{2}(L \cos 21° + D \sin 21°) = 1,177,000 \text{ lb},$
$V_z = \frac{1}{2}(-L \sin 21° + D \cos 21°) = -345,050 \text{ lb}.$

which check the table values.

Table B.9. Airload, shear, and moment distributions for example B.9.

(1) y, in	(2) c, in	(3) c_{l1}	(4) c_x	(5) c_z	(6) p_x (lb/in)	(7) p_z (lb/in)	(8) Δy, in
0	500.0	0.988	1.751	-0.511	1013	-296	
							150
150	480.0	1.006	1.782	-0.523	990	-291	
							150
300	460.0	1.020	1.806	-0.532	961	-284	
							150
450	440.0	1.029	1.822	-0.538	928	-274	
							150
600	420.0	1.033	1.829	-0.541	889	-263	
							130
730	402.7	1.029	1.822	-0.538	850	-251	
							130
860	385.3	1.019	1.805	-0.532	805	-237	
							130
990	368.0	0.998	1.769	-0.518	753	-220	
							130
1120	350.7	0.963	1.708	-0.495	694	-201	
							95
1215	338.0	0.923	1.640	-0.468	641	-183	
							95
1310	300.0	0.897	1.595	-0.451	553	-157	
							95
1405	225.0	0.879	1.564	-0.439	407	-114	
							32
1437	200.0	0.850	1.514	-0.420	350	-97	
							32
1469	160.0	0.808	1.442	-0.392	265	-73	
							31
1500	0.0	0.0	0.0	0.0	0	0	

Table B.9 (continued)

(9) y, in	(10) $(p_x)_{av}$ lb/in	(11) $(p_z)_{av}$ lb/in	(12) $\frac{\Delta V_x}{1000}$	(13) $\frac{V_x}{1000}$	(14) $\frac{(V_x)_{av}}{1000}$	(15) $\frac{\Delta M_x}{10^5}$	(16) $\frac{M_x}{10^5}$
0				1182.5			7675.9
	1002	-294	150.3		1107.3	1660.9	
150				1032.2			6015.0
	975	-288	146.5		958.9	1438.3	
300				885.7			4576.7
	944	-279	141.8		814.8	1222.2	
450				743.9			3354.5
	908	-268	136.3		675.8	1013.7	
600				607.6			2340.8
	870	-257	113.0		551.2	716.5	
730				494.6			1624.3
	828	-244	107.6		440.8	573.1	
860				387.0			1051.2
	779	-228	101.2		336.5	437.4	
990				285.9			613.8
	723	-210	94.1		238.9	310.5	
1120				191.8			303.3
	667	-190	63.4		160.0	152.0	
1285				128.3			151.3
	597	-170	56.7		99.9	95.0	

External forces on flight vehicles

Table B.9 (continued)

y, in	(9) $(p_x)_{av}$ lb/in	(10) $(p_z)_{av}$ lb/in	(11) $\frac{\Delta V_x}{1000}$	(12) $\frac{V_x}{1000}$	(13) $\frac{(V_x)_{av}}{1000}$	(14) $\frac{\Delta M_x}{10^5}$	(15) $\frac{M_x}{10^5}$
1310				71.6			56.3
	480	-136	45.6		48.9	46.4	
1405				26.1			9.9
	378	-106	12.1		20.1	6.4	
1437				13.9			3.5
	307	- 85	9.8		9.0	2.9	
1469				4.1			0.6
	132	- 36	4.1		2.1	0.6	
1500				0			0

Table B.9. (continued)

y, in	(16) $\frac{\Delta V_z}{1000}$	(17) $\frac{V_z}{1000}$	(18) $\frac{(V_z)_{av}}{1000}$	(19) $\frac{\Delta M_z}{10^5}$	(20) $\frac{M_z}{10^5}$	(21) $(x_a)_{av}$	(22) $\frac{\Delta M_y}{10^5}$	(23) $\frac{M_y}{10^5}$
0		-346.1			2233.3			964.7
	-43.9		-324.1	486.1		-98.0	147.2	
150		-302.1			1747.2			817.5
	-43.1		-280.6	420.9		-94.0	137.6	
300		-259.0			1326.3			679.9
	-42.0		-238.0	357.0		-90.0	127.6	
450		-217.0			969.3			552.3
	-40.3		-196.9	295.3		-86.0	117.2	
600		-176.7			674.0			435.1
	-33.4		-160.0	208.0		-82.3	93.0	
730		-143.3			466.0			342.1
	-31.8		-127.4	165.6		-78.8	84.8	
860		-111.5			300.4			257.3
	-29.7		- 96.7	125.7		-75.4	76.3	
990		- 81.9			174.7			181.0
	-27.3		- 68.2	88.7		-71.9	67.7	
1120		- 54.6			86.0			113.3
	-18.2		- 45.5	43.3		-68.9	43.7	
1215		- 36.4			42.7			69.6
	-16.2		- 28.2	26.8		-63.8	36.2	
1310		- 20.2			15.9			33.4
	-12.9		- 13.8	13.1		-52.5	23.9	
1405		- 7.3			2.8			9.5
	- 3.4		- 5.6	1.8		-42.5	5.2	
1437		- 3.9			1.0			4.3
	- 2.7		- 2.5	0.8		-36.0	3.6	
1469		- 1.1			0.2			0.7
	- 1.1		- 0.5	0.2		-16.0	0.7	
1500		0			0			0

B.9. Distribution of inertia forces on wing and fuselage

As shown in Section B.4, the flight vehicle must be in equilibrium with the air forces balanced by the inertia forces. These inertia forces are distributed throughout the vehicle and are determined by the mass and acceleration at any point. In general, the component inertia forces acting on any element of weight

Δw, with the gravity force included, are

$$\Delta F_{Ix} = n_x \Delta w, \quad \Delta F_{Iy} = n_y \Delta w, \quad \Delta F_{Iz} = n_z \Delta w, \tag{B.61}$$

where n_x, n_y, n_z are the component load factors acting on the element. For motion in the x-z plane, n_x and n_z are given by Equation (B.27).

For most airplanes with the major weight items in the fuselage, the inertia forces on the wing are small as compared to the air forces on the wing. However, the primary forces on the fuselage are the inertia forces. Both the wing and the fuselage are considered in this section. Although the inertia forces are body forces they can usually be treated as surface forces except for local attachments for items separate from the structure. At any wing or fuselage cross section, the weight of the cross section per unit length is regarded as acting at the center of mass of the cross section. The weight of large items such as engines, landing gear, external fuel tanks, etc. is regarded as applied at the attachments to the main structure.

If $w(y)$ is the weight per unit span at section y on the wing, then the inertia forces per unit span are usually taken as

$$p_{zw} = n_{za} w, \quad p_{xw} = 0, \quad p_{yw} = 0, \tag{B.62}$$

Here, $p_{xw} = 0$ and $p_{yw} = 0$ are approximations for the small effects of n_x and n_y on the wing. Although n_z varies with x and z in Equation (B.27), the variation is usually small on the wing so that n_z is taken as n_{za} for the flight condition under consideration on the V-n diagram. If the airplane has large items, such as tip tanks, and can have a large roll acceleration α_x, then it may be necessary to use

$$n_z = n_{za} + (\alpha_x y/g) \tag{B.63}$$

to allow for the roll effects on the wing loading.

On the wing, the usual procedure is to calculate the shear V_z, moments M_x and M_y, for a unit value $n_{za} = 1$ and then multiply by the n_{za} for the specified flight condition. Thus,

$$P_{zw} = w, \quad \Delta V_z = (p_{zw})_{av} \Delta y, \quad V_{zw} = \sum(\Delta V_z), \quad \text{from tip,}$$
$$\Delta M_x = (V_z)_{av} \Delta y, \quad M_{xw} = \sum(\Delta M_x), \quad \text{from tip,} \tag{B.64}$$
$$\Delta M_y = -(\Delta V_z)(x_w)_{av}, \quad M_{yw} = \sum(\Delta M_y), \quad \text{from tip,}$$

where x_w is the distance from the C.G. of the section to the reference axis y. For a given flight condition the net design shear and moments on the wing are

$$V_z = (V_z)_{\text{airload}} - N_{za} V_{zw}, \quad M_x = (M_x)_{\text{airload}} - n_{za} M_{xw},$$
$$M_y = (M_y)_{\text{airload}} - n_{za} M_{yw}. \tag{B.65}$$

To allow for any value of roll acceleration α_x, values can be calculated for $(\alpha_x/g) = 1$ in Equation (B.63) or $p_{zw} = (w)y$ in Equation (B.64).

On the fuselage, due to various payload items and large changes in the structural weight, it is usually simpler to specify the combined weights in blocks as

External forces on flight vehicles 303

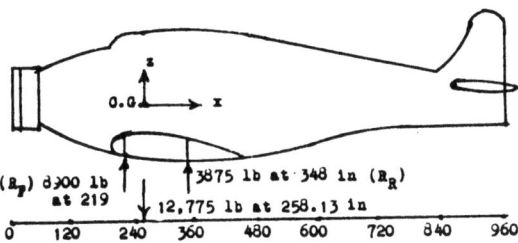

Fig. E.17. Fuselage stations to locate point weights, Table B.4.

a point weight at the center of mass of each block. Figure B.17 shows the procedure on a small airplane with two wing attach points. The reactions R_F and R_R are obtained from equilibrium of the fuselage. For $n_{za} = 1$, the shear V_z and bending moment M_y point values can be obtained directly as shown in Table B.4. For a given flight condition these values are multiplied by n_{za}.

Table B.4. Point weights, shear, and moment data on fuselage ($n_{za} = 1$), Figure B.17.

Station (in)	Weight (lb)	Shear(left) (lb)	Shear(right) (lb)	Moment (in-lb)/100
0	0	0	0	0
33	4465	0	-4465	0
150	1940	-4465	-6405	-5224
219	8900(R_F)	-6405	2495	-9643
240	1535	2495	960	-9120
348	3875(R_R)	960	4835	-8083
360	2045	4835	2790	-7503
510	1555	2790	1235	-3318
600	380	1235	855	-2200
690	50	855	805	-1436
780	105	805	700	- 712
870	590	700	110	- 82
945	110	110	0	0
960	0	0	0	0

Since the horizontal tail load varies with the flight conditions, shear and moment curves are calculated for the fuselage due to a unit tail load balanced by the wing supports. These values are then multiplied by the actual tail load for the particular flight condition.

If the pitching acceleration α_y is large and the fuselage is long, then the $(\alpha_y x/g)$ term in n_z in Equation (B.27) may be large so that V_z and M_y curves should be calculated for $\alpha_y/g = 1$ with $(-P_i x_i)$ used for the point loads P_i in Table B.4. Here x_i is measured from the airplane center of mass. If large side forces on the vertical tail produce a large yaw acceleration α_z, then for a unit $\alpha_z/g = 1$ the same curves for V_z and M_y can be used for side shear V_y and yawing moment M_z, where both cases come from the $(-P_i x_i)$ point loads.

The side tail load may also produce a torsional moment M_x on the fuselage, which is primarily balanced by the wing inertia term $I_{xx}\alpha_x$.

The P_x inertia forces on the fuselage are important for aircraft carrier planes during catapulting and arrested landing, and for tie downs on cargo and seats on transport planes.

Large inertia forces can occur on airplanes during landing. The landing gear forces for the maximum design landing load factors n_{zL} are applied as concentrated loads to the fuselage or wing and are balanced by the inertia forces. A two wheel landing may also produce a large pitching acceleration α_y and the consequent large inertia forces for large x_i values. Shear and moment curves are usually calculated for the wing and fuselage for several maximum design landing conditions such as one wheel, two wheel, three wheel, with and without pitching α_y. See Mil-A-8862 in Reference 2 for landing design conditions.

The forces, axial loads, shear forces, and bending moments acting on a typical landing gear structure are discussed in the next section.

B.10. Forces and moments on landing gear structures

When the landing gear wheels touch the runway, forces are applied in any direction so that in general the three component forces P_x, P_y, and P_z must be considered in the design of the landing structure. Figure B.18 shows the component parts, the applied component forces, and the reaction forces on a typical determinate landing gear structure. The supports for the landing gear are points (A), (B), and (D) on the wing or fuselage structure. Axially loaded member (1), or the drag brace, must be in two parts to allow the landing gear to be retracted with rotation about axis (A)-(B). Axially loaded member (2) is the side brace, which allows the joint at the junction of members (3) and (8) to have a zero bending moment. However, the cylinder (3) must take M_x and M_y bending moments and must transmit the torsion moment M_z into member (8). To transmit the torque M_z from the inner cylinder (4), which slides inside cylinder (3) to provide the shock absorber action of the landing gear, to the outer cylinder (3) it is necessary to use the torque links (5). As shown in Figure B.18, each torque link (5) acts as a simple cantilever beam to transfer the M_z moment. The cylinder (4) must take M_x and M_y bending as well as axial load P_z.

For any given dimensions in Figure B.18 the component reactions can be calculated directly from the applied component forces. Moments about a y-axis along (A)-(B) gives the axial force in member (1) and hence the reactions R_{Dx} and R_{Dz} as well as R_{Ex} and R_{Ez}. Moments about a x-axis through point (F) gives the axial force in member (2) and the component reactions at points (C) and (G). Summation of forces on members (3), (4), and (6) combined as a free body gives the component forces R_{Fx}, R_{Fy}, and R_{Fz} at point (F). From the forces R_{Fx}, R_{Fy}, R_{Fz}, R_{Gy}, and R_{Gz} on member (8) as a free body, the reactions R_{Ax}, R_{Ay}, R_{Az}, R_{Bx}, and R_{Bz} can be calculated.

Example B.10. In Figure B.18, (a) find the component reactions at points (A), (B), and (D) for $P_x = 25$ Klb, $P_y = 20$ Klb, $P_z = 100$ Klb. The location of the points by (x_R, y_R, z_R) in (in) from point (F) as reference is A(0, 10, 0), B(0, -50, 0), C(0, -10, -60), D(70, 0, 8), E(10, 0, -60), F(0, 0, 0), G(0, -45, -8), H(0, -30, -140), I(-10, 0, -80), J(-10, 0, -120), K(-40, 0, -100). (b) Draw shear and moment curves and discuss the axial load and torque distributions for beam members (3) and (4).

External forces on flight vehicles 305

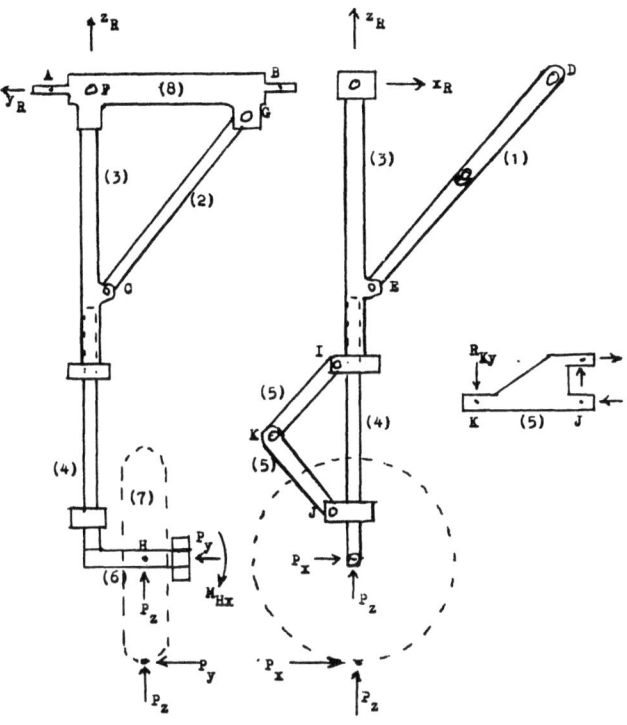

Fig. B.18. Landing gear components with landing forces.

Solution (a) From the geometry of the axially loaded member (1)

$$R_{Dx} = R_{Dz}\left[\frac{8-(-60)}{70-10}\right] = (17/15)R_{Dz},$$

whence moments about the axis (A)-(F)-(B) gives (use directions of coordinate axes for reactions and right hand rule for moments)

$$-25(140) + 8R_{Dz} - 70(17R_{Dz}/15) = 0,$$
$$R_{Dz} = -49.07 \text{ Klb} = -R_{Ex}, \quad R_{Dz} = -55.61 \text{ Klb} = -R_{Ez}.$$

Moments about the x_R-axis through point (F) with $M_{Hx} = 400$ Klb-in gives

$$400 + 20(140) - 100(30) + 8R_{Gy} - 45(-52R_{Gy}/35) = 0,$$
$$R_{Gy} = -2.67 \text{ Klb}, \quad R_{Gz} = 3.97 \text{ Klb}, \quad R_{Cy} = 2.67 \text{ Klb}, \quad R_{Cz} = -3.97 \text{ Klb}.$$

Summation of forces on members (3), (4), and (6) as a free body now gives

$$R_{Fx} + 25 - 49.07 = 0, \quad R_{Fx} = 24.07 \text{ Klb},$$
$$R_{Fy} + 20 - 2.67 = 0, \quad R_{Fy} = -17.33 \text{ Klb},$$
$$R_{Fz} + 100 - 55.61 + 3.97 = 0, \quad R_{Fz} = -48.36 \text{ Klb},$$
$$M_{Fz} + 30(25) = 0, \quad M_{Fz} = -750 \text{ Klb-in}.$$

The above results are shown on a free body of member (8) in Figure B.19, whence the reactions at (A) and (B) can be calculated as $R_{Ay} = -20.000$ Klb, $R_{Bz} = -8.49$ Klb, $R_{Ax} = 32.56$ Klb, $R_{Bz} = -10.73$ Klb, $R_{Az} = -33.66$ Klb.

306 *Appendix B*

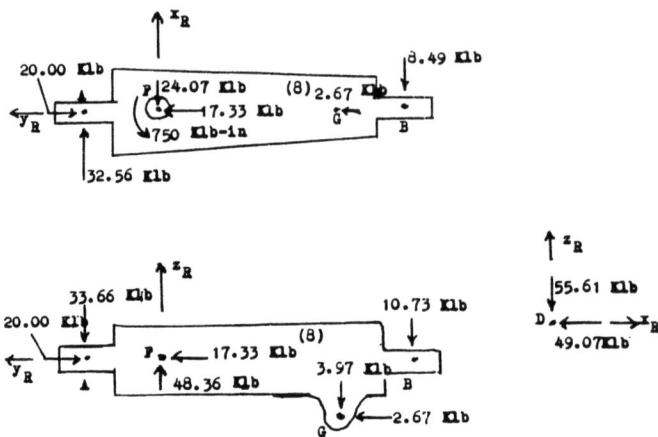

Fig. B.19. Free body of member (8) in Fig. B.18 for Example B.10.

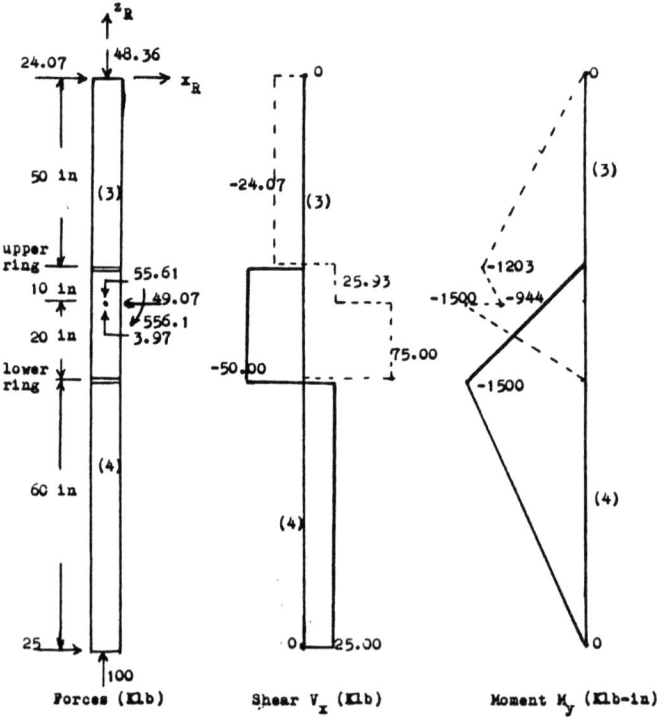

Fig. B.20. Shear and moment diagrams in x-z plane.

External forces on flight vehicles

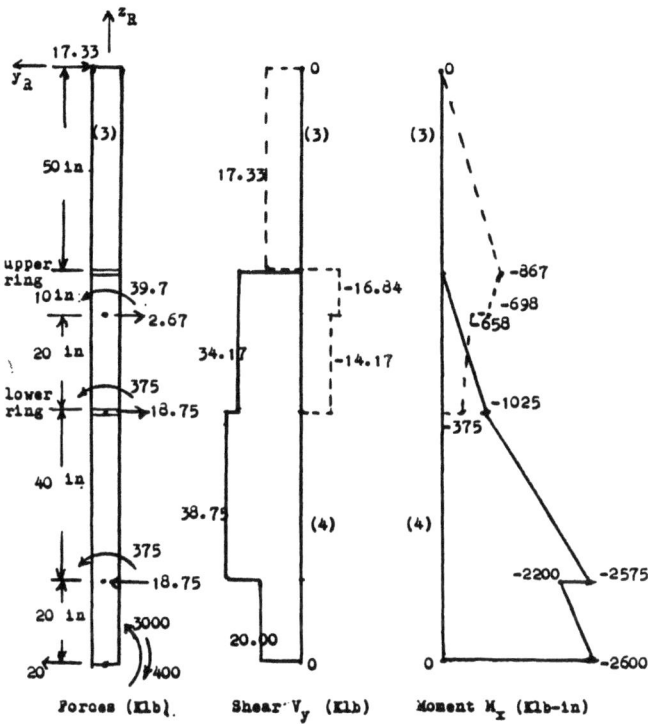

Fig. B.21. Shear and moment diagrams in y-z plane.

(b) The shear and moment curves for members (3) and (4) depend upon the location of the two ring contact points between the inner piston (4) and the outer cylinder (3). Figure B.20 shows the forces in the x-z plane, the shear distribution V_x, and the bending moment distribution M_y, for an assumed location of the contact rings. The moment is zero for member (3) at the lower ring and zero for member (4) at the upper ring so that the reactions at the rings are given by equilibrium of member (4), or

$$30R^x_{\text{upper}} - 60(25) = 0, \quad R^x_{\text{upper}} = 50 \text{ Klb}, \quad R^x_{\text{lower}} = -75 \text{ Klb}.$$

The axial load distribution depends upon the shock absorber design for members (3) and (4), with the air and oil system being involved in the transmission of the axial load through the system pressure. Usually, the air-oil system applies the 100 Klb force, which acts over a part of the length of member (4) as a compression force, to the upper end of member (3) near point (F). Thus the axial load in member (3) is the tension force of 55.61 - 3.97 = 51.64 Klb between point (E) and the cap of the cylinder. Between this cap and point (F), the axial load is a compression force of 48.36 Klb.

The torque moment M_z of 750 Klb-in applied at the lower end of member (4) at location (H) is constant in (4) up to location (J), where it is transferred into the torque links (5). Above point (J), the torque moment $M_z = 0$ in member (4), while the torque links (5) transfer the torque M_z into member (3) at location (I). From (I) to (F) in member (3), $M_z = 750$ Klb-in = constant. The shear forces and bending moments in the torque links (5) affect the shear force V_y and the bending moment M_x in members (3) and (4). The force R_{Ky} at point (K) is $R_{ky} = 750/40 = 18.75$ Klb, which produces a constant shear force and a linear bending moment in each of the torque links (5). The reactions at the locations (J) and (I) due to the

torque links are

$$R_J^y = 18.75 \text{ Klb}, \quad R_I^y = -18.75 \text{ Klb},$$
$$M_{zJ} = M_{zI} = -20(18.75) = -375 \text{ Klb-in.}$$

The reactions at the rings are

$$-30R_{\text{upper}}^y - 2600 - 375 + 60(20) + 18.75(40) = 0,$$
$$R_{\text{upper}}^y = -34.17 \text{ kN}, \quad R_{\text{lower}}^y = -(-34.17 + 20 + 18.75) = -4.58 \text{ Klb}.$$

Since the lower ring is at location (I), the combination of the torque link reactions at (I) and the ring reactions gives the values on member (3) at (I) as

$$V_{yI} = -18.75 + 4.58 = -14.17 \text{ Klb}, \quad \text{on (3)},$$
$$M_{zI} = -375 \text{ Klb-in}, \quad \text{on (3)}.$$

These values are shown in Figure B.21 and are used to draw the shear V_y and moment M_z curves in Figure B.21 for the $y - z$ plane.

See References 3 and 8 for other examples of landing gear load calculations.

B.11. Thermal forces

In high speed flight the viscous effects in the boundary layer on the surface of the vehicle heats the boundary layer to high temperatures. The heat is transmitted into the structure of the vehicle by convection, conduction, and radiation. The variable temperature in the structure can be obtained from the heat balance equations for the boundary layer and the structure using the basic procedures in texts on heat transfer. The general heat transfer equations with many applications are given in Reference 6, and some applications to flight vehicle structures are given in Reference 7.

A temperature change causes practically all materials, especially metals, to expand or contract. If the material is restrained, then this expansion or contraction will produce stresses, and hence loads, in the materials. Consider the bar of cross section A in Figure B.22 to be subjected to a uniform temperature increase ΔT. If the bar is unrestrained, it expands an amount

$$\Delta L = \alpha(\Delta T)L, \tag{B.66}$$

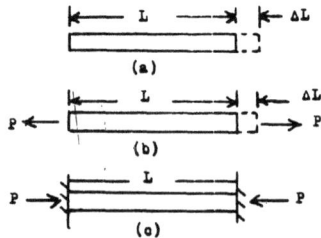

Fig. B.22. (a). ΔT expansion of bar. (b). Force P elongation of bar.
(c). Restrained bar with ΔT and $(-P)$.

External forces on flight vehicles 309

where α is the coefficient of thermal expansion per degree Celsius or Fahrenheit. If the same bar without temperature is subjected to a tension force P, then its elongation is

$$\Delta L = PL/EA, \tag{B.67}$$

where E is the modulus of elasticity for the material. In Figure B.22 it is evident that the heated bar can be returned to its original length by using a compressive force, or

$$(\Delta L)_T = (\Delta L)_P, \quad \alpha L(\Delta T) = -PL/EA,$$
$$P = -EA\alpha(\Delta T), \quad \sigma = -E\alpha(\Delta T), \tag{B.68}$$

where $\sigma = P/A$ is the stress in the bar. The minus sign indicates that the stress is compression. If the restrained bar is cooled, then the stress is tension.

Consider the case of two bars of different materials attached together, Figure B.23. Bar (1) is split in order to avoid any bending of the combination of the two bars. For simplicity, it is assumed that each bar is maintained at its uniform temperature. Since the bars will tend to expand different amounts, forces must be present to make the two bars maintain the same length. Thus from Equations (B.66)-(B.68),

$$(\Delta L)_1 = (\Delta L)_2, \quad \alpha_1 L(\Delta T)_1 + \frac{P_1 L}{E_1 A_1} = \alpha_2 L(\Delta T)_2 + \frac{P_2 L}{E_2 A_2}. \tag{B.69}$$

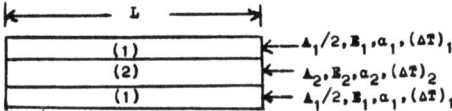

Fig. B.23. Two symmetrical heated bars joined together.

Since there are no external restraints or external forces, equilibrium requires that

$$P_1 + P_2 = 0. \tag{B.70}$$

Solve Equations (B.69) and (B.70) for P_1 and P_2 to get

$$P_1 = -KE_1 A_1 \alpha_1 (\Delta T)_1, \quad P_2 = -P_1,$$
$$K = \left[1 - \frac{\alpha_2 (\Delta T)_2}{\alpha_1 (\Delta T)_1}\right] \bigg/ \left[1 + \frac{E_1 A_1}{E_2 A_2}\right]. \tag{B.71}$$

The stresses in the bars are

$$\sigma_1 = P_1/A_1 = -KE_1 \alpha_1 (\Delta T)_1, \quad \sigma_2 = -(A_1/A_2)\sigma_1. \tag{B.72}$$

The non-dimensional coefficient K shows the effects of the material properties, the areas, and the temperature changes. If the materials are different, thermal forces are produced even though $(\Delta T)_1 = (\Delta T)_2$.

From the above discussion, it is evident that thermal forces due to temperature changes can occur in a structure with mixed materials or with either or both external restraints and internal restraints. In the flight vehicle internal restraints and mixed materials are present so that thermal forces are present whenever temperature changes occur. To allow for these thermal forces, the temperature terms are included in the basic equations for the internal stresses and internal loads in the analysis of the structure.

Example B.11. If $E_1 = 0.7(10^7)$ psi, $\alpha_1 = 25(10^{-6})/°$ C, $(\Delta T)_1 = 300°$ C, $A_1 = 5.00$ in^2, find the stresses σ_1 and σ_2 and the loads P_1 and P_2 in Figure B.23 for $E_1/E_2 = 1/3$, $\alpha_2/\alpha_1 = \frac{1}{2}$, $A_1/A_2 = 6$, $(\Delta T)_2 = 100°$ C.

Solution. The value of K in Equation (B.71) is

$$K = \frac{1 - (1/2)(1/3)}{1 + (1/3)(6)} = 5/18,$$

and from Equations (B.71) and (B.72)

$$\sigma_1 = -(5/18)(175)(300) - 14,583 \text{ psi},$$
$$\sigma_2 = -(6)(-14,583) = 87,500 \text{ psi},$$
$$P_1 = (5.00)(-14,583) = -72.917 \text{ lb}, \quad P_2 = 72,917 \text{ lb}.$$

B.12. Miscellaneous forces

Various other forces may be applied to the flight vehicle during its life. Many of these are concentrated forces applied at certain points on the structure, such as thrust and booster forces, towing forces, jacking and hoisting forces, catapult and arrested landing forces for naval carrier based airplanes, and support points for external fuel tanks, bombs, guns, etc.

A cable or push rod control system for a flight vehicle has forces in the system and applies forces at support points on the structure. Hydraulic systems or electric motors to retract the landing gears produce large forces at the support points.

For high altitude flying and for travel into space, the cabin must be designed for internal pressure forces. Also, fuel tanks and many containers for gases and liquids must be pressurized. See Chapter 4, Volume 2, for the analysis of pressurized structures.

Since the flight vehicle structure is flexible, it has various vibrations under unsteady air forces varying with time. It is possible that the motion of the vibrating structure can couple with the varying external air forces to produce dynamic loads in the structure which may add to the regular internal loads. The dynamic effects are considered in Chapter 6, Volume 2.

B.13. Deflection effects on the external forces

When external forces are applied to a flight vehicle structure, the various components of the structure deflect relative to each other. The wing bends and rotates giving vertical and horizontal deflections and a variable change in angle of attack of the wing. Since the angle of attack change is large in the outer part

External forces on flight vehicles

of the wing, the air force distribution on the wing is changed, usually increasing the load outboard toward the wing tips. If the variation of the rotation angle θ_y is known (see Chapter 5 for calculations of wing rotation) then the effect on the wing section lift coefficient distribution can be obtained by the procedure in Section B.7 for the twisted wing. After the wing is designed, its rotation can be calculated and the airload distribution recalculated. On some wings this rotation can cause a 10% to 20% increase in the rootbending moment for a given flight condition.

When the flaps and aileron are deflected, they change the wing center of pressure location. In the region of the aileron the center of aerodynamic pressure may move from about 25% chord to about 60% chord. This tends to twist the wing to reduce the angle of attack. On some wings at a certain velocity it is possible for this wing twist to completely cancel out the aileron deflection effects, or even to reverse the roll direction of the airplane. The problem of aileron reversal and wing divergence is considered in Chapter 5.

The bending deflection of the wing can produce an equivalent change in angle of attack on a swept wing. Consider Figure B.24, where the wing rotates

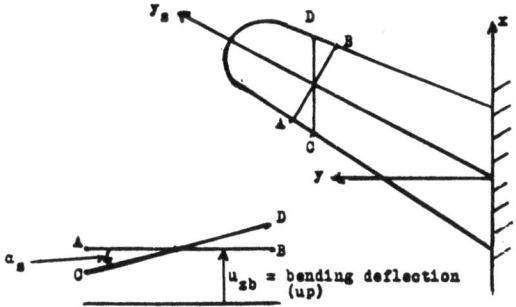

Fig. B.24. Angle of attack change due to bending deflection on swept back wing.

about the y_s-axis and the line A-B translates without rotation for a u_{zb} bending deflection of the swept back wing. It is evident that the point (C) deflects less than point (A) while point (D) deflects more than point (B). Thus the line C-D in the direction of the air flow rotates down by the angle α_s so that the angle of attack for lift on the wing is reduced, or

$$\alpha = \alpha_0 - \alpha_s. \tag{B.73}$$

This angle α_s can be combined with the rotation angle θ_{ys} of the wing about the y_s-axis, and the spanwise twist distribution $(cc_l)_0$ calculated by the procedure in Section B.7 using $\theta_{ys} - \alpha_s$.

The deflections of the wing and fuselage can affect the forces in the redundant attachments of items mounted on the wing or fuselage. If the aileron has more than two support points on the wing, then the support forces and the shear and bending moments in the aileron can be completely changed by the wing

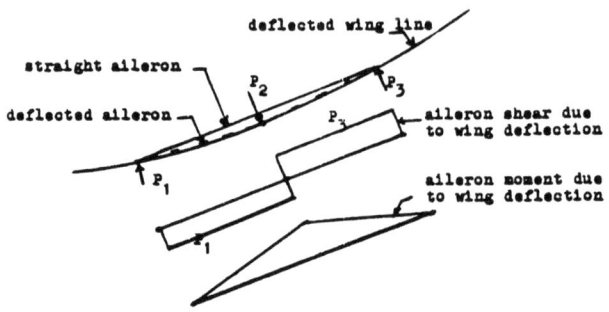

Fig. B.25. Redundant forces on aileron wing deflection.

deflection. Figure B.25 shows how the wing deflection affects an aileron with three support points. The forces, shear forces, and bending moments shown must be combined with the regular values due to the airloads.

B.14. Criteria for the structure to support the external forces

From the above discussion of the external forces acting on the flight vehicle, it is obvious that very large forces of various types are exerted on the vehicle during its lifetime. The structure of the vehicle must be designed to support these forces whenever they occur without appreciable permanent deformation and without failure. On the other hand, the weight of the structure must be lifted off the ground and transported on every flight of the vehicle. Thus, the materials used in flight vehicle structures must have very high strengths and have very low weights. In flight vehicles design, this contradiction of infinite strength and no weight must be solved in a practical way by using

(1) the most realistic V-n diagram compatible with the mission of the vehicle,

(2) the best and most accurate methods of analysis for the external force distributions on the vehicle,

(3) the best and most accurate methods of analysis for the internal loads and stresses,

(4) the most efficient type of structure consistent with the flight vehicle contour,

(5) the highest strength low density materials available,

(6) optimum design procedures for materials and structure types to obtain a minimum weight structure,

(7) criteria for material permanent deformation and failure that develop the maximum capability of the materials with a reasonable probability of safe performance for the design life of the vehicle,

(8) criteria for material fatigue and fracture that allows inspection and replacement of parts for a reasonable design life.

All the above eight items affect the primary problem in flight vehicle structures

External forces on flight vehicles 313

- to design a structure that can take the external forces safely for a specified life and that has the lowest possible weight.

In item (1) the weight of the primary structure is approximately proportional to the maximum load factor n_{za} given on the V-n diagram, Figure B.13. Although n_{za} is specified by the various agencies, References 1 and 2, for various types of airplanes, the category with the smallest n_{za} for a given type will give the lowest structural weight.

Item (2) has been the main topic of discussion in this Appendix B. Although the methods discussed are quite accurate for clean high aspect ratio wings, more refined methods must be used for low aspect wings and for wings with external engines, fuel tanks, etc., in which various interference effects change the air pressure distributions. Also, the dynamic effects of unsteady air forces must be considered. See References 10-13 and Chapter 6, Volume 2.

Items (3) and (4) are considered together so that the methods of analysis can be applied directly to the most efficient type of structure. These analysis methods together with their approximations and limitations are given in detail in this text and they constitute the major portion of the book in Volume 1 and Chapters 1-6 in Volume 2. The emphasis in Chapters 1-6 is on the very efficient thin web box beam idealized structure, which is analyzed by the elementary plane strain beam methods and by simple virtual displacement and virtual force methods for deflections and redundant beam structures. More refined methods, which take advantage of the digital computers and which can handle the more complicated real structures, are given in Chapters 5-9, Volume 2.

Items (5) and (6) must be considered together because a material must take tension, compression, and shear stresses and may have many modes of failure. It may be very good for tension but may be average for compression. The optimization of the design for least weight involves various factors that are considered in Chapters 1-4, Volume 2.

In item (7), the two basic criterions for the materials are:

(A) The stress $k_L \sigma_L$, where σ_L is produced by the largest external forces expected during the life of the vehicle, shall not exceed the yield stress F_y of the material. k_L is a non-dimensional factor to be specified for the design of a particular vehicle ($k_L = 1$ in many cases). This condition is expressed as a margin of safety (M.S.),

$$\text{M.S.} = (F_y/k_L\sigma_L) - 1 \geq 0.00, \quad (A). \tag{B.74}$$

The condition (A) is a condition to prevent permanent set of the structure during its life, where the stress σ_L is denoted as a limit stress and may be produced by the worst points on the V-n diagram.

(B) The stress $k_u \sigma_L$ shall not exceed the ultimate failing stress F_u of the material during the life of the vehicle, where k_u is a specified factor for the design. The stress σ_L is the same as in condition (A) and $k_u = 1.50$ for many airplanes, but k_u may vary from 1.20 for some guided missiles to $k_u = 2.00$ for some critical types of external forces. The condition is expressed as

$$\text{M.S.} = (F_u/k_u\sigma_L) - 1 \geq 0.00, \quad (B). \tag{B.75}$$

The condition (B) is a condition to prevent failure during the life of the vehicle.

Both conditions (A) and (B) must be satisfied at every point in the structure of the vehicle. For any particular material with F_y and F_u known and k_L and k_u specified, it is evident that one of the conditions will be more critical than the other, giving a smaller M.S., or

$$(F_u/F_y) < (k_u/k_L), \quad \text{use (B)},$$
$$(F_u/F_y) > (k_u/k_L), \quad \text{use (A)}. \tag{B.76}$$

To obtain the minimum weight of the structure, the design should give the applicable M.S. as 0.00 or small positive. It should be noted that the loads for the σ_L stress are the *worst expected loads*. This worst load may occur only a few times, or may never occur, on a large part of the structure during the vehicle life. If M.S. = 0.00 at a point in the structure for the HAA flight condition in Figure B.11, then it will be large positive for flight at V_c and $n_{za} = 1.00$. The criteria (A) and (B) are used in the design of the structural components in Chapters 2-4, Volume 2.

Item (8) involves complicated materials problems that occur for vehicles that must have a long life under varying loading conditions and in a varying environment. These problems are discussed in Chapters 1, 3, and 10, Volume 2.

B.15. Problems

B.1. If the airplane in Figure B.2. takes off at 150 ft/s after a run of 1500 ft, (a) find the constant engine thrust required during the take off run with a constant drag force of 3000 lb acting. (b) If the engine fails at the point where $V = 100$ ft/s, how long a runway is needed to stop with a braking force of 6000 lb?

B.2. In Example B.3, at touchdown the airplane has $a_z = 3.0g$ constant deceleration and $V_z = -15$ ft/s. (a) Find the distance s_z that the shock absorbers must compress to reduce the velocity V_z to 0. (b) How much time does this take?

B.3. Solve Example B.3 for the case of a concrete runway with $R_2 = 56$ Klb and $F = 4$ Klb.

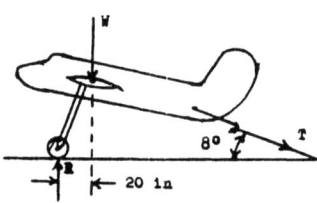

Fig B.26. Airplane for Problem B.4.

B.4. An airplane with a weight of 20 Klb makes an arrested landing, Figure B.26. (a) Find the tension T in the cable and the wheel reactions R if the decelerations are $x_{,tt} = 3.2\ g$ and $z_{,tt} = 2.0\ g$. (b) How far (above or below) from the C.G. should the line of action of T pass to produce no pitching acceleration? (c) If the landing velocity is 120 ft/s, what is the stopping distance? (d) How much time is required to stop?

B.5. Solve Example B.4 for the airplane at the position of $\theta = 60°$ on the circle.

B.6. Solve Example B.5 for the $\theta = 60°$ case of Problem B.5.

B.7. Suppose the airplane in Example B.4 encounters a gust in horizontal flight that instantaneously increases the lift to $L = 40,000$ lb. (a) What tail load is required to prevent a

External forces on flight vehicles 315

pitching acceleration? (b) If $I_{yy} = 180,000$ lb-s^2-in, what pitching acceleration α_y is produced if the tail load does not change from it's level flight value? (c) If the gust lasts 0.20s and the α_y remains constant during this time, how much does the angle of attack decrease?

B.8. An airplane is making a horizontal turn with a radius of 3000 ft and no change in altitude. Find the angle of bank (wings from horizontal) and the n_{za} load factor for velocities of (a) 200 ft/s, (b) 300 ft/s, and (c) 500 ft/s.

B.9. Draw a sea level V-n diagram for an airplane with $n_{max} = +4, -2, V_{max} = 600$ ft/s, $(C_{za})_{max} = 2.00, -1.40, m_0 = 4.5/$rad, $U = 40$ ft/s, $W/S = 60$ lb/ft^2.

B.10. Repeat Problem B.9 at an altitude of 30,000 ft.

B.11. Put column (7) of Table B.1 and column (8) of Table B.2 on the same graph to compare the twisted and untwisted unit lift coefficient distributions.

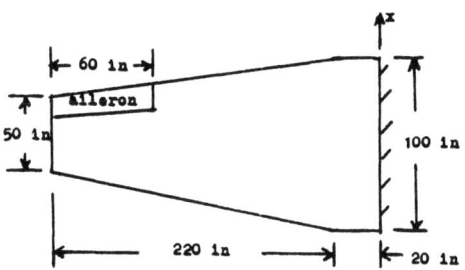

Fig. B.27. Wing for Problems B.12 and B.13.

B.12. Use Schrenk's method to calculate cc_{l1} and c_{l1} for the wing, Figure B.27. Use 20 in. intervals and show results in Table form similar to Table B.1.

B.13. When the aileron on the wing in Problem B.12 is fully deflected down to produce an effective change in the wing angle of attack of 6.0° in the region of the aileron, find the spanwise section lift coefficient for $C_L = 0$ (no lift) and $C_L = 1.00$. Take $m_0 = 0.100$ per degree and give results (with fairing) in table form similar to Table B.2.

B.14. Graph p_z, V_z, and M_z from Table B.3 on the same page to show the typical shape of the curves for the primary load, shear, and moment distributions.

B.15. Repeat Example B.9 for the low angle of attack (LAA) flight condition. Take $c_{d0} = 0.01$, $\theta = 6°$, $P_T = -10,000$ lb.

B.16. Graph the V_z shear and M_z moment curves from the data in Table B.3 and from Problem B.15 and compare for the two flight conditions.

B.17. If $w = 20 - 18(y/1500)^2$ (lb/in) is the weight distribution of the wing in Figure B.14, integrate to get the V_{zw} and M_{zw} equations for $n = 1$ in terms of $(y/1500)$. Give values of V_{zw} and M_{zw} at $y = 0$ and $y = 600$ in.

B.18. In Figure B.17 verify the C.G. location shown by calculating it from the weight and location of the blocks shown, $\bar{x} = \sum x_{Ri} w_i / \sum w_i$, Table B.4.

B.19. In Figure B.17 calculate the airplane mass moment of inertia I_{yy} about the airplane C.G., neglecting the moments of inertia of the blocks themselves and neglecting the z_i^2 term in $I_{yy} = \sum (w_i/g)(x_i^2 + z_i^2)$.

B.20. Draw the V_z shear and the M_y moment curves from the given data for the fuselage in Figure B.17, Table B.4.

B.21. In Figure B.17 calculate the point loads, point shear loads, and point moments for the case of a pitching acceleration of $\alpha_y/g = 1.00$, Table B.4.

B.22. Repeat part (a) of Example B.10 for $P_z = 40$ Klb, $P_y = 10$ Klb, $P_x = 120$ Klb.

B.23. A steel rod 50 in. long is fastened between two walls without load at 40° C. (a) What is the stress in the rod at $-50°$ C? (b) If the walls deflect 0.02 in. under the pull of the rod, what is the stress at $-50°$ C? Use $\alpha = 14(10^{-6})/°$ C and $E = 21(10^6)$ psi.

B.24. An aluminium alloy rod with a cross section area of 3.00 in.^{2} is fastened between two walls at 40° C with a preload of 10,000 lb tension. (a) What will the stress be at $-20°$ C? (b) For what temperature will the stress be zero? Use $\alpha = 25(10^{-6})/°$ C and $E = 7(10^6)$ psi.

B.25. Solve Example B.11 for the case of $(\Delta T)_1 = (\Delta T)_2 = 200°$ C and $A_1 = A_2$.

B.26. An aluminum alloy cylindrical sleeve of inner diameter 1.50 in. and wall thickness

0.30 in. is slipped over a 1.50 in. diameter steel bolt and held in place by a nut that is turned just snug. Compute the temperature rise of the assembly to produce a stress of 10,000 psi in the aluminum sleeve. Use α and E for steel and aluminum from Problems B.23, B.24.

B.27. Under the HAA flight condition the airloads produce a twist of the wing in Figure B.14 which can be represented by

$$\alpha_{aR} = 10(y/1500)^2 \text{ in degrees.}$$

Use the procedure and data in Example B.8 but with the flap neutral and calculate a new Table B.2 for this twist effect. Note that no fairing is necessary in this case.

References

Appendix B

1. Airworthiness Standards, Federal Aviation Agency Reports, various dates: Report 23, Normal, Utility, and Acrobatic Category Airplanes; Report 25, Transport Category Airplanes; Report 27, Normal Category Rotorcraft; Report 29, Transport Category Rotorcraft.
2. Airplane Strength and Rigidity, Military Specifications, various dates: MIL-A-8860 (ASG), General Specifications; MIL-A-8861, Flight Loads; MIL-A-8862, Landing and Ground Handling Loads; MIL-A-8863, Additional Loads for Carrier-Based Airplanes; MIL-A-8864, Water and Handling Loads for Seaplanes; MIL-A-8865, Miscellaneous Loads; MIL-A-8866, Reliability Requirements, Repeated Loads, and Fatigue; MIL-A-8867, Ground Test; MIL-A-8868, Data and Reports; MIL-A-8869, Nuclear Weapons Effects; MIL-A-8870, Flutter, Divergence and other Aeroelastic Instabilities; MIL-A-8871, Flight and Ground Operations Tests; MIL-A-8892, Vibration; MIL-A-8893, Sonic Fatigue.
3. D.J. Peery: *Aircraft Structures*, McGraw-Hill Book Co. (1950).
4. O. Schrenk: *A Simple Approximation Method for Obtaining the Spanwise Lift Distribution*, NACA TM 948 (1940).
5. A.M. Kuethe and J.D. Schetzer: *Foundations of Aerodynamics*, John Wiley and Sons (1950).
6. Max Jakob: *Heat Transfer*, John Wiley and Sons, Vol. I (1949), Vol. II (1957).
7. B.E. Gatewood: *Thermal Stresses*, McGraw-Hill Book Co., (1957).
8. E.F. Bruhn: *Analysis and Design of Flight Vehicle Structures*, Tri-State Offset Co., Cincinnati, Ohio (1965).
9. C.J. Wenzinger: *Pressure Distribution over an NACA 23012 Airfoil with a NACA 23012 External Airfoil Flap*, NACA TR 614 (1938).
10. V.M. Falkner: *The Calculations of Aerodynamic Loading on Surfaces of Any Shape*, R and M 1910, British Research Council (1943).
11. S.G. Hedman: *Vortex Lattice Method for Calculating Quasi-Steady State Loadings on Thin Elastic Wings*, Report 105, OCt. (1965). Aeronautical Research Institute of Sweden.
12. E. Albano and W.P. Rodden: A doublet lattice method for calculating lift distributions on oscillating surfaces in subsonic flow, *AIAA J.* **7**, No. 2, p. 269-285 (1969).
13. M.T. Landahl and V.J.E. Stark: Numerical lifting surface theory - problems and progress, *AIAA J.* **6**, No. 11, p. 2049-2060 (1968).
14. B. Etkin: *Dynamics of Flight*, John Wiley and Sons (1959).

Reference for additional reading

15. D.J. Peery and J.J. Azar: *Aircraft structures*, McGraw-Hill Book Co., New York, 2nd Ed., Chapter 2 (1982).

C

Derivation of the strain energy theorems from the virtual principles

C.1. Work and strain energy

When forces are applied to a structure, displacements, strains, and stresses are produced in the structure. As a vector force **P** is slowly applied its point of application deflects a vector distance **u** and work **W** is done, or

$$W = \int_0^u \mathbf{P} \cdot d\mathbf{u} = \int_0^u (P_x\,du_x + P_y\,du_y + P_z\,du_z). \tag{C.1}$$

Figure C.1 shows the meaning of W for **P** and **u** in the same direction. Also, from Figure C.1, complementary work W^c can be defined as

$$W^c = \int_0^P \mathbf{u} \cdot d\mathbf{P} = \int_0^P (u_x\,dP_x + u_y\,dP_y + u_z\,dP_z). \tag{C.2}$$

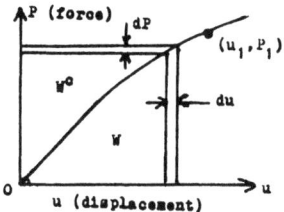

Fig. C.1. Work and complementary work.

If the structure is elastic with small displacements so that the load and displacement are proportional during the load application, then from Figure C.1,

$$W = (1/2)P_1 u_1 = W^c, \tag{C.3}$$

where P_1 and u_1 are the values at the final position. Although for this elastic case $W = W^c$, in applications below they are different in that for W the load P is specified and u is the unknown variable while in W^c the displacement u is regarded as specified and P is the unknown variable.

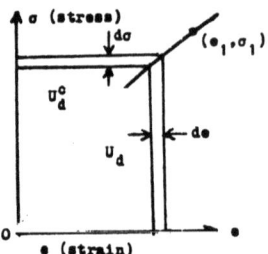

Fig. C.2. Strain energy and complementary strain energy.

For the three dimensional elastic small displacement case, Equation (C.3) becomes

$$W = W^c = (1/2) \int_V (X_x u_x + X_y u_y + X_z u_z) dV +$$
$$+ (1/2) \int_S (S_x u_x + S_y u_y + S_z u_z) dS +$$
$$+ (1/2) \sum_{m=1}^{M} (P_{mx} u_{mx} + P_{my} u_{my} + P_{mz} u_{mz}), \quad (C.4)$$

where the forces and displacements are final values and reaction forces are included.

From Figure C.2 the strain energy density U_d per unit volume is the area below the curve, whence for the structure the strain energy U and the complementary strain energy U^c are

$$U = \int_V U_d \, dV, \quad U_d = \int_0^{e_1} \sigma \, de, \quad (C.5)$$

$$U^c = \int_V U_d^c \, dV, \quad U_d^c = \int_0^{\sigma_1} e \, d\sigma. \quad (C.6)$$

For the elastic case with $\sigma = Ee$ in Figure C.2

$$U_d = (\sigma_1/2)e_1 = Ee_1^2/2, \quad (C.7)$$
$$U_d^c = (\sigma_1/2)e_1 = \sigma_1^2/2E, \quad (C.8)$$

where e_1 and σ_1 are values at the final position.

For the three dimensional small displacement case for isotropic materials

$$U = \int_V \left[\frac{\nu E(e_{xx} + e_{yy} + e_{zz})^2}{2(1+\nu)(1-2\nu)} + \frac{E}{2(1+\nu)} \{ e_{xx}^2 + e_{yy}^2 + e_{zz}^2 + \right.$$
$$\left. + \tfrac{1}{2}(e_{xy}^2 + e_{xz}^2 + e_{yz}^2) \} \right] dV, \quad (C.9)$$

$$U^c = \int_V \left[(1/2E)\{ \sigma_{xx}^2 + \sigma_{yy}^2 + \sigma_{zz}^2 + 2(1+\nu)(\sigma_{xy}^2 + \sigma_{xz}^2 + \sigma_{yz}^2) - \right.$$

$$-2\nu(\sigma_{xx}\sigma_{yy} + \sigma_{xx}\sigma_{zz} + \sigma_{yy}\sigma_{zz})\Big]dV, \tag{C.10}$$

where the strains and stresses are final values.

The kinetic energy T is defined as

$$T = (1/2)\int_V (u_{x,t}^2 + u_{y,t}^2 + u_{z,t}^2)\rho\,dV, \tag{C.11}$$

where $u_{x,t} = du_x/dt$, $u_{y,t}$ and $u_{z,t}$ are velocity components.

For the elastic conservative case, conservation of energy applies so that in the static case

$$\begin{aligned} U &= W, \quad \text{in terms of displacements,} \\ U^c &= W^c, \quad \text{in terms of stresses.} \end{aligned} \tag{C.12}$$

In this same case the total potential energy and total complimentary potential energy can be defined as

$$E_p = U + V = U - 2W = -W = -U, \tag{C.13}$$
$$E_p^c = U^c + V^c = U^c - 2W^c = -W^c = -U^c, \tag{C.14}$$

where V and V^c are the potential functions of the external forces defined by

$$V = -2W, \quad V^c = -2W^c. \tag{C.15}$$

These expressions for the potential functions of the external forces and for the total potential energy are used in the energy theorems to produce energy functions that will fit the theory of the Calculus of Variations. Because of the lack of clear definitions in much of the literature on the energy methods, this use of special expressions that will give minimum theorems in the Calculus of Variations has caused considerable confusion in the use of the energy methods.

In the following Section C.2, an effort is made to clarify the maximum and minimum cases for the strain energy and total potential energy. These cases will then be referred to in the following sections in which the major energy theorems are derived as special cases of the virtual principles.

C.2. Maximum and minimum strain energy and total potential energy

Consider Figure C.3a with the external force P specified. As pointed out in Reference 1, the work W done by P and the strain energy U in the beam due to P are *maximum* when the restraints are least. Remove restraint (2) in Figure C.3a and the deflection is larger so that P does more work and more strain energy is stored in the beam from $W = U$ for the conservation of energy. Thus, for a structure in stable equilibrium with given external forces and given restraints, the strain energy due to the given forces is a maximum. Note that any approximation for the deflection curve or any variation in the deflections which satisfies the given restraints further restrains the beam and decreases the strain energy. Under these conditions U is a maximum and $E_p = -U$ in Equation (C.13) is a minimum.

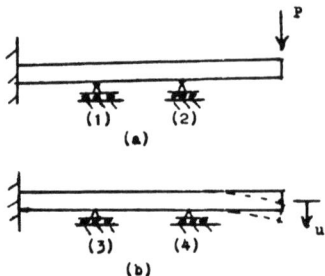

Fig. C.3. (a). External forces specified. (b) Non-zero displacements specified.

Consider Figure C.3b with the displacement u specified. From Reference 1, the work done by the application of the displacement u and the strain energy stored in the beam due to u are *minimum* when the restraints are least. If restraint (4) is removed before u is applied in Figure C.3b, then the force to produce the deflection u is smaller so that less strain energy is stored in the beam. Thus, for a structure in stable equilibrium with given applied displacements and given restraints, the strain energy U due to the given displacements is a minimum. Under these conditions, $E_p = -U$ in Equation (C.13) is a maximum.

Figure C.4 shows possible trend diagrams for the strain energy U, complementary strain U^c, total potential energy $E_p = -U$ and total complementary potential energy $E_p^c = -U^c$ as the restraints are increased to the right in the diagram. Both the external force and the non-zero displacement cases are shown. These diagrams are for the elastic conservative case so that Equations (C.12)-(C.15) apply. The internal restraints indicated in Figure C.4 may be due to redundant members in trusses and to approximations made in solutions for the stresses and displacements. If an approximate solution is made for the displacements, the internal restraints are increased so that the trend in Figure C.4 is to the right. If an approximate solution is made for the stresses, the internal restraints are decreased so that the trend in Figure C.4 is to the left.

Since the strain energy U and the total potential energy E_p are functions of displacements, and the complementary energy U^c and total complementary potential energy E_p^c are functions of the stresses, maximum and minimum values can be identified at points on the four curves in Figure C.4. Thus, for the four cases in Figure C.4,

$$\text{Curve (1)}, \quad U = U_{\max}, \quad U^c = U^c_{\min}, \tag{C.16}$$

$$\text{Curve (2)}, \quad U = U_{\min}, \quad U^c = U^c_{\max}, \tag{C.17}$$

$$\text{Curve (3)}, \quad E_p = E_{p,\min}, \quad E_p^c = E^c_{p,\max}, \tag{C.18}$$

$$\text{Curve (4)}, \quad E_p = E_{p,\max}, \quad E_p^c = E^c_{p,\min}, \tag{C.19}$$

These results for maximum and minimum values of the strain and total potential energies will be used in the discussion of the classical strain energy theorems derived in the following sections.

Derivation of the strain energy theorems

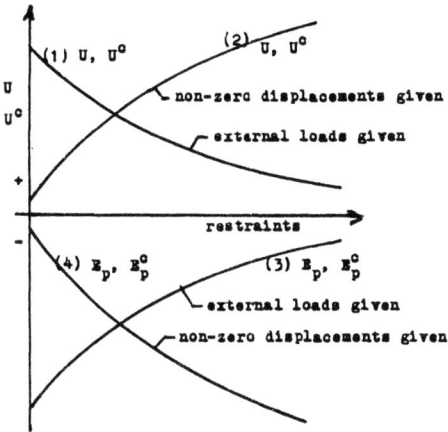

Fig. C.4. Strain energy and restraints.

Since the principle of virtual displacements involves solutions for displacements, its approximate solutions have the same trend as the strain energy U, while the principle of virtual forces, which involves solutions for stresses, has the trend of the complementary strain energy U^c. Thus, from Figure C.4 and Equations (C.16) and (C.17), approximate solutions given by the two principles are upper and lower bounds for the exact displacements and stresses. Which principle gives which bound depends upon the cases of external loads alone, Equation (C.16), non-zero specified displacements alone, Equation (C.17), or a combination of the loads and displacements.

C.3. Theorem of minimum total potential energy

If the virtual displacements u_x^V, u_y^V, u_z^V and virtual strains e_{xx}^V, e_{yy}^V, $\cdots$, e_{yz}^V in the principle of virtual displacements in Equation (2.6) are restricted to be real variations of the real displacements and real strains, then Equation (2.6) has the form

$$\int_V (\sigma_{xx}\delta e_{xx} + \cdots + \sigma_{yz}\delta e_{yz})dV - \int_V (X_x\delta u_x + X_y\delta u_y + X_z\delta u_z)dV -$$
$$- \int_S (S_x\delta u_x + S_y\delta u_y + S_z\delta u_z)dS - \sum_{m=1}^{M}(P_{mx}\delta u_{mx} + P_{my}\delta u_{my} +$$
$$+ P_{mz}\delta u_{mz}) = 0. \qquad (C.20)$$

Take the real variation of the strain energy U in Equation (C.9) and make use of Equation (1.13), or

$$\delta U = \frac{\partial U}{\partial e_{xx}}\delta e_{xx} + \frac{\partial U}{\partial e_{yy}}\delta e_{yy} + \cdots + \frac{\partial U}{\partial e_{yz}}\delta e_{yz}$$

$$= \int_V (\sigma_{xx}\delta e_{xx} + \sigma_{yy}\delta e_{yy} + \cdots + \sigma_{yz}\delta e_{yz})dV, \tag{C.21}$$

where Equation (1.13) shows that

$$\sigma_{xx} = \frac{\partial U_d}{\partial e_{xx}}, \quad \sigma_{yy} = \frac{\partial U_d}{\partial e_{yy}}, \cdots, \quad \sigma_{yz} = \frac{\partial U_d}{\partial e_{yz}}. \tag{C.22}$$

Take the variation of the real displacements in the work W in Equation (C.4) with the forces specified to get

$$2\delta W = \int_V (X_x \delta u_x + X_y \delta u_y + X_z \delta u_z)dV + \int_S (S_x \delta u_x + S_y \delta u_y + S_z \delta u_z)dS + \sum_{m=1}^M (P_{mx}\delta u_{mx} + P_{my}\delta u_{my} + P_{mz}\delta u_{mz}). \tag{C.23}$$

Compare Equations (C.21) and (C.23) to Equation (C.20) to get

$$\delta U - 2\delta W = 0, \tag{C.24}$$

whence from Equation (C.13)

$$\delta U - 2\delta W = \delta E_p = \delta(U + V) = 0, \quad \text{displacements varied.} \tag{C.25}$$

This is the *Theorem of Minimum Total Potential Energy*. It is evident that $E_p = U + V$ is a minimum in Equation (C.25) for the elastic, conservative case, provided it is applied for the given external forces on curve (3) in Figure C.4. See Equation (C.18). The specified zero or non-zero displacements for the supports are included in the restraints for the structure. Proofs of this theorem for the three dimensional case are given in Chapter 7 of Reference 5 and in Reference 4.

It should be noted that the principle of virtual displacements can be simply stated as "Virtual Strain Energy=Virtual Work", or in Equation (2.6)

$$U^V - W^V = 0. \tag{C.26}$$

This expression compares to Equation (C.24) for real variations.

C.4. Theorem of minimum strain energy

If $\delta W = 0$ in Equation (C.24) then

$$\delta U = 0, \quad \text{displacements varied,} \tag{C.27}$$

which is the *theorem of minimum strain energy* for a structure in stable equilibrium. In Equation (C.23) δW can be zero either because the applied forces are zero or because the displacements are specified so that their variations are zero. If non-zero displacements are specified, then the minimum for U follows from Equation (C.17) and curve (2) on Figure (C.4). Thus, this theorem applies to redundant structures with deflections, initial strains, or thermal strains specified.

C.5. Castigliano's theorem (Part I)

In the principle of virtual displacements, Equation (2.6), let all virtual displacements be zero except at the points m of the point forces P_m, whence Equations (C.20) and (C.21) give for real variations

$$\sum_{m=1}^{M} P_m \delta u_m = \delta U. \tag{C.28}$$

Assume the strain energy U in Equation (C.9) to be known in terms of the displacements $u_1, u_2, \cdots, u_M$, so that

$$\delta U = \sum_{m=1}^{M} \frac{\partial U}{\partial u_m} \delta u_m. \tag{C.29}$$

Thus, with δu_m arbitrary, it follows that

$$P_m = \frac{\partial U}{\partial u_m}, \quad m = 1, 2, \cdots, M, \tag{C.30}$$

which is *Castigliano's Theorem* (part I).

By using dummy or virtual displacements at the points m, it is possible to apply Equation (C.30) without knowing U in terms of $u_1, u_2, \cdots, u_M$. For example, take U for the uniaxial e_{xx} strain case in the form

$$U = (1/2) \int_V E\left(e_{xx} + e_{xxm}^1 u_m^V\right)^2 dV, \tag{C.31}$$

where e_{xxm}^1 is the virtual strain produced by a unit value of u_m^V. Thus,

$$P_m = \frac{\partial U}{\partial u_m^V} = \int_V E(e_{xx} + e_{xxm}^1 u_m^V) e_{xxm}^1 \, dV = \int_V \sigma_{xx} e_{xxm}^1 \, dV, \tag{C.32}$$

where $u_m^V = 0$ is used for the dummy value after u_m^V has served its purpose. This expression for P_m in Equation (C.32) corresponds to the unit displacement theorem, Equation (2.9). Of course, Equation (2.9) does not have the restrictions on the strains implied by Equations (C.31) and (C.32).

C.6. Hamilton's principle

Put Equation (C.26) into Equation (2.7) to get

$$\int_{t_0}^{t_1} \left[U^V - W^V + \int_V \left(u_{x,tt} u_x^V + u_{y,tt} u_y^V + u_{z,tt} u_z^V \right) \rho \, dV \right] dt = 0, \tag{C.33}$$

where $u_{x,tt} = \frac{\partial^2 u_x}{\partial t^2}$, etc. From Equation (C.11), take the virtual kinetic energy as

$$T^V = \int_V \left(u_{x,t} u_{x,t}^V + u_{y,t} u_{y,t}^V + u_{z,t} u_{z,t}^V \right) \rho \, dV, \tag{C.34}$$

whence, with integration by parts

$$\int_{t_0}^{t_1} T^V \, dt = \int_V \left[\rho \left(u_{x,t} u_x^V + u_{y,t} u_y^V + u_{z,t} u_z^V \right) \right]_{t_0}^{t_1} dV -$$
$$- \int_{t_0}^{t_1} \int_V \left(u_{x,tt} u_x^V + u_{y,tt} u_y^V + u_{z,tt} u_z^V \right) \rho \, dV \, dt. \tag{C.35}$$

Put Equation (C.35) into Equation (C.33) to get

$$\int_{t_0}^{t_1} (T^V - U^V + W^V) dt - \int_V \left[\rho \left(u_{x,t} u_x^V + u_{y,t} u_y^V + u_{z,t} u_z^V \right) \right]_{t_0}^{t_1} dV = 0, \tag{C.36}$$

which is the general form of the principle of virtual displacements for structural dynamics problems. Here U^V represents the stress terms and W^V represents the applied force terms in Equation (2.6), while T^V is defined in Equation (C.34).

If the virtual displacements are restricted to real variations of the real displacements, if the real variations are selected so that they are zero at $t = t_1$ and $t = t_0$, and if W^V is replaced by $2\delta W = -\delta V$ for a potential function of the external forces, then Equation (C.36) reduces to *Hamilton's Principle*, or

$$\delta \int_{t_0}^{t_1} (T - U - V) dt = 0. \tag{C.37}$$

For the rigid body case, $U = 0$, and

$$\delta \int_{t_0}^{t_1} (T - V) dt = 0, \tag{C.38}$$

where V is the potential of the rigid body forces.

Although some authors have stated in the literature that Hamilton's principle gives a minimum condition, it is simple to show, Reference 2, that for vibration problems the principle gives a minimum or a maximum depending upon the time interval $t_1 - t_0$. In the approximate solution the number of terms used to obtain the displacements is more important than the value of the integral in relation to a maximum or a minimum. Thus, the stationary value is the important thing.

C.7. Theorem of minimum total complementary potential energy

If the virtual forces and virtual stresses in the principle of virtual forces in Equation (2.14) are restricted to be real variations of the real forces and real stresses, then Equation (2.14) has the form

$$\int_V (e_{xx} \delta \sigma_{xx} + \cdots + e_{yz} \delta \sigma_{yz}) dV - \int_V (u_x \delta X_x + u_y \delta X_y + u_z \delta X_z) dV -$$
$$- \int_S (u_x \delta S_x + u_y \delta S_y + u_z \delta S_z) dS - \sum_{m=1}^{M} u_m \delta P_m = 0. \tag{C.39}$$

Derivation of the strain energy theorems

Take the variation of the complementary strain energy U^c in Equation (C.10) and make use of Equation (1.11), or

$$\delta U^c = \frac{\partial U^c}{\partial \sigma_{xx}} \delta \sigma_{xx} + \cdots + \frac{\partial U^c}{\partial \sigma_{yz}} \delta \sigma_{yz}$$

$$= \int_V (e_{xx} \delta \sigma_{xx} + \cdots + e_{yz} \delta \sigma_{yz}) \mathrm{d}V, \tag{C.40}$$

where Equation (1.11) shows that

$$e_{xx} = \frac{\partial U_d^c}{\partial \sigma_{xx}}, \ldots, e_{yz} = \frac{\partial U_d^c}{\partial \sigma_{yz}}. \tag{C.41}$$

Take the variation of the forces in the complementary work W^c in Equation (C.4) with the displacements specified to get

$$2\delta W^c = \int_V (u_x \delta X_x + u_y \delta X_y + u_z \delta X_z) \mathrm{d}V + \sum_{m=1}^M u_m \delta P_m +$$

$$+ \int_S (u_x \delta S_x + u_y \delta S_y + u_z \delta S_z) \mathrm{d}S. \tag{C.42}$$

Compare Equations (C.40) and (C.42) to Equation (C.39) to get

$$\delta U^c - 2\delta W^c = 0, \tag{C.43}$$

whence from Equation (C.14)

$$\delta U^c - 2\delta W^c = \delta E_p^c = \delta(U^c + V^c) = 0, \quad \text{stresses varied.} \tag{C.44}$$

This is the *Theorem of Minimum Total Complementary Potential Energy*. It is evident that E_p^c is a minimum in Equation (C.44) for the elastic conservative case, provided it is applied for the specified non-zero displacements, Equation (C.19) and curve (4) in Figure C.4. The stresses in U^c will include any applied forces acting on the structure.

It should be noted that the principle of virtual forces can be simply stated as "virtual complementary strain energy=virtual complementary work", or in Equation (2.14)

$$U^{cV} - W^{cV} = 0. \tag{C.45}$$

This expression compares to Equation (C.43) for real variations.

C.8. Theorem of minimum complementary strain energy

If $\delta W^c = 0$ in Equation (C.43) then

$$\delta U^c = 0, \quad \text{stresses varied}, \tag{C.46}$$

which is the *theorem of minimum complementary strain energy* for a structure in stable equilibrium. In Equation (C.42), δW^c can be zero because the displacements are zero or because the applied loads are specified so that their variations are zero. If external loads are specified, then the minimum for U^c follows from Equation (C.16) and curve (1) on Figure (C.4). In this case the displacements are zero at the supports.

C.9. Castigliano's theorem (Part II)

In the principle of virtual forces, Equation (2.14), let all virtual forces be zero except at the points of the desired displacements u_m, whence Equations (C.39) and (C.40) give

$$\sum_{m=1}^{M} u_m \delta P_m = \delta U^c \tag{C.47}$$

for real variations. Assume the complementary strain energy U^c in Equation (C.10) to be known in terms of the concentrated forces $P_1, P_2, \ldots, P_M$ so that, with δ as a differential operator,

$$\delta U^c = \sum_{m=1}^{M} \frac{\partial U^c}{\partial P_m} \delta P_m. \tag{C.48}$$

Thus, with δP_m arbitrary, it follows that

$$u_m = \frac{\partial U^c}{\partial P_m}, \quad m = 1, 2, \ldots, M, \tag{C.49}$$

which is *Castigliano's Theorem (part II)*. Note that the most general form of the principle of virtual forces, Equation (2.14), has been used above so that Equation (C.49) can be used at any point on the surface.

By using dummy or virtual forces at the points m, it is possible to apply Equation (C.49) without knowing U^c in terms of $P_1, \ldots, P_M$. For example, take U^c for the uniaxial σ_{xx} case in the form

$$U^c = \int_V (1/2E)(\sigma_{xx} + \sigma^1_{xxm} P^V_m)^2 dV, \tag{C.50}$$

where σ^1_{xxm} is the virtual stress produced by a unit value of P^V_m. Thus

$$u_m = \frac{\partial U^c}{\partial P^V_m} = \int_V (1/E)(\sigma_{xx} + \sigma^1_{xxm} P^V_m)\sigma^1_{xxm} dV$$

$$= \int_V e_{xx} \sigma^1_{xxm} dV, \quad P^V_m = 0, \tag{C.51}$$

which corresponds to the unit load theorem, Equation (2.16). Of course, Equation (2.16) does not have the restrictions on the strains implied by Equations (C.50) and (C.51).

C.10. Reissner's variational principle

Use Equations (C.26) and (C.45) to condense the principle of mixed virtual stresses and virtual displacements in Equation (2.20) to the form

$$U^V + U^{cV}(u) - U^{cV}(\sigma) - W^V - [W^{cV}(u) - W^{cV}(u_s)] = 0, \tag{C.52}$$

where u_s are specified displacements. From the general expressions for $U^V(\sigma)$ and $U^{cV}(u)$ it is evident that for real variations

$$\delta U + \delta U^c = \delta \int_V (\sigma_{xx} e_{xx} + \sigma_{yy} e_{yy} + \cdots + \sigma_{yz} e_{yz}) dV$$
$$= \int_V [(\sigma_{xx} \delta e_{xx} + e_{xx} \delta \sigma_{xx}) + \cdots + (\sigma_{yz} \delta e_{yz} + e_{yz} \delta \sigma_{yz})] dV. \quad \text{(C.53)}$$

With real variations for the strain energy U and complementary strain energy U^c and Equations (C.23) and (C.42) for variations of W and complementary work W^c, Equations (C.52) and (C.53) can be written in the form

$$\delta \left[\int_V \{(\sigma_{xx} e_{xx} + \cdots + \sigma_{yz} e_{yz}) - U_d^c\} dV - 2W - 2(W^c - W_s^c) \right] = 0, \quad \text{(C.54)}$$

where U_d^c is in Equation (C.10), W^c and W are in Equation (C.4). This is *Reissner's Variational Principle* in Reference 3 with body forces and specified displacements added. Since both stresses and displacements may vary in this principle, it may be a maximum or a minimum so that it is regarded simply as a stationary principle. Of course, all three virtual principles used in this text may be regarded as stationary principles.

C.11. Comparison of the virtual principles and the strain energy theorems

In the previous sections of this Appendix C the classical strain energy theorems, as well as Hamilton's principle and Reissner's principle, have been derived directly from the principles of virtual displacements, virtual forces, and mixed virtual stresses and virtual displacements. Because of the restrictions and assumptions involved in the strain energy theorems, it appears that the three virtual principles have several advantages over the strain energy theorems, such as:

(a) They avoid the confusion of defining special functions such as in Equations (C.13)-(C.15).

(b) They have no restrictions on the external forces.

(c) They are simpler to use for thermal, inelastic, and large displacement problems.

(d) They use stresses, strains, and displacements directly without any need to calculate work and strain energy.

(e) They give the Rayleigh-Ritz set of equations directly for approximate solutions without having to differentiate the energy and work with regard to each unknown constant.

(f) They apply directly to compatible substructures without any need for the strain energy in the entire structure.

(g) They require only compatible virtual displacements and only equilibrium virtual stresses.

(h) They avoid the dummy displacements, dummy loads, and differentiations in Castigliano's theorems, giving stiffness and flexibility matrices directly.

(i) They have no relation to real strain energy and real work so that expressions for real strain energy and real work are not needed and such expressions need not exist.

(j) They give approximate solutions which converge in the same manner as the strain energy theorems. This convergence is in the sense of the Galerkin and Rayleigh-Ritz approximations. These approximate solutions also provide upper and lower bounds for the exact stresses and displacements.

References

Appendix C

1. D. Williams: *An Introduction to the Theory of Aircraft Structures*, Edward Arnold (Publishers)LTD., London (1960).
2. D.R. Smith and C.V. Smith, Jr.: When is Hamilton's principle an extremum principle?, *AIAA J.* 12, No. 11, pp. 1573-1576, Nov. (1974).
3. E. Reissner: On a Variational Theorem in Elasticity, *J. of Mathematics and Physics* 29, p. 90 (1950).
4. C.L. Dym and I.H. Shames: *Solid Mechanics: A Variational Approach*, McGraw-Hill (1973).
5. I.S. Sokolnikoff: *Mathematical Theory of Elasticity*, McGraw-Hill Book Co. (1956).

Index

aerodynamic loads 284-302
  (*see* airplane loads)
aileron effectiveness 196-198
aileron redundant supports 311-312
aileron reversal speed 197-198, 311
airplane coordinate axes 280
airplane loads
  drag 287
  gust 290
  inertia 289
  landing 304-308
  lift 288-290
  pressure 288-292
  thermal 308
  V-n diagram 289-291
  weight 287
airplane wings
  divergence speed 190, 191, 195, 311
  loads 190, 193, 249, 286
  swept 189, 249, 311
  twisted 190, 249, 295
angle of twist 183, 195
apparent internal load matrix $\{S_E\}$ 74, 81, 132
apparent load matrix $\{P_E\}$ 57, 82, 131
apparent moments M_{xE}, M_{xT} 124
applied load matrix $\{P\}$ 57, 131
approximate solutions by
  finite elements 108, 119, 130-148
  Saint-Venant's principle 10-25
  thin webs 17-19
  virtual strains from virtual stresses 58
  virtual stresses from virtual strains 59
assembly of finite elements 81, 87, 131, 138
axial stresses 51, 62, 69

beam-columns 147
beam coordinate axes 11, 12, 49, 103
beam cross sections
  two dimensional stresses 16
  unsymmetrical 21-23, 175
  warping 15, 179, 180, 185

beam equations 11-21
  initial strain 21-23
  inelastic strain 21-23
  plane strain 12
  shear stresses 12-21
  temperature stresses 21-23
beam finite elements 120, 130-149
beams
  bending stresses 11-25
  box (*see* box beams)
  deflecitons 116, 174, 226
  load-strain design curves 21-32
  sequence loading 28-32
  shear deflections 178
  tapered 170
  thermal cycling 28-32
  thermal stresses 21-25, 123, 217, 227
  two elements 134
body forces 3
  potential of 8
boundary conditions 3, 9, 11, 12
box beams 18-25, 153-199
  deflections 174
  multi-cell 165
  open 160
  single cell 183
  tapered 170
brittle materials 11
buckling coefficients 147
bulkheads 210

Castigliano's theorems 37, 323, 326
characteristic values (*see* eigenvalues)
$[C]$ matrix, (*see* flexibility $[C]$ matrix)
coefficient of thermal expansion 6
column matrix 57, 277
columns
  buckling 147
  end fixity 147
combined method of solution 8
comparisons of solutions by virtual principles

329

reciprocal relations 44, 72
simple beams 117, 121, 123, 142, 146
thin web beams with stiffeners 232
trusses 73, 77, 97
compatibility equations
  beams 12-14, 117
  one dimensional 12
  plates 9
  three dimensional 5, 7
  trusses 71
  two dimensional 9
complementary potential energy 319
complementary strain energy 318
complementary work 317
concentrated point load 40
  local effect 11
coordinate axes
  airplane 280
  beam 12, 49, 103
  cylinder 9
  fuselage 303
  plate 10
critical values (*see* eigenvalues)
cut-outs
  box beams 208
  bulkheads 212
  thin-web beams 206
cylindrical structures
  plane strain 9

datum coordinates
  trusses 45, 63
deflections from principle of virtual displacements
  beam differential equations 105
  beam finite element matrices 130-135
  beam point values 109
  truss equations 65
  truss matrices 81
deflections from the mixed virtual principle
  beam finite element matrices 145, 146
  truss matrices 96
deflections from the principle of virtual forces
  beam equations 114, 115
  beam finite element matrices 136-146
  beam point values 119
  box beams 174-179
  thin-web structures 226, 233
  truss equations 70
  truss matrices 87-95
design criteria
  limit loads 313
  ultimate loads 313
design curves
  load-strain 27-31
design procedures
  load-strain curves 26-33
determinants 256
  characteristic equations 275
determinate structures 38, 44

beams 117, 127, 135
box beams 154-162
thin-web structures 202-215
trusses 65, 90
direction cosines 3
  (*see* rotation matrices)
displacement matrix $\{u\}$ 58, 88
displacements
  linear 4
  non-linear 5, 55
  virtual 40
  (*see* deflections)
displacement method of solution 6
divergence velocity 191
ductile materials 11

e_e elastic strains 22, 54
e_E strain addition to e_e 22, 54, 55
effective shear modulus 173, 174
eigenvalues 275
elastic constants 6
energy theorems, derivation of 317-327
equilibrium equations
  one dimensional 11-14
  three dimensional 3, 4
  two dimensional 9
  virtual 43
equilibrium equations for
  airplane in flight 287
  beams 11, 12
  $[p]$ and $[q]$ matrices 87-89
  trusses 64
equilibrium matrix $[s]^T$ for
  box beams 244
  simple beams 131, 136
  stiffeners 233-235
  thin web panels 233, 234
  trusses 81, 87
Euler's column formula 147
expansion, coefficient of thermal 6

finite elements for
  bars 83
  simple beam 120-146
  stiffeners 231, 232
  thin web panel 231, 232
  trusses 83-97
flexibility matrix $[C]$ 58
  bars 60
  box beams 244
  simple beams 138
  stiffeners 231
  swept back wings 249, 250
  thin web panel 231
  trusses 87
flight loading conditions 288
  V-n diagram 289
Fourier transforms 38
frame structures 118, 144
free body 1
fuselage

Index

bulkheads 212
loading 302, 303

geometric shear stresses 55, 148
Greene's first identity 39, 40
gust load factor 290, 291
gust load lines 290, 291

Hamilton's principle 323, 324
homogeneous materials 1

indeterminate structures (*see* redundant structures)
inelastic effects 21, 22, 54, 55
  axially loaded structures 23-35
  beams 21-25, 123-130, 174-176, 227
  trusses 74-80, 86, 90-94
inelastic strains e_p 21, 22, 54, 55
inertia forces 301, 302
influence coefficients
  flexibility 58
  stiffness 57
integral transforms 38
internal load matrix $\{S\}$ 59, 81, 131
isotropic materials 1
iteration for
  beams 25, 26
  inelastic axial loads 26, 31
  trusses 74-78
iteration speed-up 75

$[J]$ matrix 57
Jordan matrix transformation $[T_i]$ 89, 94, 265

$[k]$ stiffness matrix (*see* stiffness matrix $[k]$)

$[\lambda]$ rotation matrix (*see* rotation matrices)
landing gear structure 304-308
landing loads 281, 304-308
Laplace transform 38
large displacements 5, 54, 55
lift coefficient distribution for
  built-in wing twist 295
  untwisted wing 292
  wing twisted by loads 189, 311, 312
load factor 289-291
load-strain design curves 25-33

margin of safety (M.S.) 313
material property matrix $[J]$ 57
matrices
  apparent internal load $\{S_E\}$ 81, 131
  apparent load $\{P_E\}$ 57, 82, 131
  applied load $\{P\}$ 57, 131
  displacement $\{u\}$, $\{u_E\}$ 58, 88
  equilibrium $\{s\}^T$ 81, 131, 234, 235
  flexibility $[C]$ (*see* flexibility matrix $[C]$)
  internal load $\{S\}$ 59, 81, 131
  Jordan transformation $[T_i]$ 265, 266
  material property $[J]$ 57

mixed $[M]$ (*see* mixed matrix $[M]$)
redundant load $\{Q\}$ 121, 122, 236
rotation $[R]$ 132, 135, 271
shear flow 235
stiffness $[k]$ (*see* stiffness matrix $[k]$)
unit load $[p]$ 88, 136, 235, 236
unit load $[q]$ 88, 136, 235, 236
matrix displacement method 57
matrix force method 58
matrix mixed method 59
matrix operations 254-277
mixed matrix $[M]$ 59
  bars 60
  simple beams 146
  trusses 96
modulus of elasticity 6
Mohr's circle 273
moments 11, 13
  beam bending axes 12, 49
  box beam bending M_x, M_z 20-23, 160-172
  box beam torsion M_y 19-20, 158-172
  simple beam bending M_x 17, 18, 103

non-linear strains 4, 5, 55
numerical integration 176, 269

one dimensional equations 11
orthogonal matrices 270, 271

$\{P\}$ applied load matrix 88, 136
$\{P_E\}$ apparent load matrix 57, 82, 131
plane strain cross section 22
plate finite elements
  rectangular with constant shear flow 231-247
plates, plane strain 9
Poisson's ratio 6
potential energy 319
potential functions 8-10, 319
pressure loads
  distributed on beams 103-108
  wing 292-299
principle of mixed virtual stresses and virtual displacements
  beams 122, 145
  derivation 46
  derivation of Reissner's variational principle 326, 327
  matrix form 59
  one dimensional form 53
  three dimensional form 46
  trusses 96
  two dimensional form 48
  unit displacement theorem and unit load theorem 47
principle of virtual displacements
  beams 104, 105, 109, 130
  Castigliano's theorem, Part I 323
  columns 147
  derivation 39

derivation of theorem of minimum total potential energy 322
derivation of theorem of minimum strain energy 322
Hamilton's principle 323, 324
matrix form 57
one dimensional form 51
three dimensional form 40
trusses 64, 81
two dimensional form 48
unit displacement theorem 41
principle of virtual forces
   beams 114-121, 136, 233
   box beams 242-250
   Castigliano's theorem, Part II 326
   derivation 43
   derivation of theorem of minimum total complementary potential energy 325
   derivation of theorem of minimum complementary strain energy 325
   finite elements 231
   matrix form 58
   one dimensional form 52
   three dimensional form 43, 44
   trusses 70, 88
   two dimensional form 48
   unit load theorem 45
$[p]$ unit load matrix 88, 136

$\{Q\}$ redundant load matrix 88, 91
$[q]$ unit load matrix 88, 136, 137

Ramberg-Osgood equation 24-32, 125
reciprocal relations for virtual principles 44, 72
redundant structures 37, 44
   beams 114-122, 136-145, 217-226, 233-242
   box beams 242-248
   trusses 70-72, 87-95
Reissner's variational principle 326, 327
residual stresses 30
restrained box beams 242-248
ribs 210, 212
rigid body displacements 5, 6
rigid body forces 4
rotation of axes 271
rotation of beams 183, 184
rotation matrices $[R]$ and $[\lambda]$ 271, 272
   beams 132, 138, 139, 144
rotation of swept wing 189

Saint Venant's principle 10, 11, 37
Schrenk's method 293
$\{S_E\}$ apparent internal load matrix 81, 131
sequence loading 28
shear center 161, 168, 184
shear deflections 177
shear warping deflections 15, 180, 185
shear flows 156
   around cut-outs 206, 207

concentrated point loads 203
multi-cell box beams 165
open box beams 160
redundant beams 216-251
ribs and bulkheads 210
single cell box beams 163
tapered box beams 170
shear lag 202
shear modulus of elasticity 6
shear strain 4
shear stresses
   beams 12-20, 154
   buckling 173
   inelastic 172
   torsional 18, 19, 158
shear web finite elements 231-250
Simpson's rule 269
$\{S\}$ internal load matrix 59, 81, 82, 131
Sneddon, I. N. 38
spanwise airload distribution 190, 297
spherical coordinates 274
$[s]^T$ equilibrium matrix (see equilibrium matrix $[s]^T$)
stiffener finite elements 231-250
stiffness matrix $[k]$ 57
   bars 60
   simple beams 131
   trusses 82
strains 4
   additional e_E 21, 22, 55
   axial 4
   elastic 22
   geometric shear 55, 148
   inelastic 22, 23, 55, 79
   initial 22, 55, 78
   non-linear displacement 5, 54, 55
   plane 12
   shear 4
   temperature or thermal 22, 54, 55
   virtual 40
strain-displacement equations
   linear 4, 42, 62, 63
   linear geometric 55, 148
   non-linear 5, 55
   virtual 40
strain e_E effects 22, 55
   axial load P_{yE} 23, 56, 82, 131, 132
   bending moments M_{xE}, M_{zE} 23, 123, 132
   displacements $\{u_E\}$ 58, 138, 220, 232
   internal loads $\{S_E\}$ 82, 131, 132
strain energy 317
stress-displacement equations 7
stress
   axial 12
   bending 12
   shear 3
   torsion shear 18
   virtual 42
stress functions
   three dimensional 7

Index

two dimensional 9
stress method of solution 7
stress-strain equations
  elastic isotropic 6
  inelastic 22-33
  one dimensional 11, 12, 22, 104, 124
  three dimensional 6
  two dimensional 9
surface forces 3, 21
swept back wings 189, 248, 311

tail loads 288
tapered bar 69
tapered beams 170
temperature effects 6
  bars 22-33
  beams 22-25, 123, 217, 227
  cycling 28-33
  displacements 58, 74, 76, 124
  expansion coefficient 6
  inelastic 25-33, 74-80, 123-130, 174-176
  one dimensional 22-33
  plates 6-9
  strain 22, 54
  three dimensional 6-9
  trusses 74-77
  two dimensional 9, 10
theorem of minimum
  complementary strain energy 325
  strain energy 322
  total complementary potential energy 324
  total potential energy 322
thermal effects (*see* temperature effects)
thermal forces 308, 309
three dimensional equations 1-9
torsion shear stresses 18, 19, 158, 159
trapezoid rule 269
two dimensional equations 9
two element beams 134, 135

ultimate stresses 313
unit displacement theorem 41, 42
  beams 130
  Castigliano's theorem, Part I 323
  matrix form 57
  one dimensional form 51

  three dimensional form 42
  trusses 64, 80
  two dimensional form 48, 49
unit load theorem 45
  beams 114, 120, 136, 216-230, 233-250
  box beams 174, 178, 184
  Castigliano's theorem, Part II 326
  frames 118, 119
  matrix form 57
  one dimensional form 52
  shear web elements 231
  stiffener elements 231
  three dimensional form 45
  trusses 70, 87
  two dimensional form 49
unit mixed displacement and unit load theorem 49
  beams 145, 146
  matrix form 58, 59
  one dimensional 53
  three dimensional 48
  trusses 96
  two dimensional form 49

virtual displacements 39 (*see* principle of virtual displacements)
virtual forces 43 (*see* principle of virtual forces)
virtual geometric shear strain 55
virtual strains 55
  beams 104
virtual stresses 44
  beams 104
$V\text{-}n$ diagram 289, 290

warping of beam cross sections
  shear 14, 15
  thin web shear 178-180
  torsion box beam 184
weight 288
weighting numbers 269
wing loading 190, 249, 292-302
wing ribs 209, 212
wings (*see* airplane wings)
work 317

yield stress 313

MIX
Papier aus verantwortungsvollen Quellen
Paper from responsible sources
FSC® C105338

If you have any concerns about our products,
you can contact us on
ProductSafety@springernature.com

In case Publisher is established outside the EU,
the EU authorized representative is:
**Springer Nature Customer Service Center GmbH
Europaplatz 3, 69115 Heidelberg, Germany**

Printed by Libri Plureos GmbH
in Hamburg, Germany